Information Modeling and Relational Databases

Information Modeling and Relational Databases

Third Edition

Terry Halpin
Tony Morgan

ELSEVIER
MK
MORGAN KAUFMANN PUBLISHERS
AN IMPRINT OF ELSEVIER

Morgan Kaufmann is an imprint of Elsevier
50 Hampshire Street, 5th Floor, Cambridge, MA 02139, United States

ISBN: 978-0-443-23790-4

For information on all Morgan Kaufmann publications
visit our website at https://www.elsevier.com/books-and-journals

Publisher: Mara Conner
Acquisitions Editor: Chris Katsaropoulos
Editorial Project Manager: Palak Gupta
Production Project Manager: Omer Mukthar
Cover Designer: Vicky Pearson Esser

Typeset by STRAIVE, India

Contents

https://www.elsevier.com/books-and-journals/book-companion/9780443237904.

Foreword

by Dr. Enrico Franconi
Professor of Computer Science at the Free University of Bozen-Bolzano, Italy
Director of the KRDB Research Centre for Knowledge and Data.

It is with great enthusiasm and profound honour that I contribute to the foreword of the third edition of this exceptional book, a comprehensive and illuminating guide to information modeling and relational databases, underscoring its profound importance in the realm of enterprise modeling and data management. Even more in this new edition, one cannot help but acknowledge the sheer breadth and depth of knowledge included within its pages.

This book serves as an indispensable resource for comprehending the object-role modeling (ORM) language and its fact-oriented modeling methodology. ORM stands as a beacon in the realm of data modeling, offering a rigorous and yet user-friendly approach. This book meticulously dissects the semantic complexity of enterprises with strict attention to detail. The authors' dedication to exploring and evaluating various conceptual modeling notations against ORM sets this book apart, providing readers with a panoramic view of the landscape of data modeling paradigms while presenting an accurate understanding of conceptual modeling and evidence of the advantages of ORM.

The chapters clarify the nuances of conceptual modeling, emphasizing the necessary precision of a modeling language. As someone from a logic-based knowledge representation and reasoning background, I greatly appreciate the formal rigor of the ORM notation, where ambiguities and sloppiness find no place, unlike many other (widely known) conceptual modeling notations. Furthermore, the book doesn't merely describe the precise meaning of the modeling constructs but also elucidates why these constructs exist in the language. A meticulous analysis of their use cases and a methodology for their design hold equal importance in the book.

I find it fascinating how ORM offers a powerful and precise yet understandable conceptual modeling approach, and of course also how ORM is presented in this book in a precise yet understandable manner. As expected, the fact-oriented modeling methodology associated with ORM is thoroughly detailed in a clear way and serves as another key pillar of this book. It guides the understanding of data semantics and the design of data structures that authentically and correctly mirror the business reality and needs. The book focuses also quite a lot on the relationship between a conceptual model and its implementation in some database structure, emphasizing correctness and effectiveness while not forgetting the intricacies of real-world SQL.

The fact-oriented modeling methodology as explained in the previous editions of the book has influenced a lot of my research. Indeed, it was after a first reading of this book that I started working on the intersection between classical logic-based knowledge representation and ORM with its fact-oriented modeling methodology. I was also influenced by the textual version of the ORM language (FORML), also discussed in this book. I met Terry Halpin to discuss the outcome of my research, only to realize that there was much more for me to learn beforehand! Terry is a humble scientist who, after every discussion, always seems to be right.

The lifelong dedication of Terry Halpin and Tony Morgan to developing enterprise modeling concepts substantiates the significance of this work. The updated and expanded coverage, along with a plethora of exercises embedded in the chapters, enhance the book's value. The book stays current with the latest versions and trends, ensuring readers have access to the most recent developments in the field. The content encompassing ORM, ER, UML, SQL, and OWL has undergone comprehensive updates, incorporating their most recent iterations. Additional chapters addressing NoSQL technologies, such as key-value, column-oriented, and graph databases, have been integrated in this edition of the book. A chapter covering contemporary popular data file formats, such as XML and JSON, has been added. Extensive revisions have been made to all other chapters, enhancing the depth of coverage for numerous topics.

Finally, I want to emphasize the book's relevance to a diverse audience, including data modelers, data engineers, data analysts, enterprise architects, and programmers, highlighting its inclusivity and accessibility. The incorporation of exercises throughout the book offers readers an interactive and immersive learning experience, fostering a deeper understanding of conceptual modeling.

Foreword

by Dr. Herman Balsters
Partner at NorthChain, Groningen, The Netherlands.

I first became acquainted with ORM in 2005 at an ORM Workshop in Cyprus as part of the OTM Confederated International Workshops and Conference. I had just presented a paper on data federations at the OODB conference and out of pure curiosity, I paid a short visit to the ORM Workshop to learn a bit more about the ORM language. When I wandered in, Terry Halpin was offering a presentation. I was amazed with the lively discussions that went on during this presentation, and was surprised not only by the enthusiasm of the workshop participants but also by the high quality of comments and questions. Instead of paying only a short visit, I stayed at the ORM Workshop during the complete remainder of my stay in Cyprus, and I have been a great fan of ORM ever since.

During my work as a researcher and associate professor at the University of Groningen, I was heavily involved with languages such as unified modeling language (UML) and business process model and notation (BPMN), trying to capture the essence of information modeling. I had a constant struggle with these languages, which were considered state of the art, because I just could not get a full grasp of what models written in these languages really meant. I have a background in pure mathematics (in particular, mathematical logic) and computer science, and I wanted to have a clear idea of the semantics and completeness of the data and process models that I was trying to specify. I also wanted to see a clear connection between a model and how that model is implemented. In both aspects, I felt uncomfortable with the prevailing modeling languages of that time. And it was not only in research that I felt this lack of precision and completeness, but also in teaching. Nothing is harder to explain to students than something that you as a teacher do not completely understand.

All that changed the day I entered that ORM Workshop in Cyprus. It took me a while to master all the details of ORM, but after that, I was able to actually write models with complete understanding and with such precision that the models could directly be tied to implementations. Even today, I still use ORM with deep respect and satisfaction.

ORM is a so-called fact-oriented data modeling language. In ORM, you write facts about the world surrounding us. This means that you specify objects and relations between these objects, resulting in certain statements about the world. ORM does this in such a way that even nontechnical domain experts will be able to model facts about their domain of expertise. This has to do with the close resemblance that ORM has with natural language specifications about the world surrounding us. Yet ORM also does that in such a way that you can model facts with an extremely high level of detail and (mathematical) precision. ORM is a graphical language, with an intuitively understandable notation. ORM is also equipped with a powerful tool (NORMA) that allows for the graphical editing of ORM models, and validation checks. Another appealing aspect of ORM is the possibility of compiling ORM models directly into the relational data model, and after that, compilation to SQL. Hence, in ORM you can go from modeling directly to an implementation.

In 2008, authors Terry Halpin and Tony Morgan released the second edition of this wonderful book. I have used it ever since. Over the years, I have taught ORM to hundreds of students, and have supervised both Masters and Ph.D. students in their research involving ORM. The book excels in its simple and clear writing, covering almost every topic in information modeling. Also, the abundance of examples explaining the theory and practice of modeling using ORM makes this book essential for anyone wanting to understand all the details pertaining to information modeling.

This book is both technical and philosophical. It is technical because it provides for an abundance of material concerning theories, notations, examples, and tools to tackle almost anything that is thrown to an information modeler. As a bonus, we are shown how to technically implement those models as a relational database. But in a sense, the book is also philosophical because it takes you into an arena of challenging problems concerning the representation of the world surrounding us. ORM is all about using language to capture that world, and what the best language to do that job should look like.

As a university professor, I had the pleasure of using ORM to teach modeling to students, to write research papers, and to supervise Masters and PhD students in information modeling. Since leaving academia in 2018, I have been involved in information modeling on a more industrial level, where I have been confronted with some extremely challenging modeling problems in actual practice. Also in these complex situations involving modeling issues pertaining to data, business rules, and business processes, ORM provided solutions that I am sure I would otherwise not have had.

Terry Halpin is the world's leading expert on ORM, and there is no better book than this one to get an understanding of ORM and how to tackle information modeling. I therefore fully recommend the third edition of this book to anyone interested in information modeling.

Foreword

by Dr. Gordon C. Everest, Professor Emeritus of MIS and "Data" Modeling
Carlson School of Management, University of Minnesota, USA.

I am delighted and honored to again write a foreword, this time to the third edition of this book. It gives me another opportunity to convince those in the world of data modeling that there is a better way. I am absolutely convinced that object-role modeling (ORM) is a better way to do data modeling. My goal here is to sufficiently perk your interest to seriously study ORM. This book is the best resource available.

The Greek philosopher Socrates advised "Know thyself" to live a worthwhile life. Similarly, we can say "Know your world", the reality in which you live, operate, and make decisions. Because it is impossible for any one person to know everything about their world, we need a *representation* of that world, a *model* that captures information about the things, relationships, activities/processes, and rules in that world. This representation consists of two parts: the definition of the structure and format, called the *schema*, and the actual data stored consistent with its definition. A collection of "databases" of information becomes an important *surrogate* for that world.

We make decisions in and about our world based on the surrogate databases we build. That is an awesome responsibility and thus demands a high degree of integrity, quality, and availability of our data stores. That requires a data modeling scheme that can capture a rich set of semantics, and ORM is the best of anything out there today.

The use of the term "data modeling" is often misleading. It suggests we are modeling data, when we are really modeling the things and relationships (objects and roles in relationships) in some user domain of interest—modeling *in data*, using data constructs, which eventually becomes a "data" model after the model is populated with data.

Data modeling is the foundation of information systems development—if you don't get the database design "right", then the systems you build will be like a house of cards, collapsing under the weight of inevitable future forces for enhancement, integration, and quality improvement. Thus, we need a scheme to guide our data modeling efforts to produce data models that clearly and accurately represent the users' domain of discourse and facilitate human communication, a shared understanding, and validation for quality control. The people who live in and know the user domain must be directly involved in the modeling process, supported by data modeling specialists who understand the design methodology.

This book is a must for anyone who is serious about data modeling, but with a caution: you must devote enough time and effort to really understand ORM. Fortunately, I had my students as a captive audience for a whole semester—long enough for them to learn and practice ORM and become convinced that it is a better way. With ORM, you can raise your data modeling skills to a whole new level.

This book also examines record-based modeling schemes: UML, SQL based on Ted Codd's relational model, and Peter Chen's entity relationship (ER) diagrams with many variations—Barker as in Oracle, Finkelstein's information engineering (IE), and IDEF1X (as in ERwin), with new material on the data structure representations in XML, JSON, and NoSQL. Viewing these familiar modeling approaches from an ORM perspective provides an enriched understanding of their underlying nature.

Record-based modeling schemes use *three* constructs: entity, attribute, and relationship. They can directly represent only at most binary, one-to-many relationships. But it is the clustering of attributes into entity records that is the root of many of our problems in data modeling. Normalization is the test to see if we clustered too much, and record decomposition is the remedy to correct a violation of the normal forms.

Normalization is the Achilles heel of data modeling. Oh, to be able to avoid normalization altogether? The mere suggestion is intriguing to students and practitioners of data modeling. Well, with ORM you can. The problem stems from the lack of a clear and complete definition of relationships when we throw stuff (attributes) into an entity record or table, so that the intrarecord structure is implicitly defined or assumed. ORM forces you to separately consider and explicitly define all relevant relationships and constraints among the object domains in your universe.

ORM is based on only *two* constructs: objects and relationships (which correspond to the concepts of nouns as subject or object and verbs as predicates in sentences). Both entities and attributes (in the relational modeling scheme) are based on objects (not to be confused with objects in object-oriented technology) and roles in ORM. Objects play roles in relationships with other objects. Thus, objects implicitly have attributes (descriptors or properties) by virtue of the roles they play in relationships with other objects.

In record-based modeling, there are two kinds of relationships: interrecord (defined explicitly and represented using foreign keys), and intrarecord among attributes (without explicitly defining the characteristics of any of the relationships, which is why we need to enforce the normalization rules). In ORM, all relationships are represented the same way with a single construct. When the ORM model is a valid representation of the world being modeled, the functional and multivalued dependencies are explicitly defined, and hence, the generation of "records" (in a relational table) can be automated and can guarantee that the result will be normalized (to 5NF). That's good news for data modelers.

ORM does not supplant ER diagrams or relational database designs; rather, it is a stage before. It can enable, enlighten, and inform our development and understanding of ER/relational data models. We build records more for system efficiency than for human convenience or comprehension. The premature notion of a record (a cluster of attribute domains along with an identifier to represent an entity) gets in the way of good data modeling. ORM does not involve records, tables, or attributes. So consequently, we don't get bogged down in "TABLE THINK" (see https://tdan.com/recognizing-and-treating-tableitis/26060). With ORM, there is no need for an explicit normalization process.

This third edition is even more focused on the centrality of data in information systems, and on the importance of semantics. Starting with the realization that users (collectively) know more than we could ever capture in a data model, we must use a data modeling scheme that captures the widest possible range of semantics and expresses this meaning graphically. Semantics is paramount, and ORM goes way beyond any record-based modeling scheme in graphically depicting semantics.

Well, is that sufficient to pique your interest in learning more about ORM? If you are a would-be student of ORM and you take data modeling seriously, I encourage you to invest some time in reading this book. You won't regret it. You will grow to appreciate ORM and will become a better data modeler for it—just ask any of my thousands of former students. To develop effective and maintainable information systems, we need good data models, and for that we need a good data modeling scheme or methodology. ORM allows us to develop database designs at the highest conceptual level, unencumbered by things that are not of primary concern to user domain specialists. My deep desire is to see more and more database designers using ORM. The systems we build and the world we live in will be better for it. Join me in this journey and enjoy the adventure.

Preface

This book is about information systems, focusing on information modeling and database systems (especially relational database systems, but also covering nonrelational systems). It is written primarily for data modelers and database practitioners as well as students of computer science or information management. It should also be useful to anyone wishing to formulate the information structure of business domains in a way that can be readily understood by humans yet easily implemented on computers. In addition, it provides a simple conceptual framework for understanding what database systems really are as well as a thorough introduction to SQL and other key topics in data management.

A major part of this book deals with *fact-oriented modeling*, a conceptual modeling approach that views the world in terms of simple facts about objects and the roles they play. Originating in Europe, fact orientation is today used worldwide and comes in many flavors, including the Semantics of Business Vocabulary and Business Rules (SBVR). The version of fact orientation described in this book is second-generation *Object-Role Modeling* (ORM 2), and is based on extensions to the Natural-Language Information Analysis Method (NIAM).

Two other popular notations for information modeling are *Entity-Relationship* (ER) diagrams and *Unified Modeling Language* (UML) class diagrams. For conceptual information analysis, the ORM method has several advantages over the ER and UML approaches. For example, ORM models can be easily verbalized and populated for validation with domain experts, they are more stable under changes to the business domain, and they typically capture more business rules in diagram form.

However, ER diagrams and UML class diagrams are good for compact summaries, and their structures are closer to the final implementation, so they also have value. Hence, the coverage includes chapters on data modeling in ER and UML, and indicates how ER and UML data models can be easily generated from ORM models.

To make the text more approachable to the general reader with an interest in databases, the language is kept simple, and a formalized, mathematical treatment is mostly avoided. Where necessary, relevant concepts from elementary logic and set theory are discussed prior to their application. Most of the material in this book has been class tested in courses for both industry and academia, and the basic ORM method has been taught successfully even at the high school level. The content is modularized, so that instructors wishing to omit some material may make an appropriate selection for their courses.

The first chapter motivates the study of conceptual modeling, and briefly compares the ORM, ER, and UML approaches. It also includes a historical and structural overview of information systems. Chapter 2 provides a structural background, explaining the conceptual architecture of, and development frameworks for, information systems. It introduces a number of key concepts that are dealt with more thoroughly in later chapters, and should be read in full by the reader with little or no database experience.

Chapter 3 is fundamental. Following an overview of conceptual modeling language criteria and the ORM Conceptual Schema Design Procedure (CSDP), this chapter covers the first three steps of the CSDP. The first step (verbalizing familiar examples in terms of elementary facts) may seem trivial, but it should not be rushed, as it provides the foundation for the model. The rest of this chapter covers the basic graphical notation for fact types, and then offers guidance on how to classify objects into types and identify information that can be arithmetically derived.

Chapter 4 begins the task of specifying constraints on the populations of fact types. The most important kind of constraint (the uniqueness constraint) is considered in detail. Then some checks on the elementarity of the fact types are discussed. This chapter also introduces the join and projection operations at the conceptual level; the relational version of these operations is important in the later work on relational databases.

Chapter 5 covers mandatory role constraints, including a check for detecting information that can be logically derived. Reference schemes are then examined in some depth. Some of the more complex reference schemes considered here could be skipped in a short course. The CSDP steps covered so far are then reviewed by applying them in a case study, and the logical derivation check is then considered.

Chapter 6 covers value, set comparison (subset, equality, and exclusion), and subtyping constraints. Section 6.6 deals with advanced aspects of subtyping. Though important in practice, the material in this section could be skimmed over in a first reading.

Chapter 7 deals with the final step of the CSDP. Less-common constraints are considered (e.g., frequency constraints and ring constraints), and final checks are made on the design. Sections 7.3–7.5 are somewhat advanced, and could be skipped in a short course.

Chapter 8 discusses the ER approach, starting with Chen's original notation then moving on to the three most popular notations in current use: the Barker ER notation, the Information Engineering notation, and the IDEF1X notation (actually a hybrid of ER and relational notations). Comparisons with ORM are included along the way.

Chapter 9 examines the use of UML class diagrams for data modeling, including a detailed comparison with ORM. Business rule constructs in ORM with no graphic counterpart in UML are identified and then captured in UML using user-defined constraints or notes.

Chapter 10 considers several advanced aspects of information modeling, such as join constraints, historical fact types, collection types, open/closed world semantics, and higher-order types. The discussion of deontic rules and nominalization is fundamental to understanding the SBVR flavor of fact orientation. This chapter is technically the most challenging in the book, and it could be skipped in an introductory course.

Chapter 11 describes how a conceptual model may be implemented in a relational database system. The first three sections are fundamental to understanding how a conceptual schema may be mapped to a relational schema. Section 11.4 considers advanced relational mapping aspects, and it could be omitted in a short course.

Chapter 12 provides a foundational introduction to relational databases and SQL queries. Section 12.1 covers relational algebra; although not used as a practical query language, the algebra is important for understanding the basic relational operations supported by SQL. Section 12.2 provides an overview of how the relational model of data compares with data models adopted by some relational database management systems. Sections 12.3–12.14 cover the main features of SQL, focusing on versions of Standard SQL as specified by the International Standards Organization, with some attention also paid to commercial SQL systems.

Chapter 13 discusses further aspects of SQL (e.g., data definition, triggers, and stored procedures), and introduces some practical issues such as security, metadata, and concurrency.

Chapter 14 discusses how and when to transform one schema into another schema at the same level (conceptual or logical). Sections 14.1–14.4 examine the notion of conceptual schema equivalence, and ways in which conceptual schemas may be reshaped. As one application of this theory, Section 14.5 specifies a procedure for optimizing a database design by performing conceptual transformations before mapping. Section 14.6 provides concise coverage of normalization theory, including some new insights. Section 14.7 briefly considers denormalization and low-level optimization. Sections 14.8 and 14.9 illustrate the role of conceptual optimization in database reengineering, and conclude with a discussion of data migration and query transformation. Sections 14.4, 14.5, and 14.7–14.9 are of an advanced nature and may be skipped in a short course. In a very short course, the whole chapter could be skipped.

Chapter 15 broadens the treatment of information systems analysis by examining behavioral aspects of business using process and state models. The two main approaches of process modeling and state modeling are analyzed using their representations in UML. Detailed coverage of process modeling using the BPMN standard is provided, together with a methodology for linking process models and information models.

Chapter 16 examines various data file formats, including CSV, XML, JSON, YAML, and some other markup languages. In addition to their use for data interchange, XML and JSON are now fully incorporated into the SQL standard as data types. Examples of their use are provided together with a discussion on their mapping from ORM.

Chapter 17 reviews alternatives to the dominant relational database paradigm, in particular, "NoSQL" databases. The motivations for such systems are discussed and the major database types are described individually. Methods for mapping from an information model to each of these database types are introduced. The chapter concludes with some further ideas on future evolution, including the potential impact of artificial intelligence.

Chapter 18 examines other modeling aspects and trends. Topics covered include data warehousing, conceptual query languages, schema abstraction mechanisms, further design aspects, ontologies and the semantic web, and metamodeling. Though these topics are important and interesting, they could be omitted in a short course.

In line with the ORM method, this text adopts a "cookbook" approach, with plenty of diagrams and examples. Each chapter begins with a brief overview and ends with a chapter summary of the major points covered, with chapter notes to provide fine points and further references.

One of the major features of the book is its large number of exercises, which have been thoroughly class-tested. A bibliography of all cited references is included at the end of the book, where you will also find glossaries of technical symbols and terms for ORM, ER, and UML (class diagrams only). A comprehensive index provides easy access to explanations of technical topics.

For readers familiar with the second edition of this book, the major differences are now summarized. The coverage of ORM, ER, UML, SQL, OWL, and BPMN has been thoroughly updated to cover their latest versions. A significant amount of new material has been added. Various data file formats such as CSV, XML, JSON, YAML, and some other markup languages are now covered, and a more thorough treatment is provided for nonrelational databases, especially NoSQL. All previous chapters have been substantially revised, with several topics covered in greater depth. The new content has led to a larger book.

US spelling is used throughout the book, except for two forewords by European authors. US punctuation rules have also been used, except for quoted items, where item separators and sentence terminators (e.g., commas and periods) appear after, rather than just before, closing quotes.

Online resources

To reduce the size and cost of the book, supplementary material is available at the website https://www.elsevier.com/books-and-journals/book-companion/9780443237904 for downloading. There are five appendices.

Appendix A provides an overview of the evolution of computer hardware and software. Appendix B discusses two kinds of subtype matrix to determine subtype graphs from significant populations. Appendix C discusses set-comparison queries in SQL. Appendix D discusses ranking and extrema queries in SQL. Appendix E discusses ways to implement derivation rules in SQL using triggers, generated columns, stored procedures, views, user-defined functions, and common table expressions. Other appendices may be added as the need arises.

The answers to the exercises are contained in two files, one for the odd-numbered questions and one for the even-numbered questions. The answers to the odd-numbered questions are openly accessible, but the answers to the even-numbered questions are password protected to provide classroom instructors with a range of exercises for classroom discussion. Additional material on ORM and other topics is available via the Useful Websites section at the back of the book.

Further exercises and related pedagogic material are included in supplementary online resources that are available to instructors using this book as a course text. These resources and a password for exercise answers are available to classroom instructors at https://inspectioncopy.elsevier.com/.

ORM software

ORM is supported by a variety of modeling tools. At the time of writing, the Natural ORM Architect (NORMA) tool provides the most complete support for the ORM 2 notation discussed in this book. A free version of NORMA is available as an open source plug-in to Microsoft Visual Studio and may be downloaded from https://ormsolutions.com/ or https://sourceforge.net/projects/orm. For details on installing and using NORMA, see the NORMA labs downloadable from the Resources page at www.orm.net. The free, community version of Visual Studio is downloadable from https://visualstudio.microsoft.com/vs/community/.

NORMA Professional (NORMA Pro) has additional features such as multipage relational schemas and formal support for derivation rules. For details on how to get a licence for the professional version of NORMA, contact Matt Curland at mcurland@live.com.

Other tools supporting different flavors of fact-oriented modeling include CaseTalk, CogNIAM tools, ActiveFacts, and Boston. Links to these tools may be found in the Useful Websites section that is included at the back of this book.

Acknowledgments

We greatly appreciate the editorial assistance of Chris Katsaropolous and Palak Gupta at Elsevier.

Some of this book's content is based on articles written for *The Business Rules Journal*, the *Journal of Conceptual Modeling*, and material from editions of the earlier book *Conceptual Schema and Relational Database Design*, previously published by Prentice Hall Australia and WytLytPub. The first edition of that book was coauthored by Dr. Sjir Nijssen, who wrote the final four chapters (since replaced).

The ORM approach discussed in this book is based on our revisions and extensions to the NIAM method, which was largely developed in its original form by Dr. Eckhard Falkenberg, Nijssen, and Prof. Robert Meersman, with other contributions from several researchers. Some of the work discussed in this book was developed jointly with current or former colleagues, including Dr. Anthony Bloesch, Dr. Linda Bird, Dr. Peter Ritson, Dr. Erik Proper, Dr. Andy Carver, and Dr. Herman Balsters. It has been a pleasure working with these colleagues as well as with the hundreds of information modeling students and practitioners. We also gratefully acknowledge permission by The University of Queensland and Neumont University to include a small selection of our past assessment questions within the exercises.

Finally, we thank our wives, Norma and Gwen, for being so understanding and supportive while we were busily occupied in the writing task.

1

Introduction

Chapter Outline

1.1 Information Modeling

Information adds meaning to data. It's an unfortunate fact of life that names and numbers are sometimes misinterpreted. This can prove costly, as experienced by senior citizens in the United States who had their Social Security benefits cut off when government agencies incorrectly pronounced them dead because of misreading "DOD" on hospital forms as "date of death" rather than the intended "date of discharge".

In 1999, NASA's $125 million Mars Climate Orbiter burned up in the Martian atmosphere. Apparently, errors in its course settings arose from a failure to make a simple unit conversion. One team working in US customary units sent its data to a second team working in metric, but no conversion was made. If a man weighs 180, does he need to go on a drastic diet? Not if his mass is 180 lb., but yes if it's 180 kg. *Data* by itself is not enough. What we really need is *information*, the meaning or *semantics* of the data. Because computers lack common sense, we need to pay special attention to semantics when we use computers to model some aspect of reality.

This book provides a modern introduction to database systems, with the emphasis on information modeling. At its heart is a very high-level semantic approach that is *fact-oriented* in nature. If you model databases using either traditional or object-oriented approaches, you'll find that fact orientation lifts your thinking to a higher level, illuminating your current way of doing things. Even if you're a programmer rather than a database modeler, this semantic approach provides a natural and powerful way to design your data structures.

A **database** is basically a collection of related data (e.g., a company's personnel records). When interpreted by humans, a database may be viewed as a set of related *facts*—an *information base*. In the context of our semantic approach, we'll often use the popular term "database" instead of the more technical "information base". Discovering the facts that underlie a business domain, and the rules that apply to the facts, is

Information Modeling and Relational Databases. https://doi.org/10.1016/B978-0-443-23790-4.00006-2

interesting and revealing. The quality of the database *design* used to capture these facts and rules is critical. Just as a house built from a good architectural plan is more likely to be safe and convenient for living, a well-designed database simplifies the task of ensuring that its facts are correct and easy to access. Let's review some basic ideas about database systems, and then see how things can go wrong if they are poorly designed.

Each database models a *business domain*—we use this term to describe any area of interest, typically a part of the real world. Consider a library database. As changes occur in the library (e.g., a book is borrowed), the database is updated to reflect these changes. This task could be performed manually using a card catalog, or be automated with an online catalog, or both. Our focus is on automated databases. Sometimes these are implemented by means of special-purpose computer programs, coded in a general-purpose programming language (e.g., C# or Java). More often, database applications are developed using a **database management system** (DBMS). This is a software system for maintaining databases and answering queries about them (e.g., DB2, Oracle, SQL Server). The same DBMS can handle many different databases.

Typical applications use a database to house the persistent data, an in-memory object model to hold transient data, and a friendly user interface for users to enter and access data. All these structures deal with information and are best derived from an information model that clearly reveals the underlying semantics of the domain. Some tools can use information models to automatically generate not just databases, but also other kinds of models (object, logical, etc.) and user interfaces.

If an application requires maintenance and retrieval of lots of data, a DBMS offers many advantages over manual record keeping. Data may be conveniently captured via electronic interfaces (e.g., screen forms), then quickly processed and stored compactly on disk. Many data errors can be detected automatically, and access rights to data can be enforced by the system. People can spend more time on creative design rather than on routine tasks more suited to computers. Finally, developing and documenting application software can be facilitated by the use of *computer-assisted software engineering* (CASE) tool support.

In terms of the dominant employment group, the Agricultural Age was supplanted late in the 19th century by the Industrial Age, which has now been replaced by the Information Age. With the ongoing information explosion and mechanization of industry, the proportion of information workers is steadily rising. Most businesses achieve significant productivity gains by exploiting information technology. Imagine how long a newspaper firm would last if it returned to the methods used before word processing and computerized typesetting. Apart from its enabling employment opportunities, the ability to interact efficiently with information systems empowers us to exploit their information content.

Although most employees need to be familiar with information technology, there are vast differences in the amount and complexity of information management tasks required of these workers. Originally, most technical computer work was performed by computer specialists such as programmers and systems analysts. However, the advent of user-friendly software and powerful, inexpensive personal computers led to a redistribution of computing power. End users now commonly perform many information management tasks, such as spreadsheeting, with minimal reliance on professional computer experts.

This trend toward more users "driving" their own computer systems rather than relying on expert "chauffeurs" does not eliminate the need for computer specialists. There is still a need for programming in languages such as C# and Java. However, there is an increasing demand for high level skills such as modeling complex information systems.

The area of *information systems engineering* includes subdisciplines such as requirements analysis, database design, user interface design, and report writing. All these subareas deal with information. The database design phase selects the underlying structures to capture the relevant information, so it is of central importance.

To highlight the **need for good database design**, consider the task of designing a database to store movie details such as those shown in Table 1.1. The *header* of this table is shaded to help distinguish it from the *rows of data*. Even if the header is not shaded, we do not count it as a table row. The first row of data is fictitious.

Table 1.1 An output report about some motion pictures.

Movie#	*Movie Title*	*Released*	*Director*	*Stars*
1	Cosmology	2006	Lee Lafferty	
2	Kung Fu Hustle	2004	Stephen Chow	Stephen Chow
3	The Secret Garden	1987	Alan Grint	Gennie James Barret Oliver
4	The Secret Garden	1993	Agnieszka Holland	Kate Maberly Heydon Prowse
5	Crater	2023	Kyle Patrick Alvarez	Isaiah Russel-Bailey Mckenna Grace Billy Barratt

Different movies may have the same title (e.g., *The Secret Garden*). Hence, movie numbers are used to provide a simple identifier. We interpret the data in terms of facts. For example, movie 5 has the title *Crater*, was released in 2023, was directed by Kyle Patrick Alvarez, and starred Isaiah Russell-Bailey, Mckenna Grace, and Billy Barratt. Movie 1, titled *Cosmology*, had no stars (it is a documentary). This table is an *output report*. It provides one way to view the data. This might not be the same as how the data is actually stored in a database.

In Table 1.1, each *cell* (row-column slot) may contain many values. For example, Movie 3 has two stars recorded in the row 3, column 5 cell. Some databases allow a cell to contain many values like this, but in a *relational* database, each table cell may hold at most one value. Because relational database systems are dominant in the industry, our implementation discussion focuses on them. How can we design a relational database to store these facts?

Suppose we use the structure shown in Table 1.2, with one entry in each cell. Here, "?" denotes a *null* (no star is recorded for *Cosmology*). Some DBMSs display nulls differently (e.g., "<NULL>" or a blank space). To help distinguish the rows, we included lines between them. But from now on, we'll omit lines between rows.

Table 1.2 A badly designed relational database table.

Movie#	*Movie Title*	*Released*	*Director*	*Stars*
1	Cosmology	2006	Lee Lafferty	?
2	Kung Fu Hustle	2004	Stephen Chow	Stephen Chow
3	The Secret Garden	1987	Alan Grint	Gennie James
3	The Secret Garden	1987	Alan Grint	Barret Oliver
4	The Secret Garden	1993	Agnieszka Holland	Kate Maberly
4	The Secret Garden	1993	Agnieszka Holland	Heydon Prowse
5	Crater	2023	Kyle Patrick Alvarez	Isaiah Russell-Bailey
5	Crater	2023	Kyle Patrick Alvarez	Mckenna Grace
5	Crater	2023	Kyle Patrick Alvarez	Billy Barratt

Each relational table must be named. Here, we called the table "Movie". See if you can spot the problem with this design before reading on.

The table contains *redundant* information. For example, the facts that movie 5 is titled *Crater*, was released in 2023, and was directed by Kyle Patrick Alvarez are shown three times (once for each star). We might try to fix this by deleting the extra copies in the movieTitle, releaseYr, and director columns, but this artificially makes some rows special and introduces problems with nulls.

In addition to wasting space, Table 1.2 design can lead to errors. For example, there is nothing to stop us adding a row for movie 2 with a different title (e.g., *Kung Fun*), a different release year, a different director, and another star. Our database would then be inconsistent with the business domain, where a movie has only one title and release year, and only one director is to be recorded.[a]

The correct design uses two relational tables (see Fig. 1.1). The table design is shown in schematic form above the populated tables. In this example, a movie may be identified either by its movie number or by the combination of its title, release year, and director. In database terminology, each of these identifiers provides a *candidate key* for the Movie table, shown here by underlining each identifier.

[a]In rare cases, a movie may have multiple directors, but in this business domain we are interested in only one director per movie.

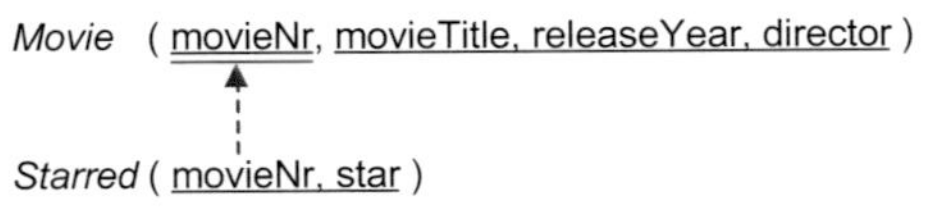

Movie:

movieNr	*movieTitle*	*releaseYear*	*director*
1	Cosmology	2006	Lee Lafferty
2	Kung Fu Hustle	2004	Stephen Chow
3	The Secret Garden	1987	Alan Grint
4	The Secret Garden	1993	Agnieszka Holland
5	Crater	2023	Kyle Patrick Alvarez

Starred:

movieNr	*star*
2	Stephen Chow
3	Gennie James
3	Barret Oliver
4	Kate Maberly
4	Heydon Prowse
5	Isaiah Russell-Bailey
5	Mckenna Grace
5	Billy Barratt

FIG. 1.1 A relational database representation of Table 1.1.

In this case, we chose movieNr as the primary way to identify movies throughout the database. This is shown here by doubly underlining the movieNr column to indicate that it is the *primary key* of the Movie table. If a table has only one candidate key, a single underline denotes the primary key.

The constraints that each movie has only one title, release year, and director are enforced by checking that each movie number occurs only once in the Movie table. The constraints that each movie must have a title, release year, and director are enforced by checking that all movies occur in the Movie table and excluding nulls from the title, release year, and director columns. In the schema, this is captured by the dotted arrow (indicating that if a movie is listed in the Starred table, it must be listed in the Movie table) and by not marking any columns as optional. In relational database terms, this arrow depicts a *foreign key constraint*, where the movieNr column in the Starred table is a foreign key referencing the primary key of the Movie table. The primary key of the Starred table is the combination of its columns, indicated here by underlining.

These concepts and notations are fully explained later in the book. Even with this simple example, care is needed for database design. With complex cases, the design problem is much more challenging. The rest of this book is largely concerned with helping you to meet such challenges.

Designing databases is both a science and an art. When supported by a good method, this design process is a stimulating and intellectually satisfying activity, with tangible benefits gained from the quality of the database applications produced. The next section explains why object-role modeling (ORM) is chosen as our first modeling method. Later sections provide historical background and highlight the essential communication skills. The chapter concludes with a summary and a supplementary note section, including references for further reading.

1.2 Information Modeling Approaches

When we design a database for a particular business domain, we create a *model* of it. Technically, the business domain being modeled is called the **universe of discourse** (UoD) because it is the universe (or world) that we are interested in discoursing (or talking) about. The UoD or business domain is typically "part" of the "real world". To build a good model requires a good understanding of the world we are modeling, and hence is a task ideally suited to people rather than machines. The main challenge is to describe the UoD clearly and precisely. Great care is required here because errors introduced here filter through to later stages in software development. The later the errors are detected, the more expensive they are to remove.

A person who models the UoD is called a *modeler*. If we are familiar with the business domain, we may do the modeling ourselves. If not, we should consult with others who, at least collectively, understand the business domain. These people are called *domain experts* or *subject matter experts*. Modeling is a collaborative activity between the modeler and the domain expert.

Because people naturally communicate (to themselves or others) with words, pictures, and examples, the best way to arrive at a clear description of the UoD is to use *natural language, intuitive diagrams*, and *examples*. To simplify the modeling task, we examine the information in the smallest units possible: *one fact at a time*.

The model should first be expressed at the *conceptual* level, in concepts that people find easy to work with. Fig. 1.1 depicted a model in terms of relational database structures. This is too far removed from natural language to be called conceptual. Instead, relational database structures are at the level of a *logical* data model. Other logical data models exist (e.g., network, XML schema, datalog, and object-oriented approaches), and each DBMS is aligned with at least one of these. However, in specifying a draft conceptual design, the modeler should be free of implementation concerns. It is a hard enough job already to develop an accurate model of the UoD without having to worry at the same time about how to translate the model into data structures specific to a chosen DBMS.

Implementation concerns are important, but they should be ignored in the early stages of modeling. Once an initial conceptual design is created, it can be mapped down to a logical design in any data model we like. This makes it easier to implement and maintain the same application on more than one kind of DBMS.

Although most applications involve processes as well as data, we'll focus first on the data because this perspective is more stable, and processes depend on the underlying data. Various information modeling approaches are discussed, such as entity-relationship (ER) modeling, fact-oriented modeling, and object-oriented modeling.

Any modeling method comprises a notation as well as a procedure for using the notation to construct models. To seed the data model in a scientific way, we need examples of the kinds of data that the system is expected to manage. We call these examples *data use cases* because they are cases of data being used by the system. They can be output reports, input screens, or forms and can present information in many ways (tables, forms, graphs, etc.). Such examples may already exist as manual or computer records. Sometimes the application is brand new, or an improved solution or adaptation is required. If needed, the modeler constructs new examples by discussing the application area with the domain expert.

As an example, suppose our information system is required to output room schedules like that shown in Table 1.3. Let's look at some different approaches to modeling this. It is not important that you understand details of the different approaches at this stage. The concepts are fully explained in later chapters.

Table 1.3 A simple data use case for room scheduling.

Room	*Time*	*Activity Code*	*Activity Name*
20	Mon 9 a.m.	ORC	ORM class
20	Tue 2 p.m.	ORC	ORM class
33	Mon 9 a.m.	XQC	XQuery class
33	Fri 5 p.m.	STP	Staff party
...	...	...	...

ER modeling was introduced by Peter Chen in 1976 and is still the most widely used approach for data modeling. It pictures the world in terms of entities that have attributes and participate in relationships. Over time, many versions of ER arose. There is no single, standard ER notation.

Different versions of ER may support different concepts and use different symbols for the same concept. Fig. 1.2 uses a popular ER notation long supported by CASE tools from Oracle Corporation. Here, entity types are shown as named, soft rectangles (rounded corners). Attributes are listed below the entity type names. An octothorpe "#" indicates that the attribute is a component of the primary identifier for the entity type, and an asterisk "*" means that the attribute is mandatory. Here, an ellipsis "..." indicates that other attributes exist but their display is suppressed.

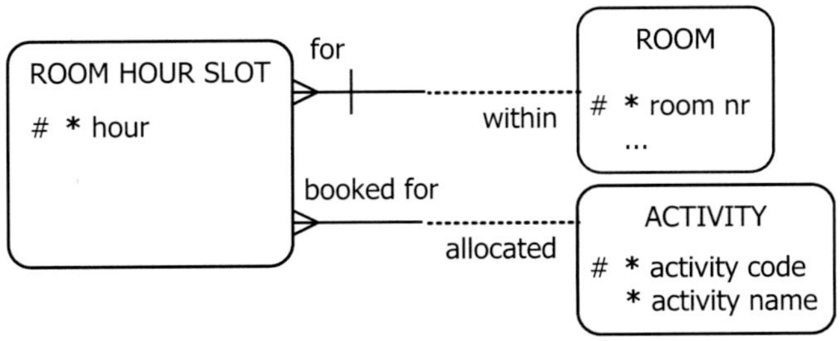

FIG. 1.2 An ER diagram for room scheduling.

Relationships are depicted as named lines connecting entity types. Only binary relationships are allowed, and each half of the relationship is shown either as a solid line (mandatory) or as a broken line (optional). For example, each RoomHourSlot must have a Room, but it is optional whether a Room is involved in a RoomHourSlot. A bar across one end of a relationship indicates that the relationship is a component of the primary identifier for the entity type at that end. For example, RoomHourSlot is identified by combining its hour and room. Room is identified by its room number, and Activity by its activity code.

A fork or "crow's foot" at one end of a relationship indicates that many instances of the entity type at that end may be associated (via that relationship) with the same entity instance at the other end of the relationship. The lack of a crow's foot indicates that at most one entity instance at that end is associated with any given entity instance at the other end. For example, an Activity may be allocated many RoomHourSlots, but each RoomHourSlot is booked for at most one Activity.

To its credit, this ER diagram portrays the domain in a way that is independent of the target software platform. For example, classifying a relationship end as mandatory is a conceptual issue. There is no attempt to specify here how this constraint is implemented (e.g., using mandatory columns, foreign key references, or object references). However, the ER diagram is incomplete (can you spot any missing constraints?). Moreover, the move from the data use case to the model is not obvious. While an experienced ER modeler might immediately see that an entity type is required for RoomHourSlot, this step might be challenging to a novice modeler.

Let's see if fact-oriented modeling can provide some help. Our treatment of fact-orientation focuses on **ORM**. ORM began in the early 1970s as a semantic modeling approach that views the world simply in terms of *objects* (things) playing *roles* (parts in relationships). For example, you are now playing the role of reading this book, and the book is playing the role of being read. ORM has appeared in a variety of forms such as the natural-language information analysis method (NIAM). The version discussed in this book is based on extensions to NIAM and is supported by industrial software tools.

Regardless of how data use cases appear, a domain expert familiar with their meaning should be able to verbalize their information content in natural language sentences. It is the modeler's responsibility to transform that informal verbalization into a formal yet natural verbalization that is clearly understood by the domain expert. These two verbalizations, one by the domain expert transformed into one by the modeler, comprise steps 1a and 1b of ORM's conceptual analysis procedure. Here, we verbalize sample data as fact instances that are then abstracted to fact types. Constraints and perhaps derivation rules are then added, and themselves validated by verbalization and sample fact populations.

To get a feeling of how this works in ORM, suppose that our system is required to output reports like Table 1.3. We ask the domain expert to read off the information contained in the table, and then we rephrase this in formal English. For example, the subject matter expert might express the facts on the top row of the table as follows: Room 20 at 9 a.m. Monday is booked for the activity "ORC", which has the name "ORM class".

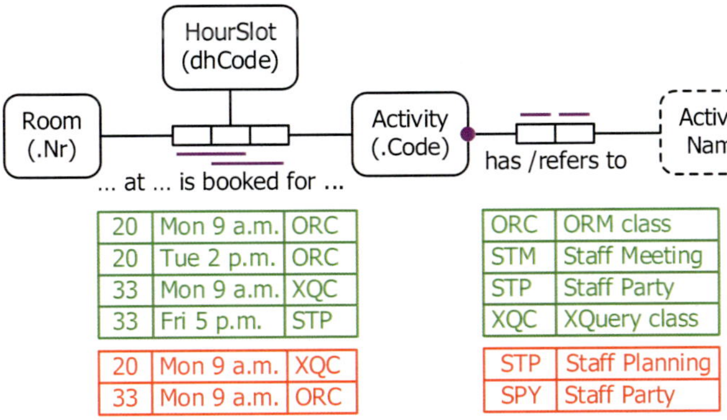

FIG. 1.3 An ORM diagram for room scheduling, with sample and counter data.

As modelers, we rephrase this into two elementary sentences, identifying each object by a definite description: the Room numbered "20" at the HourSlot with day-hour-code "Mon 9 a.m." is booked for the Activity coded "ORC"; the Activity coded "ORC" has the ActivityName "ORM class". Once the domain expert agrees with this verbalization, we abstract from the fact instances to the *fact types* (i.e., the types or kinds of fact). We might then depict this structure on an ORM diagram and populate it with sample data and counter data (explained shortly), as shown in Fig. 1.3.

By default, *entity types* are shown in ORM as named, soft rectangles (rounded corners) and must have a reference scheme, that is, a way for humans to refer to instances of that type. Simple reference schemes may be shown in parentheses (e.g., "(.Nr)" or "(.Nr)"), as an abbreviation of the relevant association, that is, Room has RoomNr. *Value types* such as types of character strings need no reference scheme and are shown as named, dashed, soft rectangles (e.g., ActivityName).

This book uses the notation of *ORM 2* (second-generation ORM), as supported by the natural ORM architect (NORMA) tool, an open-source plug-in to Microsoft VisualStudio. NET. A previous version of ORM supported by Microsoft Visio for Enterprise Architects depicts object types as ellipses, not soft rectangles.

Unless indicated otherwise, in this book the term "ORM" is understood to mean ORM 2. When specific reference is made to the previous version of ORM, the term "ORM 1" is used. The ORM glossary at the end of this book includes a side-by-side comparison of ORM 1 and ORM 2 notations.

In ORM, a *role* is a part played in a *fact type* (relationship or association). A relationship is shown as a named sequence of one or more role boxes, each connected to the object type whose instances play that role. Fig. 1.3 includes a *ternary* (three-role) association, Room at HourSlot is booked for Activity, and a *binary* (two-role) association Activity has ActivityName.

Unlike ER, ORM makes no use of attributes in its base models. All facts are represented in terms of objects (entities or values) playing roles. Although this often leads to larger diagrams, an attribute-free approach has advantages for conceptual analysis, including simplicity, stability, and ease of validation. If you are used to modeling in ER or the unified modeling language (UML) (see later), this approach may seem strange at first, but please keep an open mind about it.

ORM allows relationships of any *arity* (number of roles), so *unary* (single role) relationships are allowed. Each fact type has at least one predicate reading, corresponding to one way of traversing its roles. Any number of readings may be provided for each role ordering. For a binary association, forward and inverse predicate readings may be shown separated by a slash "/". As in logic, a predicate is a sentence with object holes in it.

Mixfix notation enables the object terms to be mixed in with the predicate reading at various positions (as required in languages such as Japanese). An object placeholder is indicated by an ellipsis "…" (e.g., the ternary predicate "… at … is booked for …"). For unary postfix predicates (e.g., "… smokes") or binary infix predicates (e.g., "… has …"), the ellipses may be omitted.

For each fact type, a *fact table* may be added with a sample population to help validate the constraints. Each column in a fact table is associated with one role. The lines beside the role boxes depict internal *uniqueness constraints*, indicating which roles or role combinations must have unique entries.

ORM schemas may be represented in diagrammatic or textual form, and some ORM tools can automatically transform between the two representations. Models are *validated* with domain experts in two main ways: *verbalization* and *population*.

For example, the uniqueness constraints on the ternary association verbalize as: **For each** Room **and** HourSlot, **that** Room at **that** HourSlot is booked for **at most one** Activity; **For each** HourSlot **and** Activity, **at most one** Room at **that** HourSlot is booked for **that** Activity.

The ternary fact table has a satisfying population (each Room-HourSlot combination and each HourSlot-Activity combination is unique). The uniqueness constraints on the binary verbalize as: **Each** Activity has **at most one** ActivityName; **Each** ActivityName refers to **at most one** Activity. The 1:1 nature of this association is illustrated by the population, where each column entry occurs only once in its column.

The solid dot on Activity is a *mandatory role* constraint, indicating that each instance in the population of Activity must play that role. This verbalizes as **Each** Activity has **some** ActivityName. A role that is not mandatory is *optional*. Because sample data are not always significant, additional data (such as STM in the binary fact type) may be needed to illustrate some rules. The optionality of the other role played by Activity is shown by the absence of STM in its population.

Because ORM schemas can be specified in unambiguous sentences backed up by illustrative examples, it is not necessary for domain experts to understand the diagram notation at all. Modelers, however, find diagrams very useful for thinking about the UoD.

To double check a constraint, a *counterexample* to that constraint may be presented. The counter-rows appended to the fact tables test the uniqueness constraints. For instance, the first row and counter-row of the ternary indicate that room 20 at 9 a.m.

Monday is booked for both the ORC and XQC activities. This challenges the constraint "**For each** Room **and** HourSlot, **that** Room at **that** HourSlot is booked for **at most one** Activity". This constraint may be recast in negative form as: **It is impossible that the same** Room at **the same** HourSlot is booked for **more than one** Activity. The counterexample provides a test case to see if this situation is actually possible.

Concrete examples help domain experts to decide whether something really is a rule. This additional validation step is very useful in cases where the domain expert's command of language suffers from imprecise or even incorrect use of logical terms (e.g., "each", "at least", "at most", "exactly", "the same", "more than", "if").

To challenge the constraint that at most one room at the same time is booked for the same activity, the first row and second counter-row of the ternary fact table in Fig. 1.3 indicate that both room 20 and room 33 are used at 9 a.m. Monday for the ORC activity. Is this kind of thing possible? If it is (and for some application domains it would be) then this constraint is not a rule, in which case the constraint should be removed and the counter-row added to the sample data.

However, if our business does not allow two rooms to be used at the same time for the same activity, then the constraint is correct and the counterexample is rejected (although it can be retained as an illustrative counterexample).

Compare Fig. 1.2 with Fig. 1.3. While ER diagrams are often more compact than ORM diagrams, ER models are *further removed from natural language* and may be harder for the domain expert to conceptualize. In this case, it was more natural to verbalize the first schedule fact as a ternary, but all popular ER notations with industrial support are restricted to binary (two-role) relationships.

Being only binary does lose expressiveness because an n-ary association ($n > 2$) may always be transformed into binaries by coreferencing or nesting (see later). However, such a transformation may introduce an object type that appears artificial to the domain expert, which can hinder communication. If possible, we should try to formulate the model in a way that appears natural to the domain expert.

ER notation is *less expressive* than ORM for capturing constraints or business rules. For example, the ER notation used for Fig. 1.2 was unable to express the constraint that activity names are unique or the constraint that it is impossible that more than one room at the same hour slot is booked for the same activity.

ER encourages decisions about *relative importance* at the conceptual analysis stage. Sometimes this may be seen as an advantage. For example, it is fairly natural to think of activity names as attributes of activities, and hence treat names as less important than activities themselves.

Sometimes, however, early distinctions on relative importance can be disadvantageous. For example, instead of using RoomHourSlot in Fig. 1.2, we could model the information using ActivityHourSlot. Which of these choices is preferable may depend on what other kind of information we might want to record. However, because we have been forced to make a decision about this without knowing what other facts need to be recorded, we may need to change this part of the model later.

In general, if you model a feature as an attribute and find out later that you need to record something about it, you are typically forced to remodel it as an entity type or relationship because attributes can't have attributes or participate in relationships.

For instance, suppose we record phone as an attribute of Room and then later discover that we want to know which phones support voice mail. Because you rarely know what all the future information requirements will be, an attribute-based model is inherently *unstable*. Moreover, applications using the model often need to be recoded when a model feature is changed. Because ORM is essentially immune to changes like this, it offers far greater *semantic stability.*

We have already seen that ORM models facilitate validation by both verbalization and population. Attributes make it awkward to use sample data populations. Moreover, populating optional attributes introduces null values, which may be a source of confusion to nontechnical people.

While an ORM model is better for conceptual analysis and model validation, you can generate an ER view from it when desired. Even if you decide to use ER throughout, ignoring the ORM notation completely, you should find that applying or adapting the modeling steps in ORM's conceptual schema design procedure to the ER notation will help you design better ER models.

Now let's consider *object-oriented* (OO) *modeling*, an approach that encapsulates both data and behavior within objects. Although used mainly for designing object-oriented program code, it can also be used for database design.

Many object-oriented approaches exist, but the most influential is **UML**, which has been adopted by the Object Management Group (OMG).

Among its many diagram types, UML includes *class diagrams* to specify static data structures. Class diagrams may be used to specify operations as well as low-level design decisions specific to object-oriented code (e.g., attribute visibility and association navigability). When stripped of such implementation detail, UML class diagrams may be regarded as an extended version of ER.

A UML class diagram for our example is shown in Fig. 1.4. To overcome some of the problems mentioned for the ER solution, a ternary association is used for the schedule information. Because of its object-oriented focus, UML does not require conceptual identification schemes for its classes. Instead, entity instances are assumed to be identified by internal object identifiers (oids).

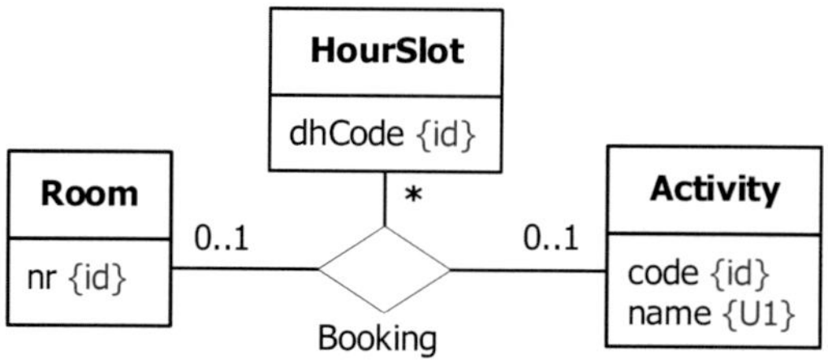

FIG. 1.4 A UML class diagram for room scheduling.

However, current versions of UML allow one to signify that an attribute value may be used as *identifier* (or part of an identifier) for entity instances in its class, by appending the attribute(s) with "{id}". This can be used in mapped implementations to denote *primary* keys in relational database tables or ID attributes in the eXtensible Markup Language (XML).

UML also allows user-defined constraints to be added in braces or notes in any language. We've added {U1} to denote an *alternate* uniqueness constraint—this symbol is not standard and hence not portable. The uniqueness constraints on the ternary are captured by the 0..1 (at most one) multiplicity constraints. Here "*" is shorthand for "0..*", meaning "0 or more". Attributes are mandatory by default.

How well does this UML model support validation with the domain expert? Let's start with verbalization. Although often less than ideal, implicit use of "has" could be used to form binary sentences from the attributes, but what about the ternary? About the best we can do is something like "Booking involves Room and HourSlot and Activity", which is of little use.

If we replaced the association name with a mixfix predicate, as we did in ORM, (e.g., "… at … is booked for …") this is no use because UML association roles (or association ends as they are now called) are not ordered. So formally we can't know if we should read the association as "Room at HourSlot is booked for Activity", or "Activity at HourSlot is booked for Room", etc. This gets worse if the same class plays more than one role in the association (e.g., Person introduced Person to Person).

UML requires association roles to have names (ORM allows role names, but does not require them), but role names don't form sentences, which are always ordered in natural language. UML's weakness for verbalization of facts applies also for verbalization of constraints and derivation rules.

The UML specification (www.omg.org/spec/UML) recommends the object constraint language (OCL) for formal expression of such rules, but OCL is simply too mathematical in nature to be used for validation by nontechnical domain experts. In principle, a higher-level language could be designed for UML that could be automatically transformed to OCL.

Because verbalization in UML has inadequate support, let's try validation with sample populations. Not much luck here either. To begin with, attribute-based notations are almost useless for multiple instantiations, and they introduce nulls into base populations, along with all their confusing properties.

UML supports object diagrams to enable talk about single instances of classes, but that doesn't help with multiple instantiations. For example, the 1:1 nature of the association between activity codes and names is transparent in the ORM fact table in Fig. 1.3, but is harder to see by scanning several activity objects.

In principle, we could introduce fact tables to instantiate binary associations in UML, but this won't work for nonbinary associations. Why not? In UML, you can't specify a reading direction for an association unless it's a binary. So there is no obvious connection between an association role and a fact column as there is in ORM.

The best we can do is to name each role and then use role names as headers to the fact table. However, the visual connection of the fact columns to the class diagram would be weak because of the nonlinear layout of the association roles, and the higher the arity of the association, the worse it gets.

In its favor, UML is far richer than ORM or ER in its ability to capture other aspects of application design (e.g., operations, activities, component packaging, and deployment). UML includes diagrams, such as state machine and activity diagrams, to capture business processes. Any full specification of a business domain needs to address these dynamic aspects. If the application is to be implemented in object-oriented code, UML enables more precise descriptions of the programming code structures to be specified (e.g., attribute visibility and association navigability).

If we restrict our attention to conceptual data modeling, however, the ORM notation is much richer than ER or UML in its capacity to express business constraints on the data as well as being far more orthogonal and less impacted by change. As a simple example, consider the output report of Table 1.4 (the reviewers and movie 3 are fictitious). You might like to try modeling this yourself before reading on.

Table 1.4 Another sample output report about Movies.

Movie		*Director*		*Reviewers*	
Nr	*Title*	*Name*	*Born*	*Name*	*Born*
1	Crater	Kyle Patrick Alvarez	US	Fred Bloggs Ann Green	US US
2	Crocodile Dundee	Peter Faiman	AU	Ann Green Ima Viewer Tom Sawme	US GB AU
3	Star Peace	Ann Green	US	?	?
...	...	...	...	...	...

One way to model this report in UML is shown in Fig. 1.5. Although the population of the sample report suggests that movie titles are unique and that a person can direct only one movie, let's assume that the domain expert confirms that this is not the case. We should adapt our sample population to illustrate this (e.g., add a new movie 4 with the same title "Star Peace" directed by Peter Faiman).

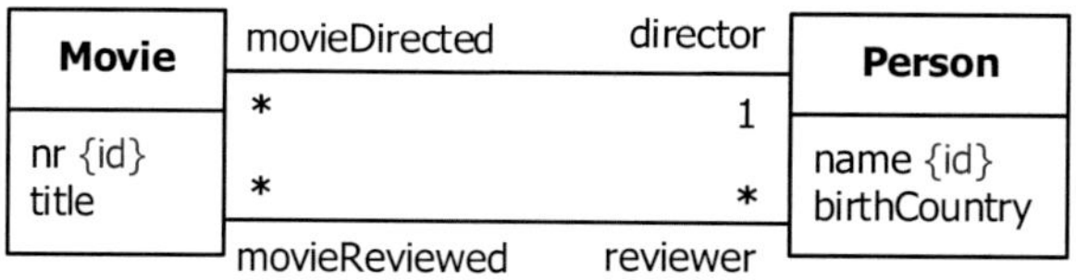

FIG. 1.5 A UML class diagram for Table 1.4.

Assuming people are identified simply by their name, Movie and Person classes may be displayed as shown. The role names "director" and "reviewer" are used here to distinguish the two roles played by Person. Similarly, role names are provided to distinguish the roles played by Movie. In this example, all four role names are required. Association names may be used as well if desired.

Unlike Chen's original ER notation, UML binary associations are typically depicted by lines without a diamond. While this is convenient, the use of diamonds in UML for longer associations is somewhat inconsistent, and the avoidance of unary relationships is unnatural. In principle, UML does allow diamonds as an alternative notation for binary associations, but in practice this is rarely seen.

In contrast, ORM's depiction of relationships as a sequence of one or more roles, where each role is associated with a fact table column, provides a uniform, general notation that facilitates validation by both verbalization and sample populations.

The multiplicity constraints indicate that each movie has exactly one director but may have many reviewers and that a person may direct or review many movies. But there is still a missing business rule. Can you spot it?

Fig. 1.6 models the same domain in ORM. Here the "◄" before "has" reverses the normal left-to-right reading direction. The rule missing from the UML model is captured graphically by the circled "X" constraint between the role-pairs comprising the "directed" and "reviewed" associations. This is called an *exclusion constraint.* The directed and reviewed fact types are populated with sample data that violate the exclusion constraint—a *counterexample* to that constraint.

This exclusion constraint verbalizes as **No** Person directed **and** reviewed **the same** Movie or, reading it the other way, **No** Movie was directed by **and** was reviewed by **the same** Person. To validate this rule with the domain expert, you should verbalize the rule and also provide a counterexample. For example, in your model is it possible for Movie 1 to be directed by Ann Green and also reviewed by Ann Green? Fig. 1.6 includes this counterexample. If the exclusion constraint really does apply, at least one of those two facts must be wrong.

Some domain experts are happy to work with diagrams and some are not. Some are good at understanding rules in natural language, and some are not. But all domain experts

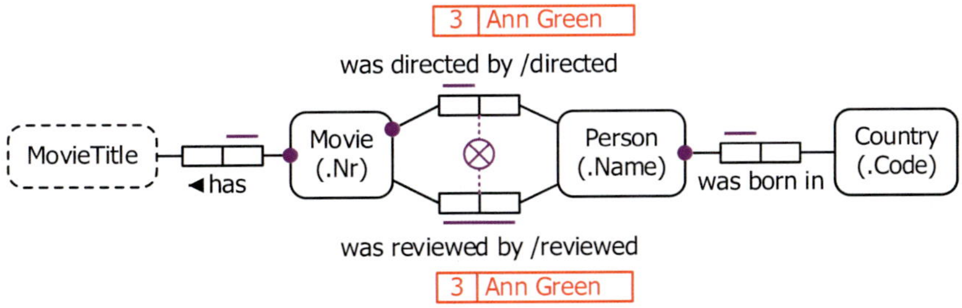

FIG. 1.6 ORM model for Table 1.4, with counterexample for the exclusion constraint.

are good at working with concrete examples. Although it is not necessary for the domain expert to see the diagram, being able to instantiate any role directly on the diagram makes it easy for you to think clearly about the rules.

UML has no graphic notation for general exclusion constraints, but does allow you to document constraints in a note attached to the relevant model elements. If a concept is already part of your modeling language, it's easier to think of it.

Because the exclusion constraint notation is not built-in to the UML language, it is easy to miss the constraint in developing the model. The same thing goes for ER. In contrast, the ORM modeling procedure prompts you to consider such a constraint and allows you to visualize and capture the rule formally. An ORM tool can then map the constraint automatically into executable code to ensure that the rule is enforced in the implementation.

ORM diagrams always display *semantic domains* as object types. The ORM diagram in Fig. 1.7(a) includes *role names* for birthdate and deathdate, shown in square brackets next to the relevant roles. These roles are clearly compatible, as they are both played by the object type Date. In ORM, role names may be used like attribute names in automatically generated attribute-views as well as in rules specified in attribute style (e.g., deathdate > birthdate).

ER diagrams typically hide attribute domains. For example, the birthdate and deathdate attributes in the Barker ER model shown in Fig. 1.7(b) should be based on the domain Date, but this is not represented visually. In ER, attribute domains can be listed in another document.

In UML class diagrams, attribute domains may be listed after the attribute name and multiplicity (if shown), as in Fig. 1.7(c). The "[0..1]" multiplicity indicates "at most one", so the attribute is optional and single-valued. However all too often in practice, only *syntactic*, or value, domains are specified (e.g., String).

An ER diagram might show population and elevation as attributes of City, and an associated table might list the domains of these attributes simply as Integer, despite the fact that it is nonsense to equate a population with an elevation.

Conceptual object types, or semantic domains, provide the conceptual "glue" that binds the various components in the application model into a coherent picture. Even at the lower level of the relational data model, E.F. Codd, the founder of the relational model, argues that "domains are the glue that holds a relational database together" (Codd, 1990, p. 45).

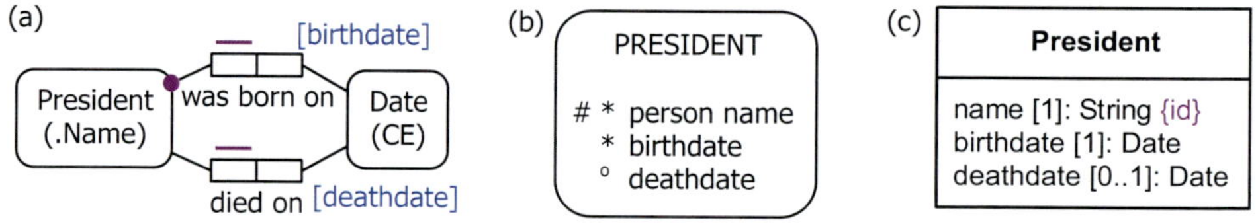

FIG. 1.7 ORM object types are semantic domains.

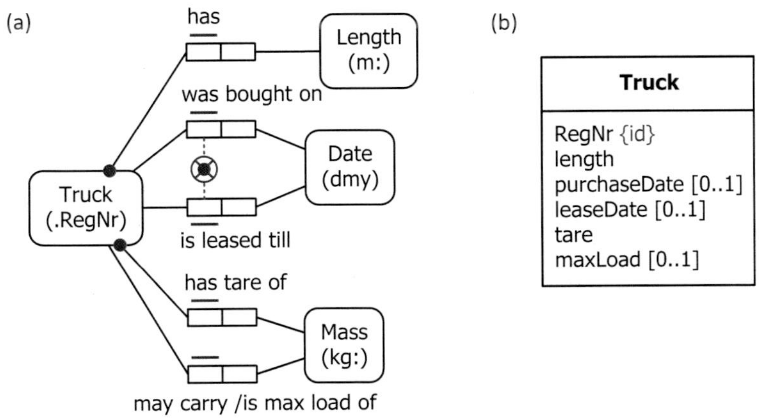

FIG. 1.8 (a) ORM model. (b) UML model, revealing less detail.

The object types in ORM diagrams are the semantic domains, so the *connectedness* of a model is transparent. This property of ORM also has significant advantages for conceptual queries because a user can query the conceptual model directly by navigating through its object types to establish the relevant connections. This notion is elaborated further in later chapters.

ER and UML diagrams often fail to express relevant *constraints on, or between, attributes.* Fig. 1.8 provides a simple example. Notice the circled dot over an "X" in the ORM model in Fig. 1.8(a). This specifies two constraints: the dot is a mandatory constraint over the disjunction of the two roles (each truck is either bought or leased) and the "X" indicates the roles are exclusive (no truck is both bought and leased). The two constraints collectively provide an *exclusive-or (xor) constraint* (each truck plays exactly one of the roles).

Unlike most versions of ER, UML does provide an xor constraint, but only between associations. Because the UML model in Fig. 1.8(b) models these two fact types as attributes instead of associations, it cannot capture the constraint graphically (other than adding a note). Notice again how the ORM diagram reveals the semantic domains. For instance, tare may be meaningfully compared with maximum load (both are masses) but not with length. In UML, this can be made explicit by appending domain names to the attributes. At various stages in the modeling process, it is helpful for the modeler to see all the relevant information in one place.

Another ORM feature is its flexible support for *subtyping,* including multiple inheritance, based on formal subtype definitions. For example, the subtype LargeUSCity may be defined as a City that is in Country "US" and has a Population > 1,000,000. As discussed in a later chapter, subtype definitions provide stronger constraints than declarations about whether subtypes are exclusive or exhaustive.

In principle, there are infinitely many kinds of constraints, so a textual constraint language may be required for completeness to supplement the diagram. This is true for ER,

ORM, and UML. However, the failure of ER and UML diagrams to include standard notations for many important ORM constraints makes it harder to develop a comprehensive model or to perform *transformations* on the model.

For example, suppose that in a movie an actor may have a starring role or a supporting role but not both. This can be modeled by two fact types: *Actor* has starring role in *Movie*; *Actor* has supporting role in *Movie*. The "but not both" condition is expressed in ORM as a pair-exclusion constraint between the fact types. Alternatively, these fact types may be replaced by a ternary: *Actor* in *Movie* has role of *RoleKind* {star, support}.

Transformations are rigorously controlled in ORM to ensure that constraints in one representation are captured in an alternative representation. For instance, the pair-exclusion constraint is transformed into the constraint that each Actor-Movie pair has only one RoleKind. The formal theory behind such transformations is easier to apply when the relevant constraints can be visualized.

Unlike UML and ER, ORM was built from a linguistic basis. To reap the benefits of verbalization and population for communication with and validation by domain experts, it's better to use a language that was designed with this in mind. The ORM notation is easy to learn and has been successfully taught to high school students.

We are not arguing here that ER and UML have no value. They do. We are simply suggesting that you consider using ORM's modeling techniques, and possibly its graphic notation, to facilitate your original conceptual analysis before using an attribute-based notation such as that of ER, UML, or relational tables.

After validating the conceptual model with the domain expert, you map it to a target such as a DBMS or program code for implementation. At this lower level, you may prefer an attribute-based model to give a compact picture of how facts are grouped into implementation structures. For database applications, you will want to see the table structures, foreign key relationships, and so on. Here, a relational or object-relational model offers a compact view, similar to an ER or UML model.

ORM models often take up much more space than an attribute-based model because they show each attribute as a relationship. This is ideal for conceptual analysis, where we should validate one fact type at a time. However, for logical design, we typically group facts into attribute-based structures such as tables or classes. At the logical design stage, attribute-based models are more useful than ORM models. For example, relational schema diagrams provide a simple, compact picture of the underlying tables and foreign key constraints between them. Also, UML is well suited for the logical and physical design of object-oriented code because it allows implementation detail on the data model (e.g., attribute visibility and association navigation) and can be used to model behavior and deployment.

Having used ER, ORM, and UML in practice, we've found that ORM often makes it easier to get the model right in the first place and to change the model as the business domain evolves. We believe in the method so strongly that we've made it the basis for much of the modeling discussion in this book. Once you understand ORM's modeling principles, you'll find it much easier to gain a proper understanding of data modeling in ER and UML.

Arguments about modeling approaches can become heated, and not everyone is as convinced of the virtues of ORM as we are. All we ask is that you look objectively at the ideas presented in this book and consider using whatever you find helpful.

Although the book focuses on ORM, it also covers data modeling in other popular notations (e.g., ER, IDEF1X, UML, and relational). These other notations have value too. Even if you decide to stay with ER or UML as your conceptual analysis approach, insight into ORM should make you a better modeler regardless.

1.3 Historical Background

This section briefly overviews the evolution of computing languages for information systems and outlines the historical development of the main kinds of logical data structures used in database systems. We begin with a simple example to illustrate how the level of a language impacts how easy it is to formulate questions.

Table 1.5 summarizes how **five generations of computing languages** might be used to request a computer to list the name, mass, and moons (if any) of each planet, assuming the information is stored in an astronomical database. The higher the generation, the closer to natural language, and usually the less you have to say. Nowadays, nobody uses machine code or assembler to access databases. Most database applications are coded using fourth-generation languages (4GLs), perhaps in combination with third-generation languages (3GLs).

Table 1.5 Five generations of computer languages.

Generation	*Language example*	*Sample code for same task*
5	ConQuer	✓Planet **that** has ✓Mass **and possibly** is orbited by ✓Moon
4	SQL	**select** X1.planetName, X1.mass, X2.moonName **from** Planet **as** X1 **left outer join** Moon **as** X2 **on** X1.planetName = X2.planetName
3	Pascal	Two pages of instructions like: **for** i := 1 **to** n **do begin** write (planetName[i], mass[i]);
2	8086 Assembler	Many pages of instructions like: ADDI AX, 1
1	8086 machine code	Many pages of instructions like: 00000101 00000001 00000000

Third-generation languages, such as C# and Java, are *procedural*, emphasizing the procedures used to carry out the task. With 3GLs, we typically need to specify how to access data one record at a time. Fourth-generation languages, such as SQL, are primarily *declarative* in nature: one declares *what* is to be done rather than *how* to do it. With a 4GL, a single statement can perform operations on whole tables, or sets of rows, at once. Hence, 4GLs are *set oriented* rather than record oriented.

Fifth-generation languages (5GLs), such as ConQuer (an ORM query language), allow you to specify queries naturally, without knowing the underlying data structures used to store the information. The widespread use of fifth-generation languages is still in the future.

The first DBMSs were developed in the early 1960s, starting with simple file managers. Various logical data architectures have been proposed as a basis for specifying the structure of databases. In the hierarchic data model, the database schema is basically a tree of linked record types, where each record type has a different structure (unlike many trees where each node is of the same type).

Records may include one or more fields, each of which can hold only a single value. Record types are related by parent-child links (e.g., using pointers), where a parent may have many children but each child has only one parent. Hence the type structure is that of a tree, or hierarchy.

For example, in Fig. 1.9 the parent record type Department has two child record types: Product and Employee. Each record type contains a sequence of named fields, shown here as boxes, and the parent-child links are shown as connecting lines. For discussion purposes, one record instance has been added below each record type. As an exercise, try reading off all the facts that are contained in this database before reading on.

To begin with, there are five facts stored in the record instances. To make these facts more obvious, Fig. 1.9 includes arcs connecting the relevant fields, one arc for each fact. Although these arcs are a useful annotation, they are not part of the schema notation. If we are familiar with the business domain, we can verbalize these arcs into relationships. For example, we might verbalize the five facts as follows:

Department 10 is located in Building 69.
Department 10 has Budget $200,000.
Product "IS2" has ProductName "InfoStar 2".
Employee 357 has EmployeeName "Jones E".
Employee 357 is of Sex "F".

Are there more facts? Yes! The parent-child links encode the following two facts:

Department 10 develops Product "IS2".
Department 10 employs Employee 357.

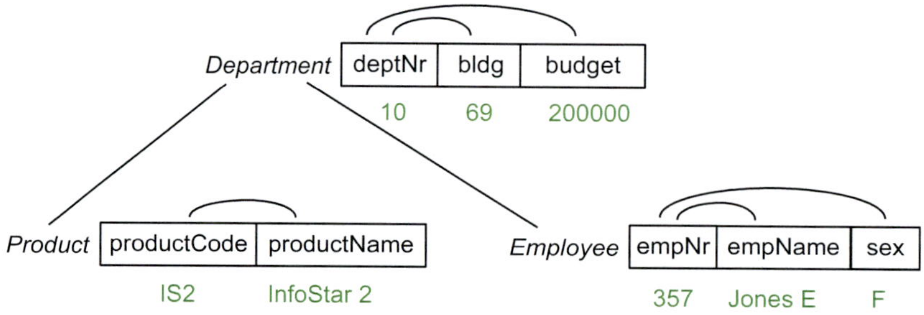

FIG. 1.9 A hierarchic database schema with sample data, and fact arcs.

Hierarchic DBMSs such as IBM's Information Management System can efficiently manage hierarchic structures (e.g., a file directory system). However, having to explicitly navigate over predefined record links to get the facts can be somewhat challenging. The complexity rises if the application is not hierarchic in nature.

Suppose that the same product may be developed by more than one department. Conceptually, the Department develops Product association is now *many:many*. Because parent-child links are always *1:many*, a workaround is needed to handle this situation.

For example, to record the facts that departments 10 and 20 both develop product "IS2", we could have the two department record instances point to separate copies of the record instance for product "IS2".

Although the type and instance link structures are trees, the fact that product "IS2" is named "InfoStar 2" now appears twice in the database, so we need to control this redundancy. Moreover, while retrieving products developed by a given department is easy, retrieving all the departments that developed a product is not so easy.

The *network data model* was developed by the Conference on Data Systems and Languages (CODASYL) Database Task Group. This model is more complex than the hierarchic model. Most of the data is stored in records, a single field of which may contain a single value, a set of values, or even a set of value groups. Record types are related by owner-member links, and the graph of these connections may be a network: a record type may have many owners and also own many record types.

As in the hierarchic model, facts are stored either in records or as record links. An owner-member link between record types is restricted to a 1:many association. To handle a many:many association, such as the case discussed earlier, we might introduce a new record type (e.g., Development) with many:1 associations to the other record types (in this case, Department and Product).

In general, encoding of facts in access paths such as interrecord links complicates the management of the application and makes it less flexible. For example, new queries will have to wait until access paths have been added for them, and internal optimization efforts can be undone as the application structure evolves.

Partly to address such problems, Dr. Edgar "Ted" Codd introduced the *relational data model*. A year after his original 1969 IBM research report on the subject, Codd published a revised version for a wider audience (Codd, 1970) where he first argued that relations should be normalized so that each data entry would be atomic; we now call this first normal form. Other normal forms were defined later.

The most significant feature of the relational model is that *all the facts are stored in tables*, treated as mathematical relations. For example, Fig. 1.10 shows the relational database for our application. Again, the database is annotated with arcs corresponding to the facts stored. Notice the extra deptNr columns in the Employee and Product tables. The facts that Department 10 employs Employee 357 and develops product "IS2" are stored in the table rows themselves. Access paths between tables are not used to specify facts (as allowed in hierarchic or network models).

To specify queries and constraints, table columns may be associated by name. This allows ad hoc queries to be specified at will and simplifies management of the application.

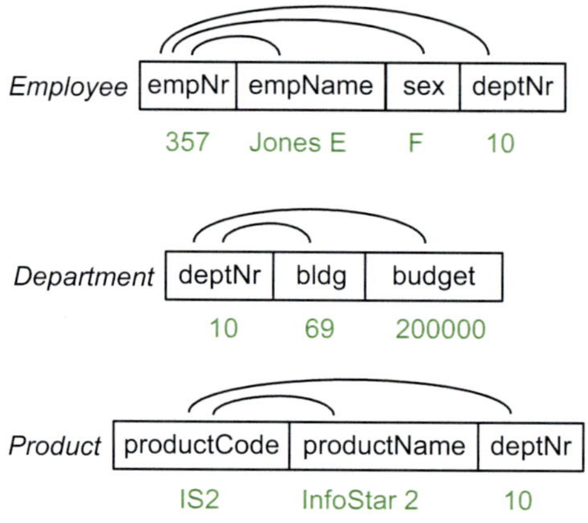

FIG. 1.10 All the facts in a relational database are stored in the tables themselves.

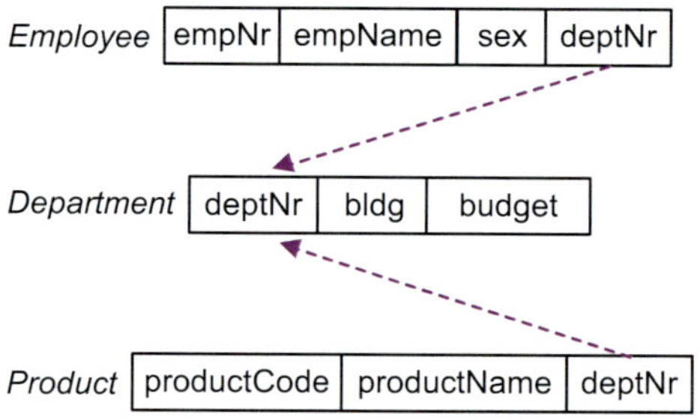

FIG. 1.11 "Links" between tables express constraints, not facts.

Note that constraints specified between tables are not the same as access paths. For example, in Fig. 1.11, arrows "link" the deptNr column of the Employee and Product tables to the deptNr column of the Department table. These "links" merely express the constraints that any value in the deptNr column of the Employee and Product tables must also occur as a value in the deptNr column of the Department table. These constraints do not express employment and product development facts.

The relational model is logically cleaner than the network and hierarchic models, but it initially had poor performance, which led to its slow acceptance. However, by the late 1980s, efficient relational systems had become commonplace. Although network and hierarchic database systems are still in use today, relational DBMSs are often the preferred choice for developing most new database applications. A DBMS should ideally provide an integrated data dictionary, dynamic optimization, data security, automatic recovery, and a user-friendly interface.

The main query languages used with relational databases are *SQL* (informally known as "structured query language") and *QBE* (query by example). Many systems support both of these. SQL has long been accepted as an international standard and is commonly used for the communication of queries between different database systems. For this reason, SQL is the main query language discussed in this book.

Although relational systems give adequate performance for most applications, they are inefficient for some applications involving complex data structures (e.g., VLSI design).

Recently the *XML* has also become widely used for communication between different systems, but this is currently focused on sharing data for purposes such as electronic commerce and web publication. Many SQL-based DBMSs now support storing of XML data as well as querying XML data directly using a query language such as XQuery.

Data architectures exist for *object-oriented databases*, *deductive databases* (such as datalog systems), and *ontological databases* in languages such as *OWL* (Web Ontology Language). Many relational systems are being enhanced with object-oriented features, leading to *object relational database systems*. *NoSQL* systems are gaining popularity for efficiently handling high volumes of less-structured data.

1.4 The Relevant Skills

Because relational database systems are still dominant, they are the main focus of our implementation discussion, with some attention also being given to XML and other systems. Although conceptual interfaces to databases are still in their infancy, developing models at this higher level is more productive and avoids wasting time acquiring knowledge and skills that will rapidly become obsolete. Consider the impact of the electronic calculator on school mathematics curricula (e.g., removal of the general square root algorithm).

Fundamentally, there are two skills that will always be relevant to interacting with an information system. Both these skills relate to *communicating* with the system about our particular application area. Recall that the application domain is technically known as the *UoD*. The two requirements are to:

- **describe** the UoD
- **query** the system about the UoD

The first skill entails describing the *structure* or *design* of the UoD and describing its *content* or *population*: the structural aspect is the only challenging part. The ability to clearly describe the application domain is critical if you wish to add a model to the system. Complex models should normally be prepared by expert modelers working with domain experts. The main aim of this text is to introduce the fundamentals of information modeling. If you master the methods discussed, you will be well on your way to becoming an expert in modeling information systems.

Issuing queries is often easy using a 4GL, but the formulation of complex queries can still be difficult. Occasionally, the ability to understand some answers given by the system requires knowledge about how the system works, especially its limitations. This book provides a conceptual basis for understanding relational structures and queries, explains the algebra behind relational query languages, and includes a solid introduction to SQL.

No matter how sophisticated the information system, if we give it the wrong picture of our UoD, we can't expect to get much sense out of it. This is one aspect of the garbage in garbage out (GIGO) principle. Most problems with database applications result from bad database design. This book shows how to model information at a very high level using natural concepts. While providing due attention to popular data modeling approaches such as ER and UML, it also provides an in-depth treatment of the higher-level ORM approach.

Conceptual schemas can be mapped onto the lower-level structures used by today's database systems. This mapping can be performed automatically using an appropriate CASE tool or manually using an appropriate procedure. This book discusses how to map a conceptual model to a relational database system and some other structures (e.g., XML schema), and how query languages may be used to retrieve data from such systems. It also discusses other methods and modern trends.

1.5 Summary

This chapter provided a motivation for conceptual modeling and presented a brief historical and structural overview of information systems. DBMSs are widely used and are a major productivity tool for information-oriented businesses. For a database to be used effectively, its data should be correct, complete, and efficiently accessed. This requires that the database be well-designed.

Designing a database involves building a formal model of the business domain or UoD. To do this properly requires a good understanding of the UoD and a means of specifying this understanding in a clear, unambiguous way. ORM simplifies the analysis and design process by using natural language, intuitive diagrams, and examples, and by examining the information in terms of simple, *elementary facts*. By expressing the model in terms of natural concepts, such as *objects* and *roles*, this fact-oriented method provides a truly *conceptual* approach to modeling.

Other valuable modeling approaches include ER modeling and object-oriented modeling. In practice, ER is still the most popular, high-level approach to the design of databases. While many popular versions of ER exist, UML is by far the most influential object-oriented approach.

Although ER and UML models are typically more compact than ORM models, they are arguably less suitable than ORM for formulating, transforming, or evolving a conceptual information model. ER models and UML class diagrams are further removed from natural language, lack the expressibility and simplicity of a role-based constraint notation, are less stable in the face of domain evolution, are harder to populate with fact instances, and may hide information about the semantic domains that glue the model together. However, ER and UML models better highlight the major features of the domain being modeled by representing currently less-important features as attributes.

In this book, ORM is used as our basic conceptual modeling method. ER models and UML class diagrams are useful as well, especially for providing compact summaries, and are best developed as views of ORM models. For database applications, conceptual models typically need to be mapped to attribute-based logical and physical models. ER models provide designs that are closer to relational database structures. For object-oriented applications, UML models can incorporate implementation details as well as behavior and deployment aspects not covered by the ORM and ER approaches.

Programming tasks are typically coded in third-generation languages such as C# and Java. Fourth-generation database languages such as SQL are declarative in nature, enabling users to declare what is to be done without the fine detail of how to do it, and are set oriented rather than record oriented.

Fifth-generation languages such as ConQuer enable users to query conceptual models directly. Hierarchic and network database systems store some facts in record types and some facts in links between record types. Relational database systems store all facts in tables.

Various postrelational systems such as object-oriented databases, object-relational databases, deductive databases, ontological databases, and NoSQL databases are gaining acceptance in industry.

No matter how "intelligent" software systems become, people are still needed to describe the UoD and to ask the relevant questions about it.

Chapter Notes

Full bibliographic entries for references are included in the bibliography at the back of this book. Codd (1969, 1970) introduced the relational model of data. For a historical discussion of these two papers, see Date (1998). Codd (1990) suggests future directions for the relational model. The classic paper that introduced ER modeling is Chen (1976). Kent (2000) is a reprint of a classic book that provides many insights into the nature of information and data.

Many papers on ORM are accessible at www.orm.net and at www.ormfoundation.org. Simsion and Witt (2005) provide a readable coverage of various data modeling topics. An overview of UML can be found in Booch et al. (1999). The latest UML specification is accessible at www.omg.org/spec/UML. Halpin (2015a) provides a comparative review of information modeling in UML and ORM.

Date (2000) provides a clear introduction to most aspects of database systems. Further background on network and hierarchic DBMSs is included in Appendices C–E of Connolly et al. (1999). Gorman (1998) provides a structural review of historical data models and argues that the modern SQL standard is a data model in its own right that is closer to some prerelational models than it is to the relational model. The latest version of the SQL standard is accessible at www.iso.org/standard/76584.html.

2

Information Levels and Frameworks

Chapter Outline

2.1 Four Information Levels

Some advanced information systems are described as "intelligent". What intelligence is and whether machines will ever be as intelligent as humans are debatable questions. In the *Turing test* of intelligence, based on a revision of a test proposed by Alan Turing in 1950,[a] an opaque screen is placed between a typical human (see *A* in Fig. 2.1) and the object being tested for intelligence. The human can communicate with the object by means of a computer (with a keyboard for input and a screen for output) in natural language about any desired topic.

If the human can't tell from the object's responses whether it is an intelligent human or a machine, perhaps the object should be classified as intelligent. Two key conditions in the test are that natural language is used and that there are no restrictions on the discussion topics. Once we restrict the language and the discussion topic, there are many examples of a computer performing at the level of a human expert (e.g., chess, diagnosis of blood diseases, mineral exploration). Such systems are called "expert systems" because they perform as well as a human expert in a specific domain. Expert systems have passed "restricted Turing tests" specific to given universes of discourse (UoD).

Recent advances in artificial intelligence (AI) systems enabled some systems to pass the unrestricted Turing test. Google's LaMDA AI chatbot passed the test in 2022. Later, OpenAI's ChatGPT also passed the test, and its March 2023 version based on GPT-4 can take images as well as text as input (see https://en.wikipedia.org/wiki/GPT-4). Microsoft's Bing chat allows both text and voice input.

[a] Turing's original test was based on the Imitation Game, requiring only that a computer be at least as good as a man in formulating responses intended to convey the idea that the man was actually a woman. For further discussion, see http://en.wikipedia.org/wiki/Turing_test.

Information Modeling and Relational Databases. https://doi.org/10.1016/B978-0-443-23790-4.00012-8

FIG. 2.1 The Turing test: can *A* distinguish between *B* and *C*?

A fifth-generation information system (5GIS) is like a "user-definable" expert system in that it allows the user to enter a description of the universe of discourse and then conduct a conversation about this, all in natural language. Just how well the system handles its end of the conversation depends on how powerful its user interface, database management, and inference capabilities are.

Although desirable, it is not necessary that a 5GIS always be able to operate at an expert level when we communicate with it. It must, however, allow us to communicate with it in a natural way. Natural languages such as English and Japanese are complex and subtle. It will be many years before an information system will be able to converse freely in unrestricted natural language. We should be content in the meantime if the system supports dialog in a formalized subset of natural language. There may be many "formal, natural languages", one for English, one for Japanese, and so on. A 5GIS should be able to respond in the same language used by the human. For example, suppose we posed the query:

What is the age of Selena?

and we received the reply

sanjusai.

This would not help unless we knew that the reply is Japanese for "30 years old". Even if we can translate Japanese to English, we might still misinterpret the reply because instead of assigning people an age of zero years when born, the Japanese assign an age of 1 year. So, an age of 30 years in the Japanese system corresponds to an age of 29 years in the Western system. In addition to the requirement for a *common language*, effective communication between two people requires that each assigns the *same meaning* to the words being used. To achieve this, they should (a) share the same context or UoD and (b) speak in sentences that are unambiguous with respect to this UoD.

With our example, the confusion over whether Selena's age is 29 or 30 years results from different age conventions, one Western and one Japanese. Natural speech abounds with examples that can be disambiguated only by context. Consider "Pluto is owned by Mickey". This is true for the world of Walt Disney's cartoon characters. But suppose someone unfamiliar with Mickey Mouse and his dog Pluto interpreted this within an astronomical context, taking "Pluto" to refer to the minor planet Pluto—a drastic communication failure! It is essential to have a clear way of describing the UoD to the information system.

An information system may be viewed from *four levels*: *conceptual, logical, physical,* and *external*. The conceptual level is the most fundamental, portraying the business domain naturally in human concepts. At this level, the blueprint of the UoD is called the **conceptual schema**. This describes the *structure* or *grammar* of the business domain (e.g., what types of objects populate it, what roles these objects play, and what constraints apply).

While the conceptual schema indicates the structure of the UoD, the **conceptual database** at any given time indicates the *content* or instances populating a specific *state* of the UoD. Because information adds meaning to the data, the term "information base" is more appropriate (van Griethuysen, 1982). However, we'll use the briefer, more popular term "database". Conceptually, the database is a set of sentences expressing propositions taken to be true of the UoD. Because sentences may be added or deleted, the database may undergo transitions from one state to another.

However, at any particular time, the sentences populating the database must individually and collectively conform to the domain-specific grammar or design plan that is the conceptual schema. To summarize:

> The conceptual schema specifies the structure for all permitted states and transitions of the conceptual database.

To enforce this law, we now introduce a third system component: the **conceptual information processor** (CIP). This supervises updates to the database by the user and answers user queries. Fig. 2.2 shows the basic conceptual architecture of an information system. This diagram assumes that the conceptual schema is already stored in the system. For each application area, a different conceptual schema is entered.

Although the diagram might suggest that the user is interacting directly with the CIP, the user's interaction with the system is external rather than conceptual. The conceptual schema is not concerned with user interfaces or the physical details of how the database

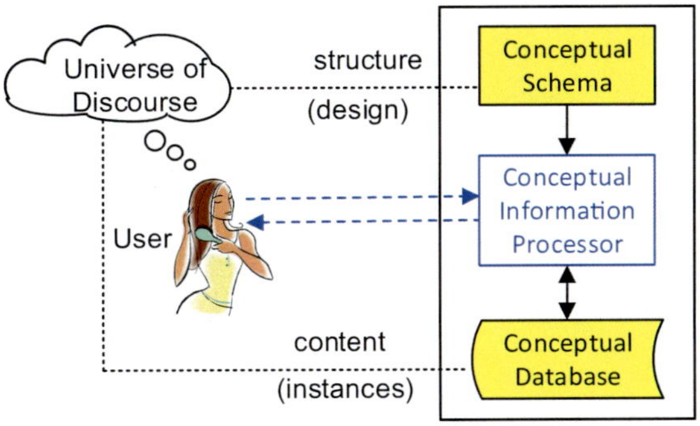

FIG. 2.2 Information system: conceptual level.

can be efficiently maintained. These concerns are catered for by including external and internal components within the overall architecture.

When interpreted by people, the conceptual schema and database both provide knowledge about the UoD. The combination of conceptual schema and (conceptual) database is called a conceptual **model** of the UoD. The model or *knowledge base* is a formal description of the UoD, and the CIP controls the flow of information between the model and humans. Some authors use the term "model" in the more restricted sense of "schema". This book uses the term "model" to include both the schema and a set of facts that populates the schema.

Recall that the *domain expert* is a person or group of people collectively familiar with the business domain. The *modeler* or conceptual designer is a person or team that specifies the conceptual schema by formalizing the domain expert's knowledge. An *end user* makes use of the implemented system. For a small system, the domain expert, modeler, and user might be the same person. A large system might have several partial domain experts, a team of analysts and designers, and thousands of end users.

The modeler inputs the conceptual schema to the system and has read/write access to it. Typically, most users have read-only capability for the schema, but read/write access to the database. Different interfaces may be created for different users so that some users have access to only part of the model. Thus, different users may access different subschemas of the global conceptual schema. This situation is summarized in Fig. 2.3. Some users may have read-only access to the database.

An **external schema** specifies the UoD design and operations accessible to a particular user or group of users. Here we specify *what* kind of facts may be read, added, or deleted, and *how* they are displayed. For *security* reasons, different user groups are often allocated different access rights (e.g., to ensure that sensitive information is not made public). For *user convenience*, views may be constructed to hide information irrelevant to a user group, or to collect and display information more efficiently. Different interfaces may be designed to cater to users with different expertise levels even when accessing the same underlying information.

Conceptual schemas are designed for clear communication, especially between modelers and domain experts. While they give a clear picture of the UoD, they are usually

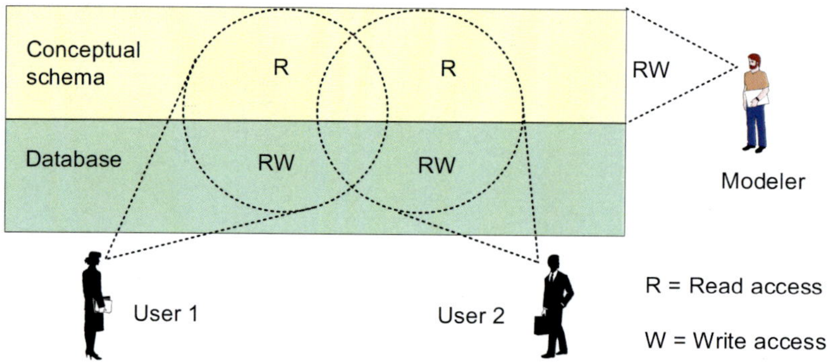

FIG. 2.3 Access to the model.

converted to a lower-level structure for efficient implementation. For a given application, an appropriate logical data model (e.g., relational, object-oriented) is chosen, and the conceptual schema is mapped to a **logical schema** expressed in terms of the abstract structures for data and operations supported in that data model. For example, in a relational schema, facts are stored in tables, and constraints are expressed using primary and foreign key declarations and so on.

The logical schema may now be realized as a **physical schema** in a specific database management system (DBMS), such as a relational schema might be implemented in Microsoft SQL Server or IBM's DB2. The physical schema includes details about the physical storage and access structures used (indexes, file clustering, etc.). Different physical choices can be made for the same DBMS, and different DBMSs often differ in what choices are possible, so different physical schemas might be chosen for the same logical schema. Operations at the external level are converted by the system into operations at the physical level. Logical and physical schemas reside at the *internal level.*

One advantage of the conceptual level is that it is the most *stable* of all the levels. It is unaffected by changes in user interfaces or storage and access techniques. Suppose a conceptual schema is implemented in a relational DBMS, and later we wish to use an object database instead. Unless the UoD has changed, the conceptual schema can remain the same. We need only apply a different mapping procedure and then migrate the data.

If a language is the object of study, it is said to be the *object language.* The language used to study it is then called the *metalanguage.* For example, you might use English as a metalanguage to study Japanese as an object language. An object language may be its own metalanguage, for example, English may be used to learn about English. Any conceptual schema may be expressed as a set of sentences, and hence may be viewed as a database in its own right.

This enables us to construct a *metaschema* for talking about conceptual schemas. This metaconceptual schema specifies the design rules that must be obeyed by any conceptual schema (e.g., each role is played by exactly one object type). Computer-aided software engineering (CASE) tools are used to assist in designing conceptual schemas of such a metaschema to ensure that schemas entered by the modeler are well formed or "grammatical".

While on the subject of grammar, let us agree to accept both "schemas" and "schemata" as plural of "schema". Although "data" is the plural of "datum", we'll adopt the common practice of allowing "data" to be used in both singular and plural senses. The first exercise in the book is shown next. Answers to exercises are included in the online supplement.

Exercise 2.1

1. Classify each of the following as A (external), B (conceptual), or C (internal).
   (a) This level deals with physical details of how to store and access data.
   (b) This level is concerned with providing a convenient and authorized view of the database for an individual user.
   (c) This level is concerned with representing information in a fundamental way.

2.2 The Conceptual Level

At the conceptual level, all communication between people and the information system is handled by the CIP. This communication may be divided into three main stages:

1. The modeler enters the conceptual schema for the UoD into the system. The CIP accepts the schema if and only if it is consistent with the metaconceptual schema.
2. The user updates the database by adding or deleting specific facts. The CIP accepts an update if and only if it is consistent with the conceptual schema.
3. The user queries the system about the UoD and is answered by the CIP. The CIP can supply information about the conceptual schema or the database, provided it has stored the information or can derive it.

In these stages, the CIP performs as a *design filter*, *data filter*, and *information supplier*, respectively. To process update and query requests, the CIP accesses the conceptual schema, which includes three main sections as shown in Fig. 2.4.

A *fact* (or fact instance) is a proposition that the business takes to be true. The **fact types** section lists the kinds of facts that may be represented in the database. Fact types indicate what types of *object* are permitted in the UoD (e.g., Employee, Country), how these are *referenced* by values in the database (e.g., SSN, CountryCode), and their *relationships* (e.g., was born in, lives in).

The **constraints** section lists the constraints or restrictions on populations of the fact types. *Static constraints* apply to every state of the database. For example, suppose a database stores information about countries and their populations. Although a country's population may change over time, at any given time each country has at most one value recorded for its current population. *Dynamic constraints* determine what transitions are allowed between states of the database. For instance, a person's age status may change from child to adult, but not vice versa.

Constraints are also known as *validation rules* or *integrity rules*. A database is said to have integrity when it is consistent with the UoD being modeled. Although most database constraints can be represented graphically on conceptual schema diagrams, some constraints may need to be represented in other ways (e.g., using conceptual sentences, formulae, tables, graphs, or program code).

The **derivation rules** section includes rules to derive new facts from other facts. A fact that is not a *derived fact* is an *asserted fact* (also known as a primitive or base fact). Derivation rules may involve *mathematical calculation* or *logical inference*. For example, an average score may be derived by summing individual scores and dividing by their number. Many operators and functions are defined for various data types. This permits a variety of possible queries without the need to document each derivation. Typical mathematical facilities include arithmetic

Fact types	*Constraints*	*Derivation rules*

FIG. 2.4 The three main sections of the conceptual schema.

operators, such as +, −, * (multiply), and / (divide); set operators such as ∪ (union); and functions for counting, summing, and computing averages, maxima, and minima.

In addition to such generic derivation facilities, specific *derived fact types* known to be required may be defined in the schema by means of derivation rules. Some mathematically computed fact types and almost all logically inferred fact types fit into this category. Fact types derived by use of logical inference typically involve rules that make use of logical operators such as **and** or **if**.

Although derivable facts could be stored in the database, it is usually better to avoid this. Apart from saving storage space, the practice of omitting derived facts from the stored data usually makes it much easier to deal with updates. For instance, suppose we want to regularly obtain individual ages (in years) of students in a class and, on occasion, the average age for the class. If we stored the average age in the database, we would have to arrange for this to be recomputed every time a student was added to or deleted from the class as well as every time the age of any individual in the class increased.

As an improvement, we might store the number of students as well as their individual ages, and have the average age derived only upon request. But this is still clumsy. Can you think of a better design?

As you probably realized, there is no need to store the number of students in a class because this can be derived using a count function. This avoids having to update the class size every time someone joins or leaves the class. Moreover, there is no need to store even the individual ages of students. Computer systems have a built-in clock that enables the current date to be accessed. If we store the birth date of the students, then we can have the system derive the age of any student upon request by using an appropriate date subtraction algorithm. This way, we don't have to worry about updating ages of students as they become a year older.

Sometimes, what is required is not the current age but the age on a certain date (e.g., entrance to a school, age grouping for a future sports competition). In such cases where a single, stable age is required for each person, it may be appropriate to store it.

Before considering a derivation example using logical inference, note the difference between *propositions* and *sentences*. Propositions are asserted by declarative sentences and are always true or false (but not both). The same proposition may be asserted by different sentences, for example, "Paris is the capital of France" and "The French capital is Paris".

While humans can deal with underlying meanings in a very sophisticated way, computers are completely literally minded: they deal in sentences rather than their meanings. If we want the computer to make connections that may be obvious to us, we have to explicitly provide it with the background or rules for making such connections.

Suppose that facts about brotherhood and parenthood are stored in the database. For simplicity, assume that the only objects we want to talk about are people identified by their first names. We might now set out facts as follows:

Alan is a brother of Betty.
Charles is a brother of Betty.
Betty is a parent of Fred.

You can look at these facts and readily see that both Alan and Charles are uncles of Fred. In doing so, you're using your understanding of the term "uncle". If we want a computer system to be able to make similar deductions, we need to provide it with a rule that expresses this understanding. For example:

Given any X, Y
X is an uncle of Y **if there is some** Z **such that** X is a brother of Z **and** Z is a parent of Y.

This may be abbreviated to

X is an uncle of Y **if** X is a brother of Z **and** Z is a parent of Y.

This rule is an example of a *Horn clause*. The head of the clause on the left of "**if**" is derivable from the conditions stated on the right side. Horn clauses are used in languages such as Prolog and Datalog, enabling many derivation rules to be set out briefly. In object-role modeling (ORM), such rules are specified using a more readable syntax, such as,

For each $Person_1$ **and** $Person_2$,
$Person_1$ is an uncle of $Person_2$ **if** $Person_1$ is a brother of **some** $Person_3$
who is a parent of $Person_2$.

To appreciate how the CIP works, let's look at an example. The notation is based on a textual ORM language called FORML (formal ORM language). A graphical version is explained in later chapters. For simplicity, some constraints are omitted. The UoD structure is set out in the conceptual schema shown.

Reference schemes:		Person(.Firstname); Country(.Code); Year(CE:)
Base fact types:	F1	Person lives in Country.
	F2	Person was born in Year.
	F3	Person is a brother of Person.
	F4	Person is a parent of/is a child of Person. *Roles:* parent, child.
Constraints:	C1	**Each** Person lives in **some** Country.
	C2	**Each** Person lives in **at most one** Country.
	C3	**Each** Person was born in **at most one** Year.
	C4	**No** Person is a brother of **itself**.
	C5	**No** Person is a parent of **itself**.
Derivation rules:	D1	$Person_1$ is an uncle of $Person_2$**if** $Person_1$ is a brother of **some** $Person_3$ **who** is a parent of $Person_2$.
	D2	nrChildren **of** Person = **count each** child **of that** Person.

The *reference schemes* section declares the kinds of *object* of interest and how they are referenced. Objects are either entities or values. *Entities* are real or abstract things referenced by relating *values* (e.g., names) to them in some way. Here, we have three *entity types*: Person, Country, and Year. In this UoD, people are identified by their first name (rarely true in practice!). Countries are identified by codes (e.g., "CU" for Cuba). Years are identified by CE (Common Era) values (e.g., World War II ended in 1945 CE). Years are time segments, so they are entities, not values.

The *fact types* section declares which kinds of facts are of interest. This indicates how object types participate in relationships. In this book, object type names in English are shown with their first letter in capitals, and role names start with a lowercase letter. Because database columns have attribute names such as "birthyear", you may feel it is better to reword "Person was born in Year" as "Person has Birthyear".

However, suppose we add the fact type "Person has Deathyear". The formal connection between Birthyear and Deathyear is now hidden. Expressing the new fact type as "Person died in Year" reveals the semantic connections to "Person was born in Year" and makes comparisons between years of birth and death meaningful (the underlying domain is Year).

Attribute names are often used in ER, UML, and relational schemas to express facts, but this can be unnatural (compare "Person has deathyear" with "Person died in Year"), and is incomplete unless domain names such as "Year" are added.

The *constraints* section declares constraints on the fact types. The examples here are all *static* constraints (i.e., each is true for each state of the database). Reserved words in the schema language are shown in bold. Constraint C1 means that for each person referenced in the database, we know at least one country where they live. C2 says nobody can live in more than one country (at the same time). C3 says nobody was born in more than one year; note that we might not know their year of birth. Constraint C4 says that nobody is his/her own brother, and C5 says that nobody is his/her own parent (brotherhood and parenthood are irreflexive).

The *derivation rules* section declares a logical rule for unclehood and a function for computing the number of children of any person. The unclehood rule is specified in relational style, using fact type readings. The nrChildren rule is specified in attribute style using a role name.

At the conceptual level, each fact in the database is a simple or *elementary fact*. Basically, this means it can't be split up into two or more simpler facts without loss of information. We may *add* or insert a fact into the database, and we may *delete* a fact from it. However, we may not modify or change just a part of a fact. By "fact" we mean "fact *instance*" not "fact type".

The operation of adding or deleting a single fact is an *elementary update* or *simple transaction*. In our sample conceptual query language, add and delete requests start with "**add**:" and "**del**:" and queries start with "**list**" or end with "?". The CIP either accepts updates or rejects them with an indication of the violation. The CIP answers legal queries and rejects illegal ones.

To explain the conceptual notions underlying database transactions, we now discuss some examples of CIP interactions. If this seems tedious, remember that we are talking about the conceptual level, not the external level where the user actually interacts with the system. At the external level, the user typically enters, deletes, or changes values in a screen version of a form or table. Conceptually, however, we may think of such an operation being translated into the appropriate delete and add operations on elementary facts before being passed on to the CIP. Suppose we now start populating our conceptual database as follows:

User:	*CIP:*
add: Person 'Terry' lives in Country 'US'.	Accepted

Because the CIP recognizes the type of this sentence and sees that no constraints are violated, it adds it to the database and issues the reply "Accepted". If an update is inconsistent with the conceptual schema, the CIP rejects it, indicating the reason for rejection by verbalizing the constraint(s) violated. You should now be able to follow the following dialog. To save space here, we list just the number of the constraint violated rather than its verbalization.

add: Person 'Norma' was born in Year 1950.	Rejected. C1 violated.
add: Person 'Norma' lives in Country 'US'.	Accepted.
add: Person 'Norma' was born in Year 1950.	Accepted.
add: Person 'Terry' is a brother of Person 'Paul'.	Rejected. C1 violated.
add: Person 'Paul' lives in Country 'AU'.	Accepted.
add: Person 'Terry' is a brother of Person 'Paul'.	Accepted.
add: Person 'Paul' is a brother of Person 'Paul'.	Rejected. C4 violated.
add: Person 'Terry' plays Sport 'Judo'.	Rejected. Unknown fact type

Now suppose that Norma moves from the United States to Australia. How would we record this change? Here are two incorrect attempts:

add: Person 'Norma' lives in Country 'AU'.	Rejected. C2 violated.
del: Person 'Norma' lives in Country 'US'.	Rejected. C1 violated.

Is there any way to change Norma's country of residence? Yes. Sometimes it is convenient, or even necessary, to submit many elementary updates in a single, *compound transaction*. To show this, we use "**begin**" and "**end**" as brackets to open and close the transaction. With compound transactions, the CIP considers only the total or *collective effect* of the transaction. It is quite possible that the CIP will accept a compound transaction even if, taken individually, some of its component simple updates would be rejected. To change Norma's location, we proceed thus:

begin
  del: Person 'Norma' lives in Country 'US'.
  add: Person 'Norma' lives in Country 'AU'.
end Accepted.

The CIP accepts a transaction if and only if its overall effect is consistent with the conceptual schema. The user entered this compound transaction as a sequence of two elementary updates. It is more natural here to specify the delete before the add operation. However, the CIP treats any compound transaction as a *set* of updates, disregarding the order of the components, and looks only at the total effect. So even if the add operation had been specified before the delete, this would have been accepted with the same result.

Note that the CIP cannot accept just part of a compound transaction: it is a case of "all or nothing". Here are a few more simple interactions:

add: Person 'Linda' lives in Country 'AU'.	Accepted.
add: Person 'Selena' lives in Country 'AU'.	Accepted.
add: Person 'Terry' is a parent of Person 'Linda'.	Accepted.
add: Person 'Terry' is a parent of Person 'Selena'.	Accepted.
nrChildren **of** Person 'Terry'?	2.
list each Person **who** is a brother of Person 'Paul'.	Terry.
list each Person **who** lives in Country 'AU'.	Linda Paul Selena
list each Person **who** is an uncle of Person 'Selena'.	No Person.
add: Person 'Paul' is a brother of Person 'Terry'.	Accepted.
list each Person **who** is an uncle of Person 'Selena'.	Paul.
list each Person **who** is an aunt of Person 'Selena'.	Rejected. Unknown fact type.

Now suppose the UoD is changed by adding the following constraint:

C6 **Each** Person was born in **some** Year.

Our current database fails to meet this constraint. Suppose we start over with an empty database and try to add the following fact. How would the CIP respond?

add: Person 'Bernard' is a brother of Person 'John'.

This update request is rejected. It violates two constraints (C1 and C6), as all people listed in the database must have both their country and birth year recorded. Chapter 5 uses the terminology "mandatory roles" to describe such constraints.

In general, the order in which constraints are listed does not matter. However, if an update request violates more than one constraint, this order may determine which constraint is reported as violated if the CIP is configured to report at most one violation. In this case, the CIP would respond thus to the previous request: "Rejected. C1 violated". If the CIP is configured to report all constraint violations, it would instead respond "Rejected. C1, C6 violated". As an exercise, convince yourself that with C6 added, the following update requests are processed as shown:

add: Person 'Ken' lives in Country 'GB'.	Rejected. C6 violated.
begin	
add: Person 'Ken' lives in Country 'GB'.	
add: Person 'Ken' was born in Year 1960.	
end	Accepted.
add: Person 'Ken' was born in Year 1959.	Rejected. C3 violated.

begin
  del: Person 'Ken' was born in Year 1960.
  add: Person 'Ken' was born in Year 1959.
  add: Person 'Erik' lives in Country 'NL'.
  add: Person 'Erik' was born in Year 1970.
  add: Person 'Ken' is a brother of Person 'Erik'.
end Accepted.

The CIP uses the conceptual schema to supervise updates of the database and to respond to questions. Think of the designer of the conceptual schema as the *"law giver"* and the schema itself as the *"law book"* containing the laws or ground rules for the UoD. The CIP is the *"law enforcer"*, as it ensures that these laws are adhered to whenever the user tries to update the database. Like any friendly police person, the CIP is also there to provide information on request.

Whenever we communicate to a person or an information system, we have in mind a particular UoD. Typically, this concerns some small part of the real universe, such as a particular business. In rare cases, we might choose a fictional UoD (e.g., one populated by comic book characters) or perhaps a fantasy world we have invented for a novel that we are writing. Fictional worlds may or may not be (logically) possible.

You can often rely on your own intuitions as to what is logically possible. For instance, a world in which the Moon is colored green is possible, but a world in which the Moon is simultaneously green all over and red all over is not. A possible world is said to be *consistent* and an impossible world is *inconsistent*.

As humans, we carry prodigious amounts of information around in our minds. It is highly likely that somewhere in our personal web of beliefs, some logical contradictions are lurking. In most cases, it does not matter if our belief web is globally inconsistent. When reasoning about a particular UoD, however, consistency is essential. It is easy to show that once you accept a logical inconsistency, it is possible to deduce anything (including rubbish) from it—an extreme case of the garbage in garbage out (GIGO) principle.

There are basically two types of garbage: logical and factual. Inconsistent designs contain logical garbage. For example, we might declare two constraints that contradict one another. A good design method supported by a CASE tool that enforces metarules can help to avoid such problems. Many factual errors can be prevented by the enforcement of constraints on the database. For example, a declared constraint that "**Each** Person was born in **at most one** Country" prevents us from assigning a person more than one birth country.

Even if the schema is consistent, and the CIP checks that the database is consistent with this world design, it is still possible to add false data into the database. For example, if we tell the CIP that Einstein was born in France it might accept this even though Einstein was actually born in Germany. If we want our knowledge base to remain factually correct, it is still our responsibility to ensure that all the sentences we enter into the database express propositions that are actually true.

The following exercise provides practice with the concepts discussed in this section and introduces some further constraint types treated formally later.

Exercise 2.2

1. (a) Assuming the conceptual schema is already stored, what are the two main functions of the CIP?
   (b) What are the three main components of the conceptual schema?
   (c) "The CIP will reject a compound transaction if any of its component update operations is inconsistent with the conceptual schema". True or false?
2. Assume the following conceptual schema is stored. Constraints apply to each database state. C1 means that each person referred to in the database must have his/her fitness rating recorded. C3 says the possible fitness values are whole numbers from 1 to 10. C4 means that no person can be recorded as an expert at more than one sport, and C5 says a person can be recorded as being an expert at a sport only if that person is also recorded as playing the same sport.

   Reference schemes: Person(.Firstname); Sport(.Name); FitnessRating(.Nr)

   Base fact types:
   F1 Person has FitnessRating.
   F2 Person plays Sport.
   F3 Person is expert at Sport.

   Constraints:
   C1 **Each** Person has **some** FitnessRating.
   C2 **Each** Person has **at most one** FitnessRating.
   C3 **The possible values of** FitnessRating **are** 1 **to** 10.
   C4 **Each** Person is expert at **at most one** Sport.
   C5 **Each** Person **who** is expert at **some** Sport **also** plays **that** Sport.

   Derivation rules:
   D1 Person is a martial artist **if**
   Person plays Sport 'judo' **or** Person plays Sport 'karatedo'.
   D2 nrPlayers **of** Sport = **count each** Person **who** plays **that** Sport.

   The database is initially empty. The user now attempts the following sequence of updates and queries. For each update, circle the letter if the update is accepted (based on the cumulative state of the database). In cases of rejection, supply a reason (e.g., state which part of the schema is violated). For queries, supply an appropriate response from the CIP.

   (a) **add**: Person 'Ann' has FitnessRating 9.
   (b) **add**: Person 'Fred' plays Sport 'tennis'.
   (c) **add**: Person 'Bob' has FitnessRating 7.
   (d) **add**: Person 'Ann' has FitnessRating 8.
   (e) **add**: Person 'Chris' has FitnessRating 7.
   (f) **add**: Person 'Fred' has FitnessRating 15.
   (g) **add**: Person 'Ann' plays Sport 'judo'.

(h) **add**: Person 'Bob' is expert at Sport 'soccer'.
(i) **add**: Person 'Ann' is expert at Sport 'judo'.
(j) **add**: Person 'Ann' programs in Language 'SQL'.
(k) **add**: Person 'Ann' plays Sport 'soccer'.
(l) **add**: Person 'Chris' plays Sport 'karatedo'.
(m) **del**: Person 'Chris' has FitnessRating 7.
(n) **begin**
  add: Person 'Bob' has FitnessRating 8;
  del: Person 'Bob' has FitnessRating 7.
 end
(o) **add**: Person 'Ann' is expert at Sport 'soccer'.
(p) **add**: Person 'Bob' plays Sport 'soccer'.
(q) Person 'Ann' plays Sport 'judo'?
(r) **list each** Person **who** plays Sport 'karatedo'.
(s) nrPlayers **of** Sport 'soccer'?
(t) **list each** Person **who** is a martial artist.
(u) **list possible values of** FitnessRating.
(v) **what** is the meaning of life?

3. In the following schema, constraints apply to the database, not necessarily to the real world. Although each student has a marital status, it is optional to record it (e.g., some students may wish to keep their marital status private). Constraint C1 combines two weaker constraints because "exactly one" means "at least one (i.e., some), and at most one".

Reference schemes: Student(.Firstname); Degree(.Code); MaritalStatus(.Name)

Base fact types: F1 Student is enrolled in Degree
F2 Student has MaritalStatus

Constraints: C1 **Each** Student is enrolled in **exactly one** Degree.
C2 **Each** Student has **at most one** MaritalStatus.
C3 **The possible values of** MaritalStatus **are** 'single', 'married', 'widowed', 'divorced'.
C4 MaritalStatus **transitions**: ("1" = "allowed")

From \ To	*single*	*married*	*widowed*	*divorced*
single	0	1	0	0
married	0	0	1	1
widowed	0	1	0	0
divorced	0	1	0	0

The database is initially empty. The user attempts the following sequence of updates and queries. For each update, circle the letter if the update is accepted; if rejected, supply a reason. Assume the questions are legal and supply an appropriate answer.

(a) **add**: Student 'Fred' is enrolled in Degree 'BSc'.
(b) **add**: Student 'Sue' has MaritalStatus 'single'.
(c) **begin**
 add: Student 'Sue' has MaritalStatus 'single'.
 add: Student 'Sue' is enrolled in Degree 'MA'.
 end
(d) **add**: Student 'Fred' is enrolled in Degree 'BA'.
(e) **add**: Student 'Fred' is studying Subject 'CS112'.
(f) **list possible values of** MaritalStatus.
(g) **add**: Student 'Bob' is enrolled in Degree 'BSc'.
(h) **add**: Student 'Sue' has MaritalStatus 'married'.
(i) **begin**
 del: Student 'Sue' has MaritalStatus 'single'.
 add: Student 'Sue' has MaritalStatus 'married'.
 end
(j) **add**: Student 'Bob' has MaritalStatus 'single'.
(k) **begin**
 del: Student 'Bob' has MaritalStatus 'single'.
 add: Student 'Bob' has MaritalStatus 'divorced'.
 end
(l) Student 'Sue' is enrolled in Degree 'BSc'?
(m) **list each** Student **who** is enrolled in Degree 'BSc'.
(n) **list each** Student **who** is enrolled in Degree 'MA'.
(o) **add**: 3 students are enrolled in Degree 'BE'.

What is the final state of the database?

4. Assume the following conceptual schema.

Reference schemes: Person(.Firstname)

Base fact types: F1 Person is male.
F2 Person is female.
F3 Person is a parent of Person.

Constraints: C1 **Each** Person is male **or** is female.
C2 **No** Person is male **and** is female.
C3 Each person has at most 2 parents
Each Person2 **instance occurs at most 2 times in** Person1 is a parent of Person2.
C4 **No** Person is a parent of **itself**.

Derivation rules: D1 Person1 is a grandparent of Person2 **if**
Person1 is a parent of **some** Person3 **who** is a parent of Person2.

Assume the database is populated with the following data. The user now attempts the following sequence of updates and queries. Indicate the CIP's response in each case.

Males: David, Paul, Terry *Females*: Alice, Chris, Linda, Norma, Selena

(a) **add**: Person 'Jim' is male.
(b) **add**: Person 'Bernie' is a parent of Person 'Terry'.
(c) **begin**
 add: Person 'Terry' is a parent of Person 'Selena'.
 add: Person 'Norma' is a parent of Person 'Selena'.
 end
(d) **add**: Person 'David' is a parent of Person 'David'.
(e) **begin**
 add: Person 'Norma' is a parent of Person 'Paul'.
 add: Person 'Alice' is a parent of Person 'Terry'.
 end
(f) **add**: Person 'Chris' is male.
(g) **add**: Person 'Chris' is a parent of Person 'Selena'.
(h) **what** Person is a grandparent of Person 'Selena'?

Formulate your own derivation rules for the following:

(i) X is a father of Y; (j) X is a daughter of Y; (k) X is a granddaughter of Y

5. Consider the following conceptual schema:

Ref. schemes: Employee(.Nr); Department(.Name); Language(.Name)

Base fact types: F1 Employee works for Department.
F2 Employee speaks Language.

Constraints: C1 **Each** Employee works for **some** Department.
C2 **Each** Employee works for **at most one** Department.
C3 **Each** Employee speaks **some** Language.

(a) Provide an update sequence to add the facts that employees 101 and 102 speak English and work for the Health department.
(b) Invent some database populations that are inconsistent with the schema.

2.3 Database Design Example

This section provides a simple overview of how conceptual schemas may be implemented in relational database systems. The topic is covered in detail in later chapters. Consider a company that is based in an English-speaking country, but carries out some of its business overseas. Some of its employees may speak languages other than English, which can help when dealing with foreign customers. The company wishes to move from a manual to an automated personnel record system. Prototype examples of some online personnel record forms are shown in Fig. 2.5. For simplicity, only a few details are displayed.

FIG. 2.5 Personnel forms for three employees.

To model this small domain, we first verbalize information on these forms. At this stage, we should ignore any user interface details, such as the geometric positioning of fields on the form or accelerator key choices (shown by underlined characters). Try this yourself for the first employee record before reading on.

Even at the conceptual level, there are often many correct ways to verbalize the information. Here is one way to read the relevant facts from the first record:

The Employee with employee number 006 has the EmployeeName "Adams, Ann".
Employee 006 is of the Gender named "Female".
Employee 006 speaks the ForeignLanguage named "Spanish".

From the sample data, it appears that we could also identify employees by their name, but this is usually not a good idea. To begin with, employees can change their name (e.g., upon marriage). Moreover, in a large company it is often possible for two employees to have the same name. Let's suppose this possibility is confirmed by the domain expert. In this case, the sample data was not significant in this regard.

We modeled the gender information of the first employee by treating Gender as an entity type with "Female" as one of its possible names and associating that employee instance with that gender instance. Instead of using names ("Male", "Female") for genders, we could use codes (e.g., "M", "F"). The second fact would then read "Employee 006 is of Gender 'F'" or more explicitly "The Employee with employee number 006 is of the Gender with code 'F'".

Because codes are more compact and their meaning is obvious, let's use them instead of names as our primary way of referring to the genders. If ever needed, we could add another two fact instances to relate the gender codes to the full gender names (actually used in the user interface).

Yet another way to model the gender information is to use unary fact types (e.g., Employee 006 is female). Later in the book, we'll discuss how to transform between different ways of conceptualizing and give you guidelines to help you choose between alternatives.

Because, unlike gender, the names of foreign languages have no obvious or common abbreviations, we'll stay with using names instead of codes to identify languages. If you verbalize the other two employee records, you'll just find different instances of the same fact types. Abstracting from the instances to the types and adding constraints, we may now set out the conceptual schema as follows:

Reference schemes: Employee(.Nr); EmployeeName(); Gender(.Code); ForeignLanguage(.Name)

Base Fact types:
F1 Employee has EmployeeName.
F2 Employee is of Gender.
F3 Employee speaks/is spoken by ForeignLanguage.

Constraints:
C1 **Each** Employee has **exactly one** EmployeeName.
C2 **Each** Employee is of **exactly one** Gender.
C3 **The possible values of** Gender **are** 'M', 'F'.

In setting out the reference schemes, we've abbreviated "number" as "nr". This works well with other languages such as German, as the first and last letters of their word for "Number" also form "Nr". Some people use "no" from the Italian "numero", but this can sometimes be confused with the negative "no". Some others use "num". The choice really is up to you, and many CASE tools allow you to declare the abbreviations you prefer to use. If there is any likelihood of confusion, you should avoid abbreviations at the conceptual level. The empty parentheses appended to "EmployeeName" in the object type reference scheme declarations indicate that EmployeeName is a value type (in this case a character string) and hence needs no reference scheme.

To ensure that terms (e.g., "Employee") used in a model are well understood, *concept definitions*, or at least clear descriptions, of each term should be included. Such definitions needn't display on the diagram, but be accessible in model reports.

Each of constraints C1 and C2 is actually a pair of elementary constraints. The phrase "exactly one" is shorthand for "at least one and at most one". The constraints do not mention the fact type Employee speaks ForeignLanguage. Hence, some employees might speak no foreign language (e.g., Cooper), some might speak many foreign languages (e.g., Bond), and many might speak the same foreign language (e.g., Spanish). This "*default*" case may be explicitly declared using the following *verbalization*:

It is possible that the same Employee speaks **more than one** Language
and that the same Language is spoken by **more than one** Employee.

The conceptual database for this schema contains three fact tables, one for each elementary fact type. This conceptual approach makes the schema easier to understand, prevents redundancy, and simplifies expression of constraints.

Once designed, a conceptual schema may be mapped to a *logical schema* described in terms of the generic logical data model (e.g., relational or hierarchic) chosen for implementation. The physical schema is then constructed by adapting the logical schema to

Employee:

empNr	*empName*	*gender*
006	Adams, Ann	F
007	Bond, James	M
008	Cooper, Susan	F

Speaks:

empNr	*foreignLanguage*
006	Spanish
007	French
007	Japanese
007	Spanish

FIG. 2.6 Relational database tables for Fig. 2.5.

the specific DBMS, improving performance by adding indexes, and so on. A schema specified in terms of the relational data model is called a *relational database schema*, or **relational schema** for short.

A relational database contains *named tables*, divided *horizontally* into *unnamed rows* and *vertically* into *named columns*. A row–column location is a *cell*. Each cell contains only one data *value*. Fig. 2.6 shows the data from Fig. 2.5 stored in a relational database. The Employee table has three rows and three columns. The Speaks table has four rows and two columns. In naming the columns, we chose to abbreviate "employee" as "emp". The number of rows of a table increases or decreases as records are added or deleted, but the number of columns stays the same.

Within a given schema, tables must have different names. Within the same table, columns must have different names. Columns are often called "fields", although this term is sometimes used for cells. Column names are also called *attribute* names. Each attribute is based on some *domain*, or pool of values from which its entries are drawn.

For example, if we added a manager column, this would have the same domain as empNr. The relational concept of domain is similar to our conceptual notion of object type, but in practice most relational DBMSs support only syntactic domains (value type declarations such assmallint), not semantic domains.

In contrast to a conceptual database, a relational database allows facts of different types to be grouped into the same table. For example, Fig. 2.6 groups the three elementary fact types into two tables. The Employee table caters for fact types F1 and F2, whereas the Speaks table caters for fact type F3.

The first row of the Employee table expresses two facts: the Employee with employee number 006 has the EmployeeName "Adams, Ann"; the Employee with employee number 006 is of the Gender with code "F". In general, *each row of a table in a relational database contains one or more elementary facts.* A rare exception to this rule may occur when a table stores only existential facts (e.g., a list of country codes). This rare case is discussed later in the context of independent object types.

With relational databases, care is needed in grouping fact types into tables. This book provides a simple procedure to map a conceptual schema onto a relational schema. This procedure shows that the choice of tables in Fig. 2.6 is correct. If instead we grouped all three fact types into just one table, this would cause problems similar to those discussed earlier with the movie star example (Table 1.2).

Relational databases are based on the relational model of data developed by E. F. Codd. The table is the only data structure. Tables with rows of data stored in the database are called *base tables.* Derived tables with only their definitions stored (not their content) are called *views.* Duplicate rows are not allowed in a base table.

The *order of the rows or the columns doesn't matter.* For example, Table 2.1 reorders the columns and rows of the Employee table but the meaning is unchanged (the same six facts are present). The order of rows doesn't matter because each fact is totally contained within a single row. The reason that the order of columns doesn't matter is that each column is named, so we know what its values mean. In a sense, the columns are "ordered by their name" rather than their position.

Table 2.1 The order of columns or rows has no impact on the meaning of the data.

Employee:

empName	*empNr*	*gender*
Cooper, Susan	008	F
Bond, James	007	M
Adams, Ann	006	F

Although the order of rows and columns is irrelevant at the logical level, at the physical level the order does exist and needs to be taken into account when tuning the database for efficient access.

Query results are also tables, but are unnamed and may contain duplicate rows. In formulating queries, you can specify the order in which the data is to be displayed, regardless of the actual order used to store the data in base tables.

We may treat a base table as a set of rows and each row as a sequence or *tuple* of data values (or, more strictly, a tuple of attribute-value pairs). In this sense, a table is a set of tuples. Because a mathematical relation is defined as a set of tuples, each table is a *relation*—hence the name "relational database".

We have seen that conceptual facts are grouped into tables in a relational schema. What about *constraints*? Relational systems typically supply only a few built-in, declarative constraint types. Columns may be specified as *mandatory* or optional for their table. In our example, all columns of the Employee table are mandatory (each entry must be an actual value, not null). This goes partway toward capturing the constraints that each employee is identified by empNr, has at least one name, and is of at least one gender. We additionally require that each employee listed in the Speaks table is also listed in the Employee table (this is a *subset* or *referential integrity* constraint).

To restrict each employee to at most one name and at most one gender, entries in the empNr column of the Employee table must be *unique* (i.e., no duplicates). This is enforced by declaring empNr as the *primary key* for the Employee table.

A *key* is a minimal combination of columns where no duplicates are allowed. In the relational model, each table has a primary key that provides the primary way of accessing its rows. Finally, the constraint on gender values may be enforced by a simple *check clause.*

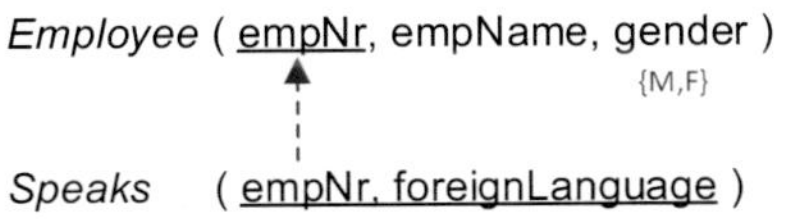

FIG. 2.7 The relational schema for Fig. 2.5.

The relational schema is set out compactly in Fig. 2.7. Column names are placed in parentheses (round brackets) after the table name, and *keys are underlined.* In this notation, columns are assumed mandatory for their table by default. The dotted arrow expresses the subset constraint mentioned earlier. The possible values for gender codes are placed in braces (curly brackets). For simplicity, domains are omitted. This notation is logical, not physical. The concrete syntax for declaring the tables and constraints depends on the relational database language used (often a dialect of structured query language, SQL).

The process of declaring a schema is often called data definition, and the language component specially designed for this task is called a *data definition language* (DDL). Our example was simple enough for the relational schema to be fully specified within the DDL component of standard SQL. Often, however, some constraints have to be coded separately.

A relational schema may contain table definitions, constraints, and derivation rules. Most derivation rules in a conceptual schema may be specified in a relational DBMS by defining views (virtual tables) or computed columns, or by coding triggers or stored procedures. If a query type is commonly used and requires derivation, it is explicitly included in the derivation rules section of the conceptual schema. In contrast, ad hoc queries are catered for by including derivation functions and operators in the query language itself, so that the user may express the required derivation rule within the query formulation itself.

In a relational database language such as SQL, it is fairly easy to express many mathematical and logical derivation rules (e.g., averages, or grandparenthood derived from parenthood). Some derivation rules that are difficult to express in SQL can be easily expressed in other languages. For example, the following Prolog rules recursively define ancestry in terms of parenthood:

X is an ancestor of Y **if** X is a parent of Y.
X is an ancestor of Y **if** X is a parent of Z **and** Z is an ancestor of Y.

The SQL standard supports recursive queries, but commercial SQLs differ in their support for this feature. Commercial SQLs that do not directly support recursion typically enable calls from SQL to other languages (e.g., C# or Prolog) to perform the needed recursion or other desired task.

Languages such as Prolog are used in some AI applications. In the past, AI systems tended to work with small populations of a large number of complex fact types, and database systems worked with large populations of fewer and simpler fact types. One of the aims of next-generation technology is to combine powerful inference capabilities with efficient database management.

Once the schema is defined, *data manipulation* may begin. The initially empty tables are populated with data (i.e., rows are *inserted*). Rows may then be *deleted* or *modified*. A row is modified if one or more of its values is changed: this is sometimes called "update" in the context of SQL, but we use the term "update" generically to include insert, delete, and modify operations. Collectively, these three operations are called *data maintenance*. If the constraints are properly specified, the DBMS automatically enforces them when a maintenance operation occurs.

Conceptually, the row modify operation of a relational DBMS simply adds or deletes elementary facts. For example, changing "Adams, Ann" to "Smith, Ann" on the third row of the Employee table of Fig. 2.6 simply deletes the fact that employee 006 is named "Adams, Ann" and adds the fact that employee 006 is named "Smith, Ann". Of course, at the external level, the user would simply modify values in a screen version of a form (as in Fig. 2.5) or a table row (as in Fig. 2.6).

Another aspect of data manipulation is *data retrieval*. This typically involves three steps: the user requests information; the system searches the database and/or the schema to locate the relevant information and uses this to obtain the required result; and the system then outputs this result. The second stage may involve sorting data and/or deriving results (e.g., an average value). Unlike maintenance, retrieval operations leave the database unaltered.

Data presentation refers to special techniques for displaying information on the screen (screen display) or for producing formatted printed output (report writing). We use the term "*output report*" generically to include screen and hard copy reports.

Typically, users of a relational database application do not interact directly with the relational tables in which the data is stored. Usually, they enter and access the data via a *forms interface*. Issues such as the design and management of screen forms, and *security* enforcement, often constitute a major part of the application development.

For reasons such as correctness, clarity, completeness, and portability, database designs should be first specified as conceptual schemas before they are mapped to actual database schemas.

2.4 Development Frameworks

Information systems development involves at least *four worlds*: the subject world, the system world, the usage world, and the development world (Jarke et al., 1992). The *subject world* is the UoD, or business domain, typically part of the real world. The *system world* is the information system's model of the UoD, and hence is a formal abstraction.

The *usage world* is the organizational world in which the system is to function and includes the community of users of the system, their user interfaces to the system, associated activities, and so on. This is sometimes called the environment of discourse. The *development world* covers the environment and processes used to develop the system, including the modelers and programmers, their design methods and rationale, project schedules, and so on.

The system world is sometimes called the *information system in the narrow sense* while the combination of the four worlds is the *information system in the broad sense* because it includes all related human factors. Specialists who focus on the system world are called "*dry*" (they deal with formal aspects), whereas those that focus on the other three worlds are called "*wet*" (they are willing to get their hands wet dealing with the informal, softer aspects). In practice, both dry and wet approaches are needed.

Determining *automation boundaries* (how much of the subject world will be managed by the system world) is a critical decision. We should first determine the services to be provided by the business (i.e., the essential business processes). Then, for each of these services we need to decide whether it will be fully automated, semiautomated, or manual.

Developing an information system for a business domain is essentially a *problem-solving* process. This general process may be broken down into four main stages: (1) defining the problem, (2) devising a plan, (3) executing the plan, and (4) evaluating what happened. Two of the most generally useful problem-solving strategies are to divide the problem into a number of *subproblems* and deal with these individually, and to try a *simpler* version of the problem first.

When the problem-solving process involves software development, it is refined into the five-stage *software life cycle* shown in Fig. 2.8: *specify* (say what the software is to do), *design* (decide how to do it), *implement* (code it), *test* (check that it works), and *maintain* (keep it working). In the naïve "waterfall" version of this process, the path through these five stages is a simple sequence. In practice, however, the process is typically *cyclic* in nature, as indicated by the dotted arrows.

If a need for change is detected at a later stage of the cycle, it is often necessary to return to one of the earlier stages. Moreover, large development projects are usually best divided into small projects, which can be further divided into components, each having its own cycle. The additional cycles at the right of Fig. 2.8 depict further iterations, each of which can be refined into a multistage cycle as shown on the left. In this *iterative* approach to development, the more essential or critical components are often developed first to minimize the risk of the project failing overall.

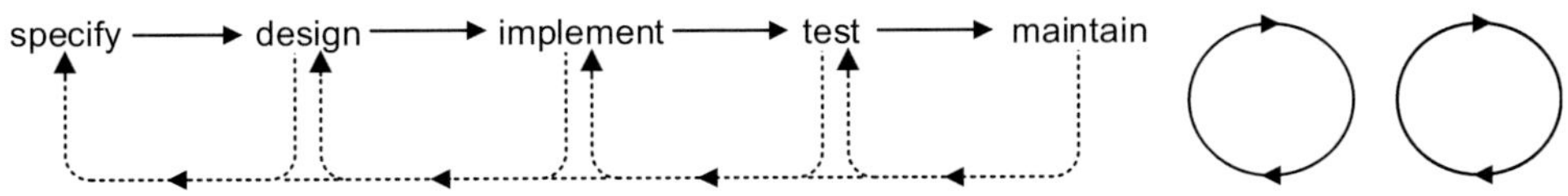

FIG. 2.8 The software life cycle, with iterations.

- Feasibility study
- Requirements analysis
- Conceptual design: data, processes
- Logical design: data, processes
- Basic physical design: data, processes
- Basic external design: data; processes
- Prototyping
- Completion of design
- Implementation of production version
- Testing and validation
- Release of software, documentation, and training
- Maintenance

In practice, these phases often overlap, feedback cycles are common, and some phases might be omitted. Many detailed procedures exist to flesh out these phases.

A **feasibility study** identifies the main objectives of the proposed information system and determines which components may be implemented with known resources (e.g., budget allocations and staff). It examines the cost-effectiveness of alternative proposals and assigns priorities to the various system components. The cost/benefit analysis might reveal that some objectives are unrealistic or that some of the objectives can be best achieved by improved manual procedures rather than by the use of computers.

Assuming the project is approved, a detailed **requirements analysis** is undertaken to determine just what the system is required to do. Components of the system are delineated, and people familiar with the relevant application areas are interviewed, including domain experts, intended users, and policy makers. Interviews may be supplemented by questionnaires. Relevant documentation is examined (e.g., forms, reports, charts, policy manuals, even code from legacy systems to be reengineered).

Where no such documentation is available, the domain experts are requested to invent examples of the kinds of information to be recorded. Simple diagrams are often used to clarify how the information system is to interact with the business environment. The main operations or transactions to be supported are identified and prioritized, and estimates are made of their expected data volumes, frequencies, and required response times.

The output of this requirements collection and analysis phase is a *requirements specifications* document detailing functional requirements, nonfunctional requirements (e.g., performance), and maintenance information (e.g., anticipated changes).

This document should be *unambiguous, complete, verifiable* (there is a way to check whether the requirements are satisfied), *consistent* (requirements do not contradict one another), *modifiable* (changes can be made easily and safely), *traceable* (requirements can be tracked to their origins, and are identifiable across different versions of the document), and *usable* (by current and future users of the document).

Various textual and graphical notations are used for different aspects of the requirements specification. Some process-oriented notations (e.g., use case diagrams, context diagrams, data flow diagrams) are discussed in a later chapter. For large, complex projects, the requirements analysis stage might take several months. As fact-oriented modeling uses the verbalization of familiar examples to clarify the UoD, its conceptual design method is also useful for the analysis of data requirements.

One way to measure the progress of a requirements activity employs the following three dimensions of requirements engineering: *specification* (opaque, fair, complete), *representation* (informal, semiformal, formal), and *agreement* (personal view, common view). The further one is along these dimensions, the better (Pohl, 1994).

With the understanding that phases may overlap, the next stage in the information systems life cycle is **conceptual design**. This is sometimes simply called "*analysis*" to distinguish it from the later stages of logical and physical design. With large applications, subproblems or components of a more manageable size might be selected, an *architecture* specified for coupling components, and a conceptual *subschema* designed for each. The subschemas may then be *integrated* within a *global* conceptual schema. In this text, the problems discussed are small enough for us to design the whole conceptual schema without needing to first design subschemas. However, the main design process itself is broken up into various stages. For example, fact types are identified before adding constraints.

Experienced modelers often notice similarities between new applications and previous ones they or others have designed. In this case, significant savings in the design effort may result from judicious **reuse** of strategies adopted in their earlier models. For example, if we have already modeled a university library system, and now have to model a video rental business, there are many features of the earlier model that may be reused. We might choose to identify an object type Loan in a similar way and either adopt or adapt several fact types from the earlier application. By abstracting similar, specific concepts (e.g., "book", "compact disc") to more general concepts (e.g., "rentable item"), it is easier to recognize a new application as related to earlier ones.

In recent years, a substantial and growing body of work on *design patterns* and *best practices* has been developed to facilitate reuse. Many resources are available that provide basic design patterns and templates for a variety of business domains. Large commercial packages for enterprise resource planning (ERP) can be adapted to handle various functions of a business (e.g., order processing, payroll management). To widen the scope of their reuse, such packages often model practices at a very high level of abstraction, which can make them harder to apply to concrete cases. The rise of component technology, design patterns, and ERP packages has seen a move from "build or buy" to "build and buy and reuse".

Over the years, many dimensions or perspectives have been suggested for capturing different aspects of information systems design. One classic survey of information systems design methods identified three design perspectives: data, process, and behavior (Olle et al., 1991).

The *data-oriented perspective* focuses on what kinds of data are stored in the database, what constraints apply to these data, and what kinds of data are derivable. The *process-oriented perspective* examines the processes or activities performed to help understand the way a particular business operates. Processes and their agents are described in addition to information flows between processes and other components.

Often, a complex process is refined into several subprocesses. The *behavior-oriented perspective* looks at how *events* trigger actions. Often, an activity analysis may be rephrased in terms of an event analysis, or vice versa.

It is important to specify the information needed and conditions sufficient for a process to execute or "fire". For example, the process of retrieving an account balance might be triggered by the event of a client requesting an account balance, and require input of information from the client (e.g., client number and account type) as well as from the relevant database tables. Because the dividing line between process and behavior is often blurred, we'll follow one common practice of bundling any discussion of processes, events, operations, activities, and so on into a single perspective, which we'll call "process". From that viewpoint, there are just *two main perspectives*: data and process. Various techniques such as *work flow models* have been developed to enable *business process modeling*, as discussed in Chapter 15.

Decades ago, systems were designed largely from the process perspective, but this proved to be inefficient and unreliable in many cases. The data perspective became dominant for information systems development because it was more fundamental and much more stable. Business processes change continually, but the underlying data tends to undergo only minor changes by comparison. Recently, object-oriented modeling has provided one way to combine data and process perspectives, with objects encapsulating data and operations. Unfortunately, this combination is often specified at a subconceptual level, and hence this integration is not as rosy as it seems.

If designed correctly, the conceptual schema provides a formal model of the structure of the UoD. Once this semantic modeling is completed, we perform a *mapping* of the conceptual design to a **logical design** expressed in terms of the generic data model of the implementation target(s). For example, we might map a conceptual schema to a relational schema. CASE tools can perform this mapping automatically.

Once the logical design is determined, a **basic physical design** can be undertaken. The logical design is adapted to the specific DBMS(s) or coding language(s) being used, and various strategies (e.g., indexes or clustering) are specified to improve the efficiency of the physical design.

The **basic external design** of an information system involves determining which kinds of data and operations will be accessible to which user groups, and designing appropriate human-computer interfaces for these groups. Typically, access rights tables are constructed for different user types, decisions are made about what functions to support on different screen forms, and the layout of screen forms and menus is decided. The external design can include cross references to the internal design (e.g., to show which forms access which tables or classes).

Except for trivial applications, the next stage of the life cycle usually involves **prototyping**. A prototype is a simplified version of the intended product that is used to gain early feedback from users on the quality of the design. It aims to cover the major functions in the requirements specification, but usually omits most of the error checking and finer details required for the final version, and uses only a small set of sample data. Because early feedback is more important than efficiency at this stage, the prototype might be coded in a higher-level language than the one(s) ultimately used. For user-interface design, the initial prototypes may simply be drawn on paper: these are called "paper prototypes". The prototypes are demonstrated to the clients, and their feedback is used to revise the requirements and designs where necessary.

Once a prototype is accepted, the (hopefully) **final design** of the actual product or component is undertaken. The **implementation** of the internal and external design is now completed by writing the actual code for the production version. The product is now subjected to extensive **testing and validation** (also known as *quality assurance*). The software is run using carefully chosen data and operations to check that it functions as expected. Selected users may then be issued with prerelease versions (e.g., alpha and one or more betas) to help find other errors. Typically, the further advanced the prerelease version, the wider the audience of test users. With some critical applications (e.g., military security systems), correctness proofs might be developed to verify that the system meets its specifications. Finally, the **release** version of the product is made available to the public, along with extensive *documentation* and *training*.

Software **maintenance** involves modifications to the software after its initial release. Maintenance may be of three main types: *corrective* (eliminate bugs); *adaptive* (alter software to cater for changes in the environment or UoD); and *perfective* (add improvements). Maintenance continues as long as the product is available. The overall development process is often *iterative*, applying the development cycle to one major component at a time rather than trying to build the whole product at once. The earlier errors are detected in the development cycle, the easier they are to fix. The cost of correcting an error at the implementation stage can be orders of magnitude greater than the cost of fixing it at the analysis stage.

Important factors in developing information systems include practicality (is automation really desirable?), correctness, clarity (designs and code should be readable and well documented), efficiency (memory requirements, speed, production costs), portability, maintainability, adaptability, user proofing, and support.

Early in the chapter, we saw how information schemas can exist at four different levels (conceptual, logical, physical, and external). The inclusion of the logical level refines the original three-schema architecture developed by an international standards group. This notion of levels provides a vertical way of partitioning an information system's architecture. It is also useful to provide a horizontal partitioning. One way of doing this is to use the "six friends" Rudyard Kipling referred to in his 1902 story "The Elephant's Child": "I keep six honest serving-men … Their names are What and Why and When and How and Where and Who". If we include a column for each of these six questions and show each of the four levels as a row, we obtain a tabular, *level/perspective framework* with 24 cells.

The most influential framework to incorporate both vertical and horizontal partitioning is John Zachman's framework for information systems architecture (Zachman, 1987). It uses six focus columns for Kipling's interrogatives, but has five perspective rows instead of four levels. Hence, the *Zachman framework* has 30 cells, as shown in Table 2.2. Any

Table 2.2 Zachman framework for enterprise architecture.

	Data (what)	*Function (how)*	*Network (where)*	*People (who)*	*Time (when)*	*Motivation (why)*
Scope						
Enterprise model						
System model						
Technology model						
Detailed representations						

systems engineering project needs to consider which of these cells is relevant as well as how to map between the levels.

Row 1 (the scope) provides a very high-level view. Row 2 (the enterprise model) corresponds roughly to the conceptual or "owner view". Row 3 (the system model) is basically the logical level, or "designer view". Row 4 (the technology model) corresponds to most of the physical level, or "builder view", with row 5 providing more detail in the "subcontractor" view. External level aspects appear in various cells.

This book focuses on column 1 (data) of rows 2 and 3, with some discussion of column 2 (function) and row 3. In practice, projects may involve other cells in this framework. For example, we may need to consider where the system will be deployed, who the stakeholders are, when system components are due, and why some business rules need to be enforced (columns 3–6).

Table 2.3 shows an *extended four-schema framework* based on both the four-schema approach and the Zachman framework. The logical and physical levels are refined into two layers. The *persistent layer* deals with data that persists between sessions: here we specify the structures to store data in a database (e.g., a relational schema). The *transient layer* deals with structures to temporarily store data in main memory during a session: the object model may be declared at the logic level as a class diagram and at the physical level as code structures in a programming language such as C# or Java.

Table 2.3 Extended four-schema framework for system development.

	Data (what)	*Function (how)*	*Network (where)*	*People (who)*	*Time (when)*	*Motivation (why)*
Conceptual						
External						
Logical-Persistent						
Logical-Transient						
Physical-Persistent						
Physical-Transient						

Connections between cells should be established, such as we should be able to map (or *forward engineer*) a conceptual model to a logical model and then a physical model, or go back the other way (*reverse engineering*).

If the application uses separate models for the transient and persistent data, then *object-relational mapping* procedures are required to map between the two layers.

Many other frameworks have been proposed in industry for architecting an information system. Four of these are now mentioned briefly. Further details on these approaches may be found in the references listed in the chapter notes.

The Object Management Group Model Driven Architecture (MDA) approach includes three levels: Computation Independent Model (CIM), Platform Independent Model (PIM), and Platform Specific Model (PSM). The CIM level is partly conceptual, the PIM level is logical, and the PSM level is physical.

The Open Group Architecture Framework (TOGAF) includes four levels or domains (Business, Application, Data, Technology) and provides a detailed Architecture Development Method (ADM) that may be tailored to an organization's needs and employed to manage the execution of architecture planning activities.

The International Standards Organization (ISO) Standard Reference Model of Open Distributed Processing (RM-ODP) incorporates five viewpoints: Enterprise, Information, Computational, Engineering, and Technology. RM-ODP consists of 14 standards covering software engineering and enterprise architecture for designing heterogeneous, interoperable IT systems. For the ISO standard, see http://committee.iso.org/sites/jtc1sc7/home/projects/flagship-standards/isoiec-10746.html.

The Framework of Information System Concepts (FRISCO) developed by an IFIP working group provides a taxonomy of concepts used in information systems work (Falkenberg et al., 1998). Although not widely adopted, the FRISCO report provides some useful insights into information system development.

2.5 Summary

An information system for a given application may be viewed from at least *four levels*: conceptual, logical, physical, and external. At each level, the formal model or knowledge base comprises a *schema* that describes the structure or design of the UoD and a database that is populated with the fact instances. Each schema determines what states and transitions are permitted for its database(s).

The *conceptual schema* does this in terms of simple, human-oriented concepts. The *logical schema* groups information into structures supported by the generic logical architecture (e.g., relational). The *physical* schema specifies the physical storage and access structures (e.g., indexes) for the specific DBMS or platform being used to implement the application. For the same global conceptual schema, different *external* schemas can be constructed for different user groups depending on what information is to be accessible and how the information is to be displayed.

A conceptual schema comprises three main sections: *fact types*, *constraints*, and *derivation rules*. Asserted or base facts are primitive (i.e., they are not derived from other facts). Derived facts are computed or inferred by applying derivation rules to other facts. In declaring fact types, we indicate the *object types*, how these are *referenced*, and what *roles* they play. Each role in a *relationship* is played by only one object type.

The simplest kind of object is a *value* (e.g., character string). *Entities* are real or abstract objects that are identified by their relationships to other objects (typically values). For example, a country may be identified by its name.

Static constraints determine the allowable populations of fact types. *Dynamic constraints* restrict transitions between these populations. For example, each city is the capital of at most one country, and no adult may become a child.

Each fact in a conceptual database is typically *elementary*. The addition or deletion of a fact is a simple update. In a *compound transaction*, several simple updates may be included; in this case, constraints apply only to the net effect of the complete update sequence, not to each individual update. Updates and queries on a conceptual database or conceptual schema are responded to by the CIP.

Each DBMS conforms to a *logical* data model (e.g., network, hierarchic, or relational). If a relational DBMS is chosen for the implementation, the conceptual schema is mapped to a *relational schema*, or relational database schema. Here, all the stored facts are placed in named *tables*, with named columns and unnamed rows. Each row of a relational table typically corresponds to one or more elementary facts. Each cell (row-column position) contains at most one value. Within a table, no row may be duplicated.

The population of a relational table may be changed by inserting, deleting, or modifying a row. Most conceptual constraints and derivation rules can be expressed within the relational DBMS language or by interfacing to another language.

The *information systems life cycle* typically involves the following stages: feasibility study; requirements analysis; conceptual design of data and processes; logical design; basic physical design; basic external design; prototyping; implementation of production version; testing and validation; release of software, documentation, and training; and maintenance. Feedback from a later stage may indicate a need to rework earlier stages.

Careful attention to earlier stages reduces the overall effort involved, as does the reuse of design strategies used previously on similar applications. For large projects, an iterative approach to development is generally recommended, where the life cycle is applied to one component before moving onto another component.

Various frameworks for information system architecture exist. The Zachman framework is composed of five rows that view the system from different perspectives (scope, enterprise model, system model, technology model, and details) and six columns with a different focus (data, function, network, people, time, and motivation). The extended four-schema framework uses the same six columns, but has six rows for conceptual, external, logical-persistent, logical-transient, physical-persistent, and physical-transient perspectives. Other well-known system architectures include TOGAF, MDA, and RM-ODP.

Chapter Notes

The classic discussion of the conceptual, internal, and external schemas of an information system is given in van Griethuysen (1982). Some people who worked on this ISO standard also worked on the FRISCO report (Falkenberg et al., 1998). Along with many other authors (e.g., Elmasri and Navathe, 1994), we consider the relational model of data to be a *logical* model rather than a conceptual model, mainly because it is too distant from natural language. From this viewpoint, a relational schema represents the logical portion of an internal schema (omitting details specific to the chosen DBMS, and storage and access details such as indexes). Codd (1990, pp. 33–34) proposes an alternative interpretation of the three-level architecture, in which the base relations of a relational database provide the conceptual schema, and views provide an external schema.

A standard reference on requirements analysis is the ANSI/IEEE standard 830-1984. A comparative review of different approaches to information system work is given in Olle et al. (1991). An informative overview of key issues in information systems modeling is provided by Mylopoulos (1998). Pohl (1994) gives a detailed treatment of the specification, representation, and agreement dimensions of requirements engineering. Bubenko (2007) provides a historical perspective on information systems modeling.

Data model patterns for various business domains are available in a number of books, including Hay (1996, 2006) and Silverston (2001a, b). These sample models are usually expressed in one of the ER notations covered later in this book.

Zachman (1987) introduces his architectural framework (see https://en.wikipedia.org/wiki/Zachman_Framework). Hay (2003) provides an extensive discussion of the Zachman framework, and argues that some of the row descriptors should be renamed.

Details on TOGAF may be accessed online at www.opengroup.org/togaf/ and at http://en.wikipedia.org/wiki/TOGAF. For resources on the MDA approach of the OMG, see http://www.omg.org/mda/. For the ISO RM-ODP standard, see http://committee.iso.org/sites/jtc1sc7/home/projects/flagship-standards/isoiec-10746.html. For further details on RM-ODP, see https://en.wikipedia.org/wiki/RM-ODP and https://rm-odp-new.lcc.uma.es/.

3

Conceptual Modeling: First Steps

Chapter Outline

3.1 Conceptual Modeling Language Criteria

Before discussing the conceptual schema design procedure (CSDP) of object-role modeling (ORM), let's review the design principles that underlie the ORM language. Some of these ideas were mentioned before, but we generalize the discussion here so that you can apply the principles to evaluate modeling languages in general. As you go through the book, consider how these principles are realized in the various languages discussed.

A modeling *method* (or modeling *process*) comprises both a *language* and a *procedure* describing how to use the language to build models. A language has associated *syntax* (marks), *semantics* (meaning), and *pragmatics* (use). Written languages may be graphical (diagrams) and/or textual. The terms "abstract syntax" and "concrete syntax" are sometimes used respectively to distinguish underlying concepts (e.g., object type) from their representation (e.g., named, soft rectangle).

Conceptual modeling portrays the business domain at a high level, using terms and concepts familiar to the business users, ignoring logical and physical level aspects (e.g., underlying database or programming structures used for implementation) and external level aspects (e.g., screen forms used for data entry). The following criteria drawn from various sources (van Griethuysen, 1982; ter Hofstede, 1993; Bloesch and Halpin, 1996) provide a basis for evaluating conceptual modeling languages.

- Expressibility
- Clarity
- Simplicity and orthogonality
- Semantic stability
- Semantic relevance

Information Modeling and Relational Databases. https://doi.org/10.1016/B978-0-443-23790-4.00003-7

- Validation mechanisms
- Abstraction mechanisms
- Formal foundation

The *expressibility* of a language is a measure of what it can be used to say. The more a language can express, the greater its expressive power. Ideally, a conceptual language should be able to completely model all details about the business domain that are conceptually relevant. This is called the *100% Principle* (van Griethuysen, 1982). ORM is a method for modeling and querying an information system at the conceptual level and for mapping between conceptual and other levels. Although ORM extensions for process modeling are being researched, the main focus of ORM is information modeling (popularly known as data modeling) because the data perspective is more stable and provides a formal foundation on which operations may be defined. Overall, unified modeling language (UML) has a wider scope than ORM because UML use case, behavior, and implementation diagrams model aspects beyond static structures. For conceptual data modeling, however, ORM's diagram notation has much greater expressive power than UML class diagrams or entity-relationship (ER) diagrams.

The *clarity* of a language is a measure of how easy it is to understand and use. To begin with, the language should be unambiguous. The more expressive a language is, the harder it is to maintain clarity. Ideally, the meaning of diagrams or textual expressions in the language should be intuitively obvious. This ideal is rarely achieved. A more realistic goal is that the language concepts and notations should be easily learned and remembered. To meet this goal, a language should exhibit *simplicity* and *orthogonality*. By avoiding attributes, ORM's role-based notation is simplified, yet easily understood by populating it with fact instances. Orthogonality allows the use of an expression wherever its meaning or value may be used. ORM's constructs were designed from the ground up to be orthogonal. For example, ORM constraints can be used and combined whenever this is meaningful. As we will see later, this is not true of languages like UML.

Semantic stability is a measure of how well models or queries expressed in the language retain their original intent in the face of changes to the business domain. The more changes we need to make to a model or query to cope with a business change, the less stable it is. Models and queries in ORM are semantically more stable than in ER or UML because they are not impacted by changes that cause attributes to be remodeled as relationships or vice versa.

Semantic relevance requires that only conceptually relevant details need be modeled. Any aspect irrelevant to the meaning (e.g., implementation choices, machine efficiency) should be avoided. This is called the *Conceptualization Principle* (van Griethuysen, 1982). ORM models and queries are purely conceptual. In contrast, UML class models include implementation aspects (e.g., attribute visibility and association navigability).

Validation mechanisms are ways in which domain experts can check whether the model matches the business domain. For example, static features may be checked by verbalization and instantiation, and dynamic features may be checked by simulation. Unlike ER and UML, ORM models are always easily verbalized and populated.

Abstraction mechanisms allow unwanted details to be removed from immediate consideration. This is very important with large models (e.g., wall-size schema diagrams). Because ORM diagrams tend to be more detailed and larger than corresponding ER or UML diagrams, abstraction mechanisms are often used. For example, a global schema may be modularized into various scopes based on span or perspective (e.g., a single page of a data model or a single page of an activity model). Successive refinement may be used to decompose higher-level views into more detailed views. Although not a language issue, tools can provide additional support such as layering and object zoom (see later). Such mechanisms can be used to hide detail, showing just that part of the model relevant to a user's immediate needs. With minor variations, these techniques can be applied to ORM, ER, and UML. ORM also includes an attribute abstraction procedure to generate ER and UML diagrams as views.

A formal foundation is needed to ensure unambiguity and executability (e.g., to automate the storage, verification, transformation, and simulation of models) and to allow formal proofs of equivalence and implication between models. Although ORM's richer, graphical constraint notation provides a more complete diagrammatic treatment of schema transformations, the use of textual constraint languages can partly offset this advantage. For their data modeling constructs, ORM, ER, and UML may be given an adequate formal foundation.

Language design often involves a number of *trade-offs* between competing criteria. One well-known trade-off is that between *expressibility* and *tractability*: the more expressive a language is, the harder it is to make it efficiently executable (Levesque, 1984). Another trade-off is between *parsimony* and *convenience*. Although *ceteris paribus*, the fewer concepts the better (cf. Occam's razor), restricting ourselves to the minimum possible number of concepts may sometimes be too inconvenient. For example, it is possible to write computer programs in pure binary using just strings of ones and zeros, but it's much easier to write programs in higher-level languages such as C#. As an example from logic, it's more convenient to use several operators such as "not", "and", "or", and "if-then" even though we could use just one (e.g., "nand"). See the chapter notes for further discussion of this example.

One basic question relevant to the parsimony-convenience trade-off is whether to use the attribute concept as a primitive modeling construct. We've already argued in favor of a negative answer to this question. ORM uses relationships instead of attributes to facilitate capturing, validating, and evolving the conceptual schema while still allowing compact attribute-views to be automatically derived for summary or implementation purposes. Traditional ER supports single-valued attributes, whereas UML supports both single-valued and multivalued attributes. We argue later that multivalued attributes are usually inappropriate for conceptual modeling, although they can be useful for logical and physical modeling.

The rest of this chapter provides an overview of ORM's CSDP and a detailed discussion of the first three steps in this procedure. The ORM approach has been used productively in industry since the mid-1970s. Details of its history can be found in the chapter notes.

3.2 Conceptual Schema Design Procedure

When developing an information system, we first specify what is required and produce a design to meet these requirements. We recommend first developing this design at the conceptual level using ORM. This entails describing the structure of the universe of discourse (UoD) formally in terms of an ORM conceptual schema. If an ORM context is understood, we'll often shorten "ORM conceptual schema" to just "conceptual schema".

The procedure for designing a conceptual schema small enough to manage as a single unit is referred to as the **Conceptual Schema Design Procedure** (CSDP). Large business domains are first divided into subareas (which may overlap). These subareas are normally prioritized to determine which ones to develop first, and a conceptual *subschema* is designed for each using the CSDP.

With a team of modelers, a consensus on terminology is reached, so the same names are used for the same concepts. Later, the subschemas are *integrated* or merged into a global conceptual schema that covers the whole UoD. This integration is often performed iteratively.

This top-down design approach may be summarized as follows:

- Divide the universe of discourse into manageable subareas.
- Apply the CSDP to each subarea.
- Integrate the subschemas into a global conceptual schema.

For each manageably sized business domain or subarea, the conceptual schema design is performed in **seven steps**, as listed below. Often, all seven steps are performed for each model component as it is discussed with the domain expert, rather than applying step 1 to all components, then step 2 and so on.

1. Transform familiar examples into elementary facts.
2. Draw the fact types, and apply a population check.
3. Check for entity types to be combined, and note any arithmetic derivations.
4. Add uniqueness constraints, and check the arity of fact types.
5. Add mandatory role constraints, and check for logical derivations.
6. Add value, set-comparison, and subtyping constraints.
7. Add other constraints and perform final checks.

A **fact** (or *fact instance*) is a proposition that is taken to be true by users in the business domain. Whether the proposition is actually true in the UoD is of no concern to the information system. Here, "taken to be true" has the sense of epistemic commitment, that is, the business users are prepared to act as if they believed the assertion is true.

A **fact type** is the set of all possible fact instances of a given kind (e.g., Person smokes, or Person was born in Country). The CSDP procedure begins with the analysis of examples of information to be output by, or input to, the information system. Basically, the first three steps are concerned with identifying the fact types.

In later steps, we add constraints to the fact types. Throughout the procedure, checks are performed to detect derived facts and to ensure that no mistakes have been made. The rest of this chapter considers the first three steps in detail.

With large business domains, the preliminary segmentation and final integration add two further stages, resulting in nine steps overall. Although the CSDP is best learned in the sequence of steps shown, in practice we might apply the steps differently. For example, we might add constraints as soon as the fact types are entered, rather than waiting for all the fact types to be entered before adding any constraints.

In the commercial world, there are many existing applications that have been implemented using lower-level approaches, resulting in database designs that may be inconsistent, incomplete, inefficient, or difficult to maintain. Such systems are often poorly documented. These problems can be overcome by **reengineering** the existing applications using conceptual modeling techniques. For example, sample populations from existing database tables can be used as input to the CSDP.

Even without sample populations, an existing database schema can be *reverse engineered* to a tentative conceptual schema by using information about constraints and data types and by making assumptions about the use of names.

The conceptual design can then be validated and completed by communicating with a domain expert. The conceptual schema can then be *forward engineered* by applying a conceptual optimization procedure and mapping to the target database system to provide an improved and maintainable implementation. This reengineering approach is discussed in a later chapter.

3.3 CSDP Step 1: From Examples to Elementary Facts

To specify what is required of an information system, we need to answer the question, *What kinds of information do we want from the system?* Clearly, any information to be output from the system must be stored in the system or be derivable by the system. Our first step is to begin with *familiar examples* of relevant information, and *express these in terms of elementary facts.* As a check on the quality of our work, we ask the following questions. *Are the entities well identified? Can the facts be split into smaller ones without losing information?* This constitutes step 1 of the CSDP.

CSDP step 1: Transform familiar examples into elementary facts.

For process modeling, it helps to begin with examples of the processes to be carried out by the system. In 1987, such process examples were coined "use cases", which suggests cases of the system being used. UML recommends such use cases to drive the modeling process. Although process use cases help with designing process models, in practice the move from such use cases to data models is often somewhat arbitrary and frequently results in data models that need substantial reworking.

Business process modeling is part of the overall process of modeling a business domain. In practice, it is often helpful to begin with a high-level view of the services to be provided by the system and then later refine these services into processes that describe how those services are provided. Because processes typically operate on data, however, we feel that a detailed description of those processes should wait until we have a clear understanding of the data. While the topic of business process modeling is the focus of a later chapter, our current and main concern is to ensure high-fidelity data models.

If you want to get the data model right, start with examples of the *data* to be delivered by the system. By analogy with the UML term, we call these "*data use cases*" because they are cases of data being used. This is just another name for the "familiar information examples" concept introduced to the ORM schema design procedure in the 1970s. If you use UML's process use cases to drive the modeling process, you should flesh them out with data samples before working on the class diagrams.

If we are modeling a business domain previously managed manually or by computer, information examples are readily available. If not, we work with the domain expert to provide examples. Three types of examples are output reports, input forms, and sample queries.

These might appear as tables, forms, diagrams, or text. To verbalize such examples in terms of elementary facts, we need to understand what an **elementary fact** is.

In everyday speech, facts are true statements about the real world. However, in computing terminology we resign ourselves to the fact that it is possible to have false "facts" in the database [just as we use the word "statement" for things that aren't really statements in languages such as structured query language (SQL)].

We may think of the UoD as a set of *objects playing roles*. Elementary facts are assertions that particular objects play particular roles. The simplest kind of elementary fact asserts that a single object plays a given role. For example, consider a small domain in which people are identified by their first names. One fact about this domain might be:

1. Ann smokes.

Here, we have one object (Ann) playing a role (being a smoker). Strictly, we should be more precise in identifying objects (e.g., expand "Ann" to "the person with firstname 'Ann'"), but let's tidy up later. With sentences such as (1), the role played by the object is sometimes called a *property* of the object. Here, an elementary fact asserts that a certain object has a certain property. This is also called a *unary relationship* because only one role is involved. Usually, however, a relationship involves at least two roles. For example:

2. Ann employs Bob.
3. Ann employs Ann.

In (2), Ann plays the role of employer and Bob plays the role of employee. In (3), Ann is self-employed so plays both roles. In general, *an elementary fact asserts that a particular object has a property, or that one or more objects participate together in a relationship.*

Here, *"elementary"* indicates that the fact *cannot be "split" into smaller units of information* (with the same objects) that collectively provide the same information as the original. Elementary facts usually do not use logical connectives (e.g., **not**, **and**, **or**, **if**) or logical quantifiers (e.g., **all**, **some**). For example, sentences (4)–(9) are not elementary facts.

4. Ann smokes **and** Bob smokes.
5. Ann smokes **or** Bob smokes.
6. Ann does **not** smoke.
7. **If** Bob smokes **then** Bob is cancer prone.
8. **All** people who smoke are cancer prone.
9. **If some** person smokes **then that** person is cancer prone.

All these sentences express information. Proposition (4) is a logical conjunction. It should be split into two elementary facts: Ann smokes; Bob smokes. Proposition (5) is a disjunction, (6) is a negation, and (7) is a conditional fact. Most database systems do not allow such information to be stored conveniently, and are incapable of making relevant inferences [e.g., deducing that Bob smokes from the combination of (5) and (6)]. For most commercial applications, there is no need to store such information.

Often, the absence of positive information (e.g., Ann smokes) is taken to imply the negative (Ann does not smoke); this is the usual "closed world" assumption. With an "open world" approach, negative information can be explicitly stored using negative predicates or status object types, in conjunction with suitable constraints. For example, the predicates "smokes" and "is a nonsmoker" are mutually exclusive, and the fact type Person has SmokerStatus {"S," "NS"} requires the constraint that **Each** Person has **at most one** SmokerStatus. Sometimes the choice of whether to store positive or negative information depends on which occupies less space, and the borderline between positive and negative may become blurred (e.g., consider Person dislikes Food and Patient is allergic to Drug).

Universally quantified conditionals such as (8) and (9) may be catered for by subset constraints or derivation rules. Derivation rules can be specified readily in SQL (e.g., by means of a view), and are also easily coded in languages such as Prolog. For example: cancerProne(X) **if** person(X) **and** smokes(X).

Elementary facts assert that objects play roles. How are these objects and roles specified? For now, we consider only basic objects: these are either *values or entities*. A **value** is a *constant* that is *self-identifying* in the sense that when you see the constant written down in some context, you always know what is being referred to. As a result, values can be referenced directly, without needing to identify them with a description.

For now, it is sufficient to recognize two kinds of value: *character string* and *number*. Character strings are shown inside quotes (e.g., "USA"). Numbers are denoted without quotes, typically using the usual Hindu-Arabic decimal notation (e.g., 37 or 5.2). Numbers are abstract objects denoted by character strings called numerals. For example, the number 37 is denoted by the numeral "37". We assume that any information system supports strings and numbers as built-in data types. Values are displayed textually, but are internally represented by bit-strings.

Conceptually, an **entity** (e.g., a particular person or car) is referenced in an information system by means of a *definite description*. For example, kangaroos hop about on an entity identified as "the Country named 'Australia'". Entities may also be called "described objects". Unlike values, some—indeed most—entities can change with time. An entity may be a tangible object (e.g., the City named "Paris") or an abstract object (e.g., the Course with CourseCode "CS114"). We consider both entities and values to be objects that exist in the UoD. Object-oriented approaches instead use the term "object" in a more restrictive sense that approximates our use of "entity". Usually, we want to talk about just the entities, but to reference them we make use of values. Sometimes we want to talk about the values themselves. Consider the following sentences:

10. Australia has six states.
11. "Australia" has nine letters.

This is a case of what logicians call the *use/mention distinction*. In (10), the word "Australia" is used to reference an entity. In (11), the word "Australia" is simply mentioned, and refers to itself. In written English language, quotes resolve this distinction. In everyday communication, entities are often referred to by a *proper name* (e.g., "Joe Biden") or by some definite description (e.g., "the 46th president of the USA", or "the president named 'Joe Biden'").

Proper names work if we know what the name refers to from the context of utterance. For example, in (10) you probably took "Australia" to refer to the country named "Australia". However, the sentence itself does not tell you this. Perhaps (10) was about a dog named "Australia" who has six moods (sleepy, playful, hungry, etc.).

Because humans may misinterpret, and information systems lack any creativity to add context, we play it safe by demanding that entities be clearly identified by definite descriptions. To begin with, the description must specify the kind of entity being referred to: the **entity type**. A *type* is the set of all possible *instances*. Each entity is an instance of a particular entity type (e.g., Person, Country). For a given UoD, the entity type Person is the set of all people we might want to talk about during the lifetime of the information system. Note that some authors use the word "entity" for "entity type". We sometimes expand "entity" to "entity instance" to avoid any confusion. Consider the sentence

12. Lee is located in 10B.

This could be talking about a horse located in a stable, or a computer in a room, and so on. By stating the entity types involved, (13) avoids this kind of referential ambiguity. Names of object types are highlighted here by starting them with a capital letter.

13. The Patient "Lee" is located in the Ward "10B".

You might recall the old joke: "*Question:* Did you hear about the man with the wooden leg named 'Smith'? *Answer:* No–What was the name of his other leg?" Here, the responder mistook "Smith" to refer to an entity of type WoodenLeg rather than of type Man. Sometimes, even stating the entity type fails to fully clarify things. Consider the sentence:

14. The Patient "Lee" has a Temperature of 37.

Now imagine that the UoD contains two patients named "Lee Jones" and "Mary Lee". There is more than one person to which the label "Lee" might apply. Worse still, there may be some confusion about the units being used to state the temperature: 37°C is normal bodily temperature, but 37°F is close to freezing!

We resolve this ambiguity by including the **reference mode** (i.e., the manner in which the value refers to the entity). Compare the following two sentences:

15. The Patient with surname "Lee" has a Temperature of 37 Celsius.
16. The Patient with firstname "Lee" has a Temperature of 37 Fahrenheit.

A common way to avoid potential confusion caused by the overlap of first names and surnames is to use fuller names (e.g., "Lee Jones", "Mary Lee"). In some cases, even these names might not be unique, and another naming convention must be employed (e.g., PatientNr). To avoid confusing "No". with the word "No", we'll use "Nr" or "#" to abbreviate "Number".

Most entity designators have three components:

- *Entity type* (e.g., Patient, Temperature)
- *Reference mode* (e.g., Surname, Celsius)
- *Value* (e.g., "Lee", 37)

This is the simplest kind of entity designation scheme. We'll restrict our discussion of reference schemes to this simple case for quite some time. Composite identification schemes are considered later.

Now that we know how to specify objects, how do we specify the roles they play? We use logical **predicates**. In logic, a predicate is basically a declarative *sentence with object holes in it*. To complete the sentence, the object holes or placeholders are filled in by *object terms*. Each object term refers to a single object in the UoD. Object terms are also called singular terms, or object designators. For us, values are designated by lexical constants (sometimes preceded by the value-type name), and entities are designated by definite descriptions that relate values to entities. Consider the following sentence:

17. The Person with firstname "Ann" *smokes*.

Here, the object term is "The Person with firstname 'Ann'". For discussion purposes, the predicate reading is shown here in *italics*. The predicate reading may be shown by itself as "… smokes", using an ellipsis "…" as a placeholder or "object hole" for an object instance. This is a *unary predicate* reading, or *sentence with one object hole* in it. It may also be called a property, or a unary relationship type.

A *binary predicate* reading is a *sentence with two object holes*. Most predicates in information models are binary. Consider this example:

18. The Person with firstname "Ann" *employs* the Person with firstname "Bob".

Here, the predicate reading may be shown by itself as "… employs …", with two object holes. Notice that the *order* in which the objects are placed here is important. For example, even if it is true that Ann employs Bob, it may be false that Bob employs Ann.

In ORM, *a predicate is identified by its fact type rather than its predicate reading*. For example, the fact types Athlete runs Race and Person runs Company have two different predicates named "runs" that have the same reading but different meanings.

A *ternary predicate* is a *sentence with three object holes*. For instance, the fact that Terry worked in the Computer Science Department for 10 years involves the predicate "… worked in … for …"

An *n-ary predicate* is a sentence with n object holes ($n > 0$). Because the order is significant, a filled-in n-ary predicate is associated with a *sequence* of n object terms, not necessarily distinct. The value of n is the **arity**, or degree, of the predicate. Predicates of arity ≥ 2 are polyadic. An elementary fact asserts a proposition of the form:

$$Ro_1 \ldots o_n$$

where R is a predicate of arity n, and $o_1 \ldots o_n$ are n object-terms, not necessarily distinct (as in *predicate logic*). Moreover, with respect to the UoD, the proposition must not be expressible as a conjunction of simpler propositions.

For naturalness, we write predicates in *mixfix* (or distfix) form, where the terms may be mixed in with the predicate reading. For example, the following ternary fact uses the predicate reading "… moved to … during … ".

19. The Scientist with surname "Einstein" *moved* to the Country with Countrycode "USA" *during* the Year 1933 CE.

Step 1 of the CSDP involves translating relevant information examples into sentences like this. As a simple example, consider the report in Table 3.1. Try to express the information in the first row as an elementary fact. For help, use the *telephone heuristic*—imagine you must convey the information over the telephone. In performing this visual to auditory transformation, you should specify each entity in terms of its entity type, reference mode, and value, and also include the predicate reading.

Table 3.1 Some languages and their designers.

Designer	*Language*
Wirth	Pascal
Kay	Smalltalk
Wirth	Modula-2

In reports like this, the column headings and table names or captions often give a clue as to the object types and predicates. The column entries provide the values. Here is one way of translating row 1 as an elementary fact:

20. The Person with surname "Wirth" designed the Language named "Pascal".

Notice that the entity types and reference modes appear as nouns, and the predicate as a verb phrase. This is fairly typical. In translating row 1 into the elementary fact (20), we read the row from left to right. If instead we read it from right to left, we might say:

21. The Language named "Pascal" *was designed by* the Person with surname "Wirth".

In reversing the order of the terms, we also reversed the predicate. We speak of "was designed by" as the *inverse* of the predicate "designed". Although semantically we might regard sentences (20) and (21) as expressing the same fact, syntactically they differ. Most logicians would describe this as a case of two different sentences expressing the same proposition. Linguists like to describe this situation by saying the two sentences have different *surface structures* but the same *deep structure*.

For example, one linguistic analysis might portray the deep structure sentence as comprising a verb phrase (Design) and various noun phrases (the object terms), each of which relates to the verb in a different case (e.g., agentive for Wirth and objective for Pascal), together with a modality (past tense). Different viewpoints exist as to the "correct" way to portray deep structures (e.g., what primitives to select), and the task of translation to deep structures is often complex. In practice, most information systems can be designed without delving further into such issues.

It is important not to treat sentences like (20) and (21) as different, unrelated facts. Our approach with binary fact types is to choose one primary way of stating the predicate, but optionally allow the inverse reading to be shown as well, such as

22. The Person with surname "Wirth" *designed/was designed by* the Language named "Pascal".

Here the predicate reading on the left of the slash "/" is used for the left-to-right reading (20). The predicate reading on the right of the slash is used for the inverse reading (21). The slash visually suggests jumping over the other predicate when reading left to right, and jumping under the other predicate when reading right to left.

Having different ways to talk about a fact type can help communication, and might improve constraint specification. For example, the constraint verbalization "Each Language was designed by some Person" is preferable to the equivalent "For each Language, some Person designed that Language".

Ternary and longer fact types have many possible orderings, but typically only one predicate reading is displayed.

Now consider Table 3.2. This is harder to verbalize because the columns don't have separate names. Assume that Terry and Tony are males, and that Norma and Gwen are females. Try to verbalize the information on the top row in terms of elementary facts before reading on.

Table 3.2 An output report about marriages.

Married couples	
Terry	Norma
Tony	Gwen

Perhaps you verbalized this as in (23). For completeness, the inverse is included.

23. The Person with firstname "Terry" *is married to/is married to* the Person with firstname "Norma".

Notice that the forward predicate is the same as the inverse. This is an example of a *symmetric* relationship. Such relationships create special problems (as discussed in a later chapter). To help avoid such problems, at the conceptual level *no asserted predicate should be the same as its inverse.*

You can always rephrase the fact to ensure this. For example, (24) does this by highlighting the different roles played by each partner.

24. The Person with firstname "Terry" *is husband of/is wife of* the Person with firstname "Norma".

As another example, consider Table 3.3. This is like Table 3.1 that was considered earlier, but it has a third column added. Try to express the information on the first row of the table in terms of elementary facts before reading the discussion about the table.

Table 3.3 Origin details about some programming languages.

Designer	*Language*	*Year*
Wirth	Pascal	1970
Kay	Smalltalk	1972
Wirth	Modula-2	1979

We might at first consider expressing this information as sentence (25), using the ternary predicate "... designed ... in ...". Do you see any problems with this?

25. The Person with surname "Wirth" *designed* the Language named "Pascal" *in* the Year 1970 CE.

Recall that an elementary fact must be simple or irreducible. It cannot be split into two or more simpler facts in the context of the UoD. The appearance of the word "and" in a sentence usually indicates that the sentence may be split into simpler facts. Here there is no "and", but "common sense" tells us that the fact can be split into the following two elementary facts with no information loss:

26. The Person with surname "Wirth" designed the Language named "Pascal".
27. The Language named "Pascal" was designed in the Year 1970 CE.

Here, "no information loss" means that if we know (26) and (27), then we also know (25). The phrase "common sense" hides some formal ideas. A sufficient reason for splitting (25) into (26) and (27) would be an implicit understanding that each language was designed in only one year. This constraint holds if we interpret "was designed in" to mean "had its design completed in". Let us agree with this interpretation.

If instead we meant "had work done on its design in", then a language may be designed in many years (e.g., Wirth did most of Pascal's design in 1968–69, but did not finalize its definition until 1970). In this case, we could still justify the split if each language had only one designer, or at least the same set of designers for each year. However, this might not be true. For example, if we include UML as a language, it had different designers in different years. This illustrates the need to be clear about the meaning of our wording and to strive for sample data that illustrate what is possible.

The ORM method encourages the use of informative terms for object types and predicates. However, sometimes it is difficult to think of a term that captures the precise meaning and is also conveniently short. In such cases, if the meaning of an object type name or predicate reading could be misinterpreted, it is important to add a *descriptive comment* to clarify its meaning. Although such comments are often suppressed on ORM diagrams, modern ORM tools allow them to be captured and displayed in various ways.

Later CSDP steps add some formal checks to detect violations of elementarity, so if our "common sense" fails us in step 1, we will normally see this error at a later stage. For now though, let's work with our intuitions. Suppose we split the ternary into the two binaries: Person designed Language; Person completed design in Year. Would this be acceptable? As an exercise, use the table's population to show that this kind of split would actually lose information.

After plenty of practice at step 1, you may wish to write the elementary facts down in abbreviated form. To start with, you can drop words such as "the" and "with" where they introduce object types and reference modes. Reference modes are placed in parentheses after the object types.

A dot before the reference mode name indicates that the object type name is prepended to it in the underlying value type name. For example, Language(.Name) and Country(.Name) have different underlying value types LanguageName and CountryName. A colon after the reference mode name, such as Height(cm:), indicates unit-based measurement. The next section discusses the various kinds of reference mode in more detail.

You may shorten some identifiers for object types, reference modes, and predicates so long as the shorter readings are still meaningful. In English, start the name of object types with a capital letter. Reference mode names usually start with a capital letter but may start with a small letter unless capitals have significance (e.g., "CE:").

For example, facts (26) and (27) may be set out more concisely as (26a) and (27a), as shown below.

26a. Person (Surname) "Wirth" designed Language (.Name) "Pascal".
27a. Language (.Name) "Pascal" was designed in Year (CE:) 1970.

If the reference schemes for entity types are declared earlier, they may be omitted in setting out the facts. For example, (26) and (27) may be specified as:

Reference schemes: Person(Surname); Language(.Name); Year(CE:)

Facts: Person "Wirth" designed Language "Pascal".
Language "Pascal" was designed in Year 1970.

Even more conveniently, a fact type may be displayed in diagram form (see next section), and fact instances may be entered into the relevant fact table simply by entering values.

The task of defining a formal grammar sufficient to capture any sentence expressed in natural language is daunting, partly because of the many ways in which objects may be referenced. For example, consider the sentence, "The next person to step on my toe will cop it".

Fortunately, such sentences don't appear in database tables, where simple value-based schemes are used to reference objects. ORM is capable of formally capturing the relevant semantics of any fact that can be represented in a database table. Structured object-terms and predicates provide the logical deep structure, independent of the natural language (English, Japanese, etc.) that is used to express the fact.

By supporting ordered, mixfix predicates, ORM enables this deep structure to be expressed in a surface structure in harmony with the ordered, mixfix nature of natural language. For example, consider Fig. 3.1.

EmployeeNr	*Department*
37	Sales

Jugyo in	*Ka*
37	Eigyo

FIG. 3.1 The same fact in English and Japanese.

Here, the two tables convey the same fact in different languages. The fact may be expressed in English as (28) and in Japanese as (29). For discussion purposes, the reference modes are underlined, and the predicates are italicized.

28. The Employee with EmployeeNr "37" *works in* the Department with name "Sales".
29. Jugyo in jugyo in bango "37" *wa* "Eigyo" to iu namae no Ka *ni shozoku suru.*

These may be parsed into the structures shown in 28a and 29a, with the predicates again italicized. The sentences have the same deep structure. The infix predicate "… works in …" corresponds to the mixfix predicate "… wa … ni shozoku suru".

28a. Employee (.Nr) "37" *works* in Department (.Name) "Sales".
29a. Jugyo in (.Bango) "37" *wa* Ka (.Namae) "Eigyo" *ni shozoku suru.*

Now consider the output reports in Fig. 3.2. Try to verbalize their top row in terms of one or more elementary facts. The data in Fig. 3.2(a) are real while the data in Fig. 3.2(b) are fictitious but possible. Assume that the following reference schemes are predeclared: Politician(.Name); Country(.Code); and Year(CE:).

(a)

Politician	*Country*	*Year*
George W. Bush	US	1946
William J. Clinton	US	1946
John Howard	AU	1939
Kevin Rudd	AU	1957

(b)

Politician	*Country*	*Year*
George W. Bush	GB	2003
George W. Bush	IE	1999
William J. Clinton	IE	1999
William J. Clinton	IE	2003
John Howard	IE	2003
John Howard	US	2003

FIG. 3.2 Two reports about politicians.

Although there are in principle infinitely many possible interpretations of the data, you can probably guess the intended meaning. We may read off the information represented in Fig. 3.2(a) and (b), respectively, as shown in the following two facts.

30. The Politician "George W. Bush" *was born in* the Country "US" *in* the Year 1946.
31. The Politician "George W. Bush" *visited* the Country "GB" *in* the Year 2003.

While these facts do provide a correct interpretation of the meaning of the data on the top rows, only one of these facts is elementary. Which is it?

Fact 30 is not elementary because it may be split into the following facts without loss of information.

32. The Politician "George W. Bush" *was born in* the Country "US".
33. The Politician "George W. Bush" *was born in* the Year 1946.

Connecting two sentences by "and" forms a logical *conjunction* of the two sentences. Intuitively, fact 30 is equivalent to the conjunction of facts 32 and 33, so fact 30 is splittable (nonelementary). Formally, this is because of a uniqueness constraint on the politician column—each politician was born in only one country and in only one year. We discuss such constraints in a later chapter and provide rules and heuristics to help determine whether a given fact is elementary. For now, however, let's just use our intuitions.

Fact 31 is elementary because you cannot treat it as a conjunction of simpler facts without losing information. For example, if we told you that George W. Bush visited the United Kingdom (the country with country code "GB") and that he travelled overseas in 2003, it does not follow that he visited the United Kingdom in that year.

In some cases, a ternary that is not splittable into two facts may be split into three facts. Here we might try to split fact 31 into the above two facts as well as the third fact that the United Kingdom was visited in 2003. However even these three facts don't guarantee that George W. Bush visited the United Kingdom in 2003.

Formally, the structural difference between reports (a) and (b) is their uniqueness constraint pattern. In report (a), the first column values must be unique (not duplicated). In report (b), only the whole rows must be unique (entries for single columns or any column pair may be duplicated). The next chapter discusses such patterns in detail.

As another example, consider the report shown in Table 3.4. Try step 1 yourself on this table before reading on.

Table 3.4 A report storing facts about lecturers.

Lecturer:

Name	*Birthyear*	*Age*	*Degrees*
Adams JB	1946	46	BSc, PhD
Leung X	1960	32	BE, MSc
O'Reilly TA	1946	46	BA, BSc, PhD

One tricky feature of this table is the final column. Entries in this column are sets of degrees. Phrase your sentences to include only one degree at a time. Step 1 applied to the first row results in four facts that may be set out as follows.

34. The Lecturer with name "Adams JB" *was born in* the Year 1946 CE.
35. The Lecturer with name "Adams JB" *has* the Age 42 years.
36. The Lecturer with name "Adams JB" *holds* the Degree with code "BSc".
37. The Lecturer with name "Adams JB" *holds* the Degree with code "PhD".

Here, the entity types and reference modes are Lecturer (.Name), Year (CE:), Age (y:), and Degree (.Code). There are three fact types: Lecturer *was born in* Year; Lecturer *has* Age; and Lecturer *holds* Degree. The entity types Year and Age are semantically different. Year involves a starting point in time, whereas Age is merely a duration of time (in this case, measured in years).

A more difficult example is shown next in Table 3.5. This is an extract of a report on tutorial groups. Perform step 1 yourself on this before reading the following discussion.

Table 3.5 An output report listing tutorial allocations.

Tute Group	*Time*	*Room*	*Student Nr*	*Student Name*
A	Mon. 3 p.m.	CS-718	302156 180064 278155 334067 200140	Bloggs FB Fletcher JB Jackson M Jones EP Kawamoto T
B1	Tue. 2 p.m.	E-B18	266010 348112	Anderson AB Bloggs FB
...	...	...	...	...

Verbalization of an example involves interpretation, so it is important that the kind of example is *familiar* to us or the domain expert who is assisting us with step 1. Because domain experts often lack technical expertise in modeling, we should not expect them to do all of step 1 themselves. It is sufficient if they verbalize the information correctly in their own terms. This is **step 1a**: *verbalize the information* (perhaps informally).

As modelers, we often need to refine their verbalization by ensuring that the facts are elementary and that the objects are well identified. This is **step 1b**: *verbalize the information as elementary facts*. For example, a domain expert might perform step 1a by verbalizing the information on the top row of this report as:

Tute group A meets at 3 p.m. Monday in Room CS-718.
Student 302156 belongs to group A and is named "Bloggs FB".

We might then perform step 1b by refining these informal sentences into the following four elementary sentences:

38. The TuteGroup with code "A" *meets at* the Time with dhcode "Mon 3 p.m.".
39. The TuteGroup with code "A" *meets in* the Room with roomNr "CS-718".
40. The Student with studentNr 302156 *belongs to* the TuteGroup with code "A".
41. The Student with studentNr 302156 *has* the StudentName "Bloggs FB".

Many features in this example rely on interpretation. For instance, we assumed that StudentNr and StudentName refer to students and that a student number and student name on the same row refer to the same student. We also filled in the associations as "meets at", "meets in", "belongs to", and "has". The report itself does not tell us this. We use our background with the situation to make such assumptions. Decisions were also made about entity types and reference schemes, for example, we treated Time as an entity type referenced by a day-hour-code rather than introducing Day and Hour entity types. A similar comment applies to Room and StudentName.

StudentNr was chosen rather than StudentName to identify Student. The report helps here because StudentNr appears first (on the left) and "Bloggs FB" appears with two different student numbers, suggesting that more than one student may have the same name. However, we are still making assumptions (e.g., that students have only one student number, or that students belong to only one group).

Another assumption is that each tutorial group meets only once a week. We need this to justify using separate facts for the time and room (rather than verbalizing this as TuteGroup meets at Time in Room). The fact that group *A* is not repeated in the report helps with this decision, but this still assumes the sample is representative in this regard. Because interpretation is always involved in the initial step, if we are not familiar with the example we should resolve any doubts by asking a domain expert, using examples familiar to them.

Although we as modelers might be expert in expressing at a formal, type level, not all domain experts can do this. By working with examples familiar to the domain experts, we can tap their implicit understanding of the UoD without forcing them to abstract and express, perhaps incorrectly, the relevant structure that we seek.

Sentence (41) expresses a relationship between an entity (a student) and a value (a name). When verbalizing facts, the **value type** is stated just before the value (e.g., StudentName "Bloggs FB"). Unlike entity terms, value terms have no reference scheme.

With many reports, it is sometimes useful to draw a connection between the relevant fields as we verbalize the corresponding fact. For example, we might add links between the columns of Table 3.5 as in Fig. 3.3. This informal summary of the fact types helps us see if some connections have been missed.

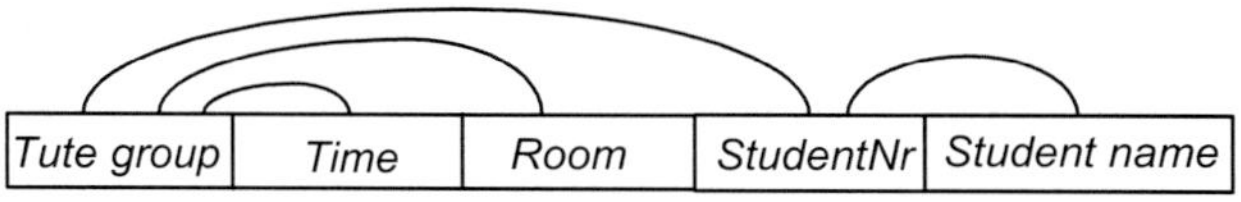

FIG. 3.3 Drawing connections for the verbalized relationships.

In addition to tables, *forms* are a common source of information examples. Consider the input form shown in Fig. 3.4. Forms are used more often for input than output, but may be used for both (e.g., the personnel forms considered in the previous chapter). Suppose students studying the subject CS114 use the form shown in Fig. 3.4 to indicate up to three preferences regarding which tutorial time is most suitable for them. This input form is used to help decide which students are assigned to which groups, and which groups are eventually used. If this information is to be taken into account in determining tutorial allocations, it must be stored.

CS114 Tutorial preferences form

Please complete this form to assist in tutorial allocation.
Tutorials are of 1 hour duration, and are available at these times:

Monday	Tuesday	Thursday
		10 a.m.
		11 a.m.
		12 noon
	2 p.m.	2 p.m.
3 p.m.		3 p.m.

Student number: ....

Student surname: .... *Initials:* ....

Tutorial preference 1: ....

Tutorial preference 2: ....

Tutorial preference 3: ....

FIG. 3.4 An input form for collecting tutorial preferences.

Let's assume that the previous output report (Table 3.5) shows the tutorial allocations for CS114 that were made after all the student preferences are considered. The preference input form lacks some of the information needed for the allocation report. For example, it does not show how groups are assigned to rooms and times. This helps prevent students from entering wrong data (they enter the times they prefer directly rather than indirectly through associated group codes) and allows flexibility in offering many tutorials at the same time. To perform step 1 here, we first fill out the form with some examples, as shown in Fig. 3.5.

CS114 Tutorial preferences form

Please complete this form to assist in tutorial allocation.
Tutorials are of 1 hour duration, and are available at these times:

Monday	Tuesday	Thursday
		10 a.m.
		11 a.m.
		12 noon
	2 p.m.	2 p.m.
3 p.m.		3 p.m.

Student number: 302156...

Student surname: Bloggs.............................. *Initials:* FB.....

Tutorial preference 1: Mon 3 p.m............

Tutorial preference 2: Thurs 11 a.m.........

Tutorial preference 3: ....

FIG. 3.5 The tutorial preference form populated with sample data.

These facts may now be verbalized as Student(.Nr) 302156 has Surname "Bloggs"; Student 302156 has Initials "FB"; Student 302156 has first preference at Time(.dhCode) "Mon 3 p.m."; Student 302156 has second preference at Time "Thurs 11 a.m.".

Taken individually, the output report and the input form reveal only partially the kinds of information needed for the system. In combination, however, they might be enough for us to arrive at the UoD structure. If so, the pair of examples is said to be significant.

In general, a set of examples is **significant** or adequate with respect to a specific UoD only if it illustrates all the relevant sorts of information and constraints. With complex UoDs, significant example sets are rare. With our current example, if a student can be allocated to only one group, then Table 3.5 is significant in this respect.

However, if more than one group can be held at the same time, Table 3.5 is not significant in this other respect. A further row is needed to show this possibility (e.g., a row indicating that group B2 meets at Tuesday 2 p.m.).

When extracting facts from information examples, we need to decide *which aspects should be modeled.* This helps determine the scope of the UoD. A useful heuristic is to ask the question, *Which parts may take on different values?* Look again at the input form in Fig. 3.5. The header section contains information that we may or may not wish to model. The first item we see is "CS114". Is it possible to have other values in place of it, within our overall application? If the only subject of interest is CS114, then the answer is "no". However, if we wish to cater for other subjects as well, we might require another form with a different value here (e.g., "CS183"). In this case, we need to introduce Subject (or some equivalent term) as an object type within our UoD.

Different universities may use different terms for a unit of study. Instead of our "Subject", terms such as "Course" or "Unit" might be used. If the domain experts all prefer the same term, use that. If different people use different terms for the same concept, get them to agree upon a *standard term*, and also note any *synonyms* in use.

Returning to the header of our preferences form in Fig. 3.5, we see the word "Tutorial". Could this change (e.g., to "Lecture")? If we wish to capture preferences for lectures as well as tutorials, the answer is "yes", and we could model this as data. But let's assume that this is not the case. The rest of the form header contains other information (e.g., duration of tutorials) but let's assume this doesn't need to be modeled.

The middle section of the form contains information about the tutorial times. If our UoD has only one subject (CS114), we could model this information as unary facts (e.g., Time "Mon 3 p.m." is available). If we need to cater for other subjects as well, then we need to treat the "CS114" at the top of the form as data, and hence verbalize the schedule as binary facts (e.g., Subject "CS114" has a tutorial slot at Time "Mon 3 p.m."). A completed tutorial preference form for a different subject is shown in Fig. 3.6.

Here, the layout of the five weekdays into columns makes it more obvious that the times are to be treated as data. The first fact from this section reads: Subject "CS183" has a tutorial slot at Time "Tues 2 p.m.".

CS183 Tutorial preferences form

Monday	Tuesday	Wednesday	Thursday	Friday
			9 a.m.	
			11. a.m.	
	2 p.m.		2 p.m.	
	3 p.m.	3 p.m.		

Student number: 211780...

Student surname: Smith................................ *Initials:* JA.....

Tutorial preference 1: Tues 2 p.m.............

Tutorial preference 2: Thurs 11 a.m.........

Tutorial preference 3: Tues 3 p.m.............

FIG. 3.6 A completed tutorial preference form for a different subject.

If you reformat the structure of the earlier CS114 example (Fig. 3.5) to agree with this structure and place the two forms side by side, you can see what aspects are to be modeled as data by looking at what data changes between the two forms (subject code, times, student details).

In this larger UoD, assuming that students may enroll in many subjects, the facts about student preferences now need to take the subject into account. Instead of binaries, the preference facts are verbalized as ternaries. For example: Student 302156 has first tutorial preference for Subject "CS114" at Time "Mon 3 p.m."; Student 211780 has third tutorial preference for Subject "CS183" at Time "Tues 3 p.m.".

Tutorial allocations would also need to indicate the relevant subject (e.g., Table 3.5 would need a header showing the subject). Tutorial groups would then need a composite reference scheme including both the subject code and the group code. For example, the first fact from the CS114-headed version of Table 3.5 would read as: TuteGroup "A" of Subject "CS114" meets at Time "Mon 3 p.m.". Composite reference schemes are discussed in detail later in the book.

It is seen later that no set of examples can be significant with respect to derivation rules or subtype definitions. In such cases, the use of a domain expert is essential. With the current domain, we made no mention of the rules used to arrive at the tutorial allocations.

If in addition to storing information about preferences and allocations, the information system has to compute the allocations in a nearly optimal way, respecting preferences and other practical constraints (e.g., size of groups), the design of the derivation rules becomes the challenging aspect of the schema. While this can be automated, an alternative is to divide the task between the human expert and the system. High-level languages facilitate such cooperative solutions.

Often, information examples used for verbalization in step 1 apply to the way the business or application domain currently works. From these examples, we can build the *as-is model* to reflect the current practice (as it is now). Some changes may also be needed to expand or improve the way the business operates. For example, we might start with separate applications to administer tutorials for just one subject and then realize it is better to integrate these into a single application capable of handling all the subjects (as discussed earlier).

By including examples of the new data requirements, we are then able to build the *to-be model*, which reflects the way we want the business to be in the future. A proper understanding of the as-is model assists in designing the to-be model.

As you gain more experience as a modeler, you will be able to draw upon lessons learned from prior modeling projects to help spot ways to improve things on future projects. Modeling is not just a science. It's an art as well, and that makes it more fun.

A comprehensive set of output reports (covering intermediate stages) may include all the information on input forms. Output reports tend to be easier to interpret, especially if the input forms have been poorly designed. Care is needed in the design of the input forms to make them clear and simple for users.

Information can appear in many ways. Apart from various tables and forms, information may be expressed graphically in various *diagrams, charts, maps,* and *graphs.* Harris (1996) discusses several hundred different ways of presenting information graphically.

Regardless of how it's presented, information can always be verbalized as facts. Because practice helps with verbalization skills, we've included many varied examples in the book to prepare you for performing step 1 in practical situations.

As a simple graphical example, Fig. 3.7 might be used to display information about nonstop flight connections provided by a particular airline, with the arrowheads indicating the direction of the flights. As an exercise, perform step 1 for this graph before reading on.

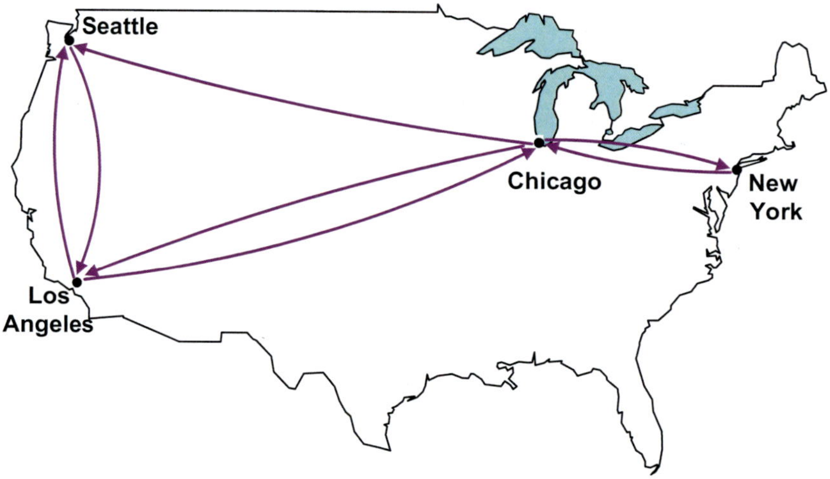

FIG. 3.7 A graph showing nonstop flight connections between cities.

Now let's see how well you performed. There is only one entity type: City (.Name). There is also only one fact type: City has a flight to City. For instance, the arrow from Chicago to Seattle may be verbalized as: City "Chicago" has a flight to City "Seattle". The "to" in the predicate is important because it conveys direction and avoids the symmetry problem with the earlier marriage example.

If this was an as-is model, and we wanted also to talk about the flight connections (e.g., state their duration) or to include many airlines, we should add flight numbers to the arrows on the graph. This to-be model leads to a different verbalization, which you might like to try for yourself. A later exercise returns to this example.

In practice, it is usually possible for a city to have more than one airport, so that connections apply directly between airports rather than cities. Airports are identified by airport codes (e.g., "JFK" and "LGA", respectively, denote John F. Kennedy and LaGuardia airports that service New York).

Airports can service a city without being located in that city. For example, Newark airport (EWR) located in New Jersey also services New York, and Sea-Tac airport (SEA) located between Seattle and Tacoma services Seattle. We need to determine which specific kinds of facts are of interest in the business domain and verbalize them precisely (e.g., Airport services City is a different fact type from Airport is located in City).

By now, you may have some sense of the power of verbalizing examples in terms of elementary facts. No matter what kind of example you start with, if you or an assistant understands the example then you should be able to express the information in simple facts. This does require practice at the technique, but this is fun anyway, isn't it? If you can't do step 1, there is little point in proceeding with the design—either you don't understand the UoD or you can't communicate it clearly.

Although it might sound hard to believe, if you have performed step 1 properly, you have completed the hardest part of the CSDP. The remaining steps consist of diagramming and constraining the fact types. Apart from the problem of detecting unusual constraints and derivation rules, once you learn the techniques you can often carry out those steps almost automatically. With step 1, however, you always need to draw upon your interpretation skills.

Exercise 3.3

1. Assuming suitable entity types and reference modes are understood, which of the following sentences fit our standard pattern for expressing exactly one elementary fact?
   (a) Adam likes Eve.
   (b) Bob does not like John.
   (c) Tom visited Los Angeles and New York.
   (d) Tom visited Los Angeles or New York.
   (e) If Tom visited Los Angeles then he visited New York.
   (f) Sue is funny.
   (g) All people are funny.
   (h) Some people in New York have toured Australia.
   (i) Brisbane and Sydney are in Australia.
   (j) Brisbane and Sydney are in the same country.
   (k) Who does Adam like?

2. Indicate at least two different meanings for each of the following sentences by including names for object types and reference modes.
   (a) Pluto is owned by Mickey.
   (b) Dallas is smaller than Sydney.
   (c) Arnold can lift 300.

   Perform step 1 of the CSDP for the following output reports. In verbalizing the facts, you may restrict yourself to the top row unless another row reveals a different kind of fact.

3.

Athlete	*Height (cm)*
Jones EM	166
Pie QT	166
Smith JA	175

4.

Athlete	*Height (cm)*
Jones EM	400
Pie QT	450
Smith JA	550

5.

Person	*Height (cm)*	*Year*
Jones EM	166	1955
Pie QT	160	1970
Smith JA	175	1955

6.

Person	*Height (cm)*	*Year*
Jones EM	160	1970
	166	1980
	166	1990

7.

Advisory panel	*Internal member*	*External member*
Databases	Codd	Ienshtein
	Kowalski	Spock
Logic programming	Colmerauer	Robinson
	Kowalski	
	Spock	

8.

Student Nr	*Course*	*Rating*
1001	CS112	7
1001	CS100	6
1002	CS112	4
1003	CS100	7
1003	CS112	4
1003	MP104	4

9.

Parents	*Children*
Ann, Bill	Colin, David, Eve
David, Fiona	Gus

10.

Country	*Friends*	*Enemies*
Disland	Oz	
Hades		Oz
		Wundrland
Wundrland	Oz	Hades
Oz	Disland	Hades
	Wundrland	

11.

Apple	*Australia*	June, July, August
	America	Oct, Dec, Jan
	Ireland	Oct, Dec
Mango	*Australia*	Nov, Dec, Jan, Feb
Pineapple	*America*	June, July
	Australia	Oct, Nov, Dec. Jan

3.4 CSDP Step 2: Draw Fact Types and Populate

Once we have translated the information examples into elementary facts, and performed quality checks, we perform the next step in the CSDP. First, we *draw* a conceptual schema diagram that shows all the *fact types*. This illustrates the relevant object types, predicates, and reference schemes. Once the diagram is drawn, we check it with a *sample population*.

CSDP step 2: Draw the fact types and apply a population check.

Consider the sample output report of Table 3.6.

Table 3.6 A relational table listing who drives which cars.

Drives:

Person	*Car*
Adams B	235PZN
Jones E	235PZN
Jones E	108AAQ

Let us agree that the information in this report can be expressed by the following three elementary facts, using "RegNr" to abbreviate "registration number":

The Person named "Adams B" *drives* the Car with RegNr "235PZN".
The Person named "Jones E" *drives* the Car with RegNr "235PZN".
The Person named "Jones E" *drives* the Car with RegNr "108AAQ".

Before looking at the relevant conceptual schema diagram, it may help to explain things if we first view an *instance diagram* for this example (see Fig. 3.8).

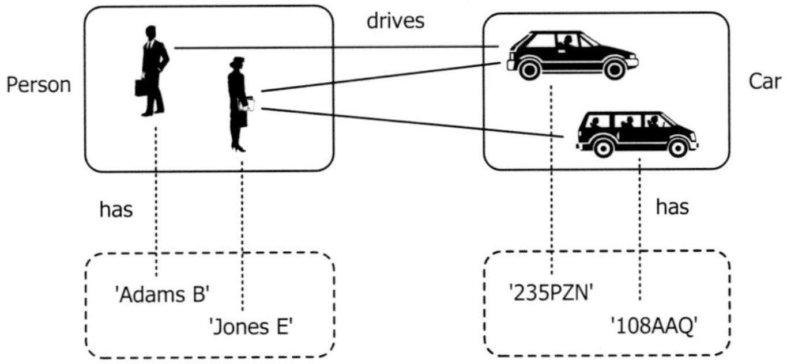

FIG. 3.8 An instance diagram.

Instance diagrams illustrate particular instances of objects and relationships. Taking advantage of the concrete nature of the entities in this example, cartoon drawings denote the actual people and cars. The values are shown as character strings. A particular fact or relationship between a person and a car is shown as a solid line. A particular reference between a value and an entity is shown as a broken line.

Fig. 3.9(a) shows a *conceptual schema diagram* for the same example, adding inverse readings for the two fact types. Instance diagrams and conceptual schema diagrams depict an *entity type* as a *named, solid, soft rectangle* (rectangle with rounded corners). A *value type* is shown as a *named, broken, soft rectangle.* The object type's name is written inside. Some versions of ORM allow an ellipse (Fig. 3.9(b)) or a hard rectangle (Fig. 3.9(c)) instead of a soft rectangle.

On an instance diagram, individual objects are portrayed explicitly (by icons or text). However, on a conceptual schema diagram, individual objects are omitted (unless we reference them in associated tables). Recalling that a type is the set of permitted instances, we may imagine that objects are represented as points inside the type shape.

On a conceptual schema diagram, the *roles* (relationship parts) played by objects are shown explicitly as *boxes.* Optionally, roles may be given a name, in which case the *role name* is displayed in square brackets next to the role box, for example, "[driver]" in Fig. 3.9(a). Each *predicate* is depicted as a *named, contiguous sequence of one or more role boxes* ("contiguous" means the boxes are adjacent, with no gaps in between).

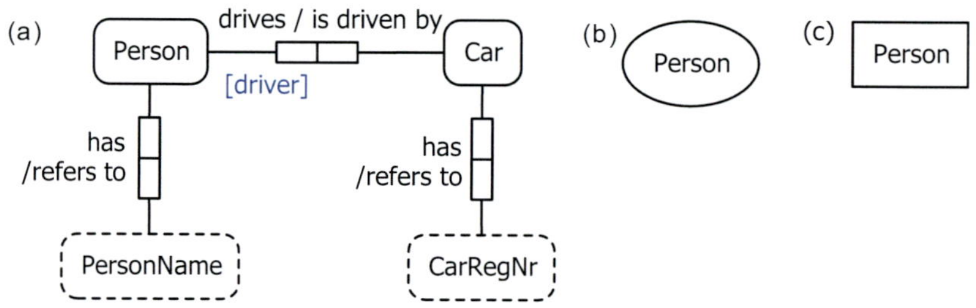

FIG. 3.9 (a) A conceptual schema (minus constraints) and (b and c) alternate shapes.

Predicates are typically ordered from one end to the other, with their reading displayed next to the predicate shape. For binary relationships (two roles), both forward and inverse predicate readings may be shown, separated by a slash "/", as in Fig. 3.9(a).

Each predicate must have at least one reading displayed. If just one predicate reading is displayed, by default this is read left to right on a horizontal predicate and downward on a vertical predicate. *Arrow tips* may be added to a predicate reading to indicate the reading direction. For example, "◄" indicates the predicate reading is right to left, and "▲" indicates the reading direction is upward.

Each role is connected to exactly one object type by a *line,* indicating that the role is played only by objects of that type. A complete conceptual schema diagram includes the relevant constraints. We'll see how to add these later.

A relationship used within the preferred identification scheme for an entity is called a *reference* or *existential fact* (e.g., There exists a car that has the car registration number "235PZN").

All other conceptual schema relationships are called *elementary facts* (e.g., The person named "Adams B" drives the car with car registration number "235PZN").

Typically, elementary facts are relationships between entities, and references are relationships between entities and values. References provide the bridge between the world of entities and the world of values. Simple references as well as elementary facts are both examples of *atomic facts.*

This is clearly seen in Fig. 3.10(a), where an instance has been added to populate each relationship. The relationship between the person and the car (depicted with icons) is an elementary fact. The relationship between the name "Adams B" and the person is a reference or existential fact, as is the relationship between the registration number "235PZN" and the car.

Although both reference predicates have the reading "has", they are different predicates. Internally, a computer-aided software engineering (CASE) tool may identify a predicate by its fact type or by using a surrogate (e.g., "P2", "P3") or an expanded name (e.g., "PersonHasPersonName", "CarHasRegNr").

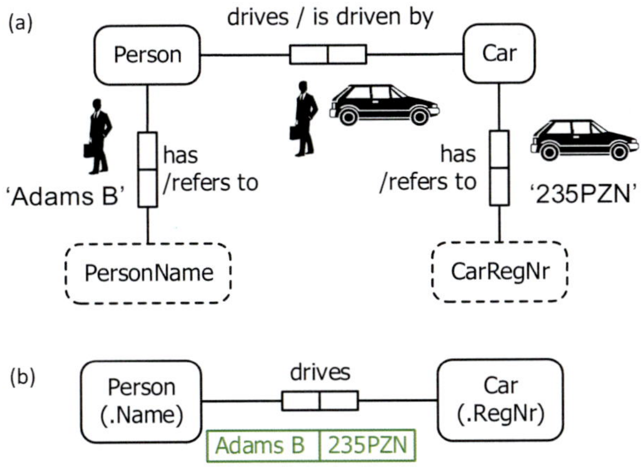

FIG. 3.10 Using reference modes for 1:1 reference: (a) model and (b) abbreviation.

Although the predicate reading "has" may be used with fact types, it is best avoided if there is a more descriptive, natural alternative. For example, "Person drives Car", if accurate, is better than "Person has Car", which could mean many things (e.g., Person owns Car).

For this example, each person has exactly one name, and each person name refers to at most one person. Each car has exactly one car registration number, and each car registration number refers to at most one car. This situation is seen clearly in the earlier instance diagram, Fig. 3.8. Each of the two reference types is said to provide a *simple 1:1 reference scheme.* We read "1:1" as "one to one". Later, we'll see how to specify this on a conceptual schema diagram using uniqueness and mandatory role constraints.

When a simple 1:1 naming convention exists, we indicate the *reference mode* by appending its name in *parentheses* to the name of the entity type, and then use the values themselves to depict object instances in associated *fact tables.* Ignoring other constraints, the populated schema of Fig. 3.10(a) may be displayed concisely by Fig. 3.10(b). Unless we want to illustrate the reference schemes explicitly, this concise form is preferred because it's closer to the way we verbalize facts, and it simplifies the diagram. The fact table is omitted if we wish to display just the schema.

Reference modes indicate the mode or manner in which objects (typically values) refer to entities. Using reference modes reduces the need to display value types explicitly. However, to understand the abbreviation scheme, we need to know how to translate between reference modes and value types. Different versions of ORM have different approaches to this. The method used in the NORMA tool for ORM 2 is now outlined.

Let the notation "*E (r)* → *V*" mean "Entity type *E* with reference mode *r* generates the Value type *V*". Reference modes may be popular, unit-based (measurement), or general. An initial list of *popular reference modes* is predefined in NORMA, including Name, name, Code, code, Nr, nr, Id, id and #. When parenthesized, popular reference modes are preceded by a *dot* ".". To obtain the value type name, a popular reference mode name is appended to the name of its entity type, for example, Product(.Name) → ProductName; Country(.Code) → CountryCode; Song(.Title) → SongTitle; Country(.code) → Countrycode. NORMA's verbalization browser automatically displays the reference type, for example, Product has ProductName. Finer control is possible using format strings.

The former VisioModeler tool for ORM upshifts the first letter (if any) of the reference mode name before appending it to the entity type name, for example, Country(.code) → CountryCode. The Visio stencil for drawing ORM diagrams in Microsoft Visio allows you to choose any name for the refmode, but it does not verbalize the reference type.

A value type corresponding to a popular reference mode identifies only one entity type. For example, CountryCode identifies countries only, and GenderCode identifies genders only.

Unit-based (or *measurement*) *reference modes* include a built-in list of physical units (e.g., cm, kg, mile) and monetary units (e.g., USD, EUR), which may be extended by the user. When parenthesized, unit-based reference modes are appended by a *colon* ":". Value type names are generated from unit-based reference modes by replacing the colon by "Value". For example: (kg:) → kgValue; (USD:) → USDValue.

Each unit is based on a *unit type* (or *dimension*), which may optionally be displayed after the colon. Sometimes, different units are based on the same dimension. For example: (kg: Mass), (lb: Mass), (USD: Money), (EUR: Money).

General reference modes have no added punctuation, and their names are the same as their value type names. For example: (ISBN) → ISBN; (URL) → URL. These value types may identify many entity types (e.g., URL could be used to identify Link and Website).

For the *Date* entity type, a basic date format is used for the refence mode. For example, "Date(mdy)" indicates that date values are displayed with the month before the day before the year, which is common in the United States, unlike Australia (dmy) and Japan (ymd).

As a check, we should populate each fact type on the diagram with some fact instance by adding a **fact table** for each fact type and entering the values in the relevant columns. In ORM, a fact table is simply a table for displaying instances of a fact type. The term "fact table" is used differently in data warehousing (see later). A diagram that includes both a schema and a sample database is called a *knowledge base diagram*.

Consider the report in Table 3.7. LicenseNr entries identify the person's driver's license. Fig. 3.11 shows the conceptual schema populated with the original data, in two fact tables. When the object types are simply identified (as here), the population of a fact role appears in a column of the fact table, with the column positioned next to the role.

Table 3.7 Driver details.

Person	*LicenseNr*	*Cars driven*
Adams B	A3050	235PZN
Jones E	A2245	235PZN, 108AAQ

The diagram omits some constraints. At least one fact from each fact table should be verbalized to ensure that the diagram makes sense. Populating the schema diagram is useful for detecting errors in the diagram, and also for clarifying constraints.

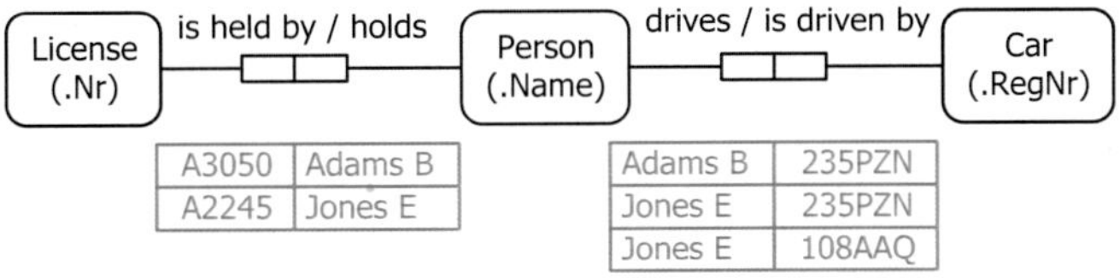

FIG. 3.11 A knowledge base diagram for Table 3.7 (constraints omitted).

Nowadays, most nonsmokers prefer a smoke-free environment in which to work, travel, eat, and so on. Hence, for some applications a report such as Table 3.8 is relevant. Please perform step 1 on this table before reading on.

Table 3.8 Smokers and nonsmokers.

Smokers	*Non-smokers*
Arnie	Lee
Sherlock	Norma
	Terry

One way to express the facts on row 1 is: Person(.Firstname) "Arnie" smokes; Person(.Firstname) "Lee" is a nonsmoker. Each of these facts is an instance of a different *unary fact type*. A unary fact type has only one role. The knowledge base diagram is shown in Fig. 3.12. Here the two roles belong to different fact types. This is shown by separating the role boxes.

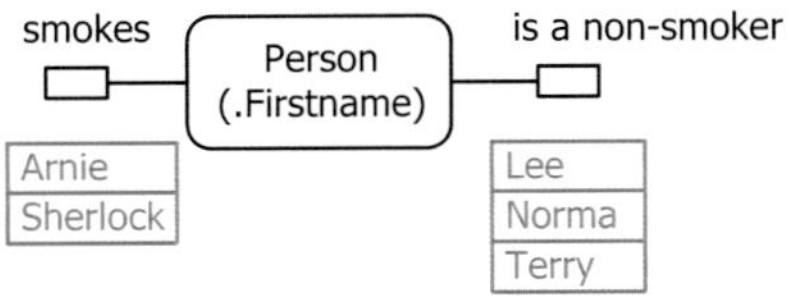

FIG. 3.12 A knowledge base diagram for Table 3.8 (unary version).

If desired, the two unaries may be transformed into a single binary by introducing SmokingStatus as another entity type, with codes "S" for smoker and "N" for nonsmoker. As a result, the first row of Table 3.8 could be rephrased as Person(.Firstname) "Arnie" has SmokingStatus(.Code) "S"; Person "Lee" has SmokingStatus "N". This approach is shown in Fig. 3.13.

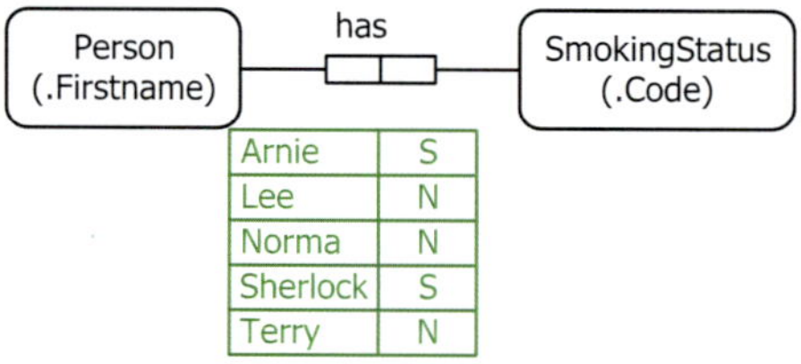

FIG. 3.13 A knowledge base diagram for Table 3.8 (binary version).

Each of the binary examples discussed had two different entity types. Fact types involving different entity types are said to be *heterogeneous fact types*. Most fact types are of this kind.

However, the fact type displayed in Fig. 3.14 has one entity type-Person. If each role in a fact type is played by the same object type, we have a *homogeneous fact type*. The binary, homogeneous fact type is called a *ring fact type* because the path from the object type through the predicate loops back to the same object type, forming a ring. Here, the forward predicate (is husband of) and inverse predicate (is wife of) are shown.

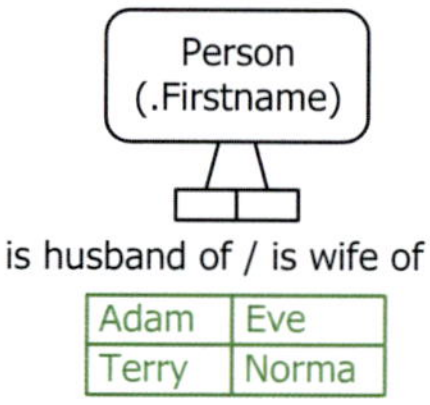

FIG. 3.14 A ring fact type with sample population.

Some versions of ORM set predicates out in gerundive form (e.g., husband of, wife of). Although useful for some styles of queries, it is generally better to use full predicate readings because this leads to better verbalization of facts and constraints.

To make diagrams more compact, you may abbreviate names for predicates, object types, and reference modes if their expanded versions are obvious to users. But this relaxed policy should be used with care. ORM tools often allow you to attach descriptions of model elements, which can compensate for such name shortening.

Apart from communication with humans, conceptual schemas provide a formal specification of the structure of the UoD, so that the model may be processed by a computer. Hence, the schema diagrams we draw must conform to the formation rules for legal schemas. They are not informal cartoons.

Now consider the report in Table 3.9. Here, we have a *ternary fact type*. The reference schemes are Student(.Nr), Course(.Code), and Rating(.Nr). We may express the fact on the first row as "Student '1001' for Course 'CS100' scored Rating 4".

Table 3.9 A relational table storing student results.

Result:

StudentNr	*Course*	*Rating*
1001	CS100	4
1002	CS100	4
1002	CS114	5

On a conceptual schema diagram, a ternary fact type appears as a sequence of three role boxes, each attached to an object type, as shown in Fig. 3.15. When names for ternary and longer predicates are written on a diagram, the place-holders are included, each depicted by an *ellipsis* "…"

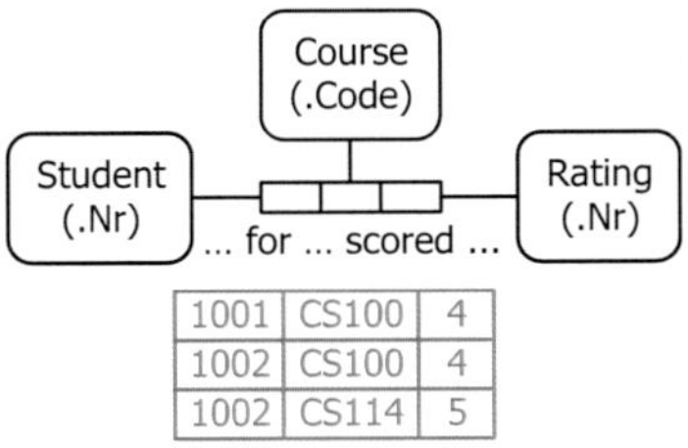

FIG. 3.15 A populated ternary fact type for Table 3.9.

Fig. 3.15 includes a sample population. No matter how high the *arity* (number of roles) of the fact type, we can easily populate it for checking purposes. Each column in the fact table is associated with one role in the predicate.

An earlier example transformed unaries into a binary. Another kind of schema transformation is **objectification** or **nesting**. This effectively *treats a relationship between objects as an object itself*. Consider again the top row of Table 3.9. Instead of expressing this as a single sentence, we might convey the information in two sentences:

Student "1001" enrolled in Course "CS100".
This Enrollment resulted in Rating 4.

Here, "This Enrollment" refers back to the enrollment relationship between the specific person and the specific subject mentioned in the first sentence. Any such enrollment may be treated as an object in its own right.

The act of making an object out of a relationship is called *objectification* or *reification* and corresponds to the linguistic act of nominalization (making a noun out of a verb phrase). An object formed by objectification is called an *objectified relationship*. The type of object so formed is called an *objectified association*, or an *objectification type*. Recall that "association" means "relationship type". Section 10.5 refines the notion of objectification to distinguish two varieties, propositional and situational.

An objectified association is depicted by a named, soft-rectangle around the predicate being objectified (see Fig. 3.16). The name of the objectified association is placed in *double quotes*. An objectified association usually has two roles, but may have one or more.

Entries in fact columns for nested objects may be shown as bracketed pairs (triples, etc.) of values. For example, the enrollment of student 1001 in CS100 appears as "(1001, CS100)" in the fact table for the resulted in predicate. Note that *nesting is not the same as splitting*. Fig. 3.16 does not show two independent binaries. The resulted in predicate cannot be shown without the enrolled in predicate. The ternary in Fig. 3.15 is still elementary. Fig. 3.15 is said to be the *flattened*, or *unnested*, version.

Chapter 14 deals with schema equivalence in detail. The nested and flattened versions are not equivalent unless the role played by the objectified association is mandatory. With our current example, this means that a rating must be known for each enrollment. In this case, the flattened version is preferred because it is simpler to diagram and populate. As discussed later, nesting is often preferred if the objectified association has an optional role, or more than one role to play.

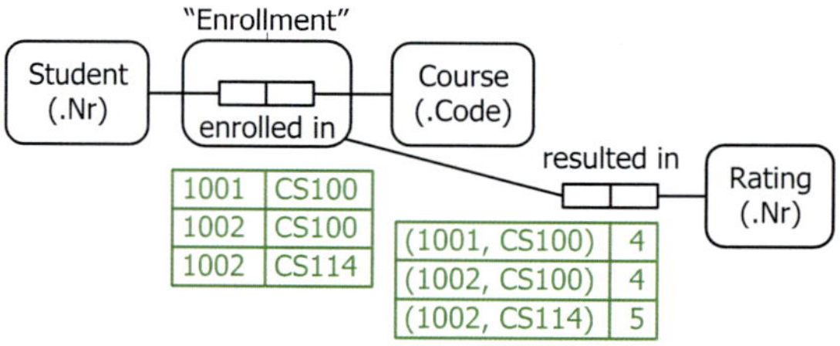

FIG. 3.16 Knowledge base for Table 3.9 (nested version).

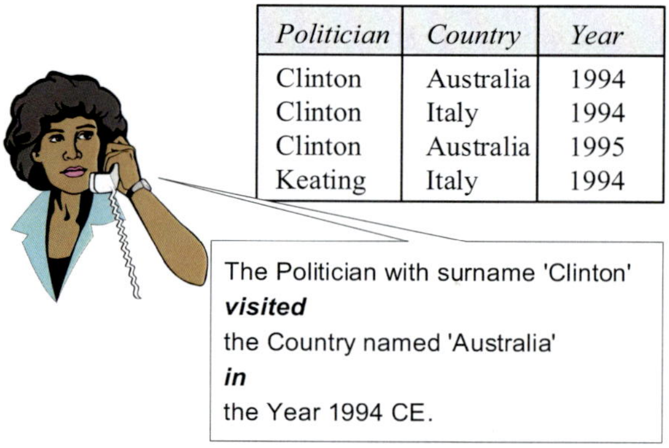

Politician	*Country*	*Year*
Clinton	Australia	1994
Clinton	Italy	1994
Clinton	Australia	1995
Keating	Italy	1994

FIG. 3.17 One way of verbalizing the first row.

For example, suppose we widen the UoD to include enrollment dates. The flat solution needs another ternary: Student enrolled in Course on Date. With the nested solution, we simply add the binary: Enrollment occurred on Date. As shown later, the nested solution also simplifies constraint specification in this case, and hence is preferred.

Now consider the travel record example depicted in Fig. 3.17. Using the telephone heuristic, the modeler verbalizes the first row of data as an instance of the ternary fact type: Politician visited Country in Year. The sample data indicates that many politicians may visit many countries in many years, allowing many nesting options. We could objectify Politician visited Country as Visit, and add Visit occurred in Year. We might instead objectify Politician traveled in Year as Travel, and add Travel visited Country, or objectify Country was visited in Year as Visit, and add Visit was by Politician. With no strong reason for one nesting choice over the other, it is simpler to leave it as a ternary.

Although we usually display only one reading for ternary and longer fact types, there are many possible readings depending on the order in which we traverse the roles. An n-ary predicate has $n!$ (factorial n, i.e., $n \times n{-}1 \times \ldots \times 1$) reading orders. Hence, a ternary has six possible reading orders, a quaternary has 24, and so on.

For example, the ternary fact type given earlier could be specified by any of the following six readings: Politician visited Country in Year; Politician in Year visited Country; Country was visited by Politician in Year; Country in Year was visited by Politician; Year included visit by Politician to Country; Year included visit to Country by Politician. Some ORM tools also support alias readings for any given reading order.

In practice, one reading is usually enough. However, if we wish to query the schema directly, it is handy to be able to navigate from any given role. To cater for this, we would need to supply for each role a reading that starts at that role (the order of the later roles doesn't matter). So, we never need any more than n readings for any n-ary fact type. This is a lot fewer than factorial n. Supplying predicate readings for more than one reading order of a fact type can also help with better verbalization of constraints and derivation rules. In practice, one often provides forward and inverse readings for binary fact types, but usually provides alternate readings for n-ary fact types only on an as-needed basis.

Now suppose we need to design a database for storing sales data that can be displayed graphically as shown in Fig. 3.18. This three-dimensional bar chart shows the sales figures for two computer-aided drafting products codenamed "ACAD" and "BCAD". As an exercise in steps 1 and 2, try to verbalize the sales information and then schematize it (on a conceptual schema diagram) before reading on.

Consider the first Q1 bar. A person familiar with the domain might verbalize the fact as: "BCAD in the first quarter had sales of one million dollars". This completes step 1a. As modelers, we complete step 1b by refining this into one or more elementary facts. In this case, we may verbalize it as a single ternary: "The Product with code 'BCAD' in the Quarter numbered 1 had sales of MoneyAmount 1000000 USD".

We chose to identify quarters using numbers (e.g., 1) but you can use codes (e.g., "Q1") if you like. We used the object type "MoneyAmount" instead of "Sales" to make the underlying domain explicit. This makes it clear that we can compare sales figures with other monetary figures (e.g., costs and profits).

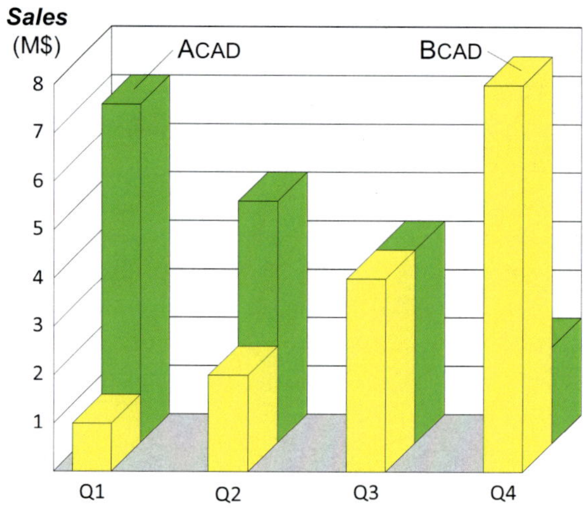

FIG. 3.18 Can you verbalize this bar chart?

Assuming the chart applies to the United States, we chose USD (United States Dollar) for the monetary unit. This distinguishes it from other dollars (e.g., AUD for Australian Dollar). This is good practice, but if there is no danger of confusion you could simply show the unit as "$".

Instead of directly using MoneyAmount as the object type, we could do this indirectly by using Sales with the expanded reference mode (USD: Money), which abbreviates to (USD:). Each other bar may be verbalized similarly. This completes step 1. In preparation for step 2, we could set the fact type out with reference modes in parentheses as follows: Product(.Code) in Quarter(.nr) had sales of MoneyAmount(USD:).

It is now a simple task to draw the conceptual schema diagram. As a check, we populate it with some sample fact instances. Fig. 3.19 shows both direct and indirect ways of typing sales to money amounts. As a display option, (USD:) may be expanded to (USD: Money).

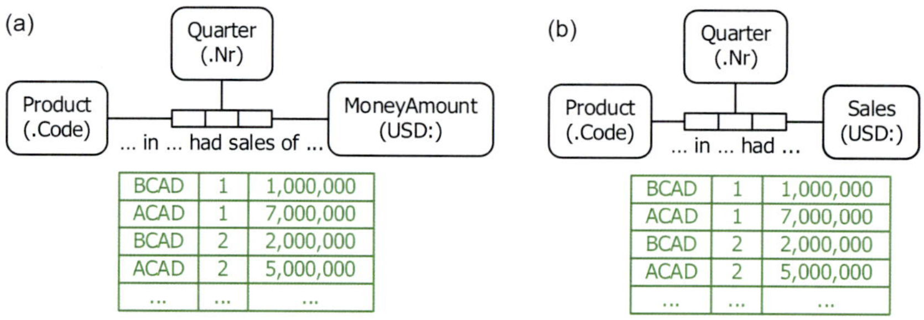

FIG. 3.19 A conceptual schema for the sales data, with a sample population.

Another common way for presenting numeric data is the pie chart. A legend is often provided next to the chart to indicate the items denoted by each slice. Each slice of the pie indicates the portion of the whole taken up by that particular item. An example is given in Fig. 3.20. Try to schematize this yourself before reading on.

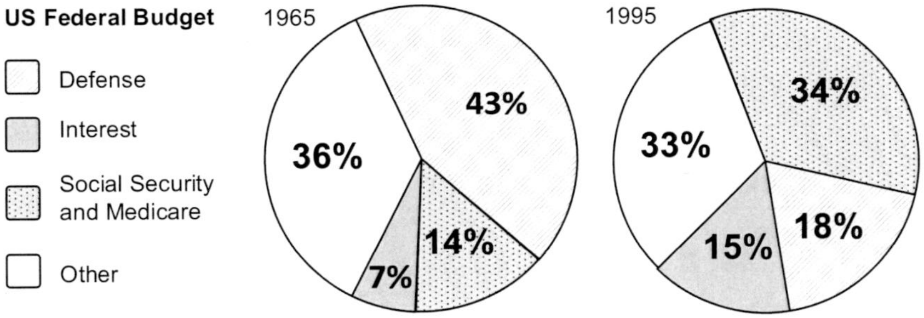

FIG. 3.20 Can you schematize this pie chart?

Applying step 1a to the defense slice of the first pie, we could verbalize the fact as "In 1965 defense consumed 43% of the budget". To complete step 1, we refine this to "In Year 1965 CE the BudgetItem named 'Defense' consumed Portion 43% of the budget". The other slices denote the same kind of fact, so we may generalize to the ternary fact type: in Year (CE:) BudgetItem(.Name) consumed Portion (%) of the budget. Here, the predicate "in … … consumed … of the budget" has front text before the first placeholder and two placeholders adjacent to one another. Although fairly rare, this is a legal mixfix predicate.

This flexibility makes verbalization easier than it would have been otherwise. To complete step 2, the resulting schema and sample data are shown in Fig. 3.21.

To conclude this section, let's review some terminology. Three terms for objects have now been introduced. Entities are the objects in the UoD that we reference by means of descriptions. Values (e.g., characters strings or numbers) appear as entries in database tables and are often used to refer to entities. Finally, relationships between objects may be treated as objects themselves: these are objectified relationships (or nested objects).

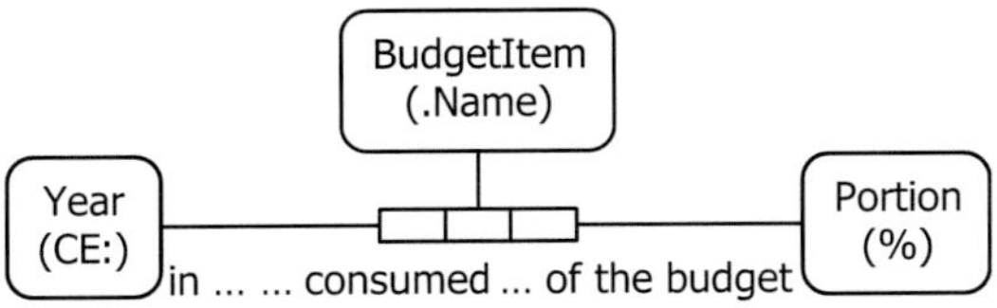

1965	Defense	43
1965	Interest	7
…	…	…

FIG. 3.21 A conceptual schema for the budget data, with a sample population.

There are two common notations for the arity or "length" of a predicate. In Table 3.10, the preferred Latin-based notation, shown as the main descriptor, is set out for the first nine cases. The alternate Greek-based descriptor tends to be restricted to the first five cases, as shown. In practice, an elementary predicate rarely exceeds five roles.

Although we should populate conceptual schema diagrams for checking purposes, fact populations do not form part of the conceptual schema diagram itself. In the following exercise, population checks are not requested. However, we strongly suggest that you populate each fact type with at least one row as a check on your work.

Table 3.10 Classification of predicates according to number of roles.

Nr roles	*Main descriptor*	*Alternate descriptor*
1	Unary	Monadic
2	Binary	Dyadic
3	Ternary	Triadic
4	Quaternary	Tetradic
5	Quinary	Pentadic
6	Senary	...
7	Septenary	
8	Octanary	
9	Nonary	

Exercise 3.4

1. The names and genders of various people are indicated below:

    Male: Fred, Tom

    Female: Ann, Mary, Sue

    (a) Express the information about Fred and Ann in unary facts.
    (b) Draw a conceptual schema diagram based on this choice.
    (c) Express the same information in terms of binary elementary facts.
    (d) Draw a conceptual schema diagram based on this choice.

    Note: For the rest of this exercise, avoid using unary facts.

2. Draw a conceptual schema for the fact types in the following questions of Exercise 3.3: (a) Q. 3; (b) Q. 4; (c) Q. 5; (d) Q. 6; (e) Q. 7; (f) Q. 8; (g) Q. 9; (h) Q. 10; (i) Q. 11. Perform steps 1 and 2 of the CSDP for the following output reports.

3.

Retailer	*Item*	*Quantity sold*
InfoWare	SQL+	330
SoftwareLand	Zappo Pascal	330
	WordLight	200
	SQL+	330
	Zappo Pascal	251

4.

Item	*Retailer*	*Quantity sold*
SQL+	InfoWare	330
	SoftwareLand	330
Zappo Pascal	InfoWare	330
	SoftwareLand	251
WordLight	InfoWare	200

5.

Tute group	*Day*	*Hour*	*Room*
A	Mon	3 p.m.	69-718
B	Tue	2 p.m.	42-B18
C1	Thu	10 a.m.	69-718
C2	Thu	10 a.m.	67-103

6. (*Hint*: Make use of nesting)

Subject	*CreditPts*	*Semester*	*Enrollment*	*Lecturer*
CS100	8	1	500	DBJ
CS102	8	2	500	EJS
CS114	8	1	300	TAH
CS115	8	2	270	TAH
CS383	16	1	50	RMC
CS383	16	2	45	PNC

7. The following interactive voting form is used to input votes by cruise club members on various motions (proposals moved by a club member). This example shows one completed form after a member has selected his/her voting choices. Although passwords are not displayed, they are captured by the information system. Perform CSDP steps 1 and 2 to schematize this UoD.

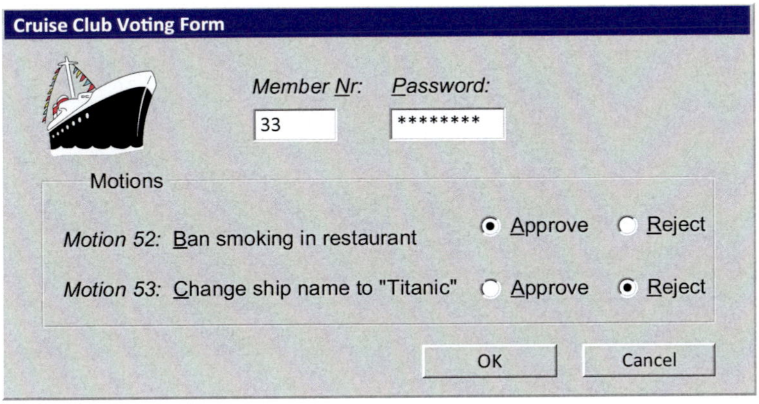

8. Assuming that appropriate names are supplied for entity types, reference modes, and predicates and that appropriate constraints are added, which of the following conceptual schema diagrams are legal? Where illegal, briefly explain the error.

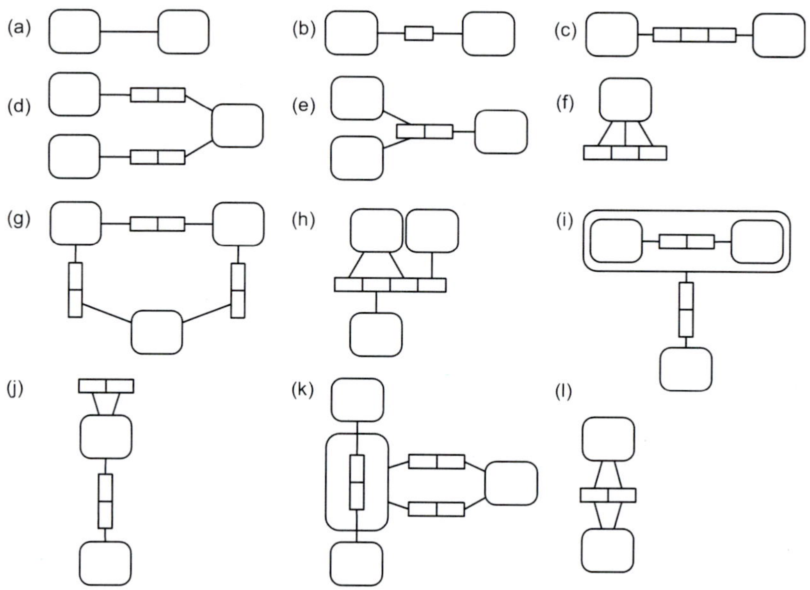

9. The following chart indicates changes in the value of the Euro and the Australian dollar relative to the US dollar over a given period. Figures may be recorded at the start of each month (only some figures are shown here). Schematize this domain.

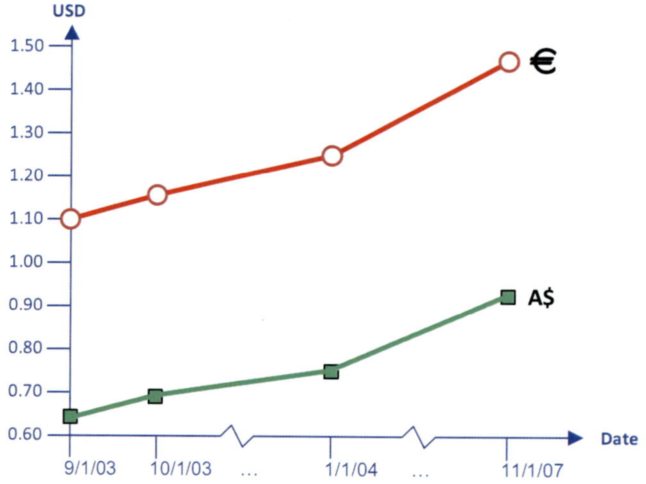

3.5 CSDP Step 3: Trim Schema; Note Basic Derivations

Having drawn the fact types and performed a population check, we now move on to step 3 of our design procedure. Here, we check to see if there are some entity types that should be combined. We also check to see if some fact types can be derived from the others by arithmetic computation.

> *CSDP step 3: Check for entity types that should be combined;*
> *note any arithmetic derivations.*

UoD are classified into types. Fig. 3.22 shows the basic division of objects into entities (nonlexical objects) and values (lexical objects). *Entities* are identified by definite descriptions and may typically change their state, whereas *values* are constants (typically character strings or numbers). Values might include other objects directly representable on a medium (e.g., sounds), but such possibilities are ignored in this book.

These subdivisions are **mutually exclusive** (i.e., they have no instance in common). For example, no character string can be an entity. The division of a whole into exclusive parts is called a **partition**. You may think of it as cutting a pie up into slices. The slices are *exclusive* (they don't overlap) and *exhaustive* (together they make up the whole pie).

Fig. 3.22 gives an example of how the object types might be divided into entity types and value types for a particular business domain. Which kinds of objects exist depends on the UoD. Basically, objects are grouped into the same type if we want to record similar information about them.

For any UoD, there is always a top-level partitioning of its entities into exclusive types: these are called **primitive entity types**. We may introduce *subtypes* of these primitive types, especially if they have some specific roles to play. In ORM, subtypes are shown connected by an arrow to their supertype. Although shown separately, it is possible that subtypes of a given entity type may overlap.

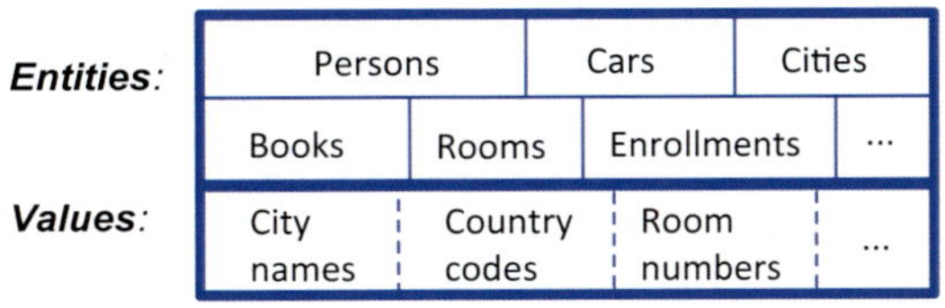

FIG. 3.22 Partitioning objects into types of entities and values.

However, *primitive entity types never overlap*. For example, no person can be a city. On a conceptual schema diagram, the visual separation of primitive entity types indicates that these types are mutually exclusive. The same is not true of subtypes. In Fig. 3.23, for example, Person and City are mutually exclusive but Manager and Woman need not be. Subtypes are discussed in detail in Chapter 6.

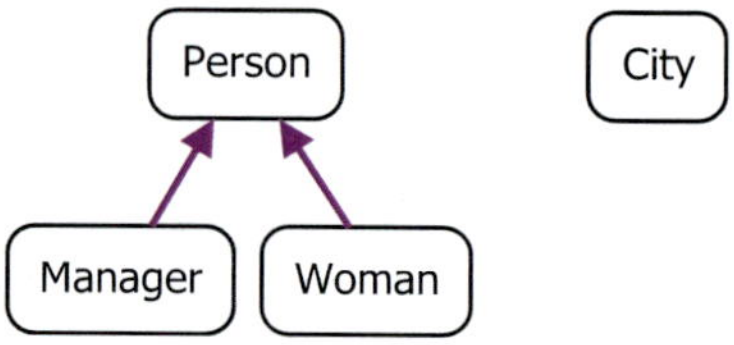

FIG. 3.23 Person and City are primitive entity types and hence mutually exclusive.

Fig. 3.24(a) shows two value types: Surname and Cityname. Even if these appear as top-level types in an ORM schema diagram, it is possible that they overlap, as shown in the Euler diagram of Fig. 3.24(b). Here, the character strings "Denver" and "Columbus" are both surnames and city names. Although subtype relationships between value types are rarely displayed, they may exist implicitly. In rare cases where the supertype plays a role of interest, we might display the connections explicitly, as in Fig. 3.24(c).

Step 3 of the design procedure begins with a check to see if some entity types should be combined. At this stage we are concerned only with primitive entity types, not entity subtypes. So, if you spot some entity types that do overlap, you should combine them into a single entity type, or at least introduce a common supertype (see Chapter 6).

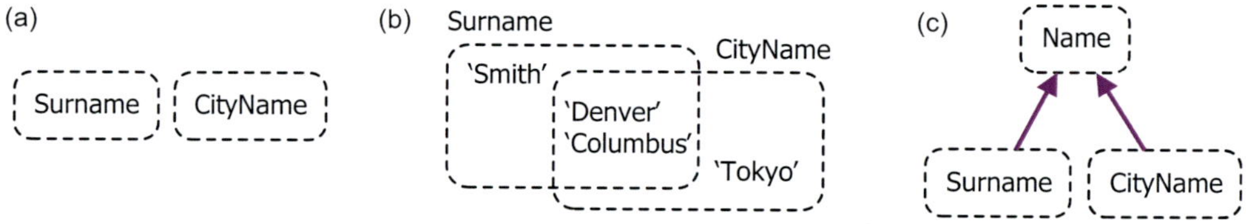

FIG. 3.24 Overlapping value types have an implicit or explicit supertype.

For example, consider Table 3.11, which is based on the movie application discussed in Chapter 1. Suppose that as a result of applying steps 1 and 2 we arrived at the diagram shown in Fig. 3.25. Do you see what's wrong with this diagram?

Table 3.11 Some motion pictures.

Movie#	*Movie Title*	*Director*	*Stars*
1	Cosmology	Lee Lafferty	
2	Kung Fu Hustle	Stephen Chow	Stephen Chow
3	The Secret Garden	Alan Grint	Gennie James, Barret Oliver
...	...	...	...

Fig. 3.25 displays MovieStar and Director as separate, primitive entity types. This implies that these types are mutually exclusive (i.e., no movie star can be a director). But is this the case? Our sample population lists the value "Stephen Chow" in both the Director column and the Stars column. Does this refer to the same person?

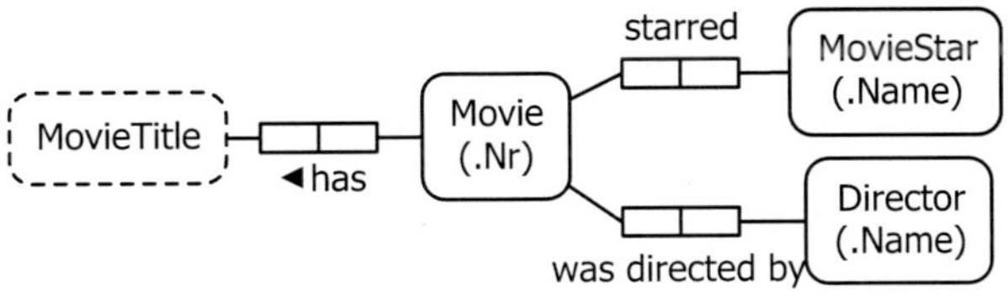

FIG. 3.25 A faulty conceptual schema.

If in doubt, you can ask a domain expert, who would respond "Yes" to the above question. As a result, we must combine the Movie Star and Director entity types into a single entity type as shown in Fig. 3.26.

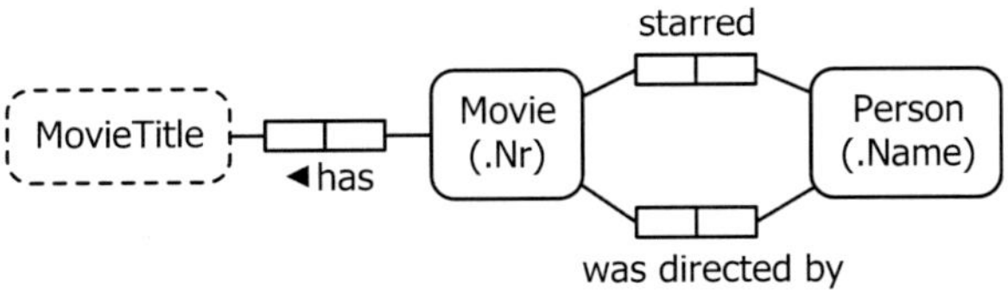

FIG. 3.26 The result of applying step 3 to Fig. 3.25.

This shows that it is possible to be both a star and a director. Of course, it does not imply that every movie star is a director. A later chapter discusses how to add subtypes later if necessary. For example, if some other facts are to be recorded only for directors, we form a Director subtype of Person for those additional facts.

One case where entity types may always be combined, if desired, is when they have the *same unit-based reference mode*. Here, the entity is envisaged as a quantity of so many units (e.g., kilograms or US dollars). Let's look at a few examples.

Consider the output report of Table 3.12, and assume that "$" means US dollar. Here, $50 is a wholesale price and a markup. In both cases, the $50 denotes the same amount of money, so wholesale prices and markups overlap. If the population is significant, the set of retail prices does not overlap the set of markups. Nevertheless, it is *meaningful to compare* retail prices and markups because they have the same unit (US dollars). For instance, article A1 has a retail price that is three times its markup.

Table 3.12 Monetary details about articles on sale.

Article	*Wholesale Price ($)*	*Retail price ($)*	*Markup*
A1	50	75	25
A2	80	130	50
A3	50	70	20
A4	100	130	30

In modeling this report, we must reveal that the entries in the final three columns are compatible because they all represent amounts of money. One way to schematize this UoD is shown in Fig. 3.27(a). By assigning WholesalePrice, RetailPrice, and Markup the same unit-based reference mode (USD:), we implicitly declare that they are all based on the same unit dimension.

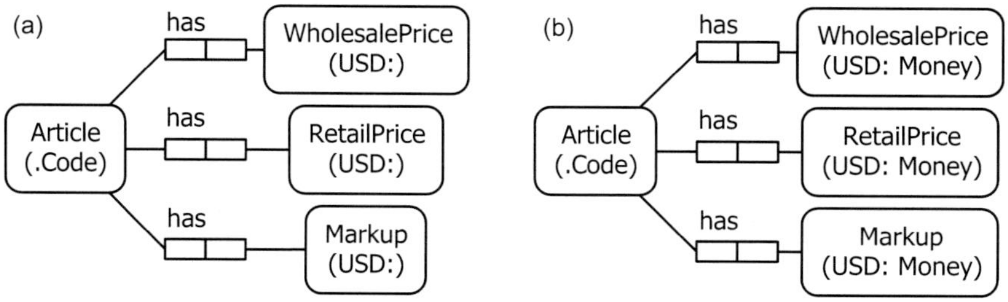

FIG. 3.27 Unit-based types are implicitly subtypes of their unit dimension.

The colon in "(USD:)" means the reference mode is unit-based. When entering this in an ORM 2 tool, we declare the unit dimension (by selecting from a predefined list or adding our own). Here, the unit dimension is MoneyAmount (or simply Money). As a display option, the unit dimension may be shown explicitly, as in Fig. 3.27(b).

WholesalePrice, RetailPrice, and Markup are not primitive types. They are implicitly subtypes of MoneyAmount. Instead, we may collapse these entity types into MoneyAmount and move their distinguishing semantics into the predicate readings, as in Fig. 3.28. We may retain the names of the former entity types as role names, as shown.

The output report satisfies the following mathematical relationship between the values in the last three columns: markup = retail price − wholesale price. Assuming this is significant, the markup value may be **derived** from the wholesale and retail values by means of this rule. To minimize the chance of human error, we have the system derive this value rather than have humans compute it.

To indicate that a fact type is derived, an ***asterisk*** *is appended to its predicate reading.* Whether or not the derived fact type is displayed on the diagram, its *derivation rule* should be supplied. In this book, derivation rules display as textual annotations.

A derivation rule may be specified in either attribute style or relational style. *Attribute style* uses role names to refer to attributes of an object type, typically cited in a for-each clause, as in Fig. 3.28. For binary associations, role names may be treated as naming attributes of the object type at the opposite end of the association.

If the *context* of a rule is declared in a for-each clause, it may be assumed thereafter. If the context is not predeclared, it may be provided *in situ* by qualified names using either the *dot notation* familiar from programming and UML (e.g., Article.markup) or the *of-notation* (e.g., markup of Article). When validating rules with nontechnical domain experts, the of-notation is usually preferable to the dot notation.

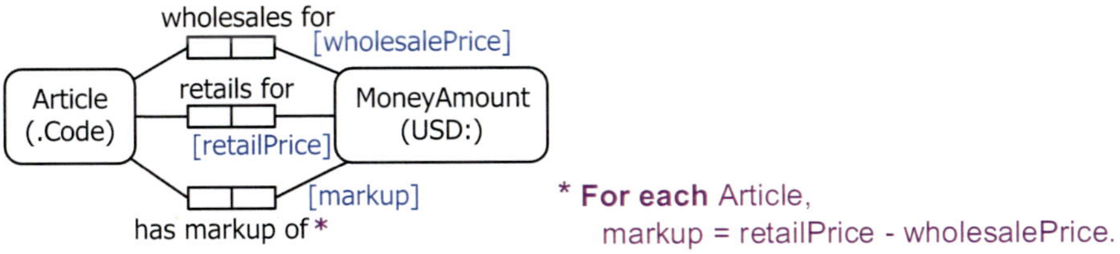

FIG. 3.28 An alternative to Fig. 3.27, with derivation rule added.

Relational style instead uses predicate readings. For example, using "**iff**" for "if and only if", the markup rule may be declared in relational style as shown below. Subscripts distinguish variables introduced in the rule body (after "iff") from variables of the same type (here MoneyAmount) that occur in the rule head (before "iff").

Article has markup of MoneyAmount **iff**
Article retails for MoneyAmount_1 **and** Article wholesales for MoneyAmount_2 **and**
MoneyAmount = MoneyAmount_1 – MoneyAmount_2.

For brevity, relational style assumes that variables in the rule head are universally quantified and that variables introduced in the body are existentially quantified. So the rule may be stated explicitly thus: **For each** Article **and** MoneyAmount, Article has markup of MoneyAmount **iff** Article retails for **some** MoneyAmount_1 **and** Article wholesales for **some** MoneyAmount_2 **and** MoneyAmount = MoneyAmount_1 – MoneyAmount_2.

Strictly, the equation markup = retailPrice – wholesalePrice is a *constraint* among the markup, retail price, and wholesale price fact types, with one degree of freedom. Any one of the three could be derived from the other two. Sometimes we might wish to enter retail and wholesale prices and have the system derive the markup. At other times, we might wish to enter the wholesale price and markup to derive the retail price. We could also enter the retail price and markup to derive the wholesale price.

If we wish the system to support all these choices, then each of the three fact types is *semiderived* (instances may be asserted or derived) and its predicate is marked with a "⁺" (intuitively, half an asterisk) instead of "*". This is sometimes useful in practice (e.g., exploring loan options based on interest rates and repayment periods).

In most derivation cases, however, we decide beforehand that one specific fact type will always be the derived one. In this case, the derivation rule may be clearly specified as a *definition* (rather than merely an equation or biconditional) where the derived fact type is defined in terms of the others using "define … as …" For example:

For each Article, **define** markup **as** retailPrice – wholesalePrice.

In this definition, the derived fact type is the *definiendum* (what is required to be defined). A fact type that is *primitive* (i.e., not defined in terms of others) is said to be an *asserted* fact type or *base* fact type. *Derived* fact types are defined in terms of other fact types (asserted or derived). For arithmetic derivation rules, such as our example, attribute style is typically more compact and readable and is preferred. For logical derivation rules (see Section 5.5), relational style is often preferable.

Now consider the output report of Table 3.13, which shows details about windows on a computer screen. Try to schematize this before reading on.

Table 3.13 Window sizes.

Window	*Height (cm)*	*Width (cm)*	*Area* (cm^2)
1	4	5	20
2	6	20	120
3	10	15	150
4	5	5	25

The values in the last three columns are numbers. Heights and widths may be meaningfully compared because both are lengths: a length of 5 cm may be an instance of a height or a width. So, Height and Width may be collapsed into the entity type Length. Do Width and Area overlap because each includes the value 20? No! Area is measured in square centimeters, a different unit from centimeters. However, a length of 20 cm is not the same thing as an area of 20 cm^2. If our final column was headed "Perimeter(cm)", we could collapse three headings into one entity type as previously. But because Area is fundamentally a different type of quantity, we must keep it separate, as in Fig. 3.29(a).

As an alternative, the height and width fact types could be modeled using the fact types Window(.nr) has Height(cm:) and Window(.nr) has Width(cm:), where the centimeter unit is based on the unit dimension Length.

The derivation rule for area in Fig. 3.29(a) is prepended by a single asterisk to indicate it is derived and not stored. The rule body includes another asterisk, which in this context means multiplication. In principle, any one of area, height, and width could be derived from the other two. In many cases, however, there simply is no choice as to which fact type is derived. For example, facts about sums and averages are derivable from facts about individual cases, but except for trivial cases we cannot derive the individual facts from such summaries.

A derived fact type that is not stored is *derived-on-query* (*lazy evaluation*), so the derived information is computed only when the information is requested. For example, if our Window schema is mapped to a relational database, no column for area is included in the base table for Window. The rule for computing area may be included in a view definition or stored query and is invoked only when the view is queried or the stored query is executed. In most cases, lazy evaluation is preferred (e.g., computing a person's age from their birth date and current date).

A fully derived fact type that is also stored is *derived-on-update* (*eager evaluation*). Sometimes eager evaluation is used because it offers better performance (e.g., computing account balances). In this case, the information is stored as soon as the defining facts are entered and is updated whenever they are updated. Fact types that are fully derived and stored are marked with a *double-asterisk* "**", as in Fig. 3.29(b). When mapping to a relational database, a column is created for the derived fact type, and the computation rule is specified as a trigger or as a computed column rule that is fired whenever the defining columns are updated (including inserts or deletes).

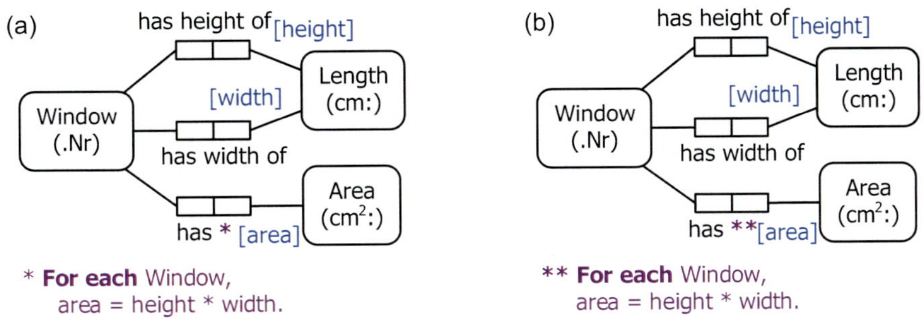

FIG. 3.29 Schema for Table 3.13, with area simply derived (a) or derived and stored (b).

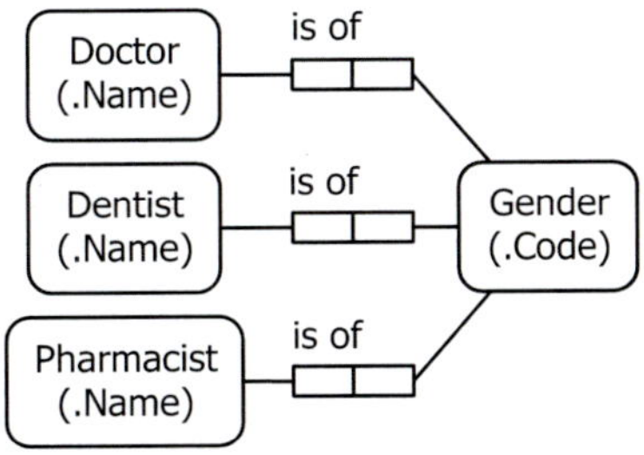

FIG. 3.30 Should Doctor, Dentist, and Pharmacist be combined?

Now consider Fig. 3.30. This might describe part of a UoD concerning practitioners in a medical clinic. Here, the entity types Doctor, Dentist, and Pharmacist have a similar reference mode (name), but because this is not unit based, this is no reason to combine the types. If somebody could hold more than one of these three jobs, then the overlap of the entity types would normally force a combination.

However, suppose that the entity types are mutually exclusive (i.e., nobody can hold more than one of these jobs). In this case, we still need to consider whether to combine the entity types because the *same kind of information* (their gender) is recorded for each.

In such cases, ask this question. *Do we ever want to list the same kind of information for the different entity types in the same query?* For example, do we want to make the request "List all the practitioners and their gender?" If we do, then we should normally combine the entity types as shown in Fig. 3.31. If we don't, there may be grounds for leaving the schema unchanged. Section 6.6 examines this issue in more detail.

Even if no doctor can be a dentist, the schema of Fig. 3.30 permits a doctor and a dentist to have the same name (e.g., "Jones E"). In Fig. 3.31, the use of "(.Name)" with Practitioner implies that each PractitionerName instance refers to only one Practitioner.

Suppose we add the constraint: **Each** Practitioner holds **at most one** Job. A graphical notation for this kind of constraint is discussed in the next chapter. With this constraint added, Fig. 3.31 would forbid any doctor from having the same name as a dentist. If the original names did overlap, we would now need to rename some practitioners to ensure that their new names are distinct (e.g., "Jones, E1" and "Jones, E2").

Alternatively, we might choose a new simple identification scheme (e.g., PractitionerNr), or identify practitioners by the combination of their original name and job. Reference schemes are discussed in detail in Section 5.3.

To preserve the distinction between the different kinds of practitioner, we introduced the entity type Job and constrained job names to the set {"doctor", "dentist", "pharmacist"}. Such "value constraints" are discussed in detail in Section 6.3.

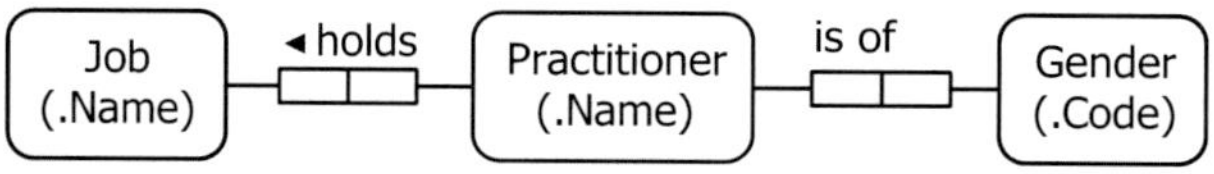

FIG. 3.31 An alternative schema for the UoD of Fig. 3.30.

The new schema is simpler because it replaced three binary fact types with two binaries. If we had even more kinds of practitioners (e.g., acupuncturist, herbalist), the savings would be even more worthwhile. If we have only two kinds of practitioner (e.g., doctor and pharmacist) both schemas would have the same number of fact types. But even in this case, the new version is generally favored. If additional information is required for specific kinds of practitioners, subtyping should be added, as discussed in Chapter 6.

In rare cases, entity types might overlap but we are not interested in knowing this, and collapsing the types is awkward. We may then leave the types separate so long as we declare that our model differs from the real world in this respect.

In performing CSDP step 3, the relevant questions to ask are as follows. Here, the derivation rules that concern us are of an arithmetic nature. Arithmetic derivations are usually fairly obvious. Logical derivations can be harder to spot, and are considered in a later step.

1. *Can the same entity belong to two entity types?* If so, combine the entity types into one (unless such identities are not of interest).
2. *Can entities of two different types be meaningfully compared (e.g., to compute ratios)? Do they have the same unit or dimension?* If so, either combine the entity types into one, or assign the units the same unit dimension.
3. *Is the same kind of information recorded for different entity types, and will you ever need to list the entities together for this information?* If so, combine the entity types, and if needed add a fact type to preserve the original distinction.
4. *Is a fact type arithmetically derivable from others?* If so, add a derivation rule.
   If you include the fact type on the diagram, mark it with "*" if fully derived, or "+" if semiderived. If the derived facts are also stored, append a final asterisk.

In addition to verbalization and population, a third way to validate a model is to see whether it enables *sample queries* to be answered, either directly from or by derivation on the fact populations. If you know what kinds of questions the system must be able to answer, navigate around the ORM model to see if you can answer them. If you can't, your schema is incomplete and you should add the fact types and/or derivation rules needed to answer the queries.

Exercise 3.5

Perform steps 1–3 of the CSDP for the following output reports. In setting out derivation rules, you may use any convenient notation.

1.

Software	*Distributor*	*Retailer*
Blossom 1234	InfoWare	PCland
		SoftKing
SQL++	TechSource	PCland
WordLight	TechSource	InfoWare
		SoftKing

2.

Project	*Manager*	*Budget*	*Salary*	*Birth year*
P1	Smith J	38000	50000	1946
P2	Jones E	42000	55000	1935
P3	Brown C	20000	38000	1946
P4	Smith J	36000	50000	1946
P5	Collins A	36000	38000	1956

3.

Dept	*Budget*	*NrStaff*	*EmpNr*	*Salary*	*Salary total*
Admin	800000	2	E01	50000	
			E02	35000	85000
Sales	900000	3	E03	40000	
			E04	35000	
			E05	40000	115000
Service	900000	2	E06	55000	
			E07	35000	90000

4.

Employee	*Project*	*Hours*	*Expenses*
E4	P8	24	400
E4	P9	26	300
E5	P8	14	200
E5	P9	16	220
E6	P8	16	240
E6	P9	14	220

5.

Female staff		*Male staff*	
Name	*Dept*	*Name*	*Dept*
Sue Bright	Admin	John Jones	Sales
Eve Jones	Admin	Bob Smith	Admin
Ann Smith	Sales		

6. The following excerpt is from the final medal tally for the 1996 Olympics. Only the top five countries are listed here.

	G	*S*	*B*	*Total*
United States	44	32	25	101
Germany	20	18	27	65
Russia	26	21	16	63
China	16	22	12	50
Australia	9	9	23	41
...	...	...	...	...

(a) Schematize this using binary fact types only.
(b) Schematize this using a ternary fact type.

7. Perform CSDP steps 1–3 for this report (13 cm = 5.25 in.; 9 cm = 3.5 inch).

Medium	*Capacity*	*Year introduced*	*Disk price (USD)*	*Cost per MB (USD)*
13 cm floppy	160 KB	1981	2.60	16.25
9 cm floppy	720 KB	1985	3.50	4.86
Zip	100 MB	1995	16.65	0.17
CD-R	650 MB	1996	1.79	0.003
DVD-R	9.4 GB	2002	7.89	0.0008

3.6 Summary

The following criteria are desirable characteristics for any language used for conceptual modeling: expressibility, clarity, simplicity, orthogonality, semantic stability, semantic relevance, validation mechanisms, abstraction mechanisms, and formal foundation.

ORM was designed with these criteria in mind. In particular, its omission of the attribute concept from base models leads to greater stability and simplicity while facilitating validation through verbalization and population.

With large-scale business domains, the UoD is divided into convenient modules, the *CSDP* is applied to each, and the resulting subschemas are integrated into the global conceptual schema. The CSDP itself has seven steps, of which the first three were discussed in this chapter.

Step
1a. Verbalize familiar information examples as facts.
1b. Refine these into formal, elementary facts, and apply quality checks.
2. Draw the fact types, and apply a population check.
3. Check for entity types that should be combined, and note any arithmetic derivations.

Step 1 is the most important. This is seeded by *data use cases*, which are relevant information *examples* (e.g., tables, forms, diagrams) that are familiar to the domain expert. These are *verbalized in terms of elementary facts*. The domain expert verbalizes the examples informally as facts in natural language sentences (step 1a).

The modeler completes step 1 by refining these into elementary facts (step 1b). An elementary fact is a simple assertion that an object has some property or that one or more objects participate together in some relationship. With respect to the UoD, an elementary fact cannot be split into smaller facts without information loss.

Elementary facts are expressed as instantiated logical predicates. A logical *predicate* is a declarative sentence with object holes in it. To complete the sentence, the placeholders are filled in by object terms. With simple reference schemes, an entity term is a definite description that includes the name of the *entity type*, *reference mode*, and *value* (e.g., "the Scientist with surname 'Einstein'"). A value term includes the name of the value type and a literal value (e.g., "the Surname 'Einstein'").

Each object hole corresponds to a *role* (part in a relationship). A predicate with one role is *unary*, with two roles is *binary*, with three roles is *ternary*, with four roles is *quaternary*, and with *n* roles is *n*-ary. The value *n* is the *arity* of the predicate.

In CSDP *step 2*, we *draw the fact types and apply a population check*. Fig. 3.32 summarizes the graphic notations met so far. Object types are depicted as named, soft rectangles (solid lines for entity types and dashed lines for value types). An *n*-ary predicate is shown as a named, contiguous sequence of *n* role-boxes. Predicates are ordered. The predicate reading is placed next to the predicate shape. Each role is played by exactly one object type, as shown by a connecting line.

A *simple 1:1 reference scheme* involves a reference predicate between an entity type and a value type, where each entity is associated with exactly one value, and each value is associated with only one entity. This kind of scheme may be abbreviated by enclosing the *reference mode in parentheses* next to the entity type name.

Popular reference modes are preceded by a dot, such as Country (.Code). *Unit-based reference modes* are appended by a colon, such as Height (cm:), possibly followed by the unit's dimension, such as (cm: Length). *General reference modes* may share their value types with other entity types, such as (URL).

Once a fact type is drawn, it should be checked by populating it with at least one fact and reading it back in natural language. Each column in the associated *fact table* corresponds to a specific role. To talk about an object corresponding to a relationship, an *objectified association* is formed from the fact type. This nominalization process is also known as *nesting*.

In *step 3* of the CSDP, we check for entity types that should be combined and note any arithmetic derivations. For any given UoD, each entity belongs to exactly one of the *primitive entity types* that have been selected (e.g., Person, Car). Hence, if we have drawn two entity types that may have a common instance, we should normally combine them.

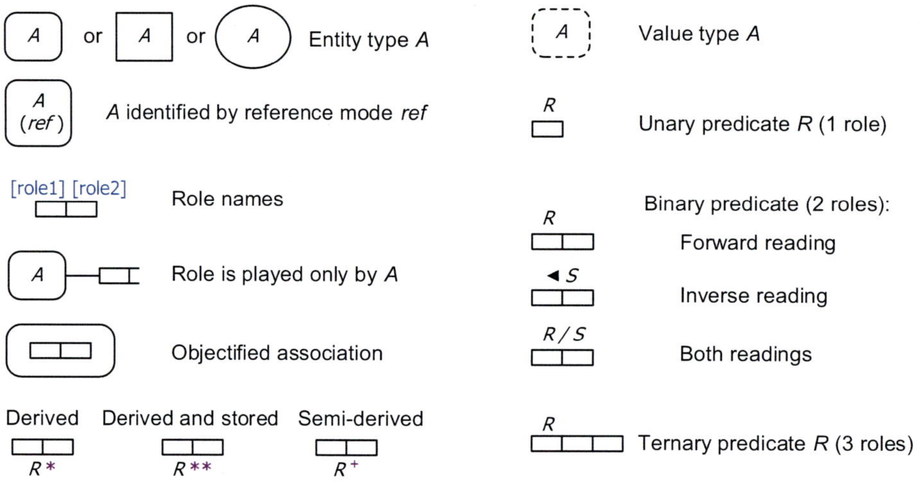

FIG. 3.32 Some basic symbols used in conceptual schema diagrams.

Even if they don't overlap, entity types that can be meaningfully compared (e.g., to compute ratios) may often be combined. If unit-based, either collapse the entity types into one or ensure that their units have the same unit dimension.

If the same kind of information is to be recorded for different entity types, these should often be combined, and if needed a fact type should be added to preserve the original distinction.

If a fact type is arithmetically derivable from others, an appropriate *derivation rule* should be declared in relational style (using predicates) or in attribute style (using attributes derived from role names). The derived fact type may be omitted from the diagram, but if included its derivation status should be marked "*" (simply derived) or "+" (semi-derived). If derived facts are also stored, append a final asterisk.

Chapter Notes

Bentley (1998) suggests the following alternative yardsticks for language design: orthogonality, generality, parsimony, completeness, similarity, extensibility, and openness. Some of these criteria (e.g., completeness, generality, extensibility) may be subsumed under expressibility. Parsimony may be treated as one aspect of simplicity. Another criterion sometimes mentioned is convenience (how convenient, suitable, or appropriate a language feature is to the user). We treat convenience as another aspect of simplicity.

The parsimony-convenience trade-off is nicely illustrated by two-valued propositional calculus, which allows for 4 monadic and 16 dyadic logical operators. All 20 of these operators can be expressed in terms of "a single logical operator (e.g., nand)", but while this might be useful in building electronic components, it is too inconvenient for direct human communication. For example, "not *p*" is far more convenient than "*p* nand *p*". In practice, we use several operators (e.g., not, and, or, if-then) because their convenience far outweighs the parsimonious benefits of having to learn only one operator such as nand.

When it comes to proving metatheorems about a given logic, it is often convenient to adopt a somewhat parsimonious stance regarding the base constructs (e.g., treat "not" and "or" as the only primitive logical operators) while introducing other constructs as derived (e.g., define "if *p* then *q*" as "not *p* or *q*"). Similar considerations apply to modeling languages.

A classic paper on linguistics that influenced the early development of a number of modeling methods is Fillmore (1968). Use case specification was central to the former Objectory approach of Ivar Jacobson, one of the main contributors to UML. For an overview of logicians' approaches to proper names and descriptions, see Chapter 5 of Haack (1978). Some heuristics to help with step 1 in interpreting common forms used in business are discussed in Choobineh et al. (1992).

Although ordered, mixfix predicates are preferred for naturalness; you could also treat a fact as a named set of (object, role) pairs: $F\{(o_1,r_1), \ldots ,(o_n,r_n)\}$. Here, each object o_i is paired with the role r_i that it plays in the fact *F*. For example, the fact that Wirth designed Pascal might be specified as Design{ (The Person with surname "Wirth", agentive), (The Language named "Pascal", objective) }. Instead of the case-adjectives "agentive" and "objective", other role names could be used (e.g., "designer" and "language", or "designing" and "being designed by"). By pairing objects with their roles, the order in which the pairs are listed is irrelevant. This approach is used in reference and idea language (RIDL) (Meersman, 1982).

In this book, the term "line" is used informally to mean "line segment" or "edge" (in geometry, lines have no beginning or end). The following notes provide a brief history of ORM.

In the 1970s, especially in Europe, substantial research was carried out to provide high-level semantics for modeling information systems. Abrial (1974), Senko (1975), and others modeled binary relationships. In 1973, Eckhard Falkenberg generalized their work to *n*-ary relationships and decided that attributes should not be used at the conceptual level because they involved "fuzzy" distinctions and they also complicated schema evolution. Later, Falkenberg (1976) proposed the fundamental ORM framework, which he called the "object-role model". This framework allowed *n*-ary and nested relationships, but depicted roles with arrowed lines.

Nijssen (1976) adapted this framework by introducing the circle-box notation for objects and roles and adding a linguistic orientation and design procedure to provide a modeling method called ENALIM (Evolving NAtural Language Information Model) (Nijssen, 1977). A major reason for the role-box notation was to facilitate validation using sample populations. Nijssen led a group of researchers at Control Data in Belgium who developed the method further, including Franz van Assche, who classified object types into lexical object types (LOTs) and nonlexical object types (NOLOTs). Today, LOTs are often called "value types" and NOLOTs are called "entity types". Kent (1977) provided several semantic insights and clarified many conceptual issues.

Robert Meersman added subtypes and made major contributions to the RIDL query language with Falkenberg and Nijssen. The method was renamed "aN Information Analysis Method" (NIAM) and was summarized in a paper by Verheijen and van Bekkum (1982). In later years, the acronym "NIAM" was given different expansions, and is now known as the "natural language information analysis method". Two matrix methods for subtypes were developed: one (the role-role matrix) by Dirk Vermeir (1983) and another by Falkenberg and others.

In the 1980s, Falkenberg and Nijssen worked jointly on the design procedure and moved to the University of Queensland, where the method was enhanced further. It was there that Terry Halpin provided the first full formalization of the method (Halpin 1989), including schema equivalence proofs, and made several refinements and extensions. In 1989, Nijssen and Halpin coauthored a book on the method. Another early book on the method was written by Wintraecken (1990).

In the early 1990s, Halpin developed an extended version of NIAM called Formal ORM (FORM), initially supported in the InfoDesigner modeling tool from ServerWare. This product later evolved to InfoModeler (at Asymetrix Corp. and then InfoModelers Inc.) and then VisioModeler (at Visio Corp.). This ORM technology was then recoded to work on the Visio engine, first appearing in Visio Enterprise. Anthony Bloesch and Terry Halpin designed an associated query language called ConQuer (Bloesch and Halpin, 1997), supported in the ActiveQuery tool. Microsoft acquired Visio in 2000 and modified its ORM solution for use in Visual Studio.

Many researchers have contributed to the ORM method, and a full history would include many not listed here. Today, various versions of the method exist but all adhere to the fundamental object-role framework. Although most ORM proponents favor *n*-ary relationships, some prefer binary-relationship modeling (BRM) such as Shoval and Shreiber (1993). Habrias (1993) developed an object-oriented version called the normalized object-oriented method (MOON). The predicator set model (PSM) was developed mainly by Arthur ter Hofstede, Erik Proper, and Theo van der Weide (ter Hofstede et al., 1993), and includes complex object constructors. De Troyer and Meersman (1995) developed another version with constructors called the natural object-relationship model (NORM). Harm van der Lek and others (Bakema et al., 1994) allowed entity types to be treated as objectified predicates, to produce fully communication-oriented information modeling (FCO-IM).

Independently of the main ORM movement, Dave Embley and others decided that using attributes in conceptual modeling was a bad idea, so they developed object-oriented systems analysis (OSA), which includes an attribute-free "object-relationship model" component that has much in common with ORM (Embley et al., 1992; Embley, 1998). The Semantics of Business Vocabulary and Business Rules (SBVR) approach adopted in 2007 by the Object Management Group is a recent addition to the family of fact-orientation, and has much in common with ORM 2, with its support of deontic as well as alethic rules. The latest version of the SBVR specification at the time of writing is accessible at www.omg.org/spec/SBVR/1.5.

Various *software tools* have been developed to support different flavors of the fact-oriented approach. The earliest tools developed by Nijssen, Meersman, and others at Control Data (e.g., IAST, RIDL*) are no longer available. Bloesch, Halpin, and others developed VisioModeler (discontinued but freely available), ActiveQuery (currently unavailable), and the ORM modeling solution in Microsoft Visio for Enterprise Architects (Halpin et al., 2003). The FCO-IM version is supported by the commercial tool CaseTalk (*www.casetalk.com*). Nijssen's group at PNA developed tools based on CogNIAM (cognition-enhanced NIAM), including various iKnow tools (www.pna-group.com). Meersman and others developed academic ORM-based tools for specifying ontologies, such as Dogma Modeler and T-Lex (Trog et al., 2006).

NORMA (Curland and Halpin, 2010) is an open-source plug-in to Microsoft Visual Studio that provides deep support for ORM 2, the next generation of ORM, and may be downloaded from https://ormsolutions.com/ or https://sourceforge.net/projects/orm. For details on installing and using NORMA, see www.orm.net/resources.

Clifford Heath developed ActiveFacts (http://dataconstellation.com/ActiveFacts/), another suite of tools designed to support ORM 2.

Victor Morgante (https://victormorgante.medium.com/) is the main developer of a variety of tools such as Boston and FactEngine that include support for ORM 2 and other modeling approaches (https://www.factengine.ai/).

4

Uniqueness Constraints

Chapter Outline

4.1 Introduction to CSDP Step 4

So far, the conceptual schema design procedure (CSDP) has focused on specifying the elementary fact types, both asserted and derived. The rest of the CSDP is concerned mostly with specifying **constraints**. Constraints apply to the database and are either static or dynamic. *Static constraints* apply to each individual state of the database and may usually be specified on a schema diagram, as discussed in CSDP steps 4–7. Examples include uniqueness, mandatory role, set comparison, value, subtyping, frequency, and ring constraints. *Dynamic constraints* restrict transitions between states, are often expressed in other ways (e.g., state charts), and are considered in step 7, along with other constraints. In practice, we usually capture all constraints relevant to the fact types being discussed before moving on to another part of the model.

This chapter discusses **uniqueness constraints** (UCs). Once UCs are added to a fact type, checks are made to see whether the fact type is of the right arity or length. There is a simple uniqueness check that shows that certain kinds of fact types are not elementary and hence should be split.

CSDP step 4: Add uniqueness constraints and check the arity of fact types.

Static constraints are often described as "constraints on the fact types". More accurately, *static constraints apply to every possible state of the database*. Here, "every state" means "*each and every state, taken one at a time*". So, static constraints apply to all possible *populations* of the fact types. For example, given the fact type *Person*(Surname) has Weight(kg:), let's add the constraint **Each** Person has **at most one** Weight. A sample population for two states is shown in Fig. 4.1.

Information Modeling and Relational Databases. https://doi.org/10.1016/B978-0-443-23790-4.00023-2

During the lifetime of the information system, the database goes through a sequence of states. In one state, Smith is recorded to weigh 90 kg, and Jones is recorded to weigh 100 kg (Fig. 4.1(a)). In a later state, Smith is still the same weight but Jones has lost weight by exercising or dieting (Fig. 4.1(b)). Over time, Jones has different weights, but in each database state at most one weight is recorded as the current weight for Jones, for Smith, or for any other person.

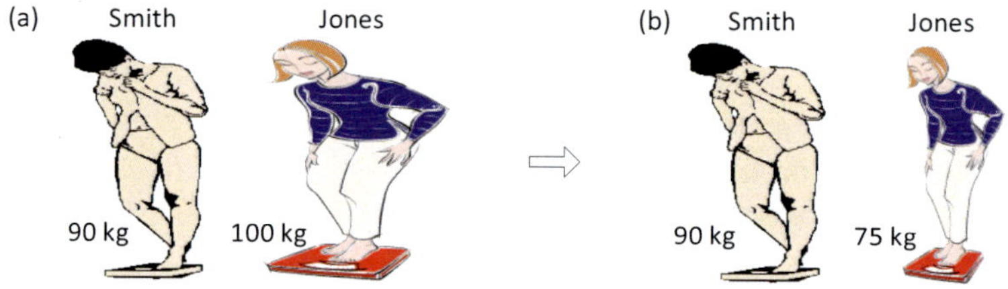

FIG. 4.1 For each state taken individually, each person has at most one weight.

This is how to interpret the static constraint **Each** Person has **at most one** Weight. Fact types such as Person has Weight are "*snapshot fact types*"—each instance applies to a single state, or snapshot, of the database. One way to record history is to use a "*historical fact type*" with reference to time, for example, Person had Weight on Date.

This chapter discusses how to specify UCs on a conceptual schema diagram. Each base fact type must be assigned at least one UC. Given the constraints on the base fact types, constraints on derived fact types are often implied by the derivation rules. For example, if a window has at most one height and at most one width, then the derivation rule "area = height * width" implies that each window has at most one area (which is the UC on the derived fact type Window has Area).

The rest of the chapter discusses CSDP step 4. First, we discuss how to mark UCs on unary and binary predicates. Then we consider UCs on longer predicates. After that, we examine external UCs (these apply between different predicates). Finally, we discuss ways to check that our fact types have the right arity (number of roles).

4.2 Uniqueness Constraints on Unaries and Binaries

Unary fact types are the easiest, so let's look at them first. Suppose that as part of a fitness application, we record which people are joggers. This can be handled with a unary fact type, as shown in Fig. 4.2.

For simplicity, let's again assume we can identify people just by their surname. A sample population is included. From the conceptual viewpoint, the fact population is the set containing the following facts:

The Person with surname 'Adams' jogs.
The Person with surname 'Brown' jogs.
The Person with surname 'Collins' jogs.

Suppose that after recording these facts, we see Adams run by again. Forgetting that we already recorded his jogging, we now try the following update:

add: The Person with surname 'Adams' jogs.

FIG. 4.2 A unary fact type with sample population.

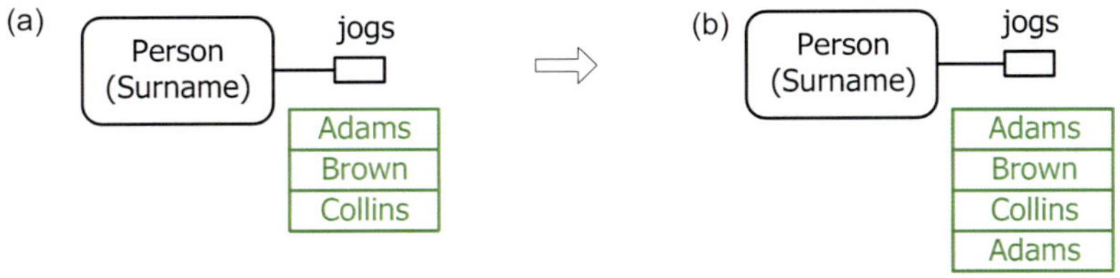

FIG. 4.3 Is this update allowed?

If accepted, this update may be pictured as shown in Fig. 4.3. The database is a variable whose population at any state is a *set* of facts. Because the fact that Adams jogs was already present, if this fact were added again the population would remain unaltered because sets are insensitive to repetition. For example, the set {Adams jogs, Adams jogs} is identical to the set {Adams jogs}. Sets are defined simply by their membership, and don't change if we display a member more than once (even if this is unusual). So, the two fact tables shown in Fig. 4.3 are equivalent if we look at each as a set of instances. Hence, there is no problem with accepting the update.

From the internal viewpoint, however, when a fact is added to a database it is typically stored in a previously unallocated space. So, accepting this update would cause the fact that Adams jogs to be stored twice, in separate locations. This is an example of **redundancy**. If redundancy occurs, the database is a *bag* of facts rather than a *set* of facts. A bag or *multiset* is just like a set, except repetition of its members is significant. For example, although the sets {1} and {1, 1} are equal, the bags [1] and [1, 1] are different.

Avoiding redundancy in a database helps maintain its integrity by simplifying the correct handling of updates. For instance, suppose the fact that Adams jogs is stored twice in the database. If we later wanted to delete or change this fact, then we would need to take care to do this for both its recorded instances. Controlling redundancy in this way can be a headache, and uncontrolled redundancy can lead to inconsistency.

A second reason for avoiding redundancy is to save space (computers have finite storage capabilities).

In practice, to retrieve information more efficiently we sometimes allow controlled redundancy at the *logical* schema level. However, for *conceptual* schema design, *no redundancy may occur in the conceptual database.* Redundancy is the repetition of an elementary or existential fact. Because each row of a conceptual fact table corresponds to one fact, *no row of a conceptual fact table may be repeated.* So, the fact table in Fig. 4.3(b) is illegal because the Adams row is repeated.

Here, for any given state of the database, each person can be recorded as being a jogger at most once. As a result, each entry in the column of the fact table is *unique* (no duplicates may appear in this column). We may represent this *UC* explicitly by placing a *bar* above or below the fact role, as shown in Fig. 4.4.

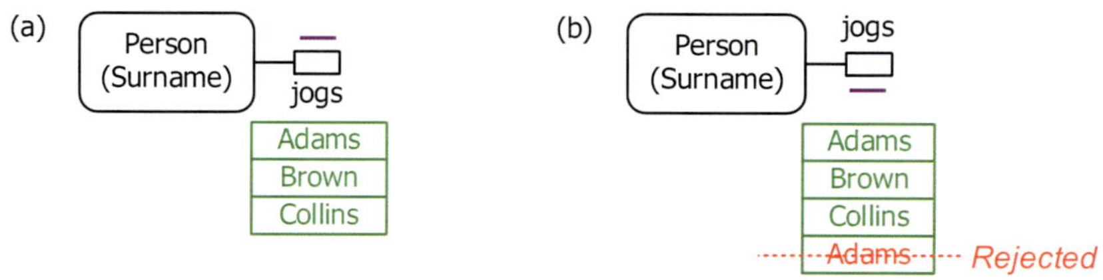

FIG. 4.4 The uniqueness constraint ensures that entries are unique (no duplicates).

Because the fact role corresponds to the fact column, a UC on the role implies that entries in the fact column are unique. We can show the constraint on the conceptual schema diagram alone by omitting the fact table.

Every object-role modeling (ORM) asserted fact type has an implied UC spanning it because for each state of the conceptual database, each asserted fact instance is unique. Because each unary predicate has just one role, only one internal UC can be applied to it. So, the UC bar on an asserted unary is implied, and hence may be omitted as a display option, as shown in Fig. 4.5(b). However, because a binary or longer predicate has more than one role, there is more than one UC pattern that could apply to the predicate. Hence, for nonunary predicates, the UCs need to be stated explicitly.

For *binary* predicates, the most common case has a UC over just one role, as in Fig. 4.6. This indicates that *each entry in that role's fact column must be unique.* Duplicate entries are allowed in the other column. Here, the politicians of interest are identified by their name, and countries are identified by their two-letter ISO codes (e.g., "AU" denotes Australia).

In the sample fact table, each value in the politician column occurs only once, but the "US" value in the country column is repeated. The constraint indicates that for all possible populations, the values in the politician column are unique. Because each politician is referenced at most once in that column, he or she is associated with at most one country in this fact type.

FIG. 4.5 Implied uniqueness constraint on a unary base predicate may be omitted.

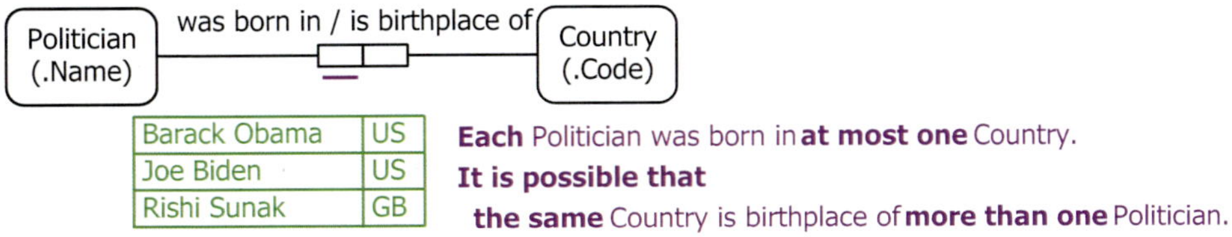

FIG. 4.6 Uniqueness constraint on first role: entries for that column must be unique.

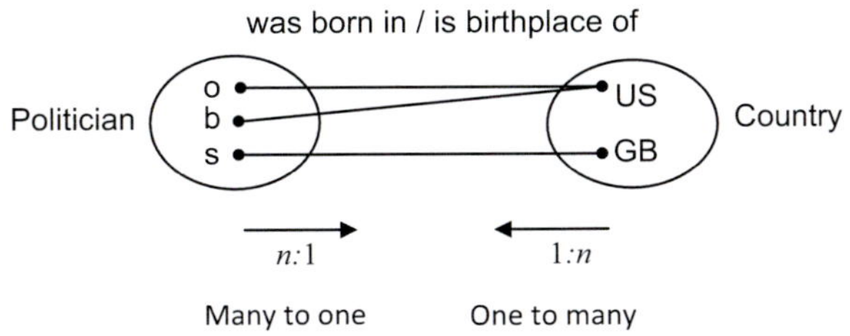

FIG. 4.7 The forward predicate is n:1; the inverse predicate is 1:n.

In other words: **Each** Politician was born in **at most one** Country. This is how the constraint is formally *verbalized* in *positive form*. The same constraint may be verbalized in *negative form* thus: **It is impossible that the same** Politician was born in **more than one** Country. The absence of a UC on the other role may be verbalized thus: **It is possible that the same** Country is birthplace of **more than one** Politician.

Fig. 4.7 shows an instance diagram for the example. Entities are depicted as dots inside the entity types, and fact instances are shown as connecting lines. Many politicians may be born in one country, but the same country may be the birthplace of many politicians. So, the predicate "… was born in …" is said to be *many to one* (*n:1*). The inverse predicate "… is birthplace of …" is *one to many* (*1:n*).

Instance diagrams such as Fig. 4.7 are often used in logic or mathematics to discuss the theory of relations. Although instance diagrams and fact tables are useful for understanding and validating constraints, they are not part of the schema itself, which is concerned only with the structure, not the population. On a conceptual schema diagram, a UC over a single role is shown by placing it next to the role (on either side), as shown in Fig. 4.8.

The role-based notation of ORM facilitates validation with sample populations because each role corresponds to a column in the associated fact table. The UC on the first role means that its entries must be unique. The absence of a UC on the second role indicates that its entries may be duplicated.

Although this notation is easy to understand, and schema diagrams reveal how things are connected, the domain expert assisting you in the modeling task should not be required to master the diagram notation. To validate a constraint, it is more important to use verbalization and sample populations.

For a binary fact type, three rows are enough for a *significant population*, so long as you pick the data carefully. If a column can have duplicates, give it two values that are the same and one different. If a column is unique, make each value different.

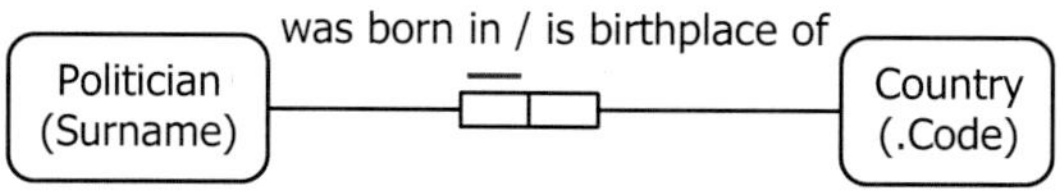

FIG. 4.8 Each person was born in at most one country.

Looking back at Fig. 4.6, note that its population is significant. Using an ORM tool, you can enter sample populations like this and have them printed out along with automatic verbalizations of the constraints for the domain expert to validate.

Now consider the example shown in Fig. 4.9, which concerns heads of government (e.g., Presidents or Prime Ministers). At any point in time, a politician can head the government of at most one country, and each country has at most one head of government. Hence, the entries in the politician column must be unique, and the entries in the country column must be unique also. So, UCs are added above each of these columns. These constraints may be verbalized as shown. The conceptual schema for this fact type can be obtained by removing the fact table.

Note the use of the *hyphen* in the inverse predicate. This binds the word with the hyphen to the object type when the constraint is verbalized. So, the right-hand UC is verbalized as "**Each** Country has **at most one** head Politician", not "**Each** Country has head **at most one** Politician". Some ORM tools interpret hyphens this way when automatically generating the constraint verbalization.

If you don't include an inverse reading, the right-hand UC verbalizes as "**At most one** Politician governs **each** Country". Usually, a constraint on a role verbalizes better if you include a predicate reading that starts at that role. We generally recommend that you include both predicate readings for binaries, but this is especially advisable if each role has a constraint that applies to it only.

For any state of the database, any one politician can be recorded as the government head of only one country, and vice versa. Hence, each population of this fact type is said to be a *one-to-one* (or *1:1*) relation. The instance diagram of Fig. 4.10 best illustrates the idea, abbreviating the surnames to their first letter.

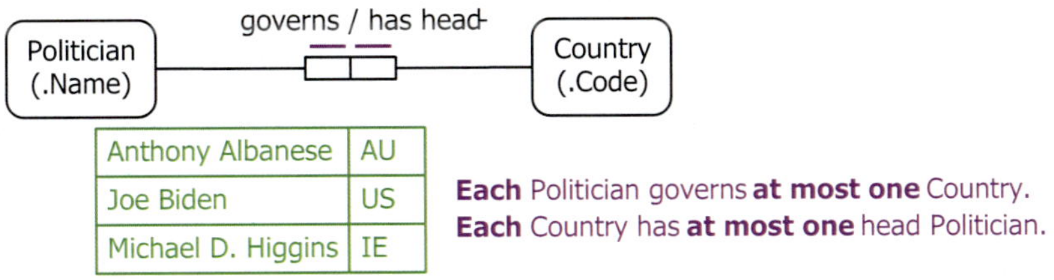

FIG. 4.9 Entries in each column must be unique.

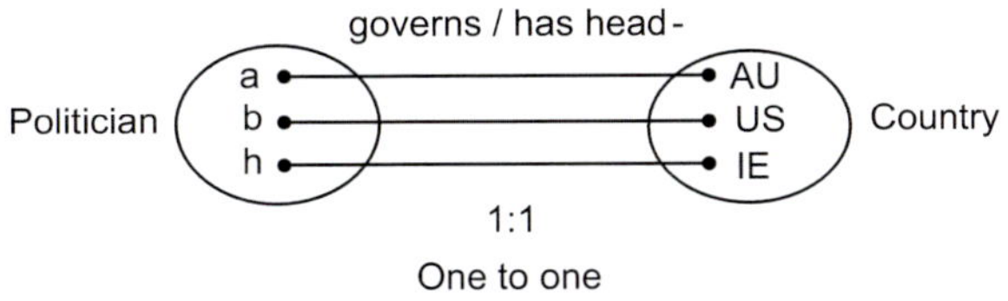

FIG. 4.10 A one-to-one relation.

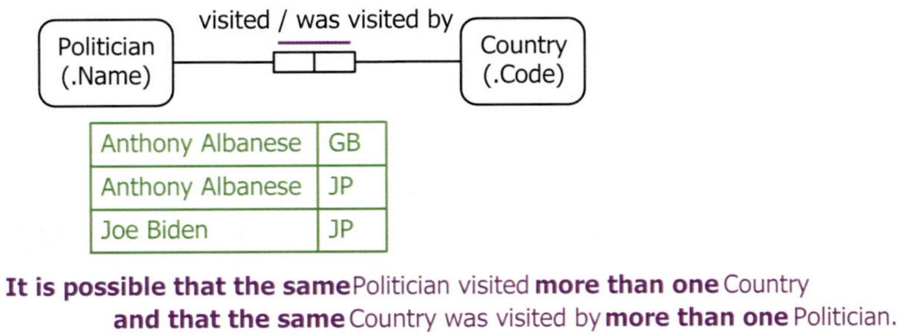

FIG. 4.11 A politician may visit many countries, and vice versa.

Now consider the populated binary fact type shown in Fig. 4.11. Note that no whole row of the fact table is repeated. This must be the case because each row corresponds to an elementary fact, and we do not duplicate facts in our database.

Although each row is unique, we cannot say the same for each politician. For example, "Anthony Albanese" appears twice in the first column. Nor is the country unique. For example, "JP" (Japan) occurs twice in the second column. For this fact type then, the only UC that applies to every one of its possible populations is that the *combination* of politician and country is unique. This UC is depicted by a uniqueness bar that spans both columns of the fact table. The constraint verbalizes as shown.

The sample data show that a politician may visit many (i.e., at least two) countries, and a country may be visited by many politicians. The fact population is a *many-to-many* (or *m:n*) relation. Fig. 4.12 shows the instance diagram.

A UC that spans all roles in a fact type is called a *spanning UC*. In addition to expressing the *m:n* nature of the binary fact type, the UC in Fig. 4.11 ensures that the fact type is populated by a set of facts (no duplicate rows). This may be verbalized as "**Each** Politician, Country **combination occurs at most once in the population of** Politician visited Country".

Whenever you have a spanning UC, you should check with the domain expert that a bag (multiset) of facts is to be excluded. For example, a politician may visit the same country many times. If we wish to distinguish between visiting a country at all and visiting it many times, we need to remodel the situation (e.g., by expanding the binary to a ternary that includes a visit date). We'll discuss such temporal aspects of modeling in detail in Section 10.3.

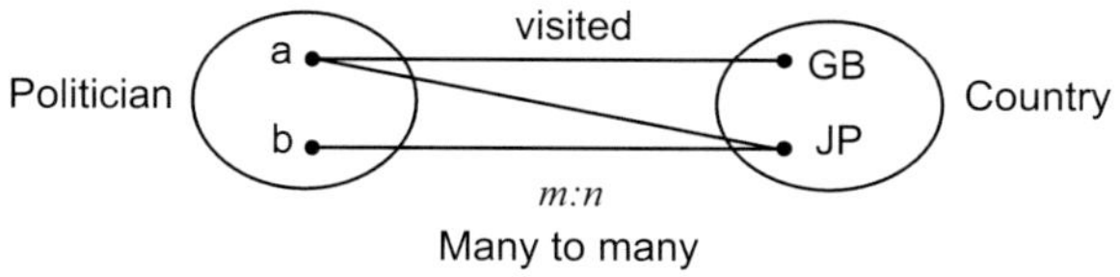

FIG. 4.12 A many-to-many relation.

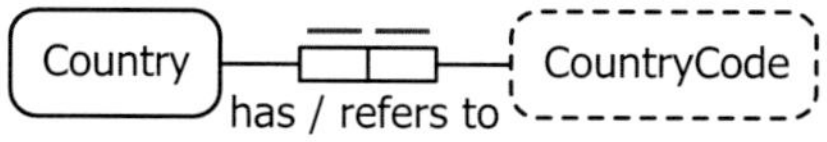

FIG. 4.13 A simple 1:1 naming convention shown explicitly.

Simple 1:1 reference schemes are usually abbreviated by placing the reference mode name in parentheses. However, we may set them out fully as in Fig. 4.13. This is an *existential fact type* (also called a *reference type*): instances are used to simply assert the existence of an entity. This diagram is still incomplete; the next chapter shows how to add the constraint that each country has a country code.

If there is only one means of referring to an entity type, the UCs on the reference type are the responsibility of the user to enforce rather than the system. For example, we must ensure that there is only one country with the code "US". If two 1:1 naming conventions are used (e.g., code or name), choose one as our preferred reference scheme, and treat the other like any other elementary fact type.

The four possible patterns of UCs for a binary predicate are: many to many; many to one; one to many; and one to one. Each of our examples so far has involved a heterogeneous predicate (different object types). Let's look now at some ring predicates (both roles played by the same object type). To clarify the different roles being played, inverse predicate names are included. Our first example concerns parenthood (see Fig. 4.14). For simplicity, assume that we can identify people by their first names.

As the sample fact table in Fig. 4.14 reveals, a person may have many children. Moreover, a person may have many parents. The column entries need not be unique, but the whole row must be. Hence, the UC spans the whole row. Contrast this with motherhood (see Fig. 4.15).

Tom	Bill
Tom	Mary
Sue	Bill
Sue	Mary

FIG. 4.14 Parenthood is many-to-many.

Although a mother may have many children, each person has only one mother, so we mark the constraint as shown. Entries in the second column of the fact table must be unique within that column (not necessarily unique within the table; e.g., after she becomes an adult, Mary could also become a mother).

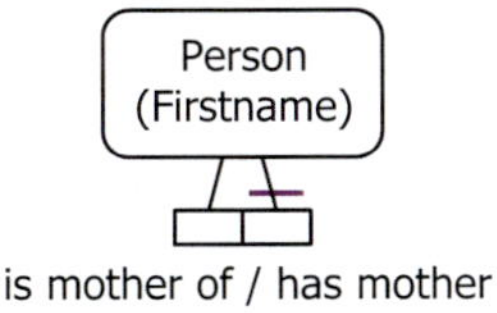

Sue	Bill
Sue	Mary
Mary	John

FIG. 4.15 Each person has only one mother.

The next example shown in Fig. 4.16 is consistent with monogamy: at any time, each man has at most one wife, and each woman has at most one husband. So, entries in each column must be unique.

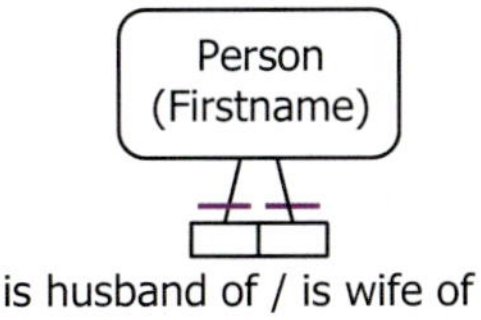

Adam	Eve
Pat	Linda
Terry	Norma

FIG. 4.16 Monogamy is one to one.

As an exercise, draw a model for the following marriage conventions: polyandry (a woman may have many husbands but not vice versa); polygyny (a man may have many wives but not vice versa); and polygamy (a man may have many wives, and a woman may have many husbands).

UCs are best understood in the context of a populated schema diagram, as discussed in this section. With this understanding, for any given fact type *each role is associated with a corresponding column of the fact table.*

In terms of logical predicates, the role boxes are the "object holes". For a set of relationship instances of this type, the holes expand to columns. Marking a *single role with a uniqueness bar* means that entries in the associated column must be unique in that column (i.e., *no duplicates are allowed in that column*).

A spanning UC requires that each fact row be unique in the table. This whole row constraint applies to any asserted fact type. If we have a stronger UC, then the whole row constraint is implied by this and hence is redundant. For this reason, we never mark the whole row constraint across a fact type unless that is the only UC that applies.

With this understanding, we must choose exactly one of the four constraint patterns shown in Fig. 4.17 for any binary predicate. For brevity, *a* and *b* denote object variables, and "*a R*'s *b*" means "*a* bears the relation *R* to *b*".

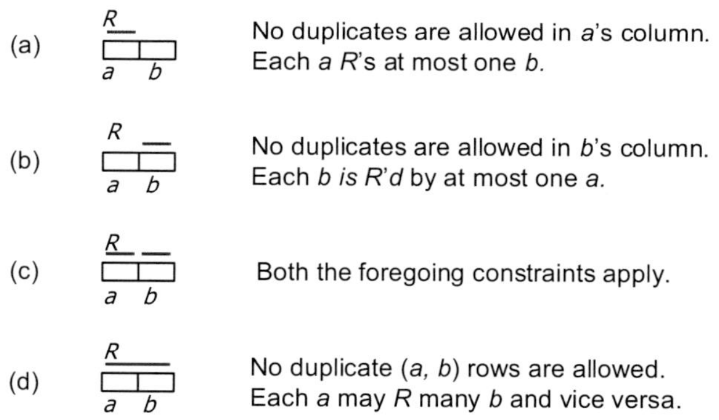

FIG. 4.17 The four uniqueness constraint patterns for a binary.

Provided a significant example is supplied, the UCs on a predicate can be determined simply by looking for duplicates in its fact table. If we are not familiar with the application, we may be unsure as to whether the population is significant. Consider for example the output report of Table 4.1.

Table 4.1 An output report of doubtful significance.

Reviewer	*Paper Nr*
Jones E	1
Smith JB	2

Suppose this is a sample output report for a conference management system. One task involves getting qualified people to review (assess the suitability of) papers submitted by people who hope to present a paper at the conference. Let us agree that the information in the table may be modeled using the fact type Person(.name) reviewed Paper(.nr). Which of the four binary UC patterns should be specified? If the population of Table 4.1 is significant, then we have a 1:1 situation, and constraints should be marked on each role, as shown in Fig. 4.18.

This means that each person reviews at most one paper (good news for the reviewers), and each paper is reviewed by at most one person (bad news for the people submitting the papers).

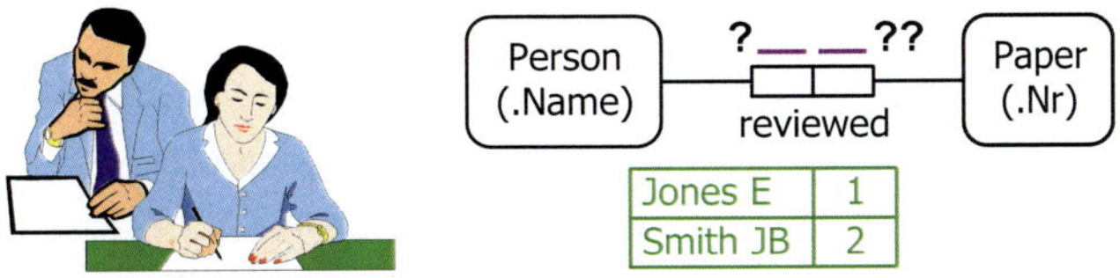

FIG. 4.18 Is the population significant?

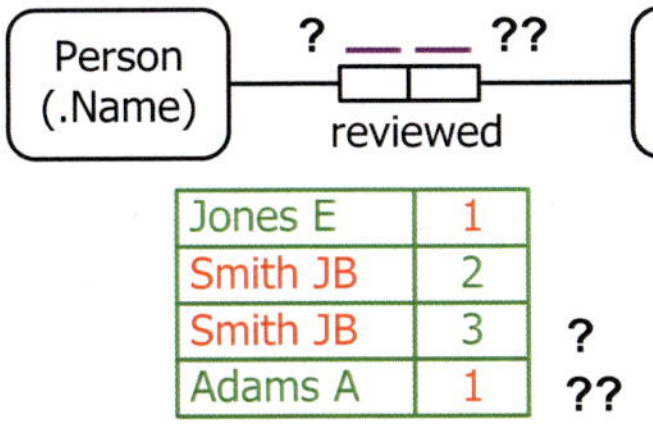

FIG. 4.19 Adding counterexamples to test the constraints.

While this arrangement is possible, we would probably doubt whether this constraint pattern is really intended. To resolve our doubts, we should *question each constraint in English, backed up by counterexamples.*

Consider the UC on the first column. Its positive verbalization is: **Each** Person reviewed **at most one** Paper. Its equivalent, negative verbalization is: **It is impossible that the same** Person reviewed **more than one** Paper. To test this constraint, we ask "Is it possible for a person to review more than one paper? For instance, could Smith review papers 2 and 3?" This counterexample appears as the second and third rows of the expanded fact table in Fig. 4.19.

To test the UC on the second column, ask "Is it possible for a paper to have more than one reviewer, for example, could paper 1 be reviewed by Jones and Adams?" This counterexample appears as the first and fourth rows of the expanded fact table in Fig. 4.19. If the domain expert replies yes to both these tests, the fact type is many to many, and we modify the constraint as shown in Fig. 4.20.

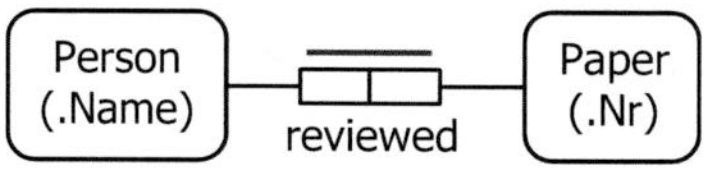

FIG. 4.20 If the test examples are accepted, the predicate is many to many.

If desired, a more compact population may be provided for the *m:n* case by using the tuple ("Smith JB", 1) instead of the last two rows. This tests both constraints at once. For testing purposes, however, it's normally better to test just one constraint at a time, as discussed. If just one of the two test cases were accepted, the final constraint pattern would be *n:1* or *1:n*. If both test cases were rejected, we would stay with the 1:1 pattern.

Remember that all constraints on fact types are to be interpreted as applying to the database, not necessarily to the real world. An information system can only enforce constraints on its formal model of the domain. The constraint schema for the "real world" need not be the same as for the model (or "recorded world").

For example, suppose the model includes the schema of Fig. 4.21(a). The UCs assert that for each state of the database, each patient is recorded as having at most one gender and at most one phone. Now it may be the case in the real world that some patients have more than one phone (e.g., a home phone and a business phone), as shown in the "real world schema" of Fig. 4.21(b).

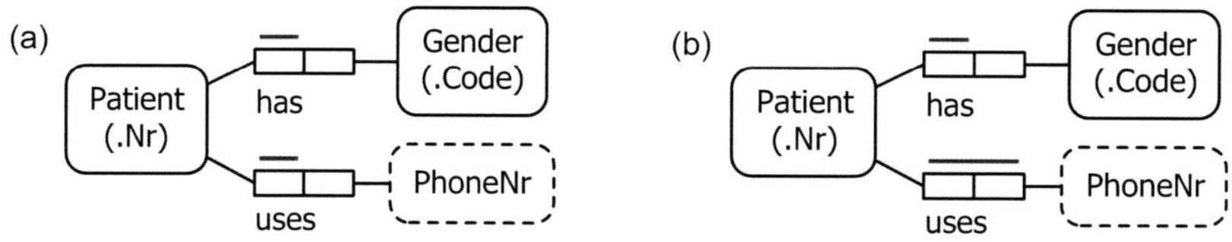

FIG. 4.21 A UC in the application model (a) may be stronger than in the real world (b).

The designer of the information system decided that no patient will be recorded to have more than one phone. Hence the constraints in the first schema (Fig. 4.21(a)) fit the intended model that is to be implemented in the information system. Unless otherwise indicated, *when we speak of a conceptual schema, we mean the structure of the conceptual model (not necessarily the real world).*

Nevertheless, we often use our background knowledge of the "real world schema" when designing the conceptual schema. In any case, *the UCs should be at least as strong as those that apply in the real world.* For the case being discussed, we know that in the real world each person has at most one gender. So, we should enforce this in our schema. With the phone fact type, we need to consciously decide whether a stronger constraint than the real-world constraint is required.

Exercise 4.2

1. For a given fact type, a sample population is significant with respect to UCs if all the relevant UCs can be deduced from the sample. A template is shown for a binary fact type. The names of the predicate and the reference modes are omitted for simplicity. For each of the fact tables, add the UCs to the template assuming the population provided is significant.

(a)

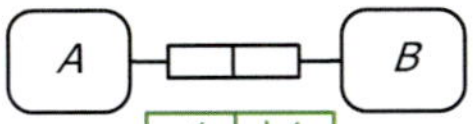

a1	b1
a2	b2
a1	b3

(b)

a1	b1
a1	b2
a2	b1

(c)

a1	b1
a2	b2
a3	b3

(d)

a1	b1
a2	b2
a3	b1

2. In a given department, employees are identified by employee numbers "e1", "e2", and so on, and projects are identified by project numbers "p1", "p2", and so on. Draw a schema diagram for the fact type Employee works on Project, and provide populations that are significant with respect to the following constraint patterns:
   (a) 1:many (b) many:1 (c) many:many (d) 1:1

3. Add the relevant UCs to the conceptual schema diagrams for the Exercise questions listed below. For some of these questions, the output report provided in the question might not be significant with respect to UCs. Using common sense, you should be able to avoid adding UCs that are likely to be violated by a larger population.
   (a) Exercise 3.4 Q.1b (b) Exercise 3.4 Q.1d (c) Exercise 3.4 Q.2
   (d) Exercise 3.4 Q.5 (e) Exercise 3.5 Q.2 (f) Exercise 3.5 Q.3

4. By now, you must be feeling like you need a challenge. The following report is an extract from an actual computer science department handbook. Verbalize this report in terms of *binaries*, draw the fact types, and add UCs.

   The first column headed "Subject title" lists the codes and titles of postgraduate topics. A topic is not the same as a subject. A postgraduate student first enrolls in a subject (identified by subject code, e.g., CS451) and then later chooses a topic to study for this subject. Subject enrollments are not part of this question. A topic may be offered in semester 1 only, in semester 2 only, or over the whole year (at half the pace).

   The fact that different students might be enrolled in the same subject (as identified by subject code) and yet be studying different topics is a source of possible confusion. As information modelers, one of the most significant contributions we can make is to suggest improvements to the universe of discourse (UoD) itself. In this example, we might argue that it would simplify things if topics were treated as subjects. The department involved eventually accepted this change. For this exercise, however, you are to model things as they were.

Postgraduate coursework topics in computer science.

Subject title	*Staff*	*When*	*Prerequisites*
AIT: Advanced Information Technology	MEO, PNC, TAH	1st sem	CS315
CLVS: Computational Logic and Verification Systems	JS, PJR	Year	CS260 preferred
DBMS: Advanced topics in Database Management	MEO, PNC, RC, TAH	2nd	CS315
FP: Functional Programming	EJS, PAB	Year	CS225 CS220 preferred
GA: Geometric Algorithms	PDE	1st	CS340

N.B. Topics are offered subject to the availability of staff and to there being sufficient demand for a topic. In each case the first-mentioned member of staff is lecturer-in-charge of the subject. Preferred prerequisites may become mandatory in later years.

4.3 Uniqueness Constraints on Longer Fact Types

This section shows how to specify UCs on fact types of arity 3 and beyond, including nested versions. Let's begin with an example. Fig. 4.22 shows a populated ternary fact type of the form Person scored Rating for Subject. For simplicity, assume that persons are identified by their surname. Even if we are unfamiliar with the example, sample populations supplied by the domain expert can reveal that certain constraints don't apply.

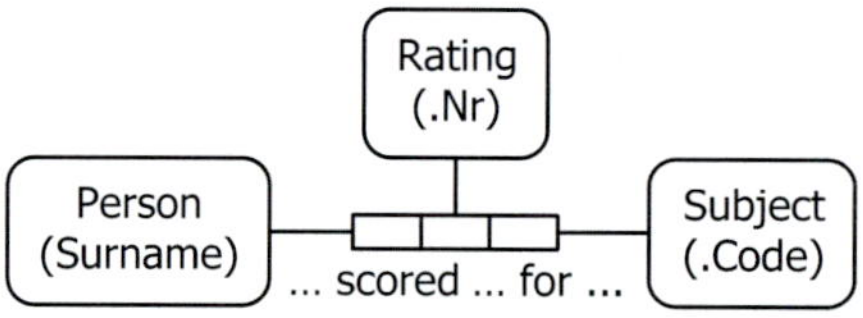

Adams	7	CS112
Adams	6	CS100
Adams	7	PD102
Brown	6	CS112
Brown	7	PD102

FIG. 4.22 What are the uniqueness constraints?

Looking at the fact table, first consider each column individually. Each column has at least one value that is repeated. So, no column by itself has a UC. As shown later, this is true of any ternary fact type that is elementary.

Next, let's look at pairs of columns. For ease of reference here, the columns are numbered 1, 2, 3 in the order shown in the diagram. Beginning with columns 1 and 2, note that the pair (Adams, 7) is repeated. With columns 1 and 3, each pair is unique. That is, each (Person, Subject) combination occurs on only one row of the table. With columns 2 and 3, the pair (7, PD102) is repeated.

There are only three ways to pair columns in a ternary. If the population of the fact table is significant, the only pair-unique constraint is that each (Person, Subject) pair must be unique to the table. If familiar with the domain, we can usually decide whether the population is significant in this regard simply by using common sense. With this example, this constraint means that each (Person, Subject) combination has at most one rating. This agrees with our background knowledge about the UoD and so the UC suggested by the table is accepted.

Sample fact tables may be obtained from output reports (which often represent a combination of separate fact tables) or from simple factual knowledge about the UoD. Unless the population is large, or well chosen, it is unlikely to be completely significant with respect to all UCs. For instance, suppose row 3 was deleted from the table in Fig. 4.22. For the smaller sample, both the column 1–2 pair and the column 2–3 pair would then show no duplicates. The table would then suggest a UC for all three column pairs. Apart from *insufficient data*, a population may fail to be significant because of *bad data* (i.e., some incorrect values have mistakenly been entered as data in the example).

So, we need to be wary of relying on just a sample fact table to determine UCs. We should *ask whether any suggested constraint really makes sense*. A UC spanning columns 1 and 2 would mean that a (Person, Rating) combination could occur for at most one subject. This would forbid somebody obtaining the same rating for two subjects–an unrealistic restriction! Similarly, the suggested constraint on the column 2–3 pair must be rejected as unrealistic (it would prevent two people from scoring the same rating for a particular subject).

UCs may often be determined simply by background knowledge of the UoD. Real world UCs usually determine the weakest UCs to be considered for the actual database. If we know the fact table is significant, then we can generate the constraints from it.

Usually, however, we won't know in advance that the table is significant, and we will have to use some of our background knowledge. If we lack such knowledge, we should consult the domain expert about any doubtful constraints, posing the relevant questions while adding concrete examples of facts to violate the constraints and asking whether such updates may be accepted.

With a ternary fact type, a spanning constraint is always implied because no whole row may be repeated. This constraint should be specified if and only if no shorter UC holds. So, for our current example, we mark just the constraint for the column 1–3 pair, indicating that each (Person, Subject) pair is assigned at most one rating.

Because the two role boxes involved are not adjacent, a *divided constraint bar* is used, as shown in Fig. 4.23. A *dotted line* connects the two active parts of the constraint bar, indicating that the constraint excludes the role under the dotted line. If we omitted the dotted line, the bars would instead denote two separate constraints, one for each end role. If we reordered the fact type instead as "Person for Subject scored Rating", there would be two contiguous roles spanned by an undivided constraint bar.

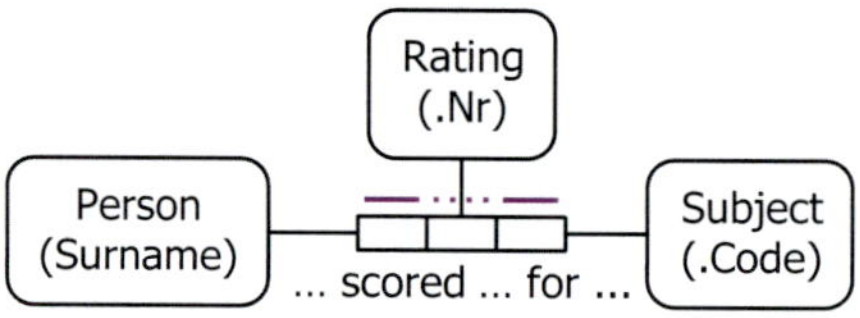

FIG. 4.23 Each (Person, Subject) combination is unique.

The term *n-role constraint* denotes a constraint that spans *n* roles. A UC that spans just one role is a **simple** UC. The constraint in Fig. 4.23 is a two-role constraint because it spans two of the three roles involved in the fact type. A binary fact type has three possible basic constraints (two one-role and one two-role), which give rise to four possible cases. A later section demonstrates that *no (elementary) ternary fact type can have a simple UC.* So a ternary fact type has four basic UCs to be considered: three two-role constraints and one three-role constraint, as shown in Fig. 4.24.

FIG. 4.24 Allowed basic uniqueness constraints for a ternary.

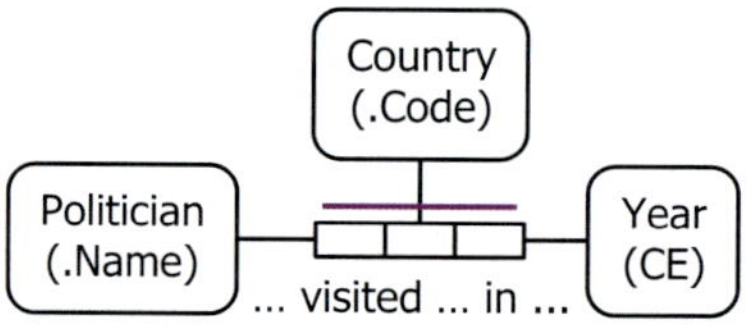

Joe Biden	GB	2022
Joe Biden	GB	2023
Joe Biden	JP	2023
Rishi Sunak	JP	2023

FIG. 4.25 A ternary with a spanning uniqueness constraint.

In practice, ternaries usually have just a single UC either over one role pair or over all three roles. As an example of the latter, a politician travel example is modeled in Fig. 4.25. Here, many politicians may visit many countries in many years. Such a fact type is said to be "many to many to many". Because no whole row can be repeated, we cannot use this fact type to record multiple visits to the same country by the same politician in the same year. If you want to record this, use a different model (e.g., replace Year by Date, or add Quantity).

With a ternary fact type, we should systematically test each of the three two-role UCs to see which ones hold. Only if no two-role constraints hold should the three-role constraint be specified. As an exercise, check for yourself that the sample data is significant for this constraint.

For a ternary association, there is only one way to have a three-role UC. There are three ways of having precisely one two-role constraint, and three ways of having exactly two two-role constraints. Finally, there is one way of having three two-role constraints. Thus, there are eight different UC patterns that may arise with a ternary fact type. Four of these involve just one constraint (Fig. 4.24) and four involve combinations (see Fig. 4.26).

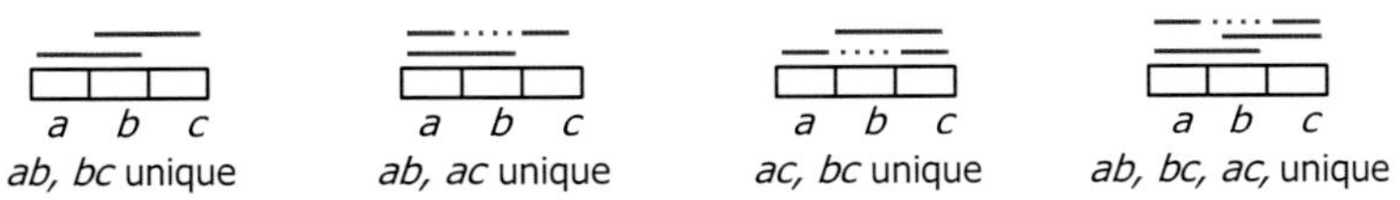

FIG. 4.26 Allowed uniqueness constraint combinations for a ternary.

All other constraint patterns for a ternary are illegal (i.e., disallowed). Some examples of these are shown in Fig. 4.27. The first two of these are fundamentally wrong because a ternary fact type with a simple UC cannot be elementary. The third example is wrong because the three-role constraint is implied by the two-role constraint and hence should not be displayed.

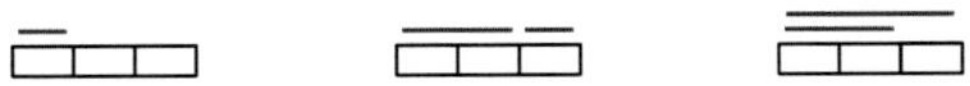

FIG. 4.27 Some illegal constraint patterns.

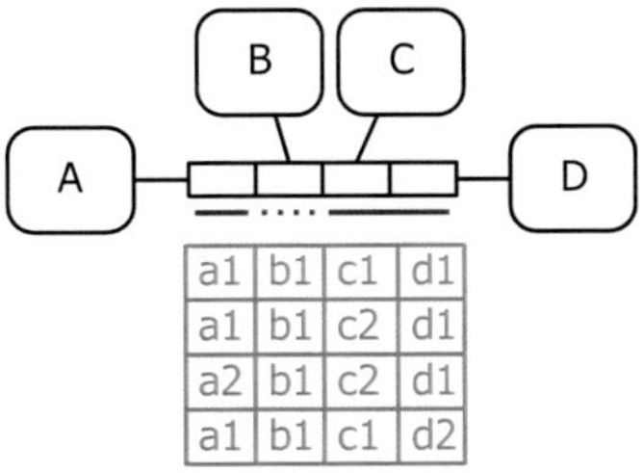

FIG. 4.28 Each *acd* combination occurs on at most one row.

This general approach for specifying UCs applies to fact types of any arity. Fig. 4.28 shows a template for a quaternary predicate with a UC spanning roles 1, 3, and 4. This indicates that each *acd* combination must be unique to the fact table, where *a*, *c*, and *d* occur in the first, third, and fourth columns, respectively, on the same row of this table. A small but significant population of the fact table is shown. Verify for yourself that this is consistent with the indicated UC.

As an example of a quaternary, consider Fig. 4.29. The UC spans the first, second, and fourth roles, and verbalizes as: **For each** Person_1, Person_2 **and** Language, **that** Person_1 judged **that** Person_2 to have **at most one** SkillLevel in **that** Language. Note the subscripts to distinguish the different roles played by Person.

It is shown later that any UCs on a quaternary must span at least three roles. As an exercise, you might like to explore all the possible cases. Or maybe not! Because elementary fact types are rarely longer than quaternaries, no examples of quinaries or longer fact types are discussed here.

To visualize the connection between roles and columns of the fact table, think of the role boxes as the holes in which the objects get placed to complete the fact. No matter what the arity of the fact type, a UC across a combination of role boxes means that any instance of that column combination that does appear in the table must occur on one row only, and hence can be used to identify that row.

Because a role combination governed by a UC provides a "key to unlock" any row of the fact table, this combination is referred to as a *key* for that table. This notion will be elaborated later in the context of the (often nonelementary) tables used in relational databases.

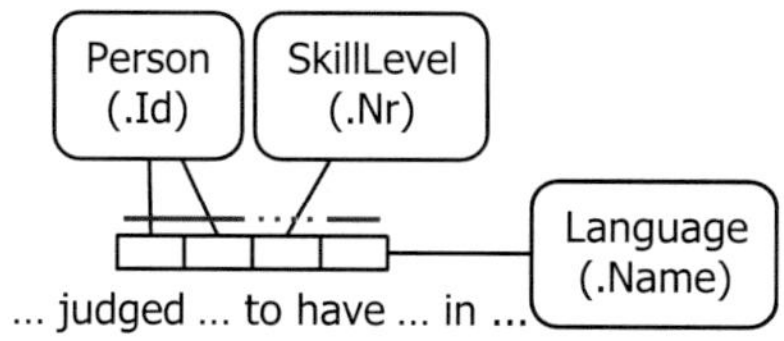

FIG. 4.29 A uniqueness constraint over roles 1, 2, and 4.

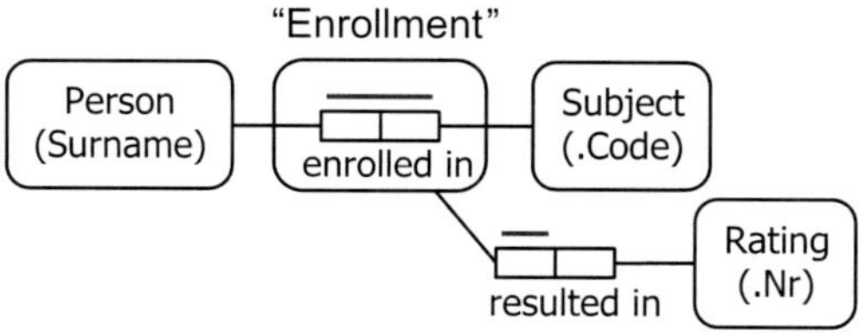

FIG. 4.30 The enrollment association is objectified as a nested object type.

Now, let's consider *nesting*. In Fig. 4.23, the subject ratings example was set out in flattened form. Suppose instead we use a nested approach, as shown in Fig. 4.30. Here each (Person, Subject) enrollment is treated as an object that may result in a rating. For example, the enrollment of Adams in CS112 results in a 7.

Fig. 4.23 is not actually equivalent to Fig. 4.30 unless a further constraint is added that a rating must be known for each enrollment. Such mandatory role constraints are discussed in the next chapter.

The Enrollment entity type in Fig. 4.30 is called an objectified association, an objectified relationship type, an objectified predicate, a nested entity type, or a nested object type. Most objectified associations are spanned by a UC (as in Fig. 4.30). This spanning UC indicates that each person may enroll in many subjects, and the same subject may be enrolled in by many people. The simple UC on the binary predicate indicates that each (Person, Subject) enrollment resulted in at most one rating.

Section 10.5 includes a discussion of objectification of associations with nonspanning UCs (e.g., 1:1 associations and *n*:1 associations). While currently allowed in ORM, such cases can violate elementarity, and are better modeled in other ways.

Exercise 4.3

1. A template for a ternary fact type is shown. For each of the fact tables provided, add the UCs, assuming that the population is significant in this regard.

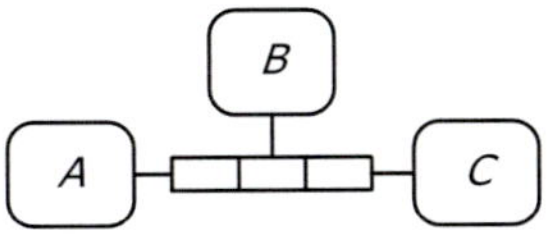

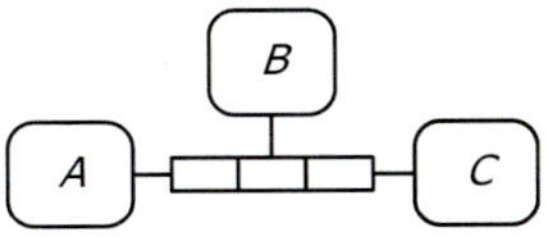

(a)

a1	b1	c1
a2	b1	c1
a1	b2	c2

(b)

a1	b1	c1
a1	b2	c2
a2	b1	c2

(c)

a1	b1	c1
a2	b2	c1
a1	b1	c2
a1	b2	c1

(d)

a1	b1	c1
a2	b2	c2
a1	b2	c2
a2	b2	c1

2. Verbalize and then schematize the following report, marking UCs.

Department	*Staff category*	*Number*
CS	Professor	3
CS	Senior Lecturer	6
CS	Lecturer	10
EE	Professor	3
EE	Senior Lecturer	7
EE	Lecturer	7

3.

Year	*Branch*	*Profit ($)*	*Total profit ($)*
2007	New York Paris	200,000 90,000	290,000
2008	New York Paris	300,000 150,000	450,000

(a) Verbalize and then schematize this report without nesting. Include UCs.
(b) As for (a) but instead use a nested approach.

4.4 External Uniqueness Constraints

The UCs discussed so far are *internal* (intrapredicate) UCs, as each applies to one or more roles in a single predicate. **External** (interpredicate) **uniqueness constraints** *apply to roles from different predicates.* To understand these constraints, the *conceptual join* operation is first discussed.

Consider the schema and sample population in Fig. 4.31. The left-most fact table indicates that employee 15 drives cars with registration numbers PKJ123 and ABC000. The right-most fact table indicates that car PKJ123 is imported from Italy. The ternary fact table indicates that employee 15 drives the car PKJ123 imported from Italy. The asterisk after the mixfix predicate reading indicates that the ternary fact type is *derived.* The derivation rule is preceded by an asterisk.

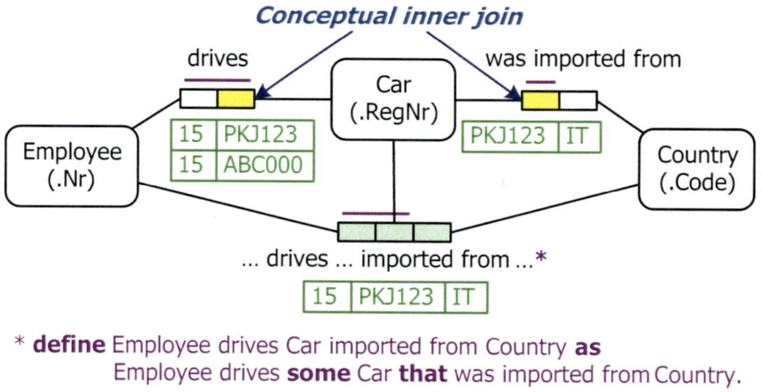

FIG. 4.31 A compound ternary derived from a conceptual inner join of the binaries.

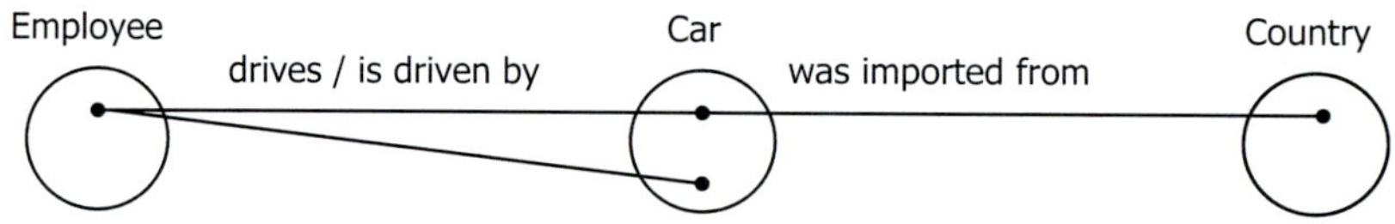

FIG. 4.32 An instance diagram for the model shown in Fig. 4.31.

The instance diagram in Fig. 4.32 may help clarify things. The binary facts appear as lines connecting two dots. The ternary fact corresponds to the continuous path from Employee to Country. Starting at the Car type, this path may also be verbalized as the conjunction Car is driven by Employee **and** was imported from Country.

The derivation rule shows how the ternary is derived by combining the binaries, with Car as the connecting or join object type. The defining expression "Employee drives **some** Car **that** was imported from Country" is a formal verbalization of the conceptual path from Employee, through the drives predicate, Car type, and import predicate, to Country. The keyword "**that**" declares that the driven car must be *the same* as the imported car. In other words, if an employee drives a car, we must use the very same car to continue the path to Country.

A fact type resulting from a join is a *compound fact type,* not an elementary fact type because it is essentially the conjunction of at least two simpler fact types. For this reason, *join fact types are normally excluded from ORM schema diagrams.* If included for discussion purposes, they may be *shaded,* as in Fig. 4.31, to indicate that they are just views. Unlike elementary derived fact types, their compound nature can lead to redundancy unless each join role is covered by a simple UC.

For example, if we add the row (16, PKJ123) to the drives table in Fig. 4.31, then row (16, PKJ123, IT) is added to the ternary and is now contained twice in the ternary. Although join fact types are never used as asserted fact types, join paths are often involved in constraints and queries, as discussed later.

In an ORM schema, to navigate from one predicate to another, we must pass through an object type, performing a *conceptual join* (or *object join*) on that object type. By default, the join condition is that the object instance remains the same as we pass through. This is called a *conceptual inner join*. This is similar to a relational natural inner join (see later chapter), except that conceptual joins require the conceptual objects to be the same, instead of matching attribute names and values. ORM object types are conceptual domains, not attributes. Conceptual joins still apply if we declare different role names (cf. attribute names) for the join roles.

In terms of the fact tables, the column value of the role entering the join object type must equal the column value of the role that exits the object type. Fig. 4.33 depicts two predicates, *R* and *S*, sharing a common object type, *B*. From a fact table perspective, the conceptual inner join of *R* and *S* is obtained by pairing rows of *R* with rows of *S* where the *B* column values match, and arranging for the final *B* column to appear just once in the result.

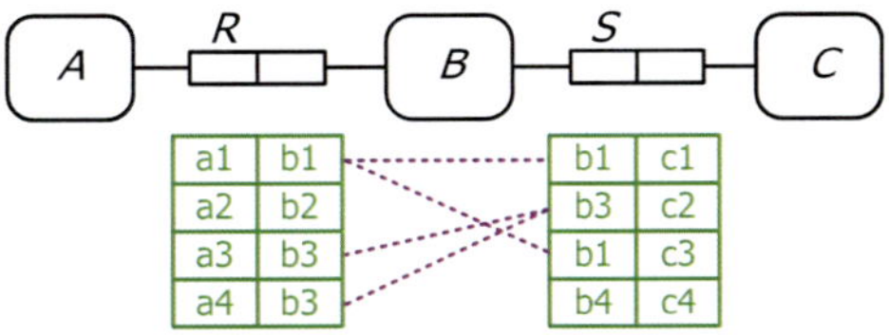

FIG. 4.33 Joining on *B* requires the values in the two *B* columns to match.

The dotted lines show how the rows are paired together. For example, row (a1, b1) is paired with rows (b1, c1) and (b1, c3) to give rows (a1, b1, c1) and rows (a1, b1, c3). Fig. 4.34 shows the result of this join operation. The final *B* column is the intersection of the original *B* columns. The natural inner join of any tables *R* and *S* is denoted by "*R* ⋈ *S*" or "*R* natural join *S*". The "inner" is often omitted and assumed by default.

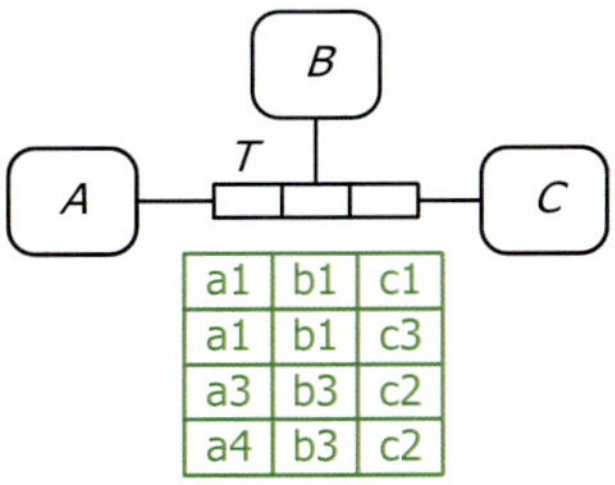

FIG. 4.34 Table resulting from the natural inner join of the tables in Fig. 4.33.

Other kinds of joins are sometimes discussed (e.g., joins based on operators other than equality, and outer joins). A left (right, full) *outer join* is obtained by adding to the inner join those rows, if any, where the join column value occurs in just the left (just the right, just one of) the tables, and padding the missing entries of such rows with nulls. For example, the left outer join of the binary fact tables in Fig. 4.31 adds the row (15, ABC000, ?) where "?" denotes a null.

The full outer join of the tables in Fig. 4.33 includes rows (a2, b2, ?) and (?, b4, c4). Outer joins are often used in queries at both conceptual and relational levels. Because asserted fact types in ORM are elementary, their fact tables cannot have nulls. We postpone further discussion of outer joins until Chapter 10.

Consider the report concerning high school students shown in Table 4.2. As an exercise, perform step 1 for the top row of this report before reading on.

Table 4.2 An output report about high school students.

Student Nr	*Name*	*Class*
001	Adams J	11A
002	Brown C	12B
003	Brown C	11A

The information for row 1 may be verbalized as two elementary facts:

The Student with studentNr '001' has the StudentName 'Adams J'.
The Student with studentNr '001' is in the Class with code '11A'.

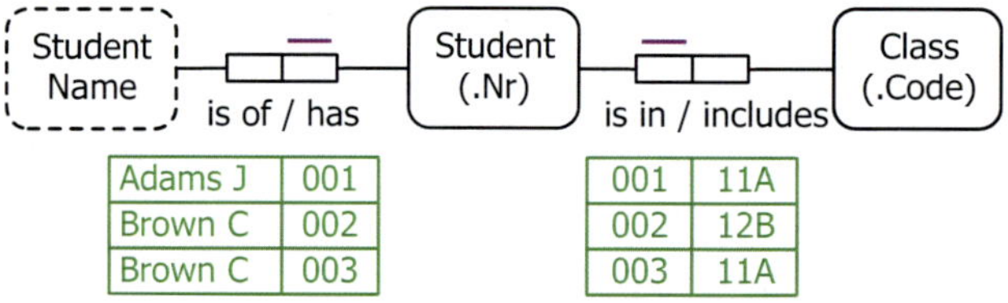

FIG. 4.35 A populated schema for Table 4.2 (draft version).

The first fact relates an entity and a value. The second fact is a relationship between two entities. In this UoD, students are identified by their student numbers. As rows 2 and 3 show, it is possible for two different students to have the same student name (e.g., "Brown C"). A populated schema diagram is shown in Fig. 4.35.

Let us agree that the population supplied is significant. It follows that each student has at most one name, and that each student is in at most one class. These UCs have been captured on the schema diagram. However, there is another UC that is missing. What is it?

The output report is reproduced in Table 4.3, with UCs marked above the relevant columns. Notice that the combination of name and class is unique. Let's suppose that this is significant. Although a student's name need not be unique in the high school, it is unique in the student's class. While there are two students named "Brown C", there can be only one student with this name in class 12B, and only one student with this name in class 11A.

Table 4.3 Uniqueness constraints added to original table.

Student Nr	*Name*	*Class*
001	Adams J	11A
002	Brown C	12B
003	Brown C	11A

If another student named "Brown C" joined one of these classes, at least one of the names would need to be modified to keep them distinct within that class. For instance, we could add extra initials or numbers (e.g., "Brown CT", "Brown C2").

Each row of the report splits into two elementary facts, one for the name and one for the class. To specify the name-class UC on the schema, we need two fact types. The role boxes to which the UC applies are joined by dotted lines to a *circled uniqueness bar,* as shown in Fig. 4.36.

Because this UC involves roles from different predicates, it is an example of an *external constraint* or interpredicate constraint. A constraint on a single predicate is an *internal* or intra-predicate constraint. So, there are two kinds of UCs: internal uniqueness and external uniqueness.

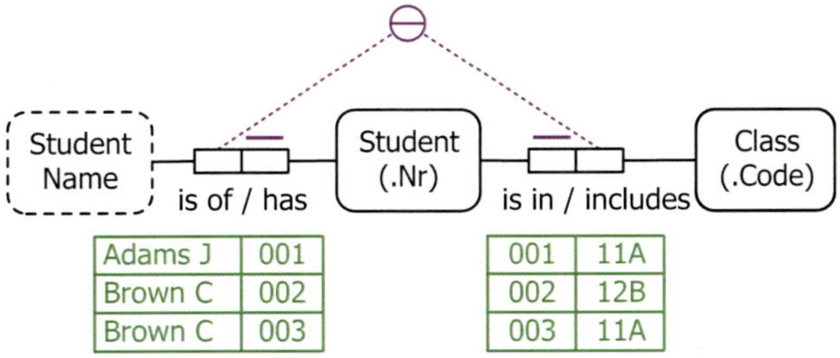

FIG. 4.36 An external uniqueness constraint has been added.

In this example, the external UC indicates that for each student, the *combination* of student name and class is unique. The constraint verbalizes as "**Given any** StudentName **and** Class, **at most one** Student has **that** StudentName **and** is in **that** Class". For instance, given "Adams J" and "11A" there is only one StudentNr sharing rows with both ("001"). Given "Adams J" and "12B" there is no StudentNr paired with both.

An easy way to understand this constraint is to say that if we perform the conceptual (inner) *join* operation on the two fact tables, then the resulting table has a UC across the (name, class) column pair. Note that when we join the two fact tables, we obtain the table in the original report (Table 4.3).

The external UC is equivalent to an internal UC on the first and last roles of the derived, ternary fact type formed from the join path: StudentName is of Student **who** is in Class. See Fig. 4.37. The object type Student is personal, so we prefer the personal pronoun "**who**" instead of "**that**".

The middle role of the join fact type has a simple UC. This is equivalent to the two simple UCs on the binaries. Although join fact types may be included on a schema diagram for discussion purposes, they are illegal in the base ORM model because they are compound, not elementary, fact types.

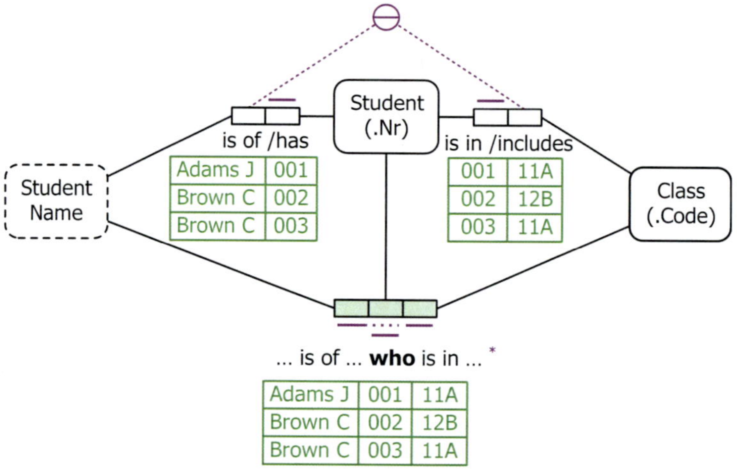

FIG. 4.37 Equivalent constraints on derived, join fact type (illegal in base model).

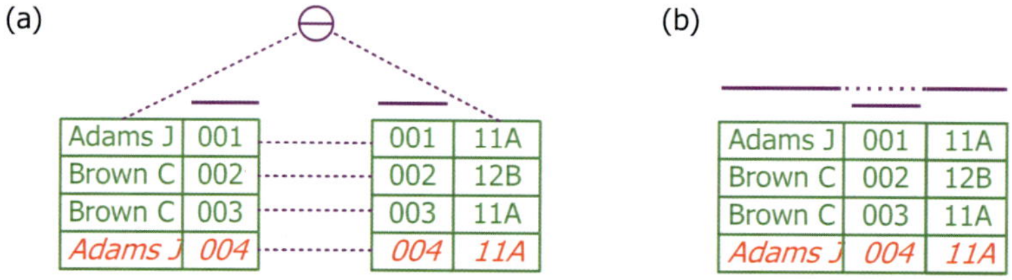

FIG. 4.38 This update would violate the external uniqueness constraint.

As discussed shortly, the presence of a simple UC on a ternary disqualifies it from being elementary. So, don't feel that it's okay to do this in your normal model.

Suppose we added the following facts to the current populations: Student '004' has StudentName 'Adams J'; Student '004' is in Class '11A'. If this update were accepted, the new populations would be shown in Fig. 4.38(a). To test your understanding, explain why this would violate the external UC before reading on.

The extra row added to each table provides a counterexample to the constraint because for "Adams J" and "11A" there are two StudentNr entries paired with these entries (001 and 004). This breaks the rule that student names are unique within a given class. In class 11A, two students (001, 004) have the same name ("Adams J").

If we join the tables on studentNr (the dotted lines indicate the matches for the join), we obtain the ternary table in Fig. 4.38(b), which clearly violates the equivalent compound UC over the StudentName and Class columns.

The general case is summarized in Fig. 4.39, where *A*, *B*, and *C* are any object types, and *R* and *S* are predicates. External UCs may apply to roles from reference types, not just fact types. This is the basis for composite reference schemes, as discussed in a later chapter.

As a more complex case, suppose persons enroll in subjects but are given subject positions (1st, 2nd, 3rd, etc.) rather than subject ratings. Each person achieves at most one position in any given subject. Moreover, no ties may occur (i.e., for each position in a subject there is only one student).

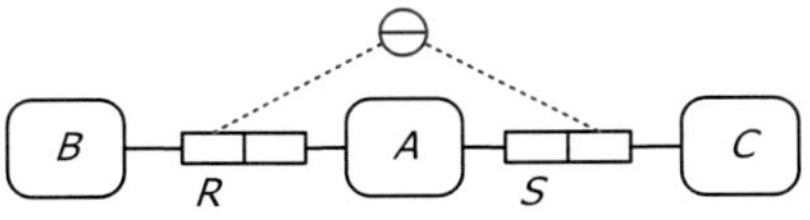

FIG. 4.39 An external uniqueness constraint.

Table 4.4 Student ranking within subjects (no ties allowed).

Person	*Subject*	*Position*
Adams J	CS114	3
Adams J	CS100	10
Adams J	PD102	3
Brown C	CS114	10
Brown C	PD102	5

Table 4.4 shows a sample report for this UoD. Performing step 1 on the first row, we may express the information as the ternary: Person (name) "Adams J" is placed in Subject (.Code) "CS114" at Position (.Nr) 3. This leads to the schema of Fig. 4.40. Notice the overlapping UCs.

We could instead express the information on row 1 thus: Person (name) "Adams J" enrolled in Subject (.Code) "CS114"; this Enrollment achieved Position (.Nr) 3. This leads to the nested version in Fig. 4.40(b). For this to be equivalent to the ternary, the role played by the objectified association must be mandatory.

The simple UC on the achievement predicate captures the UC spanning the first two roles in the flat version of Fig. 4.40(a). The external UC corresponds to the constraint spanning the last two roles in the flat version–each (Subject, Position) combination is unique.

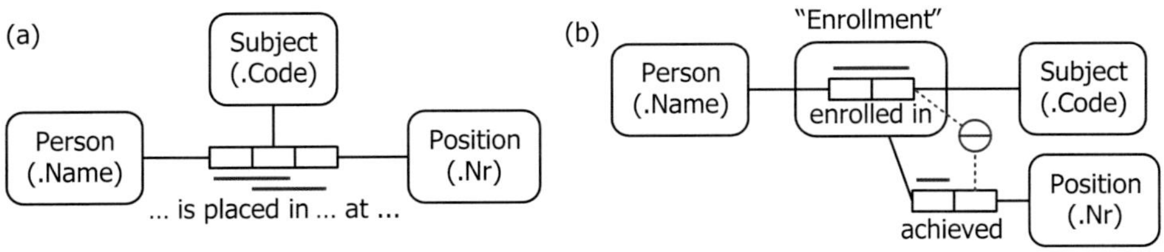

FIG. 4.40 Schemas for Table 4.4: (a) flatted version; (b) nested version.

So, if we *flatten* the nested version into a ternary, then any (Subject, Position) pair occurs on at most one row of the ternary table. To help understand this, we suggest that you add the fact tables to the diagram. The fact table for the outer predicate matches the output report.

The external UC in Fig. 4.41 is a *unique-where-true constraint*, where the uniqueness applies only to those politicians who are government heads. This verbalizes as **For each** Country, **at most one** Politician is a government head **and** serves on the government of **that** Country.

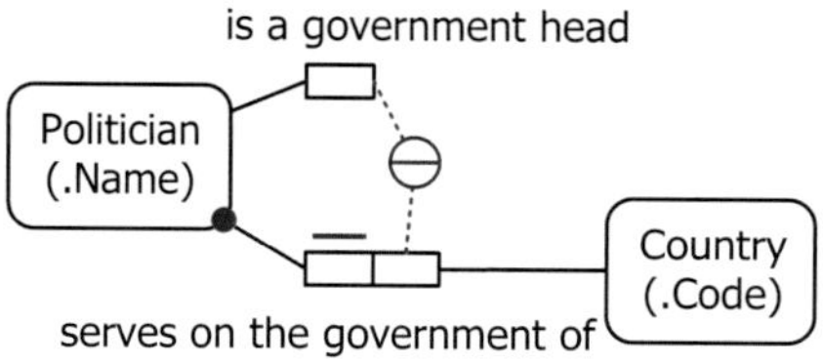

FIG. 4.41 A unique-where-true external uniqueness constraint.

Table 4.5 Students are assigned both ratings and positions.

Person	*Subject*	*Rating*	*Position*
Adams J	CS114	7	3
Adams J	CS100	6	10
Adams J	PD102	7	3
Brown C	CS114	6	10
Brown C	PD102	7	5

External UCs sometimes connect more than two roles. For example, a point in three-dimensional space might be associated with a unique combination of *x*, *y*, and *z* coordinates.

As you may gather from these examples, nesting tends to produce a more complex constraint picture. For this example, the flattened version is preferred. However, if the Position information is optional (e.g., to be added later), then the nested approach is preferred. Such modeling choices are covered in more detail later.

Another reason for nesting is to avoid embedding the same association within more than one fact type. Suppose we have to record both a rating and a unique subject position for each student taking any given subject. A sample report for this situation is shown in Table 4.5.

We might describe this UoD using two ternaries, as in Fig. 4.42. Note, however, that all (Person, Subject) combinations appearing in one fact table must also appear in the other. A later chapter shows how such a constraint may be added.

Alternatively, we may objectify the enrollment association between Person and Subject, and attach rating and position predicates to this (see Fig. 4.43). The nested approach corresponds to reading the information on row 1 of the table as Person "Adams J" enrolled

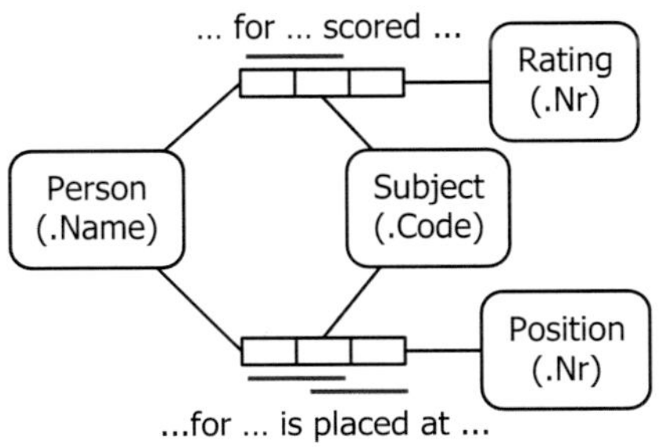

FIG. 4.42 A schema for Table 4.5 (flattened version).

in Subject "CS114"; this Enrollment scored Rating 7; this Enrollment achieved Position 3. To indicate that rating and position must be recorded, additional constraints are needed (see next chapter).

If a role has a simple, internal UC, it should not be included in an external UC. Even if a join path exists to enable an external UC to be declared, the external constraint would be implied by the stronger, simple UC; it is generally preferable to omit implied constraints.

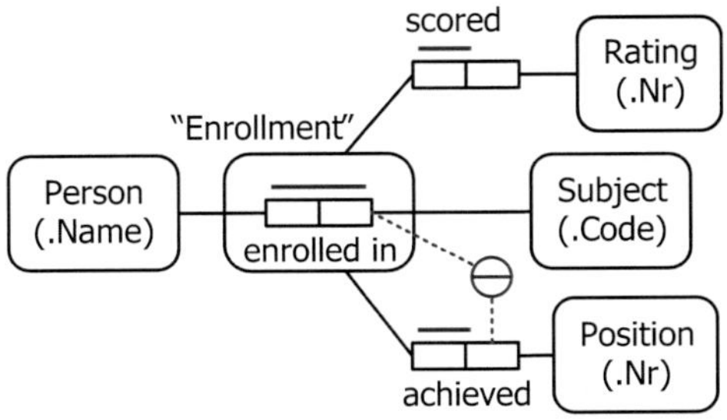

FIG. 4.43 Another schema for Table 4.5 (nested version).

As an exercise to illustrate this point, you may wish to modify Fig. 4.36 so that student names are unique, and then show that the external constraint is implied.

Before you start the section exercise, there's one point you should be clear about. When you examine an information sample, you are seeing it at the external level. *There are often many aspects of the presentation that may not be relevant to your application.* In a tabular report, for example, the order of the columns or rows is not normally something that you need to model. If the information is presented in nontabular ways (e.g., diagrammatic) there are always topological and metric aspects, and possibly other aspects (e.g., color), that are represented along with basic data. Whether these additional aspects of the presentation need to be modeled is something that only the domain expert can decide. To fully appreciate this point, make sure that you attempt at least Question 2 from the following exercise.

Exercise 4.4

1. Add the UCs to the conceptual schema diagrams for:
   (a) Exercise 3.4 Question 6 (b) Exercise 3.5 Question 4

2. Many manufactured products contain parts that may themselves be products of even smaller parts. The structure of one such product is shown:

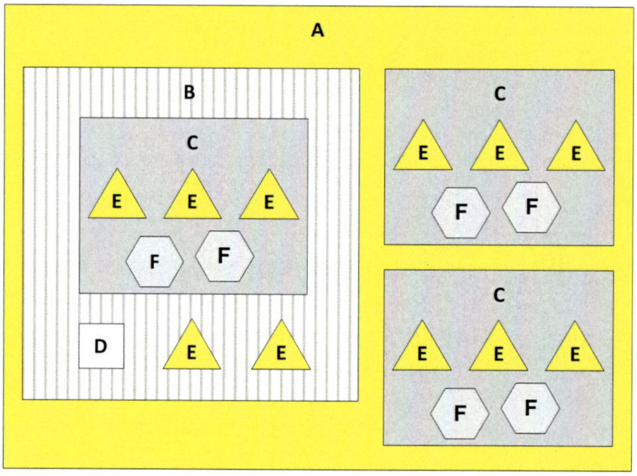

   (a) List at least three different kinds of facts that are captured in this diagram.
   (b) Assume that we are interested in modeling only containment facts. These facts may also be displayed as a labeled tree (hierarchy) as shown. Draw a conceptual schema diagram for this UoD, including UCs, but make no use of nesting.

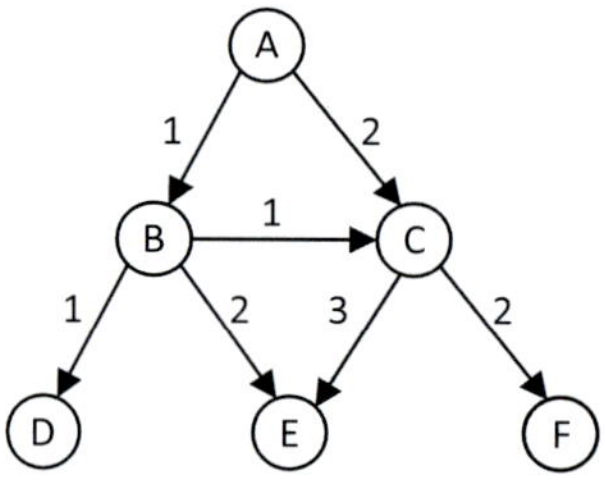

   (c) Draw an equivalent conceptual schema diagram that does make use of nesting.

3. A car dealer maintains a database on all the cars in stock. Each car is identified by the vehicle identification number displayed on a plate attached to the car (e.g., on its dashboard). For each car, the dealer records the model (e.g., Saturn SW2), the year of manufacture (e.g., 2006), the retail price (e.g., $18,000), and the color (e.g., dark green). Because of space limitations the dealer will never have in stock more than

one car of the same model, year, and color at the same time. The dealer also keeps figures on the number of cars of a particular model and color that are sold in any given year. For example, in 2007, five dark green Saturn SW2s were sold. Draw the conceptual schema diagram, including all UCs.

4.5 Arity Checks

In step 1 of the CSDP, we try to express information examples in terms of elementary facts. At that stage we rely on familiarity with the UoD to determine whether a fact type is simple or compound (splitable). Once UCs are added, a formal check can be conducted. This section discusses a check based on UCs, and the next section discusses the notions of projection and join, which may sometimes be used to perform a projection-join check.

Until we are experienced at conceptual schema design, we might include some fact types that are too long or too short. "Too long" means that the arity of the fact type is higher than it should be—the predicate has too many roles. In this case, we must split the compound fact type up into two or more simple fact types. For example, the fact type Scientist was born in Country during Year should be split into two fact types: Scientist was born in Country; Scientist was born during Year.

"Too short" means the arity of some fact types is too small, resulting in a loss of information—a fact type has been split even though it was elementary. In this case, the fact types resulting from the split need to be recombined into a fact type of higher arity.

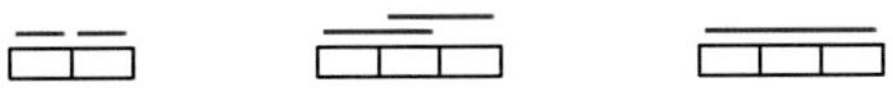

FIG. 4.44 A fact type has one or more keys.

For example, suppose that scientists may lecture in many countries in the same year. It would then be wrong to split the fact type Scientist lectured in Country during Year into the fact types Scientist lectured in Country; Scientist lectured during Year. This kind of error is rare, but serious nevertheless.

For a given fact type, a minimal combination of roles spanned by a UC is called a "*key*" for that fact type. "Minimal" means the UC has no smaller UC inside it; if it did, it is implied by the smaller (but stronger) constraint and hence should not be shown. Role combinations for such longer, implied UCs are called "proper superkeys". However, as proper superkeys are not minimal, we do not count them as keys in our discussion.

Each role is associated with a column of the fact table for the fact type. Unless the fact type is unary, it is possible for it to have more than one key. Consider the cases shown in Fig. 4.44. Here, the binary has two one-role keys, the middle ternary has two two-role keys, and the other ternary has one three-role key.

The first two predicates also have implied UCs across their whole length, but these are not counted as keys for these predicates. The *length* of a role sequence is the number of roles it contains. A key of length 1 is a *simple key*. All other keys are *composite*.

We can imagine unaries or binaries that should be split, but in practice not even a novice designer is likely to verbalize one (e.g., "The Barrister with surname 'Rumpole' smokes and drinks"). So, the question of splitability is really an issue only with ternaries and longer fact types.

To help decide when to split in such cases, the so-called "*n* – 1 *rule*" may be used: *each n-ary fact type has a key length of at least n* – 1. If this rule is violated, the fact type is not elementary, and hence should be split. Various cases of this rule are set out in Table 4.6. A proof is sketched later in the section.

If a fact type is elementary, all its keys must be of the same length. Either there is exactly one key, which spans the whole fact type, or there are one or more keys and each is one role shorter than the fact type. So, if two or more roles in a predicate are not part of a key, the fact type is compound and must be split.

Table 4.6 Key length check for ternaries and longer fact types.

Arity of fact type	*Minimum key length*	*Illegal key lengths*
3	2	1
4	3	1, 2
5	4	1, 2, 3
n	*n* – 1	1, ..., *n* – 2

Table 4.7 Country data.

Country	*Population*	*Area (sq. km)*
Australia	26,477,504	7,686,850
Germany	83,294,633	356,910
United States	340,265,309	9,372,610

To start with the simplest case, *a ternary fact type can be split if it has a simple key*. Consider the sample extract of geographic data shown in Table 4.7. The first row of data may be verbalized: "The Country with name 'Australia' has a Population of 26,477,504 people and an Area of 7,686,850 square kilometers". The presence of "and" suggests that the fact is splitable. Suppose, however, that we ignore this linguistic clue and model using the ternary Country has Population and Area.

Our familiarity with the UoD tells us that, at any given time, each country has at most one population and at most one area. So, there is a UC on the Country column of the fact table. Although the populations and areas in Table 4.7 are unique, our common sense indicates that this is not significant. For example, the population of India is projected to soon equal China's population, and it is certainly possible to have two countries of the same size. The combination of population and area need not be unique either. This leaves us with the simple UC on the Country role as shown in Fig. 4.45.

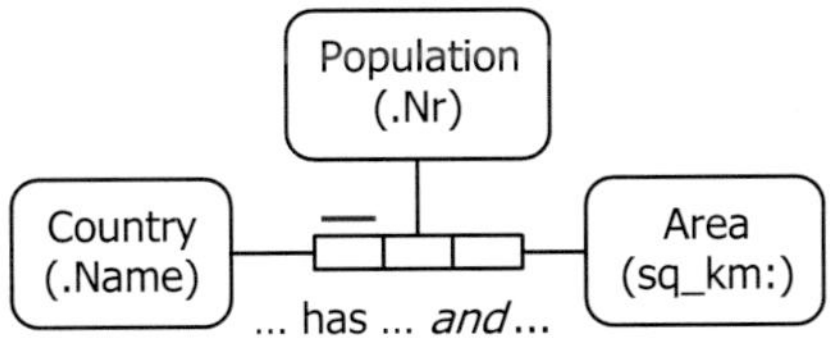

FIG. 4.45 This ternary is a compound fact type, and hence splitable.

Let's apply the $n-1$ rule. Here the key has length 1, which is 2 less than the length of the fact type. So, the fact type is not elementary and must be split. But *how do we split it?* Our verbalization of row 1 is a conjunction of two facts about the same country. Assuming our interpretation of the report is correct, we must split the fact type into two fact types as shown in Fig. 4.46. Because the common entity type is Country, the original ternary fact type is "split on Country". If we captured the semantics of the report in step 1 correctly, it would be wrong to split this fact type in any other way.

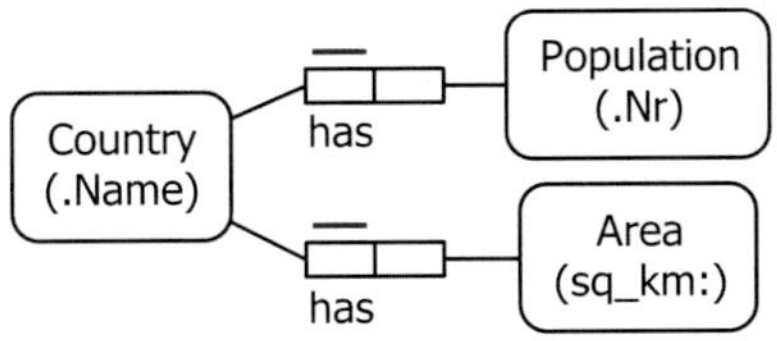

FIG. 4.46 The previous ternary splits into these two elementary fact types.

To see why the splitability rule works in this case, look back at Table 4.7. A UC on the Country column means that, given any state of the database, no country can be listed more than once in this column. Each country has at most one population and each country has at most one area. So, for any given country, if we are given as separate pieces of information that country's population and that country's area, we can reconstruct that country's row in the output report. Because the ternary fact type of the output report can be split and recombined without information loss, it is not elementary.

If a fact type violates the $n-1$ rule, it is a conjunction of smaller fact types even if its predicate has no conjunctive operator such as "and". Suppose we verbalized the previous ternary as Country with Population has Area. This still means the same, so is really a conjunction in disguise. Similarly, the ternary Invoice was issued to Customer on Date has a UC on its first role and can be split into two fact types: Invoice was issued to Customer; Invoice was issued on Date. In general, *if a fact type can be rephrased as a conjunction, it should be split* because it is not elementary.

Before considering other examples, let's review some terminology used in the normalization approach to database design. For a given fact table, let X denote a single column or combinations of columns, and let Y denote a single column. Then we say that Y is *functionally dependent on* X if and only if, for each value of X there is at most one value of Y. In other words, Y is a *function of* X. A function is just a many:1 relation (including 1:1 as a special case). If *Y* is a function of *X*, we say that *X* functionally determines *Y*, written $X \rightarrow Y$. This constraint is a **functional dependency (FD)**. Although FD arrows are not used directly in ORM schemas, the theory of FDs provides another way to understand UCs.

Now consider the sample report shown in Table 4.8, where for simplicity we assume people are identified by their surname (often unrealistic in practice). If this population is significant, any functional dependencies should be exposed here. Degree is functionally dependent on Person because for any given person, there is only one degree. Moreover, Gender is a function of Person because for each person, there is only one gender. Can you spot any other functional dependencies?

If the population is significant, Gender is a function of Degree, and Degree is a function of Gender! Does this make sense? To answer this, we need to know more about the semantics of the UoD. Spotting a functional dependency within a population is a formal game. Knowing whether this dependency reflects an actual constraint in the UoD is not. To assume the population is significant begs the question. Only someone who understands the UoD, a domain expert, can resolve the issue.

Table 4.8 Is the population significant?

Person	*Degree*	*Gender*
Adams	BS	M
Brown	BA	F
Collins	BS	M

In Section 14.6, we extend the traditional treatment of dependencies on individual table populations to dependencies representing actual constraints in the business domain (and hence applying to all possible populations of the table schemes). Some of the following discussion anticipates this extension.

If one column functionally determines another column, this reflects a many:1 (or 1: many) association that can be given a meaningful name by the domain expert. Normally, we would interpret the table in terms of degree and gender facts about specific persons. Suppose instead that the domain expert says row 1 of the output report should be read as "The Person with surname 'Adams' seeks the Degree with code 'BS', and the Degree with code 'BS' is sought by people of Gender with code 'M'". With this unusual interpretation, we have two facts about the same degree. Hence, we split on Degree rather than Person (see Fig. 4.47).

This by itself does not guarantee that Gender should be related to Person rather than Degree. For example, the schema of Fig. 4.47 with the 1:1 fact type changed to a *m:n* fact type is still a remote possibility, but it would be silly to have a report with so much redundancy (e.g., the fact that the BS is sought by males would be shown twice already). The only way to be sure is to perform step 1 properly, verbalizing the facts and having them confirmed by the domain expert.

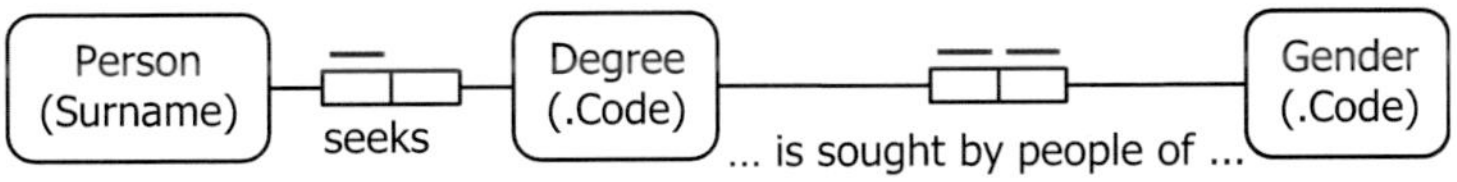

FIG. 4.47 A weird UoD where Degree and Gender are functions of each other.

If Gender really is a function of Degree, then any given degree is restricted to one gender (e.g., only males can seek a BS). If Degree really is a function of Gender, then all people of a given gender seek the same degree (e.g., all females seek just a BA). These two FDs are captured in Fig. 4.47 by the two UCs on the relationship type between Degree and Gender.

While such a UoD is possible, our knowledge about degree-awarding institutions makes this interpretation highly unlikely. To resolve our doubts, we could add test rows to the population. Although this is best done on conceptual fact tables, we could also test this example using the original output report. For example, if the domain expert accepts the population of Table 4.9 as being consistent with the UoD, then no FDs occur between Degree and Gender.

Table 4.9 A counterexample to any FD between degree and gender.

Person	*Degree*	*Gender*
Adams	BS	M
Brown	BA	F
Collins	BS	M
Davis	BS	F

It is impractical to treat fact tables in a purely formal way, hunting for FDs. The number of dependencies to check increases rapidly with the fact type length. Although our conceptual focus on elementary fact types constrains this length, the search for FDs can still be laborious. While an automated FD checker can be of use, this kind of exhaustive checking is not fit for humans. We can short-circuit this work by taking advantage of human knowledge of the UoD.

As another simple example, consider the output report of Table 4.10. Suppose that, with the help of the domain expert, we express row 1 thus: the Lecturer with surname "Halpin" works for the Department coded "CS", which is located in the Building numbered "69". Is this fact splitable, and if so, how?

Table 4.10 University data.

Lecturer	*Department*	*Building*
Halpin	CS	69
Okimura	JA	1
Orlowska	CS	69
Wang	CN	1

Here, the pronoun "which" introduces a nonrestrictive clause about the department that may be stated separately. So, this can be rephrased as two facts: Halpin works for the CS department, and the CS department is located in building 69. So, we should split on Department to get two fact types (Fig. 4.48).

As a challenge question, would it also be acceptable to split on Lecturer (e.g., Lecturer works for Department, Lecturer works in Building)? The logical derivation theory discussed in the next chapter shows that it is unwise to split on Lecturer. If the population is significant, then each lecturer works for only one department, and each department is located in only one building. If these constraints are confirmed by the domain expert, they are added to the schema as shown.

An outline proof of the $n-1$ rule is now sketched. An n-ary fact type with a key of length less than $n-1$ has at least two columns in its fact table that are functionally dependent on this key. Split the table by pairing the key with exactly one of these columns in turn. Recombining by join on this key must generate the original table because only one combined row is possible (otherwise the common key portion would not be a key). Although the n-ary fact table thus formally splits in this way, in practice the splitting might need to be made on part of the key or even a nonkey column if an FD applies there (consider the previous examples).

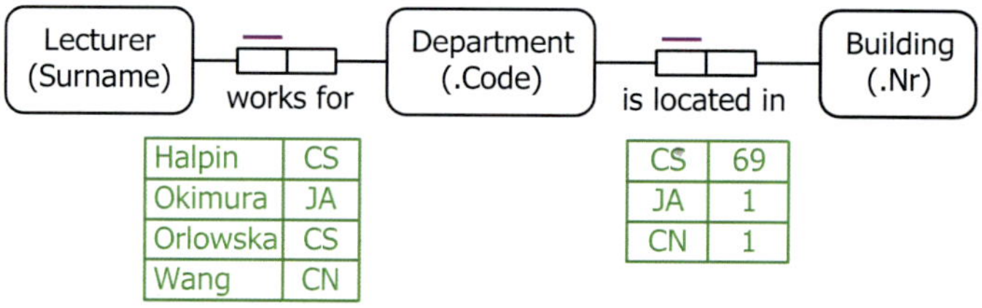

FIG. 4.48 Populated conceptual schema for Table 4.10.

Table 4.11 Student data.

Person	*Degree*	*Subject*	*Rating*
Adams	BS	CS112 CS110 PD102	7 6 7
Brown	BA	CS112 PD102	6 7
Collins	BS	CS112	7

As an example of a *quaternary* case, consider the report shown in Table 4.11. Suppose we did a bad job at step 1 and schematized this situation as in Fig. 4.49. Notice the UC. Here, we have a fact type of length 4 and a key of length 2. *Because two roles are excluded from this key, we must split the fact type.* How do we split it? Each person seeks at most one degree, but this constraint is not captured by the two-role UC. Hence, we split on Person. In terms of FD theory, the Person role functionally determines the Degree role (shown in Fig. 4.49 as a solid arrow).

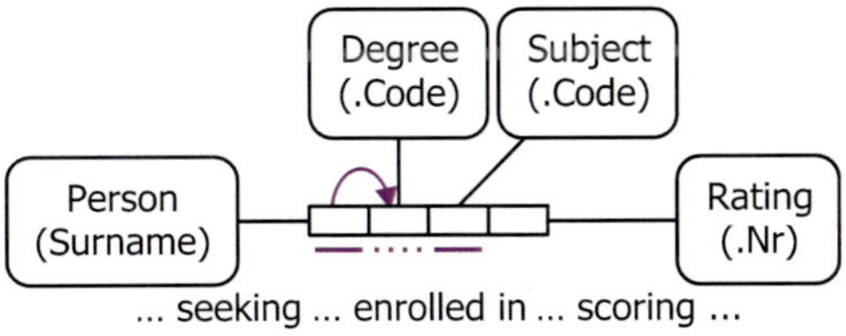

FIG. 4.49 A quaternary fact type that is splitable (FD added).

If an FD like this is involved, splitting takes place on its source. So, the quaternary splits on Person into a binary (Person seeks Degree) and a ternary (Person for Subject scored Rating). As an exercise, draw these, being sure to include all constraints. In any correct ORM schema, *all FDs should be captured by UCs*. So, FD arrows are not used on final conceptual schema diagrams. Any quaternary splits if it has a key of length 1 or 2.

Now consider the populated schema in Fig. 4.50. Suppose this interpretation of the UoD is correct. For example, suppose the first row of the output report behind this schema really does express the information "The Person with surname 'Adams' seeking the Degree with code 'BS' studies the Subject with code 'CS112'". Should the ternary be split?

Unlike the quaternary example, this ternary satisfies the $n-1$ rule. *Violating the $n-1$ rule is a sufficient condition for splitability, but it is not a necessary condition.* The ternary in this example does actually split. One way to see this is to realize that there is an *embedded FD* (i.e., an FD not captured by the UC(s)). In this case, there is an embedded FD from the Person role to the Degree role, shown as an arrow in Fig. 4.50.

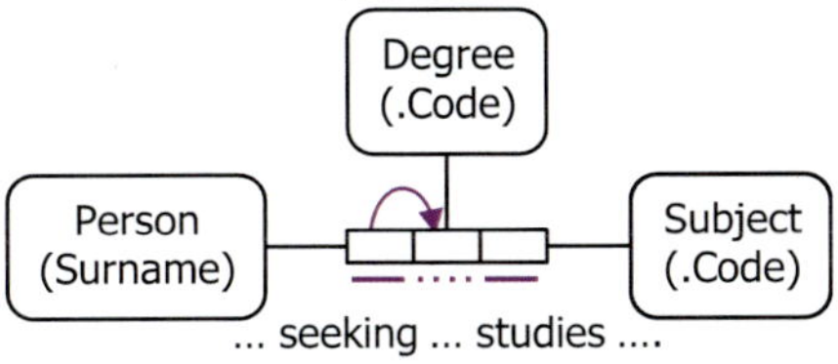

Adams	BS	CS112
Adams	BS	CS110
Adams	BS	PD102
Brown	BA	CS112
Brown	BA	PD102
Collins	BS	CS112

FIG. 4.50 Is this ternary splitable?

However, *every FD corresponds to a UC on some predicate*, and it is better to verbalize this. Here the fact type behind the embedded FD is Person seeks Degree. The remaining information can then be expressed with another fact type. Here, that is Person studies Subject.

We can achieve the same effect without talking about FDs by simply answering the question: *Can you rephrase the information in terms of a conjunction?* This approach was used earlier with the Country example. With our current example, instead of using the present participle "seeking", we can rephrase the information on the first row as:

The Person with surname 'Adams' seeks the Degree with code 'BS'
and studies the Subject with code 'CS112'.

which is equivalent to:

The Person with surname 'Adams' seeks the Degree with code 'BS'
and
the Person with surname 'Adams' studies the Subject with code 'CS112'.

This is obviously a conjunction with Person common to each conjunct. So, the ternary should be split on Person into two binaries (see Fig. 4.51).

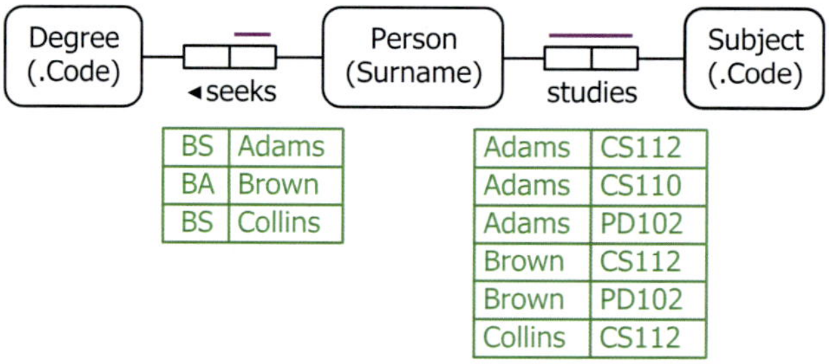

FIG. 4.51 The ternary in Fig. 4.50 should be split like this.

In general, given a significant fact table, the fact type is splitable if a column is functionally dependent on only some of the other columns. The phrase "only some" means "some but not all". This result corresponds to a basic rule in normalization theory (discussed in Section 12.6). With our present example, because the Degree column is functionally dependent on the Person column only, it should not be combined in a fact type with Subject information.

Another way to test for splitability is to *perform a redundancy check* by determining whether some fact is (unnecessarily) duplicated in the fact table. For example, in Fig. 4.50 the pair (Adams, BS) occurs three times. Using our semantic insight, we see that this pair corresponds to a fact of interest (Adams seeks a BS degree). Because this fact has been duplicated, the ternary is not elementary.

Redundancy is not a necessary requirement for splitability. For example, the ternary in Table 4.9 has no redundancy but is splitable.

Whichever method is used to spot splitability, we still need to phrase the information as a conjunction of simpler facts. Notice how the redundancy within the ternary of Fig. 4.50 has been eliminated by splitting it into two binaries in Fig. 4.51. Although we may allow redundancy in output reports, we should avoid redundancy in the actual tables stored in the database.

Our examples so far have avoided nested fact types. Recall that nesting is not the same as splitting. If nesting is used, the following *spanning rule* can be used to avoid splitable fact types: *the UC on an objectified association should normally span all its roles.*

This rule simply adapts earlier results to the nested case. Suppose the objectified predicate has a UC over only some of its roles. If an attached, outside role has a UC on it, when the outer predicate is flattened we have at least two roles not spanned by a UC.

If an attached role has no UC, then flattening generates a predicate with an embedded FD not expressed as a UC (consider the cases of Figs. 4.49 and 4.50). In either case, the flattened version must split because it is not elementary. Hence, to be elementary in this sense, the nested version must obey the spanning rule.

In the previous version of ORM, the only exception to the spanning rule was to allow 1:1 binaries to be objectified (e.g., Person is husband of / is wife of Person may be objectified as "Marriage").

The second-generation ORM discussed in this book allows any fact type to be objectified regardless of its UC pattern, and is similar to UML in this aspect (e.g., *n*:1 associations may be objectified). Although allowed for pragmatic reasons, such exceptions violate elementarity and require special treatment when mapped to logical models. Moreover, objectification without spanning uniqueness often makes the model more complicated than needed.

Consider the output report of Table 4.12. Here, the credit for a given subject is measured in points. The "?" on the second row is a null, indicating that a real value is not recorded (e.g., a lecturer for CS109 might not yet have been assigned).

Table 4.12 Subject details.

Subject	*Credit*	*Lecturer*
CS100	8	DBJ
CS109	5	?
CS112	8	TAH
CS113	8	TAH

Recall that elementary fact tables can't contain nulls (because you can't record half a fact). However, because output reports can group facts of different kinds in the one table, they often do contain nulls (indicating absence of a fact instance).

Now, let's play the part of an inexperienced–and not very clever–schema designer, so that we can learn by the mistakes made. Looking just at the first row, we might be tempted to treat the information as a ternary:

Subject (.Code) 'CS100' worth Credit (.Points) 8 is taught by Lecturer (.Initials) 'DBJ'.

If we drew the schema diagram now and populated it, we would discover an error because the second row doesn't fit this pattern (because we don't allow nulls in our fact tables). The second row reveals the need to be able to record the credit points for a subject without indicating the lecturer. So, we then rephrase the first row as follows:

> Subject (.Code) 'CS100' is worth Credit (.Points) 8.
> This SubjectCredit is taught by Lecturer (.Initials) 'DBJ'.

This overcomes the problem with the second row because we can now express the information there simply as "Subject 'CS109' is worth Credit 5". Using this approach, we can now develop the schema shown in Fig. 4.52.

Assuming that the population of the report is significant, we mark the UCs as shown. A simple UC has been added to the objectified association. This breaks the spanning rule: an objectified association has no key shorter than itself. So, something may be wrong. Because the schema follows from our handling of step 1, this means that we may have made a mistake at step 1.

Have another look at the way we set out the information for row 1. The second sentence here is the problem. In this UoD, each subject has exactly one credit point value, so this is independent of who teaches the subject. As a result, there is no need to mention the credit point value of a subject when we indicate the lecturer. In FD terminology, the Credit and Lecturer columns are each functionally dependent on the Subject column alone.

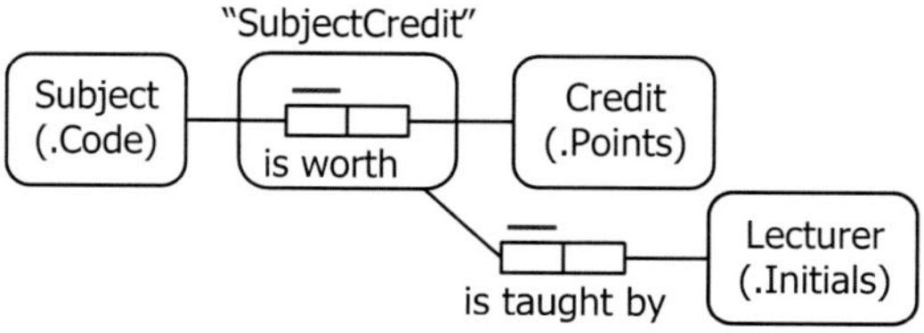

FIG. 4.52 A faulty schema.

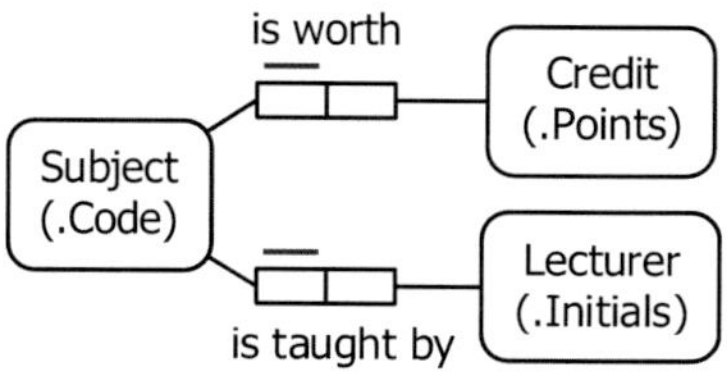

FIG. 4.53 The corrected schema for Fig. 4.52.

The information on the first row can be set out in terms of two simple binaries:

> Subject (.Code) 'CS100' is worth Credit (.Points) 8.
> Subject (.Code) 'CS100' is taught by Lecturer (.Initials) 'DBJ'.

This leads to the correct schema shown in Fig. 4.53. Note that Lecturer depends only on Subject, not Credit. The nested schema suggested incorrectly that Lecturer was dependent on both by relating Lecturer to the objectified association. With this example, Lecturer is a function of Subject, but this has no bearing on the splitability. We should use two binaries even if the same subject can have many lecturers. It is the functional dependency of Credit on Subject that disqualifies the nested approach.

In a case like this where it is difficult to come up with a natural name for the objectified association, such nesting errors are unlikely to occur in practice. When a natural name for the nested object type is available, however, it is easier to make the error. For example, we might objectify the fact type Person was born in Country as "Birth", and then add the fact type Birth occurred on Date. Although it sounds reasonable, this nesting violates the spanning rule, and should normally be replaced by the unnested binary: Person was born on Date. We discuss further cases of objectification in Chapter 10 and provide heuristics for choosing when to objectify.

One lesson to be learned from this section is to look out for UCs even when performing step 1. Consider the output report of Table 4.13. If told the population is significant, how would you model this?

If you used either a ternary or a nested fact type, you made a mistake! Although similar to an earlier example, the UoD described by this table is more restricted. Given that the population is significant, the lack of duplicates in the Person column indicates that, at any given time, each person can be enrolled in only one subject.

Table 4.13 Student results.

Person	*Subject*	*Rating*
Adams	CS112	7
Brown	CS112	5
Collins	PD102	5

For example, these people might be your employees, and you are funding their studies and want to ensure that they don't take on so much study that it interferes with their work performance.

Because we store information only about the current enrollments, and each person is enrolled in only one subject, it follows that each person can get only one rating. So, if we know the subject in which a person is enrolled and we know the rating obtained by the person, we do know the subject in which this rating is obtained by that person. To begin with, we might express the information on the first row of the table in terms of the ternary "Person (surname) "Adams" scores Rating (.nr) 7 for Subject (.code) "CS112". However, because of the UC on Person we should rephrase this information as the conjunction:

Person (Surname) 'Adams' studies Subject (.Code) 'CS112'
and
Person (Surname) 'Adams' scores Rating (.Nr) 7.

This leads to a conceptual schema with two binary fact types. Because of this unusual UC, the ternary has the same truth value as the conjunction of binaries for all possible states of this UoD.

Exercise 4.5

1. The keys for certain fact types are as shown. On this basis, which of these fact types are definitely splitable?

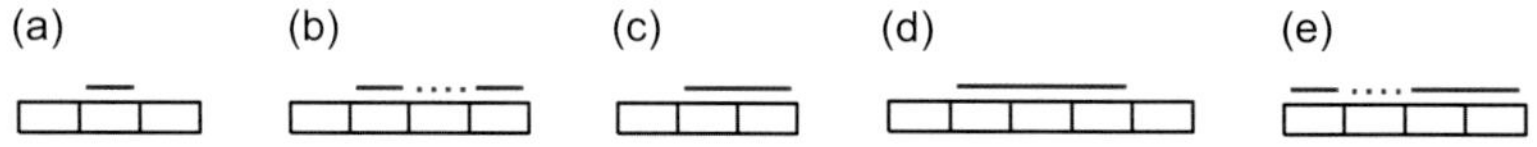

2. The following map provides information on nonstop flights between cities. The population is significant with respect to UCs.

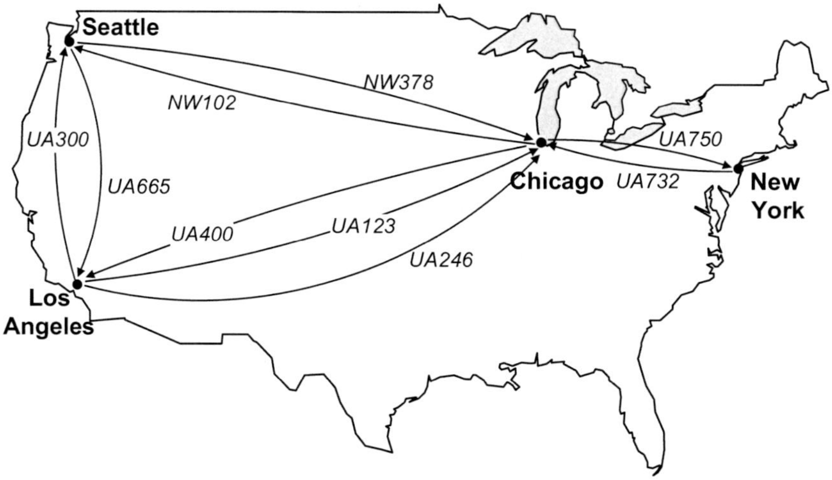

   (a) In step 1A of the design procedure, a domain expert verbalizes the information on the NW378 arrow as: Flight "NW378" goes from Seattle to Chicago. Assuming this interpretation is correct, complete steps 1 and 2 to draw a schema diagram for facts of this type and populate it with rows of data for the flights NW378, NW102, UA123, and UA246.
   (b) Now complete the first part of step 4 by adding any UCs.
   (c) Is the fact type elementary? How do you know?
   (d) Split the fact type into elementary fact types, and populate them with the data for the same four flights mentioned earlier.

3. Executives may be contacted at work on one or more phone numbers and at home on one phone number. The following report provides sample data. A novice modeler notes that this ternary table lacks a simple key, and on this basis decides to

schematize with a single ternary fact type. Evaluate this approach, and draw a correct conceptual schema.

Executive	*Work phone*	*Home phone*
Adams A	4235402	5837900
Adams A	4235444	
Adams S	4235444	5837900
Brown T	4235300	7578051

4. The following table contains data about some moons in our solar system.

Moon	*Planet*	*Orbital period (days)*
Callisto	Jupiter	16.7
Deimos	Mars	30.0
Phobos	Mars	0.3

(a) Schematize this as a ternary. Is this okay?
(b) Schematize this using nesting. Is this okay?
(c) Schematize this using only simple binaries. Is this okay?

5. The following table contains information about movie exports. Suppose we model this with the quaternary Movie exported from Country to Country achieved SalesLevel. Draw the schema for this approach. Is this acceptable? If not, explain why.

Movie	*Origin*	*Export performance*	
		Country	*Sales level*
Backdraft	US	AU	High
		NZ	Low
		GB	Medium
Crocodile Dundee	AU	NZ	Low
		US	High
Terminator 2	US	?	?

4.6 Projections and Joins

The previous section showed that a *sufficient* condition for splitting an unnested fact type is that it has a UC spanning fewer than $n-1$ roles. Because this is not a *necessary* condition for splitting, in some cases further analysis is required to make a definite decision. At the heart of such an analysis is the question, *Can the fact type be rephrased as a conjunction of smaller fact types?*

With a bit of experience behind us, we can usually answer this question fairly quickly. This section discusses a formal procedure for addressing this question in a systematic way. The procedure makes use of two operations known as "projection" and "join", which are important in relational database work. The *join* operation has already been discussed, and involves combining tables by matching values referencing the same object to form a new table (review Section 4.4).

The **projection** operation also produces a table, but is performed on a single table. To project on one or more columns of a table, we *choose just the columns of interest* (removing all other columns) and then *ensure each row in the result appears just once* (removing any duplicates).

Notationally, the columns on which a projection is made are listed in *italic square brackets* after the table name, separated by commas. For example, *T [a, c]* is a projection on columns *a* and *c* of table *T*. Academics often write this instead as $\pi_{a,c}(T)$, where "π" (pi) is the Greek "p" (first letter of "projection"). We prefer the square bracket notation because it encourages a top-down view of the operation (first find the table, then choose the columns). Visually, we can picture the operation as returning a vertical subset of the whole table (see Fig. 4.54).

For example, a ternary fact type met earlier is reproduced in Fig. 4.55(a). Its fact table T1 has six rows. Projecting on the Person and Degree roles, we obtain the fact type Person seeks Degree, with its fact table T2 as shown in Fig. 4.55(b). Notice that T2 has only three rows because projection eliminates duplicates. Projecting on the Person and Subject roles yields the fact type Person studies Subject, and its fact table T3. Because T3 is a projection on a key of T1, it must have the same number of rows as T1 (Why?).

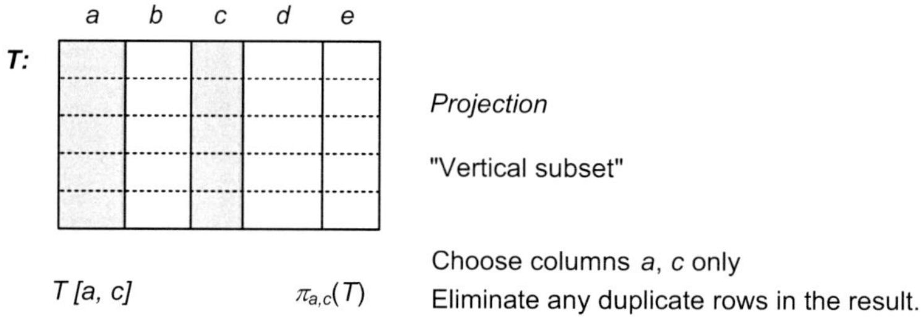

FIG. 4.54 The projection operation.

Because the roles on the ternary are played by different object types, by default the roles are named using lowercase versions of the object type names. Using these role names as attribute names, the fact tables may be set out in relational terms as T1 (person, degree, subject), T2 (person, degree), and T3 (person, subject). The two projections may now be expressed thus: T2 = T1*[*person, degree*]* and T3 = T1*[*person, subject*]*.

In an earlier section, the ternary Person seeking Degree studies Subject was reworded as the conjunction Person seeks Degree **and** studies Subject. Here, the logical "**and**" operator performs a conceptual join on Person of the two binary fact types: Person seeks Degree; Person studies Subject. Because the ternary is equivalent to the conjunction of the binaries, no information is lost if we transform the ternary into the two binaries or vice versa. So, no matter what the fact populations are, the ternary fact table must contain exactly the same elementary facts as the two binary fact tables collectively contain. In relational terms, the ternary relation is equal to the (natural inner) join of the two binary projections. So, T1 = T2 ⋈ T3. Check this out for yourself using the populations shown in Fig. 4.55.

Conceptually, the situation is summarized by the following equivalence:

Person seeking Degree studies Subject **iff** Person seeks Degree
and studies Subject.

Here "iff" abbreviates "if and only if". The projection corresponds to the conditional: *if* the ternary is true *then* the conjunction is true. The join corresponds to the conditional: *if* the conjunction is true *then* the ternary is true. Any compound fact can always be formulated as a logical conjunction (using "and").

In principle (although of little use in practice, as discussed later), the following *projection-join check* can be applied to test whether a fact type is compound (and hence splitable).

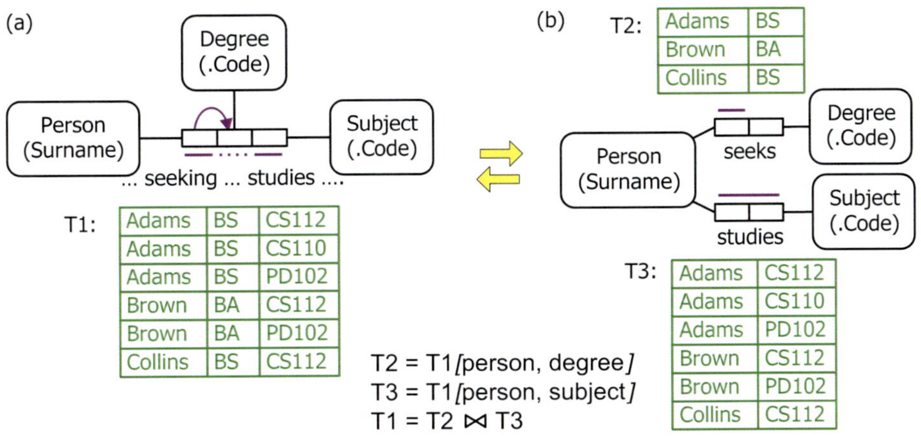

FIG. 4.55 Compound fact types can be split by projection and restored by join.

Provide a significant fact table for the fact type.
Split this table into two or more projections.
Recombine by natural (inner) join.
The fact type is splitable in this way **iff** the result is the same as the original.

An inner join is acceptable because the conceptual fact table (and hence its projections) cannot have nulls. We chose a significant population for the ternary example just discussed, so the test confirms that ternary's splitability.

Now consider our familiar ternary example about people scoring ratings for subjects. A significant fact table is provided for it in Fig. 4.56(a). Suppose we suspect that this ternary is splitable on Person. For example, we might feel that in this UoD, the following equivalence holds:

Person scored Rating for Subject **iff** Person scored Rating
and studied Subject.

To test this way to split, we could form the two binary projections shown in Fig. 4.56(b), and then recombine by joining on Person. The result is shown in Fig. 4.56(c). As an exercise, please confirm this result. In forming the join, several new rows (marked in bold italics) appeared that were not in the original table. Any one of these new rows is enough to prove that the fact type cannot be split this way (i.e., on Person).

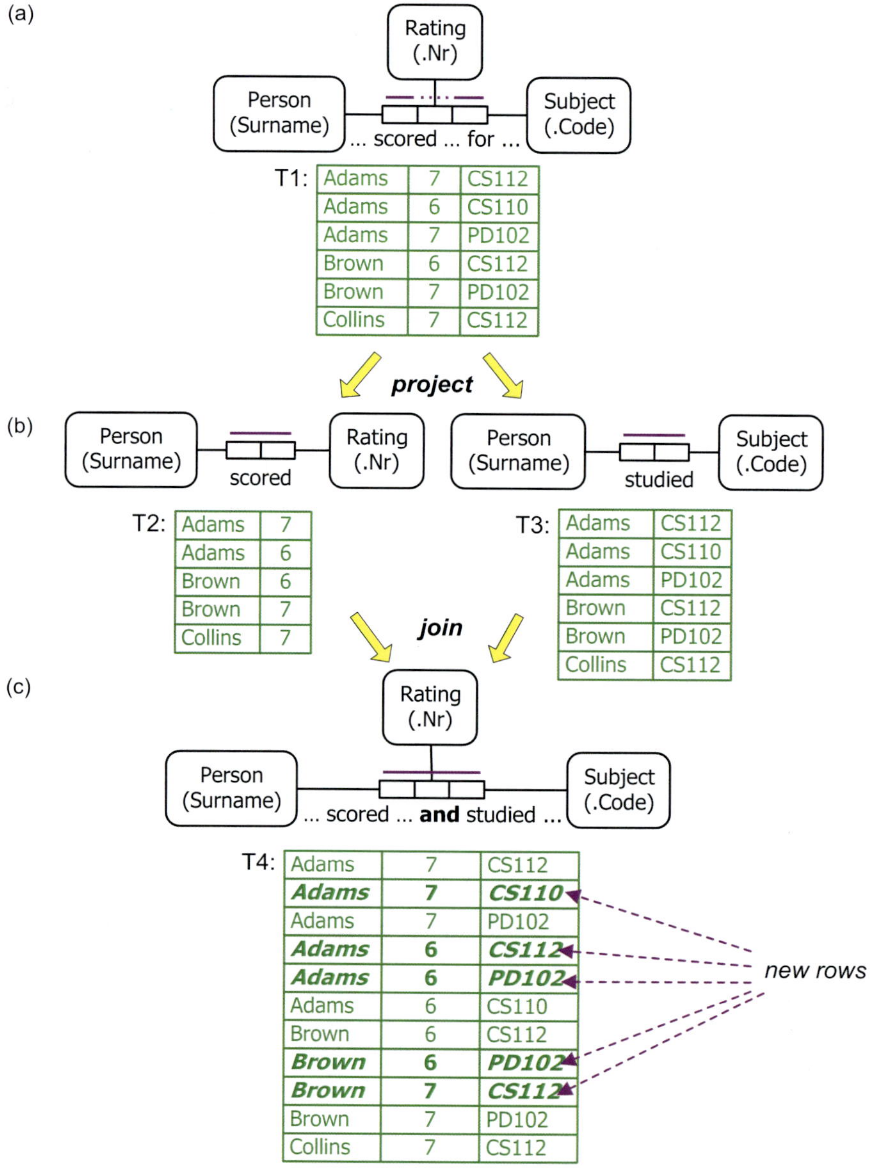

FIG. 4.56 The original ternary cannot be split in this way.

Consider for instance the fact from the first row of our original table: The person with surname "Adams" scored a rating of 7 for the subject CS112. We attempted to split this fact into the two separate facts: Adams scored a rating of 7; Adams studied CS112 (see first rows of the two projections). If this splitting is legitimate, the two separate facts must, in combination, be equivalent to the original fact. But they are not. Adams can score more than one rating and study more than one subject. Knowing that Adams scored a 7 does not tell us the subject(s) for which Adams scored this rating. It could be any of Adams's subjects (CS112, CS110, or PD102).

Joining the projections causes each rating of Adams to be paired with each subject of Adams. Some of these rows were not present in the original, and the join does not tell us which ones apply to the original ternary. Because we are unable to construct the population of the original ternary from the new ternary population, information has been lost. Hence, the new ternary fact type is not equivalent to the original fact type. So, the original ternary fact type cannot be split in this way.

This projection-join check invalidated the proposed equivalence Person scored Rating for Subject **iff** Person scored Rating **and** studied Subject. The forward conditional of the equivalence does hold (*if* ternary *then* conjunction). However, the backward conditional fails (it is not generally true that *if* conjunction *then* ternary). Simply understanding the English verbalization of the fact types should be enough for us to understand that "Person scored Rating for Subject" differs in meaning from "Person scored Rating **and** studied Subject" if we read "**and**" in a purely conjunctive sense. Another way to spot the difference is that the UC on the join fact type in Fig. 4.56(c) is weaker than that of the original in Fig. 4.56(a).

If you use some insight, it should be clear that trying to split on Subject, Rating, or all three would be pointless. In this case, you may conclude that the fact type is unsplitable and leave it as a ternary. If this is not clear, then you can use the algorithm to test all possible ways of splitting. You would find that none of them work.

In general, for a ternary A-B-C there are four ways in which it might split: A-B, A-C; B-A, B-C; C-A, C-B; and A-B, A-C, B-C. In the fourth case, the schema diagram is triangular in shape; this is referred to as three-way splitting. As an exercise, draw these four possibilities. With three-way splitting, the join should be done in two stages. For example, first join A-B and A-C (with A common); then join the A-B-C result of this join with B-C (with B and C common).

A classic example of testing for three-way splitting involves the fact type Agent sells Cartype for Company. Even if the key spans the whole fact type, the fact type can be split into three binaries if a derivation rule allows the ternary to be deduced from the binaries. Such examples are very rare in practice. For an n-ary fact type, if the n-way split fails (by generating new rows in the final join) then all other ways of splitting can be ruled out too: the fact type is unsplitable.

Although useful for understanding splitability, the projection-join test is of little practical value for determining splitability. The problem is that it only works if the population is significant. But to know the population is significant implies that we know all the constraints, and have crafted a population to illustrate these constraints. If we already know what the constraints are, it is obvious whether the fact type is elementary, so there's no need to do any projection-join test.

In Fig. 4.55(a), for example, if we already know that each person seeks at most one degree, then we need to separate out the underlying binary. In Fig. 4.56(a), there was no similar constraint because a person can score many ratings. Formally this is the only difference between the two cases, but it's enough for us to determine splitability without resorting to the tedious projection-join test. The best test for splitability is to work linguistically with the domain expert to see if a fact can be phrased as a conjunction, and to test this with carefully chosen instances. We have more to say on the semantic dependencies in Section 14.6.

In case you feel like you've just wasted a lot of time learning a projection-join check that is of theoretical interest only, please don't be upset. It's very useful to understand what the projection and join operations are at both conceptual and logical levels. Apart from their modeling impact, these operations are used extensively in both conceptual and relational query languages, as we'll discuss later.

Exercise 4.6

1. A template for a ternary fact type is outlined here. The population shown is significant.

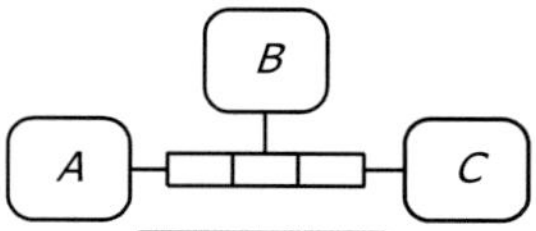

a1	b1	c1
a1	b1	c2
a2	b2	c1
a2	b2	c2

   (a) Add the UCs for this fact type.
   (b) Use the projection-join test to show that this fact type cannot be split into two binaries with *C* as the common node.
   (c) Use the projection-join test to show that this can be split on *B* into two binaries.
   (d) Draw a schema outline for this result.

4.7 Summary

In step 4 of the CSDP, we *add UCs and check the arity of fact types.* UCs come in two varieties: internal and external. An *internal* (or intrapredicate) UC applies to one or more roles of the same predicate. It is marked as a bar across the constrained role(s). If the roles are noncontiguous, the bar is divided, with a dotted line connecting its active parts.

Redundancy is repetition of an elementary fact. Because redundancy is not permitted in a conceptual fact table, each row must be unique. This is shown as a UC across the whole predicate, unless a stronger UC exists. If an internal UC spans just some of the roles spanned by another internal UC, the former UC implies the latter UC, and the implied UC should not be displayed.

Each unary predicate has a UC across its role, whether or not it is displayed. Binary predicates have four possibilities: UC on role 1 only; UC on role 2 only; UC on role 1 and another UC on role 2; and one UC spanning both roles. These cases are, respectively, described as many to one (*n*:1) , one to many (1:*n*), one to one (1,1), and many to many (*m*:*n*).

For example, if each employee works for at most one department, but many employees may work for the same department, the predicate "works for" is many to one and has a UC on its first role. In Fig. 4.57, the *R* predicates are *n*:1 and the S predicate is *m*:*n*.

Each role in a predicate is associated with a column of the predicate-table. A simple UC across just one role means no duplicates are allowed in its column. A UC spanning *n* roles means each sequence of *n* entries taken from the associated columns can occur in that column sequence on only one row.

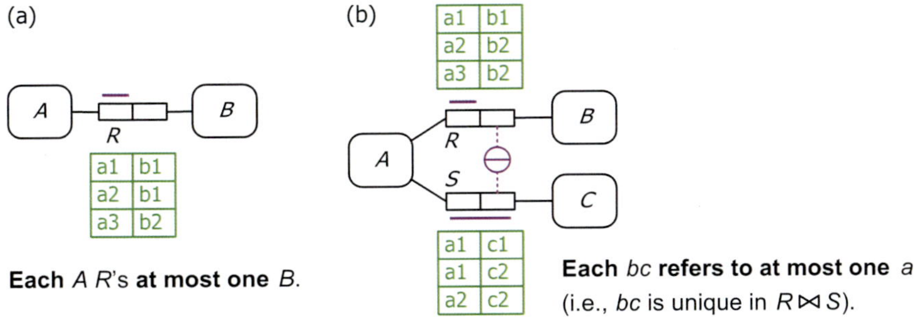

FIG. 4.57 Some uniqueness constraints.

Let *aRb* and *bSc* be two fact instances. The conjunction of these facts (*aRb* **and** *bSc*) is a compound fact instance (*aRb* **that** *Sc*) of the schema path *A*-*R*-*B*-*S*-*C* that is the conceptual (inner) *join* of *R* and *S* via the common object type *B*. The fact table for this join is obtained by pairing rows of *R* with rows of *S* where the *B* values match, and showing the *B* column(s) just once in the result. In relational terms this is a natural inner join, symbolized by "⋈". For example, *R* ⋈ *S* in Fig. 4.57(b) joins on *A* and has a population of {(a1, b1, c1), (a1, b1, c2), (a2, b2, c2)}.

An *external* (or interpredicate) UC is shown by connecting two or more roles from different predicates to a circled bar "⊖". This indicates that when the join operation is applied to the predicates, an internal UC spans these roles in the result. As an example of the constraints shown in Fig. 4.57(b), a program may appear on at most one channel, and may appear at different time slots, but for any given channel and time slot at most one program is shown.

UCs apply to the model, and must be at least as strong as those in the real world. A combination of roles spanned by an internal UC, with no smaller UC inside it, is a *key* of its predicate. A one-role key is a *simple key*.

Because conceptual predicates are elementary, a ternary fact type cannot have a simple key. It either has a three-role UC, or one or more two-role UCs. For a predicate with *n* roles, *each internal UC must span at least n – 1 roles*. This *n – 1 rule* may be applied as a *key-length check*. Violation of this rule is a sufficient but not necessary condition for splitability.

This rule also implies the following *spanning rule: a UC on an objectified association must normally span all its roles*. Exceptions to this rule are allowed for 1:1 cases and other cases discussed in Chapter 10.

For a given predicate, let *X* denote a combination of one or more roles (or columns) and *Y* denote a single role (or column). We say *X functionally determines Y*, written $X \rightarrow Y$, if and only if for each value of *X* there is at most one value of *Y*. In this case, *Y* is said to be *functionally dependent* on *X*.

The term "*FD*" abbreviates "functional dependency". If the conceptual schema is correct, all FDs corresponding to actual constraints are implied by UCs. If a nonimplied FD $X \rightarrow Y$ exists, the predicate should be split on the source *X*.

With experience, you can usually determine whether a fact type splits by using background knowledge of the UoD to answer the questions, "Can the fact type be rephrased as a conjunction? Is information lost by splitting the fact type?" The shortest key rule and the detection of a nonimplied FD provide simple ways of checking that our intuitions are correct here.

The *projection* on columns *a, c, ...* of table *T*, written *T[a, c, ...]*, is obtained by choosing those columns and removing any duplicate rows from the result. If we know we have a significant population for a fact type, the *projection-join check* can be used to determine whether the fact type is elementary. If we suspect that a predicate might split in a certain way, we can split its significant table in this way by projection and then recombine by natural join: if new instances now appear, the fact type cannot be split in this way. This test is of more theoretical than practical interest.

Chapter Notes

By depicting each role as a box associated with a table column, ORM enables UCs to be portrayed in a natural way where the impact of the constraint on populations is immediately apparent. Entity relationship (ER) modeling and UML depict internal UCs on associations using a variety of notations for "cardinality constraints" or "multiplicity constraints", as discussed later. ER and UML offer only partial graphical support for UCs on attributes. Neither ER nor UML offers any significant support for external UCs.

ORM makes minimal use of the notion of functional dependency, but this notion is central to many other methods. For a clear treatment of an ER approach as well as functional dependency theory, see Elmasri and Navathe (1994).

Although it is generally wise to objectify a predicate only if it has a spanning UC, rare cases may arise where it may be useful to remove this restriction, as discussed in Chapter 10.

5

Mandatory Roles

Chapter Outline

5.1 Introduction to CSDP Step 5

So far, we've learned how to proceed from familiar information examples to a conceptual schema diagram in which the elementary fact types are clearly set out, with the relevant uniqueness constraints marked on each. We also learned how to perform some checks on the quality of the schemas. In practice, other kinds of constraints and checks need to be considered also.

Next in importance to uniqueness constraints are *mandatory role constraints.* These indicate which roles *must* be played by the population of an object type and which roles are optional. Once mandatory role constraints are specified, a check is made to see if some fact types may be *logically derived* from others. This constitutes the next step in the design procedure.

Conceptual Schema Design Procedure (CSDP) step 5: Add mandatory role constraints, and check for logical derivations

The next two sections cover this step. This section discusses some basic concepts used in our treatment of mandatory roles and later constraints. Once mandatory roles are understood, we examine reference schemes in depth, especially composite reference—we do this later in the chapter.

Recall that a *type* may be equated with the set of all its possible instances. This is true for both object types and relationship types. For a given schema, types are fixed or unchanging. For a *given state* of the database and a given type *T,* we define *pop(T),* the **population** of *T,* to be *the set of all instances of T in that state.*

We use "{ }" to denote the *null set,* or *empty set* (the set with no members). When the database is empty, the population of each of its types = { }. As the database is updated, the

Information Modeling and Relational Databases. https://doi.org/10.1016/B978-0-443-23790-4.00009-8

population may change. For example, suppose the ternary in Fig. 5.1 stores information about medals won by countries in the next Olympic Games. The fact type and roles are numbered for easy reference.

Initially the fact table for F1 is empty because no sporting results are known, so pop(F1) = { }. Now suppose the database is to be updated after each sporting event, and in the first event the gold and bronze medals are won by the United States and the silver medal by Japan. The new state of the fact table is shown in Fig. 5.2, using "G", "S", and "B" for gold, silver, and bronze. The population of the fact type now contains three facts, and will grow further each time new results are entered.

We also define the *population of a role*. Each role of a fact type is associated with a column in its fact table. Values entered in the column refer to instances of the object type that plays that role. Given any role *r* and any state of the database:

pop(r) = population of role *r*
= set of objects referenced in the column for *r*

Typically, the objects referenced are entities, not values. For example, in Fig. 5.2, pop(r1) = {The Country with code "US", The Country with code "JP"}. The *valuation* of a role *r*, written *val(r)*, is the set of values entered in its column. For example, in Fig. 5.2, val(r1) = {"US", "JP"}.

When there is no chance of confusion, object terms may be abbreviated to constants in listing populations. With this understanding, in Fig. 5.2, pop(r1) = {US, JP}, pop(r2) = {G, S, B} and pop(r3) = {1}. Role populations are used to determine *object type populations*. With our current example, if each country referenced in the database must play r1, then

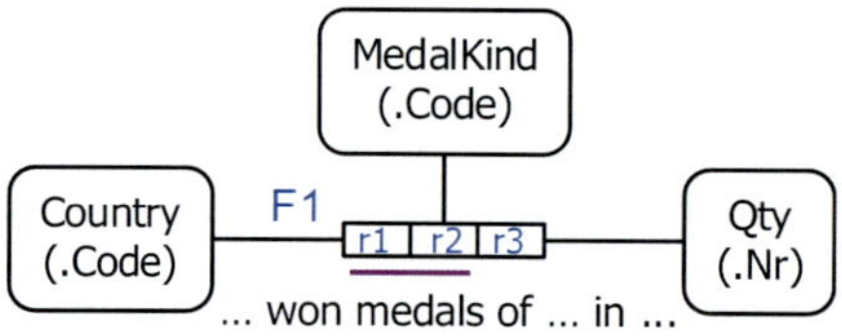

FIG. 5.1 A fact type for Olympic Games results.

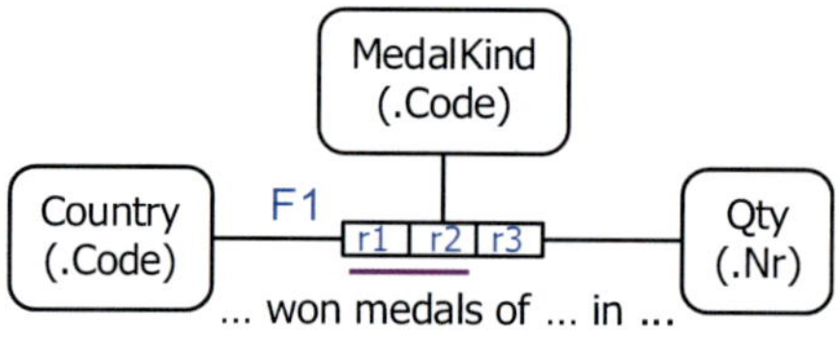

US	G	1
JP	S	1
US	B	1

FIG. 5.2 The fact type populated with results of the first event.

after the first event pop(Country) changes from { } to {US, JP}, and pop(CountryCode) changes from { } to {"US", "JP"}. Assuming the universe of discourse (UoD) is just about the Olympics, the entity type Country is the set of all countries that might possibly compete in the Olympic Games.

A predicate occurs within an elementary fact type or reference type (a.k.a. existential fact type). Reference types are usually abbreviated as a parenthesized reference mode. Roles in a reference type are called *referential roles*. Roles in an elementary fact type are called "*fact roles*". Each entity type in a completed conceptual schema plays at least one referential role and, unless it is declared independent (see Section 6.3), at least one fact role. In general, the *population of an entity type equals the union of the population of its roles*. Unless the entity type is independent, its population is the union of the populations of its fact roles.

For example, the populated schema in Fig. 5.3 indicates the size (in square kilometers) and carbon dioxide emissions (as a percentage of worldwide emissions) of some countries as measured in 2020. The Country entity *type* might include many countries (e.g., from Afghanistan to Zimbabwe), but this state of the database lists only four countries. Each instance in the database population of Country plays the role r1, r2, or both. So, the current *population* of Country is the set of all the instances referenced in either the r1 column or the r2 column.

We could set this out as:

$$\begin{aligned} \text{pop(Country)} &= \text{pop(r1)} \cup \text{pop(r2)} \\ &= \{\text{AU, FR, US}\} \cup \{\text{AU, CN, US}\} \\ &= \{\text{AU, CN, FR, US}\} \end{aligned}$$

Here, "∪" is the operator for set union. The union of two sets is the set of all the elements in either or both. For instance, $\{1,2\} \cup \{2, 3, 4\} = \{1, 2, 3, 4\}$. In the case just given, Australia and the United States occur in both role populations while China and France occur in only one.

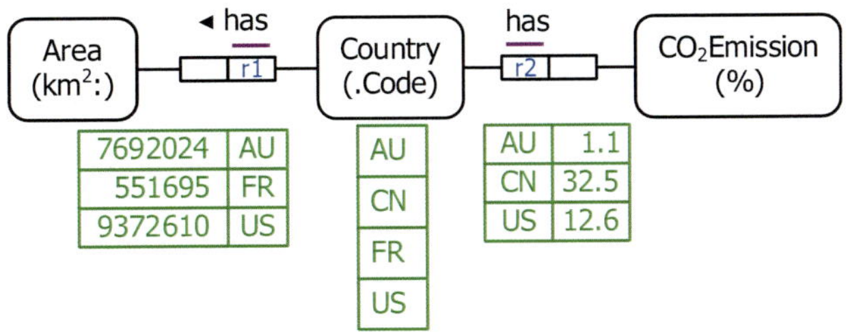

FIG. 5.3 pop(Country) = pop(r1) ∪ pop(r2).

5.2 Mandatory and Optional Roles

Consider the output report of Table 5.1. Patients are identified by a patient number. The question mark "?" denotes a *null*, indicating that an actual value is not recorded. For instance, patient 002 may have a phone but it is not recorded, or he/she may simply have no phone. Different patients may have the same name and even the same phone number, so the table population is not significant.

We must record each patient's name, but it is optional to record a phone number. Fig. 5.4 shows a preliminary conceptual model for this situation. Conceptual facts are elementary, so they cannot contain nulls. The null in Table 5.1 is catered for by omitting a fact for patient 002 in the fact type Patient has PhoneNr.

In step 5 of the design procedure, each role is classified as mandatory or optional. A **role is mandatory** if and only if, for all states of the database, the role must be played by every member of the population of its object type; otherwise, the role is **optional**. A mandatory role is also called a *total role* because it is played by the total population of its object type.

Table 5.1 Details about hospital patients.

Patient Nr	*Patient Name*	*Phone*
001	Adams C	2057642
002	Brown S	?
003	Collins T	8853020

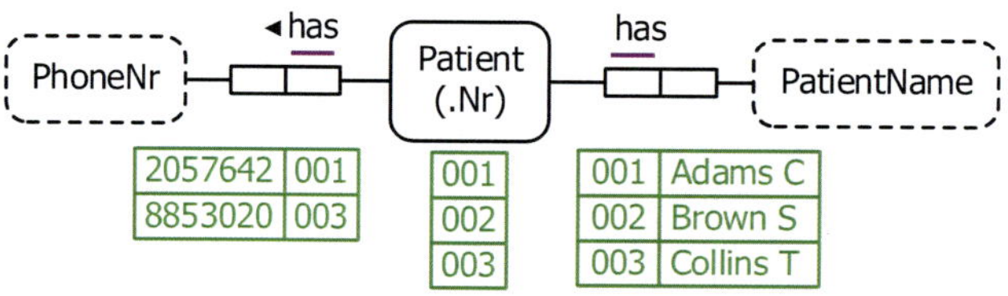

FIG. 5.4 A populated schema for Table 5.1.

Which of the four roles in Fig. 5.4 are mandatory and which are optional? If the diagram includes all the fact types for the universe of discourse (UoD), and its sample population is significant, we can easily determine which roles are mandatory.

In practice, however, we mostly work with subschemas rather than the complete, or global schema; and sample populations are rarely significant. In such cases, we should check with the domain expert as to whether the relevant information must be recorded for all instances of the object type (e.g., must we record the name of each patient?).

Consider the two fact roles played by Patient. For the database state shown, the populations of these roles are {001,003} and {001,002,003}. If these are the only fact roles of Patient, then pop(Patient) = {001, 002, 003}. The role of having a name is played by all recorded patients, and the role of having a phone number is played by only some.

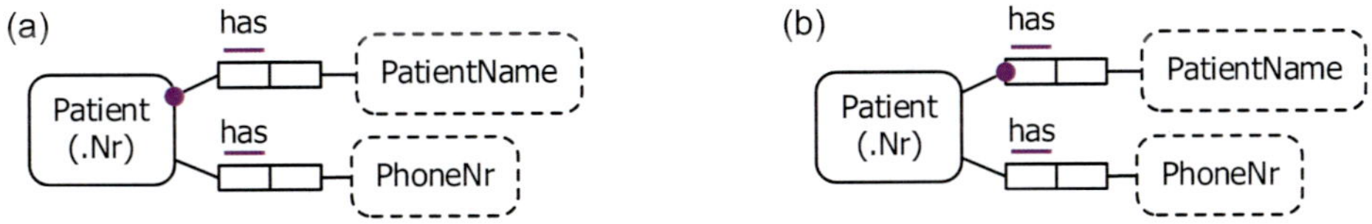

Mandatory: **Each** Patient has **some** PatientName.
Unique: **Each** Patient has **at most one** PatientName.
Each Patient has **at most one** PhoneNr.

FIG. 5.5 Mandatory role constraint shown as a dot at either end of the role line.

Assuming the population is significant in this regard, the name role is mandatory and the phone number role is optional.

To indicate explicitly that a role is mandatory, we add a **mandatory role dot** to the line that connects the role to its object type. This dot may be placed at either end of the role line. In Fig. 5.5(a), the dot is placed at the object type end. This reinforces the *global* nature of the constraint in applying to the object type's population. If we add a patient instance to the population of a mandatory Patient role, we must also include this instance in the population of each other role declared mandatory for Patient.

In this sense, a mandatory role constraint can have an impact beyond its predicate. In contrast, a uniqueness constraint is *local* in nature, constraining just the population of its role(s), with no impact on other predicates.

In Fig. 5.5(b), the mandatory dot is placed at the role end. This is best when an object type's role lines are so close together that if we added a dot at the object type end it would be unclear which role is intended. This situation can arise when an object type plays a large number of roles displayed on the same schema page.

In Fig. 5.5, the first role of the top predicate is mandatory and unique. The mandatory role constraint verbalizes as **Each** Patient has **at least one** PatientName, or as **Each** Patient has **some** PatientName. The uniqueness constraint verbalizes as **Each** Patient has **at most one** PatientName. In combination, these constraints verbalize as **Each** Patient has **exactly one** PatientName (i.e., each recorded patient has one and only one name recorded). In general, *at least one + at most one = exactly one.*

In a schema, *if two or more fact roles are played by the same object type, then individually these roles are optional unless marked mandatory.* For example, in Fig. 5.5 the first role of Patient has PhoneNr is optional. So, we may add a patient to the database population without adding a phone number for that patient.

What about the roles played by PatientName and PhoneNr? If these are the only roles played by these object types in the global schema, then these roles are mandatory. Unless declared independent (see Section 6.3), each primitive object must play some role, and each primitive entity must play some fact role. Hence *by default, if a primitive object type plays only one role, or a primitive entity type plays only one fact role* (in the global schema), *this role is mandatory.* In such cases, the implied mandatory role dot is usually

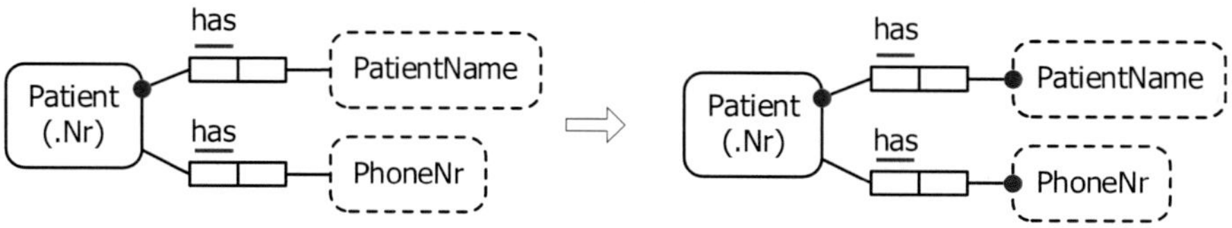

FIG. 5.6 Implied mandatory constraints specified explicitly (not recommended).

omitted. In Fig. 5.5, if no other roles are played by PatientName and PhoneNr, their roles are mandatory by implication. In this case, although not recommended, these implied constraints could be marked explicitly, as shown in on the rightmost schema in Fig. 5.6.

Explicit depiction of implied mandatory role constraints has several disadvantages. First, it complicates schema evolution. For example, suppose that tomorrow we decide to add the fact types Patient had previous—PatientName; Patient has secondary—PhoneNr. With the implicit approach, this is simply an addition to the current schema. But with the explicit approach, we need to delete the formerly implied mandatory role constraints and replace them with weaker constraints (e.g., Each PatientName is of, or was of, at least one Patient; Each PhoneNr is used by, or is a secondary number for, at least one Patient—such inclusive-or constraints are discussed shortly).

Another problem with marking implied mandatory constraints is that this de-emphasizes the important mandatory role constraints (the ones we need to enforce (e.g., we must record a name for each patient)). Implied mandatory constraints are enforced automatically, so they have no implementation impact.

If we demanded the explicit approach for simple mandatory constraints, we should do the same for disjunctive mandatory (inclusive-or) constraints. But if the roles involved in the disjunction occur on separate schema pages, there is no convenient way to mark the constraint between them.

Another problem with the explicit approach is that it complicates theorem specification. Later on, we consider some schema equivalence theorems. In this context, the implicit approach allows us to discuss schema fragments independently of other roles played in the global schema, but the explicit approach would forbid this, leading to extra complexity.

For such reasons, avoid the explicit specification unless we have some special reason to draw the attention of a human to the implied constraints. As discussed in the next chapter, the rule for implicit mandatory roles does not apply to subtypes: if they play just one fact role, this is optional unless marked mandatory.

The population of a mandatory role equals the total population of its attached object type. Fig. 5.7 depicts the general case for an object type *A* and an attached role *r*. The role *r* may occur in any position in a predicate of any arity, and *A* may participate in other predicates. Mandatory role constraints are enforced on populations rather than types. If the only fact stored is that patient 001 is named "Adams C", the name role played by Patient is still mandatory even if more instances of the Patient type are yet to be added to the population.

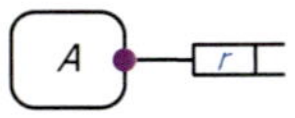

Role *r* is **mandatory** (for the population of *A*).

Each instance of type *A* that is recorded in the database is also recorded to play *r*.

pop(*r*) = pop(*A*)

FIG. 5.7 The role *r* is mandatory.

The UoD (typically part of the real world) is not the same thing as the recorded world or formal model of the UoD. Like other constraints, mandatory role constraints are assertions about our model, and do not necessarily apply to the world being modeled. With our current example, it is optional whether a patient has a phone. This simply means that we do not need to record a phone number for each patient. Maybe this is because in the real world not every patient has a phone, or perhaps patients with a phone won't supply their phone number.

The real world schema of Fig. 5.8(a) concerns applicants for an academic position. By nature, each person has exactly one gender. As a business decision, each applicant is required to have a degree. Applicants may or may not have a fax number. Fig. 5.8(b) shows the model actually used. Here, applicants may choose whether to have their gender recorded, but must provide details about their degrees. For this model, a business decision has been made (perhaps to avoid gender bias) to remove a mandatory constraint that applies in the real world. This fairly common practice of relaxing some real world mandatory constraint should always be a conscious decision.

Do *not* read a mandatory role constraint to mean "if an object plays that role in real life then we must record it". The information system can work only with the model we give it of the world—it cannot enforce real world constraints not expressed in this model.

The following checking procedure helps ensure that our mandatory role constraints are correct. For each mandatory role, is it mandatory in the real world? If not, make it optional. For each optional role, is it optional in the real world? If not, what reasons are there for making it optional? We refine this further when we discuss subtypes.

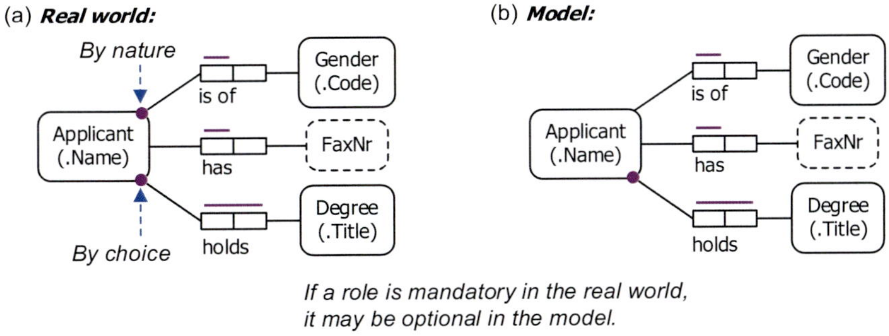

FIG. 5.8 A conscious decision to make recording of an Applicant's gender optional.

In Fig. 5.8(b), the mandatory constraint is on a *nonfunctional role* (i.e., a role not covered by a simple uniqueness constraint). In this case, it is acceptable because we wouldn't want to hire any academic without a degree, and we would be interested in all their degrees.

In general, however, you should *be wary of adding mandatory constraints to nonfunctional roles.* Do so only if you are sure the constraint is needed. Such cases can lead to complexities (e.g., referential cycles) in the implementation, which are best avoided if possible. Of course, if your application really requires such a constraint, then you should declare it, regardless of its implementation impact.

Novice modelers tend to be too heavy handed with mandatory role constraints, always making a role mandatory if it's mandatory in the real world. In practice, however, it is often best to make some of these roles optional in the model to allow for cases where for some reason we can't obtain all the information. As a general piece of advice, *make a role mandatory if and only if you need to.*

When an object type plays more than one role, special care is needed in updating the database to take account of mandatory roles. Suppose we want to add some facts from Table 5.1 into a database constrained by the conceptual schema of Fig. 5.5 (recall that it is mandatory to record a patient's name but optional to record a phone number).

Assuming the database is initially empty, an untutored user might proceed as follows.

User		*CIP*
add: Patient 001 has PatientName 'Adams C'.	→	accepted.
add: Patient 001 has PhoneNr '2057642'	→	accepted.
add: Patient 003 has PhoneNr '8853020'	→	rejected. Violates constraint: **Each** Patient has **some** PatientName.

To add the third fact into the database, we must either first record the fact that patient 003 has the name "Collins T" or include this with the phone fact in a compound transaction.

Now consider the two report extracts shown in Fig. 5.9. These list sample details maintained by a sporting club. Membership of this club is restricted to players and coaches. For simplicity, assume members may be identified by name. The term "D.O.B." means date of birth. On joining the club, each person is assigned to a team in the capacity of player or coach (possibly both).

Teams are identified by semantic codes (e.g., "MR-A", "BS-B", "WS-A" denote the men's rugby A team, the boy's soccer B team, and the women's soccer A team, respectively), but the meaning of these codes is not stored explicitly in the information system. A record is kept of the total number of points scored by each team. Each team is initially assigned zero points, even if it doesn't yet have any members.

As an exercise, you might like to model this yourself before peeking at the solution in Fig. 5.10. The uniqueness constraints assert that each member coaches at most one team, plays for at most one team, was born on at most one date, joined the club on at most one date, and each team has at most one coach and scored at most one total. The reference

Member	*D.O.B.*	*Joined*
Adams F	02-15-1995	01-02-2019
Anderson A	12-25-2009	01-02-2019
Brown C	11-02-1983	11-14-2012
Collins T	02-15-1966	05-05-2000
Crystal B	01-02-2010	11-14-2023
Downes S	11-02-1983	06-17-2015
...	...	...

Team	*Pts*	*Coach*	*Players*
MR-A	23	Downes S	Adams F Brown C Collins T ...
BS-B	0	Collins T	Anderson A Crystal B ...
WS-A	0	?	?
...	...	...	...

FIG. 5.9 Extracts of two reports from a sporting club.

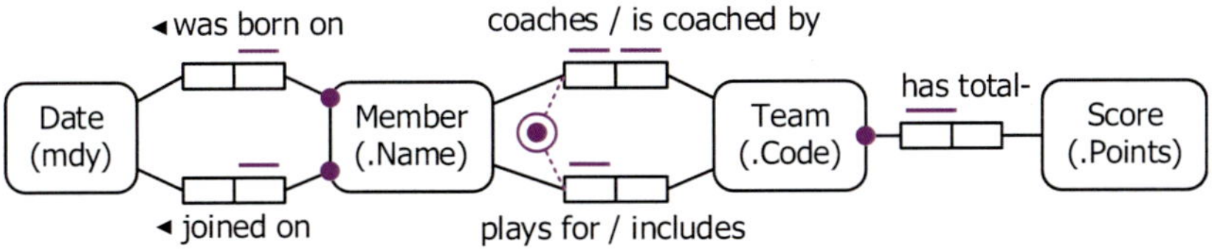

FIG. 5.10 Inclusive-or constraint: each member coaches or plays (or both).

mode "mdy" for Date indicates that for verbalization purposes, date instances are in month-day-year format—of course, this conceptual choice does not exclude other formats being specified for internal or external schemas.

The mandatory role dot on Team indicates that each team has a total score (possibly 0). The other mandatory roles on Member indicate that each member has his/her birth date and club-join date recorded. The circled mandatory role dot is linked to two roles: this is an **inclusive-or** (**disjunctive mandatory role**) **constraint**, indicating that the disjunction of these two roles is mandatory for Member. That is, each member *either* coaches *or* plays (*or both*).

For example, Adams is a player only, Downes is a coach only, and Collins is both a player and a coach. A coach of one team may even play for the same team (although the population doesn't illustrate this). The lack of an inclusive-or constraint over the "is coached by" and "includes" roles of Team indicates that it is possible to know about a team without knowing its coach or any of its players.

If Fig. 5.10 includes all the roles played by Date in the global schema, then the disjunction of Date's roles is also mandatory (each date is a birth date or a join date). This could be shown explicitly by linking these roles to an inclusive-or dot; however, this constraint is implicitly understood, and the explicit constraint would need to be changed if we added another fact type for Date, so it is better to leave the figure as is.

This Date example illustrates the following generalization of the rule mentioned earlier for single mandatory roles: *by default, the disjunction of roles played by a primitive object type is mandatory.* If the object type is an *entity type, the disjunction of its fact-roles is also mandatory by default* (this default can be overridden by declaring the object type independent; see next chapter).

Apart from highlighting the important cases where disjunctive mandatory roles need to be enforced (e.g., the coach-player disjunction) and simplifying schema evolution, this rule facilitates the drawing of schemas. For example, object types such as Date and MoneyAmount often have several roles that are disjunctively mandatory, and to display this constraint explicitly would be messy (or even impossible if the object type occurs on separate schema pages).

To avoid confusion, a mandatory role constraint (simple or disjunctive) should be shown *explicitly* if the object type it constrains is a subtype (see Chapter 6).

In the case of implied (possibly disjunctive) mandatory role constraints, it is okay to *mark the constraint explicitly if it will still apply if more roles are added to the object type* (where these extra roles are excluded from the constraint). For object types such as Date, this would almost never happen.

An inclusive-or (disjunctive mandatory role) constraint indicates that each instance in the population of the object type must play at least one of the constrained roles (and possibly all of them).

Fig. 5.11 indicates in general how to specify explicitly that a disjunction of roles r_1, r_2, ..., r_n is mandatory by linking the n roles to a circled dot. The roles may occur at any position in a predicate, and the predicates need not be distinct.

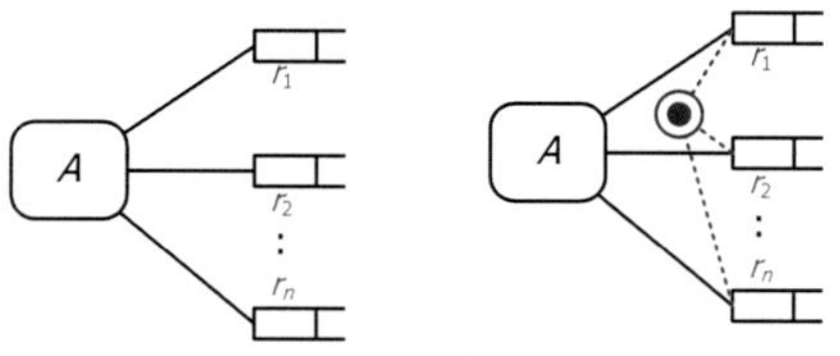

Inclusive-or (Disjunctive mandatory role) constraint

The inclusive disjunction of roles r_1 .. r_n is mandatory for *A*

i.e., each member of pop(*A*) plays r_1 **or** ... r_n (or all)

i.e., each member of pop(*A*) plays *at least one of* r_1, ..., r_n

FIG. 5.11 Inclusive-or constraint (the role disjunction is mandatory).

The inclusive "or" is used in the *verbalization* of the constraint. For example, the inclusive-or constraint in Fig. 5.10 is formally verbalized as **Each** Member coaches **some** Team **or** plays for **some** team. As in logic and computing, "or" is always interpreted in the inclusive sense unless we say otherwise.

Inclusive-or constraints sometimes apply to roles in the same predicate. Fig. 5.12 relates to a small UoD where people are identified by their first name. A sample population is shown for the ring binary; here each person plays one (or both) of the two roles. For instance, Terry is a child of Alice and Bernie, and a parent of Selena. Because Person plays another role in the schema, the disjunctive mandatory role constraint must be depicted explicitly, either by connecting the circled dot to (a) the junction of the two roles, or to (b) both of the roles.

As an example involving derivation, consider Table 5.2. Here, students sit a test and exam, and their total score is computed by adding these two scores. The schema for this, shown in Fig. 5.13, includes a derived fact type. If included on the diagram, a derived fact type must be marked with an asterisk, and its constraints are shown explicitly (even though these constraints are typically derived). The mandatory constraint on the derived fact type means the total score must be known; this does not mean that it must be stored.

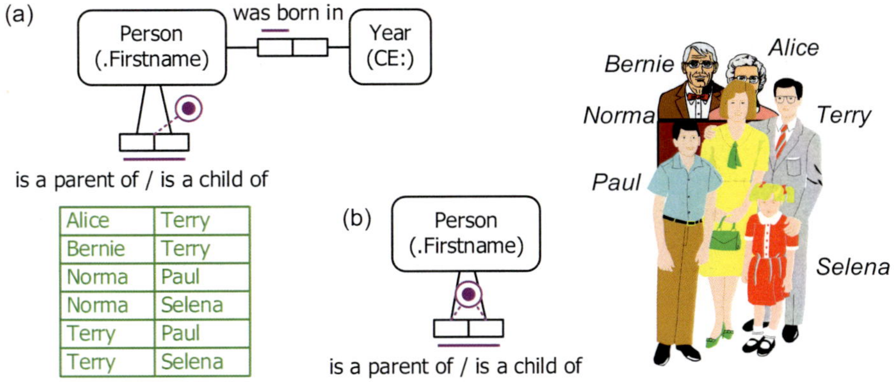

FIG. 5.12 Each person is a parent or child.

Table 5.2 Student scores.

Student Nr	*Student Name*	*Test*	*Exam*	*Total*
1001	Adams, A	15	60	75
1002	Brown, C	10	65	75
1003	Einstein, A	20	80	100

The roles played by Score are named "testScore", "examScore", and "totalScore" to enable the derivation rule to be specified in attribute style. The uniqueness constraint on the total score fact type is derivable from the derivation rule and the other uniqueness constraints (each student has only one test score and only one exam score and the total score is the sum of these). The mandatory role on the total score fact type is derivable from the derivation rule and the other mandatory roles (each student has a test score and an exam score, and the rule then provides the total score).

By default, all constraints shown on a derived fact type are derivable. This rule is discussed in more detail later. In rare cases, we may want to declare a derived fact type with a nonderivable constraint. We consider some examples much later.

In principle, a derived fact type may be drawn with no constraints if these can be derived. Typically, however, a derived fact type is included on the diagram for discussion purposes, and it is illuminating to show the constraints explicitly.

The term "mandatory role" is used in the sense of "must be *known*" (by storing it or deriving it). We may talk about the population of derived fact types as well as asserted fact types. For a given object type *A*, the population of *A*, pop(*A*), includes all members of *A* that play any of its roles (asserted, derived, or semiderived).

Derived roles may be optional. For example, if either the test or the exam role in Fig. 5.13 is optional, the total score role is also optional (but is still unique).

With large schemas, an object type may play so many roles that it is awkward to connect a single shape for the object type to all its roles. To solve this problem, *object types may be duplicated* on a schema. In this case, the rule for implicit mandatory disjunctive roles applies to the union of all the duplicate shapes for the object type.

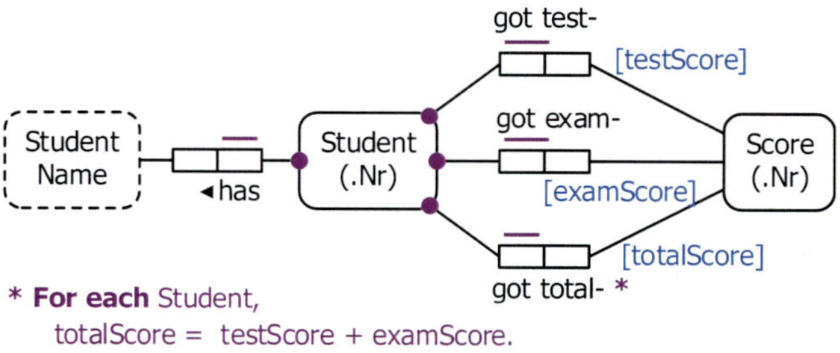

FIG. 5.13 Here, constraints on the derived fact type are derivable.

Large schemas are often divided into pages, and the same object type or fact type may appear on many pages and may also be duplicated on the same page. To indicate that an object type or fact type is duplicated in the same model, either on the same page or on a different page, the NORMA tool adds a *shadow* to the shape.

In Fig. 5.14, the object type Date is duplicated on the same page of the model. The fact type Member has MemberName is duplicated on another page (not shown here) of the global schema diagram.

With large projects involving multiple models, there may sometimes be a need to import elements such as object type or fact type into a model from another model in which the element is defined. Currently, the NORMA tool has no graphical notation to indicate that a model element is defined externally. However, various mechanisms are currently being investigated for the NORMA tool to support management of external references as well as partitioning of a model.

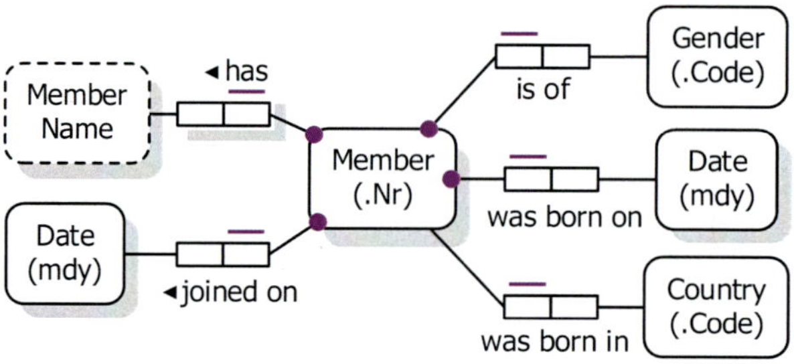

FIG. 5.14 Object types and fact types displayed more than once are shaded.

Exercise 5.2

1. Specify an ORM schema for the UoD described by the following sample output report, which is based on data accessed in 2023. The coal and iron ore reserves are measured in megatonnes (Mt), i.e., million metric tons. Include uniqueness and mandatory role constraints.

Country	*Reserves (Mt)*	
	Coal	*Iron Ore*
USA	254,197	3,000
Russia	176,770	29,000
Australia	159,634	51,000
China	149,818	20,000
India	107,726	5,500
Germany	39,802	?
Ukraine	37,891	6,500
Brazil	7,270	34,000

2. Set out the conceptual schema for the following output report. Include uniqueness and mandatory role constraints. Identify any derived fact type(s).

Subject	*Year*	*Enrollment*	*Rating*	*NrStudents*	*%*	*Lecturer*
CS121	2007	200	7	5	2.50	P.L. Cook
			6	10	5.00	
			5	75	37.50	
			4	80	40.00	
			3	10	5.00	
			2	5	2.50	
CS123	2007	150	7	4	2.67	R.V. Green
			6	8	5.33	
			5	60	40.00	
			4	70	46.67	
			1	6	4.00	
CS121	2008	250	7	10	4.00	A.B. White
			6	30	12.00	
			5	100	40.00	
			4	80	32.00	
			3	15	6.00	

3. A cricket fan maintains a record of boundaries scored by Australia, India, and New Zealand in their competition matches. In the game of cricket, a *six* is scored if the ball is hit over the field boundary on the full. If the ball reaches the boundary after landing on the ground, a *four* is scored. In either case, a *boundary* is said to have been scored.

Although it is possible to score a 4 or 6 by running between the wickets, such cases do not count as boundaries and are not included in the database. A sample output report from this information system is shown. Here, "4 s" means "number of fours" and "6 s" means "number of sixes". Schematize this UoD, including uniqueness constraints, mandatory roles, and derived fact types. Use nesting.

Year	*Australia*			*India*			*New Zealand*		
	4s	*6s*	*total*	*4s*	*6s*	*total*	*4s*	*6s*	*total*
1984	120	30	150	135	23	158	115	35	150
1985	112	33	145	110	30	140	120	25	145
1986	140	29	169	135	30	165	123	35	158

4. Report extracts are shown from an information system used for the 1990 Australian federal election for the House of Representatives (the main body of political government in Australia). The table lists codes and titles of political parties that fielded candidates in this election or the previous election (some parties competed in only one of these elections). For this exercise, *treat Independent as a party.*

PartyCode	*Title*
ALP	Australian Labor Party
AD	Australian Democrats
GRN	Greens
GRY	Grey Power
IND	Independent
LIB	Liberal Party of Australia
NDP	Nuclear Disarmament Party
NP	National Party of Australia
...	...

A snapshot of voting details is shown about two seats (i.e., voting regions) in the 1990 election. This snapshot was taken during the later stage of the vote counting. The number of votes for each politician, as well as the informal vote, is initially set to 0. During the election the voting figures are updated continually. The percentage of votes counted is calculated assuming all those on roll actually vote. For simplicity, *assume that each politician and each seat has a unique name.* Figures are maintained for all seats in all states.

An asterisk (*) preceding a politician's name indicates a sitting member (e.g., Fife is the sitting member for the seat of Hume): express this as a binary. Sitting members are recorded only if they seek re-election. Some seats may be new (these have no results for the previous election). States are identified by codes (e.g., "QLD" denotes Queensland). Each state has many seats (not shown here). Draw a conceptual schema diagram for this UoD. Include uniqueness constraints and mandatory roles. If a fact type is derived, omit it from the diagram but include a derivation rule for it (you may specify this rule informally). Do not attempt any nesting.

QLD:

FADDEN (On roll: 69110)	
CROSS (AD)	7555
FRECKLETON (NP)	2001
JULL (LIB)	24817
HEYMANN (IND)	641
WILKINSON (ALP)	21368
Informal	1404
% counted: 84	
Previous election: AD 4065; ALP 24481; LIB 21258; NP 9581	

NSW:

HUME (On roll: 70093)	
JONES (IND)	9007
* FIFE (LIB)	29553
KIRKWOOD (IND)	507
MARTIN (ALP)	20554
ROBERTS (AD)	3889
Informal	1309
% counted: 92	
Previous election: AD 2498; ALP 24516; IND (total): 2086; LIB 33687; NDP 1105	

5.3 Reference Schemes

With uniqueness and mandatory role constraints covered, it is time to consider reference schemes in more depth. With simple identification schemes, each entity is identified by associating it with a single value. For example, a country might be referenced using a code. For instance, Australia may be referenced by the definite description "The Country that has CountryCode 'AU.'"

We usually abbreviate simple reference schemes by parenthesizing the reference mode or manner in which the value relates to the entity, for example, Country(.code). In practice, more complex reference schemes are often encountered, and we need to deal with them even as early as step 1 of the CSDP, when we verbalize facts.

Before examining these more difficult cases, let's have a closer look at simple reference schemes. Table 5.3 contains data about a small UoD where people can be identified by their surnames and each city has a unique name.

Table 5.3 Personal details.

Person	*City*	*Height (cm)*	*Chest (cm)*	*Mass (kg)*	*IQ*
Adams	Brisbane	175	100	77	100
Brisbane	Sydney	182	91	80	120
Collins	Sydney	173	80	67	100
Darwin	Darwin	175	95	90	90

Fig. 5.15 shows one way to schematize this. The reference schemes are abbreviated as reference modes in parentheses. Some role names have been added for discussion purposes.

Fig. 5.16 repeats the schema, this time with the reference schemes depicted explicitly– each appears as a *mandatory, 1:1 reference type*. For example, in this schema each person has exactly one surname, and each surname refers to at most one person. Technically, this is called an *injection* (or *1:1 into* mapping) from Person to Surname.

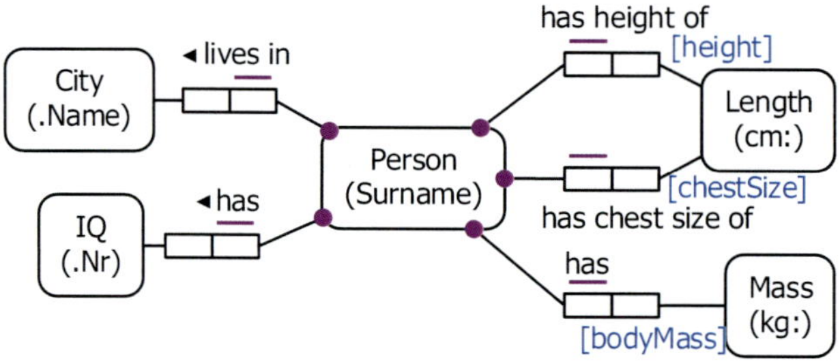

FIG. 5.15 A conceptual schema for Table 5.3 (reference schemes abbreviated).

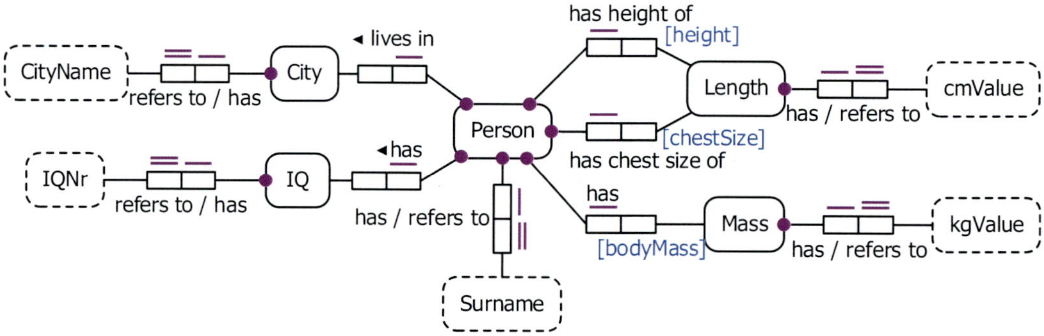

FIG. 5.16 A conceptual schema for Table 5.3 (reference schemes explicated).

A *double uniqueness bar* on a role indicates the *preferred reference scheme* (indicating the *preferred identifier* or *primary identifier*) for the entity type playing the adjacent role. In this example, surnames are the preferred way to identify persons, and centimeter values are preferred for referencing lengths.

In this case, there is only one way to reference instances of a given entity type. In general, however, object-role modeling (ORM) allows many reference schemes for the same entity type; in such cases, one of these eventually needs to be chosen as the preferred reference scheme.

The second-generation ORM considered in this book appends a colon ":" when displaying unit-based reference modes (e.g., cm:) and requires a *unit dimension* (e.g., Length) for each unit.

Because access to unit dimensions is always possible with unit-based reference modes, specific object types corresponding to roles played by the unit dimension may be used instead of the underlying unit dimension when setting out fact types. For example, our current example may be remodeled as shown in Fig. 5.17(a). The ORM tool display of unit dimensions may be toggled on/off at will. The toggle on choice is shown in Fig. 5.17(b).

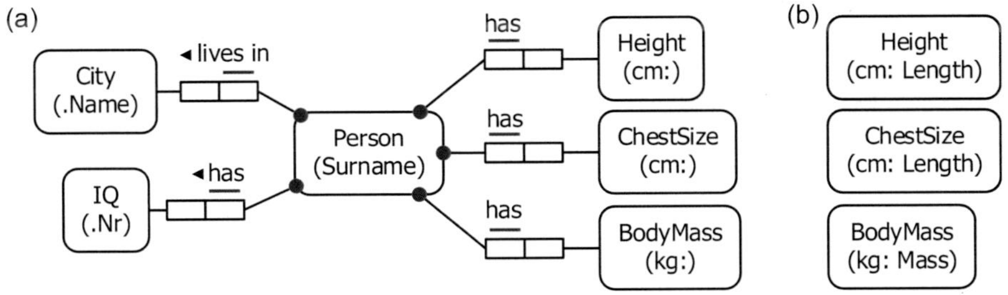

FIG. 5.17 An alternative conceptual schema for Table 5.3.

Unlike top-level entity types, value types may overlap with one another. For example, the populations of Surname and CityName overlap here because both include the character string "Brisbane".

Assuming the data in Table 5.3 and an appropriate query language, consider the following dialog between a user and the CIP.

User: Which person lives in a city with the same name as his/her surname?
CIP: Darwin.
User: Which person is a city?
CIP: None: Entity types Person and City are mutually exclusive.

It is meaningful to compare the names for Person and City because they are both character strings. But it is not sensible to compare persons and cities—they are completely different types of entities. Although the value types for IQ, mass, and length are all numeric, mass and length have different units while IQ is dimensionless (i.e., it has no units). Conceptually, two numerically referenced entities may be compared or added only if they have the same unit (or both have no unit). Hence, the following queries do not make sense:

List persons whose mass equals their IQ.
List persons whose height is more than twice their mass.
List height + mass for each person.

Of course, the numbers themselves may always be compared. Also, height and chest size are both based on the same domain and unit, so they are compatible. So the following queries are acceptable.

List persons whose mass number equals their IQ number.
List persons whose height number is more than twice their mass number.
List height + chest size for each person.

Most current database systems provide little or no support for the distinction between entities and their values. Operations specified in queries are interpreted as applying directly to the values. For example, assuming the data of Table 5.3 are stored in a relational database as the table *Athlete* (person, city, height, chest, mass, IQ), the following structured query language (SQL) query is quite legal:

```
select person from Athlete
where person=city
  and mass=iq and chest + mass>height
```

For the population in Table 5.3, this query returns "Darwin", even though taken literally the query is nonsense. The comparisons evaluated in the query are between values, not entities. We could clarify this by better choices of column names (e.g., use "Surname" and Cityname" instead of "Person" and "City").

Values are constants with a predefined interpretation, so they require no explicit reference scheme. For example, "Brisbane" denotes itself (i.e., the character string "Brisbane"), and when written without quotes, the numeral "77" denotes the number 77 (Hindu-Arabic decimal notation is assumed). It might be claimed that when written without quotes, a constant such as "Brisbane" is enough to identify a single entity. However, some context is always required (at least implicitly). Even in Table 5.3, "Brisbane" is used to refer to both a person and a city.

Earlier in the book, we saw that entities are referenced by means of definite descriptions (e.g., "The City with name 'Brisbane'"). Within the context of a reference type, a value may, somewhat loosely, be said to be an identifier for an entity if the value relates to only one entity of that type.

If only one value is used in a definite description, we have a simple 1:1 reference scheme. Sometimes, however, two or more values may be required. In this case, we have a *compound reference scheme*. As an example, consider how computer files are identified in Table 5.4. Within a given folder, a file may be identified by its local filename, for example, "flag.vsd". But when many folders are involved, we need to combine the folder name with the local filename to know which file we are talking about.

Table 5.4 Details about computer files.

File		*Size*
Folder	*Filename*	*(kB)*
MyDocs	flag.vsd	35
MyDocs	orm1.doc	1324
OzWork	flag.vsd	40

To reflect the two-part naming convention suggested by the two columns for File in Table 5.4, we could model the situation as shown in Fig. 5.18. The mandatory roles and internal uniqueness constraints declare that each file is in exactly one folder, has exactly one (local) filename, and has exactly one size. The external uniqueness constraint can be used to provide a compound reference scheme and may be verbalized as shown.

If no other identification scheme exists for File, the external uniqueness constraint is the basis for the preferred way in which humans communicate about the files. To indicate this choice of *compound, preferred reference*, a *circled double uniqueness bar* is used, as shown in Fig. 5.19.

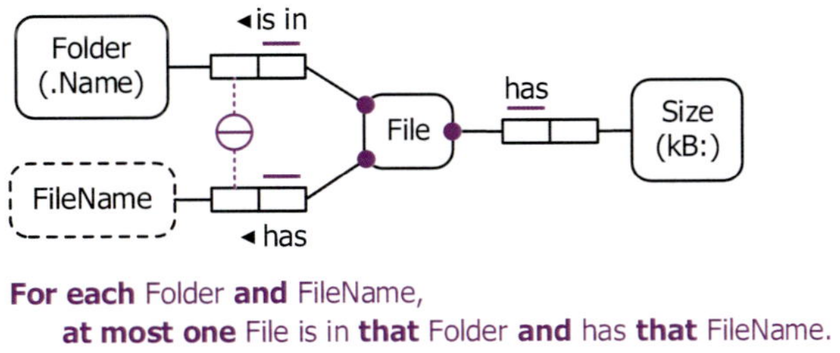

FIG. 5.18 File has a compound reference scheme.

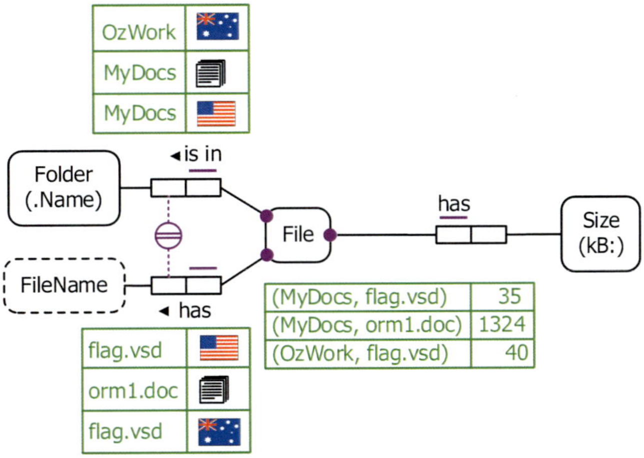

FIG. 5.19 The previous schema populated with references and elementary facts.

Using icons for entities, the reference types File is in Folder and File has FileName are populated as shown. In the fact table for File has Size, the file entries appear as value pairs. Although not apparent in the diagram, these are ordered (Folder, FileName) in the same order as the reference types were added to the schema—an ORM tool can display this order on request.

If the reference predicates are included in the definite descriptions for the files, any order can be used. For example, "the File that is in Folder 'MyDocs' and has FileName 'flag.vsd'" is equivalent to "the File that has FileName 'flag.vsd' and is in Folder 'MyDocs.'" In general, compound reference schemes involve a mandatory 1:1 map of objects to a tuple of two or more values.

Textually, a compound reference scheme for an object type *A* may be declared by listing the reference types, with the leading *A* removed, in parentheses after *A*. The relevant uniqueness and mandatory constraints on reference predicates are assumed, and if the reference schemes are declared first, the facts may be stated in shortened form. For example, the sample model may be specified textually as follows:

Reference schemes:	Folder (.name); File (is in Folder, has FileName); Size (kB:)
Fact type:	File has Size
Constraints:	**Each** File has **at most one** Size.
	Each File has **at least one** Size.
Fact instances:	File ('MyDocs', 'flag.vsd') has Size 35.
	File ('MyDocs', 'orm1.doc') has Size 1324.
	File ('OzWork', 'flag.vsd') has Size 40.

In principle, we could model Files using a simple reference scheme by concatenating the folder name and local filename into one string with a separator (e.g., "\"). For example, the files may then be named simply "MyDocs\flag.vsd", "MyDocs\orm1.doc", and "OzWork\flag.vsd". This approach is normally undesirable because it requires extra derivation work to extract and group the components to produce various reports. We'll say more about this issue later. Indeed, if we wish to issue queries about the file extension (e.g., "vsd", "doc") as well, a three-part reference scheme is desirable (using folder name, simple name, and file extension).

Identification schemes are *relative* to the particular UoD. Often, a simple scheme that works in a local context fails in a UoD of wider scope. This notion is important to bear in mind when merging schemas into a larger schema. For example, suppose that within a given university department, each subject offered by that department can be identified by its title. For instance, consider the output report in Table 5.5, which indicates which subjects are offered by the Physics Department in which semesters.

Table 5.5 Physics department offerings.

Subject	*Semester*
Electronics	1
Mechanics	1
Optics	2

If designing a schema for just the Physics Department, we have a 1:1 correspondence between subjects and their titles. So this table may be schematized as the fact type Subject (.Title) is offered in Semester (.Nr). A similar schematization could be used for just the Mathematics Department, a sample output report for which is shown in Table 5.6.

Table 5.6 Mathematics department offerings.

Subject	*Semester*
Algebra	2
Calculus	1
Mechanics	1

But now suppose we need to integrate departmental schemas into an overall schema for the whole university. Our simple identification scheme for subjects will no longer work because, in this wider UoD, different subjects may have the same title. For example, the subject Mechanics offered by the Physics Department is different from the subject Mechanics offered by the Mathematics Department. A combined output report would look like Table 5.7.

Table 5.7 University subject offerings.

Department	*Subject title*	*Semester*
Physics	Electronics	1
	Mechanics	1
	Optics	2
Mathematics	Algebra	2
	Calculus	1
	Mechanics	1
...	...	...

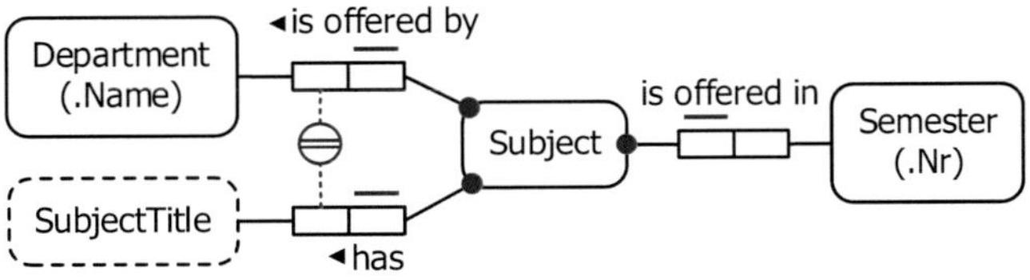

FIG. 5.20 A schema for Table 5.7 using a compound reference scheme.

The first row may be verbalized using a compound reference scheme as follows: "The Subject that is offered by the Department named 'Physics' and has the SubjectTitle 'Electronics' is offered in the Semester numbered 1". This approach is schematized in Fig. 5.20.

Although conceptually this picture is illuminating, in practice its implementation may be awkward, with two labels (one for the department and one for the title) needed to identify a subject. In such cases, a new identification scheme is often introduced to provide a simple 1:1 reference. For example, each subject may be assigned a unique subject code, as shown in Table 5.8.

Table 5.8 Subject codes provide a simple reference scheme.

Subject	*Title*	*Department*	*Semester*
PH101	Electronics	Physics	1
PH102	Mechanics	Physics	1
PH200	Optics	Physics	2
MP104	Algebra	Mathematics	2
MP210	Calculus	Mathematics	1
MA109	Mechanics	Mathematics	1

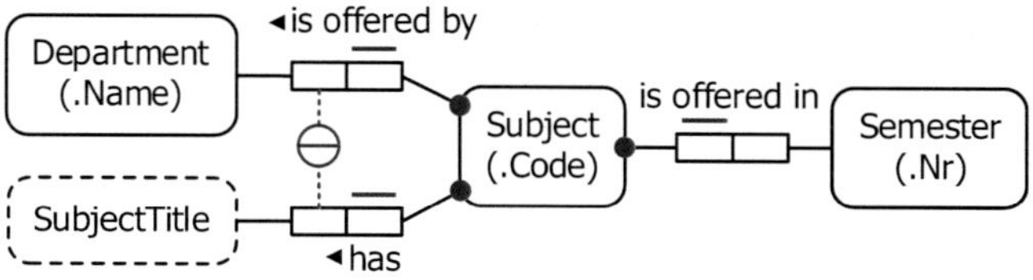

FIG. 5.21 A schema for Table 5.8 using subject code as Subject's primary identifier.

In this output report, there are two *candidate identifiers* for Subject. We could identify a subject by its code (e.g., "PH102") or by combining its department and title (e.g., "Physics", "Mechanics"). One of these is chosen as the *preferred identifier*, or standard means of referring to the entity. In this case, we would usually pick the subject code as the preferred identifier. We can indicate this choice by displaying the subject code identification scheme as reference mode in parentheses and displaying the external uniqueness constraint with a single uniqueness bar (see Fig. 5.21).

If a parenthesized reference mode is displayed, this is always taken to be the preferred reference scheme. Here, the Department and SubjectTitle predicates are treated just like any other elementary fact types; they are no longer considered to be reference types. Of course, the external uniqueness constraint between them must still be declared.

Choosing a preferred reference scheme from two or more existing schemes is not simply a conceptual issue. At the conceptual level, its main uses are to note any business decisions on preferred identification schemes and to disambiguate abbreviated references that omit the reference predicate(s), for example, "Patient 3452 has Temperature 37". This is especially useful when populating fact tables with instances. At the logical level, however, the choice of preferred reference typically has a major impact (e.g., determining the primary key of a table).

Here are some *guidelines for selecting a preferred identifier.* First, *minimize the number of components.* A single subject code is easier to enter than two values for department and title. Moreover, a compound identifier adds extra overhead in the later database implementation (joins, indexes, and integrity checks over composite columns require more effort).

Second, favor an identifier that is *more stable.* In a university environment, the same subject may undergo title changes throughout its history and sometimes the department that offers a given subject changes. It is even possible for the same department—title combination to refer to different subjects at different times. However, it is extremely rare for a subject to have its code changed, and a subject code is typically never reused for a different subject.

If a historical record is needed for an object, and its identifier changes though time, extra work is required to keep track of these changes so that we know when we are talking about the same object. This extra effort can be minimized by making the identifier as stable as possible. Ideally, the object has the same identifier throughout the lifetime of the

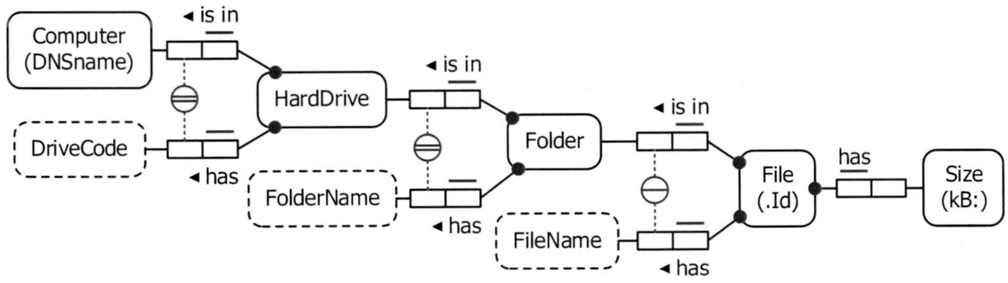

FIG. 5.22 Identifying files in a wider environment.

application; this is known as a *rigid identifier*. Most organizations choose rigid identifiers such as EmployeeNr, ClientNr, SerialNr for their employees, clients, equipment, and so on.

To explain some basic concepts in a friendly way, we've used identifiers such as PersonName, Surname, or even Firstname in some examples. However, except for trivial applications, this kind of identification is unrealistic. Often, two people may have the same name. Moreover, people may change their name—women usually change their surname when they marry, and anyone may legally change their name for other reasons. A philosophy lecturer we knew once changed his surname to "What" and his wife changed her surname to "Who"!

As another example, the schema in Fig. 5.22 expands the UoD considered in Fig. 5.19 to enable computer files to be identified across many computers. Here, an artificial FileId is introduced as the preferred identifier to provide a simple, rigid identifier that remains stable even when the file name is changed (a fairly common occurrence). In addition, a human friendly path name is provided as a multipart secondary identification scheme via the external uniqueness constraints.

Concatenating the components of this secondary identification scheme along with separators often leads to very long file names such as "www.myServer.com\c:\MyDocs \flag.vsd". In practice, many systems would also assign simple internal identifiers for other elements (e.g., folders and drives).

A third criterion sometimes used for selecting an identifier is that it be easy for users to recognize. This criterion is mainly used to minimize errors when users enter or access artificial identifiers such as subject or stock-item codes. This effectively endows codes with semantics, at least implicitly. For example, the subject code "CS114" is used in an Australian university for a particular subject in informatics. The first two characters "CS" indicate the discipline area (Computer Science), the first digit indicates the level (first), and the last two digits merely distinguish it from other computer science subjects at the same level. Such a code is less likely to be misread than an arbitrary code (e.g., "09714").

Such *"information-bearing" identifiers* are sometimes called *"semantic names"*. They should be avoided if the semantic relationships involved are unstable because the names would then need changing. In the current example, the semantics are fairly stable—the

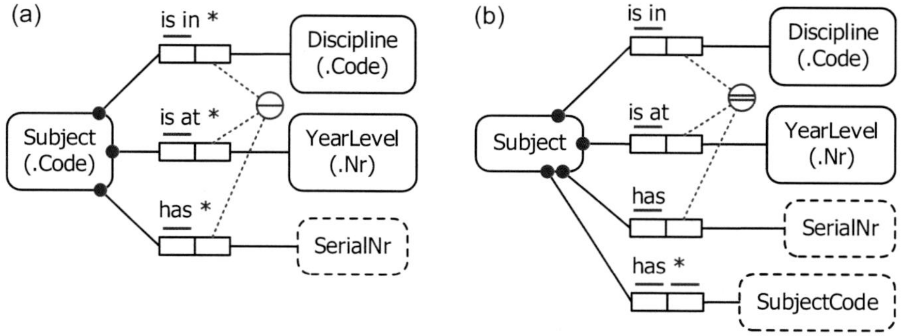

FIG. 5.23 Different ways of exposing the semantics of subject codes.

subject discipline is unlikely ever to change, and the subject level would normally be stable too. The linking of the two letters to discipline (e.g., "MP" for pure mathematics and "MA" for applied mathematics) rather than department is better because disciplines are more stable than departments.

If semantic names are used, we need to decide whether their semantics are to be explicitly modeled or instead assumed to be understood by users. The semantics of subject codes may be modeled explicitly by including three binaries: Subject is in Discipline; Subject is at YearLevel; Subject has SerialNr. An external uniqueness constraint spans the roles played here by the three components.

Fig. 5.23(a) shows the case where subject code is the preferred identifier. In this case, these three binaries would normally be derived, and appropriate derivation rules specified (e.g., the discipline code may be derived using a substring operator to select the first two characters of the subject code). The external uniqueness constraint is then derivable.

An alternative approach is shown in Fig. 5.23(b). Here, the three-part identification scheme is used for preferred reference, and the subject code is derived by concatenation.

The major advantage of storing the components separately like this is that it facilitates queries concerning the components. In particular, *if we wish to formulate queries* ***for each*** *instance of a reference component, then that component should be stored.*

The schema in Fig. 5.23(b) makes it easy to query properties for each discipline or year level (e.g., "How many subjects are offered for each discipline?"). As discussed in Section 12.10, the conceptual "for-each" construct corresponds to "group-by" in SQL.

The model in Fig. 5.23(a) is extremely awkward for such queries because groups can normally only be formed from stored columns. However, a simple code is more compact and offers much better performance for relational joins than a three-part identification scheme.

Is there a way of having your cake and eating it too? Yes, at least to some extent. In Fig. 5.24, the three components are both derived and stored (as shown by the "**"). This allows fast joins on the subject code and grouped queries on the components.

Semantic codes were often used in legacy systems, with no attempt to expose their semantics as separate components. If these legacy systems have to be retained in their

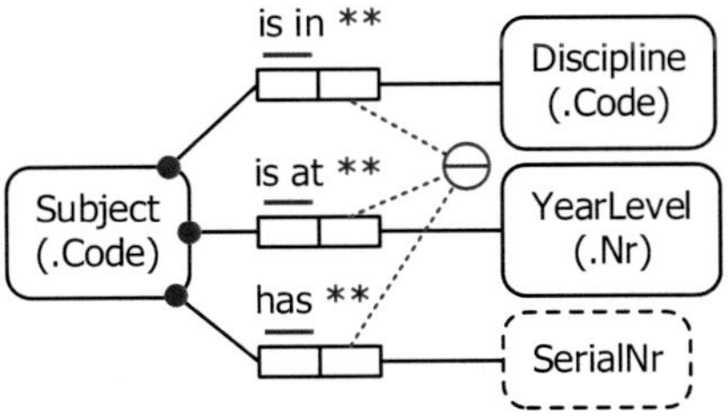

FIG. 5.24 The three components are derived and stored.

Table 5.9 Employee details.

EmpNr	*Employee Name*	*Room*
715	Adams A	69-301
139	Cantor G	67-301
503	Hagar TA	69-507
...	...	...

current form, a separate system (e.g., a data warehouse) can be constructed to better support queries on the components. In copying the data from the legacy system to the new system, the codes are transformed into separate component fields.

How much of the semantics underlying an information-bearing name should be exposed in a conceptual schema depends on the queries we wish to issue.

For example, consider the output report shown in Table 5.9. Here, room numbers have the format *dd-ddd* (where *d* denotes a digit). The first two digits provide the building number for the room's building, the third digit indicates the room's floor (in this UoD no building has more than nine floors), and the last two digits comprise a serial number to distinguish rooms on the same floor.

If we never wish to query the information system about buildings or floors, it's fine to model rooms using the simple reference scheme Room(.nr). Now suppose instead that we also expect the system to output reports such as the extract shown in Table 5.10. Here, we *are* interested in buildings, and we want to list the number of rooms *for each* building. In this case, we must expose at least the building component of the reference scheme for buildings, as in Fig. 5.25.

Table 5.10 Building details.

BuildingNr	*Building Name*	*Nr rooms*
...	...	...
67	Priestly	100
68	Chemistry	100
69	Computer Science	150
...	...	...

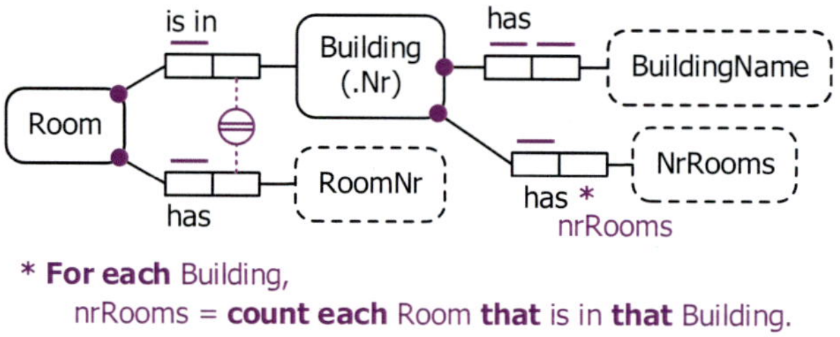

FIG. 5.25 The reference scheme for Room exposes the building component.

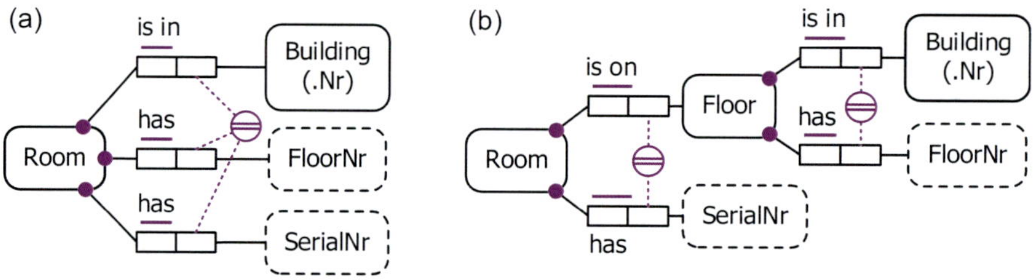

FIG. 5.26 Two ways of exposing the three-part reference scheme for rooms.

In the new model, the term "RoomNr" indicates a local room number (e.g., "301") that identifies a room within a given building. It corresponds to the last three digits in a campus-wide room number such as "67–301". If we are not interested in querying the system about floors, there's no need to expose the room number semantics any further.

Now suppose that we wish to query the system about floors; for example, "How many rooms are there on floor 3 of building 67?" To facilitate this, we could expand the reference scheme to three components to expose the floor semantics. One way to model this is shown in Fig. 5.26(a). This divides the previous room number (e.g., "301") into a FloorNr (e.g., "3") and a serial number (e.g., "01").

If we want to think about floors themselves, rather than just floor numbers, it is better to model Floor as an entity type, as shown in Fig. 5.26(b). Here, Floor plays in the composite reference scheme for Room, and also has its own composite reference scheme.

Although less convenient, it is possible to use derivation functions to formulate queries about single *instances* of a reference component without actually storing the component. For example, to list the number of rooms on floor 3 of building 67, we could use the following SQL query: **select count**(*) **from** Room **where** roomNr **like** '67–3%.' Even though this approach is possible, we still recommend exposing the semantics. If ever in doubt as to whether to expose some semantics, it's generally better to do so.

You should always store a reference component if you wish to query about a *group* of its instances; for example, "*For each* floor of building 67, how many rooms are there?" Using the schema of Fig. 5.26(b), this may be formulated in ORM by selecting the path from

Room to Building (i.e., Room is on **a** Floor **that** is in Building), adding the condition "= 67" for Building (implicitly applied to BuildingNr), and then requesting Floor and **count**(Room) **for each** Floor.

At the relational level, the query may be formulated in SQL using a group-by clause; for example, **select** floorNr, count(*) **from** Room **where** buildingNr=67 **group by** floorNr. The SQL language is discussed in detail in Chapters 12 and 13.

Choosing reference schemes for entity types is an important aspect of modeling. As a strict guideline, *an entity type's preferred identification scheme must 1:1 map each entity of that type to a tuple of one or more values* (either directly or indirectly). So, each preferred reference scheme provides an *injection* (mandatory, 1:1 mapping). For example, in Fig. 5.26, each building maps directly to a unique building number, and each room maps (indirectly via the reference predicates) to a unique triple of values (building number, floor number, serial number).

In everyday life, weaker kinds of identification schemes are used in some contexts. For example, some people have identifying nicknames (e.g., "the Great White Shark" denotes the golfer Greg Norman). However, in a typical application not everybody has a nickname, so nicknames are not used for preferred reference (they are not mandatory).

Sometimes a 1:many reference scheme is used. For example, a plant or animal type may have more than one identifying, common name. For instance, the bird species with the scientific name of *dacelo gigas* may also be identified by any of the following common names: "kookaburra", "laughing jackass", "great brown kingfisher", or "bushman's clock". As a botanical example, "gorse" and "furze" refer to the same plant.

In Fig. 5.27(a), common names are used as the preferred identifier for bird species. This choice is signified by the double uniqueness constraint bar. A sample population is included to illustrate the 1:many nature of this reference type. Do you spot any problem with this reference scheme?

This association must *not* be used for preferred reference because it is not 1:1. Imagine the problems that would arise if we used "kookaburra" to identify a bird species in

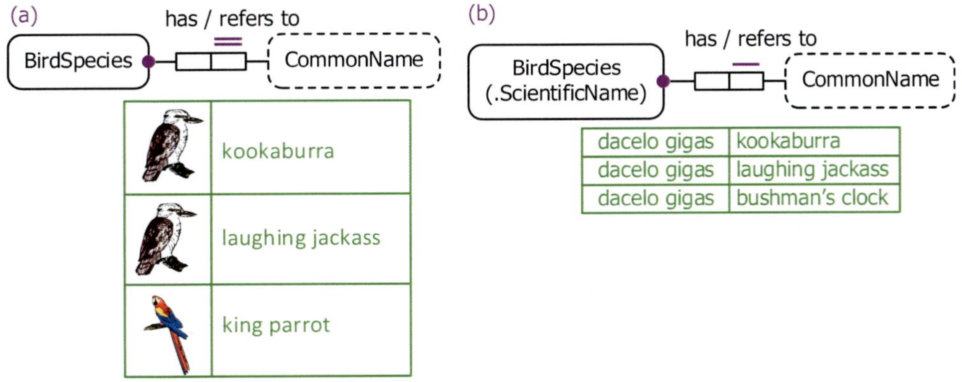

FIG. 5.27 The association (a) is not 1:1, so should *not* be used for preferred reference.

one part of a database and "laughing jackass" in another part. In the absence of any standard identifier like a scientific name, how would we know that we were talking about the same bird species?

For preferred reference, we must choose a mandatory, 1:1 scheme. Such a scheme may already exist, such as ScientificName as in Fig. 5.27(b). If not, we create one with the help of the domain expert. This might be partly artificial (e.g., standard common name) or completely artificial (e.g., BirdkindNr). Whatever we choose, we need to get the users of the information system to agree upon it.

The scientific name for a bird species is actually a semantic name with two components. For example, in *dacelo gigas*, the "dacelo" is the name of the genus, and "gigas" is a species name. By itself, a species name such as "gigas" or "americana" does not identify a species. For instance, *certhia americana* (the brown creeper) is a different bird species from *parula americana* (the northern parula).

If we wish to query about the genus or species name, we should unpack the naming semantics into a two-part reference scheme, as shown in Fig. 5.28; otherwise we can use the simple reference scheme shown in Fig. 5.27(b).

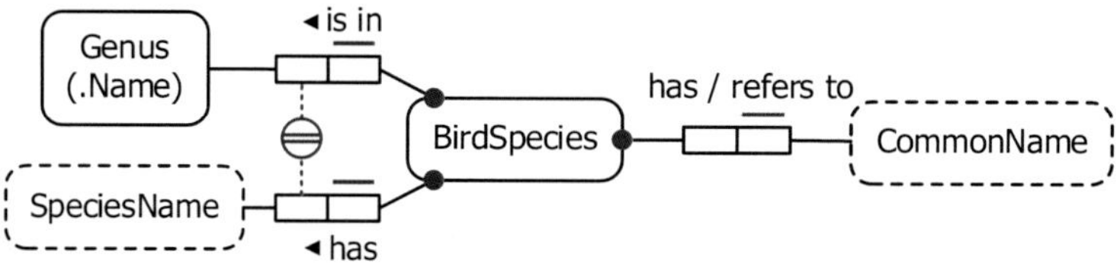

FIG. 5.28 Many common names may be used for the same species.

Notice that we can still include facts about common names. Once this schema is fully populated, we could let users interact with the system via the common names, still using the scientific naming convention to establish identity.

Simple 1:1 naming schemes relate entities directly to values (e.g., subjects to subject codes). In life, we sometimes identify entities by 1:1 relating them to other entities (e.g., "the Olympics that were held in the Year 1992", "the Warehouse that is located in the City 'Brisbane,'" "the Director who heads the Department 'Sales'"). Such identification schemes may be chosen for primary reference, and depicted as mandatory 1:1 binaries (e.g., Olympics was held in Year (CE:)). These binaries are then regarded as reference types, not elementary fact types.

This practice is rare. Instead, such binary associations are usually modeled as elementary fact types, and other primary reference schemes are chosen, such as Olympics(.SequenceNr), Warehouse(.Nr). Alternatively, the relevant entity types may be removed by using suitably descriptive predicates. For example, the assertion "The Olympics of Year 2020 were located in City 'Tokyo'" may be portrayed as a binary association between the entity types Year and City. If the original referential semantics are not needed, another alternative is to use simple value reference, but this is usually a last resort. For example, "Warehouse(.Name) 'Brisbane'" loses the semantics that this warehouse is located in the city Brisbane.

In most compound reference schemes, each reference predicate is mandatory, so each entity of the same type is identified using the same number of values. In the real world, however, we sometimes encounter identification schemes where only some of the reference roles are optional, but their disjunction is still mandatory. This is called *disjunctive reference*. In these cases, different entities of the same type may be referenced by different numbers of values. For example, consider the botanical identification scheme depicted in Fig. 5.29.

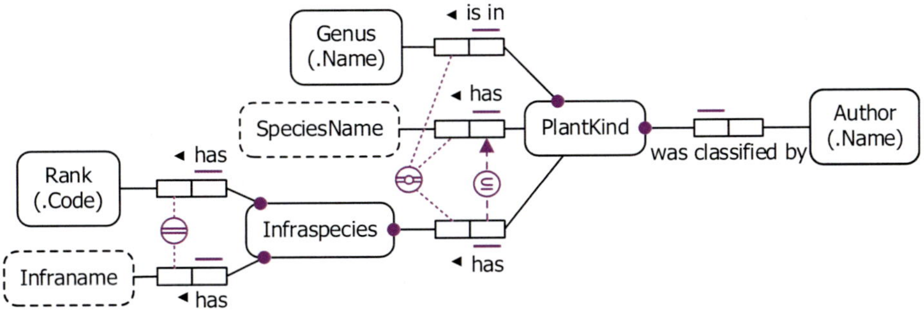

FIG. 5.29 With botanical identification, some reference roles are optional.

Some kinds of plants are identified simply by a genus (e.g., *agrostis*). Many other plant varieties are referenced by a combination of genus and species names (e.g., *acacia interior*). Still others are identified by combining genus, species, and infraspecies, where the infraspecies itself is identified by a rank and inframe (e.g., *Eucalyptus fibrosa* ssp. *nubila*). So, depending on the kind of plant, there may be one, two, or four values required to identify it.

The external uniqueness constraint over three roles indicates that each (genus, species, infraspecies) tuple refers to at most one plant kind. In PlantKind reference scheme, only the genus is mandatory. The "o" in the uniqueness constraint symbol indicates that *outer join semantics* applies (see Section 10.1), so for each genus there is at most one plant kind with only a genus name, and for each genus-species combination there is at most one plant kind with no infraspecies. In a relational database, each plant kind maps to a sequence of four values, three of which may be null, and each quadruple is unique (treating null just like any other value). The other external uniqueness constraint declares that the (rank, inframe) pair identifies infraspecies.

Years ago, when Peter Ritson and one of us formalized disjunctive reference within ORM, we added a percentage sign next to the external uniqueness marker, suggesting that "partial" sequences of components are allowed. Nowadays, the basic external uniqueness symbol is used for both ordinary and disjunctive reference. We know the reference scheme is disjunctive if at least one reference role is optional. The minimum requirement for a legal reference scheme is that it provides a mandatory 1:1 mapping to a sequence of one or more values. So, the disjunction of the reference roles must still be mandatory.

The arrow with a subset symbol running from the first role of Plant has Infraspecies to the first role of Plant has SpeciesName denotes the *subset constraint* that infraspecies is recorded only if species is. Subset constraints are discussed in detail in the next chapter. The reference scheme of Fig. 5.29 is a simplified version of the actual identification scheme used in botany, where complications such as hybrids and cultivars also need to be catered for. Life can be messy!

In practice, *try to avoid using a disjunctive reference scheme for preferred reference.* With the current example, we should consider introducing an alternative simple identifier (e.g., PlantkindNr). If the users agree to use this in their normal communication, this change can be made to the conceptual model. If not, we can still consider using it in the logical design. If we do this, we still need to model the disjunctive information as a "secondary reference" scheme. Although still messy to implement, moving the disjunction from a primary to a secondary reference simplifies and improves the performance for most implementations.

Another way to avoid disjunctive reference is to concatenate the reference components into a single name. Because this makes it difficult to issue queries about individual components, this is unlikely to be viable with our current example.

Another way to avoid disjunctive reference is to use special default values in the identification scheme. For example, if we use a symbol such as "–" or "#" for "does not exist", all the reference roles become mandatory, and this special value is treated like any other value in enforcing uniqueness. Although this approach can work in some cases, it is hardly conceptual and may be impractical if you require the user community to use the special values or if you need the same symbol to work with different data types (e.g., numbers as well as strings).

Sometimes, reference chains can get lengthy. A value is identified by a constant. Each entity is referenced directly using a sequence of one or more objects, which in turn may be values or entities. In the latter case, the reference chain continues, recursively, until finally the referencing objects are all values. In this way, each object is ultimately referenced by one or more values that appear as entries in named columns of output reports or database tables.

Candidate identifiers for the same entity are said to be *synonyms*. Some people use the term "*homonym*" for a label that refers to more than one entity. For example, the same surname "Jones" may refer to more than one person. Because "homonym" has different grammatical senses, another term such as "nonidentifying label" is preferable. At any rate, the "problem of homonyms" is solved either by augmenting the reference scheme until identification is achieved (e.g., combining surname and initials) or by using a completely different preferred identification scheme (e.g., Social Security number).

In our examples so far, an object's reference scheme is the same throughout the whole schema. The next chapter examines some cases involving subtypes where this assumption is removed; this leads to *context dependent reference*. To end this section, we consider the case of variable reference where *different units are used for the same physical quantity* in the same application domain.

In Australia, pieces of lumber have their longest length specified in meters, but their breadth and depth are measured in millimeters. Because this is a standard in the Australian building industry, a report on lumber sizes might look like Table 5.11. Suppose we want to compare the different linear measurements (e.g., Which sizes of lumber are 20 times as long as their breadth?) or compute volumes in standard units (e.g., What is the volume of lumber size C7 in cubic meters?). Fig. 5.30(a) shows one way to model this situation.

Table 5.11 Lumber details.

Lumber size	*Length (m)*	*Breadth (mm)*	*Depth (mm)*
A4	3	100	100
B7	2	100	75
C7	5	200	100

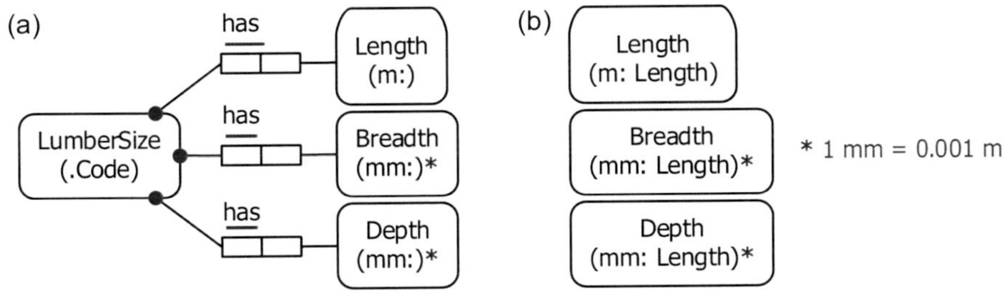

FIG. 5.30 Derived units are based on a standard unit.

The asterisk "*" after "(mm:)" indicates that millimeter (mm) is a *derived unit*, unlike meter (m), which is a *standard unit*. Fig. 5.30(b) displays the expanded form of the unit reference modes, with each unit based on the same *unit dimension* (Length). It also shows the *conversion rule* between the units, with the derived unit on the left of the "=" sign. We use the term "dimension" to include derived dimensions (e.g., Volume = Length^3). Both sides of a conversion rule equation must have the same dimension (e.g., Length) or dimension expression (e.g., Mass/Volume).

Many units might be used for the same physical quantity. To reduce the number of conversion rules, we choose a standard unit and just supply rules to convert to this standard (rather than supply conversions between each unit pair).

ORM 2 regards object types with the same unit-based reference mode to be compatible. So, if Length and Pressure occur in the same model, it would be incorrect to use the reference schemes Length(mm:) and Pressure(mm:) because this implies that Length and Pressure are based on the same semantic domain. To avoid this problem, we need to rename one of the reference modes, for example, Pressure (mmHg:).

As a more abstract approach, we could model the metadata (length, etc.) in Table 5.11 as data, as shown in the populated model in Fig. 5.31. The first fact from Table 5.11 could

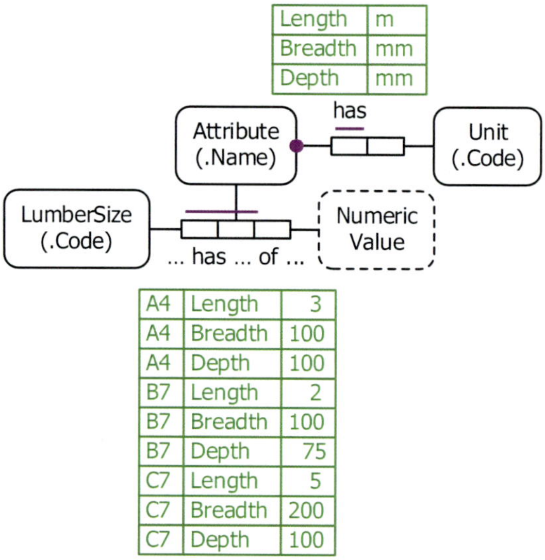

FIG. 5.31 Demoting metadata to data.

then be expressed by the facts: LumberSize "A4" has Attribute "length" of NumericValue 3; Attribute "length" has Unit "m". The unit conversion rule is still required. This technique is called *metadata demotion* because it demotes metadata (data about data) to domain data. While this approach is very powerful, it is often unnatural for most users. If used at all, it should normally be transformed into a simpler external schema for user interaction.

As a simpler example of treating different units as values, a financial application involving many currencies might model amounts of money using the compound reference scheme MoneyAmount(is in Currency(.code), has NumericValue()).

Demoting metadata (types) to data (instances) can help to minimize changes to the schema over time. An abstract model such as Fig. 5.31 might be used to permit new attributes to be added without altering the database schema. For example, a medical application might use the association Patient has Attribute of Value to anticipate new patient tests to be added later. In such cases, additional associations are often needed to control the use of units within their contexts.

Sometimes, we may need to work with multiple unit systems. For example, we may need both metric and nonmetric measures as users gradually move from one unit system to another, or merge separate applications. In addition to supplying conversion rules to transform between different units, we must take care to expose the units to humans interacting with the system (recall the fate of the Mars Climate Orbiter).

Exercise 5.3

1. It is desired to identify a warehouse by its physical location. Design an appropriate identification scheme for the each of the following contexts:

(a) UoD restricted to one suburb. The street in which the warehouse is located is identified by its name.
(b) UoD restricted to one city. Each suburb is identified by its name.
(c) UoD restricted to one country. Each city is a major city, identified by name.
(d) UoD restricted to planet Earth. Countries are identified by name.
(e) UoD restricted to Milky Way galaxy. Planets are identified by name.

2. The UoD is restricted to Earth, and we wish to store facts of the form: Warehouse has Floorspace; Warehouse contains Item in Quantity. Is the identification scheme discussed in Question 1 (d) practical from the implementation point of view? If not, suggest a better scheme and support your scheme by comparing a sample population for the two fact types, using both approaches. Comment on the relative work needed to perform a join (conceptual or relational) between the fact types.

3. (a) A triangle ABC is formed from points in a Cartesian coordinate system. The coordinates are pure numbers (no units). Assume that at any given time, only one shape can be displayed, which must be a triangle. Model the information displayed.

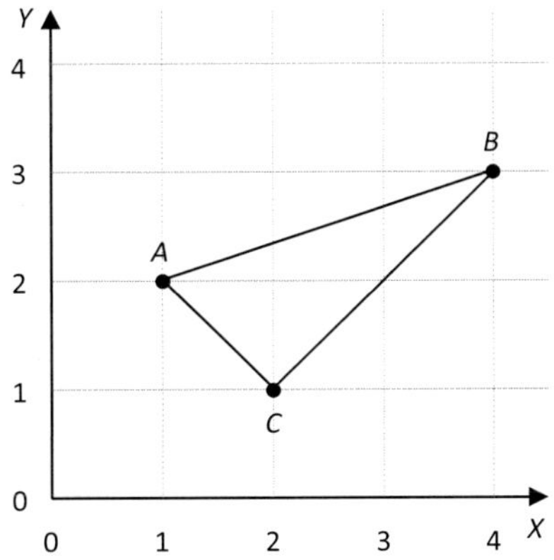

(b) The map shown here indicates the approximate location of parking stations near a famous cathedral. The squares on the grid represent city blocks. Model the parking information stored on this map.

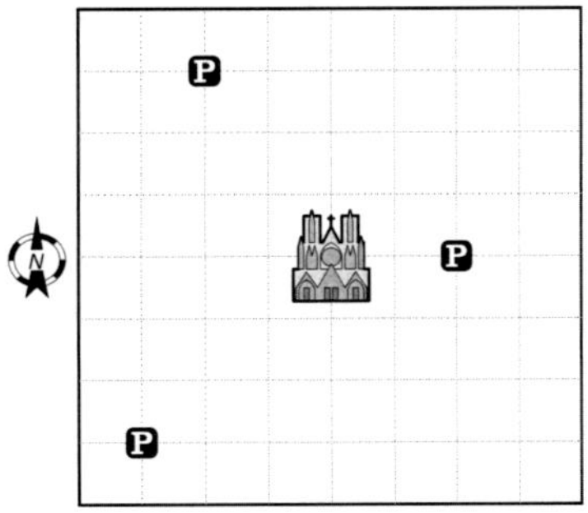

4. Members of a small social club are identified by the combination of their given names and surname. Each member has at least one and at most three given names. For example, one member is Eve Jones, another is Johann Sebastian Bach, and another is Eve Mary Elizabeth Jones. It is required that each component of their names be individually accessible. Draw a conceptual schema diagram for this situation.

5. Members of a small American gymnasium have their weight recorded in pounds (lb). The weight that each member can bench press is also recorded, but in kilograms (kg). It is desired to compare these two weights but retain the separate units. Specify a conceptual schema for this situation. Note that 1 lb = 0.454 kg and 1 kg = 2.205 lb.

6. The following table lists the common names by which beer drinks of various volumes may be ordered in hotels in the states of Australia. Volumes are measured in fluid ounces (oz). The sample data is significant. A double hyphen "–" indicates that beer drinks of that volume are not on sale in that state. For instance, in Queensland exactly three different beer drinks may be ordered. Schematize this UoD.

	4 oz	5 oz	6 oz	7 oz	8 oz	10 oz	15 oz	20 oz
Qld	--	Small beer	--	--	Glass	Pot	--	--
NSW	--	Pony	--	Seven	--	Middy	Schooner	Pint
Vic	--	Pony	Small	Glass	--	Pot	Schooner	--
SA	--	Pony	--	Butcher	--	Schooner	Pint	--
WA	Shetland Pony	Pony	--	Glass	--	Middy	Schooner	Pot
Tas	Small beer	--	Beer or Six	--	Eight	Pot or Ten	--	--
NT	--	--	--	Seven	--	Handle	Schooner	--

7. The following report is from a system that records the driving test results. A person may take as many tests as he/she wishes, but cannot take more than one test on the same day. A minimum score of 80 is required to pass the test. Sometimes a person takes another test simply to try for a better score. Schematize this UoD.

Test	*Date*	*Driver*		*Result*	
		SSN	*Name*	*Score*	*P/F*
101	1/2/00	539-31-1234	Hagar, David Paul	75	F
102	1/2/00	438-12-3456	Jones, Selena Linda	85	P
103	1/9/00	539-31-1234	Hagar, David Paul	80	P

5.4 Case Study: A Compact Disc Retailer

The following small case study is designed to consolidate the main ideas in CSDP steps 1–5. To derive maximum benefit from it, you should attempt to solve the problem yourself before looking at the solution. The business domain concerns a compact disc (CD) retailer who uses an information system to help with account and stock control, and to provide a specialized service to customers seeking information about specific musical compositions and artists.

Fig. 5.32 illustrates the kind of musical information required for compact discs (some of the data is fictitious). Each disc contains several individual musical items, referred to as "tracks". Although compact discs usually have about 20 tracks, for this example only a few tracks are listed.

cd#: 654321-2 **name**: Special Oldies
artist: **record company**: EMI

track#	*title*	*duration*	*singers*
1	Maya's Dance	225	Donovan
2	Wonderful Land	200	
3	King of the Trees	340	Cat Stevens
4	Sultans of Swing	240	Mark Knopfler
5	Seven Wonders	300	Stevie Nicks
...	...	...	...

cd#: 792542-2 **name**: The Other Side of the Mirror
artist: Stevie Nicks **record company**: EMI

track#	*title*	*duration*	*singers*
1	Rooms on Fire	300	Stevie Nicks
2	Two Kinds of Love	250	Bruce Hornsby, Stevie Nicks
...	...	...	...

cd#: 836419-2 **name**: Money for Nothing
artist: Dire Straits **record company**: Phonogram

track#	*title*	*duration*	*singers*
1	Sultans of Swing	346	Mark Knopfler
2	Walk of Life	247	Mark Knopfler
...	...	...	...

cd#: 925838-2 **name**: Fleetwood Mac Greatest Hits
artist: Fleetwood Mac **record company**: Warner Bros

track#	*title*	*duration*	*singers*
1	Say You Love Me	200	Chris McVie, Stevie Nicks
2	Seven Wonders	300	Stevie Nicks
...	...	...	...

FIG. 5.32 Musical details about four compact discs.

Before reading on, try to verbalize information from Fig. 5.32 as elementary facts. Pay particular attention to the reference schemes involved.

Each compact disc has a CD number as its preferred identifier. Although not shown here, different discs may have the same name. Here, "CD" is used in a generic sense, like a catalog stock item. The retailer may have many copies of CD 654321-2 in stock, but for our purposes these are all treated as the same CD. In some applications (e.g., car sales), you do need to distinguish between different copies of the same model, but for our application this is not a requirement.

An artist is a person or a group of persons. For a given CD, a main artist is listed if and only if most of the tracks on the disc are by this artist (this constraint is enforced by a data entry operator, not by the system). Assuming each artist has an identifying name, we use the fact type CompactDisc(CDNr) has main- Artist(.Name); otherwise we must replace Artist by ArtistName, thus CompactDisc(CDNr) has main- ArtistName(). The record company that releases the disc must be recorded.

A challenging aspect of this application is understanding what is meant by "track" and determining its identification scheme. To clarify terms, you often need to consult the domain expert. The term "track" is commonly used in two senses: the *physical track* on a compact disc and the musical composition, or *work*, recorded on this track.

Within the context of a given CD, tracks are identified by their track number, or sequential position on the disc. But there are many CDs in this domain, so we need both the CD number and the track number to identify a track. Using "track" in the physical sense, the fact involving "Maya's Dance" in the first CD report may be verbalized thus:

The Track that is on the CompactDisc with CDNr '654,321–2' and has TrackNr 1
stores a work with the Title 'Maya's Dance'.

This used the compound reference scheme: Track(is on CompactDisc, has TrackNr). Notice that the fact predicate says "*a* work" of that title, not "*the* work". Different musical works may have the same title. Although a physical track occurs on only one CD, the same work may appear on many discs. For example, track 5 of CD 654321-2 and track 2 of CD 925838-2 hold the same title, and probably the same work. There are also two tracks titled "Sultans of Swing", but these have different durations so are not the same work. In rare cases, two tracks with the same title can occur on the same CD (not shown here).

The previous verbalization did not include Work as an entity type. Should we include it? Do we need to know whether a work on one disc is exactly the same as a work on another disc? Only the domain expert can tell us whether he/she wants to know this. For our solution, let's assume the answer to this question is "no".

However, if the answer were "yes", we would need a 1:1 identification scheme for Work. A natural reference scheme for Work is hard to find (even the combination of title, artist, and duration might not be enough). We might use an artificial identifier (e.g., WorkNr), but this might be hard to enforce because humans are responsible for enforcing the constraints in preferred reference schemes.

Table 5.12 Stock quantity and recommended retail price of compact discs.

CD#	*CD name*	*Stock qty*	*RRP*
654321-2	Special Oldies	5	17.95
792542-2	The Other Side of the Mirror	100	25.00
836419-2	Money for Nothing	10	20.00
925838-2	Fleetwood Mac Greatest Hits	50	23.95
...	...	...	...

The duration of a track is the time it takes to play. Each track has exactly one duration, measured in seconds. Most tracks have one or more singers. Some tracks may have no singers (these are instrumental rather than vocal tracks).

Table 5.12 shows an extract from another report from the same application. This lists the quantity in stock and recommended retail price (in US dollars) of each compact disc. This table is updated when required, on a daily basis. The stock quantity of a disc may drop to zero. For simplicity, let's assume that only the current stock and price figures are recorded (no history).

An extract from a third report required of the information system is shown in Table 5.13. For simplicity, let's assume that sales records are kept only for the current calendar year of operation (otherwise, the year needs to be included). For each month that has passed, figures are kept of the quantity of copies sold and the net revenue (profit) accruing from sales of the compact discs in that month.

For simplicity, these figures are restricted to the four compact discs shown earlier. In reality, this table would have thousands of rows and much larger totals. This table is updated on a monthly basis. Daily sales figures are kept manually but are not part of the application.

For each month, if a disc was on the stock list and had no sales for that month, then a figure of zero is recorded for it. If a disc is a new stock item in the month, then it has no sales figures recorded for previous months (e.g., disc 792542-2 was not in stock in January).

Table 5.13 Monthly sales figures (small sample only).

Month	*CD#*	*CD name*	*Qty sold*	*Net revenue*
Jan	654321-2	Special Oldies	0	0.00
	836419-2	Money for Nothing	30	180.00
	925838-2	Fleetwood Mac Greatest Hits	50	400.00
			80	580.00
Feb	654321-2	Special Oldies	5	33.50
	792542-2	The Other Side of the Mirror	70	630.00
	836419-2	Money for Nothing	15	90.00
	925838-2	Fleetwood Mac Greatest Hits	50	350.00
			140	1103.50

This is the last report we need to model. Before reading on, try to verbalize this report and then schematize the whole application. Include all uniqueness and mandatory role constraints, and derivation rules.

How did you fare with the sales report? The reference schemes here are obvious. The totals for quantity sold and net revenue are obviously derived. You may have schematized the stored fact types thus: CompactDisc has CDname(); CompactDisc in Month sold in Quantity; CompactDisc in Month earned Profit. This is acceptable (although as shown later, an equality constraint then needs to be added).

However, it is better to nest the common part of the two ternaries, as shown in Fig. 5.33. Here, the name "Listing" is used for the objectified association that records which compact discs were listed for sale in what months. For each individual listing (of a given compact disc in a given month), the number sold (possibly zero) and profit earned (also possibly zero) are recorded. Apart from a more compact schema, the nested solution leads to a more efficient relational schema (as discussed later).

If desired, the Profit and RetailPrice entity types may be replaced by MoneyAmount (the USD unit is based on the unit dimension Money), and their specific semantics captured in predicate readings (e.g., Listing earned profit of MoneyAmount). The derivation rules are shown in attribute style, using role names.

If desired, the derived fact types defined by these rules may also be displayed, as shown on the right of Fig. 5.33. In this case, the mandatory role and uniqueness constraints on the derived fact types are colored green to indicate they are derivable.

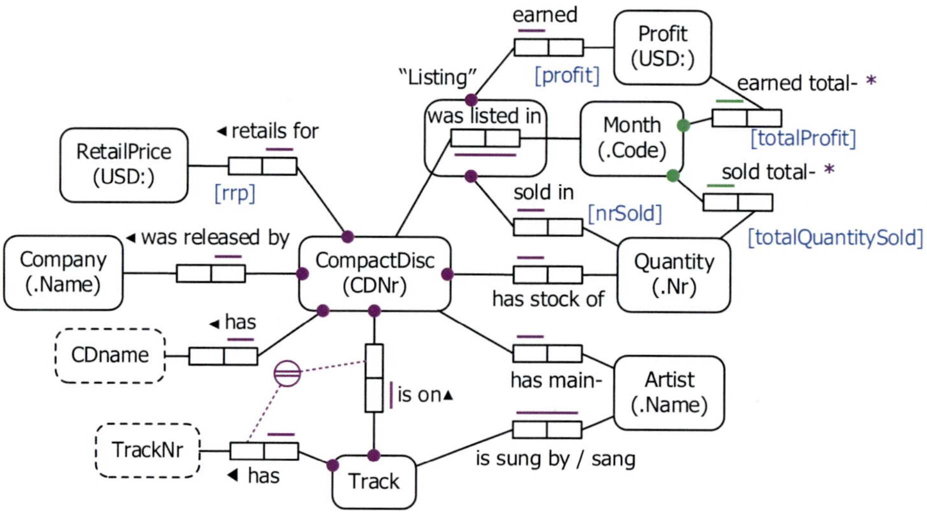

* **For each** Month,
totalQuantitySold = **sum**(listing.nrSold).

* **For each** Month,
totalProfit = **sum**(listing.profit).

FIG. 5.33 One way of schematizing the application.

Two main points arising from this case study are: care is often required in choosing and identifying object types, and sample populations are not always significant. You are now invited to try your hand at a larger schema in the following exercise, which covers all aspects of the CSDP discussed so far.

Exercise 5.4

1. The following reports are extracted from an information system used by an Australian tourism company. The first table provides details about tours offered by the company. Each tour must provide some kind of meal (snack or main meal).

Tour	*Tour Description*	*Snack*	*Main Meal*	*Duration (h)*	*Adult Fare (AUD)*	*Child Fare (AUD)*
1	Gold Coast Grand Tour		✓	11	50	30
2	Brisbane City Tour	✓		3	20	10
3	Rainforest Tour	✓	✓	10	40	22
4	Quick Brisbane Tour	✓		1	5	2
...	...	...	...	...	...	...

All tours are conducted on a daily basis. A tour may depart at more than one time. A given tour departing at a given time is called a "tour shift". The next table shows an extract from the tour shift schedule. Details about a new tour, including its duration, may be recorded before shifts are assigned for that tour (e.g., no shifts have been decided yet for the Rainforest Tour). Each tour shift is offered in at least one language. A "?" mark denotes a null.

Tour Shift	*Language(s) supported*	*Tour*	*Departs*	*Returns*	*Driver*	
					Nr	*Name*
1A	English	1	8:00	19:00	4	Smith, John
2A	English	2	8:30	11:30	1	Smith, John
2B	Japanese	2	10:00	13:00	3	Kano, Fumie
2C	English, Japanese	2	14:00	17:00	1	Smith, John
4A	English, German	4	8:00	9:00	3	Kano, Fumie
4B	English, Japanese, Spanish	4	10:00	11:00	?	?
4C	French	4	1200	13:00	2	Bell, Michelle
...	...	...	...	...	...	...

Details about tour drivers are maintained as shown below. Some drivers (e.g., driver 5) might not yet have been assigned to any tour shifts. A "?" mark denotes a null.

Driver	*Driver Name*	*Driver's Licence*	*Sex*	*Birthdate*	*Language Fluency*
1	Smith, John	17062925	M	1946-02-15	English, Japanese
2	Bell, Michelle	15534110	M	?	French
3	Kano, Fumie	20130456	F	?	English, German, Japanese
4	Smith, John	10033021	M	1965-03-26	English, Japanese, Spanish
5	Jones, Eve	17765310	F	1946-02-15	English, Japanese, Spanish

Some of the tours have a fixed sequence of sites to visit, as shown in the figure. Here the circled numbers are stop numbers, indicating the order in which the tour bus stops at a site. No tour may visit the same site more than once. Sites are primarily identified by a site code (e.g., "MW"), but also have a unique name (e.g., "Movie World"). Although site codes are not displayed in the figure, they are recorded in the information system.

A site may have zero or more facilities of interest. In the figure, these facilities are displayed as icons. In the information system, however, each facility is instead identified primarily by a facility code (e.g., "DN") and has a unique name (e.g., "Dining"). The list of facility codes and names may include facilities not yet provided by any tour site (e.g., Horseback Riding). The code, name, and facilities of other potential tour sites may be recorded even if those sites are not yet included on a tour. For example, we might record "SB" and "South Bank" for Brisbane's magnificent South Bank.

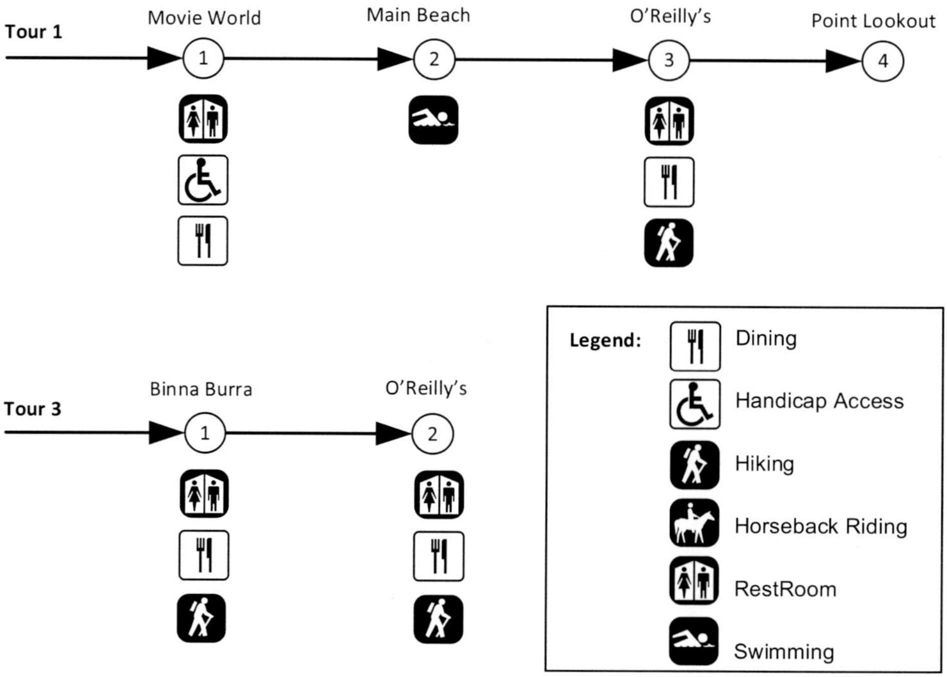

Specify an ORM conceptual schema for this UoD. Include all relevant fact types. Model the information about the sequence of sites at which a tour stops using a ternary fact type. Include all uniqueness constraints (internal or external). Include all mandatory constraints (simple or disjunctive). If a fact type is derived, provide a derivation rule for it. You may include the derived fact type on the diagram, but this is not required. Make no use of nesting. Ignore any other constraints (e.g., a driver cannot drive more than one tour shift at the same time, and drivers on a tour shift that supports a language must be fluent in that language—you'll learn how to specify such constraints in later chapters). Do not include facility icons in the conceptual model (just include their codes and names). Make no use of subtyping (even if you know how to subtype—subtyping is covered in the next chapter).

2. Provide an alternative ORM schema fragment to store data about the sequence of sites a tour visits, using *nesting* instead of a ternary. Exclude fact types for the other reports.

5.5 Logical Derivation Check

Step 3 of the design procedure included a check for basic arithmetic derivations—these are usually obvious (e.g., totals). Once uniqueness and mandatory role constraints are specified, we are in a good position to *check for logical derivations*—these can be harder to spot, especially if some important facts were missed at step 1.

As a simple example, consider the report extract shown in Table 5.14. You may recall this table from the previous chapter, where it was correctly modeled as two binaries. We'll now deliberately model it incorrectly to illustrate the logical derivation check. Suppose we verbalize the first row as the two facts: Lecturer "Halpin" works for Department "CS"; Lecturer "Halpin' works in Building "69". Assuming the table is significant with respect to uniqueness constraints and mandatory roles, this leads to the schema shown in Fig. 5.34.

Table 5.14 University data.

Lecturer	*Department*	*Building*
Halpin	CS	69
Okimura	JA	1
Orlowska	CS	69
Wang	CN	1

To begin our logical derivation check, we ask, *Are there any other associations of interest between the object types, especially functional ones?* A binary association is *functional* if at least one of its roles has a simple uniqueness constraint. Each column entry for this role functionally determines the entry for the other role. Hence, a role with a simple uniqueness constraint is said to be a *functional role.*

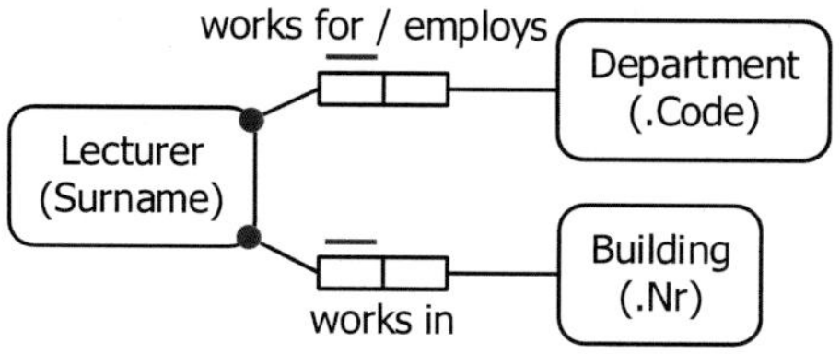

FIG. 5.34 A first draft schema for Table 5.14. Is there any other association of interest?

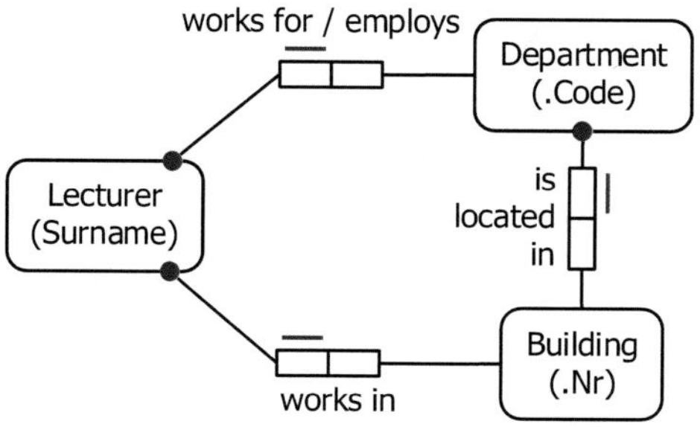

FIG. 5.35 A second draft schema for Table 5.14. Which fact type is derivable?

Looking at Fig. 5.34, we notice that the following fact type is also of interest: Department is located in Building. Suppose we interpret Table 5.14 as also giving us this information. For example, the first row of the table tells us that the computer science department is located in building 69. Is the new fact type functional?

From rows 2 and 4 we see that the same building may house more than one department. However, the population suggests that each department is located in only one building. Suppose the domain expert verifies that this is the case. We might now add this fact type to our schema to obtain Fig. 5.35.

The optional roles of Department and Building in Fig. 5.35 allow for other possibilities in the global schema (departments without lecturers, or buildings without departments) not covered by Table 5.14. We now ask ourselves the question, *Can any fact type be derived from the others?*

Looking at the three binaries in Fig. 5.35, you would probably suspect that one can be derived from the other two. Is this suspicion justified? If so, which binary should be derived? Is there a choice? What do you think?

Suppose we decide to make the new fact type about department location derivable, adding the following logical rule to our *derivation rule* section:

Department is located in Building **iff** Department employs **some** Lecturer
who works in **that** Building

This treats the department location fact type as a projection on Department and Building from the compound fact type that is the schema path from Department through Lecturer to Building. Suppose the domain expert agrees that this rule does apply. Is it now okay to make the department location derived? No! Why not? Recall the following design guideline: *by default, all constraints on a derived fact type are derivable.* Is this true with our choice of derived fact type?

With our example, the uniqueness constraint on the fact type Department is located in Building is not derivable. How can we know this? One way of seeing this is to provide a *counterexample*—in this case, a sample population that satisfies all constraints except the uniqueness constraint that each department is located in at most one building.

One such population is shown in Fig. 5.36. If we populate the horizontal binaries as shown, our derivation rule generates a population for the vertical binary that locates the computer science department in two buildings.

If we derive the vertical binary, we must specify the uniqueness on it as an additional nonderived constraint to be enforced separately—that is, when one of the two horizontal binaries is updated, we need to check that all lecturers working for the same department work in the same building.

Sometimes a design decision like this is made to speed up queries, even though updates become more expensive. However, the default design guideline should usually be followed because this simplifies updates.

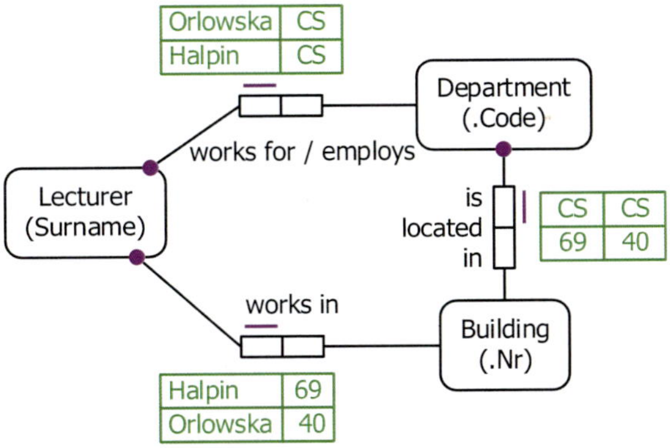

FIG. 5.36 The uniqueness constraint on the vertical binary is violated.

Another problem with deriving the department location fact type is that its mandatory role constraint can't be derived unless we add a mandatory constraint on the employs role. But perhaps some departments have no lecturers. For example, if the public relations department does not employ any lecturers, its location cannot be derived. To satisfy our default guideline, the only possible choice of derived fact type is Lecturer works in Building (see Fig. 5.37).

The derived fact type is included on the diagram for discussion purposes, but would usually be omitted, with just the derivation rule being stated. If the domain expert agrees with the derivation rule shown here, we may derive Lecturer works in Building because its constraints are implied by constraints on the asserted fact types. Because each lecturer works for exactly one department, which is located in exactly one building, the derivation

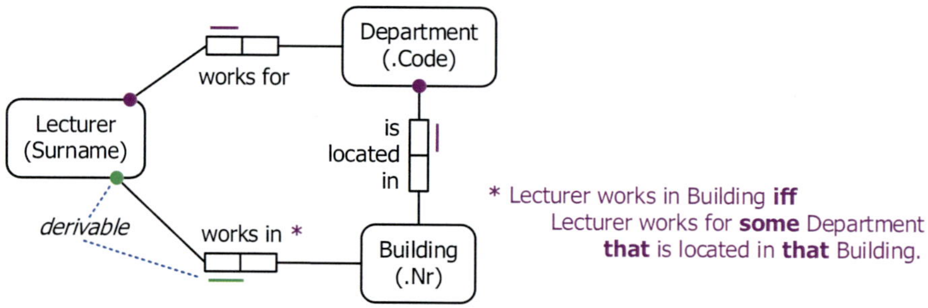

FIG. 5.37 If the rule applies, the constraints on the derived fact type are derivable.

rule implies that each lecturer works in exactly one building. So, both the uniqueness and mandatory role constraints on the derived fact type are implied, as shown by their green color.

In cases like this, we say the uniqueness constraint is *transitively implied*. You are probably familiar with the notion of transitive relations, where if the relation holds between each consecutive pair in a sequence, it holds between the first and last member of the sequence (e.g., if $x > y$ and $y > z$ then $x > z$).

Similarly, functional determinacy is transitive: *if a chain of binaries has a uniqueness constraint on the first role of each, then the binary from start to end projected from their inner join has an implied uniqueness constraint on its first role.*

In Fig. 5.38, the schema is populated with the sample facts from Table 5.14 as well as the fact that the public relations (PR) department is located in building 2. A chain of two functional binaries goes from Lecturer through Department to Building, and the derivation rule defines the [lecturer, building] projection on the conceptual join of these two binaries.

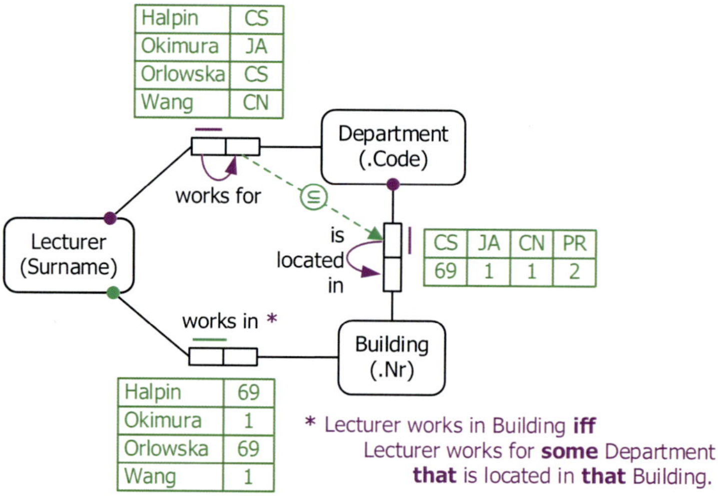

FIG. 5.38 The derived association is projected from a functional chain.

For exposition purposes, the FDs corresponding to the uniqueness constraints on the base fact types are displayed. In such cases, we may use the uniqueness constraint pattern (or FD chain) to help choose the derived fact type.

To ensure that no lecturers are lost along the chain, we also require that their departments have a location. So, each department playing the employs role must also play the role of having a location. This requirement is an example of a *subset constraint* and is depicted in Fig. 5.38 as a dotted arrow with a subset symbol from the incoming role to the outgoing role of Department.

This subset constraint is implied by the mandatory constraint on Department, so is colored green. In combination with the base mandatory constraint on Lecturer, this subset constraint implies the mandatory constraint on the derived association. Subset constraints are discussed in detail in the next chapter.

We always need to check with the domain expert whether our derivation rule is semantically correct. The mere existence of such a constraint pattern doesn't guarantee that the other fact type is derived. There are infinitely many associations that could be specified between Lecturer and Building. The other fact type cannot be derived from the work and location fact types unless within the UoD it is logically equivalent to the [lecturer, building] projection on their join path.

We still need to ensure that the following equivalence really does apply: Lecturer works in Building **iff** Lecturer works for **some** Department **that** is located in **that** Building. Is it possible for Okimura, say, who works in the Japanese department, located in building 1, to work instead in another building (e.g., building 5)? If the answer is "yes", then we have a counterexample to the equivalence, so we cannot derive the association Lecturer works in Building. If no such violation of the derivation rule is possible, then the association is derivable.

To clarify this point, note that the same constraint pattern may occur if we replace the derived fact type in Fig. 5.38 by Lecturer has lunch in Building. In most academic domains, it is common for some lecturers to have their lunch in a building different from their work building. In such a UoD, the lunch fact type is *not* derivable because the following derivation rule does not apply: Lecturer has lunch in Building **iff** Lecturer works for **some** Department **that** is located in **that** Building.

As an aside, the derivation rule in Fig. 5.38 could be replaced by a "pair-equality" constraint between the "works in" predicate and the join-path projection formed by connecting the outer roles of Department's predicates. Such "join constraints" are discussed in Chapter 10.

Sometimes, fact types may be derived even if they do not have a mandatory role. For example, suppose some lecturers (e.g., guest lecturers) don't work for a department and don't work in a building. As long as the derivation rule in Fig. 5.38 still applies, we may still derive the fact type Lecturer works in Building. The only difference is that the two mandatory role constraints on Lecturer are removed. Of course, in the global schema Lecturer would normally play some other mandatory role, but this is irrelevant to the derivation.

Now consider the case where some lecturers don't work for a department, but all lecturers do work in exactly one building. For example, say Zeno is a visiting lecturer not

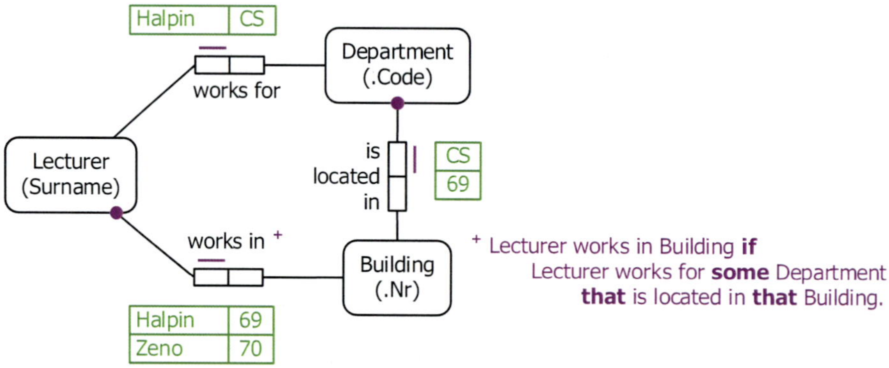

FIG. 5.39 A semiderived fact type (not all lecturers work for a department).

assigned to any department, but he works in building 70. We may handle this case with a *semiderived fact type* (a fact type that is only *partly derived*), weakening the main operator of the derivation rule from "**iff**" to "**if**".

Semiderived fact types have their predicates marked with a "+" superscript (intuitively, half a derivation asterisk). This sign is also displayed with the derivation rule, as shown in Fig. 5.39, which includes a sample schema population. This approach allows us to enter building facts directly for those lecturers without a department. The constraints on this semiderived fact type are asserted (they are no longer implied). Conceptually, semiderived fact types provide a convenient and compact way of dealing with cases where the derivation rule is one way only ("**if**"), rather than an "**iff**" equivalence.

One way to implement semiderived fact types is to replace them by *a disjunction of asserted and fully derived predicates*. For example, to cater for lecturers without a department, we use the asserted fact type Lecturer works individually in Building. For the lecturers with a department, we use the derivation rule Lecturer works departmentally in Building **iff** Lecturer works for **some** Department **that** is located in **that** Building. The overall derivation may be specified as Lecturer works in Building **iff** Lecturer works individually in **that** Building **or** works departmentally in **that** Building.

As another example often cited in logic programming, suppose our knowledge of parenthood facts is incomplete, so that we can know a person is a grandparent of somebody without knowing who the intermediate parents are. This case can be handled with an asserted parenthood fact type and a semiderived grandparenthood fact type, using the rule Person_1 is a grandparent of Person_2 **if** Person_1 is a parent of **some** Person_3 **who** is a parent of Person_2.

Semiderived fact types are also useful in cases where a constraint governs three or more quantities in an equation, and one has a choice as to which quantity will be derived. For example, in a loan application we might want to enter any two of rate, loan term, and loan amount to see what the other figure would be.

Sometimes we encounter derived associations with neither a mandatory nor a functional role. Consider the schema shown in Fig. 5.40. This has two optional, many:many

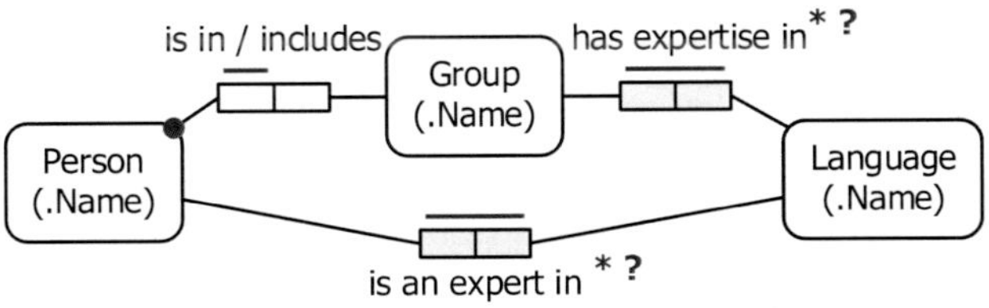

FIG. 5.40 Which (if any) of the *m:n* associations is derivable?

associations. Can you see any logical derivation possibilities? This would be easier to deal with if we had an output report or sample data. At any rate, which (if any) of the associations do you consider derivable?

Knowledge about each group member is more precise than knowledge about the group as a whole. So, if we are interested in the language expertise of each person, we should store this and derive the group expertise from the rule Group has expertise in Language **iff** Group includes **some** Person **who** is an expert in **that** Language. This approach requires a closed world approach to the expert role played by Person (if a person is an expert in a language, we must record it). Chapter 10 discusses this notion of "relative closure" in more detail.

Obviously, we cannot derive the expertise of a person from that of his/her group. But what if we are not interested in each person's expertise? In this case, the fact type Person is an expert in Language should be deleted, and the remaining binaries would both be asserted fact types.

Exercise 5.5

1. In a computer company, workers are standardly identified by their initials, but also have a unique name. Each worker has access to exactly one personal computer (PC). Each PC is accessed by at some worker. For each PC, we record its room location, the worker(s) who access it, and the language(s), if any, installed on it, as shown.

PC	*Room*	*Workers with access*		*Languages installed*
pc01	507	EFC	(Ed Codfish)	Pascal, Prolog, SQL
		TAH	(Terry Happy)	
pc02	507	NW	(Nancy Wirth)	Pascal, Modula-2
pc03	618	PAB	(Paul Boles)	Hope, Miranda
		JC	(Joan Coffee)	
pc04	508	IN	(Ima Newie)	
pc05	508	PNC	(Peter Crusoe)	C#, SQL
...	...	...	...	...

Each computer language is one of three types (declarative, functional, or procedural) and is installed on a PC *or* has an expert (or both). The PC a worker accesses must be in the room in which he/she works. The next table provides a full record of the languages, their types, who are expert at each, and each expert's room.

Language	*Type*	*Experts (rooms)*
C#	procedural	PNC (508), REK (611)
Hope	functional	?
Java	procedural	JC (618)
Modula-2	procedural	NW (507)
Miranda	functional	PAB (618), DC (708)
Pascal	procedural	NW (507), TAH (507)
Prolog	declarative	JS (407)
SQL	declarative	EFC (507), PNC (508), TAH (507)

A workshop on computer languages is to be delivered. The full workshop program, shown in the final table, indicates how many hours (h) speakers talk about each language, and the total hours for each language type. Schematize this UoD, including uniqueness and mandatory role constraints and derivation.

Declarative (6 h)		*Functional (4 h)*		*Procedural (6 h)*	
Prolog:	JS (3 h)	Hope:	PAB (1 h)	Modula-2:	NW (3 h)
SQL:	PNC (1 h)	Miranda:	PAB (3 h)	Pascal:	NW (2 h)
	TAH (2 h)				TAH (1 h)

2. The following report refers to a UoD like that discussed in this section, except that departments may be located in many buildings. Each lecturer works in exactly one building. Some departments employ no lecturers. Schematize this UoD using three binaries, and note which if any is derivable. Specify any derivation rules.

Lecturer	*Dept*	*Building*
Halpin	CS	69
Okimura	JA	1
Orlowska	CS	69
Wang	CN	1
Yamamoto	JA	3
?	PR	2

3. Consider a UoD as in Question 2, except that each department employs a lecturer, so the final row of the table in Question 2 is illegal. Schematize this UoD including derivation.

5.6 Summary

For a given state of the database, the *population* of a role *r*, *pop(r)*, is the set of objects referenced as playing that role in that state. $A \cup B$, the *union* of sets *A* and *B*, is the set of all elements in *A*, *B*, or both. The *null set* or empty set, { }, has no members.

The population of an entity type *A*, *pop(A)*, is the union of the populations of its roles. Roles of an elementary fact type are called *fact roles.*

A relationship type used purely to identify or reference some object is a reference type; its roles are called *reference roles.* For a given state, the population of a fact type (stored or derived) is its set of fact instances.

Step 5 of the conceptual schema design procedure requires us to *add mandatory role constraints and to check for logical derivations.* A role *r* is *mandatory* (or *total*) for an object type *A* **iff** each member of pop(*A*) is known to play *r* (for each state of the knowledge base); otherwise, the role is *optional.* A mandatory role is indicated by a large dot where the role connects to the object type. The dot may instead be placed at the role end.

By default, if a primitive entity type plays only one fact role in the global schema, this role is mandatory—in this case, the dot may be omitted because it is implied. Schema constraints apply to the model, not necessarily to the real world.

If a role is optional in the real world, it is optional in the model. However, a role that is mandatory in the real world may be optional in the model (e.g., the information may be unknown or omitted for privacy).

A *disjunction* of roles r_1, ..., r_n is *mandatory* for an object type *A* **iff** each member of pop(*A*) is known to play at least one of these roles (in each state). This *inclusive-or,* or *disjunctive mandatory,* constraint is shown by a circled dot ⊙ connected by a dotted line to the roles.

By default, the disjunction of fact roles played by a primitive entity type in the global schema is mandatory—in this case, the dot may be omitted because it is implied. However, a simple or disjunctive mandatory role constraint should be shown explicitly if it applies to a subtype (see next chapter).

To simplify the depiction of an object type with many roles, the object type may be *duplicated* on a schema page as well as over several pages. In this case, the rule for implicit mandatory role disjunctions applies to the union of all the duplicates. Different computer-aided software engineering (CASE) tools use different notations (e.g., shading for duplication).

Each entity type must have one or more candidate reference schemes. One of these is picked as the *preferred identification* scheme, and the others are treated as elementary fact types. Preferred reference schemes (also called primary identifiers) should be as simple and stable as possible. A *rigid identifier* identifies the same object throughout the application lifetime. The preferred identification must provide a mandatory 1:1 map of each entity to a tuple of one or more values.

A *simple reference* scheme maps each entity to a single value (e.g., Subject to Subject-Code). A *compound reference* scheme maps each entity type to two or more object types (e.g., Room to Building and RoomNr). Apart from nested cases, the identification aspect of compound reference is denoted by an external uniqueness constraint, depicted as a circled double uniqueness bar "⊜". With *disjunctive reference*, the number of values may vary for different entities of the same type (e.g., Person to Surname and Firstname and optionally Middlename).

Candidate identifiers for the same entity are called *synonyms*. When more than one candidate identification scheme exists, the preferred reference scheme is indicated on the schema as follows: if simple, use a parenthesized reference mode; if compound use "⊜". If a reference mode is expanded to an explicit binary association, the identifying role is marked with a double uniqueness bar.

If *different units* are used for the same quantity, the semantic connection between these units must be modeled (e.g., by declaring the units to be of the same *dimension*). Different units may be modeled as reference modes or by use of a Unit object type. For any given physical quantity, one unit is often picked as the standard unit; conversion rules are supplied to map the derived units to the standard unit.

Derived fact types may be omitted from the diagram, but their derivation rule must be specified. If a derived fact type is included on the diagram, all its constraints should normally be shown. *Usually, all constraints shown on a derived fact type are derivable* (from its derivation rule, and other constraints).

Once mandatory role constraints are added, a *logical derivation check* should be performed to see if some fact types are derivable using logical rather than arithmetic operations. First, check for missing fact types by asking, *Are there any other relationships of interest* between the object types, especially functional relationships? A binary relationship type is *functional* if at least one of its roles is functional (i.e., the role is a simple key). Each column entry for a functional role functionally determines the entry for the other role.

We now ask, *Can any fact type be derived* from the others? To help decide this, remember that constraints on a derived fact type should normally be derivable. If we have a chain of two or more functional fact types with uniqueness constraints on all the first roles, then a functional binary from the start to the end of this chain is derivable if it is defined by projecting on the join of these fact types–its uniqueness constraint is *transitively implied*. In this case, the first role of the derived fact type is mandatory iff the first role of the chain is mandatory and the second role of each binary subsets the first role of the next binary in the chain (e.g., if the first role of each binary in the chain is mandatory).

Derivation rules should normally be biconditionals (i.e., their main operator is "**iff**"). If their main operator is "**if**", the fact type is *semiderived*. A fact type is semiderived iff for some state of the business domain, it is possible that some of its instances are derived from other facts and some of its instances are simply asserted. Semiderived fact types are indicated by a plus "+"after their predicate reading(s).

Derived fact types do not need to have simple keys. Whether derived or not, a fact type should be excluded from the schema unless it is of interest for the application.

Chapter Notes

Some versions of ORM use a universal quantifier (∀) instead of the mandatory role dot, and place this along the role connector instead of at one end (e.g., De Troyer et al., 1988). As discussed in later chapters, entity-relationship modeling (ER) and unified modeling language (UML) support simple mandatory role constraints for most situations, but not disjunctive mandatory role constraints. Some versions of ER use a solid line for mandatory and a broken one for optional (e.g., Barker, 1990a, b), some use double lines for mandatory (e.g., Elmasri and Navathe, 1994), and some, including UML, use cardinality or multiplicity markers such as 0 for optional and 1 (or more) for mandatory. UML allows attributes to be declared mandatory or optional, but not all ER notations allow this.

The botanical example used in the discussion of disjunctive reference schemes came from Peter Ritson, who worked with one of us to develop a method for modeling such schemes. For a technical discussion of this topic, see Halpin and Ritson (1992).

6

Value, Set-Comparison, and Subtype Constraints

Chapter Outline

6.1 Introduction to CSDP Step 6

So far, we have learned how to verbalize examples in terms of elementary facts, draw the fact types, mark uniqueness constraints and mandatory roles, specify rules for derived fact types, and use simple and complex reference schemes to identify entities. The next step of the conceptual schema design procedure (CSDP) covers three kinds of constraints: value, set comparison, and subtype. Set-comparison constraints are themselves of three kinds: subset, equality, and exclusion.

CSDP step 6: Add value, subset, equality, exclusion, and subtype constraints.

This chapter covers step 6 in detail. To clarify the formal concepts underlying the constraints, some basic set theory is first reviewed. Then, we consider value constraints (i.e., restrictions on value types). For example, in modeling color monitors we might restrict the values of ColorCode to "R", "G", and "B" (for Red, Green, and Blue, respectively). After discussing the related notion of independent objects, we examine the three set-comparison constraints—these declare whether the population of one role sequence must be included in, be equal to, or be mutually exclusive with the population of another.

After that, we examine the notion of subtyping in some depth. Subtyping allows us to declare how types are related (e.g., each manager is an employee) and refines our ability

Information Modeling and Relational Databases. https://doi.org/10.1016/B978-0-443-23790-4.00002-5

to declare precisely what kinds of objects play what roles. Our treatment of subtyping proceeds from basics through to some reasonably advanced aspects.

6.2 Basic Set Theory

Because the constraints in step 6 make substantial use of sets and set operations, let's first review some basic set theory. To provide a comprehensive summary of the required background, some ideas met earlier are included.

Intuitively, a *set* is a well-defined collection of items. The items may be concrete (e.g., people, computers) or abstract (e.g., numbers, points) and are called *elements* or *members* of the set. Sets themselves are abstract. They are numerically definite in the sense that each has a definite number of elements. A *type* is a set of possible items, understood as sharing some common characteristics (possibly unstated).

The members of a set collectively constitute the *extension* of the set. While a set may *contain* members, it does not *consist* of those members. For example, the set of Martian moons is an abstraction over and above its members (Phobos and Deimos), so it has no physical properties such as mass or volume. Although sets (unlike heaps) are not to be equated with their members, they are determined by their members because two sets are *identical* or **equal** just in case they have the same extension. Using "iff" for "if and only if", the *Law of Extensionality* may be stated thus:

Given any sets *A* and *B*, *A* = *B* iff *A* and *B* have the same members.

Because sets are determined by their members, one simple way of *defining* a set is to *enumerate* or list the elements of the set. In so doing, braces are used as delimiters and commas as item separators (e.g., $A = \{3, 6\}$). Here, *A* is defined as the set containing just the elements 3 and 6.

One consequence of the Law of Extensionality is that a set is unchanged by *repeating* any of its members. For instance, if $A = \{3, 6\}$ and $B = \{3, 6, 6\}$, it follows that $A = B$ because both sets contain precisely the same members (3 and 6). The *cardinality* of a set *A*, written #*A*, is the number of different elements in *A*. For example, $\#\{3, 6\} = \#\{3, 6, 6\} = 2$.

When enumerating sets, it is usual not to repeat the members. However, it is sometimes useful to permit this. For example, when stating general results about the set variable $\{x, y\}$, it may help to include the case $x = y$. Sometimes repetition of members occurs undetected. For instance, some people do not realize that the entity set {Morning Star, Evening Star} contains just one member (the planet Venus). Of course, the value set {"Morning Star", "Evening Star"} contains two members.

If repetition is made significant, we have a *bag* or *multiset*. To distinguish between bags and sets, we use different delimiters. We use square brackets for bags and braces for sets. The cardinality of a bag equals its number of elements, including duplicates. For instance, $\#[3, 6] = 2$, but $\#[3, 6, 6] = 3$. One use of bags is in collecting values for statistical work. For example, the set {3, 6, 6} has an average of 4.5 but the bag [3, 6, 6] has an average of 5. Bags are frequently used with languages such as SQL.

Another consequence of the Law of Extensionality is that the *order* in which elements are listed is irrelevant. For example, if $A = \{3, 6\}$ and $B = \{6, 3\}$ then $A = B$ because each set contains the same members. Bags are also insensitive to order, as in $[3, 6] = [6, 3]$. If order is made significant, we have an "ordered set".

Usually, when order is made significant, so is repetition. In this case, we have a *sequence* (also called a list, permutation, or tuple). Thus, a sequence is an ordered bag. Sequences are often delimited by parentheses or angle brackets, for example, (1,2) or $\langle 1, 2\rangle$. The sequence (6, 3, 6) has three members and differs from the sequence (3, 6, 6). The cardinality of a sequence is its member count, including duplicates. For example, $\#(3, 6, 6) = 3$. In practice, several different delimiting notations are used. For example, "[", "]" are used as set delimiters in Pascal and list delimiters in Prolog.

If a set is enumerated in full, the ordering does not matter. However, a natural ordering often provides an obvious pattern; in such cases, a partial enumeration can define the set. For example, the set of decimal digits may be shown as {0..9}, where the ".." indicates the missing digits. Here, 0..9 is often called a *range*. Infinite sets may be represented with ".." at one or both ends. For example, the set of natural numbers may be shown as {1, 2, 3..}.

The preceding set definitions enumerate, wholly or partially, the extension of the set and are thus examples of an *extensional definition*. A set may also be defined by an *intensional definition*. Here, the *intension* or meaning is declared using a property that applies to just the members of the set. That is, a description is given that constitutes both a necessary and a sufficient condition (an "iff condition") for an item to belong to the set.

For example, the set *A*, defined extensionally as {1,2,3}, may be defined intensionally as $A =$ the set of natural numbers less than 4. This definition may be recast in set builder notation as {*x*: *x* is a natural number less then 4} or more briefly as $\{x: x \in \mathbb{N} \;\&\; x < 4\}$, where "$\in$" abbreviates "is a member of" and $\mathbb{N}$ is the set of natural numbers. In set builder notation, a stroke "|" may be used instead of a colon, for example, $\{x|\ x < 4\}$.

Some set operations result in propositions while others result in sets. Note the following *proposition-forming operators*: $=$, $\neq$, $\subseteq$, $\subset$, $\supseteq$, and $\supset$. These are read respectively as "equals", "is not equal to", "is a subset of", "is a proper subset of", "is a superset of", and "is a proper superset of".

Given any sets *A* and *B*, we say that *A* is a **subset** of *B* iff every member of *A* is also a member of *B*. For example, $\{1,3\} \subseteq \{1, 2, 3\}$. An equivalent definition is *A* is a subset of *B* iff *A* has no members that are not in *B*. This second definition makes it easy to see that the *null set* is a subset of every set. The null or empty set has no members and may be represented as {} or $\varnothing$.

Every set is a subset of itself. For example, $\{1, 3\} \subseteq \{1, 3\}$. We say that *A* is a *proper subset* of *B* if and only if *A* is a subset of *B* but not equal to *B*. For instance, $\{1, 3\} \subset \{1, 2, 3\}$. We say that *A* is a **superset** of *B* iff *B* is a subset of *A*, and that *A* is a *proper superset* of *B* iff *B* is a proper subset of *A*. For example, {1, 2, 3} is both a superset and a proper superset of {1, 3}.

Comparison relationships between two sets are often depicted by *Euler diagrams*. As developed by the Swiss mathematician Leonhard Euler (1707–1783 CE), these were *spatial* and *existential*. Each set is pictured as a set of points inside an ellipse. This enables the relationship between the sets to be "seen" by the spatial arrangement of the ellipses. For example, placing ellipse *A* inside *B* shows that *A* is a proper subset of *B* (see Fig. 6.1). Here, we see that every element of *A* is also an element of *B* (so *A* is a subset of *B*). The existential viewpoint implies that each of the regions in the Euler diagram contains some elements. So, *B* has some elements not in *A*. Hence *A* is a proper subset of *B*.

To show that *A* is a subset of *B* on standard Euler diagrams, we use a disjunction of two diagrams, as in Fig. 6.2. The right-hand diagram caters for the case $A=B$.

The notion of subsethood is more useful than proper subsethood. Partly to depict such relationships on a single diagram, we use *hypothetical Euler diagrams* (HEDs). In an HED, an asterisk is placed in a region to show something exists there while shading the region indicates that it is empty. If a region is unmarked, the question of whether any elements exist there is left open or hypothetical. Fig. 6.3 shows the HEDs for the six most important comparisons between two sets.

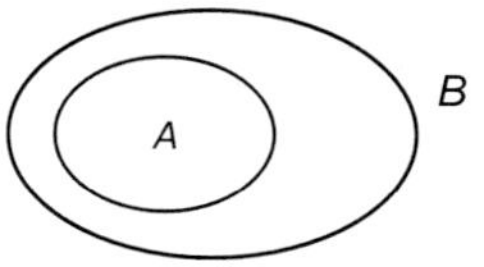

FIG. 6.1 Standard Euler diagram for *A* is proper subset of *B*.

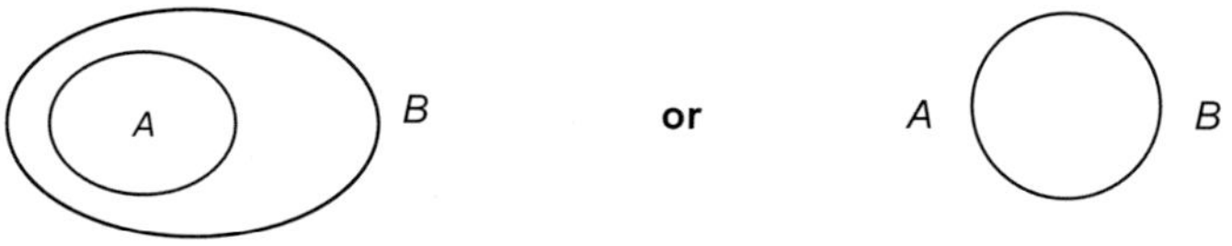

FIG. 6.2 A disjunction of standard Euler diagrams denotes A is a subset of B.

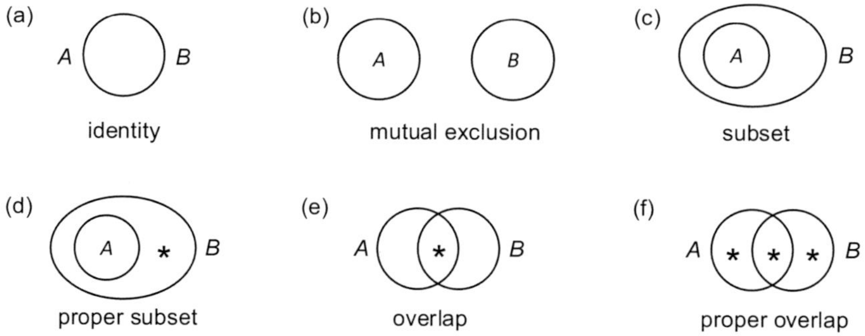

FIG. 6.3 Hypothetical Euler diagrams for set comparisons.

Fig. 6.3(a) shows *equality* or *identity* (e.g., $A = B = \{1,2\}$). In Fig. 6.3(b), A and B are *disjoint* or *mutually exclusive*; that is, they have no members in common (e.g., $A = \{1\}$, $B = \{2\}$). In Fig. 6.3(c), A is a *subset* of B (equivalently, B is a superset of A). In Fig. 6.3(d), A is a *proper subset* of B (so B is a proper superset of A). For example both $A = \{1\}$, $B = \{1, 2\}$ and $A = \{1\}$, $B = \{1\}$ are instances of subsethood, but only the former is an instance of proper subsethood. Fig. 6.3(e) shows *overlap*: the sets have some members in common. Fig. 6.3(f) shows *proper overlap*: the sets have common as well as extra members. For example, both $A = \{1, 2\}$, $B = \{2,3\}$ and $A = \{1\}$, $B = \{1, 2\}$ are instances of overlap, but only the former is a case of proper overlap.

Let's now consider *set-forming operations*, where the result is a set. Given any sets A and B, we define $A \cup B$ (i.e., *A union B*) to be the set of all elements in A or B, reading "or" in the inclusive sense. $A \cap B$ (i.e., *A intersect B*) is the set of all elements common to both A and B. Each of these operations is commutative, so the order of the operands doesn't matter. That is, $A \cup B = B \cup A$ and $A \cap B = B \cap A$.

The *set difference* (or relative complement) operation is defined thus: $A - B$ (i.e., A minus B) is the set of all elements that are in A but not in B. This operation does not commute (i.e., cases may arise where $A - B \neq B - A$). If we let U = the *universal set* (i.e., the set of all elements under consideration), we define the *complement* of A as $A' = U - A$. The *symmetric difference* between A and B is the set of elements in just one of A or B (i.e., the union minus the intersection).

The three most important set-forming operations are union, intersection, and difference. These are depicted in Fig. 6.4 by means of *Venn diagrams*, named after their inventor, the English logician John Venn (1834–1923 CE). Here, shading indicates the result of the operation.

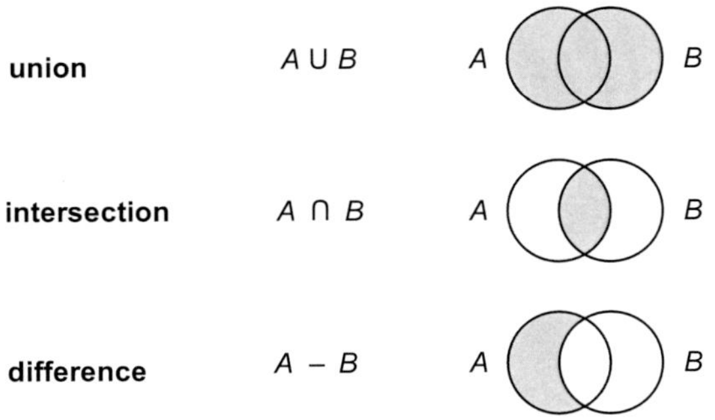

FIG. 6.4 Venn diagrams for three set-forming operations.

Unlike in Euler diagrams, Venn diagram ellipses always overlap. Like HEDs, Venn diagrams adopt the hypothetical viewpoint. As examples of these operations, if $A = \{1, 2\}$ and $B = \{2, 4\}$, then $A \cup B = \{1, 2, 4\}$, $A \cap B = \{2\}$, $A - B = \{1\}$, and $B - A = \{4\}$.

Venn diagrams are sometimes used to discuss comparisons between two sets. For example, Fig. 6.5(a) indicates that A ⊂ B, using line fill and "*", respectively, for empty and nonempty regions. However, Venn diagrams become extremely unwieldy as soon as the number of sets exceeds three—consider the Venn diagram for four sets in Fig. 6.5(b). Euler diagrams have a similar scalability problem. Section 6.5 introduces a directed graph notation for diagramming subtype connections that enables many compatible object types to be related without incurring the jumble of line crossings that Euler and Venn diagrams produce for such cases.

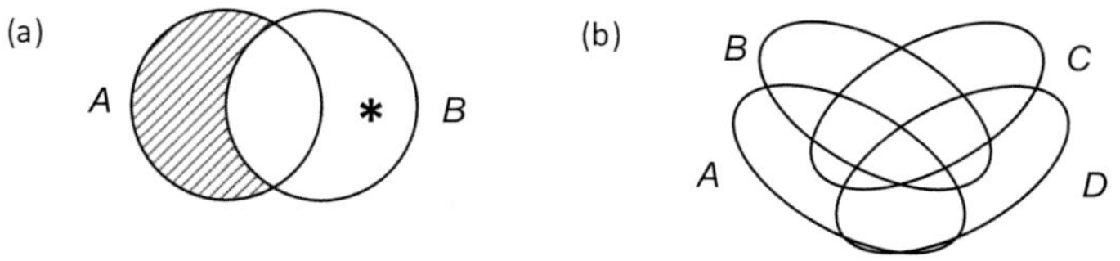

FIG. 6.5 Venn diagrams for (a) A is a proper subset of B, and (b) four sets.

6.3 Value Constraints and Independent Types

A **value constraint** indicates which values are allowed in a value type or role and is sometimes called a "domain constraint". A value constraint should be declared only if the value list is at least reasonably *stable*; otherwise, we would need to continually change the constraint, and hence the schema, as the value list changed.

A value type may be defined by declaring its extension (set of possible values) as one or more enumerations or ranges enclosed in braces (curly brackets). On a schema diagram, an *object type value constraint* is declared by displaying the extension next to either the value type itself or the entity type that is referenced by the value type.

Fig. 6.6 shows several examples. If we list or enumerate all the possible values, this is an *enumeration*, as shown in Fig. 6.6(a). If the values may be ordered in a continuous list (no gaps) from first to last, we can simply list the first and last with ".." between because we know how to fill in the intermediate values (see first two examples in Fig. 6.6(b)). This is called a *range*.

If a range is *unbounded* at one end, no value appears at that end. For example, if we measure Age as a whole number of years, the value constraint is {0..}. By default, ranges are assumed to be closed, so the end values are included. For example, {50..100} includes 50 and 100 as well as the numbers in between. Inserting *a left parenthesis "(" before a start value excludes that value*, and *appending a right parenthesis ")" after an end value excludes that value.* If desired, a square bracket may be used to explicitly indicate inclusion (the default). For example, "(0..100" and "(0..100]" each denote a range of positive (above 0) numbers up to and including 100. One may also combine enumerations and/or ranges into a single constraint, as in Fig. 6.6(c).

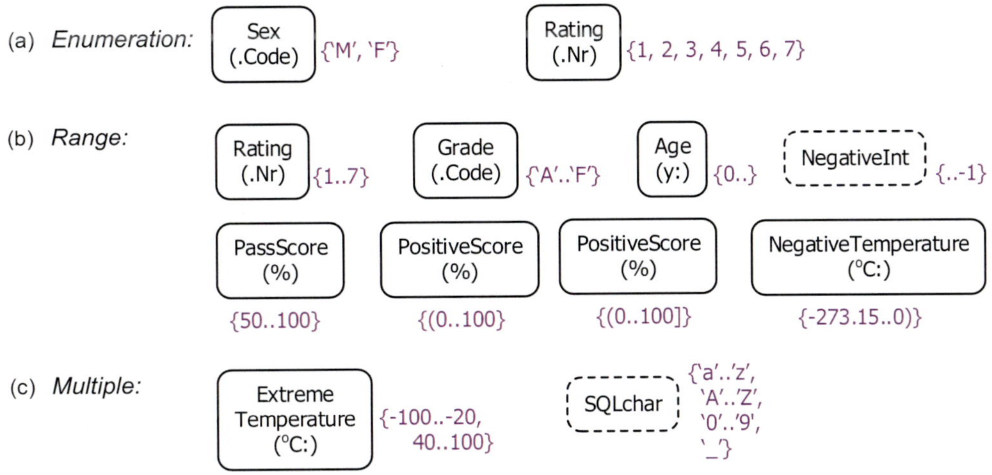

FIG. 6.6 Object value constraints list the possible values of a value type.

If the value constraint expression is short, it can be displayed in full on the diagram. If the value list is long, we can enter all the values in an object-role modeling (ORM) tool but choose to display just some leading values. A trailing ellipsis "…" then indicates that the display of later values is suppressed. For example, for the object type WeekDay we might specify the value list {"Sunday", "Monday", "Tuesday", "Wednesday", "Thursday", "Friday", "Saturday"} but choose to display only the first two (see Fig. 6.7). Although the list has been compacted on the diagram, it can be displayed in full textually when desired (e.g., by accessing a properties sheet for the object type).

Now consider Fig. 6.8, which is a modified version of an Olympics example discussed earlier, together with a sample population from the final results of the 24th summer

FIG. 6.7 The display of later values is suppressed.

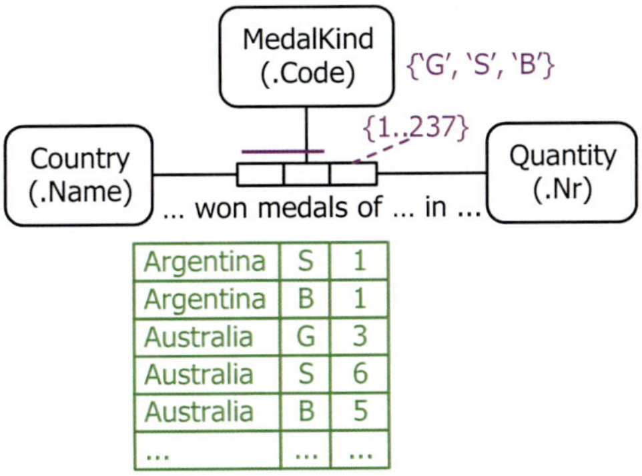

Argentina	S	1
Argentina	B	1
Australia	G	3
Australia	S	6
Australia	B	5
...	...	...

FIG. 6.8 Value constraints are declared for MedalKind and a Quantity role.

Olympics held in Seoul, South Korea, in 1988 (ordered alphabetically by Country). For this application, only three kinds of medals are allowed: gold, silver, and bronze, denoted by the codes "G", "S", and "B", respectively. This constraint is specified by listing the set of possible values {"G", "S", "B"} next to the entity type MedalKind.

In the 1988 Olympics, there were 237 events, with one gold, silver, and bronze medal to be awarded for each event (assume that there are no ties). It is possible (although highly unlikely) that the same country wins all medals of a given kind. So the role played here by Quantity (we could name this role "medal kind quantity") has a maximum possible value of 237, assuming that we exclude other Games (which could have more events) from the domain.

What is the smallest possible value of medal kind quantity? If we want to record that a country won no medals of a given kind, the lowest medal kind quantity is 0. To reduce the size of the database, we could store only facts about actual wins—this approach is suggested by our sample data. In this case, medal kind quantity has a lower limit of 1. Taking the closed world assumption, the fact that a country has won 0 gold medals could then be derived from the absence of an asserted fact about that country winning some gold medals. So, this value constraint has the value range {1..237}.

Because Quantity may well play other roles in the global schema, the {1..237} value constraint applies only to Quantity's role in this fact type rather than Quantity itself. We call this a *role value constraint* rather than an object type value constraint. Role value constraints are displayed attached by a dashed line to the role to which they apply, as shown for {1..237} in Fig. 6.8.

An alternative would be to introduce a more specific object type (e.g., MedalKindQuantity) and place the value constraint on that; however, such a move may lead to awkward verbalization, so role value constraints are quite handy.

As a fine point, an object value constraint displayed next to an entity type with a reference mode is actually a role value constraint on the referential role played by the value type. In the NORMA tool, you can display explicitly in this way by toggling ExpandRefMode to True.

In this example, "Country" denotes the set of all nations on Earth, or perhaps just nations competing in the Olympics. Should we add a value constraint for Country or Country's role? This would be impractical, because the list of country names is too long (e.g., 161 nations competed in the 1988 summer Olympics) and unstable (e.g., in 1991 the former Soviet Union fragmented into 15 nations). If the schema is reused for many games, this value constraint would need to be updated continually.

Because the list of countries is both large and unstable, we will not specify a list of values for CountryName as a *constraint* (we'll see shortly another way to supply such lists). However, we can declare a syntactic *data type* as a weak form of "value constraint". For example, we might require each country name to be a string of at most 30 characters.

Although choosing syntactic data types is not the most conceptual or exciting aspect of modeling, it's something you need to do before you implement the model. Otherwise, you'll get whatever default data type the system provides, and this won't always be what you want.

Now suppose that our Olympics information system is required to answer the following questions. *Which countries competed? Which countries didn't compete?* The first question could be addressed by adding the unary predicate "competed" to Country in Fig. 6.8. The competing countries would then all be entered in the single-column table for this unary fact type. Alternatively, we could explicitly record zero for each competing country not winning a given kind of medal.

But what about the second question? This effectively asks: Which countries exist but didn't compete? This list is usually small, but could be large (e.g., because of a boycott). So, somehow we need a way to record all the existing countries, whether or not they competed. As discussed earlier, putting this list of countries in a constraint is unwise because the list is large and unstable. Although schemas often do evolve, we usually try to design them to be as stable as possible. A classic strategy used is to model the feature so that the changes occur in the database population rather than the schema itself. How can we do this for the current example?

You might be tempted to add a unary predicate "exists" to Country. But such a predicate would apply to all object types in the schema. Moreover, there are formal problems with treating "exists" (an existential quantifier) as a predicate. In principle, we could address the problem by adding another unary predicate "did not compete" for Country and storing noncompetitors here. But such an approach is awkward—Country might play several roles (birthplace, location, etc.) and it may be arbitrary which role is negated to cater for the rest of the countries.

A cleaner approach is to list all existing countries in a *reference table* for Country. A country may appear in this table without playing any elementary fact roles in the schema. Such a country is said to be *independent* because it exists in our model independently of doing anything. An **independent entity type** is a *primitive entity type whose elementary fact roles (if any) are collectively optional.* An *independent value type* may contain some values that play in no facts (elementary or existential).

The term "independent" is not used for subtypes. If an independent object type plays some fact roles, the disjunction of its roles is not mandatory (implicitly or explicitly).

To signify that an object type is independent, an *exclamation mark* "!" is appended to its name. Fig. 6.9 depicts Country as independent. Its reference table lists all existing nations. To save space, only the first six countries are listed. The first entry may be verbalized as "The Country with name 'Afghanistan' exists"; this is a reference or existential fact.

Win counts of zero are included. The first role in the ternary is optional because some nations don't compete (the data here are fictitious). If wins of zero were excluded (as in Fig. 6.8) and the unary "competed" were added, the disjunction of this role and the wins

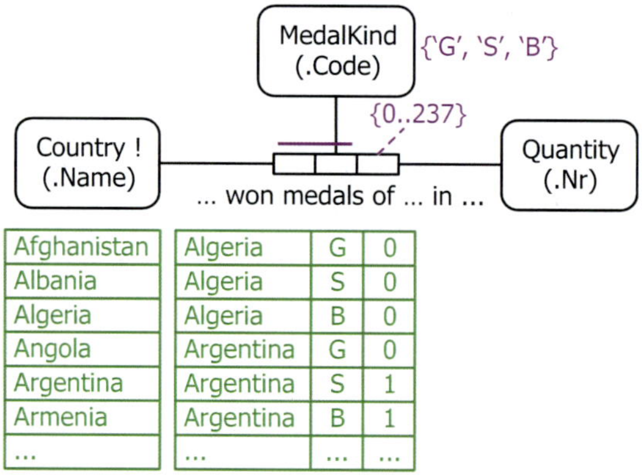

FIG. 6.9 In this schema, Country is an independent object type.

role would be optional (and a subset constraint would run from the wins to the competed role—subset constraints are treated later). So, *the implied mandatory role rule does not apply to independent object types.*

The reference scheme of an independent entity type is mandatory, so no country may be recorded to compete or win unless it is recorded in the population of Country. To fully populate Country with a list of all nations is a tedious task; however, once done it can be imported into many applications as needed. In practice, most independent object types have fewer instances. As a minor point, the value constraint on Quantity's role may need to be updated to allow for more events.

If we did not want to know about noncompeting countries, then as an alternative to adding the unary "competed", we could restrict the meaning of Country to "Competing nation", mark it as independent, and record only nonzero wins.

Objectified associations are often independent, as we often wish to record an association before recording other facts about it. For example, the schema in Fig. 6.10 allows us to record the fact that a student has enrolled in a subject before knowing the rating resulting for that enrollment (e.g., at the end of semester).

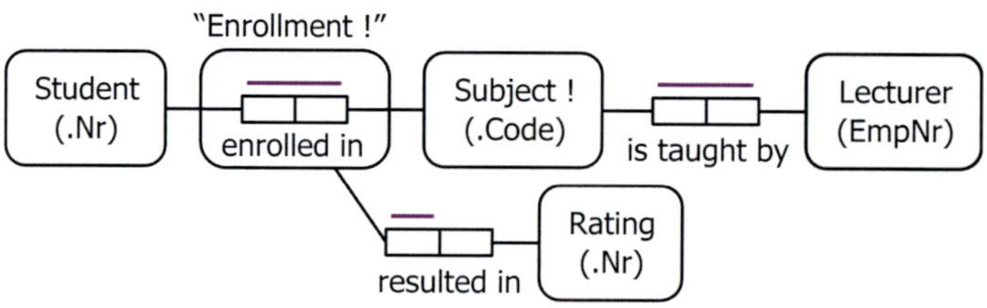

FIG. 6.10 Objectified associations are often independent.

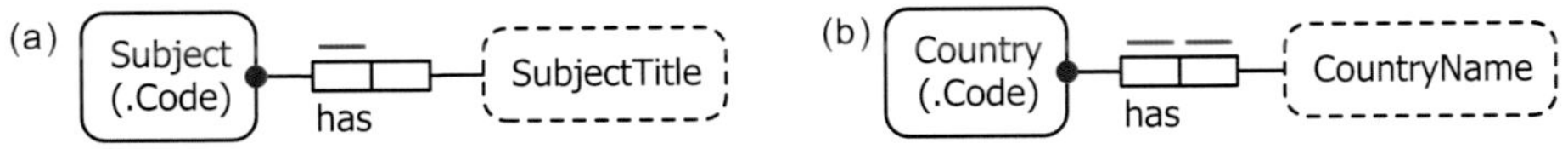

FIG. 6.11 Simple object types are rarely independent.

In this example, the simple object type Subject is also independent. In this universe of discourse (UoD), it is possible that all we know about a subject is its subject code. It is not necessary that all subjects be taught. However, *simple object types are rarely independent.*

In most academic domains, we need to know the title of any subject, not just its code. This more realistic situation is depicted in Fig. 6.11(a), where Subject is no longer independent because it has a mandatory fact role. Similarly, in our Olympics example, if all countries had codes as well as names, and we chose code for preferred reference, then Country has CountryName becomes a mandatory fact type, and Country is no longer independent (see Fig. 6.11(b)).

The use of an independent object type or unary predicates instead of a value constraint adds flexibility because the data can be changed without recompiling the schema (and relevant forms, etc.). However, the responsibility for this is now in the hands of the person entering the data rather than the schema designer. Typically, write-access to such reference tables is granted only to database administrators.

A Dutch version of ORM known as Fully Communication Oriented Information Modeling (FCO-IM) treats all entity types, independent or not, as objectified predicates. For example, in Fig. 6.12 the entity type Country is depicted as the objectification of a role played by the value type CountryName. The notion of an entity type can then be treated as a derived, rather than a base construct.

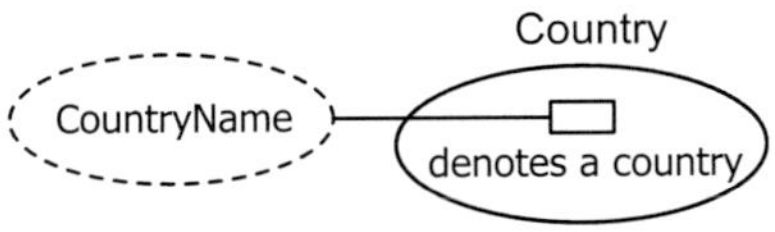

FIG. 6.12 In FCO-IM, an entity type is an objectified predicate.

Exercise 6.3

1. Schematize the following sample report about elementary particles. Include uniqueness, mandatory role, and value constraints. Set an upper limit of 2000 amu for mass

Family	*Particle*	*Charge*	*Mass (amu)*
lepton	neutrino	0	0
	electron	–	1
	positron	+	1
meson	eta	0	1074
baryon	proton	+	1836
	neutron	0	1839

2. It is desired to record a list of all sports, and for each sport, which Olympic Games (if any) included it. Some sports (e.g., running) have been included in each Olympics, some (e.g., judo) only in some, and others (e.g., surfing) never. Schematize this application. You may identify Olympics by its year or by an Olympiad number.

3. An Australian software retailer, SoftMart, maintains an information system to help with invoice and stock control. It has recently opened for business, and has made only a few sales so far. The details of the software items it has in stock are shown. The software items are standardly identified by item codes, but also have unique titles. There are exactly three software categories, identified by codes (SS=spreadsheet, DB=database, WP=wordprocesssor). The full names of these categories are not recorded. The list price of an item is the normal price at which the item is currently sold. However, SoftMart may sell an item at less than the current list price (e.g., SoftMart may give a discount for bulk orders or to favored clients, and the list price itself may change with time). There is no rule to enable the unit price (i.e., the actual price charged for a copy of an item) to be derived from the current list price.

Itemcode	*Title*	*Category*	*Stock qty*	*Listprice*
B123	Blossom 123	SS	8	799.50
DL	DataLight	DB	10	700.00
DB3	Database 3	DB	5	1999.99
Q	Quinquo	SS	6	400.00
SQL+	SQL plus	DB	4	1890.50
TS	TextStar	WP	5	500.00
WL	WordLight	WP	10	700.00

Customers are identified by a customer number, but the combination of their name and address is also unique. For simplicity, customer name is treated as a single character string, as is address. Customers have at most one phone number recorded. The next table shows customer details.

Customer#	*Name*	*Address*	*Phone*
001	Starcorp	5 Sun St, St Lucia 4067	3765000
002	Eastpac	30 Beach Rd, Sandgate 4017	2691111
003	Dr I.N. Stein	7 Sesame St, St Lucia 4067	?

Customer details may be recorded before the customer places an order. Once an order is placed, the items are issued to the customer together with an invoice. At the time the database snapshot was taken for the reports, only four invoices had been issued. When a customer pays for the items listed in an invoice, the date of payment is recorded. The following table lists the payments so far. Each invoice is identified by its invoice number. For this simple exercise, assume that invoices are paid in full or not at all.

Invoice#	*Date paid*
0501	10/07/2004
0502	20/07/2004
0503	unpaid
0504	unpaid

The four actual invoices are shown. The invoice header giving the address of SoftMart is not stored.

SoftMart, 46 Gallium Street, Brisbane 4001

invoice#: 0501 date: 03/07/04
customer#: 001 customer name: Starcorp
address: 5 Sun St, St Lucia 4067

Item code	*Title*	*Qty ordered*	*Unit price*	*Subtotal*
WL	WordLight	5	650.00	3250.00
Q	Quinquo	1	400.00	400.00
SQL+	SQL plus	5	1701.45	8507.25
B123	Blossom 123	1	799.50	799.50

⇨ total amount due: $12956.75

SoftMart, 46 Gallium Street, Brisbane 4001

invoice#: 0502 date: 03/07/04
customer#: 002 customer name: Eastpac
address: 5 Beach Rd, Sandgate 4017

Item code	*Title*	*Qty ordered*	*Unit price*	*Subtotal*
Q	Quinquo	4	400.00	1600.00
TS	TextStar	4	500.00	2000.00

⇨ total amount due: $3600.00

SoftMart, 46 Gallium Street, Brisbane 4001

invoice#: 0503 date: 10/07/04
customer#: 001 customer name: Starcorp
address: 5 Sun St, St Lucia 4067

Item code	*Title*	*Qty ordered*	*Unit price*	*Subtotal*
Q	Quinquo	4	350.00	1400.00

⇨ total amount due: $1400.00

SoftMart, 46 Gallium Street, Brisbane 4001

invoice#: 0504 date: 10/07/04
customer#: 003 customer name: Dr I.N. Stein
address: 7 Sesame St, St Lucia 4067

Item code	*Title*	*Qty ordered*	*Unit price*	*Subtotal*
B123	Blossom 123	1	799.50	799.50
DL	DataLight	1	700.00	700.00

⇨ total amount due: $1499.50

An invoice includes a table of one or more rows, called “line items” or “invoice lines”. Each invoice line lists details about the order of one or more units (copies) of a software item. For simplicity, assume that on a given invoice the same item can appear on only one invoice line. For each invoice line, the item code, title, quantity of units ordered, and unit price are listed. The total charge for the invoice line is displayed as a subtotal. The total charge for the whole invoice is displayed as the amount due.

Schematize this UoD, including uniqueness, mandatory role, and value constraints. Identify a LineItem or invoice line by using the invoice number and the item code. If a fact type is derived, omit it from the diagram but include a derivation rule for it.

4. Consider the previous question, but suppose that the same item may appear on more than one line of the same invoice. Devise an alternative identification scheme to deal with this situation. (*Hint:* Compare this with the compact disc example. Sometimes you need to identify things in terms of their position.)

5. Now suppose that a cumulative record of purchases from suppliers is also to be recorded, an extract of which is shown in the following table. Only a cumulative record is kept (individual deliveries from suppliers are not recorded in the information system). No supplier can be a customer.

ItemCode	*Supplier*	*Quantity*
B123	Macrosoft	7
	TechAtlantic	3
DL	Macrosoft	11
DB3	PacificTech	5
etc.		

   (a) Draw a conceptual schema for this table.
   (b) Assume that this subschema is integrated with your solution to Question 3, and that your fact type for stock quantity is now declared to be derivable.
      (i) Specify a derivation rule for this approach.
      (ii) In a real business, how practical would it be to use such a rule to avoid storing stock quantity? Discuss.

6.4 Subset, Equality, and Exclusion Constraints

Set-comparison constraints restrict the way the population of one role, or role sequence, relates to the population of another. In an earlier section, we considered six ways in which two sets might be related: subset, equality, exclusion, proper subset, overlap, and proper overlap. The last three of these require some objects to exist in at least one of the sets being compared. However, static constraints must apply to every state of the database, including the empty state. So, the only set-comparison constraints of interest are subset, equality, and exclusion. These three kinds of constraint are examined in this section.

Suppose a fitness club maintains an information system about its members and that Table 6.1 is an extract of a report from this system. Membership includes access to the club's normal fitness equipment (e.g., a weights gym). However, the club also has a few squash and tennis courts. To help ensure fair access to these courts, the club has a policy that members may play only one of these two racquet sports.

Table 6.1 Details about members of a fitness club.

Member	*Sex*	*Birth year*	*Sport*	*Booking*	*Reaction time (ms)*	*Heart rate (beats/min)*
Anderson PE	M	1940	tennis	Mon 5 p.m.	250	80
Bloggs F	M	1940				
Fit IM	F	1975	squash		250	70
Hume D	F	1946	squash	Tue 9 a.m.		
Jones T	M	1965			300	93

Members pay an extra fee for this right and optionally may book one regular weekly hour to use a court. For simplicity, casual bookings and court allocations are excluded from the UoD. As a service to its members, the club arranges a fitness test to measure the resting heart rate and reaction time of any member who requests it. For simplicity, assume only the latest results are kept for each member.

The determination of fact types, uniqueness constraints, mandatory roles, and value constraints for this example is straightforward (see Fig. 6.13). For simplicity, let's agree that hours may be identified by a simple code (e.g., "Mon. 5 p.m."), without needing to separate out their day and time components.

However, there are two more constraints that apply to this example. To begin with, only those members who have opted to play a sport may book a court. In terms of the schema diagram, each object that populates the booking role must also appear in the population of the playing role. In other words, the set of members booking an hour must be a subset of the set of members playing a sport. As shown in Fig. 6.13, we mark this **subset constraint** by an *arrowed subset symbol* "⊆" running from the subset role to the superset role.

Our sample data agrees with this constraint because {"Anderson PE", "Hume D"} ⊆ {"Anderson PE", "Fit IM", "Hume D"}. If we tried to add a booking for Jones without also adding a sport for him, we would violate this constraint.

In other words, the subset constraint verbalizes as **If some** Member booked **some** HourSlot **then that** Member plays **some** Sport. Fig. 6.14 summarizes the general case of a subset constraint from one role to another. For this comparison to make sense, both roles must be played by the same object type (or a supertype—see later).

In Fig. 6.14, *a* is any member of *A*. The reference to database states reminds us that the constraint applies to our domain model (not necessarily to the real world). So "*a* plays *r*" means *a* plays *r* in our model. In terms of fact tables, the set of values in the column for *r2* is always a subset of the set of values in the column for *r1*. It doesn't matter if any of these columns has duplicate values; we are comparing sets, not bags.

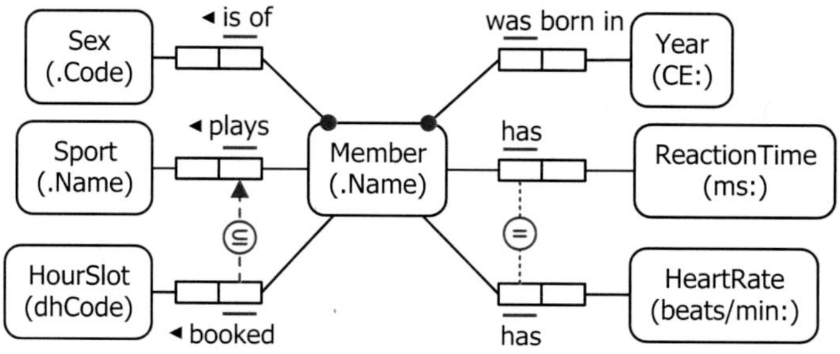

FIG. 6.13 A conceptual schema for Table 6.1.

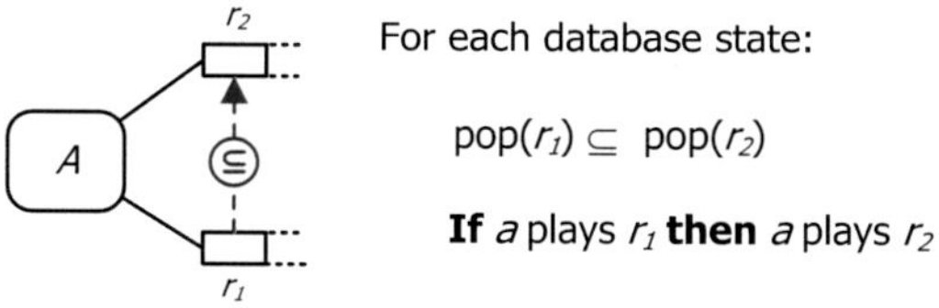

FIG. 6.14 A subset constraint between two roles.

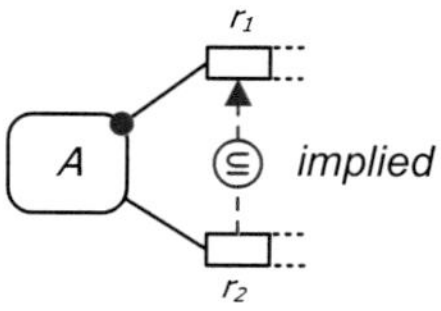

FIG. 6.15 The implied subset constraint should be omitted.

Fig. 6.14 assumes that *A* plays some other role and that both *r1* and *r2* are *optional*. If *r1* were mandatory, then a subset constraint to it would be *implied* because the total population of *A* would then play *r1* (see Fig. 6.15). In general, *A* has a mandatory role *r* only if there is a subset constraint to *r* from each of *A*'s roles.

Implied subset constraints should normally be omitted. Hence, a subset constraint between two roles is typically specified only if both these roles are optional. This rule does not extend to composite subset constraints (see later).

Our fitness club application has one more constraint. Reaction time is recorded for a member if and only if his/her heart rate is also. Each member has both fitness measures taken or neither. This might be because the club wants to provide a balanced estimate of fitness rather than risk reliance on a single figure. When the test data are entered, a compound transaction is used to enter both measures.

Such a constraint is called an **equality constraint** because for any state of the database, the set of people whose reaction time is recorded equals the set of people whose heart rate is recorded. This constraint is depicted by a *circled equality symbol "="* connected by dotted lines to the relevant roles (look back at Fig. 6.13). It asserts that the populations of these two roles must always be equal. An equality constraint is equivalent to two subset constraints, running in opposite directions.

The equality constraint in Fig. 6.13 may be expressed as **For each** Member, **that** Member has **some** ReactionTime **if and only if that** Member has **some** HeartRate. Fig. 6.16 summarizes the notion of an equality constraint between two optional roles.

If both roles are mandatory, an equality constraint is implied because each of pop(*r1*) and pop(*r2*) equals pop(*A*). Implied equality constraints should normally be omitted (see Fig. 6.17).

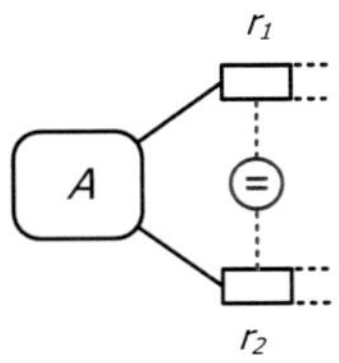

For each database state:

pop(r_1) = pop(r_2)

a plays r_1 **iff** *a* plays r_2

FIG. 6.16 An equality constraint between two roles.

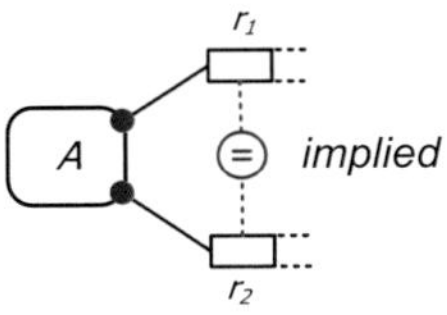

FIG. 6.17 The implied equality constraint should be omitted.

If one role is mandatory and the other is optional, then an equality constraint cannot exist between them because it is possible to have a state in which only some of *A*'s population plays the optional role. If neither role is mandatory but their disjunction is, then there is no equality constraint; otherwise, both the roles would be mandatory (why?).

In principle, all mandatory roles and equality constraints on a conceptual schema could be replaced by groups of subset constraints. In practice, however, this would lead to untidy schema diagrams, with subset constraint lines running all over them. Hence, the notions of mandatory role and equality constraint are very convenient.

Now, consider the report extract shown in Table 6.2. In this UoD, employees may request a company parking bay or a refund of their parking expenses, but not both. Employees may make neither request (e.g., they might not have a car or they might simply want more time to decide).

For this report, different kinds of null are displayed differently. The "–" value indicates "*not to be recorded*" (because the other option is chosen). The "?" value simply indicates "not recorded": an actual value might still be recorded for employee 005 later (e.g., after this employee buys a car); once such a value is recorded, however, the system must disallow the other option.

Table 6.2 Employee details.

EmpNr	*Employee name*	*Parking bay*	*Parking claim ($)*
001	Adams B	C01	–
002	Bloggs F	–	200
003	Collins T	B05	–
004	Dancer F	–	250
005	Eisai Z	?	?

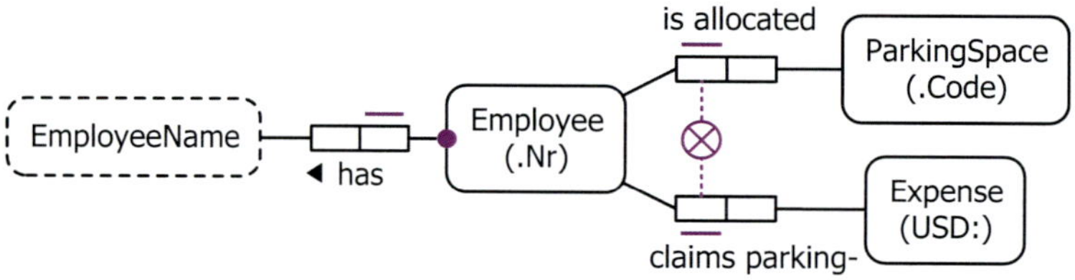

FIG. 6.18 Conceptual schema for Table 6.2.

Fig. 6.18 shows a schema for this example. A circled "X" symbol "⊗" indicates an **exclusion constraint** between the roles it connects (X for eXclusion). This asserts that for each state of the database, no employee can be recorded to play both these roles. That is, the populations of these roles are *mutually exclusive.* A textual version of this constraint is **For each** Employee **at most one of the following holds**: **that** Employee is allocated **some** ParkingBay; **that** Employee claims **some** parking Expense.

Fig. 6.19 formalizes this notion of an exclusion constraint between two roles: their populations are exclusive just in case their intersection is the null set (i.e., they have no element in common). If two roles are played by different, primitive entity types, then an implied exclusion constraint exists between them (because the entity types are mutually exclusive), and should normally be omitted.

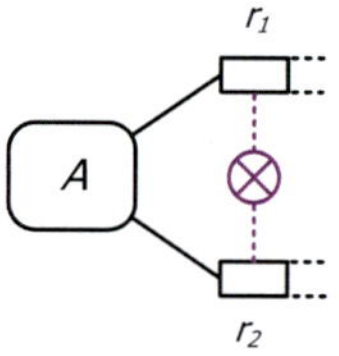

For each database state:

pop(r_1) ∩ pop(r_2) = { }

No *a* plays **both** r_1 **and** r_2

FIG. 6.19 An exclusion constraint between two roles.

An exclusion constraint may be asserted between two roles only if these roles are optional and are played by the same object type (or possibly a supertype—see later). If one of the roles were mandatory, any object playing the other role would also have to play the mandatory one, so an exclusion constraint could not apply. It is possible, however, that the disjunction of these two roles is mandatory.

Now consider Fig. 6.20. In this simple application, married partners of different genders are identified by their first names, and their country of birth must be recorded. For the ring predicate "is husband of", the inverse predicate name "is wife of" is also shown. No partner can be both a husband and a wife, so the husband and wife roles are mutually exclusive. Moreover, each partner is either a husband or a wife, so the same two roles are disjunctively mandatory.

Taken together, the exclusion and inclusive-or constraints assert that each partner is either a husband or a wife (of someone) but not both. These two constraints can be depicted together by superimposing the exclusion and mandatory symbols (⊗ and ⊙) to form the *life-buoy symbol* ⦿, connected by dotted lines to the relevant roles. If the roles are contiguous, the constraint is connected to their junction, as shown in Fig. 6.20.

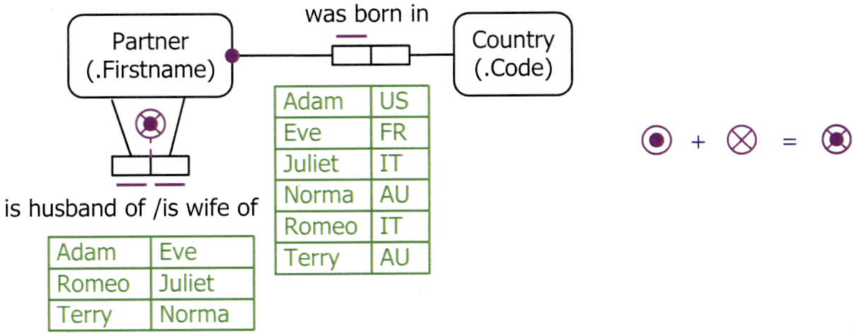

FIG. 6.20 *Exclusive or:* each partner is a husband or wife but not both.

When exclusion is combined with inclusive-or, we have an **exclusive-or constraint** (also called an **xor** constraint). Each partner plays *exactly one* of the constrained roles. Each partner appears in one column of this fact table, but no partner can appear in both columns. This exclusive-or constraint verbalizes thus: **For each** Partner, **exactly one of the following holds**: **that** Partner is husband of **some** Partner; **that** Partner is wife of **some** Partner. Contrast this with the parenthood binary in an earlier chapter, where each person is a parent or child or both—a case of inclusive or.

The life-buoy symbol is also a *partition symbol* because it partitions an object type's population into separate role populations. We return to the notion of partitions when we discuss subtypes. Although the exclusive-or constraint may be depicted as a single symbol, it is really an orthogonal combination of two constraints. As a tool option, it may be displayed and verbalized as two separate constraints.

Subset or equality constraints should not be displayed with disjunctive mandatory roles. For example, if the two constraints in the left-hand schema of Fig. 6.21 apply, then role r_1 is mandatory (as an exercise, prove this). For clarity, the mandatory role should be shown explicitly; the subset constraint is now implied by this, and hence should be omitted (see right-hand schema of Fig. 6.21).

Similarly, an equality constraint between disjunctive mandatory roles should be redrawn as two mandatory roles. Such redrawing rules are pragmatic. Formally, the left-hand schema is equivalent to the right-hand one. However, the right-hand version is preferred because it is simpler for people to work with.

As another case of redrawing, an exclusion constraint among three or more roles may be depicted by a single "⊗" connecting the roles, in preference to exhaustively marking exclusion constraints between each pair of roles. A simple example is shown in Fig. 6.22. In this UoD, each person has at most one of three vices.

FIG. 6.21 The left-hand schema should be redrawn as the right-hand schema.

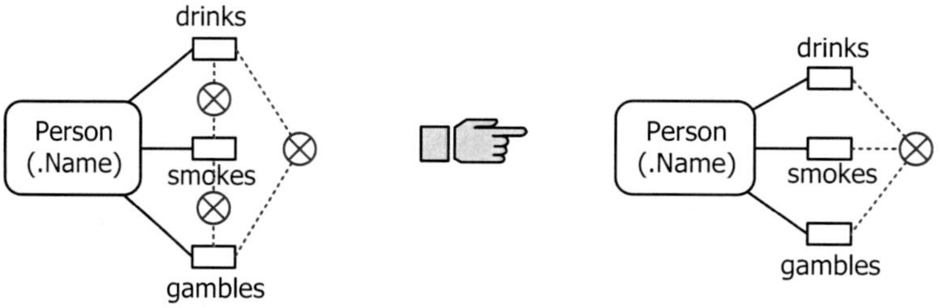

FIG. 6.22 In this UoD, each person has at most one of three vices.

In the left-hand diagram of Fig. 6.22, three exclusion constraints cover all the ways to pair the three roles. Four roles may be paired in six ways. In general, a single exclusion constraint across n roles replaces $n(n-1)/2$ separate two-role exclusion constraints. An equality constraint across n roles may also be specified using a single, circled "=" instead of multiple binary equality constraints. No notation for a subset constraint over n roles is used because direction is involved.

Although exclusion constraints between roles are common, simple subset and equality constraints are seldom used on conceptual schemas (although they are often used on relational schemas). It is fairly common to encounter "qualified" set-comparison constraints, but these are depicted by subtyping (see next section).

Domains commonly involve *set-comparison constraints between role sequences* (not just between single roles), which we cannot handle by subtyping. Let's consider some examples. Table 6.3 lists details about people and the cars they own or drive. For simplicity, assume that people are identified by their first name and that each car is identified by a registration number stamped on its license plate. If we allowed a car to change its license plate, we would need to pick some other identifier (e.g., its vehicle identification number or compliance plate number).

Fig. 6.23 shows the populated conceptual schema. The ownership association is many: many. A person may own many cars, and the same car may be owned by many people. For example, Fred and Sue are co-owners of two cars (e.g., they might be married). The drives association is also many:many.

Table 6.3 Details about car owners and drivers.

Person	*Sex*	*Cars owned*	*Cars driven*
Fred	M	272NCP, 123BOM	272NCP, 123BOM
Sue	F	272NCP, 123BOM	272NCP
Tina	F	105ABC	
Tom	M		

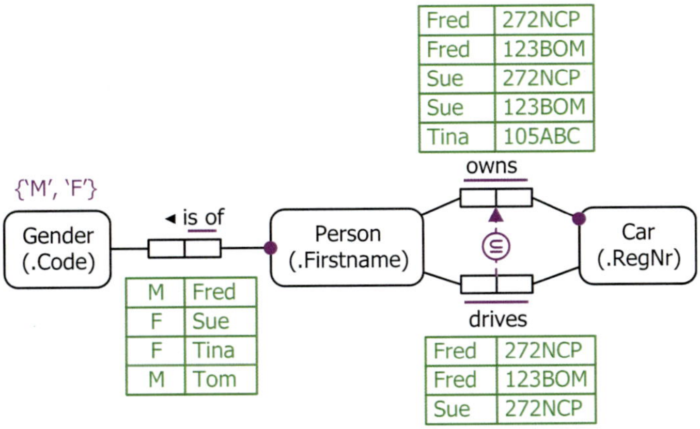

FIG. 6.23 A populated conceptual schema for Table 6.3.

Note that although Fred and Sue both own car 123BOM, only Fred drives it. Perhaps it is a "bomb" of a car and Sue refuses to drive it! Tina owns a car but doesn't drive it (maybe she has a chauffeur). Tom neither owns nor drives a car.

For the population shown, people own every car that they drive. Although unusual, for discussion purposes let us assume the sample database is significant in this respect. This rule is indicated in Fig. 6.23 by the *pair-subset constraint* running from the middle of the Drives predicate to the middle of the Owns predicate. This asserts that, for each state of the database, the set of (Person, Car) pairs referenced in the Drives fact table is a subset of the set of pairs listed in the Owns table. We may think of each pair as a tuple of values from a row of a fact table.

For the database state shown, the Drives table has three pairs: (Fred, 272NCP), (Fred, 123BOM), and (Sue, 272NCP). The Owns table has five pairs. Each pair in the Drives table is also in the Owns table; the subset constraint indicates that this is true for each state of the database. This constraint may be verbalized as **If some** Person drives **some** Car **then that** Person owns **that** Car. An alternative verbalization is **Each** Person **who** drives **some** Car **also** owns **that** Car.

As usual, the direction of the subset constraint is important. With our example, if the arrow pointed instead to the Drives fact type, this would signify that each person who owns a car also drives that car, which is quite a different constraint.

To indicate that a subset constraint applies between role pairs rather than single roles, each end of the arrow is positioned at the junction of the two roles in the pair. This notation may be used when the roles in the pair are contiguous (see Fig. 6.24).

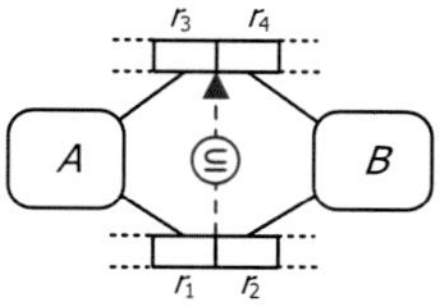

For each database state:

pop(r_1, r_2) ⊆ pop(r_3, r_4)

Each *ab* **pair in** pop(r_1, r_2) **is also in** pop(r_3, r_4)

FIG. 6.24 A pair-subset constraint.

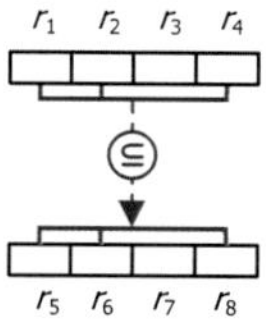

For each database state:

pop(r_1, r_2, r_4) ⊆ pop(r_5, r_6, r_8)

FIG. 6.25 An example of a tuple-subset constraint between sequences of three roles.

In terms of the associated fact tables, the projection on columns r_3, r_4 of the lower table is a subset of the projection on columns r_1, r_2 of the upper table. If the roles in the pair are not contiguous, a connecting line is used between them. Role connectors are also used when the arguments to the constraint contain more than two roles (see Fig. 6.25).

In general, a *tuple-subset constraint* may be specified between two compatible role tuples (their corresponding roles are type compatible), where each tuple is a sequence of n roles ($n \geq 1$). However, n is rarely greater than two. In rare cases, roles may need to be reordered before being compared; in this case, the constraint may be annotated by including the relevant permutation.

Now consider the case where people own a car if and only if they drive that car. Here we have subset constraints in both directions: this is a *pair-equality* constraint. In such cases where the role pairs form a whole predicate, we usually store only one fact type. For example, we might store Drives and derive Owns, or collapse both to IsOwnerDriverOf. However, if a role pair is embedded in a longer predicate, an equality constraint is not equivalent to a derivation rule (why not?). Fig. 6.26 summarizes the case of pair equality for contiguous roles.

Exclusion constraints may be specified between two or more role sequences. As a simple example, suppose we want to record information about cars that people own and cars that they want to buy. Clearly, nobody would want to buy a car that they already own. We can indicate this by the *pair-exclusion constraint* shown in Fig. 6.27.

This constraint is weaker than an exclusion constraint between just the first roles of these predicates (which would instead say that no car owner wants to buy any car, and hence would disallow Fred's appearance in both the fact tables shown). Although we must record each person's gender, both the "owns" and the "wants-to-buy" roles are optional (even disjunctively). For example, Linda neither owns nor wants to buy a car.

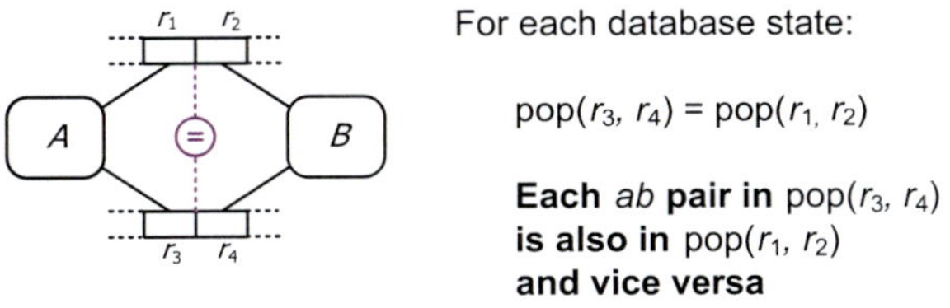

FIG. 6.26 A pair-equality constraint.

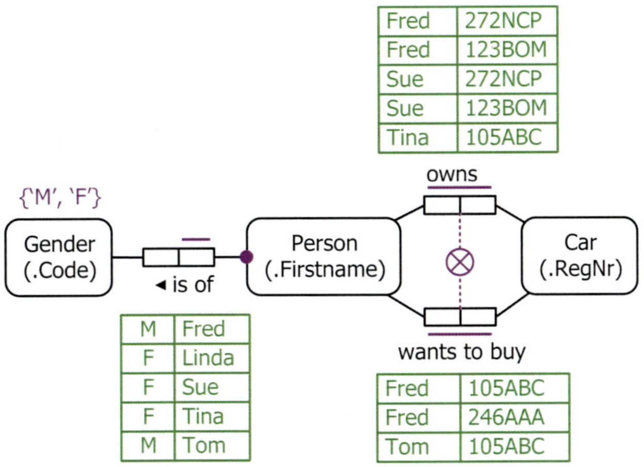

FIG. 6.27 A pair-exclusion constraint: nobody wants to buy a car that they own.

In this section, set-comparison constraints have been considered between pairs of role sequences, where each role sequence comes from a single predicate. More generally, set-comparison constraints may be defined between compatible role paths formed by joining different predicates (join constraints are covered in Section 10.1).

To conclude this section, let's look at some general results, or theorems, about set-comparison constraints. Suppose we declare a tuple-subset constraint from role sequence *rs1* to role sequence *rs2*, where each sequence has *n* roles ($n \geq 1$). Is it now possible for these role sequences to be exclusive?

If both role sequences are unpopulated, then both the subset and the exclusion constraints are trivially satisfied. But for practical reasons, we need every conceptual schema to be *strongly satisfiable* (or *population consistent*). This means that each role sequence used as a predicate or as a constraint argument can be populated in some state. It is not necessary that these role sequences be populated together in the same state.

Suppose we populate *rs1* with some sequence *a*. The subset constraint entails that *a* now populates *rs2* as well, thus violating the exclusion constraint. Swapping *rs1* with *rs2* in this reasoning shows a subset constraint in the other direction cannot hold either, if an exclusion constraint does.

This proves theorem *NXS* (*N*o e*X*clusion with a *S*ubset constraint), depicted in Fig. 6.28. Here, a long box depicts a sequence of one or more roles, and the constraints apply between the *whole* sequences (not different parts of them).

FIG. 6.28 Theorem NXS: No eXclusion with a Subset constraint.

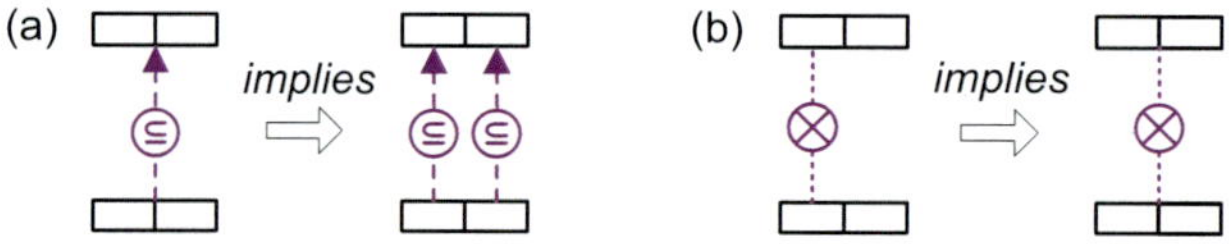

FIG. 6.29 The implied constraints should not be shown.

An equality constraint is equivalent to, and used instead of, two subset constraints. At most, one set-comparison constraint may be declared between two (whole) role sequences. However, different set-comparison constraints may be declared between different parts of role sequences, for example, a subset constraint between the first roles and an exclusion constraint between the second roles. Subset or equality constraints between two role sequences imply similar constraints between the individual roles. In contrast, an exclusion constraint between single roles implies exclusion constraints between all sequences containing these roles.

These *constraint implication* results are depicted for role pairs in Fig. 6.29, using "⇨" for "necessarily implies". In each case, the implication is in one direction only. Results for equality constraints are similar to the subset results. In the exclusion example, a similar result holds if the simple exclusion constraint is between the right-hand roles. Implied constraints should usually be omitted from a schema. As an exercise, you may wish to prove these results. If you have trouble here, try inventing some counterexamples to equivalence claims. For example, if pop(r_1, r_2) = {(a1, b1), (a2, b2)} and pop(r_3, r_4) = {(a1, b2), (a2, b1)}, then pop(r_1) = pop(r_3) and pop(r_2) = pop(r_4) but pop(r_1, r_2) ≠ pop(r_3, r_4).

Other constraint implication theorems are mentioned later in the book. One important result is the case where the target of a pair-subset constraint includes a functional role. For example, the left-hand pattern in Fig. 6.29(a) has a pair-subset constraint from the lower role pair to the top role pair. If we add a simple uniqueness constraint over the left-hand role of the top role pair, what does this imply about the left-hand role of the lower role pair? If the lower predicate is a binary, this role must also be unique! As an exercise, prove this. An important corollary of this result is that if the lower predicate is longer than a binary, then it must split in this case because there will be an implied functional dependency (FD) from the first role to the second role. This result is illustrated in Section 14.6 when we discuss normalization.

Exercise 6.4

1. A company allows some of its employees to use one or more of its company cars. The rest of its employees are given a travel allowance instead. The following report is an extract from the company's records in this regard. Schematize it.

Emp. Nr	*Emp. name*	*Cars used*	*Driver license*	*Travel allowance ($)*
001	Harding J	123ABC	A74000	–
002	Oh C	111KOR, 123ABC	A51120	–
003	Halpin T	–	–	200

2. The diagram shows the conceptual schema and current population for a given UoD. Reference schemes are omitted for simplicity. Fact tables appear below their fact types. Predicates are named **R..U**. Constraints are named *C1..C9*.

   Each of the following requests applies to the **same** database population shown on the diagram (i.e., treat each request as the *first* to be made with this population).

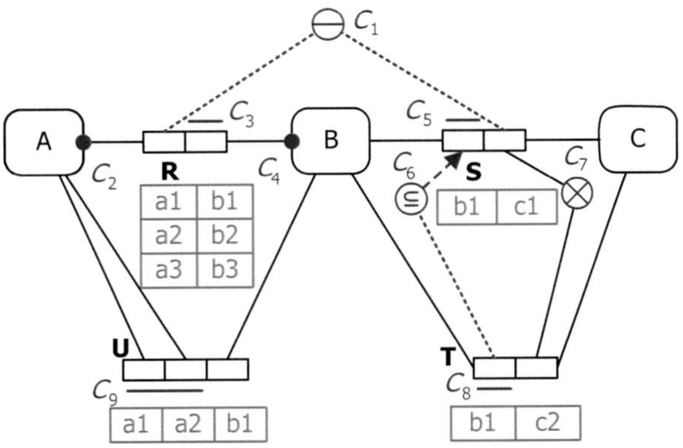

For each request, indicate the response of the conceptual information processor (CIP). If the request is legal, write "accepted". Otherwise, indicate the constraint violated (e.g., "*C2* violated").

(a) add: a1 R b2 (b) add: b4 S c3 (c) add: b1 S c3 (d) delete: b1 S c1
(e) add: b2 S c2 (f) add: b3 T c3 (g) add: U a2 a1 b2 (h) add: U a1 a2 b2
(i) begin
  add: a1 R b4
  add: b4 S c1
 end

3. Draw the conceptual schema diagram for the UoD declared in Exercise 2.2 Question 2, adding constraint labels *C1..C5* to the relevant constraints marked on the diagram.

4. The following table records details about various (fictitious) movies. Schematize this UoD. The population is not fully significant, so make some educated guesses.

Movie	*Director*	*Stars*	*Supporting cast*
Earth Song	Ima Beatle	Bing Crosby Elvis Presley	Neil Diamond Tina Turner
Me and my echo	Hugh Who	Hugh Who	
Kangaroos	Doc U. Mentary		

5. A software company has a large number of shops located in the state of Queensland and requires detailed knowledge of their distribution. The table shows some of their locations. City shops are located in suburbs, but country shops are located in towns. City and town names are unique (other states are not of interest). Schematize this UoD, revealing the reference scheme of the entity type Location in full detail. Model street addresses as values but towns, suburbs, and cities as entities.

Shop#	*Location*			
	Suburb	*City*	*Town*	*Street address*
1	St Lucia	Brisbane		Suite 5, 77 Sylvan Rd
2			Strathpine	Unit 3, 1000 Gympie Rd
3	Sandgate	Brisbane		7 Sun St
4	Clontarf	Redcliffe		25 Beach Rd
5	Sandgate	Cairns		25 Beach Rd

6. The application described here is partly based on a real-life banking system, but simplifications and changes have been made. To help you appreciate the privacy implications of a universal identification scheme, all clients and staff of the bank are identified throughout the application by their tax file number (taxNr). Thus, all bank customers and personnel are taxpayers.

An information system is required to manage accounts and staff records for Oz Bank, which has branches at various locations. Each branch is standardly identified by its branch number but also has a unique name. The first table is an extract from staff records of Oz Bank. Each employee works at exactly one branch and has at most one phone listed. The mark "?" denotes a null. The mark "…" indicates "etc." (other instances exist but are not shown here).

BranchNr	*Branch name*	*Emp. taxNr*	*Emp. name*	*Emp. phone*
1	Uni. of Qld	200	Jones E	3770000
		390	Presley E	?
		...	...	...
2	Toowong Central	377	Jones E	?
		...	...	...
3	Strathpine	222	Wong M	2051111
		...	...	...

Within the one branch, each account has a unique serial number, but different accounts in different branches may have the same serial number. Account users are identified by their taxNr, but also have a name and possibly a phone number (see the second table).

Each account is a passbook account. Five sample passbook entries are shown. For each account, transactions are numbered sequentially 1, 2, 3, etc. Dates are formatted day/month/year. For simplicity, assume that each transaction is either a deposit (DEP) or a withdrawal (WDL). In practice, other types of transaction are possible (e.g., interest and fees). The balance column shows the account balance after the transaction is executed.

Although the balance is derivable, for efficiency purposes *the balance is stored as soon as it is derived.* For example, this speeds up the production of monthly statements for the bank's customers (Oz bank has a few million customers who average several transactions each month). This *derive on update* (eager evaluation) decision contrasts with our normal *derive on query* (lazy evaluation) policy.

Account		*User*		
BranchNr	*SerialNr*	*TaxNr*	*Name*	*Phone*
1	55	200	Jones E	3770000
		311	Jones T	3770000
1	66	199	Megasoft	3771234
2	55	199	Megasoft	3771234
2	77	377	Jones E	?
3	44	300	Wong S	2051111

OZ BANK

BranchNr	SerialNr
1	55

Branch name: Uni. of Qld
Users: Jones E; Jones T

TranNr	Date	Deposit	Withdrawal	Balance
1	3/1/90	1000		1000
2	5/1/90		200	800
3	5/1/90		100	700

OZ BANK

BranchNr	SerialNr
1	66

Branch name: Uni. of Qld
Users: Megasoft

TranNr	Date	Deposit	Withdrawal	Balance
1	10/2/90	2000		2000
2	10/2/90		500	1500

OZ BANK

BranchNr	SerialNr
2	55

Branch name: Toowong central
Users: Megasoft

TranNr	Date	Deposit	Withdrawal	Balance
1	23/1/90	9000		9000
2	7/2/90	5000		14000
3	10/2/90		2000	12000
4	2/3/90		5000	7000

OZ BANK

BranchNr	SerialNr
2	77

Branch name: Toowong Central
Users: Jones E

TranNr	Date	Deposit	Withdrawal	Balance
1	3/1/90	500		500

OZ BANK

BranchNr	SerialNr
3	44

Branch name: Strathpine
Users: Wong S

TranNr	Date	Deposit	Withdrawal	Balance
1	5/1/90	100		100
2	12/1/90	600		700

(a) Draw a conceptual schema diagram for this UoD. Make use of an entity type indicating the type of transaction: TransactionType(.Code) {"DEP", "WDL"}. Include the account balance fact type on the diagram; mark it with "**" because it is derivable but it is required to be stored. Express the derivation rule as clearly as you can.

(b) Consider the fragment of your conceptual schema that captures the kind of information required for deposits and withdrawals (only two fact types are involved). Transform this subschema into an equivalent subschema that uses two different fact types, removing the entity type TransactionType altogether.

(c) In a realistic banking application, there are several kinds of transactions in addition to deposits and withdrawals. In such a situation, is it better to include Transaction Type as an object type or to extend the alternative approach proposed in (b)?

6.5 Subtyping

If instances of an object type are classified into a more specific type, this specialized type is known as a **subtype**. Consider the Person type, shown in Fig. 6.30. For discussion purposes, this is populated with eight people, depicted by clip art shapes. Each person's gender and birth country are recorded. The two instances of Gender are denoted by the biological symbols for male and female. Countries are denoted by their map shapes (only Australia, Italy, and Mexico are shown here).

Based on gender, we could classify Person into the subtypes MalePerson and FemalePerson. We've placed the males on the left and the females on the right in Fig. 6.30, so you can visualize creating these subtypes by vertically dividing the Person type in two. We could also classify persons into subtypes based on their country of birth. In Fig. 6.30, this is depicted by the horizontal division of Person into Australian, Italian, Mexican, etc. For the population shown, there are three Australians, two Italians, and three Mexicans, with other nationalities in various layers below (simply marked "etc." here). For simplicity, ignore issues about naturalization, so the term "Australian" is used here in the sense of "Australian-born", and so on.

The main reason for using a subtype in modeling is to *declare typing constraints* (in this case, to *declare that one or more specific roles are played only by a given subtype*). For example, a medical database might record whether a patient was circumcised, but this would be recorded only for male patients.

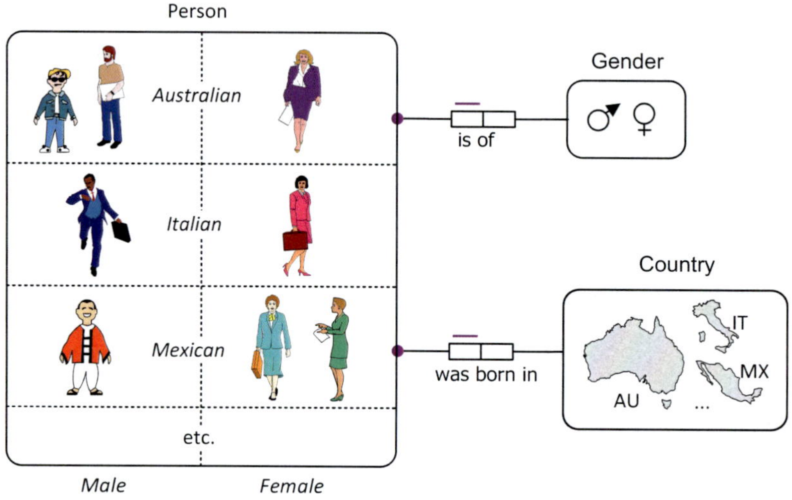

FIG. 6.30 Person classified into subtypes according to gender or birthplace.

A second reason for subtyping is to *encourage reuse of model components*. For example, if we already have an elaborate subschema describing the object type Employee and want to talk about managers, we can declare Manager as a subtype of Employee and simply add the extra details specific to managers.

If Employee is defined in a reference model, it can be imported into many different models and specialized as appropriate for each case. If an object type has behavior defined for it, this can be reused as well as the data aspects. In programming applications, this can significantly reduce code duplication.

A third reason for subtyping is to reveal a *taxonomy* or classification scheme being used in the UoD. However, in ORM, we often avoid introducing subtypes purely to expose a classification scheme. Why? First, taxonomies may often be modeled in more economical ways. For example, to indicate that there are two genders, we may simply add a value constraint such as {"M", "F"} on Gender (.code). There is no need to introduce Male and Female subtypes unless we have some specific roles for them to play in our application domain.

Second, the number of subtypes can get out of hand rapidly if we use them purely for taxonomy. For example, suppose we wish to classify people based on country of birth. All we need is the fact type: Person was born in Country. If instead we modeled this by introducing subtypes of Person, one for each country, this would lead to almost 200 subtypes in the schema!

This is crazy. If one of these subtypes plays a specific role, then we should introduce it, but if it doesn't, we have no need to. For example, suppose we wanted to record whether Australians want their country to become a republic, but had nothing specific to record about the other nationalities. In that case, Australian is the only subtype we should introduce.

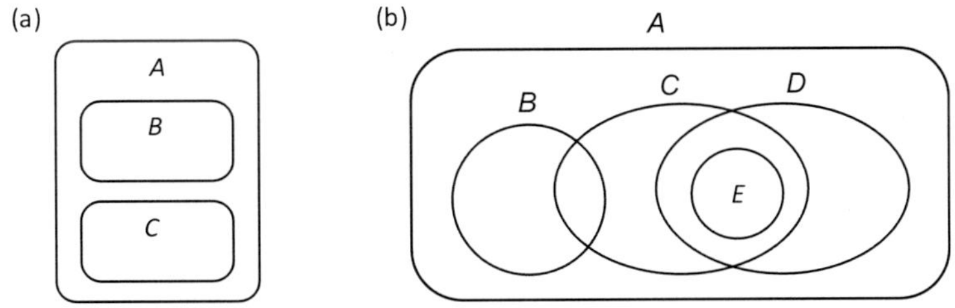

FIG. 6.31 Euler diagrams are okay for simple subtyping (a) but not complex cases (b).

Earlier in the chapter, Euler diagrams were used to depict one set as a subset of another. In designing a conceptual schema, we often need to introduce subtypes and spell out clearly what is a subtype of what. Euler diagrams may be used to do this for simple cases, such as Fig. 6.31(a) where *B* and *C* are subtypes of *A*.

Interpreting this as a hypothetical Euler diagram, the question is open as to whether *A* contains any elements not in *B* or *C*. If *A* does have more elements, this can be shown by adding an asterisk to the region outside *B* and *C*. As discussed in Section 8.2, the Barker notation for ER instead uses a special type called "Other" for this purpose.

The good thing about Euler diagrams is that they show containment intuitively by placing one set or type spatially inside another. This works well for simple cases. In practice, however, an object type may have many subtypes, which might overlap, and a subtype may have many supertypes.

For example, the subtype pattern in Fig. 6.31(b) depicts the following information: *B*, *C*, and *D* are subtypes of *A*, and *E* is a subtype of both *C* and *D*. Moreover, *B overlaps* with *C* (i.e., they may have a common instance) and *C* overlaps with *D*, but *B* and *D* are *mutually exclusive* (cannot have a common instance). For example, *A*=Person; *B*=Asian; *C*=Consultant; *D*=American; *E*=TexanConsultant.

For such cases, Euler diagrams become hopelessly complicated. Moreover, individual subtypes may have many specific details recorded for them, and there is simply no room to attach these details if the subtype nodes are crowded together inside their supertype nodes.

For such reasons, Euler diagrams are eschewed for nontrivial subtyping. What we need is a simple notation that can be used to display subtypehood and subtype roles no matter how large or complex the situation is. This is achieved by displaying subtypes *outside* their supertype(s) and depicting the subtype-supertype connections by means of an *arrow*.

The basic idea is shown in Fig. 6.32. Here, *A* and *B* are object types. Suppose *A*=Employee and *B*=Manager. The entity type Employee is the set of all employees about which facts might possibly be recorded. The subtype Manager is the set of all managers about which facts might be recorded. At any given database state, pop(Employee) and pop(Manager) are, respectively, the set of employees and set of managers actually referenced in the database.

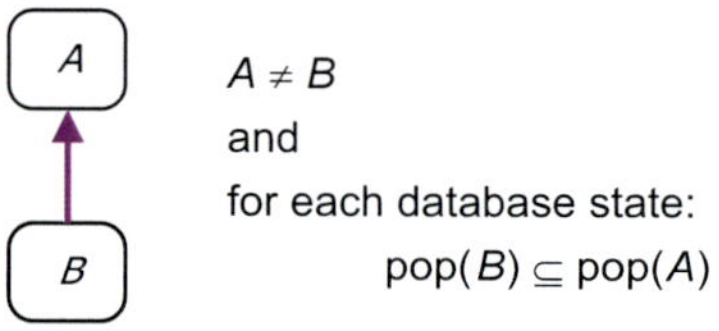

FIG. 6.32 *B* is a proper subtype of *A*.

In general, *B* is a **proper subtype** of *A* if and only if pop(*B*) is always a subset of pop(*A*), and $A \neq B$. In this case, *A* is a *proper supertype* of *B*. If $A = B$, we do not specify any subtype connection. With this understanding, we usually omit "proper" when speaking about subtypes and supertypes.

The adjective "proper" applies to the type relationship, but not necessarily to the *population* relationship. There may be a database state in which pop(*A*) = pop(*B*). For example, if no facts about *A* have yet been entered, both pop(*A*) and pop(*B*) equal {}; with our example, we may choose to enter information about managers before the other employees.

This notation is less intuitive than Euler diagrams because a subtype is shown "outside" its supertype, even though every object instance within the subtype must also be contained inside the supertype. Moreover, information about subtype overlapping is lost unless we add some way to capture it.

This is a consequence of depicting subtype connections by arrows rather than spatial containment. However, this disadvantage is more than offset by the ability to conveniently represent subtype patterns of arbitrary complexity while still allowing plenty of space around each node for details to be attached.

Using this notation, the complex subtype pattern shown in the Euler diagram of Fig. 6.31(b) may be modeled as shown in Fig. 6.33. Here, *A* is a *common supertype* of *B*, *C*, and *D*, and *E* is a *common subtype* of *C* and *D*. In general, a common supertype is at least the union of its subtypes; if it is the union of its subtypes, then its subtypes are said to *exhaust* it. In our example, *B*, *C*, and *D* do not exhaust *A*. Note that Fig. 6.33(a) depicts only the subtype connections.

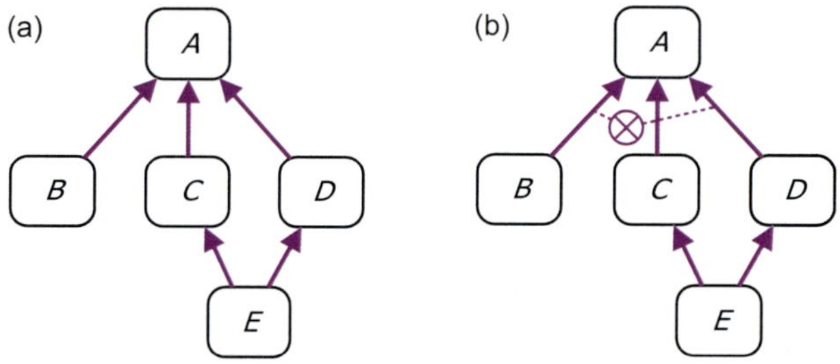

FIG. 6.33 This is another way to model the subtype connections in Fig. 6.31b.

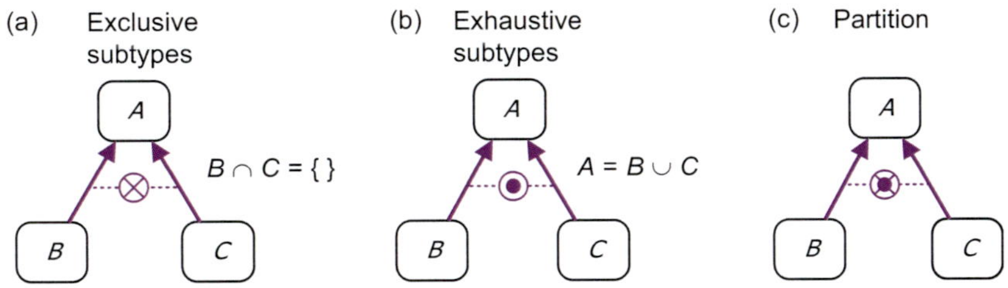

FIG. 6.34 *B* and *C* are mutually exclusive (a), exhaustive (b), or both (c).

Fig. 6.31(b) contains the additional information that *B* overlaps with *C*, *C* overlaps with *D*, and *B* and *D* are mutually exclusive. Hence, if the subtype arrow notation is used, some other means must be used to convey whether subtypes are mutually exclusive. One way to do this in ORM is to attach an *exclusion* symbol "⊗" via dotted lines to the relevant subtype links.

The absence of an exclusion symbol indicates that the populations of the types might overlap. With this added symbol, Fig. 6.33(b) now conveys the exclusion/overlap information of the Euler diagram in Fig. 6.31(b).

To indicate that two or more subtypes are *exhaustive* for their supertype (i.e., their union equals the supertype), an *inclusive-or* constraint symbol "⊙" is connected by dotted lines to the relevant subtype links, as in Fig. 6.34(b). This constraint applies to the populations as well as the types. Each instance in the population of the supertype must appear in the population of at least one of the subtypes.

If we combine both constraints, we have an *exclusive-or* constraint indicating that the supertype is *partitioned* into its subtypes, as shown in Fig. 6.34(c).

Implicitly, underlying each subtype arrow there is a derived, instance-level 1:1 binary association "is/is". The subtype exclusion and exhaustion constraints correspond to exclusion and inclusive-or (disjunctive mandatory) constraints on the roles played by the supertype in these associations—hence, the choice of symbols for these constraints.

In this light, the constraints in Fig. 6.34 may be verbalized respectively as **No** *A* is **both** *B* **and** *C*; **Each** *A* **is some** *B* **or** *C*; **Each** *A* **is exactly one of** *B* **or** *C*.

Subtypes may overlap, especially if they are based on different classification schemes. For example, in Fig. 6.35 the subtypes Australian and Female overlap. Because overlapping subtypes have elements in common, a common subtype may be formed from one or more elements in their intersection. For example, the subtype FemaleAustralian is a subtype of both Australian and Female.

The subtypehood relation is *transitive*: if *A* is a subtype of *B* and *B* is a subtype of *C*, then *A* is a subtype of *C*. In this case, *A* is said to be an *indirect subtype* of *C*. In Fig. 6.35, for example, FemaleAustralian is a direct subtype of Australian, which is a direct subtype of Person. Hence, FemaleAustralian is an indirect subtype of Person. For exposition purposes, this indirect link is marked on the diagram with an asterisk to show its derived nature. To avoid clutter, however, indirect subtype links should be omitted from diagrams because they are transitively implied.

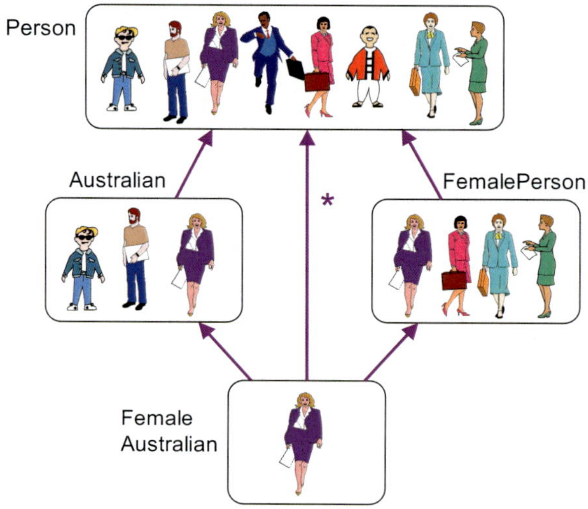

FIG. 6.35 The indirect subtype connection is implied, so should be omitted.

If a subtype has two or more direct supertypes, this is called *multiple inheritance*. For example, in Fig. 6.35, FemaleAustralian inherits from both Australian and FemalePerson. In general, a subtype is at most the intersection of its supertypes. If a subtype has only one direct supertype, this is called *single inheritance*.

A supertype may have many direct subtypes, and a subtype may have many direct supertypes, so in general we have a *subtype graph* rather than a tree. The supertype and subtypes are referred to as the *nodes* of the graph. Because the arrowheads provide direction, we have a directed graph.

Because no type can be a proper subtype of itself, no cycles or loops are permitted, no matter what the length of the cycle (Fig. 6.36 shows two illegal cycles). Thus, any pattern of type-subtype relationships forms a *directed acyclic graph*.

An entity type that is not a proper subtype of any other entity type in the schema is said to be a *primitive entity type* for the schema. In Fig. 6.30, for example, Person, Gender, and Country are primitive entity types. A single conceptual schema may have many subtype graphs (e.g., one based on Person and one based on Vehicle). Each subtype graph must stem from exactly one primitive entity type (which may be an objectified association) that is the common supertype, or *root* node (or *top*) of that graph.

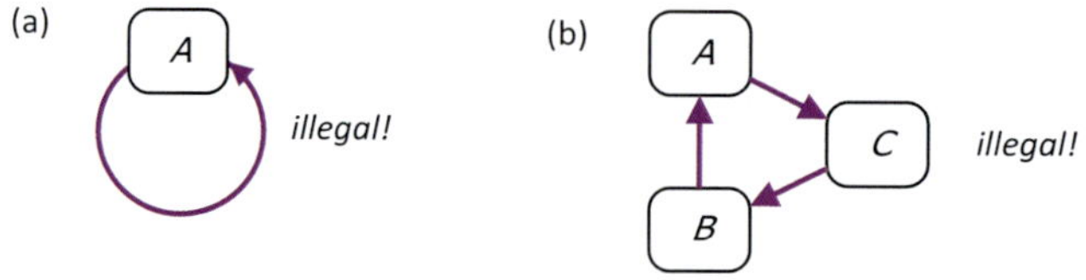

FIG. 6.36 No cycles are permitted in a subtype graph.

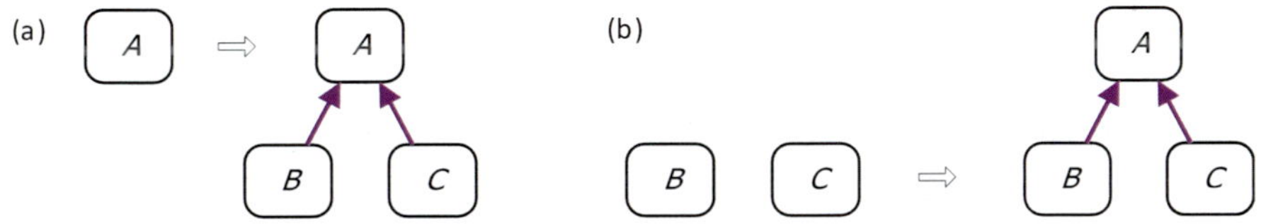

FIG. 6.37 A subtype graph may arise by (a) specialization or (b) generalization.

Subtype graphs have only one root because primitive entity types are mutually exclusive. In contrast, subtypes in a graph necessarily overlap with their supertype(s) and might even overlap with one another (e.g., Australian and Female). Because root entity types are mutually exclusive, there is no overlap between entity subtypes from different subtype graphs. To reinforce these general ideas about subtypes, you may wish to try Question 1 of the section exercises before continuing.

A subtype graph may arise in a top-down way by *specializing* an object type into subtypes (subtype introduction), as in Fig. 6.37(a). A subtype graph may also arise in a bottom-up way by *generalizing* object types to a common supertype and retaining the subtypes for specific details (supertype introduction), as in Fig. 6.37(b). Sometimes a subtype graph arises from a combination of specialization and generalization. The rest of this section discusses the specialization process, and the following section deals with the generalization process.

Now consider the extract from a hospital patient record system shown in Table 6.4. Only males have a prostate gland. In later life, this gland may cause various medical problems. Only women can become pregnant. In this table, a question mark "?" is an ordinary null, indicating that an actual value is unknown. The *minus sign* "–" is a special null, indicating that an actual value is *not to be recorded* there.

We met something like this when considering subset and exclusion constraints. But there is a difference. Here, the "–" means that an actual value is *inapplicable because of the specific value of some other entry* (or entries) for the entity involved.

On the first row, pregnancies must not be recorded for patient 101 because that patient is male ("M" entry for gender). Prostate status must not be recorded for women. The number of pregnancies must be recorded for women, even if this number is zero.

In contrast to "–", the "?" merely indicates that the information is missing. For example, patients 103 and 104 might have no phone, and patient 105 might not have had his prostate checked. Suppose we schematize this UoD as in Fig. 6.38.

Table 6.4 Details about hospital patients.

PatientNr	*Name*	*Gender*	*Phone*	*Prostate status*	*Pregnancies*
101	Adams A	M	2052061	OK	–
102	Blossom F	F	3652999	–	5
103	Jones E	F	?	–	0
104	King P	M	?	benign enlargement	–
105	Smith J	M	2057654	?	–

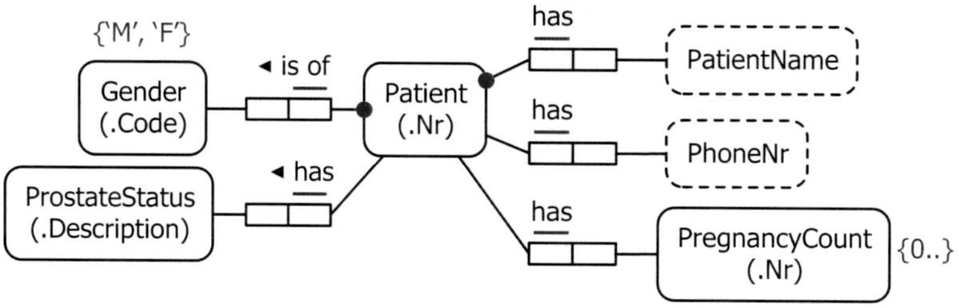

FIG. 6.38 An incomplete conceptual schema for Table 6.4.

Notice the optional roles. Although correct as far as it goes, this schema fails to express the constraints that prostate status is recorded *only for* male patients, and the number of pregnancies is recorded *just for* the females. The phrase "just for" means "for and only for" (i.e., "for all and only"). To capture these constraints, we introduce subtypes and attach their specific roles, as shown in Fig. 6.39.

The subtypes MalePatient and FemalePatient are marked with an *asterisk* "*" to indicate that they are *derived subtypes* (i.e., they are derived by applying a derivation rule to their supertype(s)).

The derivation rules appear as formal *subtype definitions* beneath the diagram. By default, subtypes inherit the identification scheme of their supertype, so there is no need to repeat it here.

Recall that a role is played only by instances of the object type to which the role is attached. Hence, the subtyping reveals that prostate status is known (to the information system) **only if** the patient is male, and pregnancy count is known **only if** the patient is female. The role attached to MalePatient is optional. Not all men need to have their prostate status recorded.

However, the role attached to FemalePatient is mandatory. So, pregnancy count is recorded **if** the patient is female. The combination of the subtype constraint and the mandatory role on FemalePatient means that a pregnancy count is recorded **if and only if** the patient is female.

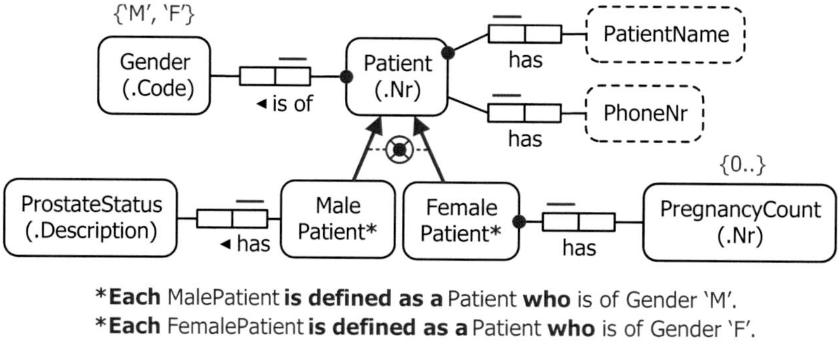

FIG. 6.39 Subtyping completes the conceptual schema for Table 6.4.

In this case, the MalePatient subtype plays only one role, and this is optional. Unlike primitive entity types, there is no default assumption of mandatory if only one role is attached to a subtype. *With subtypes, any mandatory role constraints must be explicitly shown* (including disjunctive cases).

A subtype with no specific roles attached is said to be an *inactive subtype*. For example, MalePatient and FemalePatient subtypes might be introduced merely for taxonomy purposes as a way to display that patients can be classified in this way. In ORM, we rarely use inactive subtypes, as it is usually better to simply display such a classification scheme as a binary fact type with a value constraint (e.g., Fig. 6.39 uses Patient is of Gender {"M", "F"}) or as unaries (e.g., isMan and isWoman may be attached to Patient as exclusive, disjunctively mandatory unaries). The binary form is much neater when the classification scheme involves several subtypes.

While specific fact types are attached to the relevant subtype, common fact types are attached to the supertype. In this example, name, gender, and phone details may be recorded for any patient, so are attached directly to Patient. In general, each subtype inherits all the roles of its supertype(s). Although drawn outside its supertype, the subtype is totally contained inside the supertype. In Fig. 6.39, all instances of MalePatient or FemalePatient have their name, gender, and possibly phone number recorded. Attaching a common fact type to the supertype avoids duplicating it on all the subtypes.

To determine membership of derived subtypes, *each derived subtype must be defined in terms of at least one role played by its supertype(s)*. With the present example, the gender fact type is used to determine membership in MalePatient and FemalePatient. The subtype definitions are formal—they are not just comments.

The operator "**is a**" or "**is an**" is used for "is defined as", and "**who**" or "**that**" is used after the supertype name. For persons, "who" sounds more natural than "that". The quantifier "**Each**" normally precedes the name of the subtype being defined, as shown in Fig. 6.39.

Here, each MalePatient is a Patient but because derived subtypes must be well defined, this "is a" connection must be *qualified* in some way. For example, **Each** MalePatient **is a** Patient **who** is of Gender 'M.' Subtyping provides a means of *qualifying an optional role, a simple set-comparison constraint.*

Subtyping may also be used *instead of unqualified, simple set-comparison constraints, especially if the subtype plays many specific roles.* Consider a UoD where each employee drives cars or catches buses but not both. Cars or buses are recorded for all employees, bus allowance is recorded just for the bus catchers, and driver's license numbers are recorded only for the drivers. This can be modeled without subtypes, using exclusion, equality, and subset constraints. As an exercise, draw the schema for this and then try a subtyping solution.

Different subtyping approaches are possible. One solution is to specify the fact type Employee drives Car and then introduce the subtypes Driver (Employee **who** drives **some** Car) with the license predicate attached and NonDriver (Employee **who** drives **no** Car) with both bus caught and bus allowance predicates attached. The more subtype-specific information required (e.g., total distance traveled by bus, driving violations), the tidier the subtyping portrayal becomes in comparison to the no-subtyping alternative. Note, however, that subtyping cannot replace composite set-comparison constraints (between sequences of two or more roles).

Be careful to *avoid circular definitions*—don't try to define a subtype in terms of itself. For example, suppose we defined BusCatcher as Employee **who** catches **some** Bus and then attached the bus predicate to this subtype (i.e., BusCatcher catches Bus). This specifies that the role of catching a bus is played only by those who play the role of catching a bus. Such circular reasoning is not very informative! In very rare cases, a role used to define a subtype may be a reference role. For example, we might record some property only for people whose surname begins with "Z".

For the schema of Fig. 6.39, the subtypes MalePatient and FemalePatient are mutually exclusive and collectively exhaustive. When these two conditions are met, the subtypes are said to form a partition of their supertype. We may display this situation on a hypothetical Euler diagram, as shown in Fig. 6.40(a). This is analogous to the process of partitioning or slicing a pie or pizza into pieces, as shown in Fig. 6.40(b). The pieces don't overlap (mutual exclusion), and all the pieces together make up the whole pizza (collective exhaustion).

Our understanding of the subtype names "MalePatient" and "FemalePatient" helps us to "see" a partition. But such names are only character strings to the computer system, so how is the partition formally captured in the model?

Look back at Fig. 6.39. The mutual exclusion between MalePatient and FemalePatient is implied by their definitions, together with the uniqueness constraint on the gender predicate (**Each** Patient is of **at most one** Gender). The constraint that MalePatient and FemalePatient exhaust Patient is implied by the subtype definitions, the {"M", "F"} constraint, and the mandatory role on the gender predicate (**Each** Patient is of **at least one** Gender). Because ORM demands that derived subtypes be well defined and all relevant constraints on defining predicates be declared (graphically or textually), *any exclusion or exhaustion constraints on derived subtypes are always implied.*

As humans, we can use our understanding of meaningful subtype names to "see" whether subtypes are exclusive or exhaustive. This shortcut depends on a judicious choice of subtype names. For example, if we unwisely chose the subtype names "Man" and "Woman" to partition the Patient type, the exhaustion constraint would not apply in a UoD that includes child patients or adult people who are not patients.

For clarity, it is usually best to display subtype exclusion and exhaustion constraints explicitly, even if these are implied. With complex subtype graphs, the explicit display of such implied constraints can make the diagram cluttered. Ideally, a computer-aided software engineering (CASE) tool should enable the display of such constraints to be toggled on or off.

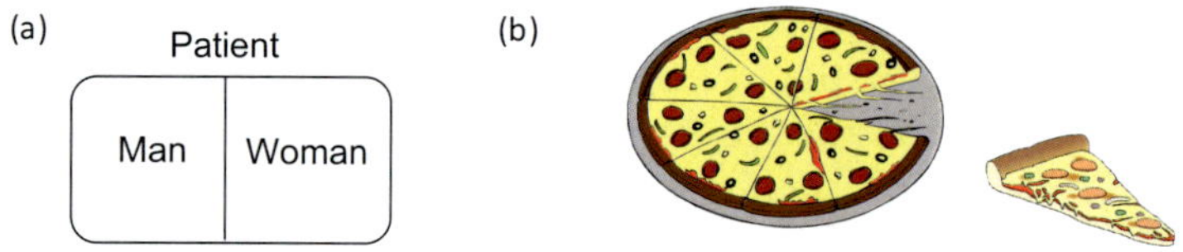

FIG. 6.40 A partition of Patient (a) and a pizza (b).

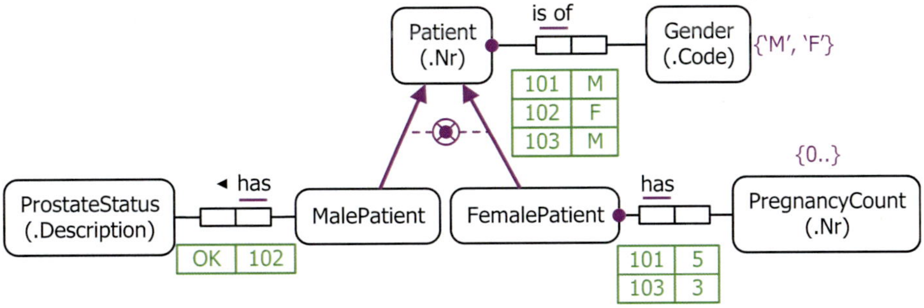

FIG. 6.41 The constraints are satisfied, but can you spot a problem?

Because MalePatient and FemalePatient are exclusive, and an object type must be populatable, no object type can be a subtype of both because such a type would always be empty (nobody is simultaneously both a man and a woman). In general, *exclusive object types cannot have a common subtype.*

In some other modeling approaches, a classification predicate or attribute (e.g., gender) may be specified, then subtypes introduced but not formally defined, and relevant exclusion and exhaustion constraints added. This alternative approach suffers from the problem that *exclusion and exhaustion constraints are weaker than subtype definitions.* For example, consider the populated schema in Fig. 6.41. Here, the partition constraint is declared but the subtype definitions are removed. The sample population satisfies all the constraints, but there is still something wrong. Can you spot the problem?

As you no doubt noted, prostate status is recorded for patient 102 (a female) and a pregnancy count is recorded for patient 101 (a male). This is nonsense.

We need to record the gender of each patient regardless of their subtype roles, but if they do play a subtype role we need to check that they are in the right subtype. The normal way to do this in ORM is to define the subtypes formally in terms of their gender and then enforce the definition. Fig. 6.42 shows the correctly populated schema for this approach.

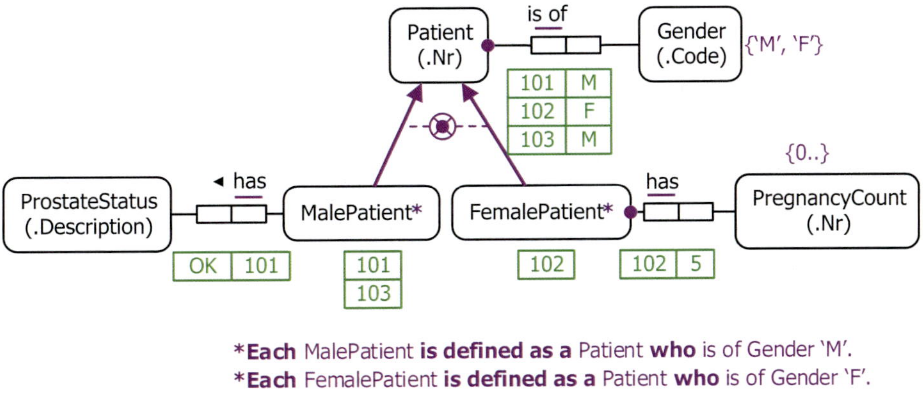

FIG. 6.42 The subtype definitions are stronger than the partition constraints.

In conjunction with the constraints on the gender fact type, the subtype definitions ensure not only that the Patient type is partitioned into the subtypes, but that patient instances appear in the correct subtype.

Note that *subtype exhaustion constraints apply to the populations of the subtypes, not necessarily to their active populations* that play a specific subtype role. Here, the prostate status role is optional for male patients, so some instances of MalePatient (in this case patient 103) need not play this role.

The problem with the model in Fig. 6.41 arose because the patient taxonomy was expressed in two ways (with a gender fact type and with subtyping) that were not kept in sync. The solution in Fig. 6.42 used subtype definitions to keep these two specifications of the taxonomy consistent. An alternative solution is to specify the taxonomy in one way only, removing the gender fact type, and simply asserting the subtyping scheme with no subtype definitions, as shown in Fig. 6.43.

A subtype that is simply asserted (no derivation rule) is called an *asserted subtype*. Unlike derived subtypes, asserted subtypes are not marked with an asterisk. Introducing an asserted subtype is equivalent to attaching a unary predicate (e.g., isMalePatient) to the supertype. With asserted subtypes, any subtype exclusion/exhaustion constraints must be explicitly declared, as there are no subtype definitions to imply them.

A third alternative is to derive a taxonomy fact type from the asserted subtypes, as shown in Fig. 6.44. Although this practice is typically closer to the way subtyping is implemented in object-oriented programming, the first solution using an asserted taxonomy fact type for gender and derived subtypes with definitions is closer to the way subtyping is best implemented in relational database systems.

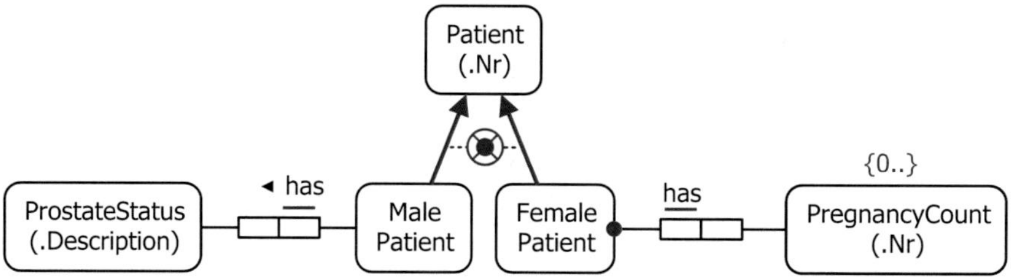

FIG. 6.43 With asserted subtypes, exclusion/exhaustion constraints are not implied.

But the main point to remember is that if a taxonomy is captured both by a classifying fact type and by subtyping, then these two ways must be kept consistent by means of derivation rules (either subtype definitions or fact type derivation rules).

Sometimes a classification predicate is required. For example, we may need to record the salary of employees and then use this to subtype employees. Employees with a salary $\geq$\$80,000 might be classified as HighPaidEmployee, for instance. Here, we could not derive the precise salary of an employee by knowing his or her membership in an asserted subtype. For such reasons, as well as uniformity, in ORM we normally prefer to derive rather than assert subtypes.

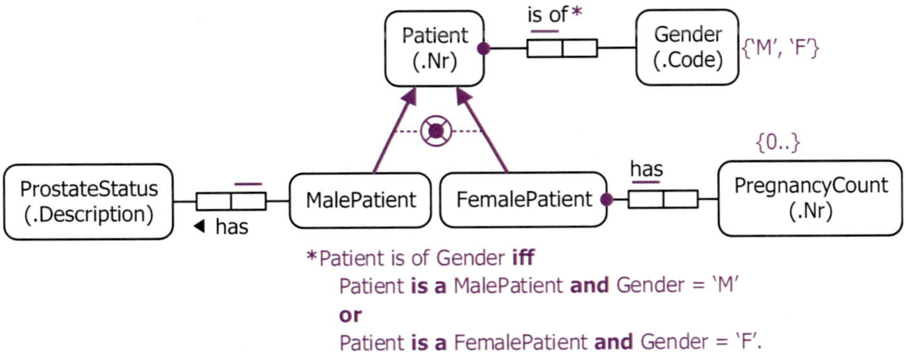

FIG. 6.44 Deriving gender from membership in asserted subtypes.

In very rare cases, a subtype may be semiderived (some instances may be derived, and some may be simply asserted). Whereas derived subtypes are fully determined by their definitions, *semiderived subtypes* are only partly determined by an implication rule. This rare situation usually occurs only when we have incomplete knowledge about the fact type used in the derivation rule.

For example, in Fig. 6.45(a) the Grandparent subtype may be derived because we have complete knowledge about parenthood facts. If instead we have incomplete knowledge of parenthood, then we might know that someone is a grandparent without knowing who their children or grandchildren are. This situation is depicted in Fig. 6.45(b) using a semiderived subtype for Grandparent.

Semiderived subtypes are marked with a "+" superscript (intuitively, half an asterisk), and the derivation rule is specified as an implication rather than an equivalence rule (either by inserting "derived" before the subtype name or by using "if" instead of "iff").

The *specialization* (subtype introduction) procedure may now be summarized as follows. Here, "known subtype" means that the subtype is known in the UoD. A subtype definition is stronger than a set-comparison constraint if it adds a restriction to a defining predicate (e.g., Gender "M", Rating > 3). The procedure is recursive: because a subtype may itself have optional roles, we apply the procedure to it to see if we need to form subtypes of it.

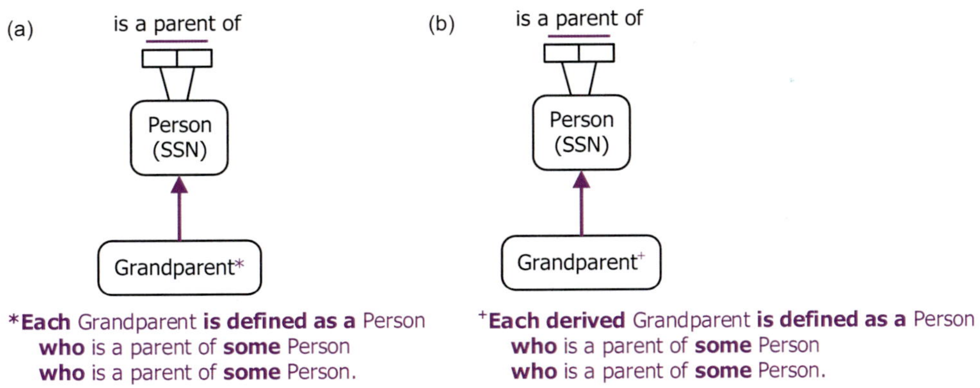

FIG. 6.45 A derived subtype (a), and a semiderived subtype (b).

Specialization Procedure (SP):

Specify all mandatory role constraints;

For each optional role:
if it's recorded only for a known subtype.
and (there is a subtype definition stronger than a set-comparison constraint
or another role is recorded only for this subtype)
then introduce the subtype;
attach its specific roles;
declare the subtype derived (*), asserted, or semiderived (⁺);
unless the subtype is asserted, add a subtype derivation rule;
apply SP to the subtype.

Note that exclusion and exhaustion constraints among subtypes are to be interpreted in a *static* sense. For example, because the subtypes MalePatient and FemalePatient are exclusive, in any given database state their populations cannot overlap. However, this does not rule out an object migrating from one subtype to another between database states. This is unlikely for our present example!

However, our next example includes the subtypes SeniorLecturer and Professor (see Fig. 6.46). SeniorLecturer and Professor are exclusive in the sense that no academic can be a member of both types simultaneously. Note that a senior lecturer in one state can be promoted to professor in a later state. The subtypes SeniorLecturer and Professor are said to be *role types* (migration allowed) rather than rigid types (migration forbidden). This distinction is discussed in more detail in Chapter 10.

In this domain, academics have exactly one rank (lecturer, senior lecturer, or professor) and have one or more degrees. Each professor holds a unique "chair" (e.g., information systems), indicating the research area that he or she manages. Each student has at most one counselor, and this must be a senior lecturer.

Nothing special is recorded for lecturers, so we don't bother introducing a subtype for them. The subtype definitions and the constraints on the rank predicate imply that the SeniorLecturer and Professor subtypes are exclusive but not exhaustive. The exclusion constraint is implied.

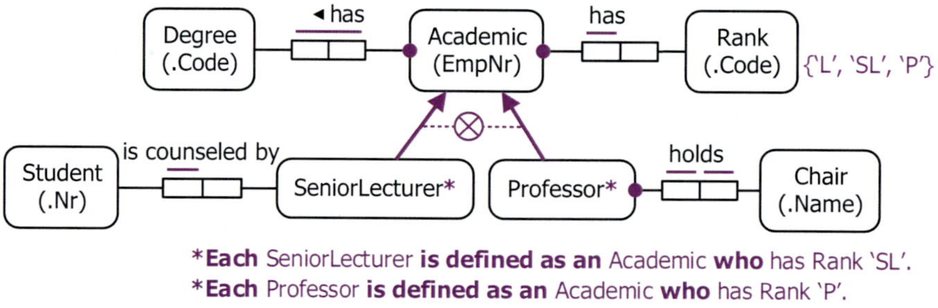

FIG. 6.46 The subtypes are exclusive but not exhaustive.

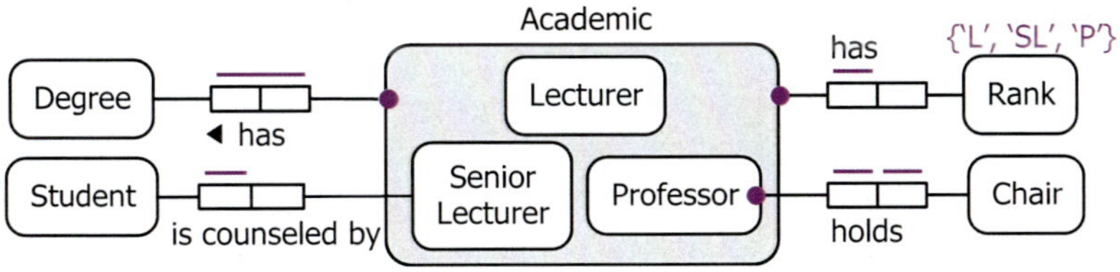

FIG. 6.47 The subtyping of Fig. 6.46 shown with a hypothetical Euler diagram.

Although subtypes are drawn outside their supertypes, every member of a subtype is also a member of its supertype(s).

To clarify this point, Fig. 6.47 depicts the subtyping of Fig. 6.46 by means of an HED; the shaded region is empty. For simplicity, reference schemes and subtype definitions are omitted. Although such diagrams may be used in simple cases for explanatory purposes, they are too awkward for complex cases and are no replacement for our subtyping notation.

Consider now a UoD peopled just by lecturers and students. Gender and PersonKind (L = Lecturer; S = Student) are recorded for all persons, salary is recorded just for lecturers, and course of study is recorded just for students (see Fig. 6.48(a)). Here, a person may be both a lecturer and a student, as indicated by the many:many constraint. So, the Student and Lecturer subtypes overlap. Although not exclusive, they are exhaustive, given the mandatory role and {"L", "S"} constraint on the personkind fact type.

Fig. 6.48(b) displays the subtyping on an HED. Because these subtypes overlap, they may have a common subtype. For example, if some further information was required only for those lecturers studying a course, we could introduce a subtype StudentLecturer (both Student and Lecturer).

Suppose we expanded the UoD to include administrative staff, who may be students but not lecturers. This classification scheme may be depicted by three unaries (studies, lectures, and administrates) with an exclusion constraint between the last two. If specific details were required for administrators, we would introduce a third subtype for this. As an exercise, draw a schema diagram and an HED for this situation.

The rest of this section focuses on the problem of providing a correct definition for subtypes. Sometimes it is not obvious from an output report what criteria have been used to decide whether some fact is to be recorded. As a simple example, consider Table 6.5. What rule determines when to record a person's favorite group?

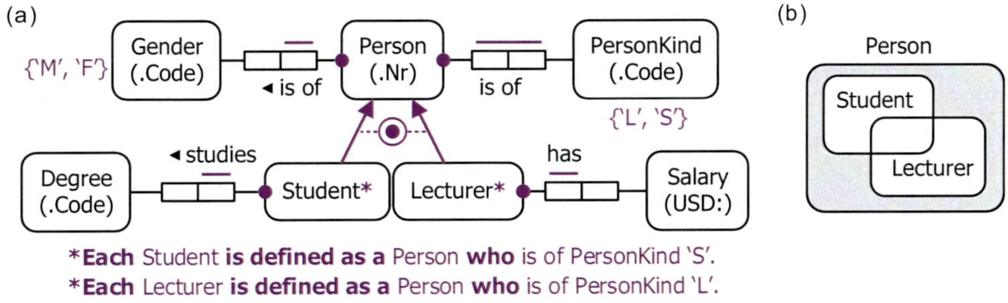

FIG. 6.48 Here, Student and Lecturer are overlapping and exhaustive subtypes.

Table 6.5 What determines whether we record a person's favorite group?

Person	*Age*	*Favorite group*
Bill	17	Abba
Fred	12	–
Mary	20	–
Sue	13	Dire Straits
Tom	19	The Beatles

One pattern that fits the sample data is to record a person's choice of favorite group if and only if the person is aged between 13 and 19 inclusive (i.e., the person is a teenager). This is not the only pattern that fits, however. Maybe favorite group is recorded just for people with odd ages (13, 17, and 19 are odd numbers).

While these might be the two most "obvious" patterns, there are infinitely many patterns that are consistent with the data. Just based on age, we could specify any set of natural numbers within the possible age range that included 13, 17, and 19 but excluded 12 and 20 (e.g., {13, 17, 19, 1}, {13, 17, 19, 5, 63} and so on).

A database population is *significant with respect to a constraint* if and only if it satisfies no alternative choice for this constraint (excluding weaker constraints implied by this choice). Obtaining a sample population that is significant with respect to uniqueness constraints can be tedious, but it can be done in a modest amount of time. Unfortunately, no decision procedure exists to automatically churn out the correct subtype definitions.

If we remove the restriction that a computer system has finite memory, then we encounter the problem that for any finite set of data, there will always remain an infinite number of possible patterns that fit the data. This is the basis of the philosophical "problem of induction".

As a scientific example, consider the graph shown in Fig. 6.49. This shows three data points determined from an experiment in which the pressure (p) and absolute temperature (T) of a gas are measured at a fixed volume. What pattern, rule, or law does this suggest to connect pressure and temperature?

One pattern that fits the data is the linear relationship between pressure and temperature shown by the straight line in Fig. 6.50(a). This shows that when the volume of gas is fixed, its pressure is directly proportional to its absolute temperature. You may recall learning this law of physics in high school. However, the curved line in Fig. 6.50(b) also fits the data. How do we know which (if any) is correct?

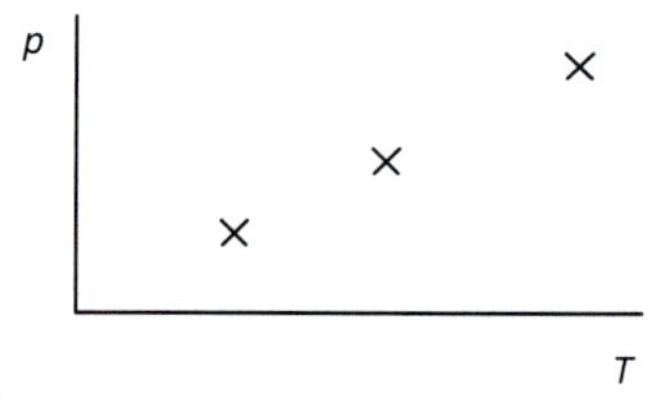

FIG. 6.49 What pattern matches the data?

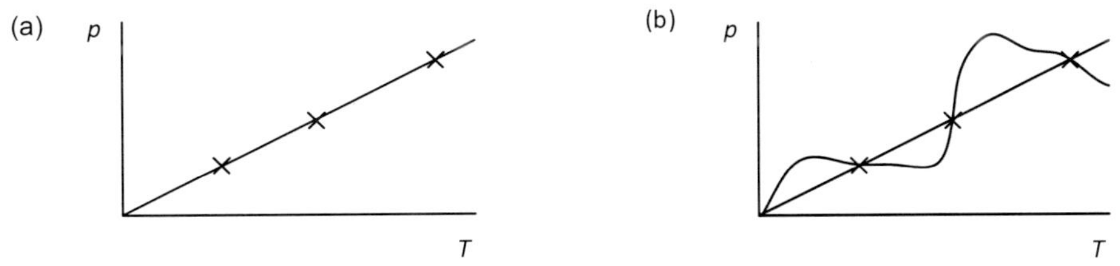

FIG. 6.50 More than one pattern fits the data.

An obvious move is to pick a temperature where the two patterns differ in their pressure reading, and then measure the actual reading. The result is shown in Fig. 6.51(a). Here, the extra data point (the third "×") fits the straight-line hypothesis, but not the curved line. This counterexample is enough to disprove that curved line. Does this prove that the straight-line law is correct? No. We can always add another curved line to fit the expanded data, as shown by the dotted line in Fig. 6.51(b). Indeed, there are infinitely many curved lines that fit the data points. You can keep this process going as long as you like. Any finite set of data is consistent with infinitely many rules. Hence, you can never really prove any scientific law.

Returning to our subtype definition problem, it is clear that any subtype data sample is consistent with infinitely many subtype definitions. Instead of getting bogged down in philosophical speculation at this point, you can use the following pragmatic approach to finalizing subtype definitions.

Look carefully at the data, use "common sense" to spot the simplest pattern, and then check with the domain expert whether this pattern is the one intended. Because the subtyping constraints reflect the decision of the domain expert on what should be recorded for what, the domain expert will always be able to resolve the matter. In a sense, it's like asking God what the rules are for the universe He/She created.

Although output reports cannot be significant with respect to subtype definitions, input forms such as tax returns or application forms often do provide a significant set of conditional instructions from which the subtyping definitions can be deduced. Typically, such forms include a number of fields to be filled in only by certain kinds of people. To let users know which questions they must answer, instructions are included that reveal the conditions under which the specific details are to be recorded. From these instructions, it is usually fairly easy to determine the subtyping.

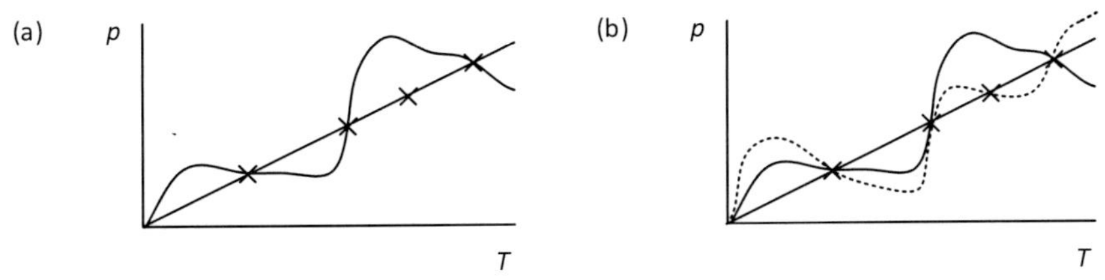

FIG. 6.51 Extra data can eliminate a pattern, but other patterns may still fit.

FormNr: 5001

1. Age (years): ....

2. Nr hours spent per week watching TV: ....

If you answered 0 then go to Question 4

3. What is your favorite TV channel? ....

---

4. Nr hours spent per week reading newspapers: ....

If you answered 0 then Stop (no more answers are required).

5. What is your favorite newspaper? ....

If you are younger than 18, or answered 0 to question 2 or 4 then Stop (no more answers are required)

---

6. Which do you prefer as a news source? (Check the box of your choice)

☐ Television

☐ Newspaper

FIG. 6.52 A sample media survey form.

As a simple example, consider the media survey form shown in Fig. 6.52. Each copy of the form has a unique form number. Each form is filled out by a different person, and each person fills out only one form. The enforcement of this 1:1 correspondence between people and forms is the responsibility of those conducting the survey rather than the information system itself. This correspondence is left implicit in the model, where people are identified directly by the form number.

The conditional instructions on this form are shown in italics. Everybody must answer questions 1, 2, and 4. Anybody who answers 0 to question 2 is told to skip question 3. Hence, favorite TV channel is recorded just for those who indicate they do watch some TV; let's call this subtype Viewer. People who answered 0 for question 4 are told to skip all later questions. Hence, favorite newspaper (question 5) is recorded just for those who read newspapers; let's call this subtype Reader.

People who are younger than 18 or are not both viewers and readers skip question 6. So, the preferred news source is recorded just for adult viewers who are also readers; let's call this subtype MediaAdult. This leads to the schema of Fig. 6.53.

Alternatively, an output report from the domain might help with the modeling (see Table 6.6). The sample data have been carefully chosen to be significant with respect to the subtype graph. Given this, and recalling that the "−" sign means "not to be recorded", we may reason as follows. Age, viewing, and reading figures are recorded for everybody. The set of people for which favorite channel is recorded properly overlaps with the set of people for which favorite paper is recorded, so these correspond to overlapping subtypes. The set of people for which preferred news source is recorded is a proper subset of the previous two sets, so this corresponds to a common subtype.

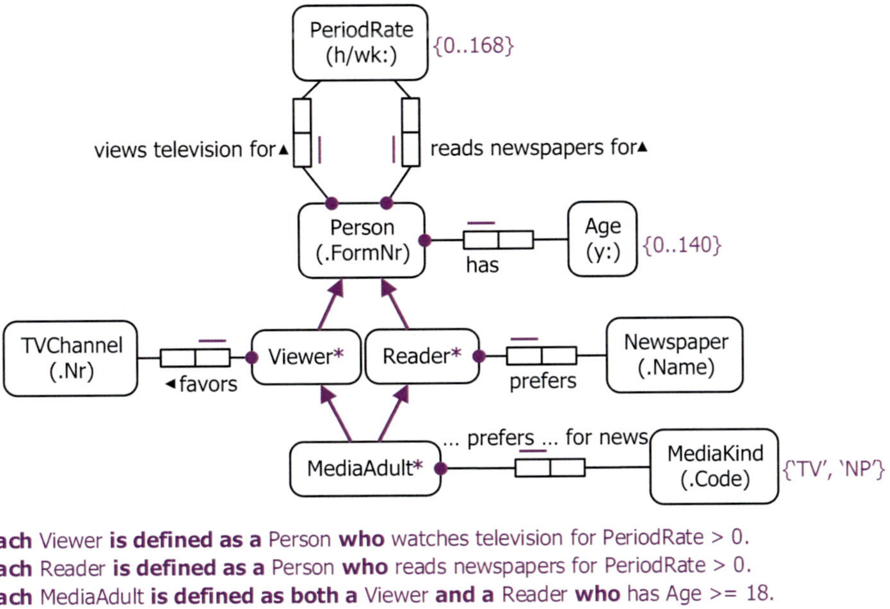

FIG. 6.53 The conceptual schema for the media survey example.

This analysis yields the "diamond-shaped" subtype graph of Fig. 6.53 as well as the information recorded for each node in the graph. However, we can only make educated guesses as to the actual subtype definitions—these should be checked with a domain expert if one is available.

In practice, reports often display different kinds of nulls in the same way. For example, a blank space might be used for both our simply unknown "?" and inapplicable "–" marks. So, you should clarify the meaning of any nulls with the domain expert before deciding how to handle them.

Sometimes, subtype-specific details appear in separate reports, and the titles of these reports may then correspond to subtypes or to subtype components (disjuncts or conjuncts).

Table 6.6 An output report from the media survey.

Person	*Age (y)*	*Television (h/week)*	*Newspaper (h/week)*	*Favorite channel*	*Favorite paper*	*Preferred news*
5001	41	0	10	–	The Times	–
5002	60	0	25	–	The Times	–
5003	16	20	2	9	The Times	–
5004	18	20	5	2	Daily Mail	TV
5005	13	35	0	7	–	–
5006	17	14	4	9	Daily Sun	–
5007	50	8	10	2	Daily Sun	NP
5008	33	0	0	–	–	–
5009	13	50	0	10	–	–

Male Patients:

PatientNr	*Name*	*Phone*	*Prostate status*
101	Adams A	2052061	OK
104	King P	?	benign enlargement
105	Smith J	2057654	?

Female Patients:

PatientNr	*Name*	*Phone*	*Pregnancies*
102	Blossom F	3652999	5
103	Jones E	?	0

FIG. 6.54 Patient data presented in subtype-specific reports.

As a simple example, the hospital patient data discussed earlier in Table 6.4 might be presented in two tabular reports as shown in Fig. 6.54. A more complex example of subtype-specific reports is included in the exercises.

To help you do the section exercises without a domain expert, the data provided are significant with respect to the subtype graph. However, this is an artificial situation. In designing real-world applications, you would rarely be handed data guaranteed to be significant in this way. Only the domain expert could provide this guarantee, and to do this he or she must know the subtyping constraints anyway. It is better to get these constraints directly from the domain expert simply by asking.

Apart from the methods discussed, an online appendix discusses two matrix algorithms to deduce the configuration of a subtype graph from a population significant in this regard. Although it is unsafe to assume that populations are significant, such procedures can at least be used to check whether a population is significant with respect to a known subtype graph.

Exercise 6.5

1. (a) For each of the following subtype graphs, assume that $A = \{1, 2, 3\}$. Provide populations for the subtypes to complete a satisfying model for each diagram.

(b) Explain what is wrong with each of the following subtype graphs.

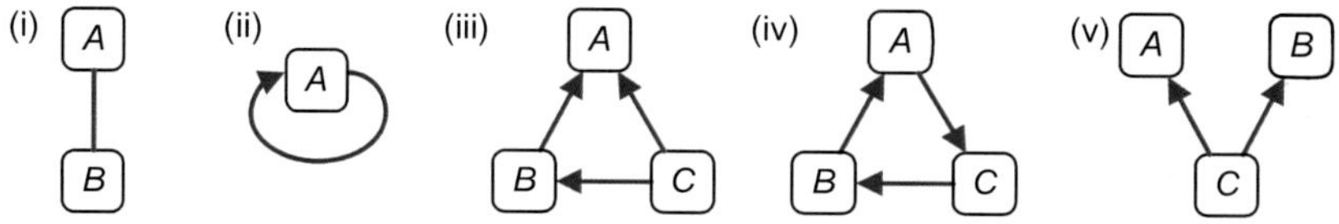

Note: *In the following questions, populations are significant with respect to the subtype graph, and "–" means "not to be recorded".*

2. The following table is a sample weekly report about animals in a certain household. Provide a conceptual schema for this UoD, including uniqueness constraints, mandatory roles, value constraints, and subtype constraints. Include a definition for each subtype.

Animal	*Animal kind*	*Sex*	*Nr cars chased*	*Nr mice caught*
Fido	dog	M	12	–
Felix	cat	M	–	1
Fluffy	dog	F	0	–
Tweetie	bird	F	–	–

3. A simplified fragment of an income tax return form is shown.
   (a) Schematize this UoD, assuming that in section 5 the same employer can appear on only one line.
   (b) Modify your solution to part (a) to handle these changes. The same employer may appear on more than one line of section 5. For each of these lines, the taxpayer attaches a wage and tax statement from his or her employer that includes the employer's name as well as the employee's tax file number, gross salary, and tax installments deducted (i.e., tax paid). Wage and tax statements are identified by serial numbers.

1. Your Tax File Number: ....

2. Your full name: ....

3. Indicate the kind(s) of income you received from an employer by checking the relevant box(es).
   If you had no employer income, skip this and go to question 7.

   ❑ salary ❑ benefit (employment related)

4. Your main employment occupation: ....

5. If you earned a salary, complete the following income details:

Employer	*Tax paid*	*Gross salary*

6. If you earned employment related benefits indicate total benefit: ....

7. etc.

4. Schematize the following table. Here, a supervisor supervises one or more persons. You may find it helpful to draw a tree showing who supervises whom. The missing value "–" means "inapplicable because of some other information". The missing value "---" means "does not exist".

Emp Nr	*Emp name*	*Supervisor*	*Smoker?*	*Home phone*	*Health bonus ($)*	*Company car*
101	Brown CA	102	N	–	300	–
102	Jones E	103	Y	2053771	–	–
103	White TA	---	N	2062050	500	597VTP
104	Mifune K	103	N	3651000	500	123KYO
105	Nelson HF	102	N	–	300	–
106	Adams PC	102	N	3719872	400	321ABC
107	Wong S	104	N	–	300	–
108	Collins T	104	Y	–	–	–
109	Smith JB	106	Y	–	–	–

5. A loan agency records information about clients borrowing money. The following report is extracted from this information system. Clients are identified by clientNr. Marital codes are "D" (divorced); "M" (married); "S" (single); "W" (widowed). Residential codes are: "B" (home buyer); "O" (home owner); "R" (home renter).

ClientNr	*Marital status*	*Residential status*	*Number of dependants*	*Home Value ($)*	*Spouse's income ($)*	*Spouse's share of home (%)*
103	M	O	3	100,000	40,000	50
220	M	R	3	–	0	–
345	S	O	–	150,000	–	–
444	W	B	2	90,000	–	–
502	D	R	0	–	–	–
600	S	R	–	–	–	–
777	M	B	0	100,000	0	40
803	D	B	1	60,000	–	–
905	W	O	0	90,000	–	–

Each client has borrowed one or more loans. For this agency, any given loan can be borrowed by only one client. The following report from the same information system provides details on some of the loans. Schematize this UoD.

ClientNr	*Loans borrowed*		
	LoanNr	*Amount ($)*	*Term (years)*
103	00508	25,000	7
220	00651	25,000	7
etc.	00652	3000	1

6. The following three reports come from an information system about these bodies in our solar system: the Sun, the naked eye planets, and their moons. Planets travel in roughly circular orbits about the Sun. A planet in inferior conjunction is lined up directly between the Earth and the Sun. A planet is in opposition if the Earth is lined up directly between it and the Sun. The missing value "---" means "does not exist", and "M_E" means one Earth mass. An astronomical unit (a.u.) equals 150,000,000 km. Schematize this.

Sun:

Name	*Mass (M_E)*	*Radius (km)*
The Sun	34,000	696,000

Planets:

Name	*Nr of moons*	*Mean distance from Sun (a.u.)*	*Mass (M_E)*	*Radius (km)*	*Orbital period (y d)*	*Next inf. conj.*	*Next oppos.*	*Atmosphere (main gases)*
Mercury	0	0.39	0.06	2440	0y 88d	Dec	–	---
Venus	0	0.72	0.81	6050	0y 224d	Jan	–	CO_2
Earth	1	1.0	1.0	6378	1y 0d	–	–	N_2,O_2
Mars	2	1.5	0.11	3095	1y 322d	–	Feb	CO_2
Jupiter	95	5.2	318.0	71,400	11y 315d	–	Sep	H_2,He
Saturn	145	9.5	95.0	60,000	29y 167d	–	Sep	H_2,He

Moons:

Name	*Planet orbited*	*Radius (km)*	*Orbital period (y d)*	*Mean apparent magnitude*
Luna	Earth	1737	0y 27.3d	-13.0
Phobos	Mars	6	0y 0.3d	11.5
Deimos	Mars	4	0y 30.0d	12.0
Io	Jupiter	1867	0y 1.7d	5.5
etc.				

6.6 Generalization of Object Types

The previous section discussed a specialization procedure in which subtypes of a more general object type are introduced to declare that specific roles are recorded only for these subtypes. Apart from this top-down procedure, a subtype graph may also arise in a bottom-up way when we need to introduce a supertype of object types that already occur in the model.

The process of introducing a supertype for object types that already exist is known as object type *generalization*. Hence, generalization is the reverse of specialization. A supertype is a more general form of its subtypes, and a subtype is a special form of its supertype(s).

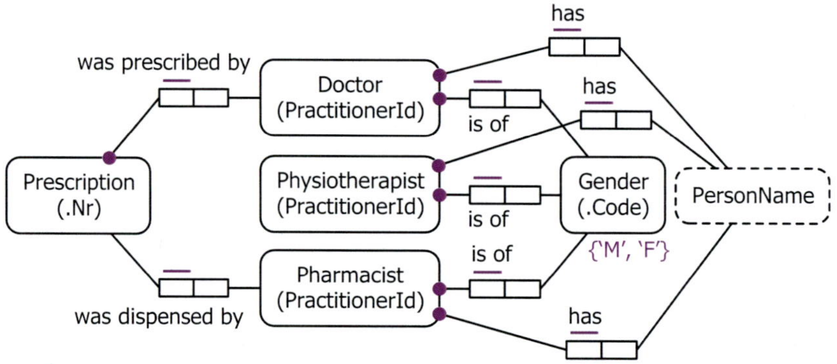

FIG. 6.55 A schema before generalization has been applied.

Step 3 of the CSDP outlined the main reasons for generalizing existing object types (e.g., Doctor, Physiotherapist, Pharmacist) into a more general type (e.g., Practitioner). If the original object types have specific roles not played by the generalized type, we need to retain them as subtypes. For example, consider Fig. 6.55. Here, name and gender are recorded for each kind of practitioner. In this case, a practitioner type is already suggested by the common identification scheme (PractitionerId), but this type has not yet been introduced.

If we want to list all the practitioners together in a query, then we should introduce a Practitioner supertype, attach the common roles to it, and add a fact type to retain the classification scheme (see Fig. 6.56). This assumes pharmacists, physiotherapists, and doctors are mutually exclusive. If a global identifier for the supertype does not already exist, (e.g., PractionerId), then we should introduce one.

The original schema shows that prescriptions are prescribed only by doctors, and dispensed only by pharmacists. These constraints are retained in the new schema by using Doctor and Pharmacist subtypes for the prescription predicate. Here, physiotherapists have no specific role, so we don't introduce a subtype for them.

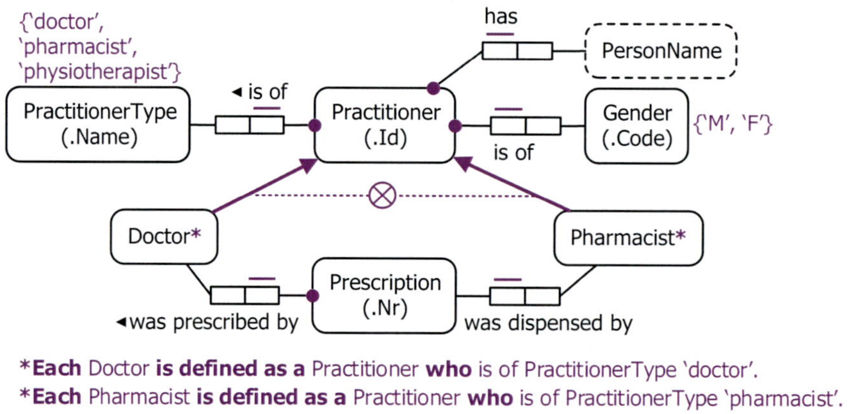

FIG. 6.56 The schema after generalization has been applied.

Doctor, pharmacist, and physiotherapist are mutually exclusive, but were generalized partly because of their common roles (having a gender and a name). In practice, further common information would normally be recorded (e.g., phone number, address, birth date), making this generalization more worthwhile.

However, as pointed out in Step 3, the sharing of a common role is not by itself sufficient to justify generalization. If the object types are exclusive, we needn't generalize them unless we want to list common details for them in the same query.

For our medical clinic example, we would usually want to issue queries such as "List the name, phone number, and address of all the practitioners". Generalization facilitates this if we have a uniform identification scheme for all the practitioners.

To clarify this point, consider a veterinary clinic that records the gender and name of its employees as well as the gender and name of the animals treated in the clinic. Employees are identified by employee numbers, but animals are identified by some other scheme (e.g., combining their name with the client number of their owner).

It is highly unlikely that we would ever list the same information about the staff and the animals together in the same query. So, Employee and Animal may be left as primitive object types without generalizing them to EmployeeOrAnimal.

As a related example, consider the schema of Fig. 6.57. Here an organization records the cars owned by its employees (e.g., to help the car park attendant check whether a car is parked legally). It also records equipment (e.g., computers, photocopiers) owned by its departments.

Cars are identified by their registration number and equipment items by an item number. An ownership predicate appears twice in the schema. Employee and Department are exclusive, but both play the role of owning. Car and Equipment are exclusive, but both play the role of being owned.

Suppose we generalize Employee and Department to Owner, and Car and Equipment to Asset. As an exercise, you might like to draw the resulting schema for yourself before peeking at the solution.

The result of this generalization is shown in Fig. 6.58. It's quite messy. The ownership predicate now appears just once, in the fact type Owner owns/is owned by Asset. But the rest of the schema is harder to conceptualize. New identifiers are required for the supertypes, and explicit fact types are needed to retain the original identification schemes as secondary reference schemes.

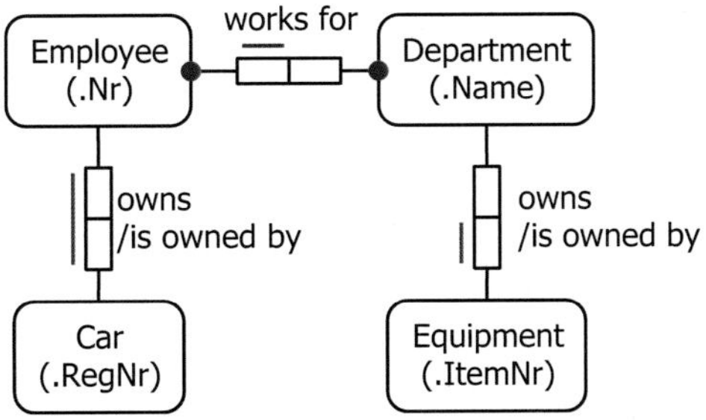

FIG. 6.57 Should the object types be generalized?

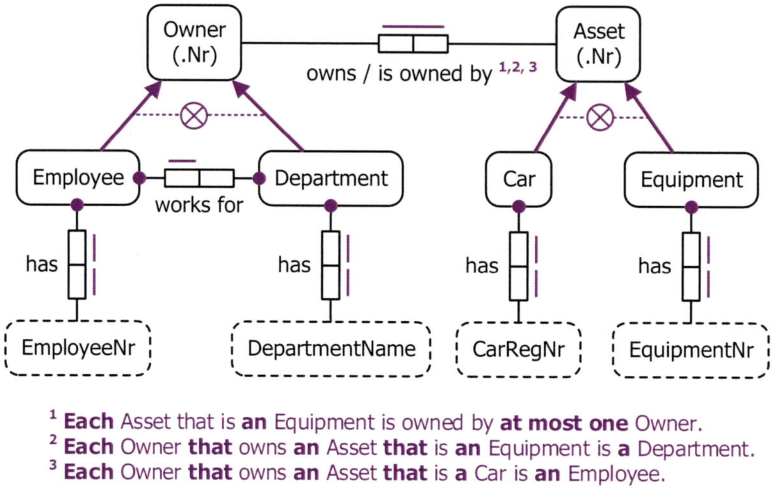

FIG. 6.58 Don't generalize the types in Fig. 6.57 like this unless you need to.

We could add the fact types Owner is of OwnerKind and Asset is of AssetKind, and use these to derive the subtypes, but it is simpler to just assert the subtypes as shown. In the schema of Fig. 6.57, Employee, Car, Department, and Equipment are implied to be mutually exclusive because they are top-level entity types. Exclusion constraints between subtypes are not implied, so in Fig. 6.58 the subtype exclusion constraints are added to ensure that no employee is a department and no car is an equipment item. To allow for other kinds of owners and assets to be added at a later time, the exclusion constraints are not strengthened to inclusive-or constraints.

An additional textual constraint is needed to ensure that each equipment item is owned by at most one Department (as shown in the footnoted constraint). To allow any kind of ownership fact types to be added later, we do not add textual constraints to ensure that only employees own cars, and that only departments own equipment. Such additional textual constraints could be added if desired.

Should we generalize exclusive object types in such a case? Not unless we want to ask a common question for all members of the general type (e.g., List all owners and their assets). If the objects are not to be listed together in the same column of an output report, then there is no need to generalize them. The more common properties there are, the more likely it is that such queries would be formulated.

For example, in many applications such as banking or insurance, a customer might be a person or a company. Because common information needs to be accessed for them in the same query, it is worthwhile generalizing Person and Company to Customer (or Party, or some term appropriate for the application). Of course we still retain Person and Company as subtypes to record details specific to them.

If two fact types have the same predicate reading and object type name sequence, they are treated as the same fact type. However, different predicates can have the same name.

This often happens with trivial predicates such as "has" and "is of". It can also happen if we don't generalize object types that share common roles. For example, in Fig. 6.57 there are two predicates called "owns". We can distinguish between these predicates by their containing fact types (e.g., Employee owns Car differs from Department owns Equipment). In a CASE tool, predicates with the same reading may be distinguished internally by surrogates or expanded names (e.g., P56, P57, or "owns1", "owns2"). Apart from obvious cases such as "has" and "is of", however, if we duplicate a predicate reading on a schema we should ask ourselves whether we should generalize.

The examples so far have exclusive subtypes. What about *overlapping types*? Recall that primitive entity types on a conceptual schema are mutually exclusive. The only way to allow different entity types to overlap is to make them subtypes. So, generally speaking, if we find two primitive entity types that do overlap, then we should introduce a supertype for them. In rare cases, however, we might portray entity types that overlap in reality to be mutually exclusive in our model because we are simply not interested in knowing about any overlap.

Consider the schema shown in Fig. 6.59. This might be part of a university schema. Students are identified by their student number, and university employees are identified by an employee number. Employees are classified as academics, administrators, or general staff. Only academics can teach subjects, and only students can study subjects. Apart from their specific roles, students and employees both have their gender and birthdate recorded. In the global schema, other specific details would be recorded for students (e.g., enrollment mode) and employees (e.g., department), and other common properties would be recorded (e.g., name, address, phone number).

If you cut Fig. 6.59 vertically down the middle, but duplicate Gender, Date, and Subject in both, the left-hand subschema deals with student records and the right-hand subschema deals with employee records. In the past, it was common practice to develop such subsystems independently of one another. Nowadays, many organizations are integrating subsystems into a general system to provide access to more information and to enforce constraints between the formally separate schemas.

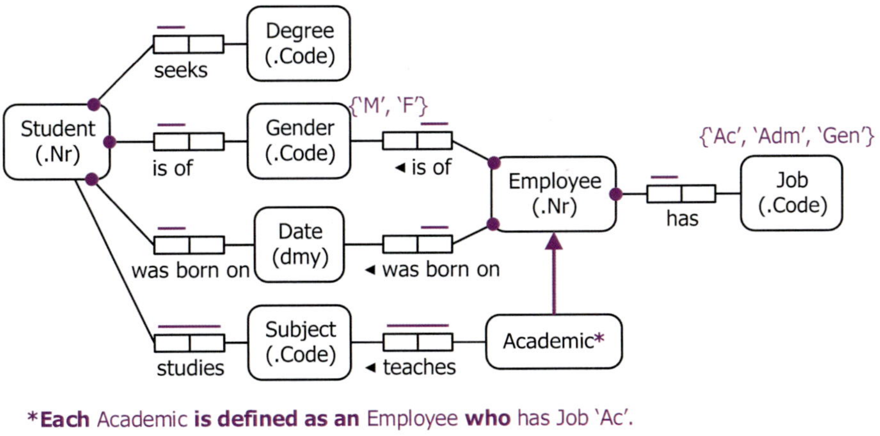

FIG. 6.59 Students and employees have some details in common.

Suppose Student and Employee are mutually exclusive. Despite their common roles, it is unlikely that we want to list such information for both students and employees in the same query. In this case, we may leave the schema as it is.

Now, suppose that in reality Student and Employee overlap, so a student may be an employee. If we want to model this possibility, we must introduce a supertype (e.g., Person). But if we are not interested in knowing whether a student is an employee, and we have no common question for the two types, then we may leave the schema as is. However, our model now differs in this respect from the real world.

By presenting Student and Employee as primitive, our model declares that these types are exclusive, whereas in reality they overlap. To avoid misunderstanding, any such disagreement between the model and reality must be clearly documented in technical and user manuals so people are adequately warned of this decision.

Such decisions should not be taken lightly, as there will often be good reasons for wanting to know whether some instance is a member of both types. For example, suppose we want to enforce the constraint that no academic can be enrolled in a subject he or she is teaching. If we want the information system to enforce this constraint for us, we have to provide it with a way of detecting whether some student is an employee. To do this, we must introduce a supertype (e.g., Person), preferably with a global identification scheme (e.g., PersonNr).

Moreover, we might want to record specific data for student employees (e.g., study leave). In this case, we also need to introduce StudentEmployee as a subtype of both Student and Employee. Fig. 6.60 depicts the subtype graph for this situation. As an exercise, expand this graph to the full schema by adding fact types and subtype definitions. The gender and birthdate fact types attach to Person. If included, the constraint about teachers should be specified textually.

In previous cases, all subtypes in a subtype graph inherited the identification scheme of their top supertype. For example, in Fig. 6.60(a), all people, whether or not they are students and/or employees, are identified by person numbers.

In practice, however, we may sometimes meet a subtype with a preferred identification scheme that differs from that of its direct supertype(s). In this case, we have a *context-dependent reference scheme* because the preferred identifier of an object depends on the context in which it is being considered.

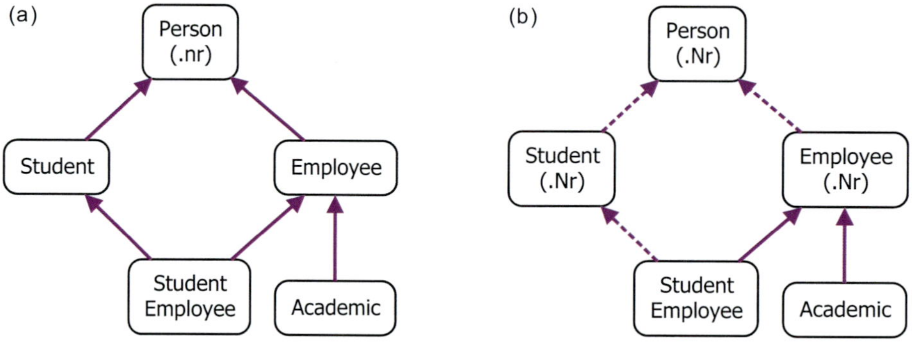

FIG. 6.60 The subtype graph after generalization and further specialization.

Suppose that within the context of student records, we want to identify students by StudentNr, and in the context of employee records, we want to identify employees by EmployeeNr. We can indicate this choice on the subtype graph by introducing the *context-dependent identifiers* on the subtype nodes themselves.

For example, in Fig. 6.60(b), Student is identified by StudentNr and Employee is identified by EmployeeNr. If a subtype does not introduce its preferred identifier, it must inherit its identifier from one of its supertypes (either direct or indirect).

A *solid subtyping arrow* provides a path to the subtype's preferred identifier, using the first identifier found on the path. A *dashed subtyping arrow* does not provide a path to the subtype's preferred identifier, so the subtype either introduces its own identifier or must inherit its identifier via another path to a supertype.

In Fig. 6.60(b), for example, Student and Employee introduce their own identifiers so they have only a dashed subtyping connection to Person. StudentEmployee has two supertypes, and is identified by EmployeeNr (shown by the solid subtyping connection) rather than by StudentNr (dashed connection).

The procedure for generalizing object types is summarized below. Here *A* and *B* are entity types that are initially depicted separately, with no common supertype.

Generalization Procedure (GP):

if *A* and *B* overlap, or can be compared (e.g., to compute ratios)
and we wish to model this

or

A and *B* are mutually exclusive.
and common information is recorded for both.
and we wish to list *A* and *B* together for this information.

then introduce their supertype $A \cup B$ with its own identification scheme;
add predicate(s) to the supertype to classify members into *A*, *B*;
attach common roles to the supertype;
if *A* (or *B*) plays some specific roles
then define it as a subtype and attach these roles

It should be emphasized that the decision to generalize or specialize is a conceptual one. This does not commit us to a particular way of mapping the resulting subtype graph to a logical database schema. A number of mapping options exist (e.g., subtypes may be absorbed into the same table as their supertype, or mapped to separate tables). Chapter 11 discusses these mapping options in detail.

Exercise 6.6

1. A taxi company records these details about its employees: employeeNr, name, address, gender, phone, salary, and weight. It records these details about its cars: regNr, model, year manufactured, cost, and weight. It also records who drives what cars (optional *m:n*).

(a) Schematize this with no subtyping. Use two predicates with the name "weighs".
(b) Generalize Employee and Car, so that "weighs" appears only once.
(c) Which of these solutions is preferable? Discuss.

2. A hospital has an information system about its employees and patients. The following report is extracted from its **employee records**. Employees are standardly identified by their employee number. Each employee has exactly one job: Administrator (admin); doctor (doc); or pharmacist (pharm). A pager is a portable electronic device that beeps when its number is rung. Some employees are assigned unique initials, which may be used as a secondary identifier. The mark "–" means "inapplicable because of other data"

EmpNr	*Name*	*Job*	*Office*	*PagerNr*	*Initials*
e10	Adams A	admin	G17	–	–
e20	Watson M	doc	302	5333	MW1
e30	Jones E	pharm	–	–	–
e40	Kent C	admin	G17	–	–
e50	Kildare J	doc	315	5400	JK
e60	Brown C	pharm	–	–	–
e70	Collins T	pharm	–	–	–
e80	Watson M	doc	315	5511	MW2

In the following extract from the **patient records**, patients are standardly identified by their patient number (PatNr) and are typed as inpatients (in) or outpatients (out), but not both. Some patients are placed in wards. Patients may be allergic to various drugs. Prescriptions are identified by their ScriptNr. Each dispensed prescription has been prescribed by a doctor, dispensed by a pharmacist, and issued to a patient. Each prescription specifies exactly one drug. Because initials are easier for humans to remember, they are used in this report; however, the system uses EmpNr as the primary identifier for employees. Unlike "–", a blank simply means "not recorded".

The population is significant with respect to mandatory roles and the subtype graph. Note, however, that any patient may be dispensed a prescription, any doctor may prescribe one, and any pharmacist may dispense one.

				Prescriptions dispensed to patients			
PatNr	*Type*	*Ward*	*Allergies*	*Script*	*Prescriber*	*Dispenser*	*Drug*
p511	in	5B	Aspirin Doxepin	7001	MW1	EJ	Warfarin
p632	out	–		7132 7250	JK MW1	CB EJ	Aspirin Paracetamol
p760	in	4C	Warfarin	8055	JK	EJ	Aspirin
p874	in	5B					

(a) In the real world, *no patient can be an employee.* Schematize this UoD.
(b) In reality, a patient may be an employee, but we are not interested in knowing such facts. What changes, if any, should be made to the model or system documentation?
(c) In reality, a patient may be an employee. Moreover, the hospital requires such facts to be known. For example, we may want to ensure that no doctor prescribes a drug for himself/herself. The hospital is willing to replace its old identification schemes for employees and patients by a new one. Modify your schema for this situation.
(d) In reality, a patient may be an employee, and we want to know this. However the hospital demands that EmpNr and PatNr be retained as preferred identifiers within their context. Modify your schema to deal with this situation.

6.7 Summary

A *set* is determined by its members, so order and repetition don't matter, for example, {3, 6} = {6, 3, 6}. A *bag* or *multiset* makes repetition significant, for example, [3, 6] = [6, 3] but [3, 6] $\neq$ [3, 6, 6]. A *sequence* is an ordered bag, for example, (3, 6) $\neq$ (6, 3). A set A is a *subset* of B (written $A \subseteq B$) iff each member of A belongs to B; in this case B is a *superset* of A. A is a *proper subset* of B iff $A \subseteq B$ and $A \neq B$. A and B are *mutually exclusive* or disjoint iff they have no common members, that is, their *intersection*, $A \cap B = \{\}$. If sets have common members they *overlap*; if each also has extra members, we have a case of *proper overlap.*

CSDP step 6 adds any value, set-comparison (subset, equality, exclusion), and subtype constraints. A *value constraint* specifies the possible values for a value type or role. It may provide a *full listing* or *enumeration* of all the values, for example, {"M", "F"}; a *subrange definition,* for example, {1..7}, {"A".."E"}; or a combination of lists and ranges.

A value constraint may also indicate a syntactic data type (e.g., variable length text of at most 20 Unicode characters). If the underlying data type is real number (e.g., float or decimal), a square bracket/parenthesis signifies inclusion/exclusion of the end value. For example, "(0..100]" denotes the positive real numbers up to and including 100.

An *independent object type* is a primitive object type whose instances may exist independently of playing any fact roles (the fact roles are collectively optional). Independent object types have "!" appended to their name. For example, we may populate a reference table for "Country!" with the names of all existing countries, even if some of these countries do not play in any elementary facts. In principle, the values that reference members of an independent object type could instead be declared in a value constraint; however, this is awkward if the values may change or there are many values that cannot be specified as a range.

Set-comparison constraints restrict the way the population of a role, or role sequence, relates to the population of another. Let *rs1* and *rs2* be role sequences (of one or more roles) played by compatible object types. A *subset constraint* from *rs1* to *rs2* is denoted by an arrow in that direction, including a circled "⊆" symbol. This indicates pop(*rs1*) ⊆ pop(*rs2*).

An *equality constraint* is equivalent to subset constraints in both directions. It is shown by a dotted line with a circled "=" symbol, demanding that pop(*rs1*) = pop(*rs2*). An *exclusion constraint* among two or more role sequences is shown by connecting them to "⊗" with dotted lines; this means their populations must be disjoint (mutually exclusive).

If a set of roles is disjunctively mandatory and mutually exclusive, this may be shown by the lifebuoy symbol. This *exclusive-or constraint* symbol is simply the inclusive-or constraint symbol ⊙ overlaid on the exclusion constraint ⊗.

If each role sequence has two roles, we talk of *pair*-subset, pair-equality, and pair-exclusion constraints. A tuple-subset constraint implies simple subset constraints. A simple exclusion constraint implies tuple exclusion. Theorem NXS states that no exclusion constraint can have exactly the same arguments as a subset constraint.

An object type *A* is a (proper) subtype of *B* iff (A ≠ *B* and) for each database state, pop(*A*) ⊆ pop(*B*). We show this by an *arrow* from *A* to *B*. If *A* is a subtype of *B*, and *B* is a subtype of *C*, then it is transitively implied that *A* is a subtype of *C*; such indirect subtype links should not normally be displayed.

An object type may have many subtypes and many supertypes. Subtype connections among a family of compatible object types are displayed in a directed, acyclic *subtype graph*. This graph has exactly one root node (or top), which must be a primitive object type (or, in rare cases, an objectified association).

In ORM, subtypes are typically introduced to declare that some roles are played only by that subtype. The process of introducing subtypes is called *specialization*. Subtypes inherit all the roles of their supertype(s) and normally have at least one specific role. By default, a subtype inherits the preferred reference scheme of a direct supertype, so the reference scheme is not displayed on the subtype. With subtypes (and objectified associations), any mandatory roles must be explicitly shown.

A subtype may be derived, asserted, or semiderived. Each *derived subtype* must be formally defined in an equivalence derivation rule using one or more roles of its supertype(s), for example, **Each** FemalePerson **is a** Person **who** is of Gender 'F'. Derived subtypes are marked "*". With derived subtypes, all relevant constraints on defining predicates must be declared; this ensures that any subtype exclusion "⊗" or totality "⊙" constraints that do exist are implied. Exclusive object types (e.g., Man, Woman) cannot have a common subtype. *Asserted subtypes* are simply asserted without any derivation rule. *Semiderived subtypes* may have both derived and asserted instances; these rare subtypes are marked "+" and have conditional derivation rules.

Output reports often have missing values or nulls. Subtyping is indicated if a missing value means "*not to be recorded*" (i.e., inapplicable because of other recorded data). This book uses a minus sign "−" for this purpose. If an output report is significant with respect to the subtype graph, we can determine the graph by examining the pattern of "−" marks. However, background knowledge is required to determine the subtype definitions, and hence meaningful subtype names. Input forms often provide instructions that indicate the conditions under which particular entries on the form are needed. Such instructions can be used to deduce the subtyping constraints. Sometimes reports are already separated on the basis of subtyping and their report titles give clues as to the subtyping scheme.

Without a full set of output reports, we might wrongly declare object types to be primitive. We then need to introduce supertypes; this process is called object type *generalization*, the reverse of specialization. We introduce a supertype if object types overlap, and we want to know this. A supertype is also introduced if the object types are exclusive but have common roles, and we want to list them together for this common information. If the original object types have specific roles, they must be retained as subtypes.

In some cases, there is a need for subtypes to have a preferred reference scheme different from that of their supertype(s). Such a *context-dependent reference* is indicated by displaying the preferred reference scheme on the relevant subtypes. A subtyping arrow appears as a *solid arrow* if it provides a path to the subtype's preferred identifier; otherwise, it is shown as a *dashed* arrow.

Chapter Notes

Not all versions of ORM allow independent entity types. Fully Communication Oriented Information Modeling (*FCO-IM*) (Bakema et al., 1994) handles the problem by treating entity types as objectifications of unaries.

Most versions of ER do not allow subset, equality, or exclusion constraints between role sequences, or attributes, to be included on the schema diagram (see Chapter 8). Unified modeling language (UML) includes graphic symbols for subset constraints only between whole associations, and exclusion constraints only in the context of an exclusive-or constraint between roles (see Chapter 9). Recent versions of entity-relationship modeling (ER) as well as UML offer basic support for subtyping, but ignore the need for formal subtype definitions or context-dependent reference (see Chapters 8 and 9).

Some versions of ORM restrict subtypes to derived, active subtypes. Context-dependent reference is not supported in some versions. The PSM variant adopts a different approach to generalization in which the supertype uses a disjunction of the identification schemes of its subtypes (e.g., see ter Hofstede et al., 1993). Although this raises additional implementation problems when a common query is executed on overlapping subtypes, this does provide another viable approach.

For early discussion of subtyping from an ORM perspective, an informal treatment can be found in Halpin (1995) and a formal treatment in Halpin and Proper (1995b). An early treatment updated for ORM 2 is provided in Halpin (2007b); this paper also distinguishes rigid subtypes from role subtypes, a distinction examined later in Section 10.3. Halpin (2009a) discusses some further enrichments to subtyping. Halpin and Curland (2013) discuss more recent enhancements to ORM, including constraints on supertype link roles, a topic discussed in Section 10.8.

7

Other Constraints and Final Checks

Chapter Outline

7.1 Introduction to CSDP Step 7

In previous steps of the *Conceptual Schema Design Procedure (CSDP)*, we verbalized familiar examples in terms of elementary facts, sketched the fact types on a diagram, and then added various constraints and derivation rules. Most applications include uniqueness, mandatory role, value, set-comparison, and subtyping constraints. Although less common, other kinds of constraint may also apply. Some of these have graphic notations. The rest are specified textually. Once all the constraints are specified, some final checks can be made to help ensure that our information model is consistent and free of redundancy. This concludes the basic conceptual schema design procedure.

CSDP step 7: Add other constraints, and perform final checks

This chapter deals with step 7 in detail. The next three sections discuss some important graphic constraints (frequency constraints, ring constraints, cardinality constraints, and value-comparison constraints). After that, we consider some constraints that must be declared textually. Finally, we examine some ways of checking that the schema is consistent and redundancy free.

Information Modeling and Relational Databases. https://doi.org/10.1016/B978-0-443-23790-4.00013-X

7.2 Frequency Constraints

Let *n* be a positive integer. To indicate that each entry in a fact column must occur there *exactly n times*, the number *n* is attached by a dashed line to the role (see Fig. 7.1). This is an example of a **frequency constraint** (a.k.a. occurrence frequency constraint). If $n=1$, this is equivalent to a uniqueness constraint—in this case, the usual bar notation for uniqueness should be used instead of a "1" mark.

Like uniqueness constraints, frequency constraints are *local* constraints on the role(s), not global constraints on the object type. The constraint in Fig. 7.1 means that for each state of the database, *if* an instance of *A* plays role *r*, it must do so *n* times. It does *not* mean that each instance of *A* must play that role *n* times. If the role is optional, some instances of *A* might not play the role at all.

One use of frequency constraints is to ensure that if details are recorded, they must be recorded for all instances of an enumerated type. Suppose a company stocks and sells three kinds of computer drives (C = CD drive, D = DVD drive, H = hard drive) and operates in just two cities (Sydney and Tokyo). Consider the conceptual model shown in Fig. 7.2. The UC on the first two roles ensures that each City-Drivekind pair appears on only one row. The frequency constraint of 3 on the first role requires each city that occurs in the first column to do so three times.

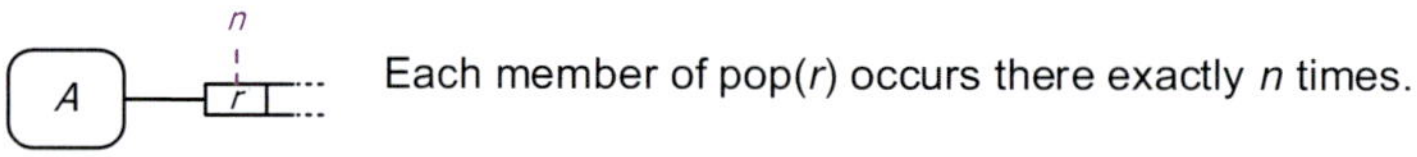

FIG. 7.1 A simple occurrence frequency constraint on role *r*.

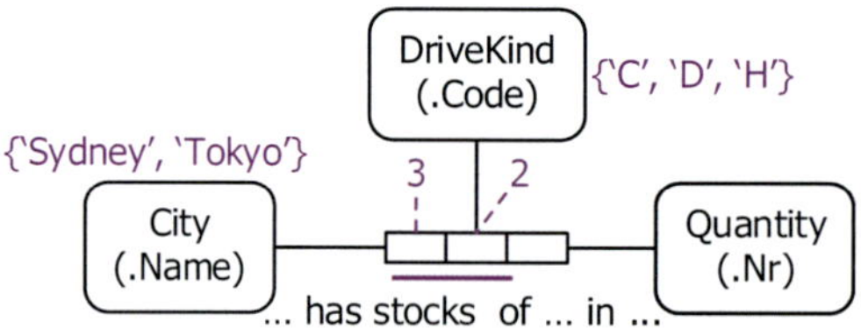

Sydney	C	3000
Sydney	D	2000
Sydney	H	600
Tokyo	C	2000
Tokyo	D	1800
Tokyo	H	300

FIG. 7.2 If stock figures are recorded, they are recorded for each city and drive kind.

Hence, if any stock figures are recorded for a city, then figures for all three kinds of drive must be recorded. Similarly, the frequency constraint of 2 on the second role ensures that each drive kind in the second column must appear there twice. Because of the uniqueness constraint, this means that if stock is recorded for a drive kind, then figures for both cities must be included.

The combination of frequency, uniqueness, and value constraints ensures that any recorded stock figures are *complete* with respect to both cities and drive kinds. A compound transaction is needed to initially populate this fact type, requiring at least six facts to be added.

A frequency constraint that spans two or more roles of a fact type is a *compound frequency constraint*. In this case, we connect the frequency by dashed lines to the relevant roles (this dotted line may be hidden if one moves the frequency number close enough to the constrained roles). If there are only two constrained roles and these are adjacent, the frequency may be attached by a dotted line to the junction of those roles. For example, in Fig. 7.3 each city-year pair in the fact table must occur there three times. Given the uniqueness constraint across the first three roles, sales must be recorded for all three drives for each city-year pair in the table. Note that this does not require that yearly sales be recorded for both cities. The sample population includes only one city, yet it satisfies all the constraints.

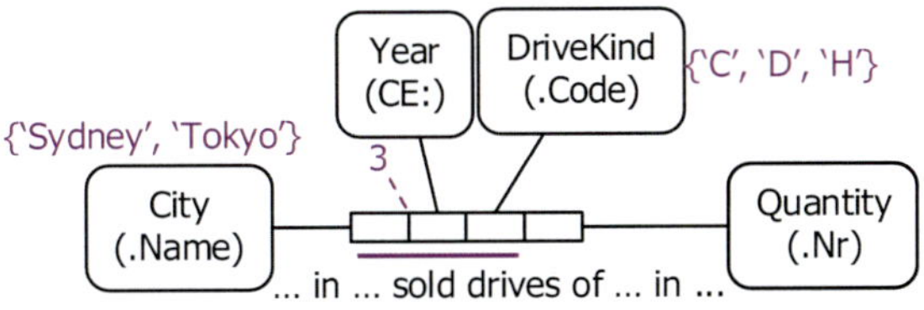

Sydney	2007	C	10000
Sydney	2007	D	7000
Sydney	2007	H	2000

FIG. 7.3 Yearly sales figures may omit a city but not a drive kind.

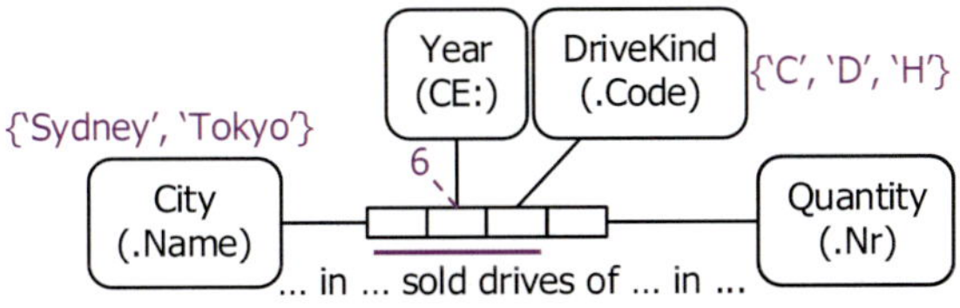

Sydney	2007	C	10000
Sydney	2007	D	7000
Sydney	2007	H	2000
Tokyo	2007	C	7000
Tokyo	2007	D	5000
Tokyo	2007	H	1000

FIG. 7.4 Yearly sales figures must cover all cities and drive kinds.

Fig. 7.4 strengthens the constraints by requiring yearly sales figures for both cities as well as the three kinds of drive. Given the uniqueness and value constraints, this is achieved by the simple occurrence frequency of 6 on the year role—as an easy exercise, prove this. The compound frequency of 3 across city-year is omitted because it is implied. A compound frequency of 2 across Year-DriveKind is also implied, and hence omitted.

Frequency ranges may also be specified. Use "$\leq n$" for *at most n*, where $n \geq 2$. Use "$\geq n$" for *at least n*, where $n \geq 1$. Use "*n* .. *m*" for *at least n and at most m*, where $2 \leq n < m$. A lower frequency of 1 is the smallest frequency allowed and is assumed by default because each entry in a column is already there once. Because the constraint applies to the population of the role(s), not the population of the object type(s), it doesn't make sense to declare an occurrence frequency of zero. An optional role indicates that some members of the object type population might not play that role.

As an example, consider Fig. 7.5. This is part of a schema for a conference application. Each expert is on at least one panel (as shown by the mandatory role) and reviews at most five papers (possibly none because the role is optional). Each panel with members has at least four and at most seven members. Because Panel is independent, we may record the existence of a panel before assigning members to it. Each reviewed paper has at least two reviewers. It is possible that some papers have not been assigned reviewers yet. In principle, frequency constraints may combine values and ranges (e.g., "1, 5..7") but in practice this is extremely rare.

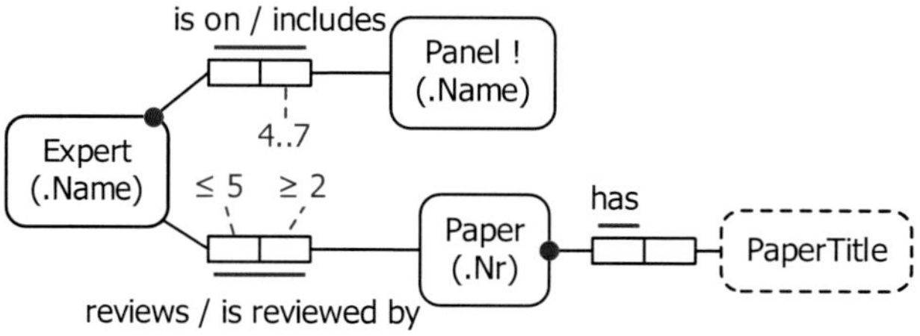

FIG. 7.5 Examples of minimum and maximum frequency constraints.

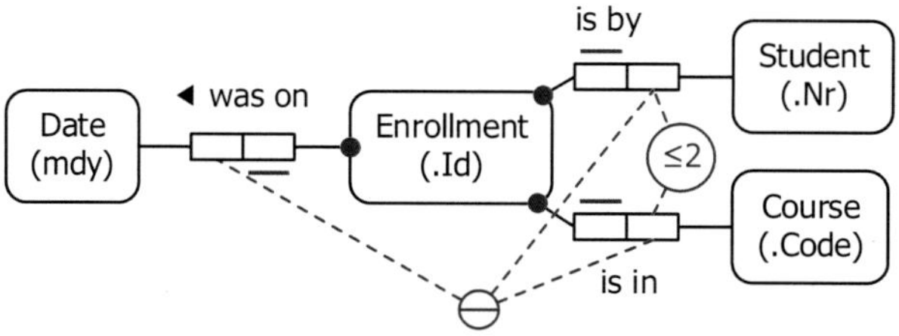

FIG. 7.6 Example of an external frequency constraint.

Just as internal uniqueness constraints can be generalized to frequency constraints, external uniqueness constraints can be generalized to *external frequency constraints*, which apply to roles from different predicates. For example, some universities allow students to enroll in the same course at most twice (if they can't pass the first time, they are allowed only one more attempt). This external frequency constraint is displayed by connecting a frequency indicator to the relevant roles, as in Fig. 7.6. In the context of these fact types, each student-course combination is associated with at most two enrollments. The external uniqueness constraint ensures at most one enrollment for each date-student-course combination.

Exercise 7.2

1. The following annual report provides information about two kinds of DBMS (R = Relational, NR = Nonrelational) for each of the four seasons. Schematize this universe of discourse (UoD), including uniqueness, mandatory role, value, and frequency constraints.

Database	*Season*	*Nr sold*
R	spring	50
NR	spring	70
R	summer	60
NR	summer	60
R	autumn	80
NR	autumn	40
R	winter	120
NR	winter	15

2. A software retailer maintains a database about various software products. Two sample output reports are shown. There is a sales tax of 20% on all software. The *ex-tax* price output reports are shown. There is a sales tax of 20% on all software. The *ex-tax* price excludes this tax; the *with-tax* price includes it. The functions of a software product are the tasks it can perform. A "✓" indicates the rating of the product for the criterion listed on that row (e.g., both products shown have a good performance). A "?" denotes missing information.

   Schematize this UoD, including uniqueness, mandatory role, value, and frequency constraints and any derivation rules. Don't nest or subtype. Model the product evaluation ratings (for performance, etc.) in terms of a single fact type.

Product: WordLight 4.0
Functions: word processor

List price: *ex tax* $500 *with tax* $600

	Poor	*OK*	*Good*	*Excellent*
Performance			✓	
Documentation				✓
Ease of learning			✓	
Ease of use			✓	
Error handling				✓
Support			✓	
Value				✓

Release date: 2023-02-15
Next upgrade: ?

Product: PCjobs 1.0
Functions: word processor
spreadsheet
database

List price: *ex tax* $1000 *with tax* $1200

	Poor	*OK*	*Good*	*Excellent*
Performance			✓	
Documentation	✓			
Ease of learning		✓		
Ease of use	✓			
Error handling		✓		
Support			✓	
Value		✓		

Release date: 2022-10-01
Next upgrade: 2024-06-30

3. In Megasoft corporation, at the start of each year, each salesperson is assigned at most three different software products to sell during that year (not necessarily the same each year). At the end of each year, each salesperson reports how many of each product he/she sold in each month of that year. Historical data is kept on who is assigned what to sell in each year and the monthly sales figures for that person (if the figures are available). No month may be omitted in these annual sales figures. The system retains information from previous years. Schematize this, including value and frequency constraints. Use nesting.

4. (Acknowledgment: This question is based on an exercise devised by Dr. E. Falkenberg and is used with permission.)

   The following tables are sample reports from a system that maintains information about communities and roads. A "–" mark means "inapplicable because of other data" (see the asterisked notes). Mayors are identified by the combination of their first name and surname. A person can be mayor of only one city, and each city has exactly one mayor. Each road connects *exactly two* communities. Schematize this UoD, including uniqueness, mandatory role, value, subtype, and frequency constraints.

Community	*Population*	*Longitude* *	*Latitude* *	*Size*** *(km²)*	*Mayor****	
					Firstname	*Surname*
Astraluna	900,000	–	–	145	Fred	Bloggs
Bradman	90,000	+120°50'	+48°45'	12	–	–
Cupidville	9,000	+120°50'	+48°40'	–	–	–

* Recorded only for villages and towns (at most 100,000 inhabitants).
** Recorded only for towns and cities (more than 10,000 inhabitants).
*** Recorded only for cities (more than 100,000 inhabitants).

Road Nr	*Road kind**	*Connection***		*Length (km)*	*Avg. travel time****	*Max. slope (%)*****
		Community	*Community*			
11,000	f	Astraluna	Bradman	25	0 h 20 m	–
11,500	h	Astraluna	Bradman	22	0 h 25 m	2
11,561	p	Bradman	Cupidville	17	0 h 20 m	2
11,562	d	Bradman	Cupidville	15	–	15

* f = freeway; h = highway; p = paved minor road; d = dirt road.
** All connections are two-way (so the order of the connection is irrelevant).
*** Recorded only for freeways, highways and paved minor roads.
**** Recorded only for dirt roads, paved minor roads and highways.

5. An award-granting institute has a rule that any given person (identified by a PersonNr) may apply for the same award (identified by an awardNr) at most three times. For each award application (identified by an ApplicationNr), its status is recorded.

   (a) Model this using an external frequency constraint.
   (b) Try to model this without using an application number, instead objectifying "Person applied for Award" as "Application". Explain why this approach fails.

7.3 Ring Constraints

When two roles in a predicate are played by the same object type, the path from the object type through the role pair and back to the object type forms a "ring". If the roles are played by subtypes with a common supertype, the path from and back to the supertype also forms a ring. The role pair typically forms a binary predicate that is then called a *ring predicate*, but may be part of a longer predicate (see the shaded role pairs in Fig. 7.7) or may even come from different predicates in a join path (join constraints are discussed in Chapter 10). A *ring constraint* may apply only to a pair of roles like this.

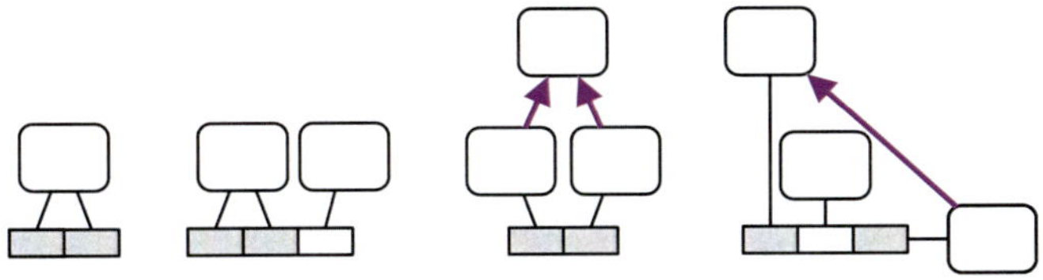

FIG. 7.7 Ring constraints may apply to the shaded role pairs.

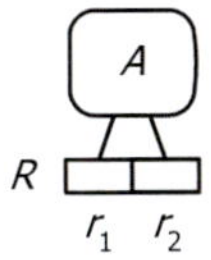

FIG. 7.8 A ring binary (possibly embedded in a longer fact type).

In Fig. 7.8, the predicate *R* might be obtained from a longer predicate by projecting on roles r_1 and r_2, and *A* might be a supertype of the object types playing r_1 and r_2. For each state of the database, pop(r_1,r_2) is a set of ordered pairs, and hence a relation, of type *R*. Derived as well as asserted relations are allowed.

Consider the model in Fig. 7.9(a). For simplicity, assume people in this UoD are identified by their first name. Here, each person likes himself/herself. Suppose that in our UoD, this is true for all possible populations—each person who plays the role of liking or of being liked must like himself/herself. In this case, the Likes predicate is said to be *locally reflexive* (or reflexive over its populations). If we delete just the third row from Fig. 7.9(a), then Bill doesn't like himself, and the relation is not reflexive. The infix notation "*xRy*" (read "*x Rs y*") means *x* bears the relationship *R* to *y*. Using *R* in the general sense discussed above, and using "→" for "*implies*" and ∀ for the *universal quantifier* (read as "for each"), we define:

R is **locally reflexive** iff $\forall x,y\,(xRy \rightarrow xRx)$
R is **purely reflexive** iff $\forall x,y\,(xRy \rightarrow x = y)$

If we picture an object as a black dot and Likes as an arrow, we can picture this reflexion as a relationship line starting and ending at the same object, as shown at the left of

Fig. 7.9(a). A *locally reflexive constraint* is displayed using a simplified version of this icon (arrow-tip removed), attached to the role pair. A *purely reflective constraint* adds a vertical stroke inside the circle, as in Fig. 7.9(b).

Many important relations are reflexive (e.g., =, ≤, is parallel to, implies). A relation that is reflexive over all things in the UoD is said to be *globally reflexive* (or *totally reflexive*), that is, for all *x*, *xRx*. Of the four examples cited, only "=" is globally reflexive. For example, "implies" is defined for propositions but not for people. When we say "reflexive", we usually mean locally reflexive. In practice, purely reflexive constraints are rare, and globally reflexive constraints are so rare that object-role modeling (ORM) does not support them.

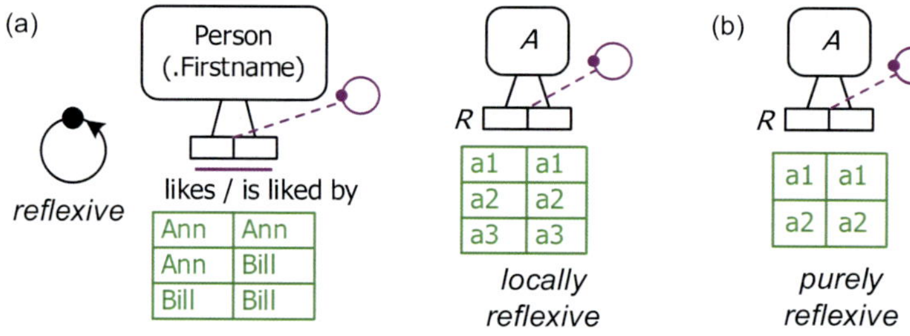

FIG. 7.9 (a) A locally reflexive constraint, and (b) a purely reflexive constraint.

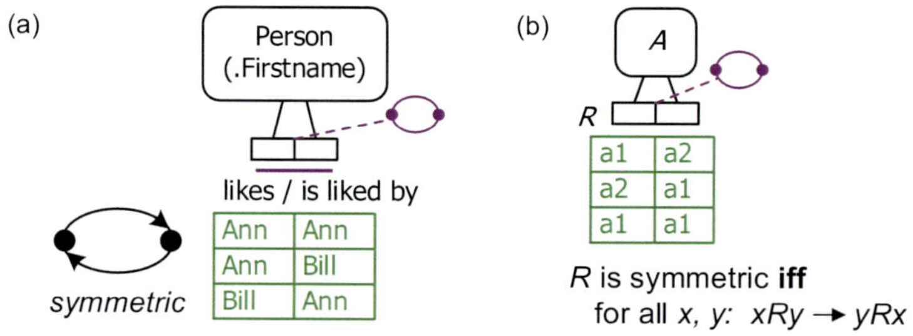

FIG. 7.10 Symmetric ring constraints.

Now, consider a UoD where anybody who likes a person is also liked by that person. One satisfying population comprises the three facts: Ann likes Bill; Bill likes Bill; Bill likes Ann (see Fig. 7.10(a)). Here, given any persons *x* and *y*, not necessarily distinct, if *x* likes *y* then *y* likes *x*. If this is always the case, Likes is *symmetric*. The dot-arrow diagram in Fig. 7.10(a) conveys the idea that if the relationship holds in one direction, it also holds in the other direction. A *symmetric constraint* is displayed using a simplified version of this icon (arrow-tips removed), attached to the role pair, as shown. We define this notion for any predicate or relation *R* as follows:

R is **symmetric** iff $\forall x,y\,(xRy \rightarrow yRx)$

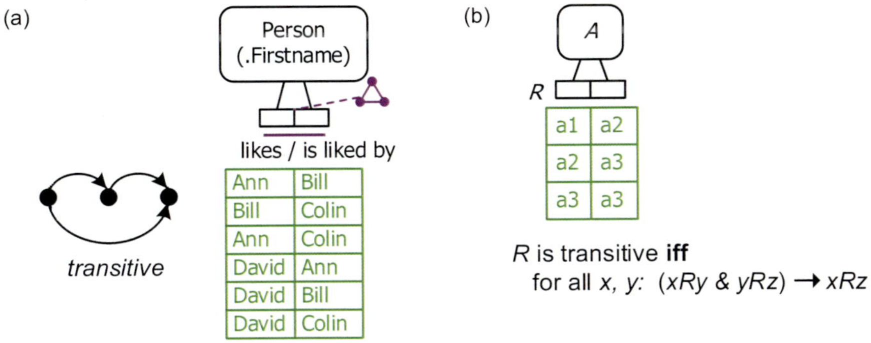

FIG. 7.11 Transitive ring constraints.

Now consider a UoD where, if one person likes a second, and the second person likes a third, then the first person must like the third. In this case, we say that likes is *transitive*. The definition includes the case where the persons are the same. The populated schema in Fig. 7.11(a) illustrates this property.

The dot-arrow diagram in Fig. 7.11(a) indicates that if a chain of relationships exists, the relationship also applies from the first to the last member of the chain. A *transitive constraint* is displayed using a simplified version of this icon (arrow-tip removed), attached to the role pair. In general, transitivity may be defined as follows. Here, we use "&" for "and" and give "&" precedence over "→" (so that the & operation is evaluated first). Note that x, y, and z need not be distinct.

R is **transitive** iff $\forall x,y,z\ [(xRy \,\&\, yRz) \rightarrow xRz]$

We leave it as an exercise to prove that a relation that is both symmetric and transitive must also be reflexive over its population. A relation that is reflexive, symmetric, and transitive is called an "RST relation" or "equivalence relation". The classic example of this is the identity relation "=". The inequality relation "≠" is symmetric only, while the less than relation "<" is transitive only.

Relational properties such as reflexivity, symmetry, and transitivity might be thought of as constraints because they limit the allowable relations. However, such properties are *positive* and are used more often to *derive* additional facts than constrain existing facts.

We now consider *negative* relational properties such as irreflexivity and asymmetry. Given the existence of some facts, such negative properties imply the nonexistence of certain other facts. Such properties are best handled as simple constraints. An example will help clarify things. Look at the populated parenthood fact type in Fig. 7.12 and see if you can spot anything wrong.

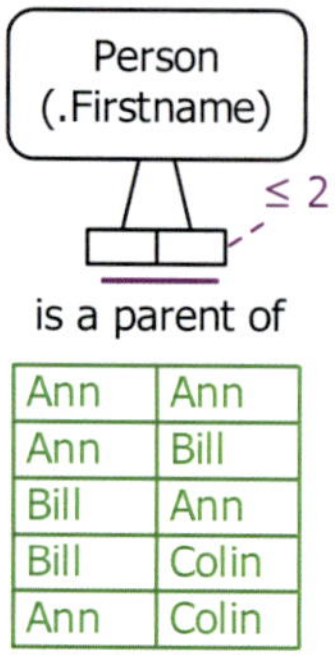

FIG. 7.12 What is wrong here?

The population shown is consistent with the constraints specified. Parenthood is many:many, and each person has at most two parents. But as you no doubt noticed, some of the facts in the table are inconsistent with the real-world concept of parenthood. We need to add more constraints so that the information system is able to reject such erroneous populations.

To begin with, the first fact (Ann is a parent of Ann) has to be rejected because nobody can be his/her own parent. We say that the parenthood relation is *irreflexive* and define this notion for any relation type *R* as follows. Here, tilde "∼" denotes the logical "**not**" (i.e., "it is not the case that") operator.

R is **irreflexive** iff $\forall x \quad {\sim}xRx$

In the dot-arrow icon in Fig. 7.13(b), the arrowed arc depicts the relationship from the object to itself, and the bar across it indicates that this is not allowed. An *irreflexive constraint* is depicted by connecting a simplified version of this icon (with the arrow-tip dropped) to the role pair (if the roles are contiguous, it connects at their junction), as shown in Fig. 7.13(c). This constraint verbalizes as "**No** Person is a parent of **itself**". An alternative notation used in ORM 1 for the irreflexive constraint is "$^{\circ}ir$".

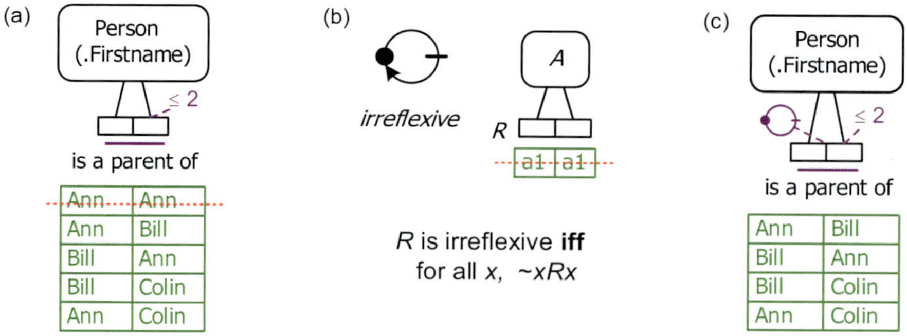

FIG. 7.13 Parenthood is declared irreflexive. But there is still something wrong.

Fig. 7.13(c) adds the irreflexive constraint and reduces the population accordingly. Note that "irreflexive" is stronger than "not reflexive". For example, if Likes is irreflexive, then nobody likes themselves. If some but not all people like themselves, then Likes is neither reflexive nor irreflexive.

Irreflexivity is an intrarow constraint (i.e., violation of the constraint by a row can be determined by examining that row only). It specifies that no entry may occur in both cells of the same row. This makes the constraint easy and inexpensive to enforce. Note that the parenthood relation is irreflexive but not exclusive. For example, Bill may appear both as a parent of Colin and as a child of Ann.

There are still some problems with Fig. 7.13(c). If we accept the first row, we should reject the second. If Ann is a parent of Bill, then it cannot be true that Bill is a parent of Ann. The parenthood relation is *asymmetric*. In general, for any predicate *R*:

R is **asymmetric** iff $\forall x,y\,(xRy \rightarrow {\sim}yRx)$

That is, if the first object *R*s the second, then the second cannot *R* the first. Here, if one person is a parent of another, then that other cannot be a parent of the first. To say that a relation is asymmetric is stronger than saying the relation is not symmetric (you may wish to prove this as an exercise). An intuitive icon for asymmetry is shown in Fig. 7.14(b).

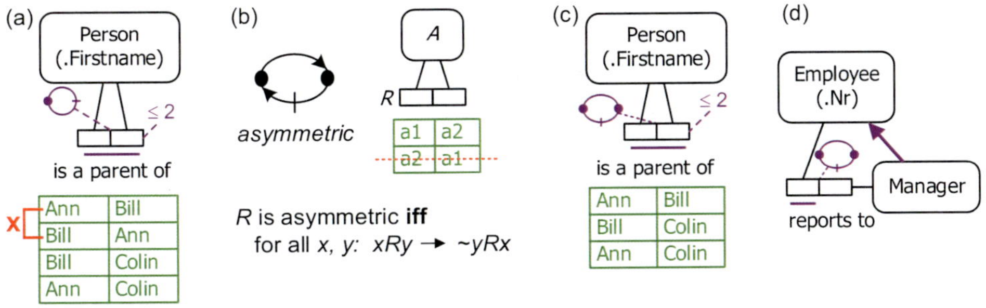

FIG. 7.14 Parenthood and reporting are asymmetric relations.

An *asymmetric constraint* is depicted graphically by a simplified version of this icon, without arrow tips, connected by a dotted line to the constrained role pair, as shown in Fig. 7.14(c). With this constraint enforced, the parenthood population is reduced as shown. Fig. 7.14(d) shows another example of the asymmetry constraint, where subtyping is involved. An alternative constraint notation for asymmetry used in ORM 1 is "$^{\circ}$*as*".

The irreflexive constraint is omitted from Fig. 7.14(c). Is this a mistake? No. Any relation that is asymmetric must be irreflexive. The proof is left as an exercise (*Hint*: *x* and *y* need not be distinct). The converse does not hold. Some irreflexive relations are not asymmetric (e.g., is a sister of). To avoid implied constraints, do not declare irreflexivity if you have already declared asymmetry.

If we remove the irreflexivity requirement, we obtain the weaker property of antisymmetry. For example, $\leq$ and $\subseteq$ are *antisymmetric* but not asymmetric. The intuitive icon in

Fig. 7.15(a) uses different-colored nodes to indicate the objects are different. A simplified version of this icon, without arrow tips, is used to depict an *antisymmetric constraint*, as shown in Fig. 7.15(b). The equivalence at the bottom of Fig. 7.15(b) indicates that asymmetry combines antisymmetry and irreflexivity.

R is **antisymmetric** iff $\forall x,y\ [(x \neq y\ \&\ xRy) \rightarrow {\sim}yRx]$

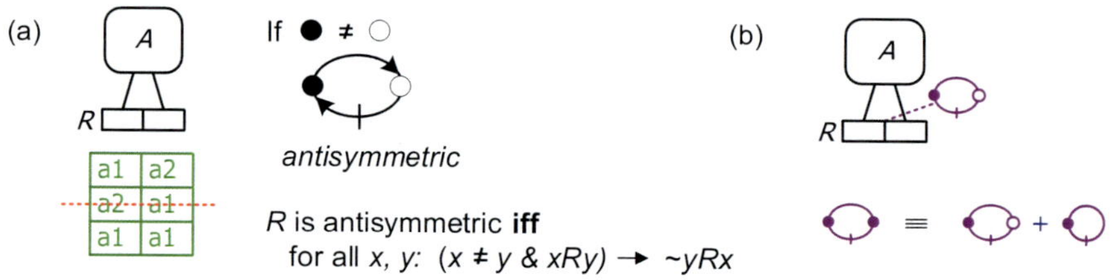

FIG. 7.15 Antisymmetry.

Before continuing with the parenthood example, let's look at some other examples to clarify the ideas discussed so far. Our first case deals with a system to record synonyms. A synonym for a word has roughly the same meaning as that word. A sample taken from a poorly designed book of synonyms is shown in Table 7.1.

Table 7.1 Extract from a poorly constructed table of synonyms.

Word	*Synonyms*
abate	diminish, lessen, reduce, decline
abbreviate	shorten, reduce, condense
:	:
condense	compress, consolidate, abridge
:	:
reduce	shorten, weaken, abate
:	:

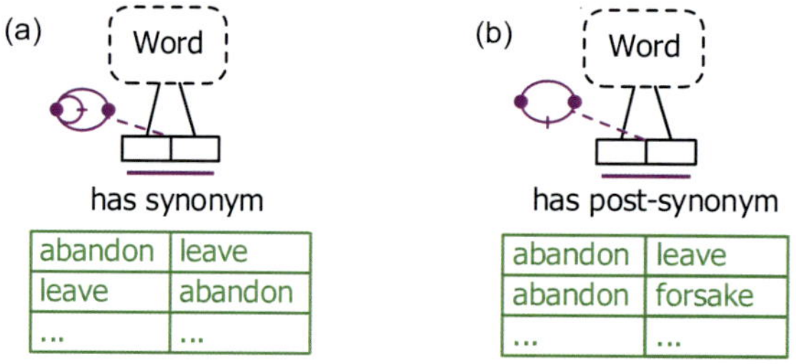

FIG. 7.16 Examples of (a) symmetric and irreflexive, and (b) asymmetric fact types.

Suppose we decide to store this information in the has-synonym binary shown in Fig. 7.16(a). Is this reflexive? Although we might argue that each word is a synonym of itself, this is a trivial result that we are not usually interested in seeing. So, as far as this relation goes, let us agree that no word is its own synonym. This makes the relation irreflexive, as shown by the inner ring constraint.

If one word is a synonym of another, then the other is a synonym for it. So has-synonym is symmetric. But Table 7.1 violates this condition. For example, the pair ("abbreviate", "reduce") appears in this order only. If we look up the word "reduce", we won't see "abbreviate" as one of its synonyms. This is a defect of the synonym book to avoid in a computerized synonym system. How can we avoid this problem?

One solution is to add the constraint that the has-synonym predicate is symmetric. This is specified on the diagram using a simplified (no arrow tips) version of the symmetry icon in Fig. 7.10, shown as the outer ring constraint in Fig. 7.16(a). The symmetric and irreflexive constraints are combined into one icon to save space. ORM 1 uses the alternative ^{O}sym notation for symmetry constraints.

Suppose the symmetric constraint is declared, the table is currently empty, and the following update is attempted: **add**: Word 'abbreviate' has synonym Word 'shorten'. A naive information processor would simply reject this for violating the symmetry constraint. A sophisticated processor might accept the fact and automatically add its converse: "shorten" has synonym "abbreviate". A consistent scheme would be adopted for the delete operation. Although this approach works, it doubles the size of the table because each pair of synonyms is stored twice, once for each ordering.

To save space, we could store each pair in one order only. The post-synonym binary in Fig. 7.16(b) has the second word alphabetically after the first. The pair ("abandon", "leave") is stored in this order only. The relation is then asymmetric, as indicated. The larger synonym relation can now be derived, using a derivation rule such as

Word$_1$ has synonym Word$_2$ **iff** Word$_1$ has post_synonym Word$_2$
or
Word$_2$ has post_synonym Word$_1$.

In SQL, the derived relation could be defined as a view on the base relation by using the **union** operator. For speed, both columns can be indexed. Users might perform updates on the base relation, but issue queries on the derived relation. Thus, we can sometimes choose whether to capture an aspect of the UoD in terms of a database constraint or as a derivation rule. Here, we distinguished the asserted and derived fact types by name. It is also possible to have the fact type semiderived.

What about transitivity of synonymy? This issue is complex. You may be familiar with the "bald-hairy paradox" based on the claim that always if we add one hair to the head of a bald man, he is still bald (where is the paradox?). A similar slippery-slope argument argues against synonym transitivity (sequences of approximations finally lead to nonapproximations). Moreover, the same word may have different meanings, leading to different synonym groups. If we still want to retain some transitivity, we may include a derivation rule for synonym transitivity, constrained by a maximum length of the transitivity chain and relativized to group meanings.

Now consider the marriage fact type in Fig. 7.17. The husband and wife roles are mutually exclusive. In contrast to an irreflexive constraint, which requires inequality between entries on the same row, an exclusion constraint applies between columns and hence is much stronger. If no entry may occur in both columns, we cannot have two rows of the form (a, b) and (b, a). So, *exclusion implies asymmetry.* Moreover, *asymmetry implies irreflexivity.* Because they are implied, neither the asymmetric nor the irreflexive constraint should be added to the schema.

Now, let's return to the populated parenthood model in Fig. 7.18(a). There is still a problem with it. Can you spot it? Let us agree that incest cannot occur in our UoD (if it can, we instead declare a deontic rule, as discussed in Chapter 10). If Ann is a parent of Bill and Bill is a parent of Colin, then Ann cannot be a parent of Colin. The parenthood relation is *intransitive* (i.e., if one person is a parent of a second and the second person is parent of a third, then the first person cannot be parent of the third). In general, for any relation type *R*:

R is **intransitive** iff $\forall x,y,z\ [(xRy\ \&\ yRz) \rightarrow {\sim}xRz]$

Fig. 7.18(b) shows an intuitive icon for intransitivity, with a bar across the forbidden jump. A simplified version of this icon (without arrow tips) is used for the intransitive constraint symbol, as shown on the right in Fig. 7.18(b). Fig. 7.18(c) signifies that parenthood is both intransitive and asymmetric by combining the constraints in a single icon. The population is changed to satisfy the constraints.

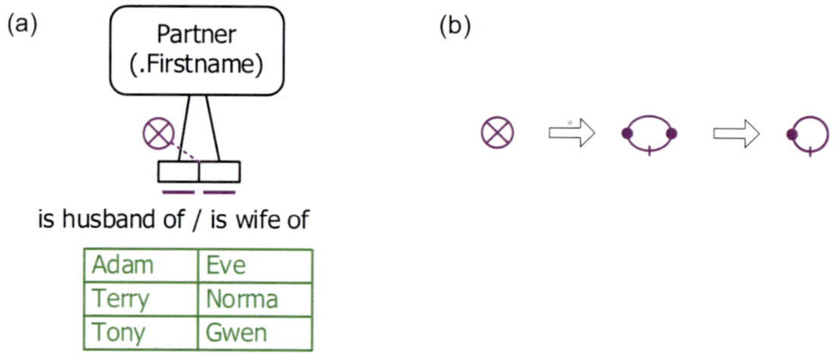

FIG. 7.17 The exclusion constraint implies asymmetry and hence irreflexivity.

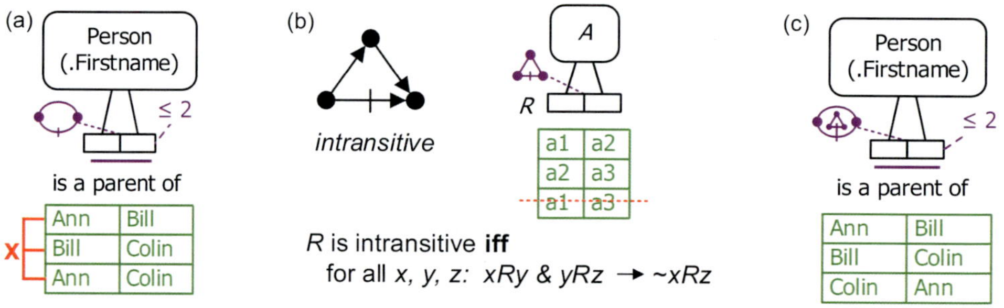

FIG. 7.18 The parenthood relation is asymmetric and intransitive.

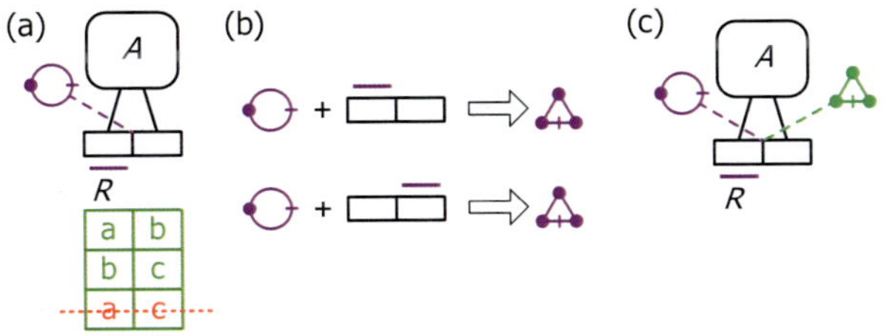

FIG. 7.19 An irreflexive, functional relation must be intransitive.

As an exercise, prove that intransitivity implies irreflexivity (*Hint: x, y,* and *z* need not be distinct). The term "atransitive" is sometimes used instead of "intransitive". Saying a relation is intransitive is stronger than saying it is not transitive. Antisymmetry, asymmetry, and intransitivity are not intrarow constraints because their enforcement requires comparing the row in question with other existing rows.

If an irreflexive relation includes a functional role (rather than being *m:n*), it must also be intransitive. Fig. 7.19 illustrates the proof of this by *reductio ad absurdum*. Let *R* be an irreflexive relation whose first role is functional (i.e., it has a simple uniqueness constraint). Now, populate *R* with the first two rows shown. Because R is irreflexive, $a \neq b$ and $b \neq c$. Assume that *R* is not intransitive. This allows us to add the third row shown. Because $b \neq c$, rows 1 and 3 are distinct rows. This means that *a* plays the first role of *R* twice. But this contradicts the uniqueness constraint. So, row 3 must be rejected (shown by the dotted line through row 3). Hence, the relation must be intransitive. The implied intransitive constraint is usually best left implicit; if shown explicitly it is colored green, as in Fig. 7.19(c). This implication also applies if the second role is functional.

Are we finished with ring constraints? Not quite. Have a look at the parenthood population in Fig. 7.18(c). All the specified constraints are satisfied, but something is still wrong. Can you spot the problem?

Here, Ann is a parent of Bill, who is a parent of Colin who is a parent of Ann. This makes Ann a great grandparent of herself. While reincarnation might make this possible, let's assume that we wish to exclude this possibility from our UoD. We can do this by declaring the relation to be *acyclic*. This means the relation has no cycles (i.e., there is no path via the relation from an object back to itself).

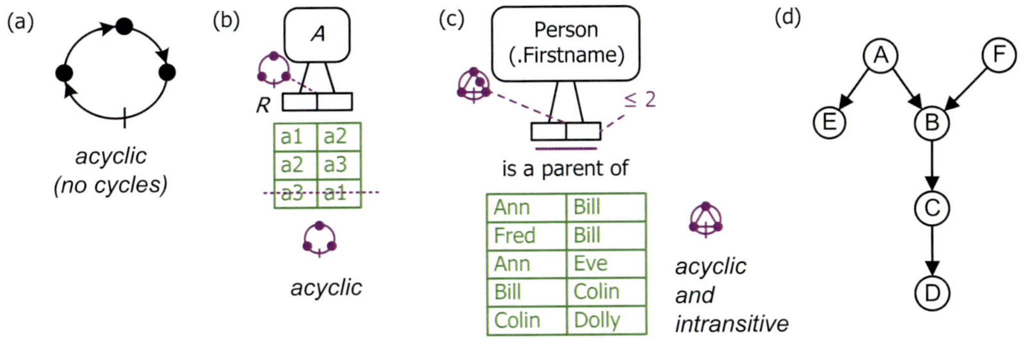

FIG. 7.20 In an acyclic relation, no loops are allowed.

Using "∃" (read "there exists some") for the *existential quantifier*, we recursively define the path predicate P (read "has a path to") to give the transitive closure of R and then use it to define the relation R as acyclic:

∀ *x,y* (*xPy* **iff** [*xRy* **or** ∃*z* (*xRz* & *zPy*)])
R is **acyclic** iff ∀*x* ∼*xPx*

Fig. 7.20(a) shows an intuitive icon for acyclicity, with the bar indicating that no cycles back are allowed. A simplified version of this icon (without arrow tips) is used for the *acyclic constraint* symbol, as shown in Fig. 7.20(b). ORM 1 uses the alternative acyclic constraint symbol ^{O}ac.

Cycles may contain one or more links. Asymmetry is just the special case of acyclicity with two links in the cycle. Because *acyclic implies asymmetric*, if a role pair is declared acyclic, then don't declare it asymmetric.

In Fig. 7.20(c), the population of the parenthood relation is both acyclic and intransitive. The two ring constraints are combined into a single icon, as shown. A homogeneous binary relation can be displayed as a graph, using named nodes for the objects and arrows for the relationship instances, as in Fig. 7.20(d). This makes it visually obvious whether a given population forms an acyclic relation.

Recall that the graph of a subtype family is directed and acyclic. Because of their recursive nature, acyclic constraints may be expensive or even impossible to enforce in some database systems. In SQL, they can be implemented using the recursive union operator, and they are easily enforceable in logic programming languages. In other systems, recursive constraint checks might be left to data entry operators to do manually.

A ring predicate R is *strongly intransitive* if and only if $\forall x,y,z[(xRy \ \& \ yPz) \rightarrow \sim xPz]$, where the path predicate P (read "has a path to") is recursively defined to give the transitive closure of R. For example, ignoring incest and reincarnation, the relation Person is an ancestor of Person is strongly intransitive. The *strong intransitivity constraint* is depicted with an extra dot to the intransitivity icon (see Fig. 7.21).

Of all the ring constraints, irreflexivity is the least expensive to enforce. With expensive constraints such as acyclicity, *incremental checking* (for violations of the current update request, assuming the previous population is consistent with the constraint) is typically much cheaper than full checking.

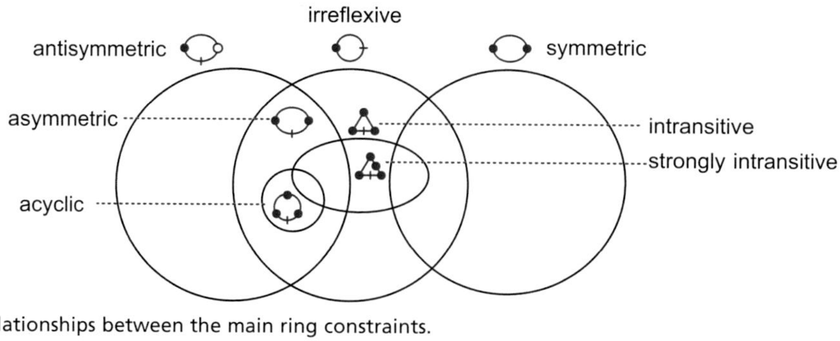

FIG. 7.21 Relationships between the main ring constraints.

The Euler diagram in Fig. 7.21 shows the relationships between the main ring constraints. The set of acyclic relations forms a proper subset of the set of *asym*metric relations, so acyclic relations must be asymmetric. An intransitive relation must be irreflexive, and an asymmetric relation is irreflexive and antisymmetric.

The symmetry constraint may be combined with intransitive or irreflexive constraints. Other properties of ring relations may apply (e.g., connectivity) but are not discussed here. A classic example of an embedded ring relation is the "Bill of Materials" or "Parts Explosion" problem (see Exercise 4.4), which involves the ternary Product contains Product in Quantity. You are invited to provide a more complete solution to this problem in the exercise that follows.

Exercise 7.3

1. For each predicate shown, state which, if any, of the following properties hold: reflexive, irreflexive, symmetric, asymmetric, transitive, intransitive.
   (a) has the same age as
   (b) is a brother of
   (c) is a sibling of
   (d) is shorter than
   (e) is an ancestor of
   (f) is at least as clever as
   (g) lives next to

2. Schematize the following report, which indicates composition of items (e.g., Item A contains two B parts and three C parts). Read "contains" as "directly contains" (e.g., A does not contain D). A more general reading for "contains" is taken in a later exercise. To help decide about ring constraints, draw a labeled graph for the containment relation, using an arrow for binary containment labeled with a number for the quantity.

Item	*Part*	*Quantity*
A	B	2
A	C	3
B	C	1
B	D	1
B	E	3
C	E	3
C	F	2

3. If you are familiar with predicate logic, prove the following theorems.
   (a) irreflex(R) & trans(R) $\rightarrow$ asym(R)
   (b) R is reflexive over its population iff $\forall x \forall y(xRy \vee yRx \rightarrow xRx)$. Show this is implied by sym(R) & trans(R).

4. The diagram shows the conceptual schema and the current population for a given UoD. Reference schemes are omitted for simplicity. Fact tables appear next to their fact types. Predicates are identified as **R**..**W**. Constraints are identified as $C_1..C_{15}$.

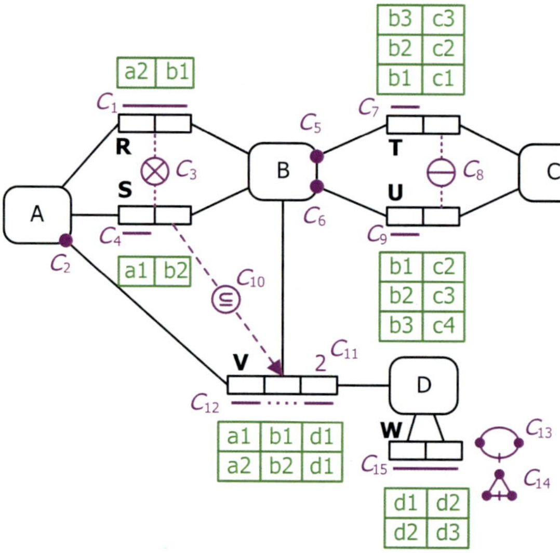

Each of the following requests applies to the **same** database population as shown (i.e., treat each request as if it was the *first* to be made on this population). For each request, indicate the response of the conceptual information processor (CIP). If the request is legal, write "accepted". Otherwise, indicate a constraint violated (e.g., "C_2 violated").

(a) add: a1 S b1
(b) add: a3 R b3
(c) add: V a1 b1 d2
(d) add: a1 R b2
(e) add: a1 R b3
(f) add: a2 S b3
(g) delete: b3 T c3
(h) begin
  add: b4 T c1
  add: b4 U c2
end
(i) begin
  add: b4 T c3
  add: b4 U c2
end
(j) begin
  add: V a1 b2 d2
  add: V a1 b3 d2
end
(k) add: d1 W d3
(l) add: d1 W d1

5. Schematize the following output report. Here, "bordering" means sharing a land border. In this sense, countries such as Australia have no bordering nations.

Nation	*Bordering nations*
Australia	
Belgium	France, Germany, Luxembourg, Netherlands
France	Belgium, Germany, Italy, Luxembourg, Spain, Switzerland
...	...

6. (a) Which, if any, of asymmetry and intransitivity are implied by acyclicity?
   (b) Convince yourself that the recursive definition of acyclicity given earlier is correct. Model relationships of type *R* graphically, and consider the fact *xRy* as a directed line from *x* to *y*.

7.4 Other Constraints and Rules

Having covered most of the main graphical constraints in ORM, we now consider three more graphical constraints: object cardinality, role cardinality, and value-comparison constraints. We then discuss basic textual constraints and derivation rules. More advanced graphical and textual rules are discussed in Chapter 10.

The hash symbol "#" denotes cardinality. *Simple cardinality constraints* are displayed by prepending "#" to the cardinality restriction, thus "#≤*n*" (**at most** *n*), "#=*n*" (**exactly** *n*), "#≥*n*" (**at least** *n*), "#*n*..*m*" (**at least** *n* **and at most** *m*).

An *object cardinality constraint* restricts the cardinality (number of members) of the population of an object type for each database state. The cardinality constraint "#=12" in Fig. 7.22(a) means there are always exactly 12 jury members. Equality settings like this are rare, as most databases start with the empty state, which would violate the constraint.

In Fig. 7.22(b), the "# ≤100" means there are at most 100 senators in each database state. Object cardinality constraints apply to the populations of the types rather than the types themselves. Over a long period, the total number of jury members or senators who ever filled these positions could be much higher.

A *role cardinality constraint* restricts the cardinality of the population of a role for each database state. The role cardinality constraint "# ≤ 1" in Fig. 7.22(c) means there is always at most one employee playing the director role. Note that ORM cardinality constraints are not the same as so-called "cardinality" constraints in entity-relationship modeling (ER), which instead correspond to frequency and/or mandatory role constraints.

Cardinality constraints are rarely included on diagrams because they are *often implied by existing value constraints or frequency constraints*. For example, an {"M", "F"} value constraint on Gender(.Code) implies "# ≤ 2" on Gender. If there is at most one senate, then a frequency constraint of ≤100 on Senate's role in Fig. 7.23(b) implies the # ≤100 cardinality constraint on Senator. Note, however, that the cardinality constraint # ≤1 on Senate is still required.

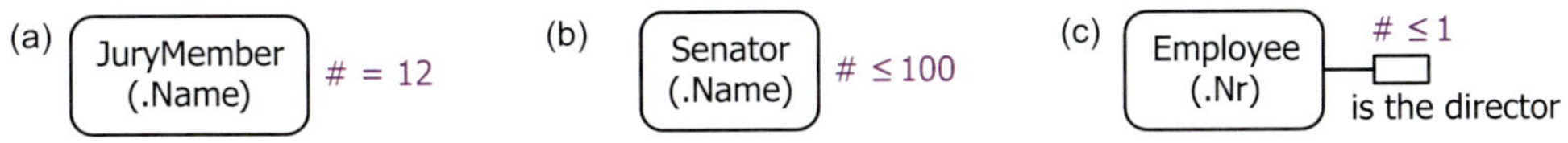

FIG. 7.22 Object cardinality constraints (a and b) and role cardinality constraints (c).

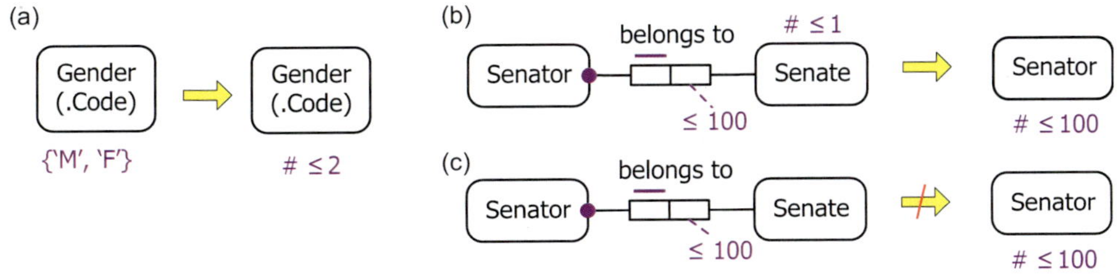

FIG. 7.23 Object cardinality constraints implied by value or frequency constraints.

In practice, however, it is often more useful to allow many senates, in which case we constrain the size of each senate to at most 100 members using a frequency constraint, even though a cardinality constraint on Senate no longer applies (see Fig. 7.23(c)).

A *disjunctive cardinality constraint* restricts the cardinality to conform to at least one of multiple, simple cardinality constraints listed as entries in a set enclosed in braces "{}" with the entries separated by commas. For example, #{0,12} means the cardinality must be 0 or 12, and #{≤2, 4, 6, 8..10} indicates the cardinality must be 0, 1, 2, 4, 6, 8, 9, or 10.

A *value-comparison constraint* compares one value with another using one of six comparison operators: $<$, $\leq$, $>$, $\geq$, $=$, $\neq$. The constraint is depicted by a circled operator, connected by a dotted arrow from the left operand role to the right operand role. Two dots are added to the circle to represent the two instances being compared (reminding us that this is a comparison between instances, not between sets).

The = and ≠ value comparison constraints are often better replaced by a ring constraint (e.g., a ≠ value comparison constraint on Country borders Country may be replaced by an irreflexive ring constraint).

The value-comparison constraints in Fig. 7.24 may be verbalized thus: **For each** Employee, birthdate < hiredate; **For each** Project, **existing** enddate ≥ startdate. For optional attributes (e.g., enddate), "**existing**" restricts the comparison to cases where attribute instances exist. So, "**existing** enddate ≥ startdate" is short for "**if** enddate **exists then** enddate ≥ startdate". Hence, if a project has no end date, the constraint is still satisfied (it maps to the SQL check clause **check** (enddate >= startdate), which is violated only when it evaluates to false).

Value-comparison constraints are a bit like ring constraints because the roles being constrained are compatible (e.g., the birthdate and hiredate being compared in Fig. 7.24(a) belong to the same employee). Like irreflexive ring constraints, value-comparison constraints are typically inexpensive to enforce, as they typically compare values on the same row of a table rather than comparing sets of values.

You have now met almost all the ORM graphical notations. Some more advanced constraints and rules are discussed in Chapter 10. Adding many graphic notations for further constraints could make the graphical language hard to learn and lead to cluttered diagrams. A conceptual schema diagram provides a human-oriented, unambiguous, but often incomplete specification of the UoD structure.

Rules not expressed using predefined graphic symbols may be specified as *textual rules*, preferably in a high-level formal language. In ORM 2, the preferred way to display these textual rules is via *footnotes*, with footnote numbers marking the involved

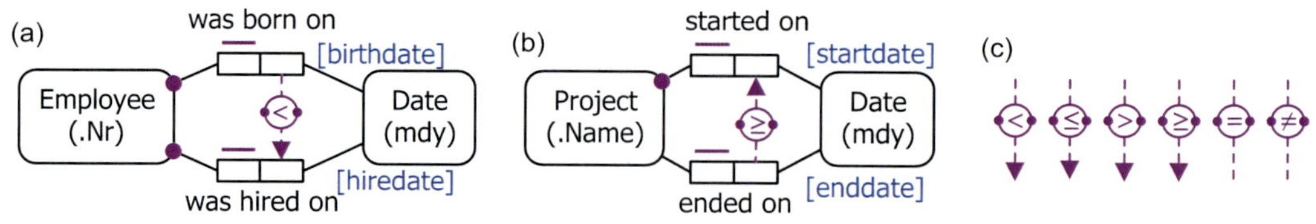

FIG. 7.24 Examples of value-comparison constraints.

model elements on the diagram. The textual rules (constraints, derivation rules, and subtype definitions) themselves may be displayed in text boxes, appearing as footnotes or elsewhere on the diagram, ideally with the option of toggling their display on or off.

Fig. 7.25(a) shows one way to display three textual rules. Rule 1 declares that employees are hired after their birth. This is the same as the value-comparison constraint in Fig. 7.24(a), but in Fig. 7.25(a) the constraint is expressed textually. The footnote number 1 is appended to the name of the Employee type. A pair-exclusion constraint also exists between the birth and hire predicates, but is omitted because it is now implied.

Rule 2 says that nonexecutives use at most one company car. This is a *restricted uniqueness constraint* because it strengthens the uniqueness constraint on Employee uses CompanyCar, but restricts this to a subtype of Employee.

Rule 3 says that each executive must have use of a company car. This is a *restricted mandatory role constraint* because it makes an optional Employee role mandatory for a subtype of Employee.

Fig. 7.25(b) shows another way to model the restricted mandatory role constraint by explicitly defining Executive as a subtype of Employee, and adding a subset constraint from the subtype link (Executive **is an** Employee) to the Employee role in Employee uses CompanyCar to declare that each executive must play the role of using a company car.

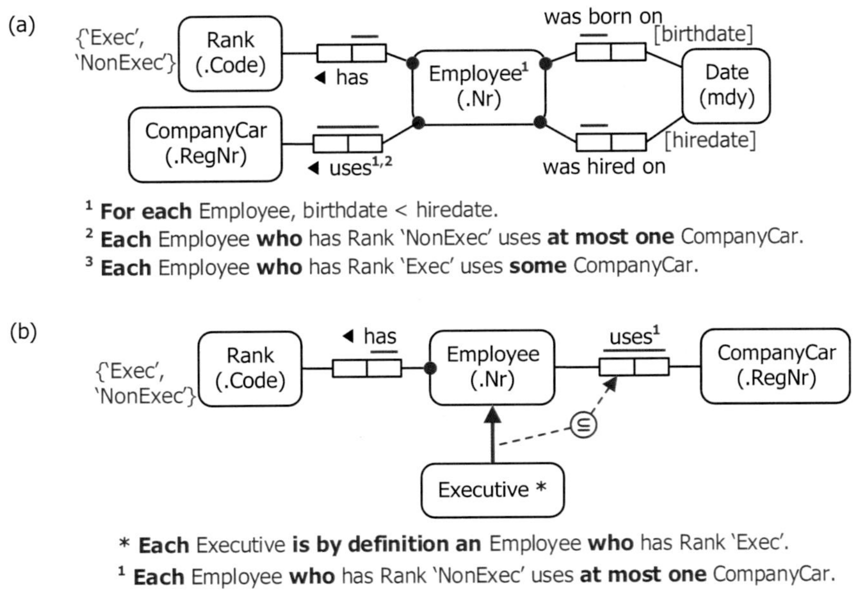

FIG. 7.25 Examples of textual rules.

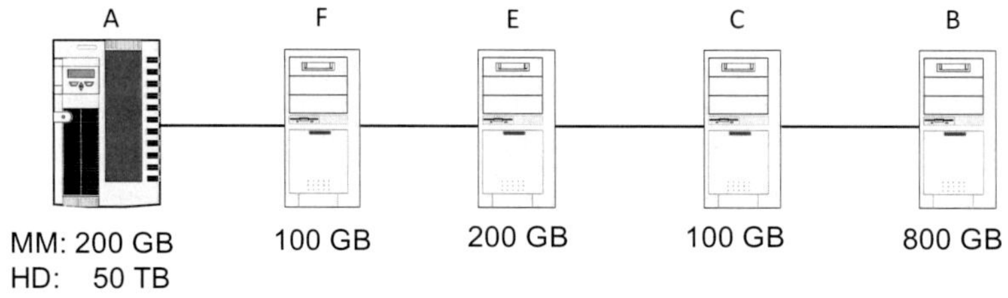

FIG. 7.26 Topology of network 1.

To conclude this section, let's consider an example that involves textual constraints and some other interesting issues. Suppose our information system is to maintain details about computer networks. In a bus network, one or more computer workstations (WS) are connected along a bus or transmission line to a file server (FS) at one end. Fig. 7.26 provides a topological picture of one such network (i.e., it shows which computers are connected to which but not their distance apart or directional pattern).

For simplicity, assume the computers are identified by a single letter. The main memory of each node and the hard disk capacity of the file server are recorded. Our task is to design a schema to record the memory and connection details, and similar details about other bus networks, so that users can extract as much information from this model as they could extract from the actual network diagrams. Try this yourself before reading on. There are many different, correct solutions.

Verbalizing the diagram in terms of elementary facts leads to several fact types, such as Node belongs to Network; Node is of NodeKind {'FS', 'WS'}; Node has main memory of Capacity; FileServer has disk of Capacity. The reference schemes are obvious. For a subtype definition, we may use **Each** FileServer **is a** Node **that** is of NodeKind 'FS'. Now comes the hard part. How do we describe how the nodes are linked together?

Depending on our choice of primitive predicates, there are various options. If we find it hard initially to express the information in terms of elementary sentences, we can begin by just jotting down samples of whatever comes to mind. For example:

1. Node "A" is directly connected to Node "F".
2. Node "F" is directly connected to Node "A".
3. Node "A" is indirectly connected to Node "E".
4. Node "A" is an end node.
5. There is at most one link between each pair of nodes.
6. No link connects a node to itself.
7. Each node is directly connected to at most two nodes.

Let's stand back from the problem for a minute. There are two basic aspects about the topology that need to be captured. First, the nodes in a given network all lie along the same *continuous line* (in contrast to other shapes such as star networks, ring networks, and fancy loop structures). Second, the nodes are *positioned in some order* on this line.

If we consider the linearity of the structure to be a constraint, how do we express the order of the nodes? Given that the nodes are positioned on a line, and that the file server is at one end of this line, let's first establish a *direction* for the line by saying the line starts at the file server. This avoids potential redundancy problems associated with an undirected, symmetric predicate (e.g., is directly connected to).

The ordering may be specified in terms of relative ordering or absolute ordering. With the first approach, we use a predicate for *directed linkage* (e.g., is just before). With the second approach, we assign each node an absolute position (e.g., 0, 1, 2, 3, 4 numbered from the file-server end).

The first approach leads to the ring fact type shown in Fig. 7.27(a). This is populated with data from the sample network. An extended version catering for multiple networks is shown in Fig. 7.27(b). This assumes that NodeId is a global identifier. The inclusive-or constraint is displayed explicitly because Node plays other roles in the complete schema. You are invited to add the relevant ring constraints as well as the other fact types. If only complete networks are considered, the kind of node can now be derived (e.g., a node is a file server if and only if it has no nodes before it).

By storing linkage information only for direct, directed links, we reduce the size of our database as well as avoiding the problems of symmetry and transitivity. If desired, information about indirect links can be derived by using a recursive rule such as:

For each Node_1 **and** Node_2,
Node_1 is before Node_2 **iff** Node_1 is just before Node_2
or
Node_1 is just before **some** Node_3 **that** is before Node_2.

Other constraints exist. For example, each network has exactly one file server. If we are using the fact type Node is in Network, this may be specified textually. We can specify it graphically if we instead use separate fact types: Node is the file server for Network; Node is a workstation for Network.

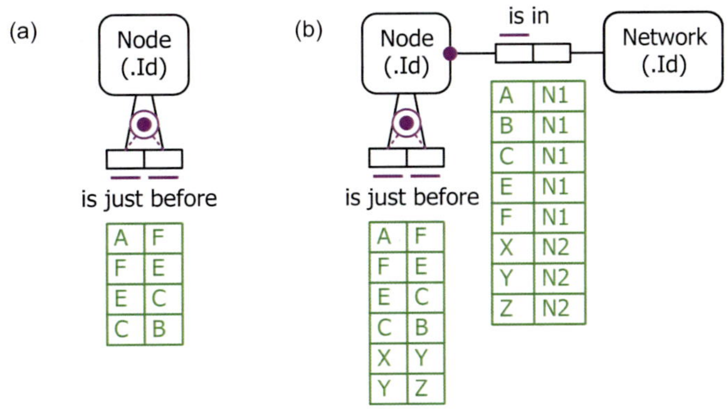

FIG. 7.27 A fact type for directed links in (a) one network and (b) many networks.

However, some constraints have to be specified textually no matter how we choose our fact types. For example, directly linked nodes belong to the same network. If each network is to be continuous (no gaps), a constraint for this also needs to be declared textually. You are invited to express such constraints (and any others) formally.

As an exercise, try to model the linkage information using node position numbers instead, and notice the impact on constraints and derivation rules. Note that this absolute position approach makes it more awkward to add or delete nodes in the network because global changes to positional information would be required.

For instance, if we add a node, say D, between F and E, then not only will D be assigned the position 2, but nodes E, C, and B will all need to have their position numbers increased by 1. With our previous approach, only a local change would be needed. Swapping nodes is simple with both approaches. As an exercise, set out the compound transactions for some sample updates.

Exercise 7.4

1. (a) Rugby union teams have 15 positions, numbered 1 through 15. Teams are identified by name. Players have names (not necessarily unique) and are identified by their position in the team. Each player is a member of exactly one team and has exactly one position. It is possible for any team that not all of its 15 positions are currently filled. Specify an ORM schema to store membership details for multiple teams.
   (b) Now restrict the UoD to a single team, whose name must still be recorded.
   (c) Now restrict the UoD to a single team, whose name is not recorded.
2. A given company always has exactly one head employee, and each employee other than the head reports to exactly one employee. The reporting relationship is acyclic. Schematize this situation.
3. Schematize the following table about the current age categories for humans. Unpack the semantics of ranges, and include any applicable graphical constraint.

Age category	*Age range (y)*
Child	0..14
Youth	15..24
Adult	25..64
Senior	65..140

4. A television survey on soap operas is conducted. Participants in this survey are identified by a number. Their gender is recorded as well as what soap operas they watch. Women may nominate all the soap operas they watch, if any, but to be included in the survey each man must specify exactly one soap opera. Model this UoD.

5. A *star network* has a centrally located file server directly linked to zero or more workstations. A sample star network is shown. For each node, the generic kind of computer is recorded (PC or Mac). Only for the file server is the processor speed recorded.

   The maximum number of computers in any network is 20. The cable distance (in meters) of each node from the file server is shown next to each link. Schematize.

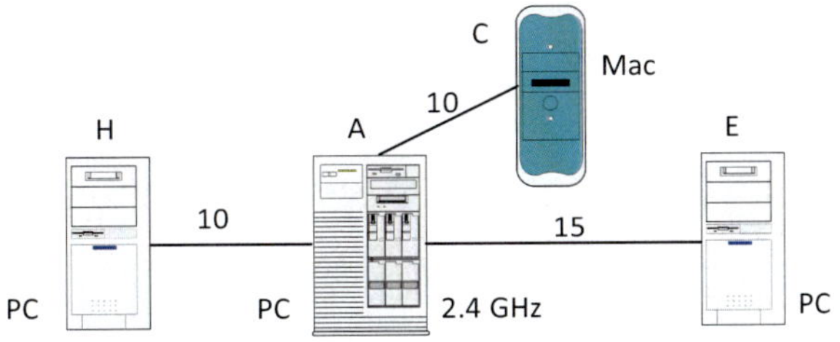

6. With reference to the Community-Roads UoD of Exercise 7.2, assume that roads are continuous, but may connect more than two communities. Discuss whether it is appropriate to use the ternary fact type Road goes from Community to Community.

7. Recall the Bill of Materials question in Exercise 7.3. A report is required to display parts contained at all levels (e.g., the fact that A contains part D is to be shown). Model this.

8. A *ring network* has a central file server with workstations arranged in a ring. A sample network is shown. Details are as for Question 5, except that no metric information is required. Specify the conceptual schema for this UoD.

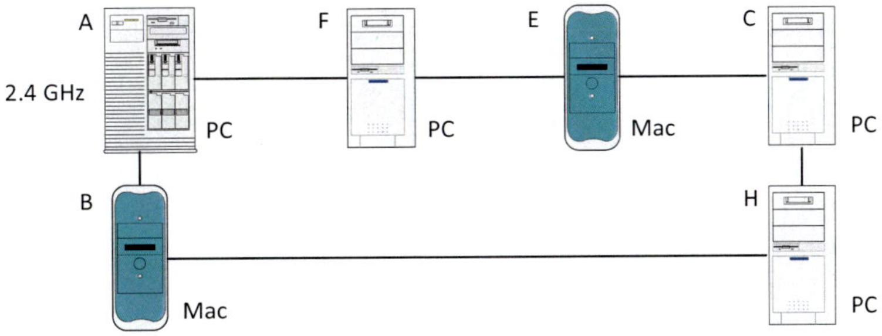

7.5 Final Checks

The conceptual schema design procedure facilitates early detection of errors by various checks, including communication with the user by way of examples. Four final checks may also be performed to help detect any errors that might have slipped through. These are designed to ensure internal consistency, external consistency, lack of redundancy, and completeness. We now consider these in order.

A conceptual schema is *internally consistent* if and only if each role sequence used as a predicate or constraint argument can be populated in some state. This topic was discussed in the previous chapter, where it was called "*population consistency*" or "*strong satisfiability*". Basically, it means that the specified constraints do not contradict one another when the system is populated. For example, the schema in Fig. 7.28 has three faults. Can you spot them?

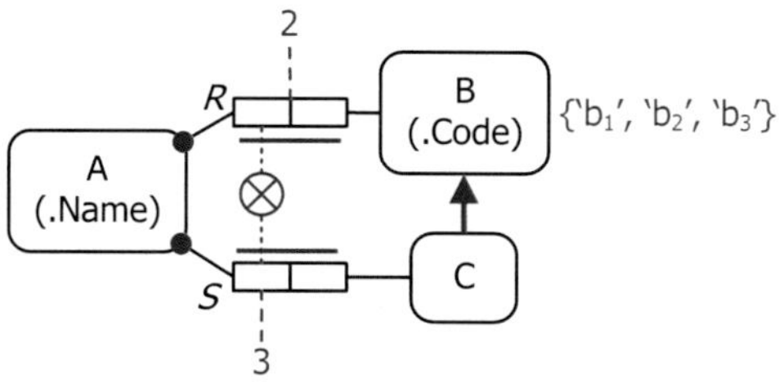

FIG. 7.28 This schema is population inconsistent on three counts.

First, the frequency constraint of 2 on predicate *R* clashes with the uniqueness constraint; moreover, any frequency above 1 should be rejected if it spans the whole predicate. Second, the mandatory roles on *A* imply an equality constraint between these roles, so if *A* is populated, the exclusion constraint cannot be satisfied (recall theorem NXS).

Finally, the frequency constraint of 3 on the first role of *S* cannot be satisfied once *A* is populated. If it were satisfied, then *C* would include at least three instances (because each row of *S* is unique), but this is impossible because *C* is a proper subtype of *B* and hence has a maximum cardinality of 2 (the value constraint shows *B* has only three possible values). With practice, it is usually easy to spot contradictory constraint patterns. Computer-aided software engineering (CASE) tools can be of assistance here. For example, the NORMA tool prevents most inconsistencies from even being entered.

A conceptual schema is *externally consistent* if it agrees with the original examples and requirements used to develop the schema. If not done earlier, we populate the schema with some of the sample data and look to see if some constraints are violated by the population. If they are, then the constraints are *too strong* in this respect because they reject legal examples. We should then modify the schema, typically by removing or softening some constraints, until the population is permitted.

As a simple illustration, consider Table 7.2. To simplify the discussion, assume that people may be identified by their first name. The null value indicates that Bob seeks a BA degree but we do not know any of Bob's subjects.

Suppose that we propose the schema in Fig. 7.29. Assuming that our fact types are correct, we might still get the constraints wrong. As well as asking the client directly about the constraints, we check by populating the fact tables with the sample data, as shown. The uniqueness constraint on studies asserts that each entry in this column is unique (i.e., each person studies at most one subject).

Table 7.2 Student details.

Person	*Course*	*Degree*
Ann	CS112	BS
Ann	CS100	BS
Bob	?	BA
Sue	CS112	BA
Tom	CS213	BS

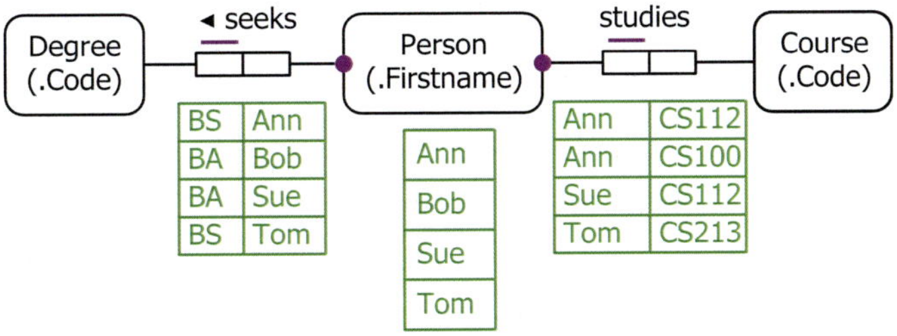

FIG. 7.29 The constraints are inconsistent with the data.

However, "Ann" appears twice here. So, this constraint is wrong. The sample population makes it clear that the uniqueness constraint should span both columns. Now consider the uniqueness constraint on Person seeks Degree. This asserts that each entry in the degree column is unique. But the entry "BS" appears twice there, as does the entry "BA". The presence of either of these cases proves that this constraint is wrong.

This does not imply that the correct uniqueness constraint should span both columns: although the population is consistent with this weaker constraint, it is also consistent with a uniqueness constraint over the Person column.

To decide between the two possibilities, we ask the domain expert an appropriate question (e.g., Can the same person seek more than one degree at the same time?). If the answer to this is No, then we should replace the old uniqueness constraint with one across the Person column for this fact type.

Finally, note that the mandatory role constraint on studies is violated by the sample data because Bob is referenced elsewhere in the database but does not play this role. So, this role should be made optional. Fig. 7.30 shows the fully corrected schema.

FIG. 7.30 The corrected conceptual schema for Table 7.2.

While populating the schema with original examples may detect constraints that shouldn't be there, this will not automatically reveal missing constraints. In other words, while this check can pick up aspects of the constraint section that are too strong, we require significant examples or background knowledge to determine whether the constraint section is *too weak* in other respects.

We now check that the schema is free of *redundancy* by ensuring that no elementary fact can appear twice. Stored redundancy was covered in step 4, in the context of arity checking. We now turn to the notion of *derived redundancy*. The most common case of this is when a recorded fact can be derived from other facts by means of specified derivation rules.

The task of identifying derived fact types was considered in earlier steps. For example, the markup of an article was derived by subtracting its wholesale price from its retail price. Here, it would be redundant to store the markup values. If markup values are not stored, then they need to be computed upon request, which has the disadvantage of adding slightly to the response time for markup queries.

If markup values are stored, this takes up extra storage. Moreover, if the wholesale or retail price of an article is changed, then failure to update markup accordingly leads to an inconsistent database. If such an update anomaly can occur, the redundancy is "*unsafe*."

You can arrange for *safe redundancy* or *controlled redundancy* in derived cases by having the derivation rule triggered by relevant updates. For example, the system can be configured to automatically "recalculate" the markup prices whenever wholesale or retail prices are updated (we assume markup prices may not be updated directly; this avoids the further overhead of a mutual recalculation among the three prices). This kind of approach is often used with spreadsheets.

Another derived case of safe redundancy occurs when the relevant facts are made non-updatable. For example, a bank account balance may be computed at the end of each transaction and then stored immediately. This is a case of "derive on update" rather than "derive on query". This "eager evaluation" of derived information can improve the performance of an application considerably.

For example, if a bank with millions of customers produces monthly statements of account for each customer, accessing stored balances can save lots of recomputation. This is safe if the bank has a policy of never overwriting any balances. If an accounting error is found, it is left there but is compensated for by a later correcting transaction. This practice can also facilitate the task of auditing.

We sometimes meet cases where a stored fact type could be handled as a *semiderived* fact type. Recall the owner-driver example from the previous chapter, where if a person drives a car then that person also owns that car. In Fig. 7.31(a), this feature is captured by the pair-subset constraint shown.

In Fig. 7.31(b), this feature is captured instead by the derivation rule: $^{++}$Person owns Car **if** Person drives Car. Some people might own a car but not drive it, so the ownership fact type must be at least partly asserted. The Person owns Car fact type is now semiderived. The "$^{++}$" superscript used here instead of the basic "$^{+}$" superscript for semiderivation indicates that the derived facts are to be stored immediately (*eager evaluation*).

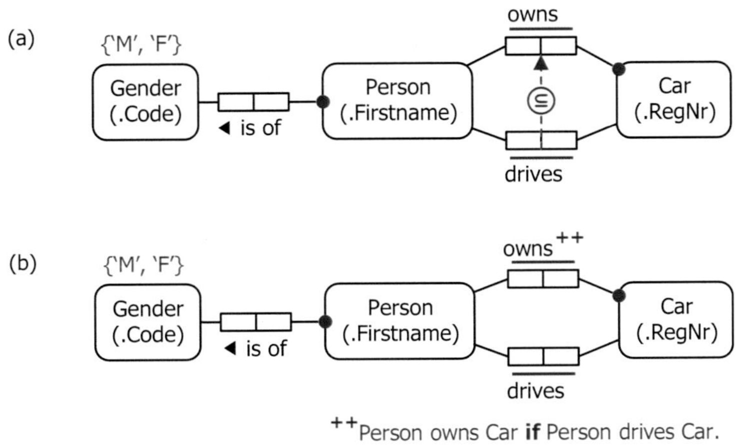

FIG. 7.31 The owns fact type is (a) asserted or (b) eagerly semiderived.

From an implementation viewpoint, it is usually better to store the whole of the ownership relation under the control of the subset constraint or the eager semiderivation rule. This may take up a bit more storage, but access to ownership facts is fast.

From the *external* viewpoint, a user interface may be constructed to automatically add the required owner fact when a driver fact is added and to delete a driver fact when its owner fact is deleted. Such an interface could inform the user of its intended action and provide the option of canceling the transaction if this is not desired.

Apart from stored and derived redundancy, redundancy can occur within the set of derivation rules. For instance, it might be possible to derive a fact in more than one way, using different rules. Moreover, some rules might be derivable from more primitive rules. This situation is fairly common in formal inference systems (e.g., computer-aided reasoning systems).

In a wider context, redundancy can sometimes be very useful from the point of view of information retrieval, understanding (e.g., human communication), and coping with partial system failure (e.g., backup systems). This book itself exhibits a great deal of redundancy (e.g., by repeating important points). Although this makes the book longer, it hopefully makes it easier to follow.

The last check in the conceptual schema design procedure is to see if it is *complete* with respect to the original requirements specified for the application. This can be done systematically by going through each requirement, one at a time, and identifying which aspects of the conceptual design cater for it. If some requirements are found to be unfulfilled, the design should be extended to cater for them.

Apart from advanced aspects covered in Chapter 10, this completes the conceptual schema design procedure. In terms of the application development life cycle, there are still several things to be done (e.g., mapping the conceptual design to a database schema;

implementing physical and external schemas; testing; and maintenance). However, the most crucial stages of the cycle have now been considered. The final exercise question is a long one, covering most aspects of the CSDP.

Exercise 7.5

1. Discuss any problems with this schema (reference schemes are omitted for simplicity)

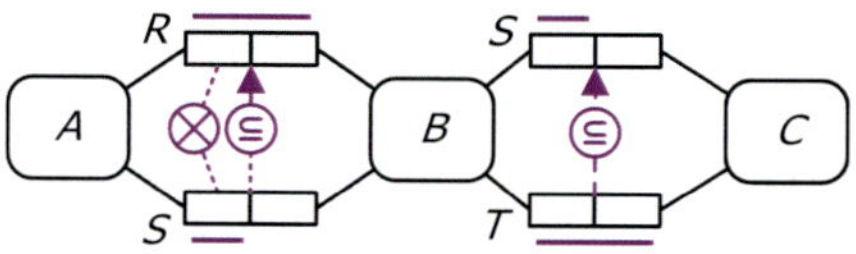

2. From the sample output report shown, a student draws the conceptual schema diagram shown. The elementary charge unit (e) is the electric charge of a single proton. Check to see if the data in the original report is a legal population of this schema. If not, modify the schema accordingly.

Particle	*Family*	*Charge (e)*
neutrino	lepton	0
electron	lepton	-1
muon	lepton	-1
pion	meson	1
kaon	meson	1
proton	baryon	1
neutron	baryon	0

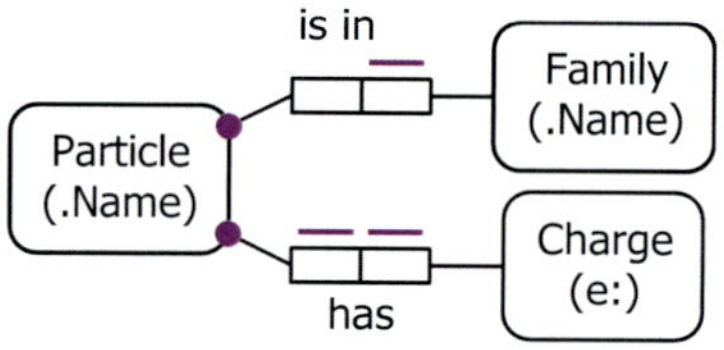

3. Is this schema guilty of redundancy? If so, correct it.

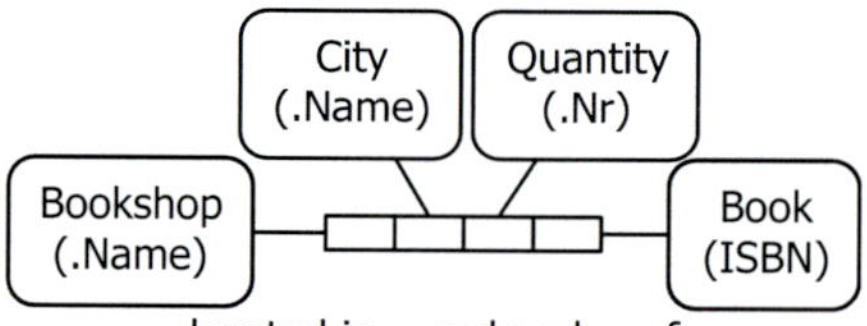

Bookland	New York	100	300705
Websters	New York	100	300705
Bookland	New York	50	123555
OKBooks	London	200	123555

4. With respect to the previous question, suppose it is now possible that bookshops in different cities have the same name. For example, the following population is legal. Draw the correct conceptual schema diagram for this UoD.

Bookland	New York	100	300705
Websters	New York	100	300705
Bookland	New York	50	123555
Bookland	London	200	123555

5. In Prolog, relations are often partly stored and partly derived. For example, at one state the knowledge base might consist of the following facts and rules. Discuss this situation making reference to the notion of derived redundancy.

   parent_of(ann, bob).
   parent_of(bob, chris).
   grandparent_of (david, chris).
   grandparent_of (X, Y) **if** parent_of(X, Z) **and** parent_of(Z, Y).

6. An information system is to deal with various colors that are classified as primary (P), secondary (S), or tertiary (T). There are three primary colors: blue, red, and yellow. A secondary color is composed of a mixture of exactly two primary colors. A tertiary color is composed of exactly three primary colors.

   Each color has a unique, identifying trade name (e.g., "forest green"). The trade names of the primary colors are "blue", "red", and "yellow". Each color has a (perhaps zero) percentage of blue, red, and yellow. This percentage is expressed as a whole number in the range 0..100. For example, forest green is 70% blue and 30% yellow, but has no red. The following extract from a sample output report indicates the sort of information that needs to be accessed.

Color	*% blue*	*% red*	*% yellow*	*Class*
forest green	70	0	30	S
mud brown	30	30	40	T
red	0	100	0	P

   It is required to reduce the size of the database as much as possible by using derivation rules. For example, the percentage of a given color that is yellow should be derived from the percentages for blue and red. Schematize this UoD, indicating all constraints. State each derivation rule—you may use obvious abbreviations (e.g., "%B", "%R", and "%Y"). As long as circularity is avoided, the rule for computing the percent yellow may be assumed in formulating other rules (i.e., other rules may use the term "%Y").

7. An information system is required to manage data about various astronomical bodies in our solar system. The first table shows an extract from a report about moons. A "?" mark indicates unknown, and "–" indicates nonexistent.

Moon	*Planet*	*Mean radius (km)*	*Discovered*	*Explored?*	*Has water ice?*	*Atmosphere*	
						Gas	*%*
The Moon	Earth	1738	?	True	True	Ne	29.0
						He	25.8
						H	22.6
						Ar	20.6
Deimos	Mars	6.3	1877	True	False	–	–
Phobos	Mars	11.1	1877	True	False	–	–
Europa	Jupiter	1561	1610	True	True	O_2	100
Ganymede*	Jupiter	2631.2	1610	True	True	O_2	100
S/2003 J2	Jupiter	1	2003	False	False	–	–

* Largest moon in our solar system

Until a moon is given a normal name, it is assigned a temporary name based on the order in which the satellite was discovered for a given year and planet (do *not* bother unpacking these semantics). Here, "Explored" means the moon has been investigated by a spacecraft (manned or unmanned) from Earth. If a moon has an atmosphere, only its main gases are listed.

If a moon has been explored, details about the missions to that moon may be recorded. A relevant extract of a report in this regard is shown in the next table. The meaning of bracketed contents in the mission entries in the first column needs to be made explicit. A "?" mark is a simple null (unknown) and a "–" mark indicates inapplicable because of other data.

Mission	*Moon*	*First Arrival*	*Crewed*	*Type*	*Astronauts*
Luna 2 (USSR)	The Moon	1959 Sep 13	No	Impact	–
Ranger 7 (US)	The Moon	1964 Jul 31	No	Impact	–
Luna 9 (USSR)	The Moon	1966 Feb 3	No	Lander	–
Apollo 11 (US)	The Moon	1969 July 20	Yes	Lander	Neil Armstrong, Buzz Aldrin, Michael Collins
Luna 17 (USSR)	The Moon	1970 Nov 17	No	Rover	–
Galileo (US)	Europa	1996 Oct 20	No	Flyby	–
Galileo (US)	Ganymede	1996 Nov 22	No	Flyby	–
Voyager 2 (US)	Europa	?	No	?	–

Details of interesting features of various moons may be recorded. The next table shows a small extract of details from this record. Feature length measures are of only two types (diameter and linear).

The type of a moon's feature does not determine its size type. For example, some grooved terrains might have a pencil-like shape and have their linear dimension measured, whereas other grooved terrains might have a roughly circular shape and have their diameter measured. Moreover, the size of a feature is recorded if and only if the size type of that feature is recorded.

Moon	*Feature*	*Feature Type*	*Age (y)*	*Size (km)*	*Size Type*
The Moon	Aristarchus Crater	Impact Crater	300,000,000	37	Diameter
	Mare Crisium	Sea	3,900,000,000	563	Diameter
	Mare Tranquillitatis	Sea	3,600,000,000	873	Diameter
	Montes Apenninus	Mountain range	3,900,000,000	?	?
Ganymede	Rupes Altai	Cliff	4,200,000,000	507	Linear
	Nicholson Regio	Dark area	?	3900	Diameter
	Uruk Sulcus	Grooved terrain	?	7	Linear
	Sippar Sulcus	Grooved terrain	?	10	Diameter

The next table shows an extract of details about astronauts. The information system identifies astronauts by numbers to allow different astronauts with the same name and to track astronauts who change their names (ignore name tracking in your solution).

Astro#	Name	Gender	Nationality	FirstSpaceTrip	Born
1	Yuri Gargarin	M	RU	1961	1934
2	Alan Shepherd	M	US	1961	1923
3	Valentina Tereshkova	F	RU	1963	1937
4	Neil Armstrong	M	US	1966	1930
5	Buzz Aldrin	M	US	1966	1930
6	Michael Collins	M	US	1966	1930
7	Sally Ride	F	US	1983	1951
8	Rodolfo Neri Vela	M	MX	1985	1952

The next table shows an extract from another report about moons and planets in our solar system. A body's name type is either normal or temp (temporary). If recorded, the origin of a body's name is treated as an atomic value (an uninterpreted character string).

Body	*Type*	*Name Type*	*Name Origin*
Mercury	Planet	Normal	God of trade, and messenger to the gods (wore winged shoes)
Venus	Planet	Normal	Goddess of love
Earth	Planet	Normal	?
Mars	Planet	Normal	God of war
Jupiter	Planet	Normal	?
The Moon	Moon	Normal	The only moon of Earth
Deimos	Moon	Normal	A son of the god Ares (Mars)
Phobos	Moon	Normal	A son of the god Ares (Mars)
Europa	Moon	Normal	Cup-bearer to the Greek gods
Ganymede	Moon	Normal	Daughter of Agenor (king of Tyre)
S/2003 J2	Moon	Temp	–

The next table shows a report of details about planets in our solar system. In 2006, the International Astronomical Union declassified Pluto as a planet, reclassifying it as a minor planet along with four other objects (Ceres, Eris, Haumea, and Makemake), so it does not appear in the table. Each planet has either its average surface temperature or its temperature range recorded. At most, four main gases may be recorded for each planet (or moon—see first table). Any data that can be derived should be derived.

Planet	*Avg. Surface Temperature*	*Temperature Range*		*Ringed?*	*Nr rings*	*Atmosphere*
		Min	*Max*			
Mercury	–	-180°C (-290°F)	430°C (810°F)	No	0	O_2 (52%), Na (39%), He (8%)
Venus	464°C (867°F)	–	–	No	0	CO_2 (96.5%), N_2 (3%)
Earth	15°C (59°F)	–	–	No	0	N_2 (78.1%), O_2 (20.9%)
Mars	–	-125°C (-195°F)	25°C (77°F)	No	0	CO_2 (95.3%), N_2 (2.7%)
Jupiter	-110°C (-160°F)	–	–	Yes	3	H_2 (89.8%), He (10%)
Saturn	-140°C (-220°F)	–	–	Yes	7	H_2 (96.3%), He (3.5%)
Uranus	-214°C (-353°F)	–	–	Yes	11	H_2 (82.5%), He (15.2%), CH_4 (2.3%)
Neptune	-200°C (-320°F)	–	–	Yes	6	H_2 (79%), He (18%), CH_4 (2.9%)

Over time, some countries may divide into other countries. For example, in 1991 the USSR (country code SU) split into 15 independent countries. The next table shows a small extract of a table to keep track of such country divisions.

Country	*Parent*
LV	SU
EE	SU
RU	SU

Specify an ORM conceptual schema for this UoD. Include all relevant uniqueness, mandatory, value, set-comparison, subtyping (including subtype definitions), frequency, ring, and value-comparison constraints. If an object type is independent, indicate this. If a fact type is derived, include it on the diagram as well as provide its derivation rule.

7.6 Summary

CSDP step 7 adds other constraints and performs final checks. A *frequency constraint* indicates that an entry in a column (or column combination) must occur there exactly n times (n), at most n times ($\leq n$), at least n times ($\geq n$), or at least n and at most m times (n .. m). A simple occurrence frequency appears next to the role. A compound occurrence frequency appears next to a line connecting the relevant roles.

A *ring constraint* can apply only to a pair of roles played by the same (or a compatible) object type. The role pair may form a binary predicate or be embedded in a longer predicate. Let R be the relation type comprising the role pair. R is (locally) *reflexive* (over its populations) iff for all x playing either role, xRx. R is *purely reflexive* iff for all x, $y(xRy \rightarrow x=y)$.

R is *symmetric* iff for all x, $y(xRy \rightarrow yRx)$. R is *transitive* iff for all x, y, $z[(xRy \;\&\; yRz) \rightarrow xRz]$. These positive properties tend to be used for derivation rather than as constraints.

The following negative properties may be marked as ring constraints next to the role pair (or role connector). R is *irreflexive* iff for all x, $\sim xRx$. R is *asymmetric* iff for all x, $y(xRy \rightarrow \sim yRx)$. R is *antisymmetric* iff for all x, $y[(x \neq y \;\&\; xRy) \rightarrow \sim yRx]$. R is *intransitive* iff for all x, y, $z[(xRy \;\&\; yRz) \rightarrow \sim xRz]$.

A relationship R is *acyclic* iff no object can cycle back to itself by applying one or more instances of the R relationship, so for all $x,y \sim xPy$ where the path predicate P is recursively defined to give the transitive closure of R.

R is *strongly intransitive* iff for all x, y, $z[(xRy \;\&\; yPz) \rightarrow \sim xRz]$ where the path predicate P is recursively defined to give the transitive closure of R.

Asymmetry and intransitivity each imply irreflexivity. Exclusion implies asymmetry (and irreflexivity). An irreflexive, functional relation must be intransitive. Symbols for all the ring constraints are summarized in the ORM symbol glossary at the end of the book.

An *object cardinality constraint* or *role cardinality constraint* limits the cardinality of each population of an object type or role, respectively. An upper limit of n is denoted by "$\# \leq n$" next to the object type or role. Cardinality constraints are rarely used because they are *often implied by existing value constraints or frequency constraints.*

A *value-comparison constraint* compares one value with another using one of the operators $<$, $\leq$, $>$, $\geq$, $=$, $\neq$. The constraint is depicted by a circled operator, connected by a dotted arrow from the left operand role to the right operand role. Two dots on the circle represent the two instances being compared. A typical example of this kind of constraint is **For each** Employee, birthdate < hiredate.

Constraints for which no graphic symbol exists may be specified as *textual rules.* The model elements involved in the rule are marked with a footnote number, and the rule itself may be displayed in a text box.

A restricted uniqueness constraint declares stronger uniqueness for a given subtype. A restricted mandatory role constraint indicates that an optional role is mandatory for a given subtype.

When modeling connections between nodes, choose directed, immediate links for the stored relation. Indirect links may then be derived recursively.

At the end of the *CSDP*, final checks are made to ensure that the conceptual schema is internally consistent (its constraint pattern can be populated without contradiction), is externally consistent (agrees with original data and conditions), is redundancy free (elementary facts can't be repeated), and is complete (covers all the requirements). Various cases of derived redundancy can be safely managed.

Chapter Notes

In contrast to our usage, frequency constraints are sometimes defined as the number of times members of an object type population must play a role—frequencies of 0 are then allowed and mean optional. Ring constraints are often discussed in logic books within the context of the theory of dyadic relations, but are rarely discussed in modeling approaches other than ORM. For further discussion of ring constraints in ORM, see Halpin and Curland (2011).

8 Entity-Relationship Modeling

Chapter Outline

8.1 Overview of ER

The *entity-relationship* (ER) modeling approach views a business domain in terms of entities that have attributes and participate in relationships. For example, the fact that an employee was born on a date is modeled by a birthdate attribute of the Employee entity type, whereas the fact that an employee works for a department is modeled as a relationship between them. This world view is quite intuitive, and despite the recent rise of unified modeling language (UML) for modeling object-oriented applications, ER is still the most popular data modeling approach for database applications.

In Chapter 1, we argued that object-role modeling (ORM) is better than ER for conceptual analysis. However, ER is widely used, and its diagrams are good for compact summaries, so you should become familiar with at least some of the mainstream ER notations. This is the main purpose of this chapter. A second purpose is to have a closer look at how ORM compares to ER. To save some explanation, we assume that you have already studied the basics of ORM in earlier chapters so that we can examine ER from an ORM perspective.

The ER approach was originally proposed by Dr. Peter Chen in the very first issue of an influential journal (Chen, 1976). Fig. 8.1 is based on examples from this journal paper. Chen's original notation uses rectangles for entity types and diamonds for relationships (binary or longer). Attributes may be defined, but are excluded from the ER diagram. As in ORM, roles are defined as parts played in relationships. Role names may optionally be shown at relationship ends (e.g., Employee plays the worker role in the Proj-Work relationship).

Chen formalized relationships in terms of ordered tuples of entities, allowing the order to be dropped if role names are used (as with attribute names in the relational model). Although not displayed on the ER diagram, relationships may have attributes, but cannot play roles in other relationships. So objectified associations are not fully supported.

Information Modeling and Relational Databases. https://doi.org/10.1016/B978-0-443-23790-4.00017-7

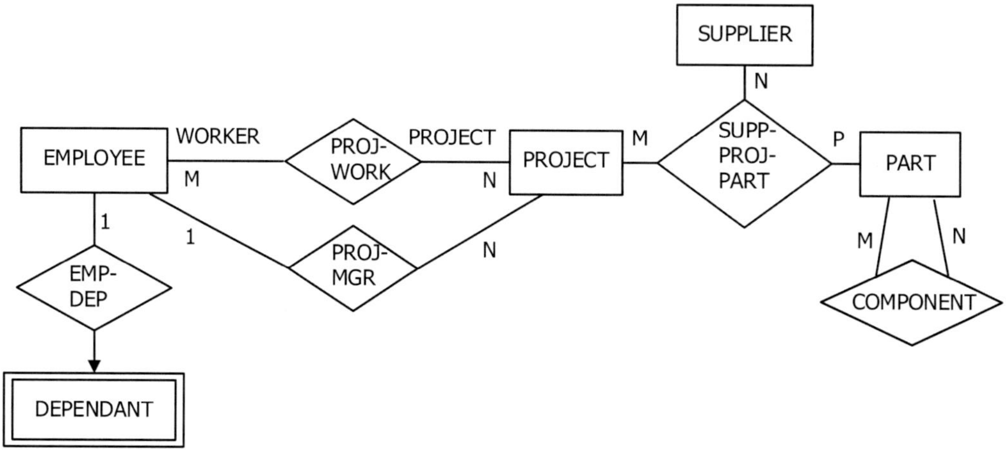

FIG. 8.1 The original ER notation used by Chen.

FIG. 8.2 An ambiguous ER schema.

Roles may be annotated with a maximum cardinality of 1 or many. For example, read left to right, the Proj-Mgr relation is one-to-many (each employee manages zero or more projects, and each project is managed by at most one employee).

As shown in Fig. 8.1, Chen used noun phrases for relationships, eliminating natural verbalization. Even if verb phrases are used, the direction in which relationship names are to be read is formally undecided, unless we add additional marks (e.g., arrows) or rules (e.g., always read from left to right and from top to bottom). For example, does the employee manage the project or does the project manage the employee? Although we can use our background knowledge to informally disambiguate this example, it is easy to find examples of relationships whose direction can only be guessed at by anybody other than the model's creator.

This problem is exacerbated if the verb phrase used to name the relationship is shortened to one word (e.g., "work"), unfortunately still a fairly common practice. As a simple example, is the ER diagram in Fig. 8.2 meant to capture the fact type Person killed Animal, or Animal killed Person? We could disambiguate this diagram by adding role names (e.g., "victim", "killer") but there is no requirement to do so.

Similarly, if we populate the Component relationship in Fig. 8.1 with the pair (*a*, *b*), we don't know whether this means *a* is a component of *b*, or vice versa. To disambiguate this, we need to add role names (e.g., "subpart", "superpart") or use a verb phrase (e.g., "is a component of") with a defined direction.

A rectangle with a double border denotes a *weak entity type*. This means that the entity type's identification scheme includes a relationship to another entity type. In Fig. 8.3(a), for example, an instance of Dependant might be identified by having the name "Eve Jones"

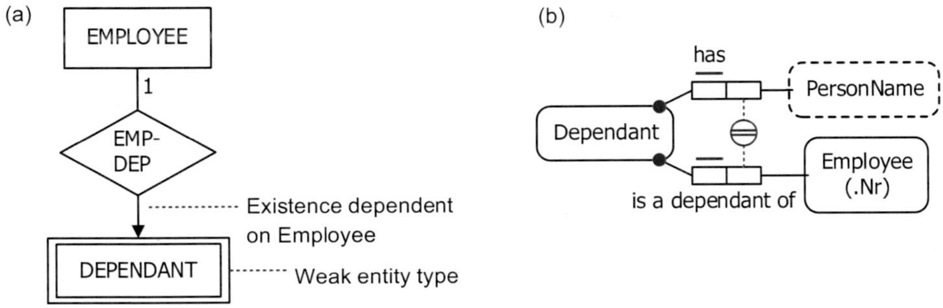

FIG. 8.3 A weak entity type in ER (a) remodeled in ORM (b).

and being related via the Emp-Dep relation to employee 007. In ORM, this is modeled by a coreferenced object type, as in Fig. 8.3(b).

The arrow tip at the Dependant end of the ER relationship indicates that Dependant is existence dependent on Employee (its existence depends on the existence of the other). Given that Dependant is weak, this is basically redundant.

A far better approach is to introduce the concept of a mandatory role, as in ORM and many other ER versions (e.g., **Each** Dependant is a dependant of **at least one** Employee). This ability to establish a minimum multiplicity of at least one for any given role was absent from Chen's original notation.

Over time, many variant notations developed. Chen's ER-Designer tool uses hexagons instead of diamonds. One problem with ER is that there are so many versions of it, with no single standard. In industry, two of the most popular versions of pure ER are the Barker and information engineering (IE) notations. These are discussed in the next two sections.

Another popular data modeling notation is IDEF1X, which is a hybrid of ER and relational notation, so it is not a pure ER notation. Nevertheless, many people talk of IDEF1X as a version of ER, so we cover it in this chapter.

The best way to develop an ER or IDEF1X model is to derive it from an ORM model; we briefly discuss mapping from ORM later in the chapter. The UML class diagram notation may be regarded as an extended version of ER, but because of the importance of UML, it is considered separately in the next chapter.

8.2 Barker Notation

We use the term "Barker notation" for the ER notation discussed in the classic treatment by Richard Barker (Barker, 1990a). Originating in the late 1980s at CACI in the United Kingdom, the notation was later adopted by Oracle Corporation in its computer-aided software engineering (CASE) design tools. Oracle now supports UML as an alternative to the Barker ER notation. For database applications, many modelers still prefer ER over UML. Embarcadero has basic support for the Barker notation in its EA/Studio product. Although dozens of ER dialects exist, we consider the Barker notation to be the best ER notation with wide support in industry.

Dave Hay, an experienced modeler and fan of Barker notation, argues that "UML is … not suitable for analyzing business requirements in cooperation with business people"

(Hay, 1999b). While UML class diagrams are less than ideal for data modeling, ER, like UML, is attribute-based. As discussed in Chapter 1, using attributes in a conceptual model adds complexity and instability, and makes it harder to validate models using verbalization and sample populations.

Attributes are great for logical design because they allow compact diagrams that directly represent the implementation data structures (e.g., tables or classes). However, for conceptual analysis, we just want to know what the *facts* and *rules* are about the business and to communicate this information in *sentences* so that the model can be understood by the domain experts. Whether some fact ends up in the design as an attribute should not be a conceptual issue.

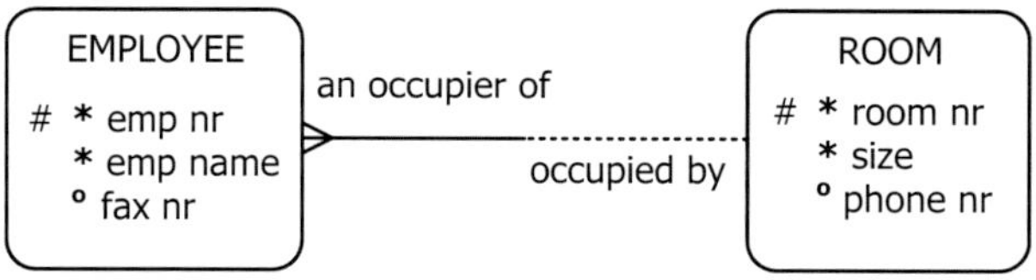

FIG. 8.4 A simple ER model in the Barker notation.

In defense of ER, it can be useful to view a conceptual schema in attribute style to gain a compact but still high-level view of the business domain. For this reason, we value ER, and the NORMA tool for ORM is currently being extended to provide ER diagrams as live views of underlying ORM schemas.

But, as Ron Ross (1998, p. 15) says, "Sponsors of business rule projects must sign off on the sentences—not on graphical data models. Most methodologies and CASE tools have this more or less backwards". ORM allows the domain expert to inspect ORM models fully verbalized into sentences with examples, making validation much easier and safer.

Now that we've stated our bias up front, let's examine the Barker notation itself. The basic conventions are illustrated in Fig. 8.4. *Entity types* are shown as soft rectangles (rounded corners) with their name in capitals. *Attributes* are written below the entity type name. Some constraint information may appear before an attribute name. An octothorpe "#" indicates that the attribute is, or is a component of, the primary identifier of the entity type.

A "*" or heavy dot "•" indicates that the attribute is mandatory (i.e., each instance in the database population of the entity type must have a nonnull value recorded for this attribute). A "○" indicates the attribute is optional. Some modelers also use a period "." to indicate that the attribute is not part of the identifier.

Relationships are restricted to binaries (no unaries, ternaries, or longer relationships) and are shown as lines with a relationship name at the end from which that relationship name is to be read. This name placement overcomes the ambiguous direction problem mentioned earlier. Both forward and inverse readings may be displayed for a binary relationship, one on either end of the line. This makes the Barker notation superior to UML for verbalizing relationships.

From an ORM perspective, each end (or half) of a relationship line corresponds to a role. Like ORM, Barker treats role optionality and cardinality as distinct, orthogonal concepts instead of lumping them together into a single concept (e.g., multiplicity in UML).

A *solid half-line denotes a mandatory role*, and a *dotted half-line indicates an optional* role. For cardinality, a *crow's foot* intuitively *indicates "many"* by its many "toes". The absence of a crow's foot intuitively indicates "one". The crow's foot notation was invented by Dr. Gordon Everest, an ORM advocate, who originally used the term "inverted arrow" (Everest, 1976) but now calls it a "fork". Fig. 8.5 shows the basic correspondence with the ORM notation for simple mandatory and uniqueness constraints.

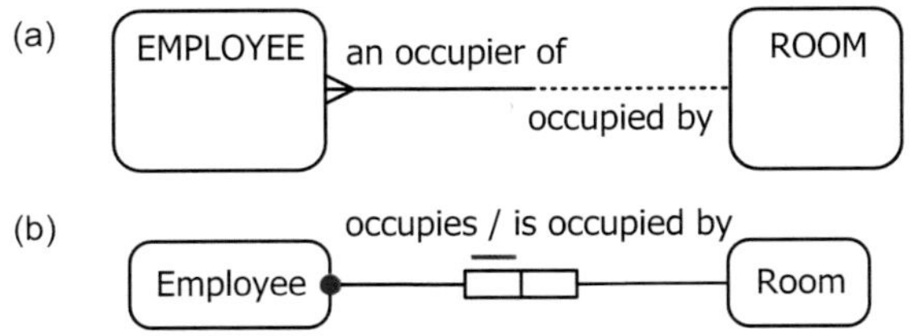

FIG. 8.5 The ER diagram (a) is equivalent to the ORM diagram (b).

To enable the optionality and cardinality settings to be verbalized, Barker (1990a, pp. 3–5) recommends the following *naming discipline for relationships*. Let *A R B* denote an infix relationship *R* from entity type *A* to entity type *B*. Name *R* in such a way that each of the following four patterns results in an English sentence:

Each A (**must** | **may**) **be** *R* (**one and only one** *B* | **one or more** *B-plural-form*)

Use "must" or "may" when the first role is mandatory or optional, respectively. Use "one and only one" or "one or more" when the cardinality on the second role is one or many, respectively. For example, the optionality/cardinality settings in Fig. 8.5(a) verbalize as **Each** Employee **must be** an occupier of **one and only one** Room; **Each** Room **may be** occupied by **one or more** Employees.

The constraints on the left-hand role in the ORM model in Fig. 8.5(b) verbalize as **Each** Employee occupies **some** Room; **Each** Employee occupies **at most one** Room. These constraints combined verbalize as **Each** Employee occupies **exactly one** Room. The lack of a uniqueness constraint on the right-hand role is verbalized as **It is possible that the same** Room is occupied by **more than one** Employee, or (if no inverse reading exists) as **It is possible that more than one** Employee occupies **the same** Room.

Regarding the lack of an explicit mandatory role constraint on the right-hand role, we don't want that verbalized explicitly because it may well be unstable. If Room plays no other fact roles, the role is mandatory by implication (Room is not declared independent), so verbalization may well confuse here. If Room does play another fact role, and we decide that some rooms may be unoccupied, we could declare this explicitly either as **It is possible that some** Room is occupied by **no** Employee or as **It is not necessary that each** Room is occupied by **some** Employee. If no inverse reading is available, it may be verbalized thus: **It is possible that no** Employee occupies **some** Room.

To its credit, the Barker verbalization convention is good for basic mandatory and uniqueness constraints on infix binaries. But it is far less general than ORM's approach, which applies to instances as well as types, for predicates of any arity, infix or mixfix, and covers many more kinds of constraint, with no need for pluralization. As a trivial example, the fact instance "Employee '101' an occupier of Room 23" is not proper English, but "Employee '101' occupies Room 23" is good English.

If each role in a binary association is assigned one of optional/mandatory and one of many/one, there are 16 patterns. The equivalent Barker ER and ORM diagrams for these cases are shown in Fig. 8.6. The last case where both roles of a many:many relationship are mandatory is considered illegal in Barker ER.

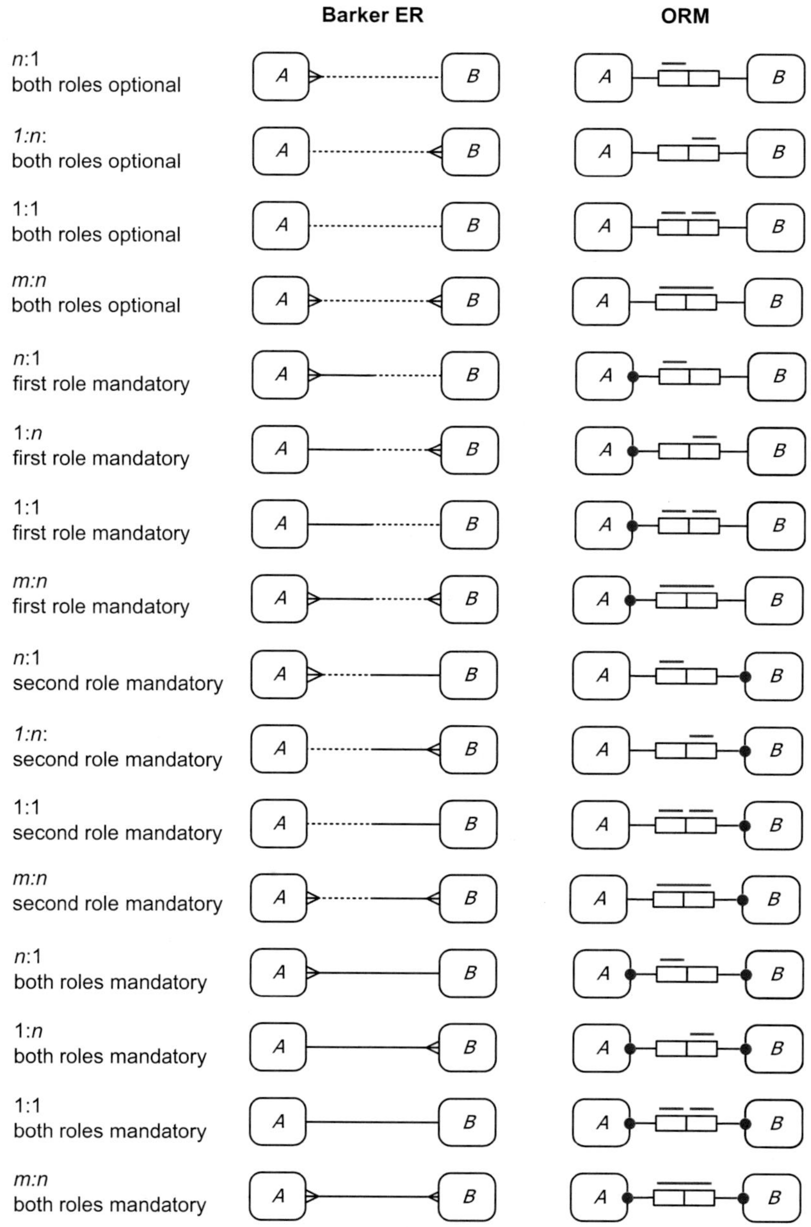

FIG. 8.6 Equivalent constraint patterns.

Ring associations considered illegal by Barker are shown in Fig. 8.7(a). Although rare, they sometimes occur in reality, so should be allowed at the conceptual level, as permitted in ORM.

As an exercise, you may wish to invent satisfying populations for the ORM associations in Fig. 8.7(b). Although considered illegal by Barker, at least some of these patterns are allowed in Oracle's CASE tools.

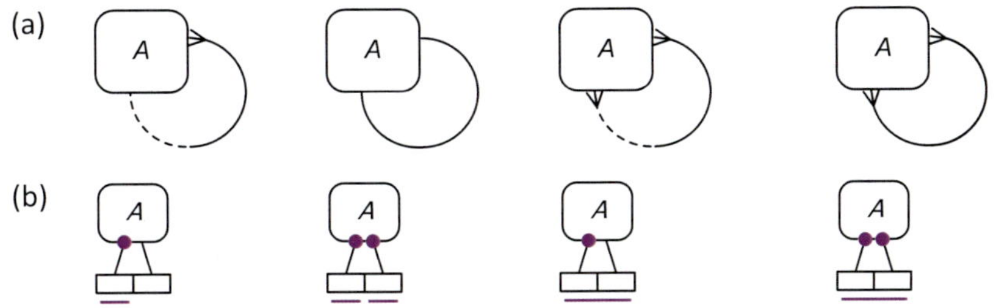

FIG. 8.7 Illegal ring associations in Barker ER (a) that are allowed in ORM (b).

In the Barker notation, a bar "|" across one end of a relationship indicates that the relationship is a component of the primary identifier for the entity type at that end. For example, in Fig. 8.8, Employee and Building have simple identifiers, but Room is compositely identified by combining its room number and building.

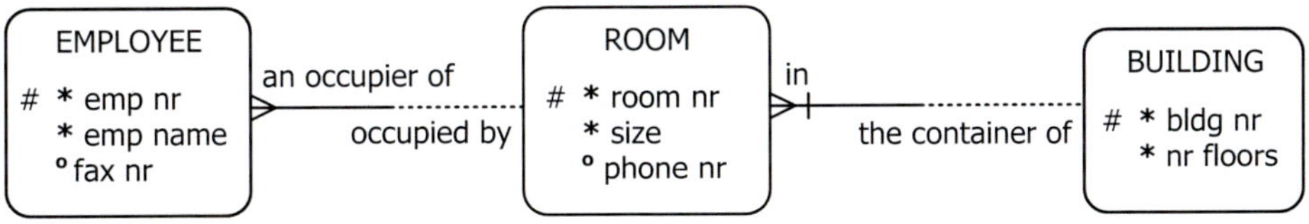

FIG. 8.8 Room is identified by combining its room nr and its Building relationship.

The use of identification bars provides some of the functionality afforded by external uniqueness constraints in ORM. For example, the schemas in Fig. 8.9 are equivalent. Any other attributes of Room and Building would be modeled in ORM as relationships. ORM's external uniqueness notation seems to us to convey more intuitively the idea that each RoomNr, Building combination is unique (i.e., refers to at most one room). But maybe we're biased. At any rate, this constraint (as well as any other graphic constraint) can be automatically verbalized in natural language.

FIG. 8.9 Composite identification in (a) Barker ER and (b) ORM.

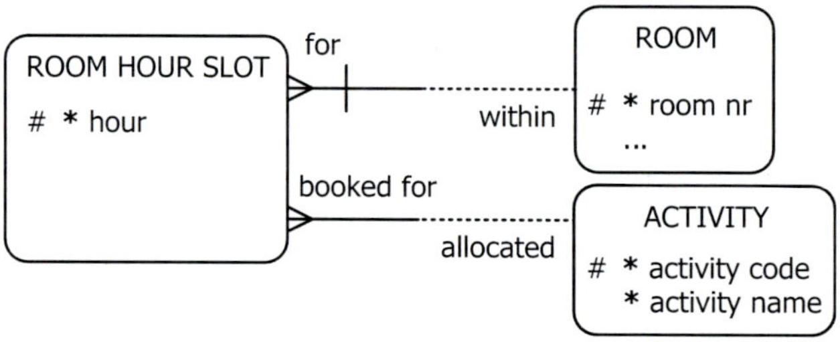

FIG. 8.10 A room schedule model in Barker notation.

Some people misread the bar notation for composite identification as a "1" because this is what the symbol means in many other ER notations. But the main problem with the bar and "#" notations is that they cannot be used to declare uniqueness constraints not used for primary reference (see later). A second problem, arising from the practice of modeling some fact types as attributes and others as relationships, is that two very different notations are used for the same fundamental concept (uniqueness).

Section 1.2 used a room scheduling example to illustrate several fundamental differences between ORM and modeling approaches such as ER and UML. You may recall that the schema shown in Fig. 8.10 was used to model the example in the Barker notation. If you skipped Section 1.2, you may wish to read it now because it discussed this example in a lot of detail. The use of attributes and the binary-only relationship restriction in this model make it hard to verbalize and populate the schema for validation purposes. Moreover, there is at least one constraint missing.

A populated ORM schema for this example is reproduced in Fig. 8.11 (minus the counterexample rows discussed in Section 1.2). Here, the facts are naturally verbalized as a ternary and a binary, and the constraints are easily checked using verbalization and sample data. With the ER model, there is no way of specifying the right-hand uniqueness constraints on either fact type because the Barker notation doesn't capture uniqueness constraints on attributes or relationships unless they are used for primary identification.

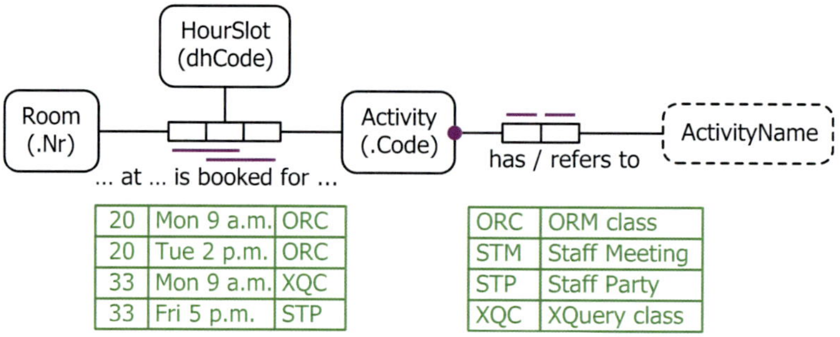

FIG. 8.11 A populated ORM model for room scheduling.

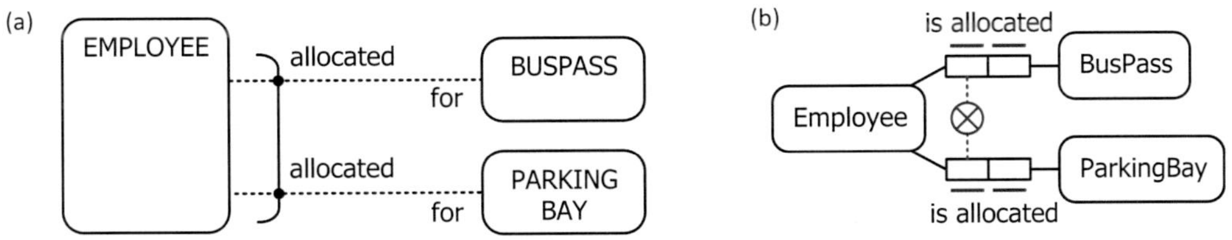

FIG. 8.12 A simple exclusion constraint in (a) Barker notation and (b) ORM.

In case it looks like we're just bashing attribute-based approaches such as ER, let us reiterate that we find attribute-based models useful for compact overviews and for getting closer to the implementation model. However, we generate these by mapping from ORM, which we use exclusively for the initial conceptual analysis. This makes it easier to get the model right in the first place and to modify it as the underlying domain evolves.

Unlike ER (and UML for that matter), ORM was built from a linguistic basis, and its graphic notation was carefully chosen to exploit the potential of sample populations. To reap the benefits of verbalization and population for communication with and validation by domain experts, it's better to use a language that was designed with this in mind.

An added benefit of ORM is that its graphic notation can capture many more business rules than popular ER notations. Although not as rich as ORM, the Barker notation is more expressive than many ER notations. The rest of this section discusses its support for advanced constraints and subtyping.

In Barker notation, an exclusion constraint over two or more roles is shown as an "*exclusive arc*" connected to the roles with a small dot or circle. For example, Fig. 8.12(a) includes the constraint that no employee may be allocated both a bus pass and a parking bay. In ORM, this constraint is depicted by connecting "⊗" to the relevant roles by a dotted line, as shown in Fig. 8.12(b).

To declare that two or more roles are mutually exclusive and disjunctively mandatory, the Barker notation uses the exclusive arc, but each role is shown as mandatory (solid line). For example, in Fig. 8.13(a) each account is owned by a person or a company, but not both.

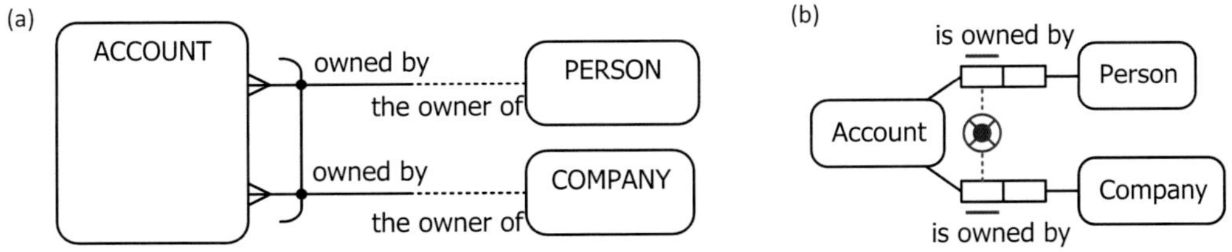

FIG. 8.13 An exclusive-or constraint in (a) Barker notation and (b) ORM.

This notation is liable to mislead because it violates the orthogonality principle in language design. Viewed by itself, the first role of the association Account owned by Person would appear to be mandatory because a solid line is used. But the role is actually optional because superimposing the exclusive arc changes the semantics of the solid line to mean the role belongs to a set of roles that is disjunctively mandatory.

Contrast this with the equivalent ORM model shown in Fig. 8.13(b). Here, an exclusion constraint ⊗ is orthogonally combined with a disjunctive mandatory (inclusive-or) constraint ⊙ to produce an exclusive-or constraint, shown here by the "lifebuoy" symbol formed by overlaying one constraint symbol on the other.

The ORM notation makes it clear that each role is individually optional and that the exclusive-or constraint is a combination of inclusive-or and exclusion constraints. Suppose we modified our business so that the same account could be owned by both a person and a company. Removing just the exclusion constraint from the model leaves us with the inclusive-or constraint ⊙ that each account is owned by a person or company. Like UML, the Barker ER notation doesn't even have a symbol for an inclusive-or constraint, so it is unable to diagram this or the many other cases of this nature that occur in practice.

In the Barker notation, a role may occur in at most one exclusive arc. ORM has no such restriction. For example, in Fig. 8.14(a) no student can be both ethnic and aboriginal, and no student can be both an aboriginal and a migrant (these rules come from a student record system in Australia). Even if the Barker notation supported unaries, this situation could not be handled by exclusive arcs. Like UML, Barker ER does not provide a graphic notation for exclusion constraints over role sequences. For instance, it cannot capture the ORM pair-exclusion constraint in Fig. 8.14(b), which declares that no person who wrote a book may review the same book. Such rules are very common. Moreover, the Barker notation cannot express any ORM subset or equality constraints at all, even over simple roles.

The Barker notation for ER allows simple *frequency constraints* to be specified. For any positive integer n, a constraint of the form $= n$, $<n$, $\leq n$, $>n$, $\geq n$ may be written next to a single role to indicate the number of instances that may be associated with an instance playing the other role. For example, the frequency constraint "≤ 2" in Fig. 8.15 indicates that each person is a child of at most two parents. In the Barker notation, this constraint is placed on the parent role, making it easy to read the constraint as a sentence starting at the other role.

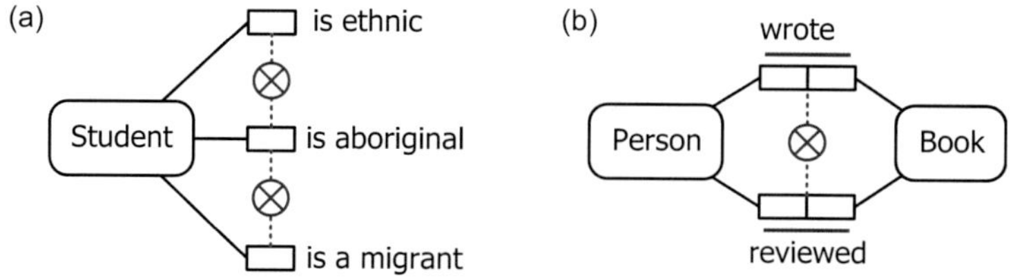

FIG. 8.14 Some ORM exclusion constraints not handled by Barker's exclusive arcs.

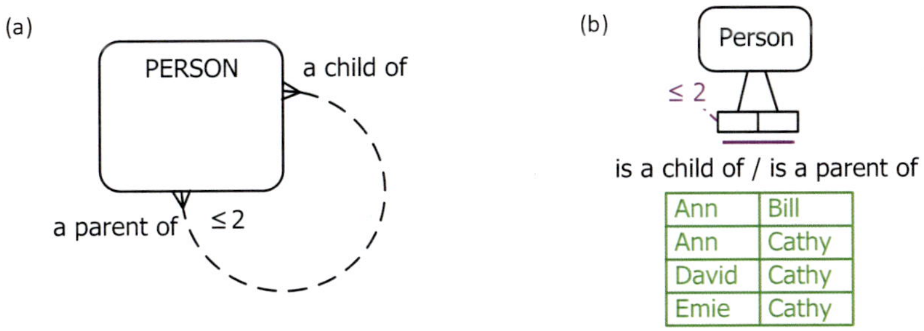

FIG. 8.15 A simple frequency constraint in (a) Barker notation and (b) ORM.

In ORM, the constraint is placed on the child role, making it easy to see the impact of the constraint on the population (each person appears at most twice in the child role population). Unlike the Barker notation, ORM allows frequency constraints to include ranges (e.g., 2..5) and to apply to role sequences.

In Barker notation, *subtyping* is depicted with a version of Euler diagrams. In effect, only partitions (exclusive and exhaustive) can be displayed. For example, Fig. 8.16(a) indicates that each patient is a male patient or female patent but not both. As discussed in Section 6.5, ORM displays subtyping using directed acyclic graphs (DAGs), and may use an exclusive-or constraint symbol to display a partition constraint (typically implied by subtype definitions and other constraints).

Euler diagrams are good for simple cases because they intuitively show the subtype inside its supertype. However, unlike DAGs, they are hopeless for complex cases (e.g., many overlapping subtypes), and they make it inconvenient to attach details to the subtypes. For the latter reason, attributes are sometimes omitted from subtypes when the Barker notation is used. Another problem with Euler diagrams is in displaying multiple partitions on a single diagram (e.g., try partitioning Patient into not just MalePatient and FemalePatient, but also InPatient and OutPatient).

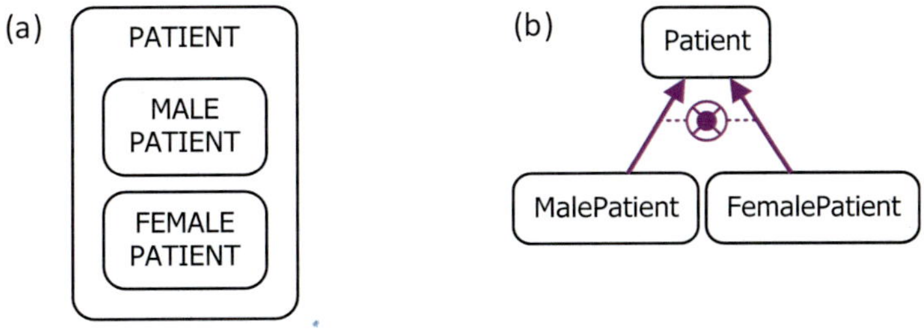

FIG. 8.16 A subtype partition in (a) Barker ER and (b) ORM

In the Barker notation, if the original subtype list is not exhaustive, an "Other" subtype is added to make it so, even if it plays no specific role. For example, in Fig. 8.17 a vehicle is a car or truck or possibly something else, and a car is a sedan or wagon or possibly something else. In ORM, there is no need to introduce subtypes for OtherCar or OtherVehicle unless they play specific roles.

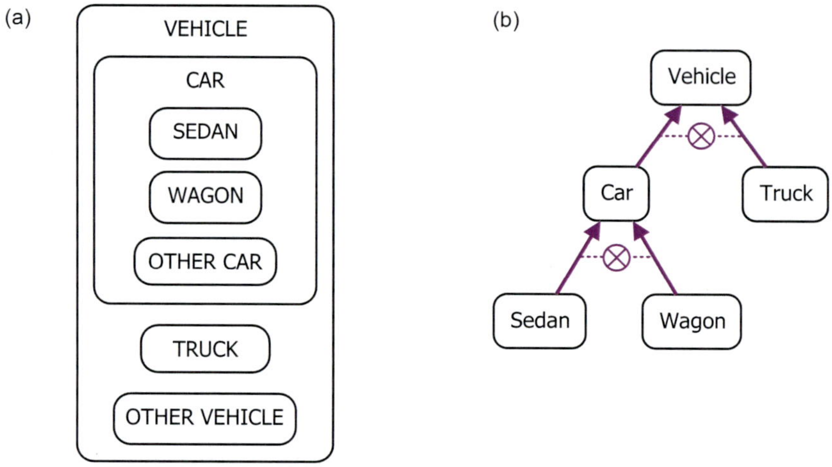

FIG. 8.17 Nonexhaustive, exclusive subtypes in (a) Barker ER and (b) ORM.

A major problem with the Barker notation for subtyping is that it does not depict overlapping subtypes (e.g., Manager and FemaleEmployee as subtypes of Employee) or multiple inheritance (e.g., FemaleManager as a subtype of FemaleEmployee and Manager). While it is possible to implement multiple inheritance in single inheritance systems (e.g., Java) by using some low-level tricks, for conceptual modeling purposes multiple inheritance should be simply modeled directly. As a final comparison point about subtyping, Barker ER lacks ORM's capability for formal subtype definitions and context-dependent identification schemes.

In addition to static constraints, Barker ER includes a dynamic "changeability constraint" for marking "*nontransferable relationships*". This constraint declares that once an instance of an entity type plays a role with an object, it cannot ever play this role with another object. This is indicated by an open diamond on the constrained role. For example, Fig. 8.18(a) declares that the birth country of a person is nontransferable.

As indicated in Fig. 8.18(b), ORM does not currently support this constraint. Adding formal support for this and other dynamic constraints in ORM is an active research area. In practice, the specification of nontransferable constraints needs to ensure that the implemented model is still open to error corrections by duly authorized users. For example, if an Australian's birth country is mistakenly entered as Austria, it should be possible to change this to Australia. We discuss this issue further in the next chapter when examining changeability properties in UML.

FIG. 8.18 Nontransferability declared in Barker, but not ORM.

Well, that pretty well covers the Barker notation for ER. As we've seen, it does a good job of expressing simple mandatory, uniqueness, exclusion, and frequency constraints; simple subtyping; and also nontransferable relationships. However, if a feature is modeled as an attribute instead of as a relationship, very few of these constraints can be specified for it.

Unlike ORM, the Barker notation does not support unary, *n*-ary, or objectified associations (nesting). Moreover, it lacks support for most of the advanced ORM constraints (e.g., subset, multirole exclusion, ring constraints, and join constraints). It does not include a formal textual language for specifying queries, other constraints, and derivation rules at the conceptual level. Nevertheless, it is better than many other ER notations and is still widely used. If you ever need to specify a model in Barker ER notation, we suggest that you first do the model in ORM, map it to the Barker notation, and make a note of any rules that can't be expressed there diagrammatically.

Rather than giving you some exercises on the Barker notation at this point, we'll wait until the end of the chapter, when we've covered the main ER notations in use as well as some techniques for mapping from ORM to ER. You can then decide which notation(s) you would like to have some practice with.

8.3 Information Engineering Notation

The IE approach began with the work of Clive Finkelstein in Australia and CACI in the United Kingdom, and was later adapted by James Martin. Different versions of IE exist, with no single standard. In one form or other, IE is supported by many data modeling tools and is one of the most popular notations for database design.

In the IE approach, *entity types* display as named rectangles, as in Fig. 8.19(a). *Attributes* are often displayed in a compartment below the entity type name, as in Fig. 8.19(b), but are sometimes displayed separately (e.g., bubble charts). Some versions support basic constraints such as Ma/Op/Unique on attributes.

Relationships are typically restricted to binary associations only, which are shown as named lines connecting the entity types. Relationship names are read left to right and top to bottom. As with the Barker notation, a half-line or line end corresponds to a role in ORM.

FIG. 8.19 Typical IE notation for (a) entity type and (b) entity type with attributes.

Optionality and cardinality settings are indicated by annotating the line ends. To indicate that a role is *optional*, a circle "○" is placed at the other end of the line, signifying a minimum participation frequency of 0. To indicate that a role is *mandatory*, a stroke "|" is placed at the other end of the line, signifying a minimum participation frequency of 1. After experimenting with different notations for a cardinality of "*many*", Finkelstein settled on the intuitive crow's foot symbol suggested by Dr. Gordon Everest.

In conjunction with a minimum frequency of 0 or 1, a stroke "|" is often used to indicate a maximum frequency of 1. With this arrangement, the combination "○|" indicates "*at most one*" and the combination "||" indicates "*exactly one*". This is the convention used in this section. However, different IE conventions exist. For example, some assume a maximum cardinality of 1 if no crows foot is used, and hence use just a single "|" for "exactly one". Clive Finkelstein uses the combination "○|" to mean "optional but will become mandatory", which is really a dynamic rather than static constraint; this can be combined with a crow's foot. Some conventions allow a crow's foot to mean that the minimum (and hence maximum) frequency is many. When using a version of IE, check which of these conventions applies.

Fig. 8.20 shows a simple IE diagram and its equivalent ORM diagram. With IE, as you read an association from left to right, you verbalize the constraint symbols at the right-hand end. Here, each employee occupies exactly one (at least 1 and at most 1) room. Inverse predicates are not always supported in IE, but you can supply them yourself to obtain a verbalization in the other direction, for example, "Each room is occupied by zero or more employees". As with the Barker notation, a plural form of the entity type name is introduced to deal with the many case.

The IE notation is similar to the Barker notation in showing the maximum frequency of a role by marking the role at the other end. But unlike the Barker notation, the IE notation shows the optionality/mandatory setting at the other end as well. In this sense, IE is like UML (even though different symbols are used). As discussed earlier, there are 16 possible constraint patterns for optionality and cardinality on binary associations. Fig. 8.21 shows these cases in IE notation together with the equivalent cases in ORM.

An example using the different notation for IE used by Finkelstein is shown in Fig. 8.22. The single bar on the left end of the association indicates that each computer is located in exactly one office. The circle, bar, and crow's foot on the right-hand role collectively indicate that each office must eventually house one or more computers. Although this "optional becoming mandatory" constraint has no counterpart in ORM and is unsupported by most IE modeling tools, it could be used to refine decisions on how to group fact types into implementation structures.

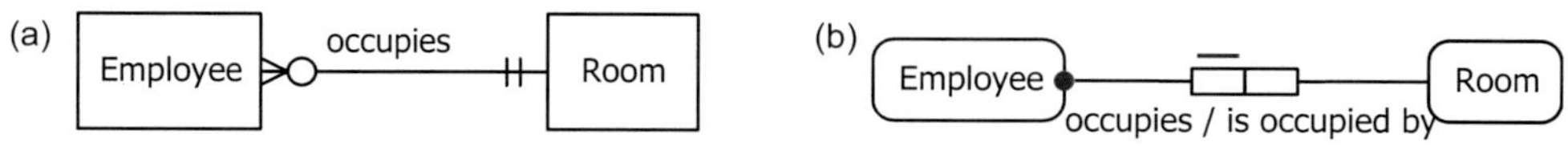

FIG. 8.20 The IE diagram (a) is equivalent to the ORM diagram (b).

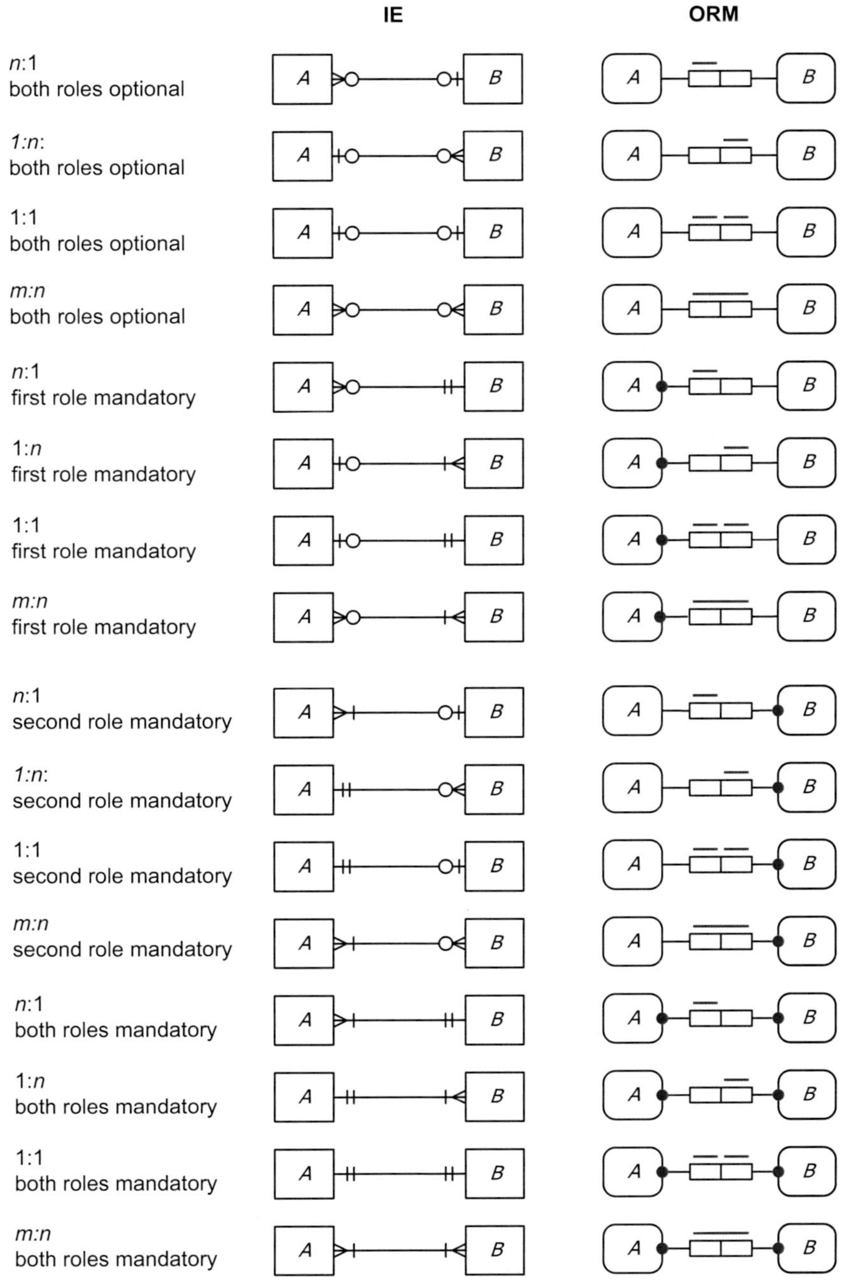

FIG. 8.21 Equivalent constraint patterns in IE and ORM.

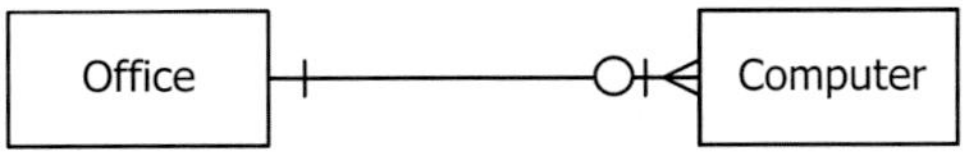

FIG. 8.22 In Finkelstein's version, circle-bar means "optional becoming mandatory".

Some modeling tools support the IE notation for *n*:1, 1:*n*, and 1:1 associations but not *m*:*n* (many to many) associations. For such tools, four of the 16 cases in Fig. 8.21 can't be directly represented. In this situation, you can model the *m*:*n* cases indirectly by introducing an "*intersection entity type*" with mandatory *n*:1 associations to the original entity types. For example, the *m*:*n* case with both roles optional may be handled by introducing the object type *C*, as shown in Fig. 8.23.

In IE, this transformation loses the constraint that each *A*-*B* combination relates to at most one *C* (unless we drop to the relational level by using two foreign keys for *C*'s primary key; see IDEF1X later). In ORM, the constraint is captured conceptually as external uniqueness, making *C* a *coreferenced* object type–the transformation is a flatten/coreference equivalence (see Chapter 14).

As an example, the *m*:*n* association Person plays Sport can be transformed into the uniquely constrained, mandatory *n*:1 associations Play is by Person; Play is of Sport. However, such a transformation is often very unnatural, especially if nothing else is recorded about the coreferenced object type. So, any truly conceptual approach must allow *m*:*n* associations to be modeled directly.

Some versions of IE support an *exclusive-or constraint*, shown as a black dot connected to the alternatives. Fig. 8.24(a) depicts the situation where each employee is allocated a bus pass or parking bay, but not both. The equivalent ORM schema is shown in Fig. 8.24(b). Unlike ORM, IE does not support an inclusive-or constraint. Nor does it support simple or compound exclusion constraints.

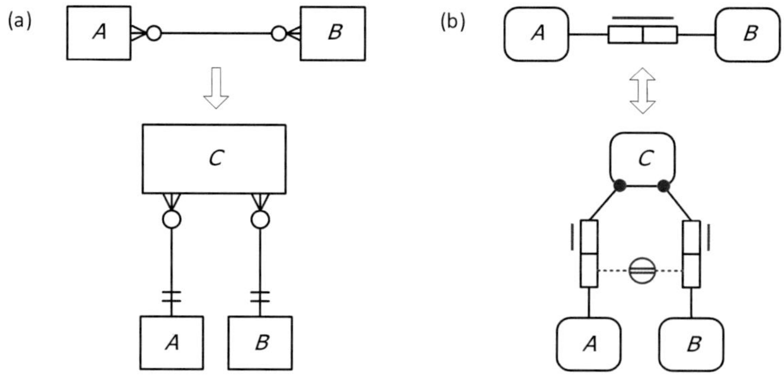

FIG. 8.23 An *m*:*n* association remodeled as an entity type with *n*:1 associations.

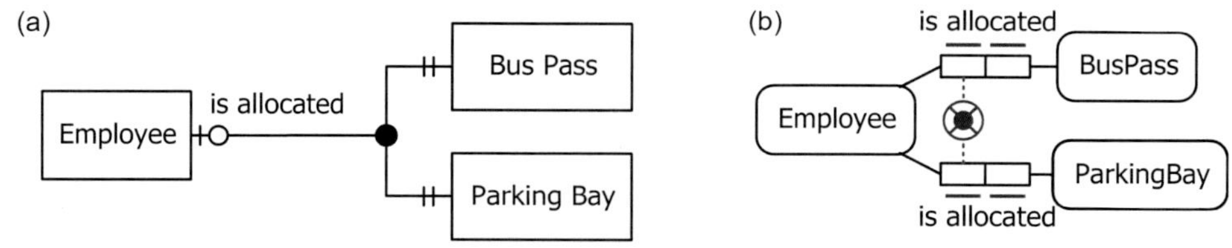

FIG. 8.24 An exclusive-or constraint in (a) IE and (b) ORM.

Subtyping schemes for IE vary. Sometimes Euler diagrams are used, adding a blank compartment if needed for "Other". Sometimes DAGs are used, possibly including subtype relationship names and optionality/cardinality constraints.

Fig. 8.25 shows three different subtyping notations for partitioning Patient into MalePatient and FemalePatient. There is no formal support for subtype definitions or context-dependent reference. Multiple inheritance may or may not be supported, depending on the version.

Although far less expressive than ORM, IE does a good job of covering basic constraints. Its founder Clive Finkelstein is an amiable Aussie who is still actively engaged in the IE discipline. He developed a set of modeling procedures to go with the notation, extended IE to enterprise engineering, and incorporated work with the extensible markup language (XML).

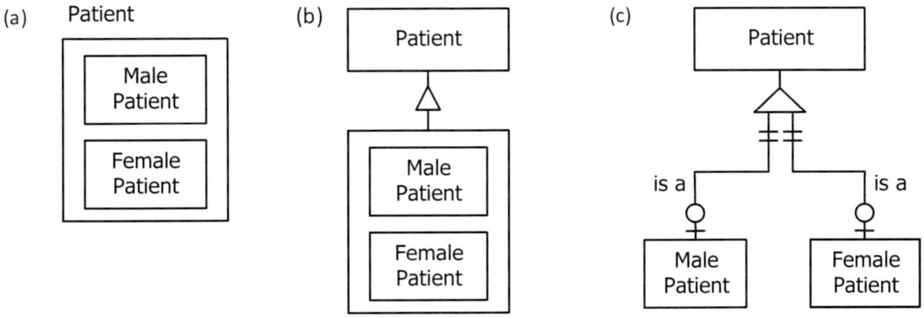

FIG. 8.25 Some different IE notations for subtyping.

8.4 IDEF1X

In the 1970s, the US Air Force began work on a program for integrated computer-aided manufacturing (ICAM). This was the genesis of a family of *IDEF modeling languages*. The acronym "IDEF" originally denoted "ICAM DEFinition", but now stands for "Integration DEFinition", reflecting its possible use for exchanging information between different modeling languages. Rather than specifying one universal modeling language, the ICAM project defined the following languages for different tasks:

- IDEF0 activity modeling
- IDEF1 conceptual data modeling
- IDEF2 simulation modeling

Later, other languages were added, including:

- IDEF1X logical data modeling
- IDEF3 process modeling
- IDEF4 object-oriented software design
- IDEF5 knowledge engineering of enterprise ontologies
- $IDEF1X_{97}$ logical data modeling with object-oriented extensions

The name "*IDEF1X*" stands for "IDEF1 eXtended". Although based on the conceptual IDEF1 language, IDEF1X was changed to focus on logical data modeling. Although regretted by some, the introduction of IDEF1X effectively spelled the end of IDEF1. Over time, IDEF3 subsumed much of IDEF2 and IDEF5. Nowadays, IDEF1X is the most popular IDEF language, and is supported by many CASE tools and is widely used in US government sectors, especially defense. The other IDEF languages are also used, although to a lesser extent.

IDEF1X is a hybrid language, combining some conceptual notions (e.g., entity, relationship) with relational database constructs (e.g., foreign keys). It was accepted as a standard by the National Institute of Standards and Technology in 1993 (NIST, 1993). A successor called IDEF1X$_{97}$ (also known as IDEF$_{object}$) was approved in 1998 by IEEE and further refined in 2012 to what is still the current version (https://www.iso.org/standard/60614.html). This extended IDEF1X with object-oriented features to facilitate implementation in object-oriented databases and programming languages while maintaining compatibility with IDEF1X for relational database implementation. This backward compatibility gives IDEF1X$_{97}$ one advantage over UML for adoption in US government sectors.

However, the object-oriented extensions of IDEF1X$_{97}$ are less encompassing than those in UML. Although UML is not yet widely used for database design, it is by far the most widely used language for designing object-oriented code, and it is supported by many CASE tools. At the time of writing, only a few CASE tools support IDEF1X$_{97}$. In contrast, the original (1993) version of the IDEF1X language continues to be very widely used for database design and is supported by many modeling tools. For these reasons, we restrict our attention to the original IDEF1X in the rest of this section.

To model facts, IDEF1X uses the three main constructs of ER: entities, attributes, and relationships. Unfortunately, the IDEF1X standard uses the term "entity" to mean "entity type", not "entity instance". Even more unfortunately, this misusage has been adopted by some practitioners. So, you may have to use the word "instance" more than you want to, simply to ensure that you are not misunderstood. Each entity type has one or more attributes that are based on domains. A domain is a named set of data values of the same data type. Relationships are restricted to binary associations between entity types.

IDEF1X allows models to be developed in phases using at least *three views*: ER view, key-based view, and fully attributed view. The essential differences among these views are summarized in Table 8.1.

Table 8.1 Three different views in IDEF1X.

ER view	Entity types	No identification scheme or keys defined No need to specify any attributes No attribute constraints (e.g., mandatory)
	Relationships:	Many:many relationships are allowed Identifying/nonidentifying distinction is not made
Key-based view	Include at least all the key-based attributes Classify connection relationships as identifying or nonidentifying	
Fully attributed view	Include all attributes	

The *ER view* may be used early in the analysis but is very inexpressive compared with true ER. No identification schemes are specified for entity types, and no attributes need be declared. If attributes are specified, no constraints (e.g., mandatory) are declared for them.

An ER view in IDEF1X is basically an incomplete sketch of an ER model. In principle, it could be refined into a true ER model. Unfortunately, instead of carrying out the refinement at the conceptual level, IDEF1X drops down to the logical level to add the extra detail. Any ER view must be resolved into a key-based view and ultimately a fully attributed view to complete the model. Key-based views and fully attributed views are similar to a relational model in many respects. Some CASE tools that support IDEF1X do not support its ER view, and we ignore the ER view from this point on.

In key-based and fully attributed views, IDEF1X entity types are basically relational tables, and the "relationships" are foreign key to primary key references. Recall that a foreign key is a set of one or more attributes whose values (possibly composite) must occur as a value of some primary key. The source or referencing entity (type) is called the *child* and the target or referenced entity (type) is the *parent*. Relationships may be assigned forward and inverse readings as if they were conceptual associations or subtype links. However, they ultimately represent subset constraints. The actual facts are all stored in attributes, as in the relational model.

An entity type is classified as identifier independent or identifier dependent. An entity type is *identifier independent* if and only if its identifier does not depend on other entity types. In other words, its identification scheme does not include a conceptual relationship to another entity type. In terms of keys, this means that *an entity type is identifier independent if and only if its primary key has no component that is a (complete) foreign key;* its primary key may, however, contain an attribute that is just part of a foreign key. Otherwise, the entity type is *identifier dependent.* The terms "identifier independent" and "identifier dependent" are often shortened to "independent" and "dependent". This notion of "independent" has nothing to do with the notion of independent object types in ORM.

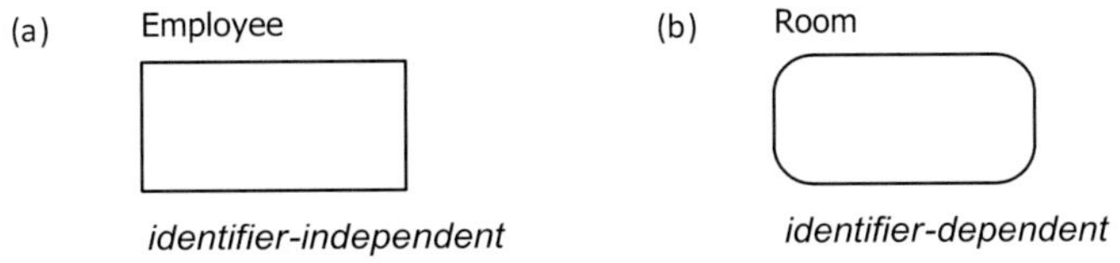

FIG. 8.26 An independent entity type (a) and a dependent entity type (b).

In IDEF1X, an independent entity (type) is depicted as a hard rectangle (square corners) with its name written above, as shown in Fig. 8.26(a). A dependent entity type is depicted as a soft rectangle (rounded corners), as shown in Fig. 8.26(b).

Attributes of an entity type are listed inside its rectangle. *Primary key* attributes appear in the top compartment. These provide the entity type's primary identification scheme. *Alternate keys* are marked by appending "(AKn)", where $n > 0$. These provide an alternate identification scheme. If an alternate key is composite, each of its components has the same value for *n*. Different alternate keys must have different values for *n*. *Foreign keys* are indicated by appending "(FK)". Attributes are *mandatory by default*. To show that an attribute is *optional*, "(O)" is appended. All attributes are single valued.

In Fig. 8.27, empNr is the primary key of Employee. The bldgNr, roomNr pair provides the composite primary key for Room. The Social Security number attribute (ssn) is an alternate key for Employee. In Employee, the pair bldgNr, roomNr is a foreign key (referencing the primary key of Room). In Room, bldgNr is a foreign key (referencing the primary key of a Building entity type not shown here). All attributes are mandatory except for fax and phoneNr, which are marked optional.

In all views, *relationships* are binary only. In the ER view, relationships are conceptual associations and may be "nonspecific" (many-to-many). In key-based and fully attributed views, each "relationship" is either a "*connection*" (*foreign key to primary key reference*) or a *categorization* (*subtype link*). Connections are "*specific*" (many-to-one) binary associations between a child entity type and parent entity type, where each child has at most one parent.

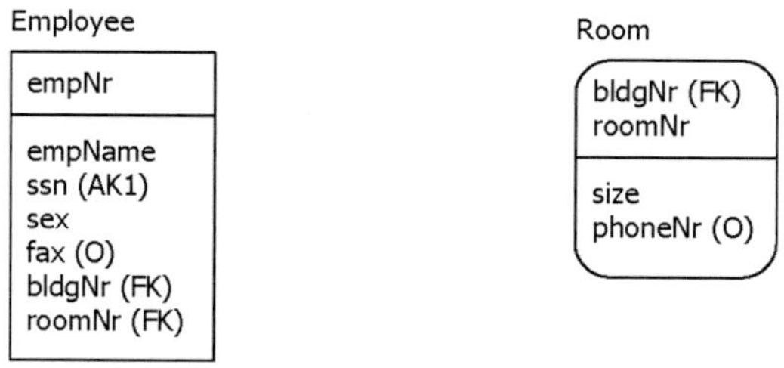

FIG. 8.27 Key (primary, alternate, and foreign) and optionality indicators.

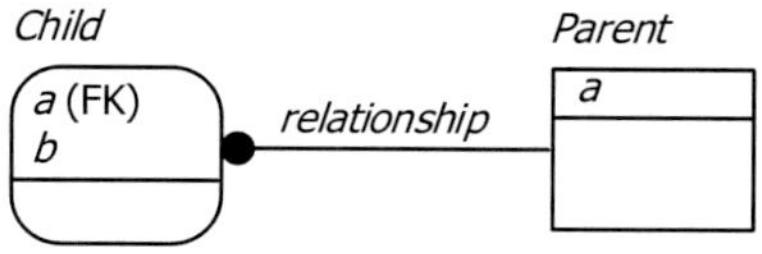

FIG. 8.28 An identifying connection "relationship."

Connection relationships are either identifying or nonidentifying. In principle, an *identifying relationship* is an association that is conceptually used in the identification scheme of the child. In actuality, it is simply a *reference from a foreign key in the child's primary key to the parent's primary key.*

In a *nonidentifying* relationship, the child *nonkey* attributes include a foreign key that references the parent's primary key.

All connection relationships are denoted as a named line with a *dot* "•" *at the child end.* An *identifying* relationship is shown as a named, *solid line*, as shown in Fig. 8.28. Here, *a* and *b* are attributes (possibly compound).

With an identifying connection, each child entity is associated with *exactly one* parent instance. This is because each child instance has a nonnull value for the foreign key (in the child's primary key) that references the parent. By definition, the child must be identifier dependent, as shown by the soft rectangle. The parent is identifier independent (as shown here) unless it is made dependent by some other relationship.

IDEF1X allows both forward and inverse predicate readings for relationships. These may be written together, as shown in Fig. 8.29(a), or on different sides of the relationship line, as shown later in Fig. 8.30(a).

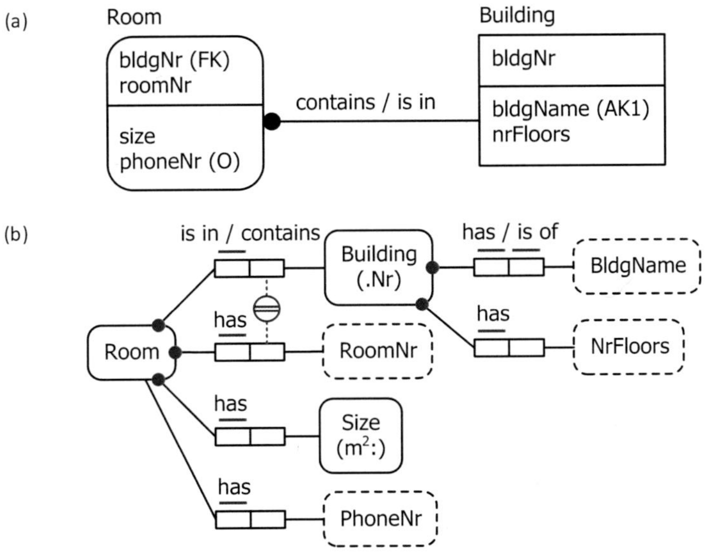

FIG. 8.29 The IDEF1X schema (a) is equivalent to the ORM schema (b).

The *forward reading of a specific connection relationship is always toward the child,* and hence *toward the dot* "●". This forward or "parent-perspective" reading must be supplied. The inverse or "child-perspective" reading, if supplied, is toward the parent (away from the dot). If both readings are supplied, a slash "/" is appended to the forward reading. In Fig. 8.29(a), the forward reading is Building contains Room and the inverse reading is Room is in Building. The child is on the left, so the forward reading is right to left. For nonspecific (*m:n*) relationships, the forward reading is always read from left to right (or top to bottom if the relationship line is vertical). Recall that *m:n* relationships exist only at the ER view level.

The equivalent ORM diagram in Fig. 8.29(b) makes it clear that the association Room is in Building is part of the composite reference scheme for Room. In the IDEF1X diagram, this association is actually depicted by including bldgNr as part of the key for Room, and the connection "relationship" shows the foreign key reference. We can partially lift the IDEF1X discussion to a conceptual level by talking of the parent's primary key *migrating* to the child (see Fig. 8.30).

Although not a legal IDEF1X diagram, Fig. 8.30(a) helps portray Room as being partly identified by its room number and partly by its relationship to Building. Compare this with the Barker notation depiction in Fig. 8.9. At this level, the association Room is in Building is conceptual.

Now, imagine a copy of Building's primary key migrating to complete the primary key of Room, as in Fig. 8.30(b). At this stage, we have dropped to the logical level. The conceptual relationship between Room and Building is now depicted by Room's bldgNr attribute. We still need to indicate that any value for this attribute must also occur as a primary key value of Building, and this foreign key reference is what the connection line now represents. We removed the "relationship" name to stress this point. You can now think of the connection line as a foreign key arrow.

For identifying relationships, each child (instance) is associated with exactly one parent (instance), as shown by a solid line with no adornments. By default, a parent is associated with zero or more children, as shown by an unadorned dot "•" at the line end. If familiar with UML, think of this as equivalent to "*". You can override this default child cardinality by adorning the dot with a *cardinality constraint* mark. The main cases are shown in Fig. 8.31 along with their ORM counterpart. Here, "P" indicates a positive number (at least 1), and "Z" indicates zero or 1 (at most 1). A single number (e.g., 3) indicates exactly that number. You can also indicate ranges. For example, "2–5" indicates at least 2 and at most 5.

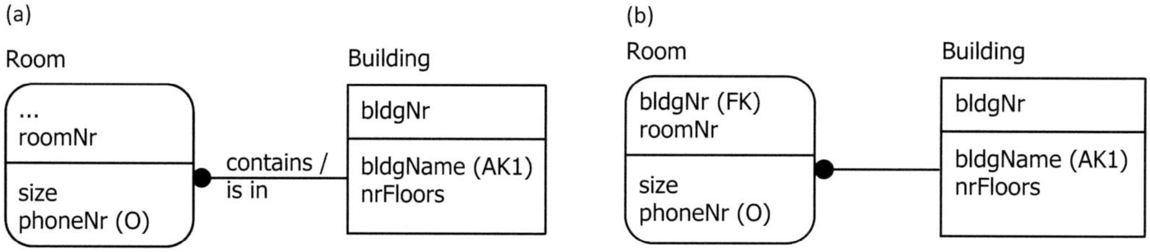

FIG. 8.30 Building's primary key "migrates" to become a foreign key in Room.

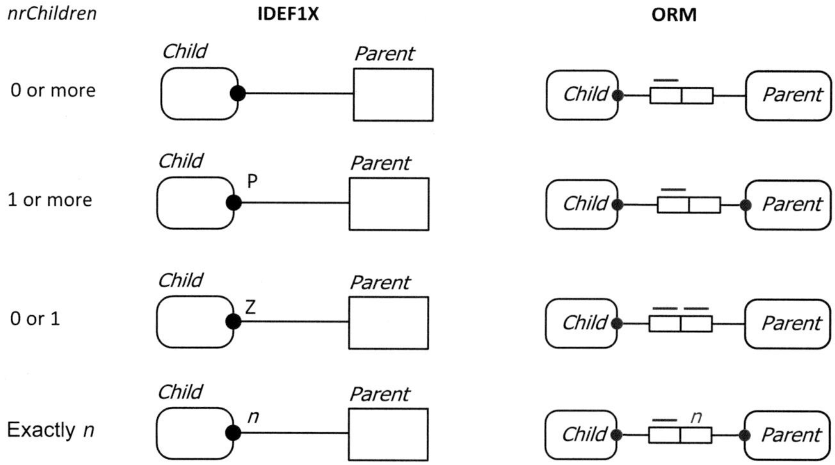

FIG. 8.31 Main child cardinality cases for identifying relationships.

To make sense of these cardinality constraints, think of them as applying to the conceptual relationship that existed before the key was migrated to the child (cardinality constraints on a foreign key reference are rather pointless). The "Z" mark is unintuitive for "at most 1", and in $IDEF1X_{97}$, an alternative notation for this constraint is to use a hollow dot "○".

A *nonidentifying relationship* is shown as a named, *dashed line.* Child and parent are independent unless made dependent by some other relationship. If the dashed line is unadorned, each child (instance) is associated with *exactly one parent* (instance), as in Fig. 8.32(a). If a *diamond* is added at the parent end, each child is associated with *at most one parent,* as in Fig. 8.32(b). The corresponding ORM diagrams are also shown.

In a nonidentifying relationship, the child *nonkey* attributes (those not in the primary key) include a foreign key that references the parent's primary key. If the foreign key is mandatory (each component is nonnull), then each child instance is associated with exactly one parent instance. This is referred to as a *mandatory* relationship, but actually means that the child role of the conceptual relationship that (before migration) gave rise to the foreign key reference is mandatory.

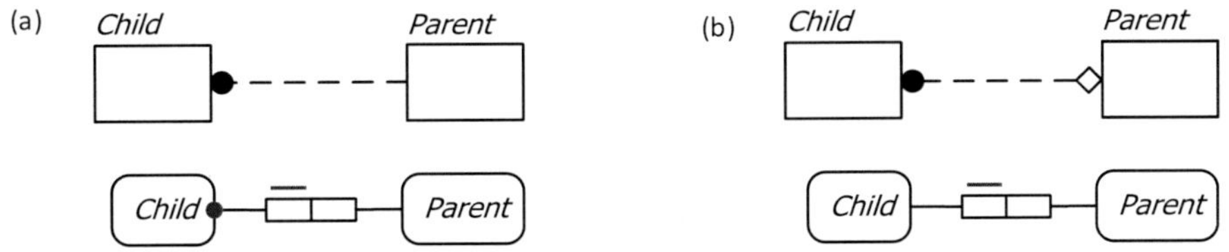

FIG. 8.32 Nonidentifying connection "relationships".

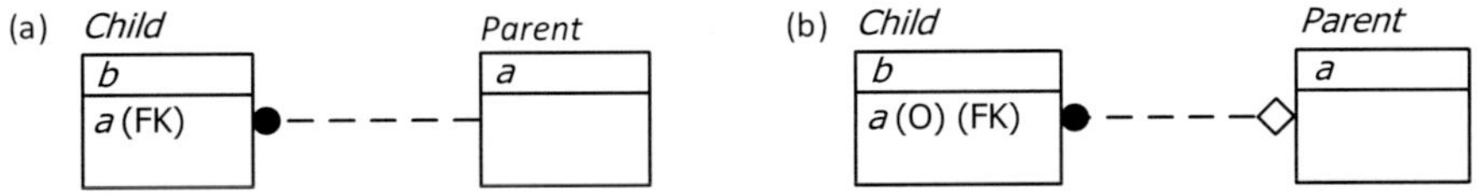

FIG. 8.33 "Mandatory" (a) and "optional" (b) nonidentifying "relationships".

Fig. 8.33(a) depicts this situation in more detail. Here, *a* and *b* are simple or composite attributes. If the foreign key is optional (its components may all be null), then each child instance is associated with at most one parent instance. Fig. 8.33(b) depicts this situation. In IDEF1X, this is referred to as an *optional* relationship.

In both mandatory and optional cases, each instance of the child's foreign key that has no null components must occur as an instance of the parent's primary key. This is the default meaning of a foreign key constraint in structured query language (SQL). For the mandatory case, of course, there can be no null components.

An example of a mandatory, nonidentifying connection relationship is shown in Fig. 8.34(a). Here, each employee is identified by an employee number, must have a name, and must work for a department. Each department is identified by its code and must have a unique name.

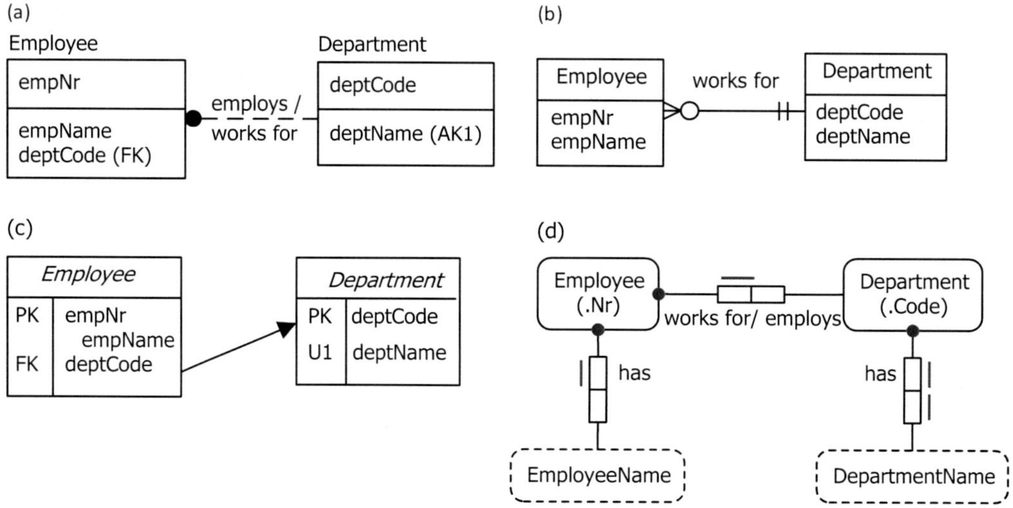

FIG. 8.34 Same example in (a) IDEF1X, (b) IE, (c) Relational, and (d) ORM.

The fact (conceptual relationship) that a given employee works for a given department is stored by instantiating the empNr and deptCode attributes of Employee (e.g., empNr = 101, deptCode = "HR"). The connection "relationship" is simply a foreign key reference indicating that each department code of an employee must be a department code of a department.

The same example is shown in IE notation in Fig. 8.34(b) and in ORM notation in Fig. 8.34(d). Here, the association Employee works for Department is depicted conceptually, rather than by including deptCode as an attribute of Employee. This would also be the case for the Barker notation. Fig. 8.34(c) depicts the situation in a relational notation. Here, facts about who works for what department are stored in the Employee table, which includes a deptCode attribute referencing the Department table. If you compare this with the IDEF1X notation in Fig. 8.34(a), it should be clear that they are just two notational variations of the same structure. In key-based and fully attributed views, the IDEF1X notation is essentially a logical notation rather than a conceptual one.

As another example of nonidentifying relationships, the room scheduling example considered in Sections 1.2 and 8.2 is modeled in IDEF1X in Fig. 8.35(a). The ORM model for this situation is reproduced in Fig. 8.35(b), without the fact populations. The primary key and alternate key constraints in the IDEF1X model correspond to the uniqueness constraints in the ORM model. Although the IDEF1X schema does capture all the constraints, it is a logical rather than a conceptual representation, and unlike ORM it does not facilitate validation by verbalization and population.

Fig. 8.36 shows a more complex IDEF1X example involving a compound foreign key relationship corresponding to a pair-subset constraint in the ORM schema.

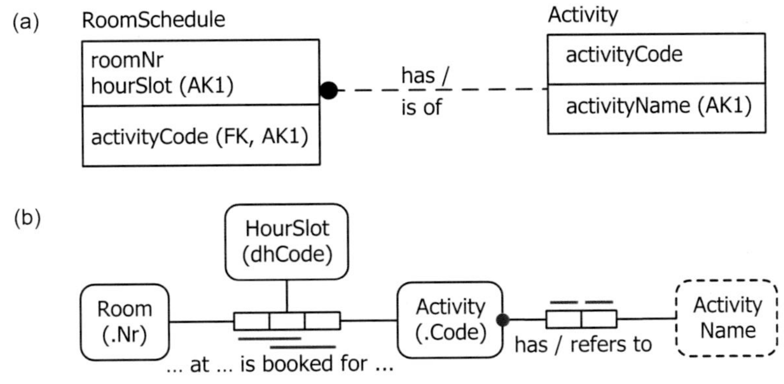

FIG. 8.35 The room schedule example in (a) IDEF1X and (b) ORM.

FIG. 8.36 A complex example in (a) IDEF1X and (b) ORM.

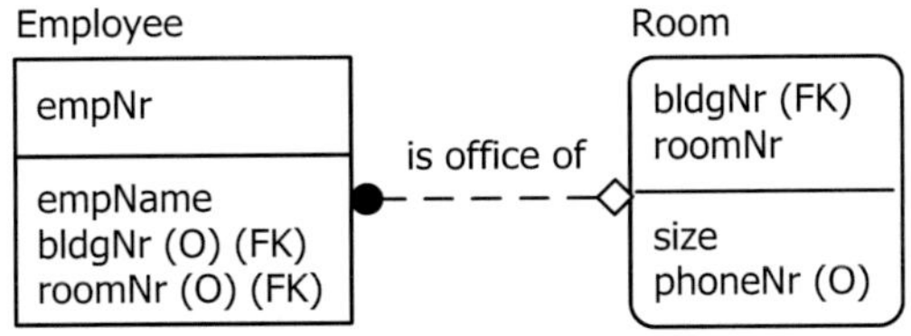

FIG. 8.37 An optional, nonidentifying "relationship."

A simple example of the optional, nonidentifying "relationship" pattern in Fig. 8.33(b) is shown in Fig. 8.37. Here, an employee optionally has an office. Each office is a room with a composite identification scheme. Room's relationship with Building is omitted here.

A more complex example is shown in Fig. 8.38. This is similar to the example in Fig. 8.36, except that it is now optional for one to like a movie or to have a favorite movie.

In an ER view or "entity level diagram", all entity types are depicted by hard rectangles, and nonspecific (*m:n*) relationships are permitted. Each nonspecific relationship line must end in a dot, possibly adorned by a cardinality mark. If unadorned, a cardinality of exactly one is assumed. In moving to a key-based or fully attributed view, each *m:n relationship* must be resolved into an "*intersection entity type*" with two *n*:1 relationships. Fig. 8.39 shows an example in both IDEF1X and ORM.

An intersection entity type is sometimes called an "*associative entity type*". Although conceptually this corresponds to a coreferenced object type in ORM, in IDEF1X the identifying attributes must be "migrated" to become foreign keys inside the primary key of the intersection entity type, which therefore must be identifier dependent, appearing as a soft rectangle. This foreign key depiction effectively lowers the representation to the relational level.

Although this resolution is often unnatural, it does allow attributes to be added to what was once a relationship. For example, we might wish to record a skill level for Play. In ORM and UML, this objective can also be achieved by simply permitting a relationship to be objectified (as an objectified association or association class), but this conceptual alternative is not allowed in IDEF1X.

If the *m:n* relationship is a ring association, role names may be prepended to the names of the migrated foreign keys to distinguish them. For example, in Fig. 8.40(a), the role names "subPartNr" and "superPartNr" are prepended to "partNr" to disambiguate the primary key attributes of Containment.

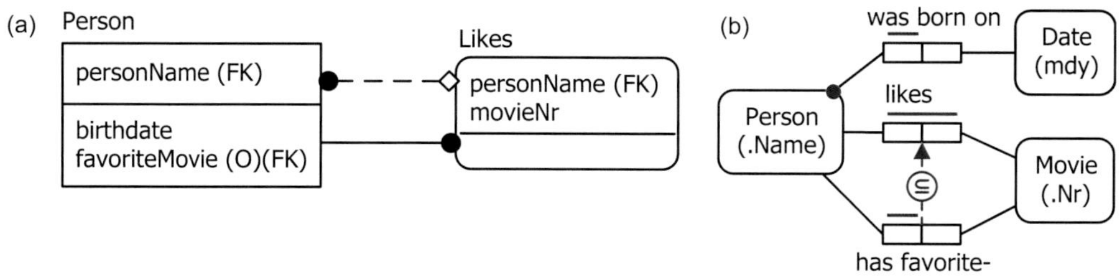

FIG. 8.38 A more complex example in (a) IDEF1X and (b) ORM.

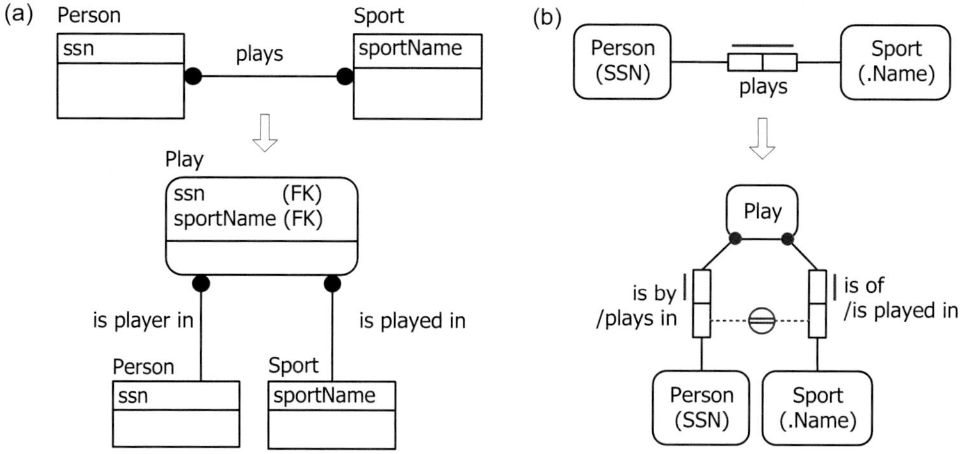

FIG. 8.39 (a) In IDEF1X, *m:n* relationships must be resolved into *n*:1 relationships. (b) The same example in ORM.

Although the IDEF1X standard uses a dot after the role name, this dot notation is not supported in some CASE tools. However, the distinction can be easily achieved without the dot by using just a role name (e.g., "subPartNr" can be used instead of "subPartNr.partNr"). In either case, the role name should be shown as a noun phrase that identifies the attribute even without appending the name of the referenced primary key. In ORM, a more natural way to model this situation is to objectify the containment association, as shown in Fig. 8.40(b).

In IDEF1X, an entity type may be classified into one or more clusters of mutually exclusive *categories* (subtypes). The supertype is called the *generic* entity (type). Each subtype inherits the primary key of its supertype, and hence is identifier dependent. The subtype linkage is called a "*categorization relationship*".

This version of subtyping is very restricted compared with ORM subtyping, which allows overlapping types within a "cluster" as well as context-dependent reference. In effect, the subtyping approach of IDEF1X treats entity types as tables, and the "categorization relationships" are simply foreign key references. In sharp contrast, ORM subtyping is purely conceptual, and different mapping strategies may be chosen at implementation time (see Chapter 11).

However, there are two aspects of the subtyping approach in IDEF1X that express more than simple foreign key references. These are the completeness and discriminator declarations. A subtype link is shown as a line with an underlined circle at the supertype end. A *single underline* means that the cluster is *incomplete* (i.e., the supertype is more than the union of its subtypes). This means that the population of the supertype may contain instances absent in any of its subtypes.

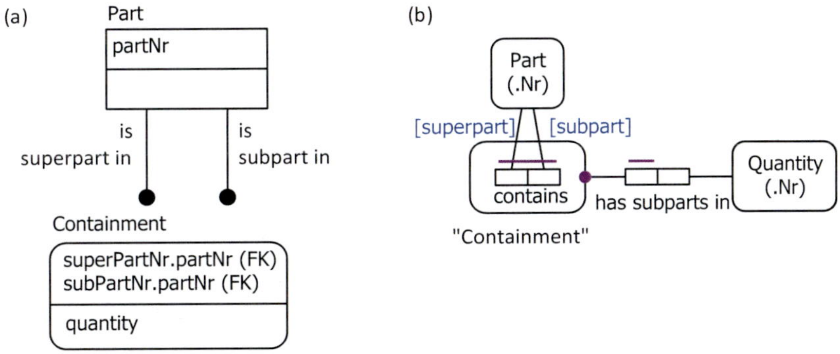

FIG. 8.40 Role names may be used in (a) IDEF1X and (b) ORM.

For example, in Fig. 8.41(a), the subtypes TallPerson and ShortPerson are mutually exclusive, but not collectively exhaustive. A person may be of medium height without being classified as either tall or short. In contrast to IDEF1X, ORM often uses derived subtypes, which must be defined in terms of roles played by their supertype(s). Fig. 8.41(b) shows one way of modeling this situation in ORM.

Here, a person's height is used to determine whether he or she is a member of TallPerson or ShortPerson. The subtype definitions and the lack of a value constraint on Height indicate that these subtypes are exclusive but not exhaustive. This can also be depicted explicitly by an implied exclusion constraint as shown. As an alternative to using height, you could attach the optional and exclusive unaries "is tall" and "is short" to Person and supply the obvious definitions.

In IDEF1X, a *double underline* at the supertype end means that the cluster is *complete*. This means that the supertype is the union of its subtypes. Because members of an IDEF1X cluster are also exclusive, this means that we have a partition. In Fig. 8.42(a), for example, Person is partitioned into MalePerson and FemalePerson.

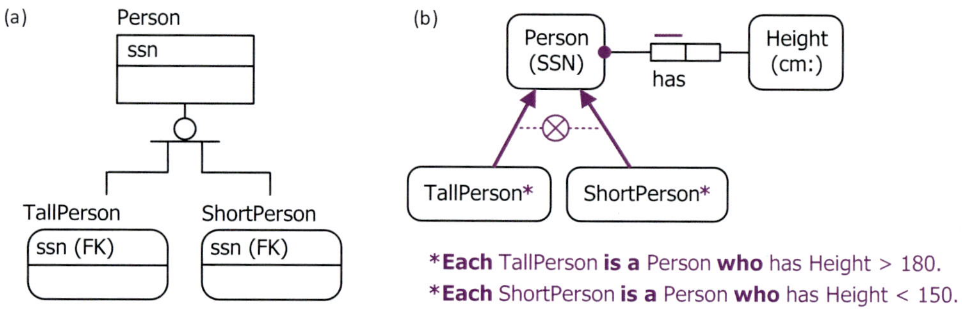

FIG. 8.41 An incomplete and exclusive subtype cluster in (a) IDEF1X and (b) ORM.

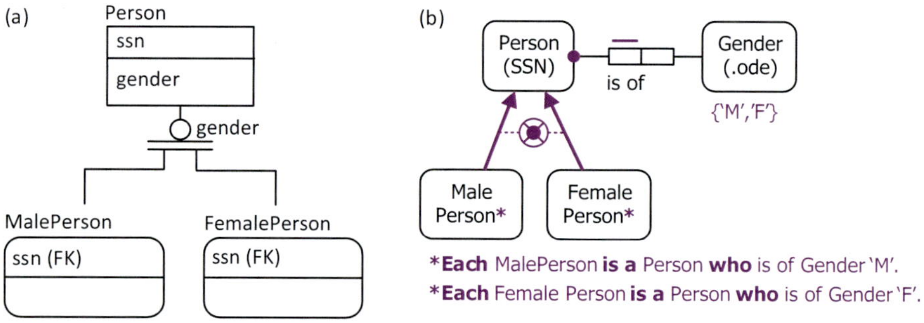

FIG. 8.42 A complete, exclusive subtype cluster in (a) IDEF1X and (b) ORM.

A mandatory attribute of a supertype may be used as a *discriminator* and written next to the circle to indicate the basis for the subtyping. In Fig. 8.42(a), for example, gender is used as a discriminator to classify Person into MalePerson and FemalePerson. The corresponding ORM schema is shown in Fig. 8.42(b). The partition constraint shown explicitly here is implied by the subtype definitions and other constraints.

Well, that pretty much covers the IDEF1X notation. Although it is a widely used standard, especially in the defense industry, we find it less suitable than the Barker or IE notations for conceptual modeling. To capture any detail or even basic constraints in IDEF1X, you have to resolve the model into what is essentially a relational model. Moreover, the

IDEF1X set of concepts and its notation are unnecessarily complex and forgettable. For conceptual analysis and validation, it also suffers from the weaknesses of an attribute-based approach.

Despite such drawbacks, IDEF1X can be used effectively by experienced modelers, who often associate connection relationships with the conceptual relationships from which they are migrated. Even for such experienced modelers, however, the best way to model in IDEF1X is to first do an ORM model and then map it to IDEF1X, expressing any additional ORM constraints as supplementary text or implementation code. Some ORM modeling tools can perform this mapping automatically.

8.5 Mapping from ORM to ER

Fully attributed IDEF1X schemas are essentially relational schemas displayed in a different notation. For this reason, Microsoft's ORM Source Model solution treats IDEF1X as simply a different relational style. If you are using that tool, you can generate IDEF1X schemas from ORM schemas by automatically generating the relational schema, setting the drawing style to IDEF1X, and then noting any extra ORM rules as textual annotations.

Table 8.2 ERmap procedure.

Step	*Action*
1	Preprocess the ORM schema (actually or mentally) 1.1 Replace objectified associations by coreferenced object types 1.2 Binarize *n*-ary associations by coreferencing 1.3 Binarize any sets of exclusive unaries
2	Model selected object types as entity types, and map a selection of their *n*:1 and 1:1 associations as attributes
3	Map remaining unary fact types to Boolean attributes or subtypes
4	Map remaining fact types to associations
5	Map ORM constraints to ER graphic constraints, textual constraints, or notes
6	Retain subtypes, but map subtype definitions to textual constraints
7	Map derived fact types to textual derivation rules, and map semiderived fact types to attributes/associations plus textual derivation rules

If instead you need to manually generate IDEF1X schemas from ORM, follow the relational mapping (Rmap) procedure discussed in Chapter 11 to obtain a relational schema and then redraw it in IDEF1X. Mapping from ORM to a true ER depends on the ER dialect (e.g., does it support *n*-ary associations, multivalued attributes, any form of objectification, or derived attributes/associations?). Industrial ER dialects typically support none of these options. The *ERmap* procedure in Table 8.2 provides basic guidelines for mapping ORM to Barker ER or IE.

We illustrate this procedure with examples using Barker ER notation. Step 1.1 replaces objectified associations by coreferenced object types. For example, if the fact type Person wrote Paper is objectified as Writing, replace this by Writing is by Person and Writing is of Paper, as shown in Fig. 8.43. Step 1.2 binarizes n-ary predicates by coreferencing. For example, the ternary Room at HourSlot is booked for Activity is replaced by forming a coreferenced entity type RoomBooking, as in Fig. 8.44.

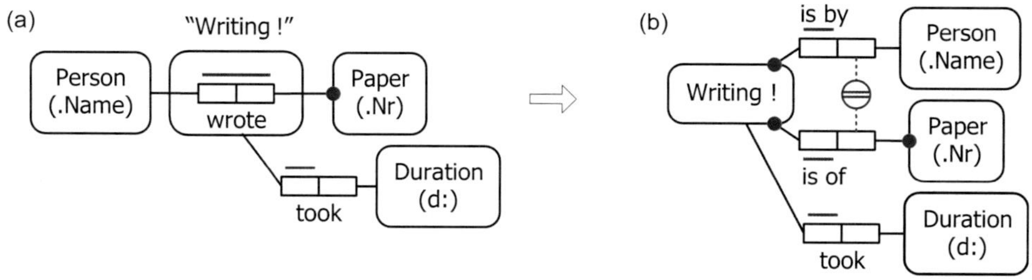

FIG. 8.43 Step 1.1: Replace objectifications by coreferenced object types.

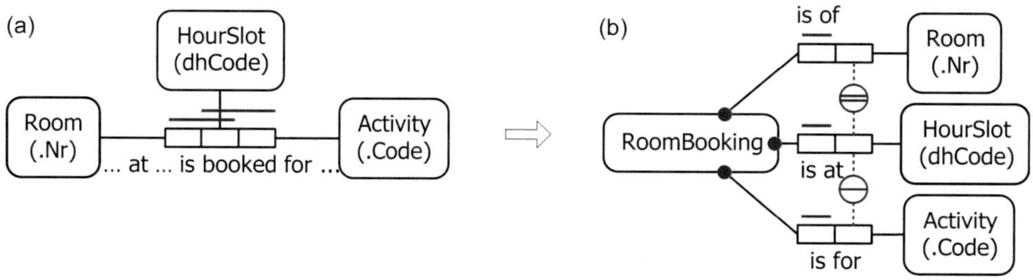

FIG. 8.44 Step 1.2: Replace *n*-ary fact types by coreferenced object types.

Step 1.3 binarizes any sets of exclusive unaries. For example, the Employee is male and Employee is female fact types are replaced by the single binary Employee is of Gender, with relevant constraint transforms, as shown in Fig. 8.45.

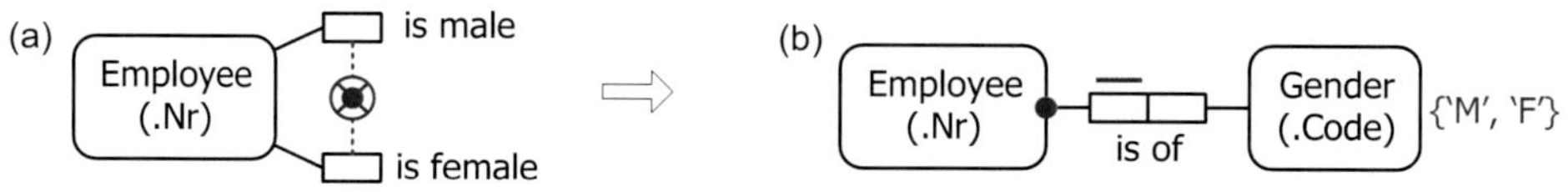

FIG. 8.45 Step 1.3: Replace any set of exclusive unaries by a binary fact type.

In step 2, we decide which object types to model as entity types and which n:1 and 1:1 ORM associations to remodel as attributes. Typically, entity types that play functional fact roles are retained as entity types. In rare cases, value types that are independent or play explicit mandatory, functional fact roles map to entity types. Functional binary (*n*:1 and 1:1) associations from an entity type *A* to a value type B, or to an entity type *B* about which you never want to record details, usually map to an attribute of *A*. If you have specified role names, these can usually be used as attribute names, with the object type name becoming the attribute's domain name.

The mapping in Fig. 8.46 illustrates several of these step 2 considerations as well as step 5 (map ORM constraints to ER graphic constraints, textual constraints, or notes).

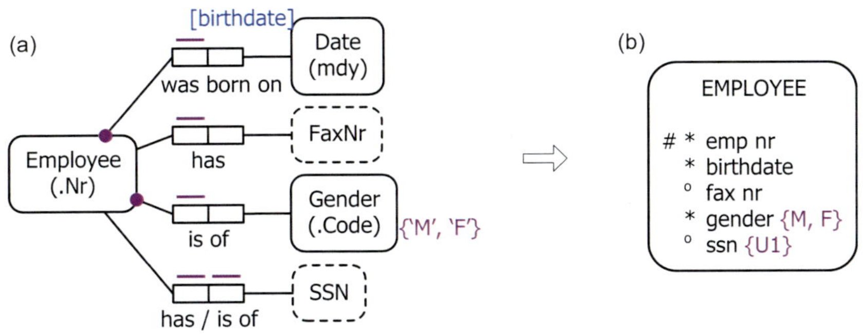

FIG. 8.46 Step 2: Map selected *n*:1 and 1:1 associations to attributes.

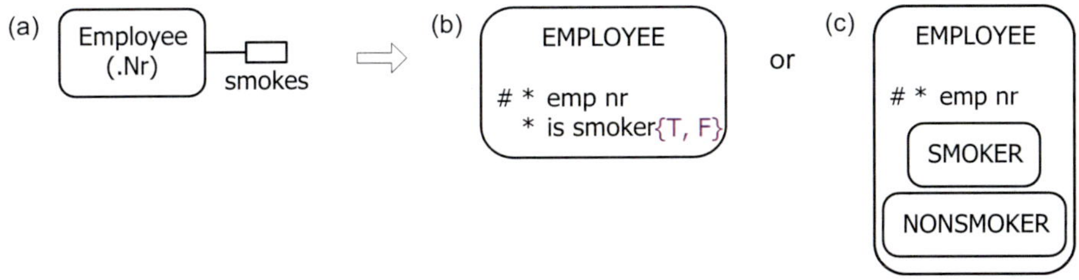

FIG. 8.47 Step 3: Map unaries to Boolean attributes or subtypes.

The "{M, F}" annotation is a textual note to capture the value constraint on gender codes. The "{U1}" annotation is a textual note to indicate a uniqueness constraint on the ssn attribute. Neither of these textual annotations is part of the Barker ER notation.

In step 3, we map unaries to Boolean attributes or to subtypes. The example in Fig. 8.47 assumes a closed-world interpretation for the unary. If this is not the case, the is-smoker attribute is made optional and an UnknownSmoking subtype is added. Open/closed world aspects are discussed in more detail in Chapter 10.

In step 4, the remaining fact types are mapped to associations. Any *m:n* associations should remain that way, unless the target notation doesn't support them (e.g., IDEF1X). In that case, you can replace the association by an intersection entity type as discussed earlier.

In Fig. 8.48, the n:1 fact type is retained as an association because it relates two entity types that will remain as such in the mapping. Even if the m:n association did not apply, we would normally retain Country as an entity type because now or later we are likely to record details for it (e.g., country name).

If an m:n association involves a value type, and the ER dialect does not support multi-valued attributes, map the value type to an entity type. Fig. 8.49 provides an example. Otherwise, where possible, transform the m:n association into multiple n:1 associations (e.g., Employee has PhoneNr1; Employee has PhoneNr2, etc.). Conceptual schema transformations like this are discussed in depth in Chapter 14.

In step 5, the simplest constraints in ORM usually map in an obvious way to multiplicity (mandatory and/or cardinality) constraints, as illustrated earlier. The more complex ORM constraints have no counterpart in ER, so you need to record these separately in textual form. Table 8.3 summarizes the main correspondences.

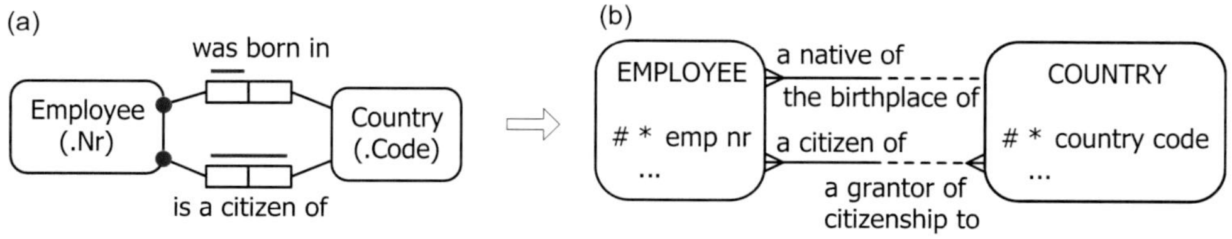

FIG. 8.48 Step 4: Map remaining fact types to associations.

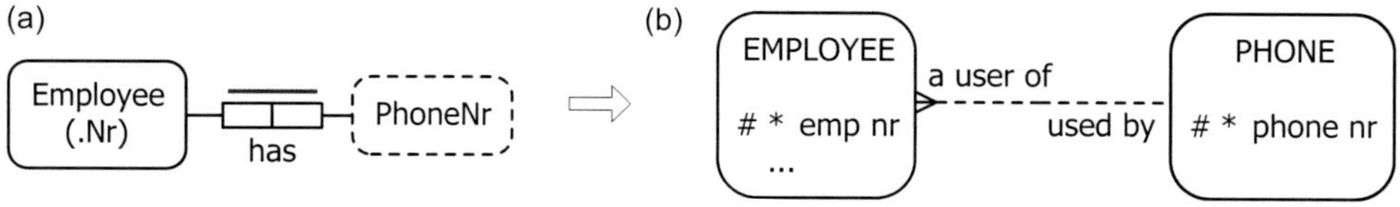

FIG. 8.49 Step 4: Map remaining fact types to associations.

Table 8.3 Mapping main ORM graphic constraints to ER (for use in step 5).

ORM constraint	*ER*
Internal UC	Maximum multiplicity of 1, or {Un}
External UC	Composite primary id (Barker) or textual constraint
Simple mandatory	Minimum multiplicity of 1, or textual constraint
Inclusive-or	Textual constraint (unless within exclusive-or)
Frequency	Multiplicity or textual constraint
Value	Textual constraint
Subset and equality	Textual constraints
Exclusion	Barker exclusive arc, or textual constraint
Ring constraints	Textual constraint

In step 6, subtypes are mapped to subtypes, adding relevant constraints if needed. Subtype definitions are handled with discriminators and/or textual constraints. If needed, introduce an "Other" subtype to complete a partition. If needed, use multiple diagrams (e.g., when multiple categorization schemes apply to the same supertype). Fig. 8.50 shows a simple example. The braced value constraint and footnoted constraints are not part of the Barker ER notation.

In step 7, we map derived and semiderived fact types. Industrial ER typically has no graphical support for derivation, so the derivation rules are captured as textual notes. Fig. 8.51 provides a simple example. With these few hints, and the examples discussed, you should have enough background to do the mapping manually. The mapping procedure is fairly boring and is best automated with a CASE tool.

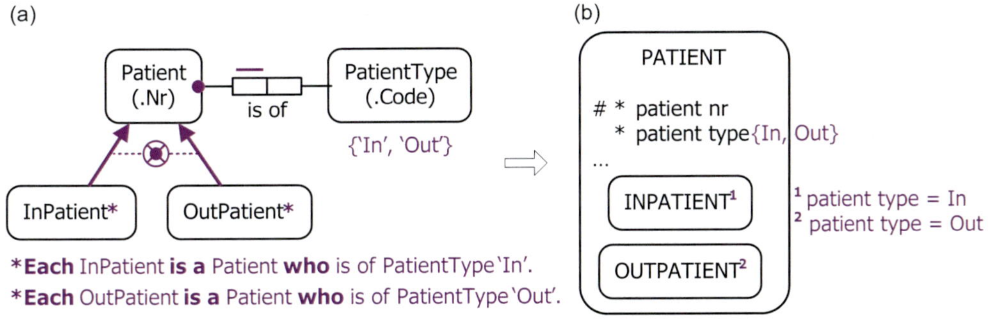

FIG. 8.50 Step 6: Map subtyping.

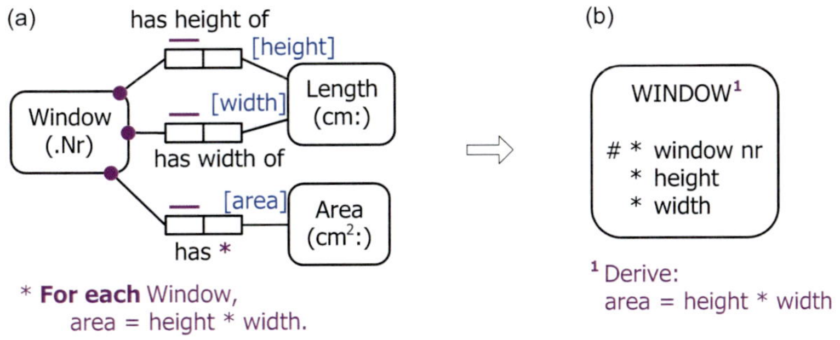

FIG. 8.51 Step 7: Map derivation rules.

For mapping to IDEF1X, the relational mapping procedure discussed in Chapter 11 may be used essentially as is, treating the relationships as foreign key references and making the relevant notational variations.

Exercise 8.5

1. Model the following application domain in your preferred version of ER or IDEF1X. We suggest that you do an ORM model first, but that's up to you.

   A video store has a library of videotapes that may be lent to customers. Six extracts are shown from the information system used by the store. Several aspects have been simplified or removed to reduce the size of this problem. For example, addresses are shown as a single entry, and financial aspects are ignored. Data about a movie may be recorded before the store obtains a videotape of it. Each videotape contains a copy of exactly one movie, and is either purchased or leased from another supplier. The status or condition (Good, OK, Poor, X) of each tape is noted on a regular basis. A status of "X" indicates that the tape is excluded from the list of tapes that may be borrowed.

Movie	*Title*	*Category*	*Copy Nr*	*Purchase date*	*Lease expiry*	*Status*
AP2K	Apocalypse 2000	R	1	01/10/2000		Good
BMF	Batman Forever	M				
CJ	City of Joy	MA	1	07/01/2000		OK
			2		01/01/2001	X
DS	Donovan Sings	G	1	01/03/1999		Good
			2	01/03/1999		OK
GHST	Ghost	M	1	06/06/1991		X
			2	07/10/1991		Poor
			3	07/10/1991		X
GQ	Galaxy Quest	PG	1	01/10/2000		OK
			2		07/01/2001	Good
MTX	The Matrix	R	1	11/11/1999		Good
			2	11/11/1999		Good
MTX2	The Matrix 2	R				

   For excluded video tapes, a record is kept to indicate the reason for their exclusion and whether they are written off for taxation purposes. Here is a sample extract.

Movie	*CopyNr*	*Comment*	*Written off?*
CJ	2	Faulty	Y
GHST	1	Faulty	N
GHST	3	Stolen	Y

   The next table records details about the categories in which movies are classified.

Category	*Description*
G	For General exhibition
PG	Parental Guidance recommended for persons under 15
M	Recommended for Mature audiences 15 years and over
MA	For Mature Adults: restrictions apply to persons under the 15 years age
R	Restricted to adults 18 years and over

For the current calendar year, a record is kept of which movies are the best sellers.

Month	*Rank*	*Movie*	*Title*
1	1	MTX	TheMatrix
	2	AP2K	Apocalypse 2000
	3	DS	Donovan Sings
	...	...	...
2	1	MTX	The Matrix
	2	GQ	Galaxy Quest
	2	DS	Donovan Sings
	4	AP2K	Apocalypse 2000
	...	...	...

Loans are charged to customers, who are identified by the number on their video store card presented at the time of borrowing. The combination of name and address is unique for a customer. Customers who indicate a phone number must indicate when they may be contacted on that phone (D = day only; N = Night only; D&N = Day and Night). Customer details may be recorded before they take out any loan.

Customer	*Name*	*Address*	*Phone*	*Call period*
1	Frog F	5 Ribbit Rd, Bellevue	425 555 7000	N
2	Jones E	3 Sun Ave, Spokane		
3	Smith J	520 Pike St, Seattle	206 555 6789	D
4	Jones E	520 Pike St, Seattle	206555 6701	D
5	Frog K	5 Ribbit Rd, Bellevue	425 555 7000	D&N

Each loan is identified by a loan number and may involve more than one tape. Customers may borrow tapes as often as they like and may even borrow the same tape more than once on the same day (after returning it). The rental of a video tape within a loan is called a LoanItem. Although return data for loan items may be derived from video tape returns and other data, to simplify the problem you can ignore this.

Loan Date	*Loan Nr*	*Customer*		*Videotape*		*Date returned*
		Nr	*Name*	*Movie*	*Copy*	
10/01/2000	9001	2	Jones E	GQ	1	?
				MTX	1	10/02/2000
	9002	3	Smith J	GQ	2	10/01/2000
10/02/2000	9003	3	Smith J	MTX	1	10/03/2000
				DS	2	10/03/2000
				GQ	2	10/03/2000
	9004	4	Jones E	GHST	2	?
10/03/2000	9005	1	Frog F	AP2K	1	10/03/2000
				MTX	1	?

2. The following ORM schema models information about university academics. As a challenge exercise, map this to your preferred ER or IDEF1X notation, including some comments to cater for advanced constraints and multiple inheritance. You may wish to delay this until you have studied the relational mapping procedure in Chapter 11.

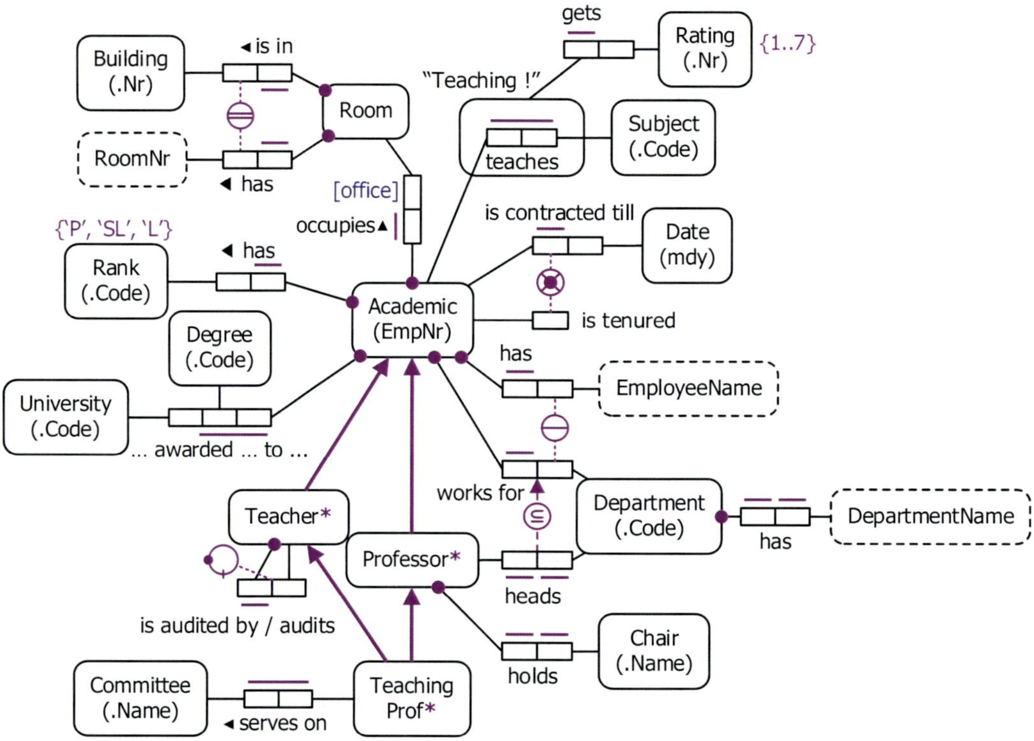

*Each Teacher is an Academic who teaches some Subject.
*Each Professor is an Academic who has Rank 'P'.
*Each TeachingProf is a Teacher and a Professor.

8.6 Summary

The ER modeling approach was originated by Peter Chen in the 1970s and allows facts to be expressed via relationships (e.g., Person was born in Country) or attributes (e.g., birthdate). Of the dozens of different ER versions in existence, the most widely used are the Barker notation and IE. The popular IDEF1X approach is often referred to as a version of ER, but is actually a mixture of ER and relational approaches, with the emphasis on relational.

The *Barker notation* represents entity types as named, soft rectangles with a list of one or more attributes. A "*" or "•" indicates that an attribute is mandatory, and a "○" indicates that the attribute is optional. All relationships are binary, and are shown as named lines. A solid half-line denotes a mandatory role, and a dotted half-line denotes an optional role. A crow's foot at the end of the line indicates the cardinality "many", and its absence indicates "one". A hash "#" in front of an attribute indicates it is, or is a part of, the primary identifier. A bar "|" across one end of a relationship indicates that the relationship is a component of the primary identifier for the entity type at that end.

An exclusive arc across roles indicates that the roles are mutually exclusive; if the role lines are solid, we have an exclusive-or constraint. Subtype partitions are denoted by Euler diagrams, placing the subtypes inside the supertype.

In the *IE* approach, entity types are displayed as named rectangles with a list of attributes. Relationships are binary and are denoted by named lines. A crow's foot at the end indicates "many", a stroke "|" indicates "one", and a circle "○" indicates optional. Two strokes "||" indicates "exactly one". Some IE versions depict an exclusive or constraint as a black dot joining relationship lines. Different subtype notations exist for IE, some using Euler diagrams and some "is a" relationship lines.

IDEF1X models may be viewed at three levels. In a high-level ER view, *m:n* (or "nonspecific") relationships may be shown directly, but these must be resolved into an intersection entity type with two *n*:1 relationships as the model is refined into key-based or fully attributed views. An entity type is identifier dependent if its primary key includes a foreign key and is shown as a named, soft rectangle. Otherwise it is identifier independent and is shown as a named, hard rectangle. Attributes are listed inside the rectangle, with the primary key in the top compartment. Alternate keys are denoted by appending "(AK*n*)" and foreign keys by appending "(FK)". An attribute is mandatory unless followed by "(O)".

Connection relationships are foreign key references from the child to the parent and are shown as a named line with a dot "•" at the child end. For a specific connection relationship, its forward name is always read toward the dot. If an inverse name is added, a slash "/" is appended to the forward reading. For nonspecific relationships, the forward reading is left to right (or top to bottom if the line is vertical).

A line end has a cardinality of 1 unless annotated. A dot indicates "0 or more" but can be strengthened by adding "P" (1 or more), "Z" (0 or 1), or "*n*" (exactly *n*).

A foreign key reference starting from a primary key attribute is an identifying relationship and is shown as a solid line. A foreign key reference starting from a nonkey attribute is a nonidentifying relationship, and is shown as a dashed line; in this case, a diamond at the parent end indicates that each child is associated with at most one parent.

An entity type may be classified into one or more clusters of mutually exclusive categories. Subtype links are depicted as categorization relationships with a circle at the supertype end. The cluster is incomplete or complete according to whether the circle has a single underline or double underline, respectively.

ER or IDEF1X models are best developed by mapping them from ORM models, and then noting any additional ORM constraints as comments. The main guidelines for mapping from ORM to ER are summarized in Table 8.2.

Chapter Notes

There are dozens of ER notations in existence, and minor variations exist in CASE tool implementations of the main notations discussed in this chapter. Dr. Peter Chen (https://en.wikipedia.org/wiki/Peter_Chen) is still active in the field, and is on the steering committee for the international ER conferences, which have been held yearly since 1981, and have been broadened in scope to include all forms of conceptual modeling. The proceedings of the ER conferences are published by Springer (*www.springer.com*).

Clive Finkelstein, the "father of information engineering", is also still active in the field. Finkelstein (1989) discusses the basic concepts and history of IE. Finkelstein (1992) provides practical advice on using IE as well as a treatment of entity dependency. For a short, authoritative overview of the IE method, see Finkelstein (1998). A treatment of his work on using XML for building corporate portals can be found in Finkelstein and Aiken (2000). His website can be accessed at *www.ies.aust.com/~ieinfo/*. For a look at the IE approach used by James Martin, see Martin (1993). Martin's more recent books tend to use the UML notation instead, for example, Martin and Odell (1998). In practice, however, IE is still used far more extensively for database design than UML, which is mostly used for object-oriented code design.

For some further details on IDEF1X, see Bruce (1992). The $IDEF1X_{97}$ standard (also known as $IDEF_{object}$) was approved in 1998 by IEEE and further refined in 2012 to the current version (https://www.iso.org/standard/60614.html).

9

Data Modeling in UML

Chapter Outline

9.1 Introduction

Although semantic approaches to information modeling appeared in the early seventies, no single approach has yet achieved universal adoption. By and large, the history of information systems modeling has been characterized by a plethora of techniques and notations, with occasional religious wars between proponents of different approaches. Each year, many new approaches would be proposed, leading to groans from academics who were charged with teaching the state of the art. This is referred to as the "YAMA" (yet another modeling approach!) or "NAMA" (not another modeling approach!) syndrome. Fig. 9.1 pictures this as a mountain of modeling methods, piled on top of one another, which nicely ties in with the Japanese meaning of "yama" (mountain), depicted as a kanji that is high in the middle and low on the ends.

While diversity is often useful, the modeling industry would benefit if practitioners used just a few standard modeling approaches, individually suited for their modeling scope, and collectively covering the tasks needed to model a wide variety of applications. This would improve communication between modelers and reduce training costs.

The rapid rise of the unified modeling language (UML) has been accompanied by claims that UML by itself is an adequate approach for modeling any software application. Some UML proponents have even been so bold as to claim that "the modeling wars are over–UML has won". This claim has been strongly rejected by several experienced data modelers, including Dave Hay, who argues that "there is no such thing as 'object-oriented analysis'" (Hay, 1999a), only object-oriented design, and that "UML is … not suitable for analyzing business requirements in cooperation with business people" (Hay, 1999b).

Information Modeling and Relational Databases. https://doi.org/10.1016/B978-0-443-23790-4.00001-3

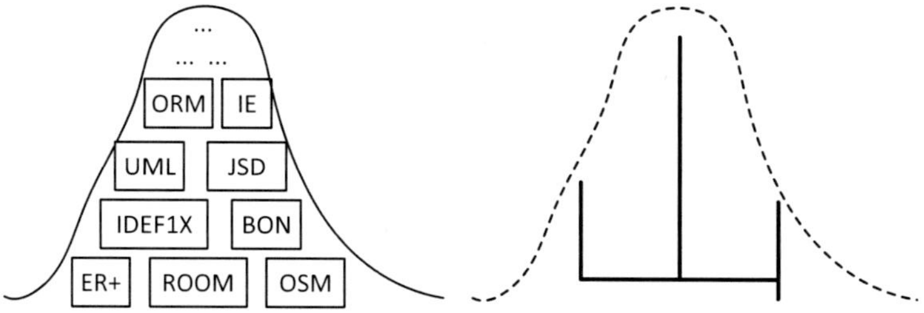

FIG. 9.1 "YAMA" (yet another modeling approach) or Japanese for "mountain".

To date, UML is mainly used in industry for designing object-oriented program code. Although it can be used for designing databases, UML has so far had little success in displacing other approaches such as entity-relationship (ER) modeling for this purpose. Given UML's object-oriented focus, and the dominance of relational database management systems (DBMSs), this is perhaps not surprising. Nevertheless, UML is a very important language that could well become popular for database design in the future.

Based on a combination of the Booch object modeling technique (OMT) and object-oriented software engineering (OOSE) methods, UML was further developed by a consortium of companies and individuals working within the Object Management Group (OMG). It includes adaptations of many other techniques (e.g., Harel's state charts).

Version 1.1 of UML was adopted in 1997 by the OMG as a language for object-oriented analysis and design, and was later accepted as a standard by the International Standards Organization (ISO). A major revision (2.0) was adopted by ISO in 2005, comprising infrastructure and superstructure specifications, plus related specifications on the object constraint language (OCL) and diagram interchange. In 2017, UML 2 was updated to version 2.5.1[a] and this is still (as of 2024) the current version. When using a UML tool, be aware that vendor support might lag behind the latest version.

As discussed later, the UML metamodel and notation have inconsistencies, with some unresolved problems being fundamental. Despite these issues, UML is the closest thing to a de facto standard in industry for object-oriented software design, and hence is worthy of study.

The UML notation is really a set of languages rather than a single language. It includes a vast number of symbols, from which various diagrams may be constructed to model different perspectives of an application. The nine main diagram types in UML 1.5 are Use Case (use case diagram); Static Structure (class diagram, object diagram); Behavior (state chart, activity diagram); Interaction (sequence diagram, collaboration diagram); and Implementation (component diagram, deployment diagram). UML 2 extended these to 13 diagram types, as set out in Table 9.1.

Some of these diagrams (e.g., collaboration diagrams) are useful only for designing object-oriented program code. Some (e.g., activity diagrams and use case diagrams)

[a]https://www.omg.org/spec/UML/2.5.1.

Table 9.1 The 13 predefined diagram types in UML 2.

Structure	Class	
	Object	
	Component	
	Deployment	
	Package	
	Composite Structure	
Behavior	Use Case	
	State Machine	
	Activity	
	Interaction	Sequence
		Collaboration
		Interaction Overview
		Timing

can be useful in requirements analysis, and some (e.g., class diagrams) have limited use for conceptual analysis and are best used for logical design.

The UML specification provides syntax and semantics for these diagram types but not yet a process for developing UML models, other than to suggest that model development should be use case driven, iterative, and architecture centric. Various companies promote their own modeling process for UML.

Although all the UML diagram types are worth studying, this book focuses mainly on information modeling for databases. This chapter addresses data modeling in UML, so considers only the static structure (class and object) diagrams. Class diagrams are used for the data schema. Object diagrams provide a limited means to discuss data populations. Some other UML diagram types are discussed in Chapter 15.

Like ER, UML uses attributes. As discussed in earlier chapters, attributes are great for logical models, but are best modeled as relationships when performing conceptual analysis because this facilitates validation and minimizes the impact of change. For such reasons, we believe the best way to develop UML data models is to first do an object-role modeling (ORM) model and then map it to UML. Because ORM will be used to clarify the data modeling concepts in UML, to gain the full benefits of this clarification you should be familiar with the ORM concepts covered in earlier chapters.

No language is perfect, ORM included. Overall, UML provides a useful suite of notations for both data and process modeling while ORM is currently focused on data modeling only.

9.2 Object-Orientation

UML facilitates object-oriented (OO) code design because it covers both data and behavioral modeling, and lets you drill down into physical design details relevant to OO code. The class diagram in Fig. 9.2 models a class whose instances are screen dialog boxes.

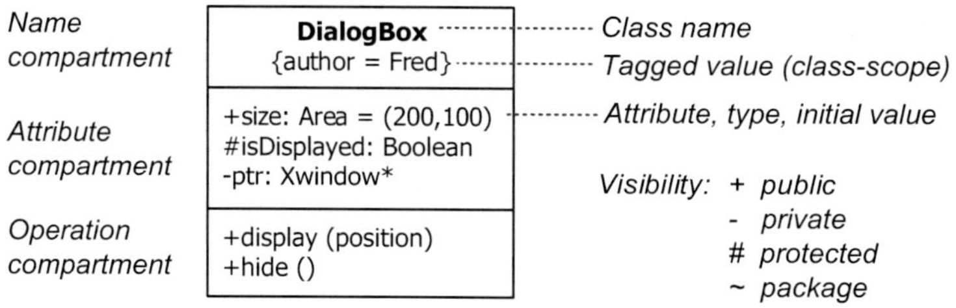

FIG. 9.2 Example of a UML class.

The class shape in Fig. 9.2 has three compartments. The name compartment includes the class name as well as a tagged value naming the author of the class. The attribute compartment lists the visibility, name, and type of each attribute. The visibility settings "+", "-", "#", and "~" indicate whether the attribute is public, private, protected, or package, respectively. These visibility settings are software, not conceptual, issues. The size attribute is initialized to a given area value. The operation compartment specifies what operations are encapsulated in instances of the class. In this example, the operations may be implemented by methods to display the dialog box at a specific position, and to hide the dialog box.

Fig. 9.3 shows another class diagram that depicts Employee and Car classes as well as an association corresponding to the ORM fact type Employee drives Car. The association is depicted by a line between the classes. The role name "driver" on the left end of the association clarifies the intended semantics (an association reading could also be supplied). The open arrow tip at the right end of the association is a navigability setting indicating that fast access is required from employee instances to their car instances. This may be implemented by including pointers from employee objects (software objects) to the car objects that model the cars that they drive. Navigability settings are implementation issues related to performance, not conceptual issues about the business domain.

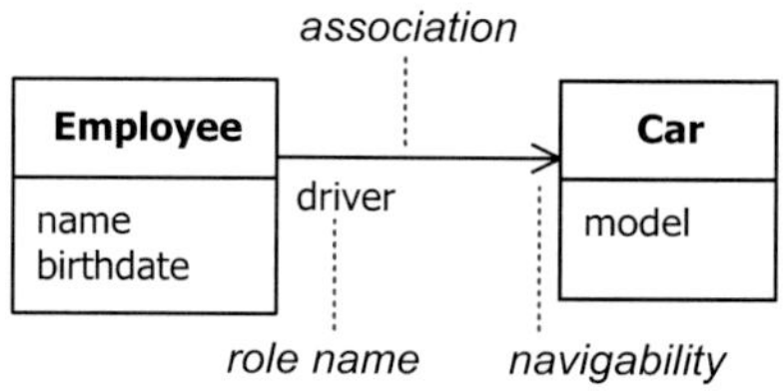

FIG. 9.3 The navigability setting demands fast access from Employee to Car.

By omitting implementation details such as attribute visibility and association navigability, class diagrams can be used for conceptual analysis. When used in this way, class diagrams are somewhat similar to ER models. But there is a significant difference arising from the OO perspective. If you look at the classes in Figs. 9.2 and 9.3, what strikes you as missing?

You guessed it! No identification schemes are provided for the classes. In object-oriented programming, objects may be identified by their memory addresses or internal object identifiers (oids), so UML does not require a value-based identification scheme for use by humans in communicating about the objects. For conceptual analysis, however, human-oriented reference schemes (e.g., dialog box numbers, employee numbers, car registration numbers) must be supplied.

To partly address this issue, UML 2.5 now allows you to mark a property (e.g., attribute or association end) as being (part of) the *identifier* for its containing class by appending "{id}" to the property name. However, UML has no standard notation for declaring a single-valued attribute to be a secondary identifier or even to be unique.

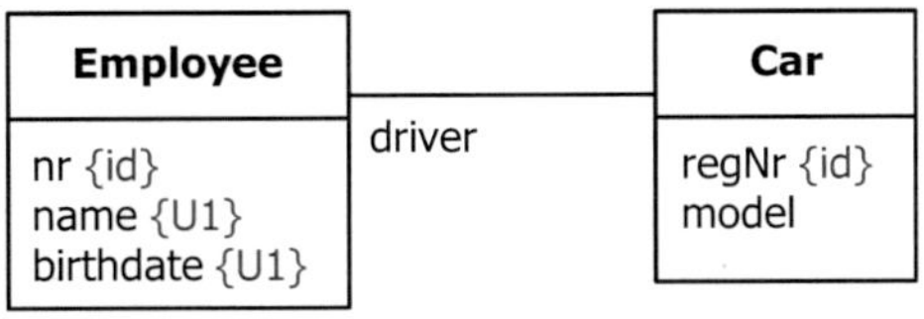

FIG. 9.4 Adding notations for identifiers and uniqueness.

This book uses "{U*n*}" for uniqueness ($n > 0$), where the n is used to disambiguate cases where the same uniqueness constraint might apply to a combination of attributes. In Fig. 9.4, for example, employee number and car registration number attributes are marked as the primary identifiers of employee and car, respectively. This entails that they are mandatory and unique. Additionally, the combination of employee name and birthdate has been declared unique. We also dropped the navigation arrow, as it is irrelevant to the business semantics.

The requirement that each class *must* have a value-based identification scheme distinguishes both ORM and ER from UML. ORM classifies objects into entities (nonlexical objects) and values (lexical objects) and requires each entity to be identified by a reference scheme used by humans to communicate about it. ORM uses "object", "entity", and "value" to mean "object instance", "entity instance", and "value instance", respectively, appending "type" for the relevant set of all possible instances. Entities may be referenced in different ways, and typically change their state over time. Glossing over some subtle points, values are constants (e.g., character strings) that denote themselves so do not require a reference scheme.

Fig. 9.5(a) depicts explicitly a simple reference scheme in ORM. If an entity type has more than one candidate reference scheme, one may be declared preferred. A preferred reference scheme for an entity type maps each instance of it onto a unique, identifying value (or a combination of values). In Fig. 9.5(a), the reference type has a sample population shown in a reference table (one column for each role). Here, icons are used to denote the real world employee entities.

Simple reference schemes may be abbreviated by enclosing the reference mode in parentheses, as in Fig. 9.5(b), and an object type's reference table includes values but

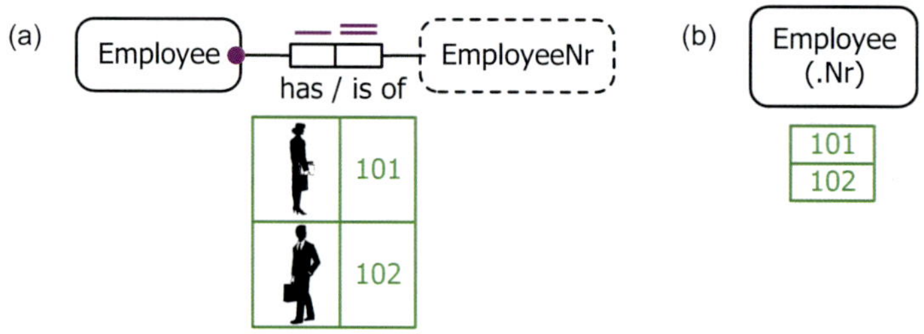

FIG. 9.5 A simple reference scheme in ORM, shown (a) explicitly and (b) implicitly.

no icons. References verbalize as existential sentences, such as "There is an Employee who has the EmployeeNr 101". Entity instances are referenced elsewhere by definite descriptions, such as, "The Employee who has the EmpNr 101".

In a relational database, we might use the preferred reference scheme to provide value-based identity or instead use system-generated row-ids. In an object-oriented implementation, we might use oids (hidden, system-generated object identifiers). Such choices can be added later as annotations to the model. For analysis and validation purposes, however, we need to ensure that humans have a way to identify objects in their normal communication. It is the responsibility of humans (not the system) to enforce constraints on preferred reference types. Assuming humans do enforce the reference type constraints, the system may be used to enforce the elementary fact type constraints.

UML classifies instances into objects and data values. UML *objects* basically correspond to ORM entities, but are assumed to be identified by oids. Although UML does not require entities to have a value-based reference scheme, we should include value-based reference in any UML class to capture all the conceptual semantics. UML *data values* basically correspond to ORM values: they are constants (e.g., character strings or numbers) and so require no oids to establish their identity. Entity types in UML are called *classes*, and value types are basically *data types*. Note that "object" means "object instance", not "object type". A relationship instance in UML is called a *link*, and a relationship type is called an *association*.

9.3 Attributes

Like other ER notations, UML allows relationships to be modeled as attributes. For instance, in Fig. 9.6(a) the Employee class has eight attributes. The corresponding ORM diagram is shown in Fig. 9.6(b).

In UML, *attributes are mandatory and single valued by default*. So the employee number, name, title, gender, and smoking status attributes are all mandatory. In the ORM model, the unary predicate "smokes" is optional (not everybody has to smoke). UML does not support unary relationships, so it models this instead as the Boolean attribute "isSmoker" with possible values True or False. In UML, the domain (i.e., type) of any attribute may optionally be displayed after it (preceded by a colon). In this example, the domain is displayed only for the isSmoker attribute. By default, ORM tools usually take a closed world approach (CWA) to unaries, which agrees with the isSmoker attribute being mandatory.

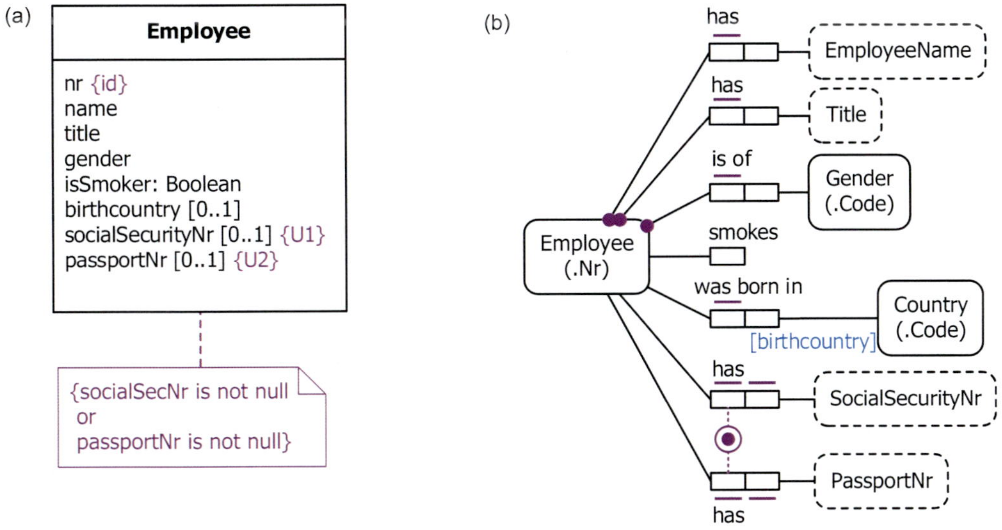

FIG. 9.6 UML attributes (a) depicted as ORM relationship types (b).

The ORM model also indicates that Gender and Country are identified by codes. We could convey some of this detail in the UML diagram by appending domain names, for example, Countrycode" could be appended to "birthcountry".

In the ORM model, it is *optional* whether we record birthcountry, Social Security number, or passport number. This is captured in UML by appending [0..1] to the attribute name (each employee has 0 or 1 birthcountry, and 0 or 1 Social Security number). This is an example of an *attribute multiplicity constraint*. The main multiplicity cases are shown in Table 9.2. If the multiplicity is not declared explicitly, it is assumed to be 1 (exactly one). If desired, we may indicate the default multiplicity explicitly by appending [1..1] or [1] to the attribute.

In the ORM model, the uniqueness constraints on the right-hand roles (including the EmployeeNr reference scheme shown explicitly earlier) indicate that each employee number, Social Security number, and passport number refer to at most one employee. UML has no standard notation for such "*attribute uniqueness constraints*", so we used our {U*n*}

Table 9.2 Multiplicities.

Multiplicity	*Abbreviation*	*Meaning*	*Note*
0..1		0 or 1 (at most one)	
0..*	*	0 to many (zero or more)	
1..1	1	Exactly 1	Assumed by default
1..*		1 or more (at least 1)	
n..*		n or more (at least n)	$n \geq 0$
n..m		at least n and at most m	$m > n \geq 0$

notation. UML 2 added the option of specifying {unique} or {nonunique} as part of a multiplicity declaration, but this is only to declare whether instances of collections for multivalued attributes or multivalued association roles may include duplicates, so it can't be used to specify that instances of single valued attributes or combinations of such attributes are unique for the class.

UML has no graphic notation for an inclusive-or constraint, so the ORM constraint that each employee has a Social Security number or passport number needs to be expressed textually in an attached *note*, as in Fig. 9.6(a). Such *textual constraints* may be expressed informally, or in some formal language interpretable by a tool. In the latter case, the constraint is placed in *braces*.

In our example, we've chosen to code the inclusive-or constraint in SQL syntax. Although UML provides OCL for this purpose, it does not mandate its use, allowing users to pick their own language (even programming code). This of course weakens the portability of the model. Moreover, the readability of the constraint is typically poor compared with the ORM verbalization.

The ORM fact type Employee was born in Country is modeled as a birthcountry attribute in the UML class diagram of Fig. 9.6(a). If we later decide to record the population of a country, then we need to introduce Country as a class, and to clarify the connection between birthcountry and Country we would probably reformulate the birthcountry attribute as an association between Employee and Country. This is a significant change to our model. Moreover, any object-based queries or code that referenced the birthcountry attribute would also need to be reformulated. ORM avoids such semantic instability by always using relationships instead of attributes.

Another reason to introduce a Country class is to enable a listing of countries to be stored without requiring all these countries to participate in a fact. To do this in ORM, we simply declare the Country type to be independent. The object type Country may be populated by a reference table that contains those country codes of interest (e.g., "AU" denotes Australia).

A typical argument in support of attributes runs like this: "Good UML modelers would declare country as a class in the first place, anticipating the need to later record something about it, or to maintain a reference list". This argument is flawed. In general, you can't be sure about what kinds of information you might want to record later, or about how important some model feature will become.

Even in the title and gender case, a complete model should include a relationship type to indicate which titles are restricted to which gender (e.g., "Mrs", "Miss", "Ms", and "Lady" apply only to females). In ORM, this kind of constraint can be captured graphically as a join-subset constraint or textually as a constraint in a formal ORM language (e.g., **If** Person_1 has **a** Title **that** is restricted to Gender_1 **then** Person_1 is of Gender_1). In contrast, attribute usage hinders expression of the relevant restriction association (try expressing and populating this rule in UML).

ORM includes algorithms for dynamically generating ER and UML diagrams as attribute views. These algorithms assign different levels of importance to object types depending on their current roles and constraints, redisplaying minor fact types as attributes of the major object types. Modeling and maintenance are iterative processes. The importance of a feature can change with time as we discover more of the global model, and the domain being modeled itself changes.

To promote semantic stability, ORM makes no commitment to relative importance in its base models, instead supporting this dynamically through views. Elementary facts are the fundamental units of information, are uniformly represented as relationships, and how they are grouped into structures is not a conceptual issue. You can have your cake and eat it too by using ORM for analysis, and if you want to work with UML class diagrams, you can use your ORM models to derive them.

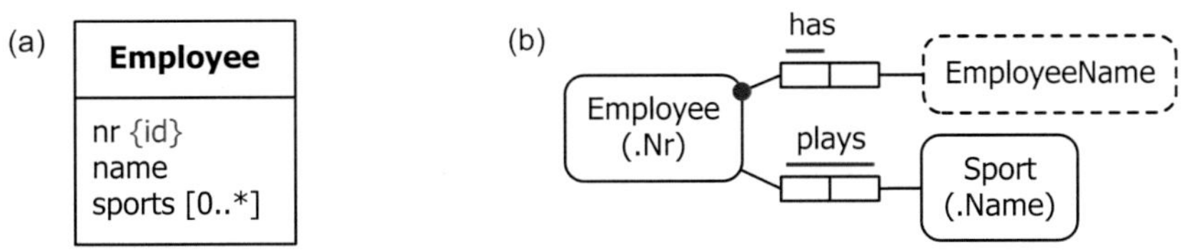

FIG. 9.7 (a) Multivalued UML sports attribute depicted as (b) ORM *m:n* fact type.

One way to model in UML that an employee may play many sports is shown in Fig. 9.7(a). Here, the information about who plays what sport is modeled as the *multivalued attribute* "sports". The "[0..*]" multiplicity constraint on this attribute indicates how many sports may be entered for each employee. The "0" indicates that possibly no sports are entered for some employee. UML uses a *null value* for this case, just like the relational model. The presence of nulls exposes users to implementation rather than conceptual issues and adds complexity to the semantics of queries. The "*" in "[0..*]" indicates there is *no upper bound* on the number of sports of a single employee. In other words, an employee may play many sports, and we don't care how many. If "*" is used without a lower bound, this is taken as an abbreviation for "0..*".

An equivalent ORM schema is shown in Fig. 9.7(b). An optional, many:many fact type is used instead of the multivalued sports attribute. As discussed in the next section, this approach may also be used in UML using an *m:n* association.

To discuss *class instance populations*, UML uses *object diagrams*. These are essentially class diagrams in which each object is shown as a separate instance of a class, with data values supplied for its attributes. As a simple example, Fig. 9.8(a) includes object diagrams to model three employee instances along with their attribute values. The ORM model in Fig. 9.8(b) displays the same sample population, using fact tables to list the fact instances.

For simple cases like this, object diagrams are useful. However, they rapidly become unwieldy if we wish to display multiple instances for more complex cases. In contrast, fact tables scale easily to handle large and complex cases.

ORM constraints are easily clarified using sample populations. For example, in Fig. 9.8(b) the absence of employee 101 in the Plays fact table shows that playing sport

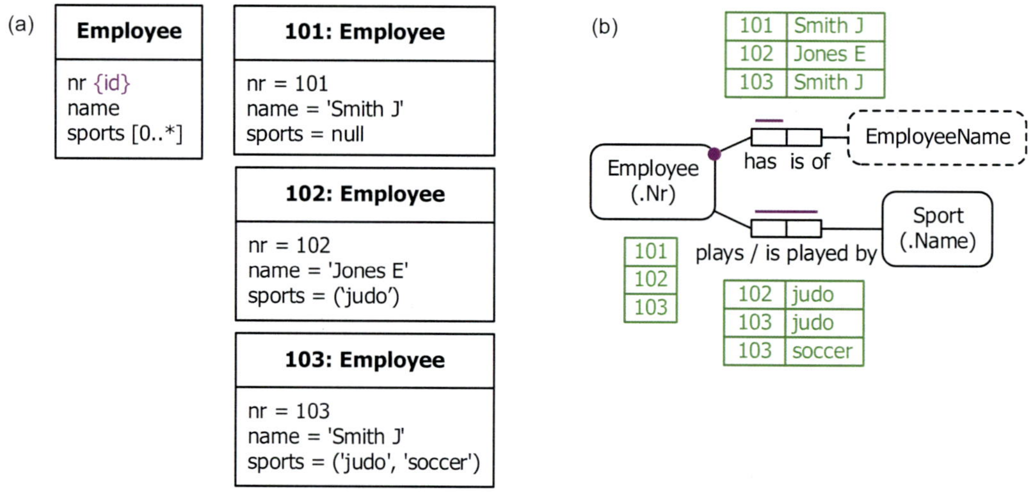

FIG. 9.8 Populated models in (a) UML and (b) ORM.

is optional, and the uniqueness constraints mark out which column or column-combination values can occur on at most one row. In the EmployeeName fact table, the first column values are unique, but the second column includes duplicates. In the Plays table, each column contains duplicates, as only the whole rows are unique. Such populations are useful for checking constraints with the subject matter experts. This validation-via-example feature of ORM holds for all its constraints, not just mandatory roles and uniqueness, because all its constraints are role-based or type-based, and each role corresponds to a fact table column.

As a final example of multivalued attributes, suppose that we wish to record the nicknames and colors of country flags. Let us agree to record at most two nicknames for any given flag and that nicknames apply to only one flag. For example, "Old Glory" and perhaps "The Star-Spangled Banner" might be used as nicknames for the US flag. Flags have at least one color.

Fig. 9.9(a) shows one way to model this in UML. The "[0..2]" indicates that each flag has at most two (from zero to two) nicknames. The ["1..*] declares that a flag has one or more colors. An additional constraint is needed to ensure that each nickname refers to at most one flag. A simple attribute uniqueness constraint (e.g., {U1}) is not enough because the nicknames attribute is set valued. Not only must each nicknames set be unique for each flag, but each element in each set must be unique (the second condition implies the former). This more complex constraint is specified informally in an attached note.

Here, the attribute domains are hidden. Nickname elements would typically have a data type domain (e.g., String). If we don't store other information about countries or colors, we might choose String as the domain for country and color as well (although this is subconceptual because real countries and colors are not character strings). However,

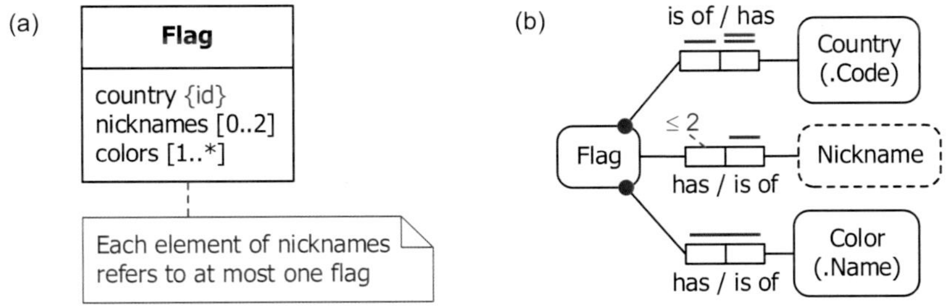

FIG. 9.9 A flag model in (a) UML and (b) ORM.

because we might want to add information about these later, it's better to use classes for their domains (e.g., Country and Color). If we do this, we need to define the classes as well.

Fig. 9.9(b) shows one way to model this in ORM. For verbalization, we identify each flag by its country. Because country is an entity type, the reference scheme is shown explicitly (reference modes may abbreviate reference schemes only when the referencing type is a value type). The "≤2" frequency constraint indicates that each flag has at most two nicknames, and the uniqueness constraint on the role of NickName indicates that each nickname refers to at most one flag.

UML gives us the choice to model a feature as an attribute or an association. For conceptual analysis and querying, explicit associations usually have many advantages over attributes, especially multivalued attributes. This choice helps us verbalize, visualize, and populate the associations and to express various constraints involving the "role played by the attribute" in standard notation, rather than resorting to a nonstandard extension. This applies not only to simple uniqueness constraints but also to other kinds of constraints (frequency, subset, exclusion, etc.) over one or more roles that include the role played by the attribute's domain (in the implicit association corresponding to the attribute).

For example, if the association Flag is of Country is depicted explicitly in UML, the constraint that each country has at most one flag can be captured by adding a multiplicity constraint of "0..1" on the left role of this association. Although country and color are naturally conceived as classes, nickname would normally be construed as a data type (e.g., a subtype of String). Although associations in UML may include data types (not just classes), this is somewhat awkward; so in UML, nicknames might best be left as a multivalued attribute.

Another reason for favoring associations over attributes is stability. If we ever want to talk about a relationship, it is possible in both ORM and UML to make an object out of it and simply attach the new details to it. If instead we modeled the feature as an attribute, we would need to first replace the attribute by an association. For example, consider the association Employee plays Sport in Fig. 9.8(b). If we need to record a skill level for this play, we can simply objectify this association as Play, and attach the fact type Play has SkillLevel. A similar move can be made in UML if the play feature has been modeled as an association. In Fig. 9.8(a), however, this feature is modeled as the sports attribute, which needs to be replaced by the equivalent association before we can add the new details about skill level. We discuss objectified relationship types and association classes later.

Another problem with multivalued attributes is that queries on them need to extract the components, and hence complicate the query process for users. As an example, compare queries Q1, Q2 expressed in ConQuer (an ORM query language) with their counterparts in object query language (OQL, proposed by the ODMG). Although this example is trivial, the use of multivalued attributes in complex structures can make it harder for users to express their requirements.

(Q1) **List each** Color **that** is of Flag 'USA'.
(Q2) **List each** Flag **that** has Color 'red'.

(Q1a) **select** x.colors **from** x **in** Flag **where** x.country = "USA"
(Q2a) **select** x.country **from** x **in** Flag **where** "red" **in** x.colors

For such reasons, multivalued attributes should normally be avoided in analysis models, especially if the attributes are based on classes rather than data types. If we avoid multivalued attributes in our conceptual model, we can still use them in the actual implementation. Some UML and ORM tools allow schemas to be annotated with instructions to override the default actions of whatever mapper is used to transform the schema to an implementation. For example, the ORM schema in Fig. 9.9 might be prepared for mapping by annotating the roles played by NickName and Color to map as sets inside the mapped Flag structure. Such annotations are not a conceptual issue, and can be postponed until mapping.

9.4 Associations

UML uses Boolean attributes instead of unary relationships, but allows relationships of all other arities. Optionally, each association may be given at most one name. Association names normally start with a capital letter. *Binary associations are depicted as lines* between classes. Association lines may include elbows to assist with layout or when needed, for example, for ring relationships. Association roles appear as line ends instead of boxes, but may be given role names. Once added, role names may not be suppressed. Verbalization into sentences is possible only for infix binaries, and then only by naming the association with a predicate reading (e.g., "Employs") and using an optional marker "►" to denote the direction.

Fig. 9.10 depicts two binary associations in UML and ORM. On the UML diagram, the association names, their directional markers, and some role names are displayed. In UML, association names are optional, but role names are mandatory. If a role name is not supplied, the role's name is assumed to be the name of its class (e.g., "employee"). *If two or more roles are played by the same class, the roles must be given different names to distinguish them* (e.g., "buyer", "acquisition"). In the ORM diagram, forward and inverse predicate readings are shown; at most one of these may be omitted. Role names are optional in ORM, and their display (in square brackets) may be toggled on or off.

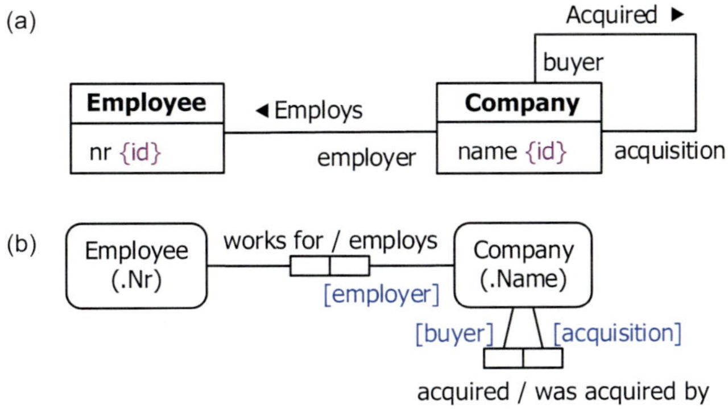

FIG. 9.10 Binary associations in (a) UML and (b) ORM.

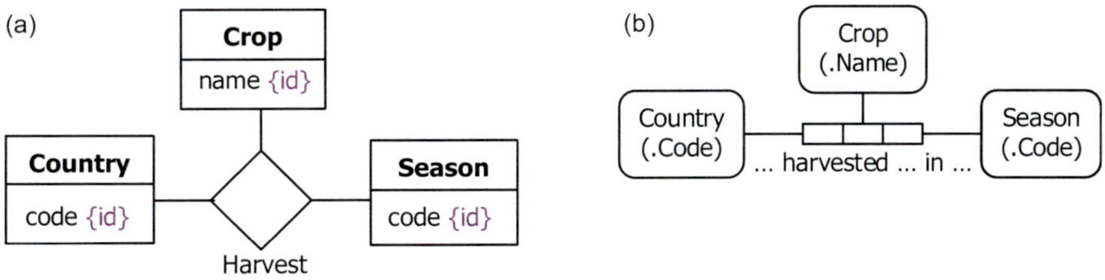

FIG. 9.11 A ternary association in (a) UML and (b) ORM.

Ternary and higher arity associations in UML are depicted as a diamond connected by lines to the classes, as shown in Fig. 9.11(a). Because many lines are used with no reading direction indicator, directional verbalization is ruled out, so the diagram can't be used to communicate in terms of sentences. This nonlinear layout also often makes it impractical to conveniently populate associations with multiple instances, unless we use role names for column names. Add to this the impracticality of displaying multiple populations of attributes, and it is clear that class diagrams are of little use for population checks.

UML does provide object diagrams for instantiation, but these are convenient only for populating associations with a *single* instance. Adding multiple instances leads to a mess (e.g., Blaha and Premerlani, 1998, p. 31). Hence, as noted in the UML Notation Guide, "the use of object diagrams is fairly limited".

The previous section discussed how UML depicts *multiplicity constraints* on attributes. A similar notation is used for associations, where the relevant multiplicities are written next to the relevant roles. Fig. 9.12(a) adds the relevant multiplicity constraints to Fig. 9.10(a). A "*" abbreviates "0..*", meaning "zero or more", "1" abbreviates "1..1", meaning "exactly one", and "0..1" means "at most one". If no multiplicity is supplied for an association role, "*" is assumed by default (unlike attributes, where 1 is the default multiplicity).

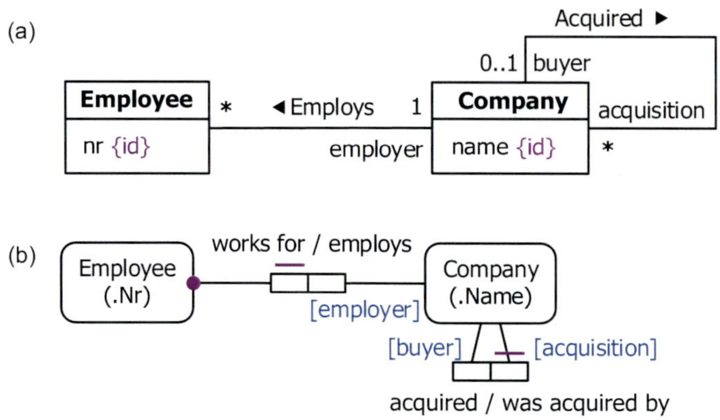

FIG. 9.12 UML multiplicity constraints (a) and equivalent ORM constraints (b).

UML places each multiplicity constraint on the "far role", in the direction in which the association is read. Hence, the constraints in this example mean that each company employs zero or more employees, each employee is employed by exactly one company, each company acquired zero or more companies, and each company was acquired by at most one company.

The corresponding ORM constraints are depicted in Fig. 9.12(b). Recall that multiplicity covers both cardinality (frequency) and optionality. Here, the mandatory role constraint indicates that each employee works for at least one company, and the uniqueness constraints indicate that each employee works for at most one company, and each company was acquired by at most one company.

For comparison purposes, Fig. 9.13 depicts the *n*:1 association Moon orbits Planet in various notations. The instance diagram in Fig. 9.13(a) includes a sample population of moons (p=Phobos, d=Deimos, c=Callisto) and planets (v=Venus, m=Mars, j=Jupiter). For illustration purposes, the ORM diagram in Fig. 9.13(e) also includes the sample object and fact populations. The population is significant with respect to multiplicity constraints. Each planet orbits exactly one moon, and the same planet may be orbited by zero or more moons.

The UML (Fig. 9.13(b)) and information engineering (Fig. 9.13(c)) approaches are similar because both express the constraints in terms of multiplicities/cardinalities. In contrast, Barker ER (Fig. 9.13(d)) and ORM (Fig. 9.13(e)) capture some constraints in terms of mandatory/optional roles and other constraints in terms of cardinality/uniqueness constraints. As shown later, the failure of UML to separate out these two kinds of constraint prevents it from graphically capturing various cases it might otherwise have handled.

For binary associations, there are four possible uniqueness constraint patterns (*n*:1, 1:*n*; 1:1, *m*:*n*) and four possible mandatory role patterns (only the left role mandatory, only the right role mandatory, both roles mandatory, both roles optional). Hence, if we restrict ourselves to a maximum frequency of one, there are 16 possible multiplicity combinations for binary associations. The 16 cases are shown in Fig. 9.14, in both UML and ORM.

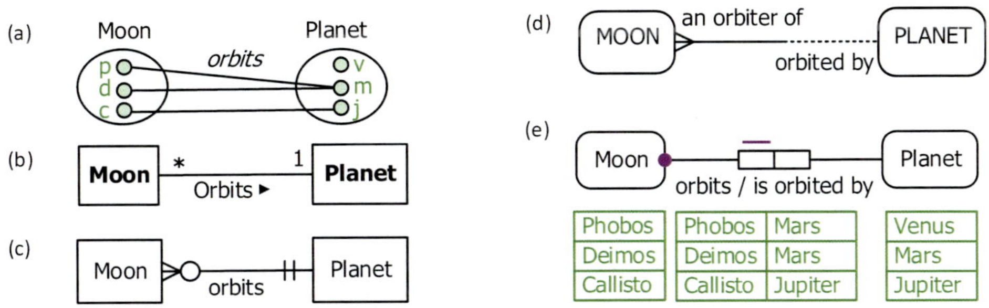

FIG. 9.13 A mandatory:optional, *n*:1 association in various notations.

UML allows multiplicity constraints with whole numbers other than zero or one, and also supports multiplicity lists or ranges (e.g., "1..7, 10"). For such cases, ORM uses frequency constraints instead of uniqueness constraints. ORM is more expressive in this regard because it can apply such constraints to arbitrary collections of roles, not just single roles.

For an elementary n-ary association, each internal uniqueness constraint in ORM must span at least $n-1$ roles. In UML, a multiplicity constraint on a role of an n-ary association effectively constrains the population of the other roles combined. For example, Fig. 9.15(a) is a UML diagram for a ternary association in which both Room-HourSlot and HourSlot-Activity pairs are unique. For simplicity, reference schemes are omitted.

An ORM depiction of the same association is shown in Fig. 9.15(b). The left-hand uniqueness constraint indicates that Room-HourSlot is unique (i.e., for any given room and hour slot, at most one activity is booked). The right-hand uniqueness constraint indicates that HourSlot-Activity is unique (i.e., for any given hour slot and activity, at most one room is booked). An extended version of this example was discussed in Section 1.2, where the ORM diagram better facilitated constraint checking by verbalization and population.

As it covers some n-ary cases like this, UML's multiplicity constraint notation is richer than the optionality/cardinality notation of typical ER. However, there are many cases with n-ary associations where the multiplicity notation of UML can't capture even a simple mandatory role constraint, or a minimum frequency constraint above 1. In contrast, the mandatory, uniqueness, and frequency constraint notations of ORM can capture any possible constraint of this nature, on roles or role sequences, on predicates of any arity. So ORM is far richer in this regard.

For example, suppose we modify our room booking example to indicate that all activities have a room-hourslot booking and also have unique names as well as their identifying codes. The modified example, including reference schemes, is shown in Fig. 9.16 in both UML and ORM. Because UML bundles both mandatory and uniqueness into a single notion of multiplicity, it cannot capture the constraint that each activity has a booking graphically. The best we can do is add a note, as shown in Fig. 9.16(a). This constraint may be expressed graphically in ORM using a mandatory role constraint, as shown in Fig. 9.16(b).

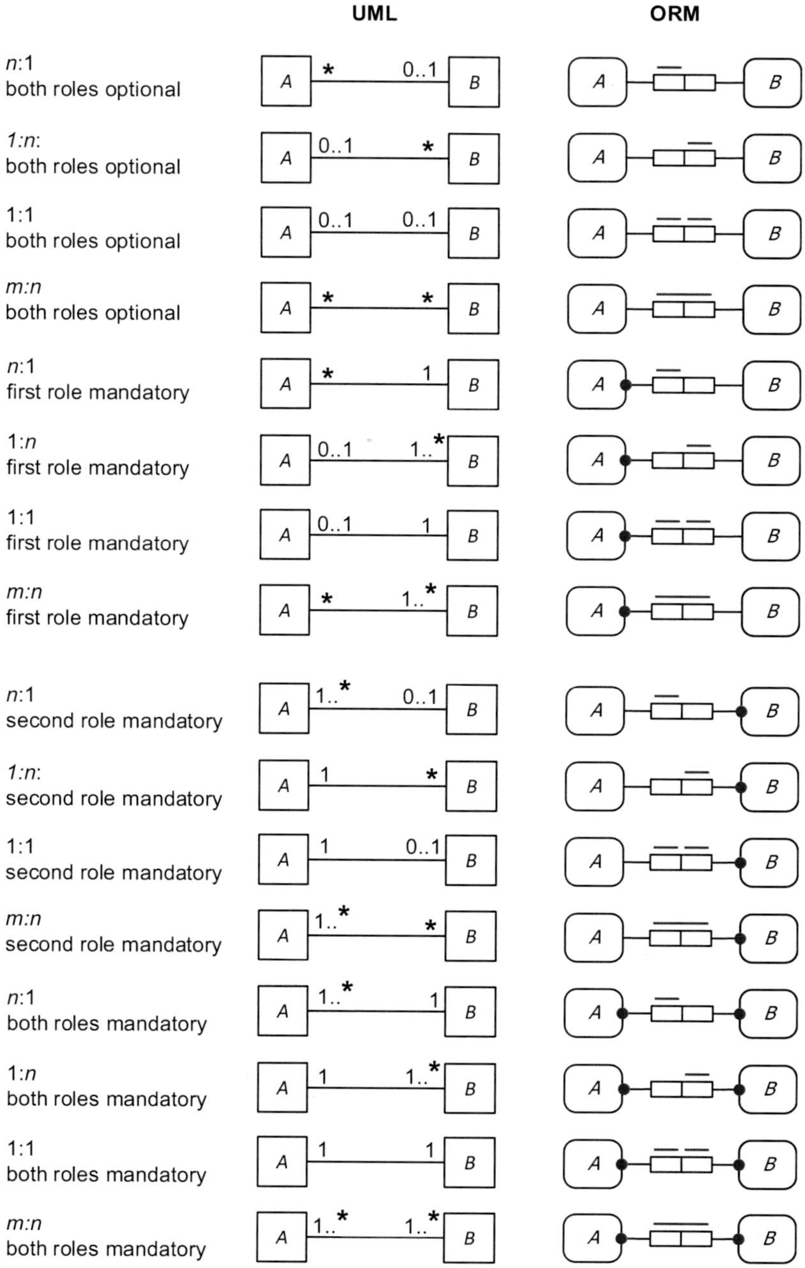

FIG. 9.14 Equivalent constraint patterns in UML and ORM.

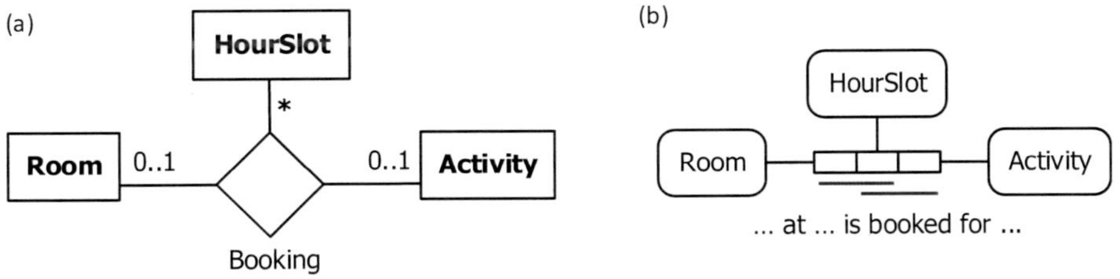

FIG. 9.15 Constraints on a ternary in (a) UML and (b) ORM.

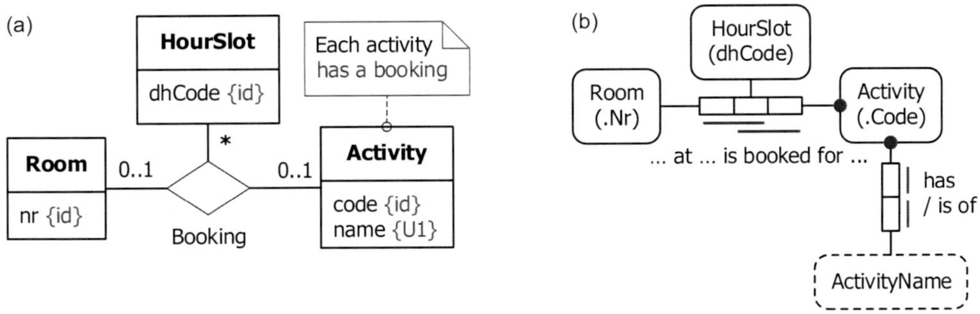

FIG. 9.16 (a) UML resorts to a note to capture (b) a mandatory constraint in ORM.

This deficiency in UML is a direct consequence of choosing to attach minimum multiplicity to a role other than the immediate role. For the same reason, UML multiplicity constraints are also unable to capture various ORM frequency constraints. In general, *given any n-ary (n > 2) association, if an ORM mandatory or frequency constraint applies to at least 1 and at most n − 2 roles, this cannot be captured by a UML multiplicity constraint.* Some examples of such cases are shown in Fig. 9.17. Further discussion on such cases may be found in Halpin (2001).

Unlike many ER versions, both UML and ORM allow associations to be objectified as first class object types, *called association classes* in UML and *objectified associations* or *nested object types* in ORM. UML requires the same name to be used for the original association and the association class, impeding natural verbalization of at least one of these constructs. In contrast, ORM nesting is based on linguistic *nominalization* (a verb phrase is objectified by a noun phrase), thus allowing both to be verbalized naturally, with different readings for each.

Although UML identifies an association class with its underlying association, it displays them separately, connected by a dashed line (see Fig. 9.18a). Each person may write many papers, and each paper is written by at least one person. Because authorship is *m:n*, the association class Writing has a primary reference scheme based on the combination of person and paper (e.g., the writing by person "Norma Jones" of paper 33). The optional period attribute stores how long that person took to write that paper.

Fig. 9.18(b) shows an ORM schema for this domain. The objectified association Writing is marked independent (by the "!") to indicate that a writing object may exist, independently of whether we record its period. ORM displays Period as an object type, not an attribute, and includes its unit.

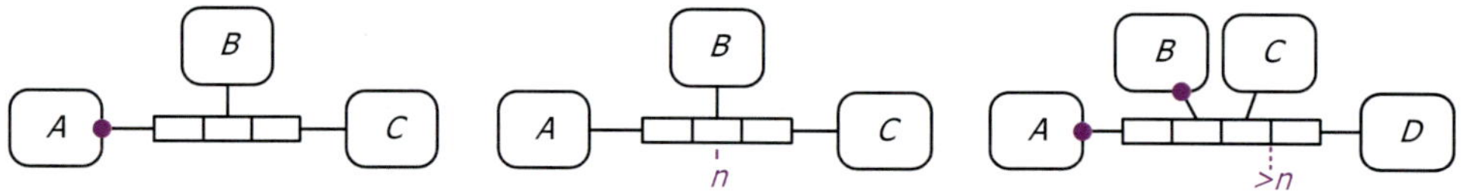

FIG. 9.17 Some ORM constraints that can't be captured by UML multiplicities.

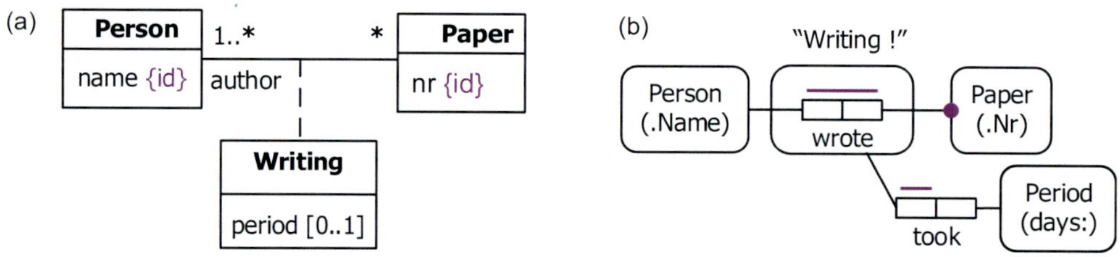

FIG. 9.18 Writing depicted as an objectified association in (a) UML and (b) ORM.

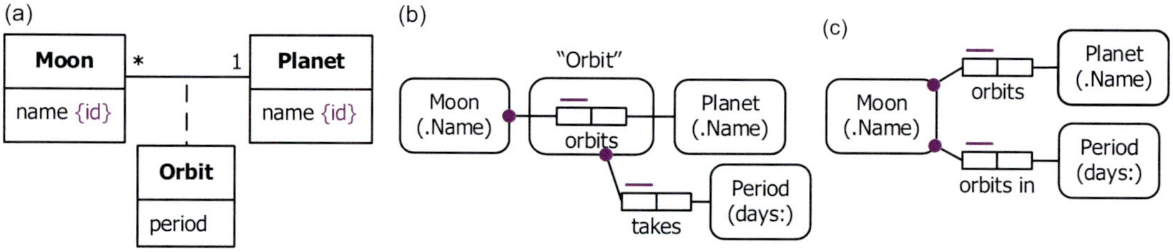

FIG. 9.19 Objectification of an *n*:1 association in (a) UML and (b) ORM.

UML allows any association (binary and above) to be objectified into a class, regardless of its multiplicity constraints. In particular, UML allows objectification of *n*:1 associations, as shown in Fig. 9.19(a). While this is allowed in ORM 2 (see Fig. 9.19(b), it is often a case of poor modeling. For example, given that a moon orbits only one planet, an orbital period may be related directly to the moon. Unless there are other things to record about Orbit (e.g., eccentricity), we could instead model the orbital period in UML as an attribute of Moon, or in ORM as an association between Moon and Period (as in Fig. 9.19c). Chapter 10 provides heuristic guidelines for applying objectification in modeling.

UML has no graphic notation to capture ORM external uniqueness constraints across roles that are all modeled as attributes in UML. However, ORM external uniqueness constraints used for *coreferencing* can be modeled in UML. Fig. 9.20(a) shows an ORM model where the Account object type is coreferenced by combining its bank with its (local) account number.

One way to model this in UML is to use a *qualified association*. Here, instead of depicting the relevant ORM roles or object types as attributes, UML uses a class, adjacent to a *qualifier*, through which connection is made to the relevant association role. A qualifier in UML is a set of one or more attributes whose values can be used to partition the class, and is depicted as a rectangular box enclosing its attributes.

Fig. 9.20(b) is based on an example from the official UML specification. Here, each bank account is used by at most one client, and each client may use many accounts. The attribute accountNr is a qualifier on the association, effectively partitioning each bank into different accounts.

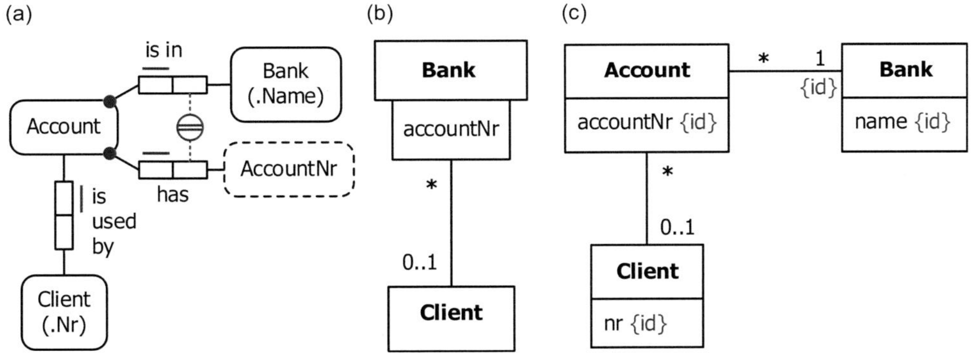

FIG. 9.20 (a) Qualified association in UML and (b) coreferenced type in ORM.

The qualified association approach in Fig. 9.20(b) is less adaptable. For example, if we now want to record something about the account (e.g., its balance) we need to introduce an Account class, and the connection to accountNr is unclear. For a similar example, see Fowler (1997, p. 92), where product is used with Order to qualify an order line association; again, this is unfortunate because we would normally introduce a Product class to record data about products, and relevant connections are then lost.

Fig. 9.20(c) shows a better way to capture coreferencing in UML. Here, both the accountNr attribute and the association end hosted by Bank are marked with "{id}", indicating that these two in combination provide an identifier for Account.

ORM's concept of an external uniqueness constraint that may be applied to a set of roles in one or more predicates provides a simple, uniform way to capture all of UML's qualified associations and unique attribute combinations as well as other cases not expressible in UML graphical notation (e.g., cases with *m:n* predicates or long join paths). As always, the ORM notation has the further advantage of facilitating validation through verbalization and multiple instantiation.

9.5 Set-Comparison Constraints

Set-comparison constraints declare a subset, equality, or exclusion relationship between the populations of role sequences. This section compares support for these constraints in UML and ORM. A detailed discussion of these constraints in ORM can be found in Section 6.4.

As an extension mechanism, UML allows subset constraints to be specified between *whole associations* by attaching the constraint label "{subset}" next to a dashed arrow between the associations. For example, the subset constraint in Fig. 9.21(a) indicates that any person who chairs a committee must be a member of that committee. Fig. 9.21(b) shows the same example in ORM.

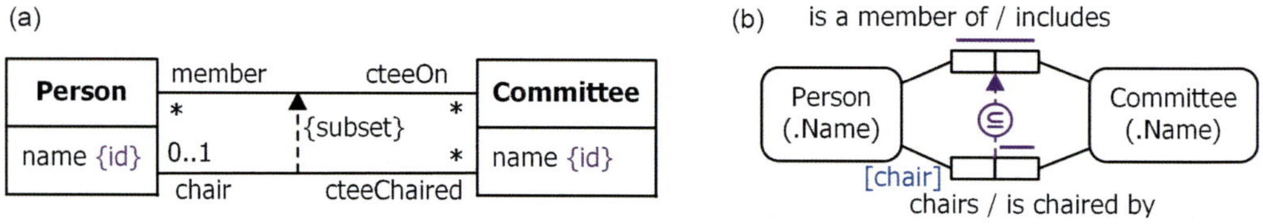

FIG. 9.21 A subset constraint in (a) UML and (b) ORM.

ORM has a mature formalization, including a rigorous theory of schema consistency, equivalence, and implication. Because formal guidelines for working with UML are somewhat immature, care is needed to avoid logical problems. For example, consider the modified version of our committee example shown in Fig. 9.22(a), which comes from an earlier version of the UML specification, with reference schemes added. Do you spot anything confusing about the constraints?

You probably noticed the problem. The multiplicity constraint of 1 on the chair association indicates that each committee must have at least one chair. The subset constraint tells us that a chair of a committee must also be a member of that committee. Taken together, these constraints imply that each committee must have a member. Hence, we would expect to see a multiplicity constraint of "1..*" (one or more) on the Person end of the membership association. However, we see a constraint of "*" (zero or more) instead, which at best is misleading. An equivalent, misleading ORM schema is shown in Fig. 9.22(b), where the upper role played by Committee appears optional when in fact it is mandatory.

One might argue that it's okay to leave these schemas unchanged, as the constraint that each committee includes at least one person is implied by other constraints. However, while display options for implied constraints may sometimes be a matter of taste, practical experience has shown that in cases like this, it is better to show implied constraints explicitly, as in Fig. 9.23, rather than expect modelers or domain experts to figure them out for themselves.

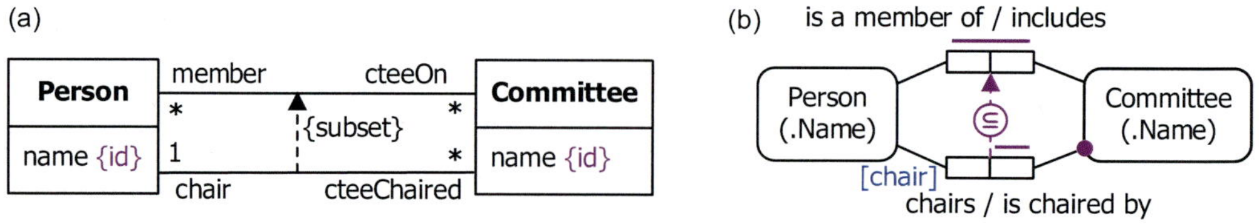

FIG. 9.22 (a) A misleading UML diagram and (b) a misleading ORM diagram.

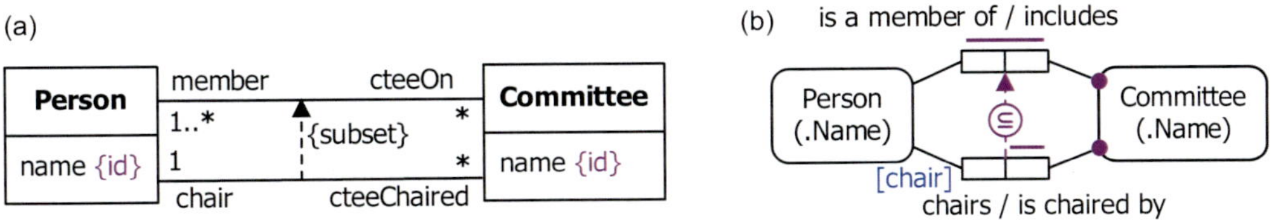

FIG. 9.23 All constraints are now shown explicitly in (a) UML and (b) ORM.

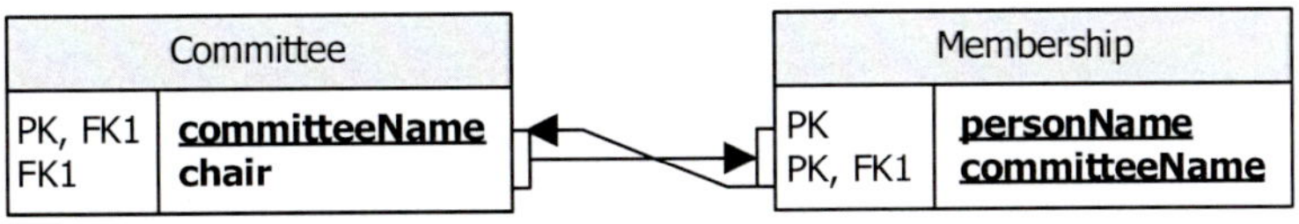

FIG. 9.24 The relational schema mapped from Fig. 9.23.

Some ORM tools can detect the misleading nature of constraint patterns like that of Fig. 9.22(b) and ask you to resolve the problem. Human interaction is the best policy here because there is more than one possible mistake (e.g., is the subset constraint correct leading to Fig. 9.23, or is the optional role correct resulting in Fig. 9.21?).

If a schema in Fig. 9.23 is mapped to a relational database, it generates a referential cycle because the mandatory fact types for Committee map to different tables (so each committee must appear in both tables). The relational schema is shown in Fig. 9.24 (arrows show the foreign key references, one simple and one composite, which correspond to the subset constraints).

Referential cycles are sometimes unavoidable, but are awkward to implement. In this case, the cycle arose because of the pair-subset constraint combined with the mandatory role constraint on is-chaired-by. Given the pair-subset constraint, the cycle would also apply if Committee is independent or plays another mandatory, functional role. Relational mapping is covered in detail in Chapter 11.

Because UML does not allow unary relationships, subset constraints between ORM unaries need to be captured textually, using a note to specify an equivalent constraint between Boolean attributes. For example, the ORM subset constraint in Fig. 9.25(b), which verbalizes as **Each** Patient **who** smokes is cancer prone, may be captured textually in UML by the note in Fig. 9.25(a).

UML 2 introduced a *subsets property* modifier to indicate that for each instance that is the context for a property (an attribute or association role), the population (extension) of that property for that instance must be a subset of the population of another compatible property for that instance. For example, adorning the citizen role in Fig. 9.26(a) with {subset resident} means that each citizen of a country is a resident of the same country. Fig. 9.26(b) shows the equivalent ORM schema.

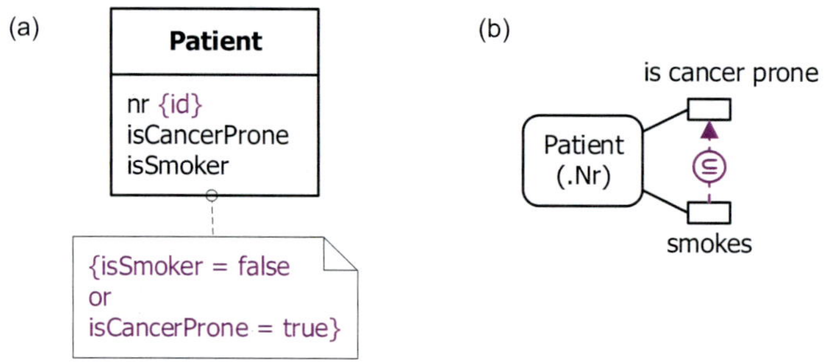

FIG. 9.25 (a) UML note for (b) ORM subset constraint between unaries.

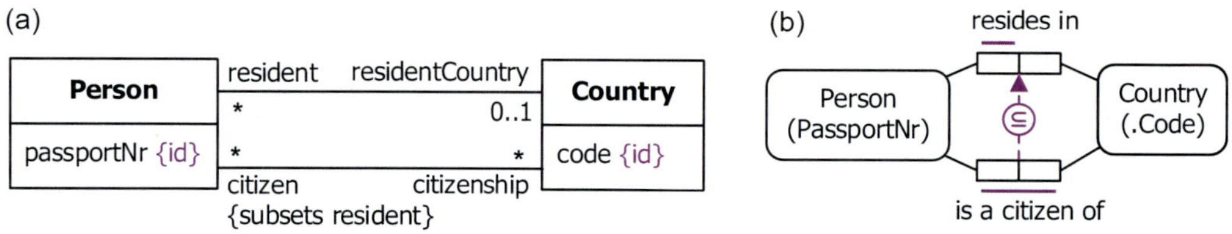

FIG. 9.26 A pair-subset constraint in (a) UML and (b) ORM.

There are still subset constraint cases in ORM that can't be represented graphically as a subset constraint in UML. For example, the subset constraint in Fig. 9.27(b) that each student with a second given name must have a first given name is captured as a note in Fig. 9.27(a) because the relevant ORM fact types are modeled as attributes in UML, and the subset constraint applies between student sets, not name sets. The subset constraint in Fig. 9.25(b) is another example.

Moreover, UML does not support subset constraints over arguments that are just parts of relationships, such as the subset constraint in Fig. 9.27(b) that students may pass tests in a course only if they enrolled in that course. Fig. 9.27(a) models this constraint in UML by transforming the ternary into a binary association class (Enrollment) that has a binary association to TestNr. Although in this situation an association class provides a good way to cater for a compound subset constraint, sometimes this nesting transformation leads to a very unnatural view of the world. Ideally, the modeler should be able to view the world naturally while having any optimization transformations that reduce the clarity of the conceptual schema performed under the covers.

As another constraint example in UML, consider Fig. 9.28, which is the UML version of an OMT diagram used in Blaha and Premerlani (1998, p. 68) to illustrate a subset constraint (if a column is a primary key column of a table, it must belong to that table). Can you spot any problems with the constraints?

One obvious problem is that the "1" on the primary key association should be "0..1" (not all columns belong to primary keys), as in Fig. 9.29(a). If we allow tables to have no columns (e.g., to cater for cases where the table is under construction), then the "*"

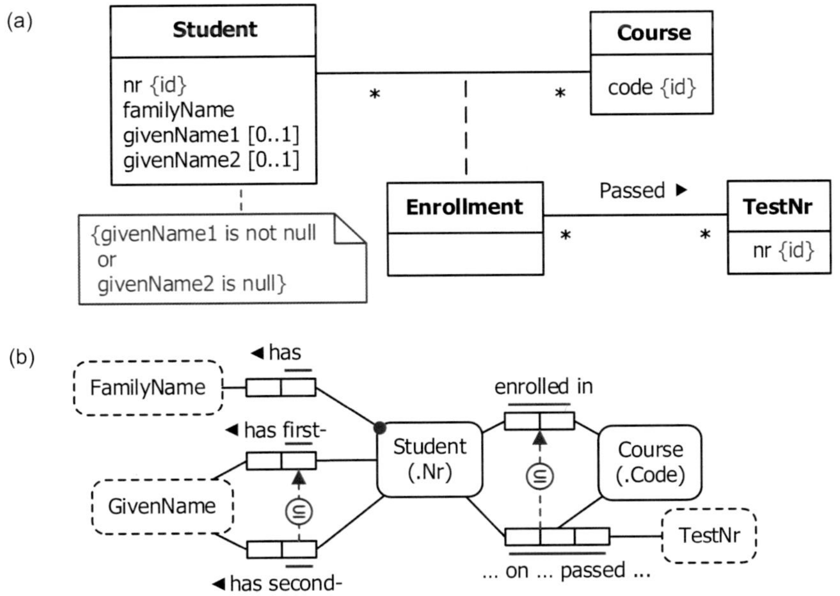

FIG. 9.27 (a) UML model capturing (b) some subset constraints in ORM.

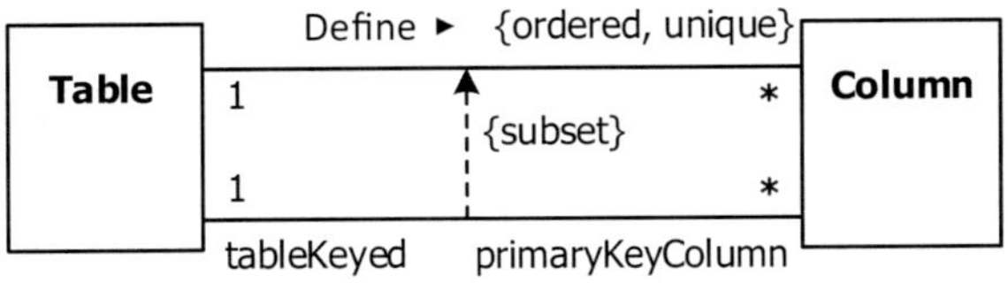

FIG. 9.28 Spot anything wrong?

on the define association is fine; otherwise it should be "1..*". Assuming that tables and columns are identified by oids or artificial identifiers, the subset constraint makes sense, but the model is arguably suboptimal because the primary key association and subset constraint could be replaced by a Boolean isaPKfield attribute on Column.

From an ORM perspective, heuristics lead us to initially model the situation using natural reference schemes, as shown in Fig. 9.29(b). Here, ColumnName denotes the local name of the column in the table. We've simplified reality by assuming that tables may be identified just by their name. The external uniqueness constraints suggest two natural reference schemes for Column: name plus table, or position plus table. We chose the first of these as preferred, but could have introduced an artificial identifier. The unary predicate indicates whether a column is, or is part of, a primary key. If desired, we could derive the association Column is a primary key field of Table from the path Column is in Table **and** Column is a primary key column (the subset constraint in the UML model is then implied).

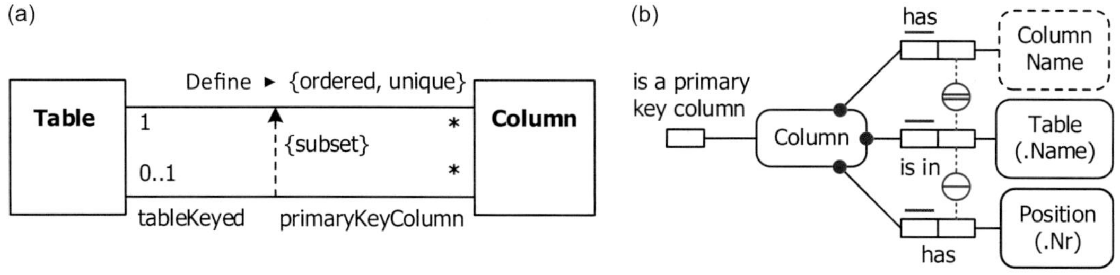

FIG. 9.29 A corrected UML schema (a) remodeled in (b) ORM.

What is interesting about this example is the difference in modeling approaches. Most UML modelers seem to assume that oids will be used as identifiers in their initial modeling, whereas ORM modelers like to expose natural reference schemes right from the start and populate their fact types accordingly. These different approaches often lead to different solutions.

The main thing is to first come up with a solution that is natural and understandable to the domain expert because here is where the most critical phase of model validation should take place. Once a correct model has been determined, optimization guidelines can be used to enhance it.

One other feature of the example is worth mentioning. The UML solution in Fig. 9.29(a) uses the annotation "{ordered, unique}" to indicate that a table is composed of an *ordered set* (i.e., a sequence with no duplicates) of columns. UML 2 allows the *unique property* to be specified with or without the *ordered property.* By default, ordered = false and unique = true. So, either of the settings {ordered} or {ordered, unique} may be used to indicate an ordered set. Either no setting, or the single setting {unique}, indicates a set (the default). If {nonunique} is allowed in this context (this in unclear in the UML specification), one could specify a bag or sequence with the settings {nonunique} or {ordered, nonunique}, respectively.

In the ORM community, a debate has been going on for many years regarding the best way to deal with constructors for collection types (e.g., set, ordered set, bag, sequence) at the conceptual level. Our view is that such constructors should not appear in the base conceptual model. Hence, the use of Position in Fig. 9.29(b) to convey column order (the uniqueness of the order is conveyed by the uniqueness constraint on Column has Position). Keeping fact types elementary has so many advantages (e.g., validation, constraint expression, flexibility, and simplicity) that it seems best to relegate constructors to derived views. We have more to say about collection types in Chapter 10.

In ORM, an *equality constraint* between two compatible role sequences is shorthand for two subset constraints (one in either direction) and is shown as circled "=". Such a constraint indicates that the populations of the role sequences must always be equal. If two roles played by an object type are mandatory, then an equality constraint between them is implied (and hence not shown). UML has no graphic notation for equality constraints. For whole associations, we could use two separate subset constraints, but this would be very messy. In general, equality constraints in UML may be specified as textual constraints in notes.

As a simple example, the equality constraint in Fig. 9.30(b) indicates that if a patient's systolic blood pressure is measured, so is the diastolic blood pressure (and vice versa). In other words, either both measurements are taken, or neither.

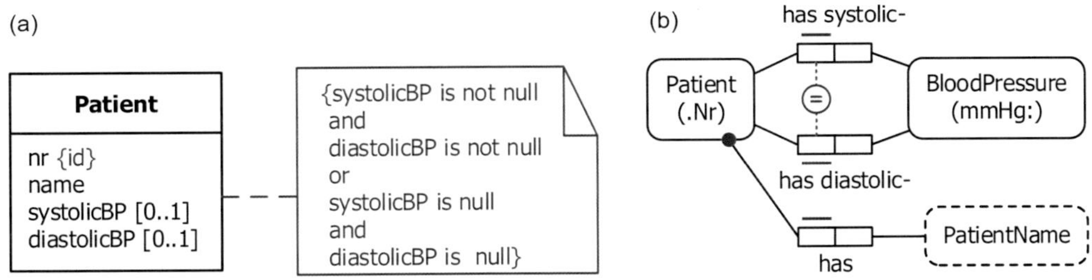

FIG. 9.30 A simple equality constraint in (a) UML and (b) ORM.

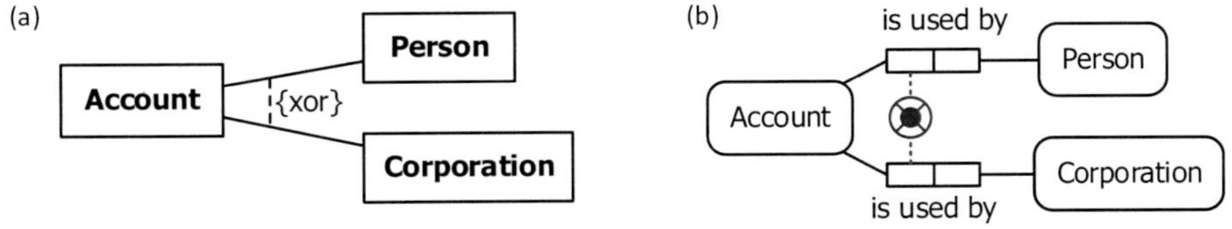

FIG. 9.31 Exclusive-or: each account is used by a person or corporation but not both.

This kind of constraint is fairly common. Less common are equality constraints between sequences of two or more roles. Fig. 9.30(a) models this in UML as a textual constraint between two attributes for blood pressure readings.

Subset and equality constraints enable various classes of schema transformations to be stated in their most general form, and ORM's more general support for these constraints allows more transformations to be easily visualized (e.g., see the equivalence theorem PSG2 in Section 14.2).

Although UML does not include a graphic notation for a pure exclusion constraint, it does include an *exclusive-or constraint* to indicate that each instance of a class plays *exactly one* association role from a specified set of alternatives. To indicate the constraint, "{xor}" is placed beside a dashed line connecting the relevant associations. Fig. 9.31(a), which is based on an example from the UML specification, indicates that each account is used by a person or corporation but not both. For simplicity, reference schemes and other constraints are omitted.

Prior to version 1.3 of UML, "{or}" was used for this constraint, which was misleading because "or" is typically interpreted in the inclusive sense. The equivalent ORM model is shown in Fig. 9.31(b), where the exclusive-or constraint is simply an orthogonal combination of a disjunctive mandatory role (inclusive-or) constraint (circled dot) and an exclusion constraint (circled "X").

Although the current UML specification describes the xor constraint as applying to a set of associations, we need to apply the constraint to a set of roles (association-ends) to avoid ambiguity in cases with multiple common classes. Visually, this could be shown by attaching the dashed line near the relevant ends of the associations, as shown in Fig. 9.32(a). Unfortunately, UML attaches no significance to such positioning, so the xor constraint could be misinterpreted to mean that each company must lease or purchase some vehicle rather than the intended constraint that each vehicle is either leased or purchased, a constraint captured unambiguously by the ORM schema in Fig. 9.32(b).

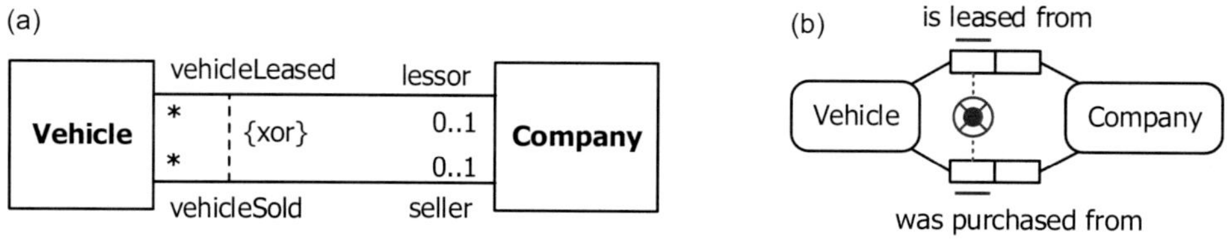

FIG. 9.32 The exclusive-or constraint should apply between association roles.

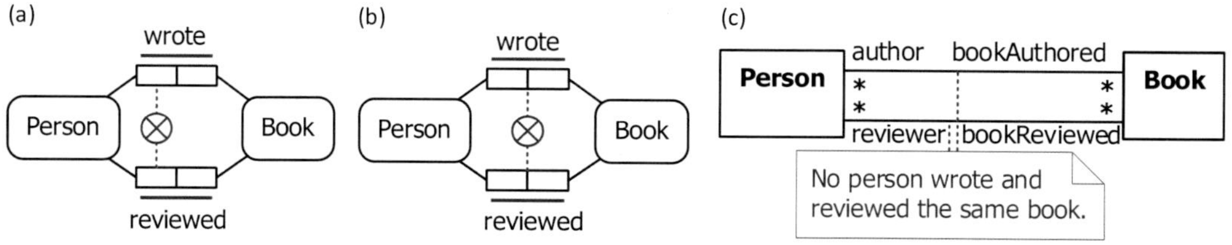

FIG. 9.33 (a) Nobody wrote and reviewed a book; (b) nobody wrote and reviewed *the same* book; (c) UML version of (b).

UML has no symbols for exclusion or inclusive-or constraints. If UML symbols for these constraints are ever considered, then "{x}" and "{or}" respectively seem appropriate–this choice also exposes the composite nature of "{xor}". UML xor-constraints are intended to apply between single roles. The UML specification seems to imply that these roles must belong to different associations. If so, UML cannot use an xor constraint between roles of a ring fact type (e.g., between the husband and the wife roles of a marriage association). ORM exclusion constraints cover this case as well as many other cases not expressible in UML graphic notation. As a trivial example, consider the difference between the following two constraints: no person both wrote a book and reviewed a book, and no person wrote and reviewed the same book. ORM clearly distinguishes these by noting the precise arguments of the constraint (compare Fig. 9.33(a) with Fig. 9.33(b)).

The pair-exclusion constraint in Fig. 9.33(b) can be expressed in UML by a note connected by dotted lines to the two associations, as shown in Fig. 9.33(c). Alternatively, one could attach a textual constraint to the Person class (e.g., "bookAuthored and bookReviewed are disjoint sets") or the Book class (e.g., "author and reviewer are disjoint sets"), but the choice of class is arbitrary.

UML has no graphic notation for exclusion between attributes, or between attributes and association roles. An exclusion constraint in such cases may often be captured as a textual constraint. For example, in Fig. 9.34(a), the exclusion constraint that each employee is either tenured or is contracted until some date may be captured by the textual constraint shown.

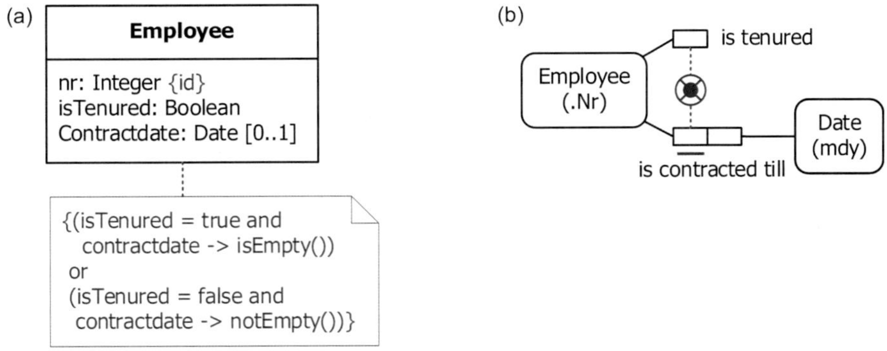

FIG. 9.34 An exclusion constraint modeled in (a) UML and (b) ORM.

Here, the constraint is specified in OCL. The expressions "->isEmpty()" and "->notEmpty()" are equivalent to "is null" and "is not null" in SQL. Fig. 9.34(b) depicts the exclusion constraint graphically in ORM. There are other ways to model this case in UML (e.g., using subtypes) that offer more chances to capture the constraints graphically.

9.6 Subtyping

Both UML and ORM support *subtyping* using "is-a" semantics, where each instance of a subtype is also an instance of its supertype(s). For example, declaring Woman to be a subtype of Person entails that each woman is a person, and hence Woman inherits all the properties of Person. Given two object types, *A* and *B*, *A* is a *subtype* of *B* if, for each database state, the population of *A* is included in the population of *B*. For data modeling, the only subtypes of interest are *proper* subtypes. We say that *A* is a proper subtype of *B* if and only if *A* is a subtype of *B*, and there is a possible state where the population of *B* includes an instance not in *A*. From now on, we'll use "subtype" as shorthand for "proper subtype".

In both UML and ORM, *specialization* is the process of introducing subtypes, and *generalization* is the inverse procedure of introducing a supertype. Both UML and ORM allow single *inheritance* as well as multiple inheritance (where a subtype has more than one direct supertype). For example, AsianWoman may be a subtype of both AsianPerson and Woman. In UML, "subclass" and "superclass" are synonyms of "subtype" and "supertype", respectively, and generalization may also be applied to things other than classes (e.g., interfaces, use case actors, and packages). This section focuses on subtyping between object types (classes).

In ORM, a subtype inherits all the roles of its supertypes. In UML, a subclass inherits all the attributes, associations, and operations/methods of its supertype(s). An operation implements a service and has a signature (name and formal parameters) and visibility, but may be realized in different ways. A method is an implementation of an operation, and includes both a signature and a body detailing an executable algorithm to perform the operation. In an inheritance graph, there may be many methods for the same operation (*polymorphism*), and scoping rules determine which method is used for a given class. If a subclass has a method with the same signature as a method of one of its supertypes, this is used instead for that subclass (*overriding*). For example, if Rectangle and Triangle are subclasses of Shape, all three classes may have different methods for display. This section focuses on data modeling, covering inheritance of static properties (attributes and associations), but ignoring inheritance of operations or methods.

Subtypes are used in data modeling to assert typing constraints, encourage reuse of model components, and show a classification scheme (taxonomy). In this context, typing constraints ensure that subtype-specific roles are played only by the relevant subtype.

Because a subtype inherits the properties of its supertype(s), only its specific roles need to be declared when it is introduced. Apart from reducing code duplication, the more generic supertypes are likely to find reuse in other applications. At the coding level, inheritance of operations/methods augments the reuse gained by inheritance of attributes and association roles. Using subtypes to show taxonomy is of limited use because taxonomy is often more efficiently captured by predicates. For example, the fact type Person is of Gender {male, female} implicitly provides the taxonomy for the subtypes MalePerson and FemalePerson.

Both UML and ORM display subtyping using *directed acyclic graphs*. A directed graph is a graph of nodes with directed connections, and "acyclic" means that there are no cycles (a consequence of proper subtyping). Fig. 9.35 shows a subtype pattern in UML and ORM. An arrow from one node to another shows that the first is a subtype of the second. UML uses a thin arrow shaft with an open arrowhead while ORM uses a solid shaft and arrowhead. UML also allows separate shafts to merge into one, with one arrowhead acting for all (e.g., *E* and *F* are subtypes of *C*). Because subtypehood is transitive, indirect connections are omitted (e.g., because *E* is a subtype of *C*, and *C* is a subtype of *A*, it follows that *E* is a subtype of *A*, so there is no need to display this implied connection).

UML includes four predefined constraints to indicate whether subtypes are exclusive or exhaustive. If subtype connections are shown with separate arrowheads, the constraints are placed in braces next to a dotted line connecting the subtype links, as in Fig. 9.35(a) (top). We assume that this line may include elbows, as shown for the disjoint

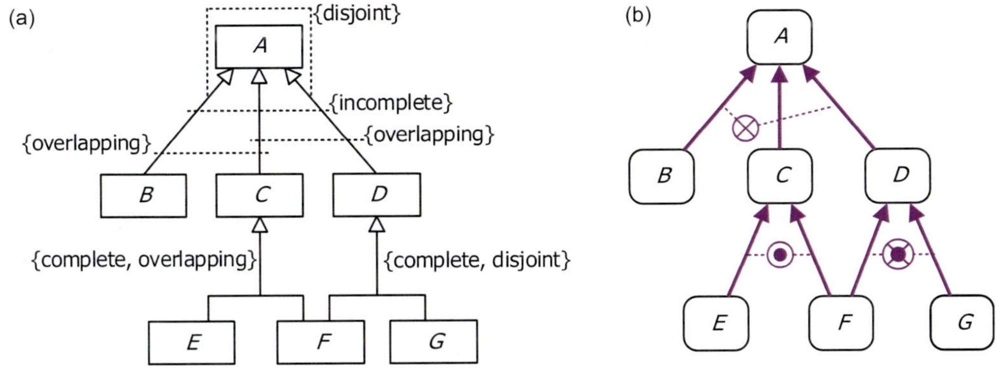

FIG. 9.35 Subtyping displayed by directed acyclic graphs in (a) UML and (b) ORM.

constraint, to enable such cases to be specified. If the subtype connections are shared, the constraints are placed near the shared arrowhead, as in Fig. 9.35(a) (bottom). The {overlapping} and {disjoint} options, respectively, indicate that the subtypes overlap or are mutually exclusive. Originally {complete} simply meant that all subtypes were shown, but it now means exhaustive (i.e., the supertype equals the union of its subtypes). The {incomplete} option means that the supertype is more than the union of its subtypes. UML 2.5 changed the default from {disjoint, incomplete} to {overlapping, incomplete}. Users may add other keywords.

By default, ORM subtypes may overlap, and subtypes need not collectively exhaust their supertype. ORM allows graphic constraints to indicate that subtypes are mutually exclusive (a circled "X" connected to the relevant subtype links), collectively exhaustive (a circled dot), or both (a circled, crossed dot), as shown in Fig. 9.35(b). ORM's approach is that exclusion and totality constraints are enforced on populations, not types. An overlapping "constraint" does not mean that the populations must overlap, just that they may overlap. From an ORM viewpoint, this is not really a constraint at all, so there is no need to depict it. In ORM, subtype exclusion and totality constraints are often implied by other constraints in conjunction with formal subtype definitions.

For any subtype graph, the top supertype is called the *root*, and the bottom subtypes (those with no descendants) are called *leaves*. In UML, this can be made explicit by adding "{root}" or "{leaf}" below the relevant class name. If the whole subtype graph is shown, there is little point in this, but if we display only part of a subtype graph, this notation clarifies whether the local top and bottom nodes are like that in the global schema. From Fig. 9.36, we know that globally Party has no supertype and that MalePerson and FemalePerson have no subtypes. Because Organization is not marked as a leaf, it may have other subtypes not shown here.

UML also allows an ellipsis "…" in place of a subclass to indicate that at least one subclass of the parent exists in the global schema, but its display is suppressed on the diagram. Currently, ORM does not include a root/leaf notation or an ellipsis notation for subtypes.

UML distinguishes between *abstract* and *concrete* classes. An abstract class cannot have any direct instances and is shown by writing its name in italics or by adding "{abstract}" below the class name. Abstract classes are realized only through their descendants. Concrete classes may be directly instantiated. This distinction seems to have little relevance at the conceptual level and is not depicted explicitly in ORM. For code design, however, the distinction is important (e.g., abstract classes provide one way of declaring interfaces, and in C++, abstract operations correspond to pure virtual operations while leaf operations map to nonvirtual operations). For further discussion of this topic, see Fowler (1997, pp. 85–88) and Booch et al. (1999, pp. 125–126).

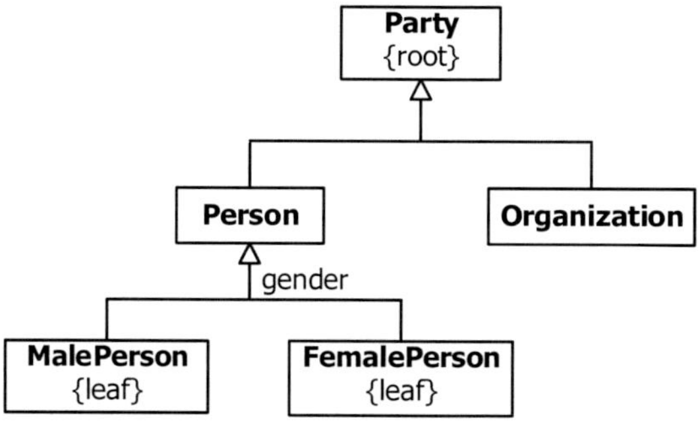

FIG. 9.36 Party may have other subtypes not shown here.

Like other ER notations, UML provides only weak support for defining subtypes. A *generalization set* (or *discriminator*) may be noted near a subtype arrow to indicate the basis for the classification. For example, Fig. 9.37 includes a "gender" discriminator to specialize Person into MalePerson and FemalePerson.

The UML specification says that the discriminator names "a partition of the subtypes of the superclass". In formal work, the term "partition" usually implies that the division is both exclusive and exhaustive. In UML, the use of a discriminator does not imply that the subtypes are exhaustive or complete, but some authors argue that they must be exclusive (Fowler, 1997, p. 782). If that is the case, there seems no way in UML to declare a discriminator for a set of overlapping subtypes.

The same discriminator name may be repeated for multiple subclass arrows to show that each subclass belongs to the same classification scheme. This repetition can be avoided by merging the arrow shafts to end in a single arrowhead, as in Fig. 9.37.

In Fig. 9.37, the gender attribute of Patient is used as a discriminator. This attribute is based on the enumerated type Gendercode, which is defined using the stereotype «enumeration», and listing its values as attributes. The notes at the bottom are needed to ensure that instances populating these subtypes have the correct gender. For example, without these notes there is nothing to stop us populating MalePatient with patients that have the value "f" for their Gendercode.

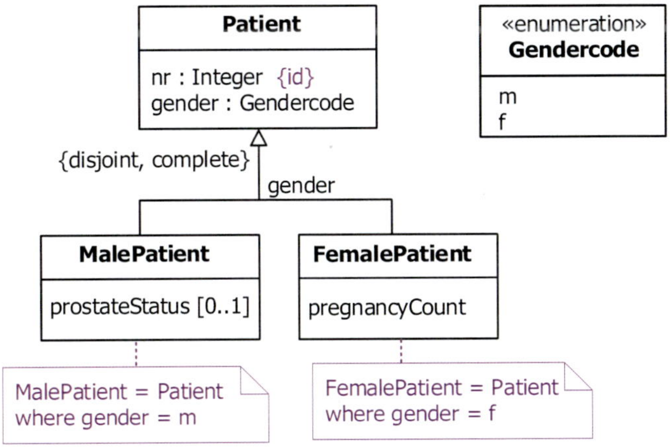

FIG. 9.37 Gender is used as a discriminator to partition patient.

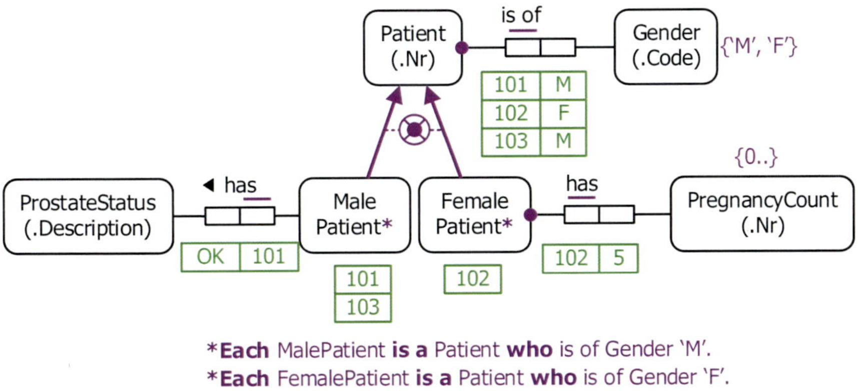

FIG. 9.38 With formal subtype definitions, subtype constraints are implied.

As discussed in Section 6.5, ORM overcomes this problem by requiring that if a taxonomy is captured both by subtyping and a classifying fact type, these two representations must be synchronized, either by deriving the subtypes from *formal subtype definitions* or by deriving the classification fact type from asserted subtypes. For example, the populated ORM schema in Fig. 9.38 adopts the first approach. The ORM partition (exclusion and totality) constraint is now implied by the combination of the subtype definitions and the three constraints on the fact type Patient is of Gender.

While the subtype definitions in Fig. 9.38 are trivial, in practice more complicated subtype definitions are sometimes required. Consider a schema with the fact types City is in Country, City has Population, and define LargeUScity as follows: **Each** LargeUScity **is a** City **that** is in Country 'US' **and** has Population > 1000000. There does not seem to be any convenient way of doing this in UML, at least not with discriminators. We could perhaps add a derived

Boolean isLarge attribute, with a derivation rule in OCL, and then add a final subtype definition in OCL, but this would be less readable than the ORM definition just given. For a more detailed discussion of subtyping in ORM, including the notion of context-dependent reference, see Sections 6.5 and 6.6. Mapping of subtypes is discussed in Chapter 11.

9.7 Other Constraints and Derivation Rules

A *value constraint* restricts the population of a value type to a finite set of values specified either in full (*enumeration*), by start and end values (*range*), or some combination of both (*mixture*). The values themselves are primitive data values, typically character strings or numbers.

In UML, enumeration types may be modeled as classes, stereotyped as enumerations, with their values listed (somewhat unintuitively) as attributes. Ranges and mixtures may be specified by declaring a textual constraint in braces, using any formal or informal language. For example, see Fig. 9.39(a).

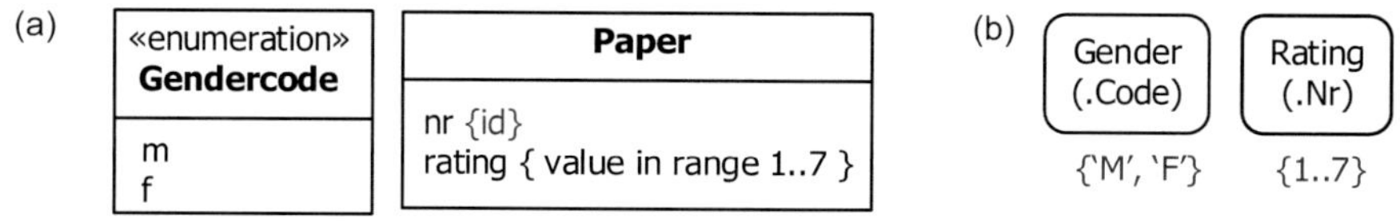

FIG. 9.39 Data value restrictions declared as enumerations or textual constraints.

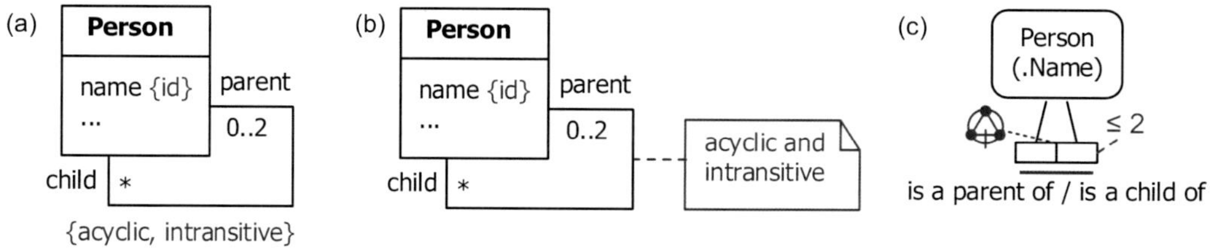

FIG. 9.40 Ring constraints expressed in (a) UML, (b) UML, and (c) ORM.

Fig. 9.39(b) depicts the same value constraints in ORM. Value constraints other than enumeration, range, and mixture may be declared in UML or ORM as textual constraints, for example, {committeeSize must be an odd number}. For further UML examples, see Rumbaugh et al. (1999, pp. 236, 268).

A ring fact type has at least two roles played by the same object type (either directly or indirectly via a supertype). A *ring constraint* applies a logical restriction on the role pair. For example, the association Person is a parent of Person might be declared acyclic and intransitive.

UML does not provide ring constraints built in, so the modeler needs to specify these as a textual constraint in some chosen language or as a note. In UML, if a textual constraint applies to just one model element (e.g., an association), it may be added in braces next to that element, as in Fig. 9.40(a). Here, the {acyclic, intransitive} notation is nonstandard but is assumed to be user supported.

It is the responsibility of the modeling tool to ensure that the constraint is linked internally to the relevant model element and to interpret any textual constraint expressions. If the tool cannot interpret the constraint, it should be placed inside a note (dog-eared rectangle), without braces, showing that it is merely a comment, and explicitly linked to the relevant model element(s), as shown in Fig. 9.40(b). Fig. 9.40(c) displays the ring constraints graphically in ORM (see Section 7.3 for a detailed discussion of ring constraints in ORM).

A *join constraint* applies to one or more role sequences, at least one of which is projected from a path from one predicate through an object type to another predicate. The act of passing from one role through an object type to another role invokes a conceptual join because the same object instance is asserted to play both the roles. Although join constraints arise frequently in real applications, UML has no graphic symbol for them. To declare them on a UML diagram, write a constraint or comment in a note attached to the model elements involved.

For example, Fig. 9.41 links a comment to three associations. This example is based on a room scheduling application at a university with built-in facilities in various lecture and tutorial rooms. Example facility codes are PA = Personal Address system, DP = Data Projection facility, INT = Internet access.

As discussed in Section 10.1, ORM provides deep support for join constraints. Role sequences featuring as arguments in set comparison constraints may arise from projections over a join path.

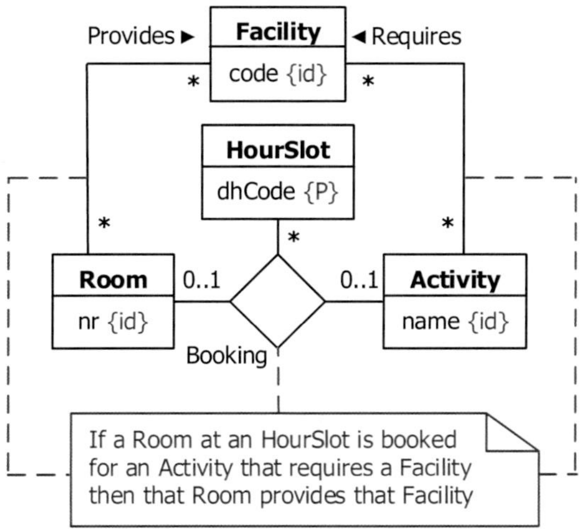

FIG. 9.41 Join constraint specified as a comment in UML.

For example, in Fig. 9.42, the subset constraint runs from the Room-Facility role-pair projected from the path Room at **an**HourSlot is booked for **an** Activity **that** requires **a** Facility. This path includes a conceptual join on Activity. The constraint may be formally verbalized as **If a** Room at **an**HourSlot is booked for **an** Activity **that** requires **a** Facility **then that** Room provides **that** Facility. Fig. 9.42 includes a satisfying population for the three fact types. This again illustrates how ORM facilitates validation constraints via sample populations. The UML associations in Fig. 9.41 are not so easily populated on the diagram. Other join constraint examples are discussed in Section 10.1.

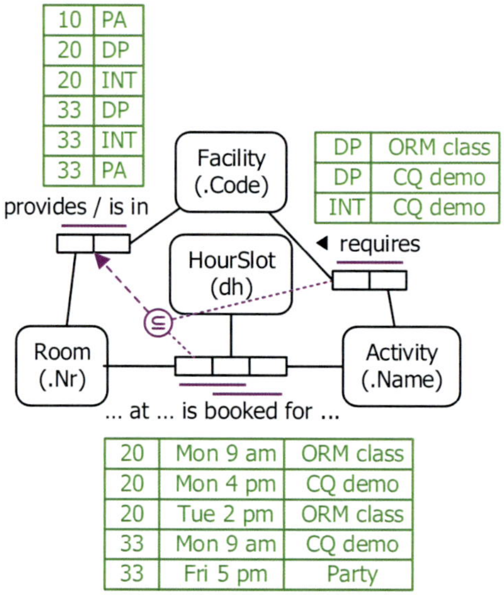

FIG. 9.42 A join-subset constraint in ORM.

In UML, the term "*aggregation*" is used to describe a *whole/part relationship*. For example, a team of people is an aggregate of its members, so this membership may be modeled as an aggregation association between Team and Person. Several different forms of aggregation might be distinguished in real-world cases. For example, Odell and Bock (Odell, 1998, pp. 137–165) discuss six varieties of aggregation (component-integral, material-object, portion-object, place-area, member-bunch, and member-partnership), and Henderson-Sellers (Barbier et al., 2000) also distinguishes several kinds of aggregation.

UML 2 associations are classified into three kinds: ordinary association (no aggregation), shared (or simple) aggregation, or composite (or strong) aggregation. Hence, UML recognizes only two varieties of aggregation: shared and composite. Some versions of ER include an aggregation symbol (typically only one kind). ORM and popular ER approaches currently include no special symbols for aggregation.

These different stances with respect to aggregation are somewhat reminiscent of the different modeling positions with respect to nulls. Although more than 20 kinds of null have been distinguished in the literature, the relational model recognizes only one kind of null. Codd's version 2 of the relational model includes two kinds of null, and ORM argues that nulls have no place in base conceptual models (because all its asserted facts are atomic). But let's return to the topic at hand.

Shared aggregation is denoted in UML as a binary association with a *hollow diamond* at the "whole" or "aggregate" end of the association. *Composition* (*composite aggregation*) is depicted with a *filled diamond*. For example, Fig. 9.43(a) depicts a composition association from Club to Team and a shared aggregation association from Team to Person.

In ORM, which currently has no special notation for aggregation, this situation would be modeled as shown in Fig. 9.43(b). Does Fig. 9.43(a) convey any extra semantics not captured in Fig. 9.43(b)? At the conceptual level, it is doubtful whether there is any additional useful semantics. At the implementation level, however, there is additional semantics. Let's discuss this in more detail.

The UML specification declares that "both kinds of aggregation define a transitive … relationship". The use of "transitive" here is somewhat misleading because it refers to indirect aggregation associations rather than base aggregation associations. For example, if Club is an aggregate of Team, and Team is an aggregate of Person, it follows that Club is an aggregate of Person.

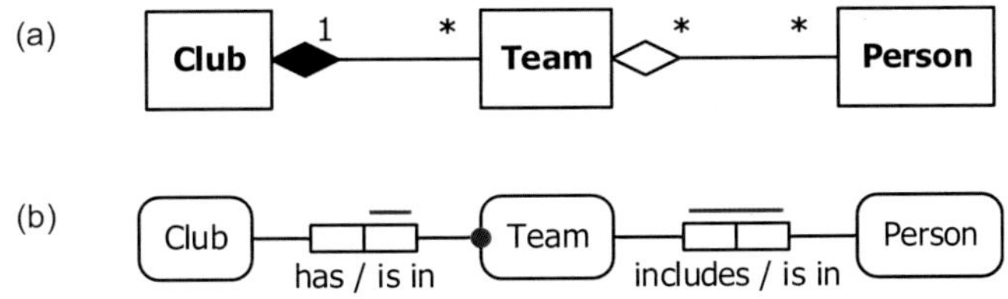

FIG. 9.43 Composition (composite aggregation) and shared aggregation in UML.

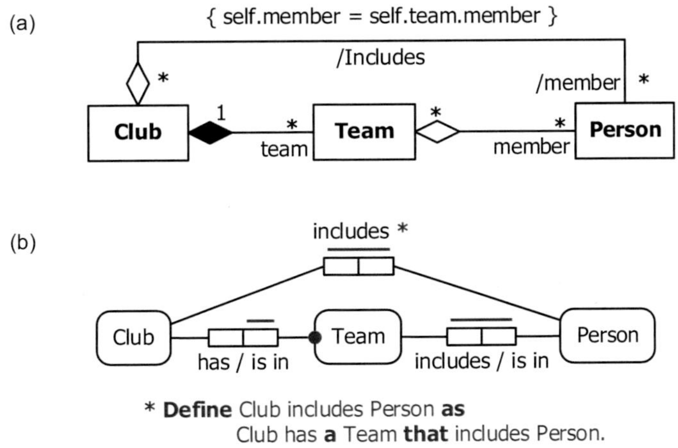

FIG. 9.44 A derived aggregation in (a) UML and (b) ORM.

However, if we wanted to discuss this result, it should be exposed as a *derived association*. In UML, derived associations are indicated by prefixing their names (association names or role names) with a *slash* "/". The *derivation rule* can be expressed as a constraint, either connected to the association by a dependency arrow or simply placed beside the association as in Fig. 9.44(a). *Derived attributes* are also indicated by prefixing their names with "/".

In ORM, derived fact types are marked with a trailing asterisk, with their derivation rules specified in an ORM textual language (see Fig. 9.44(b)). In many cases, derivation rules may also be diagrammed as a join-subset or join-equality constraint. As this example illustrates, the derived transitivity of aggregations can be captured in ORM without needing a special notation for aggregation.

Early versions of the UML specification declared that "both kinds of aggregation define a transitive, antisymmetric relationship (i.e., the instances form a directed, noncyclic graph)". But antisymmetry does not imply irreflexivity, much less acyclicity. Moreover, transitive closure is normally best achieved by derivation rather than assertion. Such property declarations for aggregation appear to be absent in UML 2.2 and later versions.

At any rate, some restriction on aggregations is needed to stop cases like that shown in Fig. 9.45. If a person is part of a team, and a team is part of a club, it doesn't make sense to say that a club is part of a person. The example should be remodeled, as shown in Fig. 9.44(a).

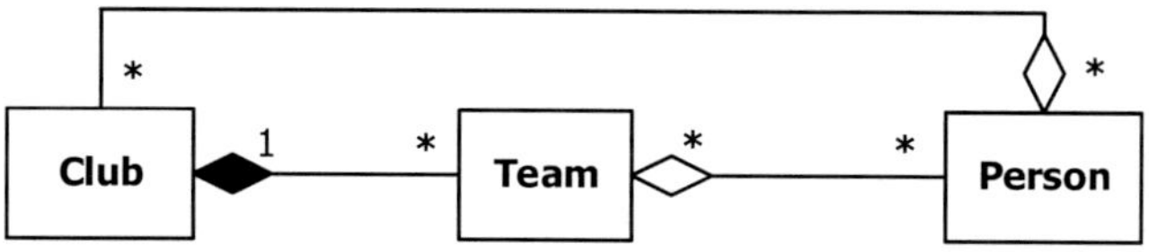

FIG. 9.45 Illegal UML model. Aggregation association chains should not form a cycle.

Of course, it is possible for an ORM modeler to make a silly mistake by adding an association such as Club is part of Person, where "is part of" is informally understood in the aggregation sense, and this would not be formally detectable. But avoidance of such a bizarre occurrence doesn't seem to be a compelling reason to add aggregation to ORM's formal notation. There are plenty of associations between Club and Person that do make sense, and plenty that don't. In some cases, however, it is important to assert constraints such as acyclicity, and this is handled in ORM by ring constraints. That said, there have been some recent proposals to add formal semantics for various forms of the part-of relationship to ORM. For example, Keet (2006) proposes adding several different mereological part-of predicates as well as four kinds of meronymic relations.

Composition adds some semantics to shared aggregation. To begin with, it requires that each part belongs to at most one whole at a time. ORM captures this by adding a uniqueness constraint to the role played by the part (e.g., see the role played by Team in Fig. 9.43(b)). In UML, the multiplicity at the whole end of the association must be 1 or 0..1. If the multiplicity is 1, as in Fig. 9.43(a), the role played by the part is both unique and mandatory, as in Fig. 9.43(b).

As an example where the multiplicity is 0..1 (i.e., where a part optionally belongs to a whole), consider the ring fact type of Fig. 9.46: Package contains Package. Here, "contains" is used in the sense of "directly contains". The ORM model is populated to capture the package containment example in Fig. 9.46(a).

The UML specification notes that "composition instances form a strict tree (or rather a forest)". This strengthening from directed acyclic graph to tree is an immediate consequence of the functional nature of the association (each part belongs to at most one whole), and hence ORM requires no additional notation for this. In this example, the ORM schema explicitly includes an acyclic constraint. This direct containment association is intransitive by implication (acyclicity implies irreflexivity, and any functional, irreflexive association is intransitive).

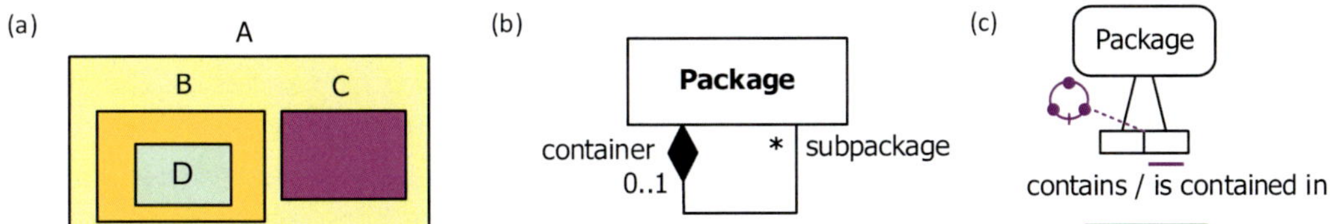

A	B
A	C
B	D

FIG. 9.46 (a) Direct containment modeled in (b) UML and (c) ORM.

UML allows some alternative notations for aggregation. If a class is an aggregate of more than one class, the association lines may be shown joined to a single diamond, as in Fig. 9.47(a). For composition, the part classes may be shown nested inside the whole by using role names, and multiplicities of components may be shown in the top right-hand corners, as in Fig. 9.47(b).

Some authors list kinds of associations that are easily confused with aggregation but should not be modeled as such [e.g., topological inclusion, classification inclusion, attribution, attachment, and ownership (Martin and Odell, 1998; Odell, 1998)]. For example, Finger belongs to Hand is an aggregation, but Ring belongs to Finger is not. There is some disagreement among authors about what should be included on this list. For example, some treat attribution as a special case of aggregation, namely, a composition between a class and the classes of its attributes (Rumbaugh et al., 1999).

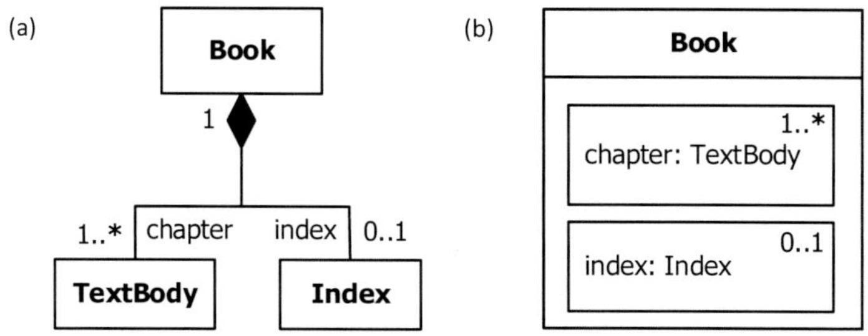

FIG. 9.47 Alternative UML notations for aggregation.

For conceptual modeling purposes, agonizing over such distinctions might not be worth the trouble. There are different stances that you could take about how, if at all, aggregation should be included in the conceptual modeling phase. You can decide what's best for you. The chapter notes provide further discussion on this.

Let's now look at the notion of *initial values*. The basic syntax of an attribute specification in UML includes six components as shown. Square and curly brackets are used literally here as delimiters (not as BNF symbols to indicate optional components).

visibililty name [*multiplicity*]: *type-expression* = *initial-value* {*property string*}

If an attribute is displayed at all, its name is the only thing that must be shown. The visibility marker (+, #, −, ~ denote public, protected, private, package, respectively) is an implementation concern and will be ignored in our discussion. Multiplicity has been discussed earlier and is specified for attributes in square brackets, for example, [1..*].

For attributes, the default multiplicity is 1, that is, [1..1]. The type expression indicates the domain on which the attribute is based (e.g., String, Date). Initial value and property string declarations may optionally be declared. Property strings may be used to specify aspects such as changeability (see later).

An attribute may be assigned an initial value by including the value in the attribute declaration after "=" (e.g., diskSize=9; country=USA; priority=normal). The language in which the value is written is an implementation concern.

In Fig. 9.48(a), the nrColors attribute is based on a simple domain (e.g., PositiveInteger) and has been given an initial value of 1. The resolution attribute is based on a composite domain (e.g., PixelArea) and is assigned an initial value of (640, 480).

Unless overridden by another initialization procedure (e.g., a constructor), declared initial values are assigned when an object of that class is created. This is similar to the database notion of *default values*, where during the insertion of a tuple an attribute may be assigned a predeclared default value if a value is not supplied by the user. However, UML uses the term "default value" in other contexts only (e.g., template and operation parameters), and some authors claim that default values are not part of UML models (Rumbaugh et al., 1999, p. 249).

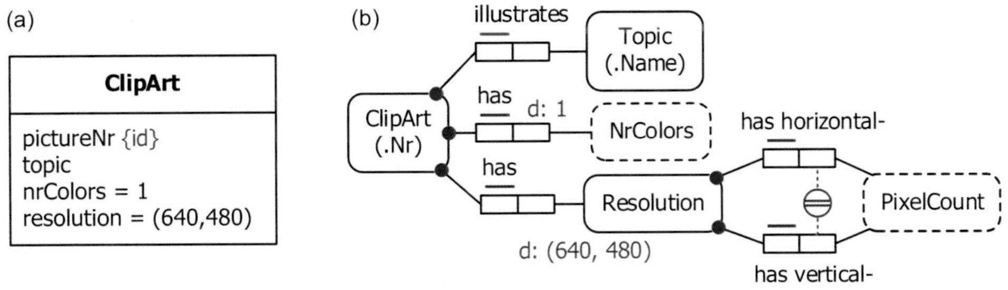

FIG. 9.48 Attributes assigned initial values in (a) UML and (b) ORM extension.

The SQL standard treats **null** as a special instance of a default value, and this is supported in UML because the specification notes that "a multiplicity of 0..1 provides for the possibility of null values: the absence of a value". So, an optional attribute in UML can be used to model a feature that will appear as a column with the default value of null when mapped to a relational database. Presumably, a multiplicity of [0..*] or [0..*n*] for any *n*> 1 also allows nulls for multivalued attributes, even though an empty collection could be used instead.

Currently, ORM has no explicit support for initial/default values. A possible extension could allow default values to be specified for ORM roles. At the metalevel, we add the fact type Role has default- Value. At the external level, instances of this could be specified on a predicate properties sheet, or entered on the diagram (e.g., by attaching an annotation such as d: *value* to the role, and preferably allowing this display to be toggled on/off). For example, the role played by NrColors in Fig. 9.48(b) is allocated a default value of 1. When mapped to SQL, this should add the declaration "default 1" to the column definition for ClipArt.nrColors.

To support the composite initial values allowed in UML, composite default values could be specified for ORM roles played by compositely identified object types (coreferenced or nested). When coreferencing involves at least two roles played by the same or compatible object types, an order is needed to disambiguate the meaning of the composite value. For example, in Fig. 9.48(b) the role played by Resolution is assigned a default composite value of (640, 480). To ensure that the 640 applies to the horizontal pixel count and the 480 applies to the vertical pixel count (rather than the other way round), this ordering needs to be applied to the defining roles of the external uniqueness constraint. ORM tools often determine this ordering from the order in which the roles are selected when entering this constraint.

If all or most roles played by an object type have the same default, it may be useful to allow a default value to be specified for the object type itself. This could be supported in ORM by adding the metafacttype ObjectType has default- Value and providing some notation for instantiating it (e.g., by an entry in an Object Type Properties sheet or by annotating the object type shape with d: *value*). This corresponds to the default clause permitted in a create-domain statement in the SQL standard. Note that an object-type default can always be expressed instead by role-based defaults, but not conversely (because the default may vary with the role).

Specification of default values does not cover all the cases that can arise with regard to default information in general. A proposal for providing greater support for default information in ORM is discussed in Halpin and Vermeir (1997), but this goes beyond the built-in support for defaults in either UML or SQL. Default information can be modeled informally by using a predicate to convey this intention to a human. For example, we might specify default medium (e.g., "CD", "DVD") preferences for delivery of soft products (e.g., music, video, software) using the 1:n fact type Medium is default preference for SoftProduct.

In cases like this where default values overlap with actual values, we may also wish to classify instances of relevant fact types as actual or default (e.g., Shipment used Medium). For the typical case where the uniqueness constraint on the fact type spans $n-1$ roles, this can be achieved by including fact types to indicate the default status (e.g., Shipment was based on Choice {actual, default}), resulting in extra columns in the database to record the status. While this approach is generic, it requires the modeler and user to take full responsibility for distinguishing between actual and default values.

In UML, restrictions may be placed on the *changeability* of attributes as well as the roles (ends) of binary associations. It is unclear whether changeability may be applied to the ends of *n*-ary associations. UML 2 recognizes the following four values for changeability, only one of which can apply at a given time:

- unrestricted
- readOnly
- addOnly
- removeOnly

The default changeability is "unrestricted" (any change is permitted). The value "unrestricted" was formerly called "changeable", which itself was formerly called "none". The other settings may be explicitly declared in braces. For an attribute, the braces are placed at the end of the attribute declaration. For an association, the braces are placed at the opposite end of the association from the object instance to which the constraint applies.

Recall that in UML, a "link" is an instance of an association. The term "*readOnly*" (formerly called "frozen") means that once an attribute value or link has been inserted, it cannot be updated or deleted, and no additional values/links may be added to the attribute/association (for the constrained object instance).

The term "*addOnly*" means that although the original value/link cannot be deleted or updated, other values/links may be added to the attribute/association (for the constrained object instance). Clearly, addOnly is only meaningful if the maximum multiplicity of the attribute/association-role exceeds its minimum multiplicity. The term "*removeOnly*" means that the only change permitted for an existing attribute value or link is to delete it.

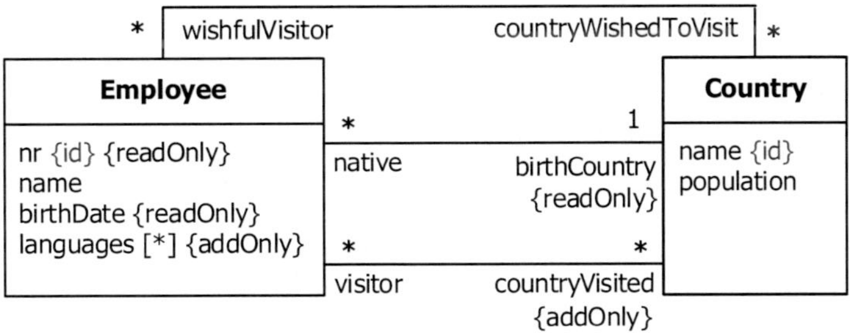

FIG. 9.49 Changeability of attributes and association roles.

As a simple if unrealistic example, see Fig. 9.49. Here employee number, birthdate, and country of birth are readOnly for Employee, so they cannot be changed from their original value. For instance, if we assign an employee the employee number 007, and enter his/her birthdate as 02/15/1946 and birth country as "Australia", then we can never make any changes or additions to that.

For each employee, the set of languages and the set of countries visited are addOnly. Suppose that for employee 007, we initially set his/her languages to {Latin, Japanese} and countries visited to {Japan}. As long as employee 007 is referenced in the database, these facts can't be deleted. However, we may add to these (e.g., later we might add the facts that employee 007 speaks German and visited India).

By default, the other properties are changeable. For example, employee 007 might legally change his name from "Terry Hagar" to "Hari Seldon", and the countries he wants to visit might change from {Ireland, USA} to {Greece, Ireland}.

Some traditional data modeling approaches also note some restrictions on changeability. The Barker ER notation includes a diamond to mark a relationship as nontransferable (once an instance of an entity type plays a role with an object, it cannot play this role with another object). Although changeability restrictions can be useful, in practice their application in database settings is limited.

One reason for this is that facts entered into a database might need to be changed. With snapshot data, this is the norm, but even with historical data changes can occur. The most common occurrence is to allow for corrections of mistakes, which might be because we were told the wrong information originally or because we carelessly made a misspelling or typo when entering the data.

In exceptional cases, we might require that mistakes of a certain kind be retained in the database (e.g., for auditing purposes) but be corrected by entering later facts to compensate for the error. This kind of approach makes sense for bank transactions (see Fig. 9.50). For example, if a deposit transaction for \$100 was mistakenly entered as \$1000, the record of this error is kept, but once the error is detected it can be compensated for by a bank withdrawal of \$900. The balance is both derived and stored, and its readOnly status is typically implied by the readOnly settings on the base attributes, together with a derivation rule.

Some authors allow changeability to be specified for a class, as an abbreviation for declaring this for all its attributes and opposite association ends (Booch et al., 1999, p. 184). For instance, all the {readOnly} constraints in Fig. 9.50 might be replaced by a single {readOnly} constraint below the name "Transaction". While this notation is neater, it would be rarely used. Even in this example, we would probably want to allow for the possibility of adding nonfrozen information later (e.g., a transaction might be audited by zero or more auditors).

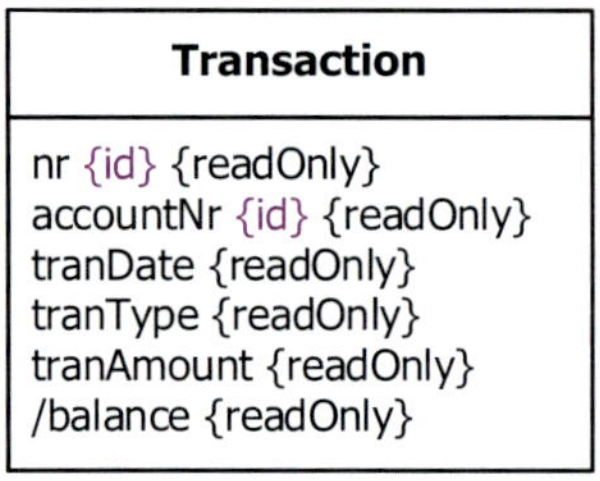

FIG. 9.50 All attributes of Transaction are readOnly.

Changeability settings are useful in the design of program code. Although changeability settings are not yet supported in ORM, which focuses on static constraints, they are being considered in extensions to support dynamic constraints (see next chapter). In the wider picture, being able to completely model security issues (e.g., who has the authority to change what) would provide extra value.

UML allows {ordered} and {unique} properties to be specified for multivalued attributes and association ends. Because {unique} is true by default, the use of {ordered} indicates an ordered set (a sequence with no duplicates). For example, Fig. 9.51(a) shows one way of modeling authorship of papers in UML. Each paper has a list of authors, each of whom may appear at most once on the list.

This may be modeled in flat ORM by using a Position object type to store the sequential position of authors on the list, as shown in Fig. 9.51(b). The uniqueness constraint on the first two roles declares that for each paper an author occupies at most one position; the constraint covering the first and third roles indicates that for any paper, each position is occupied by at most one author. The textual constraint notes that the list positions are numbered sequentially from 1.

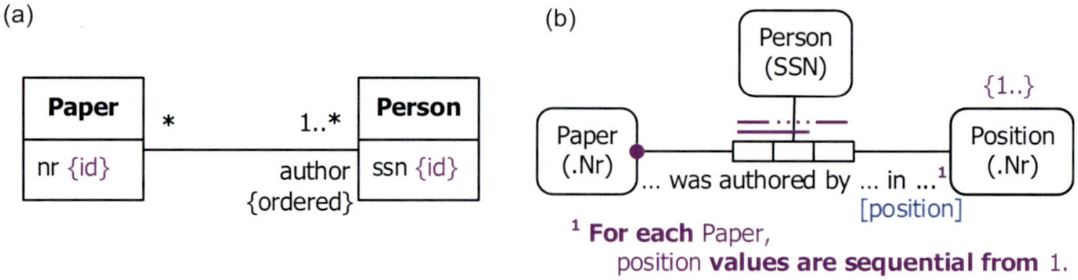

FIG. 9.51 An ordered set modeled in (a) UML and (b) ORM.

Although this ternary representation may appear awkward, it is easy to populate and it facilitates any discussion involving position (e.g., who is the second author for paper 21?). From an implementation perspective, an ordered set structure could still be chosen.

An ordered set is an example of a collection type. As discussed in Chapter 10, some versions of ORM allow collections to be specified as mapping annotations in a similar way to UML, and some ORM versions allow collections to be modeled directly as first-class objects.

UML 2 introduced the notion of *association redefinition*. This concept applies to generalizations as well as associations. One main use of it is to specify stronger constraints on an association that specializes an association involving a supertype. For example, in Fig. 9.52(a) the executiveCar role redefines the assignedCar role, applying a stronger multiplicity constraint that applies only to executives. Effectively, the association Executive is assigned CompanyCar is treated as a specialization of the Employee is assigned CompanyCar association. The stronger multiplicity constraint may be expressed in ORM by a textual constraint as in Fig. 9.52(b), or by a subset constraint involving the subtyping link (see Section 10.8).

Now let's consider *derived data*. In UML, derived elements (e.g., attributes, associations, or association-roles) are indicated by prefixing their names with "/". Optionally, a *derivation rule* may be specified as well. The derivation rule can be expressed as a constraint or note, connected to the derived element by a dashed line. This line is actually shorthand for a dependency arrow, optionally annotated with the stereotype name «derive».

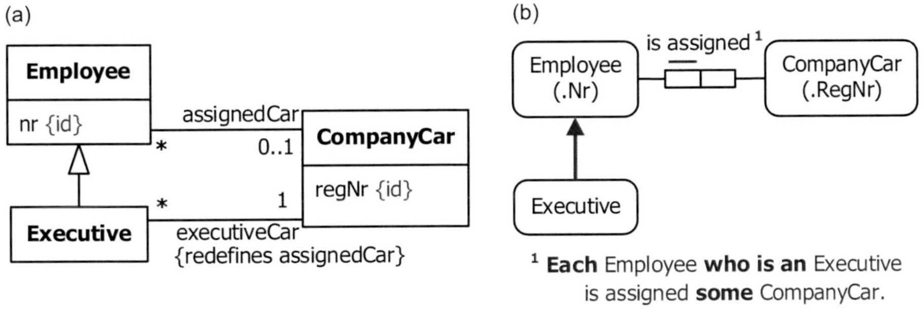

FIG. 9.52 Association redefinition in (a) UML and (b) ORM.

Because a constraint or note is involved, the arrow tip may be omitted (the constraint or note is assumed to be the source). For example, Fig. 9.53(a) includes area as a derived attribute. Fig. 9.53(b) shows the ORM schema.

The UML dependency line may also be omitted entirely, with the constraint shown in braces next to the derived element (in this case, it is the modeling tool's responsibility to maintain the graphical linkage implicitly). The example in Fig. 9.54(a) expresses unclehood as a derived association. For illustration purposes, rolenames are included for association ends. The corresponding ORM schema is shown in Fig. 9.54(b), with the derivation rule specified in relational style.

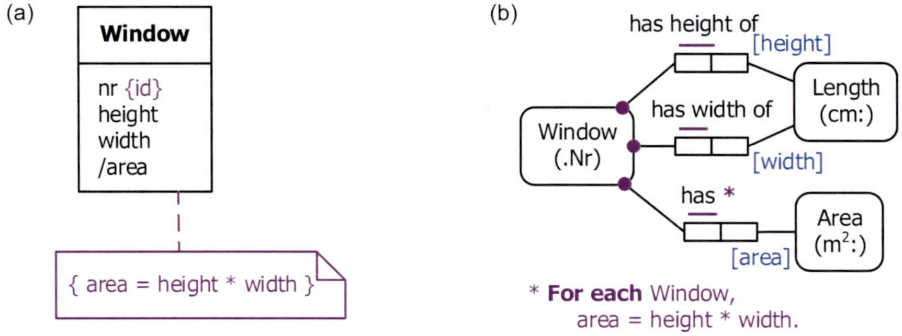

FIG. 9.53 Derivation of area in (a) UML and (b) ORM.

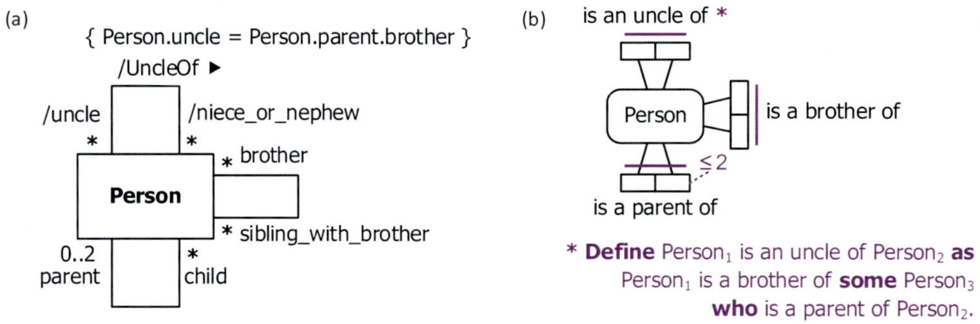

FIG. 9.54 Derived uncle association in (a) UML and (b) ORM.

Although precise role names are not always elegant, the use of role names in derivation rules involving a path projection can facilitate the concise expression of rules, as shown here in the UML model. By adding role names to the ORM schema, the derivation rule may be specified compactly in attribute style thus: * **define** uncle **of** Person **as** brother **of** parent **of** Person. More complex derivation rules can be stated informally in English or formally in a language such as OCL.

One advantage of ORM's approach to derivation rules is that it is more stable because it is not impacted by schema changes such as attributes being later remodeled as associations.

9.8 Mapping from ORM to UML

The *UML map procedure* in Table 9.3 provides basic guidelines for mapping ORM schemas to UML class diagrams. Selected entity types and value types map to object classes and data types, including attribute domains when associations are replaced by attributes. We now illustrate this procedure.

Table 9.3 UML map procedure.

Step	*Action*
1	Binarize any sets of exclusive unaries
2	Model selected object types as classes, and map a selection of their *n*:1 and 1:1 associations as attributes. To store facts about a value type, make it a class
3	Map remaining unary fact types to Boolean attributes or subclasses
4	Map other fact types to associations, association classes, or coreferenced classes. Map objectified associations to association classes or coreferenced classes
5	Map ORM constraints to UML graphic constraints, textual constraints, or notes
6	Map subtypes to subclasses, and if needed, subtype definitions to textual constraints
7	Map derived fact types to derived attributes/associations, and map semiderived fact types to attributes/associations plus derivation rules

Step 1 binarizes any sets of exclusive unaries, as shown in Fig. 9.55.

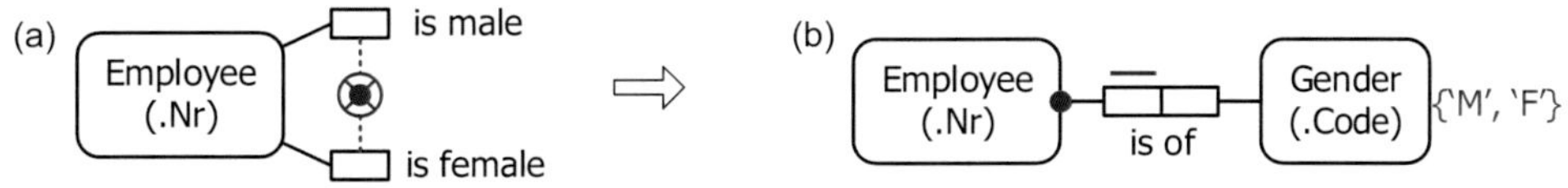

FIG. 9.55 Step 1: Replace any set of exclusive binaries by a binary fact type.

In step 2, we decide which object types to model as classes and which *n*:1 and 1:1 ORM associations to remodel as attributes. Typically, entity types that play functional fact roles become classes. Functional binary (*n*:1 and 1:1) associations from an entity type *A* to a value type B, or to an entity type *B* about which you never want to record details, usually map to an attribute of *A*. If you have specified role names, these can usually be used as attribute names, with the object type name becoming the attribute's domain name.

The mapping in Fig. 9.56 illustrates several of these step 2 considerations as well as step 6 (map ORM constraints to UML graphic constraints, textual constraints, or notes). The {U1} annotation for uniqueness is not standard UML. The value constraint on gender codes is captured using an enumeration type.

In rare cases, value types that are independent play an explicit mandatory role, or play a functional fact role in a 1:*n* fact type map to classes. The example in Fig. 9.57(a) deals with cases where we store title–gender restrictions (e.g., the title "Mr" is restricted to the male gender).

The example in Fig. 9.57(b) uses a multivalued attribute to store all the genders applicable to a title (e.g., the title "Dr." applies to both male and female genders). The Title class gives fast access from title to applicable gender, but slow access from gender to title. As discussed earlier, multivalued attributes should be used sparingly.

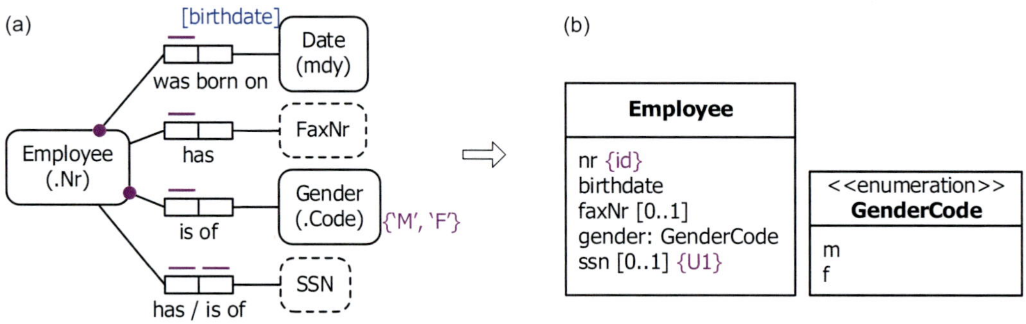

FIG. 9.56 Step 2: Map selected *n*:1 and 1:1 associations to attributes.

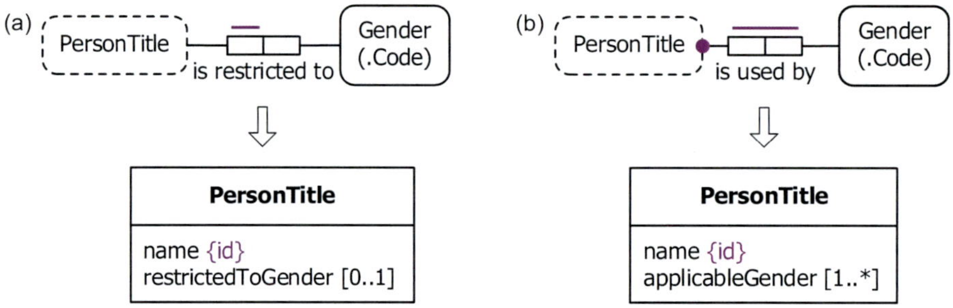

FIG. 9.57 Step 2: Rare cases of value types mapping to classes.

In step 3, we map unaries to Boolean attributes or to subclasses. The example in Fig. 9.58 assumes the usual *CWA* interpretation for the unary, so anyone not explicitly asserted to smoke is understood to be a nonsmoker. With an *open world approach* (*OWA*), failure to assert that a given person smokes means it is unknown whether that person smokes.

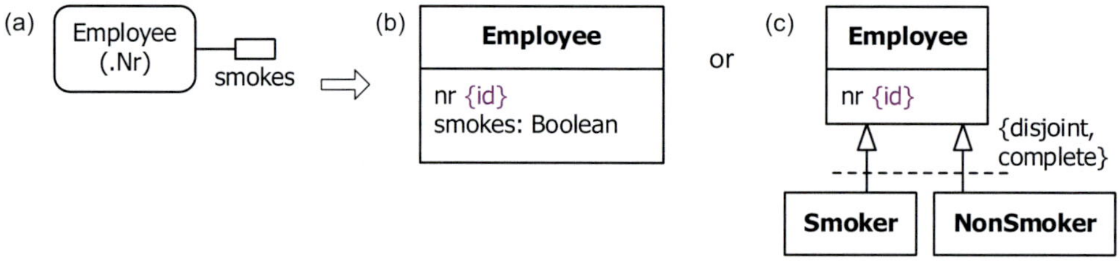

FIG. 9.58 Step 3: Map unaries to Boolean attributes or subclasses (CWA approach).

If we need to explicitly record that an employee is a smoker or is a nonsmoker, and this might be unknown for some employees, the UML class diagrams are modified so that the smokes attribute has a multiplicity of [0..1] and the {complete} constraint is changed to {incomplete}.

One way to model this situation in ORM is to replace the unary with the binary fact type Employee has SmokerStatus, where the first role has a uniqueness constraint but no mandatory role constraint, setting the data type of the SmokerStatus value type to Boolean. To do this in NORMA, edit the Properties sheet for the value type to Logical: True or False.

In step 4, the remaining fact types are mapped to associations. Any *m:n* associations should normally remain that way. In the example in Fig. 9.59, the *n*:1 fact type is retained as an association because it relates two entity types that become classes in the mapping. Even if the *m:n* association did not apply, we would normally retain Country as a class because now or later, we are likely to record details for it (e.g., country name).

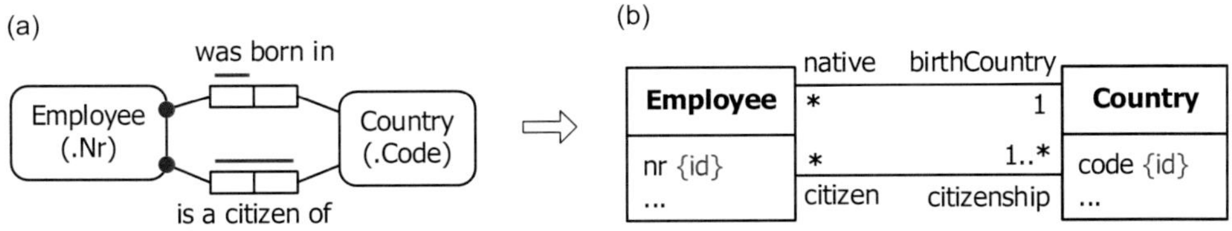

FIG. 9.59 Step 4: Map remaining fact types to associations.

If an *m:n* association involves a value type (e.g., Employee has PhoneNr), see if it is possible to transform the *m:n* association into multiple *n*:1 associations (e.g., Employee has PhoneNr1; Employee has PhoneNr2, etc.). Conceptual schema transformations like this are discussed in depth in Chapter 14. If each object type in an *n*-ary fact type should map to a class (e.g., it plays other functional roles), then map the *n*-ary fact type to an *n*-ary association. Fig. 9.60 provides an example.

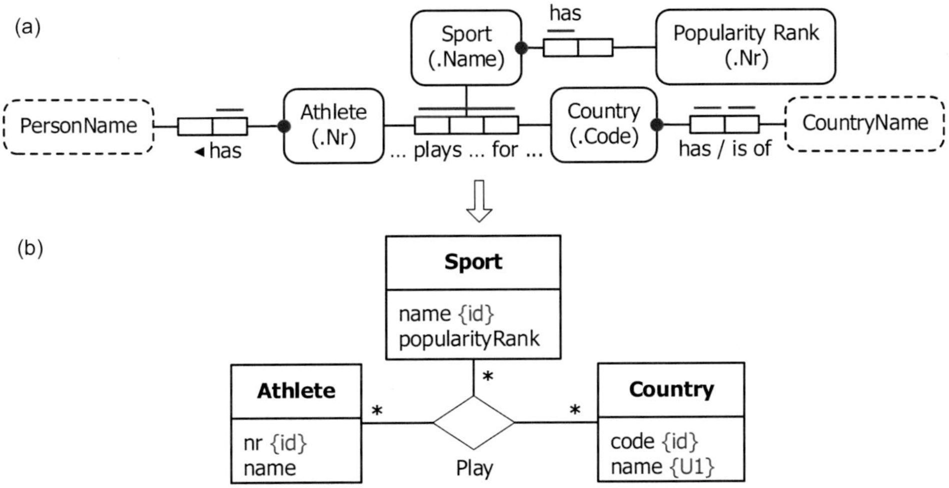

FIG. 9.60 Step 4: Map some *n*-ary fact types to *n*-ary associations.

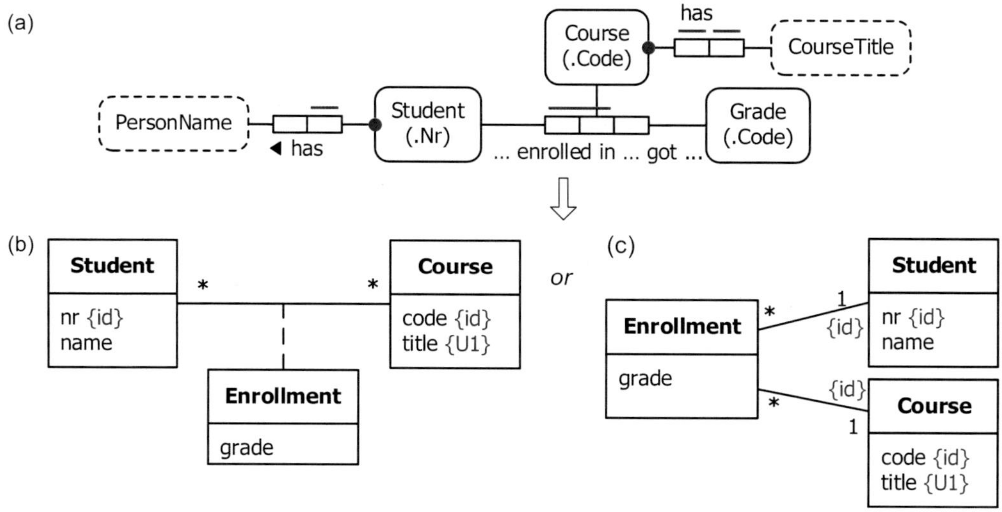

FIG. 9.61 Step 4: Map an *n*-ary fact type to association classes or coreferenced classes.

If an object type in a ternary fact type should not map to a class (typically an *m:n:1* uniqueness pattern with it outside the uniqueness constraint), then objectify the rest of the association as an association class, and map its role as an attribute. Fig. 9.61(b) shows an example. A less preferable alternative is to use a coreferenced class instead of an association class, as in Fig. 9.61(c).

Objectified associations map to association classes, as shown earlier in Fig. 9.18. Some cases of coreference could be mapped into qualified associations, but mapping to separate attributes or associations supplemented by a textual composite uniqueness constraint offers a more general solution.

In step 5, the simplest constraints in ORM usually map in an obvious way to multiplicity constraints, as illustrated earlier. The more complex ORM constraints have no graphic counterpart in UML, so you need to record these separately in textual form. Table 9.4 summarizes the main correspondences. Join constraints are considered in the next chapter.

Table 9.4 Mapping main ORM graphic constraints to UML (for use in step 5).

ORM constraint	*UML*
Internal UC	Maximum multiplicity of 1, or {Un}
External UC	Qualified association or textual constraint
Simple mandatory	Minimum multiplicity of 1, or textual constraint
Inclusive-or	Textual constraint (unless within exclusive-or)
Frequency	Multiplicity or textual constraint
Value	Enumeration or textual constraint
Subset and equality	Subset(s) or textual constraint
Exclusion	Textual constraint (unless within exclusive-or)
Ring constraints	Textual constraint
Join constraints	Textual constraint
Value-comparison	Textual constraint

In step 6, subtypes are mapped to subclasses, adding relevant subclassing constraints. Subtype definitions are handled with discriminators and/or textual constraints. For example, the ORM schema considered earlier in Fig. 9.38 maps to the UML schema in Fig. 9.37.

In step 7, we map derived and semiderived fact types. The schemas in Fig. 9.53 and Fig. 9.54 provide simple examples.

With these hints, and the examples discussed earlier, you should now have enough background to do the mapping manually for yourself. The following short exercise will give you some practice at this.

If you are keen on using UML for data modeling, you may wish to use UML to model some of the many modeling exercises in other chapters.

Exercise 9.8

1. Model the following application in UML. We suggest that you do an ORM model first, but that's up to you.

   The universe of discourse (UoD) is based on a simplified fragment of a database application used by an electrical utility to help manage the delivery of electricity to consumers in Australia. A line link is a section of an electrical feeder or mains power line that connects two adjacent nodes (e.g., power poles, pillar boxes).

   The figure shown illustrates a tiny part of the electrical power network. Here, three power poles (also called telephone poles) carry power lines to supply electricity. The house is one of many receiving electricity from the power company.

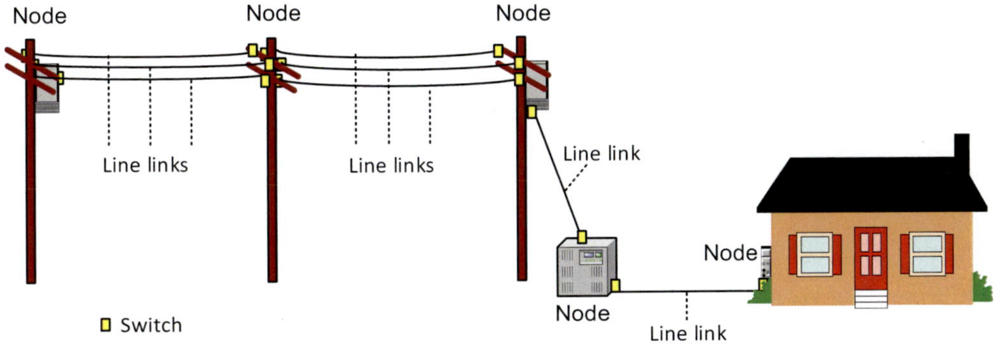

   The following output report provides an extract of sample data about line links. The link types listed are not exhaustive.

Line link	*Link type*	*Linktype description*	*Voltage (V)*	*Current (A)*	*Power (kW)*	*Length (m)*
33	OW	Open wire	11,000	400	4400	200
40	UGC	Underground cable	11,000	400	4400	150
55	ABC	Aerial bundled conductor	415	300	124.5	170
56	ABC	Aerial bundled conductor	240	200	48	180

The following report shows details about line links to help with fault detection and correction. Each power line link has a switch at both ends. The switches can be remotely toggled open (to break the circuit) or closed. Line status has only two values.

Line link	*From node*	*To node*	*Status*	*Start switch*	*End switch*
33	B	D	OK	Closed	Closed
40	D	E	OK	Closed	Open
55	F	G	Faulty	Open	Open
56	F	G	OK	Closed	Closed

The following report extract lists details about nodes and any consumers serviced with electricity by those nodes. Only nodes with a transformer (i.e., nodes for which we record a transformer type) can service a consumer. A consumer may be serviced by many nodes and has exactly one, two, or three phases supplied (regardless of which node supplies the power).

The transformer types shown here (PT = Pole Transformer, PMT = Pad-mounted transformer, GT = Ground Transformer) are not meant to be exhaustive. Codes of transformer types are stored but descriptions are not.

Note: In reality a consumer has only one node for normal service, and possibly other nodes for backup in the case of a fault; however, we ignore this refinement because its modeling requires subtypes, which we avoid in this simple exercise.

Node	*Nr fuses*	*Transformer type*	*Consumer serviced*	*Nr phases supplied*
B	3	PT	?	?
D	0	?	?	?
E	1	PMT	3001	2
			3005	3
...	...	...	...	...
F	2	GT	5678	2
			5700	1
...	...	...	...	...
G	2	GT	5678	2
...	...	...	...	...

The following report extract lists details about electrical faults suffered by consumers. Timestamps may be treated as a single value (including both date and time of day).

Consumer	Fault type	Reported	Fixed	Fault cause
3005	Dim lights	2006-04-29 09:30	2006-04-30 10:45	?
	Dim lights	2007-07-12 17:25	?	Wires down
5678	Blackout	2006-04-30 11:20	2006-05-01 6:30	Lightning strike
	Cold water	2006-04-30 11:20	?	?

The following report extract indicates electrical energy consumption (in kilowatt hours) by consumers on a quarterly basis. Some consumers (e.g., recent ones) might not yet appear in the report. Although consumption can be derived from other data not shown here, assume that it is simply stored for this exercise.

Consumer	*Quarter*		*Consumption (kW.h)*
	Year	*QtrNr*	
3001	2006	4	1622
	2007	1	1500
	2007	2	2001
	...	...	...
5678	2007	2	2001
...	...	...	...

(a) Specify a UML schema for this UoD, including constraints and derivation rules.

(b) Suppose that in the model, Fault is an association class to objectify the association Consumer reported FaultType at Instant. Suggest a new identification scheme for Fault to improve the usability of the model.

2. Model the academic schema in Exercise 8.5, Question 2 in UML.

9.9 Summary

UML has been adopted by the OMG as a method for object-oriented analysis and design. Although UML is mainly focused on the design of object-oriented programming code, it can be used for modeling database applications by supplementing its predefined notations with user-defined constraints.

UML has 13 main diagram types: six for structure (class, object, component, deployment, package, and composite structure), and seven for behavior (use case, state machine, activity, sequence, collaboration, interaction overview, and timing). Class diagrams are essentially an extended form of ER diagrams.

The basic correspondence between data structures and instances in UML and ORM is summarized in Table 9.5. Classes are basically entity types and are depicted as named rectangles, with compartments for attributes and operations, etc. In UML, facts are stored either in attributes of classes or in associations among two or more classes.

Binary associations are depicted as lines. Ternary and longer associations include a diamond. Role names may be placed at association ends, and an association may be given a name.

An association may be objectified as an association class, corresponding to nesting in ORM. Associations may be qualified to provide a weak form of co-reference.

Attributes and association ends may be annotated with multiplicity constraints to indicate both optionality and cardinality (e.g., 0..1 = at most one, 1 = exactly one, * = zero or more, 1..* = one or more). Attributes have a default multiplicity of 1, and association ends

Table 9.5 Correspondence between ORM and UML data instances and structures.

ORM	***UML***
Entity	Object
Value	Data value
Object	Object or Data value
Entity type	Class
Value type	Data type
Object type	Class or Datatype
— { use relationship type }	Attribute
Unary relationship type	— { use Boolean attribute }
2^+-ary relationship type	Association
2^+-ary relationship instance	Link
Nested object type	Association class
Co-reference	Qualified association[a]

[a]incomplete coverage of corresponding concept

have a default multiplicity of *. Refer back to Table 9.4 for the main correspondences between constraints in UML and ORM.

Subset constraints are allowed only between whole associations and are denoted by "{subset}" next to a dashed arrow. An exclusive-or constraint is depicted by "{xor}" next to a dashed line connecting the relevant associations.

Subclasses are connected to their superclasses by a line with an open arrowhead at the superclass end. Subclassing may be annotated using the keywords "{complete}", "{incomplete}", "{disjoint}", "{overlapping}", "{root}", "{leaf}". A discriminator (e.g., gender) may be used to indicate the basis for a subclass graph.

Whole/part associations may be displayed as aggregations using a small diamond at the whole end. A hollow diamond denotes shared aggregation (a part may belong to more than one whole), and a filled diamond indicates composition or composite aggregation (a part may belong to at most one whole at a time).

Attributes may be assigned initial (default) values. Derived attributes and associations are indicated by prepending "/" to their name. Attributes and binary association roles may be assigned a changeability setting: unrestricted, readOnly, addOnly, or removeOnly.

ReadOnly means that once an attribute value or link has been inserted, it cannot be updated or deleted, and no additional values/links may be added to the attribute/association (for the constrained object instance).

AddOnly means that although the original value/link cannot be deleted or updated, other values/links may be added to the attribute/association (for the constrained object instance).

A multivalued attribute or multivalued association end may be adorned with "{ordered}" to indicate implementation as an ordered set. One way of modeling this in ORM is to explicitly introduce a Position type to indicate the order.

An association may be redefined by declaring an association role to be a special case of a compatible role played by a superclass. One use of this is to strengthen the constraints on the specialized association roles.

UML models are best developed by mapping them from ORM models and noting any additional ORM constraints as comments.

Chapter Notes

The UML specification is accessible online at https://www.omg.org/spec/UML/2.5.1. For a detailed discussion of UML by "the three amigos" (Booch, Rumbaugh, and Jacobson), see Booch et al. (1999) and Rumbaugh et al. (1999). Their suggested modeling process for using the language is discussed in Jacobson et al. (1999). Martin and Odell (1998) provide a general coverage of object-oriented modeling using the UML notation. Muller (1999) provides a detailed treatment of UML for the purposes of database modeling. A thorough discussion of OMT for database applications is given in Blaha and Premerlani (1998), although its notation for multiplicity constraints differs from the UML standard. The OCL is covered in detail in Warmer and Kleppe (2003). Bennett et al. (2006) provide a detailed discussion of how to use UML 2.0 for object-oriented systems analysis and design.

On the topic of aggregation, Rumbaugh et al. (1999, p. 148) argue:

> *Aggregation conveys the thought that the aggregate is inherently the sum of its parts. In fact, the only real semantics that it adds to association is the constraint that chains of aggregate links may not form cycles ... Some authors have distinguished several kinds of aggregation, but the distinctions are fairly subtle and probably unnecessary for general modeling.*

There are plenty of other distinctions (apart from aggregation) we could make about associations, but don't feel compelled to do so. For a very detailed discussion arguing for an even more thorough treatment of aggregation in UML, see Barbier et al. (2000).

The view that security issues have priority over changeability settings is nicely captured by the following comment of John Harris, in a thread on the InConcept website:

> *Rather than talk of "immutable" data I think it is better to talk of a privilege requirement. For instance,* you *can't change your recorded salary but your boss can, whether it's because you've had a pay rise or because there's been a typing error. Privileges can be as complicated or as simple as they need to be, whereas "immutable" can only be on or off. Also, privileges can be applied to the insertion of new data and removal of old data, not just to updates.*

A collection of readings critiquing UML is contained in Siau and Halpin (2000). The Precise UML group, comprised largely of European academics, has published several papers mainly aimed at providing a more rigorous semantic basis for UML. A useful collection of their papers is accessible from their website, *www.cs.york.ac.uk/puml*.

10 Advanced Modeling Issues

Chapter Outline

10.1 Join Constraints

This chapter considers several advanced modeling topics and assumes familiarity with the basics of object-role modeling (ORM), Barker entity-relationship (ER) modeling, and unified modeling language (UML) class diagrams. While the main focus is on ORM, some attention is also given to the other approaches. Even where ER and UML fail to provide built-in support for the features discussed, many of the ideas could be applied, at least as extensions, to these approaches.

This section examines *join constraints*, first considering join constraints over a single join path, then set-comparison constraints involving joins. Although consideration of these constraints is really part of *Conceptual Schema Design Procedure (CSDP)* step 7, we postponed a detailed discussion of them until now because of their advanced nature.

The notion of conceptual joins was introduced in Section 4.4 and further illustrated in Section 9.7. With a *conceptual inner join* over two or more roles, the same object plays all the roles. In Fig. 10.1(a), there is a single join path Building contains Room **that** has RoomNr. Linguistically, the pronoun "**that**" in this path expression declares the join, indicating it is the same room playing both the join roles (shaded here for discussion purposes).

The external uniqueness constraint declares that each building, roomNr combination applies to at most one room. This is equivalent to an internal uniqueness constraint on building, roomNr in the set of (building, room, roomNr) tuples instantiating the conceptual join path.

Information Modeling and Relational Databases. https://doi.org/10.1016/B978-0-443-23790-4.00024-4

Fig. 10.1(b) shows the same schema in Barker ER notation. The external uniqueness constraint is captured by combining the "#" and "|". Fig. 10.1(c) shows the UML version, with the external uniqueness constraint captured by combing the two {id} markers.

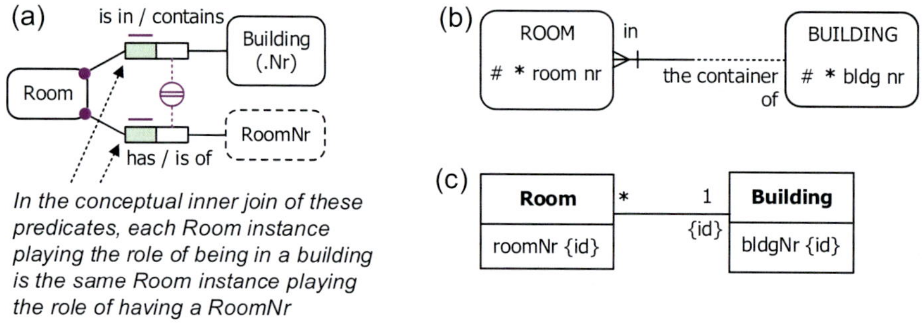

FIG. 10.1 A simple conceptual join in (a) ORM, (b) Barker ER, and (c) UML.

In the room example, the two roles participating in the conceptual join are both mandatory for the join object type (Room). This ensures that the conceptual join is an inner join. However, if one or more roles involved in the conceptual join is optional, the use of an external uniqueness constraint needs special care. For example, consider the three ORM schemas shown in Fig. 10.2.

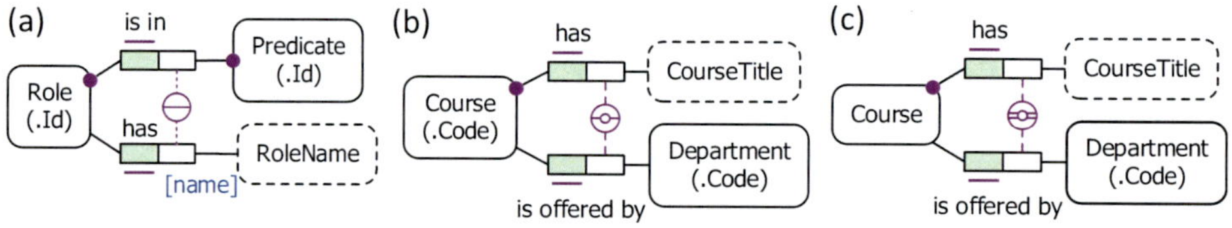

FIG. 10.2 Optional roles may involve an inner join (a) or an outer join (b), (c).

The ORM schema in Fig. 10.2(a) is a fragment of an ORM metaschema. ORM requires each predicate to have at least one reading, but it is optional whether roles are given names. In the figure, the role played by RoleName is named "name", but the other roles are not named. The external uniqueness constraint indicates that role names are unique within a given predicate. A predicate may have many unnamed roles, but if a role is named it can be identified by the combination of its name and its predicate.

This situation requires an *inner join interpretation* for the join path underlying the external uniqueness constraint. Let us identify the predicate of the association Role is in Predicate as P1 and its left and right roles as r1 and r2, respectively. Let us also identify the predicate of the association Role has RoleName as P2 and its left and right roles as r3 and r4, respectively. The inner join of the two fact types results in a compound ternary fact type equivalent to the following relational table scheme.

Role(roleId, predicateId, [roleName])
r4 P2 name

This relational notation underlines uniqueness constraints; if more than one exists, the primary key is doubly underlined. Optional columns are marked with square brackets. The external uniqueness constraint appears as a compound internal uniqueness constraint. Because we only require roles to have unique names in a predicate if they are named, the conceptual join between the roles of Role in Fig. 10.2(a) is an inner join, resulting in the single row population shown.

Using "?" to depict a null value, the outer join result would include four tuples: (r1, P1, ?), (r2, P1, ?), (r3, P2, ?), and (r4, P2, name). The first two tuples would then violate the compound uniqueness constraint, so outer join semantics should not be used with this example.

The external uniqueness constraints in Fig. 10.2(b) and (c) are depicted with an *inner "o" through the uniqueness bar.* This indicates that the external uniqueness constraint has *outer join semantics,* with the added proviso that *nulls produced in the outer join are treated as actual values.*

Hence, each course may be referenced by exactly one of the following patterns: the combination of the course title and its department (where the course is offered by a department); or its course title (where the course is offered by no department, e.g., a course offered by a visiting lecturer).

External uniqueness constraints of this nature are verbalized in two sentences, one for the inner join semantics and one for the additional semantics of the outer join. The external uniqueness constraint in Fig. 10.2(b) verbalizes thus:

For each CourseTitle **and** Department
  at most one Course has **that** CourseTitle **and** is offered by **that** Department.
For each CourseTitle
  at most one Course has **that** CourseTitle **and** is offered by **no** Department.

Consider a sample population with three courses, C1, C2, and C3, with the same title, one offered by the physics department, one by the mathematics department, and one offered by no department. The join of the title and department fact types results in a compound ternary fact type equivalent to the following relational table scheme. The external uniqueness constraint appears as the internal uniqueness constraint spanning courseTitle and departmentCode.

Course(courseCode, courseTitle, [departmentCode])

C1	Mechanics	PY	
C2	Mechanics	MA	
C3	Mechanics	?	
C4	*Mechanics*	*?*	*-- violates constraint*

The fourth row (C4, Mechanics, ?) is rejected because in the presence of row 3, it violates the constraint that courses with no department must have different titles. To determine constraint satisfaction, the nulls on rows 3 and 4 are treated like ordinary values, and thus are equated. This effectively applies *outer join semantics.*

Outer join semantics may apply even if the disjunctive reference scheme is the only means of identifying courses, as in Fig. 10.2(c).

External constraints with inner join semantics may be used for the preferred reference scheme of an entity type only if the coroles hosted by that entity type in the constrained relationships are mandatory, as in Fig. 10.2(a).

In the previous examples, the join path underlying the external uniqueness constraint involves only one join. But in practice, the *join path* may involve many inner or outer joins. Fig. 10.3 summarizes the problem of *join path ambiguity.* To find the join path underlying the external uniqueness constraint, we need to find a path connecting roles r2 and r9, moving right from r2 and left from r9.

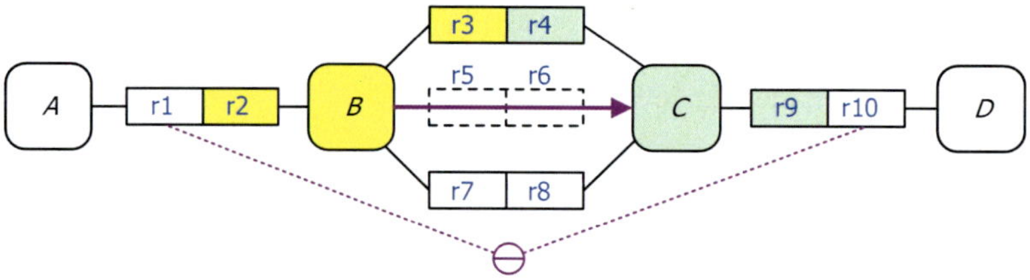

FIG. 10.3 Join paths may require explicit declaration of the join roles.

Our choice is shaded, but there are three possible paths between *B* and *C*. We could traverse the top predicate, performing the joins r2 = r3 and r4 = r9. We could traverse the middle predicate that implicitly underlies the subtype connection, performing the joins r2 = r5 and r6 = r9, or we could traverse the bottom predicate, performing the joins r2 = r7 and r8 = r9.

ORM does not graphically highlight the relevant joins (as done here by shading). In general, shading is not adequate because a path may traverse the same predicate more than once. A general graphical solution may be provided by numbering the roles used to perform the joins, but this can lead to messy diagrams. A tool could prompt the user to select the list of join roles and then toggle the display of the joins on/off under user control. An alternative is to use a textual language to formulate constraints with multiple, candidate join paths. Formal support for join paths is provided in the professional version of NORMA.

For non-ORM approaches, basic support for external uniqueness is required, and then the join paths may be disambiguated graphically as for ORM or textually if the method includes a sufficiently powerful formal query language. UML could use the object constraint language (OCL) for such specifications, although OCL expressions often appear cryptic to nontechnical domain experts.

As for external uniqueness, external frequency constraints have an underlying join path. If the joins involve optional roles, inner joins are assumed. For example, the external frequency constraint in Fig. 10.4 applies only to politicians who are coregents, so verbalizes as **For each** Country, **there are at most** 2 **instances of** Politician **such that that Politician** is a coregent **and** serves on the government of **that** Country.

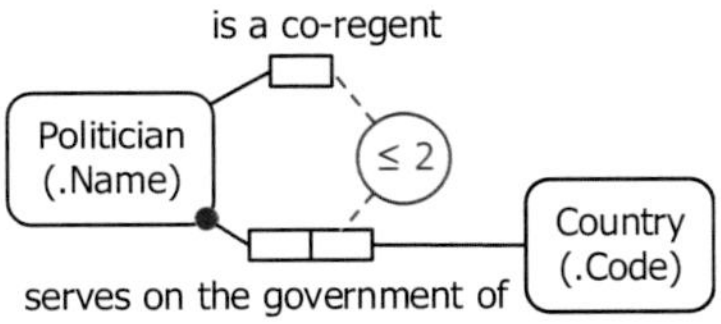

FIG. 10.4 An external frequency constraint involving a unary fact type.

Recall that ORM *set-comparison constraints* apply a subset, equality, or exclusion constraint between compatible role sequences. Sometimes one or more of these role sequences are projected from a join path. As a simple example, the join subset constraint in Fig. 10.5(a) runs from the role pair comprising the Advisor serves in Country fact type to the role pair comprising the first roles of the Advisor speaks Language and Country uses Language fact types.

The constraint means that the set of advisor–country pairs populating the serves-in predicate must be a subset of the set of advisor–country pairs obtained by projecting on the advisor and country roles of the join path Advisor speaks Language **that** is used by Country.

The join path in Fig. 10.5(a) involves a conceptual inner join on Language, equating instances playing its two roles. In other words, if an advisor serves in a country, then that advisor must speak at least one language used by that country. This constraint may be verbalized formally in ORM as **Each** Advisor **who** serves in **some** Country **also** speaks **some** Language **that** is used by **that** Country.

Like ER, UML has no graphic notation for expressing a join subset constraint, but does allow the constraint to be expressed textually in a note attached to the relevant model elements (in this case, the three associations), as shown in Fig. 10.5(b). ER could also be adapted to allow such usage of notes.

Fig. 10.5(c) shows another example discussed earlier in the book. The constraint may be verbalized in ORM thus: **If some** Employee has **some** PersonTitle **that** is restricted to **some** Gender **then that** Employee is of **that** Gender. A sample population illustrates the title-gender restriction.

The title-gender restriction can't be expressed in UML if we model personTitle as an attribute. However, it can be modeled in UML if we model PersonTitle as a class and attach a note for the restriction, as shown in Fig. 10.5(d). The note attached to the Employee class captures the title-gender restriction formally in OCL.

Join subset constraints often occur in practical applications. For a more complex join subset involving a ternary, recall the room facility example in Fig. 9.42.

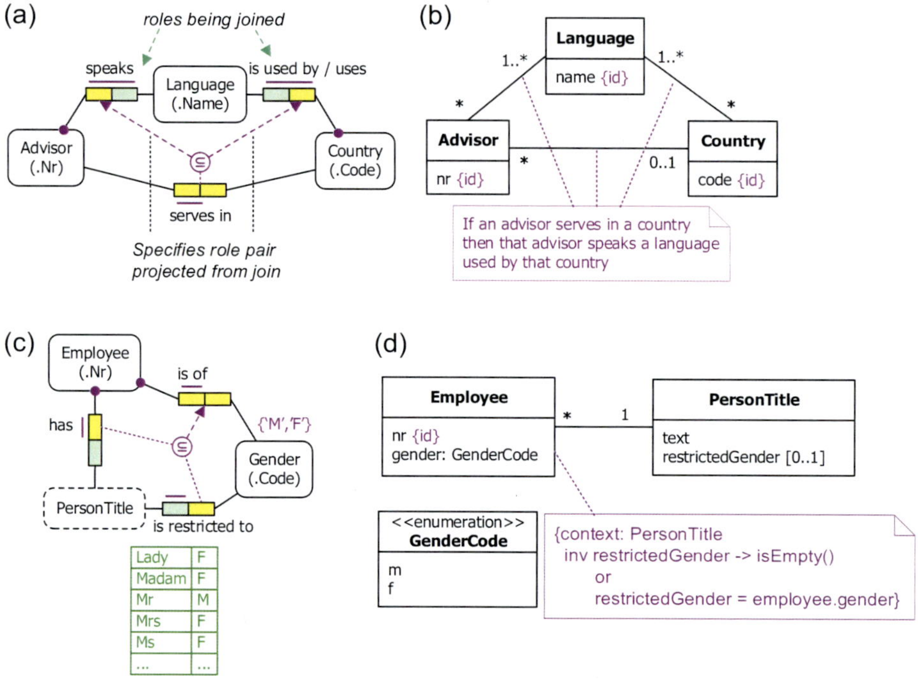

FIG. 10.5 Join subset constraints in (a), (c) ORM, and (b), (d) UML.

Join equality constraints are less common. The ORM schema in Fig. 10.6 includes an equality constraint between role triples, where the first triple involves a join on an objectified association. Role numbers are displayed here to clarify the constraint arguments (the NORMA tool displays these numbers only when the constraint is entered or selected). The constraint captures the rule that tips must be reported for, and only for, those services that earn a tip.

The join equality constraint in Fig. 10.6 captures the rule that no person can review a paper written by someone from their own institute, a rule commonly enforced in academic conferences. To avoid cluttered diagrams, the NORMA tool displays the role numbers only when the constraint is entered or selected. The constraint verbalization is also available to clarify the meaning of the constraint.

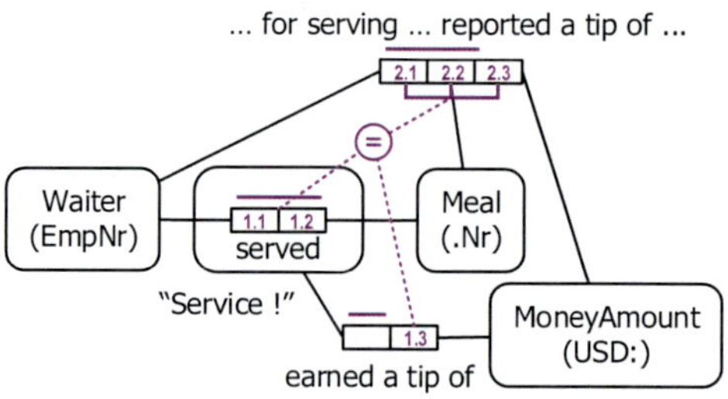

FIG. 10.6 A join equality constraint.

The ORM schema in Fig. 10.7 includes a *join exclusion constraint* between two role pairs. Each role pair is projected from a path involving a join on Person. The role pairs are numbered 1.1, 1.2 and 2.1, 2.2. Because the role played by Institute occurs in both role pairs, it is numbered twice.

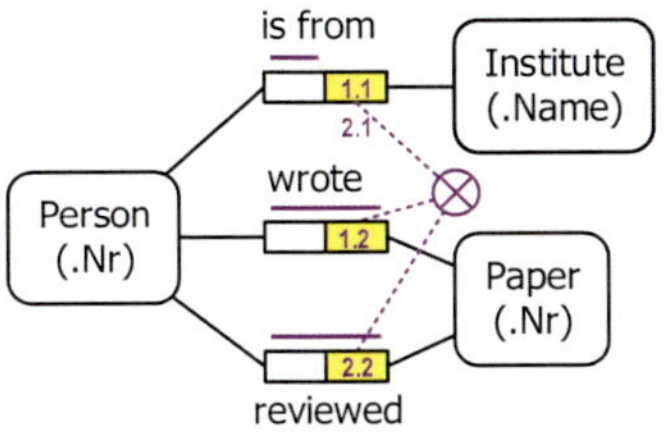

FIG. 10.7 A join exclusion constraint.

The same considerations given to join paths underlying external uniqueness and frequency constraints may be raised for each join path involved in a set-comparison constraint. Although we have modeled numerous business domains involving join set-comparison constraints, the join types have always been inner joins. External ring constraints as well as value-comparison constraints can also involve a join. For value-comparison constraints, the join type is always inner.

Rather than complicating the graphic notation for rare cases, we recommend that any set-comparison or ring constraints involving outer joins be specified textually using a formal language.

The mapping of join constraints depends on the mapping target, typically a database management system (DBMS) or third-generation programming language (3GL). Commercial structured query languages (SQLs) often diverge from standard SQL, and special care is needed when mapping uniqueness constraints over nullable columns (see Chapters 12 and 13 for further details about SQL). For example, suppose we want to implement the composite uniqueness constraint on the last two columns of *Role* (roleId, predicateId, [roleName]) in SQL Server. SQL Server allows only one null if a unique constraint is declared, but we want to allow multiple nulls. To implement inner join semantics for this composite uniqueness constraint, we need to code it up explicitly. The logic involved may be set out as follows:

```
check( not exists
    (select predicateId, roleName from Role
    where roleName is not null
    group by predicateId, roleName
    having count(*) > 1) )
```

Exercise 10.1

1. A sample population is included for each of the following ORM schemas, using surrogate identifiers to save space. In case (A) but not case (B), the following constraint applies: for each person, at most one employment end date is unknown (we have a complete record of their employments). The relational schema is also shown, using underlines to indicate uniqueness constraints.
   (a) Complete the population of the relational table for each situation.

Employment_A

enr	*pnr*	*startdate*	*enddate*

Employment_B

enr	*pnr*	*startdate*	*enddate*

   (b) What kind of join semantics (inner or outer) applies to the uniqueness constraint over (pnr, enddate) for schema A?
   (c) What kind of join semantics (inner or outer) applies to the uniqueness constraint over (pnr, enddate) for schema B?
   (d) Specify a check clause in SQL to check this uniqueness constraint for schema A.
   (e) Specify a check clause in SQL to check this uniqueness constraint for schema B.

(A)

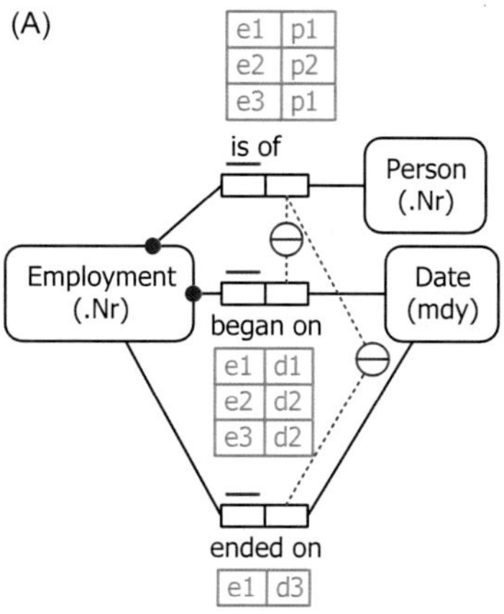

Employment (enr, pnr, startdate, [enddate])

(B)

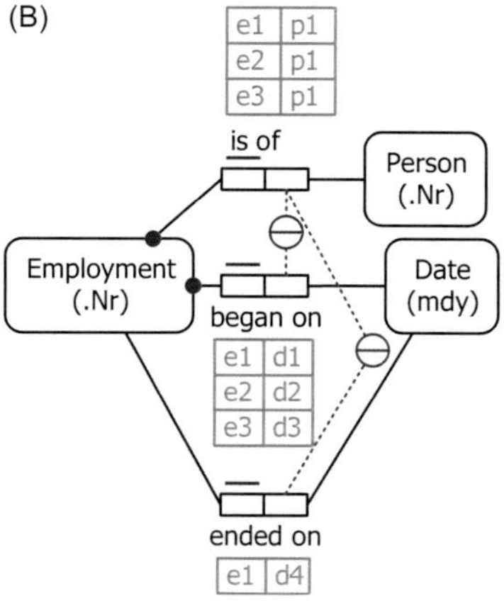

Employment (enr, pnr, startdate, [enddate])

2. Add a graphical constraint to the following ORM schema to capture the rule that if a modeling approach is used on a task that requires a specific modeling level, then the modeling approach must support that level.

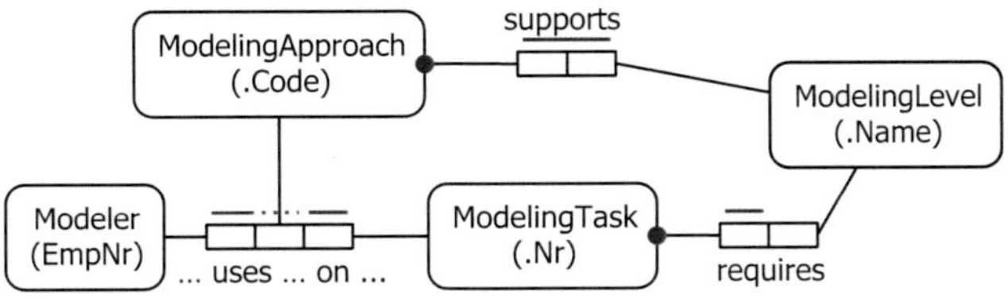

3. Add a graphical constraint to the following ORM schema to ensure that nobody reviews any book that they author.

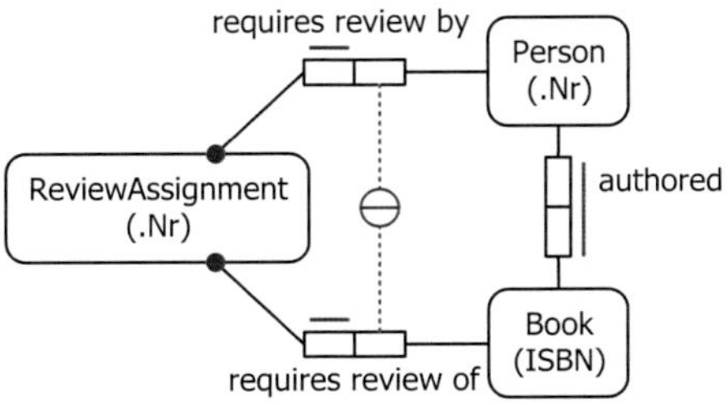

10.2 Deontic Rules

Business domains are constrained by various *business rules*, which specify required or desirable states of affairs or behavior. These rules may be of different *modalities* (e.g., alethic and deontic). *Alethic rules* impose *necessities*, which cannot be violated by the business, typically because of some physical or logical law. For example, each person was born on at most one date; no product is a part of itself.

All the constraints we've considered so far are alethic. The point of including them in an information system is not to prevent the rules being violated in the business domain (because they cannot be violated there), but to ensure that the information model accurately reflects the business domain.

Alethic rules simply prevent data representing impossibilities from being entered into the system. For example, if the information system enforces the uniqueness constraint on the employee birth fact type, we can't record two birthdates for the same employee.

Deontic rules impose *obligations*, which may be violated, even though they shouldn't. For example, in most cultures it is obligatory that each person is married to at most one person, and it is often forbidden that any person smokes in any office.

A business should have a clear understanding of all its rules, including deontic ones, whether or not it chooses to enforce these rules, or monitor violations of them by means of an automated system. In recognition of this need, both ORM 2 and the Object Management Group's Semantics of Business Vocabulary and Rules (SBVR) specification (https://www.omg.org/spec/SBVR) support the formulation of deontic rules (OMG, 2019). Although ER and UML currently confine their specification of rules to those of an alethic modality, in principle these approaches could be extended to cater for deontic rules.

Rule formulations may use any of the basic alethic or deontic modal operators from modal logic, as shown in Table 10.1. These modal operators are treated as proposition-forming operators on propositions (not actions). Equivalent readings in other concrete syntaxes may be used (e.g., "necessary" might be replaced by "required", and "is obligatory" might be replaced by "ought to be the case").

The following *modal negation rules* apply: it is not necessary that ≡ it is possible that not ($\sim\Box p \equiv \Diamond\sim p$); it is not possible that ≡ it is necessary that not ($\sim\Diamond p \equiv \Box\sim p$); it is not obligatory that ≡ it is permitted that it is not the case that ($\sim\boldsymbol{O}p \equiv \boldsymbol{P}\sim p$); it is not permitted that ≡ it is obligatory that it is not the case that ($\sim\boldsymbol{P}p \equiv \boldsymbol{O}\sim p$).

Table 10.1 Alethic and deontic modal operators.

Alethic		*Deontic*	
Reading	*Symbol*	*Reading*	*Symbol*
It is necessary that	$\Box$	It is obligatory that	***O***
It is possible that	$\Diamond$	It is permitted that	***P***
It is impossible that	$\sim\Diamond$	It is forbidden that	***F*** or **~*P***

In ORM 2, each constraint or rule has an associated modality determined by the logical modal operator that functions explicitly or implicitly as its main operator. In positive verbalizations, an alethic modality of necessity is often assumed (if no modality is explicitly specified), but may be explicitly prepended. For example, the alethic constraint **Each** Person was born in **at most one** Country may be explicitly verbalized thus: **It is necessary that each** Person was born in **at most one** Country.

We interpret this constraint in terms of possible world semantics, as introduced by Saul Kripke and other logicians in the 1950s. A proposition is necessarily true if and only if it is true in all possible worlds. With respect to a static constraint declared for a given business domain, a possible world corresponds to a state of the fact model that might exist at some point in time. The aforementioned constraint declares that for each state of the fact model, each instance in the population of Person is modeled to be born in at most one country.

A proposition is possible if and only if it is true in some possible world. A proposition is impossible if and only if it is true in no possible world (i.e., it is false in all possible worlds). The above constraint may be reformulated as the negative verbalization **It is impossible that the same** Person was born in **more than one** Country.

To avoid confusion, when declaring a deontic constraint, the deontic modality should always be explicitly included. Consider the following static, deontic constraint, verbalized in both positive and negative forms: (+ve) **It is obligatory that each** Person is a husband of **at most one** Person; (-ve) **It is forbidden that the same** Person is a husband of **more than one** Person.

This deontic rule indicates a condition that *ought* to be satisfied while recognizing that the condition *might* not be satisfied. Including the obligation operator makes the rule much weaker than a necessity claim because it allows that there could be some states of the fact model where a person is a husband of more than one wife. For such cases of polygamy, it is important to know the facts indicating that the person has multiple wives. Rather than reject this possibility, we allow it and then typically perform an action to minimize the chance of such a situation arising again (e.g., send a message to inform legal authorities about the situation).

Fig. 10.8 shows some ORM schemas with alethic and deontic constraints. When displayed in color, alethic constraints are violet, and deontic constraints are blue. Deontic constraint markers typically include a small "o" (for obligatory). Deontic ring constraints use dashed lines.

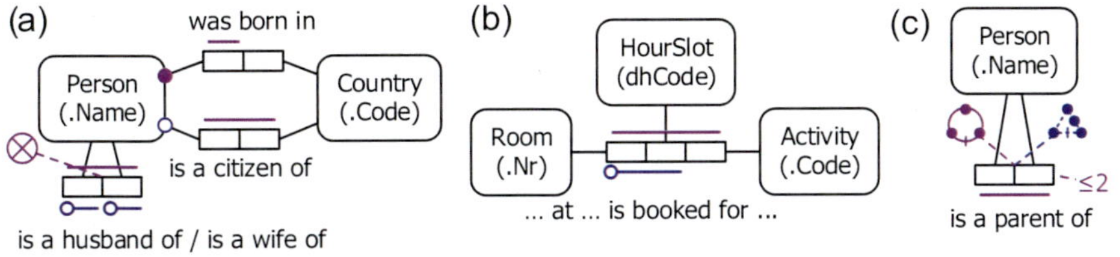

FIG. 10.8 Some ORM schemas with alethic and deontic constraints.

The citizenship and marriage fact types in Fig. 10.8(a) have spanning uniqueness constraints, and hence are alethically many-to-many associations. However, each role of the marriage association has a deontic uniqueness constraint, so it is obligatory that each person has at most one spouse.

The mandatory role dot on the citizenship role is an open "o", so that mandatory constraint is deontic, indicating that it is obligatory that each person is a citizen of some country.

Fig. 10.8(b) shows an ORM schema with a deontic uniqueness constraint spanning the first two roles of a ternary fact type. This verbalizes in positive form as **It is obligatory that for each** Room **and** HourSlot, **that** Room at **that** HourSlot is booked for **at most one** Activity. In negative form, we have **It is forbidden that the same** Room at **the same** HourSlot is booked for **more than one** Activity. Notice that in both schemas, *the deontic uniqueness constraints span a subset of the roles spanned by the relevant alethic constraints (explicit or implicit)*. If an alethic constraint already applies to a role set, a deontic constraint over the same roles is pointless, as it can't be violated.

In Fig. 10.8(c), the alethic frequency constraint ensures that each person has at most two parents, the alethic ring constraint ensures that parenthood is acyclic, and the deontic ring constraint makes it obligatory for parenthood to be strongly intransitive.

In practice, most business rules include only one modal operator, and this operator is the main operator of the whole rule expression. While ORM is restricted to rules of this nature, SBVR allows modal operators to be embedded anywhere in a rule formulation. The use of embedded modal operators, especially deontic operators, can dramatically increase the complexity of rule formalization and execution.

The only impact of tagging a rule as a necessity or obligation is on the rule enforcement policy. Enforcement of a necessity rule should never allow the rule to be violated in the model. Enforcement of an obligation rule may allow model states that do not satisfy the rule condition, and take some other remedial action: the precise action to be taken is not specified here, but a simple default is to generate a message when an update violates the rule.

At any rate, a business person should be able to specify a deontic rule first at a high level, without committing at that time to the precise action to be taken if the condition is not satisfied; of course, the action still needs to be specified later in refining the rule to make it fully operational.

The ORM glossary at the back of this book lists the symbols for all the alethic and deontic constraints supported in ORM 2.

Exercise 10.2

1. Which of the following statement pairs is/are a pair of *equivalent* rule statements?
   (a) It is necessary that each Person died on at most one Date.
   It is forbidden that the same Person died on more than one Date.
   (b) It is necessary that each Country is of at most one Area.
   It is impossible that the same Country is of more than one Area.
   (c) It is obligatory that each Person has at most one SocialSecurityNumber.
   It is forbidden that the same Person has more than one SocialSecurityNumber.
   (d) It is not permitted that any Person is a husband of more than one Person.
   It is obligatory that no Person is a husband of more than one Person.

2. Identify any errors in the following schema and explain why they are errors. Ignore the missing ring constraints.

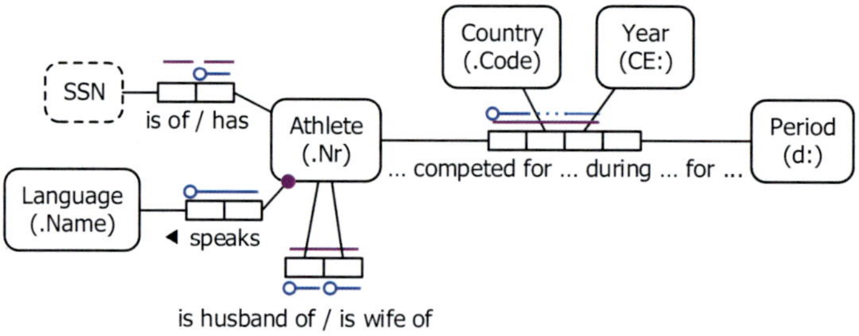

3. A given business identifies employees by employee numbers. All employees have names that are not necessarily unique. All employees should have a Social Security number, but it is possible that some do not. Safeguards make it impossible for the same employee to be issued more than one Social Security number. Although no two employees ought to have the same Social Security number, it is possible for this to happen. Model this universe of discourse (UoD).

10.3 Temporality

In the database community, many DBMS products have incorporated specific support for temporal features. Three basic *temporal data types* may be distinguished (Snodgrass, 2000):

- *Instant* (point in time, e.g., 2022 June 3, 2:50 p.m. UTC)
- *Interval* (duration of time, e.g., 3 weeks)
- *Period* (anchored duration of time, e.g., 2022 July 1 … 2023 June 30 MST)

For each of these data types, various temporal operators may be defined. For example, date subtraction (−) applied to an ordered pair of dates (e.g., today and your birthdate) results in an interval (e.g., your age), and the overlaps operator applied to two periods yields a Boolean value (True or False) indicating whether those periods overlap. Many temporal operators (e.g., before, after, during) have been defined, and are useful for specifying temporal queries and dynamic rules. Popular DBMSs typically have good support for Instant (e.g., date, time, datetime, timestamp). Interval and Period are also explicitly supported in some DBMSs.

In database work, two kinds of time may be distinguished. *Valid time* denotes the time period during which a database fact was, is, or will be valid in the business domain being modeled. In contrast, *transaction time* denotes the time period during which a database fact is/was recorded in the database. In a database table, valid time and transaction time may often be captured by extra columns (e.g., a mandatory startVT and an optional endVT column, or a mandatory startTT and an optional endTT). If the end transaction time is unknown, it is sometimes recorded as the largest time stamp supported (e.g., SQL Server uses 9999-12-31).

For example, SQL Server 2016 and later versions provide built-in support for *temporal tables*, which include two extra columns with a datetime2 datatype to record the transaction time for each row. Although these columns are typically called "ValidFrom" and "ValidTo", they measure the transaction time, not valid time in the sense defined above. A temporal table also contains a reference to a history table to store previous versions of the rows. In the absence of built-in support for periods, we may add our own support for periods by adding columns for a start and end time. Sophisticated support for temporal data is also available in various extensions (e.g., spatiotemporal data cartridges/blades).

Different facts may be recorded with different *granularities of time* (e.g., second, day). Time may be measured using different *calendars* (e.g., Gregorian, Julian) and different *time zones* (e.g., coordinated universal time (UTC), Mountain Standard Time). For most modeling tasks, calendar and time zone choices are taken to be implicitly understood.

Temporal extensions have been proposed for the relational model as well as higher-level data modeling approaches such as ER, ORM, and UML. Basic *temporal object types* (e.g., Date or Period) may be used just like other object types in the business model, with basic temporal operators (e.g., −) assumed to be predefined. *For the rest of this section, time is understood as valid time*, not transaction time.

Temporal object types may be classified into two kinds: once only and repeatable. A *once-only temporal object* is either a single *instant* or a single *period* of time, measured to a specified temporal granularity (relative to an assumed calendar and time zone). Some simple examples are depicted in Fig. 10.9(a). For example, Albert Einstein was born in the year 1879 CE. In this sense, a year is a 12-month period in the time stream. Likewise, a date is a period of 1 day in the time stream (e.g., 2023 June 5), a month is a period of approximately 30 days in the time stream (e.g., 2023 June), an hour is a period of 60 minutes (e.g., 2023 June 5, 9–10 a.m.).

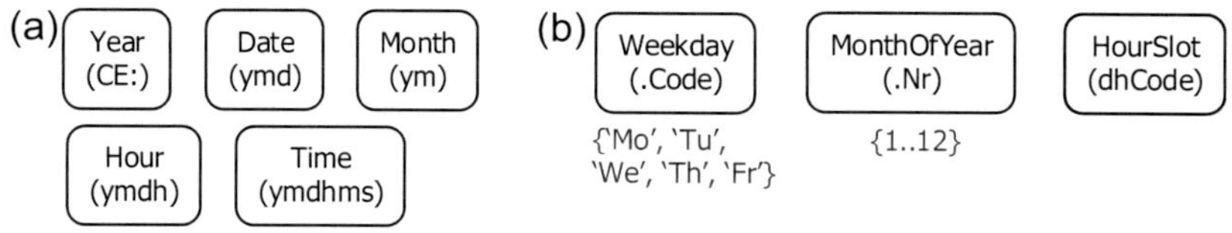

FIG. 10.9 Examples of (a) once-only and (b) repeatable temporal object types.

A time is an instant in the time stream measured to a specified granularity, such as a second, for example, 2023-06-05 14:35:27 (i.e., 2023 June 5, 14 h 35 m 27 s). Object types such as these are useful for recording when an individual event happened or will happen (e.g., the birth of Einstein, or the next Olympic Games).

Each *repeatable temporal object* corresponds to a set of instants or time periods, measured to a specified granularity (relative to an assumed time zone). Some simple examples in ORM notation are depicted in Fig. 10.9(b). Here, a weekday is a day slot of any week excluding the weekend. (e.g., Monday), a MonthOfYear instance is a month slot of any year (e.g., January, or Month 1), and an hour slot is a 1-h slot of a day. Object types such as these are useful when modeling schedules (e.g., a workout routine or a conference program).

Periods may be modeled in more detail by specifying their start and end times to the desired temporal granularity (year, month, day, hour, etc.). For example, if periods are closed (we know both start and end) and the granularity is day, then they may be modeled in ORM as shown in Fig. 10.10(a) and in UML as Fig. 10.10(b).

Intervals (unanchored durations of time) may be modeled as a simple object type with a unit for the chosen granularity. For example, if we measure ages of fossils in year units, this may be modeled in ORM as shown in Fig. 10.11(a) and in UML as shown in Fig. 10.11(b). In ORM, the colon ":" indicates that "y" is a unit of a given type (display can be expanded to "y: Interval").

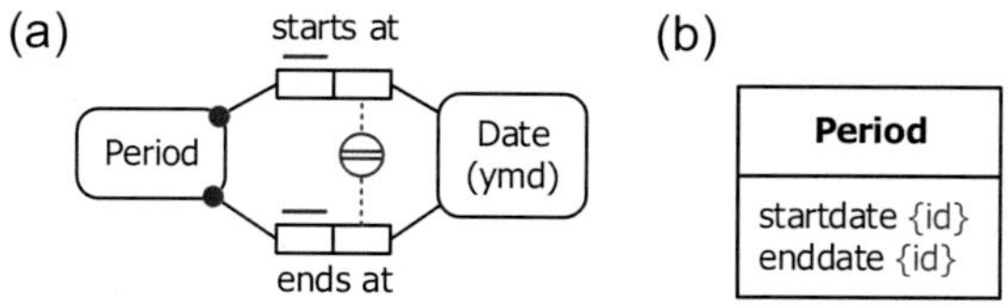

FIG. 10.10 Closed periods modeled in (a) ORM and (b) UML.

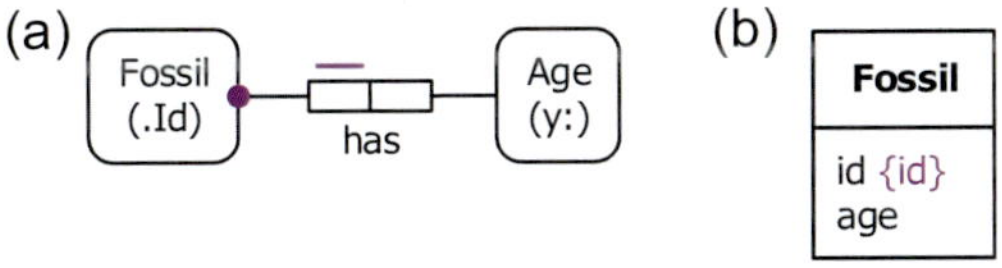

FIG. 10.11 Modeling an interval in (a) ORM and (b) UML.

As a business domain changes over time, the set of facts describing it also changes. Regardless of how we represent fact types in a model (e.g., as relationships or as attributes), for the purpose of making decisions about how they are impacted by time, it is convenient to distinguish at least *three kinds of fact types*:

- *Definitional* (truth of instances is a matter of definition, independent of events)
- *Once-only* (each instance corresponds to a single event)
- *Repeatable* (instances may correspond to multiple events)

Definitional facts have no temporal aspect. They are necessarily true by definition, rather than depending on any event in the business domain. Fig. 10.12(a) shows an ORM diagram for the definitional fact type PolygonShape has NrSides along with a sample fact population. Fig. 10.12(b) shows the same fact type as a UML class diagram.

Each *once-only fact* corresponds to a single event. Its truth derives from an event that can never be repeated in that business domain. For example, Einstein was born in Germany. This is true because of his birth event, a state of affairs that happened in 1879 CE.

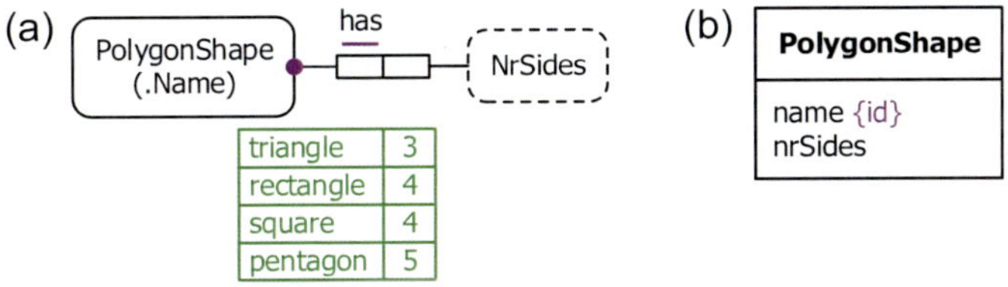

FIG. 10.12 Example of a definitional fact type in (a) ORM and (b) UML.

Whether or not reincarnation is possible, let us ignore such possibilities for the business domain modeled in Fig. 10.13, where each employee has exactly one birth event. Fig. 10.13(a) depicts the fact type Employee was born in Country in ORM, along with a sample population. Fig. 10.13(b) depicts this as a many:1 association in UML, and Fig. 10.13(c) depicts this as a single-valued attribute in UML.

Each *repeatable fact* corresponds to a set of events, in the sense that its truth is determined by any one of a set of repeatable events. For example, suppose employee 101 visited Belgium many times. The fact that he/she visited Belgium is made true by any one of those visits. Because the same employee may visit the same country repeatedly, the fact type Employee visited Country is said to be repeatable.

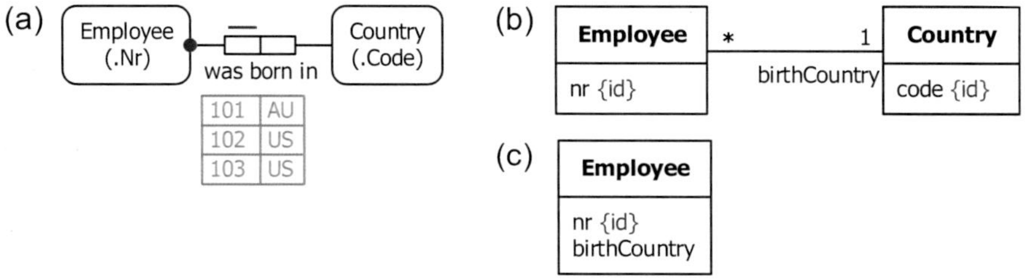

FIG. 10.13 Example of a once-only fact type in (a) ORM and (b and c) UML.

Fig. 10.14(a) depicts the fact type Employee visited Country in ORM, along with a sample population. Fig. 10.14(b) depicts this as a many:many association in UML, and Fig. 10.14(c) depicts this as a multivalued attribute in UML.

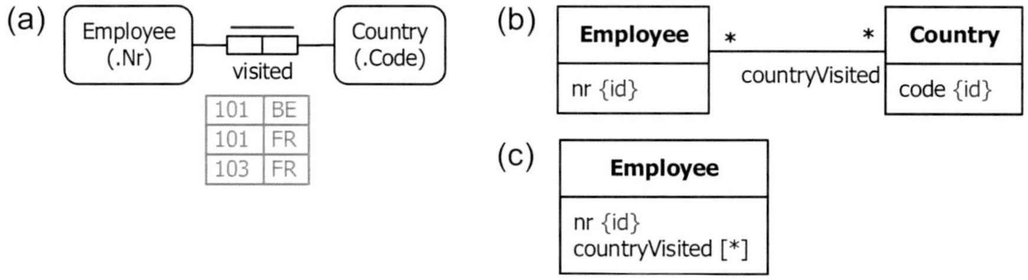

FIG. 10.14 Example of a repeatable fact type in (a) ORM and (b and c) UML.

A *proposition* (e.g., Employee 101 visited Belgium) must be true or false (and hence is a *truth bearer*). A state of affairs (e.g., Employee 101 visiting Belgium) is actual (occurs or exists in the actual world) or not. A state of affairs may be possible or impossible. Some possible states of affairs may be actual (occur in the actual world). *States of affairs* are thus *truth makers*, in that true propositions are about actual states of affairs. In sympathy with the correspondence theory of truth, we consider the relationship between propositions and states of affairs to be one of correspondence rather than identity.

Definitional facts may be viewed as corresponding to states of affairs only in a trivial sense (if at all); because they have no temporal aspect, they need no special attention in this regard. In contrast, once-only and repeatable facts correspond to states of affairs that are either single events (once only) or event sets (repeatable). *For each once-only or repeatable fact type in a model, we need to determine what (if any) temporal information is needed.*

Let's first consider once-only fact types. If a fact type is once-only and we wish to record for at least some of its fact instances when its underlying event (that made the fact instance true) occurred, then (if not already present in the model) add a functional fact type to record when the event occurred. To do this, relate a "key" of the original fact type to a temporal object type of the desired temporal granularity. Here, "key of a fact type" means an object type that plays a mandatory, functional role in that fact type. A functional role is constrained by a simple uniqueness constraint. If we wish to record the event for each of the original fact instances, the key's role in that temporal fact type should normally be mandatory.

For example, consider the once-only fact type Employee was born in Country. Suppose that for each instance of this fact type, we want to know when the underlying birth event occurred to a granularity of 1 day. The key object type is Employee (its role is mandatory and functional), so add the mandatory, functional fact type Employee was born on Date.

Fig. 10.15(a) depicts this in ORM showing a sample population. Fig. 10.15(b) depicts this in UML using a birthdate attribute. If we instead chose the temporal granularity to be a year, then we would replace Date(ymd) by Year(CE:), or birthdate by birthyear, and replace the date values by year values (1946, 1980, 1980).

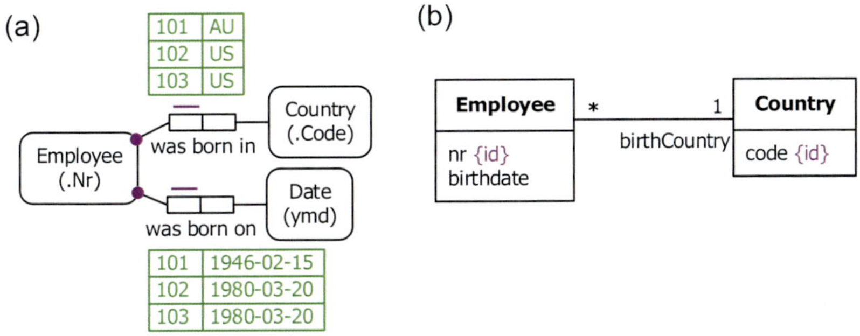

FIG. 10.15 Add a temporal fact type to record when once-only facts are "made true".

This example includes one temporal fact type (involving a temporal object type such as Date) and one nontemporal fact type (in this case a spatial fact type). Both these fact types are mandatory for employees. In other situations, either or both of these fact types might be optional for employees, and the relevant UML multiplicity constraint becomes 0..1.

If both fact types are optional for employees, but one is known only if the other is known, the relevant subset constraint should be added. For example, Fig. 10.16(a) models the case where we know an employee's birth date only if we know the employee's birth country, and Fig. 10.16(c) models the case where we know an employee's birth country only if we know the employee's birth date. The subset constraints depicted graphically in ORM are captured textually in UML using OCL, as shown in Fig. 10.16(b) and (d).

We use the term "event" to cover both *point events* (which occur at a single instant of time, e.g., a horse winning a race) and *period events* (which occur over a nonzero period of time, e.g., a rock concert). If the event described in the once-only fact is a period event, and we wish to record the start and/or end of that period, then add temporal fact types of the desired temporal granularity.

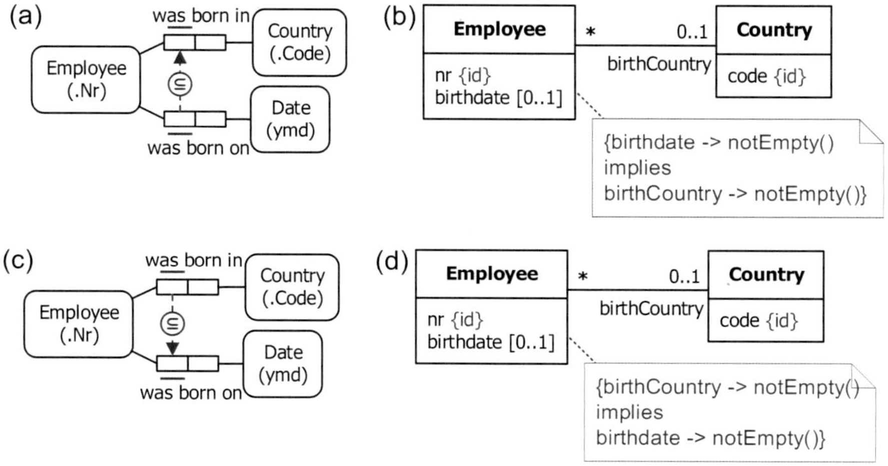

FIG. 10.16 Subset constraints between optional roles in (a, c) ORM and (b, d) UML.

As an example, suppose the fact type Employee first visited Country is used to record the event when people first visit various countries, and suppose we always know the date when such visits started but only optionally know when they ended. For example, this could record details about your first visit to Belgium, your first visit to Japan, and so on.

Fig. 10.17 shows one way to model this in ORM and UML, objectifying the *m:n* visit association and attaching fact types for the start date (mandatory) and end date (optional).

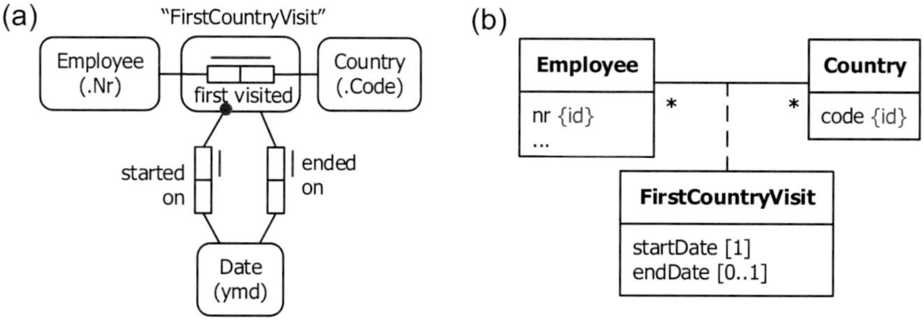

FIG. 10.17 Recording an open period for a period event in (a) ORM and (b) UML.

Alternatively, these two temporal fact types could be replaced by a single fact type linking to Period as a compositely identified object type or class. Yet another alternative is to replace the FirstCountryVisit objectified association by a coreferenced entity type or class, as in Fig. 10.18. This alternative approach is suitable for industrial versions of ER that do not support objectification.

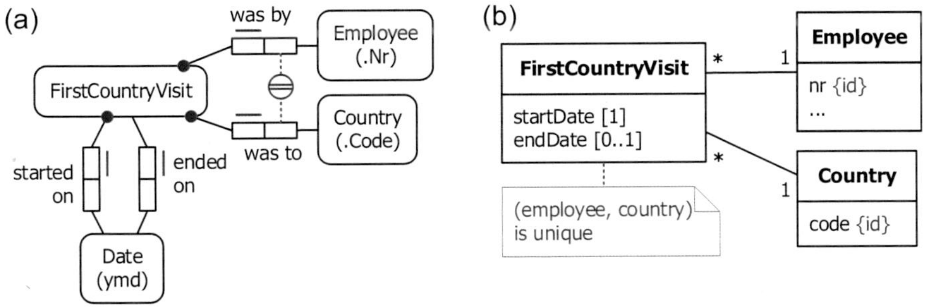

FIG. 10.18 A coreferencing alternative in (a) ORM and (b) UML.

Now suppose the fact type Employee lived first in Country is used to record where people lived first (e.g., Einstein lived first in Germany). This fact type is *n:1* rather than *m:n*. In ORM and UML, such associations may be objectified, and we may attach the temporal information as previously (see Fig. 10.19(a) and B).

ORM currently allows objectification of any fact type. This is fine for a unary fact type, as it has a spanning uniqueness constraint (asserted or implied). However, objectification of a binary or longer fact type without spanning uniqueness constraint is best avoided. For example, it might result in fact types that are not atomic, and lead to awkwardness or confusion (Halpin, 2020). Moreover, remodeling to avoid such objectifications typically results in a better schema. For example, Fig. 10.19(c) is arguably simpler than Fig. 10.19(a). This schema is also easier to emulate in modeling approaches such as Barker ER that do not allow such objectification, as shown in Fig. 10.19(d).

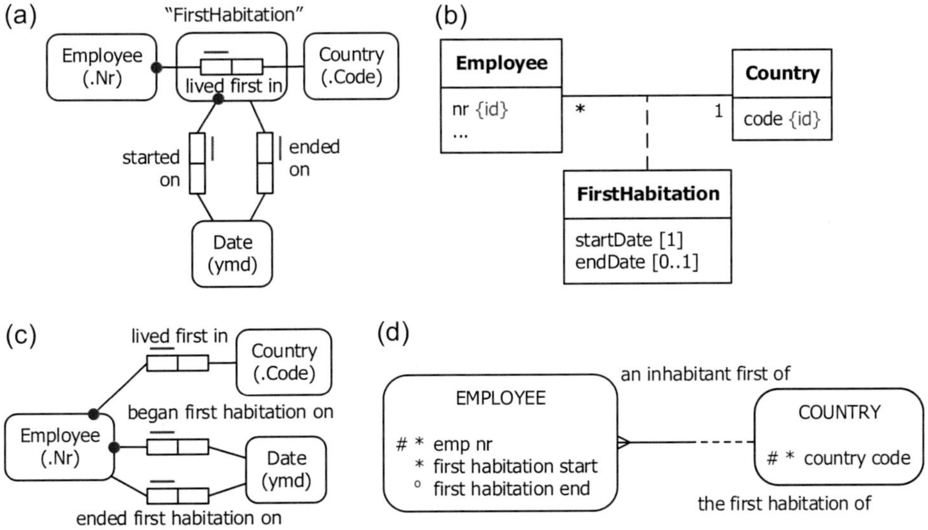

FIG. 10.19 Period events modeled by an *n:1* association.

Now, let's discuss how to record *snapshots of states that can change over time*. Consider a medical clinic where patients are identified by patient numbers, have their birthdate recorded, and optionally have their weight, temperature, and allergies (if any) recorded. Fig. 10.20 models this domain in ORM, UML, and Barker ER.

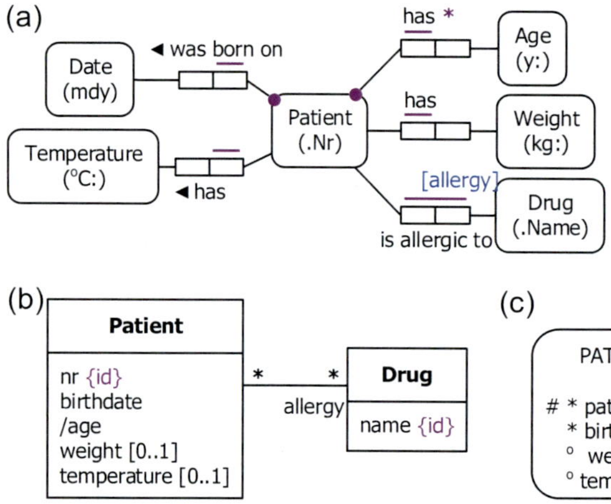

FIG. 10.20 A simple patient schema in (a) ORM, (b) UML, and (c) Barker ER.

The ORM and UML schemas include a derived fact type to compute a patient's age. A textual derivation rule may be added to compute age from birthdate and current date (in ORM, the latter is obtained from the implicit, predefined fact type Date is today).

The fact type Patient was born on Date is a once-only fact type, and hence is *unchangeable* (ignoring the possibility of error correction). Each of the other fact types is *individually changeable* (its individual instances may change over time). The fact type Patient has Weight is *functional* for Patient (at any time, a patient has at most one weight) and is repeatable (patients may gain or lose weight, and later return to their previous weight).

In physics, kg (kilogram) is a unit of mass and weight is a force. In common speech, however, "weight" is typically used to mean mass, and we use it here in that sense, using kg instead of kg wt for the unit.

The temperature fact type is also functional and repeatable. The allergy fact type is *m:n* (many-to-many) and repeatable (patients may cease to be allergic to a given drug, but later become allergic to it again). The derived age fact type is functional and changeable even though it is not repeatable.

Being functional, the age and weight fact types are changeable only at the instance level–an old instance can change only by being replaced by a new instance. The nonfunctional allergy fact type is also *collectively changeable* (its population of instances may change over time by adding new instances or deleting old ones.

The individual changeability of the age, weight, temperature, and allergy fact types follows from their *time-deictic* nature. Deixis (Greek for indicating, or point of reference) is the process whereby the referent of a linguistic structure depends on its communication (utterance or inscription) context and may be classified into five basic categories: time, place, person, discourse, and social. For example, the sentence "You are now reading this book" is person deictic ("you"), time deictic ("are", "now"), and discourse deictic ("this book").

In time (or temporal) deixis, the meaning of a communication act depends on the time in which it was performed. Temporal deictic references may appear via temporal adverbs (e.g., "now", "currently", "today", "tomorrow", "yesterday"), *tense* (e.g., "is eating", "was eating", "will eat"), combinations of deictic modifiers and time units (e.g., "last year", "this week", "next month"), or other temporal phrases (e.g., "two days ago", "within a month"). In information modeling, time-deictic aspects of predicates are usually restricted to tense and temporal adverbs.

In Fig. 10.20(a), the birthdate predicate is in past tense (was born on) while the other four predicates are in present tense (has, has, has, is allergic to). Because one cannot change the past, instances of the birthdate fact type are unchangeable. If we wish our information model to remember instances of birthdate facts, these facts also cannot be deleted. In contrast, instances of the four present tense fact types are changeable and may be replaced by other instances. Consider the following fact instance:

- Patient 101 has Weight 80 kg

This is shorthand for "Patient 101 *currently* has Weight 80 kg", which itself is shorthand for "Patient 101 has Weight 80 kg at time *t*, where *t* is the time at which the sentence was communicated". Originally, *t* is a valid time (e.g., the time at which the patient's weight was actually measured). Typically, the fact is then entered into the information model at some later transaction time and remains there until it either is updated by a more recent measurement or is deleted. From the viewpoint of the information model, the fact really means "*When last inspected*, Patient 101 had Weight 80 kg". For such reasons, the fact is known as a *snapshot* fact.

In practice, if the delay between valid and transaction times is minor, and the property (e.g., weight, allergy) is reevaluated on a suitably regular basis (which may be very infrequent if the property changes only slowly over time), then business users are willing to treat snapshot facts as if they were current.

Now, let's discuss how to record the *history of functional, changeable fact types.* If a fact type is individually changeable, and we need to maintain a *history* (full or partial) of its instances, then we should add or adapt a fact type to achieve this, using a temporal object type of the desired temporal granularity.

In ORM and UML, one simple way to do this is to insert into a "key" of the original, snapshot fact type a role that is played by the temporal object type, rephrasing the predicate as appropriate. A key of a fact type is a set of its roles that is exactly spanned by a uniqueness constraint. For example, in Fig. 10.20(a) the key of the weight fact type is the role played there by Patient.

Suppose we wish to maintain a history of patient weights where for each patient, at most one weight measurement is performed per day. Because our temporal granularity is 1 day, we choose Date as the object type and insert its relevant role into the binary fact type to produce the ternary fact type Patient on Date had Weight and expand the key uniqueness constraint to cover this role, as shown in Fig. 10.21(a).

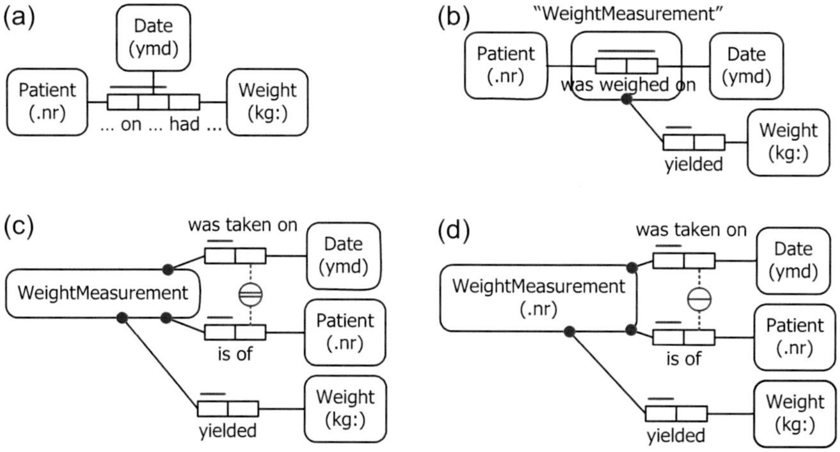

FIG. 10.21 Different options for maintaining history of patient weights in ORM.

The uniqueness constraint verbalizes as **For each** Patient **and** Date, **that** Patient on **that** Date had **at most one** Weight. If now or later you may wish to add details about the weight measurement (e.g., who performed the measurement), it is better to introduce a WeightMeasurement type either by objectification (Fig. 10.21(b)) or by coreferencing (Fig. 10.21(c)).

For easy reference, you could also introduce a measurement number to provide a simple identifier, as in Fig. 10.21(d), which includes an external uniqueness constraint to ensure that the patient and date combination provides a secondary identifier. If you wish to change the rules to allow the same patient to have more than one weight measurement on the same date, the external uniqueness constraint should be removed from Fig. 10.21(d).

In UML and Barker ER, it would be unusual to include Date as a class or entity type, so a WeightMeasurement type would normally be used. If no weight measurement number is introduced to identify weight measurements, the schemas shown in Fig. 10.22(a) and (b) may be used.

In Fig. 10.22(b), the combination of the two {id} markers on the dateTaken attribute and the association end connected to the Patient class indicate a composite identification scheme for WeightMeasurement, corresponding to the external uniqueness constraint in Fig. 10.21(c).

UML assumes that each object is identified by an internal surrogate. If desired, we could add a measurement number identifier for WeightMeasurement for Barker ER and UML. If the combination of date and patient still provides a secondary identifier for WeightMeasurement, the Barker ER identifier notation used in Fig. 10.22(a) to capture this constraint can no longer be used because it applies only to primary identifiers; so, this constraint should be noted elsewhere.

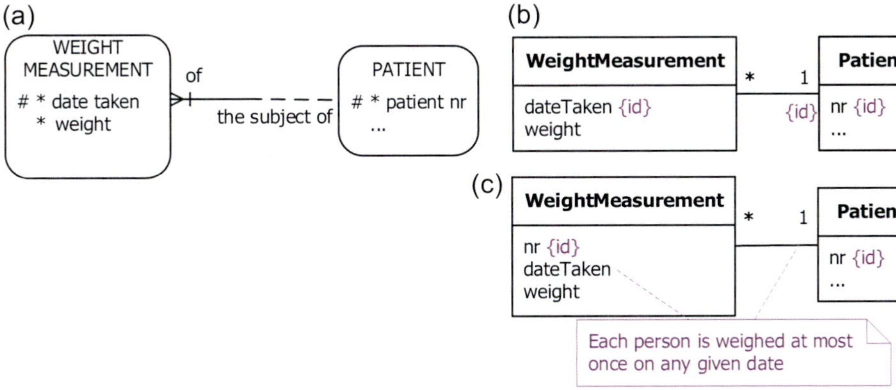

FIG. 10.22 Fig. 10.21 (c) remodeled in (a) Barker ER and (b) UML.

The UML schema in Fig. 10.22(c) adds a measurement number identifier for Weight-Measurement. The {id} marker for the WeightMeasurent.nr attribute prevents the use of the two {id} markers used in Fig. 10.22(b) to capture the composite uniqueness constraint, so this constraint is now captured textually in a note.

To maintain a history of patient temperatures, we might choose a different temporal granularity. For example, if our granularity is 1 h, we could use the fact type Patient at Hour had Temperature.

While the above approaches allow full history (at the desired granularity), they do not demand full history. For example, we are not constrained to record a patient's weight on a daily basis or a patient's temperature on an hourly basis.

For comparison purposes, we sometimes require a well-defined *partial history* (e.g., current and previous values only). The basics for such a situation can be modeled easily by using multiple fact types, as shown for patient weights in Fig. 10.23. A complete approach requires additional textual rules to assign the old current weight to the new previous weight on update.

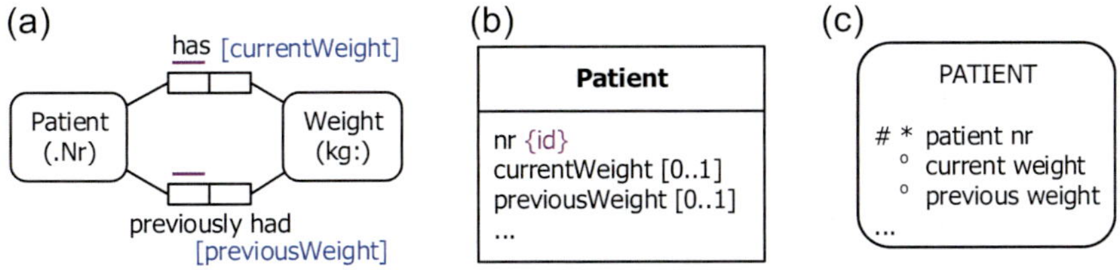

FIG. 10.23 Modeling consecutive weights in (a) ORM, (b) UML, and (c) Barker ER.

Maintaining a *history of nonfunctional fact types* is a little more complicated because they may be changeable both collectively and individually. Consider, for example, the report extract of country visits shown in Table 10.2.

Table 10.2 Extract from a record of visits to countries by employees.

EmpNr	*CountryCode*	*startdate*	*enddate*
1001	US	2022-01-01	2022-01-15
1001	NL	2023-07-11	?
1002	NL	2022-01-30	2022-02-13
1002	NL	2023-06-05	2023-06-05
1002	BE	2023-06-20	?

In this UoD, an employee may visit many countries, and a country may be visited by many employees. An employee may visit more than one country on the same date (e.g., The Netherlands and Belgium), but no employee may visit the same country more than once on the same date. For each visit, the start date is known but the end date might be unknown (e.g., the visit might still be in progress).

Fig. 10.24 shows some poor attempts to model such visits. Fig. 10.24(a) models Visit as an objectified association in ORM, and Fig. 10.24(b) takes a similar approach in UML, modeling Visit as an association class. Fig. 10.24(c) models Visit as an entity type in the Barker ER notation. All these schemas have problems. You might like to identify the main problem before reading on.

In Fig. 10.24(a), the fact type Employee visited Country is nonfunctional (in this case *m:n*) and *repeatable*. For example, the fact instance Employee "1002" visited Country "NL" is repeated, appearing on two rows. But the uniqueness constraint spanning the roles of the fact type Employee visited Country requires that such a fact instance appears at most once in any population of the fact type.

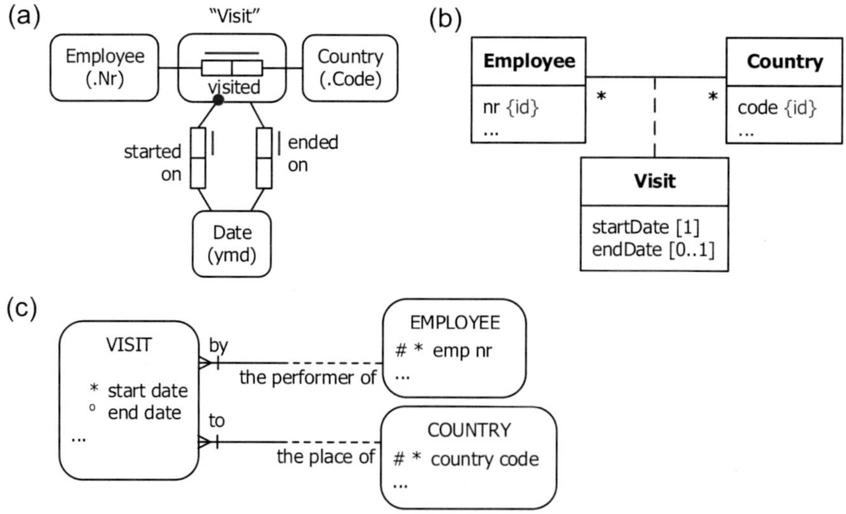

FIG. 10.24 Some attempts to model visits in (a) ORM, (b) UML, and (c) Barker ER.

In other words, the schema in Fig. 10.24(a) has *no way to record multiple visits to the same country by the same employee.* The same is true of all the other variations in Fig. 10.24—they assume that a visit may be identified simply by combining the visitor with the country visited. The schemas also lose the constraint that if a visit's end date is known, it must be on or after the start date. How *do* we record a full history of country visits? Try answering this before reading on.

There are at least three ways to resolve this problem. One approach is to *include a distinguishing temporal role as part of the identifier.* Given that each employee starts to visit any given country at most once on any given date, we may identify a visit by combining the visitor, the country visited, and the start date of the visit.

Fig. 10.25(a) and (b) shows two ORM schemas using this approach. Visit is an independent entity type, as its only role outside its preferred reference scheme is optional. The external uniqueness constraint involving the end date role indicates that when an end date exists, each combination of visitor, country visited, and end date applies to at most one visit. The value-comparison constraint ensures that if a visit's end date is known, it is on or after the start date.

UML and Barker ER versions of this approach are shown in Fig. 10.25(c) and (d). Neither has a graphic notation for the value comparison constraint nor for the external uniqueness constraint involving end date, but UML captures these constraints in a note.

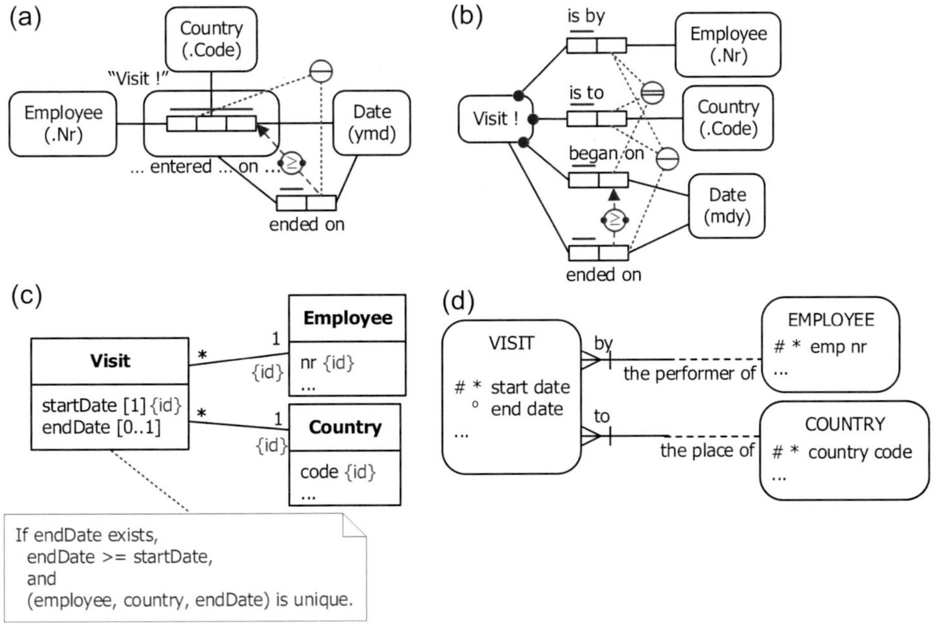

FIG. 10.25 Schemas including start date as part of the identifier for Visit.

A second approach is to *introduce a simple, visible identifier* for Visit, as shown in Fig. 10.26. Of course, both the former external uniqueness constraints still apply, but neither provides the preferred identifier.

In UML, hidden surrogate identifiers are assumed. However, we use id here as a visible identifier used by humans in the business domain to communicate about visits. This requires the addition of an id attribute to the UML diagram, as shown in Fig. 10.26(b).

In the Barker ER solution in Fig. 10.26(c), a visit-id attribute is added, with "#" marking it as the primary identifier, and the stroke is removed from the association lines. Barker ER has no notation for alternate identifiers, so now both external uniqueness constraints are lost.

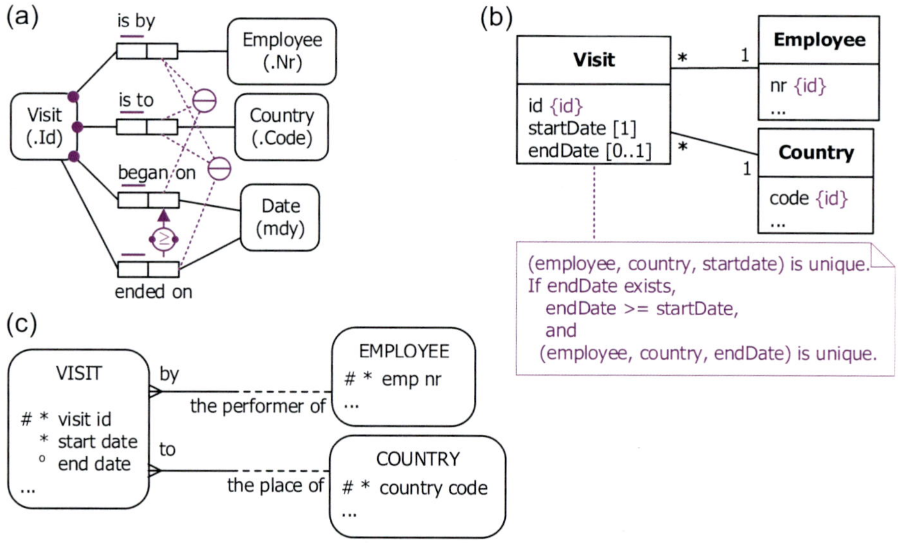

FIG. 10.26 Introducing a simple identifier in (a) ORM, (b) UML, and (c) ER.

A third approach is to *introduce an ordinal number as part of the identifier*. Here, the number is used to count the number of times the same employee visited the same country. For example, your first visit to The Netherlands is distinguished from your second visit to The Netherlands simply by including "first" and "second" in the definite descriptions. This visit number is included in the models in Fig. 10.27. For example, the first and second visits of employee 1002 to The Netherlands map to the tuples ("1002", "NL", 1) and ("1002", "NL", 2), respectively.

A fourth approach in UML is to tag the association ends in Fig. 10.24(b) with {nonunique}, so the population of the association is a bag of (employee, country) pairs. A surrogate id then provides a visit identifier. While allowed, this approach is subconceptual because the visit identifiers are no longer part of human communication.

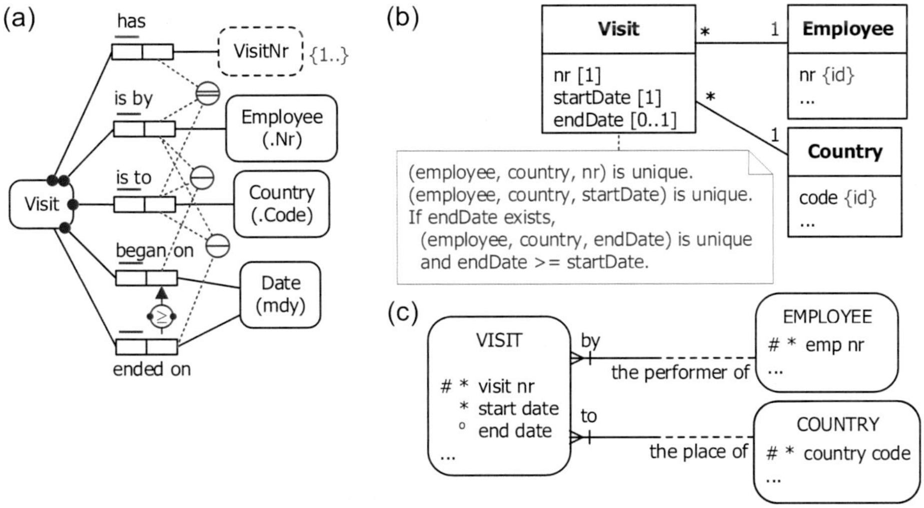

FIG. 10.27 Adding ordinals to help identify visits in (a) ORM, (b) UML, and (c) ER.

While we have now covered the most common temporal data model patterns, in practice other complexities can arise that require further analysis. So far, our constraints have been *static*—they apply to each state of the information system that models the business domain and may be checked by examining each state individually (e.g., each person was born on at most one date).

Dynamic constraints reference at least two states, which may be successive (e.g., no employee may be demoted in rank) or separated by a period (e.g., invoices ought to be paid within 30 days of issue).

UML supports dynamic constraints via both diagrams (e.g., state machines and activity diagrams, as discussed in later chapters) and textual formulation of dynamic rules in OCL. Apart from the nontransferable setting in Barker ER, industrial ER provides no support for dynamic constraints. ORM allows dynamic rules to be added as textual notes, for which a formal semantics may be provided. Various dynamic rule extensions to both ER and ORM have been proposed, but so far these have not made their way into popular modeling tools.

We now briefly sketch some ways to add dynamic rule support to ORM, focusing initially on updates to the population of a functional fact type. The fact type in Fig. 10.28 is used to store the *current* salary of each employee.

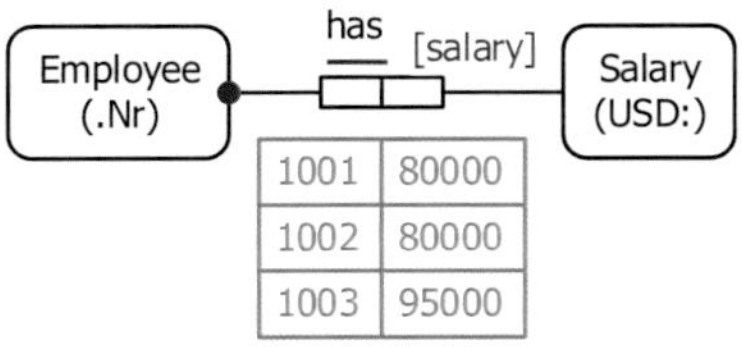

FIG. 10.28 An ORM model with a functional fact type.

At any given time, an employee has only one salary, but over time an employee may have many different salaries. Suppose the business wishes to enforce the policy that when the salary of an employee is updated, it must never decrease. Such a dynamic constraint over *successive states* is also known as a *transition constraint.*

Simple transition constraints may be formally declared in *textual rules* using reserved words with predefined semantics. For example, using "**old**" and "**new**" (or "**previous**" and "**next**"), respectively, to reference the states immediately before and after an elementary transaction, we might define a rule to prevent salary decreases in either of the following ways. The context of the constraint is the object type Employee.:

For each Employee,
  new salary >= **existing old** salary.

Context Employee
  new salary >= **existing old** salary.

The example just given involves a simple numeric comparison over an ordered quantity. When the relationship between successive states is more complicated, it may often be specified using a *state transition table or a state transition diagram.*

Fig. 10.29(a) shows a state transition table for the marital state example considered earlier in Chapter 2. Here, "1" indicates a possible transition and "0" an impossible transition.

Fig. 10.29(b) and (c) list the possible marital state changes in a state transition diagram. Two commonly used notations are shown. In Fig. 10.29(a), each state is shown as a named, horizontal bar and allowed transitions are depicted as vertical arrows. In Fig. 10.29(b), each state is shown as a named, soft rectangle.

(a)

From \ To	*Single*	*Married*	*Widowed*	*Divorced*
Single	0	1	0	0
Married	0	0	1	1
Widowed	0	1	0	0
Divorced	0	1	0	0

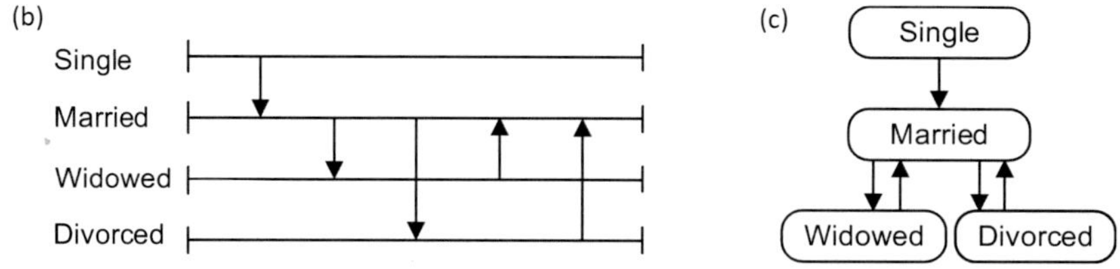

FIG. 10.29 Three state transition notations indicating possible transitions.

Given the ORM fact type Employee has MaritalStatus, the transition constraint may also be specified textually as a dynamic rule:

Context: Employee
in case old maritalStatus =
    'single':   **new** maritalStatus = 'married'

```
    'married':  new maritalStatus in ('widowed', 'divorced')
    'widowed':  new maritalStatus = 'married'
    'divorced': new maritalStatus = 'married'
end cases
```

Another possible approach is to allow ORM extensions using *reserved predicates* with predefined semantics. For example, the reserved predicate "may change to" could be applied to various object types to constrain their changes from one state to the next. Using this idea, the fact table in Fig. 10.30 could be used to declare the possible marital state transitions. This effectively demotes metadata (in this case a constraint) to object data, thus simplifying schema evolution because changes to the transition rule are now data changes rather than schema changes. Notice that the salary rule considered earlier cannot be specified in this way because its formula involves an operator, not just values.

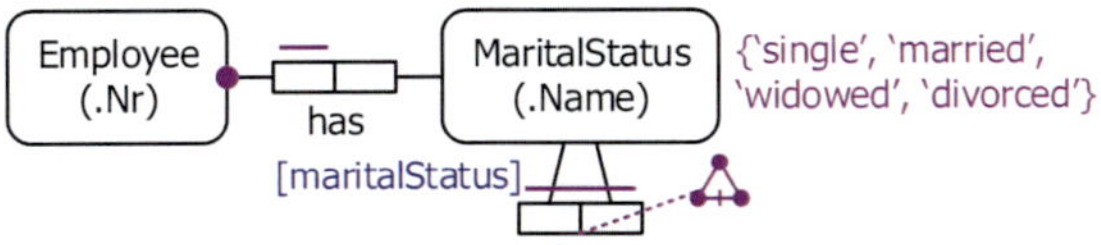

may change to	
single	married
married	widowed
married	divorced
widowed	married
divorced	married

FIG. 10.30 A transition constraint captured as data using a reserved predicate.

Now consider *adding fact instances* to a *nonfunctional fact type,* such as the Seating used Table association in Fig. 10.31, which shows a model fragment from a restaurant application. A seating is an allocation of a party (of one or more customers) to one or more vacant tables.

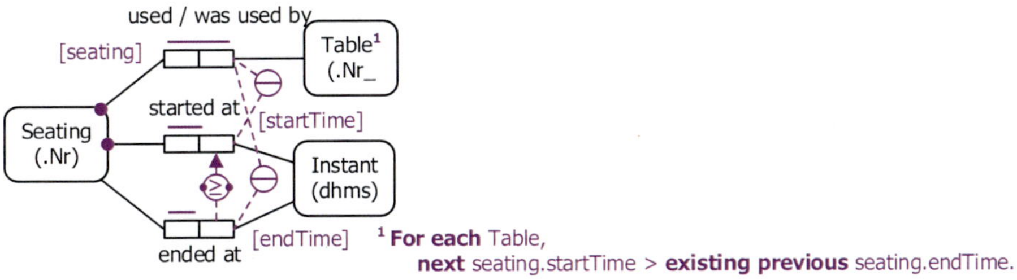

FIG. 10.31 A dynamic constraint on additions to a nonfunctional, historical fact type.

The constraint that *no two seatings that overlap in time used the same table* may be specified by the dynamic rule shown as the footnote in Fig. 10.31. This dynamic constraint is equivalent to the following static constraint:

For each $Seating_1$ **and** $Seating_2$,
  if $Seating_1$ <> $Seating_2$
    and $Seating_1$.startTime <= $Seating_2$.startTime
    and ($Seating_2$.startTime <= **some** $Seating_1$.endTime
      or not exists $Seating_1$.endtTime)
  then no Table was used by $Seating_1$ **and** was used by $Seating_2$.

For a detailed discussion of more complex examples of dynamic rules for state transitions expressed in a proposed dynamic rule language for ORM as well as a formalization in terms of logging semantics, see Balsters and Halpin (2016).

A *guarded transition diagram* allows a guard condition expressed as a formula to be applied to a transition. In practice, many different diagrams and formalisms are used to specify various kinds of dynamic constraints. These include state charts, activity diagrams, state nets, colored Petri nets, process algebra-based task structures, and *Event-Condition-Action* (*ECA*) declarations. ECA languages allow statements of the form **If** *event* **then on** *condition* **then** *action1* [**else** *action2*]. This provides a general framework for *controlling the action to be taken on violation of a constraint.* An update is an example of an event, and a constraint violation is an example of a condition.

An extreme version of a transition constraint is to make a fact *immutable* (unchangeable). In other words, once an elementary fact is entered it cannot be deleted. This kind of constraint is best considered in connection with dynamic aspects of subtyping, which we soon discuss.

As mentioned before, some dynamic rules reference states that are separated by a specified period rather than being successive. In practice, many of these rules are *deontic* in nature. For example, consider the rule that invoices *ought* to be paid within 30 days of being issued. This rule may be formally expressed as follows, assuming that the relevant fact types are declared in the schema.

It is obligatory that
  each Invoice **that** was issued on **some** $Date_1$
  is paid on **some** $Date_2$
  where $Date_2$ <= $Date_1$ + 30 days.

As the final segment of this section on temporal issues in modeling, we now turn to *dynamic aspects of subtyping.* This assumes that you are familiar with the static aspects of subtyping in ORM and UML discussed in Sections 6.5 and 6.6.

Some proposals from the ontology engineering community have employed type metaproperties to ensure that subtyping schemes are well formed from an ontological perspective. For example, Guarino and Welty (2002) argue that every property in an ontology should be labeled as rigid, nonrigid, or antirigid.

Rigid properties (e.g., being a person) necessarily apply to all their instances for their entire existence. *Nonrigid* properties (e.g., being hard) necessarily apply to some but not all their instances. *Antirigid* properties (e.g., being a patient) apply contingently to all their instances. One may then apply a meta constraint (e.g., antirigid properties cannot subsume rigid properties) to impose restrictions on subtyping (e.g., Patient cannot be a supertype of Person).

Guizzardi et al. (2004) proposed a UML profile to stereotype classes into kinds, subkinds, phases, roles, categories, roleMixins, and mixins, together with a set of meta constraints, to help ensure that UML class models are ontologically well formed. Guizzardi (2005) used this modeling profile in his doctoral thesis on ontological foundations for conceptual information models.

While the aforementioned research provides valuable contributions, we have some reservations about its use in industrial information systems modeling. Our experience with industrial data modelers suggests that the seven-classes scheme would seem overly burdensome to the majority of them.

A simpler alternative that we are currently considering for ORM 2 *classifies each subtype as either a rigid subtype or a role subtype*. A *type is rigid* if and only if each instance of that type must belong to that type for its whole lifetime (in the business domain being modeled). Examples include Person, Animal, and Book.

In contrast, an instance of a *role type* need not always be an instance of that type during its lifetime (in the business domain). Here, we use "role" liberally to include a role played by an object (e.g., Manager, Student, Patient—assuming that these are changeable in the business domain) as well as a phase or state of the object (e.g., Child, Adult, FaultyProduct—assuming changeability).

Although this rigid/role classification scheme applies to any type, we typically require this distinction to be made only for subtypes (to control subtype migration and reduce the classification burden for modelers). As an example, Fig. 10.32 depicts Dog and Cat as rigid subtypes in ORM and UML.

The rigidity notation suggested for ORM is square bracketing of the subtype name (violet for alethic; blue with "o" for deontic, e.g., changing from male to female might be possible but forbidden).

The NORMA tool does not yet support this notation to classify subtypes, but it may be informally noted in the subtype's Note property. For UML, we chose a rigid stereotype. Rigidity is a *dynamic constraint* rather than a static constraint because it restricts state changes (e.g., no dog may change into a cat).

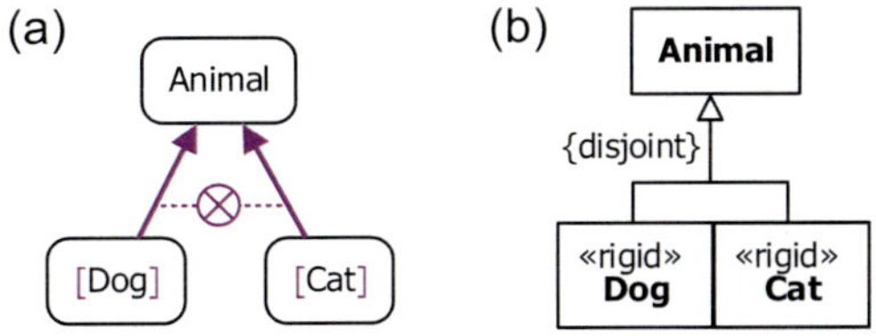

FIG. 10.32 Rigid subtypes depicted in (a) ORM and (b) UML.

In the previous, the subtypes are asserted. If we add the fact type Animal is of AnimalKind, we can use that to derive the Dog and Cat subtypes. If the fact type Animal is of AnimalKind is made *unchangeable* (an animal can't change its kind), the rigidity of the derived subtypes is now derivable. ORM does not yet support changeability status of fact types. UML recognizes four changeability settings: unrestricted, readOnly, addOnly, and removeOnly. If we derive Dog and Cat subclasses of Animal from its Animal.kind attribute, and declare that attribute read-only, the rigidity of the subclasses is implied.

To avoid explicitly declaring role subtypes as such, ORM subtypes may be assumed to be role subtypes by default. Unlike rigid subtypes, *migration between role subtypes is often permitted*. As a simple example, a person may play the role of child, teenager, and adult at different times in his/her lifetime (see Fig. 10.33).

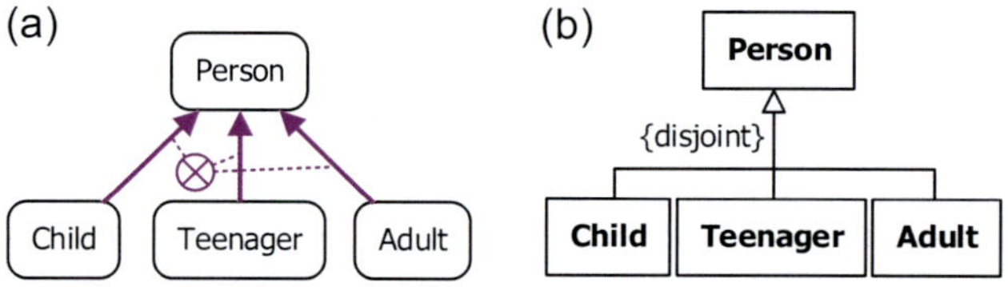

FIG. 10.33 Migration between role subtypes is allowed.

Various data model patterns cater for such *historical subtype migration*. The simplest pattern deals with linear state transitions. For example, in Fig. 10.34(a) each role has specific details, and we wish to maintain these details of a person (e.g., favorite toy, favorite pop group) as he/she passes from one role to another.

An appropriate pattern for this case is to start with a supertype that disjoins all the roles and then successively subtype to smaller disjunctions, as shown in Fig. 10.34(c). We call this the *decreasing disjunctions pattern*. Depending on the business domain, a simple name may be available for the top supertype (e.g., Person).

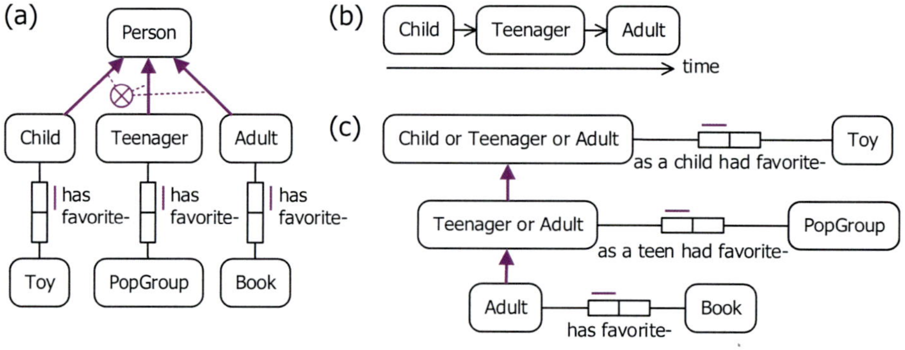

FIG. 10.34 Retaining history of subtype-specific details as a person changes roles.

An alternative approach using what we call the *once-only role-playing pattern* creates subtypes for the actual playing of the roles, as shown in Fig. 10.35. Here, the subtypes PersonAsChild, PersonAsTeenager, and PersonAsAdult are subtypes not of Person but of the objectified association LifeRolePlaying, instances of which correspond to the playing of one of those roles by a person. The temporal order can be added in a number of ways, for example, by adding a dynamic constraint to constrain the possible role transitions, as shown in Fig. 10.35.

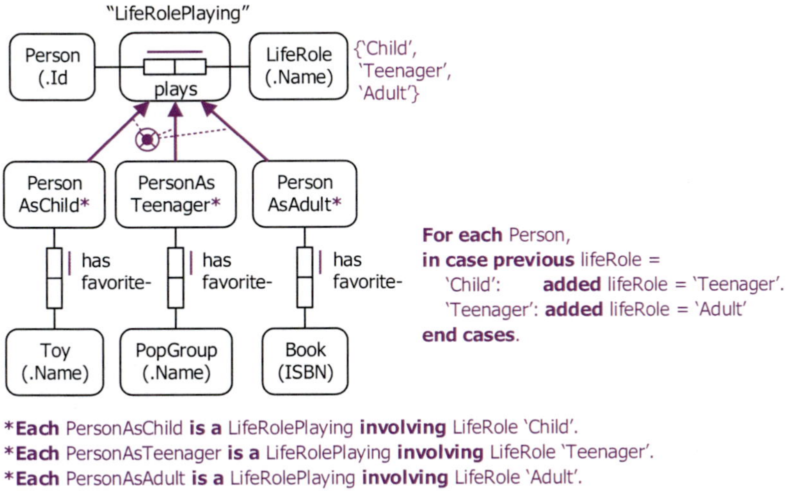

FIG. 10.35 An alternative approach using subtypes for role playing.

This role-playing pattern can be adapted to a *repeated role-playing pattern* that allows cycles in the role state graph, for example by including a start time or an ordinal number in the identifier for the role-playing object type, as in Fig. 10.36.

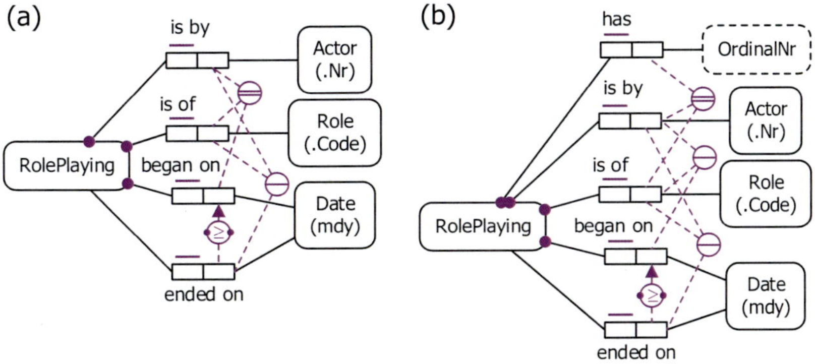

FIG. 10.36 Two versions of the repeatable role-playing pattern.

Fig. 10.37 shows a concrete example of the repeatable role-playing pattern in Fig. 10.36(a) together with a dynamic transition constraint on a person's marital status. To avoid clutter, the display of the maritalRolePlaying and maritalRole role names is suppressed.

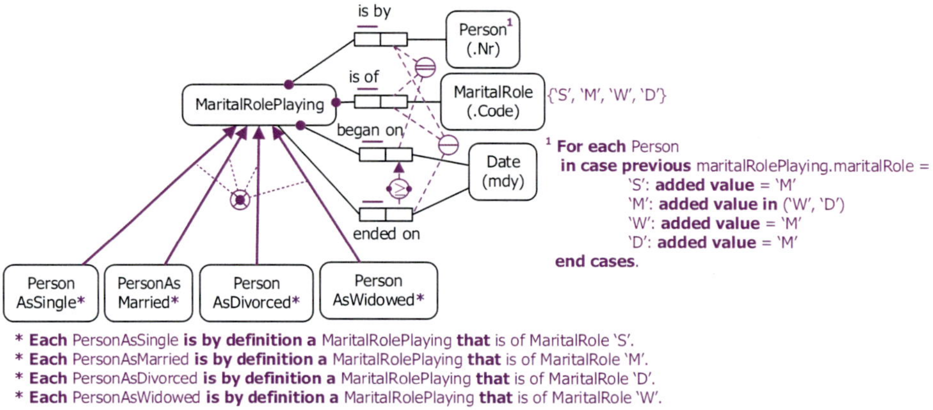

FIG. 10.37 Example of a repeatable role-playing pattern with a dynamic rule.

Exercise 10.3

1. (a) The following tiny extract is from a report about patient blood pressures (BPs). BP is determined from two readings, systolic pressure and diastolic pressure, each measured in millimeters of mercury (mmHg). For example, a BP of 150/95 has a systolic pressure of 150 mmHg and a diastolic pressure of 95 mmHg. The information system is required to explicitly capture the semantics of BP readings. Schematize this UoD in ORM, including any derivation rules. Do *not* introduce a BP measurement object type.

PatientNr	*Patient Name*	*BP*	*Time*
101	John Smith	150/95	2007-05-01 08:30:00
101	John Smith	130/90[a]	2007-05-01 15:30:00
102	Ann Jones	120/80[a]	2007-05-01 08:30:00

[a] = latest BP (i.e., most recently measured BP for that patient)

(b) Remodel this situation using the object type BP measurement, introducing a simple identifier for such measurements.

2. For each of the following sentences, state which parts are time deictic.
   (a) George W. Bush is our current president.
   (b) We will graduate next quarter.
   (c) Today's invoice should be paid within 30 days.
   (d) Australia just now won the World Cup in cricket.

3. The following ORM schema is designed to capture details about when people lived in various countries. A person may live in the same country many times. We know each time a person starts to live in a country but might not know when they end that habitation (e.g., they might still live there).
   (a) Explain why this design is poor.
   (b) Provide an improved ORM schema that removes the problem.

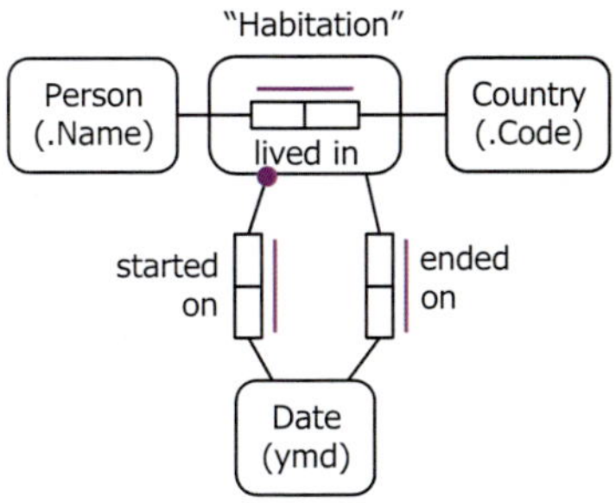

4. Each academic at an Australian university has exactly one of the following possible ranks listed in ascending order: AL=Associate Lecturer, L=Lecturer, SL=Senior Lecturer, R=Reader, P=Professor. Academics at any of the four lower ranks may be promoted to the next highest rank. Senior lecturers may also be promoted directly to Professor. No other rank changes are possible. Specify the relevant transition constraint using (a) a state chart, and (b) a transition table.

5. (a) Explain what is wrong with each of the following subtyping arrangements.

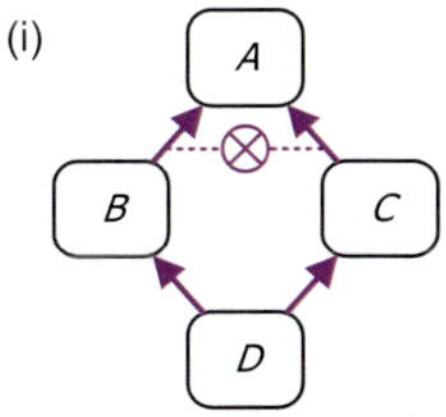

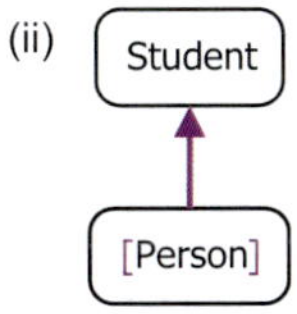

   (b) The following schema was designed to record details about people who apply for an employment position at a given university, are employed at the university, or are past employees of the university.

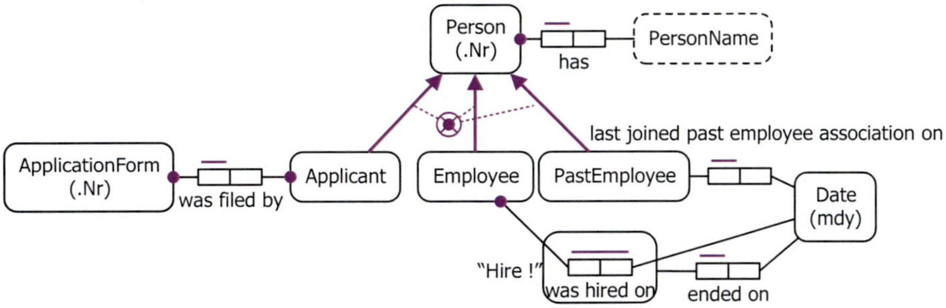

A person may be employed more than once at the university. Past employees may optionally join the past employee association; although they may do this more than once, only their most recent joining is of interest. However, the university wishes to retain a history of such details about people (e.g., employee and applicant details are retained for past employees). Modify the schema to enable this using (a) the decreasing disjunctions pattern and (b) the role-playing pattern.

10.4 Collection Types

Data modeling approaches vary considerably in their support for *collection types*. While implementation targets often support such types (e.g., SQL:2003 supports arrays, row types, and bags), high-level data modeling approaches typically provide either no direct support or only very limited direct support. Even when no direct support for fact collections is provided, it is possible to model the same information in terms of elementary facts, so the issue of collection type support is one of convenience rather than expressibility. The following collection types are the most commonly encountered in conceptual modeling:

- Set (unordered, no duplicates)
- Bag or multiset (unordered, duplicates allowed)
- Ordered set or unique sequence (ordered, no duplicates)
- Sequence (ordered bag)
- Array (indexed sequence)
- Schema

Industrial ER provides no direct support for these collections. UML enables many of these collection types to be specified as stereotyped classes, multivalued attributes, or constrained association ends. Although most versions of ORM, including ORM 2, are "flat" in the sense that they provide no direct support for collections, the Predicator Set Model (PSM) variant of ORM provides deep support for them as first-class types (ter Hofstede et al., 1993).

While the use of collection types in logical and physical data structures makes sense, whether collection types should be used at all in conceptual analysis is still a disputed issue. The addition of constructors for building various collection types requires additional language features to not only declare such structures, but also manipulate them. With collection options, the modeler now has many more choices as to how to formulate a model, update, or query.

Our experience indicates that collection types should be used with extreme care, if at all, in developing a conceptual model. Undisciplined use of constructors often leads to incorrect or incomplete solutions or to overlooking a simpler solution that does not use collections. Recall our earlier advice in Section 9.3 against using multivalued attributes in conceptual models.

The following points have been proposed in favor of conceptual constructor usage: collections lead to more compact models; they better reveal the connection between conceptual models and physical models that use collections; and they allow unnamed structures to be modeled directly.

The following points have been proposed against using collection types as first-class types in conceptual models: collections complicate verbalization and population; collections often complicate constraints and hide other details; choosing the best constructor is often difficult; unnamed structures should be avoided because they are awkward to reference; and one can still transform to collections in the physical model simply by annotating the mapping process.

The main point of disputation is whether it is a good idea to include *unnamed collections* in conceptual models. As a classic example (Hammer and McLeod, 1981), consider a convoy as an *unnamed set* of ships. This may be modeled in PSM using a set collection type, as shown in Fig. 10.38(a). While various notations exist for collection types, here we depict them as soft rectangles, with the collection kind (set, bag, etc.) indicated in bold.

Fig. 10.38(b) shows an alternative flat schema in both ORM and UML using an association, and a name to identify each convoy. In practice, convoy names are very useful, making it easy to verbalize facts about convoys and to check constraints. The uniqueness constraint on the ORM model asserts that each Ship is in at most one Convoy. This constraint is captured in the UML diagram also because the solid diamond indicates composition (strong aggregation). This constraint is missing from Fig. 10.38(a). In most cases, the use of collection types makes it harder to express constraints on collection members. The association in the flat model can also be populated conveniently with fact instances to help validate the constraint.

For snapshot purposes, it is convenient to identify a convoy by its name (or by its flagship). The convoy membership fact type can be temporalized to maintain a historical record of membership either by using the ternaries Ship joined Convoy on Date and Ship left Convoy on Date or by nesting the historical *m:n* association Ship was in Convoy as ConvoyMembership, and adding ConvoyMembership began on Date; ConvoyMembership ended on Date. In UML, this history may be modeled using ConvoyMembership as an association class with appropriate date attributes (possibly multivalued) or associations. Whether it is philosophically correct to use the word "convoy" to allow a convoy to be the same convoy after a ship leaves is of little practical interest.

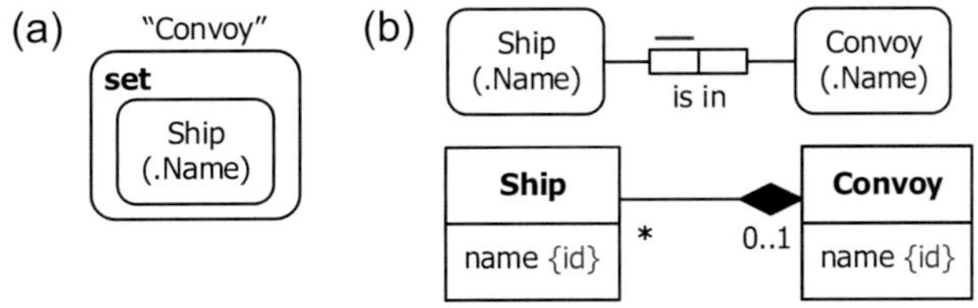

FIG. 10.38 Convoy modeled as (a) an unnamed set, and (b) a named simple object.

As a related example, consider the notion of team membership, where a person may be a member of many teams and vice versa. Fig. 10.39 shows how to model this in ORM with and without the set constructor. In Fig. 10.39(b), the circled "{_}" depicts an *extensional uniqueness constraint* (ter Hofstede and van der Weide, 1994) to declare that teams may be defined by their membership (i.e., different teams cannot have exactly the same set of members). A sample population for two teams is included. The underline (like a uniqueness bar) inside the set braces intuitively suggests that the set is unique. An alternative notation for this constraint is a circled "eu". This constraint is implicit in Fig. 10.39(a). Its use in B allows unnamed structures without introducing collection types.

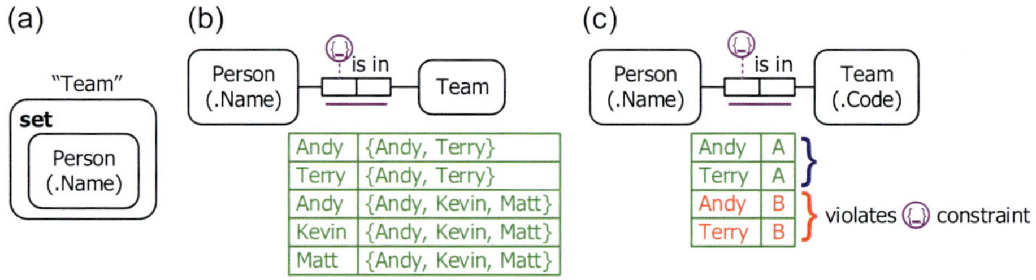

FIG. 10.39 Three ways of modeling team membership.

Bakema et al. (1994) provide another notation for this constraint, and argue that if teams are unnamed in reality, this must be reflected in the conceptual model to obey the conceptualization principle to model only conceptually relevant features. We adopt a relaxed reading of this principle, encouraging the introduction of identifiers that improve model usability. In Fig. 10.39(c), we add a team code as a simple identifier for teams to facilitate talk about teams.

The extensional uniqueness constraint is still required if we definitely don't want two different teams to have the same membership, as illustrated by the counterexample provided. This constraint is very expensive to enforce, so should be used only when necessary.

A complex example cited in the ORM literature to justify the use of constructors deals with chemical reactions (Falkenberg, 1993). This models reactions such as $2H_2 + O_2 \leftrightarrows 2H_2O$. One way to model this with a set collection type is shown in Fig. 10.40(a). This schema, based on a solution by ter Hofstede et al. (1992), treats each side of a reaction equation as a set of chemical quantities, for example, $\{(H_2, 2), (O_2, 1)\}$.

The reaction itself is then captured in the binary input/output association. An alternative flat model is shown in Fig. 10.40(b). Here, a reaction number is introduced as a convenient way of referring to reactions. This model also shows a complex example of the extensional uniqueness constraint.

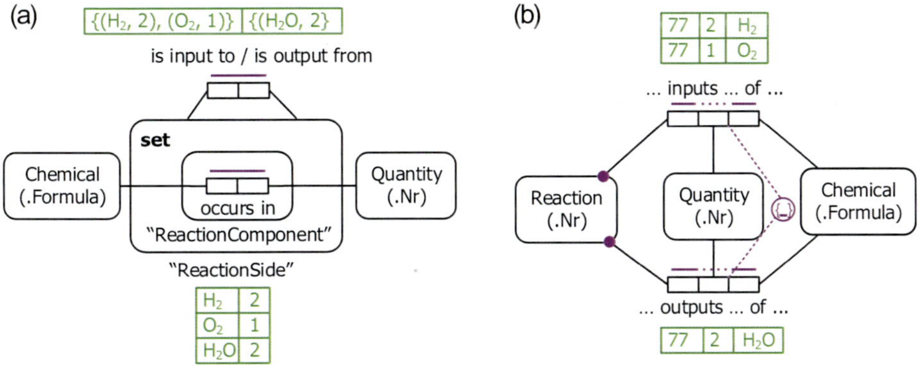

FIG. 10.40 Chemical reactions modeled (a) with and (b) without unnamed sets.

Although Fig. 10.40(a) is more compact, its use of a set constructor makes it harder to think about. For example, it lacks a simple constraint captured in Fig. 10.40(b). Can you spot it? The uniqueness constraints in Fig. 10.40(b) ensure that each chemical occurs at most once in each reaction side (e.g., to exclude equations such as 5H ⇆ 3H+2H). This is not enforced in Fig. 10.40(a). For example, we may populate the input/output association with the tuple ⟨{(H, 5)}, {(H, 3), (H, 2)}⟩. The set-based model makes this constraint harder to conceive and express.

In Fig. 10.40(a), the reaction association is *m:n* (presumably other conditions such as catalysts and temperature could be varied to have the same input reagents produce different results). If we instead demand that the same input always gives the same output, this association becomes 1:1. This is easy to model in Fig. 10.40(a), but harder to model in Fig. 10.40(b). This is one of those rare cases when we encounter a constraint on a collection that cannot be modeled easily as a constraint on the members. The opposite situation (a constraint on members that is hard to model on collections) is far more common.

Falkenberg (1993) models a chemical reaction instead as a set of sets of object-role pairs. Although the structure of this model is difficult to comprehend, even for modelers much less domain experts, Falkenberg argues that it is the only correct way to model the situation, based on a deterministic principle that one transition instance must correspond to exactly one object instance. Our viewpoint is that communication is more important than determinism, and there are often many correct ways to model the same business domain.

While there seems to be no compelling case to introduce collection types as first-class citizens in the conceptual model, we still need some way to specify collections for implementation in targets that support such structures. While this might be done directly on the logical schema, it is desirable that there be a two-way transformation between conceptual and logical levels so that changes in one can be communicated to the other to keep the models synchronized.

One simple way to do this is to specify collection type mappings as *annotations* to the pure conceptual schema, resulting in an annotated conceptual schema used only by the developer, not the domain expert, for the sole purpose of controlling the mapping between conceptual and lower levels. Different annotations could be chosen for different implementation targets. At the time of writing, no standard annotations for this purpose have yet been finalized.

We conclude this section by showing how to model a variety of collection types indirectly using a flat approach. Fig. 10.41(a) uses a unique sequence constructor to model an *ordered set* of authors. Fig. 10.41(b) models this in UML applying the {ordered} property to an association end ({unique} is assumed by default for this collection). This may be modeled in flat ORM by introducing a Position object type to store the sequential position of any author on the list, as shown in Fig. 10.41(c). The uniqueness constraint on the first two roles declares that an author occupies at most one position for each paper; the constraint on the first and third roles indicates that for any paper, each position is occupied by at most one author. The textual constraint says that list positions are numbered sequentially from 1.

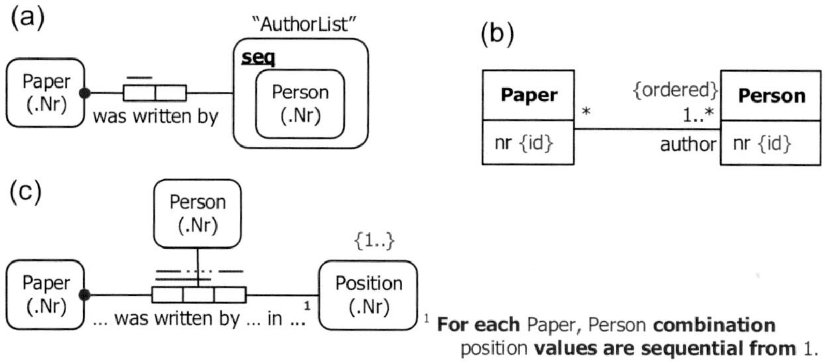

FIG. 10.41 Some different ways to model ordered sets.

Although this ternary representation may appear awkward, it is easy to populate and it facilitates any discussion involving position (e.g., who is the second author for paper 21?).

From an implementation perspective, a sequence structure could still be chosen; this can simplify updates by localizing their impact. However, the update overhead of the positional structure is not onerous, given set-at-a-time processing (e.g., to delete author n, simply set position to position–1 for position $>n$).

Care is required when modeling collection types. Fig. 10.42(a) associates each fact type with an ordered set of roles, and Fig. 10.42(b) flattens this into a ternary. See if you can spot something wrong with these schemas before reading on.

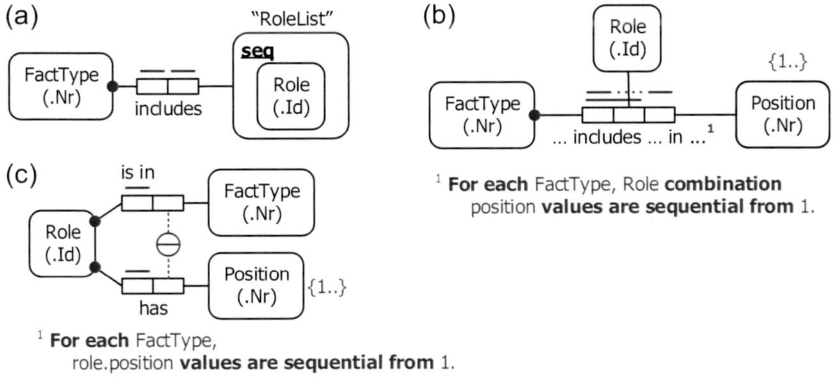

FIG. 10.42 Can you spot what is wrong with schemas (a) and (b)?

Both schemas omit the constraint that each role belongs to only one fact type. Fig. 10.42(a) requires each role list to belong to at most one fact type, but that's not strong enough; the use of a collection type here prevents us from capturing the required uniqueness constraint graphically. The ternary in Fig. 10.42(b) is not elementary (a uniqueness constraint applies to its second role). Fig. 10.42(c) provides a correct solution.

Recall that *bags* (multisets) are unordered collections with duplicates allowed. These may be implemented by ternary associations using a frequency object type to count the number of occurrences for each entry.

For example, suppose student 1001 obtains the bag of grades [A, A, B, B, B] in five tests, and student 1002 obtains the bag of grades [B, B, B] in three tests. This may be modeled in flat ORM as shown in Fig. 10.43(a). By default, multivalued association ends in UML denote sets (unique and unordered). Assuming that the nonunique setting is allowed here (the UML specification is unclear on this issue), a bag setting may be specified by the {nonunique} property, as shown in Fig. 10.43(b).

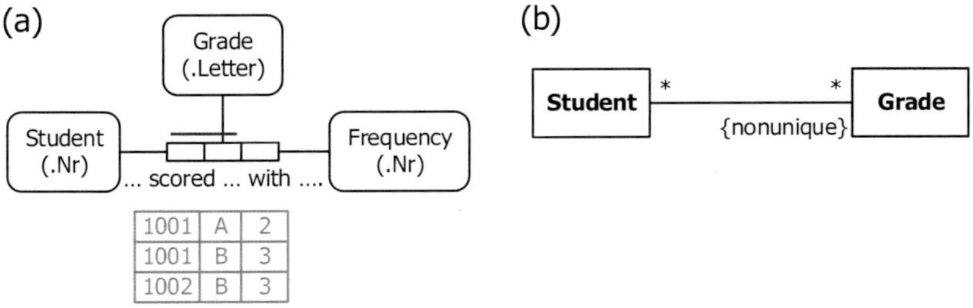

FIG. 10.43 Modeling a bag of grades for each student in (a) ORM and (b) UML.

Recall that a *sequence* is an ordered bag (ordered collection with duplicates allowed). For example, a paragraph is a sequence of sentences. Sequences may be modeled in flat ORM using a ternary association with a "Position" object type in the key. Suppose a lottery presents prizes to a sequence of people, in descending order of prize value, with no ties.

For example, the prizes for lottery L1 might be awarded to the sequence (Ann Smith, Tom Jones, Ann Smith). This may be modeled in flat ORM using Rank as a position type, as shown in Fig. 10.44(a), along with a sample population. Fig. 10.44(b) models this in UML using the {ordered, nonunique} setting, assuming this is allowed.

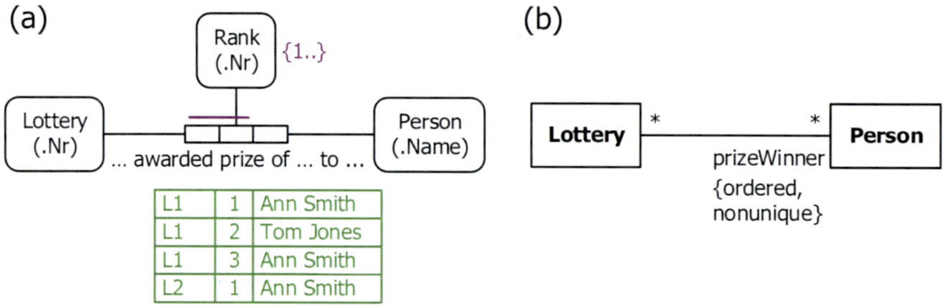

FIG. 10.44 Modeling a sequence in (a) flat ORM and (b) UML.

Arrays are indexed sequences. They may be modeled in flat ORM using a ternary with a "Position" type for the index as part of the key. For example, the ORM model in Fig. 10.45(a) uses Place as a position type to record the first three places in sporting events, assuming *no ties* are allowed.

Using a suitable mapping annotation, this could be mapped to an indexed array of countries inside sport events, leading to an object-relational database model such as

SportEvent (eventName, results; array[1..3] of Country)
'Mens 100 m freestyle' [1: 'US'
2: 'AU'
3: 'US']

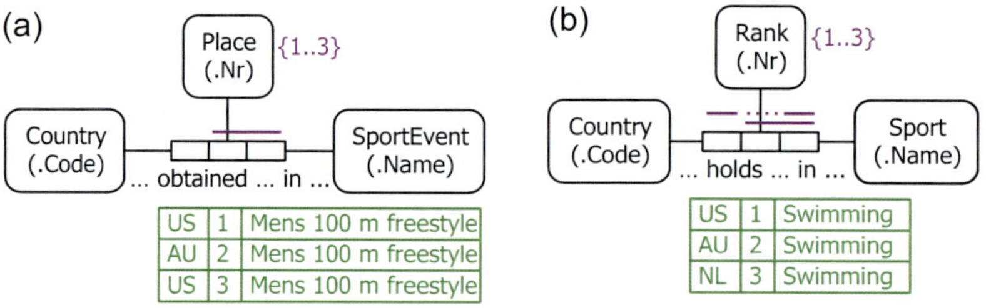

FIG. 10.45 Modeling arrays of sporting event places with no ties in flat ORM.

Fig. 10.45(b) shows a similar example involving an additional uniqueness constraint (each country has at most one rank in any given sport). Again, no ties are allowed. This could be mapped to a unique array structure (with uniqueness indicated by underlining) as follows:

Sport (sportName, results; array[1..3] of Country)
'Swimming' [1: 'US'
2: 'AU'
3: 'CN']

Although direct use of collection types in a conceptual model can lead to compactness, an alternative flat model (no constructors) of comparable convenience is generally available. The use of constructors often makes it harder to express constraints (which typically occur on members, not collections).

Because constructors involve compound facts, they are awkward to populate or check for validation, and they require more sophisticated modeling skills. Constructors are sometimes used even when there is a simpler modeling solution without them. It seems better to relegate constructor usage to annotations that guide the later mapping phase.

Exercise 10.4

1. In Terry's Railway, each train consists of exactly one locomotive and exactly one carriage train (ordered set of railway cars). Each carriage is classified as first class, second class, or third class. It is possible that a carriage is not currently part of a train. Trains, carriages, and locomotives are identified by numbers. The same locomotive may power many trains.
   (a) The same carriage may belong to more than one train (e.g., a carriage used on a morning train may be used on a different train in the afternoon). However, the same carriage train can be used in only one train. Model this in ORM using a collection type constructor.
   (b) Model the domain described in (a) without using a collection type constructor.
   (c) Suppose now that each train is uniquely determined by its composition (i.e., its locomotive and carriage train). Is any change needed to your model for (a) to capture this? Explain.
   (d) Can this additional constraint be captured graphically in your solution to (b)? Answer yes or no.
   (e) The domain is changed so that each carriage belongs to at most one train. Can this rule be added as a graphic constraint (without adding a fact type) to your solution to (a)? Answer yes or no.
   (f) Remodel your solution to (b) to ensure that the rule in (e) is expressed graphically.
   (g) If the composition constraint in (c) also applies to your solution to (f), does any further constraint need to be added to that solution? Answer yes or no.

2. In cricket, a player may score runs in singles, fours, or sixes. Some players may score a duck (no runs). The following report extract lists a bag of runs (ignore the order) for each cricketer. Each cricketer belongs to exactly one team. Model this in flat ORM.

Cricketer	*Team*	*Runs*
Don Superman	A	6, 6, 6, 4, 4, 6, 4, 1, 6, 1, 6
Fred Bloggs	A	
Mark Hitter	B	1, 1, 1, 1, 4, 1, 1

10.5 Nominalization and Objectification

Roughly speaking, objectification or nesting treats a relationship instance as an object in its own right, so it may be talked about like any other object. In UML this is called reification, where instances of an association are objectified as instances of an association class. The basic notion of objectification in ORM and UML was discussed in Chapters 4 and 9, respectively. Industrial ER typically provides no support for objectification. Some academic versions of ER support objectification, but typically with restrictions (e.g., objectified relationships may have attributes but cannot play in other relationships).

This section examines objectification in greater depth, first distinguishing two kinds of nominalization and then considering objectification of unary fact types and binary or longer associations that have a uniqueness constraint that doesn't span all its roles.

We treat *nominalization* as the recasting of a declarative sentence using a noun phrase morphologically related to a corresponding verb in the original sentence. Declarative sentences may be nominalized in different ways.

One common way is to use a gerund (verbal noun) derived from the original verb or verb phrase. For example, "Elvis sang the song 'Hound Dog'" may be nominalized as "Elvis's singing of the song 'Hound Dog'". Another way is to introduce a pronoun or description to refer back to the original (e.g., "that Elvis sang the song 'Hound Dog'" or "the fact that Elvis sang the song 'Hound Dog'").

In philosophy, it is usual to interpret the resulting nominalizations as naming either corresponding *states of affairs* or corresponding *propositions*. In linguistics, further alternatives are sometimes included. For example, states of affairs might be distinguished into events and situations.

For information modeling purposes, we adopt the philosophical approach, ignoring finer linguistic distinctions, and classify all nominalizations into just two categories.

A *situational nominalization* refers to a state of affairs or situation in the business domain being modeled. A *propositional nominalization* refers to a proposition. We treat events (instantaneous) and activities (of short or long duration) to be special cases of a state of affairs.

The relationships among states of affairs, propositions, sentences, and communication acts have long been matters of philosophical dispute. At one extreme, states of affairs and propositions are sometimes argued to be identical. Some view logic as essentially concerned with the connection between sentences and states of affairs while others view its focus to be propositions as abstract structures. Our viewpoint on some of these issues is motivated pragmatically by the need to model information systems, and is now summarized.

We define a proposition as that which is asserted when a declarative sentence is uttered or inscribed. A proposition (e.g., Elvis sang Hound Dog) must be true or false (and hence is a *truth bearer*).

Intuitively, it seems wrong to say that a state of affairs (e.g., Elvis's singing of Hound Dog) is true or false. Rather, a state of affairs is actual (occurs in the actual world, is the case) or not. A state of affairs may be possible or impossible. Some possible states of affairs may be actual (occur in the actual world).

States of affairs are thus *truth makers*, in that true propositions are about actual states of affairs. In sympathy with the correspondence theory of truth, we treat the relationship between propositions and states of affairs as one of *correspondence* rather than identity.

Although natural language may be ambiguous as to what a given usage of a nominalization phrase denotes (a state of affairs or a proposition), the intended meaning can usually be determined from the context in which the nominalization is used (i.e., the logical predicate applied to talk about it).

Consider the following examples:

```
Elvis sang the song "Hound Dog"                          -- original proposition
Elvis's singing of the song "Hound Dog" is popular       -- actual state of affairs
That Elvis sang the song "Hound Dog" is well known       -- true proposition
That Elvis sang the song "Hound Dog" is a false belief   -- false proposition

It's snowing outside                                     -- original proposition
It's true that it's snowing outside                      -- proposition
That snowing is beautiful                                -- state of affairs
```

The first three uses of the demonstrative pronoun "that" result in propositional nominalization. In the final example, "that" is used in combination with the gerund "snowing" to refer to a state of affairs (propositions aren't beautiful). In the previous two sentences, "snowing" is a present participle, not a gerund.

As indicated earlier, states of affairs may be actual or not, and propositions may be true or false. In ordinary speech, the term "fact" is often used to mean a true proposition, but when modeling information in ORM, the term "fact" is used in the weaker sense of "proposition taken to be true" (epistemic commitment) by users of the business domain.

Fig. 10.46(a) displays a typical case of objectification in information modeling in ORM. Here, the fact type Country plays Sport is objectified as the object type Playing, which itself plays in another fact type Playing is at Rank. The exclamation indicates that Playing is independent, so Playing instances may exist without participating in other fact types.

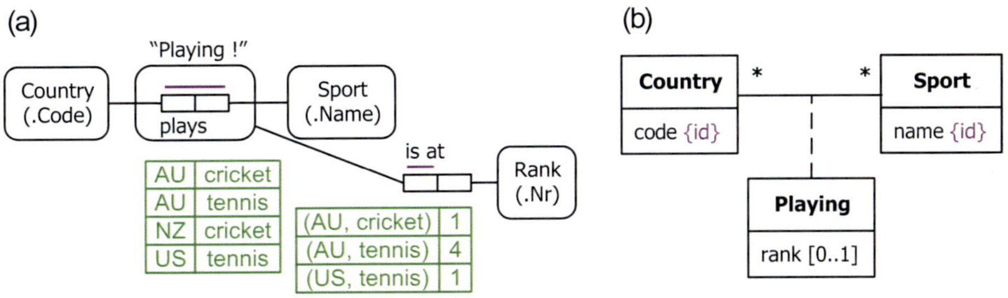

FIG. 10.46 Objectification of country plays sport as playing in (a) ORM and (b) UML.

Fig. 10.46(b) depicts the same example in UML, with the association between Country and Sport reified into the association class Playing, and the rank fact type modeled as an optional attribute.

Now, consider the question, *Are the objects resulting from objectification identical to the relationships that they objectify?* The UML metamodel answers yes to this question by treating AssociationClass as a subclass of both Association and Class. Because relationships are typically formalized in terms of propositions, this affirmative choice may be appropriate for propositional nominalization.

However, we believe that *the objectification process used in modeling information systems is typically situational nominalization*, where the object described by the nominalization is a state of affairs rather than a proposition. For situational nominalization, we answer this question in the negative, treating fact instances and the object instances resulting from their objectification as nonidentical. An intuitive argument in support of this position follows, based on the information model in Fig. 10.46.

Consider the relationship instance expressed by the sentence: "Australia plays Cricket". Clearly this relationship is a proposition, which is either true or false. Now, consider the object described by the definite description: "The Playing by Australia of Cricket" or, more strictly, "The Playing by the Country that has CountryCode 'AU' of the Sport named 'Cricket'". Clearly, this Playing object is a state of affairs (e.g., an activity). It makes sense to say that Australia's playing of cricket is at rank 1, but it makes no sense to say that Australia's playing of cricket is true or false.

Therefore, the Playing instance (The Playing by Australia of Cricket) is an object that is ontologically distinct from the fact/relationship that Australia plays Cricket. Our experience is that the same may be said of any typical objectification example that one finds in information system models. In this case, so-called *"objectified relationships" are in 1:1 correspondence with the relationships that they objectify, but they are not identical to those relationships.* Compare this with first-order logic, where predicate formulae are often tested for equivalence (≡) but not identity (=). Terms or individuals may be identical, but not equivalent.

In information models, one may sometimes encounter propositional nominalizations, where the noun phrase refers to a proposition rather than a state of affairs (e.g., [the fact] that Australia plays cricket is well known; the proposition that Australia plays cricket is true). A related although different case is where the noun phrase refers to a communication act rather than a proposition (e.g., the assertion that Australia plays cricket was made by Don Bradman). We delay discussion of such cases until the end of this section.

To facilitate high-level declaration of rules and queries on information models that make use of objectification, ORM includes (implicitly or explicitly) *link fact types* that link or relate the objectification result to the objects in the relationship that has been objectified. For example, the definite description "**The** Playing **that**: is by **the** Country **that** has CountryCode 'AU'; **and** is of the Sport **that** has SportName 'Cricket'" makes use of the *linking fact types* Playing is by Country and Playing is of Sport, as shown in Fig. 10.47(a).

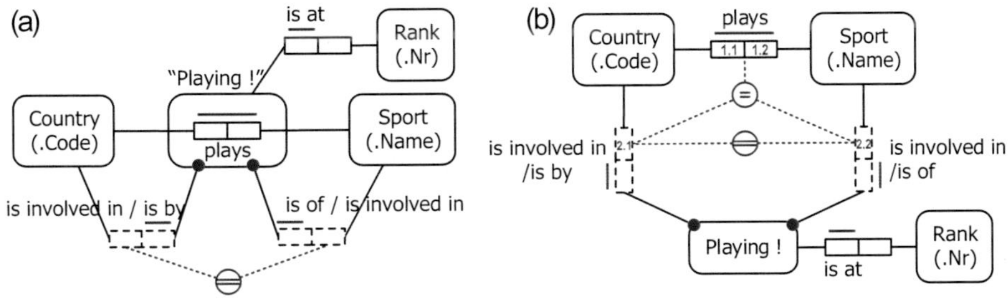

FIG. 10.47 Objectification in ORM uses linking fact types for relational navigation.

Internally, link fact types connect objectified associations to their component object types. By default, display of link fact types is suppressed. If displayed, *dashed lines* are used. The external uniqueness constraint indicates that each (Country, Sport) pair projected from the attached roles relates to at most one Playing object.

Default predicate readings for the linking predicates are assigned, such as "involves", appended by numbers if needed to distinguish linking fact types that link back to the same object type (which plays more than one role in the fact type being objectified). The modeler may edit these default readings to provide more meaningful readings for the link fact types, as shown in Fig. 10.47.

The diagram in Fig. 10.47(a) is best understood as an abbreviation of the diagram in Fig. 10.47(b). Here, Playing is a normal object type that is related back to the Country and Sport object types using the linking fact types mentioned earlier. The equality constraint depicted by a circled "=" indicates that the (Country, Sport) pairs in the population of the Country plays Sport fact type must be identical to the population of the (Country, Sport) pairs projected from the Country and Sport roles in the join path Playing is by Country **and** is of Sport. For clarification purposes, the role sequence numbers for the equality constraint are displayed here.

A similar analysis applies to the objectification of ternary and longer facts. For example, we might objectify Country plays Sport in Year as an object type Playing that has a third link fact type Playing is in Year whose year role adds a third component to the composite reference scheme for Playing.

With respect to our example in Fig. 10.47, this equivalence might be introduced to the model in one of three ways: (1) Start with the fact type Country plays Sport, and then objectify it as Playing; (2) start with the fact types Playing is by Country and Playing is of Sport and then define Country plays Sport as a fully derived fact type in terms of them; or (3) start with the three fact types Country plays Sport, Playing is by Country, and Playing is of Sport and then assert the equality constraint (corresponding to the equivalence) between them. Regardless of which way is used, the model fragment is internally stored in terms of the structure shown in Fig. 10.47(b).

Although ORM treats objectification in this way internally, by default the link fact types and 1:1 correspondence are not displayed on the diagram. So, diagrammatically objectification normally appears simply as in Fig. 10.46(a).

Like any binary fact type, link fact types may be given predicate readings in both directions and role names. For discussion purposes, the example in Fig. 10.48 includes inverse predicate readings, role names, and a sample population. Note that the role names (acquirer, target) on the acquisition fact type provide role names for the Company roles in the link fact types.

ORM schemas may be navigated in relational style (using predicate names), attribute style (using role names), or a mixture of both. To navigate from Company to Acquisition, it matters which link fact type we use. For example, from the company Visio we may navigate via the left link to its acquisition of InfoModelers or via the right link to its acquisition by Microsoft.

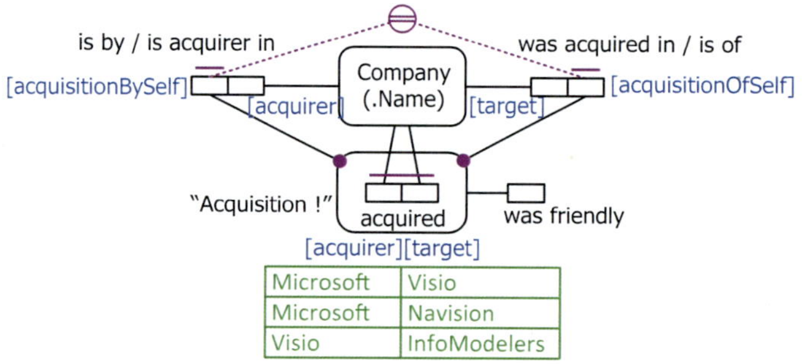

FIG. 10.48 Predicate readings and role names support full navigation.

Navigating via the left link, the schema path may be verbalized as "Company **that** is acquirer in Acquisition". Navigating via the right link we have "Company **that** was acquired in Acquisition". To navigate from Acquisition to company, it also matters which link we use. Navigating via the links, the schema paths may be verbalized in relational style as Acquisition **that** is by Company (navigation via left link); Acquisition **that** is of Company (navigation via right link).

Role paths may also be specified in attribute style, using role names for "attributes". For example, to navigate from Company to Acquisition, we have the following two different options: Company.acquisitionBySelf (navigation via left link); Company.acquisitionOfSelf (navigation via right link). To navigate from Acquisition to Company, we have the following two different options: Acquisition.acquirer (navigation via left link); Acquisition.target (navigation via right link).

Although such expressions could be used to specify projections, here we use them simply to indicate a path obtained by jumping from an object type to one of its far roles. If the dot notation is replaced by the "of-notation", then the component order is reversed (e.g., instead of Company.acquisitionBySelf we have acquisitionBySelf **of** Company).

In addition to purely relational and purely attribute styles, role paths may be specified using a mix of both styles (e.g., Company.acquisitionBySelf **that** occurred on Date). From an ORM perspective, it is all relational underneath.

UML provides no direct support for unary relationships, instead requiring their remodeling in terms of attributes or subclasses. ORM allows unary relationships to be objectified. Consider the unary fact: The President named "Abraham Lincoln" died. We may objectify this death using the nominalization "that death" and declare the following additional fact: That death occurred in the Country with country code "US". This natural way of communicating may be supported in a similar way to objectification of nonunary facts.

An ORM model for this situation is shown in Fig. 10.49(a). Sample populations are included for the object types and fact types. Here, the unary fact type President died is objectified by the object type Death. This unary objectification is internalized using the expanded schema shown in Fig. 10.49(b).

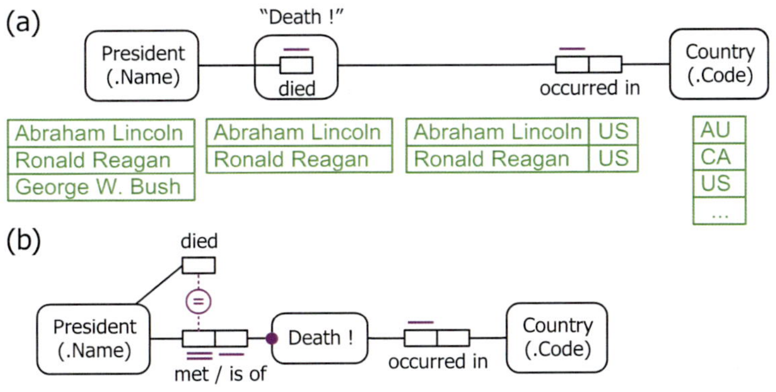

FIG. 10.49 Objectification of unary facts in ORM.

Here, Death is a normal entity type with a simple reference scheme provided by its injective relationship to President (e.g., Abraham Lincoln's death may be referenced by the definite description "**The** Death **that** is of **the** President **who** has **the** PresidentName 'Abraham Lincoln'"). Hence, the result of objectifying a relationship type may be viewed as an entity type that has a reference scheme whose reference projection bears an equality constraint to the fact type being objectified.

This interpretation does not assume that the Death is of President relationship provides the only or even preferred way of referring to deaths (e.g., we may introduce a death number to reference deaths, without impacting this analysis).

The objectification of unaries as a case of situational nominalization was introduced in ORM 2 (objectification of unaries was not supported at all in ORM 1). The fact-oriented version known as fully communication oriented information modeling (FCO-IM) introduced objectification of unaries to provide an alternative way to support existential facts and to treat all base objects as lexical in nature. With this approach, an entity (nonlexical object) is an objectification of a role played by a value (lexical object). In Fig. 10.50, for example, the entity type Country is derived by objectifying the unary fact type CountryCode refers to a country.

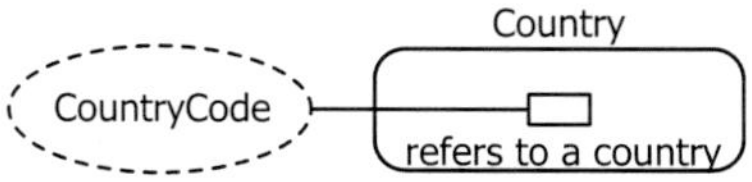

FIG. 10.50 In FCO-IM, nonlexical types objectify roles of lexical object types.

While the FCO-IM approach to objectified unaries encourages the use of natural reference schemes in modeling, and has tool support, we personally find it unintuitive and awkward in dealing with multiple inheritance, context-dependent reference schemes as well as changes to reference schemes. Nevertheless, some modelers favor this approach, so perhaps this is a subjective issue.

ORM 1 allowed objectification of 1:1 fact types as well as binary and longer fact types with a spanning uniqueness constraint. ORM 2 and UML also allow objectification of an n:1 (or 1:n) binary association, and a ternary or longer association whose longest uniqueness constraint spans exactly $n-1$ roles.

We exclude from consideration any n-ary association with a uniqueness constraint spanning fewer than $n-1$ roles because such an association is compound rather than elementary.

While NORMA allows objectification of fact types without a spanning UC, we feel this practice should be deprecated, as it can lead to confusion when populating fact types or declaring independence, and is typically better modeled in another way (Halpin, 2020).

We now recommend that *objectification of a fact type be restricted to fact types that are unary or have a spanning uniqueness constraint.*

The second edition of this book argued for objectification of a fact type with multiple uniqueness constraints to avoid forcing one to choose one of the UCs as the preferred way to identify the objectification. The simplest case is *objectification of a 1:1 fact type*, for example, the current marriage example shown in Fig. 10.51(a). Fig. 10.51(b) shows the UML version.

However, the 1:1 nature of the association means that an instance of CurrentMarriage may be identified by just the husband or just the wife, rather than the combination of husband and wife. Moreover, when mapping to SQL, each relational table must have a primary key, so one of the UCs in the binary fact type needs to be chosen as preferred for this purpose.

Entering the schema in Fig. 10.51(a) into NORMA generates two *errors*. NORMA automatically creates two *implicit 1:1 fact types* linking CurrentMarriage with Person using the fact type reading "Person involves / is involved in CurrentMarriage", thus violating the rule that fact type reading must be unique within a given model.

The only way to fix this error is to drag the implicit fact types from the Model Browser onto the drawing window, and then edit the readings in the Fact Editor or Reading Editor. Fig. 10.51(c) does this as well as adding reference modes for Person and Date and the exclusion constraint that no person can be husband and wife.

The second error is that none of the UCs in the 1:1 fact types is marked as preferred, so CurrentMarriage has no preferred reference scheme. To fix this, we select one of the UCs in the Person is husband of /is wife of Person fact type and toggle its isPreferredIdentifier property from False to True.

The selected UC and the corresponding UC on an implicit fact type are then redisplayed with a double uniqueness bar. Choosing the left UC of the marriage fact type for this results in Fig. 10.51(c), which shows a screen shot from the NORMA tool.

Formally, there is an equality constraint between the pair of husband and wife roles of Person and the pair of roles in the marriage fact type in Fig. 10.51(c). However, instead of displaying this equality constraint, purely for performance reasons the NORMA tool internally treats the implicit roles of Person as "mirror roles" or role proxies whose population always matches the population of the roles they mirror (in this case, the roles in the marriage fact type).

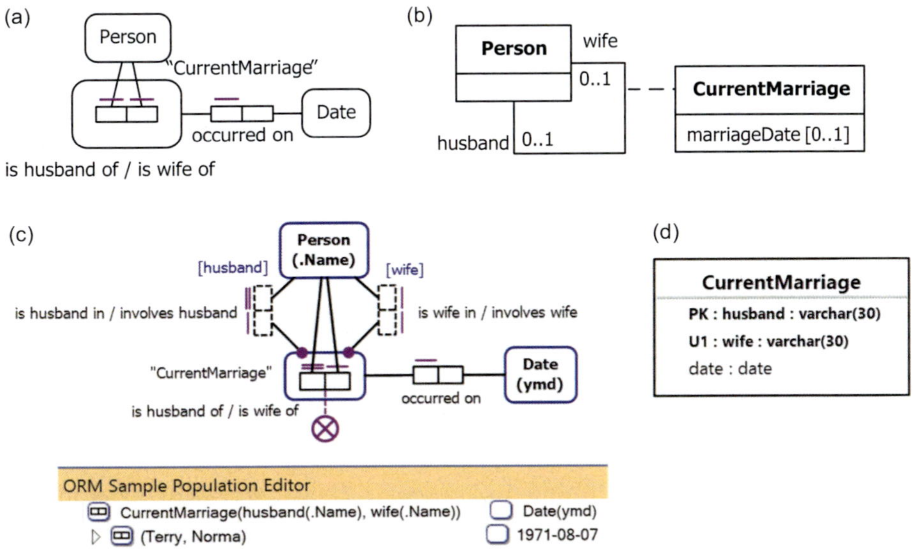

FIG. 10.51 Objectifying a 1:1 association might be confusing.

Although CurrentMarriage is now primarily identified by the husband, NORMA requires the husband, wife combination to identify it when populating the fact type CurrentMarriage occurred on Date, as shown at the bottom of Fig. 10.51(c). This might be confusing, as the fact entry is not atomic and is inconsistent with the way the sample population editor works with normal fact types.

In this example, it is possible to record a current marriage without knowing the date of the marriage. This might confuse users who, by ignoring the implicit link fact types as in Fig. 10.51(a), think that CurrentMarriage is an independent object type because instances of it can exist without playing in the marriage date fact type. However, *no objectification of a fact type with multiple UCs can be independent* because in the link fact types, it plays a mandatory role in addition to the role in its preferred reference scheme.

Fig. 10.51(d) is a screenshot of the relational schema generated by NORMA. NORMA lists the column names vertically, with the primary key columns marked "PK". Mandatory column names are displayed in bold type, and optional column names are in plain (unbolded) type. The husband and wife role names added in Fig. 10.51(c) improve the column names generated for the relational mapping.

To avoid confusion and avoid nonatomic fact entries, we think it is better to remodel without objectification of 1:1 fact types.

For the current example, Fig. 10.52(a) identifies a marriage by the husband. Fig. 10.52(b) identifies a marriage by a marriage number (e.g., a marriage certificate number).

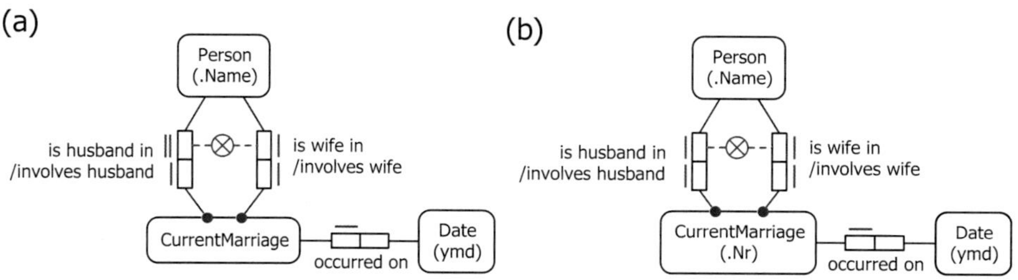

FIG. 10.52 Alternative schemas for modeling current marriages.

Similar comments apply to *objectification of an n-ary* (ternary or longer) *fact type with multiple UCs spanning n−1 roles*. For example, the ternary association in Fig. 10.53(a) has overlapping uniqueness constraints, each spanning two roles. The constraint on the first two roles indicates that within each sport, each athlete has at most one current rank. The constraint on the last two roles indicates that for each sport, each rank is assigned to at most one athlete (no ties are allowed).

For each current ranking of an athlete in a given sport, we record the date on which this ranking was awarded, and optionally the achievements on which this award was based. Fig. 10.53(b) depicts the same business domain in UML, where the ternary association is objectified as an association class, the award date is modeled as an attribute, and achievements are modeled via an association.

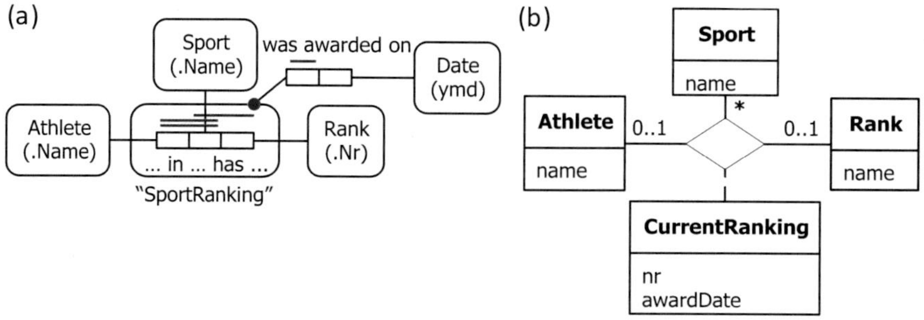

FIG. 10.53 Objectifying a ternary association with two UCs in (a) ORM and (b) UML.

Once you objectify the ternary, the NORMA tool requires one of the two UCs to be preferred, and displays the preferred UC with a double uniqueness bar. Here, SportRanking is primarily identified by the athlete, sport combination, but NORMA requires the athlete, sport, and rank to identify each sport ranking so the fact entry is not atomic. To avoid confusion and nonatomic fact entries, we prefer to model such cases objectification of a ternary with a nonspanning UC, using one of the approaches shown in Fig. 10.54.

Now consider *objectification of an n:1 fact type with multiple roles played by the objectification*. Suppose we need to record for each moon the planet it orbits and the period of that orbit. Optionally, we also record one or both of the axes (minor and major) of that elliptical orbit.

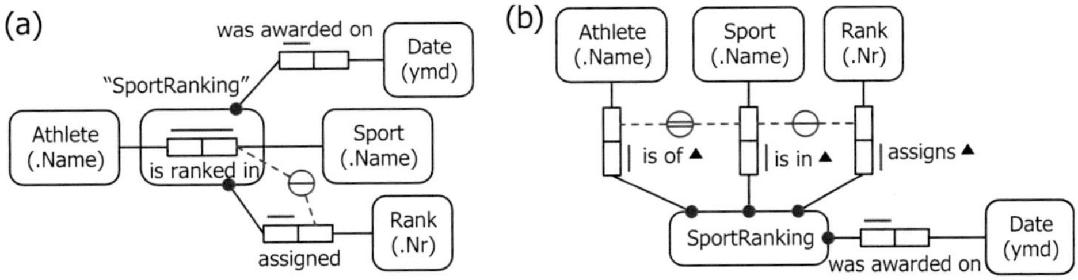

FIG. 10.54 Remodeling the sport ranking example.

Fig. 10.55(a) models this case in ORM by treating MoonOrbit as an objectification of the Moon orbits Planet association. Fig. 10.55(b) shows an equivalent model in UML.

If you enter the ORM schema in Fig. 10.55(a) in NORMA, it internally creates implicit fact types linking MoonOrbit to Moon and Planet, using the 1:1 link fact type MoonOrbit involves Moon as the preferred identification scheme for MoonOrbit. There is no need to drag the link fact types onto the drawing window. Although a moon orbit is identified by just its moon, populating the period and axis fact types requires both the moon and planet to identify each moon orbit.

Fig. 10.55(c) shows the relational schema view NORMA generates from the ORM model in Fig. 10.55(a). The moonName primary key clearly shows that NORMA identifies a moon's orbit just by the moon. Personally, we do not like the practice of hiding the preferred identification scheme at the conceptual level, and we prefer to model such cases in ORM without objectification, using an alternative schema such as that shown in Fig. 10.55(c) or (d).

However, some users might still prefer the objectification approach over such alternatives. Objectification of a binary or longer fact with just one UC spanning all but one role where the objectification plays other multiple roles is the only kind of case where one might possibly have a legitimate reason to ignore our advice to never objectify an association without a spanning UC.

So far, we have discussed objectification in the sense of situational nominalization, where the object being discussed is a state of affairs (event, activity, etc.). Although

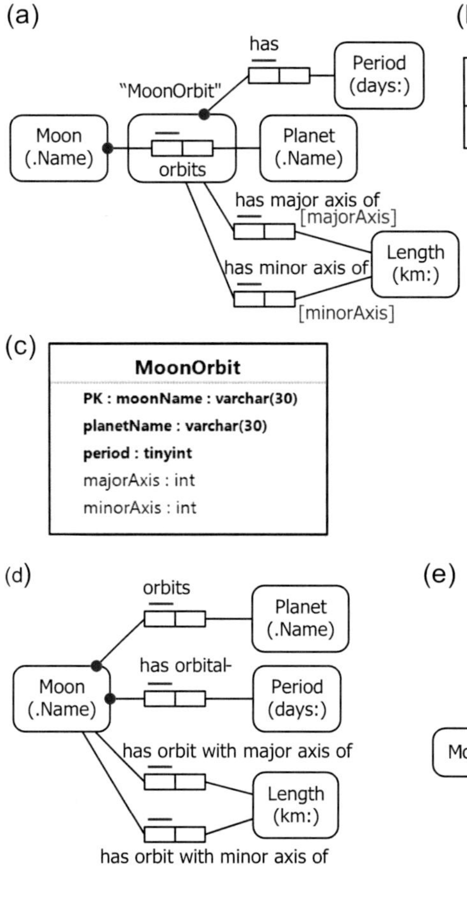

FIG. 10.55 Should we objectify an *n*:1 association with multiple roles?

comparatively rare in information modeling, one may also encounter cases where the nominalized object is either a proposition (resulting from *propositional nominalization*) or a *communication act* (e.g., an utterance or inscription act by some speaker/writer).

The OMG's SBVR specification includes support for propositional nominalization. One of the SBVR rule examples involving propositional nominalization may be verbalized as follows: If a waiter earns an amount of money as a tip from serving a meal, the waiter must report that fact.

While one may interpret this as a case of propositional nominalization (reporting the fact rather than the act), there is no need to do so for information modeling purposes, as the rule may easily be declared using situational nominalization (reporting the act rather than the fact), as shown in compact form in Fig. 10.56. The equality constraint is between the three roles numbered 1.1, 1.2, 1.3 and the three roles numbered 2.1, 2.2, 2.3.

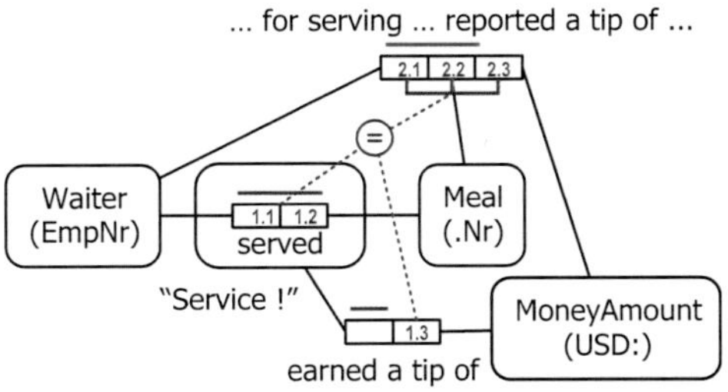

FIG. 10.56 Replacing propositional nominalization by situational nominalization.

If the rule is modified to require reporting after the service is performed, a time limit for reporting must be declared to make the rule operational. In this case, the relevant temporal object type may be added to the model to cater for the extended rule in an obvious way. For simplicity, *we recommend modeling all propositional nominalizations instead by their corresponding situational nominalizations.*

As regards modeling of communication acts, when it is of interest to model these acts, they are best modeled directly like any other business domain objects. For example, in a genealogy model we might be interested in not just descriptions of states of affairs, but assertion acts made by researchers about states of affairs.

Such a model might include fact types such as AssertionAct reported Proposition, AssertionAct was made by Researcher with ConfidenceLevel, etc. These comments relate to the information model only.

For modeling communication processes, the information model should be supplemented by other kinds of models (e.g., workflow models) that provide a more intuitive and direct way of understanding essential business processes/services.

Exercise 10.5

1. With respect to the fact informally stated as "Tolkien authored *The Lord of the Rings*", classify each of the following nominalizations as situational or propositional.
   (a) I know that Tolkien authored *The Lord of the Rings.*
   (b) Tolkien's authoring of *The Lord of the Rings* took years.
   (c) It's true that Tolkien authored *The Lord of the Rings.*

2. Expand these ORM schemas to show how objectification is implemented underneath.

(a)

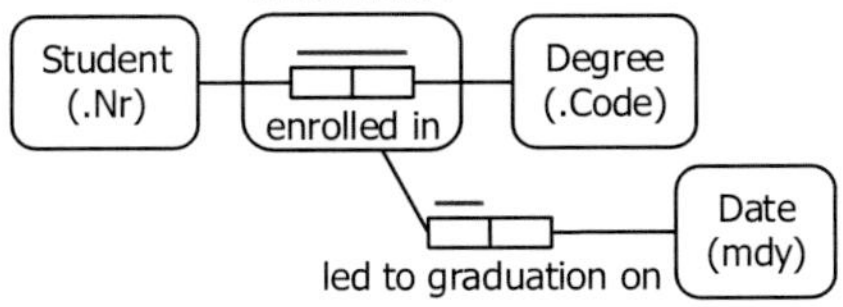

(b)

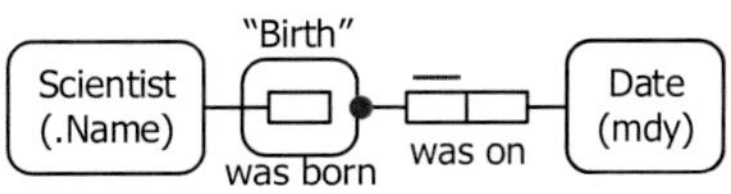

3. Remodel the following ORM schema to remove objectification. Include all constraints.

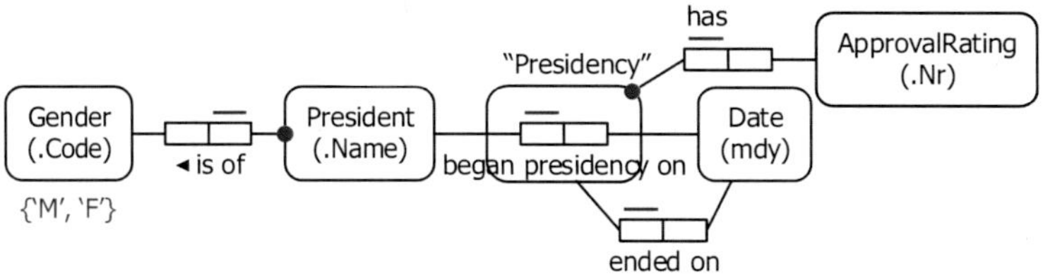

10.6 Open/Closed World Semantics

When modeling a business domain, one may take different positions with respect to the completeness of the knowledge captured in the model. The *closed world assumption* (CWA) is the assumption that all relevant facts are known (i.e., appear in the model, either as asserted facts or derived facts) and that all the relevant business rules are known. This is the usual assumption for databases and implies *negation as failure* (the failure to find or prove a proposition implies that it is false).

Consider, for example, the populated ORM model in Fig. 10.57(a). There are four students (1001–1004), but only two of these (1001 and 1003) are modeled as intelligent. Using CWA, we may deduce that students 1002 and 1004 are not intelligent. Similarly, we may deduce that students 1003 and 1004 do not work hard. Using the asserted facts and the derivation rule shown, we infer that three students (1001, 1002, and 1003) are expected to pass. If there are no other derivation rules for the fact type Student is expected to pass, the **if** operator is interpreted under CWA to mean **iff**. Hence, student 1004 is not expected to pass.

(a)

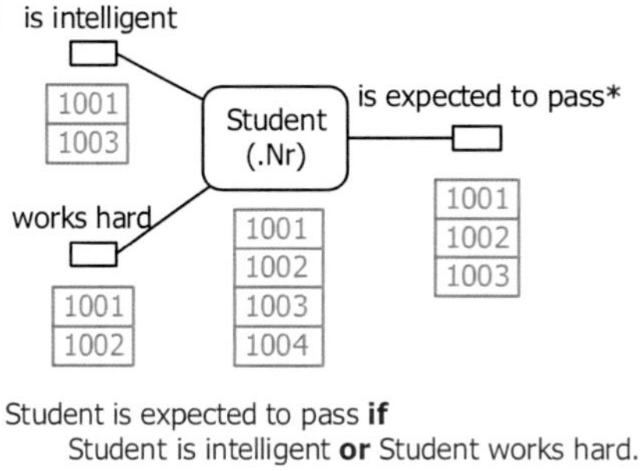

(b)

Student (<u>studentNr</u>, isIntelligent, worksHard)

1001	true	true
1002	false	true
1003	true	false
1004	false	false

(c) *Student_View* (<u>studentNr</u>, isIntelligent, worksHard, isExpectedToPass*)

1001	true	true	true
1002	false	true	true
1003	true	false	true
1004	false	false	false

FIG. 10.57 The closed world assumption requires all facts to be modeled.

Under closed world semantics, when mapped to a relational database the asserted positive facts and the implied negative facts should be grouped into the Student table shown in Fig. 10.57(b), using mandatory Boolean attributes for isIntelligent and worksHard.

The derived facts about pass expectations can be mapped to an additional isExpected-ToPass column in the Student_View table shown in Fig. 10.57(c). Instead of creating this view, a computed column for these derived facts could be added to the base Student table. Instead of mapping unaries to mandatory Booleans, the NORMA tool currently maps them to Boolean columns that are optional, so null values are allowed there.

The relational schema notation used in this book underlines the primary key name(s) and adds square brackets around the names of optional columns. Hence, NORMA maps the ORM schema in Fig. 10.57(a) to the following relational schema: *Student*(studentNr, [isIntelligent], [worksHard]).

There are plans to extend the NORMA tool with control over closed and open world semantics for unaries, and to add support for negated unaries. In the meantime, to use NORMA to map to a mandatory Boolean, replace the unary by a mandatory *n*:1 fact type that targets a value type with a Boolean data type. For example, replace the Student is intelligent fact type with the mandatory *n*:1 fact type Student has IsIntelligentStatus where IsIntelligentStatus has a Boolean data type.

A second way to remodel model CWA semantics for unaries is to replace the unary with a mandatory *n*:1 fact type that targets an entity type or value type with exactly two named values, for example, Student has IntelligenceStatus(.Name) {'Intelligent', 'Unintelligent'}.

A third way to model CWA semantics for unaries is to add an *opposite unary* with the negated meaning (i.e., *its instances are equivalent to negated instances of the original unary*), for example, for the unary fact type Student is intelligent add Student is unintelligent. An exclusive-or constraint applies between the two unaries.

There are plans for the NORMA tool to add a notation to declare the negated meaning connection between the unaries. In that case, the exclusive-or constraint is implied and the schema maps to a relational table with a single Boolean column; otherwise, this approach maps to a relational table with two columns.

As another example, Fig. 10.58(a) shows a closed world unary fact type Country is monolingual (has exactly one official language) together with sample data. Canada (CA) and Cyprus (CY) are not recorded as being monolingual, so the CWA approach implies that they are not monolingual. Fig. 10.58(b) replaces the unary by a mandatory *n*:1 fact type.

Fig. 10.58(c) adds the unary fact type Country is multilingual whose meaning is the *opposite* of the monolingual unary. If we add a way to formally declare that the unaries are opposite in meaning (different proposals are under discussion), the exclusion aspect of the exclusive-or constraint is implied, and both fact types map to a single Boolean column. Otherwise, this approach maps to a relational table with two columns.

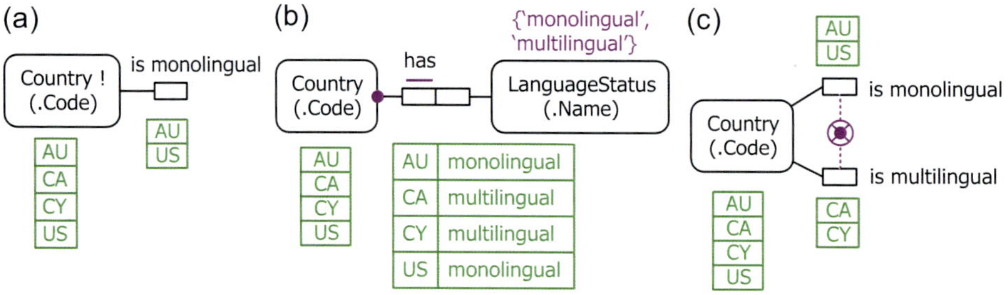

FIG. 10.58 Some CWA modeling options.

The *open world assumption* (OWA) allows that some relevant facts may be unknown (as asserted or derivable facts). Fig. 10.59 remodels our student example under OWA semantics.

In Fig. 10.59(a), the role boxes for the unary predicates have dashed lines suggesting openness and hence OWA semantics. At the time of writing, this OWA notation for ORM is not supported by NORMA.

The OWA implies *classical negation* (failure to find or prove a proposition or its negation implies that its truth value is *unknown*). For example, as indicated in the base Student table in Fig. 10.59(b), the failure to record the facts that students 1002 and 1004 are intelligent simply means that we do not know whether they are intelligent. Here we record the entry "unknown" to indicate this. In most databases, this entry would appear as a null. Check for yourself that the other "unknown" entries follow from the OWA interpretation.

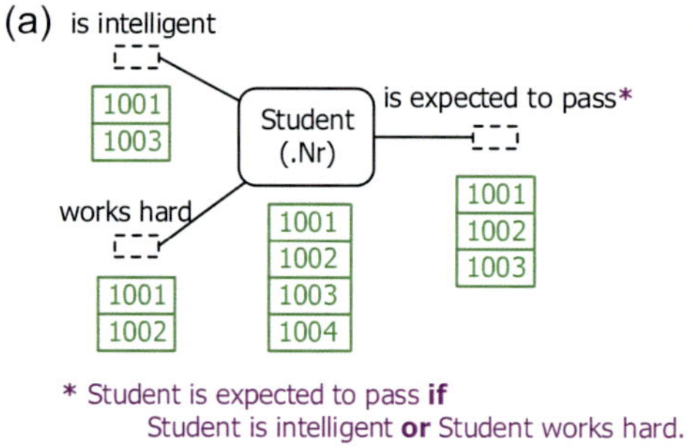

(b)

Student (<u>studentNr</u>, [isIntelligent], [worksHard])

1001	true	true
1002	unknown	true
1003	true	unknown
1004	unknown	unknown

(c) *Student_View* (<u>studentNr</u>, [isIntelligent], [worksHard], [isExpectedToPass]*)

1001	true	true	true
1002	unknown	true	true
1003	true	unknown	true
1004	unknown	unknown	unknown

FIG. 10.59 The open world assumption allows some facts to be unknown.

The relational schemas shown in Fig. 10.59(b) and (c) indicate that an open world unary is mapped to an optional Boolean whose truth value is restricted to True.

One way to obtain this relational schema result in NORMA is to replace each unary by an optional *n*:1 fact type that targets a value type with a Boolean data type restricted to

True. For example, replace Student is intelligent by the fact type Student has IsIntelligentStatus {true} where the role of IsIntelligentStatus is named "isIntelligent".

Another way to model OWA semantics for unaries is to add a unary with the negated meaning as well as with an exclusion constraint between its role and the role of the original unary. For example, add Student isUnintelligent with an exclusion-constraint between its role and the role of the Student isIntelligent fact type. However, without a way to formally declare that one unary is the negation of the other, this way maps to a relational table with two columns rather than one column.

As a halfway position between CWA and OWA, the *semiclosed world assumption* (SWA) allows that some relevant facts may be unknown as derived facts, but not as asserted facts. This implies *semipositive negation* (given any asserted fact type, failure to find a fact instance in it implies that instance is false; and failure to prove an instance of a derived fact type or its negation implies that its truth value is unknown). This approach can be used only when the fact types are asserted or fully derived (not semiderived).

The *open world with negation* (OWN) assumption allows some relevant facts to be unknown (as asserted or derived facts), but also allows negations to be asserted. Like OWA, the OWN approach also leads to classical negation, which is the usual approach in classical logic. Although it is rare to assert negative facts in databases, it is common in logic systems. The OWN approach is just a special case of OWA.

One way to support OWN semantics in ORM is to allow unaries to be populated with both positive and negative instances, with negated instances prepended by a tilde "~", often used in logic for the "**not**" (**it is not the case that**) operator, as shown in Fig. 10.60(a).

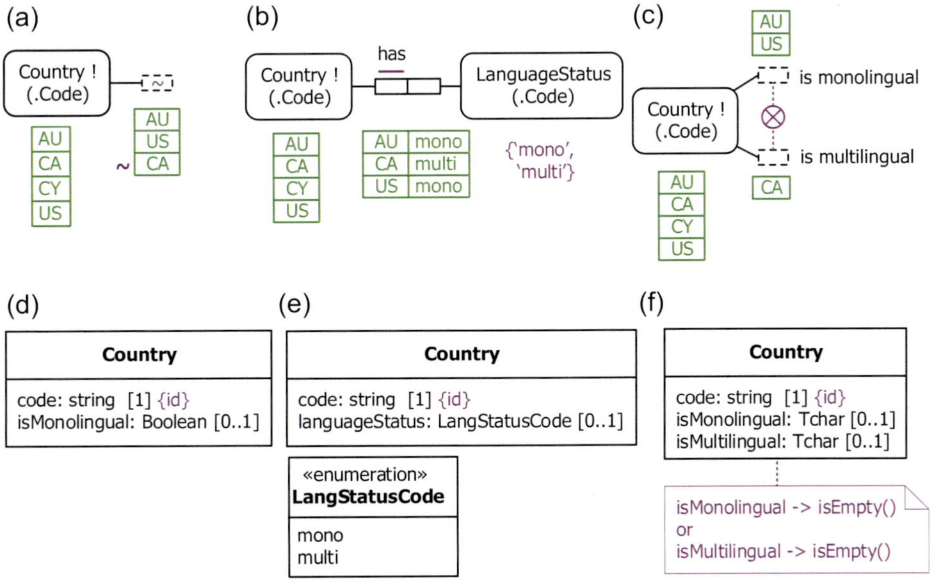

FIG. 10.60 Some OWN modeling options.

Another way to support *negated instances of an open unary is to add a separate unary fact type with the opposite meaning (its instances are equivalent to negated instances of the original unary). The opposite meaning implies an exclusion constraint between the two unaries.* For example, to record that student 1002 is not intelligent, include that student in the population of the fact type Student isUnintelligent where there is an exclusion constraint between its role and the role of the Student isIntelligent fact type.

A third option is to replace the unary with a third way to model. Suppose that we know that Australia (AU) and the United States (US) are monolingual and that Canada is multilingual, but don't know the language status of Cyprus. This corresponds to the OWN approach, as modeled directly in Fig. 10.60(a). The rest of Fig. 10.60 shows alternative ways to model this in ORM and UML.

A detailed procedure for mapping unaries for the various open/closed world approaches is discussed in Chapter 11.

In principle, support for the various closure options could be extended beyond unary fact types, but this would add considerable complexity, especially for nonfunctional fact types.

Exercise 10.6

1. With reference to the ORM model shown, complete the following table with the appropriate true, false, or unknown (UNK) entries for the two propositions about Pat, for the three cases indicated. Assume that the predicates are given the appropriate interpretation true, false or UNK entries for the two propositions about Pat, for the three cases indicated. Assume that the predicates are given the appropriate interpretation (e.g., for classical negation, the predicates are open world even if they are shown with solid lines instead of dashed lines).

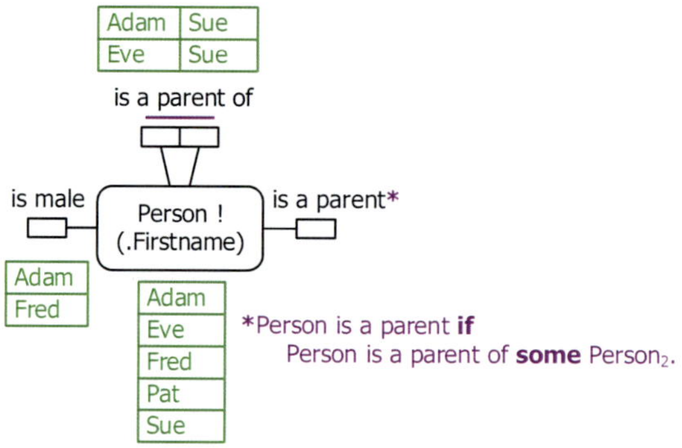

Negation approach	*Pat is male*	*Pat is a parent*
Negation as failure		
Classical negation		
Semipositive negation		

2. For the following schema, a modeler insists that the unary predicates should be declared open world rather than closed world. Does this make sense? Explain.

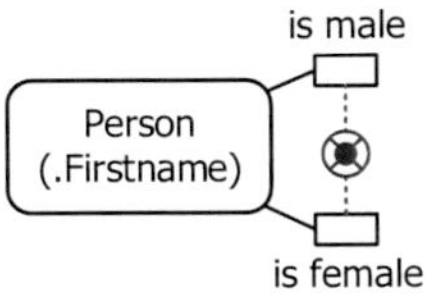

10.7 Higher-Order Types

This section evaluates the advisability of using higher-order types in information models. We first review some basic concepts from logic. *First-order logic* quantifies over individuals only, not predicates. For example, the constraint "**Each** Person was born on **at most one** Date" may be formalized in first-order typed logic thus:

$\forall x$:Person $\exists^{0..1} y$:Date x was born on y

Here, the types Person and Date appear as unary predicates, "$\forall$" is the universal quantifier ("**for each**"), and "$\exists^{0..1}$" is the "**there exists at most one**" quantifier. The quantification is over individual people and individual dates. First-order predicates may be instantiated only by individuals.

Second-order logic allows quantification over first-order predicates (as well as individuals). For example, the rule "Each asymmetric predicate is also irreflexive" may be formalized in second-order thus:

$\forall R$:Predicate (Asymmetric $R \rightarrow$ Irreflexive R)

Third-order logic allows quantification over second-order predicates, and so on. All logics whose order is above first order are called *higher-order* logics. Because types may be formalized by unary predicates, if an instance of type A is itself a type, then A is a *higher-order type.*

A *categorization type* is an object type that is used to classify another type into subtypes. For example, Gender might be used in the fact type Person is of Gender to classify Person into MalePerson and FemalePerson.

Most formalizations of ORM, ER, and the relational model are based on first-order logic. In contrast, UML introduced the notion of powertypes, whose instances may themselves be types, thus requiring higher-order semantics.

There appear to be three main arguments for using higher-order types in information modeling:

1. To allow one to think of instances of some categorization types (e.g., AccountType, CarModel) as being types themselves (as for UML powertypes).
2. To directly formalize the semantics of flexible data structures where attribute entries may themselves denote sets or general concepts (e.g., object-relational tables in nonfirst normal form).

3. To allow one to specify business rules that seem to cross levels/metalevels (or ignore level distinctions) in the same model (e.g., Only the Finance department may specify the allowed instances of AccountType).

As the move to higher-order logic may add considerable complexity to the task of formalizing and implementing a modeling approach, it is worth investigating whether the same practical modeling objectives can be achieved while staying within a first-order framework. We believe they can.

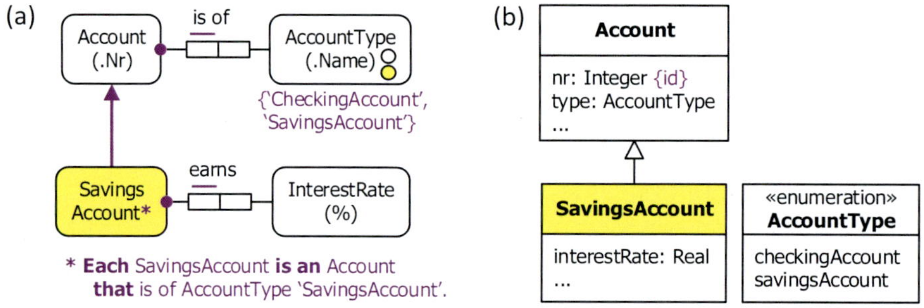

FIG. 10.61 Is an instance of AccountType identical to a subtype of Account?

Let's first address the question, May instances of categorization types be types themselves? For example, in the banking schema shown in Fig. 10.61, is the savings account instance of AccountType identical to the subtype SavingsAccount? For discussion purposes, the two AccountType instances named "CheckingAccount" and "SavingsAccount" are depicted here as a white dot and shaded dot, respectively, and the subtype has been shaded.

In practice, we would normally remove or expand the value constraint on AccountType to allow other types of accounts (e.g., LoanAccount). Regardless, we need to address the fundamental question, which may be phrased in ORM terms as, *Is the AccountType instance denoted by the shaded dot identical to the subtype SavingsAccount?* If we answer yes, then we have a case of an instance being a type in the same model, making AccountType second order.

In the UML schema, the use of an enumerated type demands a no answer because UML treats enumeration types as data types whose instances are literals. However, as discussed shortly, UML allows us to remodel the situation using a powertype for AccountType, which requires a yes answer.

For the AccountType instance denoted by the shaded dot to be identical to the subtype SavingsAccount, the semantics of the fact type Account is of AccountType should satisfy at least the following necessary conditions: (1) the "*is of*" predicate *means* "is a member of" or "is an instance of", that is, ∈ (*set membership*) and (2) o*nly Account instances may be instances of AccountType instances.*

These conditions (which in combination we call *homogeneous set membership*) follow from the *indiscernibility of identicals* (if $a = b$, then a and b have the same properties). The subtype SavingsAccount includes precisely all the possible savings accounts in the business domain—no other things can be instances of it. If the AccountType instance denoted by the shaded dot *is* actually the subtype SavingsAccount, then only accounts can bear the instance-of relationship to it.

If we allow the type=instance interpretation, we must use higher-order logic for formalization, and should apply the semantics of homogeneous set membership to categorization relationships of this kind. If we reject the type=instance interpretation, we may stay with first-order logic (at least for formalizing categorization relationships of this kind), and may optionally distinguish such categorization relationships as special and provide them with relevant formal semantics.

To note this distinction, one could adopt a special graphical or textual adornment for such categorization associations. For example, in ORM one might append a colon ":" to the forward reading of any predicate used for this purpose (based on the common use of colons to sometimes but not always introduce types). Applying this suggestion to the model in Fig. 10.61(a) would replace "is of" by "is of:".

To provide a minimal, common approach to such categorization relationships, whether or not we adopt the type=instance viewpoint, we could use the colon marker to distinguish any such relationship, and give it the semantics of an *asymmetric, intransitive*, and *locally homogeneous* relationship. A fact type of the form *A R: B* is *locally homogeneous* if and only if *B* is used as a categorization scheme for *A*, but for no other type (so no other type bears a colon relationship to *B*). For the example in Fig. 10.61, this means that only Account instances may be instances of AccountType instances. It is convenient to use the same predicate reading (e.g., "is of:", or even just ":") for all such categorization predicates, unless this makes the reading awkward. The choice of reading is language dependent.

The properties of asymmetry and intransitivity seem to be the only properties of the set-membership operator ($\in$) that are relevant here. If we always use the same reading (e.g., "is of:") for the categorization relationship, we may think of it as a predefined predicate constant that applies globally (all occurrences of this predicate have the same semantics). If we reserve the colon only for such homogeneous cases, we must not use it in cases where the classification scheme (e.g., Gender) may be applied to more than one type (e.g., Person, Dog).

Note that *any* role played by a type could be used as a basis for categorizing it. So, the main reason for marking such is-of associations as categorization relationships is to enforce the formal properties of such relationships. We may now formalize categorization relationships of this kind. Fig. 10.62(a) depicts the general case, assuming the same predicate reading "is of:" is always used (if this is not so, then without loss of generality, begin by replacing each reading by "is of:"). We use "$\sim$" for negation ("it is not the case that"), "&" for conjunction ("and"), and "$\rightarrow$" for material implication ("implies"). The asymmetric and intransitive properties may be declared independent of the object types, unlike local homogeneity.

For the *first-order logic interpretation*, all types are first order, so their instances are individuals, not types. We use lower-case letters (possibly subscripted) to range over individuals. For our *higher-order logic interpretation*, we use capital letters (in italics) to denote type variables of any order. The order (1, 2, 3, ...) of any type is implicit because it can be derived by inspecting the full schema. If it is desired to explicitly show the order of a type, a presuperscript may be used (e.g., ^{2}B indicates that B is a second-order type).

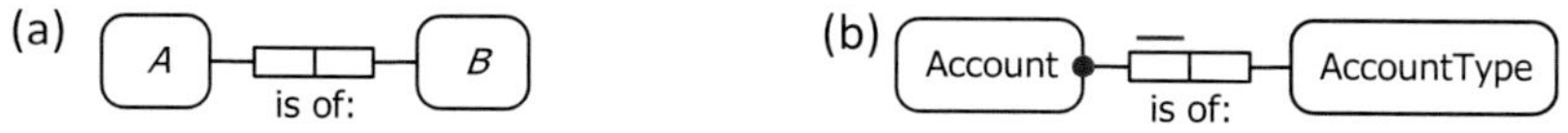

FIG. 10.62 The categorization relationship "is of:".

Postsuperscripts are typically used to denote arity, and postsubscripts are often used to distinguish variables of the same type. Ignoring the case of crossing metalevels, assign the order of a type to be 1, plus the number of relationships in a contiguous chain of zero or more categorization relationships that end at the type.

Using $\exists^1$ for the "there exists exactly one" quantifier, we may now formalize the constraints on the categorization relationship in Fig. 10.62(b), which has a mandatory and uniqueness constraint. The first-order formalization treats AccountType as a type of individuals.

First-order formalization
∀*xy* [*x* is of: *y* → ~ (*y* is of: *x*)] -- asymmetric
∀*xyz* [*x* is of: *y* & *y* is of: *z* → ~ (*x* is of: *z*)] -- intransitive
∀*xy* [AccountType *y* & *x* is of: y → Account *x*] -- local homogeneity
∀*x* [Account *x* → $\exists^1$*y* (AccountType *y* & *x* is of: *y*)] -- mandatory and unique

The higher-order formalization treats AccountType as a type of first-order types, so it replaces "*y*" and "z" in the aforementioned formulation by "*Y*" and "*Z*". Alternatively, the higher-order formalization may replace any expression of the form α is of: β, where β is a type variable, by βα because it regards α is of: β in such cases to entail that α instantiates the β predicate.

One motivation for distinguishing such categorization relationships is to facilitate transformation to or from UML models that include so-called *powertypes*, which UML includes specifically to model such categorization schemes. In UML 1.4, powertypes were declared as stereotyped classes using the «powertype» annotation, but this notation was retired in UML 2. The only way in UML 2 to know that a class is a powertype is to see it referenced in a generalization constraint.

Fig. 10.63 shows a simplified version of an example often used to illustrate the need for powertypes, using UML 2 notation. This way of classifying trees is botanically wrong, but that's irrelevant to the issue. Let's assume that trees can be classified into species such as

Oak, Elm, and Willow in this simple way. Here, TreeSpecies is a class analogous to the object type AccountType in Fig. 10.61.

If the name "powertype" derives from the notion of power set (the power set of a set *A* is the set of all subsets of *A*), the term is misleading, as the TreeSpecies powertype excludes many instances in the power set of the set of trees (e.g., the null set, the set of all trees, and many other tree sets). For this reason, the term "*higher-order type*" is more appropriate than "powertype".

At any rate, this diagram does not explicitly include a membership association such as Tree is a member of TreeSpecies (analagous to Account is of AccountType). We may treat this association as implicit here, but in practice an explicit version of this association would typically be needed because we would normally want to know the species of any given tree, and with hundreds of tree species it would be diagrammatically extravagant to introduce subtypes for all of them.

In UML 2, a colon ":" prepends the powertype name (e.g., ": TreeSpecies") to annotate the collection of displayed subtype–supertype connections that belongs to the set of all possible generalization relationships (called a *GeneralizationSet*) based on the categorization scheme provided by the association that relates the supertype to the powertype.

If the subtypes are connected to the supertype using different arrowheads, the powertype annotation is placed next to a dashed line that crosses the relevant subtype connections (see Fig. 10.63(a)).

If a common arrowhead is used, the annotation is placed next to that (see Fig. 10.63(b)).

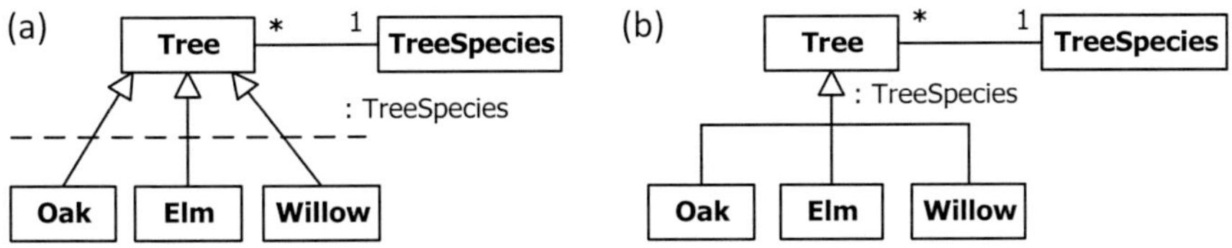

FIG. 10.63 Powertype example for TreeSpecies in UML 2 notation.

We can know that a class is a powertype if its name is used in a generalization set constraint. We may not assume that any binary association from a class to a powertype is of this nature. For example, in addition to Tree is a member of TreeSpecies we might have fact types such as Person named TreeSpecies.

Fig. 10.64 shows an example from the UML 2 specification that provides a different way to model a variation of our earlier account example. Here, AccountType is modeled as a powertype rather than an enumeration, and the fact type Account is of AccountType is modeled as an association rather than as an attribute. It is possible to include this association in the model *without explicitly introducing any subtypes* (SavingsAccount, etc.). In that case, there is *no way to formally capture the categorization semantics of the association, or to declare that AccountType is a powertype.*

While the role name "account classifier" *informally* suggests these additional semantics, this has no formal force (other examples in the UML specification use different role names such as "vehicle category", etc.). So *we need to distinguish the categorization association itself*, for example, by appending a colon to the relationship reading (e.g., "is of:") and/or the relevant association role names.

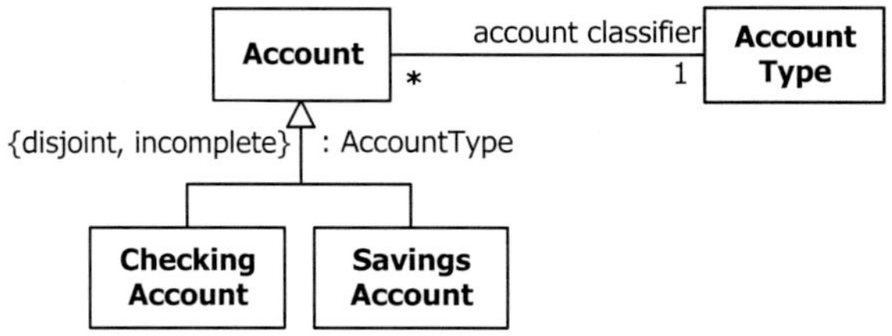

FIG. 10.64 Powertype example from the UML 2.0 Superstructure specification.

Even if generalization relationships are annotated with the relevant powertype name, this does not always formally guarantee the required semantics. For example, suppose we classify employees as part-time or full-time, and record their preferred employment status (part-time or full-time), if any. The UML model in Fig. 10.65 has two associations between Employee and EmployeeType, but there is no formal way to know which association is used as the basis for membership in the subtypes.

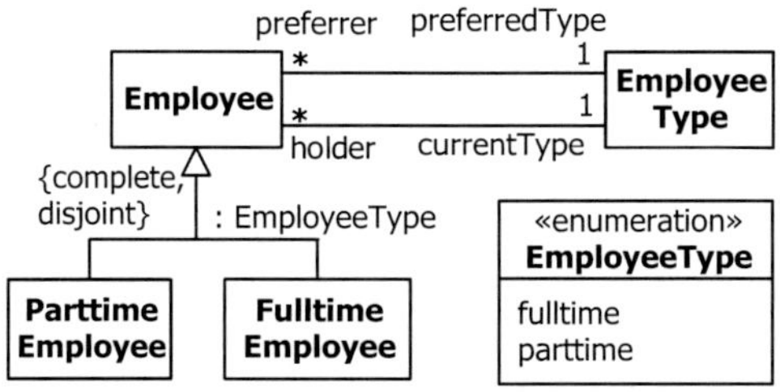

FIG. 10.65 Powertypes do not guarantee an unambiguous classification scheme.

To solve this problem, we could require role names in the annotation, or adopt the ORM practice of supplying formal subtype definitions. The types Employee, ParttimeEmployee, and FulltimeEmployee are time deictic because their sense is determined in part by the time when their terms are uttered/inscribed.

UML allows generalization annotations to include names for the generalization set (presumably prepended to powertype names when powertypes are involved, but the UML specification does not clarify this possibility). However, such names are treated as informal comments. There is no formal requirement to link these back to properties (attributes or far roles) of the supertype, as done in those database modeling techniques that use discriminators to indicate a basis for subtyping.

If we are to allow higher-order types in our models, what kind of higher-order logic is appropriate? The current ORM formalization uses only first-order logic, basic arithmetic, and bag comprehension. Once we adopt higher-order logic, we need to allow any order (not just second order) because in principle, one might always introduce new types to categorize existing types, whether or not those types are first order. If we adopt standard semantics for higher-order logic, where quantifiers may range over any imaginable predicates, we lose some useful properties of first-order logic (e.g., completeness and compactness).

Moreover, care is needed to avoid some well-known paradoxes. *Russell's paradox* considers the set of all sets that are not members of themselves: is this set a member of itself? If Yes, then No, and if No then Yes, leading to a contradiction.

Grelling's paradox deals with self-predicable or autological properties (properties that apply to themselves). For example, we might argue that the property of being nonhuman is itself nonhuman (and hence is a self-predicable property), whereas the property of being human is not itself human (and hence is a nonself-predicable property). If predicates may instantiate themselves, then using *N* and *H* for the properties of being nonhuman and human, we might formalize this as *NN* and ~*HH*, respectively. But then what about the property of being nonself-predicable? Is this self-predicable or not? If it is, then it is not, and if it is not, then it is. Either way, we have a contradiction.

To avoid such paradoxes, Bertrand Russell developed a type theory in which types are ordered in a hierarchy, and it is meaningful to say that a type is an instance of another type only if the second type is on the next level of the hierarchy. Similarly, predicates of higher order apply only to predicates or objects of lower orders. In particular, no predicate may apply to any predicate of the same order. Hence, no predicate may apply to itself. Essentially, the paradoxes are avoided by forbidding predicates to apply to themselves, by adopting a hierarchy of levels in which types can have instances at lower levels only.

While this seems a reasonable approach to adopting higher-order logic, some other versions of higher-order logic do not take this approach. For example, both the Knowledge Interchange Format (KIF) and the ISO Common Logic standard allow predicates to instantiate themselves, so expressions such as (*R R*) are allowed.

To make the implementation of higher-order logic more tractable, it seems best to adopt a nonstandard semantics, similar to Henkin semantics, to limit the range of predicates/functions over which we may quantify to retain useful properties of first-order logic such as completeness. With standard semantics, a monadic first-order predicate may range over the power set of the domain of individuals (objects: lexical or nonlexical).

To deal with categorization types (e.g., AccountType) where we wish to assert that instances are types, it seems that the only extension we need beyond first-order logic is to allow quantification over object types that are instances of a declared categorization type (whether or not these instances have been explicitly declared as a subtype).

As a separate decision, if we wish to allow crossing metalevels in the same model (see later), we should allow quantification over object types (primitive or derived), of any order, that are explicitly declared in the schema. If we allow this, there seems to be no compelling case to allow quantification over polyadic predicates, so this relaxation may be regarded as a restricted case of Henkin semantics.

We now outline an approach to categorization that remains first order. Consider the ORM schema in Fig. 10.66, which is a classic case where higher-order logic proponents would demand that CarModel is a second-order type, whose instances are subtypes of Car (e.g., Toyota Corolla 2022, Dodge Hornet 2023, Chevrolet Corvette 2025).

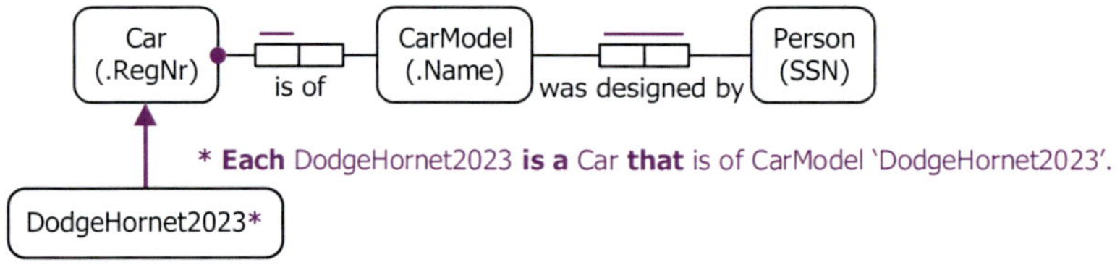

FIG. 10.66 The population of a subtype (or any type) typically varies over time.

When we think of an instance of Person and Car (concrete concepts), we think of an actual person or car. When we think of an instance of CarModel (an abstract concept), we typically think of an abstraction that is essentially a car design–a car *structure or specification* that might be denoted by a schematic diagram, for example.

For any given business domain, we define a *type* as a *set of possible instances where for any given state of the business domain, exactly one subset of the type is the population of the type in that state.*

At any given time, the *population* of a type is the set of instances of that type that exists in the business domain at that time. The temporal aspect is needed to distinguish between types that have the same *extension over time* but may differ in *extension at some time* (e.g., HumanPerson, and HumanBaby).

By the principle of extensionality, sets are defined by their extension. So unless we resort to nonwell-founded set theories, we must regard sets to be fixed—they cannot change over time. For any given business domain, the current population of a type (e.g., Person or Country) may change over time, but the type itself does not.

Given the aforementioned definition of type, it seems reasonable not to think of an instance of a car model as a set of possible cars of that model (a subtype). Each CarModel instance is in 1:1 correspondence with such a subtype (implicit or explicit), but it's not identical to a subtype. It seems to us that this distinction can always be made. If so, we can treat instances of such "categorization types" as ordinary individuals that are not types, so first-order logic is enough.

Suppose we explicitly introduce a car subtype called "DodgeHornet2023", as shown in Fig. 10.66. The subtype definition provides the formal connection between the subtype and the car model instance. Using the standard reference mode semantics for ORM, the expression "CarModel 'DodgeHornet2023'" in the definition abbreviates "CarModel **that** has CarModelName 'DodgeHornet2023'".

Suppose we add the existential fact **There exists a** CarModel **that** has CarModelName 'DodgeHornet2023' to the information base before any cars of that model are produced. At that time, the population of the car subtype DodgeHornet2023 is the null set. Suppose at a later time, 1000 cars of that model are produced. At that later time, the population of the car subtype DodgeHornet2023 includes 1000 cars.

At any time, the subtype DodgeHornet2023 includes those 1000 cars as well as all other cars of that model that will ever be produced. So, it's perfectly okay to regard an instance of CarModel (conceived of as the fundamental car structure or design to which the car instances conform) to be in 1:1 correspondence with a set of (possible) cars.

Hence, it is reasonable to think of an instance of a "categorization type" such as CarModel or AccountType as an individual (e.g., a structural pattern) that is ontologically distinct from a type (in the sense of a set of possible instances). Thus, it seems sufficient to stay with first-order logic for such cases.

Although the word "Type" in "AccountType" may suggest that its instances are themselves types, this stems more from an unfortunate naming choice for the type rather than from any fundamental intuition.

The second approach that allows one to avoid higher-order types for categorization is to *avoid uninformative categorization schemes*. The term "AccountType" is uninformative because it does not provide any basis for categorizing accounts. In principle, any object type such as Account might be categorized in many different ways, leading to different types of bank accounts. For example, we could define an AccountKind {Local, National, International}, an AccountCategory {Taxable, Nontaxable}, and so on. These are all categorization schemes, which we may wish to use in the same model, and names such as "AccountType" and "AccountKind" don't inform us at all about the criterion used by a given categorization scheme to place accounts into account categories.

One might argue that the value constraint placed on the categorization scheme provides this criterion, but this requires the modeler to induce the criterion based on his/her informal understanding of what the names for those values mean, an understanding that is not formally accessible to an automated system. More importantly, we may wish to introduce a categorization scheme without committing to a fixed set of instances.

As a pragmatic issue then, it seems reasonable to encourage the modeler to *choose informative names that reveal the basis for classification schemes*. If we adopt this approach, the type = instance issue typically disappears for the categorization case. In Fig. 10.67, for example, the subtype Savings account is defined based on its primary function. The AccountFunction instance named "Savings" and denoted by the shaded dot is clearly not identical to the subtype SavingsAccount (a function is not the same thing as a bank account). One may introduce other informative categorization schemes such as Account may be used in Region, etc. Similarly, our Car is of CarModel relationship might be renamed "Car conforms to CarModel".

Any binary fact type used for an enumerated categorization scheme may be replaced by one or more unary fact types. For example, instead of Account is of AccountType {Checking, Savings}, we may use the fact types Account is used primarily for savings and Account is used primarily for checking (the mandatory and uniqueness constraints are then captured by an xor constraint). Instead of Account is of AccountCategory {Taxable, Nontaxable}, we may use the fact type Account is taxable, applying the CWA to determine nontaxable accounts. This is yet another way to avoid higher-order types for enumerated categorization schemes.

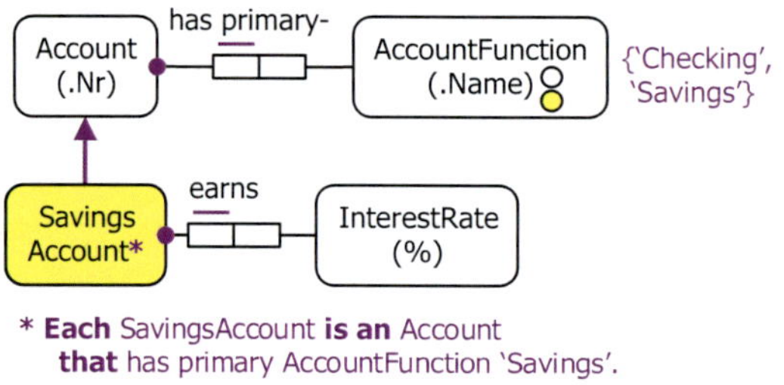

FIG. 10.67 Informative categorization explains the basis for categorization.

In many cases, *an object type may be used to categorize more than one kind of object*. For example, Fig. 10.68 includes two categorization fact types: Person is of Gender and Animal is of Gender, where Gender has two possible instances identified by the gender codes "M" (for male gender) and "F" (for female gender).

Here, the semantics of the categorization fact types does not involve homogeneous set membership. Clearly, instances of Gender (MaleGender, FemaleGender) are not identical to any of the four subtypes shown. This kind of categorization scheme is quite common and clearly excludes any type = instance identities.

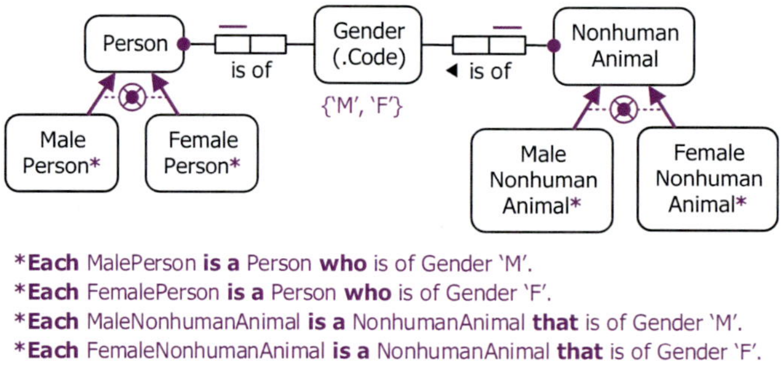

FIG. 10.68 A case where categorization instances are not identified with subtypes.

The use of Gender in this model to define the subtypes seems better than using two powertypes, PersonType {MalePerson, FemalePerson} and NonhumanAnimalType {NonhumanMaleAnimal, NonhumanFemaleAnimal}. While UML allows this case to be modeled using Gender as an attribute or nonpowertype, any use of a generalization set name (e.g., "gender") is merely an informal comment, with no formal connection to the model element used as the basis of the categorization.

We now consider the impact of higher-order types on *set structures and metalevel crossing*. Consider a categorization scheme whose instances intuitively correspond to set-like containers. Fig. 10.69 includes a model that, apart from some constraints, is structurally similar to the account example in Fig. 10.61.

In this business domain, only A-team members may earn privileges, so a subtype is created to record facts of this nature. The shaded dot denotes team A, and the white dot denotes team B. The shaded subtype A-TeamMember is the type whose population at any time is the set of A-team members at that time.

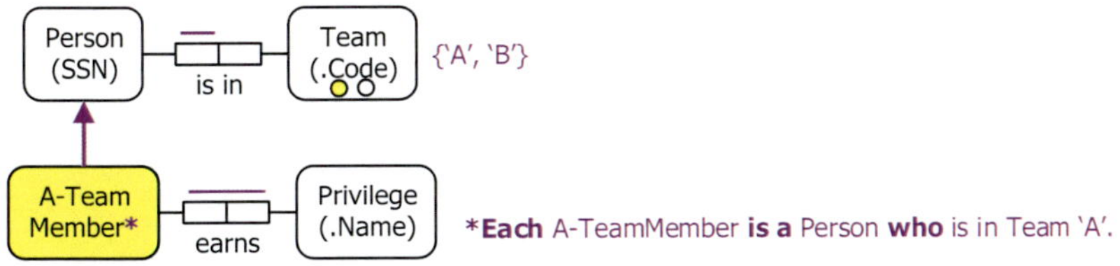

FIG. 10.69 Is Team *A* identical to the subtype A-TeamMember?

It seems natural to think of Team A at any point in time as more than just a set of people. The concept of a team brings in other semantics (a social unit whose members work together for a common purpose). While the thing denoted by the shaded dot appears to have this additional informal semantics, the subtype A-TeamMember does not—at any point in time its denotation seems to be no more than a set of people who just happen to be members of team A.

If these intuitions are correct, and team membership is considered to be a categorization scheme, here is another example where it is reasonable not to identify the subtype with the instance of the categorization scheme.

Now consider Table 10.3, which is a simplified version of an example used by Fitting (2000), to motivate a formalization of databases that uses higher-order modal logic. The table is in nonfirst normal form, allowing unnamed sets as entries (e.g., ColorChoices). This is permitted in some object-relational databases. Second, its final attribute (column) allows as entries unnamed sets whose instances appear to be attributes themselves, thus crossing levels/metalevels.

Table 10.3 Table with entries that may be sets of individuals or attributes.

Car	*CarModel*	*ColorChoices*	*AirConditioning*	*CustomerChosen*
1	Ford Escape 2003	{red, green, black}	Yes	{ColorChoices}
2	Ford Escape 2003	{red, green, black}	No	{ }
3	Mazda MPV 2004	{green, sand-mica}	Yes	{ColorChoices, AirConditioning}

Fitting's formalization of this situation is higher order, as he treats the structure directly as it stands. The price paid for this directness is deep complexity and an implementation nightmare. These disadvantages can be avoided by transformation into a first-order model that is cleaner and easier to implement.

Suppose we record the color chosen for a car. Adding the fact type Car is of Color, the situation may be modeled by the ORM schema shown in Fig. 10.70. Here, the color choice sets are handled in the usual normalized way, with a many:many association. The subset constraint ensures that the color of a car is one of those available for its car model.

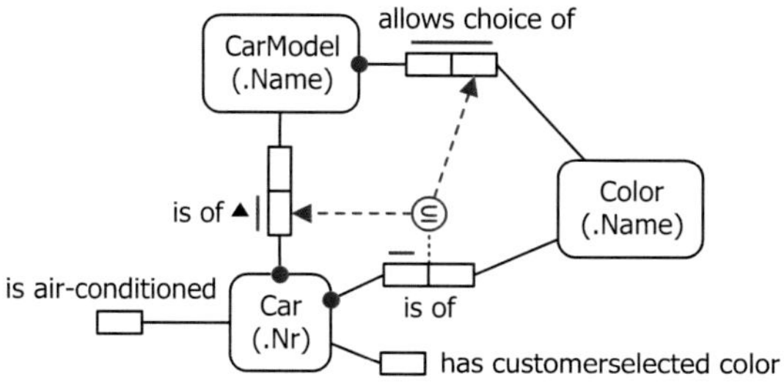

FIG. 10.70 A first-order solution for part of the unnormalized table.

In cases where there are many attributes about which information is to be recorded, and the attributes are not all known in advance, this may be modeled by introducing Attribute as a first-order type, along with fact types that record its name and value. By thus demoting metadata to ground data, we remain at first order.

A final argument for using higher-order types is to allow the expression of business rules that appear to cross levels/metalevels (or ignore level distinctions) in the same model (e.g., the Finance department is responsible for defining the possible values of AccountType).

If we really want to formalize such cases directly, then higher-order types are clearly needed. However, as a pragmatic alternative, it is usually possible to handle such rules in a first-order way, either by separating the metaaspects into a separate first-order model where the former metatypes are now ground types, or by demoting metadata to ground data, or simply ignoring the cross-level identities.

As a simple example, we might build into the core package of the metaschema such metafact types as FactRole is played by ObjectType, BusinessRule has IllocutionaryForce etc., while in the management package of the metaschema we include metafact types dealing with aspects of security and authorization, etc. (e.g., UserGroup has AccessRight to FactType). Populating the latter metafact types in the metamodel may then allow us to add rules of the desired kind without crossing metalevels.

For example, consider the simple business model in Fig. 10.71(a). Here, there are two elementary fact types (F2 and F4) and three existential fact types (F1, F3, and F5) depicted in abbreviated form using parenthesized reference modes. In this case, the current values of AccountType are stored in a reference table instead of being rigidly declared in a value-type constraint.

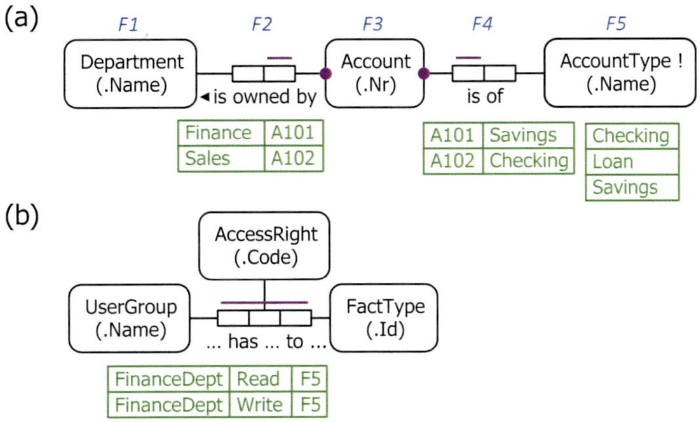

FIG. 10.71 A business model (a) and a related metamodel fragment (b).

With this approach, the one column fact table for AccountType may be treated just like any other fact table in the model, including the way in which we determine who has what kind of access to what fact type. A simplified metafragment for dealing with security is shown in Fig. 10.71(b).The metamodel in Fig. 10.71(b) exists at a level above the business model in Fig. 10.71(a). Each of these models can be formalized separately, using either first-order logic or the higher-order logic extension for categorization discussed earlier.

What is missing from this picture is the ability to identify the marketing department mentioned in the business model with the marketing department that appears as a user group in the metamodel. There does not appear to be any compelling business case to require formal support of this identity, as businesses seem to run perfectly well without such support (e.g., in SQL systems, the application tables and metatables are typically accessed separately).

Exercise 10.7

1. (a) Represent the ORM schema shown as a UML 2 class diagram, using PersonType as a "powertype". Introduce other modeling elements if needed.
   (b) Explain the disadvantages of your solution to (a).
   (c) Revise the ORM schema to improve its informativeness, adopting first-order semantics rather than second-order semantics.

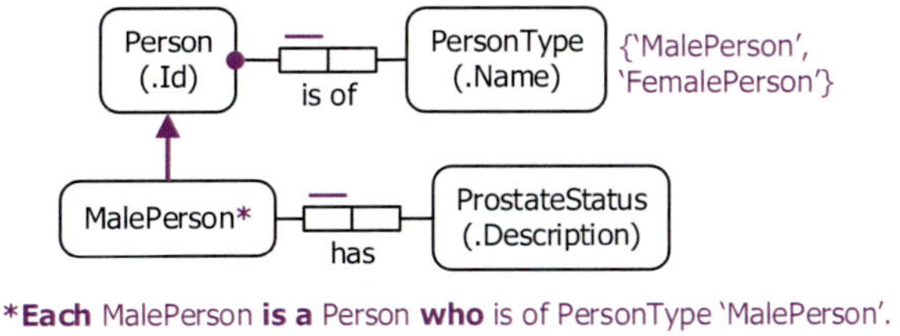

2. The UML class diagram shown is intended to model the situation where each student is currently enrolled as an undergraduate or graduate (but not both) and plans to finally be an undergraduate or graduate (but not both). For example, some current undergraduate students might plan to go on to do a graduate degree while some other current undergraduate students might plan to never progress to a graduate degree.

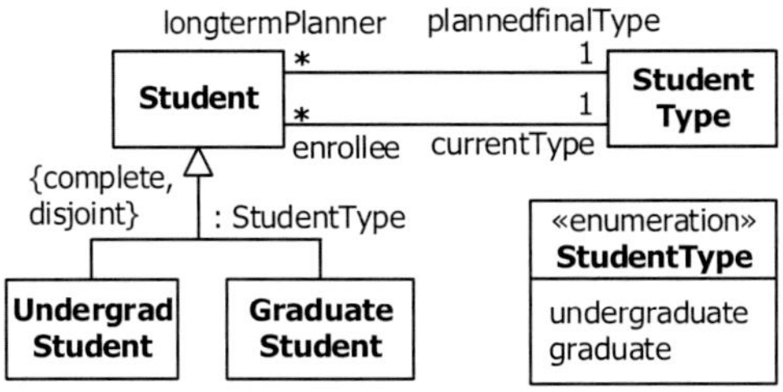

   (a) Explain what is wrong with the use of StudentType as a “powertype” in this schema.
   (b) The schema also lacks a textual constraint. Remodel the situation in ORM (include a textual constraint if needed) to address both these deficiencies.

3. Explain clearly in no more than 100 words why Russell’s Paradox really is a paradox.

4. Consider the following ORM schema.

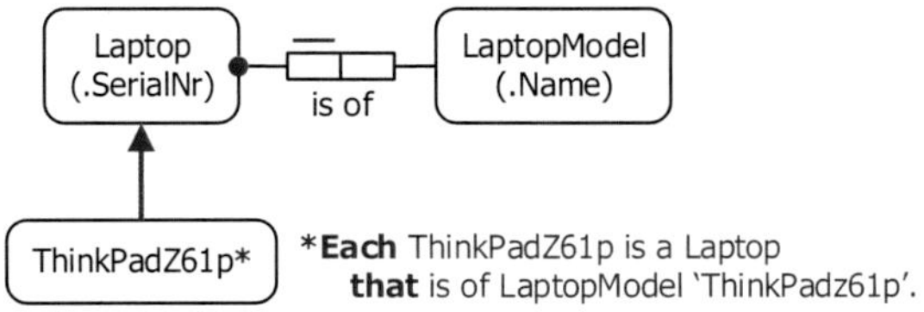

   (a) If the subtype is an instance of LaptopModel, is LaptopModel a higher-order type? Answer yes or no.
   (b) Provide in no more than 50 words an interpretation that allows one to think of LaptopModel as a first-order type.

10.8 Further Constraints Involving Subtyping

In Fig. 10.72(a), the exclusive-or constraint connects the upper parts of the subtyping arrows. In ORM, an arrow from a subtype to a supertype displays a type-level relationship that implies instance-level, identity relationships linking subtype and supertype instances. For example, the exclusive-or constraint in Fig. 10.72(a) is understood to apply to the *supertype roles* of the *instance-level link fact types* shown with dashed lines in Fig. 10.72(b).

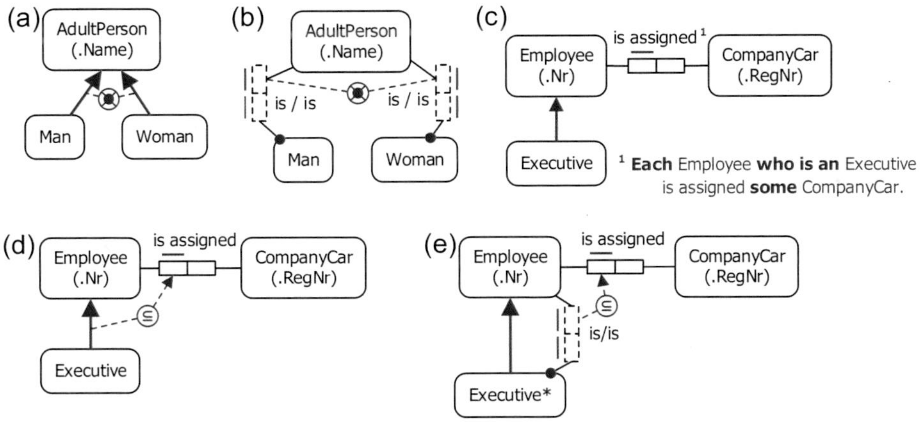

FIG. 10.72 Constraints on subtyping link roles.

In this example, the subtypes inherit their preferred reference scheme from their supertype, so these identity ("=" or "is") relationships form part of the reference scheme for subtype instances, for example, the Man who **is** the AdultPerson named "Tom Smith".

The role of using a company car is optional for employees, but mandatory for executives. Fig. 10.72(c) displays this *restricted mandatory role constraint* as a textual constraint. Fig. 10.72(c) reproduces an example discussed earlier in Section 9.7.

In 2013, ORM and NORMA were extended to allow other kinds of constraints to apply to supertype link roles. Among other things, this allows restricted mandatory role constraints to be expressed graphically, as shown by the subset constraint in Fig. 10.72(d), which is interpreted as in Fig. 10.72(e). NORMA automatically verbalizes this subset constraint as follows:

If some Employee is **some** Executive
then that Employee uses **some** CompanyCar.

Typically, an inclusive-or constraint applies to two more roles hosted by the same object type. However, *inclusive-or constraints may apply to roles hosted by different types*, so long as those types are compatible (and hence may share instances). For such cases, we define the inclusive-or constraint as follows: for each state of the model, *each instance in the population of the minimal common supertype of the types hosting the constrained role must play at least one of the constrained roles.* The minimal common supertype is typically the root of the subtype graph.

An example is shown in Fig. 10.73(a), where one constrained role is hosted by Employee and the other by Manager. Employee and Manager are compatible because Manager is a subtype of Employee. Here, the inclusive-or constraint means that each employee either reports to a manager or is a manager who manages some employee. For a one-person company, this is satisfiable only if we allow self-management.

Fig. 10.73(b) shows the same example with the implicit link fact type displayed. The inclusive-or constraint may be verbalized thus:

Each Employee reports to **some** Manager
or is **some** Manager **who** manages **some** Employee.

Fig. 10.73(c) shows a more realistic inclusive-or constraint involving different types. In this case, one of the constrained roles is a supertype link role, as shown explicitly in Fig. 10.73(d). The constraint may be verbalized simply thus:

Each Employee reports to **some** Manager **or** is **some** Manager.

If the constraint has deontic rather than alethic modality, an appropriate deontic operator is prepended (e.g., "**It is obligatory that**" or "**It is forbidden that**" for positive and negative forms of the verbalization, respectively).

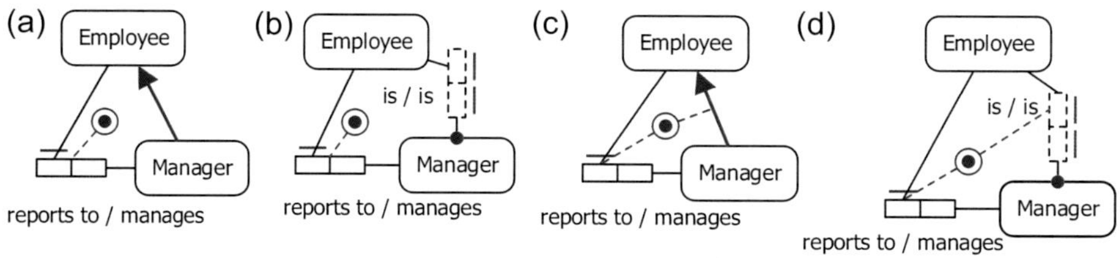

FIG. 10.73 An inclusive-or constraint on roles hosted by different types.

10.9 Summary

Join constraints apply to roles projected from an ORM path that involves at least one conceptual join. The simplest kind of join constraint applies to a single join path. External uniqueness constraints are of this kind. *Join subset constraints* are subset constraints between two role sequences, where at least one role sequence is projected from a join path. *Join equality constraints* and *join exclusion constraints* also include at least one role sequence argument projected from a join path.

Conceptual joins may be *inner* or *outer*. If there are multiple paths connecting roles in a role projection, role sequence numbers are needed to disambiguate the intended role path.

Alethic rules impose necessities that cannot possibly be violated, and hence any attempt to record a violation of an alethic rule in the information model will be rejected (e.g., each person was born on at most one date). *Deontic rules* specify obligations that ought to be obeyed in the business domain but may be violated (e.g., in most Western cultures, it is obligatory that each person is husband of at most one person). Violations of deontic rules may be recorded in the information system (e.g., to enable notification to relevant authorities who might be able to take some appropriate action concerning the violations).

Positive verbalization of deontic rules begin with "It is obligatory that" (instead of the implicit alethic "It is necessary that"), and negative verbalizations of deontic rules begin with "It is forbidden that" (instead of the alethic "It is impossible that"). The expression "It is forbidden that *p*" is equivalent to "It is obligatory that not *p*" and "It is not permitted that *p*".

There are three basic *temporal data types*: *instant* (point in time), *interval* (duration of time), and *period* (anchored duration of time). *Valid time* indicates when an event occurred in the business domain, whereas *transaction time* indicates when it was recorded in the information system. Time may be measured with different *temporal granularities* (e.g., second, minute, day) and with respect to different *calendars* (e.g., Gregorian, Julian) and different *time zones* (e.g., UTC, MDT).

Fact types may be *definitional* (truth of instances is a matter of definition, independent of events), *once only* (instances correspond to a single event), or *repeatable* (instances may be made true by any one of a set of events). For each once-only or repeatable fact type, we need to decide what (if any) temporal information or *history* is needed. Temporal data about once-only fact types (e.g., Person was born in Country, Person wrote Book) may be maintained by adding temporal fact types of appropriate temporal granularity to the key object type (e.g., Person was born in Year, Writing (objectified) began on Date, Writing ended on Date).

Fact types may be *unchangeable, individually changeable*, or *collectively changeable.* Many sentences are *time deictic* (their meaning and hence truth value depends on when they were uttered or inscribed). To keep *history of changeable fact types* (e.g., Person has Weight), insert a temporal object type into the key (e.g., Person on Date had Weight) or introduce a relevant event type (e.g., WeightMeasurement).

To keep *history of multiple events relating to a fact type with a spanning uniqueness constraint* (e.g., Person visited Country), do one of three things: include a distinguishing temporal role (e.g., Visit began on Date), introduce a simple identifier (e.g., VisitId), or introduce an ordinal number as part of the identifier (e.g., VisitNr).

Dynamic constraints apply either to transitions between successive states (e.g., changes in marital status) or over a period of time (e.g., invoices should be paid within 30 days of issue). Transition constraints can be specified textually or by state transition tables or diagrams.

Instances of a *rigid type* (e.g., Person, Car) must remain in that type for their lifetime. Instances of a *role type* (e.g., Child, Employee) may move into or out of that type at different times. No rigid type may be a subtype of a role type. Migration between subtypes (e.g., from Child to Adult) is possible. To record history of an object as it migrates between role subtypes, use either the decreasing disjunctions pattern (e.g., Fig. 10.34) or the role-playing pattern (e.g., Fig. 10.35).

Various implementation targets include data structures for various *collection types* (e.g., sets, bags, ordered sets, sequences, and arrays). Although such collections could be modeled directly at the conceptual level by using constructors, it seems preferable to model them indirectly and invoke a collection type simply as a mapping annotation. It also seems more convenient to introduce names for unnamed collections.

An *extensional uniqueness constraint* {_} may be used to declare that a set type is defined by its membership. *Ordered sets* (order significant, no duplicates) such as an author list may be modeled in ORM using a position type (e.g., Book was authored by Person in Position) and in UML using the {ordered} property. *Bags* (order not significant, duplicates allowed) such as a grade list may be modeled in ORM using a frequency type (e.g., Student scored Grade with Frequency) and in UML using the {nonunique} property. Bags are also called multisets.

Sequences (ordered bags) such as an award list may be modeled in ORM using a position type in the key (e.g., Lottery awarded prize of Rank to Person) and in UML using the {ordered, nonunique} setting. *Arrays* (indexed sequences) are modeled similarly.

Nominalization recasts a verb as a noun phrase. In information modeling, objectification of a relationship is typically *situational nominalization*, where the resulting object is a state of affairs (e.g., an event or activity) in one-to-one correspondence with, but not identical to, the relationship being objectified. For example, Person sings Song; **that** Singing is delightful.

In cases of objectification, ORM creates *linking fact types* underneath (e.g., Singing is by Person, Singing is of Song). In rare cases, objectification is *propositional nominalization*, where the object being discussed is a proposition; this is how association classes are defined in UML. In information modeling, propositional nominalization can typically be replaced by situational nominalization.

ORM 2 allows unary fact types (e.g., President died) to be objectified (e.g., Death) and includes objectification guidelines to help decide when to objectify (e.g., don't objectify an n:1 association unless it might evolve to an m:n association, or the objectification type has multiple fact types that relate directly to it). Fact types with a spanning uniqueness constraint may also be objectified (e.g., Person speaks Language, and Person visited Country in Year). ORM 2 also allows objectification of a fact type with exactly one a uniqueness constraint that spans $n-1$ roles ($n \geq 1$), but such cases are best modeled in other ways.

CWA assumes that all relevant facts are known (appear in the model either as asserted facts or as derived facts) and implies *negation as failure* (failure to find or prove a proposition implies that it is false). OWA allows that some relevant facts are unknown, and implies *classical negation* (failure to find or prove a proposition implies that its truth value is unknown).

SWA allows that some relevant facts are unknown as derived facts but not as asserted facts, and implies *semipositive negation* (missing instances of asserted fact types are false, but failure to derive a fact instance or its negation implies that its truth value is unknown). OWN allows that some relevant facts are unknown, but also allows negations to be asserted; as a special case of OWA, it implies classical negation.

Graphical notations to distinguish different closed/open world flavors for unary fact types are currently being considered for ORM 2 (e.g., dashed line for open). Even without such notations, the different approaches can be modeled, either by introducing unaries for the negated predicates or by converting to a binary. In practice, CWA and OWN are the most useful approaches, and may be captured in UML and relational schemas using mandatory Boolean attributes and optional Boolean attributes, respectively.

First-order logic quantifies over individuals only, not predicates. Higher-order logics quantify over both individuals and predicates. A *higher-order type* includes instances that are themselves types. A *categorization type* (e.g., Gender) is used to classify another type (e.g., Person) into subtypes. UML allows higher-order types called "powertypes" to be used as categorization types.

For example, in UML the subtype Oak may be treated as identical to an instance of the powertype TreeSpecies, which is used to classify the type Tree. Higher-order types are sometimes also used to directly model metastructures and to allow rules that cross levels/metalevels.

Higher-order logics are vastly more complex to implement efficiently than first-order logics. In practice, it is typically possible (and, we suggest, advisable) to avoid higher-order constructs in information modeling. For example, instances of categorization types can be thought of as individuals, informative fact types may be used for categorization, and different levels may be handled in different models. However, if higher-order logic is used in information modeling, we suggest that restricted Henkin semantics be used.

An arrow from a subtype to a supertype displays a type-level relationship that implies instance-level, identity relationships linking subtype and supertype instances. Inclusive-or constraints, exclusion constraints, and subset subsets may apply to *supertype roles* of the *instance-level link fact types.*

Typically, an inclusive-or constraint applies to two or more roles hosted by the same object type. However, *inclusive-or constraints may apply to roles hosted by different types,* so long as those types are compatible.

Chapter Notes

This chapter covered a lot of territory, and is technically the most challenging chapter in this book. Some further discussion can be found in the following research papers, upon which much of the material in this chapter is based. These papers also contain detailed references to related work.

For conceptual joins, see Halpin (2002a, 2005c). For deontic rules, see Halpin (2006a, 2007a). For temporal issues in data modeling, see Halpin's series of online articles on this topic in *Business Rules Journal* (www.brcommunity.com/nphp). For collection types, see Halpin (2000b). For nominalization and objectification, see Halpin (2005a, 2008). For higher-order types, see Halpin (2004a, 2005d). For further constraints on supertype roles and inclusive-or constraints over roles hosted by different types, see Halpin and Curland (2013).

11

Relational Mapping

Chapter Outline

11.1 Implementing a Conceptual Schema

Most database modeling tools allow you to enter a data model in one or more high level notations (e.g., ER, IDEF1X, ORM, or UML), as well as a logical level notation (e.g., relational). Typically, a high-level (conceptual or semiconceptual) schema must be mapped down to a logical and then physical schema for the database to be populated and queried. Assuming you model first at the conceptual level, the main steps in implementing your data model are as follows.

- Design the conceptual schema
- Annotate the conceptual schema with mapping choices as needed
- Map the design to a logical schema (e.g., relational or object-relational)
- Finesse the logical schema as needed (e.g., rename or reorder some columns)
- Generate the physical schema [e.g., in Microsoft SQL Server or IBM DB2]
- Create external schema(s) (e.g., forms, reports)
- Enforce security levels as needed
- Populate the database
- Issue queries and updates
- Update the schemas as needed

Earlier chapters discussed how to design the conceptual schema. This chapter focuses on mapping conceptual models to logical models. Because relational database systems are widely used, the relational model is used for the logical schema. Basic ideas about relational databases were introduced in Chapters 1 and 2. Section 11.2 summarizes these points and expands on them briefly. Section 11.3 discusses the basic procedure for

Information Modeling and Relational Databases. https://doi.org/10.1016/B978-0-443-23790-4.00014-1

mapping an ORM schema onto a relational schema. This can be adapted easily to cover mapping from other notations, such as ER or UML. Mapping from IDEF1X to relational is trivial, as discussed in Chapter 8.

Section 11.4 discusses advanced aspects of relational mapping, including the use of conceptual annotations to override default mapping choices (e.g., to control how subtypes are mapped). After automated mapping, some finessing of table and column names may be needed to meet naming standards, and columns may be reordered to improve performance.

Once the logical schema is determined, if not before, the target database management system (DBMS) is selected. The physical schema is then generated for this specific platform. Some aspects of tuning the physical model to improve performance are covered later in Chapter 14 (e.g., index selection and denormalization).

Updates and queries tend to be carried out either in a logical query language (e.g., SQL or QBE) or via an external forms and reporting interface defined on top of the logical schema. Some aspects of external schema design are covered in a later chapter. With multiuser applications, different user groups' access rights to the tables, forms, and reports should be enforced by the database administrator. Chapters 12 and 13 indicate how table updates, queries, and access rights can be declared in SQL.

All the major relational DBMSs either support SQL directly or provide translation facilities to/from SQL. Some higher-level query languages allow users to formulate queries directly in terms of dimensions (for data warehousing), or even conceptual schema constructs. Some of these are discussed in later chapters.

The next section summarizes the fundamental constructs in relational database schemas using a generic notation. The following two sections then discuss basic and advanced mapping of conceptual to relational schemas. The final section provides the chapter summary.

11.2 Relational Schemas

A *relational schema* (or relational database schema) is a set of relational table definitions, constraints, and perhaps derivation rules. You may wish to review Sections 1.3 and 2.3, where the basic ideas were discussed. The structure of a single relational table is called a *table scheme* or relation variable. This is basically a named set of *attributes* (columns) that draw their values from data *domains*. Each table scheme may be populated by a set of unnamed *tuples* (rows) of data, but the population is not part of the table scheme itself.

Many *notations* exist for table schemes. The NORMA tool uses a vertical layout, where the column names are listed vertically. To save space and to facilitate discussion of populations, this book typically uses a *horizontal layout*, where the table name precedes a parenthesized list of column names, separated by commas. If desired, domain names may be displayed after the column names, using a colon separator.

For example, ignoring constraints, the ORM conceptual schema in Fig. 11.1(a) maps to the table scheme *Employee*(empNr: EmployeeNr, salary: Money, tax: Money). Here the table name is "Employee", the attribute empNr is based on the domain EmployeeNr, and the attributes salary and tax are both based on the domain Money.

Because the unit-based reference mode USD is based on the unit dimension Money, the ORM schema could also have been displayed as in Fig. 11.1(b).

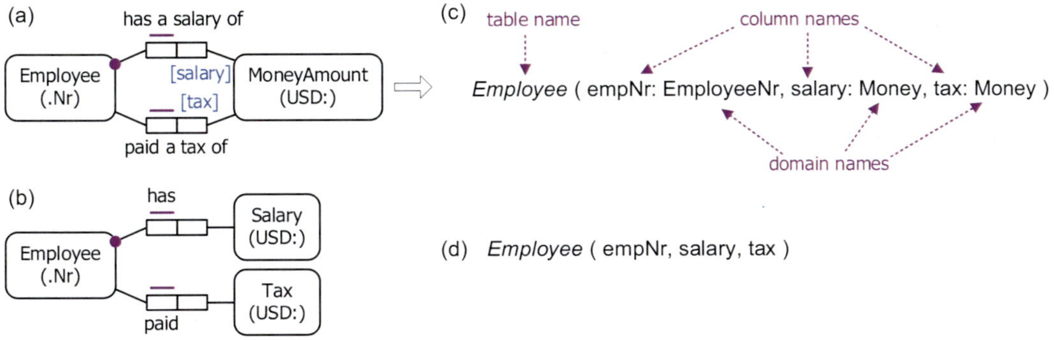

FIG. 11.1 Mapping an ORM schema to a relational schema (constraints omitted).

We adopt the convention of starting table and domain names with a capital letter and attribute names with a lowercase letter. This case convention is similar to that of UML and is fairly popular. Whichever case convention you choose, using it consistently will help make your schemas more readable.

When using horizontal layout, we usually write the table name in italics. We also prefer to concatenate name components, with later components capitalized (e.g., empNr) rather than using a separator such as an underscore or space (e.g., emp_nr or "emp nr"). As discussed in the next chapter, the SQL standard requires double quotes around a name that includes a space character.

Modeling tools usually give you control over how such names are generated. For example, the NORMA tool allows you to declare that "Employee" is to be abbreviated to "Emp" for column names but not for table names, and it can make use of role names in determining column names. You can pick your own style and stick with it. For compactness, it is usual to omit the domain names in setting out table schemes, for example, *Employee*(empNr, salary, tax).

In theory, the relational model supports semantic domains, which basically correspond to ORM conceptual object types. Prior to 1992, the SQL standard required that each attribute be defined directly over a numeric or character string data type. For example:

Person (surname: **varchar**(20), city: **varchar**(20), height: **smallint**, weight: **smallint**)

This can lead to semantic nonsense, such as comparing people and cities or height and weight. The SQL standard now allows attributes to be defined over user-defined syntactic domains that specify an underlying data type and optionally a value list and/or default value. The underlying data types in standard SQL support character strings, numbers, date, time, and bit strings, but not Money, which is instead defined in terms of a numeric data type, for example, **decimal**(9,2). SQL:1999 introduced user-defined types, which can be used as semantic domains. However, many commercial SQLs still provide no support for user-defined domains or types, so we'll typically ignore domain details from this point on.

Let's now consider the main notations used for specifying *constraints* in a relational schema. Later sections discuss how to group fact types into table schemes and how to map constraints and rules in more detail.

Using horizontal layout, **uniqueness constraints** on relational columns are shown by underlining. A *superkey* of a table is a set of one or more columns whose entries must be unique, thus providing a way to identify rows in the table. A *candidate key* is a *minimal* superkey. In other words, if an attribute is removed from a compound key, the remaining attributes are not spanned by a uniqueness constraint. If there is only one candidate key, this is automatically the *primary key*. For example, we may indicate that empNr is the primary key of the Employee table thus:

Employee (empNr, salary, tax)

If more than one candidate key exists, one of these must be selected as the primary key. The others are then called "*alternate keys*" or "*secondary keys*".

Primary keys are doubly underlined if alternate keys exist. For example, suppose we also record the name and department of each employee, and the combination of employee name and department is unique. The primary and secondary keys may be shown thus:

Employee (empNr, empName, deptCode, salary, tax)

The order in which the columns are listed is semantically irrelevant because each column has a name unique to its table. If columns in a composite key are not listed consecutively, dotted lines connect the underlines to show that a single, composite uniqueness constraint applies rather than multiple simple constraints. For example, the previous Employee table scheme may also be displayed thus:

Employee (empNr, empName, salary, deptCode, tax)

A column that does not allow null values is said to be **mandatory** *(for its table)*. A column that does allow null values is said to be *optional*. For horizontal layout, this book adopts the following conventions: optional columns are enclosed in *square brackets*; a column is mandatory in its table unless it is marked optional. This practice is consistent with the well-known Backus-Naur Form (BNF) notation. For example, because paying tax is optional, the previous ORM schema maps to the following table scheme, with the tax column optional (null values allowed):

Employee (empNr, salary, [tax])

This mapping is illustrated with a sample population in Fig. 11.2. The conceptual population includes three elementary facts: employee 101 has a salary of 80,000 US dollars (USD), employee 102 has a salary of 30,000 USD, and employee 101 paid a tax of 30,000 USD. No tax payment is yet recorded for employee 102.

Each row of a relational table expresses one or more elementary or existential facts. Because both fact types are grouped into the same table, each row expresses one or two facts.

The first row records the salary and tax facts about employee 101. The second row records the salary fact about employee 102. The second row also contains a null (shown here as "?"), indicating the absence of a tax fact for employee 102. The order in which the rows are displayed is not semantically relevant.

Fig. 11.3(a) adds two fact types to Fig. 11.1. Each employee's gender is now recorded as well as any cars driven by the employee. A role *r* is mandatory for an object type *O* only if

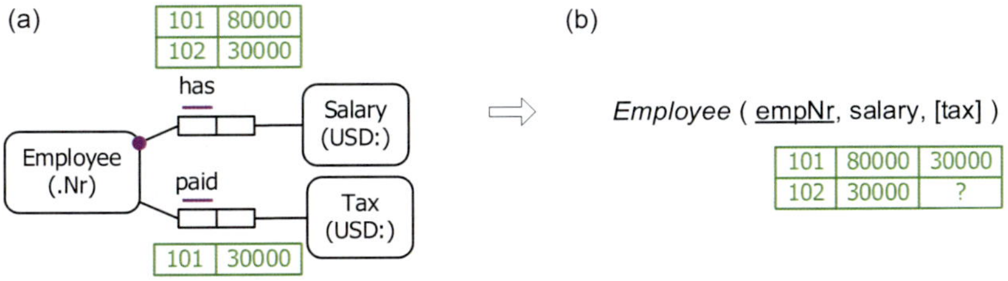

FIG. 11.2 The populated ORM model maps to the populated table scheme shown.

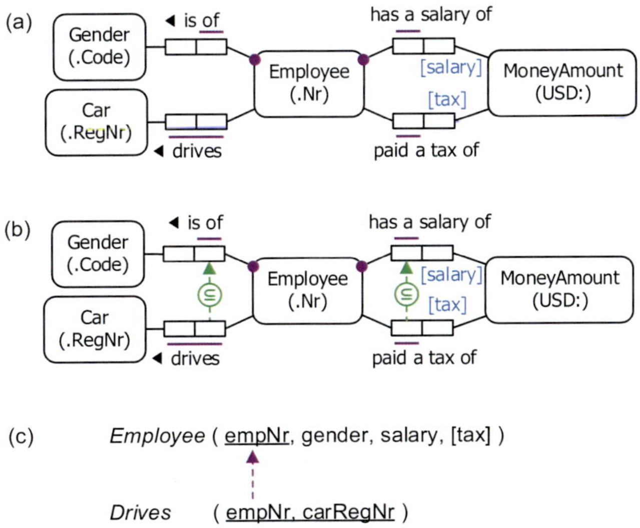

FIG. 11.3 The subset constraints are implied by the mandatory role constraints.

the population of each other role played by *O* must be a subset of the population of *r*. Hence, each optional role of an object type has an *implied subset constraint* to each mandatory role of that object type. Fig. 11.3(b) shows an implied subset constraint (colored green) from each optional role of Employee to a mandatory role of Employee. Equality constraints are implied between the mandatory roles, but are irrelevant to our discussion.

If all roles played by an object type map to the same table, its mandatory roles can be specified simply as mandatory columns. However, the relational model often requires different facts about the same object to be stored in different tables. In general, *mandatory role constraints are captured by making their columns mandatory in their table, and running a subset constraint from other tables (if any) that contain facts about that object type.*

As discussed in the next section, the *m:n* nature of the drives fact type requires it to be mapped to a different relational table from the other three fact types. So, information about Employee is spread over two relational tables shown in Fig. 11.3(c). Here, the mandatory role constraints are only partly captured—there is nothing to stop us entering employee numbers in the Drives table that do not occur in the Employee table.

How then do we map the conceptual mandatory role constraints?

The three *n*:1 fact types for salary, gender, and tax are grouped into the Employee table. The mandatory salary and gender predicates map to mandatory columns, and the subset constraint on the tax role is captured by having tax as the only optional column in the Employee table. The drives fact type maps to a separate table.

To capture the subset constraint from the drives role, we add to the relational schema a subset constraint from the empNr column of the Drives table to the empNr column of the Employee table. In horizontal layout, this subset constraint is depicted as a dotted arrow, as shown in Fig. 11.3(c). This ensures that any employee listed as driving a car is also listed in the Employee table, where gender and salary are mandatorily recorded.

In relational jargon, this subset constraint is a *referential integrity constraint*, and the empNr attribute of the Drives table is a *foreign key* that *references* the empNr attribute of the Employee table. In a standard relational model, foreign keys can reference only primary keys, but SQL also allows foreign keys to target alternate keys.

The relational model has two basic integrity rules. The *entity integrity rule* demands that primary keys contain no nulls (i.e., each column in a primary key is a mandatory column for its table). The *referential integrity rule* basically says that each nonnull value of a foreign key must match the value of some primary key. The relational model also allows for user-defined constraints and derivation rules.

To get a feel for how a relational schema might be implemented in practice, let's extend our example a little and map it to SQL. Fig. 11.4 adds fact types to record the department and name of each employee. It also adds three constraints other than internal uniqueness and simple mandatory constraints. Employee names are unique within a given department, the possible gender codes are "M" and "F", and each employee drives at most three cars.

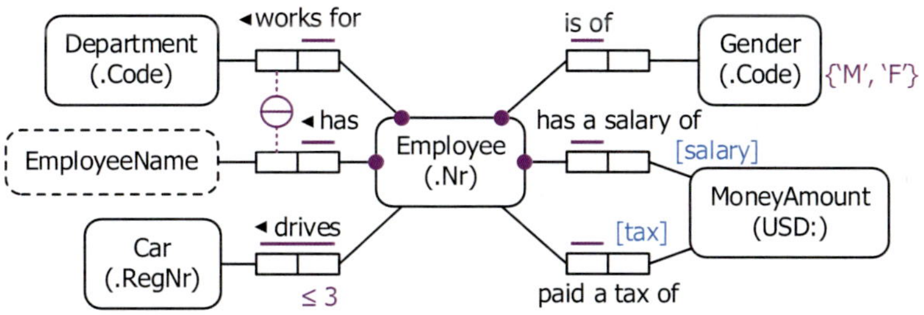

FIG. 11.4 A more detailed ORM schema for employees.

The external uniqueness constraint on employee name and department maps to a composite uniqueness constraint spanning two attributes. In the horizontal layout, this is denoted by underlining the relevant columns as discussed before. The value constraint on gender codes and the frequency constraint that each employee drives at most three cars are specified in the horizontal layout by annotating the table schemes with an ORM-like notation, as shown in Fig. 11.5(a).

Most ORM constraint notations may be used in a similar way in setting out relational schemas. Sometimes we adapt these notations or introduce new ones (see later). While our semigraphical notation is compact and fairly intuitive, commercial tools usually depict relational schemas diagrammatically using a *vertical layout*, supplemented by textual rules stored in property sheets or code. Although it takes up more space, vertical layout is convenient for tables with many columns.

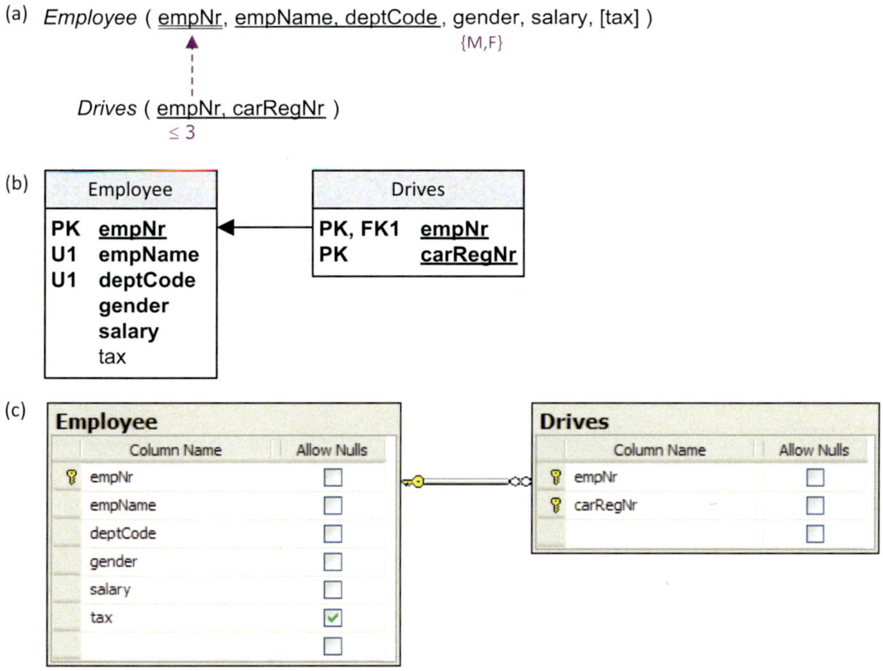

FIG. 11.5 A relational schema in (a) horizontal layout and (b, c) some vertical layouts.

Fig. 11.5(b) shows one of the ways in which Visio diagrams the same example. Here, the table name is listed in the top compartment with the column names listed vertically below. Mandatory columns are shown in bold, so tax is the only optional column. Primary key columns are underlined and annotated with "PK".

If desired, primary keys can also be displayed in a separate compartment. Foreign keys are annotated with "FK*n*" ($n > 0$), and an arrow depicts the subset constraint from the foreign key to the target table. Nonprimary uniqueness constraints are denoted by "U*n*". The value constraint on gender and the frequency constraint on carRegNr are not displayed on the diagram but can be inspected in the associated code.

Fig. 11.5(c) shows one way in which Microsoft SQL Server diagrams the same example. Here, the primary key columns are marked with a key symbol, and the foreign key constraint is depicted as a relationship line with a key symbol at the target end and "∞" (infinity, for a cardinality of "many") at the source end. The mandatory or optional nature of a column is indicated, respectively, by the absence or presence of an Allow Nulls check mark. Other constraints may be inspected in the associated code.

There is no industry standard that is uniformly adopted for diagramming relational schemas. When using a particular computer-aided software engineering (CASE) tool or DBMS, you need to familiarize yourself with its specific relational notation.

The same relational schema may be specified in SQL as set out later. Reserved words are shown in bold, but this is not required. To save some writing, domains on which more than one column are based are declared initially (EmpNr and Money), and constraint names are omitted. Ever since 1999, the ISO standard for SQL allows types to be declared instead of domains.

```
create domain EmpNr smallint;
create domain Money decimal(9,2);

create table Employee (
 empNr          EmpNr          not null   primary key,
 empName        varchar(30)    not null,
 deptCode varchar(5)        not null,
 gender         char(1)        not null   check( gender in ('M','F') ),
 salary         Money          not null,
 tax            Money,
 unique ( deptCode, empName ) );

create table Drives (
 empNr EmpNr                not null  references Employee,
 carRegNr   char(6)   not null,
 primary key ( empNr, carRegNr ),
 check( not exists
      (select empNr from Drives
       group by empNr
       having count(*) > 3) );
```

The create-table statements declare the schemes for the Employee and Drives tables. A not-null clause indicates that a column is mandatory for its table. Primary keys are declared with primary-key clauses. A unique clause specifies the alternate compound key. The intertable subset constraint is declared with a references clause. Check clauses are used to declare the value and frequency constraints. Further explanation of SQL syntax is given in the next chapter.

Although the above SQL syntax has been legal for years, some commercial SQL dialects do not yet support all this syntax. In practice, some features of a relational schema may need to be specified as procedural code rather than declaratively.

A conceptual schema comprises three sections: fact types, constraints, and *derivation rules*. Conceptual derivation rules may be specified in an appropriate language, using either relational or attribute style notation.

For example, in the schema shown in Fig. 11.6(a), an employee's net pay is derived using the ORM rule: **For each** Employee, **if** tax **exists then** netpay = salary – tax **else** netpay = salary.

As shown in Fig. 11.6(b), the derivation rule may be simplified to a simple subtraction if we make tax mandatory with a default of 0 (append "**not null default** = 0" to the SQL code above for the tax column) or assume that nulls are treated as zero for subtraction. The next chapter discusses in detail how SQL handles nulls.

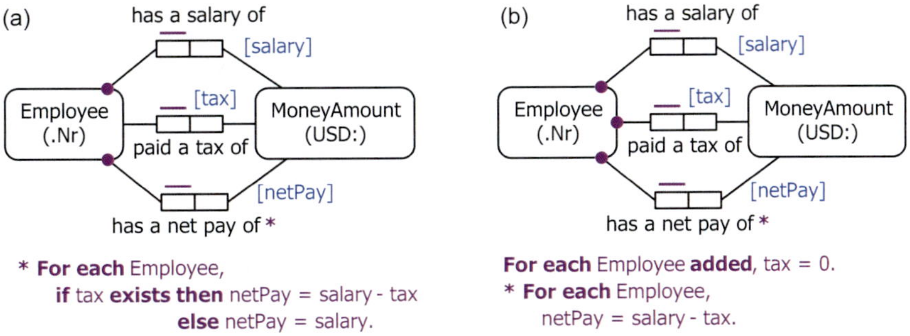

FIG. 11.6 Making tax paid optional (a) or mandatory with a zero default (b).

Various options exist for mapping derivation rules to the relational level. One option is to declare a derivation rule within a *view* (virtual table), for example, **create view** NetSalary(empNr, netPay) **as select** empNr, salary – tax **from** Employee.

A second option is to include a *generated column* in the relevant base table. In the SQL standard, this could be done by including the following definition in the table definition given earlier: netPay **generated always as** (salary – tax).

Not all DBMSs support generated columns, and some that do support this feature use a different syntax. For example, in SQL Server we may simply declare netPay **as** (salary – tax).

A third option is to include a *triggered column* in the relevant base table. As discussed in Chapter 13, triggers on a table specify actions taken automatically when relevant columns are changed in a given way. The syntax for triggers varies with the DBMS. Assuming that a netPay column is included in the Employee table, the following trigger may be used in SQL Server to derive the net pay:

```
create trigger DeriveNetPay on Employee
after insert, update as
if update(salary) or update(tax)
begin
    update Employee
    set netpay = salary – tax
end
```

A fourth option is to include the derivation rule in a *stored procedure* that is invoked when needed (see Chapter 13). The examples of SQL given earlier are intended simply for overview purposes. The next two chapters explain SQL in detail.

Once the relational schema has been declared, the initial data definition stage is over. We then perform data maintenance by adding or deleting elementary facts and perhaps some existential facts (by inserting, deleting, and modifying rows of data). Data retrieval to list base or derived facts of interest can then be carried out (e.g., by formulating appropriate SQL select statements).

Data presentation and data security tasks are also typically addressed by creating input screens and output reports, and assigning data access rights to various user groups. These operations are discussed in detail in the next two chapters.

Exercise 11.2

1. The following table contains details about members of a martial arts club. A null "?" indicates that a member has no phone or is unranked in any martial art. Black belt ranks are known as dan grades and lower ranks are kyu grades.

Member	*Gender*	*Phone*	*Arts and ranks*
Adams B	M	2052777	judo 3-dan; karatedo 2-kyu
Adams S	F	2052777	judo 2-kyu
Brown C	F	3579001	?
Collins T	M	?	aikido 2-dan; judo 2-dan
Dancer A	F	?	?

(a) Specify a conceptual schema for this universe of discourse (UoD), assuming that rank (e.g., 3-dan) may be stored as a single value. Many other martial arts are possible.

(b) Explain why the table shown is not a relational table.

(c) Given that any fact type with a composite uniqueness constraint must be stored as a separate table, map your conceptual schema to a relational schema. Underline keys, mark optional fields in square brackets, and show any intertable subset constraint as a dotted arrow from the subset end to the superset end. Then draw the tables including the above data as a sample population.

(d) In a "nested relational database", the data may be stored in a single table, as shown, where Ranking is a relation nested within the Member relation. Discuss any advantages or disadvantages that might result from this approach compared with the purely relational approach in part (c).

Member:

memberName	*gender*	*phone*	*Ranking*	
			art	*rank*
Adams B	M	2052777	judo karatedo	3-dan 2-kyu
Adams S	F	2052777	judo	2-kyu
Brown C	F	3579001	?	?
Collins T	M	?	aikido judo	2-dan 2-dan
Dancer A	F	?	?	?

11.3 Relational Mapping Procedure

The previous section introduced a generic notation for setting out a relational schema, and discussed simple examples of mapping from a conceptual to a relational schema. We now discuss the main steps of a general procedure for performing such a mapping. Advanced aspects of this procedure are considered in Section 11.4.

For a given conceptual schema, several different relational designs might be chosen. Ideally, the relational schema chosen should be *correct, efficient, and clear.* Correctness requires the relational schema to be equivalent to the conceptual schema (within the completeness allowed by relational structures). Efficiency means good response times to updates and queries, with reasonable demands on storage space. Clarity entails that the schema should be relatively easy to understand and work with.

Because correctness of data is usually more important than fast response times, and correctness requires adequate constraint enforcement, a high priority is normally placed on simplifying the enforcement of constraints at update time. The main way to do this is to *avoid redundancy.* This strategy can lead to more tables in the design, which can slow down queries and updates if extra table joins are now required. For efficiency, we should try to keep the number of tables down to an acceptable limit.

With these criteria in mind, the **Rmap** (*R*elational *map*ping) procedure guarantees a redundancy-free relational design and includes strategies to restrict the number of tables. Rmap extends and refines an older mapping procedure known as the "optimal normal form" (ONF) algorithm. The full version of Rmap includes details for completely mapping all graphical conceptual constraints and is beyond the scope of this book. However, the central steps of this procedure are covered in this section and the next.

As discussed in a later chapter, more efficient relational designs with fewer tables may possibly result if the conceptual schema is transformed by an optimization algorithm before Rmap is applied, and sometimes lower-level optimization using controlled redundancy may be needed to meet critical performance requirements.

Even without such further optimization, the Rmap procedure is extremely valuable because it guarantees a safe, reasonably efficient, normalized design. Happily, the basic steps of the procedure are simple.

Redundancy is repetition of an atomic fact. An atomic fact is typically elementary (e.g., The Scientist named "Einstein" was born in the Country named "Germany") but in rare cases could be existential (e.g., There exists a Country named "Germany"). If our conceptual fact types are atomic, we can easily avoid redundancy in our relational tables.

Typically, each row of a relational table stores one or more elementary facts. Otherwise, the table row stores a single reference or existential fact (e.g., a lookup table of country names). We can automatically *avoid redundancy* in tables by ensuring that *each fact type maps to only one table in such a way that its instances appear only once.*

To achieve this, there are two basic rules, as shown below. As an exercise, convince yourself that grouping facts like this makes it impossible for any fact to be duplicated.

1. Each fact type with a compound, internal uniqueness constraint maps to a separate table by itself, keyed on the uniqueness constraint.
2. Fact types with functional roles attached to the same object type are grouped into the same table, keyed on the object type's identifier.

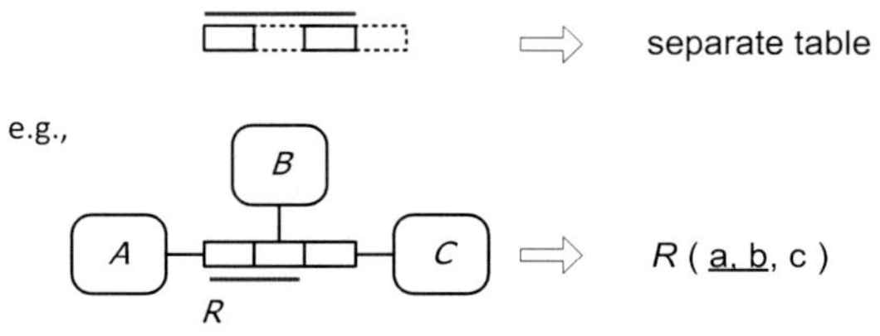

FIG. 11.7 Each fact type with a compound, internal uniqueness constraint maps to a table by itself.

These two rules show how to group fact types into table schemes. The first rule is illustrated in Fig. 11.7. Any nonobjectified predicate with a uniqueness constraint spanning two or more of its roles must map to a table all by itself. Hence, *m:n* binaries and all *n*-aries ($n \geq 3$) on the conceptual schema map to a separate table (unless they are objectified associations, as discussed later).

Each object type maps to one or more attributes depending on the number of components in its reference scheme. If the predicate has only one uniqueness constraint, the table's primary key is the attribute (or attribute set) spanned by this constraint. Otherwise, one of the uniqueness constraints spans the primary key, and the others span alternate keys. ORM tools allow you to specify which constraint you prefer for the primary key before mapping.

When mapping conceptual schemas, care should be taken to *choose meaningful table and column names*. In the case of Fig. 11.7, the table is used to store instances of the conceptual relationship type, so is often given a name similar to the conceptual predicate reading. If the object types involved are different, their names or the names of their value types, or a concatenation of the object and value type names, are often used as column names. For instance, *Drives*(empNr, carRegNr).

When information about the same object is spread over more than one table, it is usually better to consistently use the same column name for this object, unless this loses an important distinction between the semantics of the different roles involved (e.g., empNr was also used in the Employee table to refer to employees). Apart from helping the designer see the connection, this practice can simplify the formulation of natural joins in the SQL standard.

However, if different roles of the same predicate are played by the same object type, different column names must be chosen to reflect the different roles involved. For example, *Contains*(superpart, subpart, quantity).

A typical example of the second grouping rule is illustrated in Fig. 11.8. Recall that a functional role has a simple uniqueness constraint. Here, two functional roles are attached to the object type *A*. The rule also applies when there is just one functional role, or more than two.

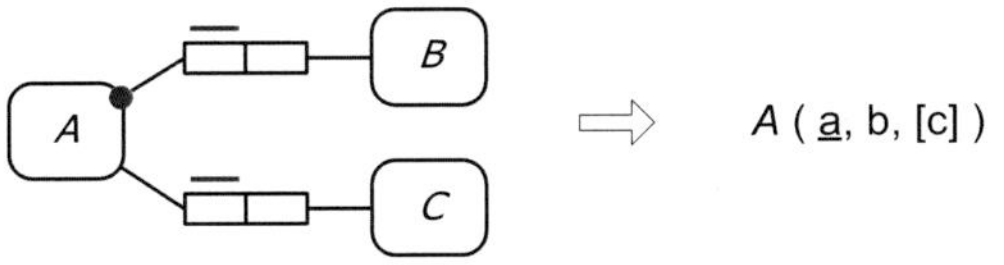

FIG. 11.8 Functional fact types of the same object type are grouped together.

Fig. 11.8 depicts mandatory and optional roles in our usual way (optional column in square brackets). The identification schemes of *A*, *B*, and *C* are not shown here, but may be simple or composite. The fact types are grouped together into a single table, with the identifier of *A* as the primary key (shown here as *a*).

The name of *A* is often chosen as the table name. To avoid confusion, it is best never to use the same name for both a table and a column, so the name of *A*'s value type is often chosen for *a*. The names of *B* and *C* (or their value types or role names) are often chosen as the other column names (here, *b* and *c*). For example, *Employee* (empNr, gender, [phone], salary, [tax]).

Once table groupings have been determined, keys underlined, optional columns marked in square brackets, and other constraints (e.g., subset, value list) mapped down, any derivation rules are also mapped, as discussed in the previous section.

To help understand the Rmap procedure, it will help to consider several examples. In our initial examples, all entity types have simple identifiers, and no subtypes or objectified associations occur.

Consider the conceptual schema in Fig. 11.9. The many:many fact type maps to a separate table (using rule 1). The unary fact type is functional because it has an implied uniqueness constraint. The functional fact types map to a table keyed on the identifier of Employee (using rule 2).

The primary keys are underlined, and the optionality of mobileNr is shown by the square brackets. Under closed world assumption (CWA) semantics, the unary fact type Employee is tenured maps to a mandatory Boolean column, Currently, NORMA instead maps all unaries to an optional Boolean; however, there are future plans for NORMA to distinguish between CWA, open world assumption (OWA), and open world with negation (OWN) semantics. The dotted arrow depicts the *subset constraint* that each employee referenced in the Drives table is also referenced in the primary key of the Employee table (referential integrity).

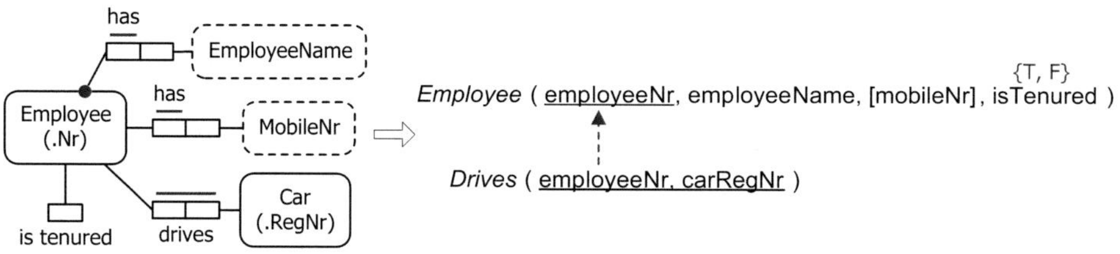

FIG. 11.9 A simple relational mapping.

Now consider the ORM schema in Fig. 11.10. Notice the *equality constraint*, shown as an arrow-tipped dotted line between the empNr fields of both tables. This is needed because both roles played by Academic are mandatory, and each maps to a different table. An equality constraint is equivalent to two subset constraints going in opposite directions. This causes a *referential cycle* because each table refers to the other.

The subset constraint from Qualification.empNr to Academic.empNr is a simple foreign key constraint because it references a primary key. But the subset constraint from Academic.empNr to Qualification.empNr is not a foreign key constraint because it targets only part of a primary key.

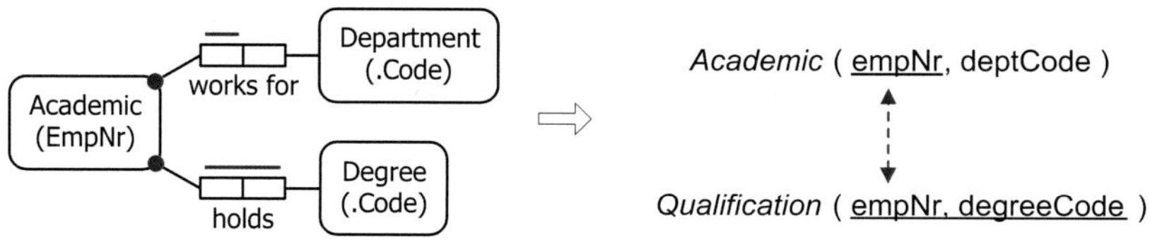

FIG. 11.10 The equality constraint entails a referential cycle.

This latter subset constraint may be enforced in various ways (e.g., by assertions, triggers, or stored procedures). Referential cycles typically result from a mandatory role that has no simple uniqueness constraint. Because referential cycles require special care to implement, this conceptual constraint pattern should be avoided if possible. In our example, however, the business rule that each academic holds at least one degree does seem reasonable, so in this case we leave it as is.

Before leaving this example, a few more comments about naming are worth making. The table name "Qualification" is one of many possible choices to convey the notion of academics being qualified by their degrees. Two other choices are to use the association name ("Holds") or to concatenate the object type names ("AcademicDegree"). The department column is named "deptCode" because we expect other departmental facts (not shown here) to be recorded in a department table, and we want to avoid using "department" to name both a column and a table. Similarly, the degree column is named "degreeCode" rather than "degree" because we might want to record other facts about degrees (e.g., their titles). In general, if the object type underlying a column will also underlie the primary key of another table, it is advisable to include the reference scheme in the column name.

Now consider Fig. 11.11. Here, each horse has its gender and weight recorded. Each race has at most one winner (we do not allow ties). There are no composite keys for rule 1 to work on. The gender and weight fact types have functional roles attached to Horse. By rule 2, these two associations must be grouped into the same table keyed on the identifier for Horse. Similarly, the time and win fact types are functions of race and are grouped into the Race table. We chose the column name "raceTime" instead of "time" mainly because "time" is a reserved word in SQL. It's best to avoid using reserved words for names because they require double quotes. We later provide links to online lists of SQL reserved words. The column name "winner" is used instead of "horseName" to convey the semantics better.

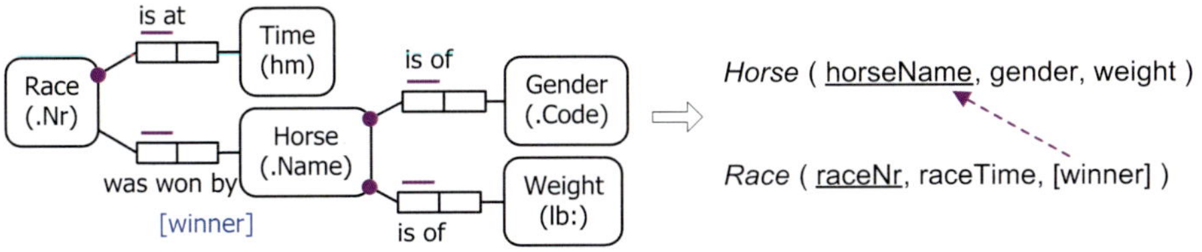

FIG. 11.11 An ORM schema for horse races and its relational map.

Mapping 1:1 Associations

Let's now consider some examples involving *1:1 fact types*. In the conceptual schema of Fig. 11.12, both Bankcard and Client share a 1:1 association. If we are certain that no other functional roles will be played by Bankcard, it is usually best to group the 1:1 fact type together with the other functional fact types into the same table, as shown.

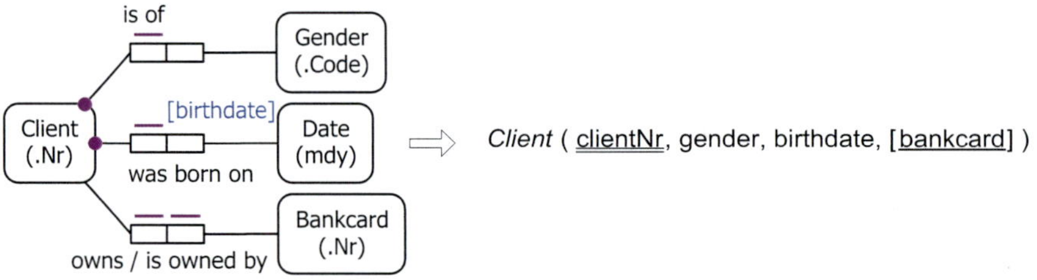

FIG. 11.12 The mapping choice if Bankcard plays no other functional roles.

The bankcard column is optional, but the disadvantage of nulls here is usually outweighed by the advantage of having all the data in one table. The underlining of bankcard indicates a uniqueness constraint for its nonnull values. Because it may contain more than one null, the bankcard column cannot be used to identify rows, and hence cannot be the primary key. With two uniqueness constraints, we doubly underline clientNr to highlight it as the primary key.

If Bankcard does play other functional roles, however, the bankcard ownership association should be grouped into a Bankcard table, as shown in Fig. 11.13. Notice that the owned role played by Bankcard is now explicitly mandatory; in the previous example, it was only implicitly mandatory (because it was the only role played by Bankcard). Notice that *only one role of this 1:1 association is mandatory.* In asymmetric cases like this, it is usually better to *group on the mandatory role side* as shown.

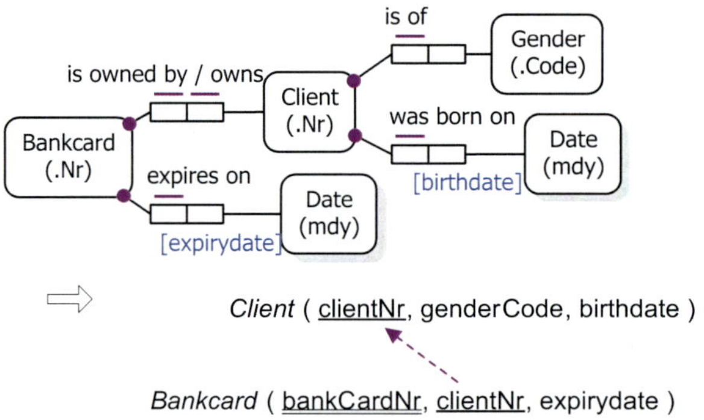

FIG. 11.13 The mapping choice if Bankcard plays another functional role.

To illustrate this idea further, consider the conceptual schema in Fig. 11.14. Each employee is identified primarily by his/her employee number, but also has a unique name (unusual, but some departments work that way by refining formerly duplicate names). Each department has one head, who heads only one department. Because not all employees are department heads, the role of heading is optional.

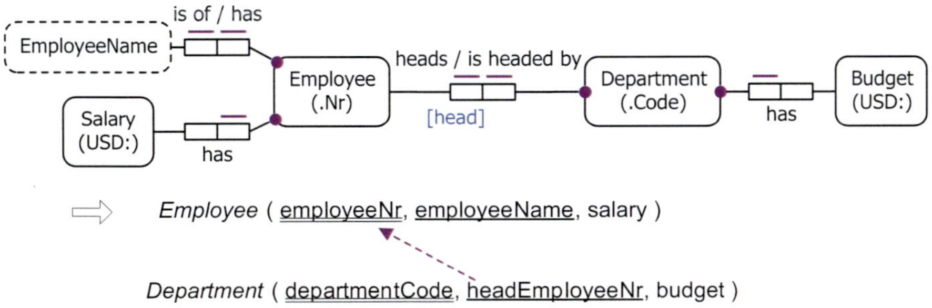

FIG. 11.14 The heads fact type is grouped on the mandatory role side.

Hence, the 1:1 association between Employee and Department is optional for Employee but mandatory for Department. Because it is usually better to group on the mandatory role side, here this means including the headEmployeeNr column in the Department table, as shown.

The NORMA tool allows you to edit the column names generated in the relational schema simply by editing their name property. For example, we may shorten “employeeNr”, “employeeName”, and “departmentCode” to “empNr”, “empName”, and “deptCode”, respectively.

Suppose we grouped the 1:1 association on the optional side instead, by adding a deptHeaded column to the Employee table. This leads to the following relational schema:

Employee (empNr, [deptHeaded], empName, salary)

Department (deptCode, budget)

This alternative has two disadvantages. First, the optional deptHeaded column permits nulls (unlike the mandatory headEmpNr column). All other things being equal, nulls should be avoided if possible. In addition to consuming storage, they are often awkward for people to work with. The second disadvantage is that we now have an equality constraint between the tables, rather than just a subset constraint.

In principle, we might group into one table all the functional predicates of both object types involved in a 1:1 association. With our example, this gives the scheme *EmployeeDept*(empNr, empName, salary, [deptHeaded, deptBudget]). Here, enclosing the last two columns in the same square brackets declares that one is null if and only if the other is.

However, apart from requiring two optional columns and a special constraint that requires them to be null together, this grouping is unnatural. For example, the primary key suggests that the whole table deals with employees rather than their departments. For such reasons, this single-table approach should normally be avoided.

Now consider Fig. 11.15(a). Here, each employee has the use of exactly one company car, and each company car is allocated to exactly one employee. Employees and cars are identified by their employee number and by their registration number, respectively. The names of employees and the car models (e.g., Mazda CX5) are also recorded, but might not be unique. Here, both roles of the 1:1 association are mandatory, and each is attached to an entity type with another functional role. Should the 1:1 fact type be grouped into an Employee table or into a Car table?

Unlike the previous example, we now have a symmetrical situation with respect to mandatory roles. An arbitrary decision could be made here. We could group to the left, as in Fig. 11.15(a) and (b), or to the right as in Fig. 11.15(c). Either of these approaches is reasonable.

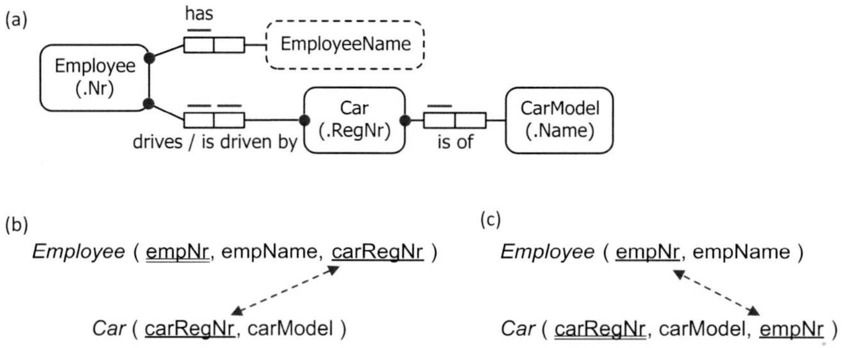

FIG. 11.15 The ORM schema (a) may map to relational schema (b) or to schema (c).

We might also try mapping to a single table: *EmployeeCar*(empNr, empName, carRegNr, carModel). Although possible, this is unnatural, requiring an arbitrary choice of primary key. This becomes more awkward if other facts about employees (e.g., gender) and cars (e.g., purchase date) are recorded in the table. Also, consider the additional update overhead to change the car used by an employee compared with the two-table approach.

Now what about the case of a 1:1 fact type with both roles optional? For example, consider a UoD identical to that just discussed except that only some employees are given company cars and only some company cars are used by employees (e.g., some may be reserved for important visitors). The two table approach is recommended.

Because of the symmetry with respect to mandatory roles, we could map the 1:1 association into the Employee table to give the schema *Employee*(empNr, empName, [carRegNr]); *Car*(carRegNr, carModel); Employee.carRegNr references Car.

Alternatively, we could map the 1:1 association to the Car table yielding the schema *Employee*(empNr, empName); *Car*(carRegNr, carModel, [empNr]); Car.empNr references Employee. The *percentage of nulls* is likely to differ in these two designs. In this case, the design with fewer nulls is usually preferable.

Yet another option is a three-table approach, in which the 1:1 association has a table to itself. This option becomes more attractive if the two-table approach yields high percentages of nulls. For example, if only 1% of employees and 1% of cars are likely to be involved in Employee drives Car, we might map this fact type into a table by itself, giving three tables overall.

A fourth option for 1:1 cases is to use two tables, but include the 1:1 association in both with a special equality constraint to control the redundancy. More detailed discussions of mapping 1:1 fact types are referenced in the chapter notes. Our default procedure is summarized in Fig. 11.16; here, "functional role" means a functional role in an elementary fact type (not a reference type).

In Fig. 11.16, the arrow "⇦" indicates that the 1:1 fact type should be grouped into the functional table of the left-hand object type. In case (a), the right-hand object type may play nonfunctional roles not shown here, and any role in the 1:1 fact type may be optional or mandatory. The third case (no other functional roles) is rare and requires a choice of primary key in the separate table. The final line refers to symmetric cases where the roles of the 1:1 predicate are both mandatory or both optional, and both object types play another functional role—here, we have a grouping choice.

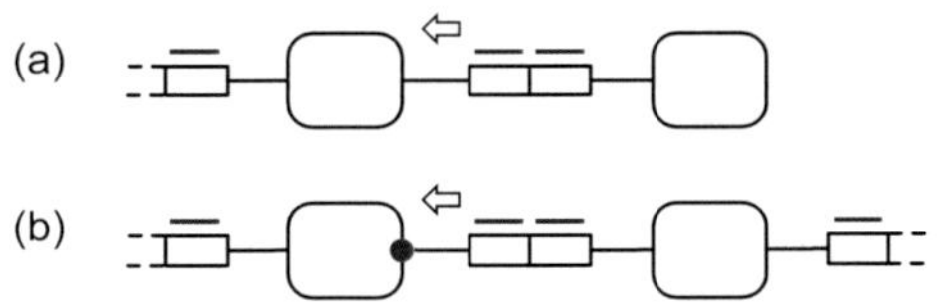

Each *1:1 fact type* maps to only one table:

if (a) only one object type in the 1:1 predicate has another functional role
then group on its side
else if (b) both object types have other functional roles
and only one role in the 1:1 is explicitly mandatory
then group on its side
else if no object type has another functional role
then map the 1:1 to a separate table
else the grouping choice is up to you

FIG. 11.16 Default procedure for mapping 1:1 fact types.

To understand some further cases, it will help to recall the *fundamental bridge between conceptual and logical levels*. This is summarized in Fig. 11.17. In the populated ORM schema in Fig. 11.17(a), the mandatory role pattern anticipates other roles being added later to Employee but not Gender.

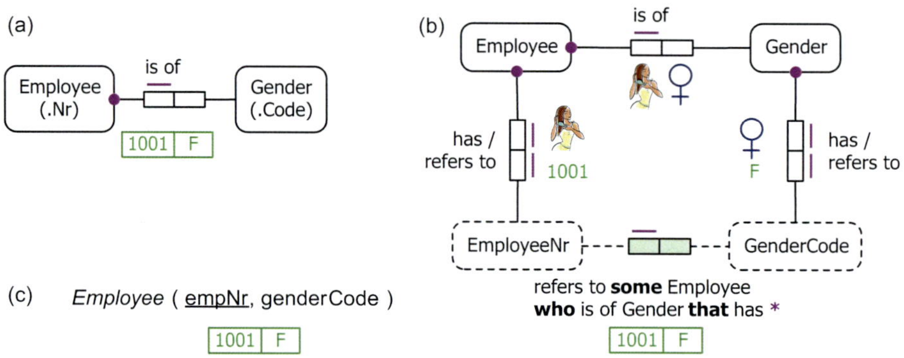

FIG. 11.17 The conceptual/logical bridge.

Fig. 11.17(a) abbreviates the preferred reference schemes shown explicitly in Fig. 11.17(b), which uses icons to directly depict a female employee and the female gender. The shaded, derived association abbreviates the conceptual path from EmployeeNr through Employee and Gender to GenderCode. This unpacks the semantics underlying the relational schema in Fig. 11.17(c).

Recall that uniqueness constraints on reference predicates are the responsibility of humans to enforce. The information system can't stop us from giving the same employee number to two different employees, or from giving the same employee two employee numbers. Hence, *uniqueness constraints on primary reference schemes are not mapped.* Instead, we must enforce them in the real world.

However, assuming we have enforced the primary reference constraints, the system can enforce constraints on the fact types. For example, it enforces the uniqueness constraint in *Employee*(empNr, genderCode) by ensuring that each employee number occurs only once in that column and hence is paired with at most one gender code. Assuming that the reference types really are 1:1, this uniqueness constraint on empNr corresponds to the uniqueness constraint on the conceptual fact type (i.e., **Each** Employee is of **at most one** Gender). It does not capture any uniqueness constraint from the reference types.

Mapping External Uniqueness Constraints

Let's now consider mapping of schemas that include *external uniqueness constraints.* Recall that such constraints appear as a circled bar, and if they are used for preferred reference they have a double bar. In the ORM schema of Fig. 11.18, employees are identified

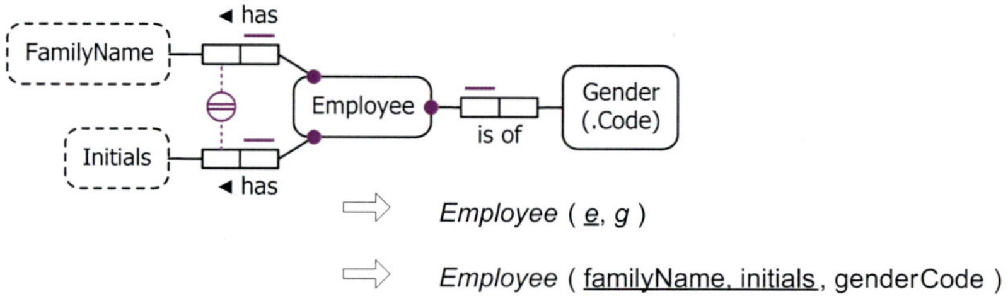

FIG. 11.18 Composite primary identifier and functional fact type.

by combining their family name and initials. Because the external uniqueness constraint in this example underlies the preferred reference scheme, it is not mapped. In the resulting table scheme *Employee*(familyName, initials, genderCode), the uniqueness constraint corresponds to the uniqueness constraint on the conceptual fact type Employee is of Gender.

A good way to visualize the mapping is as follows:

- Mentally erase the identification scheme of each object type.
- Group facts into tables, using simple surrogates for the roles of real-world objects.
- Replace each surrogate by the attribute(s) used in the table to identify the objects.

For example, the conceptual schemas of Figs. 11.17 and 11.18 each map initially to *Employee*(*e*, *g*) with the meaning Employee *e* is of Gender *g*. In both cases, *g* is then replaced by "genderCode". With Fig. 11.17, *e* is replaced by empNr, but with Fig. 11.18, *e* is unpacked into (familyName, initials) because the identification scheme is composite. Because the uniqueness constraint spans *e*, it must also span the attribute combination that replaces it.

Now consider the conceptual schema of Fig. 11.19. The structural difference here is that the fact type has a composite uniqueness constraint. We may initially think of it mapping to the table *Drives*(*e*, *c*), where *e* and *c* are surrogates for the employee and car roles. Replacing the surrogates by the real identifiers results in the table scheme *Drives*(familyName, initials, carRegNr). Because *e* was just part of a longer key, so is the composite identifier for employee that replaces it.

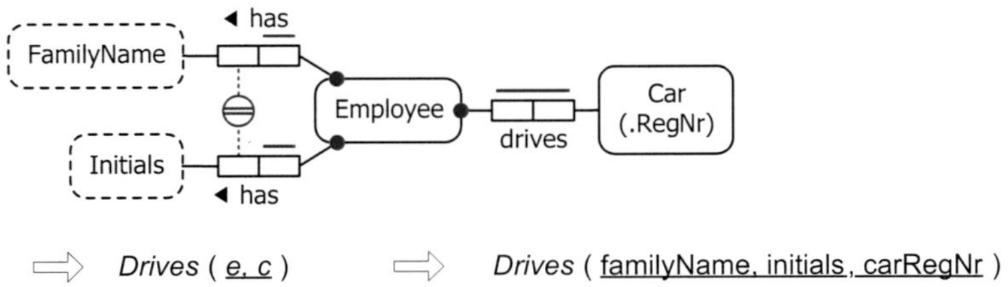

FIG. 11.19 Composite primary identifier and nonfunctional fact type.

Now consider the conceptual schema of Fig. 11.20. This is like that of Fig. 11.18 except that Employee now has empNr as its primary identifier. The external uniqueness constraint now applies to two fact types rather than reference types, and hence can be mapped and enforced by the information system.

As shown, this maps to a uniqueness constraint spanning surname and initials in the relational table (ensuring each familyName, initials combination is paired with only one employee number). The uniqueness constraint on the table's primary key, empNr, captures the three simple uniqueness constraints on the three conceptual fact types (i.e., because each empNr is unique, it is paired with only one family name, only one sequence of initials, and only one gender code).

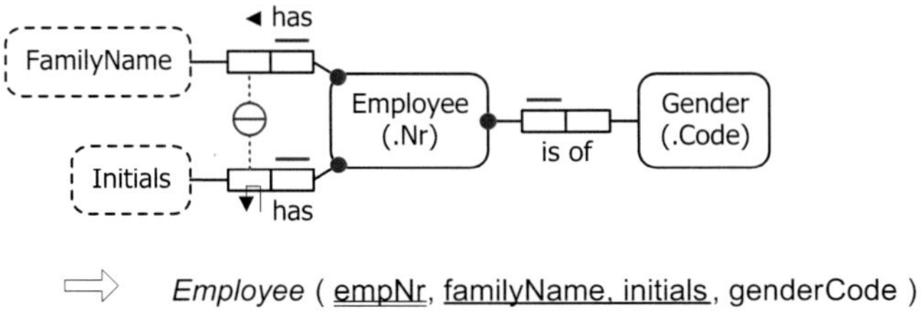

FIG. 11.20 Composite secondary identifier and functional fact type.

Now consider the conceptual schema of Fig. 11.21. Here, each laboratory session is identified primarily by a session number and is used for a particular course. Once these course bookings have been made, sessions are assigned for use by students. The external uniqueness constraint says that each student is assigned at most one session for each course (e.g., laboratory resources might be scarce). Because this constraint involves fact types rather than reference types, it can be mapped.

An unusual feature of this example is the application of the external uniqueness constraint to a set of roles that includes a role in an *m:n* fact type (LabSession is assigned to Student). Because the *m:n* fact type must map to a table by itself, the external uniqueness constraint spans two relational tables. This is equivalent to an internal uniqueness constraint spanning courseCode and studentNr in the natural join of the two tables.

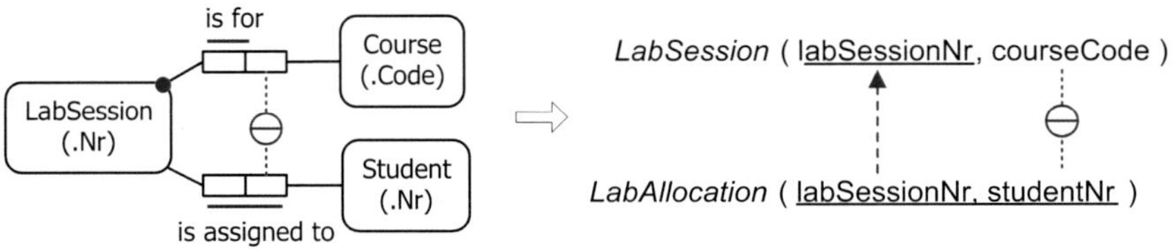

FIG. 11.21 External uniqueness constraint involving an *m:n* fact type.

Mapping Objectified Associations

Let's now discuss some cases involving *objectification* (also known as *nesting*). In Fig. 11.22, the association Employee worked on Project is objectified as Work. This objectified type plays one mandatory and one optional role, both of which are functional. As with other object types, we *initially treat the objectified association as a "black box", mentally erasing its identification scheme.*

From this viewpoint, the conceptual schema appears to have just two fact types, both having functional roles attached to the object type Work. *Fact types are now grouped in the usual way.* So, these two fact types are grouped into the same table. Visualizing the nested object type Work as a black box " ■ " results in the table *Work*(■ , startdate, [enddate]). Finally, we *unpack* ■ *into its component attributes* (empNr and projectName), giving *Work*(empNr, projectName, startdate, [enddate]).

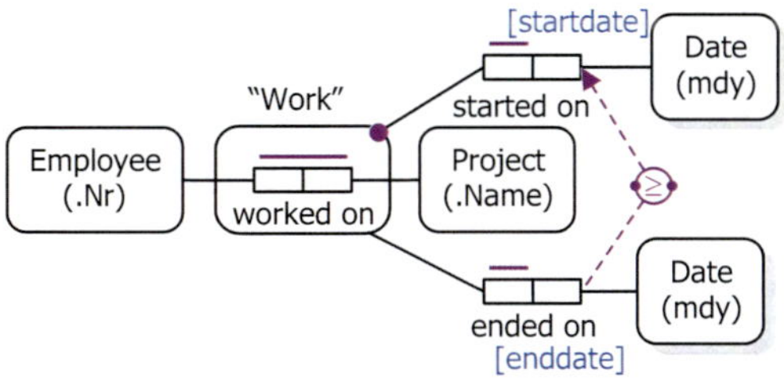

⇒ *Work* (<u>empNr, projectName</u>, startdate, [enddate])[1]

[1] enddate >= startdate

FIG. 11.22 An active, objectified association with functional roles.

The value constraint in Fig. 11.22 verbalizes as **For each** Work, **existing** enddate >= startdate. Value constraints are violated if and only if they evaluate to false (just like SQL check clauses). The footnote annotation on the relational schema here corresponds to the SQL check clause **check** (enddate >= startdate).

Now consider the ORM schema of Fig. 11.23. Here, employees might be assigned to projects before their actual starting date is known. So, some instances of the objectified Assignment association might be recorded that (in some database state) do not play either of the attached roles. In this example, because even the disjunction of the attached roles is optional, the nested object type is *independent*—this is noted by the exclamation mark in "Assignment !". Contrast this with the previous example (Fig. 11.22), where the nested object type Work is active and the startdate column in the relational schema is not optional.

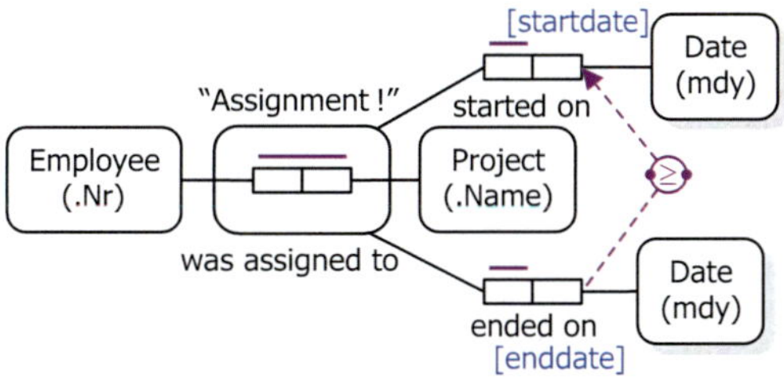

⇒ *Assignment* (<u>empNr, projectName</u>, [startdate], [enddate])[1]

[1] enddate >= startdate

FIG. 11.23 Mapping an independent, objectified association with functional roles.

Independent object types, whether nested or not, require special treatment in mapping, as follows.

- Map each independent object type and the fact types (if any) in which it plays functional roles to a separate table, with the object type's identifier as the primary key and all other attributes optional. Add a subset constraint to this primary key from each column sequence mapped from nonfunctional roles (if any) of the independent object type.

If we wish to record just the fact that an employee is assigned to a particular project, we enter values just for empNr and projectName. Details about start and end dates for work on the project can be added later when they are known. Because both start date and end date are optional, the value constraint now verbalizes as **For each** Assignment, **existing** enddate >= **existing** startdate.

Fig. 11.24 shows another way to model this UoD. The independent object type has no functional roles attached, and maps to a table all by itself. The *m:n* fact type maps to another table. A subset constraint captures its optionality. If instead we tried to map everything to one table, this would violate entity integrity (why?).

The constraint that if an assignment's start and end dates are known, the end date must come after or on the start date is now harder to formulate, both at the conceptual level and at the relational level. Moreover, the approach of Fig. 11.24 spreads the information over more tables and demands an intertable constraint. This is usually undesirable because it often slows down the execution of queries and updates. For such reasons, the approach of Fig. 11.23 is normally preferred to that of Fig. 11.24.

The schemas in Figs. 11.23 and 11.24 allow us to record an end date without a start date. In practical applications, we sometimes allow things like this because our information may be incomplete. For example, we might want to record the date somebody ended a project, but not *know* when that person started.

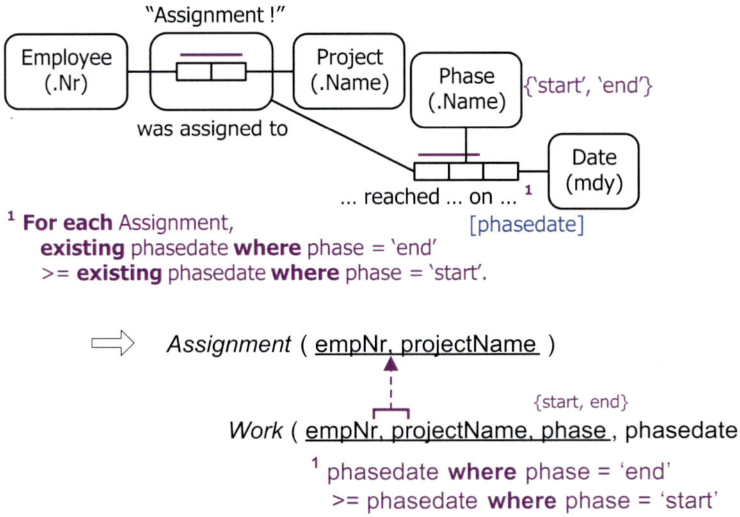

FIG. 11.24 Mapping an independent, objectified association with no functional roles.

In such cases, conceptual schemas sometimes have fewer mandatory roles than the ideal world suggests, and in consequence relational schemas may have more optional columns than complete knowledge would allow.

Suppose, however, that if we know an end date then we *do* know the starting date. To enforce this constraint in Fig. 11.23, add a subset constraint on the ORM schema from the first role of the "ended on" predicate to the first role of the "started on" predicate.

This constraint may be declared in the relational schema by using *nested option brackets*. Here, we enclose the option brackets for enddate *inside* the option brackets for startdate, giving *Assignment*(empNr, projectName, [startdate, [enddate]]). This indicates that a (nonnull) value is recorded for enddate only if a value is recorded for startdate.

To add this constraint to Fig. 11.24 is not as easy. A textual constraint is required at both conceptual and relational levels to declare that for each employee, project pair, phase "end" is recorded only if phase "start" is.

Chapter 14 examines in detail the notion of "equivalent" conceptual schemas and provides guidelines for transforming a conceptual schema to improve the efficiency of the relational schema obtained from Rmap.

If the start and end dates for an employee's work on a project are recorded, a derivation rule can be specified to compute the work period (by subtracting the start date from the end date, if known). Because each computer system has an internal clock, conceptually there is a unary fact type of the form Date is today. So, for someone still working on a project, we could also derive the time spent so far on the project by subtracting the start date from the current value for "today".

By default, derived facts are not stored. So, *by default, derived columns are excluded from the base tables* (i.e., stored tables) of the relational schema. For example, the derivation rules could be implemented using views or stored procedures.

In some cases, efficiency considerations may lead us to *derive on update* rather than at query time and to store the derived information. In such cases, the derived fact type is marked "**" on the conceptual schema diagram. During the relational mapping, the fact type is mapped to a base table, and the derivation rule is mapped to a rule that is triggered by updates to the base table(s) referenced in the derivation rule.

For example, we could implement the derivation rule by declaring appropriate generated columns or table triggers (recall the earlier net pay example). As well as derivation rules, *all* conceptual constraints should be mapped (not just uniqueness and mandatory constraints).

Mapping Subtypes

Let's now consider the *mapping of subtype constraints*. Table 11.1 was seen earlier in Chapter 6. Although we have no rule to determine when a phone number is recorded, we know that prostate status may be recorded only for male patients and pregnancies are recorded for all female patients and only for female patients. The conceptual schema is reproduced in Fig. 11.25.

Table 11.1 Patient details.

PatientNr	*Name*	*Gender*	*Phone*	*Prostate status*	*Pregnancies*
101	Adams A	M	2052061	OK	–
102	Blossom F	F	3652999	–	5
103	Jones E	F	?	–	0
104	King P	M	?	Benign enlargement	–
105	Smith J	M	2057654	?	–

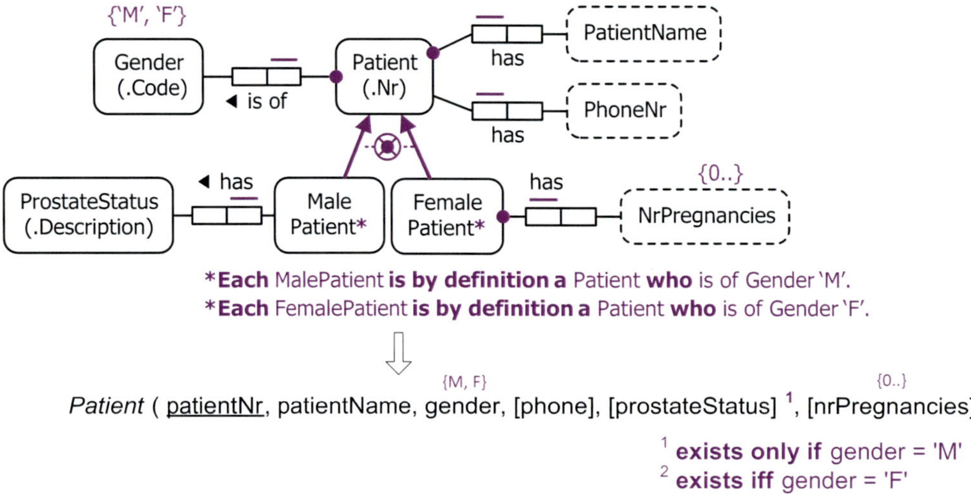

FIG. 11.25 Subtype constraints on functional roles map to qualified optionals.

Some limited support for subtyping is included in the SQL standard, but current relational systems typically do not support this concept directly. Nevertheless, there are three main ways in which subtyping can be implemented on current systems: absorption, separation, and partition. With *absorption*, we *absorb the subtypes back into the (top) supertype* (giving *qualified optional roles*), *group the fact types as usual*, and then *add the subtyping constraints as textual qualifications*, as shown in Fig. 11.25.

To best understand this example, visualize the subtypes on the conceptual schema absorbed back into the supertype Patient, with the subtype roles attached as qualified optional roles to Patient. All the roles attached to Patient are functional, so the fact types all map to the same table, as shown. The phone, prostateStatus, and nrPregnancies attributes are all optional, but the latter two are qualified.

In the qualifications, "**exists**" means the value is not null. Qualification 1 is a pure subtyping constraint, indicating that a nonnull value for prostate status is recorded **only if** the value of gender on that row is "M". Some men might have a null recorded here. Qualification 2 expresses both a subtype constraint (number of pregnancies is recorded **only if** gender is "F") and a mandatory role constraint (nrPregnancies is recorded **if** gender is "F"). Recall that "**iff**" is short for "if and only if".

Because only functional fact types are involved in this example, the absorption approach leads to a table that basically matches that of the original output report (except that most relational systems support only one kind of null). The main advantage of this approach is that it maps all the functional predicates of a subtype family into a single table. This usually makes related queries and updates more efficient. Its main disadvantage is that it generates nulls.

The second main approach, *separation*, creates separate tables for facts with subtype-specific functional roles. With this approach, the conceptual schema of Fig. 11.25 maps to three tables: one for common facts, one for male-specific facts, and one for female-specific facts.

The third main approach, *partition*, may be used when the subtypes form a partition of their supertype. Because MalePatient and FemalePatient are exclusive and exhaustive, this approach may be used here, resulting in two tables: one with the facts about the male patients and the other with facts about female patients.

The next section discusses the relative merits of these three approaches and considers mapping cases where subtypes use a preferred identification scheme different from the supertype's. Our default approach, however, is to absorb the subtypes before grouping.

Even with this approach, *any fact types with a nonfunctional role played by a subtype map to separate tables, with their subtype definitions expressed by qualified subset constraints* targeting the main supertype table.

For example, the ORM schema in Fig. 11.26 adds the optional *m:n* fact type FemalePatient attended prenatal clinic on Date. This fact type maps to a separate table with the qualified subset constraint "**only where** gender = 'F'", as shown. If the new fact type were instead mandatory for FemalePatient, the qualification would read "**exactly where**" instead of "**only where**".

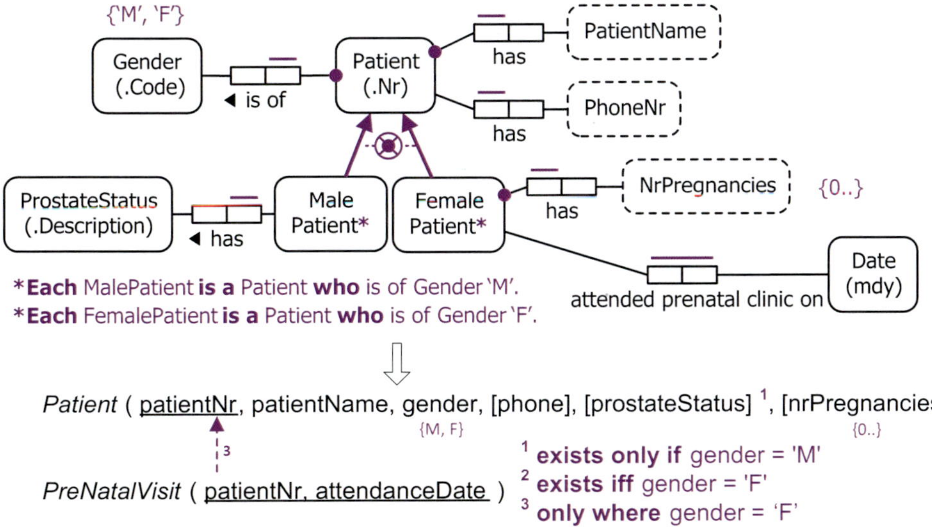

FIG. 11.26 Nonfunctional fact type of subtypes map to separate tables.

We have now covered all the basic steps in the Rmap procedure. Even if you are using a CASE tool to do the mapping for you, it's nice to understand how the mapping works. The basic steps in the relational mapping procedure are summarized below.

Basic Rmap procedure:

0 Absorb subtypes into their top supertype.
Mentally erase all explicit preferred identification schemes,
treating compositely identified object types as "black boxes".

1 Map each fact type with a compound UC to a separate table.

2 Fact types with functional roles attached to the same object type
are grouped into the same table, keyed on the object type's identifier.
Map 1:1 cases to a single table, generally favoring fewer nulls.

3 Map each independent object type with no functional roles to a separate table.

4 Unpack each "black box column" into its component attributes.

5 Map all other constraints and derivation rules.
Subtype constraints on functional roles map to qualified optional columns,
and those on nonfunctional roles map to qualified subset constraints.
Nonfunctional roles of independent object types map to column sequences
that reference the independent table.

Step 0 may be thought of as a preparatory mental exercise. Erasing all explicit preferred reference schemes (i.e., those shown other than by parenthesized reference modes) ensures that all the remaining predicates on display (as box sequences) belong to fact types (rather than reference types).

Recall that a *compositely identified object type* is either a nested object type (objectified association) or a coreferenced object type (identified via an external uniqueness constraint).

There are plenty of questions in the section exercise to give you practice at performing the mapping manually. As preparation for this exercise, some larger examples are now considered.

Fig. 11.27 shows the conceptual schema for the compact disc case study discussed in Chapter 5. If you feel confident, you might like to try mapping this to a relational schema yourself before reading on.

As there are no subtypes in this example, step 0 amounts to mentally erasing any preferred reference schemes that are shown explicitly. There are only two, each involving a compositely identified object type: Track and Listing.

Fig. 11.28 depicts this erasure by removing the predicate boxes and showing their connections to object types as dashed lines. For steps 1–3, we treat these two compositely identified entity types just like any other entity type.

We now proceed to group the asserted fact types into tables. To help visualize this, we *place a lasso around each group of predicates that maps to the same table* (see Fig. 11.28). We lasso only the predicates, not the object types. Because each fact type should map to exactly one table, *all predicates must be lassoed*, and *no lassos may overlap.*

In step 1, we look around for a predicate with a compound uniqueness constraint. Because the objectified association is now hidden, we see only one such predicate: is sung by. So, we lasso this predicate, indicating it goes to a table all by itself.

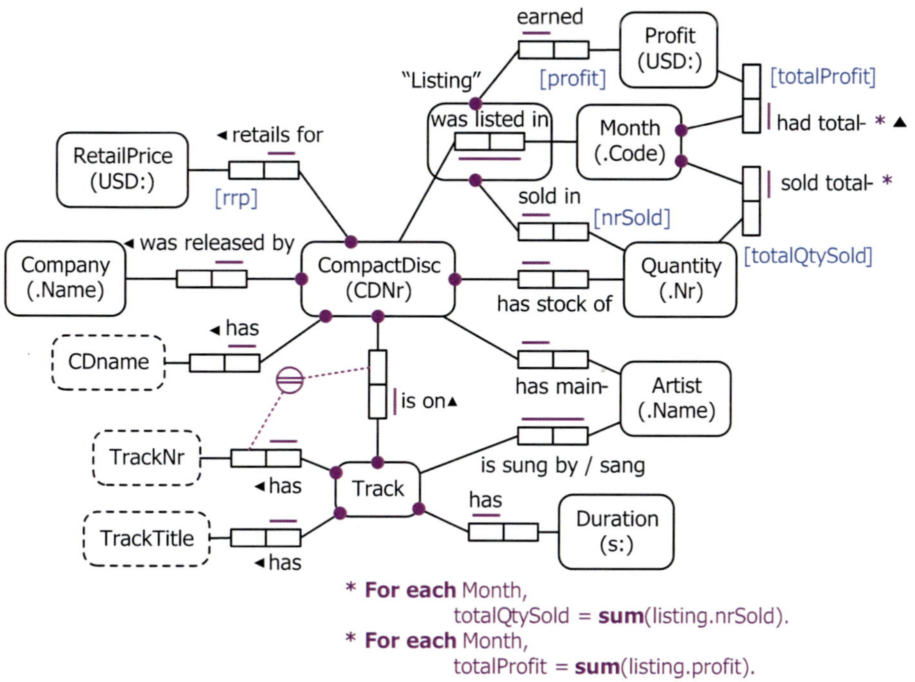

FIG. 11.27 The conceptual schema from the compact disc case study.

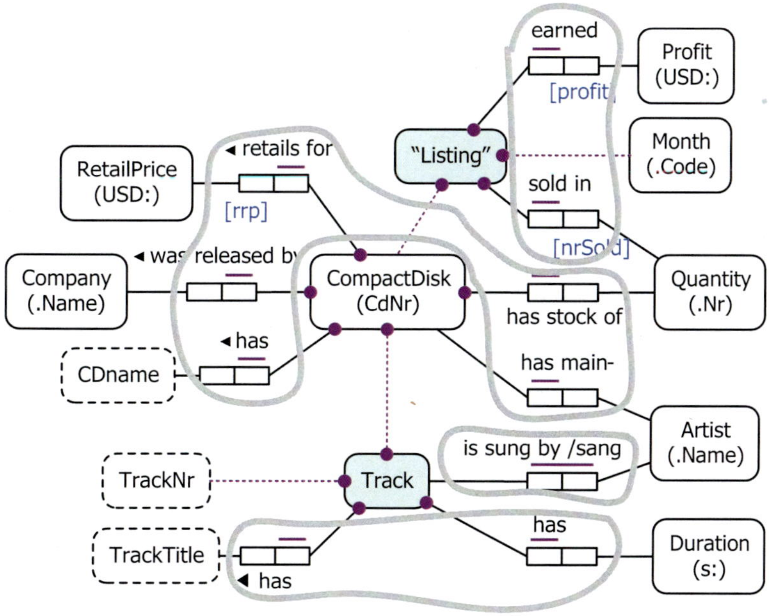

FIG. 11.28 The reference types are erased, and fact types are lassoed into groups.

In step 2, we group functional fact types of the same object type together. For example, CompactDisc has five functional roles attached to it, so these five fact types are grouped into a table keyed on the identifier for CompactDisc. Similarly, the two functional fact types for Track are lassoed together, as are the two functional fact types of the nested object type.

This example has no 1:1 cases, and no independent object types. We have now roped all the predicates, and there are four lassos, so the conceptual schema maps to four tables.

The final relational schema is shown in Fig. 11.29. Because five functional fact types map to the CD (CompactDisc) table, and all the objects involved have simple identifiers, this table has six columns (one for the key and one for each fact attribute).

The other three tables involve a compositely identified object type, which is unpacked into its component attributes (step 4).

Sales (cdNr, monthCode, qtySold, revenue)

CD (cdNr, cdName, [artist], company, stockQty, rrp)

Track (cdNr, trackNr, title, duration)

Vocals (cdNr, trackNr, singer)

* *SalesTotals*(monthCode, totalQtySold, totalRevenue) ::
monthCode, **sum**(qtySold), **sum**(revenue)
from Sales
group by monthCode

FIG. 11.29 Relational schema mapped from the conceptual schema in Fig. 11.27.

The keys of the tables are already determined, but the mandatory role constraints are enforced by mandatory columns and intertable subset/equality constraints. Note the composite subset constraint from the pair cdNr, trackNr in Vocals to the primary key of Track. Finally, the two derivation rules are mapped (step 5).

If you lasso fact types into groups before writing down the table schemes, you may mentally erase all preferred identification schemes (including reference modes) in step 0 (thus treating all object types as black boxes) and then in step 4 replace each column by its identifying attribute(s). This alternative formulation of the Rmap procedure is logically cleaner. If performing Rmap manually, you might find it convenient to photocopy the conceptual schema and use colored pencils to cross out the reference types and lasso the fact types.

As a complicated example, consider the ORM schema in Fig. 11.30. This concerns television channels. Notice that TimeSlot is modeled as a coreferenced entity type (a time slot is identified as a given hour on a given day). For variety, the object types Office and Department are modeled as objectified associations. A television channel may have different offices in different suburbs. For this UoD, a channel may have only one office in any given suburb. A given office may have many departments.

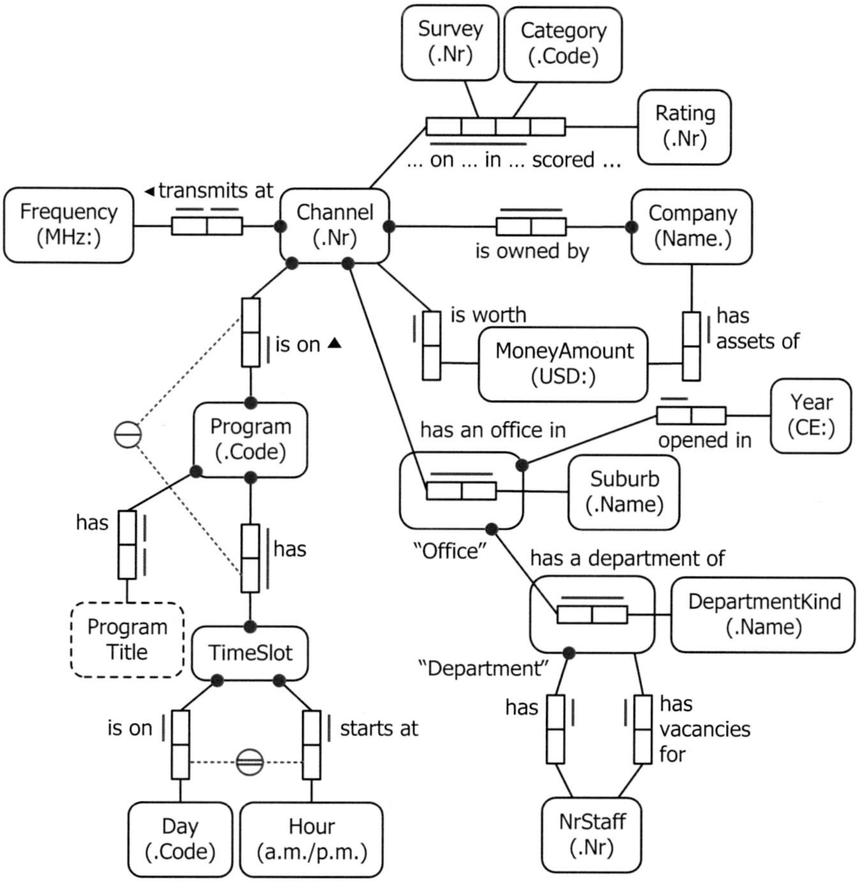

FIG. 11.30 An ORM schema for TV channels.

For example, one department might be the advertising department for the channel 9 office located in the suburb Toowong. Alternatively (and arguably more naturally), the object types Office and Department could have been modeled as coreferenced object types. However, this would not change the mapping.

The external uniqueness constraint on Program indicates that a channel can screen only one program at a given time. As an exercise, try to map this yourself before reading on. Start by mentally erasing the reference types and lassoing the fact types that should be grouped together before you write down the relational schema.

Fig. 11.31 hides the identifying predicates for Time, Office, and Department, and lassoes the fact types into groups. The 1:1 associations must be grouped with Channel and Program because these have other functional roles, but Frequency and Title do not.

The detailed relational schema is shown in Fig. 11.32. The identifier for Office unpacks into two attributes while the Department key unpacks into three (two for Office and one

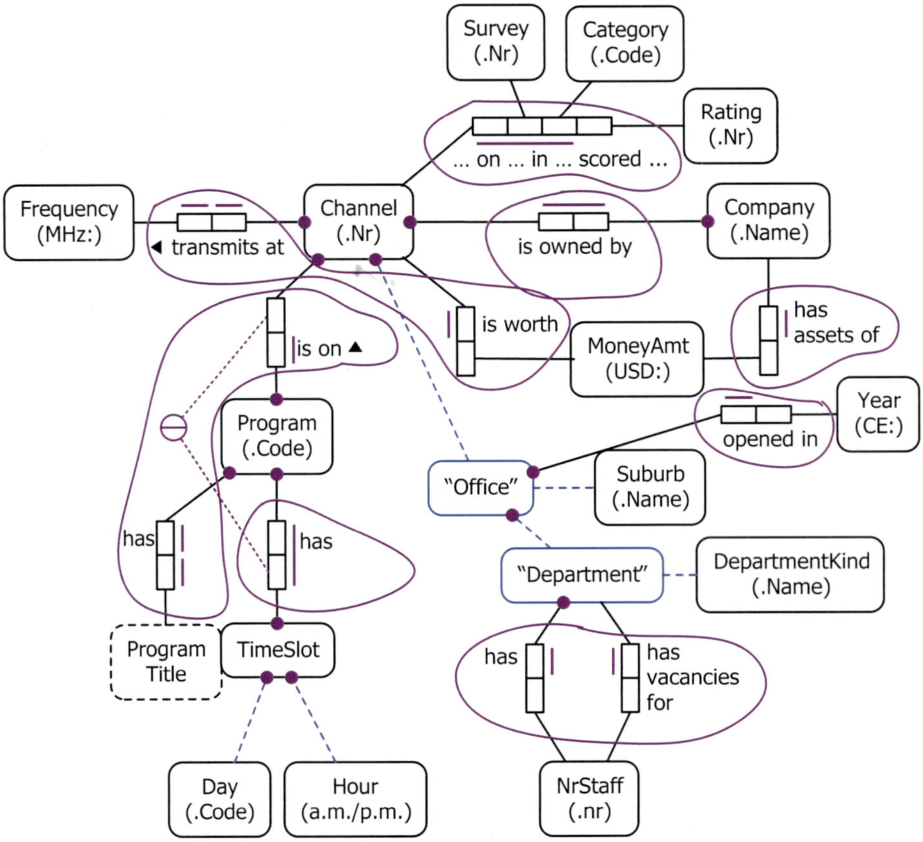

FIG. 11.31 Reference types are erased, and fact types are grouped.

for DepartmentKind). Notice also the intertable uniqueness constraint. This indicates that when a natural join is performed between the Program and ProgTime tables, there will be a uniqueness constraint spanning the three columns channelNr, progDay, and progHour. In other words, for any given channel and time, there is at most one program being shown.

Notice the many equality constraints in the relational schema. These lead to referential cycles that may be awkward to implement. The many mandatory roles also entail that a lot of data must be entered in a single transaction. As discussed earlier, to minimize referential cycles, you should not add a mandatory role constraint to a nonfunctional role unless it is absolutely required.

As an optional exercise, you may wish to consider which mandatory role constraints would best be removed from Fig. 11.30.

The following exercise contains many questions to give you practice with relational mapping. The choice of names for tables and columns is up to you, but you may wish to consider the naming guidelines discussed earlier.

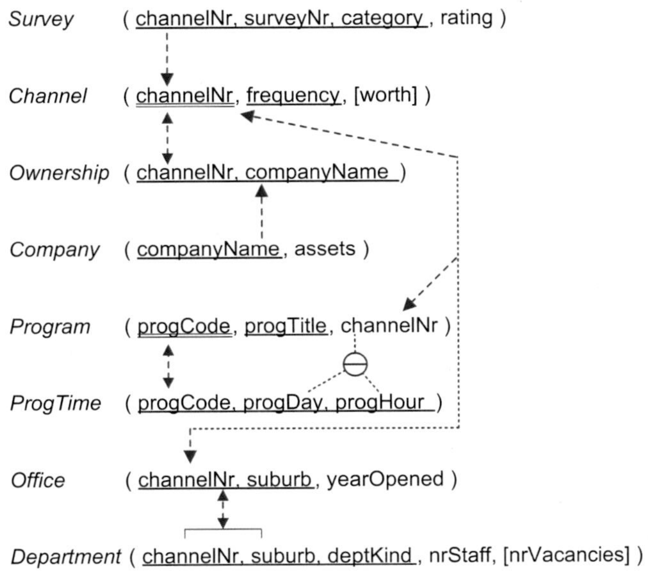

FIG. 11.32 A relational schema mapped from the model in Fig. 11.30.

Mapping from a conceptual to a relational schema is a bit like doing CSDP step 1 in reverse. Try to choose names for tables and columns that would make it easier for you to perform conceptual schema design procedure (CSDP) step 1 if presented with the relational tables. Remember that tables are basically just collections of facts.

Exercise 11.3

1. Map the following conceptual schema onto a relational schema using the Rmap procedure. Use descriptive table and column names. Underline the keys, and enclose any optional columns in square brackets. Indicate any subset or equality constraints.

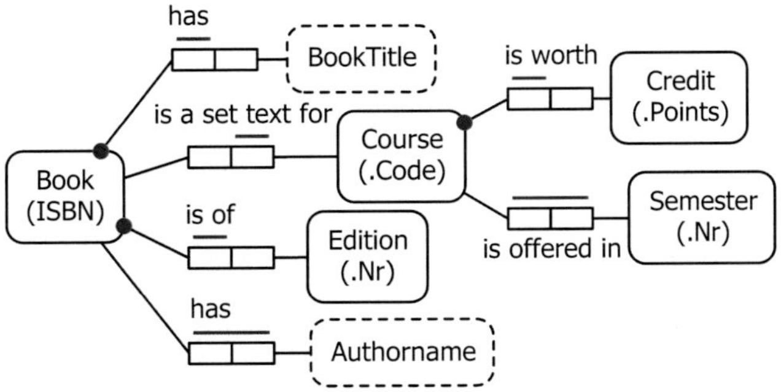

2. Rmap the following conceptual schema.

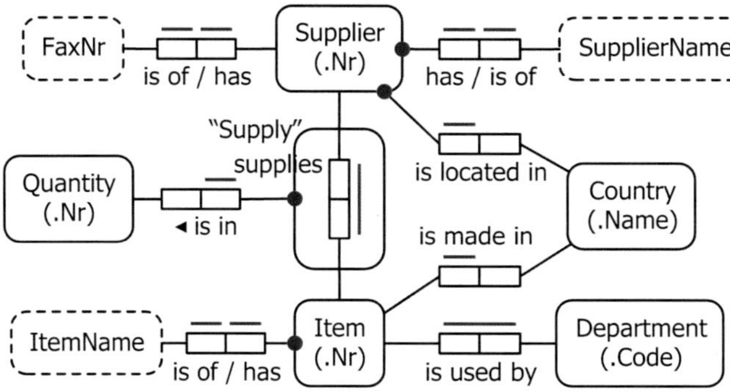

3. (a) The conceptual schema for a given UoD is shown. A novice designer maps the likes and costs fact types into a single table: *Likes* (woman, dress, cost). With the aid of a small sample population, explain why this table is badly designed. Use surrogates w1, w2, ... for women and d1, d2, ... for dresses.
   (b) Rmap the conceptual schema.

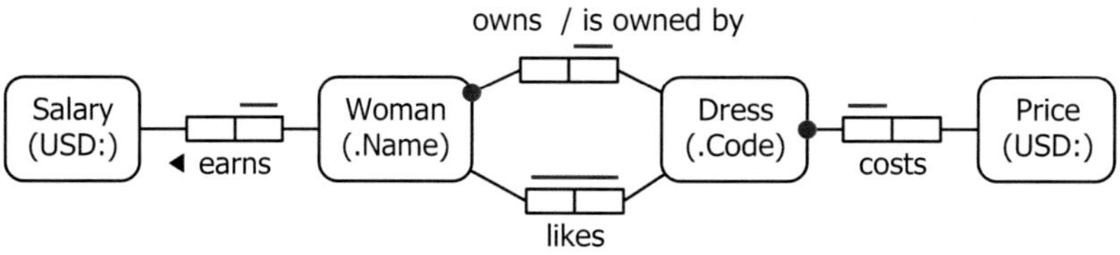

4. In a given UoD, each lecturer has at least one degree and optionally has taught at one or more institutions. Each lecturer has exactly one identifying name, and at most one nick-name. Each degree is standardly identified by its code, but also has a unique title. Some degrees might not be held by any lecturer. Each institution is identified by its name. The gender and birth year of the lecturers are recorded as well as the years in which their degrees were awarded.
   (a) A novice designer develops the following relational schema for this UoD. Explain, with the aid of sample data why these tables are badly designed.

   Lecturer (lecturerName, gender, degreeCode, degreeTitle)
   Qualification (degreeCode, yearAwarded)

   (b) Draw a conceptual schema for this UoD.
   (c) Rmap your answer to (b).

5. Add constraints to your conceptual schema for the following and then perform Rmap.
   (a) Exercise 3.5, Question 2.
   (b) Exercise 3.5, Question 3.

6. Rmap the Invoice conceptual schema for Exercise 6.3, Question 3.

7. Rmap the Oz Bank conceptual schema for Exercise 6.4, Question 6.

8. Consider a naval UoD in which sailors are identified by a sailorNr and ships by a shipNr, although both have names as well (not necessarily unique). We must record the

sex, rank, and birthdate of each sailor and the weight (in tonnes) and construction date of each ship. Each ship may have only one captain and vice versa. Specify a conceptual schema and relational schema for this UoD for the following cases.
(a) Each captain commands a ship but some ships might not have captains.
(b) Each ship has a captain but some captains might not command a ship.
(c) Each captain commands a ship, and each ship has a captain.
(d) Some captains might not command a ship, and some ships might not have captains.

9. Consider a UoD in which students enroll in subjects, and later obtain scores on one or more tests for each subject. Schematize this conceptually using nesting and then Rmap it.
   (a) Assume that scores are available for all enrollments.
   (b) Assume that enrollments are recorded before any scores become known.

10. Rmap the conceptual schema in Fig. 6.46 (absorb the subtypes before mapping).

11. Refer to the MediaSurvey conceptual schema in Fig. 6.53.
    (a) Rmap this schema (absorb subtypes).
    (b) Set out an alternative relational schema with separate tables for each node in the subtype graph. Which schema is preferable?

12. (a) Rmap the Taxpayer conceptual schema for Exercise 6.5, Question 3(a).
    (b) Rmap the Taxpayer conceptual schema for Exercise 6.5, Question 3(b).

13. Rmap the SolarSystem conceptual schema for Exercise 6.5, Question 6.

14. Rmap the conceptual schema in Fig. 7.5 (Panel is an independent object type).

15. Rmap the CountryBorders conceptual schema for Exercise 7.3, Question 5.

16. Rmap the IT Company schema shown.

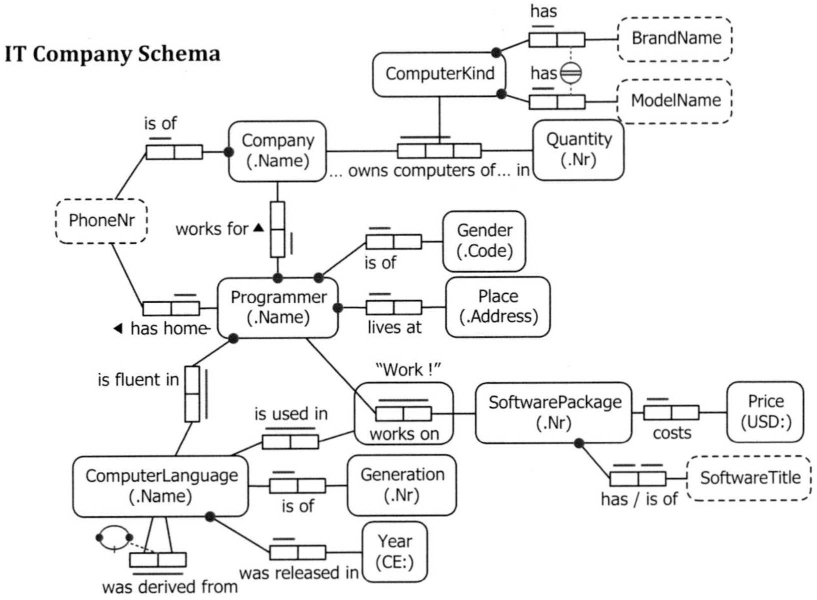

17. Rmap the University schema shown. Note that Degree is compositely identified (e.g., a Ph.D. from UCLA and a Ph.D. from MIT are treated as different degrees). Student is also compositely identified (by the time you finish the mapping, you will appreciate how much better it would be to use a StudentNr instead!).

    The codes "y", "n", "PT", "FT", "int", and "ext" abbreviate "yes", "no", "part-time", "full-time", "internal", and "external".

University Schema

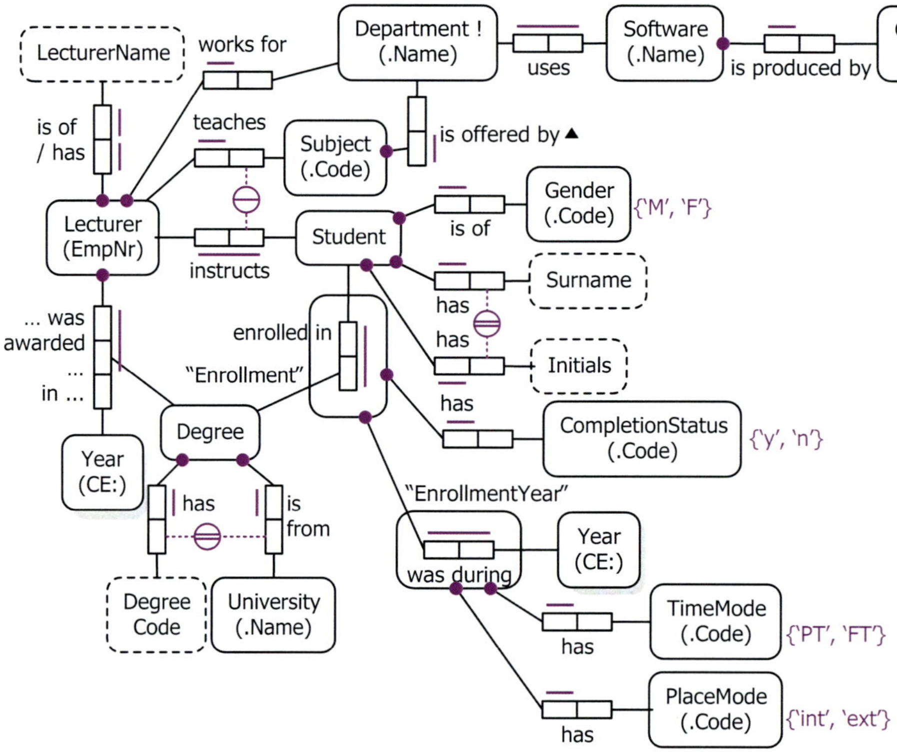

18. Rmap the following conceptual schema. Ignore the implicit semantic connection between Year and Date.

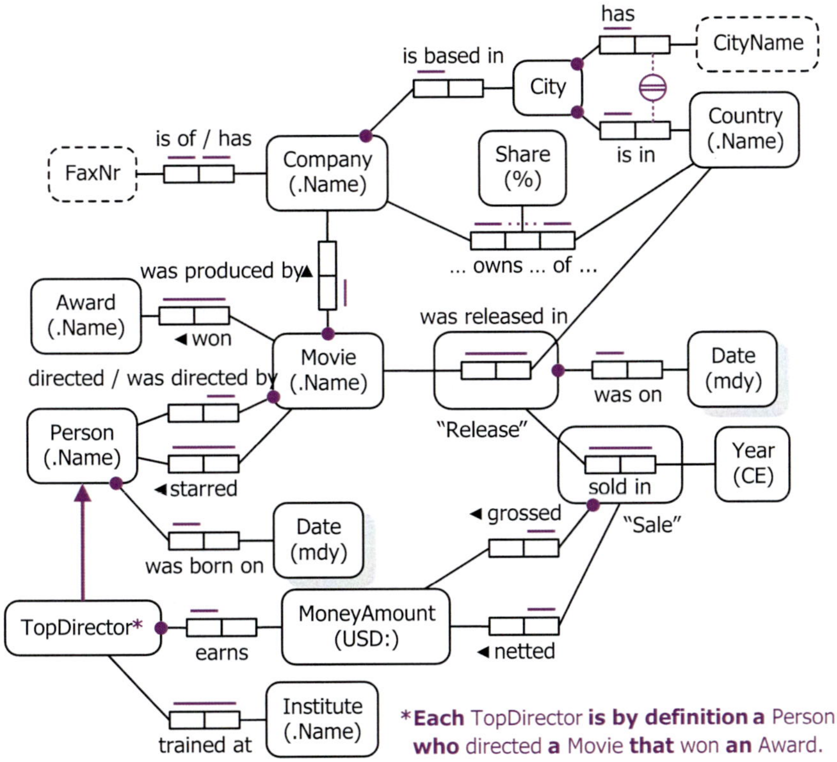

19. A life insurance company maintains an information system about its clients. The following information samples are extracted from a fragment of this system. For simplicity, several data items (e.g., client's name and address) are omitted and may be ignored for this question. Each client is identified by his/her client number (ClientNr).

The following table uses these abbreviations: Emp. = employment (EA = Employed by Another; SE = Self-Employed; NE = Not Employed); Acct = accounting; NS = Nonsmoker; S = Smoker. Some clients are referred by other clients (referrers) to take out a policy (this entitles referrers to special benefits outside our UoD). The value "?" is an ordinary null value meaning "not recorded" (e.g., nonexistent or unknown). The value "-" means "inapplicable" (because of some other data).

ClientNr	*Referrer*	*Birth date*	*Smoking status*	*Emp. status*	*Job*	*Work phone*	*Acct. firm*
101	?	15/2/46	NS	EA	lecturer	3650001	–
102	?	15/2/46	S	SE	builder	9821234	Acme
103	101	1/1/70	NS	NE	–	–	–
104	103	3/4/65	S	EA	painter	?	–
105	?	1/1/70	NS	SE	painter	2692900	?

For each client, a record of major illnesses (if any) is kept; one recovery figure and hospital is noted for each client illness. The extract for clients 101–105 is shown.

ClientNr	*Illness*	*Degree (%) of recovery*	*Hospital treated*
102	Stroke	90	Wandin Valley
	Diabetes	80	Burrigan
103	Diabetes	95	Wandin Valley

Each client selects one insurance coverage (currently $25,000, $50,000, or $100,000) and pays a monthly premium for this coverage. Premiums are determined by coverage, age, and smoking status, as shown in the following schedule. This schedule is stored in the database (it is not computed by a formula). From time to time, this schedule may change in premiums charged, coverages offered, or even age groups. Premiums for both smokers and nonsmokers are always included for each age group/coverage combination. For simplicity, only the latest version of the schedule is stored (a history of previous schedules is not kept). Moreover, payments must be for 1 year at a time and are calculated as 12 times the relevant premium current at the date of payment (using the age of the client at that date). The computer system has an internal clock, which may be viewed conceptually as providing an always up-to-date instance of the fact type Date is today. You may use "**today**" as an initialized date variable in derivation rules.

	Nonsmoker premiums ($)			*Smoker premiums ($)*		
Age (y)	25,000 coverage	50,000 coverage	100,000 coverage	25,000 coverage	50,000 coverage	100,000 coverage
21–39	5.00	7.50	12.50	6.50	10.50	18.50
40–49	6.50	10.50	18.50	10.00	17.50	32.50
50–59	13.00	23.50	34.00	24.00	45.50	88.50
60–69	36.50	70.50	138.50	57.50	112. 50	222.50

Although not shown in the schedule, age groups are identified primarily by an age group number (currently in the range 1..4). Assume that all clients have paid their 12-month fee by completing a form like that shown here (details such as name and address are omitted for this exercise). The first four fields are completed by the insurance agency and the rest by the client. Records of any previous payments are outside the UoD.

ClientNr: .... Date: ....

Insurance coverage: $ ....

Payment for next 12 months' insurance: $ ....

Method of payment (Please check one): ☐ cash ☐ check ☐ credit card

If paying by credit card, complete the following:

☐ Visa ☐ Mastercard ☐ Bankcard

Card Number: ☐☐☐☐ ☐☐☐☐ ☐☐☐☐ ☐☐☐☐

Card expiration date: ☐☐/☐☐

Choose suitable codes to abbreviate payment methods and card types. Each credit card is used by at most two clients (e.g., husband and wife). In practice, the card type could be derived from the starting digits of the card number, but ignore this possibility for this question.

(a) Specify a conceptual schema for this UoD. Include **all** graphic constraints and any noteworthy textual constraints. Include subtype definitions and derivation rules. If a derived fact type should be stored, include it on the diagram with an "**" mark to indicate it is both derived and stored

(b) Should the payment by a client be derived only, stored only, or both? Discuss the practical issues involved in making this decision.

(c) Map your conceptual schema to a relational schema, absorbing subtypes while maintaining subtype constraints. Underline keys and mark optional columns with square brackets. Include **all** constraints. As an optional exercise, map any derivation rules.

11.4 Advanced Mapping Aspects

The previous section discussed the main steps in the default relational mapping procedure (Rmap). Step 0 of this procedure may be refined as shown here. This section discusses these refinements in order and clarifies some fine points. If desired, this section may be safely skipped on a first reading.

Rmap Step 0:

0.1 Transform opposite unaries; map unaries based on their open/closed semantics.
0.2 Mentally erase all reference (preferred identification) predicates.
Treat compositely identified object types as "black boxes".
0.3 Indicate any absorption overrides (separation or partition) for subtypes.
0.4 Identify any derived fact types that must be stored.
0.5 Indicate mapping choices for symmetric 1:1 cases.
0.6 Consider replacing any disjunctive reference schemes
by using an artificial or concatenated identifier or mandatory defaults.
0.7 Indicate mapping choice where required for any objectified associations
that have no spanning uniqueness constraint.

As background to step 0.1, see the discussion of open/closed world semantics in Section 10.6. Two unaries with opposite meaning (e.g., Person is male, Person is female) should be replaced by a single closed world unary (e.g., Person is male) or a mandatory binary (e.g., Person is of Gender {'M', 'F'}).

After that, mutually exclusive unaries should be replaced by a binary. For example, in Fig. 11.33(a) the disjunction of the unaries is optional, so it is possible that some applications are pending a decision. In Fig. 11.33(b), the unaries are transformed to a single binary, where the status of the application is identified by a code (S = Succeeded, F = Failed, P = Pending). This transformation is understood to be performed automatically as a preprocessing stage to the rest of Rmap.

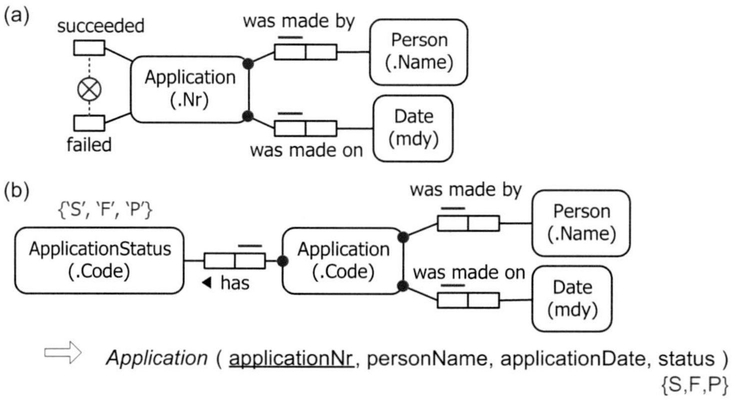

FIG. 11.33 Replacing exclusive unaries by a binary.

The final result of the mapping is shown at the bottom of Fig. 11.33(b). For analysis purposes, you may still work with the original conceptual schema.

Remaining unaries are mapped according to their open/closed world assumption. Fig. 11.34 shows a simple example, with cases (a), (b), and (c) corresponding to closed world, simple open world, and OWN, respectively. Typically, Person would play other mandatory roles and not be independent. In practice, CWA leading to a mandatory Boolean and OWN leading to an optional Boolean are the most useful cases.

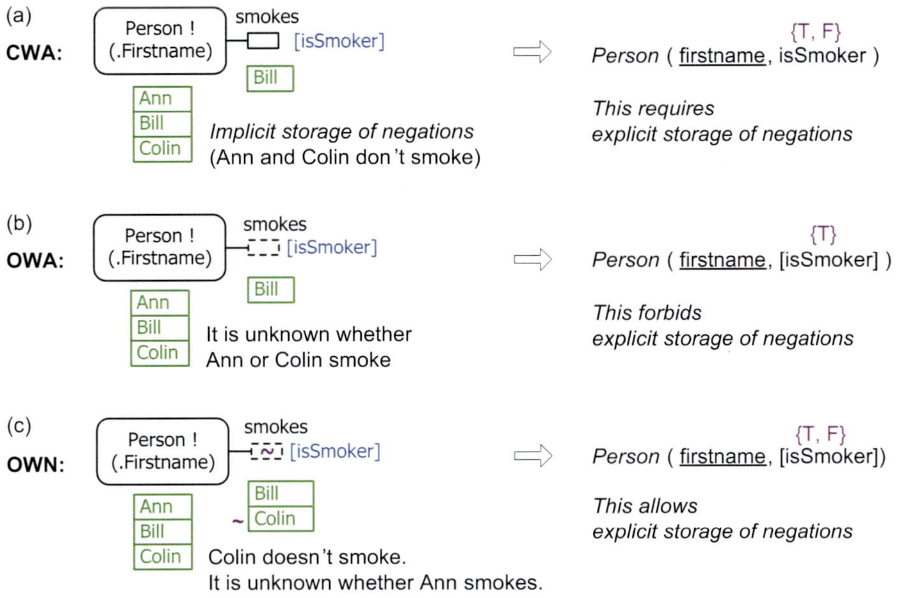

FIG. 11.34 Basic mapping patterns for unobjectified unaries.

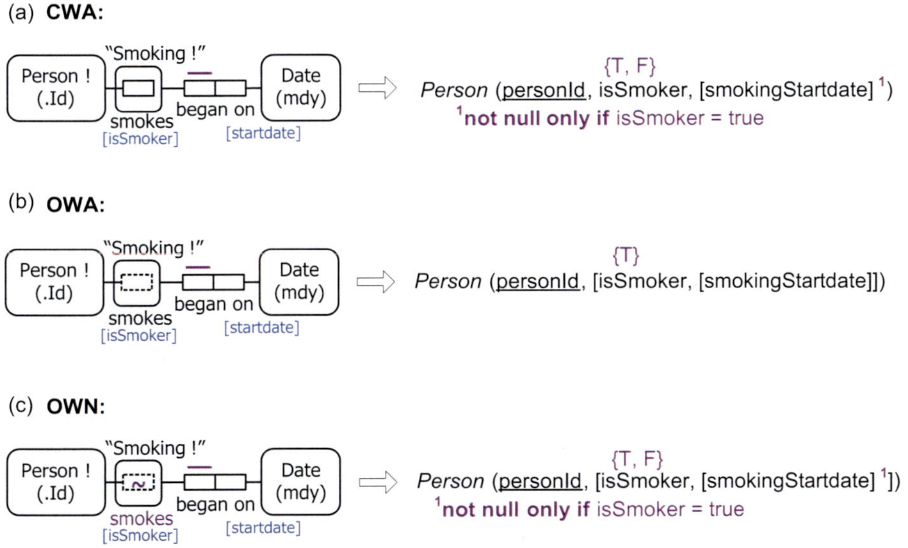

FIG. 11.35 Basic mapping patterns for objectified unaries.

Fig. 11.35 shows how to map an objectified unary, with cases (a), (b), and (c) corresponding to closed world, simple open world, and OWN, respectively. This mapping preprocesses the functional fact type attached to Smoking by attaching it directly to

Person, renaming the role name "startdate" to "smokingStartdate", and then mapping as usual, resulting in a single table rather than two. Simple variations apply if the attached role is mandatory (e.g., in the CWA case replace "**only if**" by "**iff**").

Step 0.2 involves mentally erasing any reference predicates and treating objectified associations as simple object types. The previous section discussed typical cases. Fig. 11.36 depicts a special case. Here, each Olympic Games is identified by the year in which it is held. There are three fact types (Athlete competes in Games; Athlete was born in Year; City hosts Games) and one reference type (Games is held in Year).

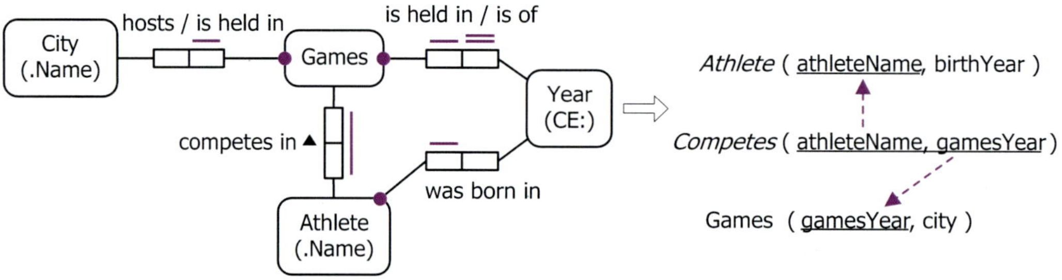

FIG. 11.36 Games is identified by a single, explicit reference type.

The reference type must be shown explicitly because Year is an entity type, not a value type. One instance of the hosts fact type is The City with name "Sydney" hosts the Games that is held in the Year 2000 CE.

It is rare to have an object type identified by means of a single, explicit reference type. To map this case, first mentally erase the reference type, leaving the three fact types to be grouped into tables. As an exercise, try the mapping yourself and then check your answer with the solution shown in Fig. 11.36.

By default, subtypes are absorbed into their root supertype for mapping purposes. Although the subtype constraints are still mapped (as qualifications on optional columns or intertable subset constraints), absorption generally causes any functional roles of subtypes to be grouped together with the functional roles of the root supertype. In step 0.3, we may wish to override this default. Let's consider some examples, starting with the usual case where the subtypes inherit the preferred reference scheme of their top supertype.

Fig. 11.37 depicts a simple subtype graph with functional roles. Reference schemes and subtype definitions are omitted for simplicity. As discussed in the previous section, there are three basic ways in which the functional fact types associated with the nodes may be grouped into tables: *absorption, separation,* and *partition*.

The first option *absorbs* the subtypes back into the supertype before grouping. For example, assume that all the roles played by *A*, *B*, and *C* are functional (i.e., they have a simple uniqueness constraint). In this case, we generate just one table (with subtype

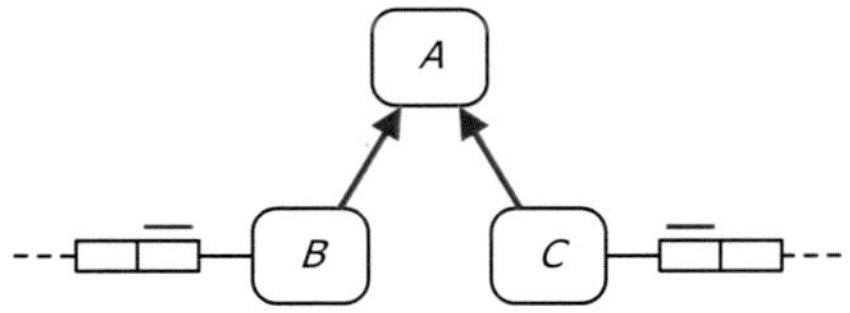

FIG. 11.37 How do we map functional fact types played by subtypes?

constraints expressed as qualifications on the optional subtype attributes). This absorption default has two main advantages: queries that require attributes from more than one node (no joins required) perform better, and subtype constraints are usually easy to specify and inexpensive to enforce (no joins).

The main disadvantages of subtype absorption are as follows: nulls are required for objects that belong to only one subtype; the supertype's functional table is larger (more columns); queries about only one subtype require a restriction; and viewing just a subtype is less convenient (projection and restriction needed). Usually, the advantages outweigh the disadvantages. Note that any nonfunctional roles of the subtypes map to separate tables, so these are not affected by our subtype mapping choice.

The second option, *separation*, groups the functional roles attached directly to each object type node into separate tables, one for each node. Here, the functional predicates of *A* map to one table (the common properties), the attributes specific to *B* map to another table, and the attributes specific to *C* map to a third table. The main advantages of separation are that it minimizes nulls, and queries about each subtype are fast. Its main weaknesses are that queries requiring attributes from more than one node are slower (joins needed) and insertions to subtype tables are slower (subtype constraints are now specified as qualified subset constraints, so access to a supertable is required to enforce them).

The third option is to horizontally *partition* the instances of *A*. This should normally be considered only if *B* and *C* form a partition of *A* (i.e., they are exclusive and exhaustive: $B \cap C = \{ \}$; $B \cup C = A$). In this case one table holds all the functional predicates of *B* (including those attached to *A*), and another holds all the functional predicates of *C*. However, if *B* and *C* do not exhaust *A*, a separate table is needed for $A - (B \cup C)$.

If *B* and *C* overlap, redundancy results for the facts about $B \cap C$, and this must be controlled. This partition option departs from the usual practice of grouping each fact type into only one table and tends to be useful only in distributed or federated database settings.

The main advantages of partitioning the supertype are as follows: it minimizes nulls; queries about all the properties of *B* (or *C*) are fast; and subtype constraints typically need not be coded (because implied) or are trivial to code (e.g., Man and Woman tables without/with gender field). Its main disadvantages are the following: it slows queries about all of *A*, or *B* and *C* (joins needed); it is very awkward unless *B* and *C* form a partition of *A*; and if *B* and *C* are exclusive, enforcement of this constraint normally requires intertable access.

The criteria discussed can help us decide whether to override the default absorption option. Override decisions may be selected from an option list in a CASE tool. For larger subtype graphs, the mapping choices increase rapidly, as mixtures of the three options might be used.

To clarify some of the previous discussion, let's look at some examples. The ORM schema in Fig. 11.38 recalls a patient example discussed in earlier sections. The three basic mapping choices are shown at the bottom of Fig. 11.38. With absorption, all the fact types are grouped into a single table, and the subtyping constraints appear as qualifications on optional columns (the **only if** condition captures the subtyping, and the **iff** condition captures both subtyping and mandatory role constraints).

With separation, the common fact types are grouped into the Patient supertable, and the subtype-specific fact types are grouped into separate subtype tables, giving three

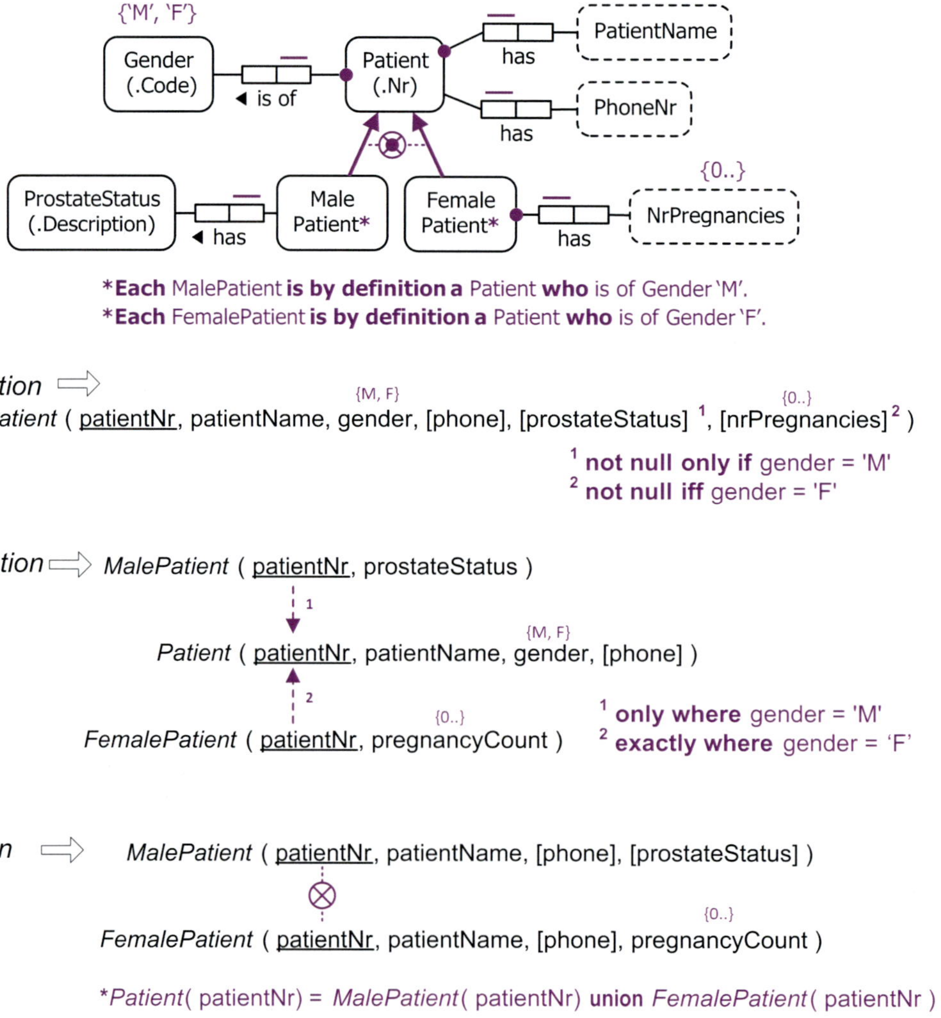

FIG. 11.38 Absorption, Separation, and Partition mapping options.

tables overall. The subtyping constraints appear as qualifications on the subset constraints (in this case, foreign key references). In the subset constraint qualifications, the **only where** clause captures just a subtyping constraint while the **exactly where** clause covers both subtyping and mandatory role constraints.

With partition, the common fact types are pushed down into subtype tables that also contain the specific fact types for those subtypes. An exclusion constraint between the keys of the tables ensures that each patient maps to only one of these tables. A view definition is included to list all the patients by unioning the keys of the two tables. The gender information is now informally captured in the table names. If desired, a view could be created to form a single table with all the Patient details, deriving gender from the table names.

While the three choices just discussed are typically the only ones used, it is possible to mix and match the options, for example, absorb the prostate fact type for male patients but keep a separate table for pregnancy count for female patients.

If we were to add the *m:n* prenatal clinic attendance fact type considered earlier in Fig. 11.26, this would map to its own separate table, regardless of which of the three subtype mapping options is chosen. Recall that the absorption choice absorbs only the fact types functional for the subtype.

Now consider the conceptual schema in Fig. 11.39. Here, academics have one of three ranks (L = Lecturer, SL = Senior Lecturer, P = Professor). Students may be counseled only by senior lecturers, and academic chairs are held only by professors. Because the subtypes do not form a partition of Academic, we would normally choose either absorption or separation to map them. As an exercise try both these options yourself before reading on.

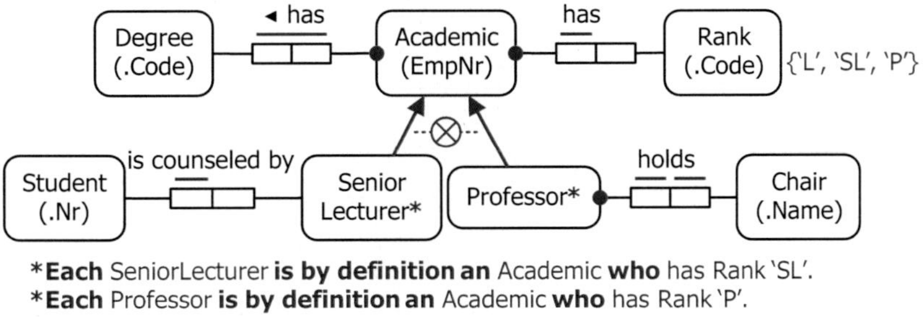

FIG. 11.39 Senior lecturers and professors play specific roles.

In the counseling predicate, the role attached to SeniorLecturer is not functional. So, this predicate maps to a separate table regardless of whether we choose absorption or separation. Also, being *m:n*, the degrees fact type maps to a separate table regardless. So, the only choice we have in the subtype mapping is whether to group the rank and chair fact types together. With absorption, we do so (see Fig. 11.40(a)). With separation we do not (see Fig. 11.40(b)). Note the different ways of specifying the subtype and mandatory role constraints for the Professor subtype.

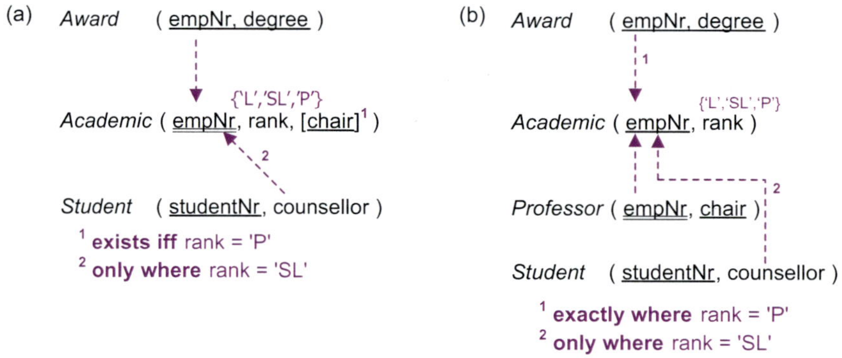

FIG. 11.40 Mapping Fig. 11.39, using subtype absorption (a) or separation (b).

Let's now restrict our attention to the mapping of functional predicates of the nodes in the subtype graph because the mapping of other predicates to separate tables with qualified subset constraints is straightforward.

As an example with multiple inheritance, consider Fig. 11.41. This figure also shows one way of displaying an ER view of an ORM diagram. Here, all attributes are single valued. For compactness, underlining an attribute indicates the 1:1 nature of the ORM binary fact type corresponding to it, e.g., "chair". The absence of a mandatory role dot on nrKids (number of children) indicates that recording of this attribute is optional for female academics.

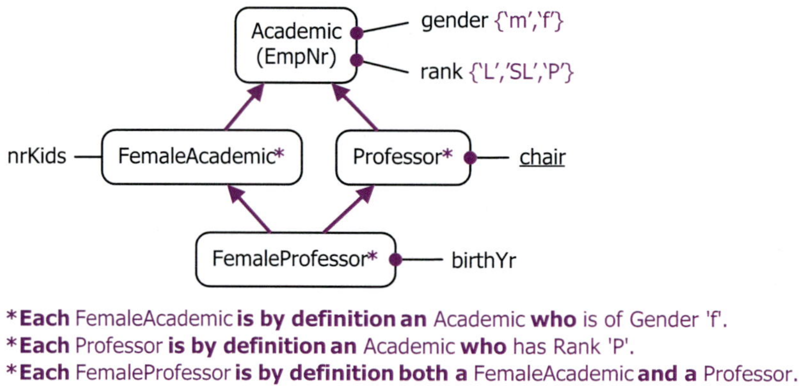

FIG. 11.41 An ER view with multiple inheritance.

Suppose we decide on separate tables for each subtype. This yields the relational schema in Fig. 11.42. Note that qualified subset constraints from FemaleProfessor.empNr to Professor.empNr and FemaleAcademic.empNr are implied.

For efficiency reasons, we adopt the default policy of specifying qualified subset constraints with respect to the root supertable where possible. The order of table creation then does not matter as long as the root table is created first, and the schema is easier to change.

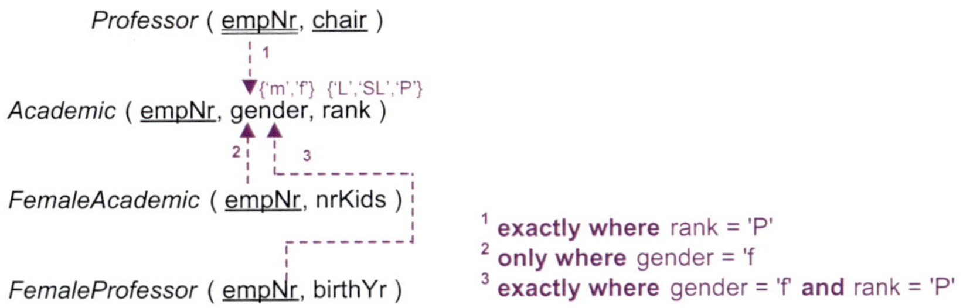

FIG. 11.42 Relational map of Fig. 11.41, choosing subtype separation.

Mixed approaches may be adopted. For example, if we absorb FemaleProfessor and Professor into Academic, but map roles specific to FemaleAcademic to a separate table, we obtain the relational schema in Fig. 11.43. Notice that qualification 2 captures two subtype constraints. Other mixtures are possible, but this gives the idea.

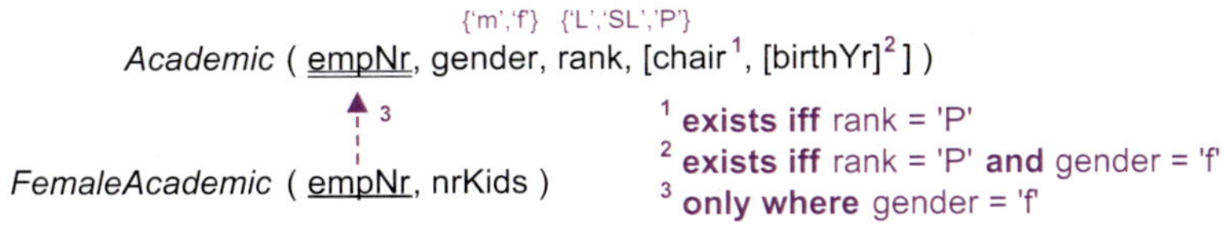

FIG. 11.43 Relational map of Fig. 11.41, separating FemaleAcademic only.

Section 6.6 discussed the awkward situation of *context-dependent reference schemes*, where a subtype may have a different preferred identification scheme from its supertype(s). Let's briefly discuss how such cases may be mapped.

Recall that a *direct* supertype of a subtype is connected directly to the subtype (i.e., with no intermediate subtypes on the connecting path). Each subtype has one or more direct supertypes. A connection between a subtype and its direct supertype(s) is shown as a *solid arrow if it provides a path to the preferred identifier of the subtype.* If a subtyping connection appears as a *dashed arrow*, the subtype either introduces its own preferred reference scheme (this is called *context-dependent reference*), or it obtains its identification scheme via a subtyping connection to another supertype.

In mapping, all specific roles attached to an entity type require the object type to be identified by its preferred reference scheme. This requirement avoids some extremely complex reference constraints that could otherwise arise in practice.

If two overlapping subtypes have the same preferred reference scheme, and one is not a subtype of the other, they must have a common supertype with this preferred reference scheme (if not, create it). This is needed to avoid redundancy later.

For example, suppose we allowed FemaleStudent(StudentNr) and PostGradStudent (StudentNr) as direct subtypes of Person(.nr). Mapping personNr down into tables for both subtypes would create redundancy of the facts that associate personNr and studentNr for the intersection of the subtypes. So this pattern is not allowed. Instead we insert the intermediate supertype Student(.nr) as a subtype of Person(.nr) and a supertype of FemaleStudent and PostGradStudent, which now inherit their identification scheme from Student (which is used for establishing the correspondence between personNr and studentNr).

We confine our discussion here to the subtype separation option (separate tables are created for functional roles specific to each subtype) and assume that all reference schemes are simple (not composite). For each subtype with a preferred reference mode different from the root's, we define its "*total table*" (if any) as set out in the following total table procedure.

Total Table Procedure:

```
if the object type has a mandatory disjunction of one or more functional roles
   all of which map to the same table
then the table to which its functional roles map is its total table
else if it has a mandatory nonfunctional role
       then select one of these roles arbitrarily;
              the table to which this role maps is its total table
       else { all roles are optional }
              if the object type is independent or is a subtype with separation mapping option
              then its table is the total table
              else if the object type plays only functional roles or only one role
                         all of which map to the same table
                     then the table for this is its total table
                     else the object type has no total table.
```

The basic idea is that the total table of an object type contains its total population (i.e., it includes all instances in the population of the object type). If an object type does have a total table, this procedure will find it. In what follows, where total tables exist we basically ignore other tables of the object type (the other tables are linked by foreign keys to the primary ones in the normal way).

The root supertype table(s) is/are computed in the normal way. Copy the preferred reference of the root supertype as an extra attribute in the total table of each subtype, which "introduces" a different preferred reference scheme to the graph; if the introducing subtype has no total table, create a reference table for it (to store just the fact connecting the two identifiers for the object type's population).

In specifying, at the relational level, subtype links between object types with different primary reference schemes, use the root reference scheme; if the object types have the same reference scheme, use this common reference scheme.

These guidelines are best understood by way of example. Consider the schema in Fig. 11.44. Here, the subtypes are all asserted rather than derived. Better students are often employed to do some tutoring. If known, the number of hours of tutoring undertaken by such a student is recorded. In practice, many other fact types might be stored about each node. Within the student records subschema, students are identified by their student number (studentNr). Within the staff records subsystem, all employees are identified by the employee number (employeeNr). A student may be an employee, and we want to know when this occurs. If it did not already exist, we introduce a person number (personNr) to enable people to be identified across the global schema. For student employees, we have three possible identifiers: personNr, studentNr, or employeeNr.

In this example, we chose employeeNr as the preferred identifier for student employees in the context of roles specific to StudentEmployee. This choice is shown by the solid subtyping arrow from StudentEmployee to Employee (indicating a path to the subtype's preferred identifier). The subtyping connection from StudentEmployee to Student is dashed because this does not provide a path to the subtype's identifier. The subtype connections from Student and Employee are also dashed because these subtypes introduce new reference schemes in preference to that of their Person supertype.

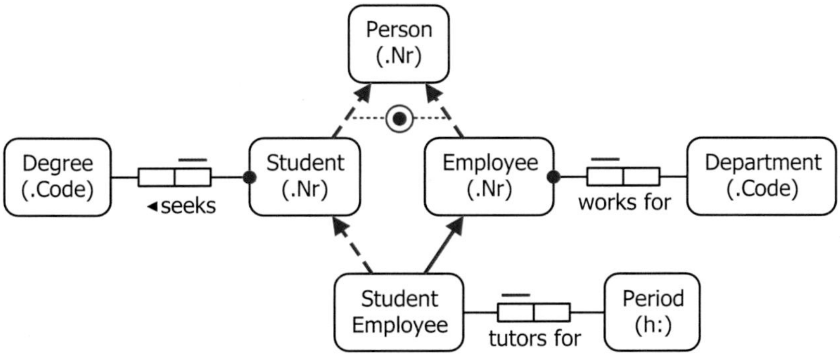

FIG. 11.44 A subtype graph with differing reference schemes.

Student (studentNr, personNr, degree, ...)

Person (personNr, ...)

Employee (empNr, personNr, deptCode, ...)

[1] **only where** personNr **in** Student.personNr

StudentEmployee (empNr, tutePeriod, ...)

FIG. 11.45 The relational map of Fig. 11.44, choosing subtype separation.

Choosing subtype separation for mapping each subtype, each of the three supertypes has a total table that is identical to its functional table. The functional fact types of the nodes in the subtype graph map to the four tables shown in Fig. 11.45. Here, an ellipsis "..." denotes any other (functional) attributes omitted in Fig. 11.44.

If there are other mandatory columns in StudentEmployee, then tutePeriod becomes optional. Notice that personNr is not included in the StudentEmployee table, as this avoids duplicating any instances of Person(.Nr) has StudentNr(). Other mapping options exist, for example, we might absorb StudentEmployee into Employee.

As explained in Section 10.8, an arrow from a subtype to a supertype displays a type-level relationship that implies instance-level, identity relationships linking subtype and supertype instances. So, the four subtype-to-supertype connections in Fig. 11.44 may be understood as implicit link fact types displayed in Fig. 11.46. For example, the subtyping link from StudentEmployee to Employee implies the link fact type StudentEmployee is Employee. The implicit is-associations between Student and Person and between Employee and Person allow us to express facts such as "The employee with EmployeeNr 23 **is** the person with PersonNr 507" and "The Person with PersonNr 507 **is** the Student with StudentNr 300".

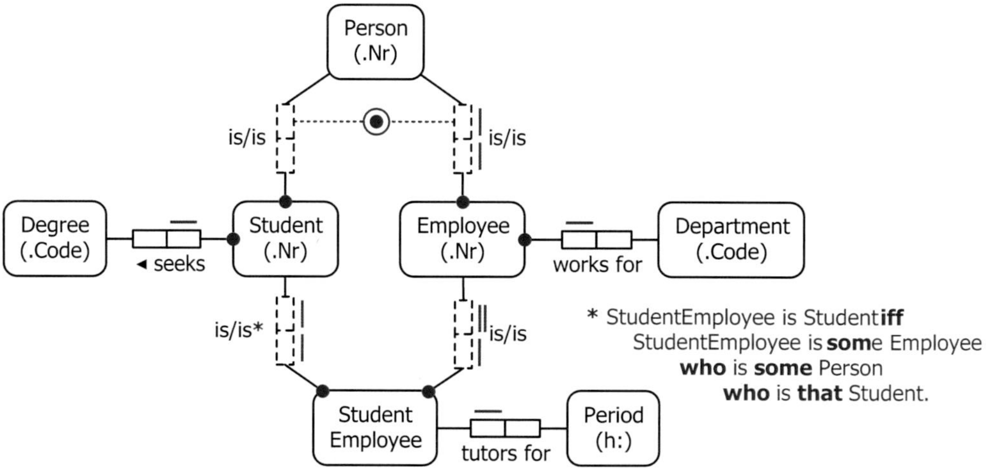

FIG. 11.46 Subtype links to new reference schemes are treated as base fact types.

The link fact type from StudentEmployee to Student may be viewed as a derived association, equivalent to the longer path from StudentEmployee through Person to Student, as shown in the derivation rule included in Fig. 11.46.

By default, derived fact types are not stored. Instead, their derivation rules are used to compute the derived values on request. Apart from saving storage space, this "derive on query" (lazy evaluation) approach ensures that every time the derived information is requested, it will be based on the latest values of the stored data. Sometimes, however, the same derivations are required many times with large volumes of stable data. In such cases, it can be much more efficient to store the derived information so that it can be accessed immediately at later times without having to be recomputed. Typically, in such cases a "derive on update" (eager evaluation) approach is used so that as soon as the base data is updated, the derived information is computed and stored. Recall the bank balance example from Exercise 6.4.

In step 0.4 of the mapping procedure, any derived fact types to be stored should be included on the schema diagram and marked "**", with the derivation rule also declared. The fact type should then be grouped into a table in the normal way, and the derivation rule mapped as well. For example, the Window schema in Fig. 11.47(a) uses the default mapping so that no column for area is included in its base table. However, the area fact type in Fig. 11.47(b) is marked "**" for storage, so it maps to a base table column.

The rule for computing area may be included in insert and update triggers, or a generated column declaration, a view, or a stored procedure. Care is required with storing derived data. For example, unless the derivation rule is fired every time the relevant base data is updated, the derived values can become outdated.

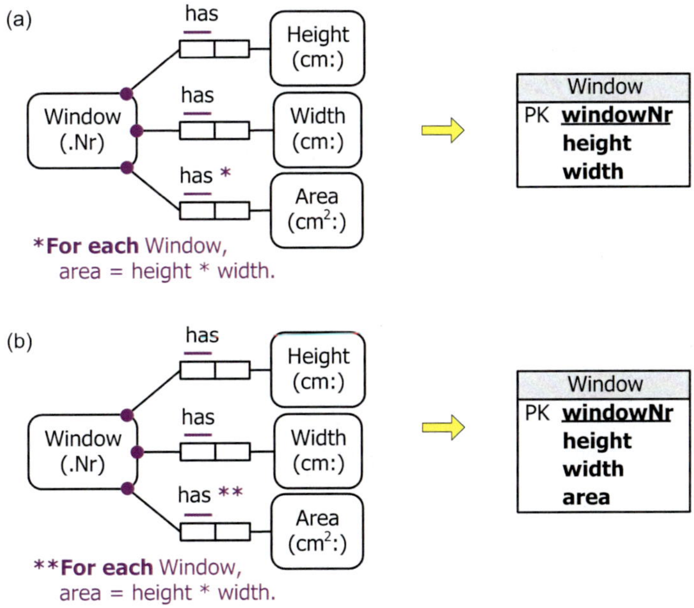

FIG. 11.47 Derived fact types may be evaluated lazily (a) or eagerly (b).

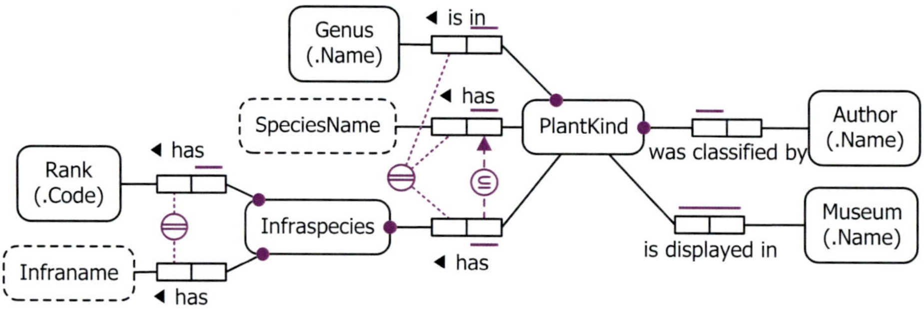

FIG. 11.48 PlantKind has a disjunctive reference scheme.

For symmetric 1:1 binaries (roles are both optional or both mandatory, and both object types have other functional roles), we have a choice as to how the 1:1 binary should be mapped. This mapping choice should be noted in step 0.5 (e.g., by annotation or option selection) and adhered to when the fact type is mapped in step 2.

In rare cases, a conceptual schema may include a disjunctive preferred reference scheme (identification by a mandatory disjunction of two or more roles, at least one of which is optional). In Fig. 11.48, PlantKind is identified in this way (this botanical naming convention was discussed in Chapter 5). When mapped to a relational schema, such schemes can prove awkward to handle.

For example, the conceptual schema in Fig. 11.48 maps to the following relational schema. This violates the relational entity integrity rule because the primary keys may contain nulls. We enclose both infraRank and infraName in the same pair of square brackets to indicate that the qualification applies to *both*, and that if one is null so is the other. Enclosing both these inside the option brackets for species indicates that they can only be given a nonnull value when species is nonnull.

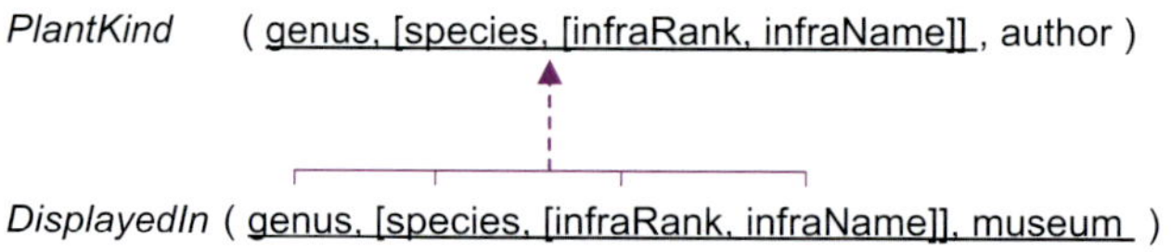

Although forbidden by the relational model, this is allowed in most relational database systems (which permit tables without primary keys), and the relevant uniqueness constraint can be enforced by a procedure or assertion. However, a simpler implementation can often be obtained by altering the conceptual schema to replace the disjunctive reference with a nondisjunctive one. In step 0.6, we consider such replacements.

There are three basic ways of replacing a disjunctive reference scheme: artificial identifiers, concatenated identifiers, or use of special default values. Let's look at these three alternatives briefly, using this botanical example. The first option is to introduce a new identifier for PlantKind (e.g., PlantKindNr), leading to the following relational schema.

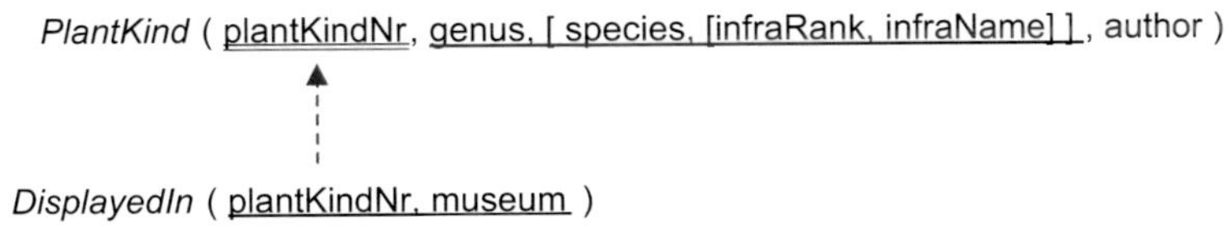

This approach has two main advantages: the primary keys have no nulls, and table joins are faster (now based just on plantKindNr). The artificial identifier option is not all good news. The secondary key in PlantKind still has optional fields. Moreover, because this secondary key uses the natural reference scheme, this is what users will normally want to see (not the artificial PlantKindNr). For example, to find what kinds of plants are on display in a museum, we now need to join the tables because we want to see the natural plant names.

A simpler solution is just to concatenate the formerly separate parts of the identifier into a single name. This leads to the very simple relational schema:

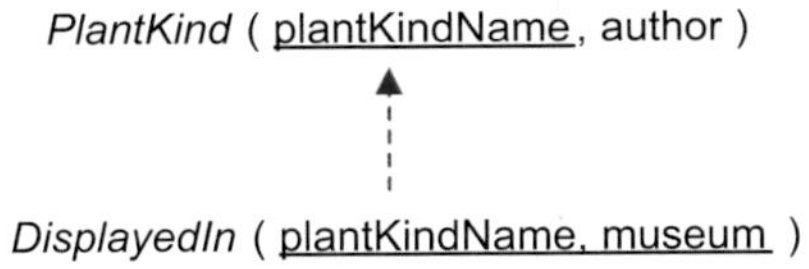

This simple schema is the best solution if we don't need to list or group the formerly separate parts of the identifier. However, if we wanted to issue a query such as "Which species has the most plant kinds?" there is now no simple way to do this.

The third option is to keep the original components of the reference scheme separate but make each of them mandatory by using *special default values* when no actual value exists. For example, the plant kind *acacia interior* could be stored as the tuple ("acacia", "interior", "nil", "nil") using "nil" as a special default value (different from null) to indicate that plant kind has no actual value for infraRank and infraName. The conceptual schema in Fig. 11.48 is modified by making the species and infraspecies predicates mandatory for PlantKind, deleting the subset constraint, and adding "nil" to any value list for Rank, e.g., {"ssp", "var", "forma", "n-ssp", "n-var", "n-forma", "nil"}. The following relational schema is obtained (value-list omitted):

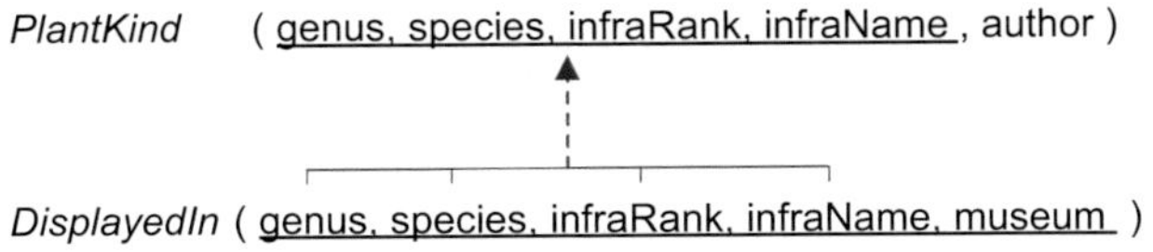

In this case, the default value "nil" is unlikely to ever be confused with an actual value. However, in some cases, such confusion might arise (e.g., a default score of 0) and the user may then be burdened with the responsibility of distinguishing default from actual values. Also, bag functions that ignore nulls will not ignore nils.

The final refinement to step 0 of the mapping procedure (step 0.7) deals with cases where an objectified predicate is not spanned by a uniqueness constraint. Some versions of ORM forbid this because it violates elementarity. As discussed in Section 10.5, we also prefer to avoid such cases.

However, if it is allowed, then some preprocessing is needed before executing the grouping part of the procedure.

In nested 1:1 cases, a decision should be made to favor one of the roles in grouping. For example, suppose the schema in Fig. 11.49 is used to model current marriages in a monogamous society. This schema violates elementarity. It can be split into two fact types: one about the marriage and one about the marriage year. If no other functional roles are played by Person, we might map this to either of the tables schemes shown in Fig. 11.49. For such 1:1 cases, a choice must be made (e.g., by choosing one of the uniqueness constraints as primary).

If only one role in the objectified predicate is functional, it is automatically chosen for the primary key. For example, in a polyandrous society where a man may be married to at most one woman but not vice versa, the husband column would be chosen as the primary key (and wife would no longer be a key).

Despite the provisions of step 0.7, objectified associations without a spanning uniqueness constraint should normally be avoided by remodeling without nesting. Recall Fig. 10.52, which showed two ways to do this for the marriage case under discussion. Figs. 10.54 and 10.55 illustrated how to remodel some other cases without using objectification.

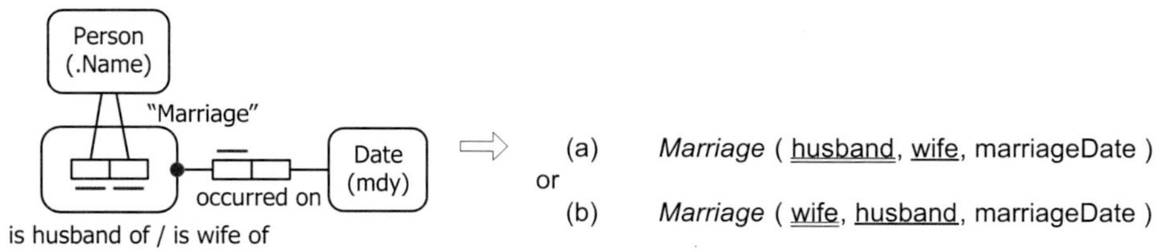

FIG. 11.49 Two possible mappings for an objectified 1:1 association.

Once the columns and constraints are determined in a relational mapping, you may wish to consider changing the *order of columns* within a table. Although the relational model treats column order as irrelevant, in practice, column order can have an impact on both readability and performance. A CASE tool will choose an ordering for you, but you can override this.

Usually, it is best to include primary key columns first, and to group columns together if they are components of a compositely identified object type or have a strong semantic affinity. In such cases, a top-down order is usually preferable (e.g., buildingNr, roomNr instead of roomNr, buildingNr). For such composite cases, placing the attributes with fewer values first can help speed up joins in some cases. A similar comment applies to composite indexes. Some systems also perform better if optional columns are placed last.

Exercise 11.4

1. Map each of the following ORM schemas to a relational schema. Use the horizontal notation. Underline keys, place square brackets around optional columns, and declare any textual constraints. A solid line unary indicates closed world, a simple dashed line indicates open world, and a dashed line around "~" indicates OWN. GraduationLevel codes have the following meaning: STD = Standard, CL = cum laude, MCL = magna cum laude, SCL = summa cum laude.

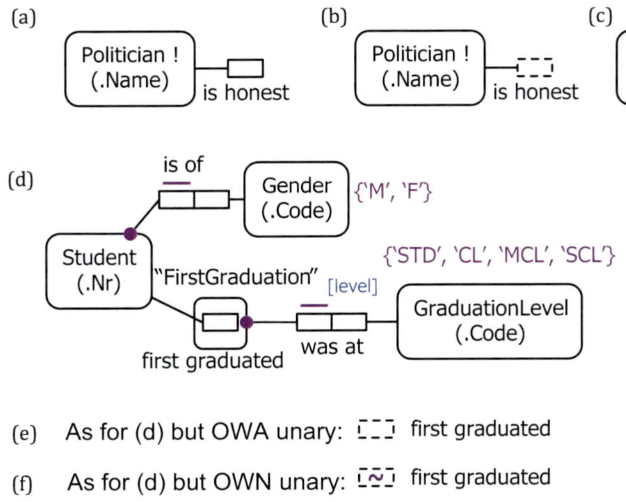

2. Suppose heads of government are identified by the country that they head, and have their salary and country of birth recorded. Schematize this and Rmap it.
3. Refer to the hospital UoD of Exercise 6.6 Question 2. Rmap the conceptual schema for part (a), then discuss any changes in the mapping for parts (b), (c), and (d).
4. Consider a UoD in which people are identified by the combination of their surname, first given name, and (if it exists) second given name. Each person's weight is recorded as well as the sports they play (if any).
   (a) Schematize this UoD.
   (b) Rmap this, using the given identification scheme for persons.
   (c) Introduce personNr as the primary identifier. Rmap the new conceptual schema.
   (d) Instead, concatenate surname and given names to a single name. Rmap this.

(e) Instead, introduce "nil" as a default for no second given name. Rmap this.
(f) Which do you prefer?

5. Consider the functional fact type: Flight(.Nr) departs Airport(.Code). Suppose that facts about the flight departure are modeled by objectifying the Flight departs Airport association as "FlightDeparture" and attaching the fact type FlightDeparture occurred at Time(hm).
   (a) Draw the conceptual schema.
   (b) Unnest the schema.
   (c) Rmap it.

11.5 Summary

Conceptual schemas are normally implemented by mapping them to a logical schema (e.g., relational), refining this, and then generating internal and external schemas (including access rights). Updates and queries may then be performed on the databases and schema(s).

A *relational (database) schema* is a set of table definitions (stored base tables or derived views), constraints, and derivation rules. A *table scheme* is a named set of *attributes* (columns) that draws values from *domains*. Each column, or column set, spanned by a minimal uniqueness constraint is a *candidate key*. Keys are underlined. Each table row is identified by its *primary key* (doubly underlined if another key exists). The *entity integrity* rule requires primary keys to have no nulls. Optional columns allow nulls and are enclosed in square brackets. For example: *Employee*(empNr, empName, address, gender, [phone]).

Mandatory roles map to nonoptional columns, with subset constraints running from any other tables that contain facts about that object type. A *referential integrity constraint* is a subset constraint from a foreign key to some primary key. Subset constraints between tables appear as dotted arrows. Other constraint notations are used.

The relational mapping procedure (Rmap) groups each fact type into a single table, using two basic ideas: each fact type with a compound uniqueness constraint (e.g., an *m:n* association or a ternary association) is mapped to a separate table; fact types with functional roles attached to the same object type are grouped into the same table, keyed on the object type's identifier.

The *basic Rmap procedure* may be summarized as follows. Absorb subtypes into their top supertype. Mentally erase all explicit preferred identification schemes, treating compositely identified object types as "black boxes". Map each fact type with a compound UC to a separate table. Group fact types with functional fact roles attached to the same object type into the same table, keyed on the object type's identifier. Map 1:1 cases to a single table, generally favoring fewer nulls. Map each independent object type with no functional fact roles to a separate table. Unpack each "black box column" into its component attributes. Map all other constraints and derivation rules. Subtype constraints on functional roles map to qualified optional columns, and those on nonfunctional roles map to qualified subset constraints.

If the total population of an object type is included in one table, this is called a total table for the object type. See the *total table procedure* for more details.

Advanced mapping aspects in the *Rmap step 0 procedure* may be summarized as follows. Transform opposite unaries and then map unaries according to their open/closed world semantics. Mentally erase all reference (preferred identification) predicates. Treat compositely identified object types as "black boxes". Indicate any absorption overrides (e.g., separation or partition) for subtypes. Identify any derived fact types that must be stored. Indicate mapping choices for symmetric 1:1 cases. Consider replacing any disjunctive reference schemes by using an artificial or concatenated identifier or mandatory defaults. Indicate mapping choice where required for any objectified associations with no spanning uniqueness constraint.

Chapter Notes

Many database textbooks discuss mapping ER schemas to relational schemas (e.g., Batini et al., 1992) and mapping UML class diagrams to relational schemas (e.g., Blaha and Premerlani, 1998). The mapping extensions to the old ONF algorithm to develop the original version of Rmap were developed jointly by Peter Ritson and Terry Halpin. The version of Rmap discussed in this chapter includes several revisions to cater for new features in ORM 2, such as further options for objectification and open/closed world semantics.

12

Relational Languages

Chapter Outline

12.1 SQL: Relational Algebra

Previous chapters discussed how to design a conceptual schema and map it to a relational schema. This chapter discusses how to implement a relational schema in a relational database management system (DBMS) and then populate its tables with data. To retrieve information from the resulting database, we need to issue queries. In practice, the most popular query languages are *SQL* (popularly but incorrectly called "structured query language") and *QBE* (query by example). Both these languages are based, at least partly, on a formal query language known as *relational algebra*, which is discussed in this section. Studying this algebra first clarifies the basic query operations without getting distracted by the specific syntax of commercial query languages.

With relational algebra under our belt, we can define what is really meant by the term "relational database system". The following sections then cover SQL in some depth, starting with the basics and moving on through intermediate to advanced concepts. Although our focus is on the SQL standard(s), some additional details will be included about commercial dialects of SQL. We begin with relational algebra.

In the relational model of data, all facts are stored in *tables* (or *relations*). New tables may be formed from existing tables by applying relational algebra operations. The tables resulting from these operations may be named and stored, using relational

Information Modeling and Relational Databases. https://doi.org/10.1016/B978-0-443-23790-4.00026-8

assignment. The original relational algebra defined by Codd contained eight *relational operators*: four based on traditional set operations (*union*, *intersection*, *difference*, and *Cartesian product*) and four special operations (*selection*, *projection*, *join*, and *division*). Relational algebra now includes a *rename* operator to define aliases for the table scheme of a query result. Each of the nine relational operators is a table-forming operator on tables. For example, the union of two tables is itself a table.

Relational algebra includes six *comparison operators* (=, <>, <, >, <=, >=). These are proposition-forming operators on terms. For example, "x <> 0" asserts that *x* is not equal to zero. It also includes three *logical operators* (**and**, **or**, **not**). These are proposition-forming operators on propositions (e.g., x > 0 **and** x < 8). Because the algebra does not include arithmetic operators (e.g., +) or functions (e.g., **count**), it is less expressive than SQL. Proposed extensions to the algebra to include these and other operators are ignored here.

Many different notations exist for expressing the relational algebra. A comparison between our notation and a common academic notation is given later. To simplify discussion, we'll often use informal terms instead of the strict relational terminology (e.g., "table" and "row" instead of "relation" and "tuple").

Union, Intersection, and Difference

Two tables are *union compatible* if and only if they have the same number of columns and their corresponding columns are based on the same domain (the columns may have different names). Treating a table as a set of rows, the traditional set operations of union, intersection, and difference may be defined between tables that are union compatible.

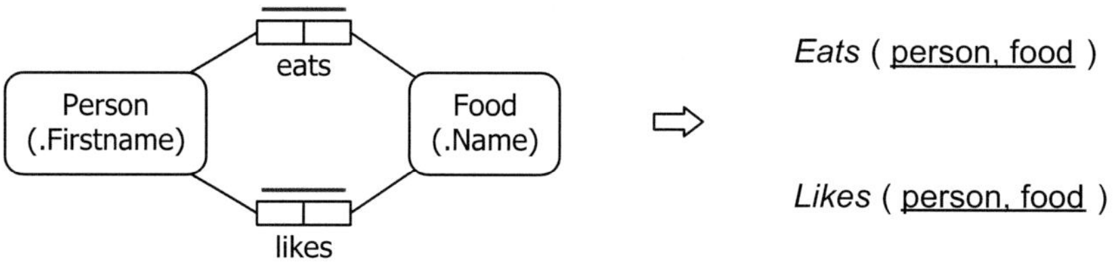

FIG. 12.1 The conceptual schema maps to two tables.

Consider the conceptual schema in Fig. 12.1. For simplicity, assume that in this small universe of discourse (UoD), people are identified by their first name. We record facts about the foods that people eat and about the foods that they like. The relational schema comprises the two tables shown. The populations of the tables may properly overlap or even be disjoint. Hence there are no intertable constraints in the relational schema—this situation is fairly unusual.

Fig. 12.2 includes sample data. The **union** of tables *A* and *B* is the set of all rows belonging to *A* or *B* (or both). We write this as "*A* ∪ *B*" or "*A* **union** *B*". Suppose that we want to pair each person up with foods that they either eat or like. This may be specified simply as the union of the tables, Eats ∪ Likes. Fig. 12.2 includes this query expression and the resulting (unnamed) table, which includes all the rows in the Eats table as well as all the rows in the Likes table. As with any table, duplicate rows are excluded, so rows in both the Eats and the Likes tables appear only once in the result.

Note that "Eats ∪ Likes" is an expression describing how the result may be derived in terms of tables stored in the database (i.e., base tables); it is not a table name. We may refer to a query result as a result table, answer table, or derived table. The order of the rows in the tables has no significance because we are dealing with sets. Actual query languages such as SQL provide ways to display rows in any preferred order.

Eats:

person	*food*
Ann	pizza
Ann	spinach
Bill	pizza

Likes:

person	*food*
Ann	chocolate
Ann	pizza
Bill	pizza
Bill	steak

Eats ∪ Likes ⇨

person	*food*
Ann	pizza
Ann	spinach
Bill	pizza
Ann	chocolate
Bill	steak

FIG. 12.2 The union of two tables is the set of rows that occur in either table (or both).

Eats ∩ Likes ⇨

person	*food*
Ann	pizza
Bill	pizza

FIG. 12.3 The intersection of two tables is the set of rows common to both.

The **intersection** of tables A and B is the set of rows common to A and B. We write this as "$A \cap B$" or "A **intersect** B". For instance, to list facts about foods that people both eat and like, we may specify this as the intersection of our two base tables, Eats ∩ Likes (see Fig. 12.3).

We define the **difference** operation between tables thus. $A - B$ is the set of rows belonging to A but not B. We may also write this as A **minus** B, or A **except** B. For example, the expression Eats –Likes returns details about people who eat foods they don't like (see Fig. 12.4(a)).

As in ordinary set theory, the union operation is commutative (i.e., the order of the operands is not significant). So, given any tables A and B it is true that $A \cup B = B \cup A$. Likewise, the intersection operation is commutative, that is, $A \cap B = B \cap A$. The difference operation, however, is not commutative so it is possible that $A - B$ does not equal $B - A$. For example, Likes – Eats returns facts about who likes a food but doesn't eat it (see Fig. 12.4 (b)). Compare this with the previous result.

(a) Eats - Likes ⇨

person	*food*
Ann	spinach

(b) Likes - Eats ⇨

person	*food*
Ann	chocolate
Bill	steak

FIG. 12.4 (a) $A - B$ is the set of rows in A that are not in B; (b) $B - A$.

Cartesian Product (Unrestricted Join)

In mathematics, the **Cartesian product** of sets A and B is defined as the set of all ordered pairs (x, y) such that x belongs to A and y belongs to B. For example, if $A = \{1,2\}$ and $B = \{3, 4, 5\}$, then the Cartesian product of A and B is {(1,3), (1, 4), (1, 5), (2,3), (2, 4), (2, 5)}. We write this product as "$A \times B$" (read "A **cross** B"). This definition applies also to tables, although the ordered pairs of rows (x, y) are considered to be single rows of the product table, which inherits the corresponding column names.

Thus, if A and B are tables, $A \times B$ is formed by *pairing each row of A with each row of B*. Here "pairing with" means "prepending to". The expression "$A \times B$" may also be written as "A **times** B". The Cartesian product is also called the "cross join" or "unrestricted join".

Because $A \times B$ pairs each row of A with all rows of B, if A has n rows and B has m rows, then the table $A \times B$ has $n \times m$ rows. So, *the number of rows in A × B is the product of the number of rows in A and B*. Because the paired rows are concatenated into single rows, *the number of columns in A × B is the sum of the number of columns in A and B.*

Let's look at a simple example. In Fig. 12.5, A is a 3×2 table (i.e., 3 rows $\times$ 2 columns) and B is a 2×1 table. So, $A \times B$ has six (3×2) rows of three ($2 + 1$) columns. The header rows that contain the column names *a1, a2, b1* are not counted in this calculation. In this example, dotted lines have been ruled between rows to aid readability. Note that the Cartesian product $A \times B$ exists for any tables A and B; it is not necessary that A and B be union compatible.

If we define Cartesian products in terms of tuples treated as ordered sequences of values, the $\times$ operator is not commutative (i.e., $A \times B$ need not equal $B \times A$). This is the usual treatment in mathematics. With relational databases, however, each column has a unique name, so a tuple may instead be treated as an (unordered) set of attribute-value pairs (each value in a row is associated with its column name). In this sense, columns are said to be "*ordered by name*". With this view of tuples, the $\times$ operator is commutative.

To clarify this, compute $B \times A$ for the example in Fig. 12.5. Although the resulting table has its columns displayed in the order *b1, a1, a2,* and the values are ordered accordingly, the information is the same as for $A \times B$ because each value can be linked to its column name. For example, the first row of $B \times A$ may be represented as {(b1, 7), (a1, 1), (a2, 2)}, which equals {(a1, 1), (a2, 2), (b1, 7)}, which represents the first row of $A \times B$. Once columns are uniquely named, their physical order has no semantic significance.

As in standard set theory, the operations of union, intersection, and Cartesian product are *associative*. For instance, $A \cup (B \cup C) = (A \cup B) \cup C$. Because no ambiguity results, parentheses may be dropped in such cases. For example, the expression "$A \cup B \cup C$" is permitted. Of the four table operations considered so far, difference is the only one that is not associative; that is, it is possible that $A - (B - C)$ is not equal to $(A - B) - C$. As an exercise, use Venn diagrams to check these claims.

A:

a1	*a2*
1	2
3	4
5	6

B:

b1
7
8

A × B ⇨

a1	*a2*	*b1*
1	2	7
1	2	8
3	4	7
3	4	8
5	6	7
5	6	8

FIG. 12.5 $A \times B$ pairs each row of A with each row of B.

Person:

firstname	*gender*
Fred	M
Sue	F

Person × Person ⇨

firstname	*gender*	*firstname*	*gender*
Fred	M	Fred	M
Fred	M	Sue	F
Sue	F	Fred	M
Sue	F	Sue	F

FIG. 12.6 What is wrong with this Cartesian product?

Recall that column names of a table must be unique. If A and B have no column names in common, then $A \times B$ uses the simple column names of A and B. If A and B are different but have some common column names, the corresponding columns in $A \times B$ are uniquely referenced by using *qualified names* (with the prefix "*A*". or "*B*".). However, what if A and B are the same (i.e., we wish to form the Cartesian product of a table with itself)?

Consider the Person × Person table in Fig. 12.6. This breaks the rule that column names in a table must be unique. In this case, simply using the qualified column names will not solve the problem either because we would have two columns named "Person.firstname" and two columns named "Person.gender".

This problem is overcome by defining **aliases** for tables. For example, we could use "Person2" as an alternative name for the table and form the product as shown in Fig. 12.7. If desired, we could use two aliases, "Person1" and "Person2", to specify the product Person1 × Person2. The need to multiply a table by itself may arise if we have to compare values on different rows. Aliases can also be used simply to save writing (by introducing shorter names). As discussed later, our **define alias** notation is an alternative syntax for trivial cases of the **rename** operator in relational algebra.

define alias Person2 **for** Person

Person × Person2 ⇨

Person.firstname	*Person.gender*	*Person2.firstname*	*Person2.gender*
Fred	M	Fred	M
Fred	M	Sue	F
Sue	F	Fred	M
Sue	F	Sue	F

FIG. 12.7 Table aliases are used to multiply a table by itself.

Relational Selection

We now consider the four table operations introduced by Codd. The first of these is selection (or *restriction*). Don't confuse this with SQL's **select** command. In relational algebra, the selection operation chooses just those rows that satisfy a specified condition. Visually, we can picture the operation as returning a *horizontal* or *row* subset because it returns zero or more rows from the original table (see Fig. 12.8).

The selection operation may be specified using an expression of the form *T* **where** *c*. Here, *T* denotes a *table expression* (i.e., an expression whose value is a table) and *c* denotes a *condition*. The selection operation returns just those rows where the condition evaluates to True, filtering out rows where the condition evaluates to False or Unknown. The "**where** *c*" part is called a *where* clause.

This notation encourages top-down thinking (find the relevant tables before worrying about the rows) and agrees with SQL syntax. The alternative notation $\sigma_c(T)$ is often used in academic journals. The "σ" is sigma, the Greek "*s*" (which is the first letter of "selection").

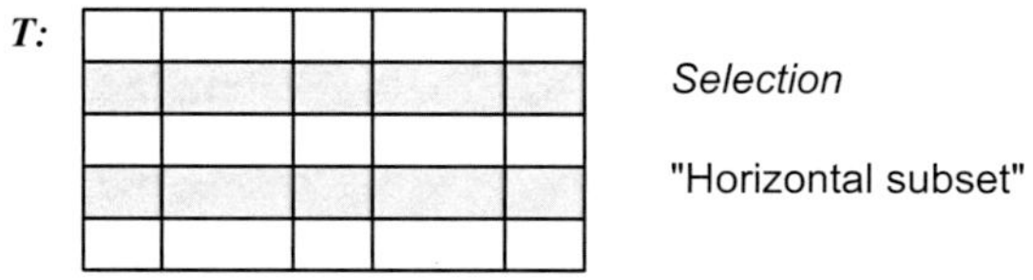

T **where** *c* or $\sigma_c(T)$

Select the rows of *T* that satisfy the condition *c*

FIG. 12.8 The selection operation in relational algebra.

In Fig. 12.9, the Player table stores details about tennis players. To list details about just the male tennis players, we could formulate this selection as indicated, giving the result shown. Here, the selection condition is just the equality: gender = 'M'. The use of single quotes with "M" indicates that this value is literally the character 'M'. Single quotes are also used when referencing character string values such as 'David'. Numeric values should not be quoted. Although the table resulting from the selection is unnamed, its columns have the same names as the original.

Player:

name	*gender*	*height*
David	M	172
Norma	F	170
Selena	F	165
Terry	M	178

Player **where** gender = 'M' ⇨

name	*gender*	*height*
David	M	172
Terry	M	178

FIG. 12.9 The selection operation picks those rows that satisfy the condition.

Player **where** (gender = 'F') **and** (height < 170)

⇨

name	*gender*	*height*
Selena	F	165

FIG. 12.10 Are the parentheses needed in this query?

The condition *c* may contain any of six *comparison operators*: = (equals); <> (is not equal to); < (is less than); > (is greater than); <= (is less than or equal to); and >= (is greater than or equal to). Sometimes, the result of a table operation is an empty table (cf. the null set). For instance, no rows satisfy the condition in Player **where** height > 180.

If conditions do not involve comparisons with null values, they are Boolean expressions–they evaluate to True or False. If conditions do include comparisons with nulls, a three-valued logic is used instead, so that the condition may evaluate to True, False, or Unknown. This is also what happens in SQL. In either case, the selection operation returns just those rows where the condition is true. Any row where the condition evaluates to false or unknown is filtered out.

In addition to comparison operators, conditions may include three *logical operators*: **and**, **or**, **not**. As usual, the **or** operator is inclusive. As an exercise, formulate a query to list details of those females who are shorter than 170 cm. Refer to the Player table and see if you can state the result, and then check your answer with Fig. 12.10. To aid readability, reserved words are in bold.

The query in Fig. 12.10 used parentheses to clarify the order in which operations are to be carried out. With complicated expressions, however, large numbers of parentheses can make things look very messy (cf. LISP). Unlike several programming languages (e.g., Pascal), most query languages (e.g., SQL) give *comparison operators higher priority than logical operators*. So, the operators =, <>, <, >, <=, >= are given precedence over **not**, **and**, and **or** for evaluation purposes. We make the same choice in relational algebra because this reduces the number of parentheses needed for complex expressions. With this understood, the query in Fig. 12.10 may be rewritten more concisely as:

Player **where** gender = 'F' **and** height < 170

To further reduce the need for parentheses, the algebra adopts the logical operator priority convention commonly used in computing languages, such as Pascal and SQL. First evaluate **not**, then **and**, and then **or**. Table 12.1 shows the evaluation order for both comparison and logical operators. This operator precedence may be overridden by the use of parentheses.

For example, suppose we want details about males who are taller than 175 cm or shorter than 170 cm. The query in Fig. 12.11(a) looks like it should work for this, but the wrong result is obtained. Why?

Table 12.1 Priority convention for the operators (1 = first).

Priority	*Operators*
1	=, <>, <, >, <=, >=
2	**not**
3	**and**
4	**or**

(a) Player **where** gender = 'M'
and height > 175 **or** height < 170 ⇨

name	*gender*	*height*
Selena	F	165
Terry	M	178

(b) Player **where** gender = 'M'
and (height > 175 **or** height < 170) ⇨

name	*gender*	*height*
Terry	M	178

FIG. 12.11 Unless parentheses are used, **and** is evaluated before **or**.

Because the **and** operator is evaluated before the **or**, this query is interpreted as Player **where** (gender='M' **and** height>175) **or** height<170. Notice that line breaks have no significance to the meaning of the query. So, the result includes details about anybody shorter than 170 cm (including females). The query in Fig. 12.11(b) corrects this by including parentheses around the height disjunction to evaluate the **or** operator before the **and** operator.

To obtain details about players who are either males over 175 cm or females at least 170 cm tall, either of the queries shown in Fig. 12.12 may be used. In the query in Fig. 12.12(b), parentheses are redundant because the **and** operators are evaluated before the **or** operator anyway. If in doubt about whether parentheses are needed, put them in.

(a) Player **where** gender = 'M' **and** height > 175
or gender = 'F' **and** height >= 170 ⇨

(b) Player **where** (gender = 'M' **and** height > 175) ⇨
or (gender = 'F' **and** height >= 170)

name	*gender*	*height*
Terry	M	178
Norma	F	170

FIG. 12.12 Two equivalent queries; if in doubt, include parentheses.

Fig. 12.13 shows three equivalent queries for listing information on players who are neither taller than 172 cm nor shorter than 168 cm. Operator precedence ensures that query (a) in Fig. 12.13 means Player **where** (**not** (height>172)) **and** (**not** (height<168)). Note that the six comparison operators may be grouped into three pairs of opposites: =, <>; <, >=; and >, <=. So, the query in Fig. 12.13(a) may be replaced by the shorter query in Fig. 12.13(b).

(a) Player **where not** height > 172 **and not** height < 168 ⇨

(b) Player **where** height <= 172 **and** height >= 168 ⇨

(c) Player **where not** (height > 172 **or** height < 168) ⇨

name	*gender*	*height*
David	M	172
Norma	F	170

FIG. 12.13 Three equivalent queries.

De Morgan's laws, named after the famous logician Augustus De Morgan, are shown below. Here, *p* and *q* denote any proposition or logical condition.

not (p **and** q) ≡ **not** p **or not** q
not (p **or** q) ≡ **not** p **and not** q

These laws hold in both two-valued and three-valued logic, so they are safe to use with null values. Using the second of these laws, it is easy to see that the query in Fig. 12.13(c) is equivalent to the query in Fig. 12.13(a).

Relational Projection

The next table operation, known as **projection**, was introduced in Chapter 4. This operation involves *choosing one or more columns from a table* and then *eliminating any duplicate rows that might result.* It is said to produce a *vertical subset* or *column subset.* Fig. 12.14 illustrates a projection on two columns.

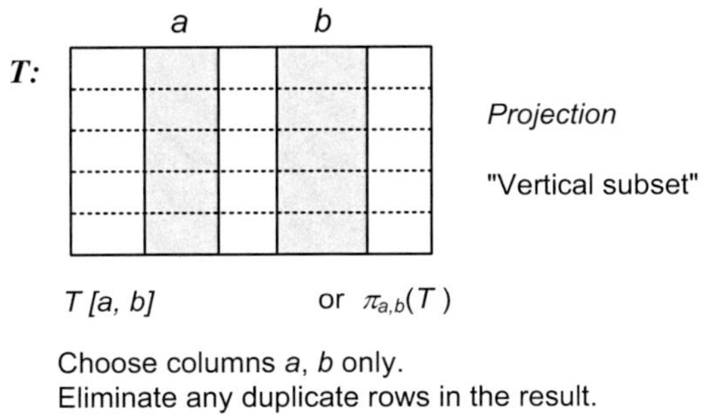

FIG. 12.14 The projection operation in relational algebra.

We may represent the projection operation as *T [a,b, ...]* where *T* is a table expression and *a, b, ...* are the names of the required columns (this column list is called the *projection list*). To delimit the projection list, we use square brackets rather than parentheses because the latter have other uses in queries (e.g., to change the evaluation order of operations). Italicizing the square brackets "*[ ]*" helps distinguish them from the use of "[ ]" to delimit optional items, or bags, and to indicate an extra task (duplicate elimination) once the columns are chosen.

The alternative notation $\pi_{a,b...}(T)$ is common in academic journals. The "π" symbol is pi, the Greek "*p*" (the first letter of "projection"). Like SQL, this notation lists the required columns before the tables. Our preferred notation instead encourages top-down thinking by identifying the relevant tables before listing the columns.

Player:

name	*gender*	*height*
David	M	172
Norma	F	170
Selena	F	165
Terry	M	178

Player *[*name, gender*]* ⇨

name	*gender*
David	M
Norma	F
Selena	F
Terry	M

Player *[*gender*]* ⇨

gender
M
F

FIG. 12.15 Projection involves picking the columns and removing any duplicates.

Fig. 12.15 gives two examples based on the table *Player* (name, gender, height). When projecting on a single base table, if the chosen columns include a candidate key then no duplicates rows can result. If not, duplicates may arise. For example, choosing just the gender column gives the bag ["M", "F", "F", "M"]. To complete the projection, the duplicate values are eliminated, giving the set {"M", "F"}. If we project on all columns, we end up with the same table. For instance, Player *[*name, gender, height*]* is the same table as Player.

The same column must not be mentioned twice in a projection. For example, Player *[*name, name*]* is illegal (why?). Projection may also be used to display columns in a different order. For example, Player *[*height, gender, name*]* reverses the order. Because unique column names are listed in the result, this doesn't change the meaning.

If the relational schema is fully normalized, it is fairly unusual to have base tables that are union compatible (the Eats and Likes tables considered earlier being an exception). So, the ∪, ∩, and − operations tend to be used almost exclusively with result tables that have become union compatible because of projections.

We now look at some examples that are a little more difficult. To facilitate a discussion of a step-by-step approach to such cases, we use the **relational assignment** operation. Because their contents may vary, table schemes may be regarded as *variables*. In fact, tables are the only kind of variable allowed in relational algebra. The relational assignment operation, which is strictly separate from the algebra itself, is similar to that in programming languages. Using the symbol ":=" for "becomes" or "is assigned the value of", we may write assignment statements of the form.

table variable := table expression

This is an instruction to first evaluate the expression on the right, and then place its value in the variable named on the left. If this variable had a previous value, the old value would simply be replaced by the new value. For example, if *X*, *A*, and *B* are table values, then the statement "*X* := *A* := *B*" means *X* is assigned the value of *A* − *B*. Some other notations use "=" or "←" instead of ":=".

Once named in a relational assignment, result tables may be used just like base tables in later expressions. For example, to store details about tall male players in a table called "TallMalePlayer", we could make the following three assignments. To list these details, we may now simply use the query TallMalePlayer.

MalePlayer := Player **where** gender = 'M'
TallPlayer := Player **where** height > 175
TallMalePlayer := MalePlayer ∩ TallPlayer

As a more difficult example, consider the UoD of Fig. 12.16. How may we formulate the following query in relational algebra: Which non-European countries speak French? Try this yourself before reading on.

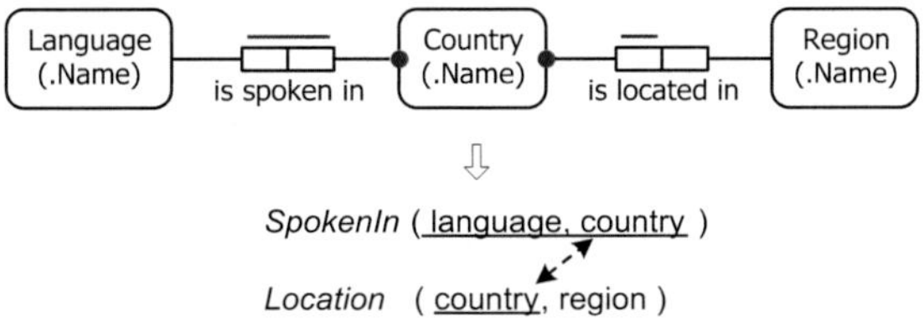

SpokenIn:

language	*country*
Dutch	Belgium
Dutch	Netherlands
English	Australia
English	Canada
French	Belgium
French	Canada
French	France

Location:

country	*region*
Australia	Australasia
Belgium	Europe
Canada	North America
France	Europe
Netherlands	Europe

FIG. 12.16 Conceptual and relational schemas, with sample population.

We could formulate the query in steps, using intermediate tables on the way to obtain our final result table. For example:

NonEuroCountry := (Location **where** region <> 'Europe') *[country]*
FrenchSpeakingCountry := (SpokenIn **where** language = 'French') *[country]*
FrenchSpeakingNonEuroCountry := NonEuroCountry ∩ FrenchSpeakingCountry

Here, our first step was to find the non-European countries: {Australia, Canada}. Then we found the French-speaking countries: {Belgium, Canada, France}. Finally, we took the intersection of these, giving the result: {Canada}.

Setting queries out this way is known as *stepwise formulation*. Doing things in stages can sometimes make it easier to formulate difficult queries. However, it saves some writing to express a query by means of a single expression. For instance, the three assignment statements just considered can be replaced by the following:

FrenchSpeakingNonEuroCountry :=
(Location **where** region <> 'Europe') *[country]* ∩
(SpokenIn **where** language = 'French') *[country]*

The lengthy table expression assigned to FrenchSpeakingNonEuroCountry is said to be a *nested formulation* because it nests one or more queries inside another. For complex queries, we might perform the stepwise formulation in our heads or scribble it down somewhere, and then convert this into the nested formulation. In general, any information capable of being extracted by a series of relational algebra queries can be specified in a single relational algebra query.

Note that projection does not distribute over intersection. In other words, given tables A and B, and a projection list p, it is possible that $(A \cap B)[p] \neq A[p] \cap B[p]$. For instance, the following query is not equivalent to the previous one (why not?).

((Location **where** region <> 'Europe') ∩ (SpokenIn **where** language = 'French')) *[country]*

This query is illegal because the table operands of ∩ are not compatible**.** Even if A and B are compatible, projection need not distribute over ∩ or over −. If A and B are compatible, projection does distribute over ∪, that is, $(A \cup B)[p] = A[p] \cup B[p]$. As an exercise, you may wish to prove these results.

Relational Joins

Object joins at the conceptual level were introduced in Section 4.4 to help clarify the meaning of external uniqueness constraints. We now consider the relational join operation between two tables (or two occurrences of the same table), which compares attribute values from the tables, using the comparison operators (=, <, >, <>, <=, >=).

There are several kinds of join operations, and we discuss only some of these here. Columns being compared in any join operation must be defined on the same domain; they do not need to have the same name. Where it is necessary to distinguish between join columns with the same local name, we use their fully qualified names.

Let Θ (theta) denote any comparison operator (=, <, etc.). Then, the **Θ-join** of tables A and B on attributes a of A and b of B equals the Cartesian product $A \times B$, restricted to those rows where $A.a \, \Theta \, B.b$. We write this as shown here. An alternative, academic notation is shown in braces.

$(A \times B)$ **where** c {or $A \bowtie_c B$}

The condition *c* used to express this comparison of attributes between tables is called the *join condition*. The join condition may be composite (e.g., a1 < b1 **and** a2 < b2). Because of the Cartesian product, the resulting table has a number of columns equal to the sum of the number of columns in *A* and *B*, but because of the selection operation, it typically has far fewer rows than the product of the numbers of rows of the joined tables.

With most joins, the comparison operator used is =. The Θ-join is then called an **equi-join**. Thus, the equijoin of *A* and *B* equating values in column *a* of *A* with column *b* of *B* is $A \times B$ restricted to the rows where $A.a = B.b$. If the join column names are different in each table, there is no need to qualify them with table names. We write this as shown here. In general, the join condition may contain many equalities.

($A \times B$) **where** $A.a = B.b$
or
($A \times B$) **where** $a = b$ { if *a*, *b* occur in only one of *A*, *B* }

Let's consider an example. Fig. 12.17 includes a simple conceptual schema and its corresponding relational schema.

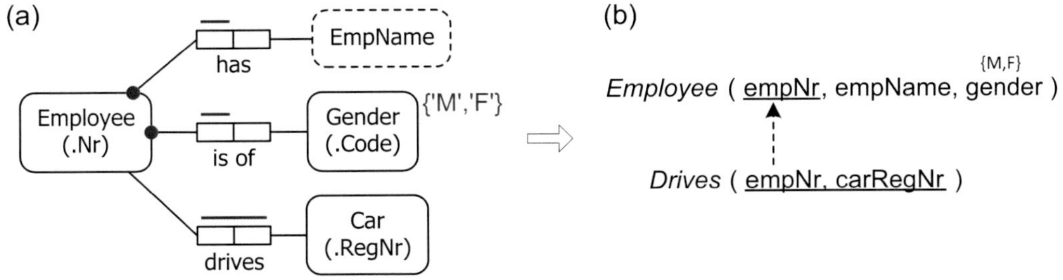

FIG. 12.17 The ORM schema (a) maps to the relational schema (b).

A sample population is given in Fig. 12.18.

Employee:

empNr	*empName*	*gender*
001	Hagar, T	M
002	Wong, S	F
003	Jones, E	F
004	Mifune, K	M

Drives:

empNr	*carRegNr*
001	ABC123
003	ABC123
003	TAH007

FIG. 12.18 A sample population for the schema in Fig. 12.17.

Before formulating queries, *familiarize yourself with the structure and contents of the database*. From the schema, we see that employees are identified by their employee number, and their name and gender must also be recorded. An employee may drive many cars, and the same car may be driven by many employees. The population is not significant because it does not include instances where the same name applies to more than one employee. However, it does illustrate that driving cars is optional (employees 002 and 004), that an employee may drive more than one car (employee 003), and that the same car may be driven by more than one employee (car ABC123). Now consider the following query: "List the employee number, name, and cars for each employee who drives a car".

We could specify this by the equijoin in Fig. 12.19, which joins the tables by equating their empNr attributes. To help explain this, dotted lines have been added between the base tables to indicate which rows match on their empNr value. Only the first and third rows from the Employee table are able to find a match. The third row (003) finds two matches. Check the result for yourself.

Because these join columns have the same local name "empNr", qualified names are used to distinguish the columns in the query and the join result. The join columns don't have to have the same (unqualified) name, but must belong to the same domain. For example, if the employee number column in the Drives table was named "empId", the join condition could be specified as "empNr = empId".

Employee:

empNr	*empName*	*gender*
001	Hagar, T	M
002	Wong, S	F
003	Jones, E	F
004	Mifune, K	M

Drives:

empNr	*carRegNr*
001	ABC123
003	ABC123
003	TAH007

(Employee Drives)
where Employee.empNr = Drives.empNr

⇨

Employee.empNr	*empName*	*gender*	*Drives.empNr*	*carRegNr*
001	Hagar, T	M	001	ABC123
003	Jones, E	F	003	ABC123
003	Jones, E	F	003	TAH007

FIG. 12.19 An equijoin.

An equijoin contains two matching columns resulting from each join attribute (note the two empNr columns). If these columns refer to the same thing in the UoD (and they typically do), then one of these columns is redundant. In this case, we lose no information if we *delete one of these matching columns* (by performing a projection on all but the column to be deleted). This is done in Fig. 12.20.

(Employee × Drives)
where Employee.empNr = Drives.empNr
*[*Employee.empNr, empName, gender, carRegNr*]*

⇨

Employee.empNr	*empName*	*gender*	*carRegNr*
001	Hagar, T	M	ABC123
003	Jones, E	F	ABC123
003	Jones, E	F	TAH007

FIG. 12.20 The duplicate column is removed by projection on the equijoin.

If the *columns used for joining have the same name in both tables*, then the unqualified name is used in the join result. The resulting table is then said to be the **natural inner join** of the original tables. Because "inner" is assumed by default, the **natural inner join** may be expressed simply as **natural join** (outer joins are discussed later). This is by far the most common join operation in practice. The natural join of tables *A* and *B* may be written as A ⋈ *B*, or in words as "*A* **natural join** *B*". This may be summarized as follows.

To compute *A* ⋈ *B*: Form *A* × *B*
For each column name *c* that occurs in both *A* and *B*
  Apply the restriction *A.c* = *B.c*
  Remove *B.c*
  Rename *A.c* to *c*

There is no need to name the join columns because these are the ones with the same names in both tables (and of course these must be based on the same domain). To help remember the bow-tie "⋈" notation, note that "⋈" looks like a cross "×" with two vertical lines added, suggesting that a natural join is a Cartesian product plus two other operations (selection of rows with equal values for the common columns, followed by projection to delete redundant columns). Fig. 12.21 shows the natural join for our driving example. Note that the empNr column appears just once and is unqualified.

Employee ⋈ Drives ⇨

empNr	*empName*	*gender*	*carRegNr*
001	Hagar, T	M	ABC123
003	Jones, E	F	ABC123
003	Jones, E	F	TAH007

FIG. 12.21 A natural join.

If the tables have no column names in common, then the natural join is simply the Cartesian product. Like the Cartesian product, natural join is associative: (*A* ⋈ *B*) ⋈ *C* = A ⋈ (*B* ⋈ *C*). So, expressions of the form *A* ⋈ *B* ⋈ *C* are unambiguous.

Tables being joined may have zero, one, or more common columns. In any case, the natural join is restricted to those rows where all the common attributes have the same value in both tables. The number of columns in *A* ⋈ *B* equals the sum of the number of columns in *A* and *B*, minus the number of columns common to both.

To illustrate a natural join on many attributes, consider the UoD in Fig. 12.22. First examine the conceptual schema. Within a bank branch, an account may be identified by its local account number. But different accounts in different branches may have the same local account number. So, globally, the bank identifies accounts by a combination of branch and account number. Clients are identified by a global client number and have a name (not necessarily unique). Not all clients need an account. The relational schema is included in Fig. 12.22.

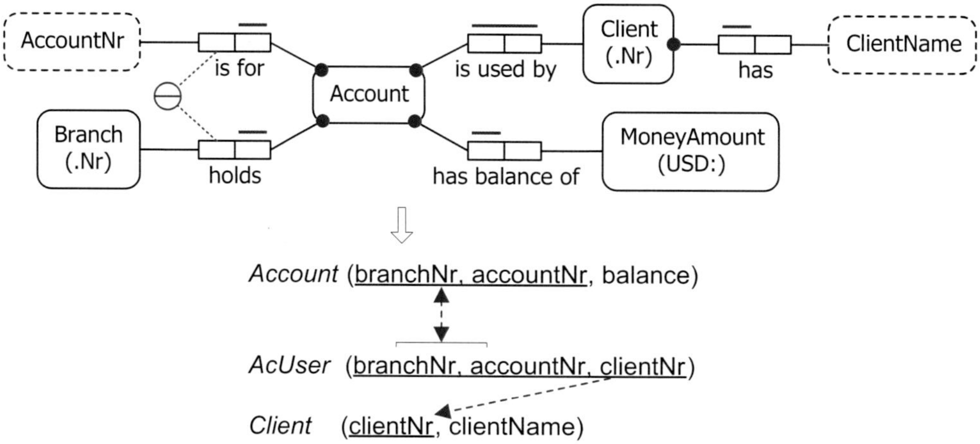

FIG. 12.22 Account has a composite identification scheme.

Notice the subset and equality constraints between the tables shown in Fig. 12.22. Joins are usually, although not always, made across such constraint links. Simple joins may be performed on clientNr, and composite joins on branchNr–accountNr pairs (which identify accounts). Fig. 12.23 shows sample data.

Account:

branchNr	*accountNr*	*balance*
10	54	3000.00
10	77	500.55
23	54	1000.00

AcUser:

branchNr	*accountNr*	*clientNr*
10	54	1001
10	54	1002
10	77	2013
23	54	7654

Client:

clientNr	*clientName*
1001	Jones, ME
1002	Jones, TA
2013	Jones, ME
7654	Seldon, H
8005	Shankara, TA

FIG. 12.23 Sample population for the bank schema.

The two queries in Fig. 12.24 use natural joins to display users and balances of accounts. The first query matches accounts by joining on both branchNr and accountNr. To add client names, the second query also matches clients by joining on clientNr.

In rare cases, comparison operators other than equality are used in joins. As a simple example, consider the Drinker and Smoker tables in Fig. 12.25. These might result from a decision to map subtypes of Patient to separate tables.

Suppose we wanted a list of drinker, smoker pairs, where the drinker and smoker are distinct persons. This can be formulated by a <>*-join* as shown. Here, the comparison operator is "<>". Notice that because drinkers and smokers overlap, some patient doubles may appear twice (in different order). Similarly, <-joins, >-joins, <=-joins and >=-joins may be defined.

(a) Account ⋈ AcUser

⇨

branchNr	*accountNr*	*balance*	*clientNr*
10	54	3000.00	1001
10	54	3000.00	1002
10	77	500.55	2013
23	54	1000.00	7654

(b) Account ⋈ AcUser ⋈ Client

⇨

branchNr	*accountNr*	*balance*	*clientNr*	*clientName*
10	54	3000.00	1001	Jones, ME
10	54	3000.00	1002	Jones, TA
10	77	500.55	2013	Jones, ME
23	54	1000.00	7654	Seldon, H

FIG. 12.24 Two queries using natural joins.

Drinker:

patient	*liver*
Bloggs, F	OK
Smith, S	poor
Stoned, IM	bad

Smoker:

patient	*lungs*
Bloggs, F	bad
Stoned, IM	bad

(Drinker × Smoker)
where Drinker.patient <> Smoker.patient
*[*Drinker.patient, Smoker.patient*]* ⇨

Drinker.patient	*Smoker.patient*
Bloggs, F	Stoned, IM
Smith, S	Bloggs, F
Smith, S	Stoned, IM
Stoned, IM	Bloggs, F

FIG. 12.25 Example of a <>-join.

The kinds of joins discussed so far are collectively known as *inner joins*. We can also have *outer joins*, which are used to include various cases with null values. An outer join is basically an inner join, with extra rows padded with nulls when the join condition is not satisfied. For example, given the data population shown in Fig. 12.23, the result of Client **left outer join** AcUser includes a row to indicate that client 8005 has the name "Shankara TA" but uses no account (branchNr and accountNr are null on this row). The left outer join includes all the clients from the left-hand table (i.e., the table on the left of the join operator), whether or not they are listed in the right-hand table. Outer joins are discussed in detail later for SQL.

Relational Division

The next relational operation we consider is relational **division**. Division of table *A* by table *B* is meaningful only if *A* has more columns than *B*. Let's assume that table *A* has two sets of columns, *X* and *Y*, and table *B* has a set of columns *Y*. In other words, *X* is the set of columns in *A* that are not present in *B*. We'll assume an ordered relation between the *Y* columns in *A* and *B* so that we know which column in *B* corresponds to which column in *A*.

The result of $A \div B$ is formed by restricting the result to those *X* entries in *A* whose associated *Y* entries in *A* include *all* the row entries of *B*. The expression $A \div B$ may also be written as *A* **divide-by** *B*. Fig. 12.26 shows a trivial example, where the attribute domains are denoted by *X1*, *X2*, *Y1*.

A:

X1	X2	Y1
a1	*a2*	*a3*
1	2	50
1	2	80
1	2	90
3	4	50

B:

Y1
b
50
80

A ÷ B ⇨

a1	*a2*
1	2

FIG. 12.26 A simple example used to explain relational division.

Although not used very often, the division operation can be useful in listing rows that are associated with at least *all* rows of another table expression (e.g., who can supply all the items on our stock list?).

As a practical example, suppose the table *Speaks* (country, language) stores facts about which countries speak (i.e., use) which languages. A sample population is shown in the large table within Fig. 12.27. Now consider the query: Which countries speak *all* the languages spoken in Canada? To answer this, we first find all the languages spoken in Canada, using the expression Speaks **where** country = 'Canada' [language]. This returns the table {English, French}. We then divide the Speaks table by this result, as shown, to obtain the final answer {Canada, Dominica}.

Speaks:

country	*language*
Australia	English
Belgium	Dutch
Belgium	French
Canada	English
Canada	French
Cuba	Spanish
Dominica	English
Dominica	French

÷

language
English
French

⇨

country
Canada
Dominica

Which countries speak
***all** the languages spoken in Canada?*

FIG. 12.27 A practical example of relational division.

Renaming

Recall Fig. 12.7, where we defined an alias for a table name. Relational assignment can also be used to provide a new table name, but it cannot be used to *change the attribute names*. For that, we need the ***rename*** operator. A relational algebra expression may include any of the relational operations discussed earlier (union, projection, selection, join, etc.) and is essentially a query that returns a table. In general, *the rename operator applied to a* *relational expression returns a table with the same rows* (same data population) *but different names for the table or columns.*

The formal syntax for the rename operator uses the Greek letter "ρ" (*rho*). If *Rexp* is a relational expression that returns a table with the table scheme $R(a_1, \ldots, a_n)$, the following expression returns a table with the table scheme $S(b_1, \ldots, b_n)$. The expression below it uses our alternative syntax.

$\rho S(b_1, \ldots, b_n)$ *Rexp*

define $S(b_1, \ldots, b_n)$ **as** *Rexp*

If the attribute list $(b_1, \ldots, b_n)$ is omitted, the operation simply provides a new table name. In that case, if *Rexp* is simply a table name, it is more natural to use the define alias notation (e.g., **define alias** Person2 **for** Person), as in Fig. 12.7.

As a simple example, consider the table schemes *Person* (personName, age, genderCode) and *Drives* (personName, carRegNr) where Drives.personName is a foreign key to the Person table. The following declaration may be used to define a WomanDriver table using simply "name" instead of "personName" for the person name attribute.

define WomanDriver (name, carRegNr) **as**
Employee ⋈ Drives
where age >= 18 **and** genderCode = 'F'
*[*personName, carRegNr*]*

Once defined, the WomanDriver table may be used in subsequent queries, just like any other table. This is similar to a *view* in SQL (see later chapter).

Query Strategies

That completes all the main operators in relational algebra. Table 12.2 indicates the operator precedence adopted in this book. The comparison operators have top (and equal) priority, so are evaluated first. The logical operators are evaluated next (first **not**, then **and**, then **or**). Then the relational selection and projection operations are evaluated (equal priority). Finally, the other six relational operators (union, intersection, difference, Cartesian product, natural join, and division) are evaluated (equal priority). Operators with equal priority are evaluated left to right. Expressions in parentheses are evaluated first. Some systems adopt a different priority convention for the relational operators.

Table 12.2 Operator priority (1 = highest).

Priority	*Operator type*	*Operator(s)*
1	Comparison	=, <>, <, >, <=, >=
2	Logical	**not**
3		**and**
4		**or**
5	Relational	selection (... **where** ...), projection ...*[* a, b*]*
6		∩ ∪ − × ⋈ ÷

To help formulate queries in relational algebra, the following *query strategies* are useful:

- Phrase the query in natural language, and understand what it means.
- Ask which tables hold the information.
- If you have table data, answer the query yourself and then retrace your mental steps.
- Divide the query up into steps or subproblems.
- If the result columns come from different tables, declare joins between the tables.
- If you need to relate two rows of the same table, use an alias or rename definition to perform a self-join.
- To change any table or column names in a result, use a rename definition.

Our first move is to formulate the query in natural language, ensuring that we understand what the query means. Next, determine what tables hold the information needed to answer our query. The information might be in a single table or be spread over two or more tables. If the columns to be listed come from different tables (or different copies of the same table), then we must normally specify joins between these tables. These joins might be of any kind: inner or outer joins, natural or theta joins, or cross joins (Cartesian products).

Apart from some cases involving ∪, ∩, or -, relational algebra requires joins whenever different tables must be accessed to determine the result, even if the result columns come from the same table. SQL introduced subqueries, which can be used instead of joins when the result columns come from the same table.

Let's look at an example using our bank account database. To reduce page turning, the database is reproduced in Fig. 12.28. Now consider the query: Which clients have an account with a balance of more than $700? Before formulating this in relational algebra, we should ensure that we understand what the natural language query means. In some cases, we may need to clarify the meaning by asking the person who originally posed the query. For example, do we want just the client number for the relevant clients or do we want their name as well? Let's assume that we want the client name as well.

Account:

branchNr	*accountNr*	*balance*
10	54	3000.00
10	77	500.55
23	54	1000.00

AcUser:

branchNr	*accountNr*	*clientNr*
10	54	1001
10	54	1002
10	77	2013
23	54	7654

Client:

clientNr	*clientName*
1001	Jones, ME
1002	Jones, TA
2013	Jones, ME
7654	Seldon, H
8005	Shankara, TA

Which clients have an account with a balance of more than $700?

FIG. 12.28 How can this query be formulated in relational algebra?

Looking at the tables, we can now see that all three tables are needed. The Account table holds the balances, the Client table holds the client names, and the AcUser table indicates who uses what account.

Because we have sample data, we can first try to answer the English query, examine what we did, and finally try to express this in the formal language of relational algebra. If we were answering the request ourselves, we might go to the Account table and select just the rows where the balance exceeded $700. This yields the account (10, 54) and the account (23, 54). We might then look at the AcUser table to see who uses these accounts. This yields the clients numbered 1001, 1002, and 7654. Now we look to the Client table to find out their names ("Jones, ME", "Jones, TA"' "Seldon, H"). So, our answer is {(1001, "Jones, ME"), (1002, "Jones, TA"), (7654, "Seldon, H")}.

Retracing and generalizing our steps, we see that we joined the Account and AcUser tables by matching the account (branchNr and accountNr), and we linked with the Client table by joining on clientNr. Because these attributes have the same names over the tables, we can use natural joins. We may set this out in relational algebra as:

(Account **where** balance > 700) *[* branchNr, accountNr *]*
⋈ AcUser
⋈ Client *[* clientNr, clientName *]*

This isn't the only way we could express the query. For instance, using a top-down approach, both joins could have been done before selecting or projecting, thus:

(Account ⋈ AcUser ⋈ Client
where balance > 700
[clientNr, clientName *]*

Notice how queries may be spread over many lines to make them more readable. A useful syntax check is to ensure that you have the same number of opening and closing brackets. Although these two queries are logically equivalent, if executed in the evaluation order shown, the second query is less efficient because it involves a larger join.

Relational algebra can be used to specify transformation rules between equivalent queries to obtain an optimally efficient, or at least a more efficient, formulation. SQL database systems include a query optimizer to translate queries into an optimized form before executing them. So, in practice we often formulate queries in ways that are easy for us to think about them rather than worrying about efficiency considerations.

Because relational algebra systems are not used in practice, you can ignore efficiency considerations when doing exercises in the algebra. However, for complex queries in SQL, practical optimizers are not perfect, and you may need to tune your queries to achieve the desired performance. In some cases, hand tuning a complex query can reduce its execution time dramatically (e.g., from hours to minutes).

As our next example, let's return to the Speaks table mentioned earlier. Consider the query: list each pair of countries that share a common language. For example, one pair is Australia and Canada because they both have English as an official language. Before looking at the solution provided, you might like to try solving this yourself. As a hint, *if you need to relate different rows of the same table, then a self-join is needed.* Also, recall that table aliases are required to join a table to itself.

The solution is shown in Fig. 12.29. First, we define two aliases, "Speaks1" and "Speaks2" for the Speaks table. You can think of this as two different copies of the table, as shown. By using the original table name, we could get by with only one alias. Check to make sure that you understand the solution. Notice that we should not use the natural join for this query (why?).

The query of Fig. 12.29 could use ">" or "<>" instead of "<". However, "<" nicely arranges for the first name of each pair to be alphabetically prior to the second name. Moreover, "<>" is inadvisable because it would result in each pair being listed twice, once for each order (recall our earlier example about drinkers and smokers).

Speaks1:

country	*language*
Australia	English
Belgium	Dutch
Belgium	French
Canada	English
Canada	French
Cuba	Spanish
Dominica	English
Dominica	French

Speaks2:

country	*language*
Australia	English
Belgium	Dutch
Belgium	French
Canada	English
Canada	French
Cuba	Spanish
Dominica	English
Dominica	French

```
define alias Speaks1 for Speaks
define alias Speaks2 for Speaks

(Speaks1    Speaks2)
where Speaks1.language = Speaks2.language
   and Speaks1.country < Speaks2.country
[Speaks1.country, Speaks2.country]
```

⇨

Speaks1.country	*Speaks2.country*
Australia	Canada
Australia	Dominica
Belgium	Canada
Belgium	Dominica
Canada	Dominica

FIG. 12.29 A self-join is needed to list pairs of countries with a common language.

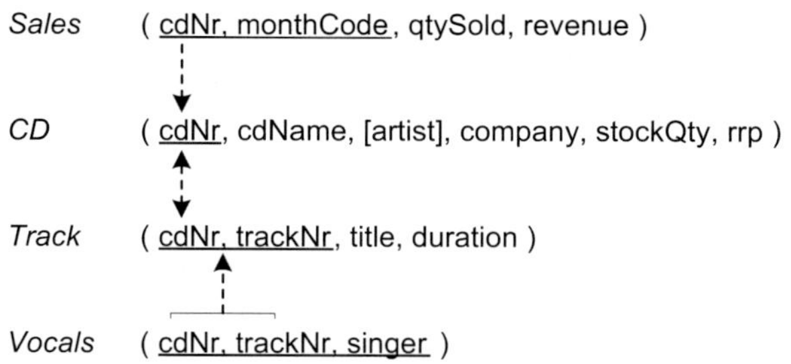

FIG. 12.30 Base table schema for compact disc retailer UoD.

For a couple of more difficult examples, we return to the compact disc retailer UoD discussed earlier in the book. The relational schema for the base tables is set out in Fig. 12.30. As an exercise, try to formulate the following English queries in relational algebra before checking the solutions provided. These queries are much harder than our earlier examples. Recall that the month codes for January and February are "Jan" and "Feb".

(a) List the CD number and name of each compact disc that *either* had sales of more than 20 copies in each of the months January and February *or* has no track with a duration longer than 300 s.
(b) Who sings a track lasting at least 250 s *and* sings on *each* compact disc that sold more copies in February than its current stock quantity?

The relational algebra query for (a) is shown in QA. Intersection is required because the quantity sold must be greater than 20 in *each* month (i.e., January *and* February). As an exercise, explain why this can't be done using "**and**" or "**or**". Union is used for the *or* because the disjuncts are not available on the same row. Subtraction is used to enforce the condition about *no* track. The join provides the cdName.

(QA) (Sales **where** monthCode = 'Jan' **and** qtySold > 20 *[* cdNr *]*
∩
Sales **where** monthCode = 'Feb' **and** qtySold > 20 *[* cdNr *]*
∪
(Track *[* cdNr *]* – Track **where** duration > 300 *[* cdNr *]*)
⋈
CD) *[* cdNr, cdName *]*

Query QB formulates the query for (b). Intersection handles the *and* operation while relational division is used to enforce the *each* requirement.

(QB) (Vocals ⋈ Track) **where** duration >= 250 *[* singer *]*
∩
(Vocals *[* singer, cdNr *]*
÷
(Sales ⋈ CD) **where** monthCode = 'Feb' **and** qtySold > stockQty *[* cdNr *]*)

Such queries are best formulated by noting the overall structure (e.g., a union) and then working on each part. These queries are unusual in requiring frequent use of ∪, ∩, and −. Remember in using these operators to ensure that the operands are compatible—this usually means a projection has to be done first.

Sometimes, a query requires several tables to be joined. In this case, if you are joining *n* tables, remember to specify $n - 1$ joins. For example, the query for Fig. 12.28 joined three tables and hence required two joins.

Of the table operations covered, only five are primitive (i.e., cannot be defined in terms of the other operations). While some choice exists as to which to consider primitive, the following list is usual: ∪; −; ×; selection; and projection. The other three (∩, ⋈, and ÷) may be defined in terms of these five operations (the proof is left as an exercise). Talking about exercises, you must be bursting to try out your relational algebra skill on some questions. So here they are.

Exercise 12.1

1. The conceptual schema for a particular UoD is shown here. "Computer" is used as an abbreviation for "kind of computer"

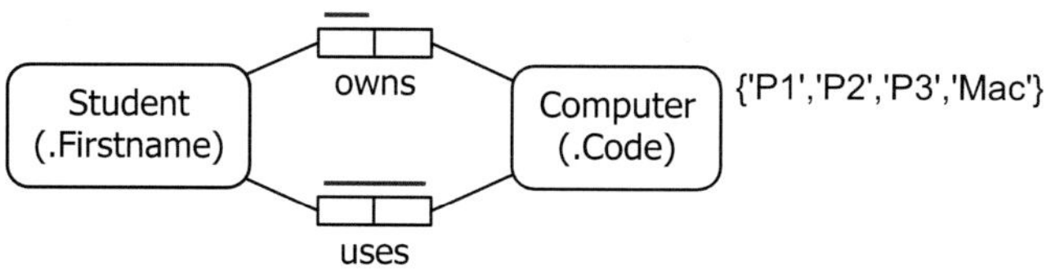

(a) Map this onto a relational schema. Then populate the tables with these data:

Ann uses a P2.
Fred uses a P2 and a Mac, and owns a P2.
Sue uses a P3, and owns a P1.
Tom owns a Mac.

Given this schema and database, formulate the following queries in relational algebra and state the result. Make use of ∪, ∩, −, and *[ ]*.

(b) List the students who own or use a computer.
(c) List the students who use a computer without owning one of that kind.
(d) List the students who use a computer but own no computer.
(e) List the students who own a computer but do not use a computer.
(f) List the students who own a computer without using one of that kind.
(g) List the students who use a computer and own a computer of that kind.
(h) List the students who use a computer and own a computer.
(i) List the computers used (by some student) but owned by no student.
(j) List the computers owned (by some student) but used by no student.

2. (a) If A is a 200 × 10 table and B is a 300 × 10 table, under what conditions (if any) are the following defined?
(i) $A \cup B$ (ii) $A \cap B$ (iii) $A - B$ (iv) $A \times B$
(b) If A is 200 × 10 and B is 100 × 5, what is the size of $A \times B$?

3. The relational schema and a sample population for a student database are shown below. Students are identified by their student number. It is possible for two students to have the same name. All students have their name, degree, gender, and

birth year recorded, and optionally subjects (e.g., a student might enroll in a degree before choosing which subjects to study). Subjects are identified by subject codes. Two subjects may have the same title. For each subject, we record its title and credit. Some subjects might not be studied (e.g., PY205 might be a newly approved subject to be introduced in the following year).

This schema applies to a one-semester period only, so we can ignore the possibility of a student repeating a subject. Subject enrollments are entered early in the semester, and ratings are assigned at the end of semester (so rating is optional). Formulate each of the following as a single relational algebra expression.

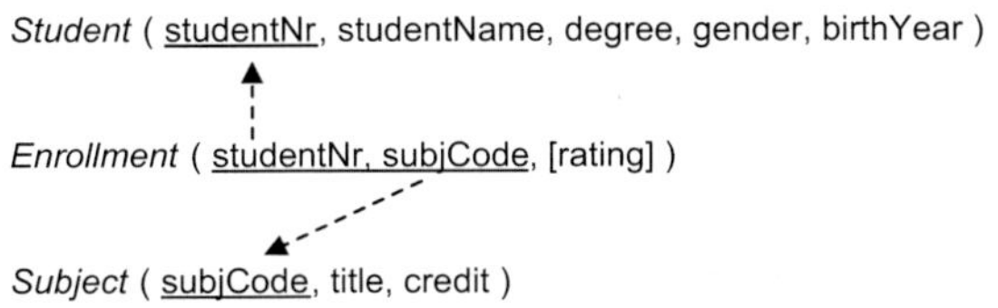

Student:

studentNr	studentName	degree	gender	birthYear
861	Smith J	BSc	M	1967
862	Jones E	BA	F	1965
863	Brown T	BSc	M	1950

Subject:

subjCode	title	credit
CS113	Databases	8
PD103	Logic	10
PY205	Xenoglossy	5

Enrollment:

studentNr	subjCode	rating
861	CS113	7
861	PD102	5
862	PD102	7
863	CS113	4
863	PD102	5

(a) List the code, title, and credit for subject CS113.
(b) List the student number, name, and degree of male students born after 1960.
(c) List the codes of the subjects studied by the student(s) named "Brown T".
(d) List the studentNr and name of those students with a 7 rating in at least one subject.
(e) List the studentNr and name of all students who obtain a 5 in a subject called Logic.
(f) List the studentNr and degree of those students who study all the subjects listed.
(g) List the studentNr, name, and gender of those students who either are enrolled in a BSc or have obtained a rating of 7 for PD102.

(h) List the studentNr, name, and birth year for male students born before 1970 who obtained at least a 5 in a subject titled "Databases".
(i) Use renaming to define the following table so it can be referenced in a later query: *TopRatedDatabaseStudent* (name, genderCode).

4. The following table contains details on students who are to debate various religious topics. For this UoD, students are identified by their first name. Each debating team is to comprise exactly two members of the same religion but opposite gender.

Phrase each of the following as a single relational algebra query.

Debater:

firstname	*gender*	*religion*
Anne	F	Buddhist
Betty	F	Christian
Cathy	F	Hindu
David	M	Christian
Ernie	M	Buddhist
Fred	M	Hindu
Gina	F	Christian
Harry	M	Buddhist
Ian	M	Christian
Jane	F	Christian
Kim	F	Hindu

(a) List the name and religion of all females who are not Buddhist.
(b) List the name and gender of those who are either male Hindus or female Christians.
(c) List all possible debating teams, mentioning females before males. For example, one team comprises Anne and Ernie.

5. (a) Map the following conceptual schema to a relational schema.

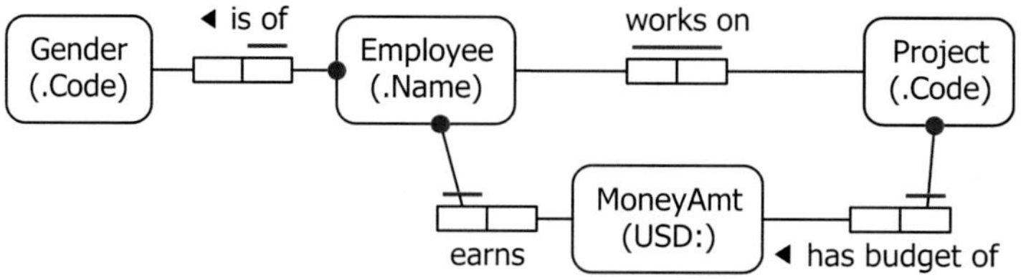

Use your schema to formulate each of the following in relational algebra.
(b) Find names and salaries of all female employees who earn more than $25,000 or work on project "5GIS".
(c) List the name and gender of those employees who work on all projects with a budget of at least $100,000.

6. The Employee table stores the employee number and name of each employee, the department they work for, and the year they joined the firm. The Department table indicates the manager and budget for each department. Each employee manages at most one department (which must be the department for which he or she works).

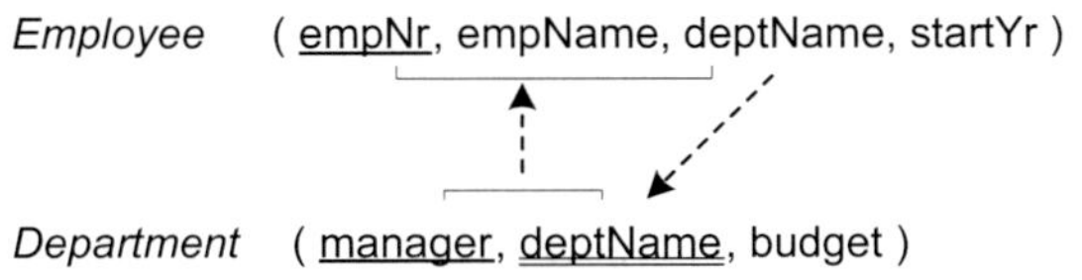

   (a) Draw a conceptual schema diagram for this UoD.
   Phrase each of the following requests as a single relational algebra query.
   (b) Who works for the Accounting department and started with the firm before 1970?
   (c) What is the budget of the department in which employee 133 works?
   (d) List the departmental managers and the year they joined the firm.
   (e) Which employees are not departmental managers?
   (f) Give the empNr, name, and year started for those managers of departments with a budget in excess of $50,000.
   (g) Which employees worked for the firm longer than their departmental managers?

7. (a) Define $A \cap B$ in terms of —.
   (b) Let relation A have attributes x, y and relation B have attributes y, z where both y attributes are defined over the same domain. Define the natural inner join operation $A \bowtie B$ in terms of $\times$, selection, and projection.
   (c) Let A have attributes x, y and B have the attribute y where both y attributes are defined over the same domain. Define $A \div B$ in terms of $\times$, $-$ and projection.

8. In the following schema, a person is identified by the combining surname and forename. Formulate each of the following as single queries in relational algebra.

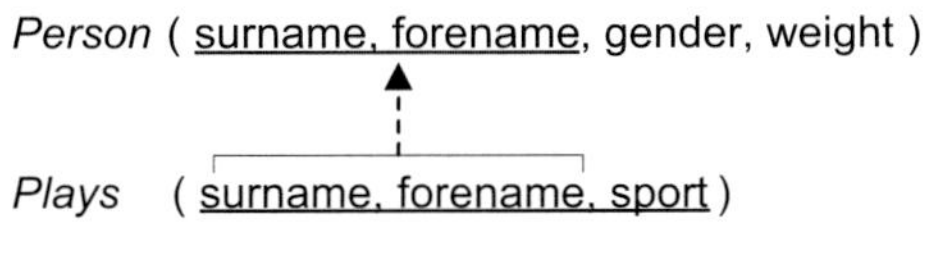

   (a) Which females play judo?
   (b) Which males in the weight range 70..80 kg play either judo or karatedo?
   (c) Which females over 50 kg play judo but not karatedo?
   (d) Who plays both judo and aikido? Do not use a join.
   (e) Who plays both judo and aikido? Do use a join.

9. The following relational schema relates to the software retailer UoD from Exercise 6.3. The category codes are: DB = Database; SS = Spreadsheet; WP = Word Processor.

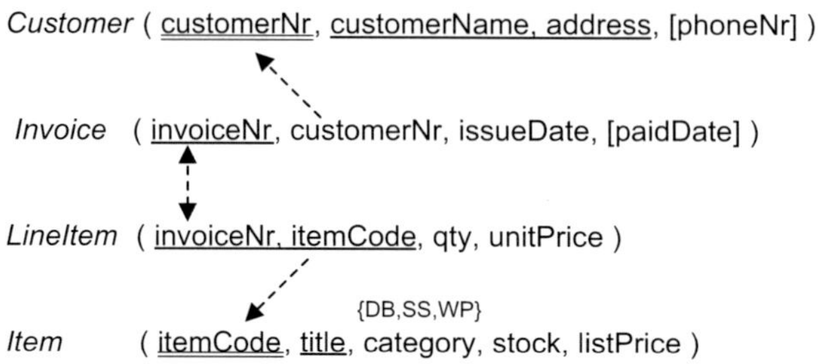

Formulate the following as single queries in relational algebra.

(a) List the customerNr of each customer who has been charged less than the list price for at least one software item, but has been sold no word processor.

(b) List the name of each customer who was sold at least one copy of all the spreadsheets on the stock list.

(c) List the customerNr of each customer who has purchased a copy of all the word processors that are in stock but who has never purchased a database.

(d) List the customerNr of those customers who purchased both a spreadsheet and word processor on the same invoice.

12.2 Relational Database Systems

We may now define a relational DBMS as a DBMS that has the relational table as its only essential data structure and supports the selection, projection, and join operations without needing specification of physical access paths. A relational system that supports all eight table operations of the relational algebra is said to be relationally complete. This doesn't entail eight distinct operators for these tasks; rather, the eight operations must be expressible in terms of the table operations provided by the system.

The two main relational languages are SQL and QBE, with SQL being the most important. Most SQL systems are relationally complete. A system that supports all aspects of the relational model, including domains and the two basic integrity rules (entity integrity and referential integrity), is sometimes said to be "fully relational". The relational model of data itself is evolving.

For Version 1 of the relational model, Codd proposed 12 basic rules to be satisfied by a relational DBMS. These may be summarized as follows:

1. All information is represented in relational tables.
2. All data is accessible using the table name, primary key value, and column name.
3. Systematic support for missing information (null values, independent of data type) must be provided.
4. The relational schema itself is represented in tables and is accessible in the same way as the database.

5. A relational query language must be provided that supports data definition, view definition, data manipulation, integrity constraints, authorization, and transaction boundaries (begin, commit, and rollback).
6. Basic support for view updatability is provided.
7. Table-at-a-time retrieval and update operations are provided.
8. Application programs are logically unaffected by changes to internal storage or access methods.
9. Application programs are logically unaffected by information-preserving changes to the base tables.
10. Integrity constraints must be definable in the query language and storable in the system tables.
11. The DBMS has distribution independence.
12. If record-at-a-time processing is supported, this cannot be used to bypass the constraints declared in the set-oriented query language.

Version 2 of the relational model as proposed by Codd (1990) includes 333 rules, including support for two kinds of missing information. While some of the proposed revisions have merit, others are debatable, and it is doubtful whether any commercial DBMS will ever try to satisfy all these rules. In practice, the SQL language itself has become more influential than the relational model in standardization efforts for relational DBMSs.

Some SQL systems are fully relational with respect to Codd's original 12 rules, but most provide only weak support for domains. For example, many require heights and weights to be defined as numeric data types, allowing comparisons such as "height = weight".

In SQL-89, the union operation is allowed only between **select** statements, not between table expressions, and separate operators for $\cap$, $-$, and $\bowtie$ (natural join) are not explicitly provided (though expressible in terms of other SQL primitives).

In 1992, the SQL standard was improved substantially, and all the table operators are explicitly included in SQL-92 (for example, $\cap$, $-$, and $\bowtie$ are called "**intersect**", "**except**", and "**natural join**"). The ability to declare constraints was also improved substantially.

Features added in SQL:1999 include triggers, user-defined data types, object identifiers, array-valued fields, recursive union, and procedural control structures.

The latest version of the SQL standard, ISO/IEC 9075-1:2023, was approved in June 2023 (see https://www.iso.org/standard/76584.html) but some of its features are still not supported in commercial systems. Further details on SQL standards are discussed in later sections.

Relational DBMSs provided higher-level query facilities than prerelational DBMSs but were originally slower. So, they were first used for decision support and low-volume online applications. Nowadays, their performance has improved greatly and they are dominant even for large-scale online transaction processing (OLTP) applications.

SQL products (e.g., DB2, Microsoft SQL Server, Oracle) dominate on larger relational systems. Initially, many smaller relational systems (e.g., Access, Paradox, FoxPro) used a version of QBE or Xbase as their main query language. Nowadays, most relational systems, small or large, provide good support for SQL, with basic QBE support offered as an alternative for simpler queries.

In addition to relational algebra operations, relational DBMSs provide further capabilities, such as sorting, arithmetic, grouping, and formatting. Most systems also provide powerful tools for creating external interfaces (e.g., screen forms), report writing, and security. Despite such productivity benefits, the success of any database application still depends critically on the database design.

12.3 SQL: Historical and Structural Overview

The rest of the chapter focuses on SQL. Full coverage would require a large book itself, so many advanced features of the language are omitted. The treatment assumes familiarity with relational schemas and the main operations of relational algebra, as discussed earlier. This section provides a brief history and structural overview of SQL.

After the publication of Dr. Codd's classic paper on the relational model of data (Codd, 1970), some early prototypes were developed to provide a relational DBMS, including a language for querying and updating relational databases. In 1974, Don Chamberlin and Raymond Boyce published a paper on a language called "SEQUEL" (Structured English Query Language) being implemented by a team at the IBM San Jose Research Laboratory as an interface to its System R relational prototype, within a project also called System R. By 1977, a revised version of this language (SEQUEL/2) had been defined and largely implemented by IBM. In the late 1970s it was discovered that "SEQUEL" was an existing trademark, so the language was renamed "SQL". Although "SQL" is often pronounced "SEQUEL", officially it is pronounced simply "ess-cue-ell".

The System R project ran from 1971 to 1979 and later evolved into a distributed database project (System R*). Using the experience gained from its System R project, IBM built its first commercial relational DBMS, known as SQL/DS, which was released in 1981. Its second and highly influential SQL product, known as DB2, was released in 1983.

As the SQL language had been widely publicized in the 1970s in research papers, other firms began developing their own systems. Relational Software Inc. (later renamed Oracle Corporation) actually beat IBM to the market by releasing its commercial SQL product in 1979. In 1982, the American National Standards Institute (ANSI) began standardization work on the language. The first SQL standard was completed in 1986 and is known as SQL-86.

In 1987, the ANSI SQL-86 standard was adopted by the International Organization for Standardization (ISO). In 1989, a more comprehensive standard, SQL-89, was adopted by both ANSI and ISO. This defined a basic Level 1 version, a richer Level 2 version, and included an addendum on integrity enhancements (including declarative entity integrity and referential integrity).

In 1992, the next standard, known as SQL-92, was approved by both ANSI and ISO. Because of its size (more than 600 pages) it was defined at three levels: entry, intermediate, and full. SQL:1999 was approved on December 8, 1999. SQL:1999 represented a significant departure from earlier versions in its inclusion of object-oriented features (e.g., object identifiers and references, array-valued fields, and procedural control structures). SQL is now object-relational rather than purely relational. Because of its size, the SQL:1999 standard (ANSI 1999) was divided into five separate parts: Framework; Foundation; Call Level Interface; Persistent Stored Modules; and Host Language Bindings. The Foundation part alone comprises 1151 pages.

The next version of the standard, SQL:2003, appeared in 2003. A significant addition was the first appearance of support for XML, but enhancements to many other areas were also included. By this time, the standard had grown to nine parts, with the Foundation part alone weighing in at more than 1300 pages. During the development of SQL:2003, the world of XML had also been evolving. To reflect the latest XML concepts, the part of the SQL standard dealing with XML alone (sometimes referred to as "SQL/XML") was revised in 2006 to allow users to import, store, and manipulate XML data; it also supported the use of the XML query language Xquery, with other parts remaining at their 2003 revisions.

In 2008, the SQL standard was revised to support **order by** clauses outside cursor definitions **instead of** triggers, **truncate** statements, and **fetch** clauses. A further revision in 2011 added support for temporal data (e.g., **period**), and enhancements to windows functions and fetch clauses. In 2016, the SQL standard was revised to support row pattern matching, polymorphic table functions, and operations on JSON data stored in character string fields. In 2019, the SQL standard was revised to add Part 15 on multidimensional arrays (MDarray type and operators).

The latest version of the SQL standard (ISO/IEC 9075:2023) was formally adopted in June 2023 (https://www.iso.org/standard/76584.html). This revision added many optional features (null values in unique constraints, string padding functions, etc.), deeper support for JSON, and Part 16, Property Graph Queries (SQL/PGQ). For further details on SQL:2023, see also SQL:2023 - Wikipedia.

Currently, most commercial DBMSs have implemented many of the features added in SQL:1999 (e.g., triggers, row types, and large objects) as well as some features added to later versions. However, it is highly unlikely that any commercial system will ever implement all of SQL:2023. Commercial dialects use their own, nonstandard syntax for some features that they do implement from the standard, and also have their own extensions that are not included in the standard. This overlap is portrayed in Fig. 12.31, where dialects A and B denote commercial versions of SQL (e.g., DB2 and Microsoft SQL Server).

Commercial SQL systems are slowly replacing much of their nonstandard syntax with standard syntax while retaining most of their alternative syntax for backward compatibility. When a feature is supported in both standard and nonstandard syntax, you should use the standard syntax. This makes your SQL code more portable and more understandable to users of other SQLs.

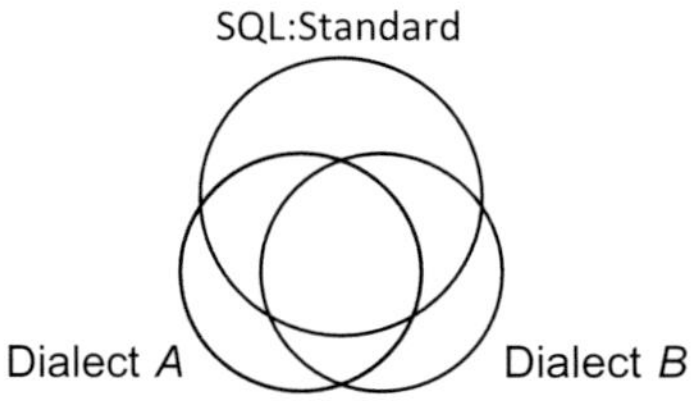

FIG. 12.31 Proper overlap between standard SQL and commercial dialects.

This chapter focuses on the SQL standard versions, but also includes some details about commercial implementations.

The SQL language is vast. To help come to terms with it, SQL is sometimes informally classified into the following sublanguages: data definition language (*DDL*); data manipulation language (*DML*); data control language (*DCL*); and data procedural language (*DPL*). The DDL includes statements for creating, altering, and dropping various kinds of database objects, such as tables and views (e.g., **create table, alter table**, **drop table**, **create view**, **drop view**). The DML includes statements for querying and updating the database (e.g., **select**, **insert**, **update**, and **delete** statements). The DCL is used for security, controlling who has what kind of access to database objects (e.g., **grant** and **revoke** statements).

The three previous sublanguages are essentially declarative in that statements are used to declare what is to be done rather than how it is to be done. However, the DPL is a procedural language that supports branching, looping, and other programming constructs (e.g., **if** and **while** statements). This classification is only approximate (e.g., a **create view** statement includes a **select** query, and a **create trigger** statement may include procedural code). Our main focus is on the DML (in this chapter) and the DDL (in the following chapter).

12.4 SQL: Identifiers and Data Types

For a given schema, some database objects such as base tables, views, domains, triggers, rules, and constraints are identified by name. A column is identified by appending its (local) name to the name of its table, using a "dot notation". This distinguishes columns with the same local name in different tables. For example, the second columns of the tables *Subject* (subjectCode, title, credit) and *Book* (isbn, title) are identified as "Subject.title" and "Book.title", respectively. In SQL, all names (including local names) are called *identifiers*, even though local names provide identification only within a limited context.

In SQL-89, identifiers were restricted to at most 18 characters, and all letters had to be in uppercase. From SQL-92 onward, an identifier is either regular or delimited. A *regular identifier* is a string of at most 128 characters, the first of which must be a letter ("a".."z", "A",.."Z"). Each later character must be a letter, digit ("0".."9"), or underscore ("_"). Moreover, *no reserved word may be used as a regular identifier*. You may, however, use a reserved word as part of an identifier. For example, "note" is fine as an identifier even though it includes the reserved word "**not**".

Microsoft SQL Server allows identifiers of up to 128 characters, but also allows the first character to be "_", "#", or "@", and later characters to be "#", "@", or "$". It uses "#" to start names of temporary objects and "@" to start variable names. Some commercial systems allow identifiers of at most 255 characters.

Some examples are shown in Table 12.3. Names in the third row cannot be used as regular identifiers because they start with a digit or include an illegal character (e.g., a space or parenthesis). Names in the fourth to sixth rows were first introduced as reserved words in SQL-92, SQL:1999, and SQL:2003, respectively.

Table 12.3 Which names may be used as regular identifiers?

Allowed as (regular) identifiers?	*SQL-89*	*SQL-92*	*SQL:1999*	*SQL:2003*
A, R2D2, CUSTOMER_NR	Yes	Yes	Yes	Yes
a, This_is_a_long_identifier, CustomerNr	No	Yes	Yes	Yes
2B, CUSTOMER NR, SPEED(MPH)	No	No	No	No
date, first, level	Yes	No	No	No
before, row, trigger	Yes	Yes	No	No
multiset, xml, bigint	Yes	Yes	Yes	No
bit	Yes	No	No	Yes

Because the list of reserved words grows with each new standard, identifiers in existing applications might become illegal at a later stage. For example, Table 12.3 shows that many words became reserved for the first time in SQL-92, SQL:1999, and SQL:2003. Different SQL dialects may omit some of these words from their reserved word lists while adding others, which makes portability between dialects even harder. Partly to avoid such problems, the SQL-92 standard introduced delimited identifiers.

A *delimited identifier* is a string of at most 128 characters, delimited by (i.e., enclosed in) double quotes. Any character at all may be used, as well as reserved words, within the double quotes. For example, the following are legal delimited identifiers: "customer nr", "speed (km/h)", "&^%!!", "date", "group". Unlike regular identifiers, delimited identifiers are case sensitive (i.e., uppercase letters are not equated to lowercase letters). For example, the delimited identifiers "invoiceNr", "InvoiceNr", and "INVOICENR" are unequal, but the regular identifiers InvoiceNr and INVOICENR are equal.

In SQL, a string constant is delimited by *single* quotes. For example, 'USA' might be a value in a countryCode column. Some commercial SQLs allow double quotes instead, for example, "USA". This practice should be discouraged because it conflicts with the now standard use of double quotes for delimited identifiers.

In Microsoft SQL Server, the command "**set quoted_identifer on**" ensures that double quotes are used only for delimited identifiers, forbidding their use to delimit string constants. SQL Server also allows square brackets ([...]) as an alternative to double quotes ("...").

Words that were reserved may cease to be so in later versions. For example, the names **avg**, **between**, **exists**, and **sum** were downgraded from reserved words to nonreserved keywords in SQL:1999. *Keywords* are words that have predefined meanings. *Reserved words* are keywords that cannot be used as regular identifiers.

As mentioned earlier, a reserved word may be embedded within an identifier. Suppose a column is to store the names of tutorial groups. Because **group** is a reserved word, we cannot use this (unquoted) for the column name. However, we may add quotes to make it a delimited identifier, or include "group" in a longer regular identifier (e.g., "group", TuteGroup, GroupName).

A first impression of how a given SQL dialect compares with the standards may be gained by inspecting its list of keywords, and especially its reserved words. Where we deal with SQL in this book, keywords are usually distinguished by displaying them in bold. Other systems often use uppercase, e.g., "SELECT".

Values entered in a table column belong to the *data type* declared for that column. Table 12.4 lists the standard data types for some older versions of the SQL standard. Here, square brackets indicate optional components. All SQL dialects support at least character string and numeric types.

Table 12.4 Standard data types.

Standard		*Data types*	
SQL-89	introduced:	fixed-length string	**char, character** (*n*), **char** (*n*)
		exact numeric	**smallint**
			int
			numeric(*p* [,*s*])
			decimal (*p* [,*s*])
		approximate numeric	**float** [*p*]
			real
			double precision
SQL-92	introduced:	variable-length string	**varchar** (*n*), **character varying** (*n*)
		national character strings	**national character** (*n*), **national character varying** (*n*) and short forms such as **nchar**(*n*)
		bit string	**bit** (*n*) **bit varying** (*n*)
		datetime	**date** {year, month, day} **time** {hour, minute, second} **timestamp** {date and time} (time and timestamp can be with or without time zone)
		interval	year-month periods day-time periods
SQL:1999	introduced:	large object	**clob** (*n*) {character large object} **nclob** (*n*) {national character large object} **blob** (*n*) {binary large object} and various lengthier alternatives
		boolean type	**boolean**
		row type	**row** (field-definition [,...]
		collection type	data-type **array**(*n*)
		user defined type (UDT)	
		reference type	**ref** (UDT)
SQL:2003	introduced	exact numeric	**bigint**
		collection type	data-type **multiset**
		xml data	**xml**
	withdrawn		**bit** (*n*) **bit varying** (*n*)

SQL:2016 and SQL:2023 added more data types, such as, **tinyint** (0.0.255), **datetime** (yyyy-mm-dd hh:mi:ss format), **text** (variable length strings up 2 GB in size), **ntext** (up to 1 GB in size), and **JSON** (Javascript Object Notation) data.

In the standard, "**char**(n)" means the value is stored as a fixed length string of n characters. If the value has fewer than n characters, blanks are appended to fill out the length. If no size (n) is specified, this is treated as a string with only one character. A value of type **varchar**(n) is stored as a string of at most n characters. If the value is shorter, it is stored without adding extra blanks. A **varchar**(n) string also has an "end-of-string" marker, the size of which varies between different implementations. This has to be taken into account when estimating the storage space required. For short strings with little variation in length, **char**(n) may be the preferred choice.

SQL-92 allows various national character sets to be declared. The use of "**nchar**" or equivalent syntactic variations indicates that characters are selected from the designated national character set. This implies that the character encoding uses *Unicode,* which typically requires two bytes of storage per character instead of the one byte per character that was commonplace for encoding the standard English characters and symbols (often referred to as the *ASCII* code, from the name of the standards body).

With the **numeric** data type, the precision p is the maximum number of digits included in the number, and the scale s is the number of digits after the decimal point. For example, columns declared **numeric**(6, 2) allow values in the range -9999.99 to +9999.99. The **decimal** type is like the numeric type except that an implementation may sometimes provide a precision greater than p. Many systems implement numeric and decimal as the same type. The **integer** (including **smallint** and **bigint**) data types allow integers only (no fractions). The three approximate numeric data types allow very large or very small numbers to be stored to a specified precision as a mantissa times an exponent of 10. The abbreviations **char**, **dec**, **int**, and **nchar** may be expanded to **character**, **decimal**, **integer**, and **national character**.

SQL-92 and later standards provide direct support for bit strings as well as time points and time intervals. Times are local unless the **with time zone** option is specified. This option includes the offset from UTC (Universal Time, Coordinated, formerly called Greenwich Mean Time (GMT)). Various temporal operators are provided (e.g., to compute intervals by subtracting one time point from another).

In SQL:1999, all the previous data types are called *predefined types.* As Table 12.4 indicates, SQL:1999 adds four more predefined types: character large object, national character large object, binary large object, and Boolean (true or false). In addition, SQL:1999 allows row types, user defined types (UDT), *reference types,* and *collection types.* A row type is basically a sequence of fields and can be used as the basis for defining a table. A UDT is identified by its name and may include a list of method specifications. If based on a single, predefined type, it is called a *distinct type.* If specified as a list of attribute definitions, it is called a *structured type.* The only collection type constructor allowed in SQL:1999 was array. The SQL:2003 standard added support for **multiset** (bags) as a collection type, large integers (**bigint**), and, significantly, **xml** as a recognized data type within SQL (we'll return to the **xml** type in a later chapter).

Currently, most commercial SQLs support all the SQL-89 data types; almost all the additional SQL-92 data types; some of the extra SQL:1999, SQL:2003, SQL:2016 and SQL:2023 data types; and some nonstandard data types.

Many differences exist. For example, Oracle treats **varchar** as **char**, and uses **varchar2** to denote the standard **varchar**. SQL Server includes many extra data types such as **money**, **smallmoney**, and **uniqueidentifier**. The range of most standard types is usually implementation defined rather than standardized. For example, the range for **smallint** is often, but not always, −32,768..32,767.

Exercise 12.4

1. Which of the following are legal identifiers in which SQL standard, if shifted to uppercase?

(a) Payroll#	(b) PayrollNr	(c) "Payroll#"
(d) 1994Tax	(e) Tax in 1994	(f) "Tax in 1994"
(g) Tax_in_1994	(h) Deposit_in_$	(i) Mass_(kg)
(j) Order	(k) WorldWideWebIdentifier	(l) count

12.5 SQL: Choosing Columns, Rows, and Order

Recall that relational algebra includes the following eight table operations: union, intersection, difference, Cartesian product, selection, projection, join, and division. All these operations (as well as others) can be expressed using SQL's powerful **select** statement. This section discusses how SQL is used to perform relational projection and selection as well as bag projection and row ordering.

First, let's see how to *choose columns*. Consider a small UoD where people are identified by their firstname. Table 12.5 provides sample data for the table scheme *Person* (firstname, gender, starsign, birthyr).

The whole table may be retrieved by projecting on all its columns. In relational algebra, this may be formulated as Person or as Person [firstname, gender, starsign, birthyr]. In SQL, this is expressed as follows:

```
select * from Person
```

Here, the asterisk "*" means "all columns" and may be read as "everything" or "all". The table named after **from** indicates the table from which the data is to be retrieved. When this command is executed, the result is an *unnamed* table with the same column names and contents as the original Person table.

If only some of the columns are required, then instead of "*" the relevant columns should be listed (separated by commas). If the columns include a key, no duplicate rows can occur in the result, so the result corresponds to a relational projection on those columns (e.g., see Fig. 12.32). To improve readability, an SQL query is usually spread over

Table 12.5 A relational table storing personal details

Priority	*Operators*
1	=, <>, <, >, <=, >=
2	**not**
3	**and**
4	**or**

```
select firstname, starsign
from Person
```
⇨

firstname	*starsign*
Bob	Gemini
Eve	Aquarius
Fred	Gemini
Norma	Aries
Selena	Taurus
Terry	Aquarius

FIG. 12.32 Projecting on columns that include a key.

(a)

```
select gender, starsign
from Person
```
⇨

```
select all gender, starsign
from Person
```
⇨

gender	*starsign*
M	**Gemini**
F	Aquarius
M	**Gemini**
F	Aries
F	Taurus
M	Aquarius

(b)

```
select distinct gender, starsign
from Person
```
⇨

gender	*starsign*
M	**Gemini**
F	Aquarius
F	Aries
F	Taurus
M	Aquarius

FIG. 12.33 The distinct qualifier may be used to remove duplicate rows.

many lines (e.g., a new line is used to start the **from** clause). Except for delimited (quoted) identifiers or string constants, the case (upper or lower) in which letters are written has no significance. We'll use lowercase most of the time, but start table names with a capital letter. Many organizations have their own detailed standards on the use of uppercase or lowercase.

If the chosen columns do not include a key, then duplicate rows may occur in the result. For example, the query in Fig. 12.33(a) returns a bag (multiset) rather than a set because ("M", "Gemini") appears twice. This operation may be called a *bag projection.*

It is sometimes useful to include duplicates (e.g., listing all the scores awarded by judges for a gymnastics event). You can explicitly request inclusion of duplicate rows by using the qualifier **all** before the **select** list, but this is what happens by default in SQL anyway. So, a query of the form **select all** a,b,... **from** T is equivalent to **select** a,b,... **from** T.

To eliminate duplicates, include the qualifier **distinct** before the **select** list. This ensures that rows displayed will be distinct, thus providing a relational projection. Hence, the **distinct** qualifier converts a bag of rows to a set of rows. For example, the query in Fig. 12.33(b) eliminates the duplicate ("M", "Gemini") row.

Columns are displayed in the order in which they are specified in the **select** list. If "*" is used, all the columns are displayed in the same order as the original table. So the command "select * from **Person**" is merely shorthand for "**select** firstname, gender, starsign, birthyr **from** Person".

In general, the relational projection *T[a, b,...]* of relational algebra may be expressed by an SQL statement of the form:

select distinct a,b,... **from** T

We may summarize that portion of the **select** statement covered so far in the following Extended Backus Naur Form (EBNF) notation. The symbol "::=" may be read "is defined as". Here, items in square brackets are optional. The expression "[,...]" means that the previous construct may be repeated any number of times, with its occurrences separated by commas. A stroke "|" is used to separate alternatives. By default, "|" has minimum scope (i.e., it applies just to the terms immediately next to it). Keywords are shown in bold. The final line of the **select** query shown here is called the from clause. From a logical point of view, it would be better to specify the **from** clause before the select list, but the syntax of the language does not allow this.

select-list ::= columnname [,...]

select-query ::= **select** [**all** | **distinct**] select-list | *
from tablename

Now, let's see how to *choose rows* in SQL. A relational *selection* (or restriction) is the operation of selecting those rows that satisfy a specified condition. In SQL, the syntax for this operation is essentially the same as the relational algebra syntax used earlier in this chapter. A **where** clause is added to the **select** statement just after the **from** clause, and the specified condition is known as the *search condition*.

Do not confuse the relational operation of selection with the **select** statement itself, which is used to perform other operations as well. Formulations in relational algebra and SQL listed in Table 12.6 are equivalent. For example, the query in Fig. 12.34 may be used to list details about the Aquarians mentioned in our earlier Person table.

Table 12.6 Equivalent formulations in relational algebra and SQL.

Relational Algebra	*SQL*
T **where** *c*	**select** * **from** *T* **where** *c*
T **where** *c* *[a, b, ...]*	**select distinct** *a, b, ...* **from** *T* **where** *c*

select *
from Person
where starsign = 'Aquarius' ⇨

firstname	*gender*	*starsign*	*birthyr*
Eve	F	Aquarius	1967
Terry	M	Aquarius	1946

FIG. 12.34 A **where** clause is used to select just the Aquarians.

Notice that the character string "Aquarius" contains a mixture of uppercase and lowercase characters. Suppose this is the way it is stored in the table, but the query has the starsign typed in uppercase, that is,

select * from Person
where starsign = 'AQUARIUS'

Unless the system has been configured to convert characters strings into the same case before comparing them, no match would be made because the string "AQUARIUS" is different from "Aquarius". To make string comparisons case insensitive in SQL-92 onward, either declare a collation set to act this way or use string functions to convert the case. For example, given a string expression *s*, the fold functions **upper**(*s*) and **lower**(*s*) return the uppercase and lowercase versions of *s*. Hence, the following query will retrieve the Aquarians regardless of the case used for the letters in "Aquarius" in the column entries:

select * from Person
where upper(starsign) = 'AQUARIUS'

Although many commercial dialects of SQL, such as SQL Server, support the standard upper and lower string functions, some dialects provide other ways of controlling case sensitivity of string constants.

The example in Fig. 12.35 illustrates three new features. The sample data for the table scheme *Actor* (surname, firstname, gender) includes the surname "D'Abot", which includes a single quote as one of its characters. As discussed, string constants are delimited by single quotes. To embed a single quote within a string like this, we need to inform SQL that the quote is part of the string rather than ending it. To do this, we *use two adjacent single quotes to denote an embedded single quote.*

Actor:

surname	*firstName*	*gender*
Bardot	Brigitte	F
D'Abot	Maryam	F
D'Abot	Tony	M

```
select *, 'This isn''t hard' as comment
from Actor
where surname - 'D''Abot'
```

⇨

surname	*firstName*	*gender*	*comment*
D'Abot	Maryam	F	This isn't hard
D'Abot	Tony	M	This isn't hard

FIG. 12.35 Two single quotes embed a single quote, and "as" introduces an alias.

The following **insert** statement may be used to add the second row of data to the table: **insert into** Actor **values** ('D''Abot', 'Maryam', 'F'). Notice that this uses two single quotes in "D''Abot". Although this may display simply as "D'Abot" when queried, we must use the two single quotes when using the value in an SQL statement. For example, note the search condition: surname = 'D''Abot'.

The second thing to note about the query is that *we may include a constant as* **select** *item.* Here the **select** list includes the string constant "This isn''t hard". Note again the use of two single quotes to embed a single quote. Any constant included as an item in the **select** list will be included in each row returned by the query. In our example, two rows were returned.

The third aspect illustrated by the example is that the keyword "**as**" *may be used to introduce an alias for a* **select** *item.* This alias will appear as the name of the column for that item in the result table. The example introduces the alias "comment" for the string constant that was added to the **select** list. In practice, aliases for **select** items are used mainly to provide simple names for more complex expressions, as discussed later.

Search conditions may include terms, *comparison operators* (=, <>, <, >, <=, >=), and *logical operators* (**not**, **and**, **or**). Some SQL dialects allow symbols other than "<>" for "is not equal to" (e.g., "^=", "!="). SQL Server also includes "!<" for "is not less than" and "!>" for "is not greater than", which are equivalent to ">=" and "<=", respectively. For portability, you should avoid such nonstandard symbols. The same priority convention as for relational algebra is adopted. So, unless brackets determine otherwise, *comparison operators are evaluated before logical operators,* which are evaluated in the order **not**; then **and**; then **or**.

Consider the query in Fig. 12.36. This lists the name and starsign of people born after 1950 who are either Aquarians or Geminis. If the parentheses were omitted, the condition would be interpreted as starsign = 'Aquarius' **or** (starsign = 'Gemini' **and** birthyr >1950). This is different because it would result in the older Aquarians (in this case Terry) being listed as well.

The selection operation becomes a bit trickier if nulls are present. Consider the relation scheme *Employee* (empNr, empName, dept, [carType]). Here, the carType column is optional. Some employees might not drive a car, and even if they do perhaps it is not recorded. Fig. 12.37 provides a sample population for this table as well as two queries and their results. Suppose we want to know who drives a Ford and who doesn't. We might formulate these questions as the queries in Fig. 12.37. Although each employee in real life either does or does not drive a Ford, employees 1002 and 1005 are absent from both the query results. Can you make sense of this?

```
select firstname, starsign
from Person
where (starsign = 'Aquarius' or starsign = 'Gemini')
        and birthyr > 1950                              ⇨
```

firstname	*starsign*
Bob	Gemini
Eve	Aquarius
Fred	Gemini

FIG. 12.36 Parentheses are needed to evaluate **or** before **and**.

Employee:

empNr	*empName*	*dept*	*carType*
1001	Thompson, E.	Sales	Ford
1002	Jones, E.	Sales	?
1003	Smith, F.	R&D	Toyota
1004	Adams, A.	Sales	Ford
1005	Dennis, A.	Admin	?

(a)

```
select empNr, empName
from Employee
where carType = 'Ford'          ⇨
```

empNr	*empName*
1001	Thompson, E.
1004	Adams, A.

(b)

```
select empNr, empName
from Employee
where carType <> 'Ford'         ⇨
```

empNr	*empName*
1003	Smith, F.

FIG. 12.37 Rows are selected when the condition is true (not false and not unknown).

Where nulls are concerned, SQL operates on a *three-valued logic*. A condition may evaluate to true, false, or unknown. More correctly, a condition is known-to-be-true, known-to-be-false, or unknown. *A comparison between terms, one of which is null, always evaluates to unknown.* This holds for any comparison operator ($=$, $<>$, $<$, etc.). Consider the row for employee 1002, where a null value (displayed here as "?") is recorded for the carType. For the query in Fig. 12.37(a), the condition is "null $=$ Ford", and for the query in Fig. 12.37(b), the condition is "null $<>$ Ford". In both cases, the condition evaluates to unknown.

The selection operation performed by the **where** clause *returns just those rows that evaluate to true* (i.e., known to be true). Rows that evaluate to false or unknown are filtered out. So, the rows for employees 1002 and 1005 (Jones and Dennis) are filtered out in both queries. Filtering out the unknown helps to avoid unwarranted assumptions. As discussed later, SQL provides a special function for detecting nulls.

Now, let's discuss how to control the *ordering* of columns and rows in a result table. Columns are displayed in the order in which they appear in the **select** list. So if you want a different column order, simply reorder the **select** list. However, the order in which *rows* are displayed may be controlled by means of an **order by** clause. If used, the **order by** clause must come at the end of the **select** statement. One or more columns in the **select** list (identified by name or by position in the **select** list) may be used as ordering criteria.

Ordering for a criterion is ascending by default. You can also explicitly specify ascending order by appending **asc** to the column identifier. To specify descending order, you must append **desc**. For numeric values, "ascending" means that smaller numbers are listed first. For example, $1 < 3 < 7$, so 1, 3, and 7 are in ascending order. For character string values, "ascending" means strings that come earlier in the collating sequence are listed first. As a rough guide, words that come earlier in alphabetical order usually come first.

```
select empNr, empName
from Employee
order by empName        ⇨

select empNr, empName
from Employee
order by empName asc    ⇨
```

empNr	*empName*
1004	Adams, A.
1005	Dennis, A.
1002	Jones, E.
1003	Smith, F.
1001	Thompson, E.

FIG. 12.38 Listing employees in ascending order of employee names.

As a simple example, each of the queries in Fig. 12.38 may be used to list the employees in alphabetical order. Although highly undesirable, "2" may be used here instead of "empName" in the **order by** clause because empName is the second item in the **select** list (it doesn't have to be the second column of the base table). This use of position numbers creates problems if you later modify the **select** list, makes the query harder to understand, and is a deprecated feature of the SQL standard (i.e., it may be removed in a later version). Because **as** can be used if needed to introduce a result-column name, there is never any need to resort to position numbers for ordering. So don't use them.

Character strings are ordered according to the character collating sequence. For example, if ASCII is used, then space (" ") precedes digits (0..9), which precede uppercase letters (A..Z), which precede lowercase letters (a..z). For instance, in ASCII, "M2" < "MY" < "Ma" < "Ma Kettle" < "MacTavish" < "Zen" < "apple". However, other collating sequences may differ from this. For example, in EBCDIC, lowercase letters precede uppercase letters, which precede digits.

The **order by** clause may be thought of as a way of converting a bag of rows to a sequence of rows. For ordering purposes, null values are treated as equal and, depending on the implementation, are either less than all nonnull values (as in SQL Server) or greater than all nonnull values. Consider the following query:

```
select carType from Employee
order by carType
```

This returns the sequence "Ford", "Ford", "Toyota" either preceded or followed by the two nulls, depending on the implementation.

If the column chosen for ordering does not have a uniqueness constraint, then further ordering can be obtained within its duplicate values by specifying more columns in the order-by clause. Criteria listed first in the **order by** clause are ordered first. For any particular column, the ascending or descending option may be applied. For instance, a query of the form.

```
select * from T order by a, b desc
```

first sorts the rows in ascending order of *a*, and then each set of rows with the same *a* value is sorted in descending order of *b*. For example, see Fig. 12.39. Here the females are listed first because "F" < "M" and within each gender, the names are shown in reverse alphabetic order.

```
select firstname, gender
from Person
order by gender, firstname desc
```

⇨

firstname	*gender*
Selena	F
Norma	F
Eve	F
Terry	M
Fred	M
Bob	M

FIG. 12.39 Ordering on two criteria.

Notice in Fig. 12.39 that **desc** has *minimum back-scope*, applying just to firstname, not to gender. If descending order of both were required, we would use gender **desc**, firstname **desc**. If an item in an **order by** list is not qualified by either an **asc** or **desc** option, then **asc** is always assumed by default. Hence, to list in ascending order of gender and then name, we may use just gender, firstname because this is taken to abbreviate gender **asc**, firstname **asc**. To test your understanding, see if you can predict the result of the query in Fig. 12.40 and then check your answer with the table shown.

```
select starsign, firstname, gender, birthyr
from Person
where starsign <> 'Taurus'
  and starsign <> 'Aries'
order by starsign, gender desc, birthyr
```

⇨

starsign	*firstname*	*gender*	*birthyr*
Aquarius	Terry	M	1946
Aquarius	Eve	F	1967
Gemini	Bob	M	1967
Gemini	Fred	M	1970

FIG. 12.40 If no order is specified after an order-by item, **asc** is assumed.

Did you get it right? The rows are first sorted on starsign in ascending order because **asc** is assumed after starsign (see column 1). Then males of the same starsign as females are listed before those females as a second criterion (see rows 1, 2) because this is descending order of gender. Finally, for those of the same starsign and gender, the older appear first using the third criterion of ascending order by birth year (see rows 3, 4). The **order by** clause may be specified explicitly by including the implied ordering options:

```
order by starsign asc, gender desc, birthyr asc
```

Let us use the term *colname* to indicate a column name, and *col* to indicate a column specification (either by name or number). The syntax of that portion of the **select** statement so far covered may be summarized in EBNF as follows. Although a column alias can be introduced immediately after a **select** item without using **as**, it is better to include **as** in such cases because it aids readability.

```
select [all | distinct] * | colname | constant [[as] column-alias] [,...]
from tablename
where condition
order by col [asc | desc] [,...]
```

Exercise 12.5

1. The following table concerns statements provided in Modula 2. In the column “extra”, the values “Y” and “N” indicate, respectively, whether the statement is extra to or already included in Pascal. Formulate each of the following requests as a single SQL query.

Statement:

kind	*composition*	*extra*
assignment	simple	N
procedure call	simple	N
if	structured	N
case	structured	N
for	structured	N
while	structured	N
repeat	structured	N
loop	structured	Y
with	structured	N
exit	simple	Y
return	simple	Y
empty	simple	N

(a) List the kind of each statement.
(b) List all information in the table.
(c) List the kind and composition of all statements.
(d) List the kind of each statement, and whether it is extra.
(e) List the possible composition values of the statements. (Avoid duplicates)
(f) Which statements are structured?
(g) Which simple statements are extra to Pascal?
(h) List the kind and composition of those statements that are extra to Pascal.
(i) List, in alphabetical order, the kinds of those statements already found in Pascal.
(j) List the kind and composition of all the statements, starting with the simple statements.
(k) List the kind and composition of all the statements, starting with the structured statements and listing statements of the same composition in alphabetic order.

(l) List the kind, composition, and extra status of all the statements, giving precedence to statements extra to Pascal. Statements with the same extra status should be listed starting with those of simple composition, with those of similar composition being shown in reverse alphabetic order.
(m) List alphabetically the kinds of statements that are neither structured nor extra to Pascal.
(n) List in reverse alphabetic order the kinds of those statements that are either structured and extra to Pascal or simple.
(o) As for (n) but exclude the exit statement from consideration and include the composition of each. Show all the structured statements first, in reverse alphabetic order, followed by the simple statements, in reverse alphabetic order.

12.6 SQL: Joins

We have seen how to choose, and order, columns and rows from a single table. But suppose the information we need from a single query is spread over many tables. SQL provides two main methods to access such information. The first of these uses relational **joins** and is discussed in this section. The second involves subqueries and is treated in a later section.

A *cross join* (Cartesian product) of tables pairs all the rows of one with all the rows of the other. In SQL-89, a cross join is specified by listing the tables in the **from** clause, using a *comma* "", for the Cartesian product operator (× in relational algebra). A *conditional join* (Θ-join) selects only those rows from the Cartesian product that satisfy a specified condition. In SQL-89, the condition is specified in a **where** clause. So, the query expressions in Table 12.7 are equivalent.

Table 12.7 SQL-89 syntax for joins.

Join type	*Relational algebra*	*SQL-89*
Cross join	*A* × *B*	**select** * **from** *A*, *B*
Conditional join	*A* × *B* **where** A.a Θ B.b	**select** * **from** *A*, *B* **where** A.a Θ B.b

The SQL-89 syntax has drawbacks. First, it fails to distinguish join conditions (intertable comparisons) from nonjoin conditions (intratable comparisons on the same base row), instead lumping them together in a single **where** clause.

Second, it uses a comma for the cross join operation. This is unintuitive and also overloads the comma with different meanings. Third, it does not provide any direct support for operations such as outer joins, which can occur in practical queries.

For such reasons, SQL-92 (and later) uses special syntax for various kinds of join. In addition to supporting the SQL-89 syntax, the newer standards include special notations for the following types of join (here, any text after "–" is a comment):

```
cross join
qualified join:    conditional join     -- on clause
                   column-list join     -- using clause
natural join
union join
```

Qualified and natural joins may be further classified into the following types:

```
inner                    -- this is the default
outer:     left
           right
           full
```

The new and old syntax for these is summarized in Table 12.8. In this summary, if tables *A* and *B* have any common columns (with the same local name), these columns are collectively referred to as "*c*". For the column-list join, "c_1, …" denotes one or more of these common columns. Some entries for SQL-89 syntax include unions and subqueries and are explained in later sections.

Table 12.8 Joins in SQL-92 onward, compared with SQL-89.

Join type	*New syntax from SQL-92 onward*	*SQL-89 syntax*
cross	**select** * **from** A **cross join** B	**select** * **from** A, B
conditional	**select** * **from** A **join** B **on** *condition*	**select** * **from** A, B **where** *condition*
column-list	**select** * **from** A **join** B **using** (c_1, ...) -- c_1, ... are unqualified	**select** A.c_1, ... , ... -- omit B.c_1, ... **from** A, B **where** A.c_1 = B.c_1 **and** ... -- join columns are qualified
natural inner	**select** * **from** A **natural** [**inner**] **join** B -- join column in result is c	**select** A.c,... -- omit B.c **from** A, B **where** A.c = B.c -- join column in result is A.c
left outer	**select** * **from** A **natural left** [**outer**] **join** B -- join cols are unqualified; -- nulls generated for non-matches -- to join on fewer cols use: A **left** [**outer**] **join** B **using** (c_1, ...) -- to join cols with different names: A **left** [**outer**] **join** B **on** *condition*	**select** A.c,... omit B.c **from** A, B **where** A.c = B.c **union all** **select** c, ..., ' ? ', ... **from** A **where** c **not in** (**select** c **from** B) -- for composite c, use **exists** with a -- correlated subquery. -- fewer or different cols cases -- are not shown here

right outer	**select** * **from** A **natural right** [**outer**] **join** B -- other cases: cf. left join	**select** B.c, ... -- omit A.c ... -- rest as for left join, -- but swap A and B
full outer	**select** * **from** A **natural full** [**outer**] **join** B -- other cases: cf. left join; -- rarely used	union of left and right outer joins
union	**select** * **from** A **union join** B -- rarely used	**select** A.*cols*, ' ', ... -- ' ' ∀ B col **from** A **union all** **select** ' ', ..., B.*cols* -- ' ' ∀ A col **from** B

Male:

firstname	*height*
Dick	72
Necito	70
Pat	68
Scot	72

Female:

firstname	*height*
Linda	60
Mary	62

```
select Male.firstname, Female.firstname
from Male cross join Female
```

⇨

Male.firstname	*Female.firstname*
Dick	Linda
Dick	Mary
Necito	Linda
Necito	Mary
Pat	Linda
Pat	Mary
Scot	Linda
Scot	Mary

FIG. 12.41 Listing all possible Male-Female pairs.

Most dialects of SQL now support the new syntax for cross joins and qualified joins (including outer joins), but some do not yet support the newer syntax for natural joins. Let's now consider each of the new join notations.

For *cross join*, the **cross join** syntax may be used instead of a comma. This syntax is best reserved for unrestricted joins (full Cartesian products). As an example, Fig. 12.41 shows details of Male and Female tennis players. The query lists possible male-female pairs for mixed-doubles teams. Cross joins are fairly rare in practice.

Much more common are conditional joins. The new syntax includes the join operator in the **from** clause, and the join condition in an **on** clause. Fig. 12.42 shows a relational schema, a sample population, and a conditional join query to retrieve the name of each female employee and her department. To aid readability, it's a good idea to indent the **on** clause, as shown. Each column mentioned in the query occurs in just one of the tables, so there is no need to qualify column names with table names. In this case, the join takes place along the foreign key reference shown in Fig. 12.42. Although joins often involve foreign key references, they don't have to. All we need is that the columns being joined are based on the same domain.

Column-list joins match values in the specified columns with the same local name in both tables. The new syntax uses the word **join** in the **from** clause, with the relevant

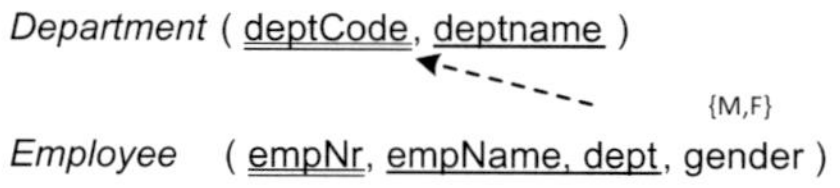

Employee:

empNr	*empName*	*dept*	*gender*
001	Hagar T	DV	M
002	Wong S	DV	F
003	Jones E	MK	F
004	Mifune K	SL	M

Department:

deptCode	*deptName*
DV	Development
MK	Marketing
RS	Research
SL	Sales

```
select empName, deptName
from  Employee join Department
      on dept = deptCode
where gender = 'F'
```

⇨

empName	*deptName*
Wong S	Development
Jones E	Marketing

FIG. 12.42 A conditional join specifies the join condition in an on clause.

columns listed in a **using** clause. As an example, consider the following schema. This is semantically equivalent to the previous schema, but different choices are made for some column names. The department code is denoted by "deptCode" in both tables, and the local identifier "name" is used for the department name as well as the employee name.

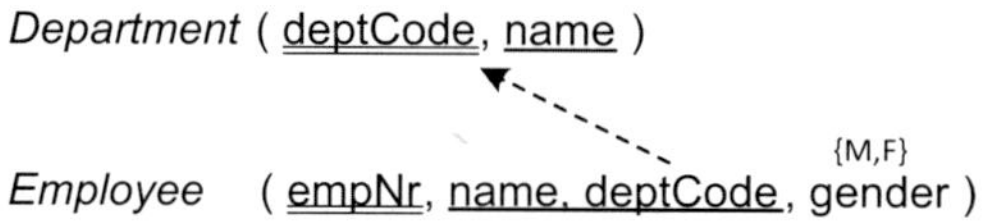

Now, consider the previous query: list the name of each female as well as her department name. How do we formulate this query for this new schema? The tables have two local column names in common: "deptCode" and "name". Here, we want to perform an equi-join between only some of the columns with common names (just "deptcode", not "name"). The common column names used for the join are listed in a **using** clause, as shown here. Only one column is used for the join. To disambiguate the query, the column names in the **select** list are qualified by their table name.

```
select Employee.name, Department.name
from   Employee join Department
   using (deptCode)
where gender = 'F'
```

As you may have realized, the column naming choices in the previous two schemas are less than ideal. If two columns in different tables are intended to signify the same thing, it is better to give the columns the same name in those tables, wherever possible. Although informal, this makes it easier to see intended semantic connections between the tables, especially those without foreign key references. For the schema under discussion, "deptCode" may be uniformly used for the department code, and the names of employees and departments may be distinguished as "empName" and "deptName", as follows:

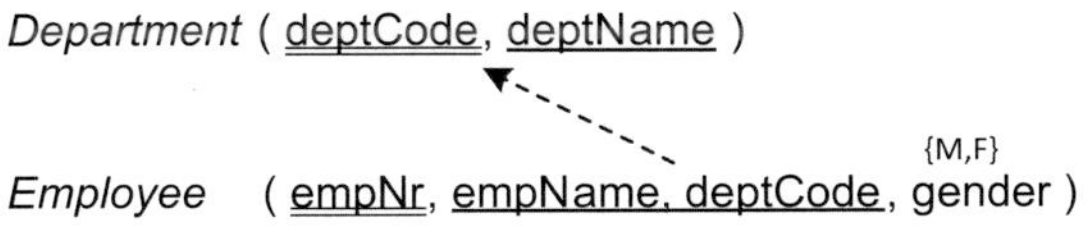

With this new schema, the conditional join query considered earlier needs to qualify the column names used in the join because their unqualified names occur in both tables. This gives.

```
select empName, deptName
from  Employee join Department
  on Employee.deptCode = Department.deptCode
where gender = 'F'
```

For the new schema, the column-list query does not require any qualifications to the column names, as shown here.

```
select empName, deptName
from  Employee join Department
  using (deptCode)
where gender = 'F'
```

As you might have noticed, the underlying join operation in the previous two queries is now a *natural join* because we are equijoining on *all* the columns with the same name (in this case there is only one such column: deptCode).

As discussed in relational algebra, *natural inner joins* perform an equijoin to match all columns with the same local names and then remove the extra copies of these columns. In SQL-89, natural inner joins are specified by first forming the Cartesian product (of the tables listed in the **from** clause) and then specifying any join conditions in a **where** clause, using the **select** list to filter out unwanted duplicate columns. In SQL-92 (and later), natural inner joins may be declared simply by inserting **natural** or **natural inner** before **join** in the **from** clause (qualified and natural joins are inner by default).

The new notation has three main advantages: it is more descriptive; the **natural join** operator gives direct support for the **natural inner join** operator of relational algebra ($\bowtie$); and it is often more concise. For example, the query just considered can be reformulated more concisely as.

```
select empName, deptName
from  Employee natural join Department
where gender = 'F'
```

There is no need to explicitly declare the join condition Employee.deptCode = Department.deptCode because this is implied by the use of **natural** and the fact that "deptCode" is the only common column name for the tables listed in the **from** clause. The compactness of this new notation is more obvious as the number of natural joins in a query grows. For example, consider the following relational schema.

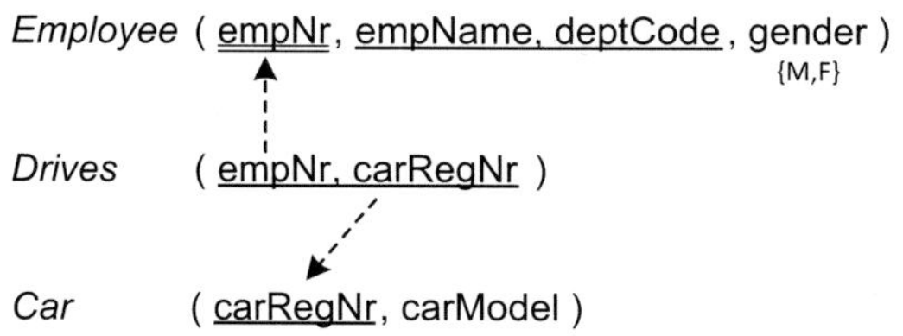

Suppose we want to list the employee number and name of each employee who drives as well as the model of their cars. Looking at the schema, the information is spread over three tables. As indicated in Fig. 12.43, we can join the Employee and Drives tables by matching empNr, and join with the Car table by matching carRegNr (join values are shown in bold). Using the special syntax for natural joins, this can be formulated simply as shown.

Employee:

empNr	*empName*	*deptCode*	*gender*
001	Hagar T	DV	M
002	Wong S	DV	F
003	Jones E	MK	F
004	Mifune K	SL	M

Drives:

empNr	*carRegNr*
001	**ABC123**
003	**ABC123**
003	**TAH007**

Car:

carRegNr	*carModel*
ABC123	Ford
PKJ900	Ford
TAH007	Saturn

```
select empNr, empName, carModel
from  Employee natural join Drives
natural join Car
```

⇨

empNr	*empName*	*carModel*
001	Hagar T	Ford
003	Jones E	Ford
003	Jones E	Saturn

FIG. 12.43 A projection on the natural join of three tables.

Currently, most SQL dialects (including SQL Server) do not support the natural join syntax. Instead, the conditional join syntax is often used to handle natural joins. For example, the query in Fig. 12.43 may be reformulated, thus:

```
select Employee.empNr, empName, carModel
from  Employee join Drives
   on Employee.empNr = Drives.empNr
  join Car
   on Drives.carRegNr = Car.carRegNr
```

Alternatively, SQL-89 syntax can be used as follows:

```
select Employee.empNr, empName, carModel
from   Employee, Drives, Car
where Employee.empNr = Drives.empNr
  and Drives.carRegNr = Car.carRegNr
```

When the natural join operator is used, any join columns in the **select** list must be denoted by their local (unqualified) names. If a natural join is specified using other syntax (e.g., conditional join or SQL-89 syntax), the join columns must be qualified.

Although most joins in practice are natural inner joins, sometimes natural joins should be avoided. The natural join operation cannot be used to join columns with different names. For example, it can't join dept. and deptCode in the query of Fig. 12.42. Moreover, the natural join operation must not be used if different tables in the **from** clause include columns with the same name that should not be equated. For example, consider the table schemes for our column-list join example: *Department* (deptCode, name); *Employee* (empNr, name, deptCode, gender).

A natural join between Department and Employee would equate not just deptCode but name as well. Any row in the result would equate the employee's name with the department's name, which is nonsense.

Sometimes, we do wish to naturally join two tables on more than one column. For example, recall the following schema discussed in the relational algebra section.

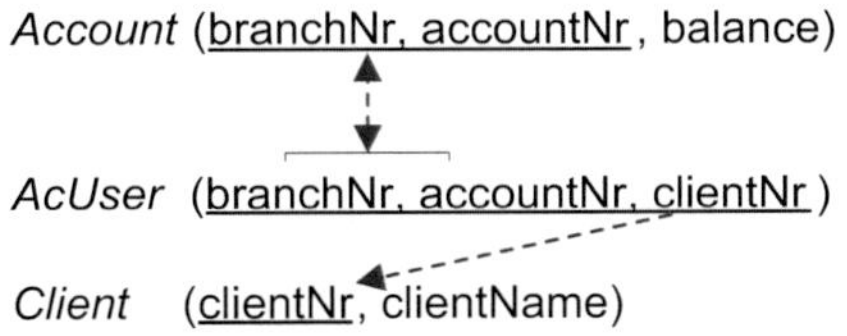

To list the balance and client details for all the accounts requires a composite join on account (compound identifier) and a simple join on client. This may be specified with the natural join syntax as:

```
select * from Account natural join AcUser natural join Client
```

or as a conditional join:

```
select Account.branchNr, Account.accountNr, balance,
Client.clientNr, clientName
from  Account join AcUser
   on Account.branchNr = AcUser.branchNr
      and Account.accountNr = AcUser.accountNr
   join Client
   on AcUser.clientNr = Client.clientNr
```

For conditional, column-list, and natural joins, **join** is assumed to mean inner join unless an outer join is explicitly specified (see later). In these cases, **inner** may be explicitly declared (e.g., **inner join** or **natural inner join**). If **natural** is declared, an **on** clause or a **using** clause must not be. If three or more tables are included in a join expression, the joins are normally evaluated in a left-to-right order. The join order can also be controlled by inserting parentheses. Because natural inner joins are associative, their order doesn't affect the actual result.

As a more complex example, consider the relational schema in Fig. 12.44. In the first table, "mgrEmpNr" denotes the employee number of the department manager (where there is a manager). The pair-subset constraint indicates that an employee manages a department only if he or she works in it. The (2,1) item-list marker on this constraint is used here to indicate the reordering of the pair of values to (mgrEmpNr, deptCode) before comparing with (empNr, deptCode).

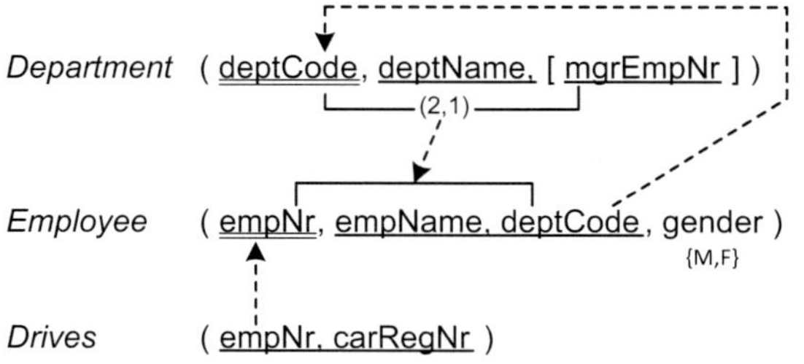

FIG. 12.44 A relational schema about employees.

Now consider the following query: for each department with a manager, list its code and name, and the employee number and name of its manager.

We must match employee numbers between Department and Employee tables. The department codes must also match, but do we need to specify this? No, because if the manager referenced in the department table is the same as the employee in the Employee table, the matching of department codes is implied by the constraints in the schema (why?). The query may be formulated with the following conditional join. A natural join won't work because the join columns have different names.

```
select Department.deptCode, deptName, mgrEmpNr, empName
from  Department join Employee
  on mgrEmpNr = empNr
```

So far, we have discussed four of the eight joins listed in Table 12.8. There are also three outer joins, as well as a union join. The union join is rarely used and is not discussed further. *Outer joins* may be *left, right,* or *full.* Left and right outer joins are often encountered in commercial applications. Given two tables *A* and *B*, their left/right/full outer join is formed by first computing their inner join and then adding the rows from the left/right/both table(s) that don't have a match in the inner join and padding them with nulls to fill the extra columns in the result table.

Fig. 12.45 provides a simple example of a natural left outer join. The query lists the employee number, name, and car registration numbers (if any) of all the employees (including those who don't drive cars). Each of the two formulations shown is legal in SQL-92 (and later). The first uses the natural join syntax, but this is not yet supported by most SQL dialects. The second formulation uses the conditional outer join syntax, which is typically supported.

The natural inner join results in the three rows shown for the two drivers (employees 001 and 003). Employees 002 and 004 have no match in the Drives table. The left outer join adds their rows padded with a null value for car registration number. We display a null value as "?". Some systems display it instead as a blank or as "NULL". Many systems allow you to choose how you would like nulls to display.

Employee:

empNr	*empName*	*deptCode*	*gender*
001	Hagar T	DV	M
002	Wong S	DV	F
003	Jones E	MK	F
004	Mifune K	SL	M

Drives:

empNr	*carRegNr*
001	ABC123
003	ABC123
003	TAH007

```
select empNr, empName, carRegNr
from Employee natural left outer join Drives                ⇨

select Employee.empNr, empName, carRegNr
from Employee left outer join Drives
     on Employee.empNr = Drives. empNr                      ⇨
```

empNr	*empName*	*carRegNr*
001	Hagar T	ABC123
002	Wong S	?
003	Jones E	ABC123
003	Jones E	TAH007
004	Mifune K	?

FIG. 12.45 Two equivalent formulations of a natural left outer join.

Although the conditional join syntax requires qualified column names for join columns, when the result table is displayed, some systems (e.g., SQL Server) are clever enough to omit the table qualification from the column header (as in Fig. 12.45). Some systems support an alternative outer join syntax that was introduced before outer joins became part of the SQL-92 standard. This syntax often has different semantics and should never be used unless the standard syntax is unsupported.

SQL-92 and later versions are *orthogonal.* Anywhere a value is legal, so is any expression that returns that value. This makes the language easier to use and increases the expressive power of single queries. However, many commercial systems are not fully orthogonal, and often place restrictions on outer join queries (e.g.. at most one outer join allowed, or no other joins allowed after an outer join). With such systems, it is sometimes necessary to formulate a single standard query as a series of queries using intermediate tables to store results on the way.

Right outer join is analogous to left. Full join is the union of left and right. The word **outer** is assumed for left/right/full joins, so may be omitted (e.g., **natural left join** or **left join**). Outer joins are not associative, so be careful with the join order when outer joining three or more tables. As discussed later, outer joins can be emulated in the old SQL-89 syntax by means of **union** and subqueries.

When different rows of the same table must be compared, we need to join the table to itself. Such *self-joins* were discussed in Section 12.1. Recall that this requires introducing an alias for the table. In SQL, a *temporary alias* may be declared as a *tuple variable* in the **from** clause after the table it aliases. This variable may be assigned any row from the table and is sometimes called a “correlation variable”, “range variable”, or “table label”. For clarity, **as** may be used to introduce the alias.

Fig. 12.46 provides a simple example. The base table Scientist stores the name and gender of various scientists, and the query lists pairs of scientists of opposite gender. The aliases S1 and S2 are declared in the **from** clause. To understand the self-join, think of S1 and S2 as copies of the original base table, as shown. The conditional join performs a <>-join on gender (to ensure opposite gender) and a <-join on PersonName (to ensure that each pair appears only once, rather than in both orders).

"copies"

Scientist:

personName	gender
Curie, Marie	F
Curie, Pierre	M
Edison, Thomas	M
Lovelace, Ada	F

S1:

personName	gender
Curie, Marie	F
Curie, Pierre	M
Edison, Thomas	M
Lovelace, Ada	F

S2:

personName	gender
Curie, Marie	F
Curie, Pierre	M
Edison, Thomas	M
Lovelace, Ada	F

```
select S1.personName, S2.personName
from  Scientist as S1 join Scientist as S2
      on S1.gender <> S2.gender
         and S1.personName < S2.personName
```

⇨

S1.personName	S2.personName
Curie, Marie	Curie, Pierre
Curie, Marie	Edison, Thomas
Curie, Pierre	Lovelace, Ada
Edison, Thomas	Lovelace, Ada

FIG. 12.46 A self-join is used to list pairs of scientists of opposite gender.

Some versions of SQL also provide a "create synonym" command for declaring a permanent alias definition. However, this is not part of the standard. In simplified form, the **from** clause syntax of SQL queries may be summarized in BNF as follows (for alternatives in braces, exactly one is required):

```
from table [ [as ] alias ]
   [ , | cross join table [ [ as ] alias ]
     | natural [ inner | { left | right | full } [ outer ] ] join table [ [as ] alias ]
         | [ inner | { left | right | full } [ outer ] ] join table [ [ as ] alias ]
                   { on condition | using ( col-list ) }
         | union join table [ [ as ] alias ]
         [ ,... ] ]
```

In formulating an SQL query, the guidelines discussed in relational algebra usually apply. First, state the query clearly in English. Then try to solve the query yourself, watching how you do this. Then formalize your steps in SQL. This usually entails the following moves.

What tables hold the required information? Name these in the **from** clause. What columns do you want, and in what order? Name these (qualified if needed) in the **select** list. If the **select** list doesn't include a key, and you wish to avoid duplicate rows, use the **distinct** option.

If you need *n* tables, specify the *n* – *1* join conditions. What rows do you want? Specify the search condition in a **where** clause. What order do you want for the rows? Use an **order by** clause for this.

The new join syntax introduced in SQL-92 is convenient, but adds little in the way of functionality. The most useful notations are those for conditional and natural joins (both inner and outer).

If you are using a version of SQL that does not support the new syntax, you should take extra care to specify the join conditions in detail and to qualify column names when required.

Any SQL statement may include *comments*. These might be used for explanation, documentation, or to simply comment out some code for debugging (comments are ignored at execution time).

In SQL-92, a comment begins with two contiguous hyphens "--" and terminates at the end of the line. In addition to these single-line comments, SQL:1999 introduced multiline comments, starting with "/*" and ending with "*/". SQL Server supports both these comments styles. Here's an example:

```
-- This query retrieves details about the employees who drive a car
select *              -- select all the columns
from Employee natural join Drives
/* If the natural join syntax is not supported,
          then omit "natural", and specify the join condition in an on-clause */
```

Exercise 12.6

1. This question refers to the student database discussed in Exercise 12.1. Its relational schema is shown. Formulate the following queries in SQL.

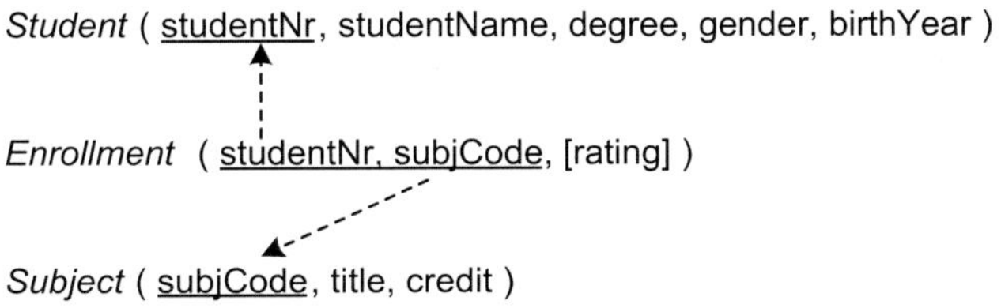

   (a) List studentNr, name, degree, and birthYr of the students in ascending order of degree. For students enrolled in the same degree, show older students first.
   (b) For each student named "Smith J", list the studentNr and the codes of subjects (being) studied.
   (c) List the studentNr, name, and gender of all students studying CS113.
   (d) List the titles of the subjects studied by the student with studentNr 863.
   (e) List the studentNr, name, and degree of those male students who obtain a rating of 5 in a subject titled "Logic". Display these in alphabetic order of name.
   (f) List the code and credit points of all subjects for which at least one male student enrolled in a BSc scores a rating of 7. Display the subjects with higher credit points first, with subjects of equal credit points listed in alphabetic order of code. Ensure that no duplicate rows occur in the result.
   (g) List the code and title of all the subjects, together with the ratings obtained for the subject (if any).

2. The relational schema shown is a fragment of an application dealing with club members and teams. Formulate the following queries in SQL.

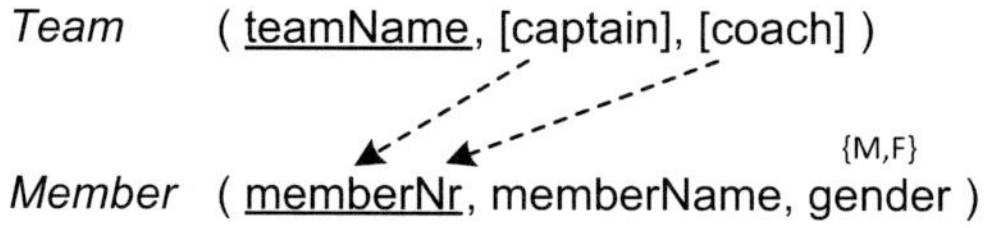

(a) List the teams as well as the member numbers and names of their captains.
(b) Who (member number) is captain and coach of the same team?
(c) Who (member number) captains some team and coaches some team?

To record who plays in what team, the following table scheme is used:

PlaysIn (memberNr, teamName)

(d) What intertable constraints apply between this table and the other two tables?

Formulate the following queries in SQL.
(e) List details of all the members as well as the teams (if any) in which they play.
(f) List details of all the members as well as the teams (if any) that they coach.
(g) Who (memberNr) plays in the judo team but is not the captain of that team?
(h) Who (memberNr and name) plays in a team with a female coach?

12.7 SQL: in, between, like, and is null Operators

This section examines four special operators (other than the comparators =, <, etc.) that SQL provides for use within search conditions: **in**; ...**between**...**and**...; **like**; and **is null.** These are sometimes called functions, and the conditions they are used to express are called "predicates" in the SQL standard.

A function is something that takes zero or more values as arguments and returns a single value as its result. You are probably familiar with functions from mathematics or programming, such as cos(x) or sqrt(x). As these examples illustrate, syntactically a function is usually represented as a function identifier preceding its arguments, which are typically included in parentheses.

If the action performed by a function is represented without bracketing all the arguments, we usually describe it as involving operators and operands rather than functions and arguments. Operators may be represented in infix, prefix, postfix, and mixfix notation according to whether the operator appears between, before, after, or mixed among the operands. For instance, the sum of 2 and 3 might be set out as:

```
sum (2, 3)         -- function
2 + 3              -- infix operator
+ 2 3              -- prefix operator
2 3 +              -- postfix operator
sum of 2 and 3     -- mixfix operator
```

The four operators we are about to discuss are used to express search conditions. The first three return the value True, False, or Unknown while the **is null** operator returns True or False. Our initial treatment focuses on the SQL-89 version of these operators. Extensions for SQL-92 onward are mentioned later. Most of our examples are based on the Person table, reproduced here as Table 12.9.

Table 12.9 A relational table storing personal details.

Person:

firstname	*gender*	*starsign*	*birthyr*
Bob	M	Gemini	1967
Eve	F	Aquarius	1967
Fred	M	Gemini	1970
Norma	F	Aries	1950
Selena	F	Taurus	1974
Terry	M	Aquarius	1946

Suppose we want the names and birth years of the people born in 1950, 1967, or 1974. One way of requesting this information is shown in the following SQL query. As an exercise, check that this results in four rows.

```
select firstname, birthyr from Person
where birthyr = 1950 or birthyr = 1967 or birthyr = 1974
```

Imagine how tedious this way of phrasing the request would be if there were a dozen or more years involved. Partly to make life easier in such situations, SQL includes an **in** operator to handle *bag membership* and hence *set membership*. Using this infix operator, the aforementioned request may be formulated more briefly as:

```
select firstname, birthyr from Person
where birthyr in (1950,1967,1974)
```

Here, the search condition is that the birthyr value is a member of the bag containing the values 1950, 1967, and 1974. The order in which these values are written does not matter, nor would it matter if any are duplicated. In general, if x is some expression (e.g., a column name) and a, b, etc., are data values (e.g., numeric or string constants), then the SQL condition shown here on the left is equivalent to the mathematical expression shown on the right:

x **in** (a, b ...) means $x \in [\, a, b \ldots]$

We use **in** instead of "∈" and parentheses "( )" instead of square brackets "[ ]" for *bag delimiters*. Because a set may be thought of as a bag with no duplicates, the **in** operator may also be used for set membership. Unlike many programming languages, SQL allows bags or sets to contain character strings (not just numbers). For example, the following query lists the Aquarians and Taureans (Eve, Selena, and Terry):

```
select firstname from Person
where starsign in ('Aquarius','Taurus')
```

To indicate that the value of an expression does *not* belong to a bag or set, the logical **not** operator may be used with the **in** operator. As with ordinary comparison operators, SQL gives the **in** operator higher priority than logical operators (unlike most programming languages). In SQL, the condition "not (*x* in *S*)" may be rendered more briefly as **not x in S**. Even better, SQL also allows **not in** for "$\notin$", so this may set out more naturally as x **not in** S.

x **not in** (*a*, *b* ...) means *x* $\notin$ [*a*, *b* ...]

For example, to obtain the names and birth years of those not born in any of the years 1950, 1967, or 1974, the following query may be used (yielding Fred and Terry):

```
select firstname, birthyr from Person
where birthyr not in (1950,1967,1974)
```

Sometimes, we wish to determine whether an expression has a value occurring in a *range* of values. In mathematics, to say that some variable *x* has a value in the range from *a* to *b*, we usually express this as $a \leq x \leq b$. This notation is illegal in SQL. Instead, we could say a<= x **and** x<= b. For the sample population, the following query returns the name and birth year of all but Terry.

```
select firstname, birthyr from Person
where 1950 <= birthyr and birthyr <= 1974
```

As another way to specify range membership, SQL provides the ternary mixfix operator "**...between...and...**" that may be defined as shown, where *x*, *a*, and *b* may be scalar expressions:

x **between** *a* **and** *b* means *a* <= *x* **and** *x* <= *b*

For example, the previous query may be reformulated as:

```
select firstname, birthyr from Person
where birthyr between 1950 and 1974
```

Note carefully that, in contrast to ordinary English, **between** in SQL is read in an *inclusive sense*. For instance, both 1950 and 1974 are included in the given range.

Notice also that here **and** is just part of the mixfix operator; it is not a logical operator. As already noted, the <= operator may be used to order strings as well as numbers. So, strings may be used as operands. For example, the following query returns the bag ("Bob", "Eve", "Fred").

```
select firstname from Person
where firstname between 'Bob' and 'Fred'
```

Nonmembership in a range may be expressed with the help of the **not** operator, which has lower priority than **between**, and may be placed just before the word **between**. Thus, each of the following conditions is equivalent to *x* < *a* **or** *x* > *b*.

x **not between** *a* **and** *b* **not** (*x* **between** *a* **and** *b*) **not** *x* **between** *a* **and** *b*

The **not between** formulation is easier to read. For example, the following query returns just the tuple ('Terry', 1946).

select firstname, birthyr **from** Person
where birthyr **not between** 1950 and 1974

In the pure relational model, column values are atomic. However, SQL provides a number of ways of accessing substrings within character string values. In particular, the **like** operator may be used for *pattern matching* with character strings. The general form of the condition in SQL-89 may be set out in BNF thus:

char-col [**not**] **like** *quoted-string* [**escape** *quoted-char*]

Here, char-col is the name of a column based on a character string data type. The quoted string is a character string, surrounded by single quotes, which may contain *wildcard characters.* You may be familiar with the use of wildcards for matching filenames at the operating system level (e.g., "*" in Microsoft Windows).

In the absence of an escape clause, SQL gives the percentage character "%" and the underscore character "_" the following special meanings if included in a quoted string operated on by **like**:

%	means	zero or more characters
_	means	any single character

The "%" is generally more useful, but "_" is needed if the character's position in the string is important. Fig. 12.47 provides a few examples based on Table 12.9. The **like** operator has priority over logical operators, so "*x* **not like** *s*" means "**not** (*x* **like** *s*)".

select distinct starsign **from** Person
where starsign **like** 'A%' ⇨

starsign
Aquarius
Aries

select distinct starsign **from** Person
where starsign **like** '%a%' ⇨

starsign
Aquarius
Taurus

select distinct starsign **from** Person
where starsign **like** '_a%' ⇨

starsign
Taurus

select firstname **from** Person
where name **not like** '%e%' ⇨

firstname
Bob
Norma

FIG. 12.47 "%" and "_" wildcards for pattern matching with the **like** operator.

Note that the "%" and "_" are interpreted as wildcard characters only in the context of a **like** condition. For example, the following query returns the null set. There are no rows that satisfy the condition because if "=" is used instead of **like**, the expression "A%" is taken literally.

select starsign **from** Person
where starsign = 'A%'

As discussed earlier, to include a single quote in a string, we use two single quotes. For example, for the table scheme *Actor* (surname, firstname, gender) considered earlier, the following query returns each actor whose surname begins with "D'": **select** surname **from** Actor **where** surname **like** 'D''%'.

To see a few examples where the **like** operator is quite useful, consider Table 12.10, which indicates which subjects are offered in which semester at a given university. The sample population is small to save space.

At this university, subject codes have the following meaning: the first two characters indicate the discipline area (e.g., "CS" denotes Computer Science and "PD" denotes Philosophy) and the third character denotes the year level (e.g., "1" for first-year level).

Table 12.10 A relational table showing subjects offered and semesters.

Offering:

subject	*semester*
CS112	1
CS112	2
PD102	1
PD102	2
CS225	2
CS314	1

Try to formulate the following queries yourself before peeking at the answers.

List the computer science subjects.
List the first-year level subjects.
List subjects higher than first level that are offered in semester 2.

The queries in order are:

select distinct subject **from** Offering **where** subject **like** 'CS%'

select distinct subject **from** Offering **where** subject **like** '_ _1%'

select distinct subject **from** Offering
where subject **not like** '_ _1%' **and** semester = 2

Another common example of a string that is often interpreted as having structure is a person name. Usually, a person's family name as well as either a first given name or initials is recorded (e.g., "Smith, James" or "Smith JB"). In Western culture, the terms "surname" and "first name" are often used for family name and first given name. In some countries (e.g., Japan), the family name is stated first.

Suppose we wish to distinguish between these two parts of a person's name. One way to implement this is to include two columns, one for the family name and another for the first given name (or initials). The structure is then known to the system, and each of these two parts of the name can be accessed individually (this is especially handy for SQL's **group by** facility).

Alternatively, we might use just one column to store the whole name and then use the **like** operator to distinguish the two parts—the name structure then becomes derived rather than stored.

Table 12.11 A relational table concerning tennis players.

Player:

personName	*gender*	*height*
Smith, James	M	180
Smith, Sue	F	?
Smithers, James	M	175
James, Susan	F	170

For example, consider Table 12.11, which provides details of players in a mixed doubles tennis match. Try the following queries and then check your answers.

List details of people with family name "Smith"
List details of people with first given name "James"

In order, the queries are:

select * **from** Player **where** personName **like** 'Smith, %'

select * **from** Player **where** personName **like** '%, James'

Conditions using the **like** operator may optionally include an *escape character*. If chosen, it may be used as a lead-in character to have "%" and "_" interpreted literally rather than as wildcards. For example, the query in Fig. 12.48 lists starships with an underscore character in their name. Here, "\" is used as the lead-in character; any character that doesn't occur in the string being investigated could be used. Some SQL dialects extend the capability of the **like** operator with additional wildcards. For example, SQL Server uses [ ] to include a range of characters, and [^] to exclude characters.

Although nulls may be displayed (e.g., as "?"), they cannot be used with the usual comparison operators. Instead, a special postfix Boolean ...**is null** operator is used. This operator is placed after the name of the column on which it operates. It returns True if the column value is null, and False otherwise. For example, the following query on the Player table returns "Smith, Sue" because her height is unknown:

Starship:

shipName	*maxSpeed*
Alpha_1	warp 5
Enterprise	warp 8
Epsilon_33	warp 7
Galactica	warp 5

select shipName **from** Starship
where shipName **like** '%_%' **escape** '\' ⇨

shipName
Alpha_1
Epsilon_33

FIG. 12.48 Here, "\" is used as an escape character to detect an underscore.

select personName **from** Player
where height **is null**

Neither of the expressions "height = null" nor "height = '?'" is allowed. To specify that a value is not null, use the logical **not** operator before the column name or the word **null**. For instance, either query shown here returns the names of the three other players.

select personName **from** Player **where not** height **is null**
select personName **from** Player **where** height **is not null**

Using "col" to denote a column, the BNF syntax for null comparisons in SQL-89 is:

col **is** [**not**] **null**

While the **is null** operator always returns True or False, other comparison operators return unknown when one of the arguments is the null value. Suppose we want the names of the players who are between 170 and 175 cm tall (inclusive). The following query returns just two names: "Smithers, James", "James, Susan".

select personName **from** Player
where height **between** 170 **and** 175

Although Sue Smith's height might be in this range in the real world, she is excluded from this result. The system does not know her height, so will not evaluate the condition to True in her case. Suppose now we want the names of those players whose height is not in the range 170..175. The following query returns just one name: "Smith, James".

select personName **from** Player
where height **not between** 170 **and** 175

Notice that Sue Smith is excluded from this result as well. As far as the system is concerned, in her case the truth value of the condition "height between 170 and 175" is neither true nor false; rather, it is unknown.

A row is included in a result only if it satisfies the search condition, that is, the condition is (known by the system to be) True for that row. If the condition is either False or Unknown, the row is excluded from the result.

As an extreme example, the following query returns all the players except Sue Smith.

```
select personName from Player
where height = 170 or height <> 170
```

p	**not** *p*
1	0
0	1
?	?

p	*q*	*p* **and** *q*
1	1	1
1	0	0
1	?	?
0	1	0
0	0	0
0	?	0
?	1	?
?	0	0
?	?	?

p	*q*	*p* **or** *q*
1	1	1
1	0	1
1	?	1
0	1	1
0	0	0
0	?	?
?	1	1
?	0	?
?	?	?

FIG. 12.49 Truth tables for three-valued logic (1 = True, 0 = False, ? = Unknown).

For Sue Smith's row, each comparison in this condition evaluates to unknown, and applying the **or** operator to unknowns returns unknown. The truth value of any condition is evaluated according to a *three-valued logic*, as set out in Fig. 12.49. Here, "1", "0", and "?" denote the values "True", "False", and "Unknown", respectively.

Although SQL-89 and SQL-92 allow only one kind of null value, Codd (1990) proposed a four-valued logic in "version 2" of the relational model to allow for two kinds of unknown (applicable and inapplicable), but this has not received popular approval.

A summary of the SQL-89 syntax of the four operators introduced in this section is shown. Here, "expn" denotes an expression such as a column name or constant (or a combination of these connected by arithmetic operators +, −, *, / ... see later).

expn [**not**] **in** (*constant-list*)
expn [**not**] **between** *expn* **and** *expn*
char-col [**not**] **like** *quoted-string* [escape *quoted-char*]
col **is** [**not**] **null**

In addition, SQL-89 allows the **in** operator to be used with subqueries, and the use of quantified comparisons and the **exists** quantifier in search conditions (see later). SQL-92 (and later) significantly extended the kinds of expressions and operators allowed in search conditions.

For example, SQL-92 allows a *row value expression* (e.g., a tuple of values) as the left argument of any comparator (e.g., =, **in**) and includes **unique**, **match**, and **overlaps** predicates. **Unique** is used to check whether each null-free row returned by a subquery is unique, **match** is used to check whether a row matches a row returned by a subquery, and **overlaps** is used to test whether two datetime periods overlap.

SQL:1999 goes further, adding **similar**, **distinct**, and **type** predicates for use in search conditions. **Similar** allows character strings to be compared by means of a regular expression,

distinct tests whether two row values are distinct, and **type** tests whether a user defined type value expression conforms to a user defined type specification. Dialects of SQL differ in their support for these standard predicates. However, all SQL dialects now support the SQL-89 syntax discussed in this section.

Exercise 12.7

1. The table shown below contains data about computer languages. Phrase the following queries in SQL.
   (a) Which languages were released in the years 1959, 1975, or 1979?
   (b) Which languages were released in the period 1959–1979?
   (c) Which languages do not have their release year recorded?
   (d) List the name and release year of any Goodo language.
   (e) List the name of any language ending with "OL".
   (f) List the name and release year of any Pascal language.
   (g) List language names that are five characters long.
   (h) List language names with "o" as the second character.
   (i) List the name of any language containing the letter "O" or "o".
   (j) List the name and release year of languages not starting with "P" that were released after 1959 and before 1979.
   (k) List language names that either are six characters long and have "a" as the last or second last character or are five characters long and end with "OL".

CompLanguage:

title	*releaseYr*
FORTRAN	1957
ALGOL	1958
COBOL	1959
Logo	?
Pascal	1971
Prolog	1972
SQL	?
Modula	1975
Modula 2	1979
Goodo Pascal 1.0	1983
Goodo Pascal 2.0	1984
Goodo Pascal 3.0	1985
Goodo Prolog	1986

2. Suppose that in later years, computer languages with the following titles were released: Modula_3; Oberon_2; Ada%93; COBOL%93. Write an SQL query to list all languages whose title includes an underscore or a percentage sign.

12.8 SQL: Union and Simple Subqueries

This section discusses two ways in which **select** queries may be used as components within a larger **select** query. First, we examine how to specify the operations of union, intersection, and difference in SQL. Next, we look at basic subqueries in SQL. Let's begin with the union operation. The basic syntax is shown next. The reserved word "union" is allowed only between **select** queries, not between table names.

select-query
union [**all**]
select-query
...

The **select** queries must be union compatible (i.e., same number of columns, and corresponding columns are based on compatible data types). By default, a set is returned that is the union of the tables returned from the individual **select** queries, with any duplicate rows eliminated. To return a bag, specify **all** to include duplicate rows. Note that this is the opposite of the SQL **select** list syntax, which returns a bag by default and uses **distinct** to remove duplicates.

SQL:1999 allows the **distinct** option to be explicitly declared after **union**; if omitted it is assumed by default. SQL Server supports the "**union all**" syntax, but not the "**union distinct**" syntax because it always removes duplicates from **union** unless **all** is added (so adding "**distinct**" is redundant even if it might clarify the default).

As an example, the query in Fig. 12.50 lists each person who is an American or drives a Ford. The first **select** finds the Americans ("JimO", "Lance"), the second **select** finds the Ford drivers ("Colleen", "Lance"), and the **union** includes any person in either or both of these intermediate results. Although Lance appears in both tables, he appears just once in the final union because duplicates are removed by default.

If *duplicates* are desired, the **all** option must be specified after **union**, yielding a bag. For example, **union all** in the previous query would cause Lance to appear twice in the result. In general, if a row appears m times in one result table and n times in another, then it appears $m + n$ times in the bag union of those results.

Because column headings in the final result are chosen from the first query in the union, the heading "shortName" appears here. Although this query may be formulated in other ways (see subqueries later), the union operation provides the most natural method of merging results from compatible result tables.

Because base tables are rarely union compatible, the union operator is mostly used on derived tables that are made union compatible by projection, as in this example. The **union** operation may be applied several times in a query. If an **order by** clause is used, it must come after the last query in the union. For example, appending the clause **order by** shortName sorts the result into "Colleen", "JimO", "Lance", "Walter".

Person:

shortName	*nationality*
Colleen	Irish
JimO	American
Lance	American
Norma	Aussie
Terry	Aussie
Walter	American

Drives:

driver	*carModel*
Colleen	Ford
JimO	Honda
Lance	Ford
Lance	Honda
Terry	Saturn

Who is an American or drives a Ford?

select shortName **from** Person
where nationality = 'American'
union
select driver **from** Drives
where carModel = 'Ford' ⇨

shortName
Colleen
Lance
Walter
JimO

FIG. 12.50 By default, union eliminates duplicates.

In SQL-92 (and later), a **corresponding** option may be added to specify the common columns used for the union. For example, if the driver column in Fig. 12.50 were instead named “shortName”, the query could be declared as **select** * **from** Person **union corresponding** Drives or as **select * from** Person **union corresponding by** (shortName) Drives. This option is not yet supported by most SQLs, including SQL Server.

Now consider the query, Who drives a Ford, Holden, or Honda (i.e., who drives at least one of these car models)? Although we could formulate this using two unions, the query can be answered with one scan of the rows from the Drives table. So, it is simpler (and more efficient) to use the **or** operator, thus:

select driver **from** Drives
where carModel = 'Ford' **or** carModel = Holden' **or** carModel = 'Honda'

When the operands of a **union** operator are disjoint (i.e., mutually exclusive), the **union** operator is sometimes used to add descriptions to the output even when only one base table is involved. As an example, consider the Expert and Novice tables in Fig. 12.51. Suppose that we want to list the names of the judo players as well as a label to indicate who are the experts and who are the novices. To obtain the names, we can union a **select** query for the judo experts with another **select** query for the judo novices. We could then identify the expert and novice judoka by including the strings “expert judoka” and “novice judoka” in the relevant **select** lists.

We can improve on this by introducing “judo level” as an alias for the labels. This provides a meaningful column header for the labels and allows us to shorten the labels to just “expert” and “novice”, as shown. Notice the use of double quotes in “judo level” to allow an embedded space.

Expert:

person	*sport*
Ann	aikido
Ann	judo
Bill	judo
John	bojutsu

Novice:

person	*sport*
Ann	karatedo
Bill	karatedo
Cathy	aikido
Cathy	judo
David	kendo

```
select person, 'expert' as "judo level"
from Expert
where sport = 'judo'
union
select person, 'novice' as "judo level"
from Novice
where sport = 'judo'
order by "judo level", person   ⇨
```

person	*judo level*
Ann	expert
Bill	expert
Cathy	novice

FIG. 12.51 Constants may be used with union to indicate the source of the rows.

Although this technique of adding descriptive strings in unions is handy when the operands are disjoint, it is not of much use if they overlap. Any originally overlapping rows become disjoint when the string values are added into the rows (because those values differ). Hence, such rows would appear more than once in the union.

In SQL-92 (but not SQL-89), the **intersect** operator is provided for *set intersection* and the **except** operator handles *set difference*. These work analogously to the **union** operator. They are allowed only between **select** queries that return union-compatible tables. Duplicates are removed from the result unless the **all** option is specified.

Refer back to the tables *Person*(shortName, nationality) and *Drives*(driver, carModel) in Fig. 12.50. Now consider the query, Which Americans drive a Ford? This means, Who is an American and drives a Ford? Because the nationality facts and carModel facts are in separate tables, this logical "**and**" operator can naturally be implemented by a table intersection operator, as follows, yielding the result {"Lance"}.

```
select shortName from Person
where nationality = 'American'
intersect
select driver from Drives
where carModel = 'Ford'
```

Similarly, the query "Which Americans do not drive a Ford?" can be formulated as a set difference as follows, yielding the result {"JimO", "Walter"}.

```
select shortName from Person
where nationality = 'American'
except
select driver from Drives
where carModel = 'Ford'
```

Unlike union and intersection, the difference operator is not commutative. For example, reversing the order of the operands in the previous query results in {"Colleen"} because the reversed query has the different meaning, "Which Ford drivers are not American?"

If the **all** option is used after **intersect** or **except**, it returns a bag intersection or bag difference, respectively. In general, if a row appears *m* times in one table and *n* times in another, then it appears min(*m, n*) times in the bag intersection and max(*m* - *n*, 0) times in the bag difference.

Although all the main commercial SQL dialects support the **union** operator, several do not yet support the **intersect** and **except** operators. As we now discuss, these two operators can be emulated using subqueries, so their absence does not entail a loss of functionality.

SQL *subqueries* are like component queries within nested formulations in relational algebra. Although SQL subqueries are queries in their own right, they must be enclosed in parentheses (unlike **select** queries that are unioned), and they have some limitations on their use. Ideally, a language ought to be orthogonal—wherever a value or result is legal, an expression that returns that value or result should be legal. In SQL-89, however, subqueries are allowed only in a few places where their result is legal. Although SQL-92 removed these restrictions, few commercial SQLs are fully orthogonal.

A subquery is a parenthesized **select** query embedded in another query. Let Θ denote one of the comparison operators (=, <>, <, >, <=, >=). In SQL-89, a subquery is legal only in the following three contexts:

```
expression [not] in (subquery)                    -- membership subquery
expression Θ [all | some | any] (subquery)        -- [quantified] comparison subquery
exists (subquery)                                 -- existential subquery
```

A *membership subquery* follows an **in** operator. In SQL-89, a membership subquery must be a *singleton query* (i.e., it must return a single column). This is consistent with the use of the **in** operator with a bag of values (a column is a just bag of values, displayed vertically). In SQL-92, the **in** operator may be preceded by a row value expression, in which case the subquery may return a multicolumn table.

A *comparison subquery* follows a comparator (e.g., =). If the comparator is modified by one of the quantifiers **all**, **some**, or **any**, this is a *quantified comparison*. In SQL-89, a comparison subquery must return a single column; if the comparison is not quantified, the subquery must return a single value. This is consistent with the normal use of comparison operators. In SQL-92, a comparator may be preceded by a row value expression, in which case the subquery may return a multicolumn table or, if unquantified, a row value. An *existential subquery* appears after the **exists** quantifier and may return any table.

In SQL-92, subqueries may be used in other places (e.g., **from** clauses and **check** clauses), as long as their result table makes sense there (orthogonality). **From** clause subqueries require an alias for the derived table even if not needed. The syntax is:

```
... from (subquery) [as] alias
```

SQL Server supports subqueries in a **from** clause, but not in a **check** clause. Some SQL dialects limit their subquery support to the three SQL-89 cases, each of which embeds the subquery in a where clause of an outer query, thus:

```
select ...         -- outer query
from ...
where ...
   (select ...     -- subquery
     from ... )
```

If the subquery is self-contained (i.e., it makes no reference to the outer query), it is said to be *uncorrelated*. In this case, it is evaluated first and replaced by its result (e.g., a bag of values) before the outer query is processed. In contrast, *correlated subqueries* refer to their outer query. This section considers only simple, uncorrelated membership and comparison subqueries. Existential and correlated subqueries are covered later.

Country:

countryName	*population*
Australia	26,477,504
India	1,428,627,663
Netherlands	17,618 299
USA	340,265,309

Mountain:

mtName	*countryName*
Kosciusko	Australia
Mana	India
McKinley	USA
Rainier	USA

FIG. 12.52 Some countries and some of their mountains (if any).

Let's now consider *membership subqueries.* The condition *expression* **in** (*subquery*) returns True if and only if the value of the expression is included in the subquery result. If the subquery returns the null set, the condition evaluates to unknown. The outer query result includes only the rows where the condition evaluates to True.

To discuss some examples, consider the tables shown in Fig. 12.52. Assume that there is a subset constraint from Mountain.countryName to Country.countryName. The opposite constraint does not apply (e.g., The Netherlands has no mountains). To save space, only some mountains are shown for the countries.

Consider the query, Which mountains are located in a country with more than 100 million people? We could begin by finding the countries with the relevant populations, and then finding the mountains in those countries. This may be set out in SQL as follows:

```
select mtName from Mountain
where countryName in
      (select countryName from Country
      where population > 100000000)
```

The subquery is evaluated first, resulting in the bag ("India", "USA"). After the subquery has been replaced by this intermediate result, the outer query is evaluated, finally resulting in a list of three mountains: Mana, McKinley, and Rainier. Of course, this query could have been phrased instead using a join. As an exercise, you may wish to do this.

Subqueries may be nested (i.e., their search conditions may include subqueries). To improve readability, subqueries should always be indented. The deeper the level of nesting, the greater the indentation should be.

Often, queries using subqueries may be replaced by **join** queries, and vice versa. Some joins cannot be reformulated in terms of subqueries. In particular, *if the* **select** *list includes columns from more than one table, a join must be used* (self-joins must be used if the select columns are from different copies of the same table).

For example, the following query cannot be reformulated as a subquery (what columns would the result table contain?).

```
select * from Country natural join Mountain
```

Now consider the relational schema for our software retailer UoD, which was discussed in Exercise 6.3 and is reproduced in Fig. 12.53.

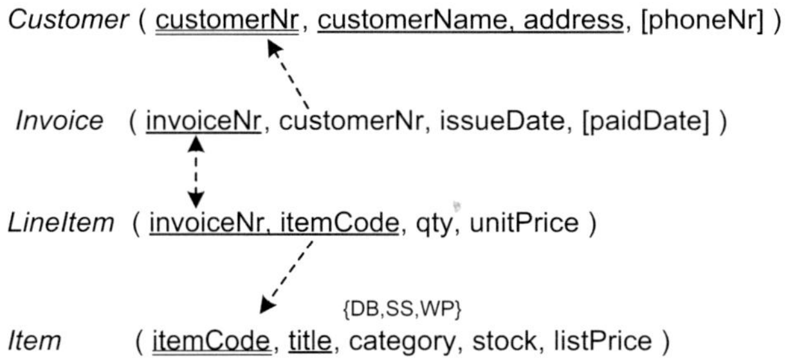

FIG. 12.53 The relational schema for an invoice application.

Suppose we wish to issue the query, List the customer number, customer name, and invoice number of each invoice that records a shipment of the product titled "Quinquo". You might like to try this yourself in SQL before reading on.

The customer name, customer number, and invoice number are to be selected, but do not all occur in the same table. Hence, we must include a join in the query to bring these together on the same row of the result table.

Looking at the schema, the minimum join we require is to join the Customer and Invoice tables together on customerNr. The rest of the navigation between tables can be done with subqueries because we don't need to include any other columns from those tables in the **select** list of the final result. This approach leads to a formulation with one join and nested subqueries as follows:

```
select customerNr, customerName, invoiceNr
from  Customer natural join Invoice
where invoiceNr in
   (select invoiceNr from Lineitem
   where itemCode in
       (select itemCode from Item
       where title = 'Quinquo')
```

Alternatively, we could replace any of the subqueries by a join. For example, the following formulation uses three joins and no subqueries:

```
select customerNr, customerName, invoiceNr
from   Customer natural join Invoice
       natural join Lineitem
       natural join Item
```

If your SQL dialect doesn't support the natural join syntax, use either the conditional join syntax or the old SQL-89 join syntax instead.

As a negated example of subqueries, the following query lists the countries with no mountains. For the given population, this results in just one country, The Netherlands.

```
select countryName from Country
where countryName not in
      (select countryName from Mountain)
```

Can you replace a negated subquery with an inner join? No! However, you can replace it with an outer join. For example, the aforementioned query may be reformulated as:

```
select countryName
from Country natural left outer join Mountain
where mtName is null
```

The query may also be formulated using a set difference operation, thus:

```
select countryName from Country
except
select countryName from Mountain
```

To list those countries that do have mountains, the following query could be used:

```
select countryName from Country
where countryName in
      (select countryName from Mountain)
```

This performs a set intersection and is equivalent to:

```
select countryName from Country
intersect
select countryName from Mountain
```

So, when just single columns are required, *the use of* **in** *and* **not in** *with subqueries enables set intersection and set difference operations to be formulated* even in SQL-89. To discuss this further, recall the *Person* (shortName, nationality) and *Drives* (driver, carModel) tables from the previous section. For convenience, the sample population for these tables is repeated in Fig. 12.54.

Person:

shortName	*nationality*
Colleen	Irish
JimO	American
Lance	American
Norma	Aussie
Terry	Aussie
Walter	American

Drives:

driver	*carModel*
Colleen	Ford
JimO	Honda
Lance	Ford
Lance	Honda
Terry	Saturn

FIG. 12.54 Two relational tables.

Recall that the following queries were used to list the Americans who drive a Ford or do not drive a Ford, respectively:

```
select shortName from Person
where nationality = 'American'
intersect
   select driver from Drives
   where carModel = 'Ford'
```

```
select shortName from Person
where nationality = 'American'
except
   select driver from Drives
   where carModel = 'Ford'
```

For a dialect that does not support the **intersect** or **except** operators, these queries may be reformulated using membership subqueries, thus:

```
select shortName from Person
where nationality = 'American'
  and shortName in
   (select driver from Drives
   where carModel = 'Ford')
```

```
select shortName from Person
where nationality = 'American'
and shortName not in
  (select driver from Drives
   where carModel = 'Ford')
```

This technique will not work in SQL-89 if we want to obtain the intersection or difference of tables with more than one column. However, the exists quantifier may be used with subqueries to handle such cases, as discussed later. In SQL-92, the multiple column case can be handled with membership subqueries because the **in** operator works with row value expressions.

Now, suppose that we tried to list the Americans who did not drive a Ford by using the following query. Is this correct? Work out the result table for yourself before reading on.

```
select shortName
from Person join Drives on shortName = driver
where nationality = 'American' and carModel <> 'Ford'
```

The result from this query lists both JimO and Lance. The query is incorrect for two reasons. As formulated, the query actually asks for the Americans who drive a car other than a Ford. This is very different from the intended query. Lance is wrongly included because he drives a Honda (even though he does drive a Ford). Walter is wrongly eliminated because he drives no car. The condition of never matching a specified value (on any row) is quite different from the condition of matching a different value on some row.

Subqueries may be used with the **union** operator to emulate *outer joins*. For example, consider the query, "List the empNr and name of each employee as well as their car models (if any)". In SQL-92 syntax, this may be set out as:

```
select Employee.empNr, empName, carModel
from   Employee left outer join Drives on Employee.empNr = Drives.empNr
```

In SQL-89 syntax, this may be rephrased as follows. Here, a string constant "?" (padded with blanks if needed) is used to emulate a null value because the standard forbids **null** as a **select** item. However, many dialects, including SQL Server, do allow us to say "**select null** ...", which can be handy at times.

```
select Employee.empNr, empName, carModel
from   Employee, Drives
where Employee.empNr = Drives.empNr
union
select Employee.empNr, empName, ' ? '
from  Employee
where empNr not in
   (select empNr from Drives)
```

The remainder of this section discusses *quantified comparison subqueries*. Recall that comparison subqueries may be used to formulate conditions of the form:

expression Θ [**all** | **some** | **any**] (*subquery*)

where Θ is a comparison operator (e.g., =, >) and **all**, **some**, and **any** are quantifiers. In the absence of a quantifier, the subquery must return a single value, which can be guaranteed by using a function (see next section). If such a quantifier is used, we have a quantified comparison subquery. The quantifiers have the following meanings:

some	means	*at least one*
any	means	*at least one*
all	means	*each* (taken one at a time)

In consequence, the following equivalences hold:

in	≡	= **some**	≡	= **any**
not in	≡	<> **all**		

To avoid confusion, **in** and **not in** should be used instead of their equivalent quantified comparisons, and **some** should always be used instead of **any**. In SQL, **some** and **any** are treated as synonyms. However, this is not always the case in English. For example, "<> any" means "not in" in English, but not in SQL! To appreciate this point, compute the result of the following query for yourself.

```
select countryName from Country
where population <> any (select population from Country)
```

After evaluating the subquery, we may express the condition as "population <> **any** (190000000, 1000000000, 16000000, 270000000)". In English, "<> any" means "=none". Because each population equals itself, we would expect no rows to be returned. But the query returns all the populations because each is not equal to some (i.e., at least one) of the populations, and SQL treats "<> any" to mean "<> some".

As a proper use of a quantified comparison, consider the query: Which country has the largest population? This may be formulated as follows.

```
select countryName
from   Country
where population >= all (select population from Country)
```

Remember that **all** is interpreted distributively as "each, taken individually" rather than collectively. The query result correctly lists India. As discussed in the next section, a max function could alternatively be used to reformulate the search condition as population=(**select max**(population) **from** Country). As discussed later, however, quantified comparisons can also handle some cases not amenable to functions (e.g., group extrema).

Exercise 12.8

1. A database is used to store information on diets. The relational schema and a sample population are provided. Weight (or, more correctly, mass) is measured in kilograms.

Eater (person, weight, gender)

Eats (person, foodName)

Food (foodName, foodClass)

Eater:

person	*weight*	*gender*
Ann	70	F
Bill	70	M
Humphrey	150	M
Sue	60	F

Food:

foodName	*foodClass*
apple	fruit
beef	meat
chicken	meat
orange	fruit
peas	vegetable
potato	vegetable

Eats:

person	*foodName*
Ann	apple
Ann	beef
Ann	potato
Bill	apple
Bill	potato
Humphrey	apple
Humphrey	beef
Humphrey	chicken
Humphrey	orange
Humphrey	peas
Humphrey	potato
Sue	apple
Sue	chicken
Sue	orange
Sue	peas

For this database, formulate each of the following queries in SQL.

(a) Who is either a male or a person weighing over 60 kg?
(b) Who is either a male or a person who eats peas?
(c) Who is female and weighs over 60 kg?
(d) Who is female and eats potatoes?
(e) Who weighs over 60 kg but does not eat beef?
(f) List all pairs of people who are of the same weight but opposite gender.
(g) Who eats vegetables?
(h) List the name and weight of those males who eat at least one kind of meat.
(i) Who eats some food other than meat?
(j) Who are vegetarians (i.e., who does not eat meat)?
(k) List the name, weight, and gender of all eaters, identifying each as either a vegetarian or meat eater, and listing vegetarians first, then ordering by gender (females first), then by weight, and finally by name.

2. Consider the following relational schema and sample population (for simplicity, assume that academics are identified by their surname).

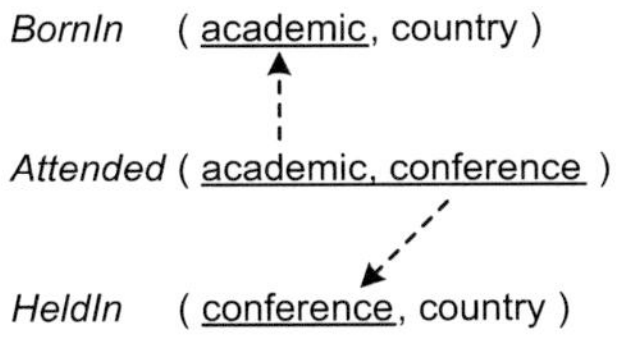

BornIn:

academic	*country*
Everest	America
Smith	America
Meersman	Belgium

Attended:

academic	*conference*
Everest	ER-99
Everest	ORM-1
Meersman	ORM-1
Meersman	CAiSE-00

HeldIn:

conference	*country*
CAiSE-00	Sweden
DAMA-00	America
ER-99	France
ER-00	America
ORM-1	Australia

Formulate queries (a)–(f) in SQL-89:

(a) For conference attendees, list their name, birth country, and conferences attended.
(b) As for (a), but include all academics.
(c) For those academics who attended conferences, list their surname and the name and place of their conferences.
(d) As for (c), but include the conferences not attended by the listed academics.
(e) List the name and birth country of academics born in a country that held a conference as well as the conferences held there.
(f) As for (e), but include all countries (each additional country will have birth or conference details, but not both).

(g)–(l) Formulate queries (a)–(f) in SQL:1999, making use of its extra join operators.

12.9 SQL: Scalar Operators and Bag Functions

A *scalar* is a data value (e.g., a number, string, or date). This section provides a brief discussion of the scalar operators and bag functions available in SQL-89 as well as some of the further operators and functions included in later versions.

In SQL-89, the scalar operators are the four *arithmetic operators* +, −, *, and /. As unary operators, + and − provide the sign of the number (e.g., +3 or −3). If omitted, + is assumed (e.g., 3 = +3). As binary operators, + and − perform *addition* and *subtraction*. The binary operators * and / perform *multiplication* and *division*; they have no unary version. Each is a number-forming operator on numbers. Their operands may appear as column names, constants, function calls, or expressions formed from these by the use of arithmetic operators and perhaps parentheses. Scalar expressions may be used as items in a **select** list or a search condition.

Let's begin by using these operators to provide a simple calculator facility in SQL to compute the value of an arithmetic expression. We simply include in our database a dummy table with just one row and then include the expression as the sole item in a **select** list for this table. Some DBMSs have dummy tables built in, but let's create our own by issuing the command **create table** C (c **char**) and then populating this table with the dummy value "c", thus: **insert into** C **values** ("c"). We can now use this table to do simple calculations. As a trivial example, to compute the sum obtained by adding 2 to the product of 3 and 4, we can issue the following query.

select 2 + 3 * 4 **from** C → 14

For such calculations, the contents of C are not accessed, and hence are irrelevant (except that by confining C to one row, the desired value displayed just once). In SQL Server, the **from** clause is optional, so you can simply use **select** 2+3 * 4 without having to create a dummy table. The arithmetic operators obey the usual priority convention. Multiplication and division have top priority, with addition and subtraction second. Operators on the same level are left associative (i.e., are evaluated left to right in the order in which they appear). Parentheses may be used to override this evaluation order (parenthesized expressions are evaluated before being operated on from outside). For example: 2 + 8/2 * 4 = 18.

For the division operation on exact numeric data, the SQL standard leaves it implementation-defined as to the scale (number of digits after the decimal point) of the result. In many SQL dialects (including SQL Server), this scale is 0, so that the division operator performs integer division if its operands are integers, for example,

select 14/3 **from** C → 4

Real division can be obtained by making one of the operands real. For example:

select 14.0/3 **from** C → 4.666666

In some dialects, this displays as 4.666666E+000. Here, "E" is read "times 10 to the power", so the aforementioned value is read "4.666666 times 10 to the power 0". Because $10^0 = 1$, this result boils down to just 4.666666.

SQL-92 introduced the **cast** function to enable data to be recast or converted into data of another type (with some restrictions). This can be used to control the scale of a division result. For example, in SQL Server the following query results in 4.67. Whether results are rounded (as here) or truncated is implementation dependent.

```
select cast(14.0/3 as numeric(9,2)) from C    →  4.67
```

When arithmetic operands are of different types, the result is typically coerced into the "greater" type. For example, the result of the division was coerced into real by including a real operand. When column names rather than constants are involved, coercion to real is achieved by multiplying by 1.0 (e.g., **select** 1.0 * quantity/2).

The result of an arithmetic operation between two numbers of the same type (e.g., **smallint**) may be coerced into a larger type (e.g., **integer**) to avoid possible overflow problems. As not all SQL dialects behave in the same way, you should check the numeric computation rules for your own dialect.

As a practical example using scalar operators, consider the table scheme *LineItem*(invoiceNr, itemCode, qty., unitPrice). The items and line totals in Australian dollars for invoice 502 where the unit price is above 100 AUD may be requested as follows, assuming unit prices are stored in USD and a conversion rate of 1 AUD = 0.70 USD.

```
select itemCode, qty * unitPrice/0.70 as lineTotalAUD
from LineItem
 where invoiceNr = 502
   and unitPrice/0.70 > 100        -- unitPrice > 100 AUD
```

SQL-92 introduced one more scalar operator, "||", for *concatenating strings* (character strings or bit strings). For example, suppose that we use the table scheme *Pupil* (pupilNr, surname, firstname, gender, [iq]) to store data about school students. Fig. 12.55 shows a sample population ("?" denotes null), and a query to list the full name of each student in the normal way as well as how much above average their IQ is.

Pupil:

pupilNr	*surname*	*firstname*	*gender*	*iq*
103	Adams	Ann	F	120
101	Brown	Tom	M	120
106	Brown	Chris	F	100
102	Collins	Don	M	?
104	Dancer	Ernie	M	95
105	Evans	Ann	F	115

```
select firstname || ' ' || surname as fullname,
       iq - 100 as iqDelta
from  Pupil
order by fullname                              ⇨
```

fullname	*iqDelta*
Ann Adams	20
Ann Evans	15
Chris Brown	0
Don Collins	?
Ernie Dancer	-5
Tom Brown	20

FIG. 12.55 String concatenation using the "||" operator (sometimes rendered as "+").

Some dialects, including SQL Server, use "+" instead of "||" for the concatenate operator. In the result, each fullname appears as the concatenation of firstname, a single space, and surname. This assumes that the firstname and surname columns are defined using a **varchar** data type. If instead these columns used a fixed length **char** type, extra spaces would appear between the fullname components, aligning the surnames. The IQ delta for Don Collins is unknown because his IQ is unknown, and *a scalar operation returns unknown if any of its operands is null.*

SQL-92 extended the use of the +, –, *, and / operators to dates, times, and intervals (e.g., one date minus another yields an interval). SQL:1999 extended the use of the string concatenation operator to work with large object strings (**clobs**, **nclobs**, and **blobs**) and arrays. Table 12.12 shows the priority order for the main operators.

SQL-89 provides the following five *bag functions:* **count**, **sum**, **avg**, **max**, and **min**. Each operates on a bag of rows and returns a single, scalar value. In SQL-89, the return value must be a number in the case of **count**, **sum**, and **avg** and is either a number or a character string in the case of **max** and **min**. Recall that a bag or multiset is a set in which repetition (although not order) is significant. Each function takes a bag as its argument. For **count**, this may be a table (bag of rows) or a column (bag of data values); for the others, it must be a column.

Table 12.13 summarizes the cases. The **distinct** option may also be applied with the **max** and **min** functions, but obviously would have no effect.

Table 12.12 Operator priority (1 = highest).

Priority	*Operator type*	*Operator(s)*
1	Arithmetic/temporal	* /
2		+ –
	String/array	||
3	Comparison	= <> < > <= >= **in between like null**
4	Logical	**not**
5		**and**
6		**or**
7	Relational	**union intersect except**

Table 12.13 The five bag functions in SQL-89.

Bag function	*Result*
count (*)	number of rows in bag
count ([**distinct**] *colname*)	number of [distinct] values in column
sum (*numeric-expn*)	sum of *numeric-expn* values in bag
sum ([**distinct**] *colname*)	sum of [distinct] values in column
avg (*numeric-expn*)	average of *numeric-expn* values in bag
avg ([**distinct**] *colname*)	average of [distinct] values in column
max (*expn*)	maximum of *expn* values in bag
min (*expn*)	minimum of *expn* values in bag

SQL-92 and SQL:1999 introduced no new bag functions, but more were added in SQL:2003. Commercial dialects often support further bag functions (e.g., SQL Server includes **stdev** and **stdevp** functions to compute standard deviations).

A function is called simply by naming it and placing its argument in parentheses after its name. In SQL-89, the argument must be an expression of the following kind: a column name, a constant, or an arithmetic expression formed from column names and constants with the use of arithmetic operators and perhaps parentheses. Any duplicate values are included in the computation unless the keyword **distinct** is placed inside the parentheses just before the expression.

Function calls may be included in a **select** list. In this case, every other item in the **select** list must also include a function call, unless the item is a constant or grouping is being performed (see later).

The **sum** and **avg** functions require numeric expressions, whereas **max** and **min** accept number or string expressions. **Count** (*) treats nulls like actual values. All other function calls remove the nulls before computing the result. If the argument bag is empty (either initially or because nulls were removed), **count** returns 0 and the other functions return null.

The count function may be used in two ways. **Count** (*) returns the number of rows in the specified table, whereas **count** (*colname*) returns the number of values in the named column. If **distinct** is specified with a bag function, any duplicates are excluded. If duplicates are wanted, the keyword **all** may be used; however, because this is the default, it is often omitted. For obvious reasons, adding **distinct** to **max** or **min** has no effect.

A few examples based on the Pupil table in Fig. 12.55 are shown in the following queries. With the third example, a **where** clause is used to filter out unwanted rows before the function is called.

```
select count ( * ) from Pupil                      →  6

select count ( distinct gender ) from Pupil        →  2

select count ( * ) from Pupil
where iq > 100                                     →  3
```

The **sum** function returns the sum of the values in the column, and **avg** returns the average of these values. For example:

```
select sum ( iq ) from Pupil                 →  550
select sum ( distinct iq ) from Pupil        →  430
select avg ( iq ) from Pupil                 →  110
```

Note the exclusion of nulls here. Because Don Collins's IQ is unknown, there are only five IQ values to be considered. Because these total 550, their average is 110.

If the argument of the **avg** function is exact numeric, the SQL standard leaves the scale of the result implementation defined. For example, the following query returns different results depending on the SQL dialect used.

```
select avg ( distinct iq ) from Pupil   →  107.50   – in some SQLs
                                        →  107      – in SQL Server
```

In SQL Server, the scale here is 0, so any fraction in the average result is lost. To include the fraction, you can multiply the argument by 1.0 to coerce it to real. For example:

select avg (**distinct** 1.0 *iq) **from** Pupil → 107.500000

The functions **max** and **min** return, respectively, maximum and minimum values. If the data type is character string rather than numeric, these values are computed according to ordinal positions in the character collating. For example:

```
select max ( iq ) from Pupil        →  120
select min ( iq ) from Pupil        →  95
select max ( surname ) from Pupil   →  Evans
```

In SQL-89, the bag functions may be called only in a **select** list (or in a **having** clause—see later). For example, consider the request, Who has an IQ above the average pupil IQ? The following formulation is illegal:

```
select firstname, surname from Pupil
where iq > avg (iq)                  →  Error!
```

Instead, we need to embed the function call within a subquery as follows. This query correctly lists Ann Adams, Tom Brown, and Ann Evans in its result.

```
select firstname, surname from Pupil
where iq >
     (select avg( iq ) from Pupil )
```

A few more examples are given in the following SQL queries. The first query computes the difference between the highest and the lowest IQ. The second query determines the ratio of highest to lowest IQ (note the multiplication by 1.0 to ensure that the decimal fraction is included). Similarly, in the third example, to compute the mean of the highest and lowest IQ, the divisor is 2.0 rather than 2. The fourth example lists those pupils where 80% of their IQ is greater than 90% of the minimum IQ. As an exercise, explain why it would be wrong in SQL-89 to express the search condition as "0.8 * IQ > 0.9 * (**select min**(IQ) **from** Pupil)".

```
select max( iq) – min(iq) from Pupil  →  25

select 1.0 * max(iq) / min iq )
from Pupil                             →  1.263157

select ( max(iq) + min(iq) ) / 2.0
from Pupil                             →  107.500000

select firstname, surname from Pupil
where 0.8 * iq >
```

```
( select 0.9 * min(iq) from Pupil )    →    Ann   Adams
                                            Tom   Brown
                                            Ann   Evans
```

This example illustrates the fact that in SQL-89, unlike a function call, an arithmetic expression may be used as a term to be compared in a search condition.

Our next example additionally illustrates the fact that built-in functions may take an arithmetic expression as an argument. Consider the following relation scheme: *Window* (windowNr, height, width). The following query may be used to list details about those windows whose height exceeds their width by the greatest amount.

```
select * from Window
where height - width =
    ( select max( height – width ) from Window)
```

SQL-92 and SQL:1999 extended the range of expressions that may feature as arguments of bag functions. For example, **sum** and **avg** may apply to interval types, and **max** and **min** may be applied to user-defined types. Although SQL-92 and SQL:1999 did not add further bag functions, they did add many scalar functions. For example, SQL-92 introduced the following scalar functions, each of which takes zero or more arguments and returns a scalar: **bit_length**, **char_length**, **octet_length**, **case**, **cast**, **collate**, **convert**, **translate**, **current_date**, **current_time**, **current_timestamp**, **current_user**, **session_user**, **system_user**, **user**, **extract**, **lower**, **upper**, **position**, **substring**, and **trim**. Most dialects support some of these as well as other scalar functions of their own.

The **lower** and **upper** functions return uppercase and lowercase versions of the argument while the **cast** function converts the argument to another data type. Examples of these were given earlier. The **substring** function returns a part of a string, using the syntax **substring** (*string-expn* **from** *start-posn* [**for** *length*]). For example:

```
select substring(surname from 2 for 3)
from Pupil
where surname = 'Adams'    →  dam
```

Some dialects use a nonstandard syntax for this function. For example, SQL Server uses commas instead of the **from** and **for** keywords, for example, **substring** (surname, 2, 3).

The **case** function is quite useful for arranging different output for different cases. Conditions are placed after **when**, with their return option after **then**, and an **else** clause is used to specify the return option if all conditions fail. For example:

Query:

```
select firstname, surname,
   case
      when iq > 100 then 'above average IQ'
      when iq = 100 then 'average IQ'
      when iq < 100 then 'below average IQ'
        else 'IQ is unknown'
   end
from Pupil
```

Result:

```
Ann    Adams     above average IQ
Tom    Brown     above average IQ
Chris  Brown     average IQ
Don    Collins   IQ is unknown
Ernie  Dancer    below average IQ
Ann    Evans     above average IQ
```

That covers the most important scalar operators and functions in SQL. A discussion of other scalar functions can be found in references cited in the chapter notes.

Exercise 12.9

1. This question refers to the Diet database used in Exercise 12.8. The table schemes are: Eater(person, weight, gender), Eats(person, foodName), and Food(foodName, foodClass). Formulate SQL queries for the following
   (a) How many males are there above 100 kg in weight?
   (b) How many different weights are there?
   (c) What is the sum of the weights of the males?
   (d) What is the average weight of the females?
   (e) What is the heaviest weight of those who eat beef?
   (f) For each of the next two questions, give two equivalent solutions, one of which uses a function while the other uses an "all" or "some" quantifier.
   (g) Which females are lighter than every male?
   (h) State the name and weight of those females who are as heavy as at least one male.
2. The Log table shown is used by a hardware store to record details about wooden logs. The schema is *Log* (code, diameter, len, mass, cost_price, retail_price). The diameter, length (here called "len"), mass, and price of the logs are measured in cm, m, kg, and $, respectively. Formulate the following in SQL.

Log:

code	*diameter*	*len*	*mass*	*cost_price*	*retail_price*
2A	15	2	10	4.00	7.00
3B	20	3	20	6.00	9.50
5C	20	5	30	9.00	13.00
5D	20	5	25	8.00	2.00

   (a) List the absolute mark-up (i.e., retail price−cost price) for all the logs.
   (b) List the volume of each log in cubic meters, with the column heading "Volume (cubic m)". Use the formula $V = \pi D^2 L/4$ for the volume, approximating π as 3.14. Express each volume as a fixed-point number, truncated to three decimal places.
   (c) List the density (i.e., mass/volume), in kg/m^3, of those logs that are 5 m long. Use an appropriate heading and show each density as a fixed-point number truncated to two decimal places.
   (d) Which log has the highest relative mark-up (use ratio R.P./C.P.)?
   (e) Which log has a less than average ratio of length to diameter? Include this ratio (dimensionless, two decimal places) in the output.
3. A library uses the following database schema to record details of books. Authors are identified by name, and books are identified by their International Standard Book Number (ISBN). A sample population is shown.

Book (<u>isbn</u>, title, publisher, yrPublished, nrCopies)

WrittenBy (<u>isbn, authorName</u>)

Author (<u>authorName</u>, gender, nationality)

Book:

isbn	*title*	*publisher*	*yrPublished*	*nrCopies*
101	Databases	Possum	1985	4
202	SQL Primer	Hall	1984	4
246	Databases	West	1980	1
345	dBase III	West	1986	1
400	Modula 2	Possum	1986	4
444	Advanced SQL	Hall	1986	2
500	Informatics	Hall	1986	2

Formulate the following queries in SQL.

(a) Who are the Australian male authors?
(b) List the titles of books published in the period 1984.0.1986, ordered by title. (Explain the position of "dBaseIII" in the result.)
(c) List all details of books with "SQL" as part of the title, showing the most recently published ones first.
(d) Who wrote a book titled "Databases" published in 1980?
(e) List the ISBN and title of all books with at least one Australian author.
(f) List the name and nationality of author(s) of any book(s) titled "Informatics" published by Hall in 1986.
(g) List, with suitable descriptors, names of the Australian male authors and the American female authors, with the former shown first.

WrittenBy:

isbn	*authorName*
101	Brown J
101	Collins T
202	Adams A
246	Smith JB
345	Jones S
400	Smith JA
444	Adams A
500	Brown J
500	Smith JA

Author:

authorName	*gender*	*nationality*
Adams A	F	Aussie
Brown J	M	Aussie
Collins T	M	Kiwi
Jones S	F	American
Smith JA	M	Aussie
Smith JB	M	American

4. Specify the following SQL queries for the Pupil table discussed in this section.
   (a) List the names of the female pupils in a single column, with surname first, followed by a comma, space, and firstname, for example, "Adams, Ann".
   (b) List the number of pupils as well as the number of pupils whose IQ is known.
   (c) List the pupilNr, surname, firstname, and gender of each pupil whose IQ is known, specifying gender as a name ("Female" or "Male") instead of a code ("F" or "M").

12.10 SQL: Grouping

Sometimes, we wish to partition the rows of a table into a number of *groups* and display properties that apply to each group as a whole. SQL provides a **group by** clause that may be included in the **select** statement to retrieve such grouped data. *Each group is a bag of rows with the same value(s) for the grouping criteria* listed in the **group by** clause. *The final query result has at most one row for each group.*

Fig. 12.56 pictures grouping by column *a*. Here, there are three groups, one for each of the *a* values. The result may include some group property or function of *a*, depicted here as *f(a)*.

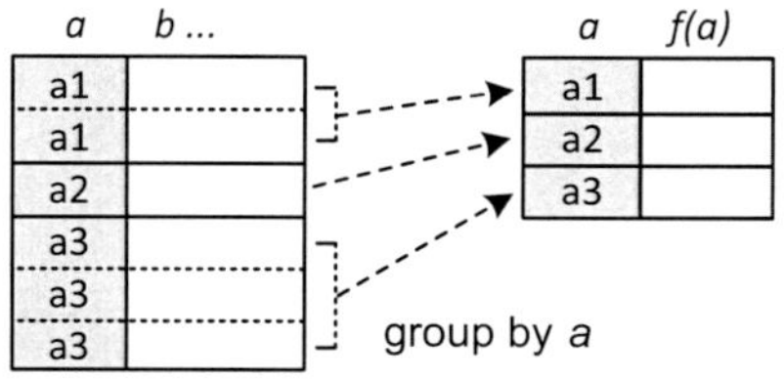

FIG. 12.56 Partitioning rows into groups with the same value for column *a*.

Because the final result includes just one row for each group, the **select** list may include group properties only.

In other words, for each group, each **select** item has only one value. To ensure this, each group property must be a grouping criterion (a column name used for grouping), a function call (e.g., **count**(*)), a constant, or an expression formed from these terms (e.g., 2 * groupcol). The basic syntax for **group by** queries in SQL is set out below:

select *group-property1*, ...
from ...
[**where** ...]
group by *group-criterion1*, ...

Table 12.14 will be used to help explain most of these ideas. The relation scheme is *Particle* (<u>particleName</u>, family, charge, mass). To help you visualize the grouping into families, the rows are sorted by family, and a blank line is added between the groups. The charge of each particle is given in elementary charges. An elementary charge is the charge of a proton, and equals 1.6×10^{-19} C. The mass of an electron is 9.1×10^{-31} kg. The mass of each particle is expressed as a multiple of the electron mass.

Table 12.14 Some atomic particles.

Particle:

particleName	*family*	*charge*	*mass*
neutrino	lepton	0	0
mu neutrino	lepton	0	0
electron	lepton	-1	1
mu –	lepton	-1	207
pi 0	meson	0	264
pi +	meson	1	273
pi –	meson	-1	273
K +	meson	1	966
K–	meson	-1	966
K 0	meson	0	974
eta	meson	0	1074
proton	baryon	1	1836
neutron	baryon	0	1839
lambda	baryon	0	2183
sigma +	baryon	1	2328
sigma 0	baryon	0	2334
sigma –	baryon	-1	2343
xi –	baryon	-1	2585
xi 0	baryon	0	2573

Consider the following query: List the families and the number of particles in each. This may be formulated in SQL as shown. For this section, the result tables are listed with the column headers underlined by hyphens (this is typical of most SQLs).

```
select family, count(*)    →    family   count(*)     -- function header
from Particle                   ------------------    -- may be omitted
group by family                 baryon    8           -- in some SQLs
                                lepton    4
                                meson     7
```

Here, the **group by** clause indicates that the particles are to be *grouped* into families (i.e., particles with the same family name are placed in the same group). The *items in the* **select** *list are then applied to each group as a whole.*

If **count** (*) is used, it returns the number of rows in the group. In this case, we have eight baryons, four leptons, and seven mesons. If an expression other than a column name is used in the **select** list, some SQLs use the expression as a column header in the result (as shown here), whereas others, including SQL Server, omit the column header. It's normally best to use an **as** clause after the expression to provide your own header.

In many versions of SQL, the use of **group by** causes a default ordering equivalent to a matching **order by** clause (e.g., **order by** family in this case). However, this is not part of the SQL standard. The syntax of the **group by** clause is:

group by *col-name* [, ...]

Members of the same group must have matching values for all the columns named in the **group by** clause.

The next example places particles into groups having both the same family name and the same charge.

```
select family, charge, count(*) as tally    →    family    charge    tally
from Particle                                    --------------------------
group by family, charge                          baryon        -1        2
order by family, charge                          baryon         0        4
                                                 baryon         1        2
                                                 lepton        -1        2
                                                 lepton         0        2
                                                 meson         -1        2
                                                 meson          0        3
                                                 meson          1        2
```

If more than one column is used for grouping, the order in which these columns are specified is irrelevant to the content of the result. This is because the same set of groups must be formed. For example, if we group by attributes *a, b,* or *b, a* then the group with values a=a1, b=b1 is the same as the group with values b=b1, a=a1, and so on.

However, if the row order is based on the group order, the order of rows in the result may be different. For example, the following query returns the same rows as the previous query, in a different order. You can adjust the **order by** clause to change the order:

```
select family, charge, count(*) as tally    →    family    charge    tally
from Particle                                    --------------------------
group by charge, family                          baryon        -1        2
order by charge, family                          lepton        -1        2
                                                 meson         -1        2
                                                 baryon         0        4
                                                 lepton         0        2
                                                 meson          0        3
                                                 baryon         1        2
                                                 meson          1        2
```

If a function call in the **select** list is used as an ordering criterion, you should introduce an alias for it so that you can reference it by name in the **order by** clause. For example, the following query lists each family and the mass of its lightest particle, ordering by that minimum mass. As an undesirable alternative, you can reference it by its position in the **select** list (e.g., **order by** 2).

```
select family, min (mass) as minMass   →   family   minMass
from Particle                              -----------------
group by family                            lepton         0
order by minMass                           meson        264
                                           baryon      1836
```

Use of a **group by** clause implies that items in the **select** list are to be applied to each group as a whole, returning just one value for each group. So, any columns mentioned in the **select** list must either be grouped columns or must appear there as an argument to a built-in function (e.g., family and mass in the aforementioned query). So, when grouping is being performed it makes no sense, and is in fact illegal, to include a nongrouped column as a whole item in the **select** list. For example:

```
select family, mass from Particle   →   Error!
group by family
```

Because the same family may include several particles of different mass, this request is plain silly (which mass would we choose?). We can, however, obtain a single grouped value related to mass by using a built-in function, such as **min** (mass) or **avg** (mass), as we did with the previous query.

Although the **select** list must not include any nongrouped columns, it is legal to group by columns that are not mentioned in the **select** list. For example, the following query lists the minimum mass, but not the name, of each family.

```
select min (mass) as minMass   →   minMass
from Particle                      ---------
group by family                            0
                                         264
                                        1836
```

Within the **select** statement, *a* **where** *clause may be included to filter out unwanted rows before the groups are formed and functions are applied.* Sensibly, the **where** clause must be placed before the **group by** clause. For example, the following query lists each family and the average mass of its uncharged particles. To cater for dialects such as SQL Server that truncate averages, the mass has been coerced to real by multiplying by 1.0, and the **cast** function is used to display just two digits after the decimal point. Note that it's legal to apply a scalar function to a bag function, for example, **cast** (**avg** (...)). It is, however, illegal to apply a bag function to a bag function, for example, **min** (**avg** (...)).

```
select family, cast (avg (1.0*mass) as decimal (6,2)) as avgUnchargedMass
from Particle
where charge = 0                        →   family    avgUnchargedMass
group by family                             ----------------------------
                                            baryon             2232.25
                                            lepton                0.00
                                            meson               770.65
```

Sometimes, we wish to restrict our attention to only those groups that satisfy a certain condition. Just as a **where** clause may be used to impose a search condition on individual rows, a **having** clause may be used to impose a search condition on individual groups. In this case, the **having** clause must be placed straight after the **group by** clause to which it is logically connected. The syntax is:

group by *col-name* [, ...]
having *group-condition*

The usual rules for search conditions apply. Logical and comparison operators may be included in the condition. Each comparison compares some property of the group, such as **min** (mass), with a constant or another group property. (Although rare, it is possible to include a **having** clause without an associated **group by** clause; in this case, the entire table is treated as one group.) As a simple example, the following query lists the families with more than four particles, and the number of particles for each. Although it makes sense to allow aliases (e.g., tally) in the **having** clause, this is not allowed in many SQLs (e.g., SQL Server).

```
select family, count (*) as tally        →    family   tally
from Particle                                  ------------
group by family                                baryon    8
having count (*) > 4                           meson     7
```

The next example is harder. To help clarify how it works, the numbers 1 through 6 have been added to indicate the order in which the query lines are processed. These numbers are not part of the query. Trace through the steps yourself to make sure that you follow this order. With respect to the **having** clause, note that the charged leptons are included (uncharged leptons with zero mass are filtered out by the **where** clause) and that the charged baryons are excluded (their maximum mass exceeds 2000).

Considering only the charged particles, form groups with the same family name, where the group's lightest particle has a mass above 0 and the group's heaviest particle has a mass below 2000. List the family name, number of particles, and largest mass for each group, showing groups with more members first.

```
5   select family, count (*) as chargedTally, max(mass) as maxChargedMass
1   from Particle
2   where charge <> 0
3   group by family
4   having min (mass) > 0 and max (mass) < 2000
6   order by chargedTally desc
```

```
→ family     chargedTally    maxChargedMass
  ------------------------------------------
  meson           4               966
  lepton          2               207
```

One trivial use of grouping is simply to avoid duplicates. For example, the following query may be used as an alternative to **select distinct** family **from** Particle:

```
select family from Particle   →   family
group by family                   --------
                                  baryon
                                  lepton
                                  meson
```

Now suppose we issue the following query. What will be the result?

```
select family from Particle
group by family
having count (*) = 4
```

Comparing this with the previous query, you may feel that the count of each family will be 1 and hence no rows will be returned (i.e., the result is the null set). No! When the **group by** clause does its work, each group still has all its members (which of course all have the same group value). The **having** clause now filters out those groups that do not satisfy its search condition. Finally, the **select** list determines which group properties are listed. The groups are first sorted into three families, two of these families are eliminated (the baryons and mesons), and finally just the family name "lepton" of the remaining group (the four leptons) is listed.

As a further application of grouping, let's consider how simple cases of *relational division* can be handled. Fig. 12.57 shows an SQL query for an example handled earlier in Fig. 12.27 by relational division. Here, the column name "language" is double quoted because it is a reserved word.

Although SQL provides no operator for relational division, we achieve the same effect by grouping as shown. The first subquery yields ("English," "French"), so the **where** clause filters out the Dutch and Spanish rows before the groups are formed.

Speaks:

country	*language*
Australia	English
Belgium	Dutch
Belgium	French
Canada	English
Canada	French
Cuba	Spanish
Dominica	English
Dominica	French

Which countries speak
***all** the languages spoken in Canada?*

```
select country from Speaks
where "language" in
  (select "language" from Speaks
   where country = 'Canada')
group by country                          ⇨
having count(*) =
          (select count(*) from Speaks
           where country = 'Canada')
```

country
Canada
Dominica

FIG. 12.57 Grouping can be used to perform relational division.

Because of the uniqueness constraint on Speaks, the number of Canadian languages spoken by a country is now the number of rows in the group for that country. So, the left **count** (*) of the **having** condition returns the values 1, 1, 2, and 2 for Australia, Belgium, Canada, and Dominica, respectively. The second subquery evaluates to 2, so the overall query returns Canada and Dominica.

This approach can be adapted to handle cases when the relevant column association is not unique by using the **distinct** option. For example, consider the relation schemes *Localization* (productCode, "language") and *LanguageUse* ("language", country, nrSpeakers). The query "Which products are localized into all languages spoken by a country where at least 10 million speak that language?" requires **count** (**distinct** ("language")) instead of **count** (*) in the subquery of the **having** clause. As discussed later, this grouping approach can be extended to cater for queries that involve set comparisons other than those covered by relational division.

We have now covered most of the basic working of SQL's **select** statement. The BNF syntax covered so far may be summarized as shown. Unlike the syntax shown here, the SQL standard actually allows a **having** clause without a **group by** clause (effectively meaning "group by no columns") but the semantics are problematic, so we'll ignore this option.

```
select [all | distinct ] * | expression [ [ as ] column-alias ] [,...]
from table [ [ as ] alias ]
        [,| cross join table [ [as ] alias ]
          | natural [ inner | { left | right | full }[ outer ] ] join table [ [as ] alias ]
          | [ inner | { left | right | full } [ outer ] ] join table [ [as ] alias ]
              { on condition | using ( col-list ) }
          | union join table [ [as ] alias ]
                    [,...] ]
[ where col-condition ]
[ group by colname [,...]
      [having group-condition ] ]
[ union | intersect | except ... ]
[ order by col [ asc | desc] [,...] ]
```

Of the nine table operations of relational algebra, we have discussed the equivalent SQL formulation for the general case of five (projection, selection, Cartesian product, Θ-join, and union) and for special cases of three (intersection, difference, and division). The rename operation of relational algebra is handled using a view (see later chapter). The general cases of intersection, difference, and division can be handled using joins, and also with correlated and existential subqueries, as discussed in the next section.

Exercise 12.10

1. This question refers to the library database discussed in Exercise 12.9. The relational schema is repeated here.

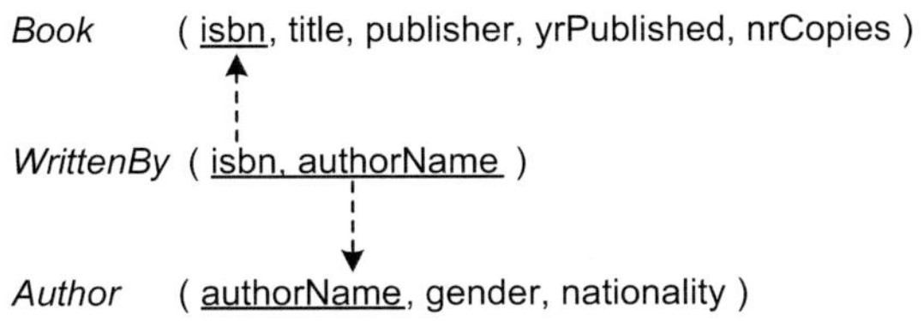

2. Assume ISBN is based on a character data type. Formulate SQL queries for the following requests.
   (a) How many authors of each gender are there?
   (b) Place authors into groups of the same gender and nationality, indicating how many there are in each group, with the larger groups shown first.
   (c) Considering only the books published after 1980, list the publishers and the total number of copies of their books in the library.
   (d) For each publisher having an average number of library copies per book above two, show the earliest publication year.
   (e) List the ISBN and the number of authors for each book published before 1986 that has more than one author.
   (f) Restricting attention to male authors of a book published by Hall or Possum, list the number of such authors for each nationality that has at least as many authors of this kind as there are copies in the library of the book with ISBN "444".
3. The relational schema for a dietary database is shown (for a sample population, see Exercise 12.8). Formulate the following queries in SQL

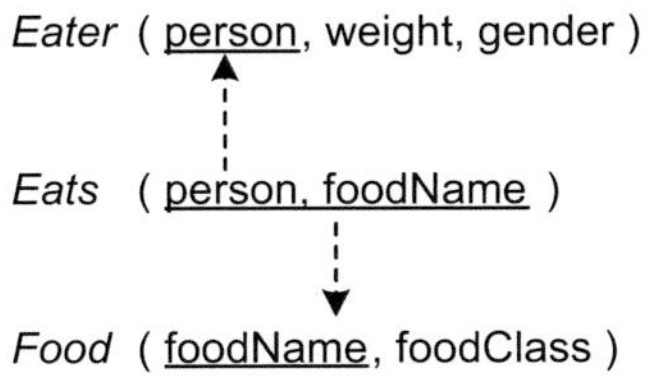

   (a) Who eats all the foods?
   (b) Which foods are eaten by all eaters?

12.11 SQL: Correlated and Existential Subqueries

This section extends the subquery work in Section 12.8 by considering correlated and existential subqueries. The subqueries discussed earlier were simple, uncorrelated subqueries. An uncorrelated subquery is computed once, replaced by its result, and the outer query is run. In contrast, a *correlated subquery* relates its search condition to each row of a table named in the outer query and is effectively recomputed for each outer row (see Fig. 12.58).

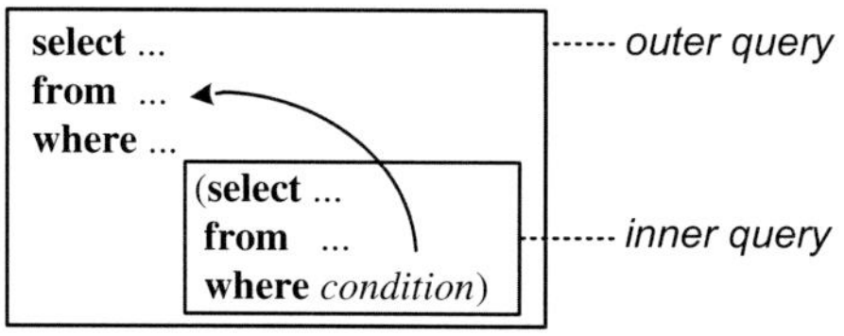

FIG. 12.58 A correlated subquery relates its condition to each row of the outer query.

Because of repeated computation, correlated subqueries can sometimes be slow to run. However, they significantly extend the range of queries that can be expressed as a single SQL query and are well worth mastering. Let us refer to the condition inside the correlated subquery as the correlation condition. This condition typically takes the form:

$$a \, \Theta \, b$$

where a is an *inner column* (listed in the **from** clause of the subquery), b is an *outer column* (listed in the **from** clause of the outer query), and Θ is a *comparator* (e.g., =, >).

To avoid ambiguity, the column names for a or b may need to be qualified by a table name. The basic cases are summarized in Table 12.15. Local scope is assumed by default, so we may normally leave a unqualified (unless it occurs in two inner tables). However, if a and b have the same unqualified name, we need to distinguish b by qualifying it with its outer table name. If the outer table has the same name as the inner table, introduce an alias for it to distinguish it from the inner table.

Table 12.15 Correlation between an inner column *a* and an outer column *b*.

Case	*Situation*	*Declaration*
1	a is the name of a column in inner table only b is the name of a column in outer table only	$a \, \Theta \, b$
2	a, b are the same column name but in different tables T is the base name of the outer table	$a \, \Theta \, T.b$
3	inner and outer tables have the same base name X is an alias introduced for the outer table	$a \, \Theta \, X.b$

For example, consider the table scheme *Human* (firstname, gender, height). For simplicity, assume that people in this UoD can be identified by their firstname. A sample population is provided in Fig. 12.59, along with a correlated subquery to retrieve the name and height (in cm) of those who are above average height for *their* gender. In English, correlations are often made with a *pronoun*, such as "their", "his/her", "its", and "that". In this context, pronouns effectively function as object variables.

Human:

firstname	*gender*	*height*
Ann	F	160
Bill	M	180
Carol	F	170
David	M	175
Emily	F	165
Fred	M	170

*List the name and height of those above average height for **their** gender*

correlation pronoun

```
select firstname, height
from Human as X
where height >
   (select avg(height)
    from Human
    where gender = X.gender)
```

⇩

firstname	*height*
Bill	180
Carol	170

FIG. 12.59 Correlation between inner and outer tables with the same base name.

Here, the same table name "Human" appears in the **from** clause of both inner and outer queries. So, to distinguish the gender column in the inner table from the gender column in the outer table, the alias *X* is introduced for the outer table. This allows us to specify the correlation condition as "gender = *X*.gender", rather than "gender = gender" (which, of course, would always be true). Here, *X* is said to be a *correlation variable*. You can think of *X* as an extra copy of the Human table.

Alternatively, you can think of it as a tuple variable that ranges over the rows of the Human table, as depicted in Fig. 12.60. Despite its set-oriented syntax, SQL queries are processed one row at a time. Here, *X* is successively assigned each row of the Human table, passing through six states (in this example), in each of which it holds one row. For each of these states, the inner query is evaluated, computing the average height for the gender currently assigned to *X*. If you're familiar with programming languages, this should remind you of a nested **for** loop.

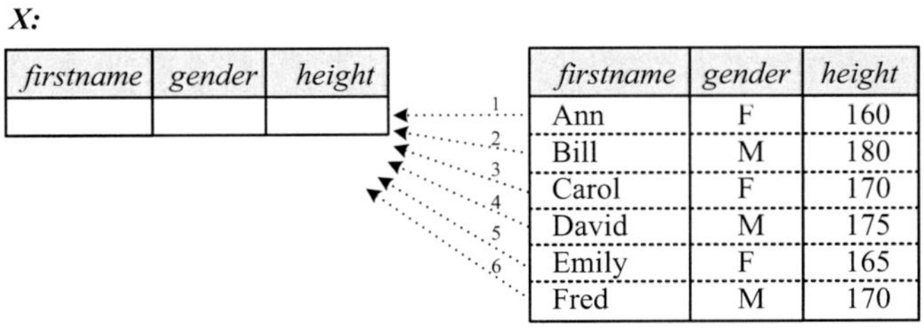

FIG. 12.60 In the outer query, X is successively assigned one of the six rows.

Let's walk through the execution of the query. At the outer level, we start at the top row, so *X* is assigned the tuple ("Ann", "F", 160). Hence, in the outer level, gender = "F" and height = 160. In other words, *X*.gender = "F" and *X*.height = 160. In the subquery, the correlation condition "gender = *X*.gender" now becomes "gender = 'F'". So, the subquery returns the average height of the females, which for the given data is 165.

The condition in the outer query now becomes "160 > 165", which is false. Because the top row does not satisfy the condition, it is filtered out, so Ann is excluded from the result.

At the outer level, we now move on to the second row, so *X* is assigned the tuple ("Bill", "M", 180). Hence, *X*.gender= "M" and height=180. In the subquery, the condition "gender = *X*.gender" becomes "gender = 'M'". So, the subquery returns the average height of the males, which is 175. The condition in the outer query now becomes "180 > 175", which is true. So, Bill's firstname and height ("Bill", 180) are included in the result.

Similarly, we move through rows 3, 4, 5, and 6 at the outer level, recomputing the subquery each time. Carol is accepted (170 > 165 is true) but the others are rejected (because the conditions 175 > 175, 165 > 165, and 170 > 175 are all false). So, the final result lists the firstname and height of just Bill and Carol as shown earlier in Fig. 12.59.

Now consider the table scheme *Employee* (empNr, empName, pay, [bossNr]), where pay stores the weekly pay and bossNr is the employee number of the boss of the employee on that row. The foreign key constraint bossNr **references** empNr also applies. A sample population is shown in Table 12.16.

Table 12.16 Who earns at least 80% of his/her boss's pay?

Employee:

empNr	*empName*	*pay*	*bossNr*
1	Davis, J	1000	?
2	Jones, E	900	1
3	Smith, J	700	1
4	Brown, T	780	2

Consider the query: Who earns at least 80% of his/her boss's pay? Let's try answering the question first, using the sample data: Employee 1 is eliminated because he/she has no boss. Employee 2 earns 900/1000 or 90% of his/her boss's pay, employee 3 earns 70% of his/her boss's pay, and employee 4 earns 780/900 or 86.67% of his/her boss's pay. So, the result should list employees 2 and 4.

How can we formulate the query in SQL to work with any set of data? You might like to try it yourself before reading on.

The inclusion of the pronoun "his/her" in the question is a clue that the query can be formulated with a correlated subquery, as set out later. *E* is used to denote the employee at the outer level. Although not needed, a second alias *B* has been used to help indicate that the inner employee is a boss.

```
select empNr
from Employee as E
where pay >=
   (select 0.8 * pay
    from Employee as B
    where B.empNr = E.bossNr)
```

As an alternative, you can formulate the query using a self-join as follows:

```
select E.empNr
from Employee as E join Employee as B
  on E.bossNr = B.empNr
    and E.pay >= 0.8 * B.pay
```

To help you understand the join approach, the following list shows rows formed from the join before the final projection is made on E.empNr.

E				B			
empNr	empName	pay	bossNr	empNr	empName	pay	bossNr
2	Jones, E	900	1	1	Davis, J	1000	?
4	Brown, T	780	2	2	Jones, E	900	1

As this example suggests, computing a correlated subquery is basically equivalent to computing a join, so some care is needed in the formulation of correlated subqueries to avoid unacceptable performance.

Now, let's consider *existential subqueries*. In predicate logic, the proposition "There is a frog" may be formulated as "$\exists x\, Fx$". Here, "$\exists$" is the existential quantifier, with the meaning "there exists at least one", "x" is an individual variable standing for some object, and "F ..." abbreviates the predicate "... is a frog".

An existential subquery in SQL is somewhat similar. It appears in an *existential condition* of the form:

exists (*subquery*)

Here, **exists** is used instead of "$\exists$" for the existential quantifier, and subquery is effectively a tuple variable ranging over rows returned by the subquery. The **exists** quantifier may be thought of as a Boolean function, *returning True if the subquery returns a row and False otherwise.*

The **exists** quantifier may be preceded by the logical **not** operator. The possible cases are shown here. The term "a row" means "at least one row". Because the system always knows whether or not a row is returned, a simple two-valued logic applies—existential conditions evaluate to True or False, never Unknown.

exists (*subquery*)	→	True	-- a row is returned (by the subquery)
	→	False	-- no row is returned
not exists (*subquery*)	→	True	-- no row is returned
	→	False	-- a row is returned

Existential subqueries may be used within search conditions in a **where** clause, leading to overall queries of the following form:

```
select ...
from ...
where [ not ] exists
   ( select * ...)
```

Because the mere existence of a row is all that counts in the subquery result, it is normal to simply use "*" instead of a column list in the **select** list for the existential subquery.

Prior to SQL-92, existential subqueries were the only subqueries that could return multiple columns because in SQL-89, both membership and comparison subqueries had to return a single column. Some SQL dialects still have this restriction.

Let's look at some examples using the populated schema in Fig. 12.61. The Member table stores details about members of a martial arts club. The Ranked table indicates what rank (if any) members have achieved in what martial arts. Below black belt level, *kyu* grades are used, whereas *dan* grades are used for black belt ranks.

Member (<u>memberNr</u>, memberName, gender)

↑

Ranked (<u>memberNr, art</u>, rank)

Member:

memberNr	*memberName*	*gender*
1	Adams, Ann	F
2	Brown, Ann	F
3	Halpin, John	M
4	Halpin, Terry	M

Ranked:

memberNr	*art*	*rank*
2	kendo	3 kyu
3	judo	2 dan
3	karatedo	6 dan
4	judo	1 dan
4	karatedo	3 kyu

FIG. 12.61 Martial arts club members and their ranks (if any).

Consider the query, Who (member number and name) is a black belt in at least one art? We can formulate this in SQL with an existential subquery as shown here. This is also a correlated subquery because the subquery condition refers back to the outer table (Member). Because the outer table has a different name from the inner table (Ranked), there is no need to introduce an alias for it. The existential condition checks to see if the member has a row in the Ranked table with a dan rank.

```
select memberNr, memberName
from Member                               →    memberNr    memberName
where exists                                   ---------------------------
   (select * from Ranked                          3        Halpin, John
     where memberNr = Member.memberNr             4        Halpin, Terry
       and rank like '% dan')
```

The same query could have been formulated using a membership query instead, for example,

```
select memberNr, memberName
from Member
```

```
where memberNr in
    (select memberNr from Ranked
     where rank like '% dan')
```

To list the members who do *not* have a black belt, we simply substitute **not exists** for **exists** and **not in** for **in** in the previous queries. Would the following query also work for this?

```
select memberNr, memberName
from Member natural join Ranked
where rank not like '% dan'
```

As you no doubt realized, this fails because it also returns members with both a dan rank and a nondan rank. As a more complex example, suppose that club members are identified instead by combining their surname and firstname. The populated schema for this is shown in Fig. 12.62.

Member (<u>surname, firstname</u>, gender)

Ranked (<u>surname, firstname, art</u>, rank)

Member:

surname	*firstname*	*gender*
Adams	Ann	F
Brown	Ann	F
Halpin	John	M
Halpin	Terry	M

Ranked:

surname	*firstname*	*art*	*rank*
Brown	Ann	kendo	3 kyu
Halpin	John	judo	2 dan
Halpin	John	karatedo	6 dan
Halpin	Terry	judo	1 dan
Halpin	Terry	karatedo	3 kyu

FIG. 12.62 Members have a composite reference scheme.

Now consider the query, Who is not ranked in judo? The existential subquery solution is straightforward, matching on both surname and firstname:

```
select surname, firstname from Member
where not exists
  (select * from Ranked                  →  surname      firstname
   where surname = Member.surname            ------------------------------
     and firstname = Member.firstname        Adams        Ann
         and art = 'judo')                   Brown        Ann
```

Although we can do this with a membership subquery, in SQL-89 the logic becomes more complex because we cannot use a column-name pair (here, surname, firstname) as an argument for the **in** operator. In the following query, the surname column is compared using the **in** operator and the firstname is matched using a correlation:

```
select surname, firstname from Member
where surname not in
      (select surname from Ranked
      where firstname = Member.firstname
                and art = 'judo')
```

For such cases, the existential solution is easier to understand.

Yet another solution to this query is to use the concatenate operator to combine the name parts into a single string (recall that some dialects use "+" instead of "||"):

```
select surname, firstname from Member
where surname || firstname not in
   (select surname || firstname from Ranked
    where art = 'judo')
```

Further examples of correlated and existential subqueries are discussed in the context of set-comparison queries in the advanced SQL supplement (Appendix C).

Exercise 12.11

1. A population for the table scheme *Pupil* (pupilNr, surname, firstname, gender, [iq]) was provided in Fig. 12.55. Formulate the following queries in SQL:
   (a) Which person has the highest IQ?
   (b) Who has the highest IQ for his/her gender?
   (c) List each male, female pair where both genders have the same IQ.
2. For the Member and Ranked tables discussed in Fig. 12.62, formulate the following in SQL.
   (a) Who holds a kyu rank?
   (b) Which genders hold a kyu rank?
   (c) Who is not ranked?
   (d) Who does not hold a kyu rank?
3. The following schema refers to part of the student database considered in Exercise 12.1. Specify the following queries in SQL.

   Student (studentNr, studentName, degree, gender, birthYr)

   Result (studentNr, subjCode, [rating])

   (a) List the student number and name of each student who did not score any 7 s. (Use a subquery.)
   (b) Same question, but use a join instead of a subquery. (*Hint*: 7 is the highest rating.)

12.12 SQL: Recursive Queries

Hierarchical data structures are found quite often in business. Some common examples are organizational structures with departments and subdepartments, and bill-of-material information involving assemblies and subassemblies. Recursion provides a powerful way to query such structures.

In SQL, recursive queries involve a three-phase process:

1. Create an initial result set (sometimes known as the *anchor*).
2. Recursively add results from further queries to the initial result set.
3. Carry out a final query on the accumulated results.

Recursion generally requires some termination condition. Here, the termination is implicit: the recursion ends when there are no more results to add in Phase 2. Particular SQL implementations may provide facilities for controlling the maximum depth of recursion to avoid infinite loops.

Creating the initial result set involves the concept of a *common table expression* (CTE), first introduced in SQL:1999. A common table expression can be imagined as a view definition with a scope that is limited to the query that immediately follows. The next chapter discusses *views*, which are basically named **select** statements stored in the system. The result of a view can be visualized as a table, although such a table is not normally stored in any physical sense. The basic syntax for a CTE is:

```
with cte-name [ (col-list) ] as cte-query-definition
query-statement
```

The col-list is not required as the cte-query-definition returns distinct columns. The query-statement immediately following the **with** clause uses the results of the cte-query-definition, referencing the cte-name. Attempts by any other query to refer to this CTE will meet with failure, unlike a view definition. This construct has other uses in SQL, but we're interested here in the role it plays in forming a recursive query. The anchor part of the recursive query can involve multiple subqueries with results combined using the standard SQL set operations (**union**, **intersect**, and **except**), but for simplicity we'll assume here that we have a single query.

Once the anchor results are formed, results from the recursion are added using SQL's **union all** operation. At each stage, the recursive part of the query takes the accumulated results as input, runs the query, and adds the results on to the result set. This continues until no more results are produced. Typically, this happens when there are no more unexplored nodes in the hierarchy.

A skeleton of the recursive CTE looks like the following:

```
with cte-name [ (col-list) ] as anchor-query-definition
union all
recursive-query-definition
query-statement
```

To make this more concrete, we'll look at a simple example. Fig. 12.63 shows a schema for a table that holds hierarchical information about departments in an organization, together with a sample data population.

As shown in Fig. 12.63(b), this organization has three departmental levels. We may want to write various queries that exploit the intrinsic parent-child relationships in the hierarchy, and recursive queries are a natural way to do this. Given the simple nature of the scenario in Fig. 12.63, using recursive queries may seem overkill, but we can easily demonstrate the essential features.

(a)

Department (<u>deptId</u>, <u>deptName</u>, totalStaff, [parentDept])

Department:

deptId	*deptName*	*totalStaff*	*parentDept*
101	Management	3	?
201	Sales	2	101
202	Production	6	101
301	Sales - West	2	201
302	Sales - Central	3	201
303	Sales - East	4	201

(b)

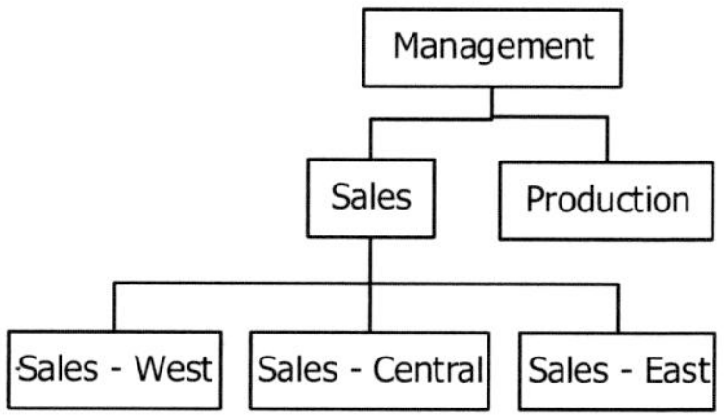

FIG. 12.63 (a) Table schema and data (b) Departmental hierarchy.

We'll begin with a simple level-by-level listing of the departments to illustrate how the result is produced. Here is the query:

```
with myCTE (deptId, deptName, parentDept, "level") as
(  select anchor.deptId, anchor.deptName, anchor.parentDept, 1 as "level"
   from Dept as anchor
   where parentDept is null
         union all
              select child.deptId, child.deptName, child.parentDept, "level" + 1
              from Dept as child join myCTE as parent
              on child.parentDept = parent.deptId
)
select deptId, deptName, "level" from myCTE
```

When executed, the query produces the result:

deptId	*deptName*	*level*
101	Management	1
201	Sales	2
202	Production	2
301	Sales - West	3
302	Sales - Central	3
303	Sales - East	3

Here, the "level" column indicates the relative depth in the hierarchy, taking 1 as the top level ("level" is a delimited identifier here because it's a recognized keyword in SQL).

We've taken as our anchor the topmost level by specifying **where** parentDept **is null**, so the anchor tuple is <101, 'Management', null, 1>. At the next level, we're now looking for departments that have a deptId of 101. The **union all** operation adds the tuples <201, 'Sales', 101, 2> and <202, 'Production', 101, 2>.

The join in the recursive part of the CTE can now add the departments under Sales. Production has no subdepartments and so the join produces no results. After adding the three Sales subdepartments there are no other possibilities for the join (because no department has a parent deptId of 301, 302, or 303), so the recursion ends.

When the **select** statement after the CTE is executed, it creates the result set for myCTE and then applies its own query, which in this case just extracts three of the columns produced by the recursive CTE.

Of course, the final result depends on the query that operates on the CTE results and it does not have to list all the CTE rows.

For instance, the following recursive query finds the total number of staff below a particular point in the hierarchy—again, we've taken the top node, but with a simple change to the query defining the anchor, we could have taken any point in the hierarchy.

```
with myCTE (deptId, totalStaff, parentDept) as
(  select anchor.deptId, anchor.totalStaff, anchor.parentDept
   from Dept as anchor
   where parentDept is null
          union all
                select child.deptId, child.totalStaff, child.parentDept
                from Dept as child join myCTE as parent
                on child.parentDept = parent.deptId
)
select sum(totalStaff) as totalStaff from myCTE
```

which produces the expected result:

totalStaff
20

Clearly, we could have produced this particular result using a much simpler query, but in general, recursion allows us to frame queries on hierarchical structures that would be difficult (or clumsy) to achieve in other ways. Most commercial dialects of SQL support CTEs and recursion, but the details may vary from the outline we've given here.

12.13 SQL: Updating Table Populations

Generically, an "update" of the population of the database tables may involve any of three relational operations: insert a row into a table, delete a row from a table, or modify one or more values on a row. In SQL, the modify operation is called "update" so the term is sometimes used in this restricted sense.

Let's first consider how to **insert rows** into a table. For this, SQL provides an **insert** statement in two basic forms. The first adds a *single row of data* to a table. Its SQL-89 syntax is:

insert into *tablename* [(*col-list*)]
  values (*constant-list*)

If no columns are listed, values for all columns must be given. Character strings must be delimited by single quotes. A null value may be entered as **null**. SQL-92 introduced the following form of the **insert** statement: **insert into** *tablename* **default values**. This adds a single row where each column is given its default value (possibly **null**). An error results if any column has no default. SQL:1999 added options for using constructors and overrides.

As a simple example, consider the table scheme *Employee* (empNr, surname, firstname, gender, [phone], [email]). Suppose a new female employee named "Eve Jones" is assigned the employee number 715. To add these details of the new employee to the table, we could use either of the following commands.

insert into Employee
  values (715, 'Jones', 'Eve', 'F', null, null)

or

insert into Employee (empNr, surname, firstname, gender)
  values (715, 'Jones', 'Eve', 'F')

Commercial systems also provide commands for fast bulk load of data from files (e.g., in delimited ASCII format). Data entry from users is typically via screen forms rather than directly to tables. Because of constraints on forms or between tables, a single row insert might require other data to be inserted in the same transaction. A transaction is a series of SQL statements, terminated by a **commit** statement:

commit [work]

If not committed, a transaction may be cancelled, thus:

rollback [work]

SQL:1999 allows inclusion of one or more *savepoints* within a transaction. Work can then be rolled back to a specific savepoint rather than rolling back the whole transaction. Releasing a savepoint acts like a tentative commit of the part of the transaction preceding the savepoint. Many SQL vendors do not yet support savepoints.

The most powerful form of the **insert** statement allows us to insert *multiple rows* into a table in one go. Its basic syntax is:

insert into *tablename* [(*col-list*)]
  select-query

This inserts the set of rows returned from the subquery. This has many uses, such as making copies of tables, or temporarily storing join results to speed up later queries. For example, consider the software retailer database schematized earlier in Fig. 12.63. Now, suppose we want to run many queries about sales of the database products. To facilitate this, we could create an appropriate table called "DBsales" and then populate it, thus:

```
insert into DBsales
  select *
  from Invoice natural join LineItem natural join Item
  where category = 'DB'
```

For analysis purposes for a given period (e.g., 1 week), it may be acceptable to ignore updates to the underlying base tables. Queries can now be formulated on DBsales without the performance overhead of the original joins. This approach is used in data warehousing. Now, let's consider how to **delete rows**. In SQL, the **delete** statement is used to delete some or all rows from a table. After deletion, the table structure still exists (until the relevant **drop table** command). The basic syntax of the **delete** statement is:

```
delete from tablename
[where condition]
```

Here are two examples, based on our retailer database:

```
delete from Item        -- deletes all rows from the Item table

delete from Item
where category = 'DB' -- deletes all the database item rows from Item
```

Next, let's consider how to **update rows** (i.e., change one or more values in existing rows). In SQL, the **insert** statement is used to modify rows. Its basic syntax is shown. In SQL-89, expression must be a scalar expression, or **null** or **default**. A scalar expression may include column names, constants, and scalar operators.

```
update tablename
     set colname = expression   [, ...]
     [where condition]
```

For example, consider the table scheme *Employee* (empNr, ..., job, salary). The following statement gives all modelers earning less than $50,000 a 5% increase in salary:

```
update Employee
  set salary = salary * 1.05
  where job = 'Modeler'
    and salary < 50000
```

Because of the bulk updating power of **insert**, **delete**, and **update** commands, it is critical to have facilities for undo (rollback) and *sec*urity (e.g., granting relevant access rights).

12.14 Summary

The data manipulation aspect of the relational model includes relational algebra and relational assignment (:=). Apart from the use of logical and comparison operators in expressing conditions, the algebra includes nine basic table operations. Comparison operators have precedence over logical operators, which have precedence over table operators.

Table 12.3 listed the six priority levels of the operators: =, <>, <, >, <=, >=; **not**; **and**; **or**; selection (**where** ...) and projection (*[ ]*); and finally **union** ∪, **intersection** ∩, difference − (termed **except** in SQL), Cartesian product ×, natural join ⋈, and division ÷. Parentheses may be used to override this order. A notation common in academic journals is shown on the left, with our notation on the right.

$$\sigma_c(T) \equiv T \textbf{ where } c$$
$$\pi_{a,b,\ldots}(T) \equiv T\ [a,b,\ldots]$$
$$A \bowtie_c B \equiv (A \times B) \textbf{ where } c$$

Operators on the same precedence level are left associative (evaluated in left to right order as they appear in an expression). Our priority convention is designed to minimize the use of parentheses and is stronger than some other conventions in use. If in doubt, or purely for the sake of clarity, feel free to add extra parentheses.

A *relational DBMS* has the relational table as its only essential data structure and supports relational selection, projection, and join without needing to specify access paths.

The table operators are summarized visually in Fig. 12.64. Many different notations exist for selection, projection, and joins.

The laws for the comparison and logical operators are well known. Table 12.17 sets out most of the main laws for the table operators. Some of these have not been discussed; their proof is left as an easy exercise. Here *A*, *B*, and *C* are tables, *c* is a condition, and *p* is a projection list of attributes.

Various other distributive laws could be stated; recall, however, that *[ ]* does not distribute over ∩ or −.

SQL is the dominant language used in relational DBMSs, and various versions have been standardized over the years, with the most recent being SQL:2023. Most commercial SQL dialects support all of SQL-89, some of the extra features in SQL-92 and SQL:1999, and some of their own special features.

Database objects such as tables and columns may be named using standard identifiers or delimited identifiers (allowing any characters inside double quotes). Database fields may be defined in terms of predefined data types (e.g., **varchar**(20)).

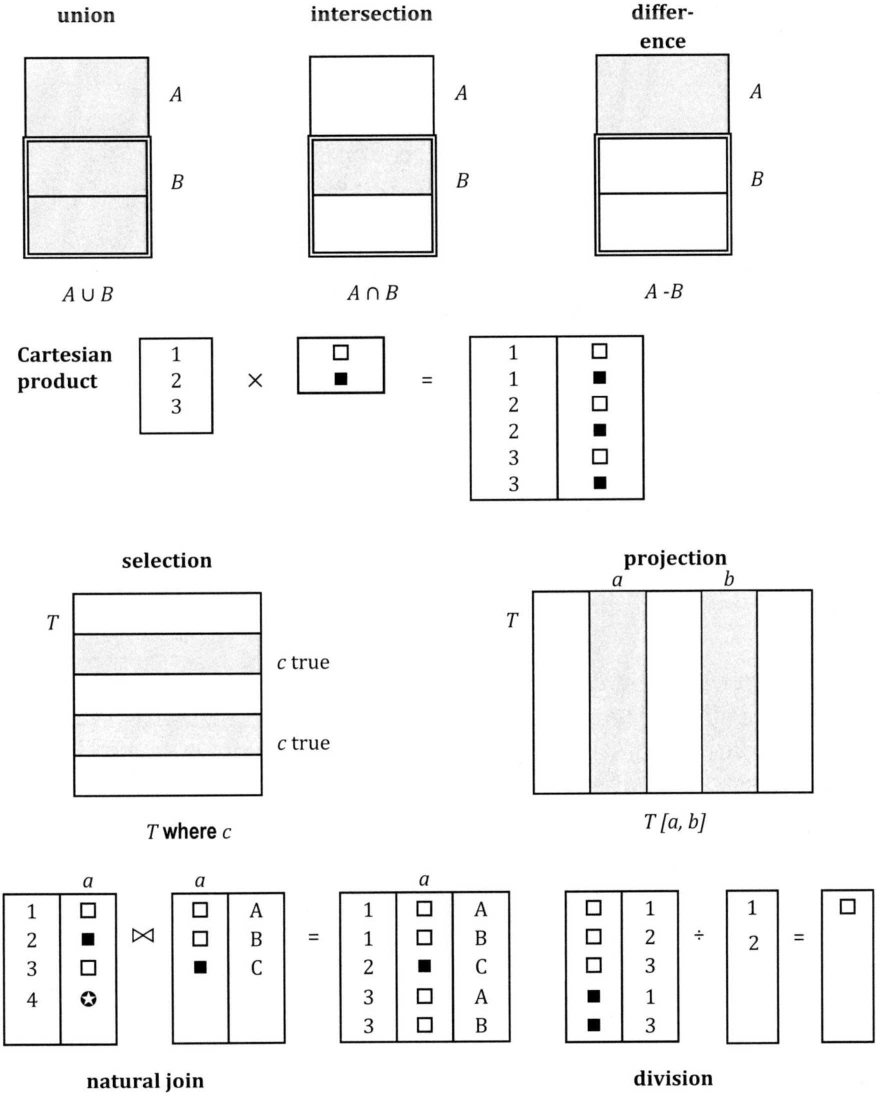

FIG. 12.64 The eight basic table operations of relational algebra.

Table 12.17 Main laws for the relational operators.

Commutative laws:	$A \cup B$	=	$B \cup A$
	$A \cap B$	=	$B \cap A$
Associative laws:	$A \cup (B \cup C)$	=	$(A \cup B) \cup C$
	$A \cap (B \cap C)$	=	$(A \cap B) \cap C$
	$A \times (B \times C)$	=	$(A \times B) \times C$
	$A \bowtie (B \bowtie C)$	=	$(A \bowtie B) \bowtie C$
Distributive laws:	$A \cup (B \cap C)$	=	$(A \cup B) \cap (A \cup C)$
	$A \cap (B \cup C)$	=	$(A \cap B) \cup (A \cap C)$
	$(A \cup B)$ **where** c	=	A **where** $c \cup B$ **where** c
	$(A \cap B)$ **where** c	=	A **where** $c \cap B$ **where** c
If A, B are compatible:	$(A \cup B)\ [p]$	=	$A\ [p] \cup B\ [p]$ -- false for $\cap$
Exportation:	A **where** $c1$ **where** $c2$	=	A **where** $c1$ **and** $c2$
Delayed Projection:	$A\ [p]$ **where** c	=	$(A$ **where** $c)\ [p]$

SQL:1999 added user-defined types, row types, reference types, and arrays. SQL:2003 added support for **multiset** (bags) as a collection type, and **bigint** and **xml** data types.

SQL:2016 added support for row pattern matching, polymorphic table functions, and operations on JSON data stored in character string fields. SQL:2019 added support for multidimensional arrays (MDarray type and operators).

The SQL:2023 (https://www.iso.org/standard/76584.html) revision added many optional features (null values in unique constraints, string padding functions, etc.), deeper support for JSON, and support for property graph queries (SQL/PGQ).

The SQL **select** statement used for queries has the following basic syntax.

```
select [all | distinct] * | expression [[as] column-alias ] [,...]
        from table [[as] alias]
        [, | cross join table [[as] alias]
           | natural [inner | {left | right | full} [outer]] join table [[as] alias]
           | [inner | {left | right | full} [outer]] join table [[as] alias]
               { on condition | using (col-list) }
           | union join table [[as] alias]
             [,...]]
[where col-condition]
[group by colname [,...]
    [having group-condition]]
[union | intersect | except ...]
[order by col [asc | desc] [,...] ]
```

The **select** list chooses the result columns, the **from** clause declares the source tables and joins, the **where** clause is used to filter out rows, the **group by** clause forms groups of rows, the **having** clause is used to filter out groups, and the **order by** clause determines the order of rows.

A *subquery* is a parenthesized **select** query inside an outer query. If the subquery has a condition that refers to the outer query, it is a *correlated* subquery. A subquery after an **exists** quantifier is an existential subquery. An **exists** condition is true if the subquery returns a row, and false otherwise. The three most common forms of conditions with an embedded subquery are as shown, where Θ is a comparison operator.

```
expression [not] in (subquery)                -- membership subquery
expression Θ [all | some | any] (subquery)    -- [quantified] comparison subquery
exists (subquery)                             -- existential subquery
```

Table populations may be changed using **insert**, **update**, and **delete** statements, with the following basic syntax.

```
insert into tablename [ (col-list) ]
  {values (constant-list) | select-query}

update tablename
  set colname = expression   [, ...]
  [where condition]

delete from tablename
[ where condition ]
```

Chapter Notes

Relational algebra is covered in most database texts, e.g., Elmasri and Navathe (1994), Date (2000). Arithmetic operators, set functions, semijoin and other operators have been proposed to extend the relational algebra, e.g., Date (2000), sec. 6.7. As an alternative to the relational algebra, the relational calculus may be used, e.g., Date (2000), ch. 7. Codd (1990) proposed version 2 of the relational model, with 333 rules, but this has not achieved any significant adoption.

The SQL standard is published by the International Standards Organization (*www.iso.org*), At the time this book was published, the current version (SQL:2023) is ISO/IEC 9075-2:2023); see https://www.iso.org/standard/76584.html, and press the Preview button to browse the publicly available informative sections. For a clear, thorough account of the SQL-92 standard, see Date and Darwen (1997), and for SQL:1999 see Melton and Simon (2002).

The original SEQUEL language, described in Chamberlin and Boyce (1974), was itself based on an earlier relational language called "SQUARE". Chamberlin (1998) provides very readable coverage of DB2, with Section 1.3 sketching an authoritative, early history of SQL, including an explanation for why SQL permits duplicate rows in base tables.

For a readable and practical discussion of many kinds of advanced queries in SQL, see the series of books by Celko (1999, 2000, 2004, 2005). An in-depth critique of SQL and an alternative theoretical foundation for object-relational databases are provided by Date and Darwen (1998).

For recent coverage of SQL fundamentals as well as how to use SQL with different DBMSs such as Oracle, MySQL, PostgreSQL, and SQL Server, see Taylor (2018) and Zhao (2021). For comprehensive coverage of Oracle up to version 11g, see Greenwald et al. (2013).

SQL Server is covered in many books, such as Syverson and Murach (2023). *SQL Server Pro Magazine* (formerly called *"SQL Server Magazine"*) is a good source of technical articles on SQL Server (www.sqlmag.com).

13

Other Database Features

Chapter Outline

13.1 SQL: The Bigger Picture

In the previous chapter, we looked at structured query language (SQL) statements that retrieve data (**select**...) and modify data (**insert**..., **update**..., **delete**...). This set of statements is traditionally known as the *data manipulation language* (DML). We now turn our attention to other aspects of SQL. We'll look at statements used to define the various elements of a database schema, which uses a subset of SQL that is traditionally known as the *data definition language* (DDL).

Tables are, of course, fundamentally important to any schema, but we'll also discuss several other types of database objects and the roles they play in the realization of a conceptual schema. In particular, database objects such as *views* and *triggers* are often used to implement constraints and derivation rules that may have been defined initially at the conceptual level.

The widespread adoption of extensible markup language (XML) has been fueled primarily by its use for the interchange of structured data. However, this has created a corresponding need for better integration between relational and XML data. A later chapter discusses the ways in which the SQL standard has been adapted and extended to add XML-related features and how non-SQL data manipulation (such as in XQuery) can be accommodated in a relational framework.

Information Modeling and Relational Databases. https://doi.org/10.1016/B978-0-443-23790-4.00004-9

13.2 SQL: Defining Tables

Creating Tables

The population of a *base table* resides in the database (unlike result tables and views, which we'll look at shortly). Base tables must be created before they are populated. In SQL-92 onward, base tables (and other aspects of the database) are defined within the scope of a *database schema* that is created using a **create schema** statement. The schema may be deleted using a **drop schema** statement. Commercial SQLs often provide a different syntax for creating a new database (e.g., **create database**).

In the SQL standard, tables are created using a **create table** statement. In addition to this, most database management systems (DBMSs) provide a graphical interface for users to create tables by entering text and selecting options. The main part of a table definition is a list of columns, identifying for each column the column name, its data type, and whether *nulls* are prohibited in the column (i.e., the column is mandatory). Constraints may also be declared for some of the columns.

The SQL-89 syntax of the **create table** statement may be summarized as shown. Here, *literal* denotes a numeric or string constant, *col-list* denotes a list of one or more column names separated by commas, and *unique-col-list* denotes a column list that has collectively been declared to be the primary key or to be unique. If a constraint applies to a single column, it may be declared at the end of the column definition. Multicolumn constraints must be declared at the end of the table definition. **Check** clauses are restricted to conditions that involve only one row of the table.

```
create table tablename (
  colname   data-type  [ not null ]
                       [ default { literal | null | user } ]
                       [ primary key | unique ]
                       [ references tablename ]
                       [ check (col-condition-on-same-row) ]
  [,...]
  [, primary key (col-list) ]
  [, unique (col-list) ]
  [, foreign key (col-list) references tablename [ (unique-col-list) ] [,...] ]
  [, check (table-condition-on-same-row) [,...] ] )
```

SQL-92 extended this syntax. Constraints may be named by inserting **constraint** *constraint-name* at the start of their definition. This facilitates meaningful error messages and dropping of constraints. **Check** clause conditions may range over many rows or tables. General constraints may be declared using **create assertion** (mainly for intertable constraints).

Foreign key constraints may include full/partial match conditions and referential actions. Constraints may be declared immediate or deferred. Macros for constrained data types may be declared with a **create domain** statement.

SQL:1999 went much further, for example, an **of** clause may be used to declare a table to be of a user-defined type, previously declared as a row type with a **create type** statement. Subtables may be defined using an **under** clause to reference the supertable.

Commercial SQL dialects typically support all the SQL-89 functionality, possibly with some different syntax. They tend to support only some of the new syntax in recent versions of the standard, such as SQL:2019 and SQL:2023. Constraint names and referential actions are widely supported, but assertions and unrestricted **check** clauses are rarely supported. Most vendors also provide extra features and syntax of their own.

As an example, let's see how to declare the relational schema for our software-retailer database. The graphical version of this schema is reproduced in Fig. 13.1. The schema declaration in SQL-92 syntax is shown next. You can adapt this where needed to your SQL dialect, and also add constraint names as you see fit.

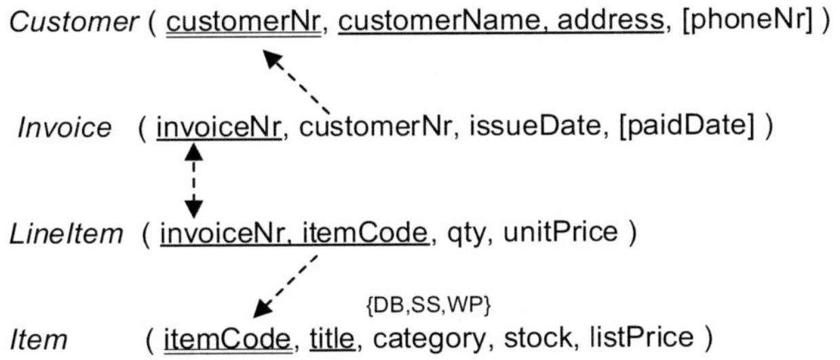

FIG. 13.1 The relational schema for the software-retailer database.

```
create schema SoftwareRetailer;

create table Customer (
  customerNr     smallint      not null  primary key,
  customerName   varchar(20)   not null,
  address        varchar(40)   not null,
  phonenr        varchar(10),
        unique (customerName, address) );

create table Item (
  itemCode       char(4)       not null  primary key,
  title          varchar(20)   not null  unique,
  category       char(2)       not null
                               check (category in ('DB','SS','WP')),
  stock          smallint      not null  default 0,
  listPrice      decimal(6,2)  not null );

create table Invoice (
  invoiceNr      smallint      not null  primary key,
  customerNr     smallint      not null  references Customer,
```

```
  issueDate        date             not null,
  paidDatepaid     date );

create table LineItem (
  invoiceNr        smallint         not null  references Invoice,
  itemCode         char(4)          not null  references Item,
  qty              smallint         not null,
  unitPrice        decimal(6,2)     not null,
      primary key ( invoiceNr, itemCode ) );

create assertion each_invoice_has_a_line_item
  check ( not exists
                   (select * from Invoice
                   where invoiceNr not in
                       (select invoiceNr from LineItem)) );
```

Notice the four intertable subset constraints (the equality constraint shows two of these). Unless we add these intertable constraints later, using an **alter table** or **create assertion** statement, the *order* in which we create the tables is important. To minimize the need for these later additions, you should wherever possible create the referenced table before the referring table. In our example, this entails creating the Customer table before the Invoice table and the Item table before the LineItem table.

Columns are optional by default, but may be declared mandatory using **not null**. SQL requires **primary key** columns to be declared not null. A **unique** constraint is assumed for primary keys, but may be explicitly declared for other columns or column lists, with the meaning that nonnull values are unique. For example, if the optional phoneNr column had instead been declared **unique**, an actual phone number may appear at most once in that column, but there may be many nulls in that column. Some commercial systems may consider null to be a value under these circumstances, in which case there could be at most one null value in the phoneNr column.

Foreign keys are declared with a **references** clause, indicating the referenced table, optionally followed by the referenced column list, which must already be declared to be the primary key or unique. If no referenced column list is given, the primary key is assumed to be the target.

The equality constraint in the graphical version of the software-retailer schema is equivalent to two subset constraints in either direction between LineItem.invoiceNr and Invoice.invoiceNr—a referential cycle. The subset constraint from LineItem.invoiceNr to Invoice.invoiceNr is a foreign key constraint because the target is a primary key. So, this may be declared using a **references** clause as shown. However, the subset constraint from Invoice.invoiceNr to LineItem.invoiceNr is *not* a foreign key constraint because its target is just one part of a composite primary key.

In SQL-92 onward, this constraint may be declared using a **create assertion** statement as shown. In SQL dialects that do not support assertions (such as SQL Server), assertions can be coded in other ways (e.g., by means of check constraints, triggers, or stored procedures). We'll look at referential integrity in more detail shortly.

A **check** clause specifies a condition to be satisfied. Value constraints are often implemented using **check** constraints. A **check** constraint is violated if and only if its condition evaluates to false (not true and not unknown). Some people find this unintuitive when compared with the **where** clause in a **select** statement, which must evaluate to true (not false and not unknown) for a row to be included in the result set.

Default options may be declared with a **default** clause. If a user inserts a row without specifying a value for a column, then the system automatically inserts the default value defined for that column. In SQL-89, the only allowed defaults were a literal value, **user** or **null**. In the absence of a default clause for an optional column, **null** is assumed to be the default. SQL-92 expanded the possible defaults to include the nullary functions **current_user**, **session_user**, **system_user**, **current_date**, **current_time**, **current_timestamp**. SQL:1999 added **current_role**, **current_path** and any implicitly typed value specification (e.g., to deal with collection types).

Here is a suggested procedure for deciding where to declare constraints in SQL that you may find useful if you want to minimize keystrokes:

```
If a constraint applies to only one table
                or is a foreign key constraint
then if it applies to only one column
      then declare it with the column
      else declare it at the end of the table definition
else declare it with an assertion (or trigger etc.)
```

SQL:2003 added another column option to deal with the creation of *identity columns*, for which the column syntax is

```
colname data-type    [ generated { always | by default } as identity ]
                       [ ( start with start-value | increment by increment-value |
                         { maxvalue value | no maxvalue } | { minvalue value | no minvalue } |
                         { cycle | no cycle } [ ,... ] ) ]
```

There can be at most one identity column in a table and its data type must be numeric with a scale of zero (e.g., integer). When a row is inserted into the table, the system generates the next value in sequence, as defined by the rather complicated set of options. If **always** is specified, it is an error to try to insert a value into the identity column. If **by default** is specified, the system will generate a value if none is supplied; otherwise, it will use the supplied value. If a row is subsequently deleted, its identity value is not reused.

The default value for both **start with** and **increment by** is 1, unless explicitly specified otherwise. The increment value can be positive or negative. The maximum and minimum values possible for a sequence are implementation dependent, and these will be adopted unless a specific **maxvalue** or **minvalue** is defined. The sequence stops and no more values will be generated once **maxvalue** (counting up) or **minvalue** (counting down) is reached or exceeded. Depending on the relationship between the start value and increment size, the sequence may or may not land exactly on a **maxvalue** or **minvalue**.

A sequence will not cycle back and duplicate values already generated unless the **cycle** option is specified. If **cycle** is specified, the sequence will stop at **maxvalue** and reset to **minvalue** if counting up, and the converse if counting down. The start value can be anywhere in the range defined by these two values.

Identity columns are supported in most implementations, although not necessarily with the SQL standard syntax. The guarantee that the generated value will be unique within a table makes them attractive to use as a primary key. In cases where the natural primary key for the table is clumsy, for example, a combination of multiple **varchar** columns, an identity column is often introduced as a surrogate key.

The SQL:1999 standard allowed the creation of tables based on existing table definitions. This capability was refined in SQL:2003 to provide more precise control over the columns that should be included or excluded in the new table. The basic syntax is as follows.

create table new-tablename **like** existing-tablename
[{ **including identity** | **excluding identity** } |
{ **including defaults** | **excluding defaults** } |
{ **including generated** | **excluding generated** } [,..]]

The result of the **create table** ... **like** ... statement is an empty table with the same column names, data types, and nullability as the original table. No dependency is established between existing and new tables. The options specify what action should be taken for "special" columns.

Maintaining Referential Integrity

Let's now discuss the notion of *referential integrity* in more detail. Recall that referential integrity is a special case of a subset constraint. The following relational schema outline shows a typical case, where *S.a* is a foreign key referencing *R.a*. Basically, the subset constraint is satisfied if each value in *S.a* either occurs as a value of *R.a* or is null.

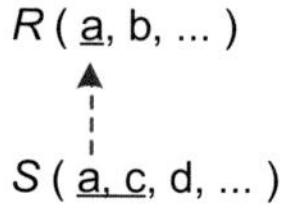

The subset constraint may be violated in any of the following four ways:

(1) insert a row into *S* with a value for *S.a* that is not a value in *R.a*
(2) update a row of *S* to replace the value for *S.a* with a value that is not in *R.a*

(3) delete a row from *R* where the value of *R.a* is also a value in *S.a*
(4) update a row of *R* to replace a value of *R.a* that is also a value in *S.a*

Instead of declaring the subset constraint as "*S.a* **references** *R.a*", we could enforce the constraint by writing trigger code to handle each of these four cases. A *trigger* is basically a chunk of procedural code that is defined for a single table to carry out some action whenever a specified kind of event (insert, delete, or update) occurs for that table. We'll look at triggers in more detail a little later in this chapter. SQL:1999 includes a full procedural trigger language with the power of a typical third-generation programming language.

Although triggers can be very useful for enforcing rules (such as constraints or derivation rules), if a declarative version of the rule is supported by the system, then the declarative approach is normally preferred. First, the declarative version of the rule is more easily understood, and hence easier to validate with domain experts. Second, the declarative version is shorter and easier to formulate than the procedural version. For example, the declaration "*S.a* **references** *R.a*" is trivial compared to the code for the corresponding triggers. Finally, the system can often do a better job of optimizing declarative code than procedural code. Having said that, there are still many cases where triggers or stored procedures provide the only way of getting the system to carry out some action. Until recently, referential actions fitted into this category.

A *referential action* is an action to be taken in response to a violation of a referential integrity constraint. In SQL-89, if an attempted update (insert/delete/update) violates a foreign key constraint, the update is simply rejected; in other words, no action is taken. In SQL-92, this "no action" default may be replaced for update and delete operations by other options by declaring the appropriate referential action(s) at the end of the **references** clause. SQL-92 also allows a **match** clause, with **full** or **partial** options for finer control over null matching. SQL:1999 added the **simple** option to allow the default option for matching to be explicitly declared, but otherwise uses the same syntax for the **references** clause, the full syntax of which is

references *table* [(*col-list*)]
[**match** { **full** | **partial** | **simple** }]
[*delete-action* [*update-action*] | *update-action* [*delete-action*]]

where *delete-action* has the syntax:

on delete { **no action** | **cascade** | **set null** | **set default** }

and *update-action* has the syntax:

on update { **no action** | **cascade** | **set null** | **set default** }

If an attempt is made to change the database with an insert, delete, or update operation that would violate the referential constraint, the default **no action** option rejects the change. The other three referential action options accept the change, but also take some compensating action to keep the database consistent. The **cascade** option propagates the change to all referencing foreign key values; **set null** sets the relevant foreign key values to null; and

set default sets the relevant foreign key values to the column default. For example, consider the following relational schema:

CarModel (model, manufacturer)

Car (carRegNr, model)

Drives (empNr, carRegNr)

Here, we have a chain of two subset constraints. Suppose "**on delete cascade**" is declared for both constraints. If we attempt to delete the model "Ford T" from the CarModel table, the delete is accepted. But if there are any rows in the Car table with the model "Ford T", they will be deleted also, which in turn causes any rows in the Drives table with matching registration number to be deleted also.

SQL-92 introduced the **match** operator as a Boolean operator to test whether a tuple belongs to a set of tuples, that is, does $(a_1, \ldots, a_n) \in \{(b_1, \ldots, b_n), \ldots\}$? Except for its Boolean nature, it is a generalization of the **in** operator and may be used in search conditions of the form $(a_1, \ldots, a_n)$ **match [unique] [full | partial | simple]** (*subquery*). The **unique** option requires the tuple to match exactly one row. By default, or if **simple** is declared, the match operation returns True if all values $a_1..a_n$ match or *some* are null; otherwise, it returns False. If full is specified, the match operation returns True if all values $a_1..a_n$ match or *all* are null; otherwise, it returns False. If partial is specified, the match operation returns True if all *non-null* values match; otherwise, it returns False. The match operator may also be used in a **references** clause in a similar way. Although the **references** clause syntax for SQL-89 is widely supported, most SQL dialects have less support for the further referential syntax in SQL-92 or SQL:1999.

Changing Table Definitions

After a base table is created and even populated, we may wish to change its definition. This was impossible in SQL-89, but SQL-92 allows a table scheme to be changed using an **alter table** statement. SQL:1999 extended this slightly to cater for column scoping. Each change requires a separate statement. The basic syntax is as follows:

alter table *tablename*
add [**column**] *col-defn*
| **alter** [**column**] *colname* **set** *default-defn* | **drop default**
| **drop** [**column**] *colname* **restrict** | **cascade**
| **add** *table-constraint-defn*
| **drop constraint** *constraint-name* **restrict** | **cascade**

Dropping a column is not allowed if **restrict** is specified and the column is referenced in constraint or view definitions. The **cascade** option causes the other definitions to be dropped too. Commercial SQLs typically do not support **restrict** or **cascade** options for dropping columns or constraints.

Although the standard forbids the data type of a column to be altered, some dialects allow a column's data type to be changed to a "larger type", such as change char(2) to char(4) for a roomNr column. Some alter actions are not supported by many vendors, such as **drop column**. To simplify automated table creation, some vendors prefer to add most table constraints in **alter table** statements. The following examples are self-explanatory:

```
alter table Person
  add column email_address varchar(20)

alter table Person
  add foreign key starsign references starsignPeriod
```

SQL-92 allows a table (including its population) to be deleted from the database using a **drop table** statement. Although this is not allowed in SQL-89, all vendors support it. The syntax is:

drop table *tablename* **cascade** | **restrict**

Dropping a table is not allowed if **restrict** is specified and the table is referenced in constraint or view definitions. The **cascade** option causes the other definitions to be dropped too. Most commercial SQLs do not require cascade/restrict here.

As a first step on the way to user-defined types, SQL-92 introduced statements for creating, altering, and dropping syntactic *domains*. Once a domain has been defined, it can be used in column definitions. For example:

```
create domain Money as decimal ( 9, 2 )
create domain Gendercode as char
constraint ValidGendercode check ( value in ( 'F' , 'M' ) )
      create domain Centimeters as smallint check ( value > 0 )
      create domain Kilograms as smallint check ( value > 0 )

create table Employee (
      empNr        smallint          not null primary key,
      gender       Gendercode        not null,
      salary       Money             not null,
      tax          Money,
      height       Centimeters,
      weight       Kilograms )
```

These domains are merely macros or abbreviations for their definitions. So, nonsense queries such as "**select** * **from** Employee **where** height = weight" are still legal. However, SQL:1999 does support true user-defined types, with strong type checking. For example, if we replace the previous domain definitions for Centimeters and Kilograms by the following type definitions, the condition "height = weight" generates an error. If you ever want to compare just the numbers rather than the height and weight entities, you can use the **cast** function to convert to compatible data types. The declaration of "**final**" means the type cannot have proper subtypes.

```
create type Centimeters   as smallint   final
create type Kilograms     as smallint   final
```

Centimeters and Kilograms are defined in terms of a predefined scalar type, but are treated as distinct from their base type, and are said to be *"distinct types"*. SQL:1999 also allows user-defined *structured types* involving many attributes. In SQL:1999, *array types* and *row types* may be introduced within a table definition. The type definition features in SQL:1999 onward are vast, but are not yet widely supported. There is no space here to discuss them in detail, but here is an example:

```
create table Voter (
    voterId       smallint                     primary key,
    preferences   smallint array[3],
    address       row( streetAddress    varchar(20),
                       cityname         varchar(20),
                       zip              row( main char(5), ext char(4))))
```

13.3 SQL: Views

In addition to defining base tables, SQL also allows us to define **views**, or "virtual tables", which are basically *named, derived tables.* Their definition is stored, but their contents are not. By means of the view definitions, queries and updates on views are automatically translated into equivalent queries/updates on base tables. Hence, unlike working tables that you might create to temporarily store results, views are always up to date.

Views have many uses. They may be used for security reasons by allowing a particular group of users access to views but not to their underlying base tables. They can also be used to hide unwanted details, simplify queries or updates, implement derivation rules, and allow base table reconstruction with minimum impact. Their main limitations are that some views cannot be updated, and queries on views sometimes lower performance. Further discussion on the implementation of derivation rules using views and generated columns can be found on the website that accompanies this book.

A view is created using the **create view** statement. The SQL-89 syntax is as follows:

```
create view viewname [ (col-list) ] as
    select-query
    [ with check option ]
```

Because the result of a view is a derived table, the **select** statement in a view can reference other views as well as base tables. In fact, from the perspective of a **select** statement, there is little difference between a query on a table and a query on a view.

SQL-92 allows the check option to be qualified as local or cascaded, and SQL:1999 allows views to be recursive and to be based on user-defined types. Although SQL-89 does not allow views to be dropped, SQL-92 does, using the syntax:

```
drop view viewname { restrict | cascade }
```

If the **restrict** option is specified, the view is not dropped if it is referenced in another view or a constraint. If the **cascade** option is specified, the view is dropped as well as any views or constraints that reference it. Many dialects, including SQL Server, do not yet support these two options. Although the standard does not provide an **alter view** statement, some dialects, including SQL Server's T-SQL, do. One advantage of altering a view, instead of dropping it and recreating it, is that any privileges associated with the view are preserved.

Some restrictions apply to the **select** query used to define a view, although details depend on the SQL dialect. Typically, no **order-by** clause or any of the **union**, **intersect**, or **except** operators are allowed. An *updatable view* can be used to insert, delete, or modify rows. Many more restrictions apply to updatable views (see later).

Consider the table scheme *Academic* (empNr, empName, dept, [gender], grade) used to store details about academics. Table 13.1 includes some data. The dept column stores the academic's department code (e.g., CS = Computer Science, MA = Mathematics), and the grade column is the academic's rank (e.g., Lec = Lecturer, Prof = Professor). For example, to retrieve the female professors of computer science, we request:

```
select empNr, name from FemaleCompScientist
where rank = 'Prof'                              →    Rolland, C
```

Table 13.1 A base table about academics.

Academic:

empNr	*empName*	*dept*	*gender*	*grade*
101	Aldave, L	CS	F	Lec
102	Jones, E	MA	?	Lec
103	Codd, E	CS	M	Prof
104	Rolland, C	CS	F	Prof

The following statement may be used to create a view about just the female computer scientists. With this definition stored, we may now act as if a table corresponding to the view really exists, as shown in Table 13.2.

```
create view FemaleCompScientist ( empNr, name, rank ) as
    select empNr, empName, grade from Academic
    where dept = 'CS' and gender = 'F'
```

Table 13.2 A view.

FemaleCompScientist:

empNr	*name*	*rank*
101	Aldave, L	Lec
104	Rolland, C	Prof

When this query is processed, the view definition is accessed and used to automatically translate the query into the following equivalent query on the underlying base table. The base table query is then processed to produce the result:

```
select empNr, empName
from Academic
where dept = 'CS' and gender = 'F'
and grade = 'Prof'                 → Rolland, C
```

That view was based on just a projection and selection from a single table. Let's look at a harder case based on a join. Consider the table scheme *ParentOf* (parent, child). A sample population is shown in Fig. 13.2. For simplicity, assume people are identified by their first name.

Now consider the query, Who are the grandparents of Fred? To make it easy to answer the question, Fig. 13.2 includes a graph where a dotted arrow from one node to another means the first is the parent of the second. Starting at Fred, we can easily find Chris as Fred's only recorded parent, and then move up to Chris's parents. So, the answer is the set {Ann, Bill}. We can also follow this procedure using the table. The query is easy to formulate in SQL using a membership subquery, thus:

```
select parent from ParentOf
where child in
     ( select parent from ParentOf
     where child = 'Fred' )
```

But now suppose we often wanted to list the grandparents of different people on different occasions. We could save ourselves some work by defining a view for the *GrandparentOf* (grandparent, grandchild) relation thus:

```
create view GrandparentOf ( grandparent, grandchild ) as
     select A.parent, B.child
     from ParentOf as A join ParentOf as B
     on A.child = B.parent
```

ParentOf:

parent	*child*
Ann	Chris
Ann	David
Bill	Chris
Bill	David
Chris	Ernie
Chris	Fred
David	Gina
Gina	Helen

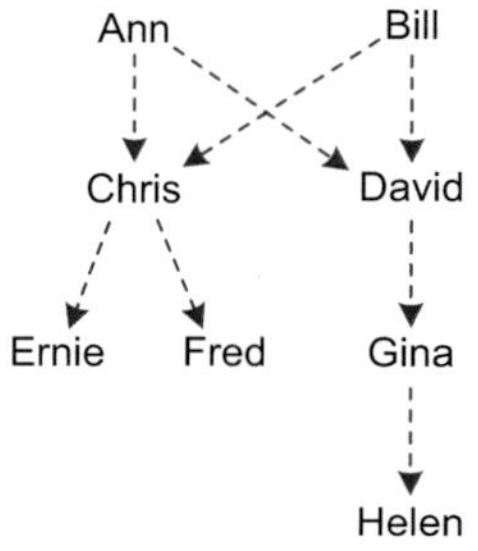

FIG. 13.2 Who are the grandparents of Fred?

Table 13.3 A view based on a join.

GrandparentOf:

grandparent	*grandchild*
Ann	Ernie
Bill	Ernie
Ann	Fred
Bill	Fred
Ann	Gina
Bill	Gina
David	Helen

This view definition involves a projection on a self-join, where we match the child of the grandparent to the parent of the grandchild. Table 13.3 shows the virtual population, which could be listed using the command **select** * **from** GrandParentOf.

The grandparents of Fred can be listed using **select** * **from** GrandParentOf **where** grandchild = 'Fred'. The grandparents of any other person can be requested by using their name in place of "Fred". Although the view is obviously useful for queries, can it also be used for updates?

As discussed in the previous chapter, rows can be inserted into base tables using an **insert** command. Suppose we try to insert a row into the view, thus:

```
insert into GrandparentOf
   values ( 'Bernie', 'Selena' )
```

An update on a view must be translated into an equivalent update on the base table(s). But because we don't know the intermediate parents (i.e., Selena's parents), the best we could hope for in this case is to add the following two rows into the ParentOf table:

```
ParentOf:  parent   child
           -------------
           Bernie     ?
           ?          Selena
```

But this is unacceptable because it doesn't really tell us that Bernie is a grandparent of Selena. In fact, the two rows would not satisfy the view definition anyway because they fail the join condition A.child = B.parent. Because both values are null here, the condition evaluates to unknown, and the row ("Bernie", "Selena") is excluded from the view. So, the GrandparentOf view is *nonupdatable*.

SQL-92 places many restrictions (more than logically required) on what kinds of views are updatable. In particular, an updatable view must contain no joins, no union (or intersect or except), no **distinct** option, and no **group by** clause. SQL Server gets around many such restrictions by use of **instead-of** triggers on views (see later).

Even with no joins, similar problems with nulls arise if the view excludes a column used in its search condition. For example, suppose we try the following insertion on our FemaleCompScientist view:

```
insert into FemaleCompScientist
    values ( 105, 'Bird, L', 'Lec' )
```

If we allow this update, is "Bird, L" included in the result? Given our earlier view definition, if the update were accepted it would translate into the following row being inserted into the Academic table: (105, "Bird, L", **null**, **null**, "Lec"). The new row fails the view definition because the condition (dept = 'CS' **and** gender = 'F') evaluates to unknown. So, if the update is accepted, it is not part of the view!

It is better to reject any update that fails to satisfy the view definition. This is done by appending "**with check option**" to the view definition. For example, the following view may be used for computer science professors:

```
create view CSProfessor ( empNr, name, dept, rank ) as
    select empNr, empName, dept, grade
    from Academic
    where grade = 'Prof' and dept = 'CS'
        with check option
```

The following attempts to update this view would now have the results shown.

```
insert into CSProfessor
  values ( 106,'Wirth N','CS','Prof' )     →   accepted

insert into CSProfessor
  values ( 107,'Russell B','MA','Prof' )   →   rejected

update CSProfessor
  set rank = 'Lec'
  where empNr = 103                        →   rejected
```

The type of view we have considered so far is basically a predefined query on one or more base tables. The result of the query is not stored in the view, only the query definition. A subsequent query on the view is combined with the query in the view definition, and the composite query is run against the base table(s). This ensures that the result is always up to date with the content of the base tables, but may be inefficient if a view is queried frequently.

Another option is to execute the query defined for the view and cache the result. Now, the view represents an actual data structure held in physical storage, not just the definition of a data structure. This is known as a *materialized* view. Such views have obvious advantages in efficiency, but subsequent changes to data in the base table(s) can lead to the view becoming increasingly unrepresentative of the "real" data. Commercial systems that support materialized views normally provide some mechanism for periodically refreshing them.

SQL:2003 added a new option to the **create table** statement to allow for the creation of tables based on the result of a **select** statement. This provides functionality somewhat similar to a materialized view. The basic syntax is:

create table tablename **as** select-statement
{ **with data** | **with no data** }

If the intention is to populate the table with data at the time of its creation, the **with data** option must be specified. If it is not, an empty table will be created with the appropriate columns but no data. A better description of the result is a *materialized query table* because no dependency is established between the new table and the underlying query expression. After the table is initially populated, updates to the tables in the query expression will not automatically be reflected in the new table.

Exercise 13.3

1. Specify SQL DDL statements to create the following relational schema.

2. A conceptual subschema, its relational tables, and a sample population are shown.

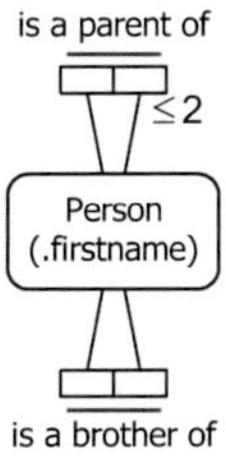

ParentOf: (≤2)

parent	*child*
Ann	Chris
Ann	David
Bill	Chris
Bill	David
Chis	Ernie
Chris	Fred
David	Gina
Gina	Helen

BrotherOf:

brother	*sibling*
Ian	Ann
John	Ann
Ken	Bill
Bill	Ken
David	Chris
Ernie	Fred
Fred	Ernie

(a) Complete the following derivation rule: *Person₁* is an uncle of *Person₂* **if** …
(b) Specify this rule as a view *UncleOf* (uncle, child) in SQL.
(c) Formulate the query "Who is an uncle of whom?" in SQL, using the view.
(d) Formulate (c) in SQL without using the view.
(e) Formulate in SQL: Who is recorded as having a parent but not an uncle?

3. The following schema shows part of a database that holds information about repair technicians and the types of assemblies that they are competent to repair. Different technicians may be competent at repairing different assemblies. Assemblies may be parts of other assemblies, unless, of course, they are a top-level assembly.

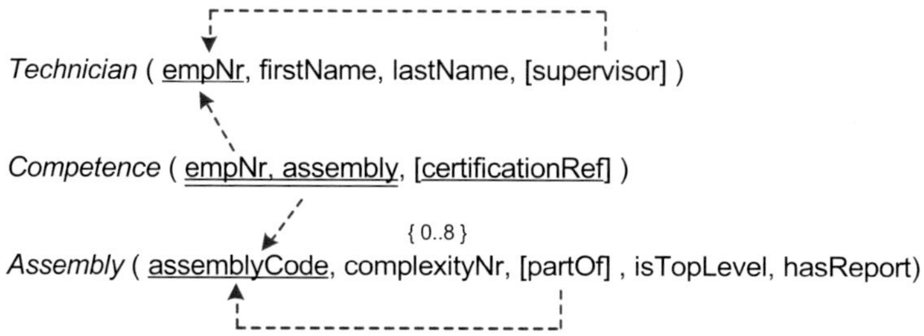

(a) Create the tables for this schema in the most convenient SQL form. Assume that empNr is some unique integer, assemblyCode is some unique 10-character string, complexityNr is an integer, and firstName, lastName, and certificationRef are varchar strings of appropriate length. isTopLevel and hasReport are flags that reflect a Boolean condition. For this exercise, specify certificationRef simply as a unique column.
(b) Is the order of creation of the tables important?
(c) Both of the Technician and Assembly tables could represent hierarchies (how can you tell?). What would be the distinguishing characteristics of a node at the top of a Technician hierarchy and a node at the top of an Assembly hierarchy?

4. Given the schema defined in Q1, we want to identify technicians with high-level skills.
(a) Define a view for all technicians that are competent to work on an assembly of complexity 6 or higher. This should list only the Technician's firstName and lastName, assemblyCode, and complexityNr.
(b) Is this view updateable? Explain your reasoning.

5. Do you agree or disagree with the following statements about ordinary SQL views? Explain your reasoning.
(a) Views create temporary tables that can be queried using select statements.
(b) A select statement associated with a view may not reference other views.
(c) Defining a large number of views will create additional demands for data storage.
(d) A view defined on exactly one base table is always materialized.
(e) A view can be queried in much the same way as a table, but updates are more restricted.

13.4 SQL: Triggers

Triggers provide yet another way to change table populations. Although not included in the standard until SQL:1999, triggers have been used by commercial dialects for many years. As well as enforcing constraints, triggers may be used to perform other actions, including computation of derived columns, maintaining audit or summary data in other tables, and initiating external actions such as sending email.

A trigger specifies a set of SQL statements that will be executed automatically when a given event takes place on a table (or, in some circumstances that we'll discuss shortly, a view). The event causing the execution of the trigger can be an attempt to *insert* new rows, *delete* existing rows, or *update* existing rows in the table. The trigger actions can be specified to take place either immediately *before* or immediately *after* the related operation (insert, update, or delete) takes place.

Multiple triggers can be defined on a table. All **before** triggers will execute before any operation is carried out on the table, including constraint checking. This provides an opportunity to examine the action that is about to take place and to modify it if necessary. All **after** triggers will execute after constraint checking has been performed and the action has taken place, but before any changes are committed. This provides an opportunity to examine changes that have been made and to roll back the associated transaction if required. Different vendors may provide facilities for specifying the trigger sequence if there are multiple **before** triggers, and similarly for **after** triggers.

Within the body of a trigger, it's possible to reference the contents of one or two *transition tables*. These contain rows that are being affected by the trigger event. The columns of the transition tables match the column definitions of the table on which the trigger is based. For a delete trigger, there is an *old* transition table containing the rows being deleted. For an insert trigger, there is a *new* transition table containing the rows being inserted. An update trigger has both an *old* transition table containing the row data prior to the update and a *new* transition table containing the row data after the update. In the update case, every row in the *old* transition table will have a corresponding row in the *new* transition table. Transition tables only exist for the duration of the trigger and cannot be accessed from outside the trigger.

The trigger action can be specified to take place either at the *row level* or at the *statement level*. In the former case, the action takes place for each row of the transition table. In the latter case, the action occurs just once, when the trigger event takes place. Depending on whether the trigger has been defined at the row level or the statement level, the trigger code can access *transition variables* that are either individual rows from the old/new transition tables or the complete transition tables.

In SQL:1999, the statements for creating and dropping triggers have the following syntax:

```
create trigger triggername
    { before | after } { insert | delete | update [ of col-list ] } on tablename
    [ referencing   { old | new } [row] [ as ] correlation-name |
                    { old | new } table [ as ] table-alias       [, ...] ]
    [ for each { row | statement } ]
          [ when ( condition ) ]
          triggered-SQL-statement

drop trigger triggername
```

Although most SQL dialects support triggers, they do not necessarily conform to the standard syntax just shown for creating triggers. For example, SQL Server does not currently support **before** triggers.

In addition to **before** and **after** triggers, SQL:2003 introduced a third kind of trigger: the **instead of** trigger. As the name suggests, the trigger action is carried out *instead of* the triggering event. An important application for **instead of** triggers is to intercept an attempt to modify a view and replace it with an appropriate set of modifications to the table(s) on which the view is based.

As an example, we can return to the *FemaleCompScientist* view defined earlier.

```
create view FemaleCompScientist ( empNr, name, rank ) as
    select empNr, empName, grade
    from Academic
    where dept = 'CS' and gender = 'F'
```

Recall that the base table, Academic, had columns (empNr, empName, dept, [gender], grade). An attempt to insert a new row into the view, such as

```
insert into FemaleCompScientist
    values ( 105, 'Bird, L', 'Lec' )
```

would fail because we have no values for the dept and gender columns; null entries would violate the view definition. However, we can use an **instead of** trigger to intercept the attempted insertion, and instead make the correct insertion in the base table, as follows:

```
create trigger AddFCS
instead of insert on FemaleCompScientist
referencing new row as Inserted
    for each row
    insert into Academic ( empNr, empName, grade, dept, gender )
        select empNr, empName, rank, 'CS', 'F'
        from Inserted
```

We obtain the appropriate values for empNr, empName, and grade from the empNr, empName, and rank columns in the *new transition variable*. We already know from the view definition that the values for dept and gender must be "CS" and "F", respectively. This allows us to correctly insert a new row into the Academic base table. A subsequent **select** command on the view would show that the insert has apparently succeeded, with the new row now appearing in the view.

Exercise 13.4

The following schema shows part of a database that holds information about repair technicians and the types of assemblies that they are competent to repair. Different technicians may be competent at repairing different assemblies.

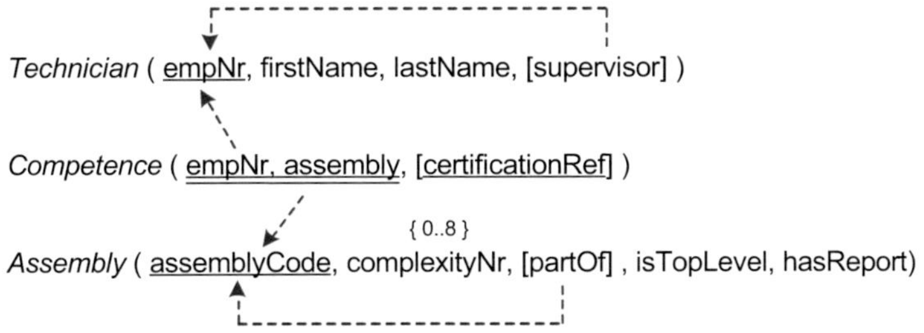

1. The foreign key reference from Competence to Technician would normally be implemented by a **references** clause. Assume that this is omitted, and create the equivalent functionality using SQL triggers to prevent violations of referential integrity (only **restrict** is required).
2. Values for certificationRef must be unique if they exist, but some implementations of SQL do not allow multiple nulls in a unique column. Instead of specifying the certificationRef column as unique, write a trigger that ensures that nonnull values are unique but allows multiple nulls.
3. From time to time, the complexityNr of an assembly may be adjusted up or down, but only by a single level. For example, a complexityNr of 6 may only be changed to 5 or 7. Write an update trigger to limit changes to complexityNr to at most one level on a single update.
4. Reports are produced detailing faults that have been found on assemblies in the field. These are stored in a table with the schema *Report* (reportId, assemblyCode, reportDoc), where reportId and reportDoc have the data types **int** and **clob**, respectively. A report can be deleted if it is subsequently found that it does not reflect a genuine fault. For efficiency, we want to set a flag in the Assembly table for any assembly that has a report. Write a trigger for the Report table that will maintain the hasReport flags in the Assembly table.

13.5 SQL: Routines

The SQL standard uses the generic term *routine* to describe a body of code that can be invoked from some other code—this covers functions, procedures, and the methods of user-defined data types.

User-Defined Functions (UDFs)

We'll look first at functions, which, in simple terms, are bodies of code that return a result. The SQL standard defines a large number of built-in functions for calculating logarithms, generating random numbers, transforming character strings into all uppercase or all lowercase, and so on. These can be used wherever appropriate in SQL code. It's also convenient to be able to define new functions to encapsulate pieces of code that are likely to be reused. In SQL, functions can return scalar (single) values of a given data type. The SQL:2003 standard extended the concept to include functions that return tables.

The basic SQL:2003 syntax for creating a function is as follows.

```
create function functionname ( [ parameter-list ] )
    returns { data-type | table col-name data-type [,...] }
    function-body
```

The parameter-list is a (possibly empty) comma-separated list of parameter name, parameter type pairs. The data-types for the returned value can be any legitimate SQL type. The function body can be composed of any valid combination of SQL statements, but must include a **return** clause that returns a parameter of the type specified in the **returns** clause. UDFs can be removed by using **drop function** *function-name*.

Commercial DBMSs may place other restrictions. For example, in SQL Server, UDFs cannot use SQL statements that would produce different results if the function was invoked repeatedly with the same parameters. Examples of such statements include getting the current time and generating random numbers. SQL Server also prohibits side effects in UDFs so, for instance, a function cannot be used to update a base table.

The following example shows a scalar UDF for SQL Server that takes a currency amount (defined as a decimal value) and a currency symbol as parameters, and returns a string consisting of the currency symbol concatenated with a string version of the amount. SQL Server uses "+" for string concatenation, as opposed to the standard "||".

```
create function moneyString ( @amount decimal( 9, 2 ), @symbol char ( 1 ) )
    returns varchar ( 11 )
    return @symbol + cast ( @amount as varchar(10) )
```

Once defined, functions can generally be used at any point where a value of the return type could be used. So, taking the example just given, we could use a call to moneyString( ) at any point where a varchar(11) value would be legitimate.

Stored Procedures

Stored procedures, commonly referred to as "sprocs", are bodies of code that are explicitly called from other code. In this section, we'll assume that the body of the procedure is written in SQL, but it's possible for the procedure actions to be written in some other language, such as "C", and then called from SQL. Stored procedures can (and often do) have side effects, such as updating tables during their execution. Encapsulating code in a stored procedure offers several potential benefits.

Modularity. Code in a stored procedure is readily reusable in other code. A well-maintained library of stored procedures can be a valuable resource when developing new applications. Modularity also helps to ensure consistency. For example, a business rule can be implemented once in a stored procedure and called as needed, instead of having the same rule implemented in perhaps slightly different ways in different places. The same argument applies to maintainability. If a change is needed, it need only be applied in one place. As long as the previous interface is maintained, other programs that call the stored procedure need not be aware of any changes that have been made.

Security. Users of a stored procedure are offered only the functionality that the procedure provides. Limiting access in this way drastically reduces the possibility of damage to the database, whether deliberate or accidental. The authority given to users of a stored procedure can be different from the authority to access any underlying tables, so any actions can only be taken in the context of the procedure. For example, all modifications to an "Orders" table can be channeled through a stored procedure that imposes business rules and other constraints on what changes can be made, instead of giving users direct access to the base table.

Performance. Stored procedures are typically compiled, optimized, and cached ready for use when needed. The details vary from one implementation to another, and extensions to the basic syntax are usually provided to address issues such as the need to redo compilation and optimization if the usage circumstances change. Stored procedures can also reduce network traffic: instead of constructing a query in an application and passing the resulting SQL across a network, only the parameters need be passed to a stored procedure.

The basic syntax for creating and dropping a stored procedure is as follows.

```
create procedure procedurename
    ( [ { in | out | inout } parmname parmtype [,...] ] )
        procedure-body

drop procedure procedurename
```

A stored procedure takes zero or more parameters. Each parameter can be specified as **in** (external values are provided to the procedure), **out** (the procedure makes its results available externally), or **inout** (the input parameter may be modified by the procedure for external access). As usual, commercial systems may vary from the standard syntax. For example, in SQL Server any parameters of a stored procedure are considered as **in** parameters unless they are explicitly denoted as **out**, and there is no **inout** option.

The procedure body is made up of any legal combination of SQL statements. Stored procedures may make calls to other procedures, and a procedure may call itself recursively. An implementation may place restrictions on the number of levels of nesting permitted, for example, SQL Server allows nesting up to 32 levels deep. Stored procedures can also be called from within a trigger, which gives both the advantage of automatic invocation (when the relevant trigger event occurs) and the modularity offered by a stored procedure. For instance, we may want to apply the same constraint under several different circumstances, which we could do by having several different triggers call the same stored procedure.

One SQL statement that requires attention in a stored procedure is the **return** statement. Once a **return** is encountered, the stored procedure is exited immediately, so a **return** is often associated with a branch of a test condition, along the lines of "*if x is the case then return, otherwise, …*". Although not necessarily enforced by an implementation, a recommended practice is to use a value associated with the **return** statement to indicate status. Typically, a return code of zero indicates "all is well", and a nonzero value is used to indicate a specific error condition that had been encountered. The results of any computation carried out by the stored procedure should be returned through **out** parameters rather than using the **return** value.

Exercise 13.5

The following schema shows part of a database that holds information about repair technicians and the types of assemblies that they are competent to repair. Different technicians may be competent at repairing different assemblies.

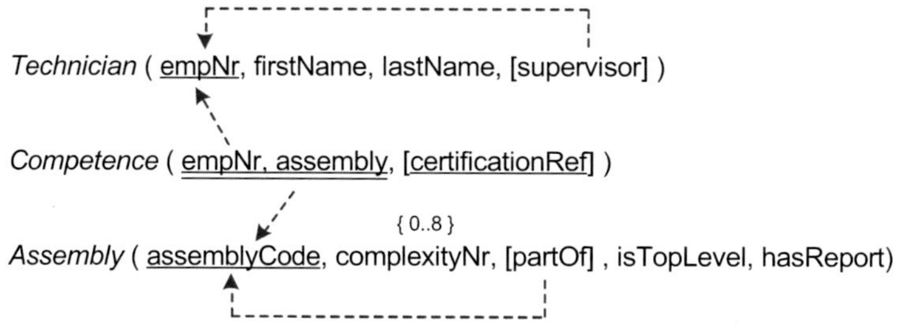

1. Write a stored procedure that inserts a new row into the Technician table and returns a result of zero if it succeeds. If the new row would create a loop in the technician—supervisor hierarchy, the insert should not be performed and the procedure should return the result code 99. Use recursion to check for loops in the hierarchy. You can assume that no hierarchy will be more than 10 levels deep.

2. Most commercial systems provide a function that returns the last identity value used in the update of a table. The empNr column in the Technician table would be a typical application for an identity column. Write a stored procedure that adds a new row to

the Technician table and returns as an **out** parameter the identity value that was generated for the new row.

3. Assume that we have a table with the schema *Currency* (code, symbol, rate), where code identifies a particular currency, symbol is a single Unicode character, and rate is a conversion rate. Write a function that takes a currency code and a decimal number as parameters and returns a string consisting of the number provided multiplied by the rate for the currency code appended to the appropriate currency symbol. For example, if the Currency table contains a row ("USD", "$", 2.0), given the parameters "USD" and 100.00 the function should return "$200.00". The returned amount should have two decimal digits.

13.6 More Database Objects

Sequences

In SQL, a sequence is a database object that automatically generates a succession of values; note that this is a completely different meaning of the word "sequence" from its usage in XQuery, as we'll see later. Sequence objects were first introduced in SQL:2003 and are useful whenever it's necessary to produce a unique key value. Having the unique value generated inside the database, instead of in an external application, reduces the possibility of issues arising in areas such as performance and concurrency. Some SQL implementations may precompute the next values in a sequence and cache these ready for immediate use.

In some respects, a sequence is similar to an identity column in a table. However, unlike an identity column, a sequence is not tied to one specific table, and sequence values are unique within a database, rather than being unique within a table. Sequences can be created using the following syntax.

```
create sequence sequence-name
  [ as data-type | start with start-value |
  increment by increment-value |
  maxvalue value | no maxvalue
  minvalue value | no minvalue |
  cycle | no cycle [ ,... ] ]
```

The data type must be a numeric type with a scale of zero (e.g., integer). The options are similar to those for identity columns, which were discussed earlier. One additional feature of sequences is that some implementations also allow a zero increment, so that the sequence effectively becomes a global constant value.

The standard syntax for obtaining the next value from a sequence is:

```
next value for sequence-name
```

but some implementations may have variations on this. They may also provide some means of accessing the last sequence number generated without generating a new one, such as:

previous value for sequence-name

SQL also defines **drop sequence** and **alter sequence** commands, but because these work in the obvious way, we won't discuss them here.

Indexes

An *index* is typically an ordered binary tree of key values, with pointers to the locations of the full records used to store the data. Because indexes are relevant only for physical modeling, not logical modeling, the SQL standard will never mention them. However, most commercial systems allow users to explicitly create and drop indexes, using syntax such as **create** [**unique**] **index** indexname **on** tablename (index-column [**asc** | **desc**] [,...]) and **drop index** indexname. Optimizers use indexes to improve performance and may create and drop indexes themselves. If supported, the unique index option provides one way of enforcing uniqueness constraints. However, most uniqueness constraints are better declared using **primary key** or **unique** declarations, although uniqueness constraints on optional columns usually require triggers. Indexes are often created automatically for primary and foreign keys.

Indexing can dramatically improve query performance, especially for large tables, because indexes often allow random access in main memory rather than sequential disk access. Some commercial systems may also allow a chosen index to be specified as *clustered*. This creates an alignment between the logical sequence of values in the indexed column(s) and the order in which rows are physically stored, potentially offering improved performance in retrieving row data. Obviously, if a table has multiple indexes, only one can be selected as the basis for clustering; typically, this will be the index on the primary key column(s).

Although query efficiency (and hence perhaps constraint enforcement performance) can be improved by indexing, there is a price to be paid. Indexing may slow down the updating of a table because any change to the data will involve not only a change to the base table, but also to any indexes that have been defined. However, this trade-off may still be acceptable, depending on the situation. Individual DBMSs may offer many different kinds of indexes, and strategies for choosing indexes may also vary from one DBMS to another. Because this book is primary concerned with conceptual and logical modeling, index strategies will not be considered further.

Cursors

A cursor is essentially a mechanism to provide row-at-a-time access to the results of a pre-specified **select** statement. The result is retained for as long as the cursor is being used (during which time the cursor is said to be *open*). Using a cursor is a four-stage process.

1. The cursor is *declared,* which defines the **select** statement and the mode of operation of the cursor.
2. The cursor is *opened,* at which time the associated **select** statement is executed and the temporary result table is formed.
3. The application navigates through the result table, typically row by row, to *fetch* the data needed for any required operations.
4. When the application has completed its work, the cursor is *closed,* which results in deletion of the results of the original **select** statement.

It's fairly common practice for these steps to be controlled directly from an application program. This is one reason why cursors have a reputation for sluggish operation, which is discussed in more detail later.

Most vendors support the basic SQL-92 syntax for cursor declaration, which is shown here. Some details were added in later versions of the standard, but these did not affect the basic SQL cursor concept.

```
declare cursor-name [ sensitive | insensitive ] [ scroll | no scroll ] cursor
   for select_statement
      [ for { read only | update [ of column_name [ ,... ] ] } ]
```

Optionally, the cursor can be declared as **for read only** (which improves performance if updating is not required) or **for update**, possibly restricted to specific columns. A cursor that is declared as **insensitive** makes a temporary copy of the data to be used, and all data requests are answered from this temporary table. Modifications therefore cannot be made to base tables. If **insensitive** is omitted, or the sensitive option is chosen, changes made to the underlying tables may be reflected in subsequent fetches. The **scroll** option specifies that a range of navigation operations should be enabled for the cursor. If omitted, or if the **no scroll** option is chosen, the cursor is limited to just fetching the next row of the result. Once a cursor is defined, it can be opened. The syntax could hardly be simpler:

```
open cursor-name
```

There's no option for the open clause to change the **select** statement; it simply executes the one already defined in the associated cursor declaration. It also positions the cursor to point just before the first row of the result set. Given an open cursor, data can be fetched from the results produced by the **select** statement.

```
fetch [ [ next | prior | first | last | { absolute | relative } ] from ] cursor-name
  into target-specification [ ,... ]
```

In the navigation options, **next** fetches the next row from the current cursor position; **prior** fetches the previous row; **first** and **last**, respectively, fetch the first and last rows of the result set; **absolute** retrieves a row by its row number in the result set; and **relative** moves the cursor the specified number of rows backward or forward in the result set. If none of the navigation options is specified in the **fetch** clause, **next** is assumed. If the cursor declaration did not include **scroll**, the only option available is **next**. Under these circumstances, if we want to move to a previous row we have to close the cursor, reopen it, and then **fetch** repeatedly until we reach the required row.

The target-specification varies somewhat depending on the environment that the data is being fetched into. The main restriction is that the number of variables waiting to receive values must be the same as the number of columns produced by the **select** statement associated with the cursor. Closing a cursor is as simple as opening it.

```
close cursor-name
```

Once the cursor is closed, its result set becomes undefined. Cursors are automatically closed at the end of the transaction they are contained within, even if a **close** command is not explicitly used. The following code fragment illustrates a typical series of cursor activities. The host language here is "C".

```
/* variables dnum etc. will have been declared previously */
exec sql declare c1 cursor for
      select deptno, deptname, mgrno from tdept
      where admrdept = 'A00';

exec sql open c1;

while ( SQLCODE==0 ) {
      exec sql fetch c1 into :dnum, :dname, :mnum;
      /*  --
          -- do something interesting with the values
          --
      */
}

exec sql close c1;

/* carry on */
```

Cursors have a reputation for poor performance. One reason for this is that they are often used to retrieve data for use in an application program running on a different server to the database system. This requires some processing effort on the application server to construct the cursor and submit it to the databases server. More importantly, the network overhead (serializing, transmission, and deserializing, both ways) introduces a significant lag in operations. This can be partly addressed by moving the cursor to a stored procedure in the database server. A better solution is to rethink the data requirements in terms of set operations rather than row-at-a-time operations, using the inherent relational power of the database server.

13.7 Transactions and Concurrency

Operations affecting data in a DBMS are contained in transactions. A transaction may be composed of many individual actions, which, in some sense, can be considered together as an atomic unit of activity. To preserve the integrity of the database, we can make positive decisions about whether to accept the results of a unit of activity or not. The corresponding database operations are **commit**, which makes the results permanent, and **rollback**, which aborts the activity and removes the results of the transaction, leaving the database in the state it was in before the transaction began. In some situations, the **commit** or **rollback** is carried out implicitly, depending on whether a program terminates normally or abnormally.

If database systems handled transactions one at a time, this mechanism would be sufficient to maintain integrity. We could call this idealistic situation *serial* execution. However, a realistic database system could be expected to have many concurrent users, each carrying out their own individual transactions. It's therefore entirely likely that several users could be working in the same part of the database, for example, writing and reading data to and from the same table. Although we may expect integrity within a transaction, some potential problems arise when transactions overlap. The following situations are possible.

Dirty reads (also known as uncommitted reads). Transaction1 modifies a row in a table. Transaction2 reads that row before Transaction1 performs a commit. If Transaction1 subsequently performs a rollback, Transaction2 will have read a row that never really existed.

Nonrepeatable reads. Transaction1 reads a row from a table. Transaction2 modifies the row, or perhaps deletes it, and then performs a commit. If Transaction1 attempts to read the row again, it will find that it has changed or perhaps even disappeared completely!

Phantoms. Transaction1 reads some rows that satisfy a search condition. Transaction2 then creates some rows that would match the search condition used by Transaction1. If Transaction1 attempts to read the rows again, using the same search condition, it will obtain the original result plus some additional rows that have now mysteriously appeared.

Interactions between transactions can be controlled using *locks* on resources. A transaction that wants to use a resource, such as a row or a table, can acquire a lock that makes it the temporary owner of the resource. Other transactions are obliged to wait until the first transaction has relinquished the lock before they can compete for the same resource. A typical sequence is shown in Fig. 13.3.

There are two main kinds of lock: *shared* and *exclusive*. A shared lock is typically acquired by a transaction when it wants to read some data but does not want another transaction to change it. The resource can be shared with other transactions that have the same intentions. An exclusive lock is typically acquired when a transaction wants to modify the resource, and wants to prevent other transactions from either writing to or reading from the resource until the operation is complete. Commercial systems may have many subtly different variations on these two main kinds of locks.

1. Applications *A* and *B* connect to the database
2. Application *A* starts to update Table 1
   update Table1 **set** num = 0
   where alpha = 'R'
3. The application acquires an *exclusive lock* (**X**) on the relevant row
4. Application *B* tries to retrieve all rows
   select * **from** Table1
5. Application *B* can retrieve the first two rows but is forced to wait for the row with alpha = R
6. Application *A* commits, and the lock is removed
7. Application *B* can now retrieve the row for alpha = R and beyond

Table1:

	alpha	num
	P	1
	Q	2
X	R	3
	S	4
	T	5

Table1:

alpha	num
P	1
Q	2
R	0
S	4
T	5

FIG. 13.3 A typical locking sequence.

Obviously, locking is likely to have an impact on how quickly transactions can complete their business as they compete for resources. We have a potential trade-off here between better isolation between transactions but reduced performance, or reduced isolation but better performance, as shown in Fig. 13.4.

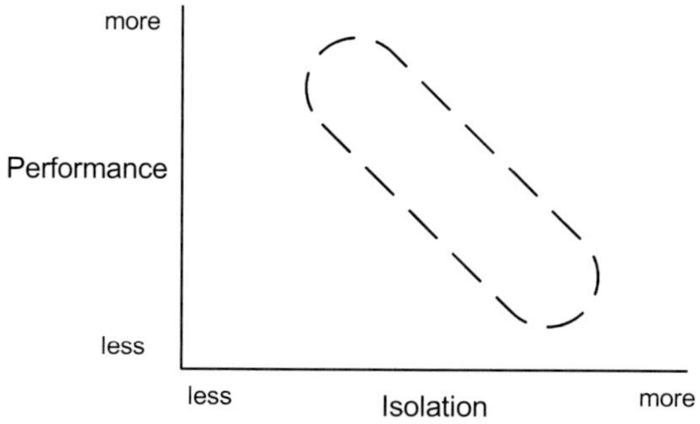

FIG. 13.4 Isolation versus performance.

The choice will be dependent on the nature of the application, and so the SQL standard defines several *isolation levels* that represent different points along this spectrum. The standard isolation levels are summarized in Table 13.4. At the highest level, SERIALIZABLE, we can treat transactions as though they are running in a one-at-a-time serial mode, albeit with a certain cost in terms of performance.

Table 13.4 Standard isolation levels.

Isolation level	*Dirty reads*	*Nonrepeatable reads*	*Phantoms*
READ UNCOMMITTED	Possible	Possible	Possible
READ COMMITTED	Not possible	Possible	Possible
REPEATABLE READ	Not possible	Not possible	Possible
SERIALIZABLE	Not possible	Not possible	Not possible

Most database vendors support at least some of these isolation levels, but not necessarily under the same names. For example, DB2 refers to these same four levels as UNCOMMITTED READ, CURSOR STABILITY, READ STABILITY, and REPEATABLE READ (the last of these being particularly confusing when compared to the standard). Other vendors may offer additional isolation levels.

An application can specify the isolation level required for a connection. If no level is specified, the DBMS will assume a default, such as READ COMMITTED, that represents a reasonable compromise between isolation and performance. It may also be appropriate to switch temporarily to a different isolation level to improve either isolation or performance. For example, an application reading data from a table that is read only could temporarily drop down to the READ UNCOMMITTED level to improve performance—in this case without compromising isolation because dirty reads cannot occur on a table that is not being updated concurrently.

A practical DBMS also has to deal with a number of side effects of locking. One is the *deadlock* situation. Transaction1 requires resources *A* and *B*, and successfully acquires a lock on *A*. Meanwhile, Transaction2, which requires the same two resources, has acquired a lock on *B*. The result is that Transaction1 is unable to proceed because it is waiting for *B*, and Transaction2 is unable to proceed because it is waiting for *A*. Most DBMSs have a built-in background process to detect deadlock situations. Typically, deadlock situations are resolved automatically by rolling back one of the transactions. Another potential problem is a poorly designed application that acquires a lock and retains it for an unduly long period. Again, an automatic process can be included to detect such conditions and roll back the offending transactions.

13.8 Security and Metadata

This section looks briefly at two topics that any database administrator needs to be familiar with: security and metadata in SQL. Let's begin with security. A database is *secure* if and only if operations on it can be performed only by users authorized to do so. Each system user is assigned a user identifier (AuthID). The user who "owns" or creates a database has all access privileges on it and can grant and revoke access privileges on it to other users.

SQL provides a *grant statement* for granting various kinds of privileges to users. The SQL-92 syntax is:

> **grant all privileges** |
> **select** | **insert** [(*col*)] | **update** [(*col*)] | **delete** | **usage** | **references** [(*col*)] [,...]
> **on** *object*
>   **to** *user-list* [**with grant option**]

where *object* is one of [**table**] tablename, **domain** domainname, etc. The user list may include **public**, meaning all users. SQL:1999 extended the statement to cater for "roles" (user groups). If **with grant option** is included, the user list has the power to grant the same privilege to others. For example, the following statement grants all users read access to the Stock table: **grant select on** Stock **to public**. The following grants the user with authorization id "Mgr" read access to the Stock table as well as the right to update values in the price column: **grant select, update**(price) **on** Stock **to** Mgr. Privileges may be removed with the *revoke statement*:

> **revoke** [**grant option for**] *privilege-list*
> **on** *object*
> **from** *user-list* [**restrict** | **cascade**]

For example, to revoke the update right granted earlier, use **revoke update** (price) **on** Stock **from** Mgr. The table name in **grant** or **revoke** statements may be the name of a base table or view. Granting access to views but not the base tables is one convenient way of enforcing security. The **restrict** and **cascade** options are not yet widely supported.

In addition to database privileges, commercial systems also provide controls over operations related to database servers and instances of a DBMS. One server may contain multiple instances and one instance may support many individual databases. Typically, roles are arranged hierarchically, with lower levels of the hierarchy having fewer privileges than higher levels. For example, a lower-level administrator may be able to carry out maintenance and utility operations but not access the data stored in the databases.

Now, let's move on to *metadata* (data about data). SQL systems automatically maintain a set of *system tables* holding information about the database schema itself (e.g., base tables, views, domains, constraints, and privileges). In commercial DBMSs, such tables are collectively referred to under various names such as "system catalog". SQL-92 uses the term "catalog" in a different sense (for a collection of schemas within an environment). System tables of commercial systems also include physical information (e.g., about indexes) that is used by the optimizer. Users with access to the system tables may query them in SQL (just like the application tables).

From SQL-92 onward, the SQL standard defines metadata in terms of tables and assertions in a special schema named DEFINITION_SCHEMA. This provides a reference model for metadata, but there is no intention that this should be implemented in this way in any commercial realization—it's sufficient for an implementation to simulate its existence.

The intended access to metadata is though *views* defined in another schema, named INFORMATION_SCHEMA. Views in the INFORMATION_SCHEMA are defined in terms of the model in the DEFINITION_SCHEMA. Most of the views are defined with a **select** privilege granted to **public**, but no other privileges are granted, so the views cannot be updated. These schemas have been extended as new features of SQL, such as triggers, have been added in subsequent versions of the standard.

To conform to the standard, a DBMS should support these views, and many popular DBMSs are at least in the process of doing so. Commercial systems may still use their own specific names (e.g., SysTables, SysCatalog, etc.) and structures for the metadata. These generally provide equivalent features to the standard schemas, but with some variation in the detailed syntax.

As a simple example, the following query uses the standard Columns view to request the number of columns in the Customer table:

```
select count ( * ) from Columns
where table_name = 'CUSTOMER'
```

To give an idea of the variation to be found among commercial systems, the equivalent query in SQL Server might look like:

```
select count ( * ) from sys.Columns
where object_id = object_id ( 'CUSTOMER', 'U' )
```

whereas in DB2 we might see:

```
select count ( * ) from sysibm.Columns
where table_name = 'CUSTOMER'
```

13.9 Summary

Base table schemes may be defined using a *create-table statement*, whose main syntax is:

```
create table tablename (
  colname    data-type        [ not null ]
                              [ default literal | null | user ]
                              [ primary key | unique ]
                              [ references tablename ]
                              [ check (col-condition-on-same-row) ]
   [,...]
   [, primary key ( col-list ) ]
   [, unique ( col-list ) ]
   [, foreign key ( col-list ) references tablename [ ( unique-col-list ) ] [,...] ]
   [, check (table-condition-on-same-row) [,...] ] )
```

In addition to creating tables by definition, as just shown, tables can also be created **like** another table or **as** the result of a select query. Other SQL statements exist for tasks such as altering and dropping tables, and equivalent alter or drop statements are available for other types of database objects.

Views may be created using a *create-view statement*, with the syntax:

```
create view viewname [ ( col-list ) ] as
    select-query
        [ with check option ]
```

View definitions are used to translate queries over views into equivalent queries on base tables. Some views are nonupdatable (e.g., views involving joins).

Triggers are used to respond to specific events in the context of a given table or view. When the specified event is detected, the body of code in the trigger is executed. The data leading to the event can be examined by the trigger to determine the appropriate course of action. The basic syntax for creating a trigger is:

```
create trigger triggername
    { before | after } { insert | delete | update [ of col-list ] } on tablename
    [ referencing  { old | new } [row] [ as ] correlation-name |
                   { old | new } table [ as ] table-alias     [, ...] ]
    [ for each { row | statement } ]
        [ when ( condition ) ]
        triggered-SQL-statement
```

Routines are bodies of code that are explicitly *called*, in contrast to triggers, which are inherently event driven. UDFs return either a scalar or a table result. The basic syntax for creating a function is:

```
create function functionname ( [ parameter-list ] )
    returns { data-type | table col-name data-type [,...] }
    function-body
```

Stored procedures are able to carry out a greater range of activities than functions, such as updating tables, and are the standard way of encapsulating SQL code. Stored procedures are often used to "insulate" the database from user actions that could prove damaging. The basic syntax for creating a stored procedure is:

```
create procedure procedurename
    ( [ { in | out | inout } parmname parmtype [,...] ] )
        procedure-body
```

A database schema may contain other types of objects. *Sequences* provide for the automatic generation of unique identifiers. *Indexes* can be used to improve query performance. *Cursors* can be used to provide row-at-a-time processing.

Practical database systems include measures to deal with concurrency, typically based on the selective locking of resources to avoid interference between transactions. The preferred position between more isolation and more throughput can be specified at the statement or the connection level.

Database security revolves around two main concepts. Users are authenticated against specified credentials to confirm their identity. Privileges—the rights to carry out certain types of action—can be given to, or withheld from, specific individuals or groups.

Metadata—information about the database system itself—is stored in various tables for the system's internal use. This information can also be queried, although standard practice is to discourage direct access to these tables and instead use system views.

This chapter completes our basic coverage of the major feature of modern relational database systems. Some other aspects of SQL are considered elsewhere in the book in the context of the topic under consideration. More advanced SQL aspects are included in the online Appendix C, and the chapter notes provide further detail. While knowledge of the relevant SQL standards provides an excellent basis, the easiest way to learn the language is by using it, which entails discovering how your chosen SQL dialect differs from the standard. Fortunately, most commercial DBMSs provide extensive online resources to assist you in this regard.

Chapter Notes

The SQL standard has been adopted by the International Standards Organization (ISO) as ISO/IEC 9075. At the time this book was published, the current version (SQL:2023) is ISO/IEC 9075-2:2023); see https://www.iso.org/standard/76584.html, and press the Preview button for some further details. Because of its size (several thousand pages in total), the standard is split into a number of parts. Part 2 (Foundations) is the most relevant to this chapter.

The SQL syntax used by most commercial database systems tends to differ somewhat from the SQL standard syntax. For recent coverage of SQL using different DBMSs such as Oracle, MySQL, PostgreSQL, and SQL Server, see Taylor (2018) and Zhao (2021). SQL Server is covered in many books, such as Syverson and Murach (2023). *SQL Server Pro Magazine* is a good source of technical articles on SQL Server (www.sqlmag.com). Most commercial systems provide online help that can be searched in various ways.

14

Schema Transformations

Chapter Outline

14.1 Schema Equivalence and Optimization

Previous chapters discussed how to model the structure of an application in terms of a conceptual schema, and then map it to a logical schema for implementation in a relational database system. Although much of the design and mapping can be automated, humans are required to perform conceptual schema design procedure (CSDP) step 1 because verbalizing the relevant facts about the real world involves human understanding. Given the informal nature of this initial step in modeling the universe of discourse (UoD), it is not surprising that people often come up with different ways of describing the same reality.

Hence, the same business domain may be modeled by more than one conceptual schema. Moreover, the same conceptual schema may be mapped in different ways to a logical schema, either by choosing a different type of logical model (e.g., object relational) or by overriding the default mapping choices (e.g., choosing separate subtype tables or denormalizing). Finally, the same logical schema may be implemented in more than one way in the physical schema (e.g., different index choices).

To help you choose from such a variety of different designs, this chapter examines various ways in which schemas may be transformed to provide alternatives and then presents guidelines for optimizing your choice. On the way, we'll cover some theory on schema transformations and normalization to ensure a proper understanding of the underlying concepts. We'll start at the conceptual level and then move down to the logical and physical levels. Examples will illustrate how the ideas can also be used to reengineer existing applications that proved unsatisfactory.

Information Modeling and Relational Databases. https://doi.org/10.1016/B978-0-443-23790-4.00018-9

The rest of this section provides a simple introduction to schema transformation and optimization, using an example. Earlier chapters introduced some modeling choices. For example, we may model a ternary fact type in either flattened or nested form. Basically, two conceptual schemas are **equivalent** if and only if whatever UoD state or transition can be modeled in one can also be modeled in the other.

As a simple example, consider the medical report shown in Table 14.1. Here, a check mark in the appropriate column indicates if a patient smokes or drinks. Because these "vices" can impair health, doctors are often interested in this information. Try to schematize Table 14.1 for yourself before looking at the solutions provided.

Fig. 14.1(a) shows one conceptual schema for this UoD, together with the sample population. Here, two optional unaries are used for the smoker-drinker facts. In Table 14.1, the absence of a check mark might mean that the patient doesn't smoke/drink (closed-world approach) or simply that we don't know (open-world approach). You would need to check with the domain expert as to what is intended here. For our discussion, we'll assume that closed-world semantics are intended.

Table 14.1 A report on vices of medical patients.

PatientNr	*Patient name*	*Smokes?*	*Drinks?*
1001	Adams, A		✓
1002	Bloggs, F	✓	✓
1003	Collins, T		

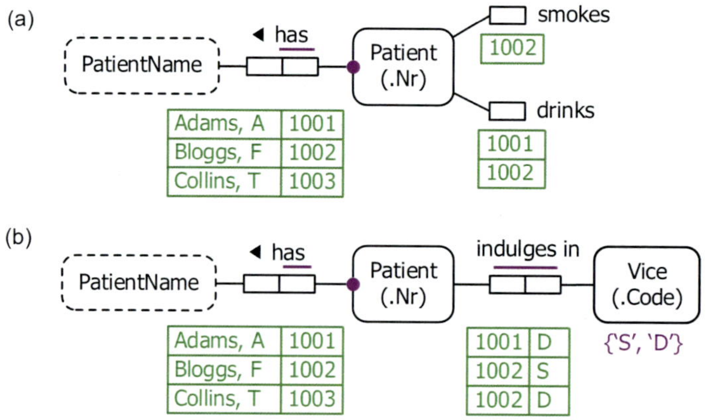

FIG. 14.1 Two ways of modeling (Table 14.1).

Another way to model this domain is to generalize the smoking and drinking predicates into a single binary, introducing the object type Vice {S = Smoking, D = Drinking} to maintain the distinction, as in Fig. 14.1(b). Intuitively, most people would consider the two schemas in Fig. 14.1 to be equivalent. Formally, this intuition can be backed up by introducing Vice as a "virtual object type" to the schema of Fig. 14.1(a), and by specifying how the predicates of each are translated into the predicates of the other. For example, facts expressed in the first model may be expressed in the second model using the following translations:

Patient smokes **iff** Patient indulges in Vice 'S'
Patient drinks **iff** Patient indulges in Vice 'D'

and facts in the second model may be expressed in the first model using the following translation. Like the schemas, these translations can be mapped into formulae of predicate logic. Formal logic may then be used to prove schema equivalence.

Patient indulges in Vice **iff** Patient smokes **and** Vice has ViceCode 'S'
or
Patient drinks **and** Vice has ViceCode 'D'

The equivalence just considered is an example of *contextual equivalence* because it adds definitional context to the original schemas. It also requires formal recognition of "virtual object types" (e.g., Vice exists implicitly in the first schema). This notion is called "*object relativity*" because it effectively allows that "objects lie in the eye of the beholder". This permits classical logic proof techniques to be applied to such cases.

The next few sections discuss some theorems that can help us decide whether two schemas are equivalent. These theorems can also help us to transform one schema into an alternative schema that is either equivalent, or at least acceptably close to being equivalent (sometimes we may wish to strengthen or weaken our schema a little by adding or deleting information). The act of reshaping a schema like this is said to be a **conceptual schema transformation**.

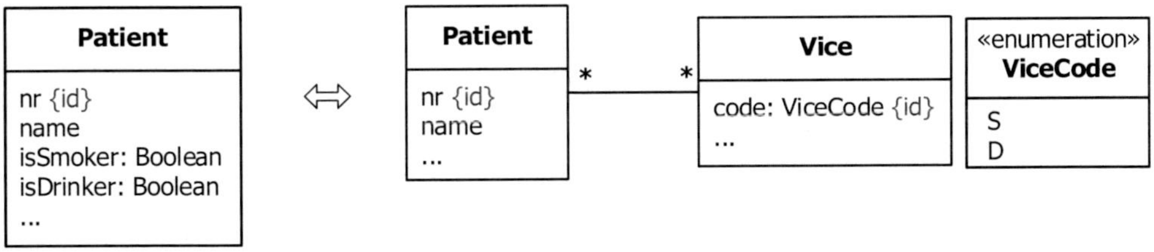

FIG. 14.2 The schema equivalence example in unified modeling language (UML).

Although the conceptual transformations are specified in object-role modeling (ORM), they can be translated into other modeling notations. For example, the previous equivalence (minus definitional context) may be set out in unified modeling language (UML), as shown in Fig. 14.2, using "⇔" for "is equivalent to". Using ORM enables many more complex transformations to be captured graphically, without needing to supplement the diagram with textual constraints. But if you want to work in a non-ORM notation, you should be able to translate the transformation theorems into your preferred notation without much trouble.

Knowledge of schema transformations helps us to see what different design choices are possible. Moreover, if two independently developed schemas are to be either fully or partly integrated, we often need to resolve the differences in the ways that each schema models common UoD features. To do this, we need to know whether one representation can be transformed into the other, and if so how.

Another use of conceptual schema transformations is to reshape the original conceptual schema into one that maps directly to a more efficient implementation. This process is known as **conceptual schema optimization**. For example, the conceptual schema in Fig. 14.1(a) maps to a single table. Taking a closed-world interpretation, we may use the table scheme shown in Fig. 14.3(a). For discussion purposes, the population for the sample data is included. However, because the indulges fact type is *m:n*, the schema in Fig. 14.1(b) maps to two tables, as shown in Fig. 14.3(b). Again, the sample population is included.

(a)

{Y, N} {Y, N}

Patient (patientNr, patientName, isSmoker, isDrinker)

1001	Adams, A	N	Y
1002	Bloggs, F	Y	Y
1003	Collins, T	N	N

(b)

{S, D}

Patient (patientNr, patientName) *Indulges* (patientNr, vice)

1001	Adams, A
1002	Bloggs, F
1003	Collins, T

1001	D
1002	S
1002	D

FIG. 14.3 Relational models based on the conceptual models in Fig. 14.1.

For many applications, the single-table solution is more efficient. It enables us to list all the information about the patients without requiring a table join. It also avoids any inter-table accesses required to enforce the subset constraint (e.g., when adding a vice fact or deleting a patient). So, the schema of Fig. 14.1(b) might be optimized by transforming it to that of Fig. 14.1(a) before passing it to the normal Rmap procedure.

However, if the set of vices we wish to record often changes, the two-table solution may be preferable because it caters for such changes without requiring the schema itself to change. Guidelines for performing such optimizations are given later in the chapter. If the optimized conceptual schema still fails to give an efficient map, some denormalization may then be required, as discussed in Section 14.7.

In reengineering, optimization is applied to an existing database. So, after the schema itself has been transformed, we have the additional task of *data migration* to populate the new schema with the information stored in the original database. This task can be lengthy and tedious without automated assistance.

Here is one basic way of transforming the populated two-table model in Fig. 14.3(b) into the single-table model in Fig. 14.3(a), assuming that a structured query language (SQL) database is used. First, add an optional isSmoker column to the Patient table, using the following statement:

```
alter table Patient
  add column isSmoker char check (isSmoker in ('Y','N'))
```

At this stage, the isSmoker column is populated with nulls. The check constraint is not violated because it evaluates to unknown (instead of false). Now, add the appropriate "Y" or "N" values to indicate who smokes:

```
update Patient
  set isSmoker = case
                   when patientNr in
                          (select patientNr from Indulges where vice = 'S')
                   then 'Y'
                   else 'N'
                 end
```

Now, make the isSmoker column mandatory:

```
alter table Patient
  add constraint isSmokerNonNull check (isSmoker is not null)
```

Similarly, add the isDrinker column to the Patient table, update its population, and finally drop the Indulges table. In practice, the most efficient way to migrate data may depend on the database management system (DBMS) used. In contrast to the simple example just discussed, it is more typical to leave the original database unaltered, create a new schema, migrate data to it, and then drop the original database (see Section 14.9).

That completes our high-level introduction to the basic concepts of schema equivalence and optimization. The following sections develop these concepts further while also providing a concise account of normalization theory.

14.2 Predicate Specialization and Generalization

The previous section noted how the same UoD structure may be described by different, equivalent schemas and discussed transformations to reshape one schema into another. This section considers a class of schema transformations known as **predicate specialization** as well as its inverse, **predicate generalization**.

If two or more predicates may be thought of as special cases of a more general predicate, then we may replace them by the more general predicate, as long as the original distinction can be preserved in some way. For example, if we transform the schema of Fig. 14.1(a) into that of Fig. 14.1(b), we *generalize* smoking and drinking into indulging in a vice, where vice has two specific cases. If we transform in the opposite direction, we *specialize* indulging in a vice into two predicates, one for each case.

Predicate specialization and generalization are similar notions to object type specialization and generalization, except that it is rare to specify any subtype connections between predicates (the predicates must be objectified for this).

A predicate may be specialized if a *value constraint or a frequency constraint* indicates that it has a finite number of cases. Examples with value constraints are more common, so we examine these first. The drinker-smoker example provides one illustration, where Vice has the value constraint {"S", "D"}. As another example, recall the Olympic Games schema reproduced in Fig. 14.4(a).

There are exactly three kinds of medals, so the ternary may be specialized into three binaries, one for each medal kind, as in Fig. 14.4(b). You may visualize the transformation from schema (a) into schema (b) as *absorbing* the object type MedalKind into the ternary predicate to divide (or specialize) it into the three binaries. Hence, this transformation is also known as *object type absorption*.

The reverse transformation from (b) to (a) generalizes the three binaries into the ternary by extracting the object type MedalKind—this may be called *object type extraction*. In the vices example, a binary is specialized into unaries. With the games example, a ternary is specialized into binaries. In general, when an n-valued object type is absorbed into a predicate, the n specialized predicates that result each have one less role than the original (because the object type has been absorbed).

Although the equivalence is set out diagrammatically to aid understanding, there is a formal mapping of the diagrams to predicate logic where the actual equivalence proofs were made (Halpin, 1989b). In this sense, the results may be called theorems. This general result is set out in Fig. 14.5.

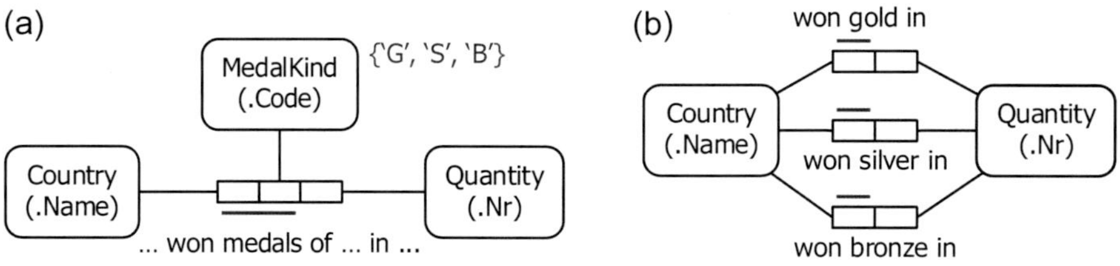

FIG. 14.4 The ternary is specialized into three binaries by absorbing MedalKind.

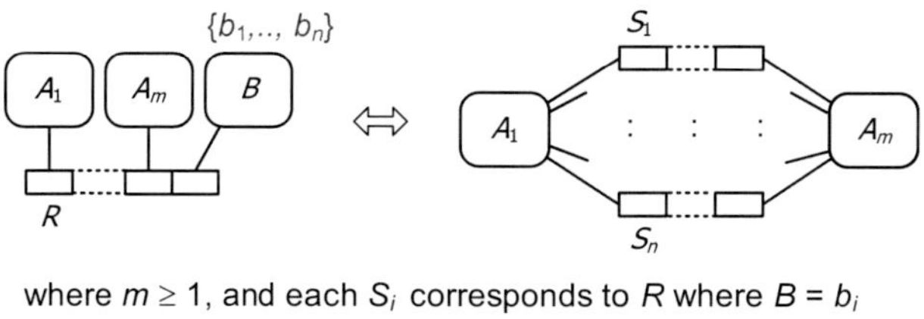

FIG. 14.5 PSG1: R may be specialized into $S_1 \ldots S_n$ by absorbing B.

The schema equivalence in Fig. 14.5 is called *predicate specialization/generalization theorem 1* (PSG1). The predicates R and S, respectively, have $m+1$ roles and m roles, where $m \geq 1$. The object types $A_1 \ldots A_m$ and B are not necessarily distinct. The value type for B has n values $b_1, \ldots, b_n$. If $m = 1$, we have conversion between a binary and n unaries; if $m = 2$, the conversion is between a ternary and n binaries; and so on. Transforming from left to right specializes the predicate R into n predicates $S_1 \ldots S_n$ by absorbing the object type B. The reverse transformation from right to left generalizes the shorter predicates into the longer one by extracting B. As an exercise, draw the diagrams for the cases $m = 1$, 2, and 3.

The theorem PSG1 holds regardless of whatever additional constraints are added. However, *any constraint added to one of the schemas must be translated into an equivalent, additional constraint on the other schema*. For example, the uniqueness constraint in Fig. 14.4(a) translates into the three shorter uniqueness constraints in Fig. 14.4(b). This is an instance of the following corollary to PSG1 (using "UC" for "uniqueness constraint" and reading "spans" as "exactly spans").

> If a UC in R spans a combination of B's role and other roles,
> then a UC spans the specialization of these other roles in $S_1 \ldots S_n$; and conversely.

Fig. 14.6 illustrates the most common case, where R is a ternary with a UC spanning B's role and one other. However, the result applies for longer predicates too. The games example of Fig. 14.4 is an instance of the equivalence in Fig. 14.6, where B = MedalKind, $n = 3$, and B's role is included in a compound uniqueness constraint.

What happens, however, if B's role in the general predicate R is not included in a uniqueness constraint? Because R is elementary, its other roles must be spanned by a uniqueness constraint; this constraint is transformed into a mutual *exclusion* constraint over the specialized predicates.

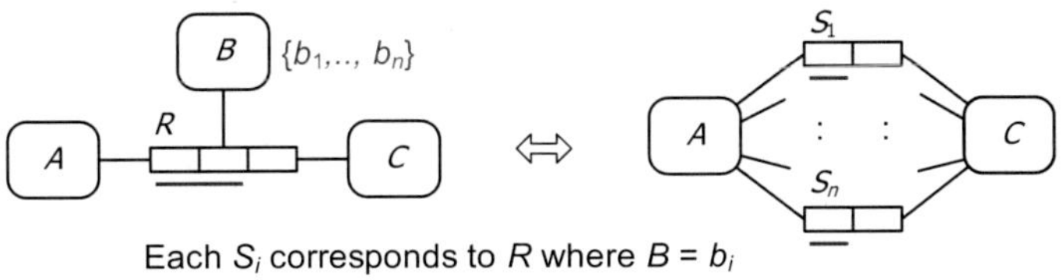

FIG. 14.6 The UC on the left is equivalent to the UCs on the right.

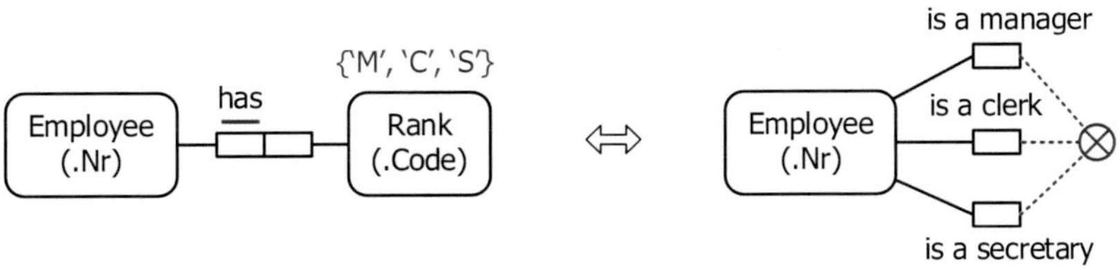

FIG. 14.7 The UC on the left is equivalent to the exclusion constraint on the right.

Exclusive unaries provide the simplest case of this. For example, suppose that workers in a company hold at most one position (manager, clerk, or secretary). This may be rephrased using exclusive unaries, as shown in Fig. 14.7, assuming appropriate translations between the predicates. For implementation, the binary version is usually preferred (e.g., its relational schema simplifies both updates to a worker's position, and schema updates to the list of allowable positions). The larger the number of unaries, the worse the unary solution becomes (why?).

Another example, this time with exclusive binaries, is shown in Fig. 14.8. For simplicity, reference schemes are omitted. Here, the task codes 'A' and 'R' denote authoring and reviewing, respectively. Note the two alternative ways of saying that a person cannot act both as an author and as a reviewer of the *same* book.

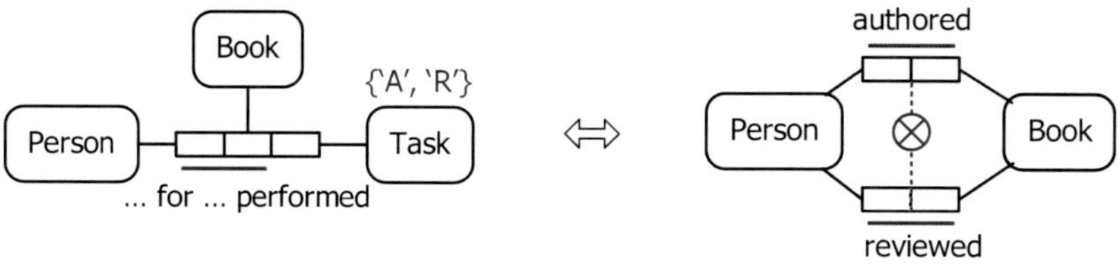

FIG. 14.8 The UC on the left is equivalent to the exclusion constraint on the right.

These two examples illustrate unary ($m = 1$) and binary ($m = 2$) cases of the following second corollary to theorem PSG1. This result applies to longer predicates as well (see Fig. 14.9). Here, the exclusion constraint means that no row of values may appear in more than one of the S_i fact tables.

If a UC spans all roles of R except for B's role,
then $S_1 \ldots S_n$ are mutually exclusive; and conversely.

In rare cases of schema transformations, a graphical constraint in one schema may have no corresponding graphical constraint in the other. In this case, the constraint should be expressed as a *textual constraint* in the other version.

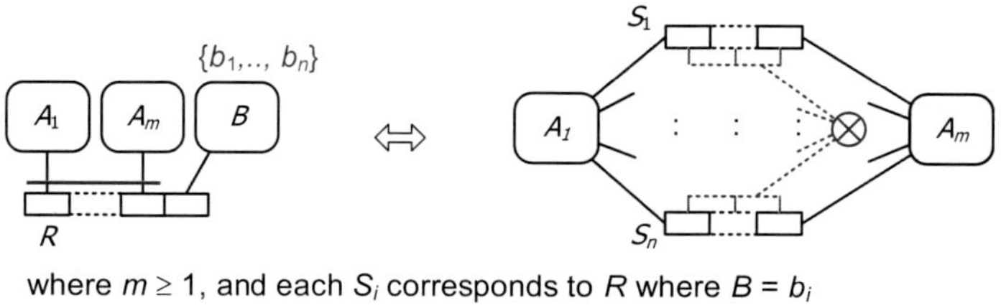

FIG. 14.9 The UC on the left is equivalent to the exclusion constraint on the right.

For example, consider the two schemas in Fig. 14.10. The codes "A" and "S" indicate "assistant" and "supervisor", respectively. Reference schemes are omitted for simplicity. A person might supervise one project and assist on another.

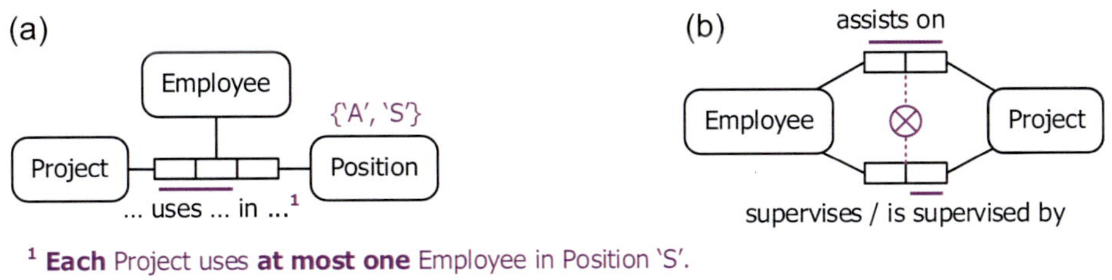

FIG. 14.10 A textual constraint in schema (a) appears as a graphical constraint in (b).

Assuming appropriate translations between the predicates, the uniqueness constraint in schema (a) translates to the exclusion constraint in schema (b), and the top uniqueness constraint in (b) is implied.

The lower uniqueness constraint in (b) declares that each project has at most one supervisor. This cannot be expressed as a graphical constraint on the ternary because it is a *restricted uniqueness constraint* (as discussed in Section 7.4). To ensure equivalence, this is added to the left-hand schema as the textual constraint **Each** Project uses **at most one** Employee in Position 'S'.

If Project plays other functional roles, the binary approach of the right-hand schema would normally be preferred because its relational version simplifies the enforcement of this supervision constraint. If Project has no other functional roles, the ternary approach might be preferred because it maps to a single table.

Because schema transformation theorems are typically applied to subschemas within some global schema, care must be taken to preserve any *mandatory role constraints* (implicit or explicit).

For example, in Fig. 14.4, if Country plays no other role, then its role in schema (a) and the disjunction of its roles in (b) are implicitly mandatory. If Country plays another role in the global schema, and at least one medal result must be recorded for it, then these simple and disjunctive mandatory role (inclusive-or) constraints must be explicitly shown (see Fig. 14.11).

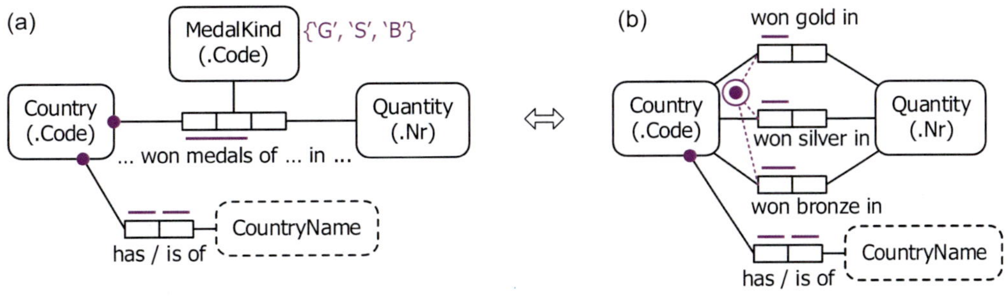

FIG. 14.11 A mandatory constraint in (a) maps to an inclusive-or constraint in (b).

This illustrates our third corollary to theorem PSG1:

If *Ai*'s role (or role disjunction) in *R* is mandatory,
then the disjunction of its specialized roles is mandatory, and conversely ($1 \leq i \leq m$).

The inclusion of "(or role disjunction)" covers the rare case when A_i plays more than one role in R ($A_1 \ldots A_m$ are not necessarily distinct). Note also how our convention for implicit mandatory role constraints enables theorems such as PSG1 to be specified without any assumption about other roles played in the global schema.

Now consider Fig. 14.12(a). Here, Country's medal-winning role is optional (this implies Country plays another role in a unary or longer fact type, but it has a *frequency constraint* of 3). So, if any medal results are recorded for a country, all three medal results (gold, silver, and bronze) are required. To express this in Fig. 14.12(b), we add an *equality constraint* between the medal-winning roles played by Country. This is an instance of the following fourth corollary to theorem PSG:

If *R* is a ternary with a UC spanning just *B*'s role and one other role, then adding a frequency constraint of *n* to this other role is equivalent to adding an equality constraint over the specialized versions of that role.

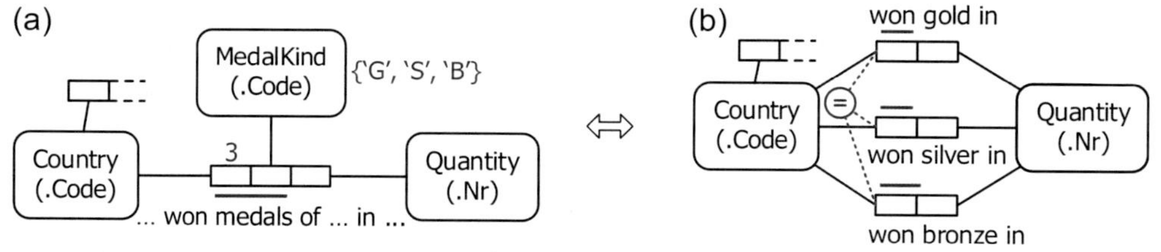

FIG. 14.12 A frequency constraint in (a) maps to an equality constraint in (b).

Now consider Fig. 14.13(a), where again Country plays another role in the global schema, but now its medal-winning role is mandatory. The mandatory, frequency, uniqueness, and value constraints together ensure that each country must have its medal tally recorded for each medal kind, so Fig. 14.13(b) has three mandatory roles. This example illustrates the following fifth corollary to theorem PSG1:

If *R* is a ternary with a UC spanning just *B*'s role and one other role, then adding a mandatory role constraint and frequency constraint of *n* (the number of possible values for *B*) to this other role is equivalent to making each specialized version of that role mandatory.

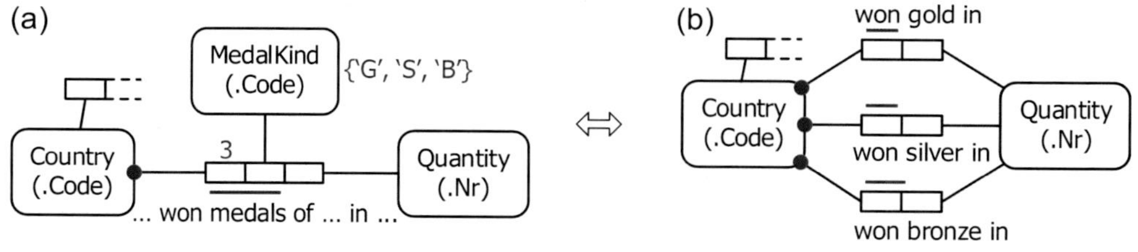

FIG. 14.13 The impact of adding mandatory role and frequency constraints.

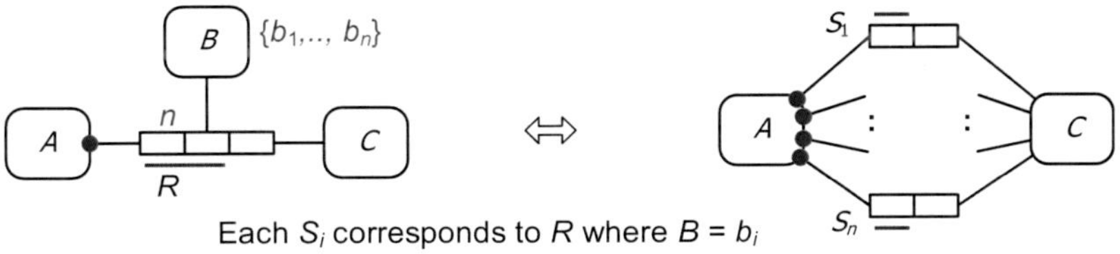

FIG. 14.14 Fifth corollary to theorem PSG1.

This corollary is illustrated in Fig. 14.14. It is implied by the previous two corollaries because an equality constraint across a set of disjunctively mandatory roles implies that each of these roles is mandatory.

Other Kinds of Predicate Specialization/Generalization

As an introduction to a second equivalence theorem, consider Fig. 14.15. The two schemas provide alternative models for a fragment of a car rally application. Each car in the rally has two drivers (a main driver and a backup driver), and each person drives exactly one car. The schema in Fig. 14.15(a) is transformed into the schema in Fig. 14.15(b) by absorbing the object type DriverStatus into the drives predicate, specializing this into the main driver and backup driver predicates. The reverse transformation generalizes the specific driver predicates into the general one by extracting the object type DriverStatus. Because this object type appears in a different fact type, this equivalence does not fit the pattern of PSG1.

Note how the constraints are transformed. The external uniqueness constraint in (a) says that each car has at most one main driver and at most one backup driver. This is captured in Fig. 14.15(b) by the uniqueness constraints on the roles of Car. The uniqueness constraint on the drives predicate in Fig. 14.15(a) corresponds in Fig. 14.15(b) to the uniqueness constraints on the roles of Driver. The uniqueness constraint on the driver status predicate in Fig. 14.15(a) is captured by the exclusion constraint in Fig. 14.15(b).

The mandatory and frequency constraints on Car's role in Fig. 14.15(a) require the two mandatory role constraints on Car in Fig. 14.15(b). Finally, the mandatory role constraints on Driver in Fig. 14.15(a) are catered for in Fig. 14.15(b) by the inclusive-or constraint (shown explicitly here as part of the exclusive-or constraint).

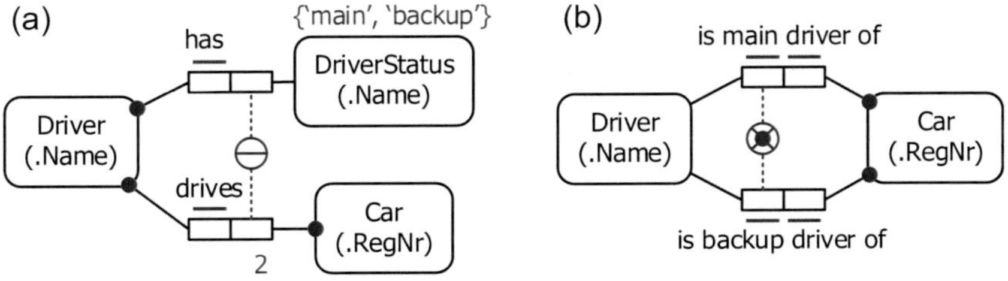

FIG. 14.15 The drives predicate is specialized by absorbing DriverStatus.

This example illustrates our *second predicate specialization/generalization theorem* (PSG2) as well as four of its corollaries (see Fig. 14.16). The terms "LHS" and "RHS" abbreviate "left-hand schema" and "right-hand schema". In our example, *A*, *B*, and *C* correspond to Driver, DriverStatus, and Car, respectively; the equality constraint is implied by the two mandatory role constraints on Driver.

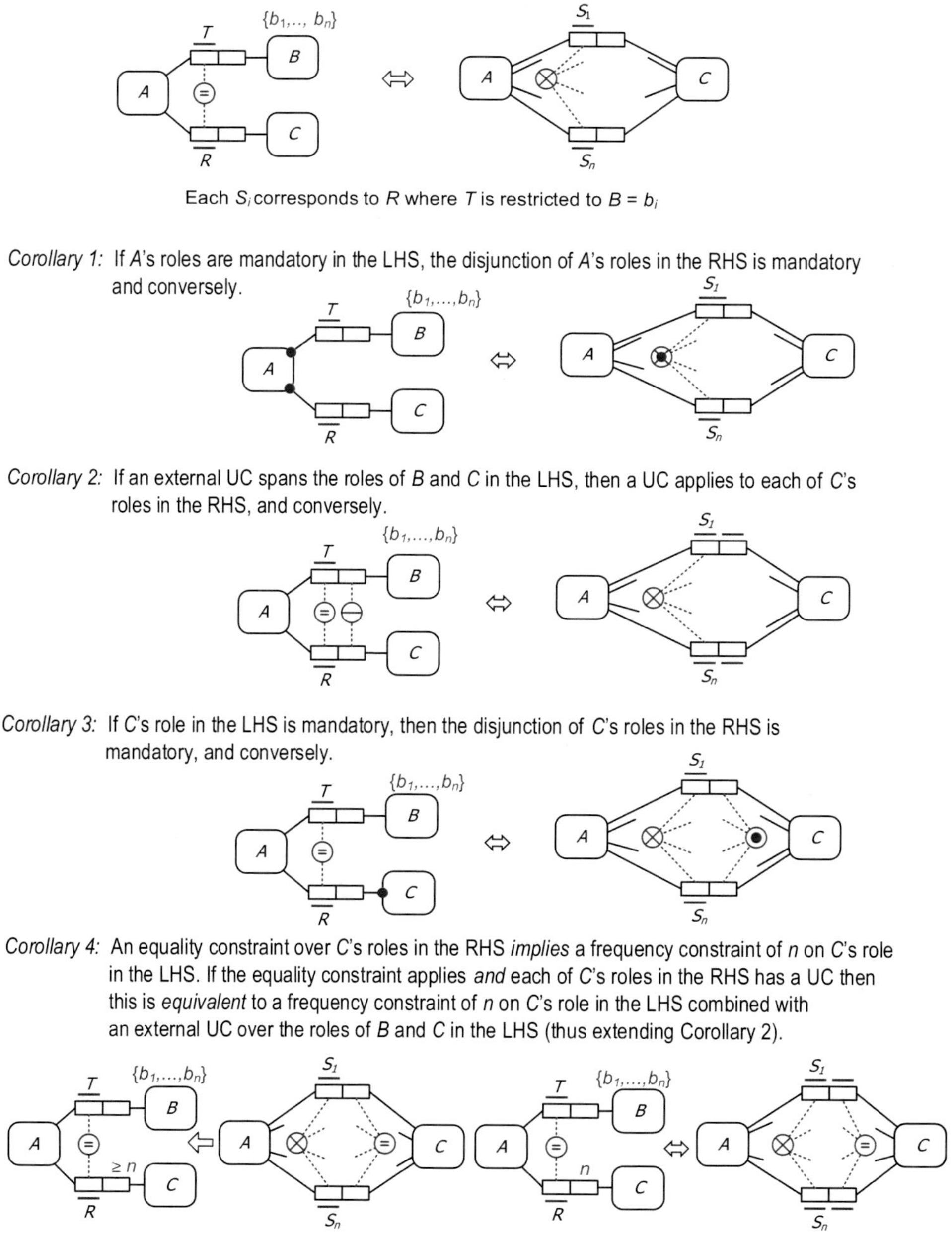

FIG. 14.16 PSG2: *R* may be specialized into $S_1 \ldots S_n$ by absorbing *B*.

Theorem PSG2 can be derived from earlier results. For example, adding an exclusion constraint across *A*'s roles in the RHS of Fig. 14.6 makes *A*'s role unique in the LHS, causing this compound fact type to split into two binaries, as in the LHS of Fig. 14.16. Section 14.4 considers two more general versions of this theorem.

Sometimes, we may wish to transform a schema into another that is not quite equivalent. For example, suppose that in our car rally application, we limit each car to at most two drivers, but do not classify the drivers in any meaningful way (e.g., as main or backup drivers). Let us also remove the constraint that drivers may drive only one car. This situation is schematized in Fig. 14.17.

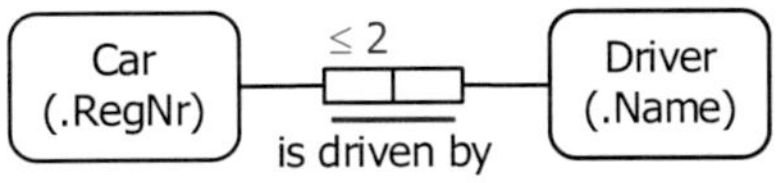

FIG. 14.17 Can the predicate be specialized?

Although no DriverStatus object type is present, the frequency constraint in Fig. 14.17 says that each car has at most two drivers. This enables us to introduce an artificial distinction to specialize the predicate into two cases, as in Fig. 14.18.

Because this distinction is not present in the original schema, the alternatives shown in Fig. 14.18 are *not equivalent* to the original. They are, in fact, *stronger*—each implies the original schema, but is not implied by it.

Notice the use of *hyphens* in the predicates of Fig. 14.18 to bind adjectival phrases to their object terms. This improves verbalization of the fact types and constraints. The four fact types are Car has Driver-A, Car has Driver-B, Car has first-Driver, and Car has second-Driver. When the NORMA tool generates constraint verbalization, it binds the adjective to the object noun. For example, the upper uniqueness constraint in Fig. 14.18(b) verbalizes as "**Each** Car has **at most one** first Driver". Without the hyphen, the verbalization defaults to "**Each** Car has first **at most one** Driver".

The schema in Fig. 14.18(b) is actually stronger than the schema in Fig. 14.18(a). If the Car role in Fig. 14.17 is mandatory, then the Car roles in Fig. 14.18(a) are disjunctively mandatory, and the top Car role in Fig. 14.18(b) is mandatory (which then implies the subset constraint).

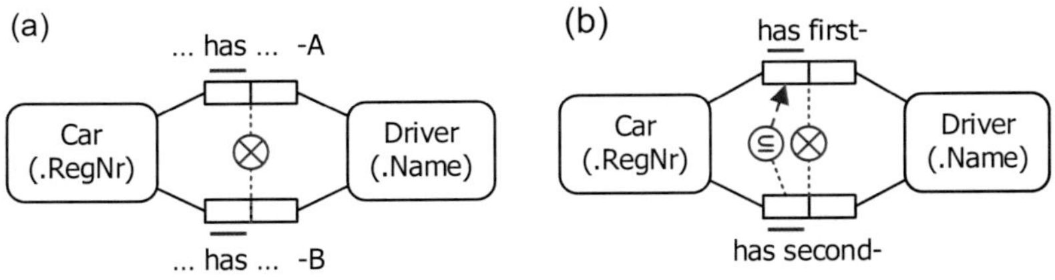

FIG. 14.18 Two ways of strengthening the schema in Fig. 14.17.

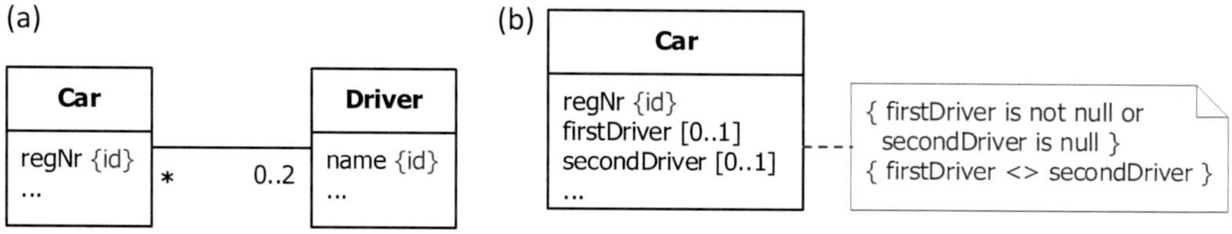

FIG. 14.19 UML versions of the schemas in Figs. 14.17 and 14.18(b).

In a business domain where other facts are stored about cars but not about drivers, one of these alternatives may well be chosen to avoid a separate table being generated for car-driver facts when the schema is mapped (the specialized predicates are functional rather than *m:n*). In practice, the schema in Fig. 14.18(b) would normally be chosen.

Fig. 14.19(b) shows the UML version of this choice, using a note for the subset and exclusion constraints. Fig. 14.19(a) shows the UML version of Fig. 14.17.

Transforming from the original schema in Fig. 14.17 to one of those in Fig. 14.18 *strengthens* the schema by adding information. Transforming in the opposite direction *weakens* the schema by losing information. Any such transformations that add or lose information should be the result of conscious decisions that are acceptable to the client (for which the business domain is being modeled).

This example illustrates our *third specialization/generalization theorem* (PSG3), as shown in Fig. 14.20. Note the use of "⇨" for "implies" because this result is a schema implication rather than an equivalence. In practice, the transformation is usually

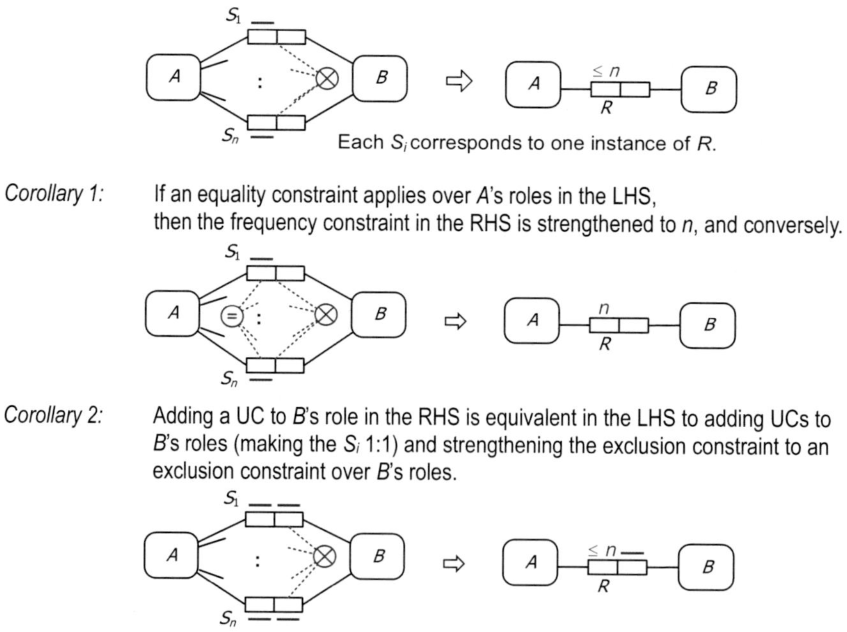

FIG. 14.20 PSG3: The left-hand schema implies the right-hand schema.

performed from right to left (strengthening the schema), with a subset constraint added from the first role of S_2 to that of S_1 if $n = 2$. Some corollaries for this result are also specified.

That covers the most important results concerning predicate specialization and generalization. Note that the theorems require that the appropriate translations hold between general and special predicates. Humans are needed to provide natural readings for the new predicate(s) and to confirm that the translation holds.

Exercise 14.2

1. Schematize the following table using (a) unaries and (b) a binary.

Male staff	*Female staff*
Halpin, T	Jones, E
Morgan, T	Smith, A

2. A company committee has to decide whether to increase its budget on staff training. The following table indicates the current views of the committee members on this issue. Schematize this using (a) unaries and (b) a binary.

For	*Against*	*Undecided*
Alan	Betty	Chris
David	Eve	Fred
Gearty		

3. The following table is an extract from an output report indicating quarterly sales figures for software products marketed by a particular company.
   (a) Schematize this using a ternary.
   (b) Transform your solution into an equivalent one using binaries.

Software	*Quarter*	*Sales ($)*
DataModeler	1	200,000
	2	500,000
	3	500,000
	4	700,000
WordLight	1	90,000
	2	150,000
	3	155,000
	4	200,000

4. An embassy maintains details about how well its employees speak foreign languages. The following table is an extract from this system.
   (a) Schematize this using binaries.
   (b) Transform this into a ternary.

Language	*Expert*	*Novice*
Arabic		Smith, J
Dutch	Bruza, P Proper, HA	
French	Rolland, C	Bruza, P

5. University debating teams are to be selected, with four students in each team, with one student from each year level (1...4). No student may be in more than one team. Students are identified by their student number, but their name is also recorded. Debating teams are identified by codes.
   (a) Schematize this UoD using YearLevel as an object type.
   (b) Transform your schema by absorbing this object type.

6. Employee records are kept that show the employee number, name, and up to two phone numbers for each employee.
   (a) Schematize this in the natural way.
   (b) Map this to a relational schema.
   (c) It is now required to map all the information into a single relational table. Transform the conceptual schema to enable this to happen.
   (d) Rmap your revised conceptual schema.

   Each phone is now to be classified as a work phone or a home phone. At most one work phone and at most one home phone may be recorded for an employee.
   (e) Modify your solution to (a) accordingly.
   (f) Rmap this.
   (g) Modify your solution to (c) accordingly.
   (h) Rmap this.

7. The following conceptual schemas are meant to describe the same UoD, but each fails to capture some constraint in the other. Add a textual constraint to schema (1) and a graphical constraint to schema (2) to obtain equivalence. The codes "chr", "sec", and "ord" abbreviate "chairperson", "secretary", and "ordinary member", respectively.

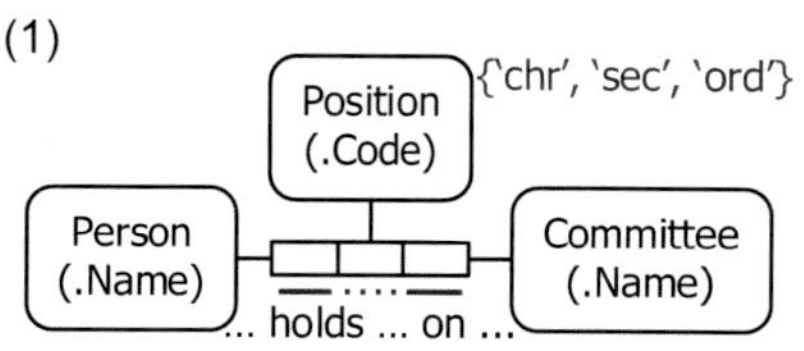

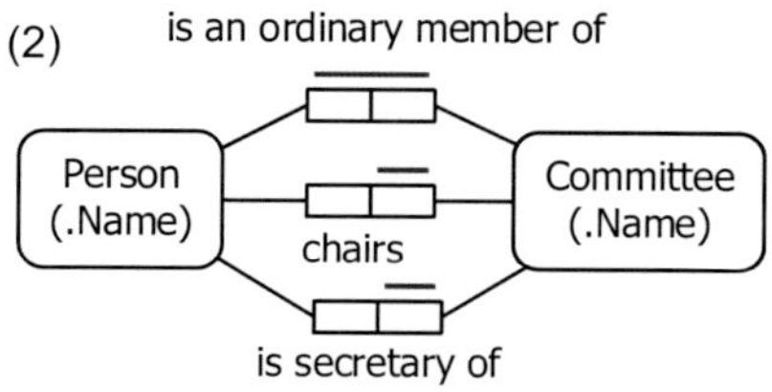

8. Consider the following academic UoD. Each subject is identified by its code but also has a unique title. Each subject has at most three assignments (possibly none),

numbered 1, 2, and 3. Each subject has a second assignment only if it has a first assignment, and it has a third assignment only if it has a first and a second assignment. Each assignment has exactly one due date. Although assignments for different subjects may be due on the same date, no subject has more than one assignment due on the same date.

It is possible that within a subject, the chronological order of due dates differs from the numerical order of the assignments (e.g., because of software problems the due date for CS400 assignment 1 is postponed until after the due date for CS400 assignment 2).

(a) Model this by adding constraints (graphic, and textual if needed) to the schema.

Reference schemes:	Subject(.Code)
	Assignment (has AssignmentNr(), is for Subject)
Fact types:	Subject has SubjectTitle()
	Assignment is due on Date(mdy)

(b) Map this to a relational schema, including all constraints.
(c) Specialize the is-due-on predicate in (a) by absorbing AssignmentNr into it.
(d) Map this to a relational schema, including all constraints.
(e) Suppose that within each subject the chronological order of assignment due dates must match their numeric order. How does this affect answers to (c) and (d)?

14.3 Nesting, Coreferencing, and Flattening

Humans may think of the universe in different ways. One modeling choice is whether to represent some feature in terms of an object type, and if so, how. The previous section discussed how to absorb or extract object types by predicate specialization and generalization. An object type with a composite identification scheme may be portrayed explicitly as a nested object type or as a coreferenced object type. If we don't want to think of this feature in terms of an object type, we can model it with predicates using a flattened approach. This section considers transformations among nested, coreferenced, and flattened approaches.

Let's start with a familiar example. The report in Table 14.2 may be modeled in three ways, as shown in Fig. 14.21. The schema in Fig. 14.21(a) takes the flattened approach, using a ternary fact type. In the schema in Fig. 14.21(b), Enrollment is a coreferenced object

Table 14.2 A relational table of student results.

Result:

studentNr	*subject*	*rating*
1001	CS100	4
1002	CS100	4
1002	CS114	5

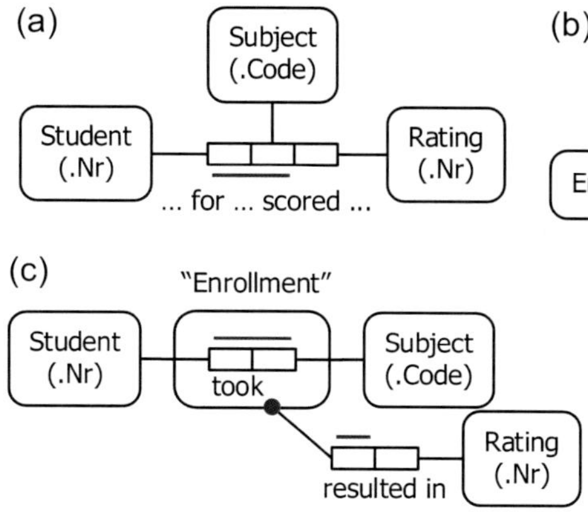

FIG. 14.21 Modeling Table 14.2 by (a) flattening, (b) coreferencing, and (c) nesting.

type, where its primary identification scheme combines two reference types. In the schema in Fig. 14.21(c), Enrollment is modeled as a nested object type or objectified association. As discussed in Chapter 9, UML uses association classes for nesting, and has a limited form of coreference.

In this example, the nested or coreferenced object type plays just one fact role, and this role is mandatory. In such a situation, the relational mapping is the same whether we nest, coreference, or flatten. In this case, each maps to the table scheme *Result* (studentNr, subjectcode, rating). So, we may choose which of the three approaches appeals most to us. Although this choice is partly subjective, there are a few guidelines that may be helpful.

First, let's consider choosing between nesting and coreferencing. Both approaches always have the same relational mapping, no matter what the situation. Choose what seems most "natural". If in thinking about the real world, you first "see" a relationship and later want to talk about it, then nesting is probably the best choice. On this basis, more people would probably nest Enrollment rather than coreference it. However, it often seems unnatural to think of a visible object as an objectified relationship. For example, consider Table 14.3, which provides details about the length of annual reports by departments within a company.

Table 14.3 Length of departmental annual reports.

Year	*Department*	*Report length (pp.)*
1993	Research	40
1993	Sales	25
1993	Support	40
1994	Research	67
...	...	...

Because you can hold a report (or at least a copy of a report) in your hand, it is natural to think of it first as an object and later ask yourself how to identify it. In this case, we use both the year and the department to coreference it, as in Fig. 14.22(a). The nested approach in Fig. 14.22(b) treats a report as a relationship between Department and Year. The authors find this unnatural, but because both map the same way, you may choose which one suits your taste.

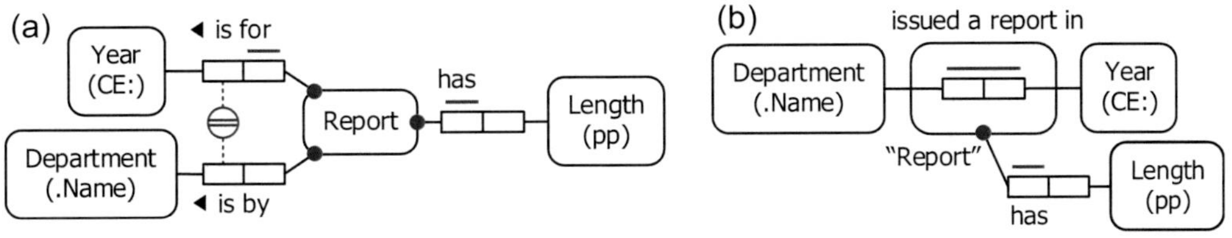

FIG. 14.22 Because a report is a physical object, is (a) more natural than (b)?

If at least one of the object types used for the composite identification is a value type, then coreferencing is often more natural than nesting. Recall the following coreferenced example: Subject (is offered by Department, has SubjectTitle). We could model this instead by objectifying the association Department offers a subject with SubjectTitle, but this seems unnatural. As another example, consider the composite reference scheme Account (is at Branch, has AccountNr).

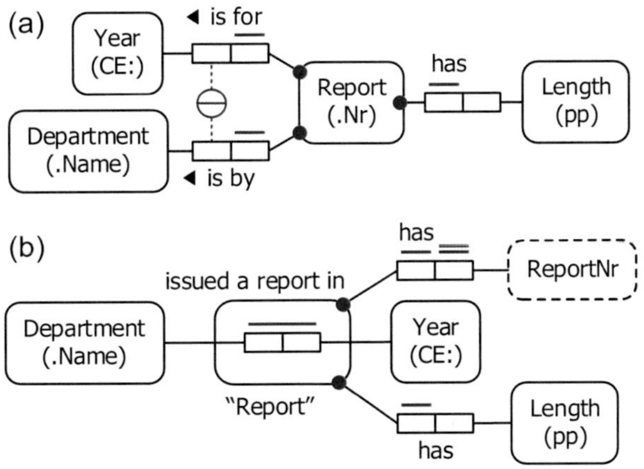

FIG. 14.23 Annual departmental reports are now identified by a report number.

For efficiency, we sometimes introduce a new, simple identifier to replace a composite one, such as Subject(.code). In this case, the former reference types become elementary fact types. For example, in Fig. 14.23(a) a report number is introduced to identify annual reports. We could use a plain serial number (e.g., 1, 2, …) or a coded number for humans to derive the semantics (e.g., “RES-2008”, “SAL-2008”, …). Though unnatural, a simple identifier may be introduced for Report even though it remains objectified, as shown in Fig. 14.23(b).

Fig. 14.23(a) includes the relationship types Report is for Year; Report is by Department. These were reference types in Fig. 14.22(a), but are now elementary fact types. For example, two fact instances might be Report "RES-2008" is for Year 2008; Report "RES-2008" is by Department "Research". If the semantic basis of such introduced identifiers is unstable, then plain serial numbers might be preferred (e.g., suppose departments are often split, merged, or renamed). However, if the semantic basis is stable, such "information-bearing" identifiers can lead to efficiency gains by reducing the need to perform relational joins when information about the object is spread over many tables.

For example, suppose one table is used to store the data in Table 14.3 as well as an extra column for reportNr, and a separate table is used to store the *m:n* fact type Employee(.nr) authored Report(.nr). To find out who authored what reports, most users would be content with a listing of this second table, if meaningful report identifiers such as "RES-2008" are used, because this would be enough for them to deduce the year and department. However, if plain serial numbers are used, most users would want the year and department listed too, requiring a join to the first table. As another example, consider the use of subject codes (e.g., "CS115") in listing results.

If a new identifier is introduced, the new schema is no longer strictly equivalent to the original one. For example, the schemas in Fig. 14.23 are stronger than the schemas in Fig. 14.22. Any schema strengthening or weakening should result from a conscious decision of the modeler. However, shifting between a coreferenced and a nested approach (as in Fig. 14.22) may be treated as an equivalence transformation.

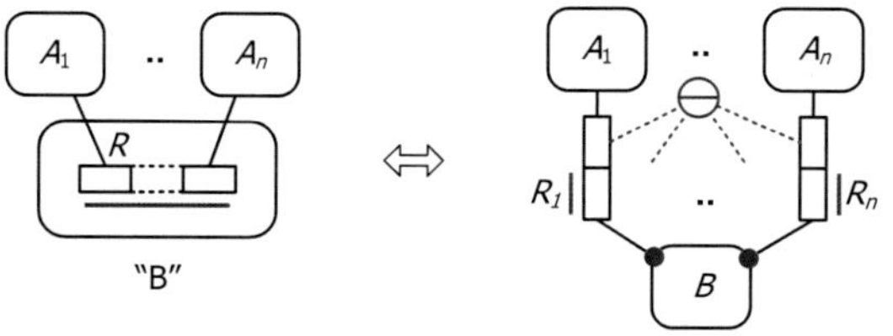

Where $n > 1$, and $Ra_1 \ldots a_n$ iff there is a b such that bR_1a_1 & .. & bR_na_n.

Note: If *B* is independent, show it as *B* ! in the coreferenced version.

FIG. 14.24 N/CR: The nested version is equivalent to the coreferenced version.

The general theorem, called **N/CR** (*nest/coreference*), is set out in Fig. 14.24. Here, the objectified predicate *R* has *n* roles, where *n* is at least 2. The object types $A_1 \ldots A_n$ are not necessarily distinct. Although not shown in the figure, we assume the nested and coreferenced object types have at least one role attached.

An object type is independent if the disjunction of its fact roles is optional. If independent, the object type must be explicitly marked so by appending "!" to its name. For example, look back at Fig. 14.21(b) and (c), and suppose that enrollments are recorded before any results are known. The mandatory role constraint must now be removed from the rating association, and the Enrollment object type renamed "Enrollment !". As shown shortly, a more drastic change is needed to the schema in Fig. 14.21(a) to maintain a flattened approach.

As long as the nested or coreferenced object type plays only one fact role, and this role is mandatory, we may transform to a single, flattened fact type. Recall the example in Fig. 14.21. In such a case, the flattened approach is usually recommended because it gives a simpler diagram, is easier to verbalize, and it sometimes avoids arbitrary decisions about which part of the predicate to objectify.

When a ternary or longer predicate has just one uniqueness constraint, and this spans all but one of its roles, the subpredicate spanned by this UC provides the best choice for objectifying. This was our choice in Fig. 14.21. In principle, we could pick any subpredicate to objectify. For example, suppose in Fig. 14.25(a) that we objectify the subpredicate comprising the first and third roles, instead of the first two roles. This yields the awkward schema shown in Fig. 14.25(b).

The external UC is needed to capture the UC in the flattened version (i.e., each student obtains at most one rating for each subject). Notice also that the UC on the outer predicate is compound. Although it maps to the same relational schema as the others, Fig. 14.25(b) is more difficult for a human to understand.

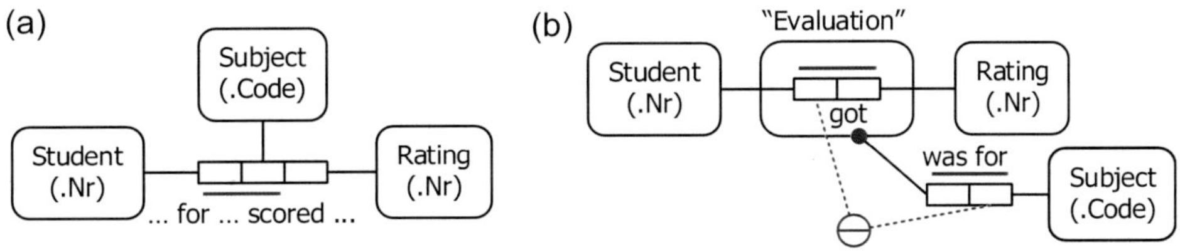

FIG. 14.25 A poor choice for nesting (the UC spans other roles).

If a ternary or longer predicate has overlapping UCs, then any objectification is best based on one of these. Each other UC in the flattened version is then captured by an external UC. See Figs. 4.40 and 4.41 for an earlier example.

With overlapping UCs, there is more than one "natural" choice of a subpredicate to objectify. This is also true if a ternary or longer predicate has a spanning UC, such as Lecturer visits Country in Year. Here, we could objectify any of three role pairs: roles 1 and 2; roles 1 and 3; or roles 2 and 3. Unless there are additional facts about the subpredicates, the flattened fact type provides the simpler approach.

Fig. 14.26 shows the ternary version of the **N/F** (*nest/flatten*) equivalence theorem. Here *A*, *B*, and *C* need not be distinct. A spanning UC over the roles in the nested object type *S* is assumed. Typically, a UC spans the *A* and *B* roles in *R*; this is equivalent to a UC over the first role of *T*. For equivalence, the predicates must be formally related by the condition shown in the where clause.

In Fig. 14.26, if a UC spans all the roles of *R*, then a UC spans both roles of *T*. We do not bother stating this corollary in the figure because each predicate has an implied UC spanning its full length. Recall that for generality, schemas used to depict theorems may omit constraints that are not relevant to the transformation.

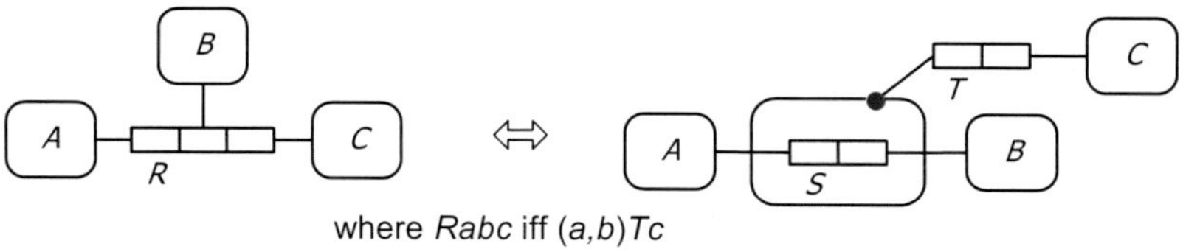

Corollaries: A UC exactly spans the *A*, *B* roles of *R* iff a simple UC spans the first role of *T*.
A binary UC over the *A*, *C* (or *B*, *C*) roles of *R* is equivalent to an external UC over the *A*, *C* (or *B*, *C*) roles, respectively, in the nested version.

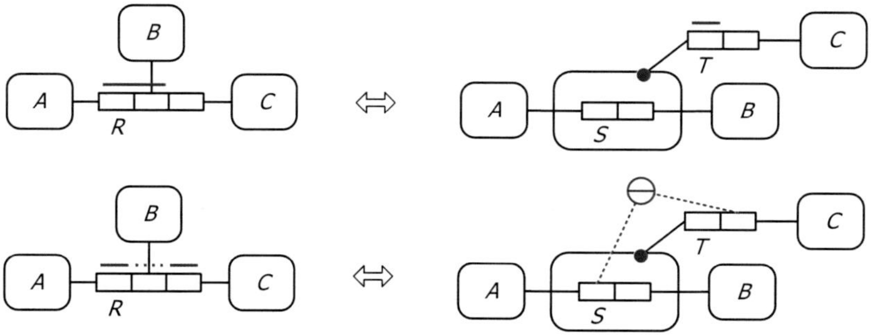

FIG. 14.26 N/F: The nested version is equivalent to the flattened version.

Our equivalence theorems state formal connections between predicates in the different versions, but do not specify how predicate names used in one schema might help choose the different predicate names used in the other. A computer-aided software engineering (CASE) tool could generate suggested names, but humans can often provide more natural ones (cf. generation of table names in relational mapping).

In choosing identifiers, we should ensure that their background meaning agrees with the formal connection required by the transformation rule. This still leaves many possibilities. In Fig. 14.25(a), for example, instead of "… for … scored …" we might use "… enrolled in … obtaining …".

Note that because nesting can always be replaced by coreferencing, we could specify **CR/F** (*coreference/flatten*) equivalence theorems analogous to any N/F theorem. For example, consider transforming between the two schemas in Fig. 14.21. To save space, however, we limit our unflattening discussion to nesting.

The ternary nest/flatten equivalence may be generalized to flattened fact types of any arity above two. Note that nesting always introduces an extra role (to be played by the objectified association). If we objectify m roles of an n-ary predicate, the outer predicate will have $n-m+1$ roles. This is shown in Fig. 14.27(a), where other object types and role connectors are omitted for simplicity.

Recall that each UC of an elementary predicate must span at least all but one of its roles. If a UC exactly spans all but one role of a predicate, and we objectify on this UC, the outer predicate in the nested version has two roles, with a simple UC on its first role (Fig. 14.27(b)).

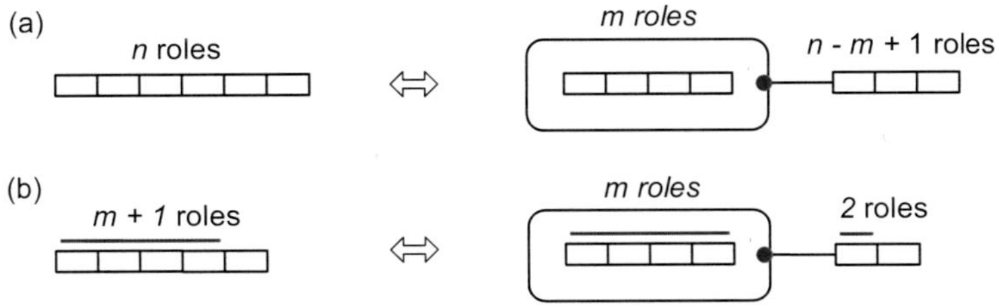

FIG. 14.27 (a) Nesting ternaries and beyond; (b) nesting on a UC.

With long predicates, nesting may be applied more than once, as shown in Fig. 14.28, where a quaternary is binarized in two stages. For simplicity, other object types and role links are omitted. We saw earlier how unaries may be converted into binaries. Nesting or coreferencing may be used to convert n-ary predicates into binaries. So, in principle, any domain may be modeled with binaries only. However, nonbinary predicates often enable an application to be modeled in a more natural and convenient way.

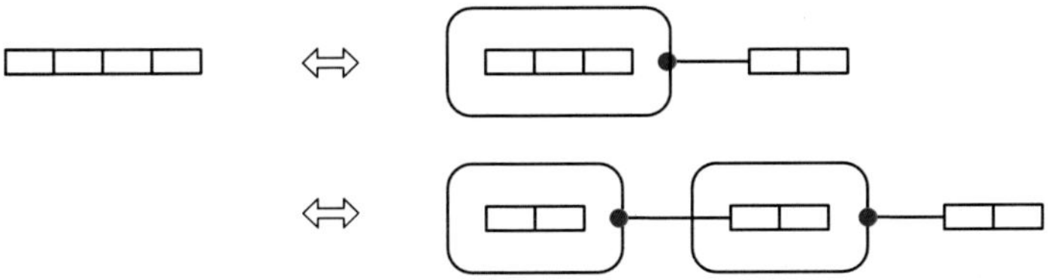

FIG. 14.28 Binarizing a quaternary by nesting.

In our examples so far, the objectified association played just a single, mandatory role. In this case, flattening is generally preferable. But what if this role is optional? For example, suppose we modify our earlier Table 14.2 by allowing nulls in the rating column (see Table 14.4). In this population, no ratings appear for the subject CS100. Perhaps the final exam for this subject is still to be held, or marking for it has not finished. In this UoD, it is possible to record the fact that a person enrolls in a subject before knowing what rating the student gets for that subject.

Table 14.4 Enrollments may be recorded before ratings are finalized.

Result:	*studentNr*	*subject*	*rating*
	1001	CS100	?
	1002	CS100	?
	1002	CS114	5

Because nulls are not allowed at the conceptual level, we cannot model this as a single ternary. We may, however, use two fact types–one for enrollments and one for ratings. Fig. 14.29(a) shows this flattened approach, together with the sample population.

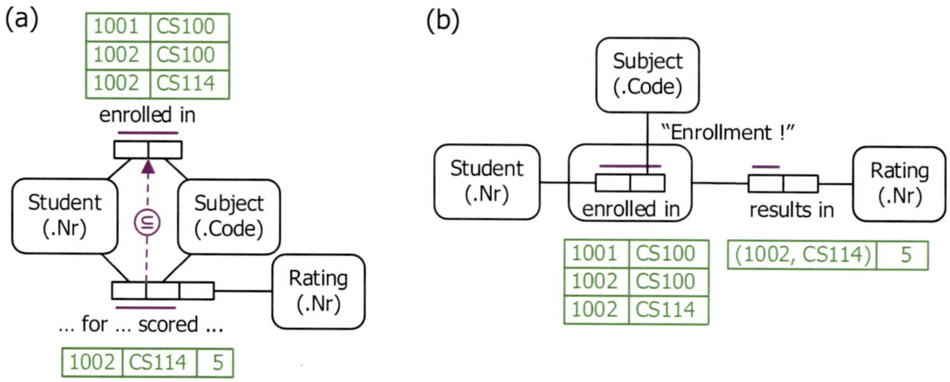

FIG. 14.29 Modeling Table 14.4 with (a) flat fact types and (b) nesting.

The *pair-subset constraint* in Fig. 14.29(a) indicates that students can score ratings only in subjects in which they enrolled. If this figure is part of a model with other roles for Student and Subject, the roles in the binary fact type may be optional; if the figure is not, they should be marked mandatory.

This UoD may also be modeled using a nested or coreferenced object type, as shown in Fig. 14.29(b). The role played by the objectified association is now *optional*. The objectified predicate corresponds to the binary predicate in Fig. 14.29(a).

If no other role is played by Enrollment in the global schema, it is independent, as shown by the appended "!". This is not unusual with nesting. Any independent object type, whether simple, nested, or coreferenced, should be explicitly declared independent by appending "!" to its name. Fig. 14.30 remodels the example using a coreferenced object type.

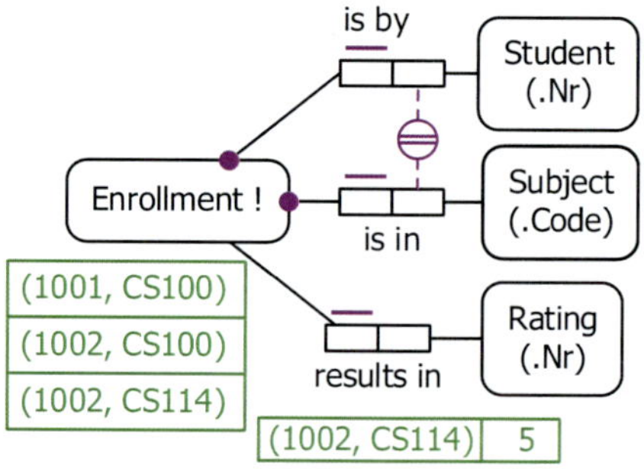

FIG. 14.30 Modeling Table 14.4 with a coreferenced object type.

When two or more flat fact types are involved, the equivalence between nested and flattened versions is called **N/Fm** (*nest/flatten into many fact types*). The subset pattern illustrated by our example is only one of many cases for this theorem, but because it is the most common case, we set it out in Fig. 14.31.

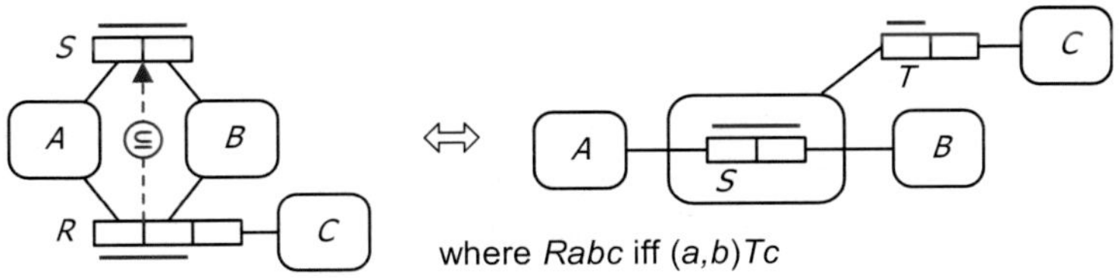

FIG. 14.31 Another nest/flatten equivalence.

We generally favor the nested version in this case because it is more compact and it Rmaps directly to just one table. In contrast, if the flattened version is passed to our standard Rmap procedure, two tables result. Nevertheless, the flattened approach is quite natural, and it verbalizes easily. If you prefer the flattened approach for these reasons, it would be nice if your CASE tool displayed it this way, but gave you the option of having it mapped to a single table in the same way as the nested version. However, such flexibility is typically not provided by current CASE tools. So, the nested approach is generally preferable for this case.

In the flattened version in Fig. 14.31, the *S* predicate is compatible with the subpredicate comprising the first two roles of *R* (they are played by the same object types *A* and *B*). This compatibility enables the nesting to occur. For this subset case, any (*a*, *b*) pair in the population of either must belong to the population of *S*. So, we objectify *S* and add an optional connection to *C* to handle the *R* facts.

Recall that in our CSDP, whenever different predicates have compatible role sequences (of two or more roles), we should ask ourselves to what extent the populations of these [sub-] predicates must overlap. If a set-comparison (subset, equality or exclusion) constraint exists between them, it must be declared.

The kind of nesting transformation that may now occur depends partly on what kind of set-comparison constraint exists (if any) and on whether other roles occur in the full predicates. Let's look at some more examples before summarizing the overall procedure.

Consider the output report shown in Table 14.5. This may be modeled with two flat fact types, as shown in Fig. 14.32(a). Notice the *equality constraint* between the two role-pairs. An assignment mark is recorded for a student in a subject if and only if an exam mark is recorded for that student in that subject.

Table 14.5 Student assignment and exam marks.

Subject:	CS102		CS115	
StudentNr	*Assignt*	*Exam*	*Assignt*	*Exam*
1001	17	65	15	58
1002	20	79	20	67
1003	15	60	12	55

The compatible subpredicates might be read as "did assignment in" and "did exam in". Since their populations must be equal, we may objectify their conjunction did assignment and exam in. The nested version in Fig. 14.32(b) uses the wording "was assessed in". The specific marks are catered for by attaching two specific predicates as shown. Because of the equality constraint, these are both mandatory.

When passed to the standard Rmap procedure, the schema in Fig. 14.32(a) maps to two tables, whereas the nested version in Fig. 14.32(b) maps to a single table, so the nested version is generally preferred whenever two or more functional roles are attached to the objectification.

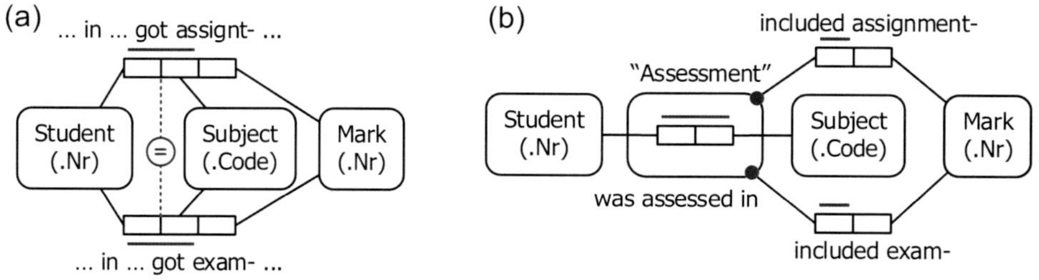

FIG. 14.32 ORM schemas for Table 14.5: (a) flattened version and (b) nested version.

A more general N/Fm equivalence result is shown at the top of Fig. 14.33, with four corollaries listed and illustrated below it. This may be generalized further to n predicates ($n \geq 2$) and compatible subpredicates of any arity above 1. The object types $A \ldots D$ need not be distinct. As long as [sub-]predicates are compatible, nesting may always be performed by objectifying their *disjunction*.

The predicate T in Fig. 14.33 is a logical disjunction, so may be verbalized using the "**or**" to connect verbalizations of the compatible [sub-]predicates. In terms of populations, the objectified relation T in Fig. 14.33 is the *union* of the compatible [sub-]relations., that is, $T = R[a,b] \cup S[a,b]$.

Suppose we remove the equality constraint in Fig. 14.32(a). If no other set-comparison constraint exists, we replace the two mandatory role constraints in Fig. 14.32(b) by an inclusive-or constraint: **Each** assessment included **an** assignment **or an** exam mark **or both.** The uniqueness constraints in the flat version correspond to the simple UCs in the nested version (see corollary 1 in Fig. 14.33).

Adding a pair-subset constraint from the first two roles of S to the first two roles of R is equivalent to adding a subset constraint from the first role of V to the first role of U. In the context of the disjunctive mandatory constraint, this means that U becomes mandatory (see corollary 2). In terms of populations, it also means that the T relation becomes $R[a,b]$ because if one set is a subset of a second, then their union is just the second set. Compare this with the result in Fig. 14.31.

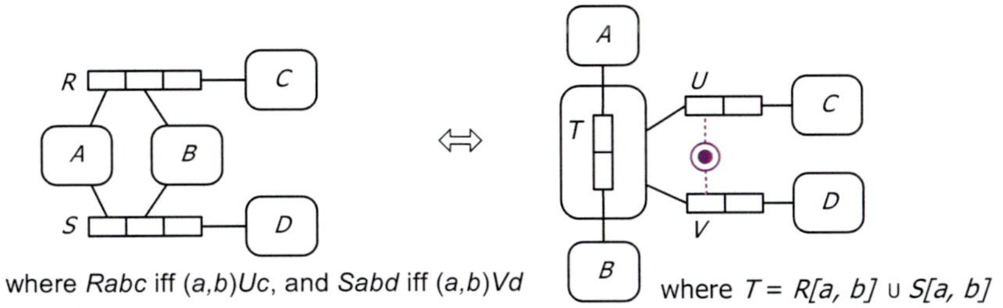

Corollaries:

(1) A binary UC over *R[a,b]* is equivalent to a UC on the first role of *U*, and a binary UC over *S[a,b]* is equivalent to a UC on the first role of *V*.

(2) A pair-subset constraint from *S[a,b]* to *R[a,b]* makes *U* mandatory.

(3) A pair-equality constraint between *R[a,b]* and *S[a,b]* makes *U* and *V* mandatory.

(4) A pair-exclusion constraint between *R[a,b]* and *S[a,b]* adds an exclusion constraint between the attached roles in the nested version.

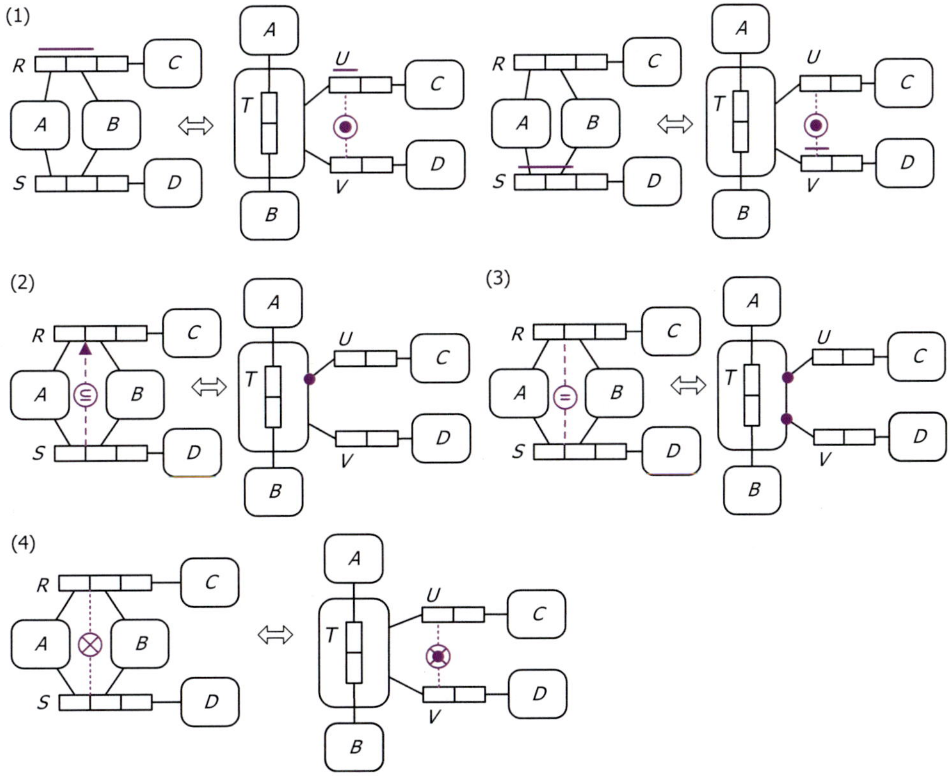

FIG. 14.33 A more general nest/flatten equivalence.

A subset constraint in both directions is an equality constraint. So adding an equality constraint between *R[a,b]* and *S[a,b]* in Fig. 14.33 makes both *U* and *V* mandatory in the nested version. Fig. 14.32 provides an example.

If an exclusion constraint exists between *R[a,b]* and *S[a,b]* in Fig. 14.33, and we decide to nest, we must add an exclusion constraint between the attached roles in the nested version. For example, suppose that for any given subject, a student can be awarded either an actual grade or a notional grade, but not both (a notional grade might be awarded on the basis of performance in a similar subject from another university). As an exercise, schematize this in both flat and nested versions.

A relation is said to be *partial* if it is a projection of a longer relation; otherwise, it is said to be *whole*. Sometimes, we run into a case where a whole relation must be a subset of another. Consider Table 14.6, for instance. Here, "?" is an ordinary null. In this UoD, we want to know what subjects students have passed, but are not interested in their actual ratings.

Table 14.6 Student enrollments and passes (if known).

Student	*Subjects enrolled in*	*Subjects passed*
3001	CS100, CS114, MP105	CS100, CS114
3002	CS114, MP105	CS114
3003	CS100, CS114	?

Fig. 14.34(a) schematizes this using a flattened approach. We can nest this by objectifying enrollment, and then using a unary to indicate which enrollments resulted in a pass, as in Fig. 14.34(b). By default, the flat version maps to two relational tables while the nested version maps to one. So, the nested version is usually preferred in such a case. You can decide for yourself whether to take an open- or closed-world approach when mapping the unary.

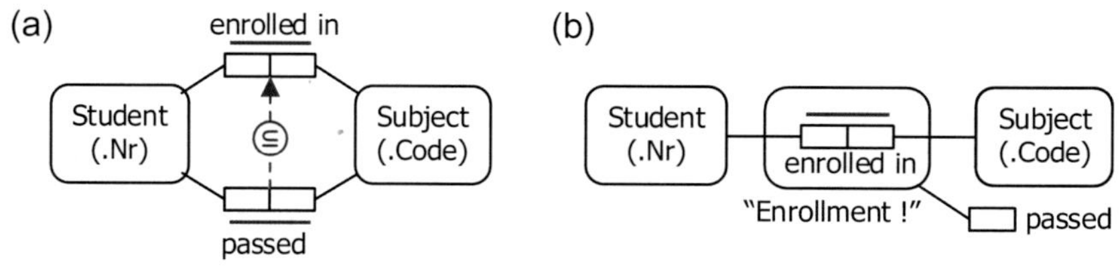

FIG. 14.34 ORM schemas for Table 14.6: (a) (flat version) and (b) nested version.

The **overlap algorithm** in Fig. 14.35 summarizes the main cases of the N/Fm equivalence. This shows how to nest when two or more fact types include compatible role sequences, each with at least two roles. For simplicity, the diagram shows just two predicates with compatible role pairs (shaded). The predicates may have additional roles.

Overlap algorithm:

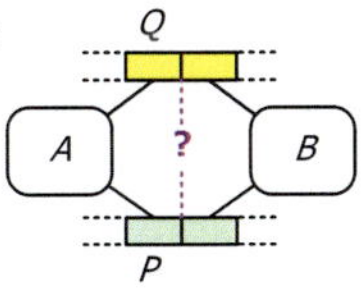

case ? constraint **of**

↑ : -- *P*'s population must be a subset of *Q*'s, but need not equal *Q*'s
objectify *Q*;
if *Q* is partial **then** attach rest of its predicate via mandatory extra role;
if *P* is partial **then** attach rest of its predicate via optional extra role
else attach optional unary "is-*P*"

↓ : reverse of previous case

↕ : -- *P*'s population must equal *Q*'s
if *P* or *Q* is partial
then objectify the single predicate "*P* and *Q*";
attach rest of their predicates via mandatory extra role(s)
else replace both by the single predicate "*P* and *Q*"

⊗ : -- *P*'s population is mutually exclusive with *Q*'s
objectify "*P* or *Q*";
if *P* is partial **then** attach rest of its predicate via extra role
else attach as unary "is-*P*";
similarly for *Q*;
mark the attached roles as exclusive and disjunctively mandatory

: -- no constraint; proper overlap is possible
as for ⊗ but omit exclusion constraint

end

FIG. 14.35 Overlap algorithm for nesting when compatible subpredicates exist.

Let *P* and *Q* be the compatible [sub-]predicates. The nesting action depends on the amount of overlap allowed between the populations of *P* and *Q*. In any given case, this overlap condition is specified by replacing the pair connection marked "?" by a pair-subset (either direction), pair-equality or pair-exclusion constraint, or by no constraint (proper overlap is possible).

In Fig. 14.35, a subset constraint upward from *P* to *Q* is denoted "↑", a downward subset constraint is shown as "↓", and an equality constraint appears as "↕". Recall that a predicate is partial if it is embedded in a longer predicate.

The five cases are based on which set-comparison constraint is explicitly *displayed*. An equality constraint must be specified if it exists. Hence, in this context, display of a subset constraint is taken to imply that an equality constraint does not exist. In any situation, exactly one of the five cases listed will apply.

As an example of the "↑" case, consider the transformation from Fig. 14.29 to Fig. 14.30. Here, *Q* is the whole enrollment predicate while *P* is the partial predicate comprising the first two roles of the fact type Student for Subject scored Rating.

As an example of the "↑" case where Q is partial, replace the enrollment predicate in Fig. 14.29 by the ternary: Student enrolled in Subject on Date. The nested version in Fig. 14.30 must now be modified by adding the mandatory binary: Enrollment occurred on Date. As an example when both P and Q are whole, consider the transformation of Fig. 14.34(a) to (b).

As an example of the "↕" case, consider reshaping Fig. 14.32(a) to (b). Here, P and Q are partial, and their conjunction is reworded as "was assessed in". As an example of the "↕" case where neither P nor Q is partial, remove the mark information from Fig. 14.32(a) to give two binaries: Student did assignment in Subject; Student did exam in Subject. The equality constraint enables these to be replaced by a single binary: Student did assignment and exam in Subject (or Student was assessed in Subject). Derivation rules may be specified for the former binaries if desired.

As an example of the "⊗" case where both P and Q are partial, recall this exclusive disjunction instance of Fig. 14.33: Student for Subject obtained actual Grade; Student for Subject obtained notional Grade. This may be modeled by objectifying Student was assessed in Subject and attaching the mandatory, exclusive disjunction: resulted in actual Grade; resulted in notional Grade.

As an example where both P and Q are whole, consider the exclusive disjunction: Student passed Subject; Student failed Subject. Suppose we objectify the disjunction of these predicates (i.e., Student passed or failed Subject) as Student was assessed in Subject. This Assessment object type may now be classified by attaching an exclusive disjunction of the following unaries: is a pass; is a fail. As an exercise, draw these examples.

While the mutual exclusion case may be transformed by nesting in this way, this is rarely the best modeling alternative. Sometimes the flat version is preferable, and sometimes a predicate generalization transformation is more appropriate. For instance, the pass/fail example just cited may be modeled instead using the ternary Student in Subject obtained Result {'pass','fail'}. Also, the nested solution to the actual/notional grade example may be transformed using PSG2 to replace the exclusive disjunction by has Grade; is of GradeType {'actual', 'notional'}.

In the final case of the overlap algorithm, no set-comparison constraint applies between P and Q, so their populations may properly overlap. For example, suppose that for any given student and subject we might have a predicted grade, an actual grade, or both. This may be modeled by two flat fact types: Student for Subject has predicted Grade; Student for Subject has actual Grade.

We may nest this by objectifying Student is assessed for Subject and attaching the mandatory, inclusive disjunction: has predicted Grade; has actual Grade. As with the previous mutual exclusion case, other modeling alternatives may be preferable.

If P and Q are keys (each is exactly spanned by a UC) and the Rmap procedure is used, applying the overlap algorithm to flattened cases before mapping reduces the number of relational tables. In this situation, especially for the subset and equality cases, nesting (or the equivalent coreferenced solution) is generally preferred.

Exercise 14.3

1. Consider the fact type City in Year has Population. Population figures are collated only once each year. Populations may go up, down, or remain the same.
   (a) Model this using a flattened approach.
   (b) Now use a nested approach.
   (c) Now use a coreferenced approach.
   (d) Which solution do you prefer?
   (e) Use a nested approach that objectifies the association between City and Population. State whether this is better or worse than your solution to (b).

2. Consider the fact type: Flower blooms in City in Month. Note that flowers may bloom for more than 1 month in the same city.
   (a) Model this as a flat fact type.
   (b) Show three alternative nested solutions.

3. Schematize the following table using (a) flat, (b) nested, and (c) coreferenced approaches. Which do you find the most natural?

Software title	*ReleaseNr*	*Size (Mb)*
DataModeler	1	3
DataModeler	2	5
DataModeler	3	5
WordLight	1	5
...	...	...

4. Suppose employees are identified by their surname and initials, and each works for exactly one Department. In principle, we could model this using the ternary fact type Surname and Initials belong to an employee who works for Department.
   (a) Draw the flat fact type. Is this natural?
   (b) Nest this instead. Is this natural?
   (c) Now use coreferencing. Is this natural?
   (d) We now decide to primarily identify employees by an employee number, but still require that surname and initials provide an alternative identifier. Model this.

5. The following conceptual schema refers to the recent Olympic Games.
   (a) Rmap this.
   (b) Transform the conceptual schema by nesting.
   (c) Rmap your solution to (b).
   (d) Which conceptual schema do you prefer?

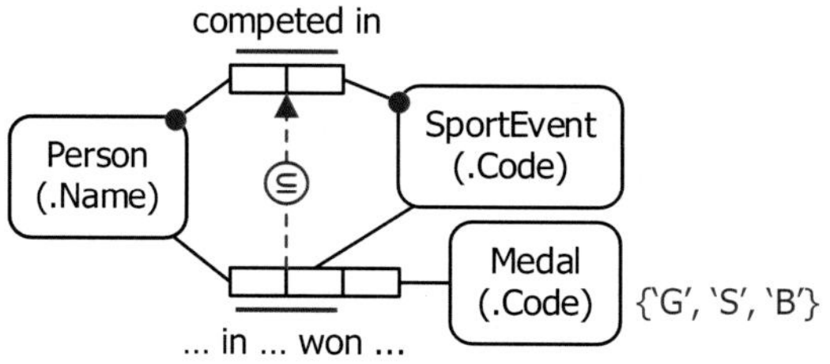

6. The following is an extract from a yearly report giving test results of reaction time (in milliseconds) and resting heart rate (in beats per minute) for members of a health club. As the table indicates, the club may gain or lose members during the year.
   (a) Schematize this using two flat ternaries.
   (b) Transform this by nesting.
   (c) Which schema do you prefer?

Month	*Member*	*Reaction time*	*Heart rate*
Jan	Jones, E	250	80
	Matthews, S	320	120
	Robinson, S	300	100
Feb	Jones E	250	75
	Matthews, S	300	100
Mar	Anderson, P	250	80
	Matthews, S	280	85
...	...	...	...

7. The following table indicates the rooms and times for lectures in various subjects. Schematize this using (a) flat, and (b) coreferenced approaches.

Subject	*Time*	*Room*
CS213	Mon 3 p.m.	B19
CS213	Wed 9 a.m.	A01
CS213	Wed 10 a.m.	A01
EN100	Mon 3 p.m.	F23
EN100	Tue 3 p.m.	G24
...	...	...

8. The report shown indicates student performance in subjects in one semester. Once a student fails a subject, the subject cannot be passed by the student in that semester. As students pass or fail subjects these results are recorded. It is possible that a student might have neither passed nor failed a subject taken (e.g., the exam is yet to be held).

(a) Schematize this UoD using three binaries.
(b) Transform this by nesting, with two attached unaries.
(c) Transform the two unaries into a binary.
(d) For this UoD, is it possible that the null for Adams AB might be updated to an actual value? What about the null for Casey J?

Student	*Subjects taken*	*Subjects passed*	*Subjects failed*
Adams AB	CS100, CS114, MP104	CS100, CS114	?
Brown SS	CS114, MP102	CS114	?
Casey J	CS100, CS114	?	CS100, CS114

9. The following table is an extract from an output report concerning the finals of a recent judo competition. For each weight division, the four clubs that made it to the finals are recorded, together with the results for first and second places. For a given weight division, a club can obtain at most one place. No ties are possible.
   (a) An information modeler schematizes this in terms of three fact types: Club is finalist in Event; Club wins Event; Club is second in Event. Set out this conceptual schema.
   (b) Transform the winner and second-place predicates into a single ternary.
   (c) Transform this schema into an equivalent nested version.
   (d) Assuming complete information, show an alternative nested solution by assigning the places "3A" and "3B" to the clubs that didn't get first or second in the event.
   (e) Transform this to a flattened fact type.
   (f) Which solution do you prefer?

Event	*Finalists*	*Winner*	*Runner-up*
Lightweight	Budokan, Judokai, Kodokan, Zendokan	Kodokan	Judokai
Middleweight	Budokan, Judokai, Kodokan, Zendokan	Kodokan	Zendokan
Heavyweight	Budokan, Kanodojo, Kodokan, Mifunekan	Mifunekan	Kodokan

10. The following examples are extracts of output reports from an information system about media channels. Each channel (TV or radio) has a unique, identifying call sign. Some radio channels broadcast using frequency modulation (FM), whereas others use amplitude modulation (AM). All TV channels are rated in the range 1…7 in three categories on two surveys. Channels are either commercial (com) or owned by the government (govt). All commercial radio channels have their audience composition assessed (see the sample pie charts). The mark "–" means "inapplicable because of other data".

Schematize this, including uniqueness, mandatory role, value, subtype, and frequency constraints. Provide meaningful names and definitions for each subtype. If a fact type is derived, omit it from the diagram but provide a derivation rule. Do not nest. Minimize the number of fact types in your schema (if necessary, use transformations).

TV channels:

		Ownership details		
Call sign	*Ownership*	*Company*	*% share*	*Head office*
ATQ8	com	MediaCo	100	Brisbane
CTQ3	com	MediaCo	50	Brisbane
		TVbaron	50	Sydney
TVQ3	govt	–	–	–

Radio channels:

		Ownership details				
Call sign	*Ownership*	*Company*	*% share*	*Head office*	*Modul-ation*	*Music played*
4BZ	com	MediaCo	100	Brisbane	FM	rock country
RB3	govt	–	–	–	AM	–
STR5	com	MediaCo OzRadio	30 70	Brisbane Cairns	AM	country
4AA	govt	–	–	–	FM	–

TV survey ratings:

		Category			
Channel	*Survey*	*News*	*Drama*	*Sport*	*Totals*
ATQ8	A	5	3	3	
	B	4	4	3	22
CTQ3	A	5	3	4	
	B	5	4	5	26
TVQ3	A	3	4	4	
	B	2	4	5	22

Audience composition of commercial radio channels:

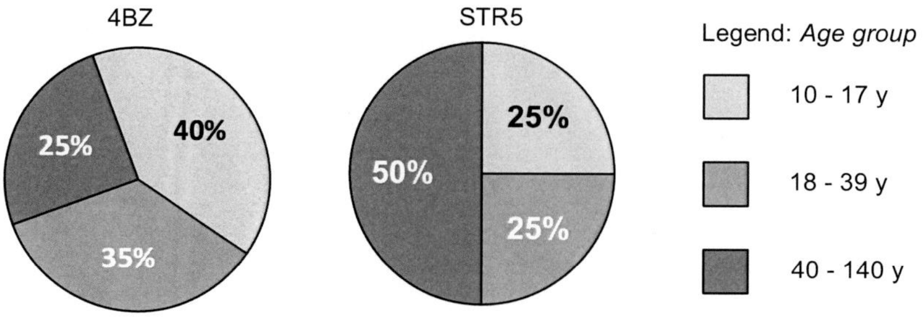

14.4 Other Transformations

The previous two sections covered the most useful conceptual schema transformations. In this section, several other transformations of lesser importance are considered briefly.

Recall PSG2, our second predicate specialization/generalization theorem (Fig. 14.16). This may be generalized further by removing constraints. First note that removing the UC on *A*'s role in *R* corresponds to removing the UC's on *A*'s roles in each of the S_i predicates (see Fig. 14.36). In the absence of additional internal UCs, the *R* and S_i binaries will be many:many. As an example, let *A* = Car, *B* = Status {"company", "private"}, *C* = Employee, *T* = has, *R* = is used by, S_1 = is provided for, and S_2 = is privately used by. Allow the same employee to use many company cars and many private cars, and vice versa.

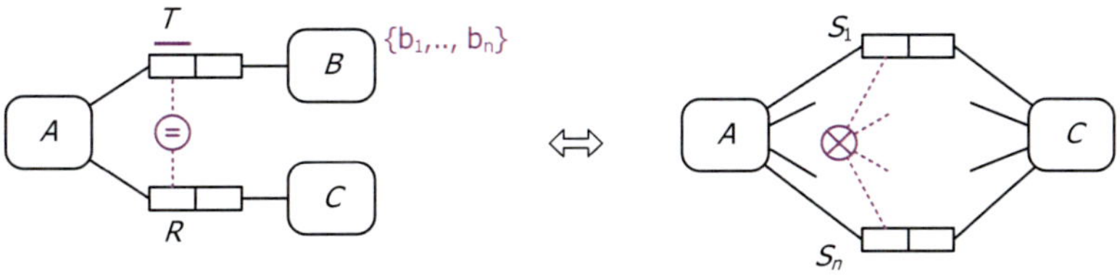

Each S_i corresponds to *R* where *T* is restricted to *B* = b_i.

FIG. 14.36 Another PSG2 case: *R* may be specialized into $S_1 \ldots S_n$ by absorbing *B*.

Now suppose the equality constraint is weakened to an upward subset constraint. The *T* predicate must be retained because there may be instances of *A* that do not play *R*, and hence *T* will not in general be derivable from the S_i. Moreover, subtypes must now be introduced for the specialized predicates (Fig. 14.37). The exclusion constraint over the subtype roles is omitted because it is implied by the subtype definitions.

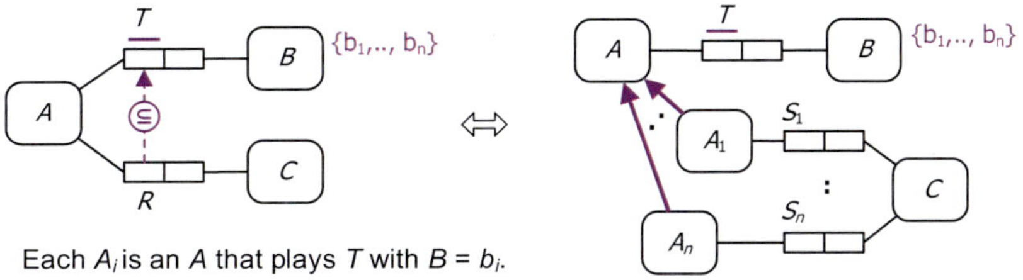

FIG. 14.37 Another PSG2 case: *R* is specialized into $S_1 \ldots S_n$, but *B* is not absorbed.

For example, consider the car UoD just discussed, where some cars might not yet be used. In this case, the subtypes are CompanyCar and PrivateCar. Cases that require one of these more general forms of PSG2 usually invoke some of the corollaries stated for the original version of the theorem (Fig. 14.16).

Suppose each employee uses at most one company car and at most one private car. An external UC is now required between the Status and Employee roles in the left-hand version, and a simple UC must be applied to each role of Employee in the right-hand version. As an exercise, draw both schema versions for this UoD and prove that a frequency constraint of ≤ 2 is implied on the role played by Employee in the left-hand version.

If an object type in an *m:n* binary has two values, the fact type can be converted to a functional one by converting the object type to one that allows concatenated values. A simple example is shown in Fig. 14.38. The transformation from left to right is called *value concatenation*, and the inverse transform is *value separation*. In this example, a physician may be trained in medicine, acupuncture, or both. Allowing the both-case to be recorded as a concatenated value (med-acu) is a "quick and dirty" way of effectively creating a set object type (in this case, DisciplineSet).

Appropriate translation rules are needed to license the transformation (e.g., Physician is trained in Discipline 'med' **iff** Physician is trained in DisciplineSet **in** {'med', 'med-acu'}). Moreover, the open/closed-world approach adopted should be the same for both. For example, if the left-hand version has a closed-world interpretation, then the code 'med' in the right-hand version means "medicine only".

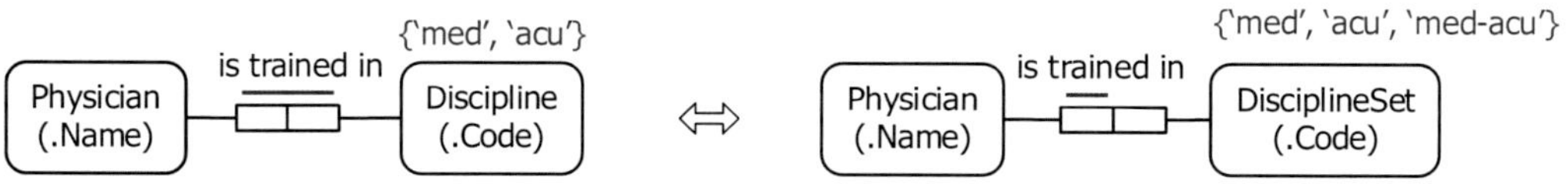

FIG. 14.38 An example of value concatenation/separation.

Although sometimes used as a quick fix, this kind of transformation rapidly becomes unwieldy as the number of atomic values increases because a list of n atomic values leads to $2^n - 1$ concatenated values (e.g., if we added herbalism and physiotherapy to give four atomic disciplines, we now have 15 concatenated disciplines). In such cases, if functional predicates are required, it is better to replace the binary with n unaries, one for each value, before mapping.

Alternatively, if the target system supports set-valued fields (e.g., a nested relational or an object-oriented database), a mapping to a set-valued structure is possible.

In the presence of a pair-subset constraint, a binary may be *contracted* to a unary; conversely, the unary may be *expanded* to a binary (see Fig. 14.39). Note that the UC on the first role of *R* is implied by the other constraints (as an exercise, prove this).

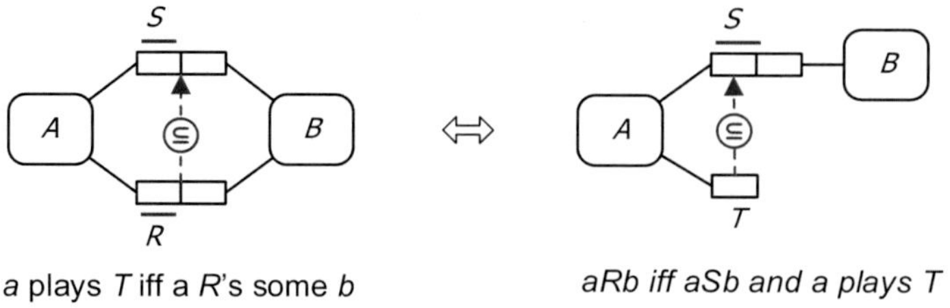

FIG. 14.39 Binary-unary contraction and unary-binary expansion.

As an example of the left-hand schema, consider the fact types (*S* listed first) Politician is a member of Party; Politician is a minister in Party. Because one can be a minister only in one's own party, a pair-subset constraint runs from the minister to the member fact type. In the right-hand version, the minister fact type is contracted to the unary: Politician is a minister. Here, a minister's party can be derived using the rule Politician is a minister in Party **iff** Politician is a member of Party **and** is a minister.

In this case, contraction is usually preferred to expansion. However, if the binary to be contracted is 1:1, its second uniqueness constraint may be captured as a textual constraint in the unary version. For example, suppose the following fact types appear in the position of *S* and *R* in Fig. 14.39: Politician is a member of / includes Party; Politician leads Party. Because each party has at most one leader, there is an extra UC on the lower binary, making it 1:1.

Now, suppose we contract this binary to the unary Politician is a leader. The extra UC in the binary version appears in the unary version as a textual constraint: **Each** Party includes **at most one** Politician **who** is a leader. At the time of writing, ORM is being extended to capture such textual constraints graphically using a "unique-where-true" constraint (see the chapter notes).

Previous sections have shown how object types can be absorbed to specialize predicates and how nested and coreferenced object types may be removed by flattening. Apart from these cases, modelers sometimes eliminate an entity type from a fact type, phrasing the association in terms of the object type(s) formerly used to reference the entity type (usually, however, it is more convenient not to do this).

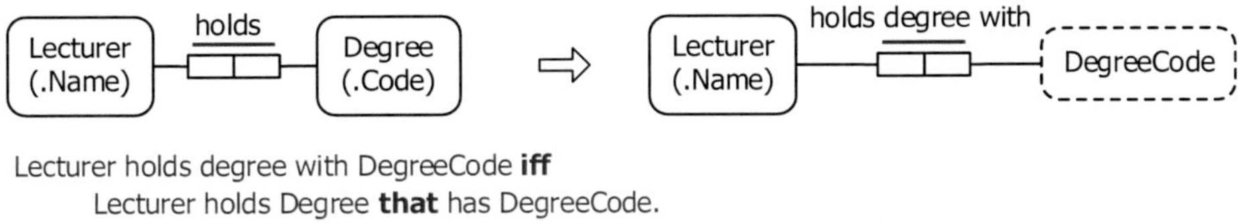

FIG. 14.40 A simple case of entity type elimination/introduction.

Fig. 14.40 provides an example with a simple reference scheme. Transforming from the left-hand to the right-hand version *eliminates* the entity type Degree. This weakens the schema because the entity type and its reference scheme are lost. The translation defines the right-hand predicate in terms of the left. This is an implication rather than an equivalence. The reverse transformation from right to left *introduces* an entity type. In almost all cases, the left-hand version is preferred because of its richer formal semantics.

This transformation is sometimes performed with a coreferenced object type to which another object type is functionally related. Fig. 14.41 provides an example. Here, a room is identified by combining its local room number and building. Note the equality constraint; this is implied if the office roles of Employee are mandatory.

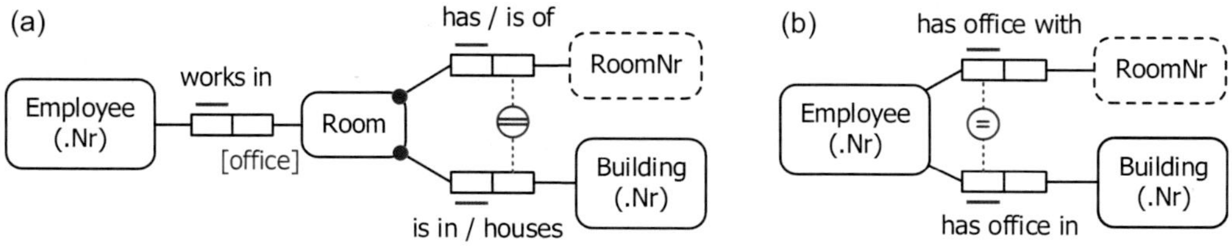

FIG. 14.41 A functional, coreferenced case; (a) is generally preferable to (b).

Fig. 14.41(a) is generally preferable to Fig. 14.41(b) for two reasons. First, it facilitates schema changes that add other facts about rooms (e.g., Room has Size(m^2:)). Second, it is natural to think in terms of rooms, and the composite identification scheme is clearly displayed. In general, don't eliminate an entity type unless you feel that no further details will be recorded for it and that no clarity is lost by its removal.

If you do decide to eliminate a compositely identified entity type, ensure that it is the target of only *functional* predicates. For example, Fig. 14.41(a) includes the fact type Employee works in Room. Suppose we now make this association *m:n* to cater for some unusual case where an employee may have more than one office. For example, if an employee has to move his/her office from one building to another, he/she may be assigned an office in each building to facilitate working efficiently in both places until the move is completed. It would now be wrong to eliminate Room as in Fig. 14.41(b) because the two associations are now *m:n*, losing the information about which room number goes with which building.

That covers the main conceptual schema transformations of practical relevance in information modeling. Other transformations exist, but we have no space to discuss them here. Some transformations that deal with compound (nonelementary) fact types are discussed later in the context of denormalization and nonrelational mappings.

Exercise 14.4

1. (a) Each employee uses one or two phone numbers, and each phone number is used by an employee. Each phone number is classified as a work phone or home phone. No employee can have two phone numbers with the same classification. Schematize this using the following entity types: Employee, PhoneNr, and PhoneType.
   (b) Provide an alternative schema by absorbing PhoneType.
   (c) Now suppose that some phones might not be used by any employee. Modify your answer to (a) accordingly.
   (d) Provide an alternative schema by specializing the uses fact type.

2. (a) Consider the fact types Person is of Gender {'m', 'f'}; Person is a parent of Person. Each person's gender is recorded. Parents of the same child must differ in gender. Some people might not be recorded as a parent or child. Schematize this.
   (b) Set out an alternative schema by specializing the parenthood fact type.

3. (a) Each person has at least one PersonType {"lecturer", "student"}. It is possible that a lecturer is also a student. Assume that all persons are identified by a PersonNr. Schematize this in terms of a many:many fact type.
   (b) Transform this to a schema comprising one functional fact type.
   (c) Provide an alternative schema using unaries.
   (d) People may now be classified as any or all of the following: lecturer, student, driver. How would you best model this?

4. (a) Consider the fact types Employee works for Department; Employee heads Department. Each employee works for exactly one department and heads at most one department (the department he/she works for). Each department has workers and at most one head. Schematize this using the two fact types given.
   (b) Set out an alternative schema that contracts the heads fact type to a unary.
   (c) Which schema is preferable? Discuss.

5. The following table indicates where certain objects are placed in three-dimensional space (*x*, *y*, and *z* are the Cartesian coordinates). Only one object can occupy the same position at any given time.
   (a) Schematize this in terms of three binaries.
   (b) Set out an alternative schema using Position as an entity type.
   (c) Set out an alternative schema using AxisType {'x', 'y', 'z'} as an object type. Why is this alternative schema inferior?

Object	*x*	*y*	*z*
A	3	1	0
B	3	1	2
C	0	1	2

14.5 Conceptual Schema Optimization

Previous sections discussed how to transform conceptual schemas to equivalent, or at least acceptable, alternatives. *Conceptual schema optimization* involves transforming a conceptual schema into an alternative conceptual schema that maps to a more efficient (ideally, most efficient) implementation. Section 14.1 gave a simple example of this. Guidelines for performing such optimizations are now discussed.

Four **main factors** to consider when optimizing a conceptual schema are the *target system*, the *query pattern*, the *update pattern*, and *clarity*. The target system is the DBMS or other platform used for implementation. Here, we assume it is a centralized, relational system. The query pattern includes the kinds of questions the system has to answer as well as statistics about the expected frequency and priority of these questions. The update pattern includes the kinds, frequencies, and priorities of the expected insertions, deletions, and modifications to the database.

Response times for queries are usually more vital than for updates, so as a first criterion, we might try to minimize the response times of the "focused queries" (i.e., those queries with high priority or frequency). Clarity here refers to the ease with which the modeler can fully grasp the semantics conveyed by the schema.

Before the optimization procedure is executed, the global conceptual schema should be completed and validated. Any mapping choices (e.g., for 1:1 cases or subtypes) should also be declared. The procedure is then run on the whole schema. The complete optimization procedure for ORM schemas is somewhat complex, so only an overview of its main components is given here.

Although the default procedure could be executed without human intervention, better results may often be obtained by allowing modelers to override a suggested transformation when their additional insights reveal other factors that make the transformation counterproductive.

Several optimization strategies were introduced earlier in the chapter. Overall, the procedure comprises two main stages:

- Transform to reduce the number of mapped tables (steps 1 and 2)
- Transform to simplify individual tables (steps 3–5)

Because slow queries tend to involve joins or subqueries, a default strategy is to try to reduce the number of focused queries that involve joins or subqueries. Steps 1 and 2 of the optimization procedure do this for the main case (two or more tables involved) by transforming to reduce the number of tables involved in these queries. Because updates that require checking constraints between tables also tend to be expensive, this strategy usually improves update performance as well.

Recall that a composite key in a predicate involves a UC spanning two or more roles. Because the mapping algorithm maps predicates with composite keys to separate tables, a basic strategy is to reduce the number of (relevant) compositely keyed fact types in the conceptual schema. There are two main situations with the potential for achieving this: compatible keys that may be unified and nonfunctional roles that may be replaced by functional roles. Let's consider these in turn.

Step 1 aims to unify compatible, composite keys (see Fig. 14.42). As preparation, step 1.1 includes two moves. As an example of its second move, consider a schema with the following fact types: Company in State has staff of Gender {'m', 'f'} in Quantity; Company in State has Budget. Compatible keys based on Company–State are formed by absorbing Gender to specialize the quaternary into two ternaries: Company in State has male staff in Quantity; Company in State has female staff in Quantity.

Steps 1.2 and 1.3 ensure that each pattern involving n compatible, composite keys is replaced by a single predicate ($n \geq 2$). Several examples were discussed in previous sections. Whereas the original pattern maps to n tables, the nested version maps to just one table.

If nesting leads to a loss of clarity in the conceptual schema, the modeler may prefer not to nest. Ideally, the modeler may then choose whether to have nesting performed automatically as an invisible, preprocessing stage to Rmap. Any nested object type may be recast as a coreferenced object type if this seems more natural.

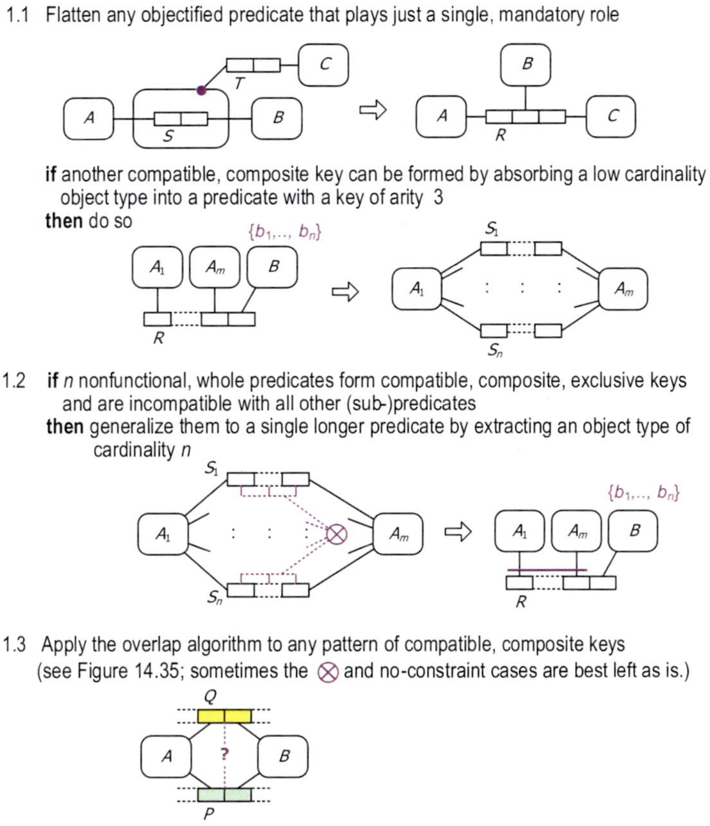

FIG. 14.42 Optimization step 1: unifying compatible, composite keys.

Step 2 examines *object types with both a functional and a nonfunctional role,* and typically attempts to replace the latter by functional roles that can be grouped into the same table as the former (see Fig. 14.43).

Most of step 2.1 amounts to predicate specialization using enumerated object types or frequency constraints. Several examples were discussed earlier. The predicate generalization case with a restricted UC is rare (e.g., recall Fig. 14.10).

The first case of step 2.2 for pattern (a) uses PSG3 with its second corollary (see Fig. 14.20). The second case for pattern (a) is extremely rare and was ignored in earlier transformations.

As an example, consider the fact type Official holds Position {'president', 'secretary', 'treasurer'}. We may transform to three unaries: Official is president; Official is secretary; Official is treasurer. Assuming that each position is held by at most one official, we also need to add a role cardinality constraint declaring $\# \leq 1$ on each unary.

The final stage of step 2.2 (pattern (b)) uses PSG2 and its second corollary. As an example, consider the main and backup driver UoD of Fig. 14.15 and add the functional fact type Car is of CarModel. Specializing the drives predicate enables all the information to be mapped to a single Car table instead of a Driver and a Car table.

2.1 For each case where an object type *A* has simple and binary keys attached, where the other role of the binary key is played by *B*:

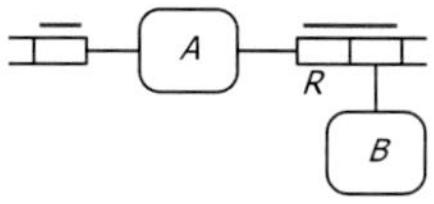

if *R* is a binary
then if *B* has values $b_1,\ldots, b_n$ (and *n* is small)
    then specialize *R* to *n* unaries **or** replace *B* by *B'* {b_1, b_2, both} (if *n* = 2) making the key(s) simple
    else if *A*'s role in *R* has a frequency constraint *n* or *n* (and *n* is small)
        then specialize *R* into *n* exclusive binaries simply keyed on *A*
        else if *A* has just one functional role
            and the predicates are compatible, pair-exclusive binaries
            then generalize both to a single ternary with a restricted UC
else -- *R* is a ternary
    if *B* has values $b_1,.., b_n$ (and *n* is small)
    then absorb *B*, specializing *R* into *n* binaries simply keyed on *A*.

2.2 For each case where an object type *A* has a functional role attached, as well as a 1:*n* binary predicate *R* connected to object type *B*:

(a)
(b)

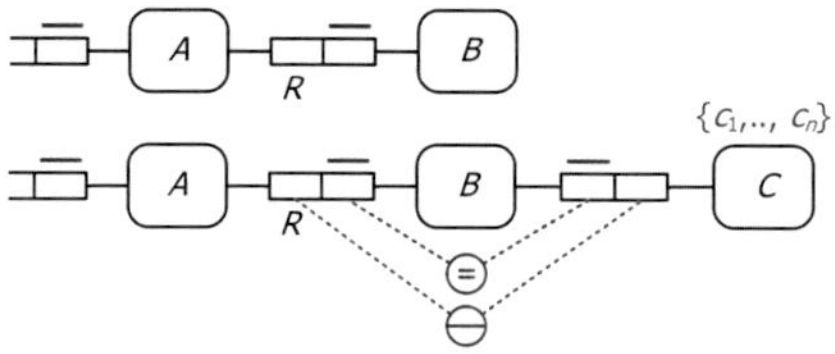

if *B* has no other functional roles (see (a))
then if *A*'s role in *R* has a frequency constraint of *n*, or *n* (and *n* is small)
    then specialize *R* into *n* 1:1 binaries with *B*'s roles mutually exclusive
    else if *B* has values $b_1,.., b_n$ (and *n* is small)
        then consider specializing *R* into *n* unaries
else if *B* has exactly one more functional predicate, linked to *C* {$c_1,.., c_n$}
    and an equality constraint spans *B*'s functional roles
    and an external UC spans *B*'s co-roles (see (b))
    then specialize *R* into *n* 1:1 binaries with *B*'s roles exclusive, by absorbing *C*
        (this is PSG2 with corollary 2: see Figure 14.16)

FIG. 14.43 Optimization step 2.

As an exercise, draw the schemas and perform the mappings. Step 2.2 completes the optimization strategy to reduce the number of mapped tables.

Before going on to step 3, let's consolidate the first two steps by considering an example for which substantial optimization is possible. Fig. 14.44 depicts a conceptual schema for postgraduate coursework in a computer science department. This UoD was first introduced in Exercise 4.2.

The pair-subset constraint reflects the fact that after students enroll in a subject, they choose a topic for it. Each topic has a lecturer in charge, and at most two (other) colecturers.

As an exercise, map the schema in Fig. 14.44 to a relational schema and then check your solution with the one in Fig. 14.45.

If you did this correctly, you should have obtained six tables. Notice the intertable constraints. The "⊙" superimposed on "⊗" in the relational schema denotes a partition of

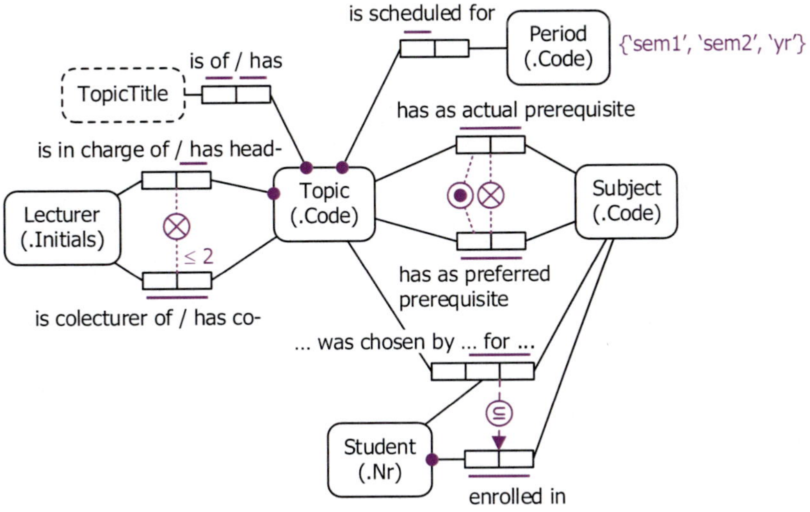

FIG. 14.44 This suboptimal schema maps to six tables.

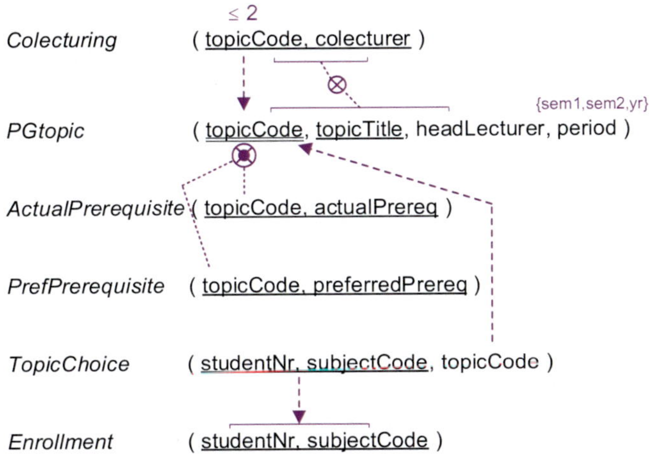

FIG. 14.45 The relational schema mapped from Fig. 14.44.

PGtopic.topicCode (i.e., PGtopic.topicCode is the disjoint union of ActualPrerequsite.topicCode and PrefPrerequisite.topicCode). The ordinary "⊗" indicates that the topic-lecturer pairs in Colecturing and PGtopic are exclusive. The other constraints are straightforward.

Suppose a focused query for this domain is the following: list all topics (code and title) together with their main lecturers, colecturers, actual prerequisites, and preferred prerequisites. Let another focused query be the following: list students, the subjects in which they are enrolled, and the topics chosen by them (if any) for those subjects. With the relational schema of Fig. 14.45, the first query involves a join of four tables, and the second query involves a composite outer join of two tables. So, these queries will be slow to run.

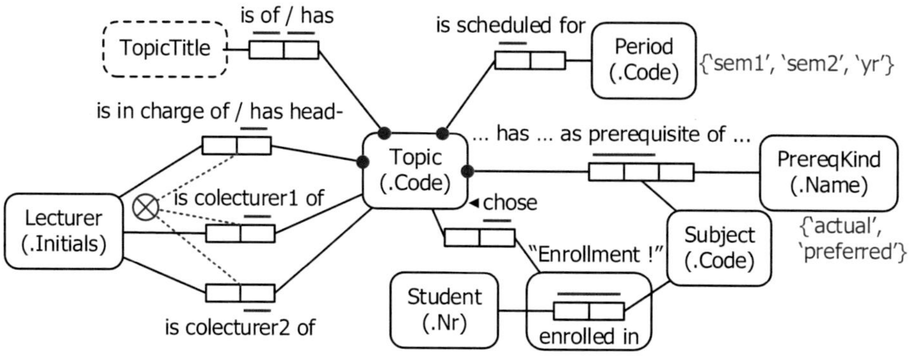

FIG. 14.46 The conceptual schema obtained by optimizing Fig. 14.44.

Applying optimization steps 1 and 2 to the conceptual schema of Fig. 14.44 results in three main changes, as shown in Fig. 14.46.

Step 1.2 generalizes the two prerequisite binaries to a ternary by extracting the object type PrereqKind {"actual", "preferred"}. Step 1.3 uses the overlap algorithm to nest the enrollment and topic fact types. Step 2.1 specializes the colecturer predicate into two exclusive binaries simply keyed on Topic. If we also demand that a topic has colecturer2 it must have a colecturer1, then this may be added as a subset constraint. As an exercise, Rmap the schema in Fig. 14.46, then check your solution with that shown in Fig. 14.47.

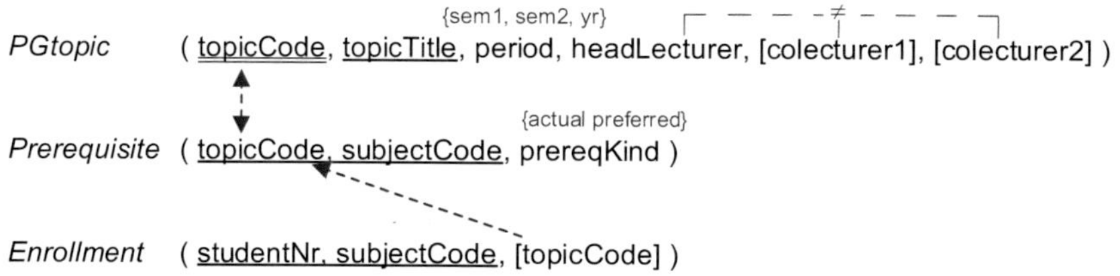

FIG. 14.47 The relational schema mapped from Fig. 14.46.

The optimized relational schema has just three tables: one for the ternary, one for the nesting, and one for the five functional binaries of Topic. The "≠" symbol means that on the same row, any nonnull values for headLecturer, colecturer1, and colecturer2 must differ. This relational schema leads to much faster execution of the focused queries as well as simpler constraint enforcement on updates.

We now consider the final steps (3, 4, 5) in the optimization procedure. Unlike the previous steps, these do not reduce the number of tables. *Step 3* is set out in Fig. 14.48. This aims to reduce self-joins and is called the *table width guideline* (TWG) because it indicates when a wider table (more columns) may be preferable.

The schema equivalence illustrated was considered earlier. If the object type *A* plays a functional role, the transformation to binaries would have already occurred in the last phase of step 2.1. Even if *A* has no functional role, the specialization into binaries may still be worthwhile if it avoids self-joins in focused queries.

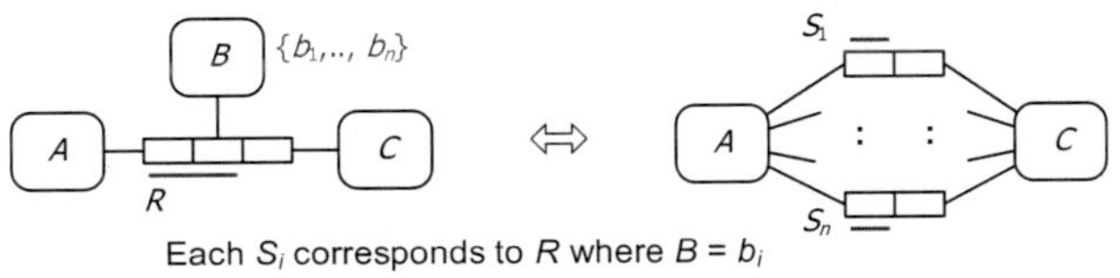

Each S_i corresponds to R where $B = b_i$

3 For the ternary pattern above, if focused queries on *R* require comparisons between different tuples, and the values $b_1, ..., b_n$ are few and fairly stable, then absorb *B*, specializing *R* into *n* functional binaries.

FIG. 14.48 Optimization step 3: the table width guideline.

As an example of step 3, suppose we wish to see how full-time university fees varied for various degrees over the triennium 2021–2023. Let's model this with the ternary Degree in Year had FulltimeFee with a uniqueness constraint spanning the first two roles, a frequency constraint of 3 on the first role, and a value constraint of {2021…2023} on Year's role, as shown in Fig. 14.49. Suppose we map this to the relational schema *FulltimeFee* (degree, feeYear, Fee), with the frequency and value constraint enforced at the relational level.

A fragment of a sample population for this table is shown in Fig. 14.49. Let a sample focused query be, How much did the BS full-time fee increase over the triennium? Answer this query yourself by inspecting the table.

In arriving at the answer ($500), you compared values from two different rows from the same table. To perform this query in a relational database requires joining the table to itself (a self-join). An SQL version of the query is included in Fig. 14.49.

FulltimeFee:

degree	*feeYear*	*fee*
BS	2021	5000
BS	2022	5000
BS	2023	5500
MBA	2021	7000
MBA	2022	7300
MBA	2023	7700

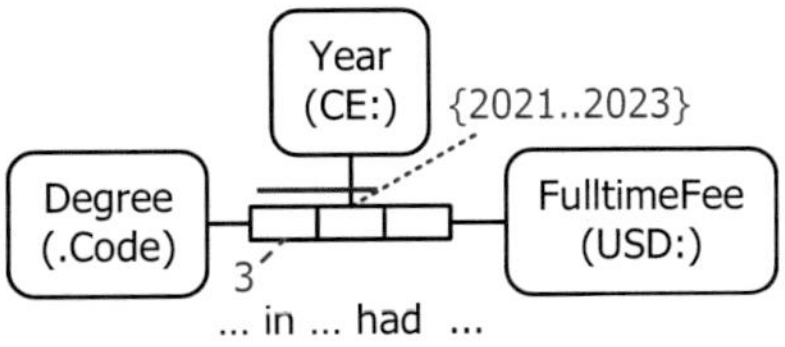

How much did the BS degree fulltime fee increase over the 3 years?

SQL code for query:

```
select New.fee - Old.fee
from   FullTimeFee as Old cross join FulltimeFee as New
where  Old.feeYear = 2021 and Old.degree = 'BS' and
       New.feeYear = 2023 and New.degree = 'BS'
```

FIG. 14.49 This query requires a self-join because it compares different rows.

Such joins can be avoided by first transforming the schema via TWG. Applying a variant of PSG1 where the value constraint applies to a *role* instead of an object type, the ternary specializes to the three conceptual binaries: Degree had 2021- FulltimeFee; Degree had 2022-FulltimeFee; Degree had 2023- FulltimeFee.

Here, the roles of Degree are functional and linked by an equality constraint. If, in the global schema, fee information is mandatory for Degree, these roles are mandatory (which implies the equality constraint). If Degree has no other functional roles, these three conceptual binaries map to a single table: *FulltimeFee* (degree, fee2021, fee2002, fee2023)

If Degree had other functional roles, the optimization would have been performed at step 2.1, with the extra functional fact types mapping to the same table.

Fig. 14.50 shows the conceptual schema, and also illustrates how the wider table enables the query to be performed simply by comparing values on the same row, which is much easier than the previous query.

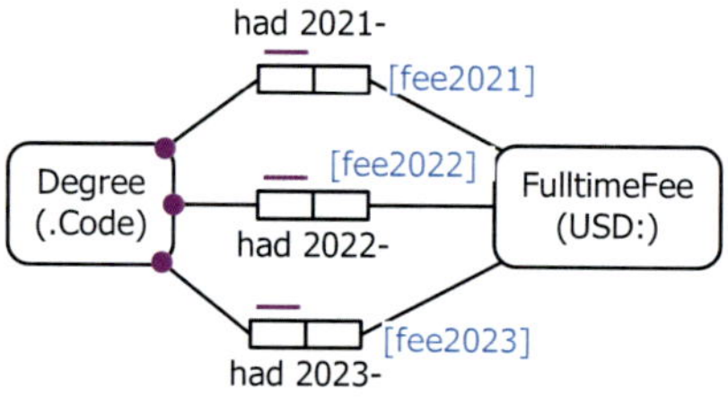

FulltimeFee:

degree	*fee2021*	*fee2022*	*fee2023*
BS	5000	5000	5500
MBA	7000	7300	7700

How much did the BS degree fulltime fee increase over the 3 years?

SQL code for query:

```
select fee2023 - fee2021
from   FullTimeFee
where degree = 'BS'
```

FIG. 14.50 The transformed schema leads to a more efficient query.

This example was trivial and clear-cut, but life is not always so simple. If the fee year range is large, the number of extra columns generated by the transformation may be too high. For example, suppose that we must record fees for the period 1984...2023. Specializing the ternary would generate 50 binaries and lead to a relational table with 51 columns: *FulltimeFee*(degree, fee1984, ..., fee2023).

Such a large number of binaries would clutter the conceptual schema (although a flexible CASE tool could be set to display the ternary with the optimization hidden in the mapping). The very wide relational table would also be awkward to view (lots of horizontal scrolling) and to print. For these reasons, once the *cardinality* of the value constraint exceeds a reasonable number (e.g., 5), the ternary fact type might well be preferred.

The *stability* of the value constraint is also important. In many cases, such a constraint is known to be stable (e.g., {"m", "f"} for gendercode). Sometimes, however, the value constraint changes with time. For example, suppose fees are to be recorded for all years from 2001 to the current year, or for just the most recent triennium. The binary approach requires changes to the fact types and relational columns each year. In this case, the ternary approach might be preferred because no structural change is needed (the year changes are made only to rows of data, not the table structure itself).

The wording of optimization step 3 reflects this trade-off between query efficiency and the ease of table display and schema evolution. Although default choices can be built into an automated optimizer, interaction with human modelers is advisable to ensure the best trade-off.

The final steps (4 and 5) of the optimization procedure deal with exclusive roles (see Fig. 14.51).

4. Generalize each group of *n* exclusive unaries ($n \geq 2$) to a functional binary by extracting an object type with *n* values.

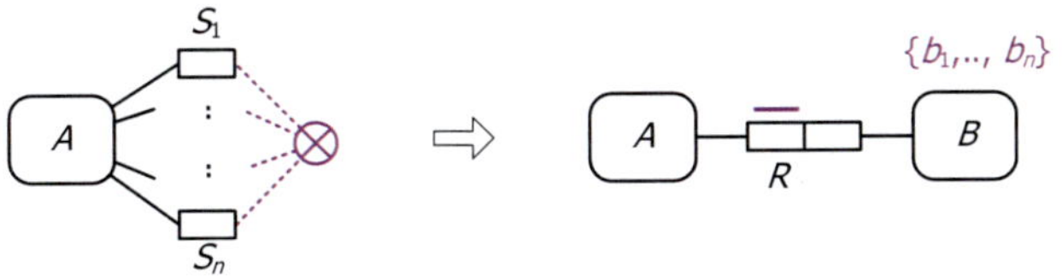

5. **if** *A* plays exclusive, functional roles in *n* *1:many* predicates with *C*,
and *C* has no other functional role
then replace these *n* predicates by two binaries, by extracting *B* $\{b_1, \ldots, b_n\}$ as per the LHS of PSG2.

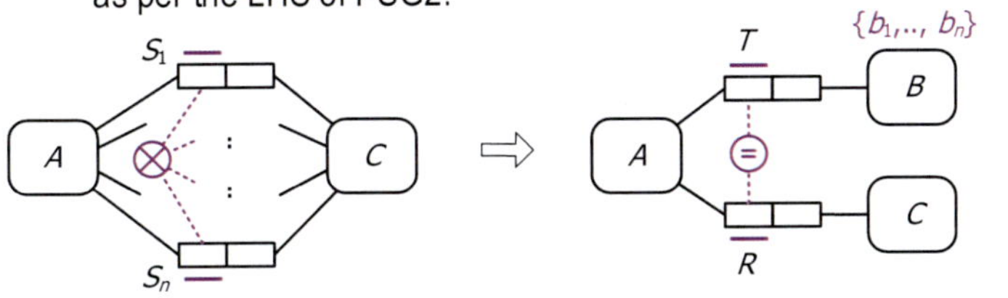

FIG. 14.51 Optimization steps 4 and 5: further aspects of exclusive roles.

Step 4 is straightforward. Different unaries attached to the same object type map to different columns of the same table. If these unaries are exclusive, we may replace them all by a functional binary that maps to just one column. Moreover, the transformation of the exclusion constraint to a uniqueness constraint enables it to be simply enforced by the primary key constraint of the table. Recall Fig. 14.7 where three exclusive unaries (Employee is a manager; Employee is a clerk; Employee is a secretary) are replaced by the functional binary Employee has Rank {'M', 'C', 'S'}.

Step 5 helps us decide when to favor predicate generalization in applying theorem PSG2. This is illustrated, along with other optimizations, by our next example. Consider the conceptual schema shown in Fig. 14.52.

The interpredicate uniqueness constraint in Fig. 14.52 asserts that a team may have only one player of each sex. This conceptual schema maps to seven tables. As an exercise, you might like to check this for yourself and then optimize the conceptual schema before reading on.

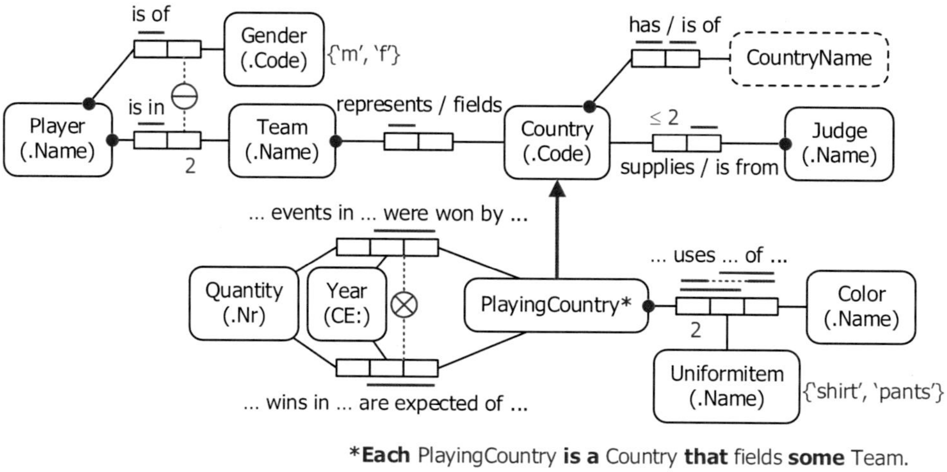

FIG. 14.52 This unoptimized conceptual schema maps to seven tables.

The overlap algorithm (step 1.3) applied to the bottom ternaries of Fig. 14.52 generates the intermediate, nested subschema in Fig. 14.53. This is then transformed (step 5) to the nested pattern included in the optimized conceptual schema (Fig. 14.54). This modification to the nesting does not change the number of tables or columns, but it does simplify the constraint pattern that has to be enforced.

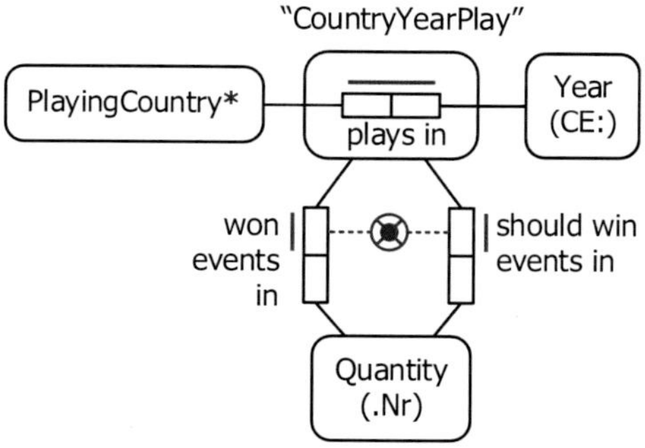

FIG. 14.53 An intermediate transform of the bottom left ternaries of Fig. 14.52.

At the final stage of step 2.1, the subtype PlayingCountry in Fig. 14.52 has a functional role attached (via its supertype) and a role within a composite UC, so the two-valued UniformItem is absorbed to specialize the ternary into two functional binaries. Note the constraint pattern arising from the application of corollaries.

At the first stage of step 2.2, the frequency constraint on the supplies predicate is used to specialize it into two functional binaries. The second half of step 2.2 is used to absorb the object type Gender, specializing the team membership predicate into binaries in which Team now has functional roles that may be mapped to the same table as Team's other functional fact type. The inclusive-or component of the exclusive-or constraints on the Player and Judge roles is implied in the global schema, so is ignored in the relational

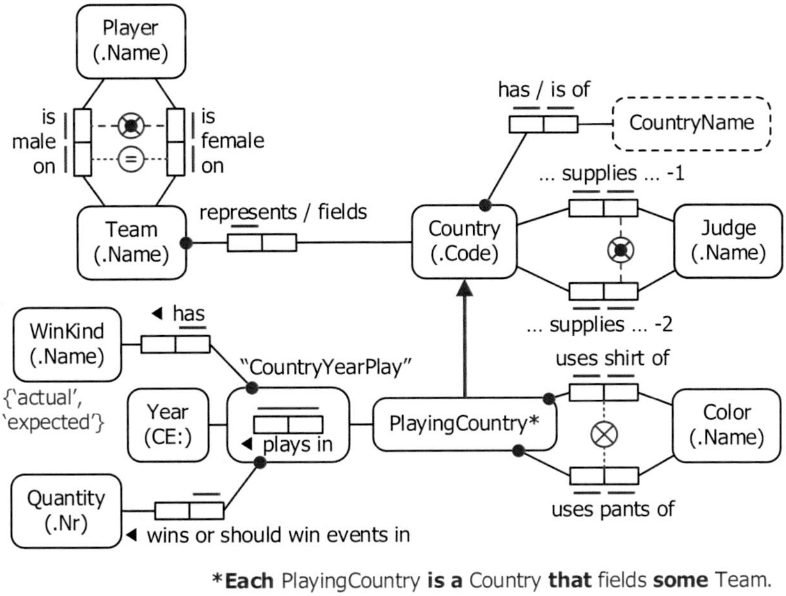

FIG. 14.54 This optimized conceptual schema maps to three tables.

mapping. Steps 3 and 4 are not invoked by this example. Step 5 transformed the original nesting to another, as discussed earlier.

This completes the basic optimization procedure. Generation of all the constraints in the optimized version is determined rigorously by the underlying transformation theorems and their corollaries. In the absence of such formal grounding, it would be easy to lose some constraints in transforming such a schema.

Although UML and commercial versions of ER are useful for summarizing schemas, they typically lack the required detail (e.g., disjunctive mandatory or pair-exclusion constraints) to illustrate many complex transformations, unless supplemented by textual constraints. The optimized conceptual schema maps to just three tables, as shown in Fig. 14.55, thus reducing the number of tables by four.

FIG. 14.55 The relational schema mapped from Fig. 14.54.

Exclusion between column sets is shown by "⊗". The weaker constraint of exclusion between values on the same row is shown by "≠". Pairing attributes in the same square brackets entails that any qualification applies to each and that if one is null so is the other (this constraint is implied for the color attributes by their qualification).

The schema optimization procedure discussed here does not include decisions about how subtypes and symmetric 1:1 fact types should be mapped because these do not change the conceptual schema itself. As discussed in Chapter 11, however, these mapping choices can have a significant impact on the efficiency of the relational schema obtained.

Recall that the optimization procedure assumes that the target system is a relational DBMS, at least in the sense that each entry in a table column is either an atomic value or a null. If multivalued fields are permitted (as in SQL:2003 onward), other designs may be used to avoid table joins, leading to further possible optimizations (see Section 14.7).

Exercise 14.5

1. Consider a conceptual schema in which the only fact type associated with Department (code) is Department at Level {UG, PG} has students in Quantity. Here, "UG" and "PG" are abbreviations for "undergraduate" and "postgraduate", respectively. The role played by Department has a frequency constraint of 2. Let a focused query be, What is the ratio of postgraduate to undergraduate enrollments for the department of computer science?
   (a) Draw the conceptual fact type, and Rmap it.
   (b) Optimize the conceptual (sub)schema, and Rmap it.
   (c) Although primarily identified by its code (e.g., "CS"), each department also has a unique name (e.g., "Computer Science") that is now mandatorily recorded.
   (d) The enrollment figures are now optional, but if any figures are recorded, both UG and PG figures are required. Draw the new conceptual schema and Rmap it.
   (e) Optimize the new conceptual schema, and Rmap it.

2. The following conceptual schema deals with applicants for positions as astronauts. Applicants are given ability tests (C=Cognitive, A=Affective, P=Psychomotor) and their performance on various tasks is also measured.

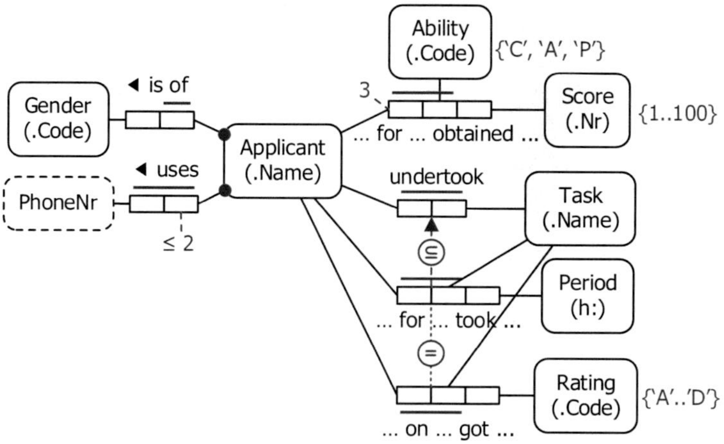

(a) Rmap this.
(b) Optimize the conceptual schema.
(c) Rmap your optimized conceptual schema.

3. Consider the following conceptual schema (see next page). Assume that the Project-Programmer-Class subschema may be treated simply in snapshot fashion (i.e., only current data for these fact types are recorded).
   (a) Rmap this.
   (b) Optimize the conceptual schema.
   (c) Rmap your answer to (b).
   (d) As indicated in the original conceptual schema, each project must normally have a junior and a senior programmer. However, suppose now that while, or after, working on a project, a programmer may be promoted or even demoted in class (junior, senior). Discuss briefly how you would deal with this situation from a practical business standpoint and what changes, if any, you would suggest for the original and optimized conceptual schemas.

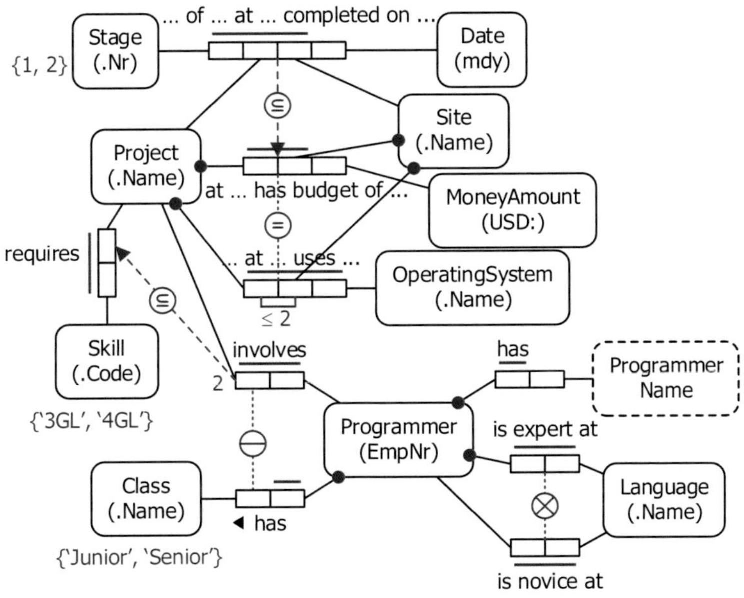

14.6 Normalization

The process of mapping from a conceptual schema expressed in atomic fact types to a relational schema involving table types is one of **deconceptualization**. This approach facilitates sound design primarily because it emphasizes working from examples, using natural language, and thinking in terms of the real-world objects being modeled.

The advantages of using *elementary facts* are clear: working in simple units helps get each one correct; constraints are easier to express and check; nulls are avoided; redundancy control is facilitated; the schema is easier to modify (one fact type at a time); and the same conceptual schema may be mapped to different logical data models (so decisions about grouping fact types together may be delayed until mapping).

Nowadays, many database designers do a conceptual design first, whether it be in ORM, entity-relationship (ER) modeling, or UML notation, and then apply a mapping procedure to it. In the past, in place of conceptual modeling, many relational database designers used a technique known as **normalization**. Some still use this, even if only to refine the logical design mapped from the conceptual one.

If an ORM conceptual schema is correct, then the table design resulting from Rmap is already fully normalized, so ORM modelers don't actually need to learn about normalization to get their designs correct.

However, a brief look at normalization is worthwhile, partly to consolidate aspects of good design and partly to facilitate communication with other modelers who use the normalization approach. The rest of this section provides an overview of the normalization procedure.

There are two techniques for normalization: *synthesis* and *decomposition* (sometimes called "analysis"). Each operates only at the relational level and begins with the complete set of relational attributes as well as a set of dependencies on these, such as functional dependencies (FDs). Recall that if *X* and *Y* are simple or composite attributes (sets of one or more column headers), we say that *X functionally determines Y* (written $X \rightarrow Y$) if and only if for each value of *X* there is at most one value of *Y*.

In any realistic business domain, it is easy to get an incorrect model if we begin by simply writing down a list of attributes and dependencies because the advantages and safeguards of conceptual modeling are removed (e.g., validation by verbalization and population). Moreover, the set of dependencies considered in normalization is usually very limited, so that many UoD constraints are simply ignored.

The output of both synthesis and decomposition techniques is a set of table schemes, each of which is guaranteed to be in a particular "normal form". We will have a closer look at normal forms presently. The synthesis algorithm basically groups attributes into tables by finding a minimum, reduced, annular cover for the original dependencies. Redundant FDs (e.g., transitive FDs) are removed, redundant source attributes (those on the left of a "→") are removed, and FDs with the same source attribute are placed in the same table. With respect to its specified class of dependencies and attributes, the synthesis algorithm guarantees a design with the minimum number of tables, where each table is in elementary key normal form (see later for a definition of this concept).

For example, given the FDs {empNr → gender, empNr → birthdate, empNr → job, job → salary, empNr → salary}, synthesis generates two tables: *R1*(empNr, gender, birthdate, job); *R2*(job, salary). While this example is trivial, for more complex cases the execution of the synthesis algorithm is arduous and is best relegated to a computer. It should be apparent that the derived fact type checks in the CSDP, and the basic fact type grouping in Rmap, have a close correspondence to the synthesis method.

Although synthesis has often been taught in academia, it is rarely used by practitioners, partly because of its highly technical nature, but mainly because it is almost useless in practice. The main problem with synthesis is that it assumes for its input a correct global set of attributes and dependencies. In practice, however, such an input is rarely available, and determining such an input is often harder than determining the correct design by other means. For such reasons, we essentially ignore the synthesis technique from this point on.

Another problem with normalization (by synthesis or decomposition) is that it can't change the set of attributes. Hence, apart from ignoring transformations that involve other kinds of constraints, normalization misses semantic optimization opportunities that might arise by allowing transformations to change attributes.

For example, the FD set {department → budget, (department, gender) → nrStaff} synthesizes to two table schemes, *R1*(department, budget) and *R2*(department, gender, nrStaff). But normalization is unable to transform the FD set into {department → budget, department → nrMaleStaff, department → nrFemaleStaff}, which would lead to the single table design *R*(department, budget, nrMaleStaff, nrFemaleStaff). Compare this with the ORM schema optimization approach just discussed.

Such considerations indicate that the use of normalization alone is inadequate as a design method. Nevertheless, the use of normalization as a check on our conceptual modeling may at times be helpful. On the positive side, by limiting their scope to a small but important class of constraints, and ignoring semantic transformations, normalization theorists have established several interesting results.

Unlike synthesis, the decomposition approach to normalization is an iterative one, progressively splitting badly designed tables into smaller ones until finally they are free of certain update anomalies, and it provides no guarantee of minimality of number of tables. However, the decomposition approach is more well known, and some modelers find its principles useful for correcting poor relational designs. The rest of this section focuses on the decomposition approach, with some comparisons to ORM. To assist the reader who may read this section out of sequence, some basic concepts treated earlier are reviewed briefly.

A relational schema obtained by Rmapping it from a correct ORM conceptual schema is already fully normalized. Suppose, however, that some error was made in either the conceptualization or the mapping, or that the relational schema was designed directly without using a conceptual schema. It is now possible that some of the table designs might be unsafe because some conceptual constraint is not enforced. If a conceptual constraint is violated when a table is updated by inserting, deleting, or modifying a row, the table design is said to contain an *update anomaly*.

Constraints on conceptual reference types are the responsibility of humans. Only people can ensure that preferred reference schemes actually do provide a correct 1:1-into map from real-world entity sets to sets of data values (or data tuples). Special care is required when an entity type plays just one fact role, this role is not functional, and an instance of this entity type changes its preferred identifier.

For example, suppose City occurs only in the *m:n* fact type RussianTour(.Nr) visits City (.Name). In the early 1990s, the city then known as "Leningrad" reverted to its original name "St. Petersburg". As an exercise, discuss some problems arising if only some instances of "Leningrad" are renamed.

Assuming preferred reference schemes are already enforced, the information system itself should normally enforce all the fact type constraints. So all constraints on (and between) conceptual fact types should be captured in the relational schema. The normalization procedure ensures that most of these constraints are so captured. In particular, it aims to remove any chance of redundancy (repetition of an elementary fact). If a fact were duplicated in the database, when the fact changed it would be necessary to update every instance of it; otherwise, the database would become inconsistent.

Starting with a possibly bad table design, the normalization procedure applies rules to successively refine it into higher "normal forms" until it is fully normalized. Let's confine our attention to the normal forms that focus on eliminating problems with a table by splitting it into smaller ones. There are many such normal forms. In increasing order of acceptability, the main ones are first normal form (1NF), second normal form (2NF), third normal form (3NF), elementary key normal form (EKNF), Boyce/Codd normal form (BCNF), fourth normal form (4NF), fifth normal form (5NF), and domain key normal form (DKNF). We'll discuss each of these in turn.

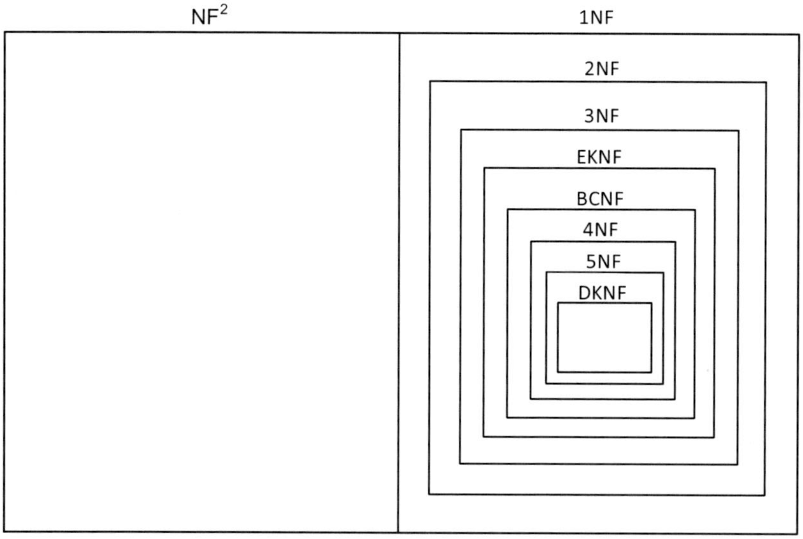

FIG. 14.56 Classification of tables based on their level of normalization.

A table that is not in first normal form is said to be "unnormalized" or in *Non-First Normal Form* (NFNF = NF^2). A table in a higher normal form is also in its lower normal forms (e.g., a 3NF table is also in 2NF and 1NF). Hence, tables may be classified as shown in Fig. 14.56. A table's *highest normal form* (hNF) is its deepest normalization level.

The first three normal forms were originally proposed by E. F. "Ted" Codd, who founded the relational model of data. The other forms were introduced later to cater for additional cases. These improvements were due to the work of several researchers, including Codd, R. Boyce, R. Fagin, A. Aho, C. Beeri, and J. Ullman. Other normal forms have also been proposed (e.g., horizontal normal form and nested normal form) but are not discussed here.

A table is normalized, or in **first normal form** (**1NF**), if and only if its attributes are single valued and fixed. In other words, a 1NF table has a fixed number of columns, and each entry in a row-column position is a simple value (possibly null). Almost every table we've seen so far in this book has been in 1NF.

A table may violate 1NF (and hence be an NF^2 table) either by having multivalued attributes (e.g., entries that are a collection such as sets, arrays, or even relations) or by allowing different rows to have different numbers of attributes.

As an example of the first kind of violation, consider Table 14.7. While the first two columns in this Student table store only atomic values, the Car column stores sets of values, and the Result column holds sets of (course, rating) pairs.

To model this in a relational database, we need to flatten the structure out into three tables: *Student*(studentNr, gender); *Drives*(studentNr, car); and *Scores*(*studentNr, course, rating*). Hence, 1NF relations are often called *flat relations*.

Table 14.7 A nested-relational structure.

Student:

studentNr	*gender*	*Car*	*Result*	
		regNr	*course*	*rating*
1001	F	ABC123	CS100	6
		PKJ155	CS114	7
			PD102	6
1002	M		CS100	6
			MA104	5

However, it is possible to model the structure directly by using a relation that is not in first normal form. For example, we might use the following *nested relation* scheme: *Student* (studentNr, gender, *Car*(regNr), *Result*(course, rating)).

Here, an entry in the third column is itself a relation (here, a set of car registration numbers). Similarly, each entry in the Result column is a relation (with course and rating attributes). Because the logical data model underlying this approach allows relations to be nested inside relations, it is called the *nested relational model*.

Because the nested table is not in 1NF, it is not a relational table (i.e., it does not conform to the relational model of data). Although pure nested relational systems have yet to make any significant commercial impact, many of the major relational DBMSs have evolved into object-relational systems with support for NF^2 tables, as discussed in the next section. While tables not in 1NF might be in some other exotic "normal form", it is common practice to refer to them as being "unnormalized".

Set-valued fields such as Car in Table 14.7 are sometimes called “repeating attributes”, and fields such as Result that hold sets of grouped values are sometimes called “repeating groups”. This term may be used to include the notion of repeating attribute by allowing degenerate groups with one value. Using this somewhat misleading terminology, we may say that 1NF tables have no repeating groups.

Another example of an unnormalized table is the “variant record type” used in languages such as Pascal and Modula. Here, different “rows” in the same record type may contain different fields. For example, the figure record type in Fig. 14.57 is designed to store details about colored geometric figures. A sample data population is included. The record type includes common fields for figureId, color, and shape (circle, rhombus, triangle).

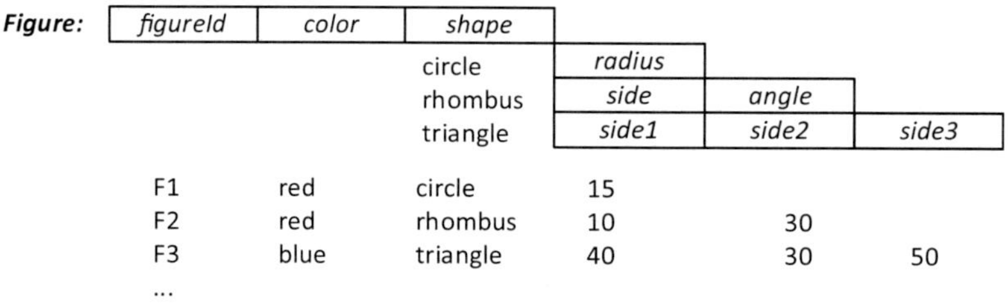

FIG. 14.57 A variant record type, with a sample population.

The record type has different remaining fields depending on the shape (radius for circle; side and angle for rhombus; three sides for triangle). This provides one way to implement a restricted notion of subtypes without using subclasses, as is common in modern object-oriented programming languages.

A *key* of a table is a set of one or more attributes spanned by an explicit uniqueness constraint. Here, “explicit” excludes any UCs that are implied by shorter UCs inside them; so, keys are spanned by “minimal” UCs. For a given relation, an attribute is a *key attribute* if and only if it belongs to some key (primary or secondary) of the relation. A *nonkey* attribute is neither a key nor a part of a composite key.

For example, consider the relation scheme *City*(cityNr, cityname, statecode, population, [mayor]). Here, cityNr is the primary key and (cityname, statecocde) is an alternate key. So cityNr, cityname, and statecode are key attributes. Although unique, the mayor attribute is not a key because it is optional. Because they don’t belong to a key, population and mayor are nonkey attributes.

Although definitions of normal form and related dependencies are often given in terms of relations (fixed sets of tuples), we should define them in term of *relation schemes* because we want the table structures to exhibit the desired quality for each possible population (this requires revised definitions for some of the higher normal forms). With this in mind, we define an FD over a relation scheme as follows. Given attributes X and Y of a relation scheme, X functionally determines Y ($X \rightarrow Y$) if and only if, *given any possible population* of the table scheme, for each value of *X* there is only one value for *Y*. Attributes *X* and *Y* may be composite. To decide whether a population is possible for a business domain requires the understanding of a domain expert (an informal semantic issue, not a syntactic issue).

A table scheme is in **second normal form** (**2NF**) if and only if it is in 1NF and every nonkey attribute is (functionally) dependent on the whole of a key (not just a part of it). If a table is normalized, it is in at least 1NF. For example, the table in Fig. 14.58(a) is in 2NF because gender and birthdate are both functions of student number. Here, the FDs from studentNr are depicted as arrows.

The table in Fig. 14.58(b) is in 1NF but not 2NF because height, a nonkey attribute, is functionally dependent on athletename, which is just part of the composite key. Note the redundancy: the fact that Jones E has a height of 180 cm is recorded twice.

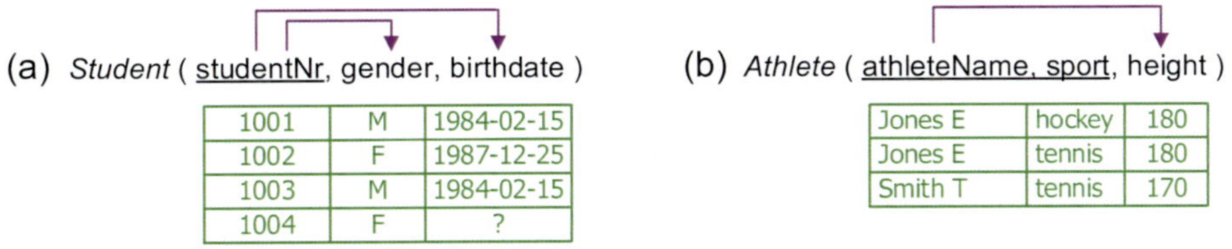

1001	M	1984-02-15
1002	F	1987-12-25
1003	M	1984-02-15
1004	F	?

Jones E	hockey	180
Jones E	tennis	180
Smith T	tennis	170

FIG. 14.58 Table (a) is in 2NF but table (b) is not.

To avoid redundancy, we split the original Athlete table into two smaller ones, as in Fig. 14.59. This figure shows both conceptual and relational schemas for the example (assuming that the information is mandatory for athletes).

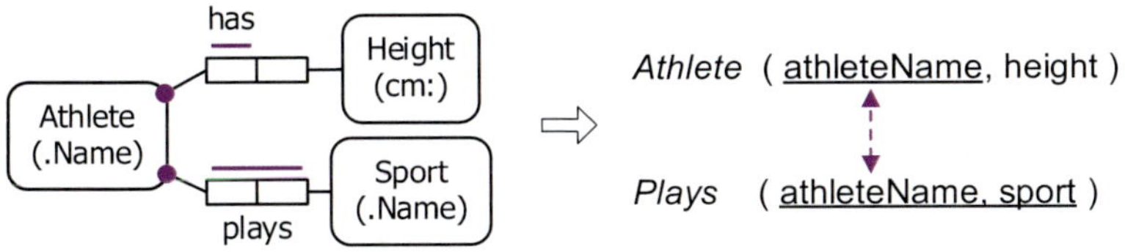

FIG. 14.59 Conceptual schema and correct relational schema for Fig. 14.58(b).

If you verbalized Fig. 14.58(b) correctly in terms of elementary facts, you would obtain this conceptual schema directly, and the relational schema would follow automatically using Rmap.

You would have to be a real novice to verbalize Fig. 14.58(b) in terms of a ternary, but suppose you did. The checks provided in the CSDP, especially steps 4 and 5, prompt you to split the ternary into two binaries.

If you were careless in applying the CSDP, you could end up with the ternary in your relational schema. At this stage, you could look for an FD coming from only part of a key, and use the 2NF normalization rule to at last correct your error. Even at this level, you should think in terms of the functional fact type behind the FD.

Recall that if fact types are elementary, all FDs are implied by UCs. This property is preserved by Rmap because it groups together only functional fact types with a common key. So, if we find an FD in a relational schema that is not implied by a UC, we must have gone wrong earlier. Because UCs imply FDs, every nonkey attribute is functionally dependent on any key of the relation. Note that to determine the hNF of a relation, we need to be told what the relevant dependencies are (e.g., keys and FDs) or have a significant population from which these may be deduced.

Although normalization to 2NF overcomes the redundancy problem in the original Athlete table, this reduces the efficiency of those queries that now have to access both the new tables. For example, if we want the name and height of all the hockey players, the 2NF design means that two tables must be searched and their athleteName fields matched. This kind of efficiency loss, which is common to each normalization refinement (each involves splitting), is usually more than offset by the higher degree of data integrity resulting from the elimination of redundancy.

For example, if we need to change the height of Jones E to 182 cm, and we record this change on only the first row of the original Athlete table, we now have two different values for the height. With more serious examples (e.g., defense, medical, business), such inconsistencies could prove disastrous.

As discussed in the next section, we sometimes denormalize by introducing controlled redundancy to speed up queries. Although the redundancy is then safe, control of the redundancy is more expensive to enforce than in fully normalized tables, where redundancy is eliminated simply by enforcing key constraints. In short, normalization tends to make updates more efficient (by making constraints easier to enforce) while slowing down queries that now require additional table joins.

Given attributes X and Y (possibly composite), if $X \rightarrow Y$, then an update of X entails a possible update of Y. Within a relation, a set of attributes is *mutually independent* if and only if none of the attributes is (functionally) dependent on any of the others. The attributes may then be updated independently of one another.

(a)

Student (studentNr, gender, birthdate)

1001	M	1984-02-15
1002	F	1987-12-25
1003	M	1984-02-15
1004	F	?

(b)

Student (studentNr, birthcountry, population)

1001	US	340265309
1002	AU	26477504
1003	US	340265309
1004	CA	38829398

FIG. 14.60 Table (a) is in 3NF; table (b) is in 2NF but not 3NF.

A table scheme is in **third normal form** (**3NF**) if and only if it is in 2NF and its *nonkey attributes are mutually independent*. Hence, in a 3NF table no FD can be transitively implied by two FDs, one of which is an FD between two nonkey attributes. For example, the table in Fig. 14.60(a) is in 3NF because there is no dependency between gender and birthdate.

However, the table in Fig. 14.60(b) is not in 3NF because population depends on birthcountry, and these are both nonkey attributes. Note that population is transitively dependent on the key (studentNr determines birthcountry, and birthcountry determines population). This table exhibits redundancy: the fact that the United States has a population of 301,139,947 is shown twice. To avoid this redundancy, the fact type underlying the FD between the nonkey attributes is split off to another table, as in Fig. 14.61, which shows both ORM and relational schemas for the example.

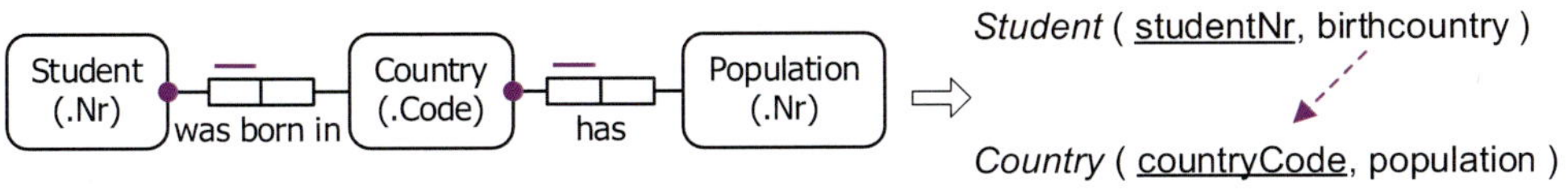

FIG. 14.61 Conceptual schema and correct relational schema for Fig. 14.60(b).

Now consider the table in Fig. 14.62, which is a modified version of an example used in Section 5.5 to illustrate the logical derivation check in the CSDP. If this CSDP step has been carried out properly, FDs from one nonkey attribute to another cannot arise in any tables obtained from Rmap. In the original Section 5.5 example, we split this table into two, one for the fact type Lecturer works for Department and one for the fact type Department is located in Building, deriving the fact type Lecturer works in Building from the rule Lecturer works in Building **iff** Lecturer works for **a** Department **that** is located in Building.

Lecturer (surname, department, building)

Halpin	CS	69
Okimura	JA	1
Orlowska	CS	69
Wang	?	1
Zelewski	?	2

FIG. 14.62 Are facts underlying the FD surname → building transitively derivable?

However, if the "**only if**" part of the "**iff**" in this biconditional is not satisfied, the two-table design loses information. For example, suppose that Wang and Zalewski don't work for any department (e.g., they might be visiting lecturers). This situation is represented by the last two table rows in Fig. 14.62, using "?" to indicate a null. If we split the table in two as described earlier, it would be impossible to derive the facts about where Wang and Zelewski work.

The original work on normalization assumed that nulls do not occur, but this assumption is often false. If we agree with Codd (1990, p. 201) that nulls should be ignored when applying rules about FDs, then the FDs shown in Fig. 14.62 still exist; however, it is wrong to decompose this table into two, so this normalization step can be unsafe. The safest way to deal with such issues is at the conceptual level.

The principle underlying the refinement to second and third normal forms has been nicely summarized by Bill Kent (1983) as follows: "A nonkey field must provide a fact about the key, the whole key, and nothing but the key". This should be refined by replacing the final occurrence of "the key" by "a key".

Codd's work on 2NF and 3NF classified attributes as "prime" or "nonprime" depending on whether they belong to a candidate key. We now use the terms "key attribute" and "nonkey attribute" for these concepts. So, a key attribute belongs to *some* key (primary or alternate), and a nonkey attribute belongs to *no* key.

Hence, in one sense the definition of 3NF does allow transitive dependencies if the intermediate attribute is a key. Consider the table in Fig. 14.63. It has two keys. We could argue that the FD countrycode → population is transitively implied by the FDs countryCode → countryName and countryName → population. Alternatively, we could argue that countryName → population is implied by countryName → countryCode and countryCode → population. However, the relation is still in 3NF as there is no FD between nonkey attributes. The ORM schema is shown at the left of the figure.

AU	Australia	26477504
CA	Canada	38829398
US	United States	340265309

FIG. 14.63 The FD countryCode → population is a harmless transitive dependency.

Let *X* and *Y* be attributes (possibly composite) of the same table. A FD *X* → *Y* is *trivial* if and only if *X* contains *Y*. For example, the table scheme *Student*(studentNr, gender) has six trivial FDs: studentNr → studentNr, (studentNr, gender) → studentNr, etc. A FD *X* → *Y* is *full* (rather than partial) if and only if there is no FD from just part of *X* to *Y*.

A full, nontrivial FD is said to be an *elementary FD*. Consider the table in Fig. 14.64 about populations of US cities (2023 figures). This has three elementary FDs that start at the primary key cityNr, and two elementary FDs that start at the alternate key (cityname, stateCode). Some nonelementary FDs in that table are cityNr → cityNr and (cityName, stateCode) → cityName. We typically avoid displaying nonelementary FDs because they are obviously implied by the elementary FDs.

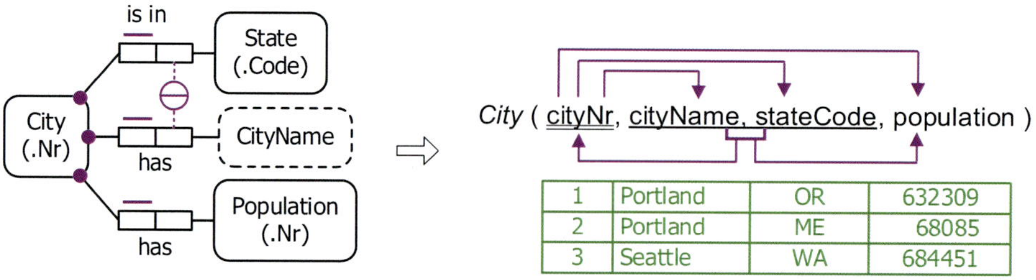

1	Portland	OR	632309
2	Portland	ME	68085
3	Seattle	WA	684451

FIG. 14.64 Five elementary FDs: three from cityNr and two from (cityName, stateCode).

A key is an *elementary key* if there is an elementary FD from it to some attribute in the table. An attribute is an elementary key attribute just in case it belongs to some elementary key. For example, the city table in Fig. 14.64 has two elementary keys, cityNr and (cityName, stateCode). We may now define the next strongest normal form, which was proposed as an improvement on 3NF (Zaniolo, 1982).

A table scheme is in **elementary key normal form** (**EKNF**) if and only if all its elementary FDs begin at whole keys or end at elementary key attributes. In other words, for every full, nontrivial FD of the form $X \rightarrow Y$, either X is a key or Y is (part of) an elementary key. For example, the City table in Fig. 14.64 is in EKNF. If the primary key of a table includes all its columns, the table is automatically in EKNF because it has no elementary FDs; for example: *Enrollment*(studentNr, subjectCode).

Now consider the OfficialLanguage table in Fig. 14.65(a), which records the languages officially used in various countries. Countries may be identified by their codes or by their names. This table has two composite keys: (countryCode, language); (countryName, language). The primary key is indicated by a double underline; because its attributes are not adjacent, a dotted line connector is added to show that this is a single, composite key rather than two simple key constraints. The secondary key is indicated by a single underline.

Because the OfficialLanguage table has no nonkey attributes, it is automatically in 3NF. However, it has obvious problems (e.g., the fact that the country with code "CA" is named "Canada" is duplicated).

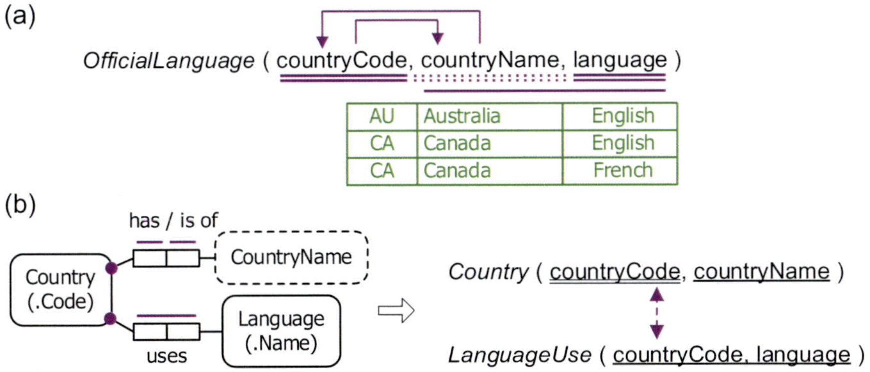

FIG. 14.65 (a) Table in 3NF but not EKNF. (b) Rmap provides a correct schema.

One way to spot this bad design is to note that the table is not in EKNF. There is an FD from countryCode to countryName, and an FD from countryName to countryCode (depicted by arrows above the table). These elementary FDs come from parts of keys rather than whole keys, and they do not end at elementary key attributes because neither of the composite keys is elementary. So, the table is not in EKNF.

The redundancy is avoided by splitting the table into two EKNF tables as in Fig. 14.65(b): *Country*(countryCode, countryName); *LanguageUse*(countryCode, language). The two FDs missed in the original table scheme are now captured by key constraints in the Country table, and the redundancy is eliminated. The same design is obtained by Rmapping the ORM schema shown. If the language fact type is not mandatory for Country, replace the equality constraint in the relational schema by a simple subset constraint from LanguageUse.countryCode to Country.countryCode.

The next normal form is named after Raymond Boyce and Edgar Codd, who proposed it as an improvement on 3NF. An equivalent definition was given earlier by Heath (1971). A table scheme is in **Boyce–Codd Normal Form** (**BCNF**) if and only if *all its elementary FDs begin at whole keys* (i.e., given any full, nontrivial FD $X \rightarrow Y$, it must be that X is a key). For example, the relation scheme *Student* (studentNr, gender, birthdate) is in EKNF because all its elementary FDs begin at the studentNr key. The only time that a relation can be in 3NF (or EKNF) but not in BCNF is when it has at least two candidate keys that overlap (Vincent and Srinivasan 1994).

Now consider the table in Fig. 14.66(a). This is like the previous table, except that the percentage of the country's population that actually uses the language is recorded (the figures here for Canada are fictitious). The keys are now elementary because they functionally determine the percentUsage attribute. Because the FDs between countryCode and countryName now end at elementary key attributes, the table is in EKNF. But the table has the same redundancy as the previous table.

The table is not in BCNF because the FDs shown above the table do not start at whole keys. For instance, the FD countryCode → countryName exists, but countryCode is only part of the (countryCode, language) key. The redundancy is avoided by splitting the table into two BCNF tables, as in Fig. 14.66(b).

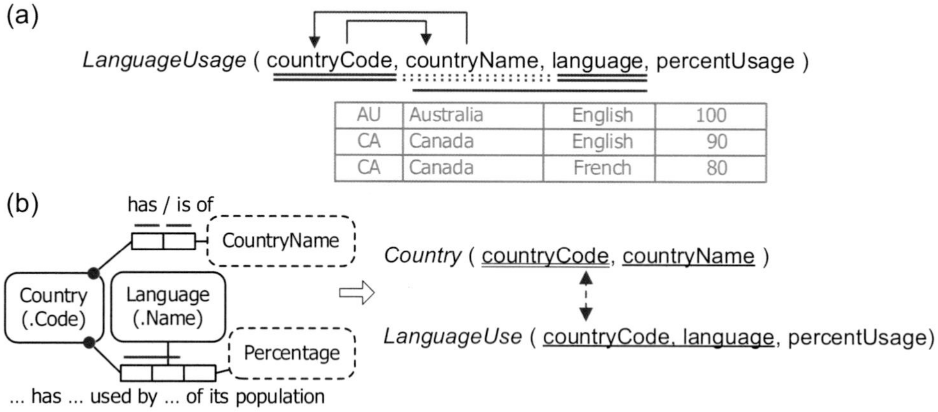

FIG. 14.66 (a) Table in EKNF but not BCNF. (b) Rmap provides a correct schema.

This design is automatically obtained by Rmapping the ORM schema shown. As before, if the language use fact type is optional, change the equality constraint to a subset constraint.

In the previous two examples, a table with overlapping keys had to be split. Sometimes, overlapping keys are permitted. Recall the example from Chapter 4 where students are assigned unique positions in subjects, with no ties. This leads to this BCNF table scheme:

Enrollment(studentNr, subjectCode, position)

The previous normal forms considered only FDs. The next normal form is due to Ronald Fagin (1977) and considers *multivalued dependencies* (MVDs). For a given relation with attributes *X*, *Y*, and *Z* (possibly composite), *X multidetermines Y* (written as $X \twoheadrightarrow Y$) if and only if the *set* of *Y* values is a function of *X* only (independent of the value of *Z*). In this case, *Y* is said to be multivalued dependent, or multidependent, on *X*; in such a case, *Z* will also be multidependent on *X*. A FD is a special case of an MVD, namely when the set of dependent values is a unit set. An MVD $X \twoheadrightarrow Y$ is trivial if and only if *X* includes *Y*, or *X* and *Y* together include all the attributes in the relation.

For example, the Athlete table considered earlier in Fig. 14.58(b) is repeated in Fig. 14.67(a), this time showing the nontrivial MVD athleteName ↠ sport. As the ORM schema in Fig. 14.67(b) indicates, MVDs usually arise from *m:n* associations. Although this poor design would be detected earlier because of its embedded FD (it is not even in 2NF), some relations with MVDs are in BCNF yet still have redundancy problems. To address such cases, we now consider the next normal form.

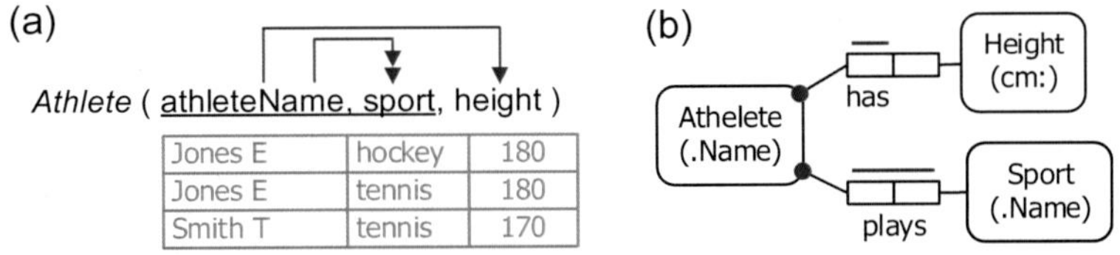

FIG. 14.67 The MVD athleteName ↠ sport in (a) is due to the *m:n* association in (b).

A relation is in **fourth normal form (4NF)** if and only if it is in BCNF and all its nontrivial dependencies are functional (single-valued) dependencies. So, a 4NF relation cannot have any nontrivial MVDs that are not FDs. Basically, each nonfunctional MVD requires a separate table for itself.

Consider the Lecturer table in Fig. 14.68(a). It is in BCNF because it is "all-key". The table is not in 4NF because language and sport are each *multidependent* (but not functionally dependent) on surname. The MVDs are depicted as double arrows above the table. Assuming that the ternary means Lecturer plays Sport and speaks Language, the table exhibits redundancy. The facts that Halpin plays judo and karatedo and speaks English and Japanese are duplicated. To avoid this problem, the table is split into two 4NF tables (as obtained by Rmapping the ORM schema) as shown in Fig. 14.68(c).

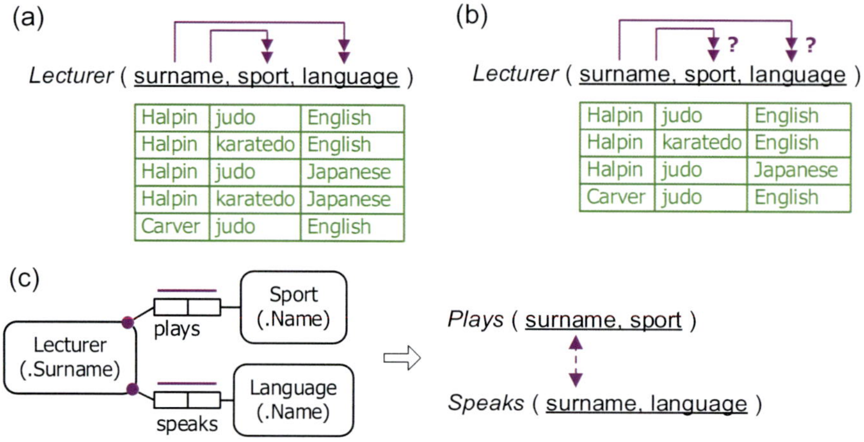

(a)

Halpin	judo	English
Halpin	karatedo	English
Halpin	judo	Japanese
Halpin	karatedo	Japanese
Carver	judo	English

(b)

Halpin	judo	English
Halpin	karatedo	English
Halpin	judo	Japanese
Carver	judo	English

FIG. 14.68 (a) Table in BCNF but not 4NF; (b) debatable MVDs; and (c) Rmap result.

Now, let's delete the fourth row of this Lecturer table to obtain the smaller Lecturer table in Fig. 14.68(b). Notice the question mark next to the MVDs. If we strictly apply Fagin's definition of MVD, the relation (here a set of four tuples) does not exhibit these MVDs. According to Fagin (1977), the MVD $X \twoheadrightarrow Y$ holds for the relation $R(X, Y, Z)$ if and only if whenever (x, y_1, z_1) and (x, y_2, z_2) are tuples of R, then so are (x, y_1, z_2) and (x, y_2, z_1). By this definition, if the MVD surname $\twoheadrightarrow$ sport did apply, then the presence of the tuples ("Halpin", "karatedo", "English") and ("Halpin", "judo", "Japanese") requires the presence of the tuples ("Halpin", "judo", "English") and ("Halpin", "karatedo", "Japanese"). But the second of these tuples is absent (we just deleted it). Similarly, the MVD surname $\twoheadrightarrow$ language does not apply because of this missing tuple. However, even without these MVDs, the table shows fact redundancy (the facts that Halpin plays judo and speaks English still appear twice).

What is important in practice is that the relation *scheme* is well designed, regardless of whether its current population satisfies some required dependency. If somebody produced the table in Fig. 14.68(b), the table would have no nontrivial MVDs according to Fagin's definition, but the design is still faulty if the ternary means Lecturer plays Sport and speaks Language because it allows fact redundancy.

For such reasons, we adopt the notion of what we call *semantic 4NF*, which relates MVDs to the *meaning* of the underlying fact types. If the ternary in Fig. 14.68(b) means Lecturer plays Sport and speaks Language, then there is a semantic MVD surname $\twoheadrightarrow$ sport because, in terms of the underlying fact type Lecturer plays Sport, the set of sports is determined by the lecturer alone (without regard to language). Similarly, there is a semantic MVD surname $\twoheadrightarrow$ language.

Suppose instead the ternary meant Lecturer plays Sport when it is conducted in Language. For example, Halpin might play judo only when it is conducted in English. In this case, the ternary is elementary so can't be split without information loss. Hence, it is not in general possible to determine whether a table is well designed simply by inspecting its population(s). You also need to know what the data means.

ORM's Rmap procedure forbids grouping two or more fact types with composite UCs (e.g., *m:n* associations) into the same table scheme, regardless of what the current populations might be. ORM's notion of an alethic, static constraint refers to *any possible population*, and in principle requires an understanding of the business domain by the domain expert who asserts that such a constraint must apply. While sample populations can reveal the absence of a constraint, and provide hints as to the presence of constraints, they cannot guarantee that such constraints really do apply—such guarantees are the responsibility of the domain experts.

It can be shown that BCNF is equivalent to no redundancy when there are FDs, and 4NF is equivalent to no redundancy when there are FDs and MVDs (Vincent and Srinivasan, 1993). Moreover, if a relation scheme is in BCNF and some key is simple, then it is also in 4NF (Date and Fagin, 1992).

The next normal form is also due to Fagin (1979) and is based on *join dependencies* (JDs). A relation has a join dependency if it can be reconstructed without information loss by taking a join of some of its projections. If one of these projections is the table itself, this is a trivial join dependency.

A table is in **fifth normal form** (**5NF**) if and only if, for each nontrivial join dependency, each projection includes a key of the original table. A table is in **project-join normal form** (**PJNF**) just in case each JD is the result of the key constraints. Any 3NF relation scheme whose keys are all simple is also in PJNF (Date and Fagin 1992). The forms 5NF and PJNF are often treated as equivalent, but some subtle differences can be distinguished to show that PJNF is a stronger notion (Orlowska and Zhang, 1992).

As an example of key-based join dependency, *Employee*(empNr, birthdate, gender) is equivalent to the join of the projections *Emp1*(empNr, birthdate); *Emp2*(empNr, gender). The original table as well as the two smaller ones are all in 5NF. Because a relational schema is in a given normal form if its table schemes are all in that form, a relational schema that includes just the Emp1 and Emp2 tables is a 5NF schema. This illustrates that 5NF of itself does not guarantee that a schema is minimal with respect to the number of tables.

Nontrivial join dependencies that are not key-based are rare, so 4NF tables are almost always in 5NF as well. The theory underlying the test for 5NF was included in the projection-join check at CSDP step 4, so we confine ourselves here to a brief discussion. A classic example used to discuss the notion is portrayed in the SellsFor table in Fig. 14.69(a). A small sample population is provided to show that the only uniqueness constraint is the weakest possible (verify this for yourself).

Conceptually, this ternary fact type may be expressed as Agent sells Cartype for Company. The fact types behind the three binary projections may be expressed as Agent sells Cartype; Agent represents Company; Company makes Cartype. Any attempt to split this ternary into *two* binaries will result in information loss. For example, from the facts that Smith is a representative for Foord, and Foord makes a four-wheel drive (4WD), it does not follow that Smith sells a 4WD.

However, the population of the SellsFor table does equal the join of the three binary projections (confirm this for yourself). So we could split this particular table population into three binary table populations without information loss.

(a)

SellsFor (agent, cartype, company)

Jones E	sedan	Foord
Jones E	4WD	Foord
Jones E	sedan	Yotsubishi
Smith J	sedan	Foord

(b)

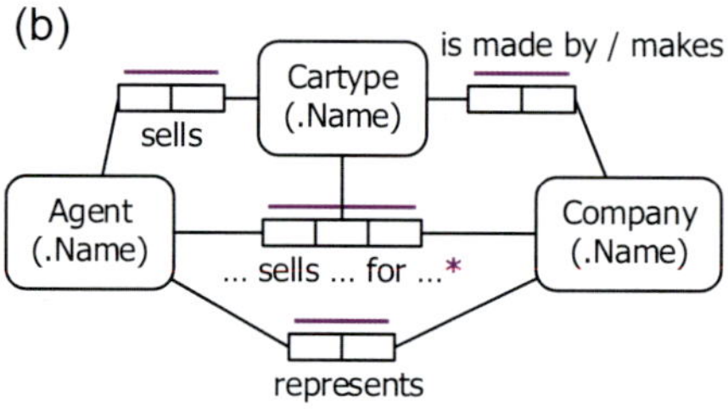

*Agent sells Cartype for Company **iff**
Agent sells Cartype **and**
Agent represents Company **and**
Company makes Cartype.

(c)

Sells (agent, cartype)

Represents (agent, company)

Makes (company, cartype)

FIG. 14.69 (a) Is this a 5NF Table? (b) Possible ORM schema that maps to schema (c).

But because database populations are updated continually, if we are to split the table scheme into three binary table schemes, we need to know that this join dependency applies to all possible, complete populations. In other words, is the sample population significant in this regard? If it is, the derivation rule shown in Fig. 14.69(b) applies.

The only way to check that this is an actual business rule is to ask the domain expert. Let's suppose that it's *not* a rule. For example, it is now acceptable to delete just the first row of the SellsFor table in Fig. 14.69(b). Check for yourself that this deletion should be rejected if the rule did apply. In this case, the ternary fact type is elementary and should be left as it is. If the derivation rule does not apply, the SellsFor table *scheme* is already in 5NF.

If the derivation rule *does* apply, as shown in Fig. 14.69(b), this rule needs to be enforced, but the key constraint in the SellsFor ternary fails to do this (e.g., it allows the first row to be deleted). To avoid this problem, the 4NF ternary should be split into three 5NF binaries, as in Fig. 14.69(c). If some roles in the ORM schema are mandatory, some intertable constraints need to be added. The original ternary may also be defined as a view derived from the join of the new tables.

It has been argued (e.g., by Kent) that when relation schemes such as SellsFor in Fig. 14.69(a) are elementary, there is a kind of "unavoidable redundancy". For example, the "fact" that agent "Jones E" sells sedans is repeated on rows 1 and 3. However, such "redundancy" is harmless because it is derived rather than stored. Although the tuple ("Jones E", "sedan") is stored twice, the fact that agent "Jones E" sells sedans can only be obtained via the derivation rule Agent sells Cartype **iff** Agent sells Cartype for Company.

The seven normal forms discussed so far are strictly ordered. Each higher normal form satisfies the lower forms. Recall that "⊂" denotes "is a proper subset of". Let "5NF-rels" denote the set of all possible relations in fifth normal form, and so on. The strict ordering may now be set out thus (in decreasing order of normality):

5NF-rels ⊂ 4NF-rels ⊂ BCNF-rels ⊂ EKNF-rels ⊂ 3NF-rels ⊂ 2NF-rels ⊂ 1NF-rels

Essentially, normalization through to 5NF simplifies table updates by ensuring that three kinds of dependency (FDs, MVDs, and JDs) will automatically be enforced by key constraints (i.e., uniqueness constraints within each table). An FD is a special case of an MVD, which in turn is a special case of a JD. As noted by Fagin, BCNF ensures that all FDs are implied by key constraints, 4NF ensures that all MVDs are implied by key constraints, and 5NF ensures that all JDs are implied by key constraints.

The next normal form we consider is **domain key normal form** (**DKNF**), which was introduced by Fagin (1981). A table scheme is in DKNF if and only if all its constraints are expressible as domain dependencies or key dependencies. A domain dependency specifies a set of possible values for a given attribute. This basically declares a data type (e.g., **varchar**(20)) or a value constraint (e.g., {"M", "F"}) for the attribute. A key dependency is a uniqueness constraint spanning one or more attributes in the table scheme. If all domains are infinite, any DKNF table scheme is also in 5NF.

For a relational database schema to be in DKNF, each of its table schemes should be in DKNF, and there should be either no intertable constraints or only "simple" intertable constraints that are easily enforced, such as inclusion dependencies (subset constraints).

DKNF designs are easy to understand, and their constraints can normally be efficiently enforced by a DBMS. Unfortunately, however, life is not so simple. Many other kinds of constraints often arise in practical applications, and need to be enforced even though their inclusion in the relational schema violates DKNF.

As a simple example, Fig. 14.70 shows a conceptual schema and its relational mapping. The table scheme Person is not in DKNF because it contains two constraints other than domain and key dependencies. First, the subset constraint from manager to personNr is an intratable inclusion dependency. Second, the "personNr ≠ manager" constraint enforces the irreflexive nature of the reporting association. The overall schema is not in DKNF because its Person scheme is not in DKNF, and because of the exclusion constraint (no person wrote and reviewed the same book).

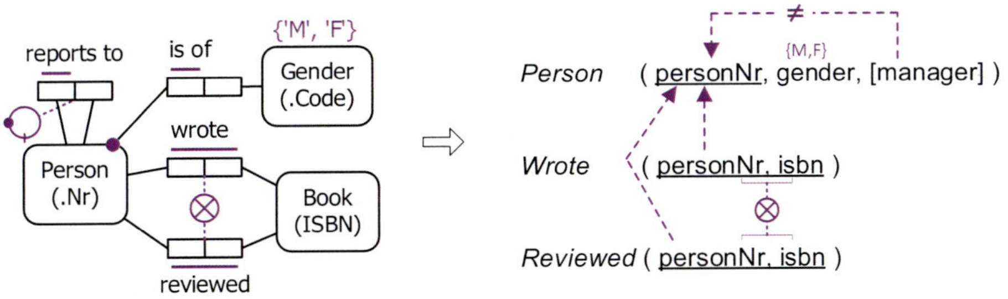

FIG. 14.70 The relational schema is not in DKNF.

Although the relational schema in Fig. 14.70 is not in DKNF, each of its table schemes is in 5NF. The additional constraints are not included in the definition of 5NF so do not violate it. Any valid ORM schema, where each asserted fact type is atomic, will result in a 5NF relational schema if mapped with our Rmap procedure.

As this example shows, ORM actually goes beyond 5NF by catering for other kinds of constraint. Although this rich support for expressing and mapping constraints may lead to schemas that violate DKNF, that's an advantage rather than a weakness because such violations are needed to capture the relevant business rules.

It is very common to encounter cases where normalization to 5NF or DKNF fails to fully capture the relevant constraints. A modified version of an awkward but classic example is set out in Table 14.8. In this UoD, lecturers are assigned exactly one course, and this is the only course they may teach.

Table 14.8 An unusual and awkward example.

Lecturer	*Course*	*Student*
Halpin	CS113	Brown A
Rose	CS102	Brown A
Nijssen	CS113	Smith J
Rose	CS102	Smith J
Halpin	CS113	Wang J
Bloggs	CS226	?

The null indicates that Bloggs does not teach the assigned course (e.g., because too few enrolled for it). For each course they take, students have exactly one lecturer. The same course may have many assigned lecturers. Because of its strange constraint pattern, this example is not easy to verbalize in elementary facts on a first attempt. You are invited to try this before reading on.

An initial conceptual schema is depicted in Fig. 14.71(a). If this is the global schema, the mandatory role constraints are implied by the pair-subset constraint. More importantly, the uniqueness constraint on the binary when combined with the pair-subset constraint has other implications. Can you spot these?

The pair-subset constraint indicates that lecturers may teach only their assigned courses, but the uniqueness constraint on the binary indicates that each lecturer is assigned at most one course. Hence, lecturers may teach at most one course. So, there is an FD from the first role of the ternary to the second role, as shown in Fig. 14.71(b). Because the ternary has an implied, spanning UC, the implied FD implies the UC across the first and third roles. Moreover, because there is a third role in the ternary, the fact type now cannot be elementary. This is an example of the general rule set out in Fig. 14.72. Two corollaries are included, one of which we will make use of in our example. You may wish to prove these corollaries as an exercise.

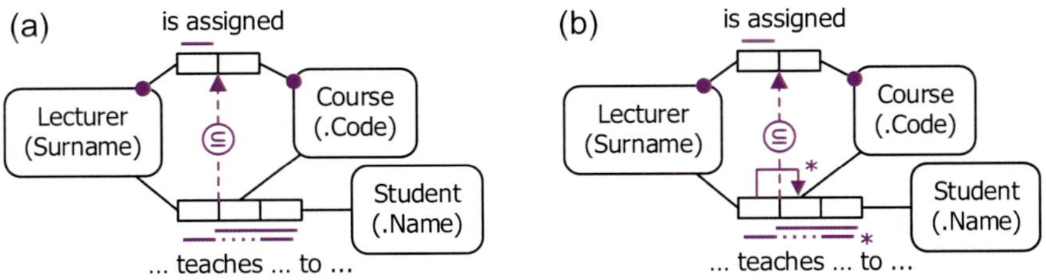

FIG. 14.71 Is the ternary fact type elementary?

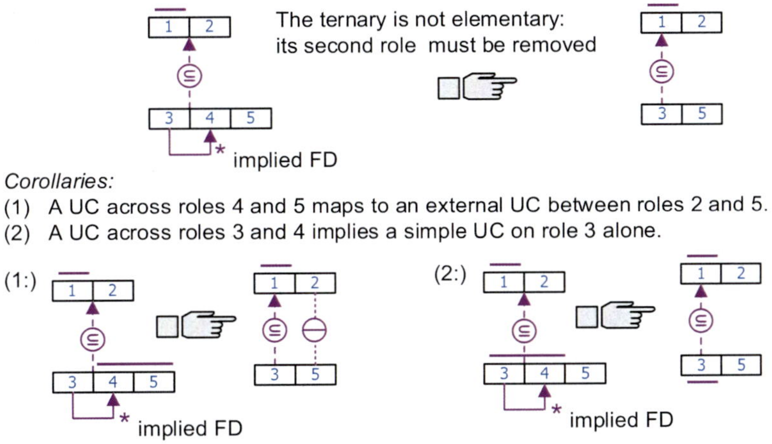

FIG. 14.72 A basic splittability check.

If a fact type is elementary, all its FDs are the result of its UCs. If another FD exists, then semantically underlying this FD is another fact type, which is the target of a pair-subset constraint from the original fact type. If this is expressed at the conceptual level, we know the original fact type is not elementary, and we should remove the relevant role (as in Fig. 14.72).

The ternary in Fig. 14.72 matches this pattern, but has an additional UC across its last two roles (for each course, a student has only one lecturer). To preserve this constraint after removing the role, an external UC must be added between the course and student roles in the new schema (see Fig. 14.73).

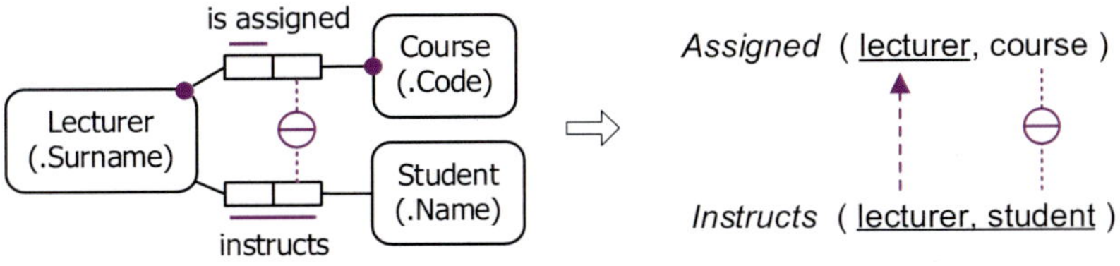

FIG. 14.73 The corrected conceptual schema and its relational map.

The general requirement for this extra move is specified as the first corollary in Fig. 14.72. When the corrected conceptual schema is Rmapped, we obtain two table schemes with two intertable constraints (Fig. 14.73).

Suppose that instead of dealing with Table 14.8 conceptually, we develop a relational schema for it using the normal form theory discussed earlier. We might begin with a single ternary: *Teaches*(lecturer, course, [student]). Suppose that someone now tells us all the constraints or cleverly constructs a significant population from which the constraints can be deduced. We note that there are two composite uniqueness constraints over (lecturer, student) and (course, student). Because student is an optional attribute, neither of these can be used as a primary key, so we know something is wrong. Indeed, we know the table is not in BCNF because we have the FD lecturer → course, and lecturer is not a key.

At this stage we split the table into two: *Assigned*(lecturer, course); *Instructs*(lecturer, student). These are the same two tables we arrived at by deconceptualization in Fig. 14.73. However, the normal form theory discussed in this section tells us nothing about the two intertable constraints. We have to figure these out for ourselves.

The final normal form we discuss is **sixth normal form** (**6NF**), which was proposed by Chris Date and others in the late 1990s mainly to simplify the implementation of temporal databases. A relation scheme is in 6NF if and only if it satisfies no nontrivial join dependencies. A 6NF table scheme is automatically in 5NF and DKNF.

Intuitively, a 6NF table scheme contains the primary key and at most one other attribute. So, each row of a 6NF table contains just one atomic fact. Recall that an atomic fact is either an elementary fact or an existential fact.

For example, the 5NF table scheme *Person*(name, gender, height, weight, maritalStatus) may be decomposed into the following 6NF table schemes: *Person*(name, gender); *Person*(name, height); *Person*(name, weight); *Person*(name, maritalStatus). This makes it easier to store temporal information such as the period during which a person has a specific height, a specific weight, or a specific maritalStatus.

As a trivial example, the table scheme *Person*(*name*) is also in 6NF because each of its rows holds just one existential fact (e.g., there is a person named "Joseph Robinette Biden Jr.").

Further normal forms could be invented to cater for further constraints, but surely it is better to focus on cleaning up design faults at the conceptual level where humans can utilize their semantic understanding of the UoD more easily.

As discussed in the next section, it is sometimes necessary to denormalize a relational schema to achieve acceptable performance. This is a possible option for the current example. Note that the intertable constraints are expensive to enforce, unlike key constraints. For example, enforcing the external uniqueness constraint effectively requires a table join each time a row is inserted into either table. Moreover, queries to list who teaches what to whom need a table join.

Another design is obtained by mapping the original ORM schema in Fig. 14.71(a) to give the relational schema shown in Fig. 14.74(a). The pair-subset constraint implies the ternary's other key constraint on (lecturer, student), which may thus be omitted.

Although the ternary is denormalized, it is safe because its problematic FD is enforced by the pair-subset constraint and the key constraint in the binary. We may now list who teaches what to whom, without a table join. If this is a focused query for the application, this design might be preferred.

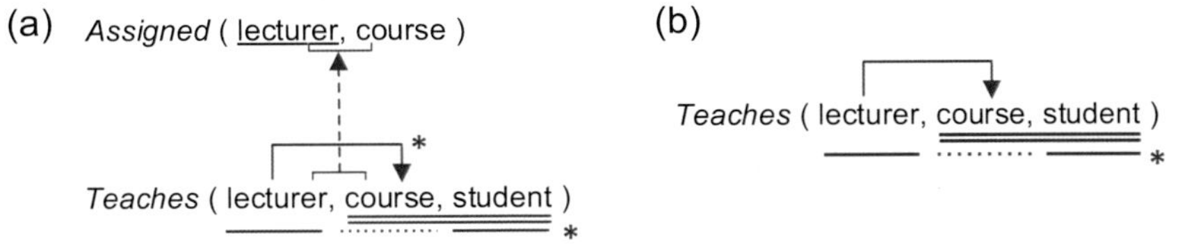

FIG. 14.74 Some denormalization options.

To modify the example somewhat, suppose lecturers must teach the course assigned to them (so the null in Table 14.8 cannot occur). Fig. 14.71(a) is changed to make both roles of Lecturer mandatory, with the subset constraint becoming an equality constraint, and Fig. 14.73 is changed to make both roles of Lecturer mandatory, and the relational subset constraint becomes an equality constraint.

If the normalized, two-table design has poor performance, we might denormalize the design to a single table that is their natural join, as shown in Fig. 14.74(b). With the subset constraint of Fig. 14.71(a) replaced by an equality constraint, assignment facts can be retrieved by projecting on the first two roles of the ternary, but the uniqueness constraint on the assignment binary still needs to be catered for. In the denormalized ternary, this constraint cannot be enforced as a uniqueness constraint, so a separate constraint is required to enforce the FD lecturer → course. As long as this additional constraint is enforced (e.g., as a user-defined constraint in SQL), the single-table design is safe. As before, the alternate key constraint is implied and hence omitted.

Advances in relational languages are making it easier to specify constraints such as FDs that are not key based. However, the overhead of enforcing such constraints needs to be factored in when choosing the final relational design. In general, we should first try a fully normalized design, using conceptual optimization where relevant to improve it. If performance is still unacceptable, controlled denormalization is an option.

Exercise 14.6

1. The conceptual schema for a given UoD is as shown. Some table schemes designed by novices for parts of this UoD are listed here. Indicate briefly, in terms of normalization theory, why each of these table schemes is not fully normalized.
   (a) *Parenthood* (parentName, gender, *Child*(childName, gender))
   (b) *Teaches* (person, subjectTutored, [subjectLectured])
   (c) *ParentOf* (parent, child, genderOfParent, genderOfChild)
   (d) *Tutor* (tutorName, subject, [child])

(e) Use Rmap to generate a correct relational schema.

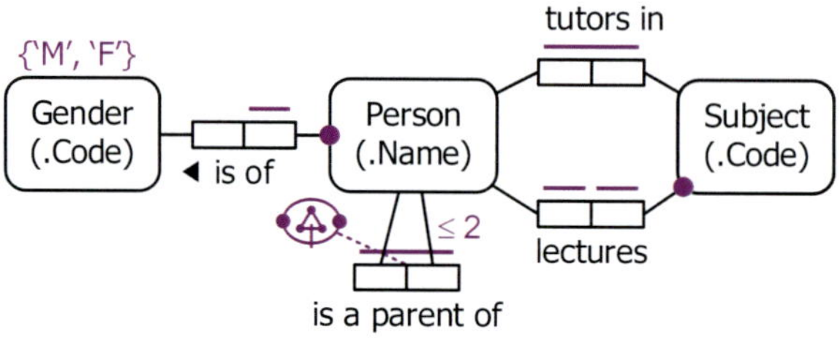

2. (a) The conceptual schema of Question 1 is incomplete with respect to real-world constraints. For example, parents of the same child must differ in gender, but this constraint is not captured. Replace the parent predicate with mother and father predicates to develop a more complete schema that includes this constraint (Hint: include subtyping).
   (b) Rmap your solution to (a).

3. Suppose that while developing a conceptual schema, you specify the following fact type: Manager supervises Trainee on Project. Assume that no manager is a trainee, and that all objects are simply identified by name. The following constraint applies: no manager may supervise the same trainee on more than one project.
   (a) Draw this as a ternary fact type, including the constraint.
   (b) You now check whether there are any other functional fact types of interest between these object types, and discover that each manager supervises at most one project. Add this fact type to your schema and then add any set-comparison constraint that applies.
   (c) Is the ternary elementary? If not, explain why not, and correct the schema.
   The following business rule is now added: each trainee on a given project has only one supervisor for that project. Add this constraint to your schema.
   (d) Rmap your answer to (d).
   (e) Because of poor performance, this relational schema is regarded as unsatisfactory. Using controlled denormalization, suggest an alternative relational schema.

14.7 Denormalization and Low-Level Optimization

Applying the conceptual optimization procedure followed by the Rmap procedure ensures a redundancy-free relational schema that should be reasonably efficient. To test the performance of this logical schema, it should be implemented as an internal schema on the target DBMS, using a realistic population. At this stage, *indexes* should normally be created for all primary and foreign keys as well as any other columns that play a significant role in focused queries.

For large tables, indexes can dramatically improve performance because they can normally be loaded, in part or whole, into main memory and their entries are sorted into structures that permit fast retrieval (e.g., binary trees). At times, indexes may slow down updates because they must be updated when relevant changes are made to the base tables, but even with updates, indexes can speed up many kinds of constraint checking (e.g., ensuring that keys are unique). There are many kinds of indexes (e.g., clustered, nonclustered), and indexing schemes often differ among vendors.

To speed up or facilitate certain complex queries, *working tables* (either predeclared or ad hoc) may be used (e.g., to hold intermediate results, perhaps obtained by joining several tables, which might be reused several times before the base values are updated). Predeclared working tables are populated at the start of a session to hold intermediate results that may be treated as stable for the session; these temporary tables are dropped at the end of the session. Ad hoc working tables are similar, but are created during a session and may be dropped before the end of the session as soon as they are no longer needed.

Another popular option is to use *materialized views*—these are views that are populated automatically with data according to their definition, and updated automatically as needed when their base tables are updated. Support for materialized views varies with the DBMS; in SQL Server, they are called indexed views.

Different DBMSs offer other means of tuning the internal schema (e.g., clustering the storage of related data). In addition, because no query optimizer is perfect, the form of the actual query itself can impact (sometimes seriously) on the performance. With some older optimizers, merely changing the order in which tables are listed in a join condition can cause significant changes in the response time. In most cases, you can use the performance tuning assistance provided by the DBMS to compare execution plans and select the best alternative.

If the data has been prechecked for integrity, consider deactivating constraint checking when performing bulk updates (e.g., loading a data warehouse from operational databases). You might also delay expensive constraint checking until after performing a bulk update. Although most database applications can be developed completely in 4GLs, consider writing some modules in a 3GL (e.g., C#) to speed up critical aspects or to achieve better control (e.g., checking acyclic constraints is usually faster in a 3GL than using recursive union in SQL).

For a few specific applications, a relational DBMS system might not be able to give the desired performance, and a different kind of DBMS might be needed (e.g., an object database).

Because realistic applications typically require data updates by users to be performed via an external interface (e.g., screen forms) rather than directly on the base tables, it needs to be decided how much constraint checking will be done at the table level and how much at the form level. With client-server networks, we need to decide how much work is done at the client end and how much at the server end. Constraints are often implemented both at the user interface and at the DBMS.

Implementing constraints on screen forms can detect data entry errors as early as possible. Implementing constraints on the DBMS ensures that they are checked regardless of the user interface. With noncentralized database systems, great care is needed in deciding how to replicate data to reduce communication overheads. A proper discussion of these issues is beyond the scope of this book.

If the performance of the system is still unacceptable after the aforementioned internal schema optimization has been carried out, it may be necessary to *denormalize* the base relational schema. Typically, this entails introducing *controlled redundancy* to reduce the number of table joins required for focused queries. This slows down updates (redundant facts must be kept consistent) and makes the system harder to work with, but it can dramatically reduce response times for queries.

Any denormalization decisions should be carefully documented, and appropriate subset or equality constraints must be enforced to control the redundancy, unless the denormalized version is used only for queries, not for updates (e.g., a data warehouse).

As a simple example, consider the following relational subschema, which is extracted from our software retailer example.

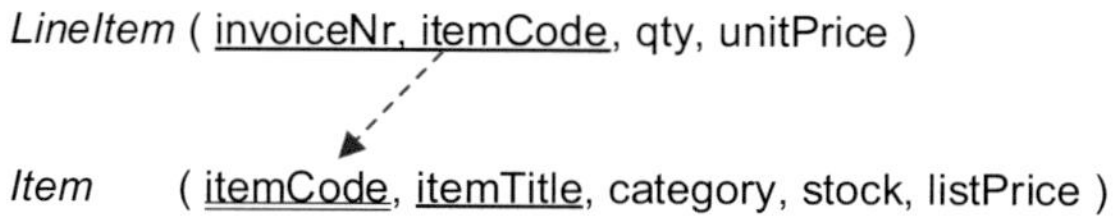

Suppose we frequently need to list the line items, including item titles, for specified invoices. With the schema given earlier, this requires a join of the two tables. If the tables are very large, accessing the second table simply to get the item title may slow things down too much. To avoid this, we might denormalize by inserting an extra copy of title in the LineItem table. This means the fact type Item has Title is now mapped to both tables. To control this *redundancy between tables,* we add a pair-subset constraint between the tables as shown:

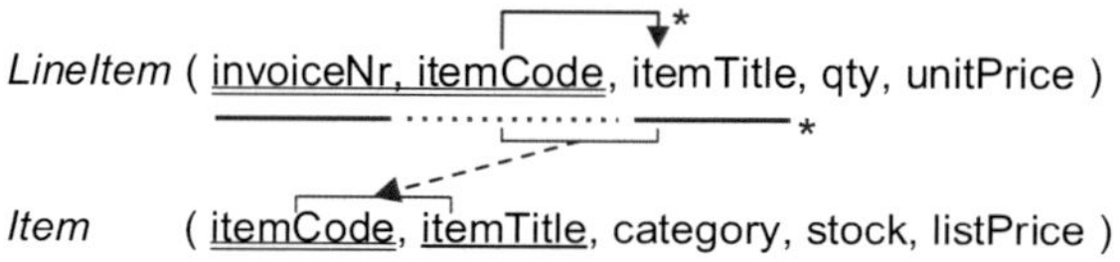

The new LineItem table is denormalized to 1NF because it includes the FD itemCode → itemTitle, and itemCode is only part of its primary key. Although the constraint pattern is more complex, the query may now be performed by accessing just the LineItem table. The asterisked uniqueness constraint and FD are both implied by the other constraints (why?), so need not be specified. The primary and alternate key constraints are inexpensive, but the pair-subset constraint may be expensive. It is not a foreign key constraint, but in the SQL standard it may be declared as the following assertion:

```
create assertion LineItem_itemCode_itemTitle_pair in_Item
  check (not exists
            (select * from LineItem
             where itemCode not in
                  (select itemCode from Item
                   where itemTitle = LineItem.itemTitle)))
```

In most DBMSs, assertions are not supported, so this constraint would normally be coded with triggers. The subset constraint needs to be checked whenever items are deleted from or renamed (code or title) in the Item table, or inserted or renamed in the LineItem table. Because items are rarely deleted or renamed, the main overhead for the subset constraint involves validating insertions to the LineItem table. For some applications, the additional update overhead may be a price well worth paying to improve the query performance.

Alternatively, explicitly declare the implied constraint **unique**(itemCode, itemTitle) on the Item table, leading to the following schema:

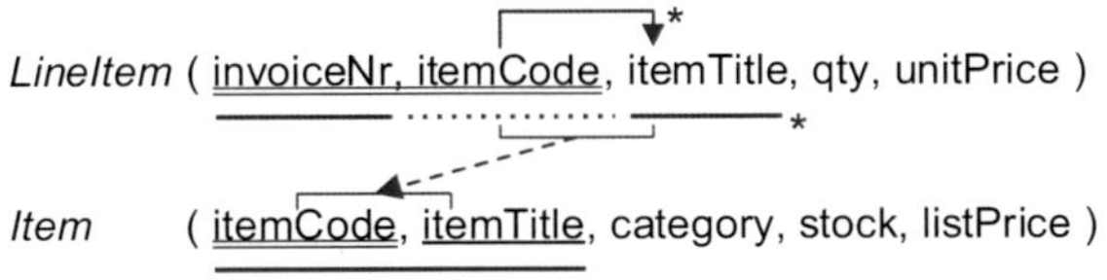

Although this adds some unneeded overhead (the extra constraint is implied by either of the candidate key constraints), it enables us to declare the intertable subset constraint simply as a foreign key constraint (as its target is declared unique).

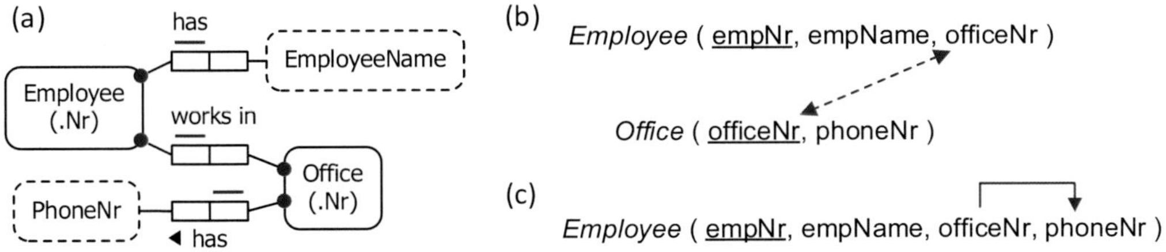

FIG. 14.75 (a) An ORM schema, and its (b) normalized and (c) denormalized mapping.

The previous example denormalized by storing item title facts twice in different tables. Another denormalization technique is to embed facts that functionally depend on a non-key attribute, hence violating 3NF. For example, the conceptual schema in Fig. 14.75(a) by default maps to the normalized relational schema in Fig. 14.75(b).

The FD from roomNr to phoneNr is efficiently enforced by the primary key constraint in the Room table, but a query to list an employee's phone number requires a join. To speed up this query, we could denormalize to 2NF by embedding the phone facts in the Employee table, as shown in Fig. 14.75(c).

Because many employees may share the same room, this denormalized schema is open to *redundancy within a table*. For example, if employees 1 and 2 share room 200 that has phoneNr 4251234567, then the fact that room 200 has this phone number is stored twice. To control this redundancy, we need to enforce the FD roomNr → phoneNr. Because this is no longer enforced by a key constraint, we need to specify it separately. In the SQL standard, the following assertion may be used:

```
create assertion eachRoomHasAtMostOnePhoneNr
 check( not exists
  (select roomNr from Employee
   group by roomNr
   having count(distinct phoneNr) > 1))
```

If assertions are not supported (which is likely), the embedded FD can be enforced by insert and update triggers on the Employee table.

As shown from these examples, denormalization can speed up queries but may slow down updates. So, there is a trade-off in performance. In general, don't denormalize unless it is the only way to obtain the desired performance. If you do decide to denormalize, and your database will be used for updates, then make sure you control the denormalization by coding the relevant constraints.

Real-world applications sometimes use very poor database designs. In particular, it is not uncommon for novice modelers to introduce uncontrolled redundancy (e.g., to ignore the pair-subset constraint and embedded FD in the above denormalized schemas). The next section discusses a reengineering example that illustrates, among other things, how such poor designs can be improved by applying conceptual modeling techniques.

If you are using an object-relational or object-oriented DBMS, you have the option of denormalizing to NF^2 using complex types. For example, SQL:1999 onward allows array types and row types to be used as columns that may themselves be nested as components of array or row structures used for outer columns. SQL:2003 onward supports multisets. Recall the Voter table example in an earlier chapter. Some DBMSs already support these collection types. This gives you a far greater range of mapping choices, which need to be used judiciously.

As a simple example, Fig. 14.76(a) models details about books in two normalized tables, using a position column to explicitly store each author's position in the book's author list. Up to 10 authors are allowed for a book, and some or all of these might not be recorded.

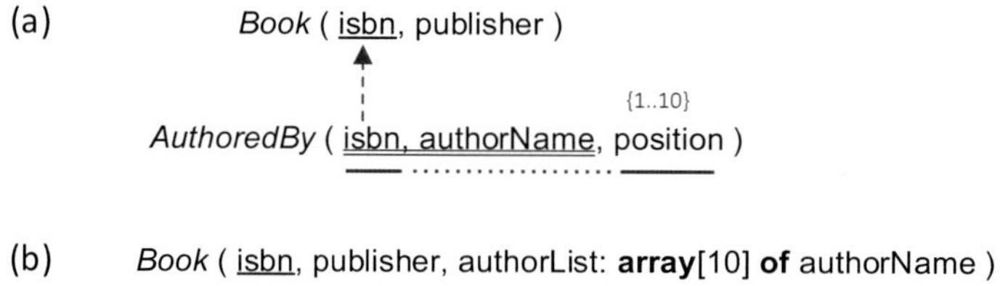

FIG. 14.76 Modeling author lists with (a) normalized tables or (b) an NF^2 table.

By using an array-valued column for the author list, the denormalized table scheme in Fig. 14.76(b) allows all the details to be captured in one table. Although this solution might seem to be preferable, extra care is needed to enforce constraints. For example, the entries in the array must be unique. The best choice depends on the query/update pattern for the application. For example, the normalized solution makes it easier to list all the books of a given author.

Considerable care is required in the use of collection types within table design, especially in constraint enforcement. We'll have a few more words to say about this in a later chapter. Given the current lack of a uniform approach to such object-relational extensions by commercial DBMSs, we'll say no more about object-relational DBMSs here.

In addition to denormalization, a technique known as *fragmentation* is commonly used to improve performance on very large tables. For example, a PhoneBook table for all the United States would contain hundreds of millions of rows. To speed things up, we can partition this table into 50 smaller tables, one for each state (if needed, we could partition further into smaller regions). Each of these smaller tables has the same structure, with the name of the table providing the state information.

If this process sounds familiar, it should be. It's simply an example of predicate specialization, as discussed early in the chapter. At any rate, if you want to query or update phone details about one state only, you can do this on the table for that state. This gives much better performance than trying to work with a table for the whole country. If you need to work with multiple states at a time, you can create a view for it based on a union of the smaller tables.

Some DBMSs enable good performance with such views. For example, if you include a column for the statecode in each table, and a check constraint to ensure the right statecode, SQL Server can use this information in the union view to consider only the base tables for the states mentioned in your query. For a good discussion of this technique, based on partitioning trades by month, see Seidman (2000).

Physical optimization techniques vary considerably among DBMSs. For example, Oracle's *materialized view* support differs somewhat from Microsoft SQL Server's *indexed views* support. Different systems support different kinds of *indexing schemes* (e.g., clustered, hash table), and Microsoft SQL Server even provides a tuning wizard to make the best index selection based on a significant sample of transactions. Detailed suggestions on physical optimization should be available in the online documentation for your specific DBMS.

Exercise 14.7

1. The following relational schema is used to store details about students and their results. A focused query is as follows: List the studentNr and name of all students who enrolled in CS114. As there are more than 50,000 students, and on average each is enrolled in more than 20 subjects, the tables are large. Denormalize the schema to improve the performance of the query, and discuss any possible disadvantages of this action.

Student (studentNr, studentName, degree)

↑

Result (studentNr, subjectCode, [rating])

2. (a) Schematize this UoD including derivation. Note that external uniqueness constraints may be declared across any join path. If any subset constraints apply (possibly from a join), specify them. Ensure that the fact types are all elementary.
   (b) Rmap your solution.
   (c) Discuss whether a denormalized relational schema may be more appropriate.

14.8 Reengineering

While conceptual modeling provides the best approach to developing new applications, it can also be used to remodel existing systems to better meet requirements. There are three main reasons for replacing or revising (perhaps drastically) an existing information system.

First, the current system might provide an *incorrect* model of the UoD, either by being *inaccurate* (wrong information) or by being *incomplete* (missing information). Wrong or missing fact types, constraints, and derivation rules might arise because of bad modeling in the first place. For example, the schema might assert that planets orbit moons, allow people to have more than one birth date, or derive ages from a simple formula that ignores the extra day in leap years.

Such errors can also arise because the model has not kept pace with changes in the UoD. For example, a system for recording academic results based on a seven-point numeric rating scale becomes outdated if the rating scheme is replaced by a five-point letter grade.

Second, the existing system might be *inefficient*. It might be too slow (e.g., too many table joins needed in queries), too large (e.g., too many nulls gobbling up space), or difficult to evolve (e.g., outdated rules hardwired into program code rather than modeled as data in tables). Perhaps the hardware platform itself is outmoded and needs to be replaced, or the external interface is awkward for users.

Finally, the existing system might be *unclear*. This also makes it hard to maintain or evolve. Many legacy systems in place today were designed at the logical, or even physical, level and have no conceptual schema formulated for them. They may have been developed completely in 3GLs, with no underlying design method, have little or poor documentation, and be riddled with misleading or cryptic identifiers.

Often, a combination of the three reasons leads to an information system that needs to be *reengineered*, or restructured, for its owner (e.g., a company) to remain competitive in today's marketplace. The company might even reengineer the very way it does business. Although the need for restructuring is sometimes used as an excuse for retrenching, a creative response to this need can lead to major enhancements in the quality and efficiency of the business and its supporting information systems. Here, we consider only the problem of reengineering a database system. This process may be divided into four stages, as shown in Fig. 14.77.

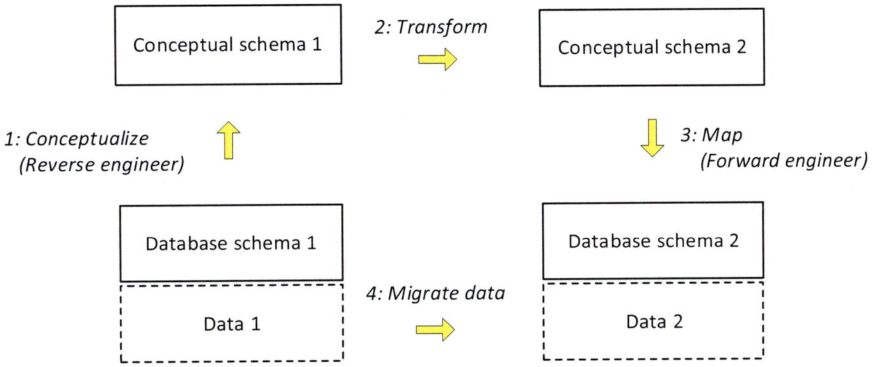

FIG. 14.77 Four steps in reengineering a database.

The first stage, *conceptualization* or *reverse engineering*, involves developing an initial conceptual schema of the application from the original database. This can be done simply by applying the CSDP to the database, treating it as a set of output reports. Most of the verbalization stage is usually straightforward because tables provide the easiest kind of output report to interpret. Usually, at least an incomplete logical schema for the database is also available, and constraints in this schema (e.g., primary and foreign key constraints) can be used to help derive some of the constraints in a fairly automatic fashion.

If the database is not relational, some of the intertable connections may need to be verbalized as fact types. Often, additional constraints that have not been specified will be identified with the assistance of the domain expert.

The second stage, *transformation*, involves modifying the original conceptual schema to cater for changes in the UoD not modeled in the original database, and/or optimizing the conceptual schema as discussed in the previous sections.

The third stage, *mapping* or *forward engineering*, takes us down to the logical schema. If a relational database system is to be used, we may use the Rmap procedure for this. The fourth stage, *data migration*, copies the facts stored in the original database to the new database. This involves conversion as well as copying because some of the fact types involved will be grouped into different table structures (recall the patient data example in Section 14.1). If the final two stages update the original database (rather than simply generating a new database), the whole reengineering process is called round-trip engineering or simply "round-tripping".

Much of the reengineering process can be automated, especially if a very detailed database schema is provided at stage 1. However, human interaction is also required to ensure a complete and elegant result. While our discussion is limited to the logical and conceptual levels, reengineering an application involves work at the external level (e.g., revising the user interface) and physical level (e.g., retuning) as well.

The four basic reengineering stages are now illustrated with a worked example. As the main input to the process, Fig. 14.78 shows a relational schema as well as two rows of data for each table (shown as tuples on the right to save space).

Relational schema (some constraints missing):		**Sample data** (not significant)
Committee	{Prog,Org} (cteeCode, chairPerson, [budget]) 2	('Prog', 'Adams, Prof. A.B.',?) ('Org', 'Bloggs, Dr F.',10000)
Person	(personName, affiliation, [email])	('Adams, Prof. A.B.','MIT',?) ('Pi, Dr Q.T.', 'UQ','pi@uq.au')
Rated	{1..10} (referee, paperNr, rating)	('Adams, Prof. A.B.',43,8) ('Pi, Dr Q.T.',5,4)
Referees	(personName, paperNr) ⊗	('Adams, Prof. A.B.',43) ('Knot, Prof. I.M.',43)
Authored	(personName, paperNr, paperTitle)	('Sea, Ms A.B.',5,'EER models') ('Pi, Dr Q.T.',43,'ORM dialects')
Presents	(author, paperNr)	('Pi, Dr Q.T.', 43) ('Knot, Dr I.M.',61)
Paper	{undec,accept,reject} (paperNr, paperTitle, status)	(5,'EER models','undec') (43,'ORM dialects','accept')
AcceptedPaper	(paperNr, [pageLength])	(43,14) (75,?)
AcPaperDiags	{tbl, fig} (paperNr, diagramKind, qty) 2	(43,'tbl',3) (43,'fig',5)
PaperSlot	{A,B} (slotNr, stream, hr, [bldgNr, roomNr], [paperNr])	(1,'A',9,69,'110',43) (2,'B',9,50,'110',?)
Room	{lab,lec,office} (bldgNr, roomNr, roomType, area)	(69,'110','lec',100) (50,'110','lab',120)
LabOrLecRm	(bldgNr, roomNr, [nrPCs], [nrSeats])	(69,'110',?,350) (50,'110',40,?)

FIG. 14.78 The original database schema and some sample data.

Suppose this design is used to record details about a 1-day computer conference, but its performance is poor mainly because of the number of table joins required for common queries.

Moreover, application developers find parts of the schema awkward or even unsafe. For example, the budget column in the Committee table applies to the committee as a whole, not to a particular chairperson of that committee (each committee has at most two chairpersons). It is suspected that this is not the only unnormalized feature of the tables.

Because of your expertise in conceptual modeling, you are hired to provide a conceptual model of the application so that a clear and complete picture of the business domain is available. Moreover, you are asked to provide a new relational design that is both normalized and efficient, if this is possible.

In practice, reengineering usually requires dialog with the domain expert to clarify the UoD, especially to identify missing constraints. To help you try the reverse engineering phase yourself, uniqueness, value, exclusion, and frequency constraints are depicted. However, other constraints are omitted (e.g., subset constraints, and nonimplied FDs due to denormalization). Some missing constraints are "obvious" and others may be identified from the brief description that follows.

Also, the sample rows of data enable you to apply CSDP step 1. These rows do not provide a significant, or even a legal, population (e.g., if the database were composed of only these data, subset constraints between the tables would be violated).

Abbreviated codes have the following meanings: Prog = Programming; Org = Organizing; undec = undecided; tbl = table; fig = figure; lec = lecture; lab = laboratory. Authors submit papers that are then sent to referees to be rated. The method for determining which papers are accepted is not modeled here.

Various statistics (number of pages, tables, and figures) may be kept about accepted papers, partly to help with publication of the conference proceedings. If statistics are kept for a paper, all three numbers must be recorded.

An accepted paper may be presented by one or more of its authors at the conference. Which authors present the paper might be unknown until after the paper is accepted. The conference program includes slots to which accepted papers are eventually allocated. Each slot is of 1 hour duration. Time is measured in military hours (e.g., 14 = 2 p.m.). To allow more papers on the same day, two streams (A and B) may be run in parallel (so two paper slots may have the same time).

A room directory is used to help assign slots to rooms; room area is in square meters. Only lecture rooms or laboratories may be used for paper presentations. The number of seats is recorded just for lecture rooms, and the number of PCs is recorded just for laboratories. Before peeking at the solution, try to provide a conceptual schema for this application. Include all constraints (including those missing from the original relational schema). In doing this, it will become apparent that some of the tables have redundancy problems. Which tables are these?

Compare your solution with the one provided in Fig. 14.79. Give yourself a pat on the back if you got all the constraints.

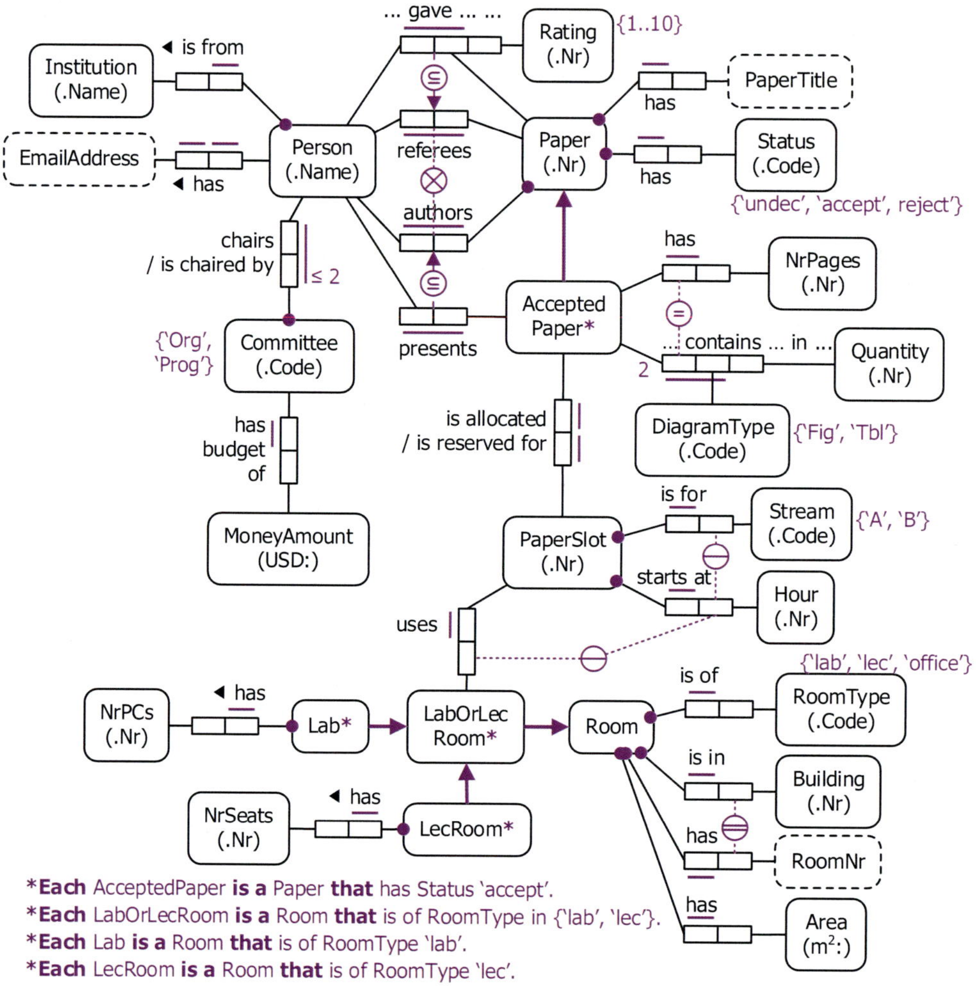

FIG. 14.79 The conceptual schema resulting from reverse engineering.

The table grouping differs from what Rmap would generate, for example, the two Committee fact types are wrongly grouped into the same table, even though one of them is nonfunctional. As a result, the original Committee table is open to redundancy problems with facts of the type Committee has budget of MoneyAmt. Similarly, the Authored table exhibits redundancy by including facts of the type Paper has PaperTitle. So, the original design denormalizes the relational schema in an uncontrolled way.

Although the ORM schema looks complicated, the conceptualization clarified the UoD, for example, the semantics of the PaperSlot table are easier to understand on the ORM schema. Now that the detailed semantics are displayed graphically, we are in a better position to explore optimization possibilities in a controlled manner.

As the second phase of reengineering, use the conceptual optimization procedure to transform the conceptual schema into one that Rmaps to a more efficient relational schema. Compare your solution with Fig. 14.80. The specialization of the chairperson and diagram predicates into functional binaries is straightforward.

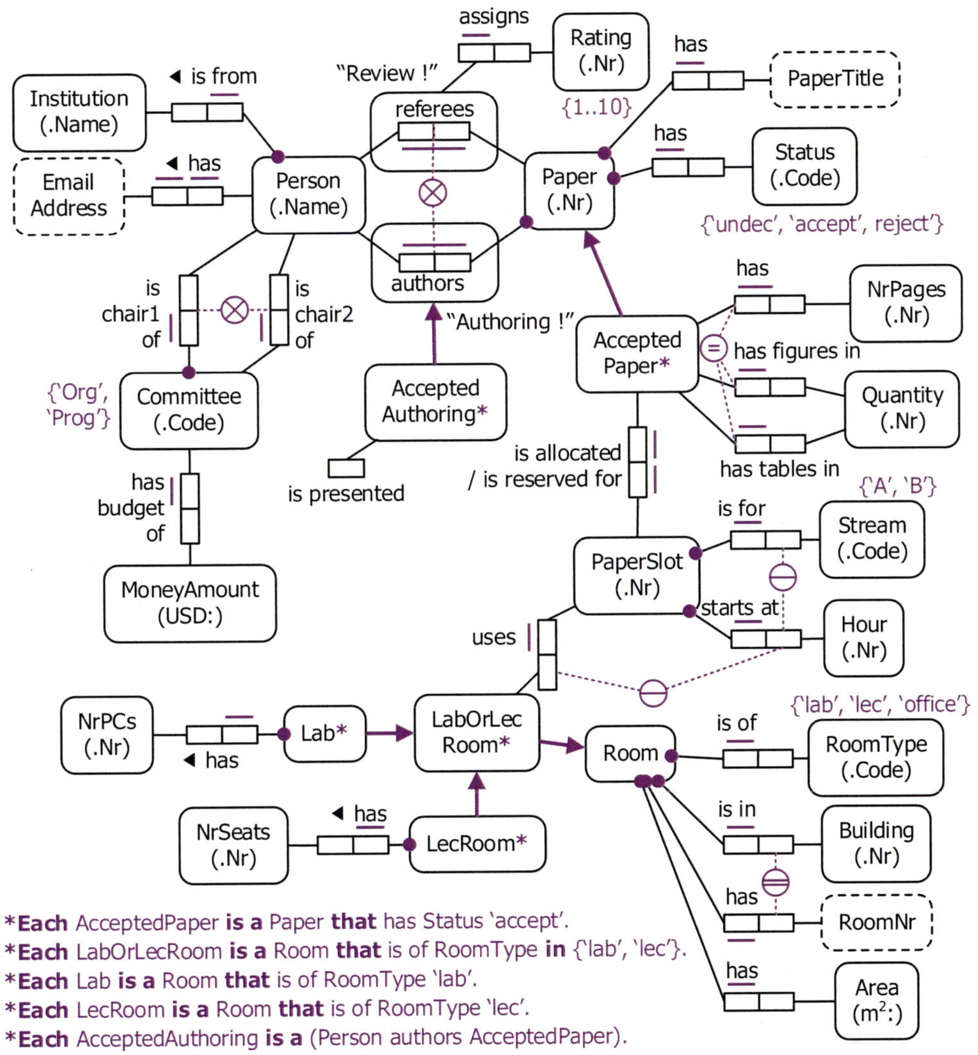

FIG. 14.80 The optimized conceptual schema.

In nesting the referee and author predicates, we could instead have united these exclusive but compatible predicates. Although this alternative move would lead to one less table, nesting these exclusive predicates would detract from the readability and lead to awkward constraint patterns.

Even the nesting in Fig. 14.80 leads to awkward verbalization. The original flattening into four predicates seems more natural than the two nestings. Ideally, a CASE tool would allow the modeler to choose the flattened display for clarity, with the option of nesting in a hidden, preprocessing stage to the standard mapping.

The next stage is to map the optimized conceptual schema to a new relational schema. As an exercise, try this yourself before looking at the solution in Fig. 14.81. Notice that the final relational schema contains just seven tables (compared with the 12 original tables), is fully normalized, and captures all the original constraints as well as those that were missing from the original.

Finally, the data migration is performed. The main aspect of this conversion involves defining the new relation types in terms of the original ones. Once this is done, the data migration can be performed automatically. Fig. 14.81 displays one row of converted data for each of the new tables.

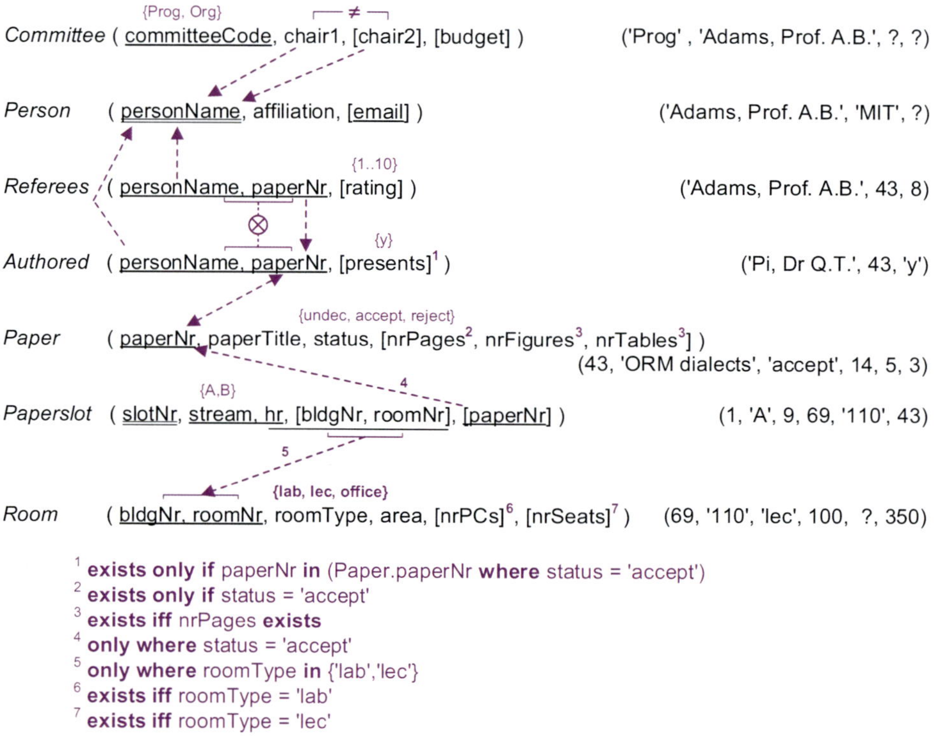

FIG. 14.81 The final relational schema.

Exercise 14.8

1. The following relational schema was designed to record details about academics and subjects, but its performance is poor, mainly because many joins are needed for common queries. Also, developers find the Academic table and its qualified

constraint to be awkward to think about. This constraint says that on each row of the Academic table, there is exactly one nonnull value in the last three columns.

The column "dept_where_lec" means the department (if any) where that academic is a lecturer, "dept_where_sen_lec" means the department (if any) where that academic is a senior lecturer, and "dept_where_prof" means the department (if any) where that academic is a professor. Here is one sample row from each table ("?" denotes a null):

Subject: ("CS115", 8)
Teaches: (30572, "CS115")
Academic: (30572, ?, "Computer Science", ?)
AwardedBy: (30572, "PhD", "UQ")
AwardedIn: (30572, "PhD", 1990)

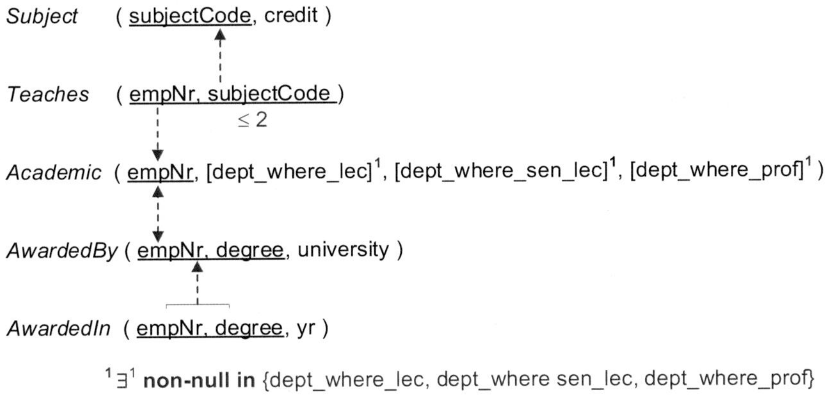

Reengineer the model to improve its performance and clarity, using the following steps.

(a) Reverse engineer the relational schema to an ORM schema. Include all constraints.
(b) Optimize the conceptual schema to prepare it for relational mapping.
(c) Forward engineer your new conceptual schema by mapping it to a relational schema. Include all constraints.
(d) Populate the new relational schema with the sample data provided.

14.9 Data Migration and Query Transformation

Data migration, the fourth phase of reengineering, can sometimes be challenging. When the original database is populated with massive amounts of data, it is typically impractical to convert and renter the data manually into the new database schema. While most DBMSs provide tools to assist with data migration, these tools usually handle only simple data transformations, so you need to write your own transforms for more complex cases. Such data transformations are normally written in SQL, although XSLT could be used if the data is available in XML form.

Once the data migration is completed, any stored queries on the old database need to be recoded to perform equivalent queries on the new databases. This section illustrates the tasks of data migration and query transformation using a small reengineering example. The exercise at the end of this section provides a larger example for you to try out yourself.

Fig. 14.82(a) shows a fragment of a relational schema about martial arts clubs. In martial arts (e.g., judo and karatedo), there are exactly two kinds of rank: dan (black belt grade) and kyu (below black belt). The ClubNrRankedPlayers table stores facts about how many dan and kyu graded players belong to certain clubs. For example, the tuples ("KDK", dan, 300) and ("KDK", kyu, 250) would mean that the club with the clubcode "KDK" has 300 dan grades and 250 kyu grades.

(a) *Club* (clubCode, clubname, cityname)

{dan, kyu}

ClubNrRankedPlayers (clubCode, rankType, nrPlayers)

(b) *Club* (clubCode, clubname, cityname, [nrDanGrades], [nrKyuGrades])

FIG. 14.82 (a) Original relational schema and (b) reengineered relational schema.

Suppose that the original database has a stored query to list the code and name of each club, together with the number of black belt players (if known). This query may be formulated in SQL as query Q1.

```
Q1: select Club.clubCode, clubName, nrPlayers as nrDanGrades
        from   Club left outer join ClubNrRankedPlayers
          on Club.clubCode = ClubNrRankedPlayers.clubCode
        and ClubNrRankedPlayers.rankType = 'dan'
```

Now suppose that in reengineering the original database, these two table schemes are replaced by the equivalent, single table scheme in Fig. 14.82(b). This optimization is based on the predicate specialization transform discussed in Section 14.2.

Suppose the original schema is called MartialArts1 and the new schema is called MartialArts2. We may now migrate the data from the original two tables into the new table using the following SQL command.

```
insert into MartialArts2.Club
select Club.clubCode, clubName, cityname, ClubDans.nrPlayers, ClubKyus.nrPlayers
from MartialArts1.Club left outer join MartialArts1.ClubNrRankedPlayers as ClubDans
                  on Club.clubCode = ClubDans.clubCode
                     and ClubDans.rankType = 'dan'
      left outer join MartialArts1.ClubNrRankedPlayers as ClubKyus
                 on Club.clubCode = ClubKyus.clubCode
                    and ClubKyus.rankType = 'kyu'
```

The original query Q1 may now be reformulated to work with the new database, as shown in Q2.

```
Q2: select clubCode, clubName, nrDanGrades
    from Club
```

Even with this simple example, you can see that writing the code to do the data migration can require a little thought. In practice, it can get much more difficult than this. The following exercise, which covers most aspects of the data reengineering process, includes several more challenging data migration tasks. The solutions to this exercise (and all other odd-numbered questions) are available online.

Exercise 14.9

1. The SQL Server diagram shown depicts the main aspects of a relational database schema CarSale1 used by a car retailer to record details about cars and sales. The developers do not fully understand the schema, and suspect that some aspects of the schema are incomplete or even unsafe. Also, the database has performance problems mainly because of the number of table joins required for common queries.

   The tables were originally thrown together with little design discipline by someone who has now left the application team. You are asked to provide a new relational design that is both normalized and efficient.

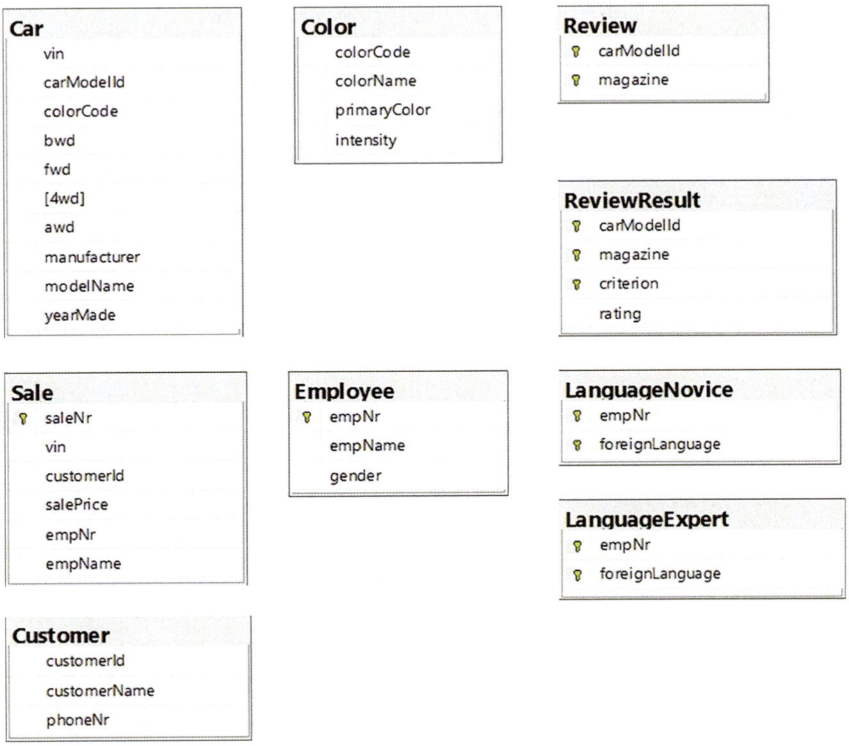

Although the schema includes some primary key declarations, it is missing several constraints that need to be enforced. Moreover, it includes some nonimplied FDs due to denormalization. Some missing constraints are "obvious" and some others may be inferred from the incomplete description that follows. In addition, some rows of data are provided for each table so you can apply ORM CSDP step 1. The sample data are not fully significant (i.e., not all constraints may be deduced from them).

For this UoD, each car is identified by a vehicle identification number (vin) and has exactly one color (identified by a code). Car models are back wheel drive (bwd), front wheel drive (fwd), four-wheel drive (4wd), or all-wheel drive (awd). Each car color has a unique name, and optionally its RGB (Red, Green, Blue) composition may be recorded. At most two phone numbers are recorded for each customer.

There is no restriction on what criteria may be used by magazines to rate car models, but each rating must be in the range 1…7 (the higher the rating the better). Whether a customer is an employee of the car retailer is not of interest. An SQL Server database with sample data for this question is available online.

(a) Which table schemes in the original schema are not fully normalized? Explain briefly what design flaws cause them to be denormalized.

(b) *Reverse engineer* the relational schema to an ORM conceptual schema. Do *not* concern yourself with efficiency or optimization at this stage. Your ORM schema, while being redundancy free with all fact types elementary, should in other respects directly reflect the design decisions made in the original relational schema. *Include all constraints* (graphical or textual) that apply to the application domain, including those constraints omitted from the original relational schema.

Formulate the following queries in SQL:

(c) For each car model that is reviewed, list its id and the magazines in which it is reviewed, as well as its economy rating(s) (if known).

(d) For each employee who has made a car sale and is a novice or an expert in Spanish, list their employee number, name, and gender as well as their sales total (i.e., the total amount they have earned for the company from car sales).

(e) For each car whose color has an equal measure of red and green, list its vehicle identification number, color code, car model id, and drive kind.

Criterion is now constrained to the values "Safety", "Economy", "Performance", and "Style".

(f) Transform the ORM schema to an "optimized conceptual schema" that minimizes the number of tables obtained if the Rmap procedure is applied.

(g) Forward engineer your optimized ORM schema to a new relational schema CarSale2. Ensure that all tables and columns are named well. Annotate the relational schema diagram to display all constraints, using footnote numbers and text boxes where needed. Value constraints should appear in braces after relevant column name(s).

(h) The external uniqueness constraint that each RGB intensity combination applies to at most one Color cannot be specified as a unique clause in most DBMSs, but can be coded using a trigger. Write appropriate trigger code to enforce this constraint.

(i) Now write SQL code to perform the data migration needed to populate the CarSale2 schema.

(j) Using your new CarSale2 schema, reformulate the queries in (c), (d), and (e).

14.10 Summary

Conceptual schemas are *equivalent* iff they model the same UoD. Various *schema transformations* may be performed to reshape a conceptual schema into one that is either equivalent, or an acceptable alternative.

Predicate specialization fragments a predicate into two or more special cases, typically by *absorbing* an enumerated object type into it. For example, the association Person is of Gender {'M','F'} may be specialized into Person is male and Person is female by absorbing Gender.

The inverse transformation, *predicate generalization*, typically involves *extracting* an enumerated object type from a predicate. Predicate specialization/generalization (PSG) may also be performed by considering the n cases of a role with a frequency constraint of n or $\leq n$. For example, the association Person is a parent of Person with a frequency constraint of 2 on its second role may be specialized into Person is parent1 of Person and Person is parent2 of Person.

Nested object types may be interchanged with *coreferenced* object types, using the Nest/Coreference transformation (N/CR). Either representation may be replaced by *flattening* into one or more fact types. For example, consider the nesting Enrollment(Student enrolled in Subject) results in Rating. If results are mandatory, this may be flattened into the ternary Student for Subject scored Rating. If results are optional, the flattened approach requires an extra binary (Student enrolled in Subject) targeted by a pair-subset constraint from the ternary.

The *overlap algorithm* indicates how nesting may be performed for compatible subpredicates, depending on the constraint between them (subset, equality, exclusion, etc.). See Fig. 14.35. Other schema transformations exist. A transformation might convert a graphical constraint to a textual constraint (or vice versa).

Although nested and coreferenced approaches have the same relational map, many other cases of equivalent conceptual schemas result in different relational schemas when subjected to the standard Rmap procedure. The process of transforming a conceptual schema into another that results in a more efficient relational map is called *conceptual schema optimization*.

Many optimization moves aim to reduce the number of relational tables, typically by specializing nonfunctional predicates into functional ones that can be grouped into the same table as another functional fact type or by using nesting (or coreferencing) to unify compatible composite keys. This is particularly useful in reducing the number of table joins required for focused queries. Predicate specialization is sometimes used to enable comparisons between rows of the same table to be replaced by comparisons between columns on the same row, thus eliminating the need for a self-join to perform the query.

Rmapping a correctly designed conceptual schema automatically results in a safe, redundancy-free, relational schema where the number of tables is reduced by grouping together functional fact types with the same key. In contrast to this deconceptualization approach, *normalization* provides a set of rules for achieving reasonable table designs by catering for dependencies between attributes.

Normalization by *synthesis* inputs a set of attributes and some basic dependencies, and groups the attributes into tables, with the aim of minimizing the number of tables required to satisfy these dependencies. Normalization by *decomposition* applies rules to split a poorly designed table into two or more tables in an effort to ensure that some basic dependencies are implied by the table key constraints.

This decomposition approach recognizes a number of *normal forms*. A table is in first normal form (*1NF*) if it has a fixed number of columns, all of whose values are atomic. A nonkey attribute is neither a key nor a part of a key. In a *2NF* table, every nonkey attribute is functionally dependent on a whole key. In a *3NF* table there is no FD between nonkey attributes.

An FD $X \rightarrow Y$ is trivial if X contains Y, and full if the FD requires all of X, not just part of it. An elementary FD is full and nontrivial. An elementary key has an elementary FD to some other attribute. In an *EKNF* (elementary key NF) table, each elementary FD (if any) comes from a whole key or ends inside an elementary key. In a *BCNF* (Boyce-Codd NF) table, each elementary FD begins at a whole key.

A *multivalued dependency* (MVD) $X \twoheadrightarrow Y$ exists if the set of Y values in the table depends only on X. An MVD $X \twoheadrightarrow Y$ is trivial if X includes Y, or X and Y together include all the attributes in the table. A *4NF* table is a BCNF table where all its nontrivial dependencies are FDs. So, each nonfunctional MVD requires a separate table for itself. To properly apply this notion to a table scheme, we strengthen it to *semantic 4NF*, where a semantic MVD $X \twoheadrightarrow Y$ applies iff the set of Y values is dependent on X alone for all possible populations of the underlying fact types.

A table has a *join dependency* (JD) if it is equivalent to the join of some of its projections. If one of these projections is the table itself, the JD is trivial. A table is in 5NF if for each nontrivial JD, each projection includes a key of the original table. The 5NF form is often called “PJNF” (Project-Join NF), although some subtle differences exist between these notions.

If a fact type is elementary, all its FDs are the result of its UCs. If another FD exists, underlying it is another fact type targeted by a tuple-subset constraint from the original fact type, which then requires role removal (see Fig. 14.72).

A table is in domain key normal form (*DKNF*) if its only constraints are domain constraints and key constraints. A relational schema is in DKNF if all its tables are in DKNF and intertable constraints are either absent or simple (e.g., inclusion dependencies, which are typically foreign key constraints). In practice, more complex constraints often exist in the UoD being modeled, so DKNF is often inadequate for complete models.

A table is in 6NF if and only if all its join dependencies are trivial. Each row of a 6NF table stores exactly one atomic fact.

These nine normal forms may be listed in increasing order of normality: 1NF, 2NF, 3NF, EKNF, BCNF, 4NF, 5NF, DKNF, and 6NF. Higher-level forms satisfy all the lower forms. In general, normalization ignores various constraints that we have considered at both conceptual and relational levels.

Once conceptual optimization and mapping have been completed, the physical schema should be *tuned* (e.g., by adding indexes, working tables, or materialized views). Various choices are also made as to the best place(s) to enforce constraints (e.g., at the database and/or in screen forms). If performance is still poor, it may be necessary to *denormalize* some of the base tables via *controlled redundancy* to reduce the need for joins.

If an existing database application is unsatisfactory, it may be necessary to *reengineer* it. This remodeling may involve four main steps: conceptualize (or *reverse engineer*) the existing database schema to a conceptual schema; *optimize* this by transformation; *map* (or *forward engineer*) it to the improved database schema; and migrate the original data to populate the new database. Once the *data migration* is complete, *query transformation* is often needed to transform stored queries based on the old schema into equivalent queries on the new schema.

Chapter Notes

Schema equivalence has been studied within the relational model (e.g., Kobayashi, 1990), the ER model (e.g., D'Atri and Sacca, 1984), extended ER models (e.g., Batini et al., 1992), and UML (e.g., Blaha and Premerlani, 1998). To keep our treatment intuitive, some technical issues have been glossed over. More formal treatments of schema transformation in ORM may be found in Halpin (1989b, b) and Halpin and Proper (1995a). For formal proofs of some conceptual schema equivalences, see Halpin (1989b). Further reduction transformations and additional ORM constraints (e.g., unique-where-true constraints) are discussed by Halpin, Carver, and Owen (2007).

Most work on schema optimization focuses on subconceptual levels. Date (2000) provides clear discussion of various optimization aspects at the relational level. For an approach to schema optimization that evaluates fitness functions on randomly generated, equivalent internal schemas, see van Bommel and van der Weide (1992).

A simple introduction to normalization is provided by Kent (1983). Classic papers include Codd (1970), Fagin (1977, 1979, 1981), Rissanen (1977), and Zaniolo (1982). In Fagin's original DKNF paper (Fagin, 1981), the term "key" is used in the sense of "superkey", so a UC may be implied by a smaller UC within it. Fagin uses the terms "relation schema" and "database schema" for what we call a "table scheme" and "relational schema", respectively. For further details on 6NF and its use for temporal databases, see Date et al. (2014).

Date and Fagin (1992) discuss simple, sufficient conditions for some of the higher normal forms. The extensions to semantic dependencies and semantic 4NF are due to Carver and Halpin (2008). A general study of join dependencies (including functional and multivalued dependencies) is provided by Aho et al. (1979). For an early paper on deductive normal forms, see Thalheim (1984). For a thorough analysis of normal forms, see Vincent (1994). An in-depth, formal treatment of normal forms and their relationship to ER modeling is given by Thalheim (1994). Chapters 11 and 12 of Date (2000) cover the main aspects of normalization and provide extensive references to the relevant literature.

15

Process and State Modeling

Chapter Outline

15.1 Modeling Dynamic Behavior

In this book, we have been mostly concerned with understanding information structures. This has very practical applications: if we want to build a business information system, it is crucial that we have an accurate definition of the information that we intend to use. We have seen how information can be modeled and how we can use the knowledge gained through modeling as the basis for storing and manipulating the information. Of course, modeling does not in itself determine what kind of information fulfills a business need. For example, should a customer information system include each customer's shoe size? For a shoe retailer, this is likely to be important information, but it would probably be irrelevant for an auto insurance company. Whatever the chosen scope, we've seen how it is possible to produce very precise definitions of the required business information.

Information models make minimal assumptions about how the information will be used. This is quite deliberate: a business may find that information originally intended for one purpose turns out to be useful for other purposes. Some implementation choices may be made below the conceptual level, such as denormalizing a database schema to improve query performance. However, it's generally desirable to have an information model that reflects the intrinsic structure of the information itself and is neutral with respect to information usage.

Information Modeling and Relational Databases. https://doi.org/10.1016/B978-0-443-23790-4.00019-0

A side effect of this is that an information model does not provide a complete specification of the information dynamics, the temporal relationships among information creation, modification, and deletion. To represent these aspects fully, we need some additional specification.

We can illustrate this point with an example. Fig. 15.1 shows a fragment of an information model. In practice, each of Data1, Data2, and Data3 could represent a complex set of data, but that's not important for our purposes here. Let us assume that our business has three activities, CollectData1, CollectData2, and CollectData3, which result in the creation of Data1, Data2, and Data3, respectively. Each of these activities takes the same amount of time and has the same cost in terms of resources required. We will further assume that our business is only interested in Data3 if both Data1 and Data2 meet certain criteria; otherwise, Data3 is not needed. On average, 90% of cases meet the Data1 criterion, and 10% of cases meet the Data2 criterion.

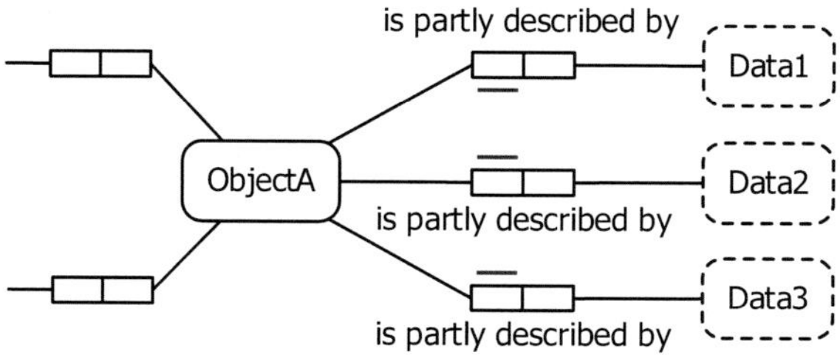

FIG. 15.1 A simple information structure.

Now, let us consider the question of how the information represented by Data1, Data2, and Data3 could evolve. Given the business scenario, it would not seem very sensible to collect Data3 before trying to collect Data1 and Data2, so we can immediately eliminate some of the possible orderings of CollectData1, CollectData2, and CollectData3. The remaining options are as follows:

(a) CollectData1 followed by CollectData2, then CollectData3 if needed.
(b) CollectData2 followed by CollectData1, then CollectData3 if needed.
(c) CollectData1 and CollectData2 simultaneously, then CollectData3 if needed.
(d) CollectData1, CollectData2, and CollectData3 in parallel, accepting that we will be carrying out some additional unnecessary work.

We can compare these alternatives by considering an average set of 100 cases. For option (a), we carry out 100 CollectData1 operations. On average, 90% of the cases meet the Data1 criterion, so we then carry out 90 CollectData2 operations. On average, 10% of the cases meet the Data2 criterion, so we finally carry out nine CollectData3 operations. The total cost will be $100+90+9=199$ units.

For option (b), we carry out 100 CollectData2 operations. On average, 10% of the cases meet the Data2 criterion, so we then carry out 10 CollectData1 operations. On average, 10% of the cases meet the Data1 criterion, so we finally carry out nine CollectData3 operations. The total cost will be $100+10+9=119$ units.

For option (c), we carry out 100 CollectData1 operations and 100 CollectData2 operations. On average, 9% of the cases meet both the Data1 and the Data2 criterion, so we finally carry out nine CollectData3 operations. The total cost will be 100 + 100 + 9 = 209 units.

Finally, for option (d) we carry out 100 CollectData1 operations, 100 CollectData2 operations, and 100 CollectData3 operations. We end up with the same nine cases, but this time at a cost of 300 units. Fig. 15.2 summarizes these results.

From this analysis, we can see that option (b) is likely to have the lowest cost. However, our business may have decided that speed of service is more important than lowest cost, and may prefer option (d), despite the higher cost. In fact, the company might choose to offer multiple options, passing the choice between cost and speed on to the customer.

Even examples as simple as this can point to some general conclusions. First, information can potentially evolve in many different ways, and the differences might be very important to a business; in the example just given, different options resulted in different costs and different processing times. Second, the relationship between information *structure* and information *usage* is one to many: a single information structure has the potential to be used in many ways, but a particular sequence of activities will always relate to one specific information structure. Third, we cannot define business activities (as in Fig. 15.2) by considering the information structure alone (as in Fig. 15.1). Although a model of information structure is necessary, it does not provide sufficient information to specify an information system fully, as we also need to know something about how the information will be used.

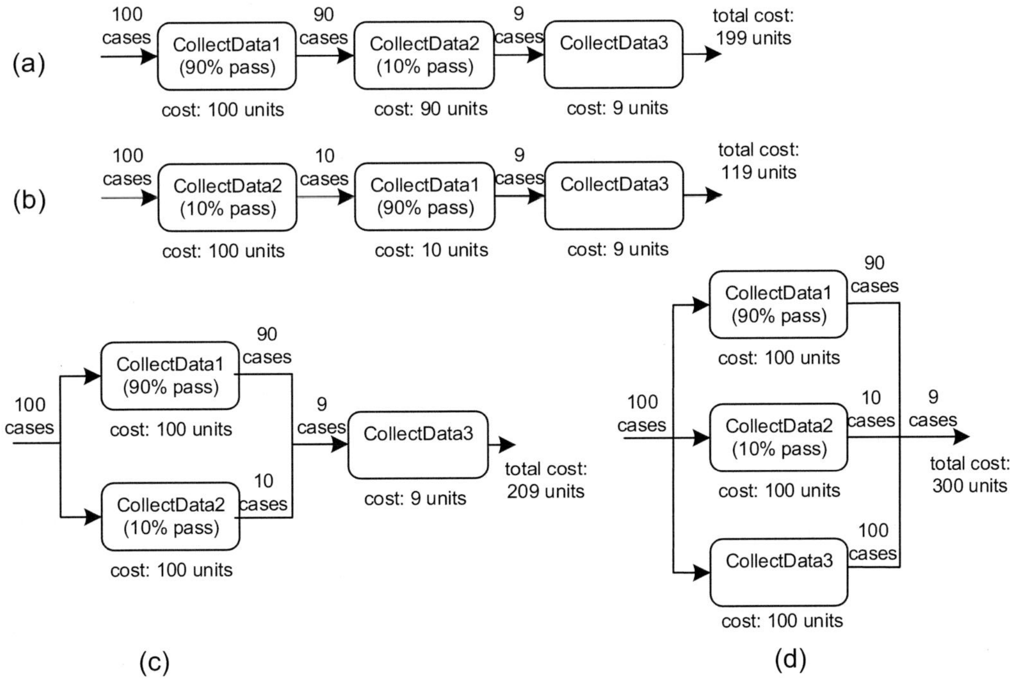

FIG. 15.2 Alternative processing strategies.

This separation of concerns turns out to be a useful aspect of information modeling. Businesses frequently review and revise their activities in light of perceived opportunities or threats. Having some degree of orthogonality between a "structural" view and a "behavioral" view minimizes the ripple effect of changes. For example, a business could switch between "minimum cost" and "minimum time" strategies without necessarily requiring any change to the information model.

Although information structure alone does not fully specify activity sequencing, it may place some constraints on the order in which information can be created. This is easy to see if we look at the information structure in the form of a relational schema. Fig. 15.3 shows an example.

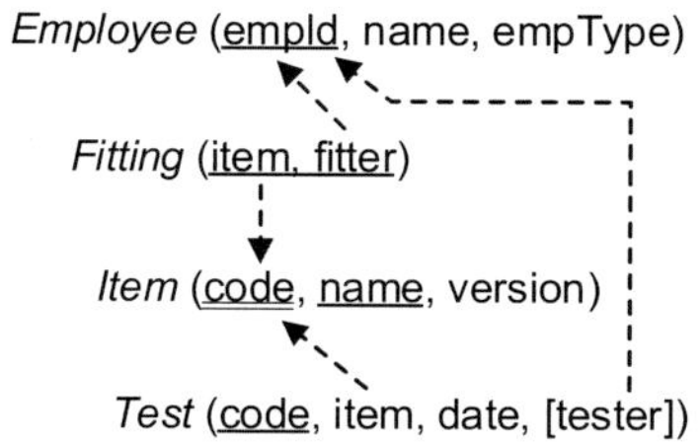

FIG. 15.3 A relational schema.

For this schema, we cannot enter information about a Fitting unless we have previously recorded the Employee and Item involved. Similarly, a Test cannot be entered unless the Item has already been recorded. These constraints are recognizable from the foreign key references between tables. However, there is a weaker ordering constraint for the tester information: this column is optional, and so we could create a Test record leaving the value for tester as null. The referenced Employee only needs to exist at the time we add the tester information, which could be some time after creating the Test record. Corresponding constraints apply to the removal of information. For instance, we would probably not want to delete an Employee while we still have Fitting records relating to that Employee. For the tester column, the foreign key reference would probably be specified as **on delete set null**, which would allow us to delete an Employee row that was still referenced in a Test, setting any existing tester references to the Employee table to null.

Even though constraints of this kind do not fully determine the sequence in which we can create, update, or delete information, the constraints must still be observed in whatever modeling approach we adopt to define information behavior. This chapter mainly focuses on two approaches used commonly to model behavioral (sometimes called *dynamic*) aspects: process modeling and state modeling.

15.2 Processes and Workflow

It has always been important for an enterprise to manage its business processes. Process change, whether dramatic or subtle, is a common mechanism employed to keep a business efficient and competitive (Hammer and Champy, 1993). Any medium to large business is likely to have a comprehensive set of processes that defines how it intends to conduct its activities. The definition is needed by companies aiming to achieve quality accreditation to a standard such as the ISO 9000 series. Among other things, these quality standards require an organization to define its processes and to maintain adequate records of their execution.

Having a defined set of business processes does not necessarily imply any commitment to automation. Until the second half of the 20th century, the opportunities for process automation were limited by the available technology, and manual processes were commonplace. Today, it's more usual to have some automation support. We can broadly distinguish several levels of automation, as shown in Table 15.1.

Table 15.1 Levels of process automation.

Level	*Characteristics*
Level 0	Process definition, execution, and record keeping are all carried out manually.
Level 1	As Level 0, but with records of process execution maintained in a computer-based information system.
Level 2	As Level 1, but with process design supported by computer-based tools.
Level 3	As Level 2, plus the automation of at least some of the process activities.

An organization may have different processes at different levels, but the general trend is for processes to migrate toward Level 3. At the extreme end of Level 3, we have processes in which all the activities are automated, a position that is facilitated by the growing availability of technology to support service-oriented architecture.

We can also classify processes based on the structure and the repeatability of the activities involved. Some processes are ad hoc: the activities involved are not predefined, and so, obviously, their order cannot be predefined either. In place of a predefined structure, ad hoc processes are usually subject to a set of rules that defines what kinds of activities are permitted, how activities can be started or ended, and so on.

Most of the current interest in process modeling relates to more structured arrangements of activities, sometimes termed *framed* processes. *Individually* framed processes are well-defined, but the process definition is executed only a few times, and perhaps just once. An example is a project plan, which usually defines a detailed set of activities and its sequencing but is unlikely to be reused in any other project without modification. *Loosely* framed processes, sometimes also known as “administrative processes”, have defined activities and coordination rules, but with the flexibility to vary the activities and their sequencing within some set limits. *Tightly* framed processes, sometimes also known as “production processes”, have highly predictable activities and stable rules for coordination that are followed consistently.

An important characteristic of loosely framed and tightly framed processes is that they are generally repeatable. This makes investment in their automation an attractive proposition, as the cost of development can be justified by the potential savings made through better control over process execution. In the remainder of this chapter, we will assume that we are dealing with processes of this kind.

A term that commonly appears in relation to process modeling is *workflow*. The original focus of workflow was the mechanization of business activities that, historically, involved passing paper documents between in trays and out trays. Consequently, workflow, and its embodiment in workflow systems, placed a great deal of emphasis on enactment. More recently, it has been realized that workflow is part of a set of larger concerns that relates to the systematic handling of business information. In many contexts, the term workflow can be used interchangeably with business process modeling, but workflow generally implies a slightly lower-level viewpoint dealing mainly with operational and technical issues.

In contrast, business process modeling places more emphasis on the definition of the process, with the assumption that the process design can ultimately be mapped to some realization. These concerns recur throughout a process lifecycle: process definition and design naturally precede its realization in a technical form and its subsequent operation, but the relation between these is cyclic as business processes are reviewed and revised in the light of changing business circumstances.

Fig. 15.4 summarizes the general picture, but it's worth emphasizing that the overlap is fuzzy and that the definitions of these terms vary a great deal from one person or organization to another.

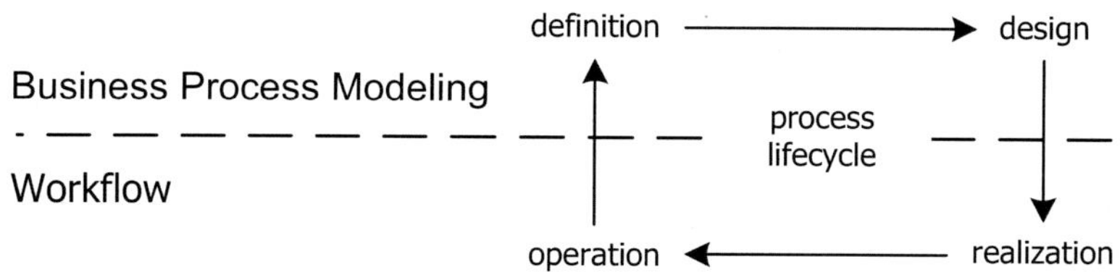

FIG. 15.4 Business process modeling and workflow.

Modeling a process usually involves breaking it down into activities that, taken together, are intended to produce a specific result of value to the business. One way of looking at process structure is as a hierarchical assembly of activities—a tree—as shown in Fig. 15.5. Loops are not permitted; a process element (at any level) cannot contain itself, but the same activity could appear at several places in the tree.

The root node of the activity tree represents the process itself. The leaf nodes represent the smallest units of activity (with no lower-level breakdown into finer-grained activities), which in most methodologies and tools have some special name, such as "activity", "task", "elementary business process", and so on. The intermediate nodes may also have some special name, such as "block" or "subprocess". In this chapter, we adopt the terms "process", "subprocess", and "activity", respectively, for the top, intermediate, and bottom levels of the process structure. In some methodologies and tools, actual work is assumed

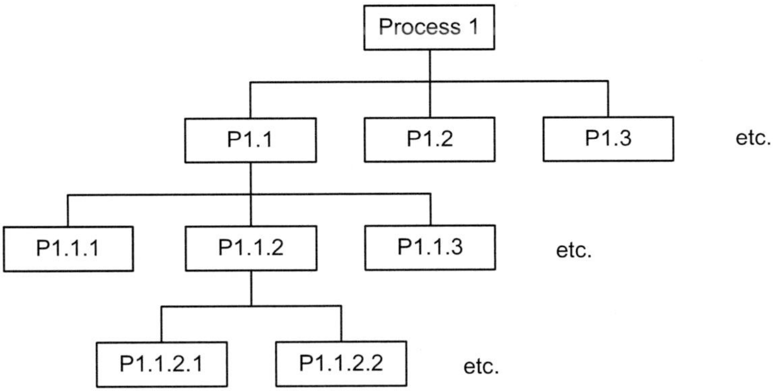

FIG. 15.5 An activity hierarchy.

to happen inside the top and intermediate nodes; in others, these nodes exist only as a mechanism for grouping the leaf nodes, which are assumed to contain all the actual work, into convenient clusters.

Clearly, we could have both a model of business activity and a model of information structure relating to the same area of a business, and therefore it should be possible to establish some kind of relationship between them. Most process modeling approaches focus mainly on flow-of-control issues and have surprisingly little to say about the information used by, produced by, or modified by the process. We look at some examples later in this chapter.

To establish a connection between process activities and the information they use, produce, or consume, we need to examine process structure in a little more detail. Specifically, we need to establish how the "activity" level of a process structure can be aligned with identifiable elements in an information model. We will return to the detailed sequencing and control of the activities later, but our focus for now will be on the *structure* of a process.

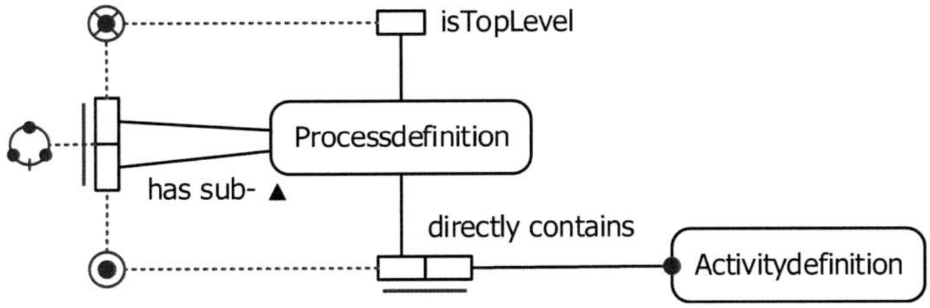

FIG. 15.6 Main process-definition structure.

A model fragment containing the main structure of a process definition is shown in Fig. 15.6. Processes can contain subprocesses to any number of levels, but loops are not allowed. We use the special term "activity" for the lowest level of the hierarchy and assume that a process definition (at any level) will contain subprocesses, activities, or both. The same type of activity could appear several times in the same process. For example, a "Charge Customer's Credit Card" activity could appear at various stages of the same process, and so could be directly contained in several parent nodes. Similar considerations apply to subprocess definitions.

Each process, subprocess, and activity can have multiple instances. Fig. 15.7 shows the relationship between the definitions of process elements and their instances. Each instance is an execution of the corresponding definition. A specific activity instance will always be running in the context of a specific process instance, even though the process instance could contain several instances of the same activity. Hence, the relationship between ProcessInstance and ActivityInstance is one-to-many instead of the many-to-many relationship between ProcessDefinition and ActivityDefinition. The current behavior of an activity can be indicated by a status value. The corresponding definitions do not have this kind of dynamic behavior, as they simply exist.

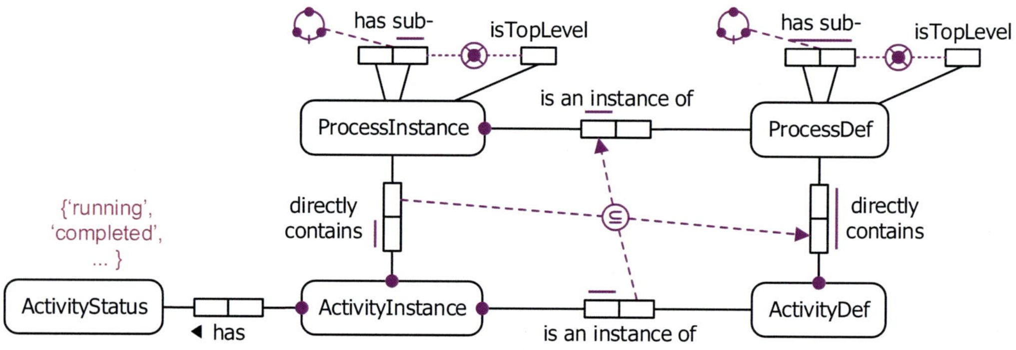

FIG. 15.7 Definitions and instances of process elements.

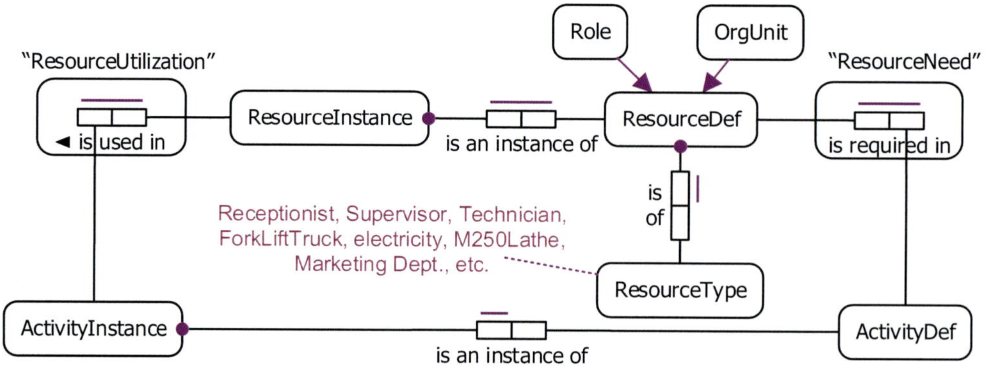

FIG. 15.8 Definitions and instances of resources.

Activities require resources to carry out useful work. The activity definition will normally identify the type of resource required. Instances of relevant resource types are associated with running instances of activities. One resource instance may meet the needs of more than one resource type. Some resources, such as electrical power, may be consumed by an activity. Other resources may be simply utilized for a while and then freed for use in other activities.

Fig. 15.8 shows how resources relate to the process structure. For simplicity, Fig. 15.8 shows resources as associated with activities. As discussed earlier, if process and subprocess definitions specify actual work, then resource definitions could also be associated with those (and, similarly, resource instances could be associated with the process and subprocess instances). "Role" and "OrgUnit" are examples of resource definitions, but there are many other possibilities. A particular resource instance may satisfy more than one resource definition. For example, an individual might be both a "Level 3 Manager" and a "Mortgage Specialist".

It's worth noting that resource allocation, just as information structure, tends to be treated lightly by modeling approaches centered on flow control. The relationship of process elements to organizational units also tends to have weak support, with the allocation of activities into "swimlanes" (which are discussed again later in this chapter) being the only mechanism provided in some approaches.

We won't dwell any further on resources and organizational relationships because we are going to be mainly concerned with the information aspects of the process. The activity definitions will specify the work to be carried out, but for our purposes, it's more important to focus on the data modified or accessed by the process. If the information structure (the business objects and their relationships) has been defined in ORM, this would correspond to references to relevant object types and fact types, as shown in Fig. 15.9. Instances of these object types and fact types are used in executing instances of the corresponding activities. An ObjectTypeInstance (or a FactTypeInstance) is, of course, simply an Object (or a Fact), respectively; "instance" has been retained in the terms to emphasize the difference between the two sides of the model. As discussed previously, object types and fact types could also be related to processes and subprocesses if these contain actual work definitions.

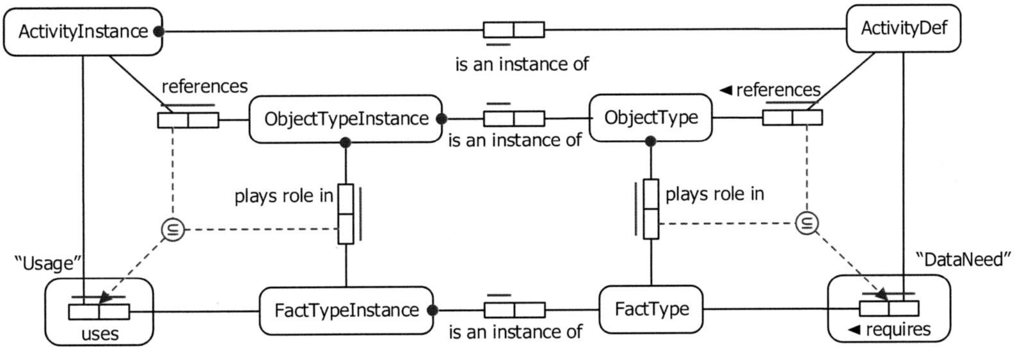

FIG. 15.9 Fact types, object types, and their instances.

At the most primitive level, information manipulation in object-role modeling (ORM) can be reduced to adding or deleting instances of facts. At a slightly higher level, we may wish to specify data usage in terms of whether the information is being Created, Read, Updated, or Deleted in the context of a particular activity. Fig. 15.10 shows the possible ways of using a fact type (or its instance) in the context of a particular activity, where "C", "R", "U", and "D" stand for Create, Read, Update, and Delete, respectively. Again, these notions could be extended to processes and subprocesses if necessary.

We may also wish to specify that a particular activity should be explicitly denied usage of a particular fact type because of security or other similar considerations. The unary isDenied fact type allows this to be specified.

The hierarchical structure of the processes can be utilized to define information scoping. This is not illustrated here for reasons of space, but we can imagine specifying usage of a fact type at a subprocess level that could "trickle down" to its successive lower-level process elements. This would avoid the need to specify data scoping only at the activity level. Mechanisms similar to the isDenied fact type could also come into play to prevent a lower-level element from inheriting its parent scope where necessary.

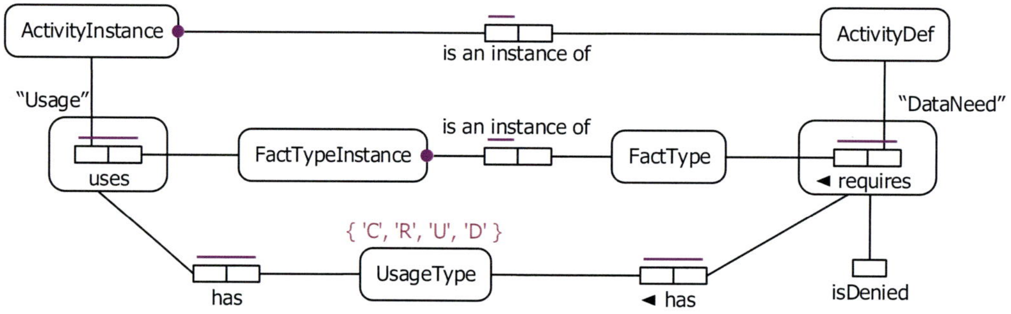

FIG. 15.10 Data usage from a fact-based perspective.

15.3 Foundations for Process Theory

Process modeling is an application that has motivated many technologists to build automated systems to support process definition, execution, management, and so on. There are probably well more than a hundred tools and techniques in current circulation, with more arriving all the time. Unfortunately, most of these approaches are based on vague intuitions about process features, with the result that their associated semantics are often a matter of guesswork. This is an unfortunate situation if we have ambitions to derive information system elements from business models because our model does not provide a definitive baseline to work from. The problems resulting from this have prompted interest in more formal ways of defining processes. The formality is not necessarily something that needs to be exposed at the business level, but we need to be sure that our business models are built on firm foundations.

The technique that has attracted the most attention in this regard is the use of Petri nets. These were first defined in Carl Adam Petri's Ph.D. thesis in 1962. They immediately attracted interest from theorists, but it took a little longer for them to find practical uses. Petri nets are now used in many fields, and the basic concepts have been extended in several ways. This section looks briefly at how they can be used to represent business processes. References to more detailed treatments of Petri nets are given in the chapter notes.

Fundamentally, a Petri net is a bipartite graph with variable markings at some of the nodes. There are two node types: *places* and *transitions*. Places and transitions are connected by directed arcs. The possible connections are place-to-transition or transition-to-place, but never place-to-place or transition-to-transition. A Petri net is used to represent a dynamic situation, but the structure of the graph does not change as the situation unfolds. Instead, the different conditions are marked by different arrangements of *tokens* at the places in the network. A particular configuration of tokens is sometimes referred to as the *state* of the network—"state" here does not have quite the same meaning that it does in the context of state machines that we discuss later in this chapter. Tokens are created at places, or removed from places, according to certain rules. A token cannot exist at a transition. Graphically, places are usually shown as large open circles; tokens as small, filled circles; and transitions as either rectangles or thick bars. Fig. 15.11 shows a simple Petri net with tokens at some of the network places.

A transition is said to be *enabled* if it has a token at each of its input places. An enabled transition can *fire*, and in doing so it consumes a token from each of its input places and produces a token at each of its output places. For the configuration shown in Fig. 15.11, transition t1 is enabled but t2 and t3 are not (t3 does not have a token at all its input places). The resulting situation after transition t1 has fired is shown in Fig. 15.12. A token has been removed from p1 and a new token added at p2.

Conceptually, tokens are consumed or created at transitions. Transitions can have input places (the arc from the place is directed toward the transition) or output places (the arc to the place is directed away from the transition). The same place can be both an input place and an output place relative to some transition. In Fig. 15.11, transition t1 has p1 as an input place and p2 as an output place. Place p3 is both an input place and an output place for t3.

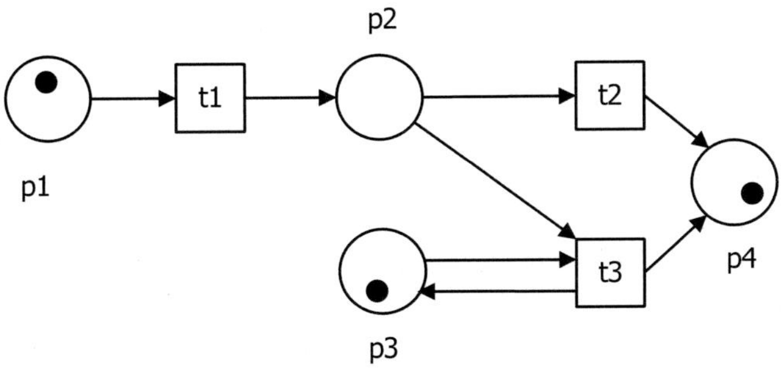

FIG. 15.11 A simple Petri net.

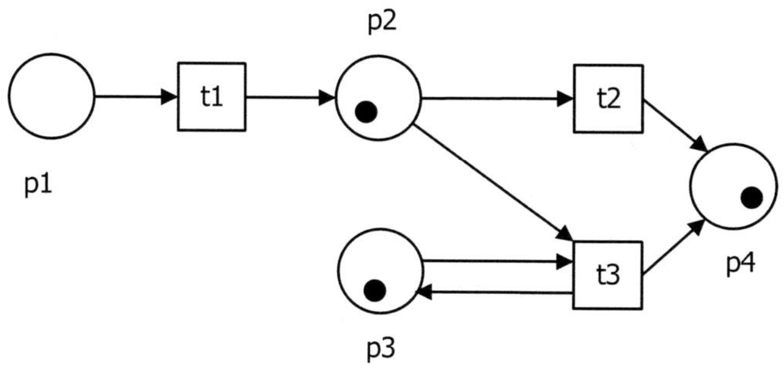

FIG. 15.12 The network of Fig. 15.11 after transition t1 has fired.

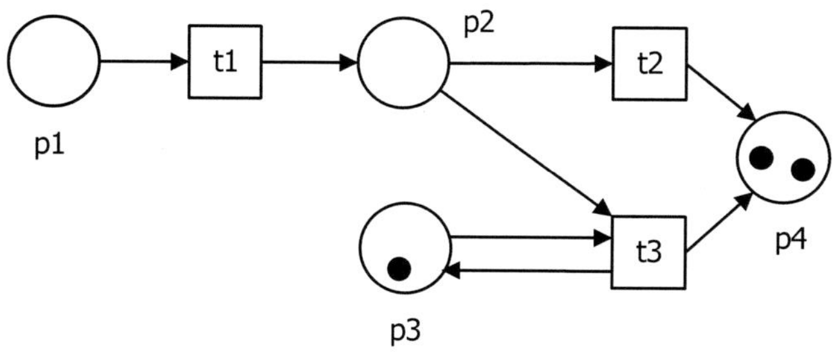

FIG. 15.13 The network of Fig. 15.12 after transition t2 or transition t3 has fired.

Now, both t2 and t3 are enabled. Because we have only one token at p2, only one of t2 and t3 can fire. In a basic Petri net, this situation is nondeterministic, as we cannot decide in advance which of the transitions will fire. Adding more tokens to place p2 would not solve the problem, as transitions t2 and t3 would still compete for each token. However, in this case, the same state results whichever transition fires. If t2 fires, it consumes the token at p2 and produces a token at p4. If t3 fires, it consumes the tokens at p2 and p3 and produces a token at p3 and p4. The final result is shown in Fig. 15.13. At this point, no further transitions are enabled and activity ceases.

To make use of a Petri net, we must map its elements onto the real world. Typically, tokens are taken to represent objects (people, goods, machines), information conditions, or object status; places represent buffers, channels, geographical locations, conditions, or situations; and transitions represent events, transportation, or transformation. In the specific case of business process modeling, transitions are usually taken to represent business activities, whereas the combination of places and tokens determines when the activity is carried out.

The basic constructs of the Petri net can be composed into more convenient forms to provide building blocks to model business processes. Some constructs appear frequently, and it's useful to be able to insert these into a model without repeating the basic Petri elements that they are built from. Processes commonly need to split into parallel branches at some point and then reconcile the branches at some later point, for example by waiting until activities on preceding parallel branches have been completed. Various forms of "Split" and "Join" (or "Merge") are commonplace in process models. Both come in two flavors: OR and AND. An OR-split sends a token along one path or another while an AND-split sends a token along multiple paths. Similarly, an OR-join expects a token to arrive on one of several branches, whereas an AND-join expects a token on all branches.

Fig. 15.14 shows the high-level elements with their equivalent constructs in a basic Petri net. The notation used for the high-level constructs is taken from YAWL: a workflow language based firmly on Petri nets. Note particularly that the construction for the OR-split is nondeterministic in its basic Petri net form. To make it more useful for business modeling, we assume the existence of a business rule (or rule set) that can steer a token in the appropriate direction.

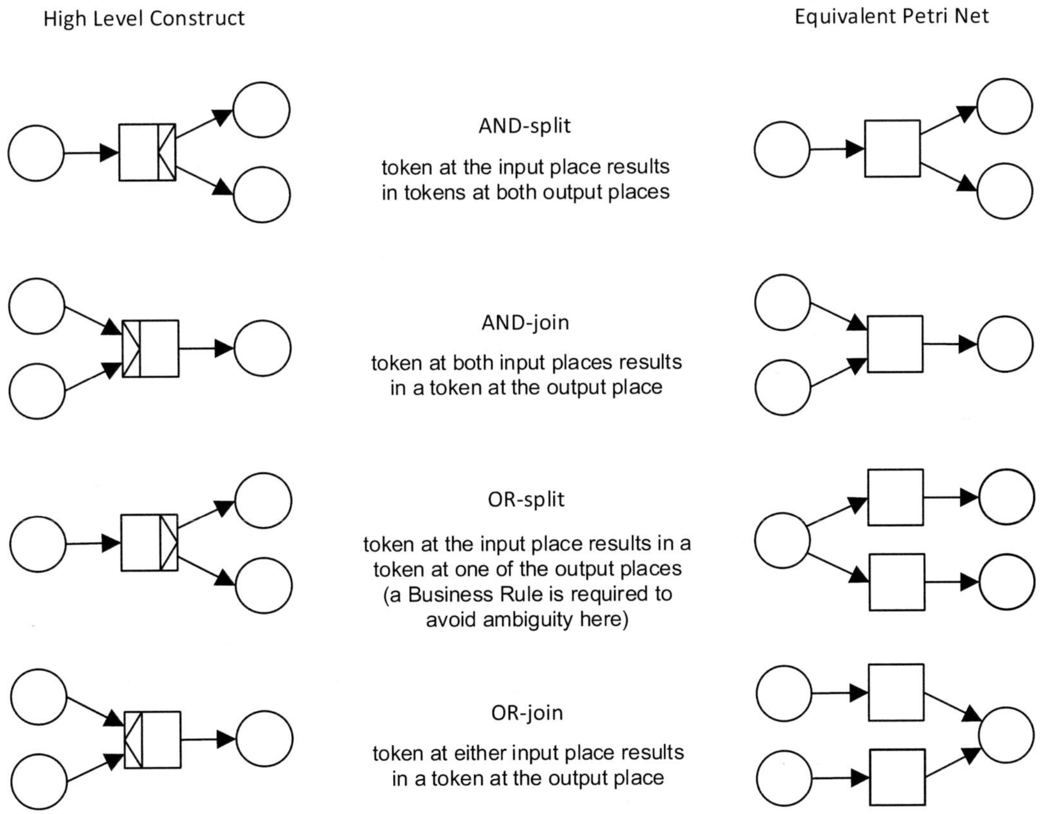

FIG. 15.14 Higher-level split and join constructs.

Given these constructs, we can begin to put together networks of increasing complexity. As an example, Fig. 15.15 shows how we can define some classical looping constructs using Petri nets.

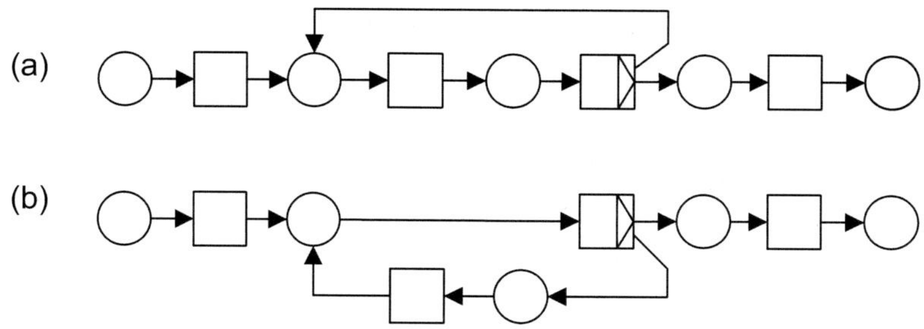

FIG. 15.15 (a) A "repeat-until" loop. (b) A "while-do" loop.

The standard Petri net allows us to construct models of dynamic behavior, but it has some shortcomings for process modeling. Nets that result from the primitive constructs become too large and clumsy for realistic cases, and it's difficult to represent fine-grained business information. Because of this, several extensions have been proposed to make Petri nets more useful in a business context.

The first idea is to use different types of tokens. In the basic Petri net, tokens are all the same kind. If we retain the place/transition network but allow multiple token types, we open some new possibilities. For example, we can make the firing of a transition dependent on token types and values. Conceptually, we can think of the different token types as being tokens of different colors, and so a network extended in this way is known as a colored Petri net.

A second important idea is to provide ways of organizing networks hierarchically. We have already seen this concept from other perspectives. In the case of Petri nets, we have networks containing subnetworks, potentially nested to several levels. At a high level, we can represent a subnetwork (which might correspond to a subprocess) as a single block, allowing us to selectively hide detail in complex networks.

A third extension introduces the dimension of time. Activities are typically modeled as transitions, but in a basic Petri net, transitions are considered instantaneous. We can provide a more realistic simulation by time-stamping tokens so that we have information about their temporal evolution, and allowing transitions to have delays that would represent the passage of time while an activity is carried out in the real world.

Fig. 15.16 shows an example of a hierarchical network for a process to handle complaints that uses different types of tokens to indicate different cases that are being progressed.

Several other candidate formalisms for business processes have been proposed. One is the use of state machines, discussed shortly. There is not a single state machine formalism, but several state machine frameworks have been formally defined, notably by David Harel (see Harel and Politi, 1998). If we begin with a state machine that has a formal description, then we can build other constructs on top that inherit the formal foundations.

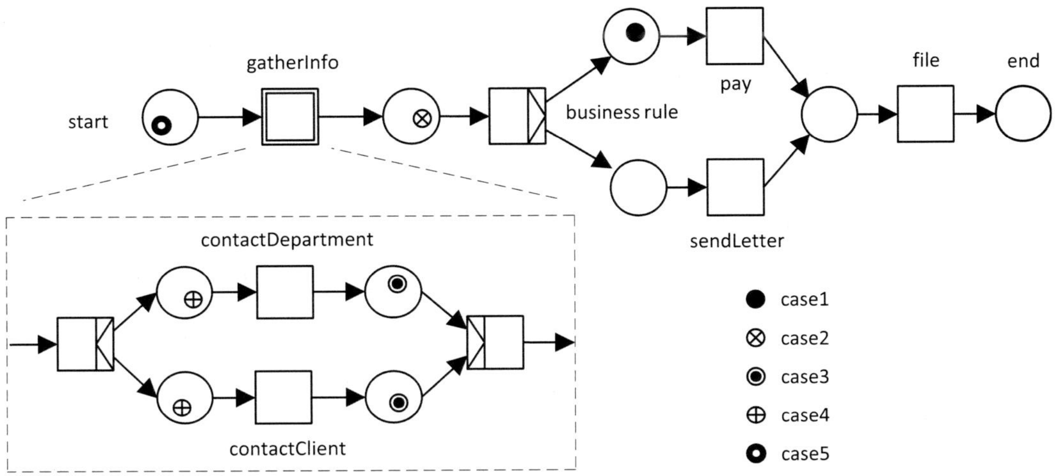

FIG. 15.16 A process for handling complaints.

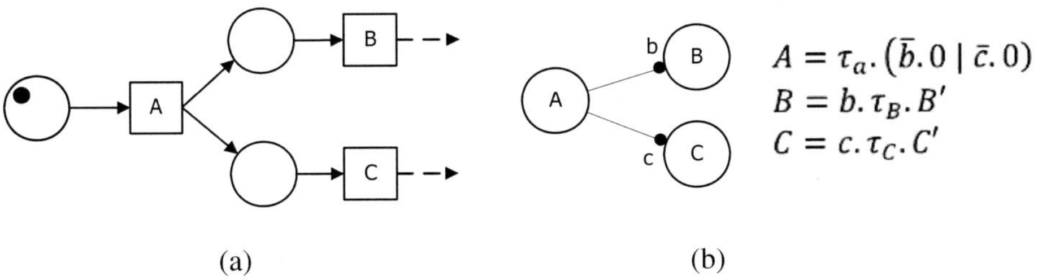

FIG. 15.17 Parallel split pattern: (a) Petri net and (b) Pi calculus.

Another approach that has received a fair amount of publicity is the use of process algebras, with the most prominent example being the *pi calculus* defined by Milner (1999). In these approaches, a series of algebraic expressions replaces a network as the basis for defining process structure. For space reasons, we won't go into detail here, but the following example provides something of the flavor of the approach. As a simple example, we can take a common workflow pattern, a parallel split. (We discuss this and other patterns in more detail later in the chapter.) Fig. 15.17 shows equivalent specifications of this pattern using both Petri nets and pi calculus.

There is currently no agreement on whether algebraic approaches offer significant advantages over other approaches such as Petri nets. They are certainly not at all useful at the business level, but their place is more likely to be hidden under the covers of a tool that presents a more user-friendly interface.

15.4 State Models

The term "state" is used in several different ways in modeling. One usage relates to the changes made to information in a particular domain. An information processor can add or delete facts or modify existing facts in a fact store. After any of these changes, we could say that the fact store is in a different *state* than previously. In this sense, the number of possible states for an information model (or its equivalent realization) would be too large to count. Individual states of this sort are not identified by name or any other designation, and so the term is usually used in a before-and-after sense in relation to some update.

A different usage of the term state relates not to changes in *information* but to changes in *behavior*. As a simple example, let's say that we have a device that counts occurrences of a specific event. When a certain number has been reached, it takes some action and then waits to be reset. An alarm clock is a practical device of this kind.

While the device is counting, its information state is continually changing as the count is updated. However, its behavior remains consistent until the limit value is reached. From the information perspective, the device would have many states (e.g., one for each count value), but from the behavioral perspective, it can only be in one of three states.

1. Counting events and checking the total against the predefined limit.
2. Carrying out defined actions.
3. Waiting for the user reset signal.

In the early days of electronic computers, it was recognized that behavioral state descriptions provided a convenient way to define and build automata with the capability to react in seemingly complex ways to changes in their environment, and these reactive systems became known by the generic term *state machine*. Two alternative formulations appeared in the 1950s, named (after their inventors) Moore machines and Mealy machines. These are arranged slightly differently but have similar capabilities–in fact, a Moore machine can be transformed into an equivalent Mealy machine, and vice versa. Fig. 15.18 gives a high-level view of the two approaches.

Both approaches store information relating to the current state of the machine. The combination of the current state and the input to the machine determines whether a state change occurs. The main difference between these two approaches is that the outputs of a Moore machine are related to the machine being *in a particular state*, whereas in a Mealy machine, the output is determined by the input in the context of that state. For a Mealy machine, changes to the output therefore tend to be associated with *changes in state*.

We can illustrate some of the differences between the two using a bubblegum vending machine as an example. The machine sells only one type of item, which costs 15 cents (15¢). The machine takes coins of value 5¢ (a nickel) or 10¢ (a dime) but gives no change if excess money is inserted.

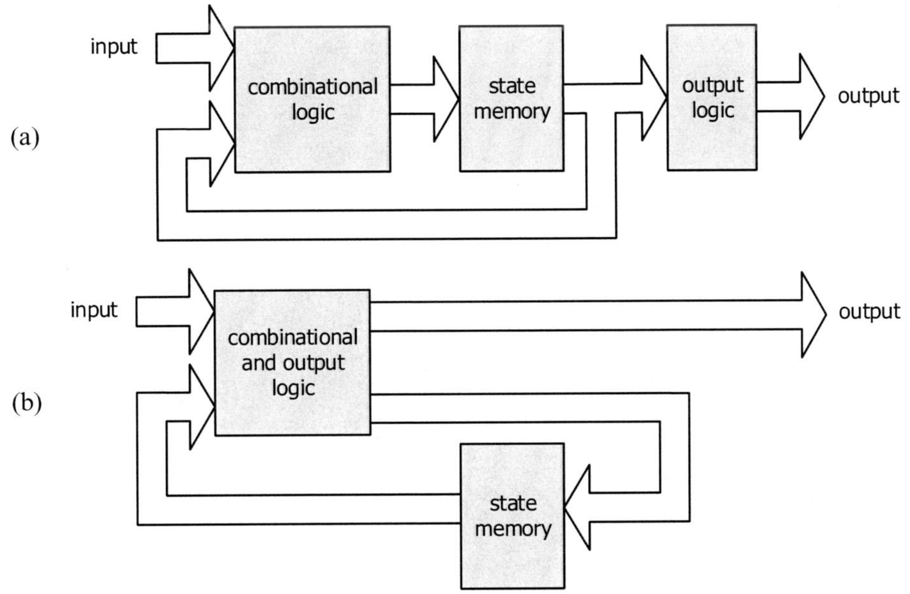

FIG. 15.18 (a) Moore machine. (b) Mealy machine.

The machine has a reset button: if this is pushed before vending, money already inserted is refunded and the machine returns to its initial condition (state 0). Once sufficient money has been inserted, the machine automatically delivers the item and resets ready for the next customer. We will ignore various practical issues such as running out of gum, jamming of the mechanism, timing out if an attempt to purchase is abandoned before completion, and so on.

Fig. 15.19 shows a Moore machine model of the vending machine. The states are shown as boxes labeled with the state name. For convenience, the state names here are chosen to represent the total amount of money that has been inserted in the current transaction, so the possible states are 0, 5, 10, and 15.

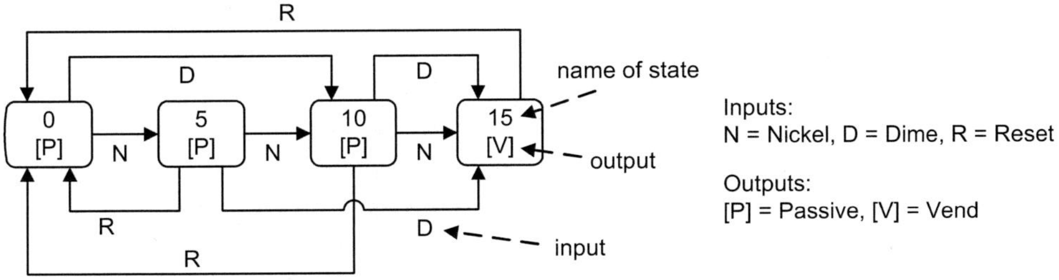

FIG. 15.19 A vending machine modeled as a Moore state machine.

We are ignoring excess money being inserted, so 15¢, the sale price, is the maximum we need to consider. The state also shows the action the machine takes in that state, which is either Passive, if it's waiting for more money, or Vend, if sufficient money has been inserted. The links between the state nodes represent transitions between states. The transitions are labeled with the event that causes the transition. The transition between state 15 and state 0 has no associated event because it happens automatically once the activity in state 15 has completed. Again, for simplicity, we're using the money amount inserted to identify the event, and so the possible inputs are either nickel (N) or dime (D).

We are taking state 0 as the initial state. The possible events in this state are either N or D, which take us to state 5 or state 10, respectively. From state 5, the event N or D takes us to state 10 or state 15, respectively. From state 10, either type of coin input (N or D) will take us to state 15. In state 5 and state 10, a reset will return the machine to state 0. In state 0, state 5, and state 10, the vending machine is passive, but once state 15 is reached it produces the output (some gum). The transition from state 15 to state 0 is not labeled with an event because the transition is taken automatically when the activity in state 15 is completed.

Fig. 15.20 shows the same vending machine as a Mealy state machine. We have the same states, but there is no output (Passive or Vend) associated with a state. Instead, the outputs are associated with the transitions between states, which are labeled both with the event causing the transition and the output produced in that transition.

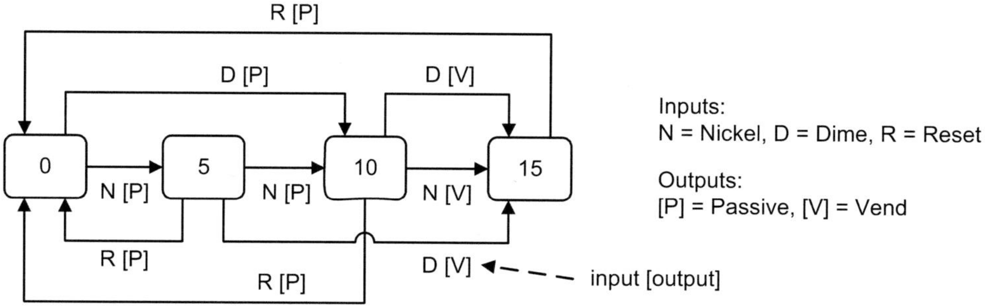

FIG. 15.20 A vending machine modeled as a Mealy state machine.

State machine descriptions have been popular for several decades as a way of specifying and implementing automata; we see many examples in everyday life in the form of digital watches, cell phones, video recorders, and so on. However, for a long time, it was believed that state machines did not scale well because of the explosion of states required to deal with all the combinations of variables. In a naïve state machine interpretation, the total number of states required is of the same order as the product of the number of different values for each involved variable. For instance, if we have three variables with 3, 4, and 5 possible values, respectively, the likely number of states is $3 \times 4 \times 5 = 60$ to cover all possible combinations of values. Of course, some combinations may actually be associated with the same state but considering the large number of variables involved in a typical business system, the idea of a state-based description seems futile.

A major contribution was made by David Harel, who pointed out that state descriptions could be composed in more efficient ways, and also arranged hierarchically so that state transition diagrams are not swamped with detail. The key notion is that states can be defined to represent the composition of other (sub-)states, allowing us to consider some states as a group rather than having to deal with them individually.

For example, we could have the situation shown in Fig. 15.21(a). States *U* and *V* both have a similar transition to another state *W*. By placing *U* and *V* in a composite state *X*, we only need to show a single transition from *X* to *W* instead of the individual *U* and *V* transitions to *W*. The small, filled circle and the line that originates from it represent the initial transition into the default state when the state machine is started. Fig. 15.21(b) shows a composite equivalent of Fig. 15.21(a). The state machine begins in state *X*. When state *X* is entered, it immediately transitions to state *U*. In both cases shown in Fig. 15.21, the state machine therefore begins in state *U*. This is known as OR composition because we can only be in state *U or* in state *V* within the composite state *X*.

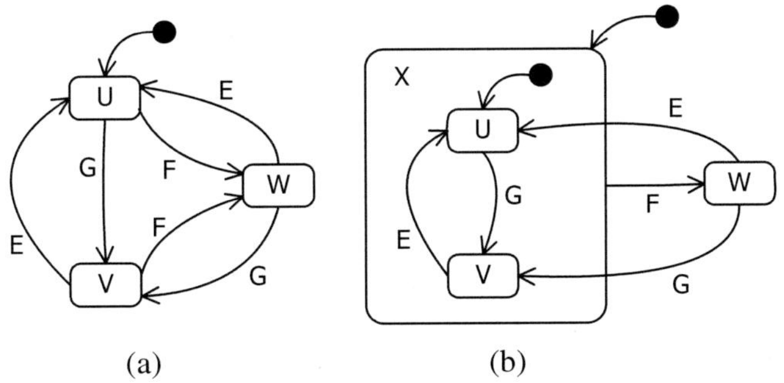

FIG. 15.21 OR composition of states: (a) decomposed and (b) composed.

A second form of composition allows a state to contain several regions. These are considered *orthogonal* because state transitions can evolve separately in each of the regions. Fig. 15.22 shows a simple example. The states in Fig. 15.22(a) have composite names to make it easier to see the relationship between the composed and the decomposed variants. In Fig. 15.22(b), a new state *S* has been introduced with two regions, *R* and *T*. Informally, *R* and *T* could be visualized as substates of *S*. The crucial difference in this case is when we enter state *S*, *both* of its regions become active, with a default transition to state *U* in region *R* and a default transition to state *X* in region *T*. The combination of the state *U* in *R* and state *X* in *T* in the composed variant is equivalent to the state *U-X* in the decomposed variant. For example, receiving event *J* in the composed variant has no effect in region *R*, but causes a transition to state *W* in region *T*. This is equivalent to the transition from state *U-X* to the state *U-W* in the decomposed variant.

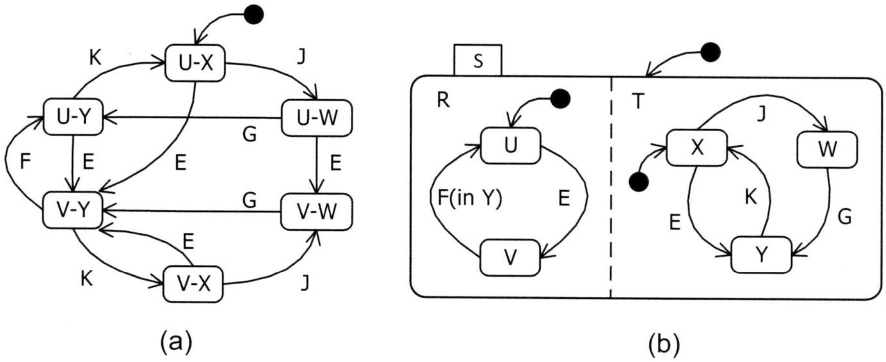

FIG. 15.22 AND composition of states (a) decomposed and (b) composed.

This approach is sometimes known as AND composition because we are in one state in region *R* *and* one state in region *T*. Another way of looking at this is to see it as an AND composition of OR compositions (because in any orthogonal region we can only be in one state at a time).

One additional feature shown in Fig. 15.22(b) is the use of a guard associated with the transition from state *V* to state *U*. The guard condition must evaluate to "True" to enable the transition. The result is that event *F* will not trigger the transition in region *R* unless we are in state *Y* in region *T*. This is consistent with the equivalent transition in Fig. 15.22(a) between state *V-Y* and state *U-Y*.

Although the savings made by OR and AND composition may appear trivial in simple examples, they can become very significant in larger models. For instance, in Fig. 15.22(b), region R has two substates and region T has three substates. If these were increased to four substates and five substates, respectively, the orthogonal version would have $(4+5)=9$ nodes, but the nonorthogonal version would have $(4 \times 5)=20$ nodes.

It's important to recognize that state models are primarily intended to define the locus of control within a given scope. A state model does not in itself define any actions that should be taken. However, an action definition can be linked to the state model to identify the (relative) time that the action should be taken.

The main possibilities are for actions to take place as part of a particular transition from one state to another (Mealy semantics) or in a particular state (Moore semantics). Within a state, we can further distinguish between actions that are taken just after the state is entered, sometime during the state, or just before the state is exited.

So far, we've looked at some basic ideas about describing information dynamics in a generic way. The features of process- and state-based models discussed so far can be found in many methodologies and tools that address this area. To provide a more concrete illustration, we turn to one of the most widely used modeling approaches: the unified modeling language (UML).

15.5 Modeling Information Dynamics in UML

The UML includes a range of different diagram types—13 types in all in UML Version 2 (an increase over the nine types available in UML Version 1). Fig. 15.23 shows the range of diagram types; boxes with dashed lines are simply summary supertypes and not actual types of diagram.

UML supports information modeling through class diagrams and, to a lesser extent, object diagrams. These have been covered in some detail elsewhere in this book and we won't discuss them again here. UML also has several diagram types that can be used to model behavior.

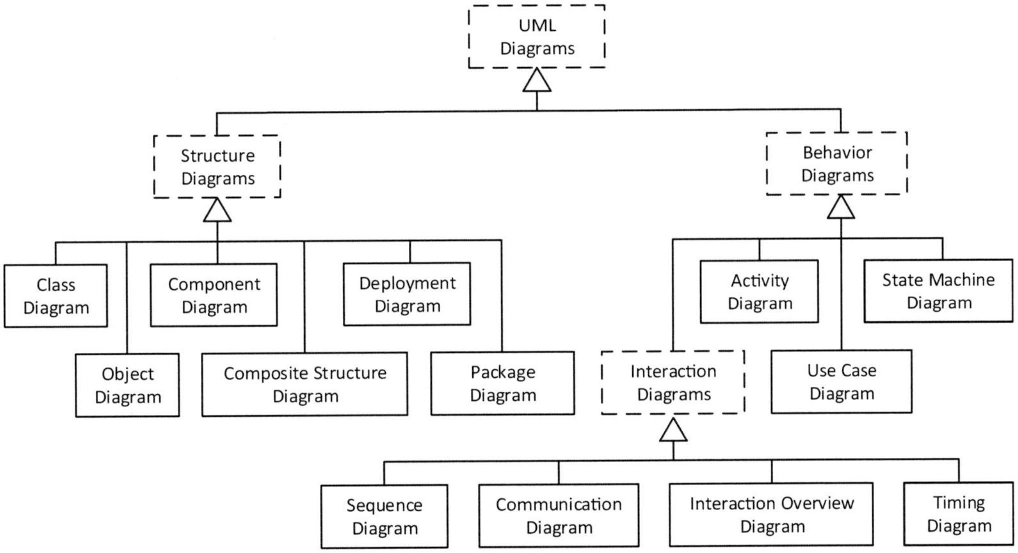

FIG. 15.23 UML 2 diagram types.

The four types of interaction diagrams shown in Fig. 15.23 are mainly of interest in software development and have limited application in business modeling. Sequence diagrams and communication diagrams describe messages passed between object instances. Interaction overview diagrams are a specialized variant of activity diagrams—we discuss activity diagrams in more detail shortly. Timing diagrams show the detailed timing of interactions between various system elements and are mostly only relevant to real-time systems.

The types most relevant to business modeling fall under the heading of Behavior Diagrams. We first take a very brief look at use case diagrams. A use case describes a connected series of actions performed by one or more actors related to the system under discussion. An actor may be human or machine, but, by convention, both are shown graphically as a stick figure icon. The use case description is usually expressed from the viewpoint of one specific actor—the principal actor. A use case diagram contains none of this description and is essentially no more than a list of the relationships between actors and use cases (and sometimes between use case and use case), depicted in a graphical rather than a tabular form. A simple example is shown in Fig. 15.24.

The oval shapes in Fig. 15.24 represent use cases. These are usually named with a verb phrase that summarizes the actions being performed in the use case. The optional rectangular boundary represents the system, which is slightly misleading because the use case describes the actor's interactions with the system (i.e., the use case is not "in" the system).

The UML standard describes various extensions to the basic concept. For instance, some shared activity that potentially appears in several use cases can be extracted out into a separate use case and referenced by an «include» relationship. Some activity that only occurs in exceptional cases can be separated out into a different use case and referenced in an «extend» relationship.

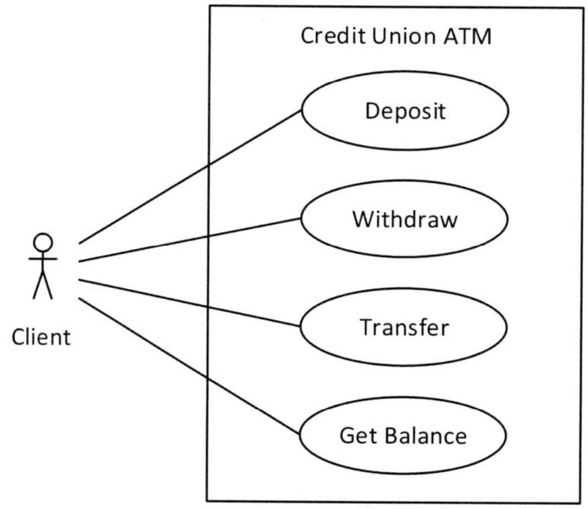

FIG. 15.24 A simple use case diagram for an automated teller system.

Actors and use cases can be treated as classes, with subtyping used to show hierarchical relationships between actors and between use cases. However, these and other elaborations often end up causing more problems than they solve and should be used with caution. The valuable part of a use case is the actual behavioral information contained in the use case description, but, strangely, this aspect is not defined in UML.

A common error is to equate a use case with a business process. This may be true in very small systems, but in most major enterprises, cross-functional business processes span many organizational units. This makes it unlikely that individual actors (from whose perspective the use case is being described) will see the whole of the process. It is more likely that a use case describes a segment of a process, as shown in Fig. 15.25 (note that this figure is *not* a UML use case diagram!).

Use cases that overlap in their descriptions (as do UC2 and UC3 in Fig. 15.25) should corroborate each other. If not, further investigation is required. Gaps in the coverage of the process (as between UC3 and UC4 in Fig. 15.34) also require investigation. Perhaps we have a missing use case? Or perhaps our understanding of the process is not correct?

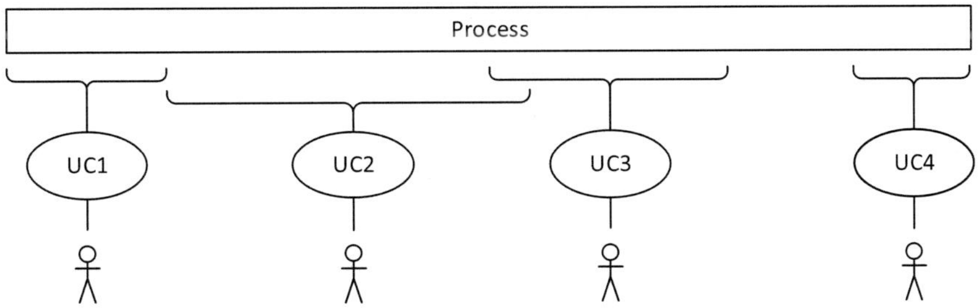

FIG. 15.25 Relationship between use cases and processes.

Although use cases can help uncover some dynamic aspects of business information, they are not sufficiently rigorous to be our primary modeling tool, and they are best seen as a step on the path to producing a more comprehensive and precise business model. The remaining two UML behavioral diagram types can be pressed into service to model dynamic business behavior, and we will look in a little more detail at some of their features.

Activity Diagrams

Activity diagrams are mainly focused on the flow of control within a set of related activities. The level of granularity is not defined by UML, so the scope of each individual activity is at the discretion of the modeler. This has the advantage that activity diagrams can be used for both high-level and low-level descriptions, but places the onus of defining activity boundaries onto some other authority. Activities in UML are considered to be composed of atomic actions. Activities can be interrupted but actions cannot, which implies that any interruption to an activity must take place on a boundary between actions. We have already seen flow-of-control chaining of activities earlier in this chapter, and it's not surprising that UML activity diagrams are seen as a candidate notation for documenting business processes. The underlying semantics of activity diagrams changed between UML versions 1 and 2. In UML 1, activity diagrams were seen as a special kind of state machine description and shared a great deal of terminology with UML statecharts. In UML 2, the semantics of activity diagrams became realigned to be closer to the semantics of Petri nets.

An activity diagram consists of several nodes connected by arrowed lines (directed edges). There are three main types of nodes: action nodes, object nodes, and control nodes. Action nodes represent some work being carried out and are shown as a rectangle with rounded corners. The rectangle typically contains the name of the action; other information may optionally be added that we won't consider here. Object nodes, shown as rectangles with square corners, represent an object type, for example, a business entity such as an invoice that may be passed from one activity to another. Object nodes are generally only shown when some special consideration needs to be given to the objects being passed; the routine use of business objects by activities is just assumed. The "flow" along the activity edges is either of control tokens, indicating the transfer of control from one activity to another, or of objects. Because the edges are directed, there is no assumed direction of flow, although diagrams are conventionally laid out so that the main flows are either left to right or top to bottom.

Control nodes are abstract activity nodes that coordinate flows in an activity diagram and come in various subtypes. An *initial* node is represented by a filled circle and represents a starting point for the activity diagram. There can be more than one initial node. There are two kinds of *final* nodes. An *activity final node* is shown as a target symbol, and an activity diagram can have more than one of these. As soon as an activity final node receives a token, the entire activity immediately terminates, even if there are other control tokens active in other paths. A *flow final node*, shown as an "X" in a circle, destroys any token that reaches it without affecting any other concurrent flows.

A *decision* node, denoted by a diamond shape, has one input edge and one or more output edges. A token arriving on the input edge will leave on only one of the output edges (i.e., exclusive-or). To control this, the output edges have guard conditions, denoted inside square brackets. The guard conditions should be complete and disjoint to ensure that token flow is not inadvertently halted or duplicated. A default branch can be provided by specifying it with the guard condition [else].

The diamond shape used for a decision node is also used for a *merge* node. This performs the complementary action to a decision node by bringing together multiple alternate flows on its input edges. Any token arriving on an input edge is offered to the output edge. There is no synchronization of tokens arriving on the incoming edges.

A *fork* node splits a flow into multiple concurrent flows. It is shown as a solid bar with one input edge and two or more output edges. Each token arriving on the input edge is duplicated and offered to each output edge. There is no sequencing implied between the outgoing flows.

A *join* node is used to combine parallel flow paths. The join notation is the same as for a fork, except that the join has multiple input edges and one output edge. An optional join expression determines how incoming tokens are combined to produce an output token. The default is "and", implying that the join waits for a token to arrive on each incoming edge before producing a token on the output edge (i.e., synchronization).

The UML 2 specification does not require that branches introduced in the diagram should be balanced. In other words, multiple paths introduced by decisions or forks do not necessarily have to be recombined by complementary merges or joins. Although this provides maximum flexibility in defining a process, it can also provide subtle traps for the unwary and can lead to unsound process definitions.

Fig. 15.26 shows an example activity diagram for a simple sales process. The diagram is based on an example given in the UML specification, modified slightly to contain examples of the features we discuss. Although the modified diagram is still syntactically legal, it represents an unsafe process design, as we will see shortly.

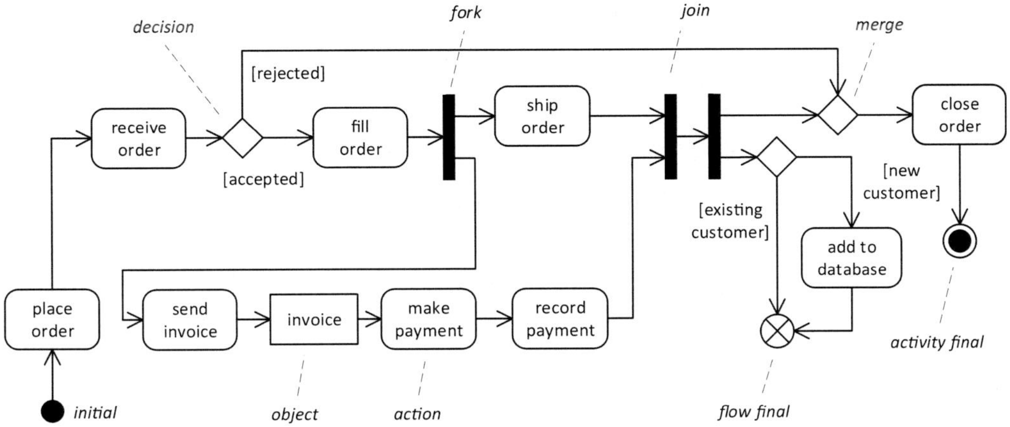

FIG. 15.26 An activity diagram for a simple sales process.

The diagram is self-explanatory, but not unique, in the sense that the same process could have been depicted in several slightly different ways. For example, the join and the fork that immediately follow it could have been combined into a join/fork construct that would have appeared as a thick bar with two edges entering and two edges leaving. Combining a join and a fork in this way can make diagrams slightly tidier but does not affect the semantics.

The execution of the process can be simulated by tracing the flow of a token in a particular instance of the process represented by the diagram. The token originates at the initial node. A customer places an order, which is received by the company's sales department. The first decision is to either accept or reject the order. The two guard conditions, [rejected] and [accepted], are mutually exclusive. If the token takes the [rejected] branch, it immediately reaches the merge and appears on its output edge. After the order is closed, the token reaches the activity final node, which immediately ends the process instance. If the token takes the [accepted] branch, the order is filled. When the token reaches the fork, two tokens are generated. The upper token causes the order to be shipped, and the lower token stimulates a series of actions to invoice the customer and eventually receive payment.

The upper and lower branches have no implied timing constraints, so either could be the first to complete. The following join waits for a token to arrive on both branches before emitting its own single token. This immediately forks into two other branches. The token on the upper branch reaches the merge that we have already discussed and causes the order to be closed. The token on the lower branch reaches a decision. If the customer is new, then they are added to the customer database, but for an existing customer, the token is immediately consumed by the flow final node.

The second fork in the diagram is not balanced (there is no corresponding join), but we eliminate the token in the lower branch because it eventually reaches a flow final node. This ensures that adding the customer to the database does not interfere with the closing of the order. However, the design is unsound because the lower branch is in a race with the upper branch. Once the order is closed, a token will reach the activity final node and immediately terminate the process, regardless of the status of the token in the other branch. This could result in a new customer *not* being added to the database, contradicting the process designer's original intentions.

One other feature of activity diagrams is support for swimlanes. These are optional but can be used to indicate features such as separation of responsibilities, usage of resources, geographical distribution, and so on. As with other features, the precise meaning of a swimlane is not defined in UML; the interpretation is left up to the user of the activity diagram. Fig. 15.27 shows the process model of Fig. 15.26 with swimlanes added. In Fig. 15.27, the swimlanes are drawn to indicate who carries out the actions (customer or sales clerk). The invoice object is used by both, and so is shown on a swimlane boundary. Multiple orthogonal swimlanes can be defined. For example, one axis could represent organizational units and the other could represent resource types. The result resembles a kind of Venn diagram with the actions placed in whichever swimlane intersection is appropriate. However, this quickly gets difficult to draw and can be confusing to read.

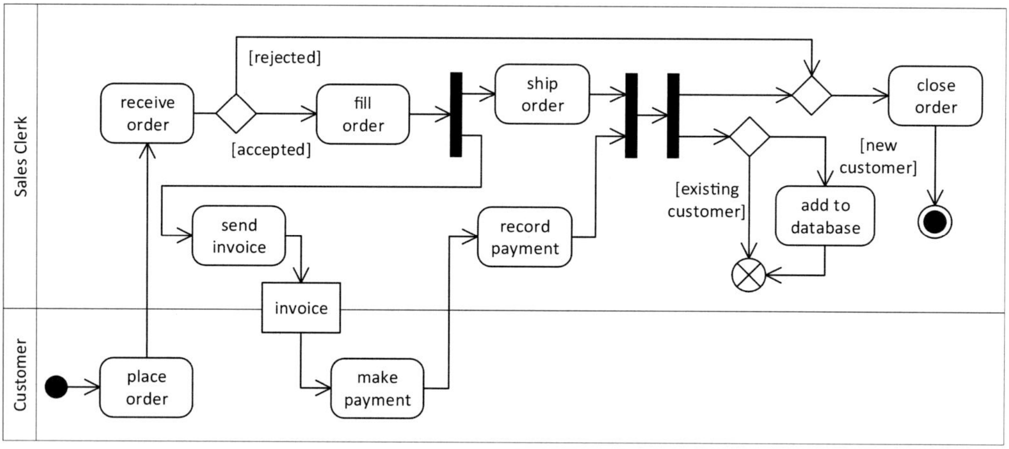

FIG. 15.27 An activity diagram with swimlanes.

Activity diagrams also provide for the hierarchical composition/decomposition of actions. A structured activity of this kind can be treated at one level as a single action but can be expanded if required to show the breakdown into its subordinate activities. Activities can be nested in this way to any level that's useful.

When considering the relationship between an activity diagram and the data that the activities may produce or consume, we need to bear in mind that the primary purpose of the diagram is to illustrate activity sequences. There is only a limited ability to show data in the form of objects being transferred between activities, as with the Invoice object in Fig. 15.27. These objects are transient and disappear when the sequence of activities terminates.

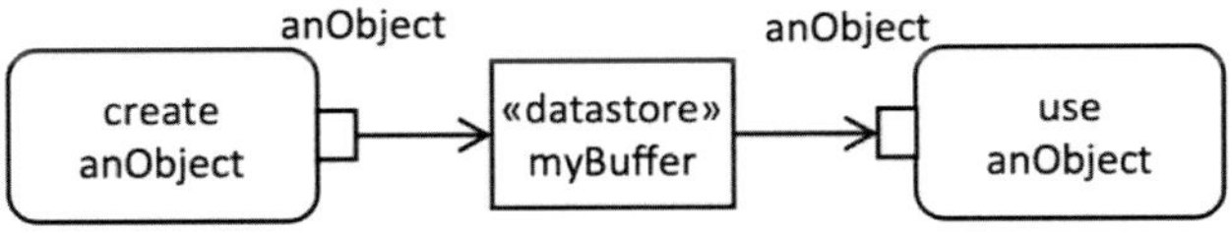

FIG. 15.28 Internal buffering of objects.

There is no direct way of showing how activities relate to any kind of persistent data. UML does provide a promising-sounding element with the stereotype «datastore», but this is used to indicate a specific type of internal buffer, as shown in Fig. 15.28, and still represents only transient data.

Fig. 15.28 shows an object transferred between activities, using the alternative "pin" notation to indicate the object. The transfer is indirect, with the object being buffered in the datastore ready for use by the succeeding activity. UML also provides another type of buffer with the stereotype «commonBuffer», the difference being that a «datastore» transfers a copy of the object so that the same object can be retrieved multiple times.

Activity diagrams have several other features that we haven't space to discuss here. Examples include signaling between system elements, events used to trigger activities, the definition of interrupts and interruptible regions of activity, and so on. Many of these features can be useful in the definition of software but are less relevant for conceptual modeling.

State Machine Diagrams

In UML version 1, state machine diagrams were known as statecharts and were closely aligned with activity diagrams, to the extent that an activity diagram could be considered as a special type of statechart. In UML version 2, state machine diagrams and activity diagrams occupy positions that are rather more distinct.

UML state machine diagrams have many of the same main constituents as other state diagramming conventions. States are normally shown as rectangles with rounded corners. The name of the state can be shown free form inside the rectangle, in a separate compartment in the rectangle, or as a name tag attached to the rectangle, as shown in Fig. 15.29.

States are interconnected by transitions, shown as directed lines. The lines typically begin on the border of one state and end on the border of another state; we discuss some special cases later. A transition can begin and end at the same state, and there can be more than one transition between a particular pair of states.

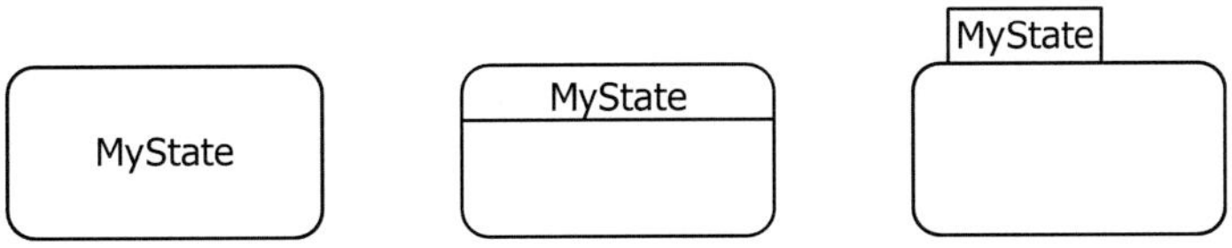

FIG. 15.29 Alternative state notations in UML.

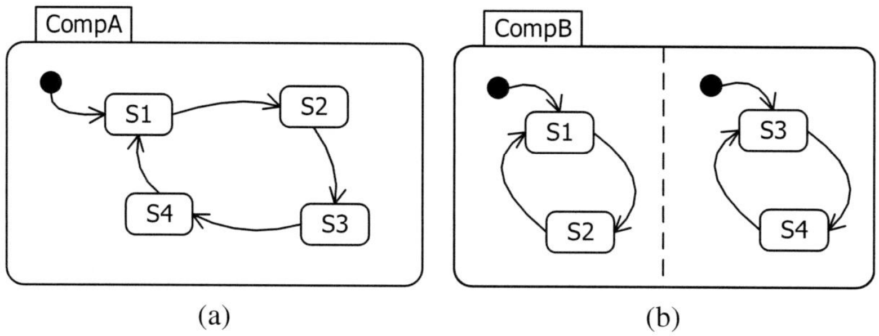

FIG. 15.30 (a) Simple composite state. (b) Orthogonal state.

A state can be simple or composite. Composite states contain other substates within them. If a state has an active substate, then the containing state is also active. A composite state that is subdivided into multiple regions is known as an orthogonal state. Fig. 15.30 shows examples of a simple composite state (containing exactly one region) and an orthogonal state (in this case, with two regions).

The small black nodes in Fig. 15.30 are special *pseudostates* that represent the initial conditions for their respective states. Within a region, only one substate can be active at any one time, so the state shown in Fig. 15.30(a) can be in state S1, S2, S3, or S4. In contrast, the state shown in Fig. 15.30(b) can be in state S1 or S2 and simultaneously in state S3 or S4. A state can also contain nested state machines—called submachine state machines—that are semantically equivalent to composite states. Submachines are a convenient way of encapsulating common behaviors for reuse.

Actions can be carried out in a state at one of three times:

(a) on entry to the state, denoted as entry / action
(b) during the state (perhaps continuously in a loop), denoted as do / action
(c) on exit from the state, denoted as exit / action.

All these are optional. The "during" actions do not begin until any entry actions are completed and will be completed or aborted before any exit actions begin. This ensures that entry actions are always carried out before anything else and that exit actions are always carried out after anything else. Actions carried out within a state correspond to the Moore machine semantics discussed earlier. If required, the actions can be shown on a state machine diagram by listing them in a compartment of the state symbol just below the compartment with the state name, as shown in Fig. 15.31.

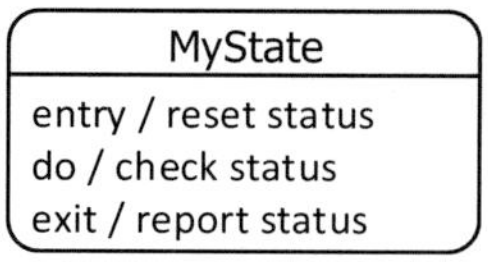

FIG. 15.31 Showing state actions.

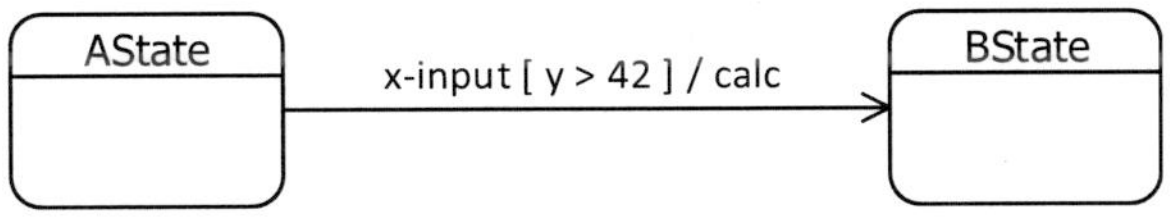

FIG. 15.32 A labeled transition.

Transitions between states can have a label that consists of up to three parts. The normal notation is `e [g] /a`, where

- `e` identifies the triggering event causing the transition (the event may have parameters).
- `g` is a guard expression, enclosed in square brackets.
- `a` identifies a list of one or more actions, preceded by a "slash" character.

All three components are optional, so zero, one, two, or all three of the components could be present. Fig. 15.32 shows a typical transition with all three components present.

If no event is specified, the transition is enabled when the actions in the preceding state have been completed (a *completion transition*). If a guard expression is specified, it must be evaluated to "True" for the transition to be taken. The guard expression is evaluated at the point in time that the transition is about to be taken. If it evaluates to "False", the transition will not be taken, even if it has been enabled by an event. If actions are specified on the transition, they are completed before entering the target state. This corresponds to the Mealy machine semantics discussed earlier. UML state machines can therefore take on the characteristics of a Moore machine, a Mealy machine, or both.

Events are handled by a state machine one at a time, with run-to-completion semantics. This means that an event is only recognized and processed after the processing of the previous event is fully completed. This simplifies the state machine because it can never get into a situation of processing an event while it is in some intermediate configuration (e.g., part of the way through a transition from one state to another). Events do not necessarily cause transitions, as an event can be recognized and processed within a state. Events can also be recognized but *deferred* in a particular state. A deferred event is saved and passed to the following state, where it must be either accepted or deferred once again. This continues until the event is no longer deferred or is where it triggers a transition.

UML defines several *pseudostates* that can be either the source or the target of a transition. The notations used for each of the pseudostates are listed in Table 15.2.

The *initial* pseudostate is the source for a single transition to the default state of a composite state. There can be at most one initial pseudostate per region. A *final* pseudostate indicates the completion of its containing region. The containing state for a region is considered completed when all its contained regions are completed. If all regions contained in a state machine are completed, then the entire state machine terminates. This is equivalent to destroying the object (in the programming sense!) with which the state machine is associated.

Table 15.2 UML pseudostates.

Symbol	*Pseudostate*	*Incoming transitions*	*Outgoing transitions*
●	initial	Not permitted.	One. Guards and triggers not permitted.
◉	final	One or more.	Not permitted.
○	entry	One or more.	One or more. Triggers not permitted.
⊗	exit	One or more.	One or more. Triggers not permitted.
•	junction	One or more.	One or more. Triggers not permitted.
◇	choice	One or more.	One or more. Triggers not permitted.
▮	fork	One.	Two or more. Guards and triggers not permitted.
▮	join	Two or more. Guards and triggers not permitted.	One. Guards and triggers not permitted.
Ⓗ	shallowHistory	One or more.	One. Triggers not permitted.
(H*)	deepHistory	One or more.	One. Triggers not permitted.
×	terminate	One or more.	Not permitted.

An *entry* point provides for the entry to a state machine or composite state that targets a specific state within the enclosing region. The entry actions associated with the enclosing state are executed before the transition to the enclosed state. A transition that enters an *exit* point implies an exit from the enclosing state machine or composite state. Any exit behavior of the enclosing state is executed after the transition from the enclosed state. Entry and exit points can be shown within the boundary of a state or on the boundary line of the state.

Junction pseudostates are used to combine multiple transitions and/or to split a transition into several branches, each with a different guard condition. The latter case is known as a static conditional branch. A *choice* pseudostate is a dynamic conditional branch—it evaluates the guards on its outgoing transitions dynamically to reflect the results of actions performed in the same run-to-completion step.

Fork pseudostates split transitions that are targeted at states in different orthogonal regions. *Join* pseudostates perform the complementary operation of combining transitions that originate from states in different orthogonal regions.

A transition entering a *deepHistory* pseudostate restores the most recently active configuration of the composite state and all its substates. A transition into a *shallowHistory* state is similar, except that the configurations of any substates are not restored.

A composite state can have at most one deepHistory pseudostate and at most one shallowHistory pseudostate. In either case, if the composite state had not been active previously (i.e., it has no history), the outgoing transition from the history state is taken.

Entering a *terminate* pseudostate implies that the execution of the containing state machine is terminated. This is equivalent to destroying the object with which the state machine is associated–again, "destruction" is intended here in the programming sense.

In a hierarchical state machine, more than one state can be active at the same time. If a nested state is active, then its parent state and all its ancestors back to the top-level state are also active. There can be at most one active state in any one region. An orthogonal state can have an active state in each of its regions. Entering an orthogonal state implies entering each of its regions (explicitly or by default). Similarly, exiting an orthogonal state implies exiting all its regions.

It's sometimes convenient to hide the decomposition of a composite state; suppressing selected details can allow attention to be focused on larger issues. A composite state with a hidden decomposition is shown graphically by adding a small icon to the state that looks like two horizontally placed states with a connecting line, illustrated in Fig. 15.33. A tool that supports state modeling would typically have some convenient mechanism for expanding back to the detailed view of the composite state. Hiding detail in this way does not change the state model, only the way that it is displayed.

FIG. 15.33 A state with a hidden decomposition.

We've glossed over some of the finer details of how state machines are defined in UML. For instance, some situations can lead to nondeterministic results. One example is two transitions exiting the same state, triggered by the same event, but with different guard conditions. If the guard conditions overlap so that both evaluate to "True", both transitions are enabled but only one of them will fire. In such cases, the algorithm for selecting which transition to fire is determined by the implementation. If guards on transition branches are framed so that it is possible for all guards to simultaneously evaluate to "False", the model is considered ill-formed. A simple cure for this latter situation is to include a branch with the guard [else], which will be chosen if the guards on all the other branches evaluate to "False". Another complex area is the processing of events in state machines that have extensive nesting of substates. A given event may be relevant to several substates, but deciding which one actually gets to process the event is not a trivial exercise.

A final point to note is that, like activity diagrams, UML state machines have no direct representation of persistent data, so it's not possible to show if a specific database item is used, modified, or created in any specific state. The only way of finding this out is by inspection of the code for the actions in each state.

State machines are very popular with some users of UML, but most of these appear to be in application areas dominated by either real-time constraints or the need to be highly reactive to external events. In a typical business information system, time constraints are less onerous, so a processing delay is more likely to be annoying than fatal. Historically, business systems were also less concerned with being reactive. In fact, early batch processing systems could hardly be said to be reactive at all. With the increasing trend toward online processing, reactivity is becoming more important.

Anecdotal reports on the use of state machines in business applications suggest that business people find state-based descriptions less intuitive than process-based descriptions. It also seems to be true that a large proportion of the developers of business software are unfamiliar with the basic concepts of state machines. However, state machines do offer some powerful capabilities, particularly the ability to generate application code from suitably constructed models. The next section discusses one approach that relies on state modeling to provide the behavioral aspects of an executable specification.

Executable UML

As the name suggests, executable UML is a dialect of UML that is focused on a style of modeling that results in a model that can be compiled and executed (Mellor and Balcer, 2002). It is based on a fusion of ideas from the earlier Shlaer-Mellor methodology and standard UML. An executable UML model is built around three core concepts: specifying structural aspects, such as data, using UML class diagrams; specifying dynamic behavior and control using state machine diagrams; and specifying actions using an action language.

The use of class diagrams and state diagrams is close to standard UML but with a few detailed differences. At present, UML does not define any notation or syntax for an action language, although there *is* a specification for the semantics that must be supported by such a language. The consequence is that each variant of Executable UML has its own language. A proposal for standardizing an action language is currently in progress within the Object Management Group (OMG).

The first stage of analysis separates different areas of concern into different *domains* to be modeled. A domain contains a set of cohesive concepts that can be considered without requiring detailed knowledge of other domains. Domains are interconnected by *bridges* that resolve intercommunications required between domains.

The assumptions made in one domain can form the requirements for another domain. For example, a business-related domain may assume the existence of a communication network with certain characteristics. These business domain assumptions contribute to the requirements for the communication network domain.

Bridges allow the separation of concerns, so, for example, the elements in a business domain can be developed and refined without considering the details of any technical domain. This also makes the system design more resilient. For instance, modification to a technical domain does not necessarily have to impact a business domain, so we could "unplug" one technical realization and "plug in" a different one without impacting the business, if we can define a suitable bridge.

The functionality required in a domain can be established in any desired way. We assume here that this is done through use case analysis. The information contained in the use case descriptions can be dissected to determine an initial set of classes. The initial class definitions will likely be extended and refined many times as the design progresses.

Associations between classes on a class diagram are handled in a slightly different way from standard UML. Each association is given a unique identifier, and the association ends are named with verb phrases instead of the noun phrases of standard UML. Classes have operations defined in the usual way. These operations typically compute and return some value, or create, delete, or modify some object (or perhaps a combination of these). Each instance of a class (i.e., an object) has a life cycle. For some classes, the behavior of their instances does not vary. An object is created, it reacts in a fixed way to any stimulus (a call to one of its operations), and eventually it goes away—not a very exciting life cycle.

Other classes have behaviors that change over time, depending on the history of events to which the object has reacted. For these classes, a state machine can be defined to model their life cycle. State machines in most implementations of Executable UML adopt a subset of the possibilities defined in the UML specification. For example, actions are typically encapsulated into procedures that are executed on entry to a state (standard UML allows other possibilities).

The procedures are defined in an action language. The language is at a higher level of abstraction than standard programming languages such as C# or Java to remove dependence on a particular software platform. Action language statements are conceptually concurrent, as decisions about how to serialize these are left to the model compiler. Similarly, such concerns as data access, persistence, distribution, and so on are not specified in the action language.

The only direct manipulation that can be specified in an action language is the manipulation of the model elements themselves. At the time the model is compiled, the action language specification can be mapped onto the desired architecture. In principle, the same model could therefore be used in an embedded system, a web service, a distributed application, and so on.

Executable UML has been applied successfully in a variety of applications. Most of these are real-time applications that lend themselves very naturally to state-based specifications. Products are available from a small number of vendors who tend to be focused on the reactive systems market. However, there seems to be no reason why the concepts should not be equally applicable to general business systems, so the current focus may just reflect historical market positioning rather than any inherent feature of the technical approach.

15.6 Business Process Standards Initiatives

The late 1990s saw rising interest in supporting the automation of business activities, aimed at replacing the various ad hoc approaches that existed with a more general and systematic way of building business-focused systems. Although UML activity diagrams could be pressed into service, their primary purpose of describing object-oriented software made them less than satisfactory for expressing business concepts. As a result, the early 2000s saw several initiatives aimed at producing a standard for defining business activity.

Table 15.3 A partial list of initiatives related to process standards.

Organization	*Initiative*
BEA	Process Definition for Java (PD4J)—later absorbed into BPEL
IBM	WSFL—later absorbed into BPEL
Microsoft	XLANG - later absorbed into BPEL Windows Workflow Foundation (WWF)
The Business Process Management Initiative (BPMI)	Business Process Modeling Notation (BPMN) (later merged with OMG – see below)
The Organization for the Advancement of Structured Information Standards (OASIS)	Business Process Execution Language (BPEL)
The Object Management Group (OMG)	Business Process Definition Metamodel (BPDM) Business Process Maturity Model (BPMM)
The Workflow Management Coalition (WfMC)	XML Process Definition Language (XPDL) Workflow XML protocol (Wf-XML) Workflow Application Program Interface (WAPI) The WfMC reference model for workflow
World Wide Web Consortium (W3C)	Web Services Choreography Description Language (WS-CDL) Web Service Choreography Interface (WSCI) Web Services Conversation Language (WSCL) Web Service Definition Language (WSDL)

Table 15.3 shows a partial list of organizations active in process-related fields and the initiatives with which they were involved. These efforts were focused in three main directions.

Metamodel. Although humans typically view a process design through some form of notation, automation of the process requires the construction of a data structure and procedures that can navigate through the structure. Transferability of process designs between tools and between organizations would be facilitated by having a common metamodel to ensure consistency. A leading proposal in this area was the Business Process Definition Metamodel (BPDM).

Execution. The rapid growth of web services has led to a demand for a consistent way of expressing business processes in an executable form. The basic web protocols are stateless, but business processes may require lengthy conversations between distributed participants in a process. One way of tackling this problem is to have a specialized programming language targeted at a business process execution engine. A leading proposal in this area was the Business Process Execution Language (BPEL).

Notation. Most process models use some form of diagrammatic presentation of the process design. This typically takes the form of a flow diagram showing sequencing, branching, and so on, perhaps with some additional annotation to indicate related features such as the use of resources and organizational structure. The graphical syntax (including symbols and connection rules) varies from one notation to another. Perhaps more importantly, the semantics also vary. For example, a diamond shape may look similar in two different notations but have subtly different implications in each. A common notational standard would simplify staff training, make process designs more readily understandable, and reduce the scope for semantic variation. The leading proposal in this area was the Business Process Modeling Notation (BPMN) which, over time, evolved to become the mainstay of process modeling. We will return to further details of BPMN later.

BPDM was aimed at representing business process elements in a way that was independent of any specific notation or process modeling methodology. The concept of a metamodel is intended to provide a kind of vocabulary of concepts related to process modeling, with a clear definition of the meanings of the concepts and how one concept can relate to another.

BPDM supported two complementary views of business processes: "Orchestration" and "Choreography". Orchestration is concerned primarily with *what* happens and *when* and is usually defined in some flow-based notation (although the notation itself was considered outside the scope of BPDM). Choreography describes how business entities collaborate in a process, considering that the entities may be independent and may have their own internal processes. The focus of choreography is on the responsibilities of the role players in a process and the interactions between them.

The BPDM Metamodel was based on the "Meta Object Facility" (MOF) that also underpins UML and some other OMG standards. As well as providing a baseline for concept definition, MOF also supports an XML syntax that can be used for transferring business process models between tools and between organizations. An obvious shortcoming of BPDM was the lack of a standard business-friendly notation, and so BPMN was proposed to fulfill this role.

BPEL aimed at extending standards for web services to add explicit support for business processes. It was conceived as a programming language using XML syntax. The original version, called BPEL4WS, was based on Microsoft's WSFL and IBM's XLANG approaches. The merged version was renamed WS-BPEL, or simply BPEL. BPEL could be applied to abstract process definition, but its main application was seen as the definition of executable business processes.

BPEL assumed the existence of one or more Web Service Definition Language (WSDL) files that define the web service interfaces. WSDL is a separate standard, which is used for other purposes besides support for BPEL.

Because of the mixed inheritance of BPEL, the implementer sometimes had a choice of alternative constructs that could potentially produce the same effect. In some cases, it was necessary to use the block-structured part of the language. In other cases, it is necessary to use the graph-based part of the language. Some of the semantics of the constructs were very subtle, and effects could be produced that may not have been what the process implementer intended.

One consequence of block-structured elements is that some process constructs that can be expressed in a graph-oriented approach cannot be expressed in block-structured form, and some remodeling may be required. Fig. 15.34 shows a nonstructured process flow and one possible way in which it could be remodeled. To get to a block-structured form, any looping constructs must be arranged so that there is only one entry point into the loop and only one exit point from it. We look at process structures in more detail in Section 15.4.

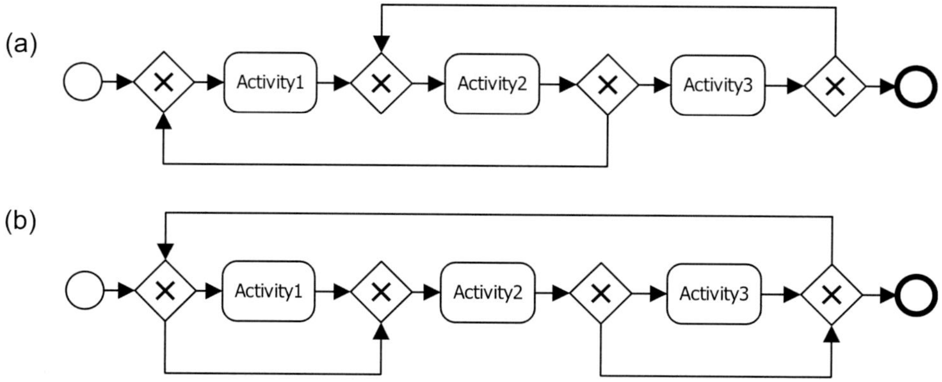

FIG. 15.34 (a) An unstructured process. (b) A structured equivalent.

15.7 Business Process Model and Notation

The original specification for the Business Process Modeling Notation was produced by BPMI, and, as the name suggests, was aimed directly at providing a systematic way of defining business processes in a way that would be accessible to a business community as well as technical experts. It defined a graphical syntax for drawing process diagrams in a flowchart-like style that would be easy to understand. This made it possible to validate a process description from a business perspective, something that could not be achieved with other technical languages that were proposed around the same time.

In 2005, the assets of BPMI were taken over by the OMG and the development of BPMN continued under their management. Because the OMG's existing BPDM standard lacked a convenient notation, a natural evolution was to merge the two standards under a slightly modified name: Business Process Model and Notation (still BPMN!). The new BPMN Version 2 was ratified in 2011 and became an ISO standard in 2013. It is now the dominant process modeling standard and has effectively incorporated useful strands from other initiatives, including the process execution concerns of BPEL. BPMN supports three different process perspectives:

Orchestration describes a sequence of activities from the point of view of the business entity owning the process. The process is assumed to be contained in a single *Pool.*

Collaboration describes the interaction between two or more process participants, shown as separate pools. The interactions are identified as message flows between pools. The internal details of a pool need not be shown.

Choreography describes interactions between participants in a different format and concentrates mainly on message flows.

Orchestration and Collaboration use a common set of diagram conventions. Choreography diagrams use a different style and are much less widely used, so they will not be considered further in this chapter.

A pool contains three main types of objects: Activities, Events, and Gateways. These are linked by connecting objects, which come in three varieties. *Sequence flows* show the order (sequence) in which activities will be performed during a process and are denoted as solid lines with a filled arrowhead. Sequence flows can also carry additional markings to indicate whether a flow is a conditional or a default flow. A sequence flow identifies the flow of control and does not necessarily imply a flow of material or information. *Message flows* show the interchange of messages between participants in different processes and are denoted as dashed lines with unfilled arrowheads. *Associations* are used to connect data, text, and so on with flow objects and are denoted as dotted lines with an optional open arrowhead. One use of these is to show the inputs and outputs of activities.

An *Activity* represents work that is performed during a process, denoted in BPMN as a rectangle with rounded corners. Activities can be nested with the details of subactivities hidden. Compound activities are identified by adding a "+" icon to the activity symbol. BPMN uses the term "Task" for atomic activities that have no lower-level breakdown. Various other icons can be added to the basic activity shape to indicate other types of activity, such as looping, transactional, compensating, ad hoc (the sequencing of any internal activities is not defined), multiple instance (several instances of the activity can be performed in parallel), and so on. With some restrictions, these can be combined to indicate such things as a looping ad hoc compensation activity.

Events represent something that happens during a process and are denoted by circular symbols. There are three main subtypes: start events have a single-line circle, intermediate events have a double-line circle, and end events have a thick-line circle. Icons can be added within the circle to give a more precise indication of the type of event, such as *message, timer, error,* and so on.

The splitting and joining of flows can be shown using a diamond-shaped *Gateway* symbol. The symbol can carry an additional icon to indicate the nature of the split or join.

Variants include *exclusive-or* (the default if no icon is shown), *and, inclusive-or,* and *complex* (where the split or join conditions are determined by expressions).

Fig. 15.35 shows a typical BPMN process diagram. The process shown starts with the receipt of a loan application (identified as a message). After the loan assessment activity, the control flow passes to one of the three following activities. Each of these is a compound activity; the internal details are not shown here. The final exclusive-or gateway joins the split paths together and the process ends. In the referral case, there is a time event associated with the applicant review. If the review is not completed within the given period, it is escalated to be dealt with by a supervisor. The escalated review process also has branching in the flow, but rather than using gateways, the subprocess uses decisions on sequence flows and implicit joining of flows. The escalated review subprocess also shows the reuse of the definition of the compound activities for loan approval and loan rejection.

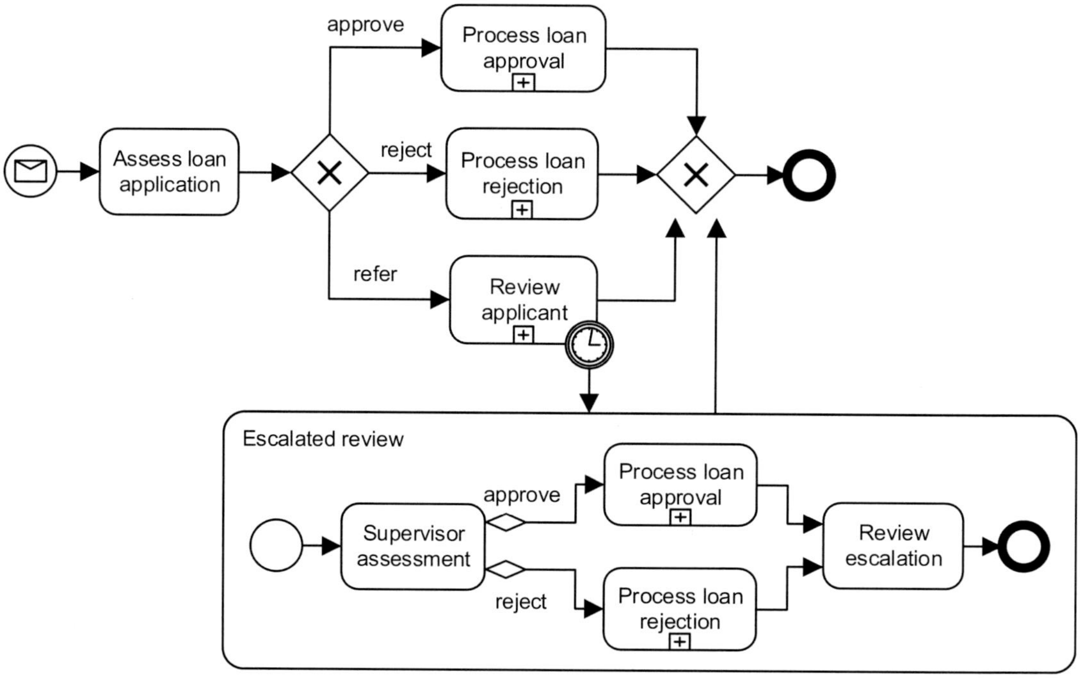

FIG. 15.35 A BPMN diagram.

15.8 Standard Process Patterns

A significant contribution to the establishment of a common foundation for business process modeling was made by the identification of 20 or so workflow patterns by van der Aalst, ter Hofstede, and others. The patterns have subsequently provided a valuable benchmark for describing the characteristics of alternative process modeling notations, methodologies, and tools. Table 15.4 provides a summary of the original workflow patterns using (mostly) BPMN notation. There is no particular significance to the numbering system used.

Table 15.4 Standard process patterns.

Nr	*Pattern*	*Description*
1	Sequence	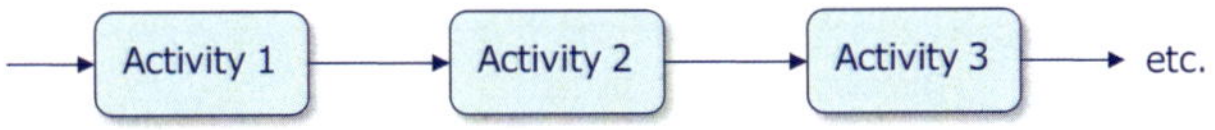 Sequence is the simplest pattern. It expresses the notion that one activity follows another, so in the diagram, Activity 2 starts when Activity 1 finishes, Activity 3 starts when Activity 2 finishes, etc.
2	Parallel split	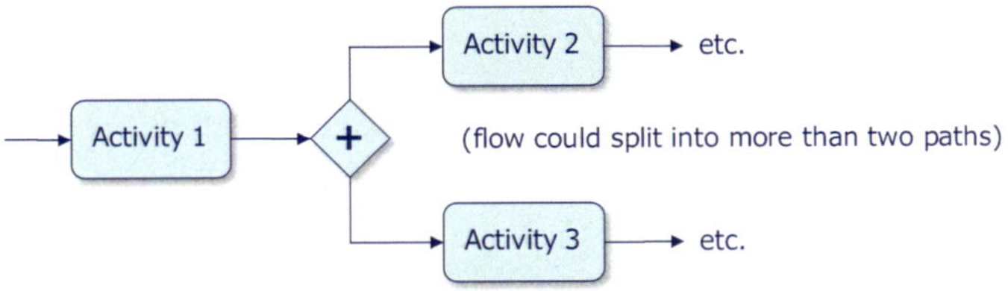 In the parallel split pattern, a token released by the completion of a preceding activity is split into multiple tokens, perhaps at an AND gateway, enabling several succeeding activities. In the diagram here, there is no defined ordering between Activity 2 and Activity 3.
3	Synchroniza-tion	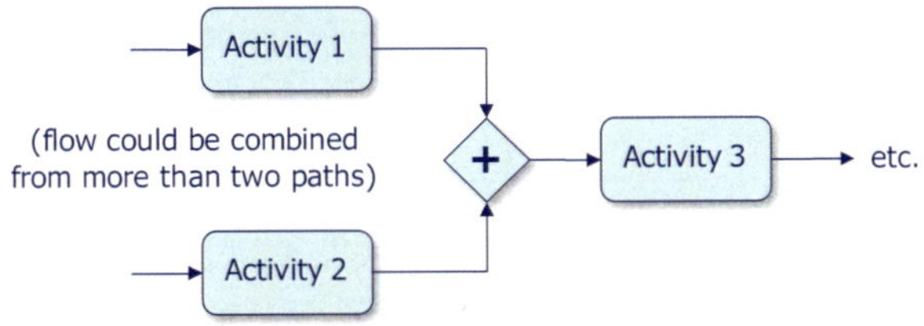 In synchronization, tokens from multiple branches merge, perhaps at a gateway, so that one token is emitted once a token has been received on all the preceding branches. Here, both Activity 1 and Activity 2 must complete before Activity 3 can be executed.
4	Exclusive choice	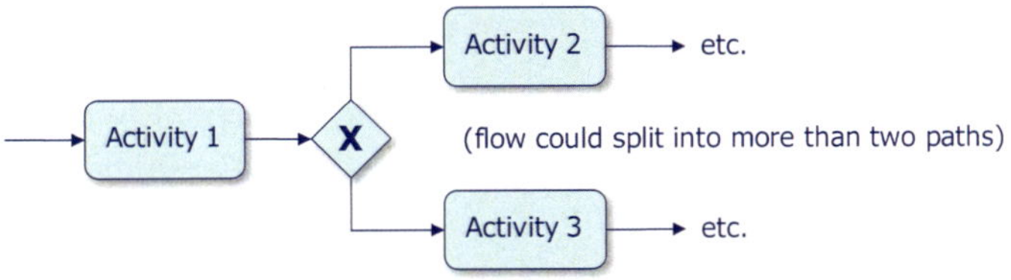 An exclusive choice is a decision point at which the thread of execution will take one of several alternative branches. The branching decision is typically defined by a business rule or rule set. One of the branches may be designated as the default. In the diagram here, Activity 1 can be followed by either Activity 2 or Activity 3, but only one of these will be executed.

Nr	*Pattern*	*Description*
5	Simple merge	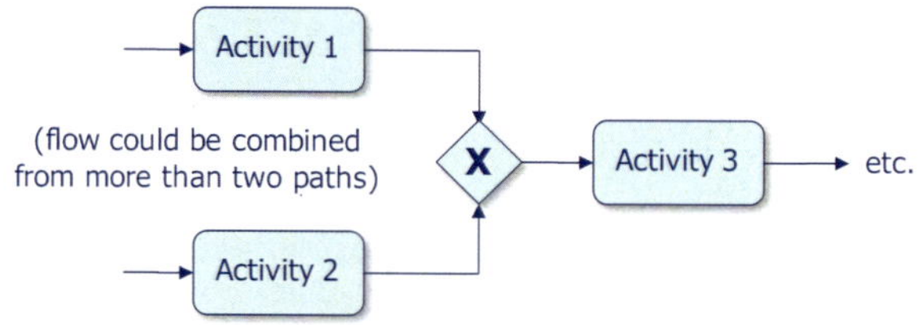 A simple merge is a point at which two or more alternative branches come together without synchronization. Only one of the preceding branches is assumed to be active (later patterns deal with merging multiple active branches). In the diagram here, either Activity 1 or Activity 2 must complete before Activity 3 can be executed.
6	Multiple choice	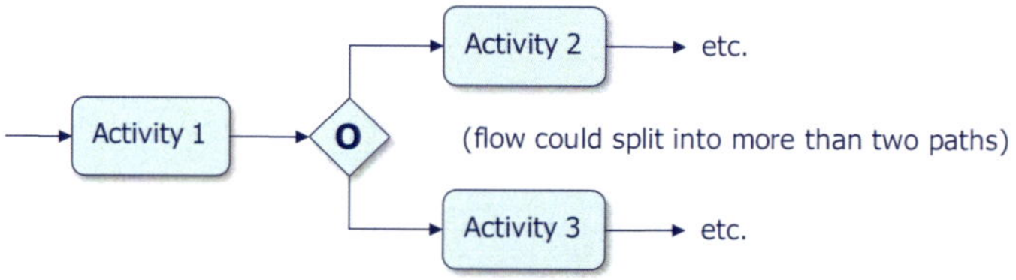 In multiple choice, the thread of control diverges into several parallel branches on a selective basis. The branching decision is typically defined by a business rule or rule set. One of the branches may be designated as the default. In the diagram here, Activity 1 can be followed by Activity 2, Activity 3, or both.
7	Synchronized merge	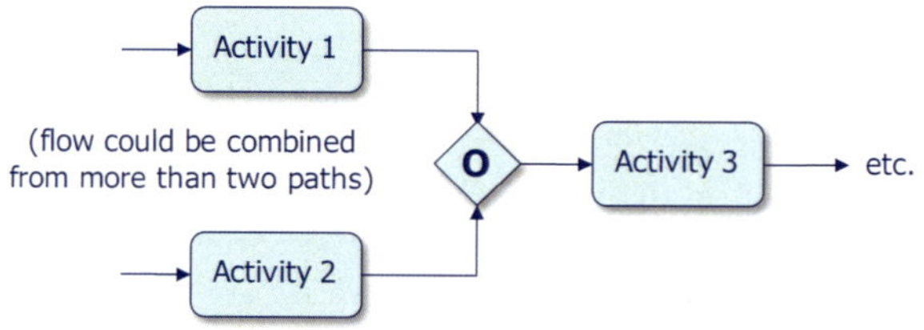 Synchronized merge represents the completion of activities that have been started on parallel branches. One or more of the preceding branches can be active. All preceding active branches must complete before the succeeding Activity 3 can be executed. This assumes that the gateway has information about which preceding activities are expected to complete (sometimes called a "structured workflow" assumption). In the diagram here, Activity 1 or Activity 2 must complete (if they were started) before Activity 3 can begin.

<table>
<tr><th>Nr</th><th>Pattern</th><th>Description</th></tr>
<tr><td>8</td><td>Multiple merge</td><td>
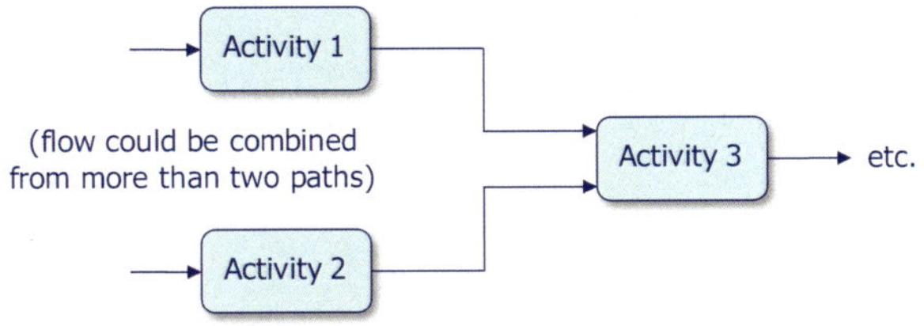

Multiple merge is the unsynchronized convergence of several alternative branches. More than one of the branches can be active. The succeeding activity will be executed when any of the preceding activities completes. If multiple branches complete, the succeeding activity is executed multiple times. In the diagram here, either Activity 1 or Activity 2 must complete before Activity 3 can be executed, and a new instance of Activity 3 will be started for each preceding branch that completes.</td></tr>
<tr><td>9</td><td>Discriminator</td><td>
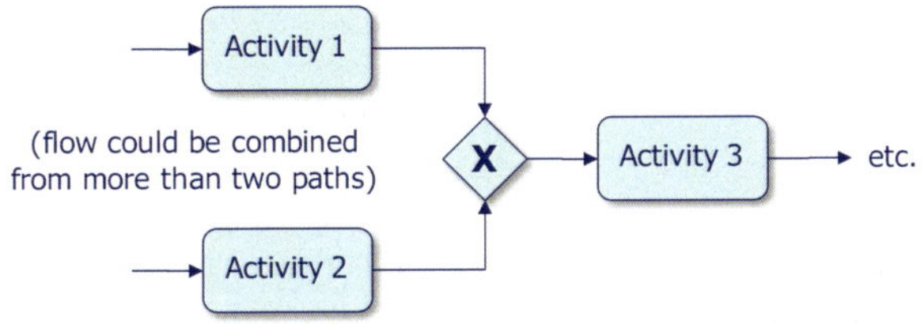

The discriminator provides for synchronized convergence after one out of m alternative branches completes. One or more preceding branches will be active. Completion of any preceding active branch will forward a token to the succeeding activity. The completion of subsequent branches is ignored. A Complex gateway in place of the XOR gateway shown here would allow n-out-of-m merges: an expression would determine which path(s) must join. Here, either Activity 1 or Activity 2 (or both) must complete before Activity 3 can be executed.</td></tr>
<tr><td>10</td><td>Arbitrary cycles</td><td>
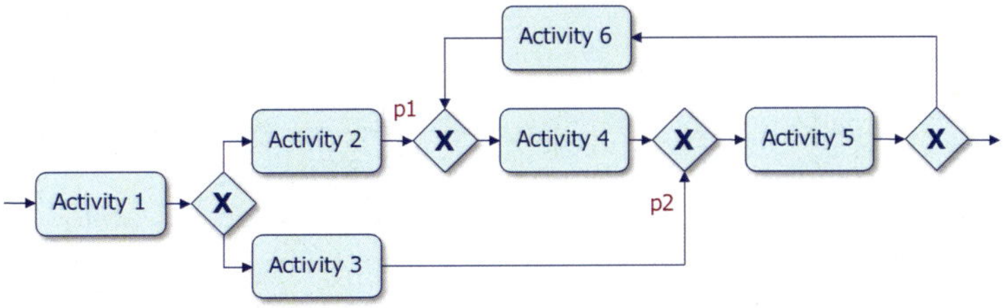

Arbitrary cycles of activity have more than one entry or exit point in a repetitive cycle of activities. This pattern supports unstructured repetition in a process without requiring special looping constructs. This is not possible in "block structured" modeling languages. Here, Activities 4, 5, and 6 form a loop that can be entered at either points p1 or p2.</td></tr>
</table>

<table>
<tr><th>Nr</th><th>Pattern</th><th>Description</th></tr>
<tr><td rowspan="2">11</td><td rowspan="2">Implicit termination</td><td>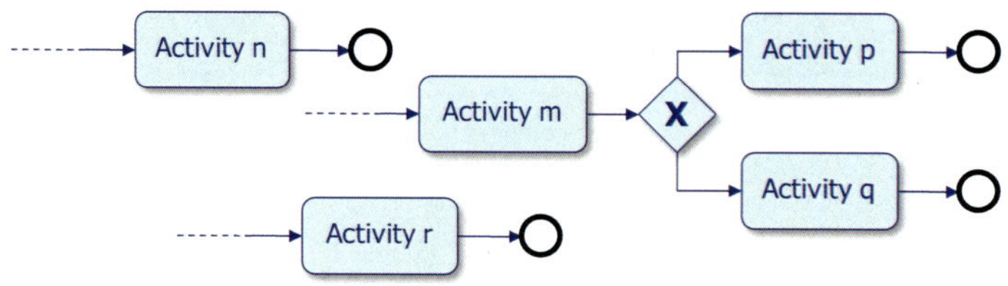
</td></tr>
<tr><td>Implicit termination specifies that when a process is terminated, any subprocesses should also be terminated. The intention is that on termination there are no active activities, there is no activity that can be made active, and the process is not deadlocked. In BPMN this is specified by ensuring that every thread of control in a process ends with an end event (as shown in the diagram here).</td></tr>
<tr><td rowspan="2">12</td><td rowspan="2">Multiple instances, without synchronization</td><td>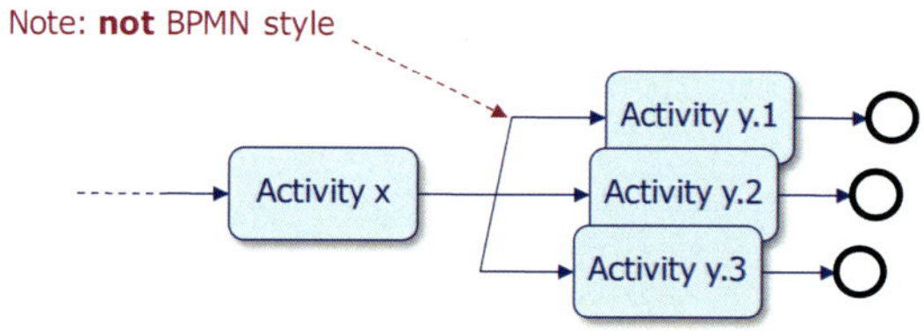
</td></tr>
<tr><td>Multiple instances of the same activity are initiated in parallel without requiring that they resynchronize on completion. Each instance of the activity runs independently of the other instances. This is not the same as a parallel split because all instances are of the same activity. Here, Activities y.1, y.2, and y.3 are all separate instances of Activity y. This diagram (and the other multiple instance diagrams) departs from the BPMN notation to make the patterns more distinct.</td></tr>
<tr><td rowspan="2">13</td><td rowspan="2">Multiple instances with a priori design-time knowledge</td><td>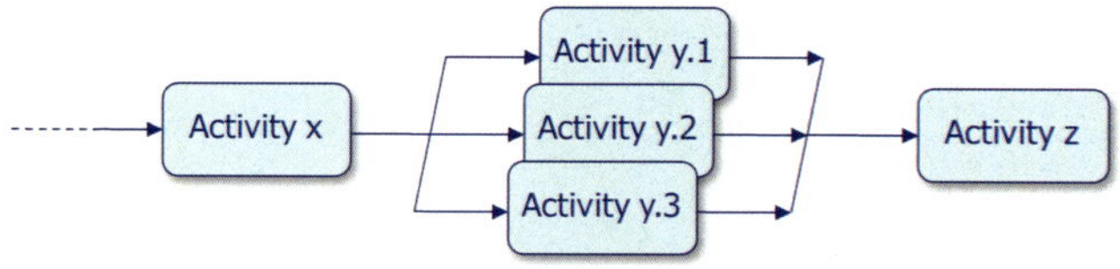
</td></tr>
<tr><td>Several instances of the same activity are initiated in parallel with intended resynchronization on completion. The number of instances is known when the process is defined. Each instance of the activity (such as Activities y.1, y.2, and y.3 shown here) runs independently of the other instances but succeeding activities (such as Activity z above) cannot be triggered until every instance has completed. Note: this is not the same as a parallel split/join because these are all instances of the same activity.</td></tr>
</table>

<table>
<tr><th>Nr</th><th>Pattern</th><th>Description</th></tr>
<tr><td rowspan="2">14</td><td rowspan="2">Multiple instances with a priori runtime knowledge</td><td>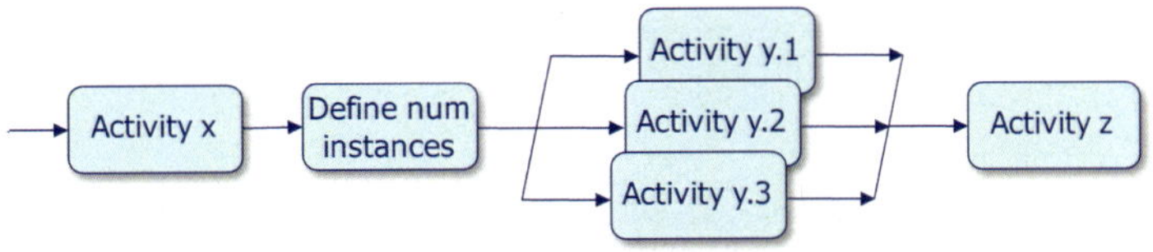
</td></tr>
<tr><td>Several instances of the same activity are initiated in parallel with intended resynchronization on completion. The number of instances required is determined in the process itself, perhaps only at the very last moment before the instances must be executed. Each instance of the activity (such as Activities y.1, y.2, and y.3 shown here) runs independently of the other instances but succeeding activities (such as Activity z) cannot be triggered until every instance has completed.</td></tr>
<tr><td rowspan="2">15</td><td rowspan="2">Multiple instances without a priori knowledge</td><td>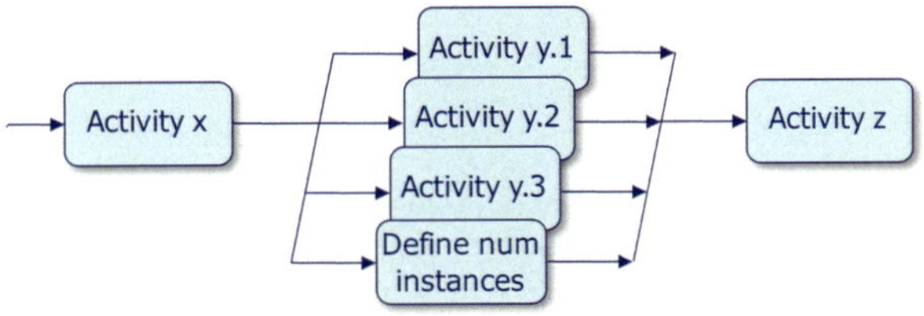
</td></tr>
<tr><td>Several instances of the same activity are initiated in parallel with intended resynchronization on completion. The number of instances required is determined in the process itself during execution of the parallel instances and is known only after the last instance finishes. Each instance of the activity (such as Activities y.1, y.2, and y.3 shown here) runs independently of the other instances but succeeding activities (such as Activity z) cannot be triggered until every instance has completed.</td></tr>
<tr><td rowspan="2">16</td><td rowspan="2">Deferred choice</td><td>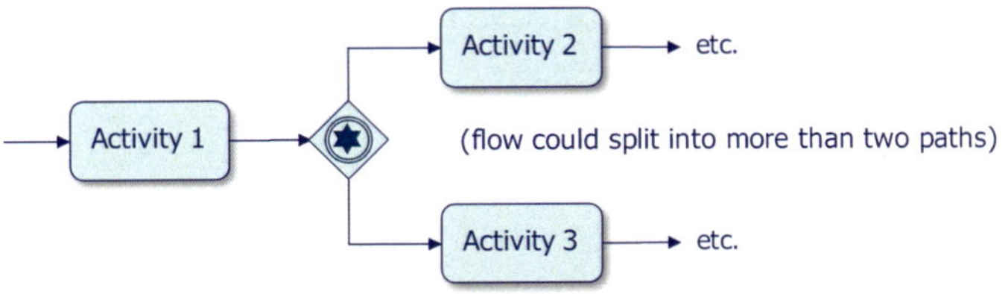
</td></tr>
<tr><td>Deferred choice involves a single thread of control that is split into two or more threads, all of which could execute concurrently. Only one branch will complete, but which one of the branches will complete will not be known until after the split. This can be considered as a "race" between the activities in the alternative branches. Once a decision is made in one of the branches the other alternatives are immediately withdrawn. The special marker in the gateway in the diagram given here is the BPMN notation for an "Event-based split".</td></tr>
</table>

<table>
<tr><th>Nr</th><th>Pattern</th><th>Description</th></tr>
<tr><td>17</td><td>Interleaved parallel routing</td><td>
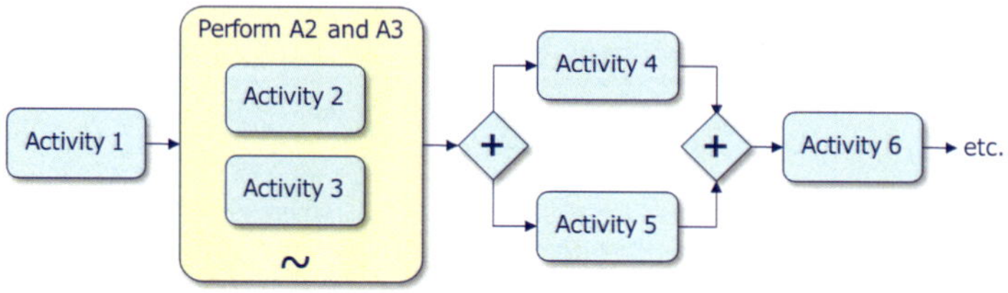

This pattern provides for a partial ordering of activities, also constraints that prevent certain activities from running in parallel. Here, Activity 1 is followed by Activities 2 and 3 in any order (but not simultaneously), followed by Activities 4 and 5 in parallel, and then Activity 6. The properties of the "Perform A2 and A3" activity must be set appropriately (e.g., its subactivities must be defined as Sequential in BPMN). The tilde symbol on "Perform A2 and A3" is the BPMN notation for an ad hoc activity.</td></tr>
<tr><td>18</td><td>Milestone</td><td>
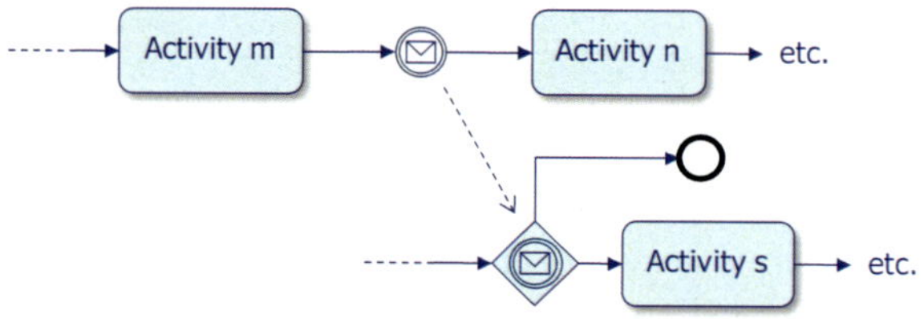

Milestones indicate that an activity in one branch of a process cannot be initiated once another branch of the process passes a certain point. Here, Activity s cannot be carried out once Activity m has completed. This is hard to model accurately in BPMN because it relies on state, rather than the triggering of activity. Some workarounds have been proposed, but these do not totally capture the intention.</td></tr>
<tr><td>19</td><td>Cancel activity</td><td>
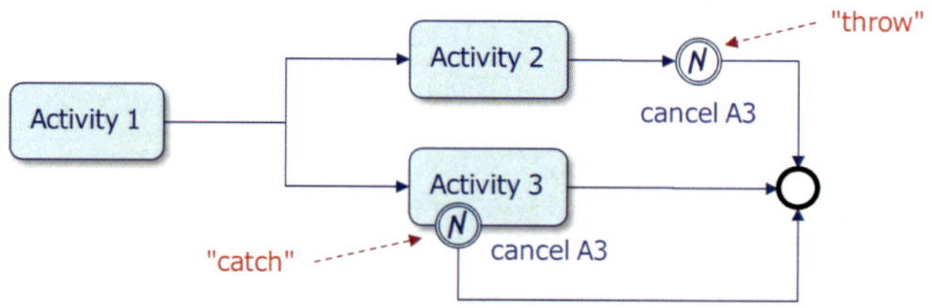

The cancel activity pattern indicates that a process has progressed to a point where it becomes clear that a particular activity should be canceled. Here, the event created after Activity 2 causes a message to be sent that causes Activity 3 to exit at the point shown. The symbol used on the event in the diagram here indicates an error.</td></tr>
</table>

Nr	*Pattern*	*Description*
20	Cancel case	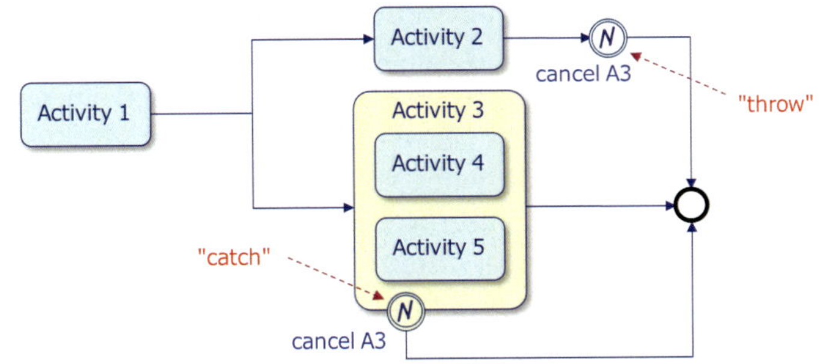
		The progression of the process instance makes it clear that a particular case in progress should be canceled. This is like the previous pattern but here we are canceling a complete case (and all the activities relating to it) instead of just an activity. If the case represents the whole process, we can just use a Terminate event, which kills every activity in progress.

The original workflow patterns focused on flow-of-control aspects. A similar analysis has been applied to workflow data patterns by Russell and colleagues (Russell et al., 2016), who define five main categories of pattern: Data Visibility (7 patterns), Internal Data Interaction (6 patterns), External Data Interaction (12 patterns), Data Transfer (7 patterns), and Data Routing (7 patterns). The patterns illustrate the way in which many data-related features are handled by current workflow tools. Although no claim is made that the list of features is exhaustive, the patterns probably represent a reasonable catalog of concepts that need to be considered in the context of process data.

Because the data patterns were gleaned by examining current practice, they effectively make statements about the way that process data is typically handled in a software implementation. However, here we are mainly concerned with modeling processes at a conceptual level. This corresponds to the distinction between an ORM information model and its implementation in (say) a relational database management system. Examination of the documented data patterns shows that several of them are primarily abstractions of implementation patterns, rather than information usage patterns.

As an example, we could consider two activities, *A* and *B*, with a dependency between them. Activity *B* needs some information created by Activity *A*, and so it would be natural for an instance of Activity *A* to precede temporally the corresponding instance of Activity *B*. This is true whether we are considering the process at a purely conceptual level or looking at the implementation.

If the activities are implemented in some object-oriented software language, a standard approach would be to create some object within the Activity *A* instance and then pass it to the corresponding Activity *B* instance; in other words, we could say that Activity *A pushes* the data to Activity *B*. If the two activities are operating in the same memory space, this could be accomplished by passing a reference to the relevant object; otherwise, an additional step of serializing the object may be required.

From a conceptual viewpoint, these last details are irrelevant. Whether Activity *A* "pushes" an object to Activity *B*, whether Activity *B* "pulls" an object from Activity *A*, or, indeed, whether any "object" is involved at all is purely an implementation issue. From a conceptual standpoint, we certainly need to be concerned about the rights and responsibilities of activities regarding data, but we explicitly do not want implementation issues to become entangled with the process definitions.

Processes elements (at any level) utilize data in three main ways: (a) to compute some result of value to the process, (b) to determine a course of action, and (c) as information to guide process activity, for example, as reference material for a human involved in the process. Our main need is to define data scoping: which activity needs which item of data and for what purpose. If we have ambitions to automate the construction of automation systems to support business processes, this is clearly crucial. Even if we have more modest ambitions, such as simply providing a checklist to implementers, this still seems to be an important facet of process definition.

From a high-level perspective, we can classify an activity's usage of a data item in terms of whether it is limited to reading the data item or whether it can also modify it. If modification is allowed, we can further distinguish between the ability to create new data, modify existing data, or delete existing data. (One could also predict other uses for such information, such as identifying the data privileges that should be given to specific user roles, but that is outside the scope of this chapter.) In contrast, the types of features that we would *not* want to specify at a conceptual level include the following.

Communication mechanisms

- between different levels (process, subprocess, activity)
- between different elements (e.g., activity to activity)
- between process elements and the environment

Data transfer mechanisms

- "push" (data explicitly sent to process element)
- "pull" (data requested by process element)
- pass by value versus pass by reference
- with or without locking
- with or without transformation
- etc.

These issues are largely outside the scope of a business process model, and so we don't consider them in detail here. Of course, a business model alone does not contain sufficient information to determine the detailed mechanisms (and deliberately so). What we do have is the ability to identify the relationships between low-level atomic activities and low-level data manipulation and query operations.

We are not limited to operating only at this level because, as we have seen, we can compose collections of low-level elements into higher-level elements until we reach a more convenient level of expression. However, this still leaves the realization underconstrained, as there could be many ways of linking data to process for the same conceptual design. The choices between the alternatives are somewhat orthogonal to the conceptual elements. For example, are we building a self-contained or distributed system operating over a network? How many users do we intend to support? Are we concerned about issues such as availability and scalability? The additional information required is what might be termed the *technical architecture* of the system, which is more concerned with the engineering aspects. In principle, given a definition of the technical architecture of a system, we should be able to generate the necessary system elements required from a business model. We have already seen how this could work in examples such as Executable UML. In the future, we can expect similar ideas to be widely adopted as a natural outgrowth of conceptual modeling.

15.9 Process Models, Databases, and ORM

BPMN provides a standard notation and semantics for process descriptions. It has support for defining the execution of business processes but can be used to define processes at a purely conceptual (implementation-independent) level. Equivalently, ORM supports information modeling at a conceptual level, so we have the capability of defining both business information and business activities in a consistent way. But although both BPMN and ORM produce valid business models, neither approach makes any reference to the other in their basic forms. In this section, we will see how these two viewpoints can be reconciled.

The BPMN standard provides two elements relating to data along with the ability to associate them with activities, summarized in Fig. 15.36.

Data Object	Represents temporary data that exists in a process instance while it is running. When the process instance terminates, the data object disappears.	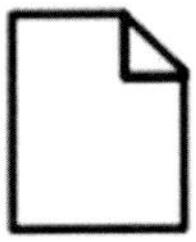
Data Store	Represents an item of information that is stored persistently outside the process. When the process instance terminates the information remains. Note that the reference is to an item of information in some data store, not an entire data store or any specific data store.	

FIG. 15.36 Data options in BPMN.

Because an ORM model generally specifies information that is intended to be stored (not temporary variables that are used then discarded), the BPMN Data Object is not relevant to a BPMN-ORM linkage. This leaves the Data Store reference as the only way of representing persistent information in a BPMN model.

A BPMN Data Store reference is linked to an Activity (usually a Task) by a Data Association. In BPMN, associations are generally used to link artifacts to process elements and are shown with a single dotted line (dashed lines have a different meaning). Associations can have an arrowhead showing the directionality of the association, but the semantics of this for a Data Store are a little ambiguous in the BPMN specification. Access to a database is usually a request-response cycle, so it's uncertain what an arrowhead would be indicating. The sensible option is therefore to use an association with no arrowhead.

The only textual notation provided for the Data Store reference is its name, which can be shown inside the icon or near it. A name alone is obviously not sufficient to define an adequate ORM linkage. A possible BPMN-compliant supplement would be the use of a Text Annotation linked to the Data Store, which could be used to add arbitrary text to a BPMN diagram. Text Annotations have no significant limitations or format requirements, and so a convention could be established for describing the linkages between Activities and Data Stores in a way that precisely locates the relevant areas of an associated ORM model. However, this would add considerable clutter to the BPMN diagram.

This exhausts the graphical options in BPMN, but a Data Association also has attributes that are not displayed on BPMN diagrams. One of these is "assignment", which identifies one or more data elements involved in the transfer from source to target. Potentially, this provides a point at which references to an ORM model could be stored, but these would only be visible in the BPMN tool being used to manage the process model and not to anyone viewing the graphical representation of the process from a business perspective. Such attributes are intended to facilitate process automation and have little relevance to modeling at a conceptual level.

An alternative approach is to introduce an intermediate stage between the BPMN model and the ORM model, forming a Process-Information Bridge (PIB). The Data Store reference can then be to a PIB element containing the relevant ORM linkage. One data item can be used in multiple activities in a process, so this also provides a location to resolve the one-to-many relationship. Fig. 15.37 shows the relevant diagram notation.

Next, we can turn to the ORM end of the relationship. In ORM (as opposed to an Entity-Attribute model), the attributes (properties) of the entity are unbundled and can be

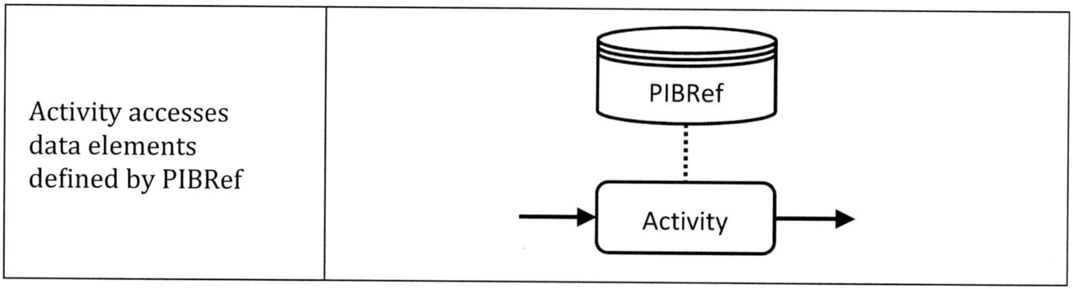

FIG. 15.37 Activity linked to PIBRef.

addressed in terms of separate fact types. "Entities" are formed from collections of fact types in a mapping process that produces a new model of a type defined by the mapping (for example, a relational model).

Although the unbundling of entities in ORM potentially offers a finer-grained definition of a relationship between activity and information, it is not necessarily convenient for cross-referencing of the kind we are considering. An activity is likely to require all information about an entity to be treated together rather than as isolated facts. For example, an agent updating stored data about personnel would probably expect to see all relevant information about a particular employee to be displayed at the same time. Instead of referring to individual fact types, it would be much simpler if the fact types were arranged in convenient bundles related to major entity types.

A straightforward way of achieving this is to group fact types into *relations*, in the sense originally expressed by Codd. Facilities already exist in ORM to produce normalized relations from an ORM model with the Rmap procedure. In a conceptual model, there is no prior commitment to a relational database solution, but this does not matter because normalized relations can be realized in other types of data structures such as XML and JSON; see Chapter 16 for more details.

A convenient data reference to a group of fact types in an information model can therefore be made by simply using the (unique) name of the relation. Accessing a particular data item during the execution of a process would require the relation name plus a key value that is unique within the relation. The same activity may access different instances of the relation in different instances (executions) of the process. However, only the relation's name is necessary to define the linkage between a process model and an information model.

This approach would create a set of hierarchical links between the process model and the information model, as outlined in Fig. 15.38.

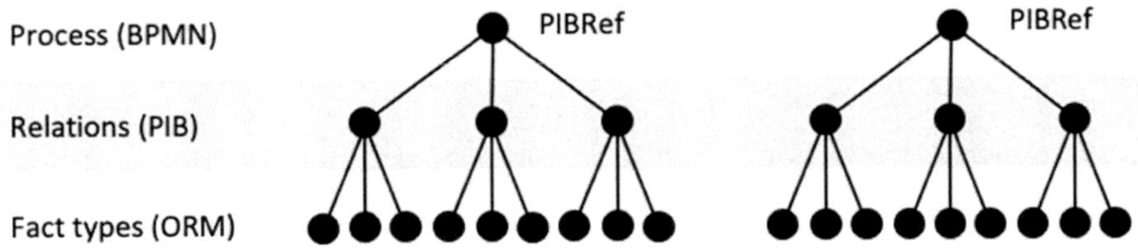

FIG. 15.38 Reference hierarchies.

Each PIBRef will contain the following:

- An indicator showing the type of CRUD access intended in this case. The usual access will be either Read-only, where the activity is not expected to change the data, or Full, where the activity can view the data and possibly also change it. These can be indicated by a single letter (R or F, where F is a shorthand for CRUD). In the unlikely event of more specificity being necessary, then the individual letters of CRUD can be used.
- The identification of the ORM elements involved in this case, using relation names. More than one relation could be used in the same PIBRef. For example, an activity might involve reading from one relation and updating another.

Each set of PIBRef access/relation values is likely to appear in one PIBRef only—the same PIBRef can be reused in the process model so there would be no need to repeat it. However, if duplication is accidentally introduced, it would not be harmful, merely slightly wasteful. The general scheme is summarized in Fig. 15.39.

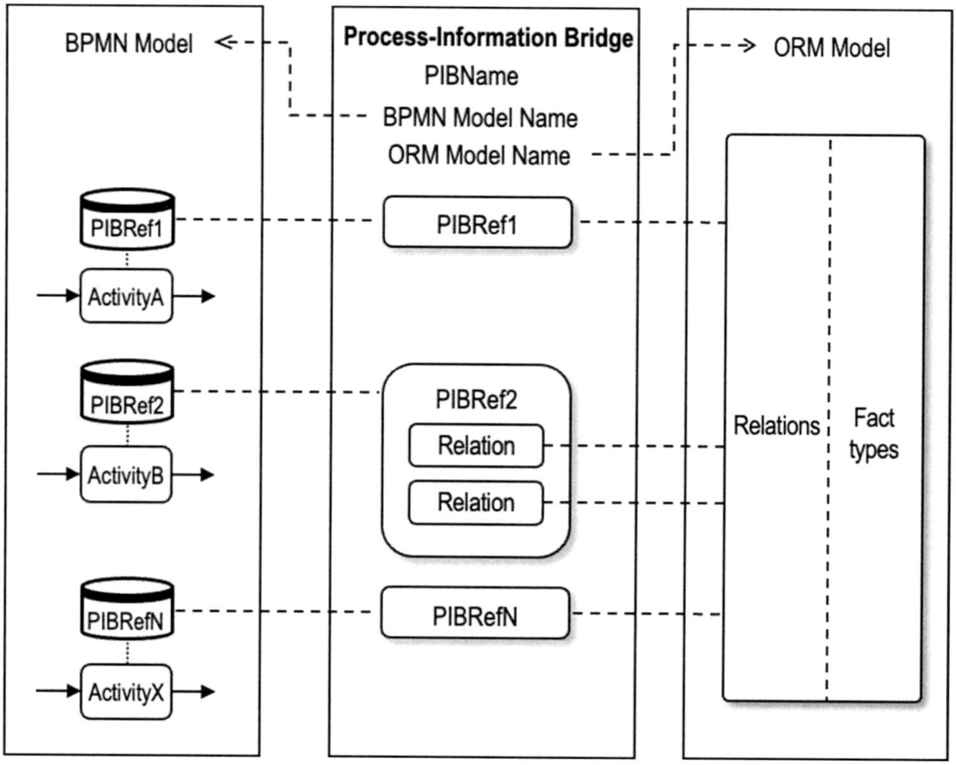

FIG. 15.39 PIB general scheme.

To illustrate some of these points, we can look at an example process model. Fig. 15.40 illustrates a simple complaints process using BPMN. After recording the complaint, the client is contacted to acknowledge receipt and the relevant department is contacted to assess the complaint. After assessment, the client is either compensated in some way or a standard letter is sent to explain the resolution. Details of the case are filed in case of any future query, and the process terminates.

To provide a concrete example of the relationship between process and information, we'll look at the SendStandardLetter activity; a corresponding picture could be drawn for any of the other activities. We will assume that appropriate resources and manual services are available to the activity, such as printing, envelope stuffing, and franking, and focus on the information usage. Fig. 15.41 shows a fragment of an information model, including the types of facts that might be relevant to the SendStandardLetter activity.

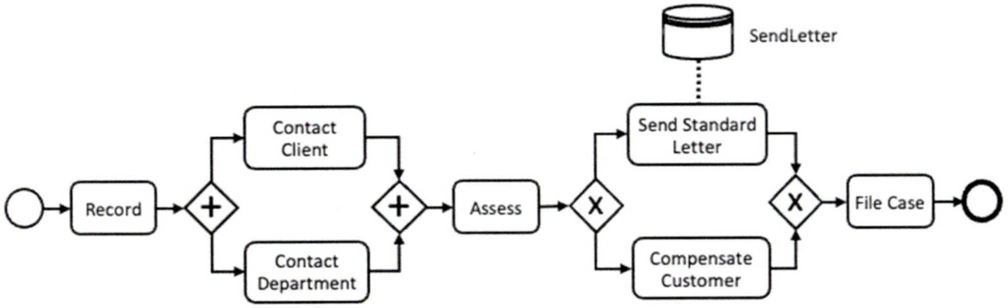

FIG. 15.40 A simple complaints process.

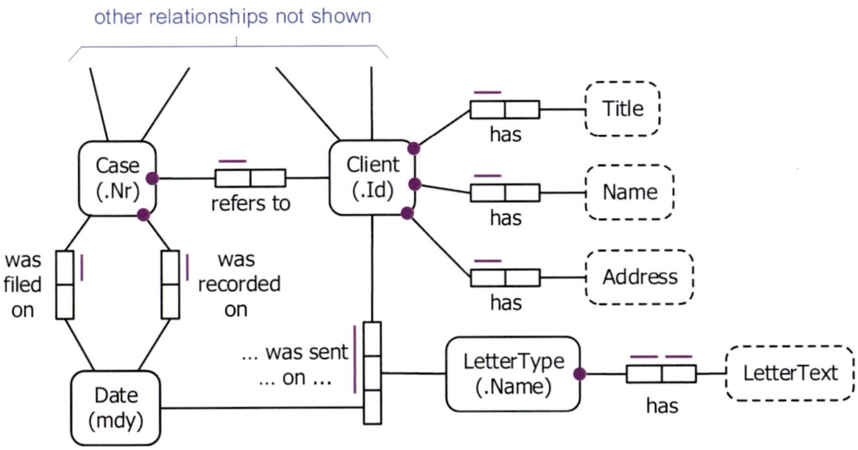

FIG. 15.41 An ORM model fragment relating to the SendStandardLetter information.

We are assuming here that the letter consists mainly of "boilerplate" text with the client information, the case reference, and dates inserted at the appropriate points. To identify the relevant facts, the activity requires two primary objects to be identified: a Case and a LetterType. From the Case, we can uniquely identify the Client and hence the Client's Title, Name, and Address. From the LetterType, we can identify the appropriate text to form the body of the letter. The current date is assumed to be available from the activity's environment. This information is sufficient for the activity to create and send the standard letter.

Fig. 15.42 shows a relational view of the information. This can be used as the basis for defining the necessary bridge between process and information.

From the process perspective, the necessary link is provided by the SendLetter Data Store reference in the process model of Fig. 15.40. The SendStandardLetter activity requires Read access (R) to the LetterType, Client, and Case relations but full access (F) to the ClientWasSentLetterTypeOnDate relation. The associated PIB can be defined in a suitable editing tool, as illustrated in Fig. 15.43. This shows the four relations, with the last relation being

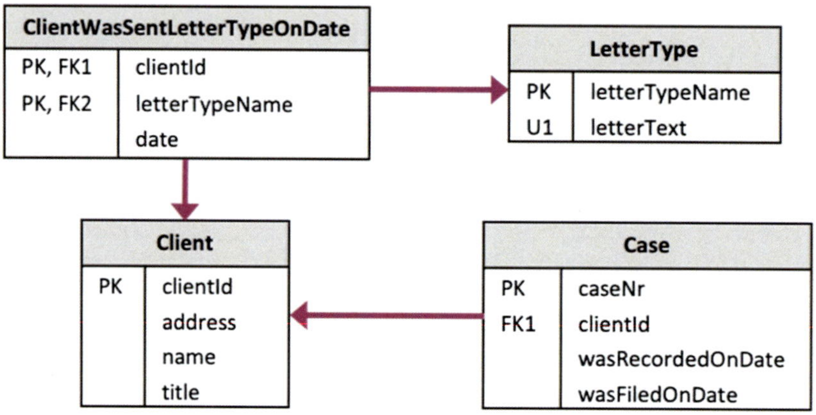

FIG. 15.42 A relational view of the SendStandardLetter information.

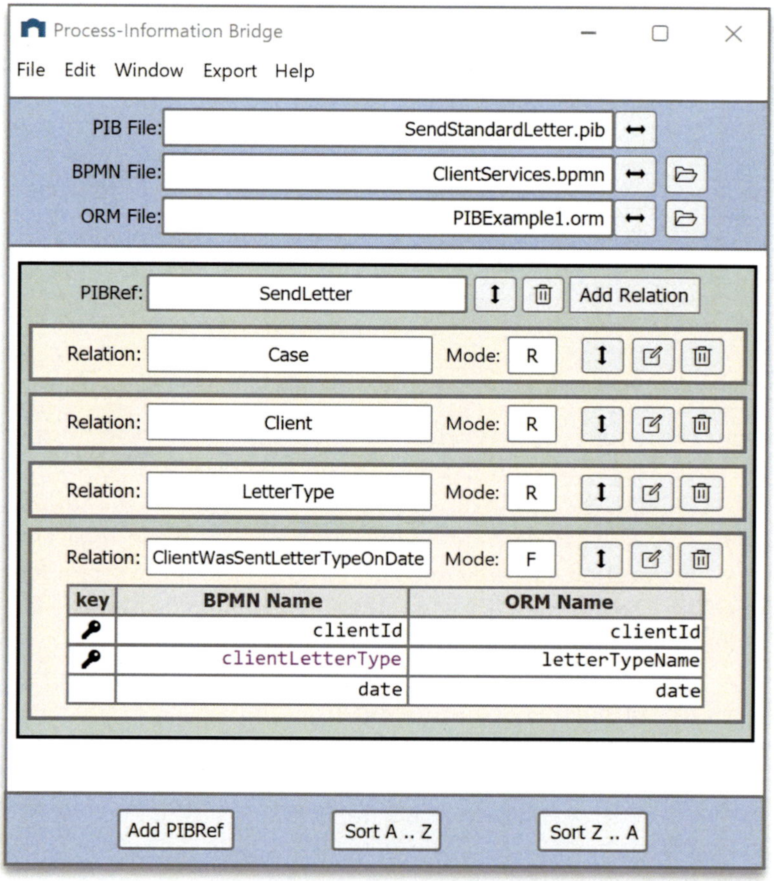

FIG. 15.43 The SendLetter PIB defined in an editing tool.

expanded to show its attributes. This specific tool also allows the use of alternative names in the process context.

A PIB file of this kind (probably expressed in XML or JSON) is a useful resource in its own right. It provides traceability between process and information at a conceptual level. The structure can be used to answer queries such as which activities rely on a given relation, or which activities create instances of a given fact type. This can help to ensure that systems designs are coherent before embarking on implementation.

15.10 Decision Model and Notation

Simple decisions can be defined in BPMN using appropriate gateway constructs, but this becomes infeasible for more complex decisions, such as evaluating the many factors influencing the approval of a loan application. To address this need, a new standard called Decision Model and Notation (DMN) was created by OMG. The first version of the standard was proposed in 2015, but it has undergone several revisions and is expected to evolve further over time. DMN is intended to be used in conjunction with BPMN in processes involving complex decisions.

The basis of a decision is the evaluation of some given input data against predefined business logic. A decision involving many factors may be broken down into a set of subdecisions combined to give the final result. DMN provides a graphical notation to show the linking of data (including intermediate results) and decision logic blocks. Here, diagrams only show how the logic is organized for human convenience. The real definition of the decision lies in the business logic contained in the blocks. DMN provides two means of expressing this logic: decision tables and a special expression language.

A decision table shows a list of input/output conditions indicating the output that results from a specific set of inputs. The table contains all possible combinations of input values. A row of the table contains columns for input values, corresponding outputs, and optional annotations. If more convenient, the orientation of the table can be changed by switching rows and columns.

As an alternative, DMN provides a language for defining business logic called Friendly Enough Expression Language (FEEL). This is claimed to be "friendly enough" for nontechnical business people to understand. It provides a collection of operators and functions that can be used to specify how a conclusion can be reached from a combination of input data. These are organized into "boxed expressions" that can be nested in complex cases.

The decision logic is invoked from BPMN through a special "Business Rule" activity type. Although the logic may be complex, the result is often very simple, for example, just a "yes" or "no" value. Activities that follow later in the process can use the decision result to influence the path taken, possibly using an appropriate gateway element.

Because DMN is an adjunct to BPMN, it requires no additional treatment from a conceptual modeling perspective. A general approach to combining ORM, BPMN, and DMN is outlined in Fig. 15.44.

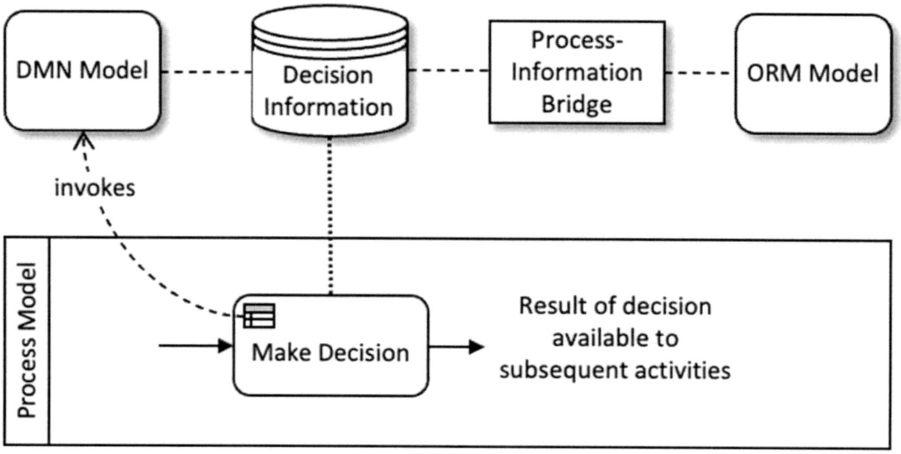

FIG. 15.44 Outline of DMN application.

Broadly speaking, the relevant information from an ORM model is marshaled together using the Process-Information bridge approach discussed above. A Business Rule activity invokes the decision at the appropriate point in the process. The decision logic to be applied at that point is expressed in DMN, referring to the available information used as inputs to the decision and defining the decision outcome (which may or may not be stored persistently). DMN is still a relatively young standard, and further revisions can be expected in the future.

Exercise 15.10

1. We are going to make a series of cabinets in a really organized way! First, we'll select a plan and get the wood required. Next, we'll lay out the parts on the pieces of wood according to the plan. We'll cut out the parts and then assemble them. This will require us to check the fitting of the joints and adjust them as necessary before gluing them together. Finally, we'll sand the cabinet smooth, paint it, and then move on to the next cabinet. We have the following resources in our workshop: saw, ruler, pencil, sandpaper, paint, screwdriver, chisel, and glue.
   (a) Draw a process structure for making our cabinets, using any suitable notation.
   (b) Identify which resources are required for each activity.
   (c) Are the resources consumed?
   (d) Are there any points where the process should be suspended?
   (e) Could we have multiple parallel instances of any of these activities?

2. The following diagrams show two workflow designs. The workflow on the left is unstructured. The workflow on the right is intended to restructure the workflow on the left into a form that is appropriate for a block-structured workflow language. Assume that the values of w and x are set by Activities D and F, respectively.
   Are these two workflows equivalent?

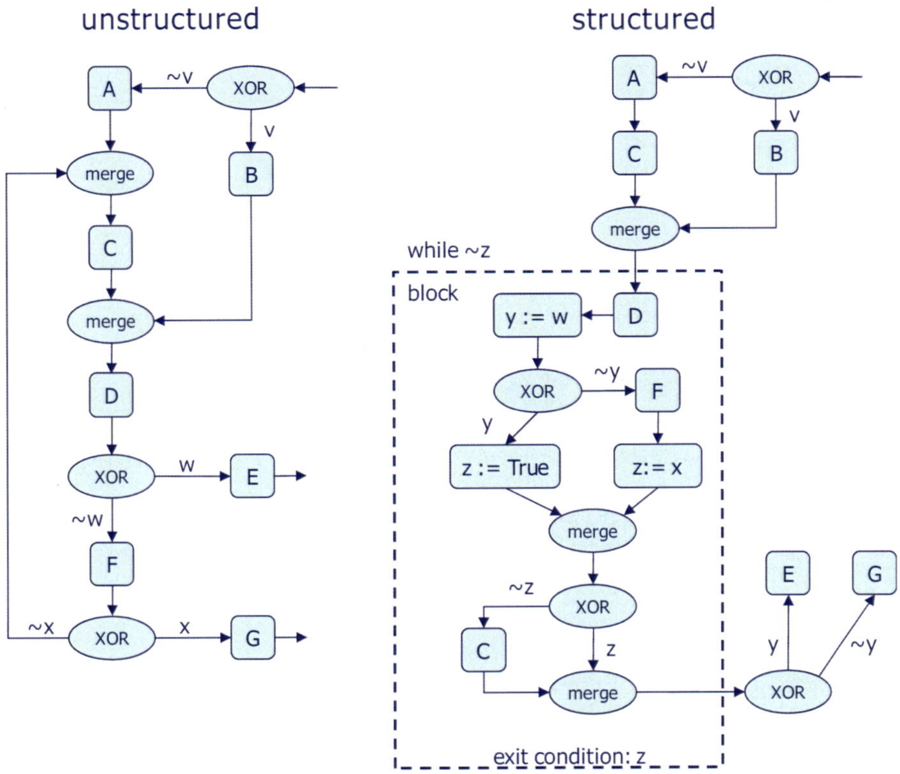

3. A company has a process for planning new products. Management sets out the product strategy. Sales identifies product needs from the market. Once these have been done, sales develops proposals for new products

After the proposal is complete, Sales carries out market testing and then estimates the likely sales volume. At the same time, Finance estimates the costings and then either reserves a budget from existing funds or sets up a loan facility with the company's bank. Once this is done, Finance produces a financial plan for the new product.

When the financial plan and the sales estimates are available, Management carries out a feasibility study. If the feasibility is rated good, Production carries out production planning and the process ends. If the feasibility is rated poor, the viability of the product concept is reassessed.

If the product concept is judged not viable, the process ends. If the product is judged still potentially viable, Sales is asked to develop new proposals.

Draw a BPMN-style process diagram for the scenario just described. Show the Sales, Management, Finance, and Production departments in lanes. Use only the following symbols:

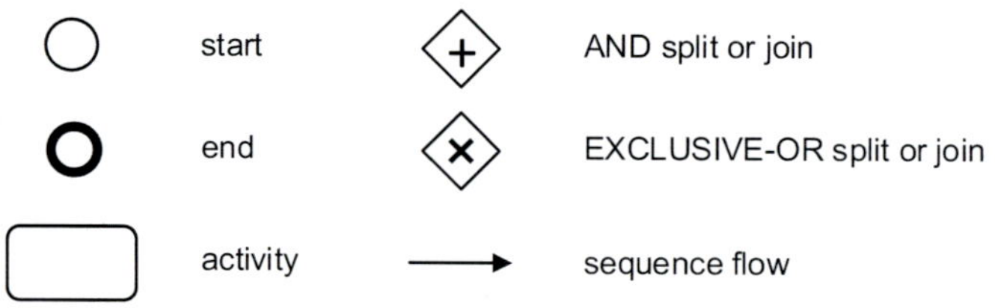

4. The following schema shows information related to the testing of manufactured items

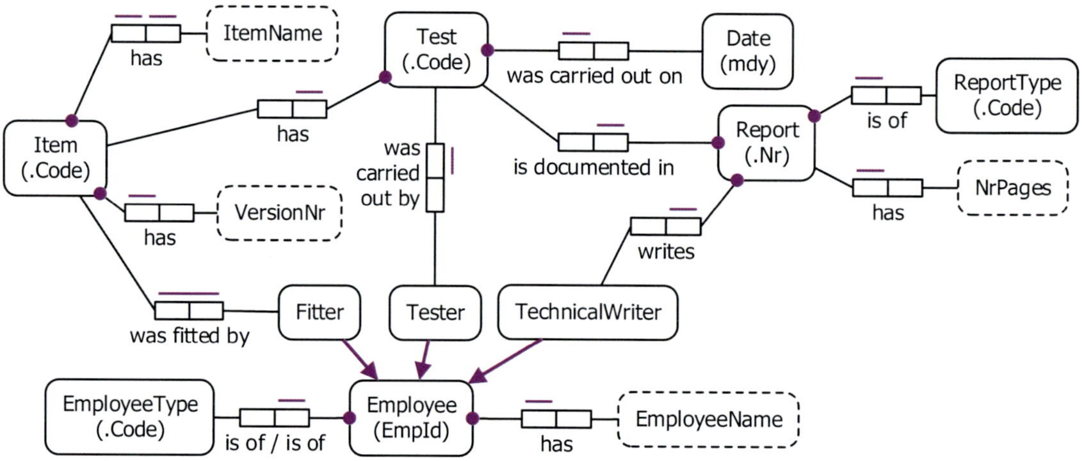

(a) Given the schema here, are there any implied constraints on the sequence in which business objects can be created?
(b) What can we say about this information model from the standpoint of business processes?

5. Assume that we have a hair dryer controlled by two switches: SW1 controls the motor speed and SW2 controls the heating element. The possible switch positions are shown here.

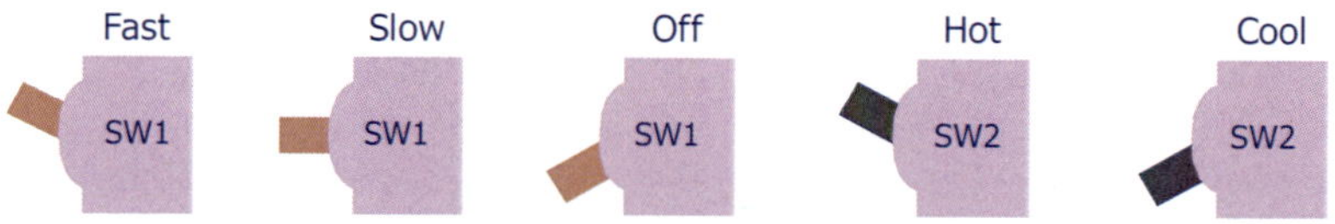

SW1 has three positions and SW2 has two positions. Assume switches can't go past their range (e.g., "Up" has no effect if the switch is already all the way up).

A possible Moore state machine is shown, using Up and Down as the input events for the two switches.

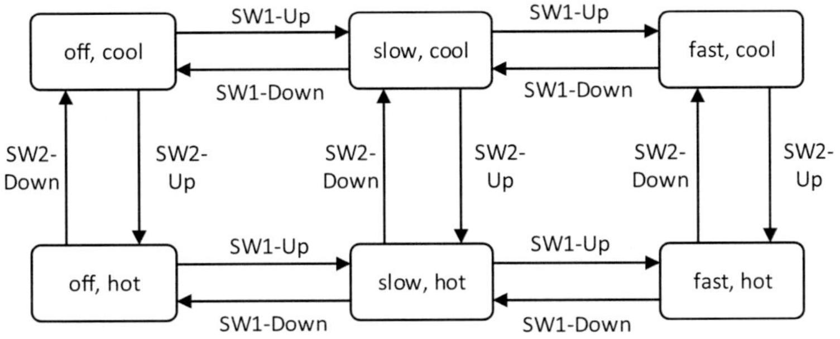

(a) From a common-sense point of view, does this model seem sound?
(b) Produce an equivalent diagram using composite states. Assume the starting position is Heat = Cool, Fan = Off.
(c) Add a "Swirl" switch to the composite state diagram, with three positions: Off, Medium, and High. Assume the starting position is Swirl = Off.
(d) What would the diagram look like without composite states?

15.11 Summary

ORM provides a powerful methodology and notation for capturing the structure of information at a conceptual (i.e., implementation-free) level. From an ORM model, one can infer some partial ordering on the dynamics of the information. For example, information on mandatory fact types associated with a primary object type must be available no later than the time at which that object type is instantiated. References between primary fact types must be ordered in a way that maintains referential integrity—in a database system, these typically appear as foreign key references.

However, these limited constraints fall far short of a definition of a process. Some other forms of expression are necessary to define explicitly the intentions of a business in executing its activities. Business process models provide one way of doing this; state models provide another. Nor does ORM provide any way of expressing other aspects of process elements, such as the resources involved. This is not a criticism of ORM: it's just that a description of a business involves many aspects in addition to the structure of the information it uses.

Clearly, it may be appropriate to consider a business from many different perspectives, but these different points of view must be reconciled because they are ultimately an expression of the same business system. Trying to model these different perspectives with ORM alone would be like trying to model an object-oriented system in UML and being limited to using only class diagrams. This implies a multi-viewpoint approach to business modeling. However, combining different perspectives does not mean using independent models, as we are really talking about different views onto one common model.

Both process models and state models can be used to express behavioral aspects. The techniques that they embody have evolved in similar ways because they have responded to the same pressures, such as the need to deal with large, complex sets of activities. UML includes both activity and state modeling, but its use in business modeling is compromised by its primary orientation toward defining software artifacts.

Standards are becoming more important in process modeling. Informal standards, such as the popularly quoted workflow patterns, have helped establish a better understanding of important features of business processes. Standards of a more official kind, such as BPMN and DMN, are helping facilitate the uptake of process modeling by businesses of all kinds.

Chapter Notes

Rummler and Brache (1995), Burlton (2001), and many others have described methodologies for capturing business process definitions, and Hammer (1997) and others have emphasized the restructuring of businesses around processes. Havey (2005) gives a broad summary of several business process modeling approaches. A good introduction to Workflow is given by van der Aalst and van Hee (2004). Many valuable contributions to state modeling have been made by David Harel; see for example Harel and Politi (1998). The systematic modeling of business processes remains an area of great interest in information technology. Several standardization initiatives continue within organizations such as the Object Management Group (OMG) and the World Wide Web Consortium (W3C), including the Business Process Model and Notation (BPMN), Business Process Definition Metamodel (BPDM), Business Process Maturity Model (BPMM), Decision Model and Notation (DMN), the Business Motivation Model (BMM), and many others, including related initiatives in the area of web services. The latest public information on OMG standards (BPMN, DMN, etc., and UML) can be found at www.omg.org. Members of OMG also have access to the latest work in progress on each of the standards. The relevant W3C standards relating to process orchestration and process choreography can be accessed at www.w3c.org. The use of UML for business modeling is discussed by Eriksson and Penker (2000). Lassen and van der Aalst (2006) and Ouyang et al. (2006) discuss techniques for translating workflow processes to BPEL. Many papers have used the "classic" workflow patterns defined by van der Aalst, ter Hofstede, and others, and there is a workflow patterns website (www.workflowpatterns.com). This provides detailed descriptions with a Petri-net orientation. Additional discussion of state modeling and ORM is given in Morgan (2006).

16

Data File Formats

Chapter Outline

16.1 External Data Structures

Previous chapters have made the implicit assumption that data will be stored in a database system of some kind, such as a relational database. In this chapter, we consider how data can be represented in a way that's independent of a particular type of data store. This may be necessary when we need to transfer data between different systems. If the transfer takes place between two systems of the same kind, we can make use of any functionality built into the system. As examples, we can use a server-to-server replication mechanism, or restore a backup made on one server onto a different server. But these approaches will not usually work between two unlike systems, even if they are of the same general kind. For instance, a backup copy made on SQL Server cannot be restored on an Oracle system. We also need to be aware of possible interchange between different kinds of systems, such as a database management system and a spreadsheet.

A widely used and long-established approach is to use comma-separated values, typically referred to as a "CSV file". This is a simple file format used to store tabular data in a plain text format. In a CSV file, each line of text represents a record, with the fields in each record being separated by commas. This structure is roughly equivalent to the values in a single table in a relational database. Tab-separated values are a close cousin; as the name suggests, the separator between fields here is a tab character instead of a comma. If the data itself contains delimiter characters (comma, tab, newline), additional measures such as quoting or escaping are needed.

The main advantage of a CSV file is its simplicity. The file can be opened and edited with any text editor, and the structure is easy to understand, just rows and columns (though the columns may not be obvious because of different field lengths in different rows). CSV files are widely supported by a variety of applications, including spreadsheets, programming languages, and databases. They are quite compact because they consist mainly of data with only a minimal set of delimiter characters. Because they contain

Information Modeling and Relational Databases. https://doi.org/10.1016/B978-0-443-23790-4.00020-7

no information about how the data might be stored, they are a popular choice for straightforward data transfer between different kinds of systems. Text files of this kind can also be used as simple databases when partnered with an appropriate query facility, such as the venerable AWK language found in the UNIX environment.

However, this same lack of structure limits the expressiveness of the format. There's no built-in way to include metadata about the data or the file itself. There's no way of defining data types. Most important of all, there's no way of expressing even moderately complex structures or relationships.

The need to address these issues has promoted interest in more sophisticated representations. Two major alternatives have emerged: the eXtensible Markup Language (XML) and JavaScript Object Notation (JSON). Both are in widespread use, especially for the transfer of structured data between systems, though they can both be used as information stores in their own right. They share some common features.

1. They are text-based. Like CSV files, XML and JSON files can be opened and edited with a simple text editor. However, the complexity of the structure can make this a daunting task. It's very easy to make editing errors that render the structure invalid, and it's difficult to visualize the structure unless it is laid out appropriately. For this reason, it's common to use special editors that are sensitive to the conventions of the notation. The result, though, is still just text, which can be stored and retrieved using standard facilities.
2. They are hierarchical. Elements can contain other elements (recursively), which results in the data being contained in a tree-like structure. This is the case even when the original data is not a natural hierarchy. An inevitable consequence is the possibility of data duplication, which will need to be managed, either in a data store or in applications that use the data.
3. They use a "bracketed" notation. Data elements are contained within a matching pair of opening and closing delimiters such as < x >, (x), [x], or { x }. Nested elements must be wholly contained, so < x < y >> may be valid, but <x < > y > is definitely not.
4. They are nondeterminative, in the sense that the same data structure could be represented in different, but equally valid, ways. For example, in XML a data item could appear as the value of an attribute or the content of an element. The choice between these can be guided by some high-level nonspecific policies or by the intended use of the data, but this is not enforced by any language constraints or tools. This is different from the application of normalization rules in a relational scenario.

Although both XML and JSON are text-based, in structured query language (SQL) terms, they can also be considered as complex data types alongside BLOB, DATETIME, etc., and both have been adopted as such within the SQL standard. XML was added as a data type in SQL:2003. JSON was introduced through text operations in SQL:2016, and as a data type in SQL:2023. Using a built-in data type instead of text has some advantages. Validation is simpler, some operations are more efficient, and the overhead of conversion can be avoided when the input or output is required to be in XML or JSON form. In this chapter, we will focus mainly on how XML and JSON can be used for the representation of structured data.

16.2 XML

XML is a key member of a family of standards produced by the World Wide Web Consortium (W3C). Version 1.0 first appeared in 1998, with Version 1.1 following in 2004. Several subsequent revisions have corrected errata and provided clarifications. The growth of WWW systems using HTML (itself introduced in 1993) exposed the need for a language that provided a stronger way of representing structured data to support seamless data exchange, transformation, and interoperability across different systems and platforms, and also to allow for extensions to customize the language for specific applications.

There are several other standards relating to XML. In addition to XML itself (and its relationship to SQL), we will limit our main discussion in this chapter to the following:

XPath. This is an expression language that provides a means of hierarchic addressing of the nodes in an XML tree. As well as modeling the tree structure of XML, the data model also includes atomic values, function items, and sequences.

XQuery. This is a query language designed to be broadly applicable across many types of XML data sources, including structured and semistructured documents, relational databases, and object repositories. XQuery uses the structure of XML intelligently to express queries across all these kinds of data, whether physically stored in XML or viewed as XML via middleware.

These both depend in turn on the *XQuery and XPath Data Model*. This defines the information contained in the input to a query processor and all permissible values of expressions in the XQuery and XPath languages. Both languages have been extended to provide support for JSON as well as XML.

The data structure for a particular universe of discourse (UoD) can be defined by a schema. The XML Schema standard provides a means of defining the structure, content, and semantics of XML documents. It replaces the earlier document type definitions (DTDs) used to define structure. An XML document can be validated against a schema to ensure that its structure is consistent. Like SQL, XML defines a variety of data types, although these do not correspond exactly. The SQL specification (part 14) defines a mapping between XML Schema and SQL datatypes, but XML Schema is a complex topic, and we will not examine it in detail in this chapter.

XML Structure

The crucial distinction between relational and XML data is that XML structure is inherently hierarchical. This is a consequence of the expression of the structure as items embedded within other items. Fig. 16.1 illustrates this with a simple list of items. XML

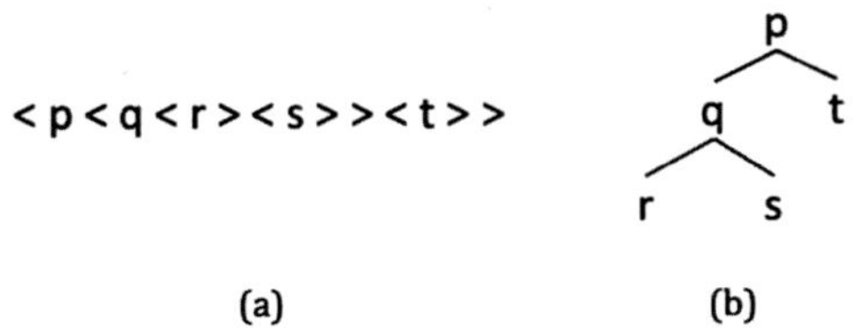

FIG. 16.1 Hierarchies: (a) List with embedded items. (b) Interpretation as a hierarchy.

has a more complex syntax, but the principle is the same. As the figure shows, a natural interpretation of embedding one item within another is that a contained item becomes a child of the containing item.

Interestingly, early hierarchical database systems were commercially successful—with IBM's IMS being a prominent example—before they became overtaken by later developments. The rigidity of these earlier hierarchical systems proved to be uncompetitive against alternatives (especially the relational model) and so they became obsolete. XML does not set out to imitate these earlier systems, but there seems to be something compelling about the hierarchical notion that just won't go away. Some differences in structure between relational and XML structure are summarized in Table 16.1.

Table 16.1 Some differences between relational and XML structured data.

Feature	Relational (SQL)	Basic XML
Primary structure	"Flat" sets	Hierarchical
Data sequencing	Row order is not significant	Nodes are ordered
Primary access mechanism	Set operations	Navigation through hierarchy
Data typing	Always typed	Typed or untyped
Missing values	NULL	Absent or empty element

The directory structure of most operating systems provides a familiar example of a hierarchy. Even in a graphical interface, we have become used to a system of "folders" that can contain a combination of other "folders" and leaf nodes called "files". The file system analogy is particularly relevant here because the XPath/XQuery expressions used to navigate the XML structures resemble the textual commands used to navigate directory structures in environments that aren't based on a graphical user interface (GUI). The basic hierarchy concepts are shown in Fig. 16.2.

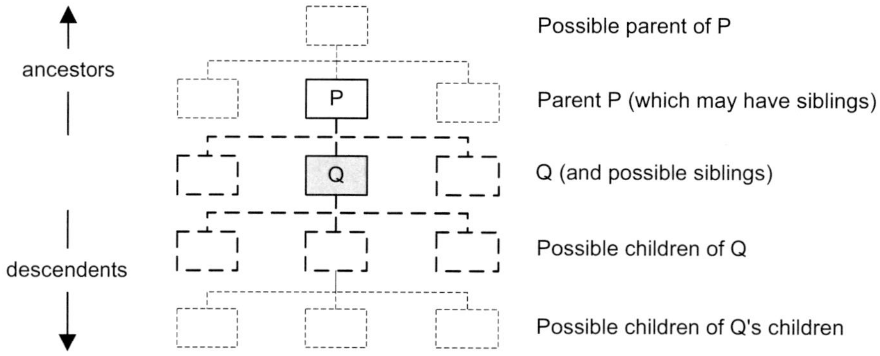

FIG. 16.2 A node in a hierarchy.

We're going to be navigating from node Q, which provides the "context" for our path expression. Node Q might be at the top of the hierarchy we're considering, in which case it can be taken to represent the whole of our XML document. Apart from that special case, we can always go up one level from Q to the node P, the *parent* of Q. If Q has *siblings* (nodes that also have the parent P), we can navigate sideways. If Q has *children*, then we can navigate down. Some combination of up/sideways/down movements can take us to a new node. Once we get there, the newly selected node can become the context for some other navigation.

Most of the time, we're going to be concerned with relationships between XML elements, so, for example, we might want to know about all elements that are "contained in" element Q (i.e., they are descendent nodes of the node Q). But we might also want to know about all elements that have some specific attribute, all elements related to some specific namespace, and so on. All these possible dimensions are known in XPath/XQuery as *axes*, and the languages provide a sophisticated notation for the precise definition of navigation along various combinations of axes. Unfortunately, the full notation is a little cumbersome. Fortunately, there are short-form notations that cover the main possibilities, and we'll concentrate on those.

Nodes or "Elements" are identified in XML with a pair of named tags (opening and closing). Nodes can have zero or more attributes, which are listed in the opening tag. Each opened tag must ultimately be closed after all the content of the tag has been listed. The following XML expression shows an example:

```
<movie movieNr="1"> <title> Cosmology </title> </movie>
```

Here, we have a tag named "movie" with one attribute, movieNr, that contains data. It contains another tag named title, which itself contains some data. The closing tag repeats the tag name preceded by a *solidus* or *slash* character. Optionally, elements that contain no other elements can be closed with just a single solidus character following any attributes, as in the following.

```
<movie movieNr="1" />
```

The guidelines for choosing between attribute or element content as the location for data are a little vague. Broadly speaking, attributes are supposed to be for intrinsic properties of the node, but this is open to interpretation. It's certainly true that the following representation is just as valid, syntactically, as the first.

```
<movie title="Cosmology"> <movieNr> 1 </movieNr> </movie>
```

In many cases, finding a path to the relevant node or nodes is all that's necessary to extract relevant data from XML. The required path can be specified using an XPath expression. To provide some concrete examples, we'll return to the *Movies* example from Chapter 1. There, we saw how the content of a report could be captured as relational tables. Listing 16.1 shows similar information (with slightly different data from the example in Chapter 1). The initial tag simply identifies this as an XML document. This listing is a lot more verbose than an equivalent rendering in, say, a CSV file, but that's the price that must be paid to achieve the expressiveness necessary to represent complex data structures.

```
<?xml version="1.0" encoding="UTF-8"?>
<movies>
        <movie movieNr="1">
          <title>Cosmology</title>
          <released>2006</released>
          <director>Lee Lafferty</director>
        </movie>
        <movie movieNr="2">
          <title>Kung Fu Hustle</title>
          <released>2004</released>
          <director>Stephen Chow</director>
          <star gender="M">Stephen Chow</star>
        </movie>
        <movie movieNr="3">
          <title>The Secret Garden</title>
          <released>1987</released>
          <director>Alan Grint</director>
          <star gender="F">Gennie James</star>
          <star gender="M">Barret Oliver</star>
        </movie>
        <movie movieNr="4">
          <title>The Secret Garden</title>
          <released>1993</released>
          <director>Agnieszka Holland</director>
          <star gender="F">Kate Maberley</star>
          <star gender="M">Heydon Prowse</star>
        </movie>
        <movie movieNr="5">
          <title>The Da Vinci Code</title>
          <released>2006</released>
          <director>Ron Howard</director>
    <star gender="M">Tom Hanks</star>
          <star gender="M">Ian McKellen</star>
          <star gender="F">Audrey Tautou</star>
        </movie>
        <movie movieNr="6">
          <title>Cast Away</title>
          <released>2000</released>
          <director>Robert Zemeckis</director>
          <star gender="M">Tom Hanks</star>
        </movie>
  </movies>
```

LISTING 16.1 Movie information in XML.

Navigation is relative to a given node in an XPath sequence. The basic navigational notation resembles that of a conventional file system. A *period* or *full stop* " . " means "the current node", " .. " means "the parent of the current node". A *slash* or *solidus* character " / " without further qualification means "go down one level"; " // " means go down any number of levels. Names are taken to refer to elements unless prefixed with "@", in which case they refer to attributes. At any step in the path, a predicate can be introduced in square brackets. This will filter the resulting nodes, so only those for which the predicate is "True" will be included. Table 16.2 shows some typical expressions that use these concepts.

Table 16.2 Example path expressions.

Expression	Result
Find all directors /movie/director	<director>Lee Lafferty</director> <director>Stephen Chow</director> <director>Alan Grint</director> <director>Agnieszka Holland</director> <director>Ron Howard</director> <director>Robert Zemeckis</director>
Find one star from each movie /movie/star [1]	<star gender="M">Stephen Chow</star> <star gender="F">Gennie James</star> <star gender="F">Kate Maberley</star> <star gender="M">Tom Hanks</star> <star gender="M">Tom Hanks</star>
Find the titles of all movies with a female star /movie/star [@gender = "F"] / .. / title	<title>The Secret Garden</title> <title>The Secret Garden</title> <title>The Da Vinci Code</title>
Find the release year of all movies starring Tom Hanks /movie [star = "Tom Hanks"] / released	<released>2006</released> <released>2000</released>

Now that we have some idea of how to move around an XML hierarchy, we can consider how to query XML data. In simple cases, the result of the path expression may be sufficient, but in general, we will need more powerful expressions. This need is met by XQuery, a language for manipulating XML values. Roughly speaking, XQuery is partnered with XML data in the same way that SQL is partnered with relational data. Like SQL, XQuery is declarative. It expresses the data sources and the processing rules to be applied, but it does not specify the explicit procedure by which the results should be produced.

XQuery is a *functional* language, so we must specify what we want to do in functional terms. Most software developers are familiar with the concept of a function. A function has a name, zero or more input parameters, and a body that computes some result to be returned. From a functional programming perspective, the arguments to a function can be constants, or the results produced by other functions (i.e., a parameter can be a call to another function). Now, we have another hierarchy: functions that provide results to functions that provide … and so on. This *function* hierarchy may be established dynamically and may bear no direct relationship to the *data* hierarchy. The top level of the function hierarchy computes the result we are seeking; the lower-level functions provide results to their parents to (ultimately) enable the top-level computation. XQuery defines more than a hundred built-in functions, but commercial systems may not yet implement all of these.

XQuery allows information to be combined from multiple sources and restructured to produce a new result. The most common expressions of this kind in XQuery are called FLWOR expressions, from the first letter of the types of clauses that may appear in the expression: **for**, **let**, **where**, **order by**, and **return**. FLWOR expressions are the XQuery equivalent of **select** statements in SQL.

The clauses in a FLWOR expression are defined in terms of *tuples*. The tuples are created by binding variables to values in **for** and/or **let** clauses, so every FLWOR expression must have at least one **for** or **let** clause. The difference between the two is that **for** iterates over the items in an expression, producing a tuple for each, and **let** produces one tuple for the entire result of an expression. As a simple example, the following FLWOR expression produces the same result as the last path example in Table 16.2. The dollar sign "$" indicates a variable in XQuery.

```
for $m in movie
  where $m/star = "Tom Hanks"
  return $m/released
```

As a slightly more complex example, let's assume that we want to invert the hierarchy of the source XML. For our Movies XML fragment, we could list the movies in which an actor starred under each actor, instead of listing the stars under each movie. An expression in XQuery to do this might look like the following:

```
for $s in distinct-values ( movie/star )
order by $s descending
return
   <starring>
         <actor> { $s } </actor>
         <movies> { for $m in movie where $m/star = $s return $m/title } </movies>
   </starring>
```

Here, we've used the distinct-values ( ) function to make sure that each actor only appears once and sorted the result in descending order of actor name. The query contains *constructors*, indicated by braces around an inner expression, which are used to build element content. Evaluating the query produces the result:

```
<starring>
    <actor>Tom Hanks</actor>
    <movies>
        <title>The Da Vinci Code</title>
        <title>Cast Away</title>
    </movies>
</starring>
<starring>
    <actor>Stephen Chow</actor>
    <movies>
        <title>Kung Fu Hustle</title>
    </movies>
</starring>
...etc.
```

Of course, we're not limited to using a single XML document or fragment. Let's say we have some XML data containing reviews of the movies, like the example shown in Listing 16.2.

The reviews are related to the movies by the unique movieNr reference. In standard SQL, if we wanted to show the movie titles along with the rating scores for each movie, we could use a join. We can do the same kind of thing in XQuery.

```
<review reviewNr="1">
   <movieNr>4</movieNr>
   <rating>4</rating>
   <comments>Classic story</comments>
</review>
<review reviewNr="2">
   <movieNr>5</movieNr>
   <rating>4</rating>
   <comments>Great movie!</comments>
</review>
<review reviewNr="3">
   <movieNr>5</movieNr>
   <rating>5</rating>
   <comments>Superb</comments>
</review>
<review reviewNr="4">
   <movieNr>5</movieNr>
   <rating>1</rating>
   <comments>Not my kind of movie</comments>
</review>
```

LISTING 16.2 Movie reviews in XML.

Here's one way of doing the equivalent of an inner join.

```
for $m in movie
where some $r in review satisfies( $m/@movieNr = $r/movieNr )
return
<reviewscore>
  { $m/title, $m/released }
  { for $r in review where $m/@movieNr = $r/movieNr return $r/rating }
</reviewscore>
```

The query includes the release year as well as the title to avoid ambiguities between movies with the same name. The "**where some ... satisfies (...)**" construct is an existential quantifier test so that we only get results for movies that have had a review. The query produces the following result:

```
<reviewscore>
    <title>The Secret Garden</title>
    <released>1993</released>
    <rating>4</rating>
</reviewscore>
<reviewscore>
    <title>The Da Vinci Code</title>
    <released>2006</released>
    <rating>4</rating>
    <rating>5</rating>
    <rating>1</rating>
</reviewscore>
```

To get the equivalent of an outer join, which would also include the movies with no reviews, we would simply need to omit the existential quantifier test in the second line of the query.

As a final XQuery example, we'll show the equivalents of the *aggregate* or *bag* functions in SQL, which include **sum( )**, **count( )**, **avg( )**, **max( )**, and **min( )**. Let's suppose we want to produce a summary element for each movie, showing the total number of reviews and the average score as attributes. The following query would be one way of producing the required result.

```
for $m in movie
return
<reviewsummary total = "{ count ( review [ movieNr = $m/@movieNr ] ) }"
   avg = "{ avg ( review [ movieNr = $m/@movieNr]/rating ) }" >
  { $m/title, $m/released }
</reviewsummary>
```

Here, we've used the XQuery functions **count( )** and **avg( )**, which operate in a similar way to their SQL counterparts. For an empty list of elements (in this case, movies that have no reviews), the **count( )** function returns zero and the **avg( )** function returns an empty result. Running the query produces the following XML fragment:

```
<reviewsummary total="0" avg="">
  <title>Cosmology</title>
  <released>2006</released>
</reviewsummary>
<reviewsummary total="0" avg="">
  <title>Kung Fu Hustle</title>
  <released>2004</released>
</reviewsummary>
<reviewsummary total="0" avg="">
  <title>The Secret Garden</title>
  <released>1987</released>
</reviewsummary>
<reviewsummary total="1" avg="4">
  <title>The Secret Garden</title>
  <released>1993</released>
</reviewsummary>
<reviewsummary total="3" avg="3.33333333333333">
  <title>The Da Vinci Code</title>
  <released>2006</released>
</reviewsummary>
<reviewsummary total="0" avg="">
  <title>Cast Away</title>
  <released>2000</released>
</reviewsummary>
```

XQuery has many other features, and a more complete description would require a book in its own right. Some references to further information sources are included in the chapter notes.

We can also work with XML values using a standard programming language. Library functions are available for most languages to allow an XML processor to be incorporated into a program. Two types are available. The first, usually known as a DOM processor, loads the complete Document Object Model into an internal program structure. From there, various functions can be used to navigate through the model to extract tags and attributes as individual items or lists within the program. Because the whole DOM is in memory, this is convenient and reasonably efficient, but there is going to be some upper limit on the size of the XML file that can be loaded.

The alternative is usually referred to as a SAX processor (Simple API for XML). This uses a different event-based strategy to process an XML file. As the XML document is read, each significant item (such as a start tag) is sent to the program where it can be stored for later use or processed immediately. The important distinction is that the entire document does not need to be stored in memory, and so there is no upper limit on document size.

The disadvantage is that it is more difficult to carry out operations that require references back and forth between various parts of the model.

SQL/XML

SQL:2003 introduced XML as a new built-in data type. XML data is strongly typed, as XML values are assumed to be tree-like data structures that are distinct from their representation in the form of a text string. The XML type can be used wherever other SQL types are allowed, such as in table columns, as parameters of a routine, as variables, and so on. Although relational data and XML data are both structured, there are some significant differences between them. Table 16.3 (repeated from Table 16.1) identifies some key points.

Table 16.3 Some differences between relational and XML structured data.

Feature	Relational (SQL)	Basic XML
Primary structure	"Flat" sets	Hierarchical
Data sequencing	Row order is not significant	Nodes are ordered
Primary access mechanism	Set operations	Navigation through hierarchy
Data typing	Always typed	Typed or untyped
Missing values	NULL	Absent or empty element

XML data can be stored in a database either in a text column (i.e., **char**, **varchar**, or **clob**) or in a column of data type **xml**. A textual representation might be preferred if it is important to retain the textual fidelity of a document (which may include text that has no relevance to the data structure) or if the XML is always entered and retrieved as a complete document. However, storing the data in an XML column allows efficient processing of the XML data in the database, rather than in a middle-tier application.

In addition to the XML data type, the SQL:2003 standard also introduced some related features, such as new XML operators, functions, and rules for mapping between SQL concepts and XML concepts. Most vendors broadly support these features or their equivalents, but the need to integrate the additional XML features into different preexisting frameworks has inevitably resulted in some differences in syntactic detail. The fact that XML support is an area of keen competition for vendors is also likely to produce a variety of nonstandard extensions. The picture is further complicated by the need for interoperability with other standards, such as XQuery. Fortunately, this competition is limited to the variety of operators, functions, and so on that are provided to integrate XML with SQL. The standards governing the structure of XML data are defined outside SQL, and so the XML values themselves are reasonably consistent from one database system to another.

In 2006, the part of the SQL standard dealing with XML (Part 14) was revised and extended to consolidate the features of SQL/XML, as it has come to be called.

(The standard is sometimes referred to as SQL:2006, although this is a little misleading because only the XML part was revised in 2006.) Here, we will use SQL/XML:2006 to specifically distinguish the more recent version where necessary.

In SQL/XML:2006, the **xml** data type is based on the XQuery data model.[a] Legal values are either **null** (indicating the absence of any XML) or any legal value of the XQuery data model. An instance of the XQuery data model is a *sequence* of zero or more items, which are either atomic values of one of the defined data types or a *node*. There are seven kinds of nodes: *document, element, attribute, text, namespace, processing instruction,* and *comment*. The **xml** data type can take optional modifiers that give a finer-grained definition of the XML structure. The general sequence structure is indicated by **xml(sequence)**.

A sequence containing exactly one document node is termed *XML content,* with the type modifier **xml(content)**. A special type of content is where the document node contains exactly one element node (plus, possibly comment and processing instruction nodes), and this is termed an *XML document,* with the type modifier **xml (document)**.

A secondary modifier can be added for **xml(content)** and **xml(document)** to indicate whether the content or document can contain both typed and untyped elements (**any**), only untyped elements (**untyped**), or elements defined by a schema (**schema** name). If a schema is defined, the element and attribute nodes are guaranteed to be valid according to the schema. These distinctions are summarized in Fig. 16.3. If the type **xml** is specified without a modifier, any default assumption about these more specialized subtypes is implementation-dependent.

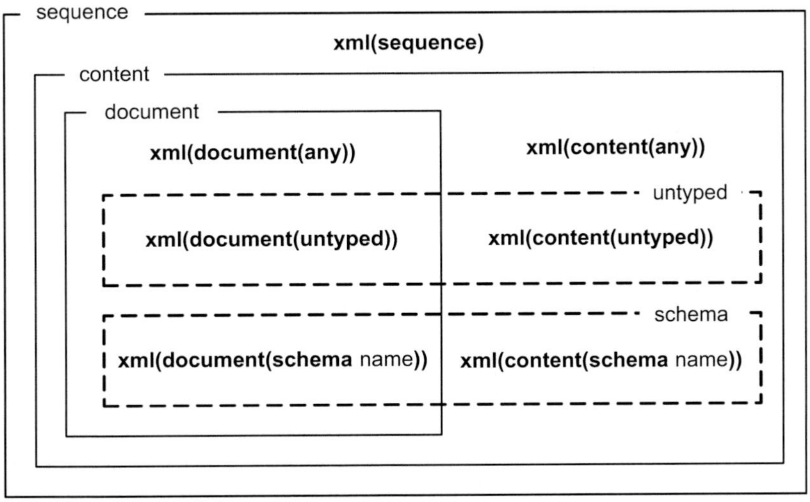

FIG. 16.3 XML data type modifiers.

[a]To be more precise, SQL/XML is based on the XQuery 1.0 and XPath 2.0 Data Model.

The SQL:2003 standard introduced the predicate **is document** to test whether an XML expression is a document. The basic syntax is:

xmlexpression **is [not] document**

Although the syntax of the predicate remained unchanged from SQL:2003, the meaning shifted somewhat because of the realignment of SQL/XML:2006 with the XQuery data model. Three more predicates were added: **is content, xmlexists,** and **is valid.** The **is content** predicate tests whether an XML expression is legitimate XML content, using the syntax:

xmlexpression **is [not] content**

To test whether an expression contains valid XML, the **xmlexists** predicate can be used. The syntax, simplified a little, is as follows:

```
xmlexists ( XQuery-expression
  [ passing value-expression
  [ as identifier ] [ { by ref | by value } ] ] [,...] )
```

If the XQuery evaluation is of a value that turns out to be null (perhaps passed as an argument from SQL), then the predicate result is unknown. For nonnull values, if the XQuery evaluation returns an empty sequence, the predicate's value is false; otherwise, the result is true. The following example shows the use of **xmlexists** in a select statement to retrieve "id" values for customers in a given city, with the customer details being stored in a column of the **xml** type.

```
select c.custId from customer as c
  where xmlexists ( ' $d/*:customerinfo/*:addr [ *:city = "Salt Lake City" ] '
  passing info as "d" )
```

The **is valid** predicate checks to see if an XML value is valid according to a given XML schema. The details of validity assessment are a little complex and are not covered here. The simplified syntax is:

```
xml-value-expression is [ not ] valid
  [ identity-constraint-option ]
  [ according to xmlschema { uri | registered-name } ]
  [ valid-element-clause ]
```

The xml-value-expression is checked according to the schema identified by the choice of namespace. If none is specified, each element is validated according to the schema associated with its namespace. If specified, the schema can be identified either by URI or by its registered name. The optional identity-constraint-option and valid-element-clause parameters provide for a more detailed specification of exactly how schema checking should be done.

Several built-in XML operators were introduced in the SQL:2003 standard. One of the operators, **xmlroot**, could be used to modify the version and standalone properties of an XML expression, but this was removed in SQL/XML:2006 so we don't cover it here.

The first two SQL/XML operations we'll look at, **xmlparse** and **xmlserialize**, convert back and forth between *structured* and *string* representations of XML. The **xmlparse** operation takes an SQL string expression and produces the equivalent XML structure. A parameter defines whether the string expression is expected to represent a document or content. The basic syntax has remained consistent since SQL:2003.

```
xmlparse ( { document | content } stringexpression
[ { preserve | strip } whitespace ] )
```

The result of the operation could be assigned to a variable of type **xml** or stored in an **xml** column. The precise way in which whitespace is handled is by **xmlparse**, which is subject to detailed rules that are not considered here.

The **xmlserialize** operation performs the inverse of **xmlparse**, producing a string version of a given XML expression. The SQL standard does not mandate the precise format of the resulting string, but does require that a subsequent **xmlparse** of a string produced by **xmlserialize** should reproduce the original XML expression. In other words, given an XML document called myXmlData, the following construct:

```
xmlparse ( document ( xmlserialize ( document myXmlData as varchar ( max ) ) ) )
```

should produce a result equal to myXmlData. The original SQL:2003 syntax of **xmlserialize** was extended in SQL/XML:2006.

```
xmlserialize ( { document | content } xmlexpression as SQLdatatype
  [ version string ]
  [ including xmldeclaration | excluding xmldeclaration ] )
```

The SQL datatype for the result could be any SQL string type (such as **char**, **varchar**), but serialized XML tends to be lengthy, so a character large object (**clob**) type is likely to be a common choice. Optionally, **including xmldeclaration** prepends the XML declaration "< ?xml version="1.0" encoding="UTF-8"? >" to the result; the default is **excluding xmldeclaration**. The XML version can also be separately specified; at present, this will be limited to "1.0".

The **xmlconcat** operation introduced in SQL:2003 was retained in SQL:2006. It simply concatenates a list of XML expressions, producing a new XML expression.

```
xmlconcat ( xmlexpression [, ...] [ returning { content | sequence } ] )
```

The optional **returning** clause was added in SQL/XML:2006. If this is omitted, it is implementation-defined whether returning content or sequence is implicit.

The SQL:2003 standard defined several "publishing" functions to generate XML expressions from SQL expressions. These were extended in SQL/XML:2006, and new functions were added. Here, we cover only the SQL/XML:2006 definitions of the functions.

The **xmlelement** function is used to create an XML element from a given set of parameters. Recall that an XML element has the basic form:

```
<tagname optionalAttributes> optionalElementContent </tagname>
```

So, for example, an element representing a customer might look something like:

```
<cust custid = "1234"> Widgets Inc </cust>
```

If required, the tagname can be further qualified by prefixing it with a namespace, in which case the namespace URI must also be specified. The **xmlelement** function constructs an individual XML element from the parameters provided and returns it as an **xml** value. The basic syntax for character strings is shown here. There is also a variant for binary strings that we won't consider.

The embedded **xmlnamespaces** and **xmlattributes** functions are optionally used in the context of a given element to add any required namespace declarations or attributes to the element. The element content is defined by a list of expressions that is optional, so it's possible to create an empty element. Also optional is the **returning** clause, which we've already seen.

```
xmlelement ( name xmlelementname
    [ , xmlnamespaces (namespace-declaration-item [,...] ) ]
    [ , xmlattributes ( attributevalue [as attributename ] [,...] )
  [ , elementcontent [,...]
  [ option { null on null | empty on null | absent on null | nil on null | nil on no content } ] ]
  [ returning { content | sequence } ] )
```

The content option requires a little more explanation. In SQL, a missing optional value is explicitly marked as **null**. This is not a value recognized in XML, although **null** is legitimate for the SQL **xml** data type. Depending on the desired result, a missing value could be indicated in XML by simply omitting the relevant element or by including the element begin/end tags with no contained value. If an explicit indicator equivalent to **null** is required, a further option is the "nil" mechanism specified in the "*XML Schema For Instances*" namespace. The namespace is typically introduced with a declaration such as:

```
xmlns:xsi = "http://www.w3.org/2001/XMLSchema-instance"
```

The xsi prefix is used by convention, but some other prefix could be substituted. Once this is defined, the **nil** attribute defined in XMLSchema-instance can be set to True.

```
<myElement xsi:nil = "true"></myElement>
```

An element attributed in this way may not have any element content but may still carry other attributes. The SQL:2003 standard acknowledged the xsi:nil option, but did not prescribe how this should be specified in conjunction with operations such as **xmlelement**. SQL/XML:2006 provided close control over these choices, as summarized in Table 16.4.

Table 16.4 Options for handling null and nil.

Option	Result when the function is applied to an SQL null
null on null	A null SQL value results in a null XML value
empty on null	A null SQL value results in an empty XML element
absent on null	A null SQL value results in no element (i.e., an empty sequence)
nil on null	A null SQL value results in a XML element marked with xsi:nil = "true"
nil on no content	An element being created with no children is marked with xsi:nil = "true"

As a simple example of the use of **xmlelement**, consider a Customers table that is purely relational (it contains no XML). We could produce a result table containing a mixture of relational data and XML representing selected customer information in the following way:

```
select c.id,
  xmlelement( name "cust", xmlattributes (c.id as "custid"), custName) as Xcust
  from Customers as c where custid = 1234
```

which would produce a result such as:

id	Xcust
1234	<cust custid = "1234"> Widgets Inc </cust>

where the data contained in the *Xcust* column is of type **xml**. Calls to **xmlelement** can be nested so that elements can be created within elements, producing the characteristic hierarchical structure of XML.

The **xmlforest** function is a convenient way of applying **xmlconcat** to a sequence of **xmlelement** results: it produces a sequence of elements from a list of expressions. The elements can be named explicitly or implicitly. As with **xmlelement**, calls to **xmlforest** can also be nested. The basic syntax is as follows:

```
xmlforest ( [ xmlnamespaces (namespace-declaration-item , ), ]
elementvalue [ as elementname ] [,...]
[ option { null on null | empty on null | absent on null | nil on null | nil on no content } ] ]
[ returning { content | sequence } ] )
```

The parameters are used in the same way as **xmlelement**. For example,

```
select o.orderId,
xmlelement ( name "order",   xmlforest ( o.custName as "customer", o.date ) ) as Xorder
from Orders as o
   where orderid > 9800
```

would produce a result such as:

orderId	Xorder
9876	<order> <customer> Widgets Inc </customer> <date> 2007-12-12 </date> </order>
9877	<order> <customer> Thingummy Co </customer> <date> 2007-08-15 </date> </order>

If preservation of the case in SQL names is important, they should be aliased as delimited identifiers. Regular identifiers are likely to be mapped to all uppercase in some implementations. For example, the *date* element in the example just given could appear as <DATE> ... </DATE> instead of <date> ... </date>.

An equivalent result table could be produced in SQL Server as follows:

```
select o.orderId,
(select o,custName 'customer', o.date 'date'
    from orders as o
    for xml path ('order'), type) as Xorder
from orders as o
where orderid > 9800
```

SQL Server uses various **for xml** ... constructs to convert relational data to XML. The **for xml path** variant gives full control over how the XML structure is to be generated on a column-by-column basis. The **type** parameter specifies that the result should be of type **xml**, as opposed to its string equivalent.

Returning to SQL/XML, the **xmlagg** function is an aggregate function, similar to SQL's **sum**, **avg**, etc. It returns a single value from a given bag of values, which must be the result of an **xml** type expression. The resulting **xml** values are concatenated to produce a single **xml** value. Optionally, an **order by** clause can be used to order the results before concatenating, using a similar syntax to SQL. Any **null** produced is dropped before concatenating. If all the results of the expression are **null**, or the bag is empty, the result of the **xmlagg** function is the **null** value. The basic syntax is as follows:

```
xmlagg ( XML-value-expression [ order by sort-specification ]
[ returning { content | sequence } ] )
```

As an example of the **xmlagg** function, suppose we have an "Orders" table and want to group by customer, showing, for each customer, an element for each order placed by that customer. For simplicity, we'll just assume that both orders and customers are represented by a simple ID value. We could construct the result as follows:

```
select xmlelement
  ( name "Customer", xmlattributes ( o.custId as "customerId" ),
  xmlagg ( xmlelement ( name "order", o.orderId )
  order by o.orderId ) ) as XcustOrders
from Orders as o
group by o.custId
```

which would give a result resembling:

XcustOrders
<Customer customerId = 1234> <order> AB123 </order> <order> CD456 </order> </Customer>
<Customer customerId = 5678> <order> BC345 </order> </Customer>
etc.

The **xmlpi** function generates a single XQuery processing instruction node, as follows:

xmlpi (**name** target [, string-expression] [**returning** { **content** | **sequence** }])

The target identifier (**name**) cannot be 'xml' (in any case combination) and the string expression cannot contain the substring '?>'. The **xmlcomment** and **xmltext** functions are also straightforward, each simply producing a node of the appropriate type. Here are some examples of the use of these functions.

xmlpi (**name** "Warning", "Turn off now")	produces	<?Warning Turn off now?>
xmlcomment ("This is my comment")	produces	<!-- This is my comment -->
xmltext ("Here is my text")	produces	Here is my text

The **xmldocument** function returns a document node (as defined by the XQuery data model) that has the given XML value as its content. This is useful for converting an arbitrary XQuery sequence into an XML document.

xmldocument (xmlvalue [**returning** { **content** | **sequence** }])

A separate casting function, **xmlcast**, is provided for casting to and from XML values. This works in the same way as the standard SQL cast function, except that at least one of the data types involved must be **xml**. It's also possible to cast between the subtypes of **xml**, subject to certain rules that we won't go into here. Note that casting backward and forward between XML values and strings is not the same as using the **xmlparse** and **xmlserialize** functions.

Like the **valid** predicate, the **xmlvalidate** function ensures that a given XML expression is valid according to the related schema definitions. As with the **valid** predicate, schemas can optionally be explicitly specified. However, unlike the **valid** predicate, the **xmlvalidate** function also persistently annotates the XML expression with type information. The syntax is:

xmlvalidate({ **document** | **content** | **sequence** } xmlexpression
[**according to xmlschema** { uri | registered-name }]
[valid-element-clause])

We saw earlier how the **xmlnamespaces** function could be used in the context of **xmlelement** and **xmlforest** functions. A more complete syntax for **xmlnamespaces** is:

xmlnamespaces ({ XML-namespace-URI **as** namespace-prefix |
default XML-namespace-URI | **no default** } [,...])

The **default** option applies the namespace definition to any nested elements, unless overridden by a lower-level **no default**. Similarly, the **no default** option can be overridden by a lower-level **default**. The **default** and the **no default** options can be specified at most once in an **xmlnamespaces** argument list.

The SQL:2003 standard introduced an extensive set of mapping rules to define how SQL entities map from SQL to XML (and the reverse). The specification of the rules is quite extensive, and some aspects are implementation-dependent, so we won't cover all the details here. Broadly speaking, SQL data types are mapped to the closest equivalent type defined in XML Schema. XML Schema facets and annotations can be used to indicate distinctions in certain areas, For example, basic XML Schema does not distinguish between **character varying** and **character large object**, which are distinct types in SQL. Delimited identifiers in SQL are mapped as defined, preserving the upper/lower case of characters. Case is not guaranteed to be preserved for regular identifiers.

Some special treatment is given to characters that would result in illegal names in XML. The general technique is to replace the character by _xNNNN_ or _xNNNNNN_, where N is a hexadecimal digit and NNNN or NNNNNN is the Unicode representation of the character. There are two variants of the identifier mapping: *partially escaped* and *fully escaped*. The key differences between these are in the treatment of noninitial colons and identifiers beginning with the letters "xml" (in any combination of uppercase or lowercase). The differences are summarized in the examples in Table 16.5.

Table 16.5 Mapping SQL identifiers to XML.

Mode	SQL identifier	XML name
Partially escaped	xmlthing XMLTHING :ab:cd	xmlthing XMLTHING _x003A_ab:cd
Fully escaped	xmlthing XMLTHING :ab:cd	_x0078_mlthing _x0058_mlthing _x003A_ab_x003A_cd

Using XQuery

One of the most important aspects introduced in SQL/XML:2006 was the ability to integrate XQuery into the established relational database framework–a key motivator in moving SQL/XML to the XQuery data model. One point of caution: if you have a background in SQL, you will need to remember that, unlike SQL, XQuery is case-sensitive, and so care is needed in specifying expressions. It's also important to keep in mind that in SQL/XML, we will be operating on XML values and not their serialized representations. So, for example, we're not trying to textually process "angle brackets", although these may be useful in denoting structures for human readability.

Finally, we'll look at some mechanisms provided in SQL/XML that help integrate XQuery into the relational framework. The most important of these is the **xmlquery** function, which allows an XQuery expression to be evaluated from within an SQL context. SQL alone cannot operate on parts of an XML document: the whole document is treated as a single value. With the **xmlquery** function, it's possible to query inside an XML document so that XML data can participate in SQL queries and SQL processing can be applied to returned XML values (for instance, using **order by** to order the results). The basic syntax of **xmlquery** is as follows. The arguments are used in a similar way to the equivalents we have already seen in other functions.

```
xmlquery ( XQuery-expression
[ passing [ { by ref | by value } ] XML-query-argument [,...] ]
[ returning { content | sequence } ]
[ { by ref | by value } ]
[ { null on empty | empty on empty } ] )
```

Another function, **xmltable**, provides a useful way to integrate XQuery results back into SQL. Normally, an XQuery expression would return a sequence, but by using **xmltable**, we can execute an XQuery expression and return values as a table instead. The returned table can contain columns of any SQL data type, including XML, although values inserted into **xml** columns in **xmltable** must be well-formed with a single root node. The basic syntax is:

```
xmltable ( [ xmlnamespaces ( namespace-declaration-item [,...] ), ]
row-pattern [ passing [ { by ref | by value } ] XMLquery-argument [,...] ]
columns { { colname for ordinality |
colname datatype [ by ref | by value ] path column-pattern} [,...] } )
```

The row-pattern defines an expression that produces a sequence of elements. Each element in this sequence will become a row in the resulting table. The **columns** clause lists the column definitions and the path expression used to find the column value within each element of the row-pattern. The **for ordinality** option creates a column with an automatic ordinal reference to the row, so the first row is 1, the second row 2, and so on. An example is shown here. This assumes that we have an **xml** value called movieXML, the contents of which are similar to the examples we used when discussing XQuery. If this **xmltable** function call is contained in a **select** statement, the value being passed could be selected from a suitable xml column in a table.

```
xmltable ('//movies'
passing movieXML
columns
"Seqno" for ordinality,
"MovieNr" int path '@movieNr',
"Movie Name" varchar (50) path 'title',
"Release Year" char(4) path 'released'
"Director" varchar(50) path 'director' )
```

The resulting table would have the following form, and could be aliased and so on, in the same way as a table produced by other means.

In this case, values in the automatically generated SeqNo column and the retrieved MovieNr column are coincidentally the same. Had the XML data been in a different order, the SeqNo column would remain unchanged, but the MovieNr column (and the other retrieved columns) would reflect the XML element ordering.

SeqNo	MovieNr	Movie Name	Release year	Director
1	1	Cosmology	2006	Lee Lafferty
2	2	Kung Fu Hustle	2004	Stephen Chow
3	3	The Secret Garden	1987	Alan Grint
4	4	The Secret Garden	1993	Agnieszka Holland
5	5	The Da Vinci Code	2006	Ron Howard

Exercise 16.2

The following schema shows part of a database that holds information about repair technicians and the types of assemblies that they are competent to repair. Different technicians may be competent at repairing different assemblies.

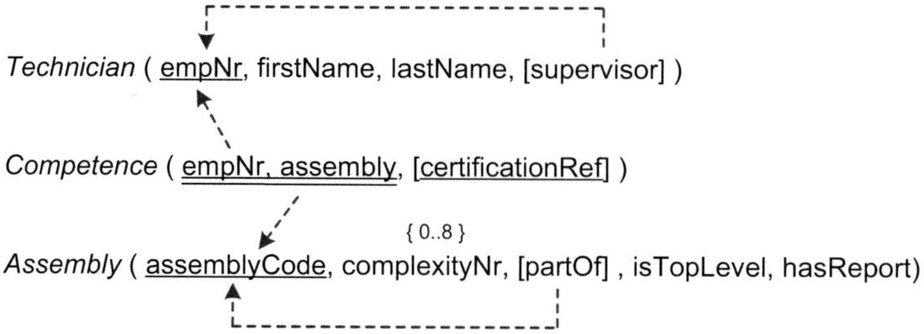

1. Using the facilities provided in a modern database system, turn the contents of the repair database into three XML documents, that is, one XML document per table.
2. Use XPath and XQuery expressions to produce the following results, using one or more of the Technician, Competence, and Assembly XML documents. Produce the results either as an XML document or as an XML fragment.
   (a) List all supervisors and under each supervisor all technicians that report to them. The Supervisor and Technician elements should have empNr as an attribute and the concatenation of 'firstName' || ' ' || 'lastName' as the element content.
   (b) List the assemblies in a hierarchic form (i.e., the top-level assemblies will have subassemblies, which will have their own subassemblies, etc.). Show the assemblyCode as the content of each element.
   (c) List all the assemblies with a complexity code of 5 or higher that have a report. Order the list by complexityNr (descending) and assemblyCode (ascending).
   (d) For each technician, list the types of assembly that they are competent to repair. Include each assembly's complexityNr as an attribute.
   (e) List all technicians with a certificationRefs element under each technician. Use xsi:nil = "true" in cases where the technician has no certificationRef. For each certificationRef that does exist, show the related assemblyCode.
   (f) List all assemblies that no technician is competent to repair.
3. Using the facilities provided in a modern database system, turn each XML result from the queries in Question 2 into a relational table.

16.3 JSON

JSON stands for JavaScript Object Notation, which describes its origin and function very clearly. Like XML, JSON is a text-based format that makes it very easy to interchange data between different systems. Each system can store the data in its own internal form for efficient processing, but these internal representations do not need to be the same. JSON was first defined by Douglas Crockford in the early 2000s as a format for data interchange, but it subsequently became widely adopted as a convenient way of expressing data structure and is supported in most programming languages. An early "informational" standard (Crockford, 2006) was followed by the first international standard (ECMA-404) in 2013.

The syntax of JSON is very simple and is a slightly modified version of JavaScript syntax. There are two main structural elements.

1. Sets of key-value pairs. A colon character separates the key and the value, as in:
   ```
   "givenName" : "Alan",
   "familyName" : "Turing"
   ```
   The key is enclosed in double quotes. Successive items are separated by a comma character and the set is enclosed in braces. The order of items in the set is not significant. Such sets are termed "objects" from their JavaScript origins.

2. Arrays containing an ordered collection of values. Values are separated by comma characters. An array is enclosed in square brackets.

Whitespace characters (space, tab, newline, ...) outside quoted values are ignored. Values can be a string, a number, a Boolean, the special value null, or either of the structured types (objects and arrays). The broad guidelines are that arrays are used for collections of similar items and objects are used for collections of properties for an individual item.

The JSON syntax is slightly different to JavaScript, which does not require quoted keys and allows other possible values, such as functions, but the general style is similar. This makes JSON processing in JavaScript very straightforward, and the language has built-in functions for handling JSON text. `JSON.parse()` converts a text representation into a JavaScript object structure and `JSON.stringify()` converts a JavaScript object structure into text. Similar functionality is provided in other object-oriented languages. One noticeable absence from JSON is any capability for expressing comments. This was a deliberate choice made by its designers but has been a source of frustration for many users.

This straightforward processing is a distinct advantage over XML, which requires a more complex parser to process text into an internal representation and provides only indirect access to data values.

The nesting of entities (e.g., objects containing objects) produces a hierarchical structure in much the same way as XML, but JSON also provides arrays, which have no direct equivalent in XML. Like XML, a JSON file can be opened and edited with nothing more than a simple text editor. Human readability can be helped by sensible use of whitespace characters, but dedicated editors, sensitive to JSON structure, are also available. JSON files are generally more compact than the equivalent XML representation of the same data.

Another difference between XML and JSON is the approach to data typing. XML, like SQL, has an extensive range of data types that can be specified in a schema. JavaScript is a dynamically typed language, and this is carried over into JSON, which has only three scalar data types: string, number, and Boolean (plus null, which, like SQL, indicates the absence of a value). Numbers can be integer or floating point, and Booleans have the value true or false.

JSON support is available in many languages, not just JavaScript. Once a JSON file is loaded into the language of choice, the standard facilities of that language can be used to carry out operations on the data in the usual way, such as iterating over the members of a collection. Extracting data from the object structure created by loading the JSON file requires code in JavaScript, Python, C#, or whatever host language is being used. Strictly speaking, JSON refers to the text format, so once it has been loaded, it is no longer JSON but a structure of the kind appropriate to the language. The syntax may vary by language, there may be different conventions for single versus double quotes, and so on. Data operations are normally carried out on this internal structure, not on the JSON text. Each language will also provide some function to carry out the reverse process to output a textual representation of the inner structure. Different languages may change the format slightly by introducing extra whitespace characters into the output. This does not change the data, but it means that results may not be identical if moving from one programming language to another.

Like XML, a JSON file can be described precisely with a schema. The JSON Schema specification describes the structure, constraints, and validation rules for JSON documents. JSON Schema is itself written in a JSON format. It uses JSON objects to describe the structure of the data it validates. This means that JSON Schema documents themselves can be validated. The expected properties of an object can be defined, each property with its own validation rules and data types. This includes maximum and minimum values for numeric properties, maximum and minimum and lengths for strings, regular expression patterns, and special formats such as email or URL. Array properties can be specified with data types and maximum and minimum numbers of items. JSON Schema includes many other features that we will not cover here. The specification is constantly reviewed and updated and is freely available. There are many tools available for validation of JSON against a specific schema, generating a JSON schema, generating code from a schema, and so on. As with XML Schema, we will not delve into the full details here.

To illustrate the general features of JSON, we can use the Movies information previously covered as an XML example in Listing 16.1. Listing 16.3 shows the same information in JSON format. The movies object contains an array (a list) of individual movies. Array items are accessed by index, so the first movie is movies[0] (array indexes are zero-based). Each movie item in the array is an object containing properties such as title, director, etc. A property can itself be an object or an array, as in the case of the *stars* property, which contains a list of *star* objects. Object properties can be accessed via the name of the key.

```
{ "movies": [
   { "movieNr": 1,
       "title": "Cosmology",
       "released": "2006",
       "director": "Lee Lafferty"
   },
   { "movieNr": 2,
       "title": "Kung Fu Hustle",
       "released": "2004",
       "director": "Stephen Chow",
       "stars": [
           { "name": "Stephen Chow", "gender": "M" }
       ]
   },
   { "movieNr": 3,
       "title": "The Secret Garden",
       "released": "1987",
       "director": "Alan Grint",
       "stars": [
           { "name": "Gennie James", "gender": "F" },
           { "name": "Barret Oliver", "gender": "M" }
       ]
   },
```

```
    { "movieNr": 4,
        "title": "The Secret Garden",
        "released": "1993",
        "director": "Agnieszka Holland",
        "stars": [
            { "name": "Kate Maberley", "gender": "F" },
            { "name": "Heydon Prowse", "gender": "M" }
        ]
    },
    { "movieNr": 5,
        "title": "The Da Vinci Code",
        "released": "2006",
        "director": "Ron Howard",
        "stars": [
            { "name": "Tom Hanks", "gender": "M" },
            { "name": "Ian McKellen", "gender": "M" },
            { "name": "Audrey Tautou", "gender": "F" }
        ]
    },
    { "movieNr": 6,
        "title": "Cast Away",
        "released": "2000",
        "director": "Robert Zemeckis",
        "stars": [
            { "name": "Tom Hanks", "gender": "M" }
        ]
    }
]}
```

LISTING 16.3 Movie information in JSON.

Unlike XML and XQuery, JSON has no special query language. Instead, the general features of a host programming language are used to process data. We need to bear in mind that once a JSON file is loaded it is no longer JSON, but instead is a structure that is meaningful in the context of the host language. To provide some concrete examples of extracting required data values, we can use the same queries as in the previous XML example, shown in Table 16.2. Here we will use Python to express the queries, but other languages follow a similar approach. The first step is to load the JSON file into a Python dictionary with the aid of a presupplied json library.

```
import json
 # Open JSON file
theFile = open('movies.json')

 # load JSON object into Python dictionary
data = json.load(theFile)
```

```
# Iterate over the json object
for item in data['movies']:
    # code fragment goes here
```

Now we can iterate over the content of the dictionary to extract (and in this case print) the required data. The code fragments in Table 16.6 are assumed to directly follow the Python code above.

Table 16.6 Example Python code fragments.

Expression	Result
Find all directors `print (item['director'])`	Lee Lafferty Stephen Chow Alan Grint Agnieszka Holland Ron Howard Robert Zemeckis

Find one star from each movie
 	
`starList = item.get('stars')`
`if starList is not None:`
   `print (starList[0])`
	{'name': 'Stephen Chow', 'gender': 'M'}
	{'name': 'Gennie James', 'gender': 'F'}
	{'name': 'Kate Maberley', 'gender': 'F'}
	{'name': 'Tom Hanks', 'gender': 'M'}
	{'name': 'Tom Hanks', 'gender': 'M'}

Find the titles of all movies with a female star
 	
`starList = item.get('stars')`
   `if starList is not None:`
      `for star in starList:`
         `if star['gender'] == "F":`
            `print (item['title'])`
	The Secret Garden
	The Secret Garden
	The Da Vinci Code

Find the release year of all starring Tom Hanks
 	
`starList = item.get('stars')`
   `if starList is not None:`
      `for star in starList:`
         `if star['name'] == "Tom Hanks":`
            `print (item['released'])`
	2006
	2000

In some respects, a JSON file (and a corresponding program to load and process the file) could be considered a small database in its own right, though perhaps lacking many features expected in a complete database system such as transactions, concurrency, and so on. However, the JSON format has proved so useful that it has been used as the basis for full database systems, commonly referred to as *Document Databases*. These are usually classified as a type of "NoSQL" database and are discussed in more detail in Chapter 17.

SQL/JSON

Given the widespread use of JSON, the SQL standard needed to be extended to accommodate it in the same way that it had previously accommodated XML. SQL:2016 provided support for JSON stored as text in a character-based column (typically **varchar**). Time pressures did not allow the full definition of JSON as a data type, but this was completed in SQL:2023, so data can be stored in a column of data type **json**. This had already been anticipated by several vendors, but their solutions are not necessarily compatible with each other or with the SQL:2023 standard. However, all vendors consistently adhere to the standard definition of JSON. There are two immediate advantages of a **json** data type.

1. Values can be automatically validated as correctly formed JSON. A data type such as **varchar** has no intrinsic understanding of JSON structure, so an additional function is required for text to establish that it represents a valid JSON value.
2. Values can be processed more efficiently by converting JSON text into a binary form for internal operations, avoiding relatively expensive string processing. The binary versions also require less storage.

The SQL/JSON data model covers SQL/JSON items and SQL/JSON sequences. A sequence is an ordered list of items. An item is any of the following:

- An SQL/JSON object: an unordered collection of members, where each member is a key-value pair, the value component being an SQL/JSON item.
- An SQL/JSON array: an ordered list of elements, where each element is an SQL/JSON item.
- An SQL/JSON scalar: a nonnull value with a type that is one of the standard JSON types (character string, numeric, or Boolean) or datetime.
- An SQL/JSON null (different from an SQL null).

Because the structural items (object and array) can be nested, JSON text can represent data with any required degree of complexity.

Values can be inserted into a **json** column in the usual way. Assume we have the following table definition:

```
create table movies (
movienr int primary key
moviedata json
)
```

The primary key column here is included simply as a convenience. We could enter JSON data using one row for each movie in a similar way to any other SQL data type.

```
insert into movies values (3,
  '{ "title": "The Secret Garden",
    "released": "1987",
    "director": "Alan Grint",
```

```
  "stars": [
    { "name": "Gennie James", "gender": "F" },
    { "name": "Barret Oliver", "gender": "M" }
  ]
}')
```

However, we don't necessarily have to split data up in this way. For example, all the movie data in Listing 16.2 could be entered as a single value in a **json** column.

SQL:2023 provides several functions to work with JSON values. These fall into two categories. SQL/JSON *retrieval* functions operate on SQL/JSON data and return values either as SQL data or SQL/JSON data. SQL/JSON *construction* functions return SQL/JSON values from either SQL data or SQL/JSON data.

Although JSON support was added to SQL:2016, it did not include the **json** data type until SQL:2023. During the intervening period, several vendors introduced individual interpretations of the data type and the corresponding functions to work with it. The result was some variance between different systems, including the following:

1. Certain SQL/JSON functions are missing completely, though it may be possible to obtain similar results in a more roundabout way.
2. Functions are provided with the same names as SQL/JSON functions but with different syntax.
3. Functions have the same main syntax as the SQL/JSON definitions but do not support some optional clauses defined in SQL:2023.

Hopefully, this situation will normalize in the future. The main SQL/JSON query functions in SQL:2023 are summarized in Table 16.7.

Table 16.7 SQL/JSON query functions.

Operator	Functionality
json_query()	Returns an object or an array from the specified JSON text
json_value()	Returns a scalar value from the specified JSON text
json_exists()	Returns true if the data is found in the specified JSON text.
json_table()	Converts identified parts of specified JSON text into a relational table.

Some systems may provide alternatives such as **json_extract()**, which offers similar functionality to **json_query()**, and **json_contains()**, which may be used as a replacement for **json_exists()**.

These query operators locate relevant parts of the JSON data using a path language. This is similar in principle to using XPath with XML data, but the JSON path language has a syntax closer to JavaScript.

A path starts from the root of the relevant JSON text. This will be an object or an array and is represented as a "dollar" symbol ($). The path shows the steps from the root to the required element (or elements) within the JSON text using *accessors*: square brackets ([ ]) for elements in an array and a period or "dot" (.) for members of an object. The ECMAScript Language Specification for JSONpath also allows bracket notation for object members, but this is not supported in SQL/JSON. An asterisk (*) is used as a wildcard character to indicate all elements in an array or all members of an object. To illustrate these concepts, assume we have the JSON text of Listing 16.3 stored in a movie column of a table called videos.

This text contains a single object called movies. The value of this object is a list of six elements. Each element of the list is an object with members such as "title", "director", and so on. We can use the following query to get a list of all movie names.

select json_query(movie, '$.movies[*].title') **from** videos

To build the path from the root, we first add the movies member (the only member at this level). Next, we add a wildcard specification for all elements of the movies array. Finally, we add the title member from each of the array elements. All path expressions work in this way, though they can be much more complex.

JSON path expressions can be used in two modes: *lax* and *strict*. Lax mode is the default, and it provides automatic handling of some structural problems. If an object accessor in a path expression finds an array instead of an object, the array is converted to a sequence and the object accessor is then applied to each element of the sequence. If an array accessor finds a value that is not an array, the value is converted to a single-element array. If an error cannot be handled by these mechanisms, an empty SQL/JSON sequence is returned.

In strict mode, a structural error is reported without any attempt at correction. This is useful for identifying why a path expression is returning unexpected results.

The purpose of **json_query** is to return an extract from the JSON text. This will be a structural item (an array or an object) containing other items. To return a scalar value, **json_value** is used. This has a similar syntax to **json_query**, but the path must identify a single item. Consider the following two queries:

select json_query(movie, '$.movies[*].movieNr') **from** videos
select json_value(movie, '$.movies[*].movieNr') **from** videos

The first will return a list of movie numbers, but the second will return **null** because the result is not a scalar value. It's possible to cast a scalar returned by **json_value** into an appropriate SQL data type. For example:

select json_value(movie, '$.movies[0].movieNr **returning decimal**(5,2)') **from** videos

returns a scalar value (because this query specifies only the first array item) with a cast to the designated SQL type. If no cast is specified, the type of the return value is a system default (typically **varchar**).

If a null is returned, we might want to know the reason. There are three possibilities: the null may have been retrieved from the JSON text being queried, the specified value does not exist, or some error has occurred. The default for all these cases is to return null, but we can distinguish between them with optional clauses to deal with the "on empty" and "on error" cases. For example, the following query will return the text "Not found" because we have specified an array index that does not exist in the data. Such optional clauses are not supported in all systems.

```
select json_value(movie, '$.movies[8].movieNr' default 'Not found' on empty) from videos
```

We can use **json_exists** to check whether a specific item is present or not. In this case, the return value is 1 if the item exists and 0 if it does not. For example:

```
select json_exists(movie, '$.movies[*].director') from videos
```

will return 1 indicating that at least one occurrence of the specified item has been found, but:

```
select json_exists(movie, '$.movies[*].producer') from videos
```

returns 0 because our movie objects do not contain a producer member.

One useful application of **json_exists** is in the creation of check constraints to ensure that JSON data contains some mandatory elements. Further predicates can be used to check whether an item is (or is not) a JSON value, array, object, or scalar, but we won't go into the details of these here.

Although **json_query** can return results from JSON documents, the results remain in a JSON format. To be consistent with other operations, we may prefer results in the form of a standard SQL table. This is the functionality provided by **json_table**, which extracts data from a JSON document and returns it as a relational table. Here's a simple example, based on our usual movie data.

```
select jt.Title, jt.Director from videos as v,
  json_table (v.movies, '$.movies[*]' columns(
    Title varchar(20) path '$.title',
    Director varchar(20) path '$.director'
  )) as jt
```

The select statement identifies columns from the aliased JSON table ("jt") created by the **json_table** function. The first argument to the function specifies the JSON source, in this case the movies column of the videos table. The second argument defines the starting point in the JSON text, in this case the movies array. The column list defines the column name, the SQL data type, and the path in the current context (in this case, the starting point defined previously). The result is a straightforward SQL table.

Title	Director
Cosmology	Lee Lafferty
Kung Fu Hustle	Stephen Chow
The Secret Garden	Alan Grint
The Secret Garden	Agnieszka Holland
The Da Vinci Code	Ron Howard
Cast Away	Robert Zemeckis

It's possible to include data from different levels of the JSON text by using "nested" columns. In our data, the names of any stars of a movie are at a level below the title of the movie. The following example shows how these levels can be combined.

```
select jt.* from videos as v,
    json_table (v.movies, '$.movies[*]' columns(
      Title varchar(20) path '$.title',
      Released int path '$.released',
        nested path '$.stars[*]'
            columns (
              Starring varchar(20) path '$.name')
  )) as jt
order by Released
```

Again, this produces a standard SQL result table.

Title	Released	Starring
The Secret Garden	1987	Gennie James
The Secret Garden	1987	Barret Oliver
The Secret Garden	1993	Kate Maberley
The Secret Garden	1993	Heydon Prowse
Cast Away	2000	Tom Hanks
Kung Fu Hustle	2004	Stephen Chow
Cosmology	2006	NULL
The Da Vinci Code	2006	Tom Hanks
The Da Vinci Code	2006	Ian McKellen
The Da Vinci Code	2006	Audrey Tautou

Tables produced in this way can participate in all the usual table operations such as join operations with other (non-JSON) tables in the database. A possible use of **json_table** is to create views that allow the JSON origins of the data to be hidden from SQL developers.

There is one more SQL/JSON retrieval function, and that is **json_serialize**. This takes a JSON value as input and returns JSON text in one of the SQL character data types (typically **varchar**). This is useful if the "raw" JSON value is in a binary form, which is not intended for human readability. Some implementations may also include a "pretty" option to produce nicely formatted text.

The operations we have considered so far assume that we have some JSON value stored in the database and want to extract some or all of it, in various ways. SQL/JSON also defines construction functions that build SQL/JSON items from non-SQL/JSON values.

The most obvious of these is **json_parse**, which takes a text input and converts it to a JSON sequence. This is essentially the reverse operation to **json_serialize**. If the input text is not well-formed JSON, an error will be returned. The **json_scalar** function is simple: it creates a JSON value from a non-JSON value. For example, an SQL value of data type **int** would be converted to a JSON number.

More interesting are the functions to construct JSON structural elements, objects and arrays. The constructors can be nested, so we can build JSON values of arbitrary complexity. These functions are summarized in Table 16.8.

Table 16.8 SQL/JSON structural construction functions.

Operator	Functionality
json_object()	Creates an SQL/JSON object from a given list of name-value pairs
json_array()	Creates an SQL/JSON array from a given list of values
json_objectagg()	Creates an SQL/JSON object from a given collection of rows
json_arrayagg()	Creates an SQL/JSON array from a given collection of rows

These functions have many options, and we'll explore some of these with examples. The relational tables in Fig. 16.4 show some data about agents and their allocation to projects. The SQL data types are **int** for employee numbers and rate and **varchar** for other columns. Blank spaces represent SQL **null**. We'll assume that we ultimately want to create a single JSON object containing an individual object for each agent, including any project allocation(s) that they may have.

Agent:

nr	*givenName*	*familyName*	*workPhone*	*homePhone*	*office*
101	Alice	Andrews	222-1234	222-9876	Boston
232	Bob	Brown	555-6789		Portland
397	Carol	Codd	777-4321	777-6789	Durham
468	Dave	Dunk			

Project:

empNr	*projName*	*rate*
101	Mallard	1
101	Grebe	2
232	Grebe	3
232	Plover	1
232	Bittern	2

FIG. 16.4 Agent and project data.

Creating objects is straightforward. The following query defines an object structure with members for employee number and office, which is then applied to each row of the Agent table.

```
select json_object('empNr' value nr, 'office' value office) from Agent
```

Result:

```
{"empNr": 101, "office": "Boston"}
{"empNr": 232, "office": "Portland"}
{"empNr": 397, "office": "Durham"}
{"empNr": 468, "office": null}
```

Some implementations of **json_object** replace the **value** keyword with a comma. An ugly feature of the result is the **null** in the last object. A more natural JSON representation would omit this member entirely. SQL/JSON defines an optional **absent on null** clause to effect this, but it's not supported in all implementations. We may be able to find a workaround by providing something more meaningful based on domain knowledge. For instance, if we know that the reason for Dave Dunk having no office recorded is that he is mobile, so his office varies, we can replace the null.

```
select json_object('empNr' , nr,
                'office', case when office is not null then office else 'varies' end)
from Agent
```

Result:
{"empNr": 101, "office": "Boston"}
{"empNr": 232, "office": "Portland"}
{"empNr": 397, "office": "Durham"}
{"empNr": 468, "office": "varies"}

It's obvious how this query could be extended to include separate members for givenName and familyName, but an alternative would be to create a name member containing objects for each of the name components. This can be done by nesting additional **json_object** constructors within the name member.

Phone information is a little more complex because the member is not single-valued. We'll assume that a list of numbers is required, each represented by an object. This can be produced by nesting object constructors inside an array constructor in the following way.

```
select json_array(  json_object('type' value 'home', 'number' value homePhone
                    json_object('type' value 'work', 'number' value workPhone) )
from Agent
```

Result:
[{"type": "home", "number": "222-9876"}, {"type": "work", "number": "222-1234"}]
[{"type": "home", "number": null}, {"type": "work", "number": "555-6789"}]
[{"type": "home", "number": "777-6789"}, {"type": "work", "number": "777-4321"}]
[{"type": "home", "number": null}, {"type": "work", "number": null}]

To integrate this information, we can introduce a phones member into our earlier object, with the array construct above nested inside it. We have the same issue as before with nulls. If the **absent on empty** clause is not available, we could create a more complex SQL query to remove the null members from the JSON result, but for our current purposes, we'll just accept the nulls.

The functions not covered so far are the object aggregate and array aggregate constructors. We can illustrate these with the Project table. The **json_objectagg** function requires a key-value pair, as in the following example:

```
select empNr, json_objectagg(projName, rate)
from Project group by empNr
```

Result:
101 {"Grebe": 2, "Mallard": 1}
232 {"Grebe": 3, "Plover": 1, "Bittern": 2}

Here, the resulting object for each group has a member for each item in the group. This differs from **json_object**, which would give a separate object for each item.

Creating an aggregating array works similarly but requires only the identification of the value to be listed. This gives an array with an element for each item in the group.

```
select empNr, json_arrayagg(projName)
from Project group by empNr
```

Result:
101 ["Mallard", "Grebe"]
232 ["Grebe", "Plover", "Bittern"]

To meet our aim of a comprehensive Agent object, the **json_arrayagg** constructor looks the most appropriate. This can be integrated into the larger query by nesting it within a projects member of the main object structure.

To create the required final JSON structure, the rows containing the above elements can be aggregated together to form a list of agents within an enveloping object, as shown in the following query.

```
select json_object ('Agents', json_arrayagg(
   json_object(
      'empNr' value nr,
      'name', json_object('given' value givenName, 'family' value familyName),
      'phone', json_array(
            json_object('work' value workPhone),
            json_object('home' value homePhone)),
      'office', case when office is not null then office else 'varies' end,
      'projects', (select json_arrayagg(projName)
            from Project p where p.empNr = a.nr group by p.empNr))))
from Agent a
```

The following listing shows the combined result, formatted slightly to make the structure more obvious.

```
{"Agents": [
  {"empNr": 101,
  "name": {"given": "Alice", "family": "Andrews"},
  "phone": [{"work": "222-1234"}, {"home": "222-9876"}],
  "office": "Boston",
  "projects": ["Mallard", "Grebe"]
  },
  {"empNr": 232,
  "name": {"given": "Bob", "family": "Brown"},
  "phone": [{"work": "555-6789"}, {"home": null}],
  "office": "Portland",
  "projects": ["Grebe", "Plover", "Bittern"]
```

```
    },
    {"empNr": 397,
    "name": {"given": "Carol", "family": "Codd"},
    "phone": [{"work": "777-4321"}, {"home": "777-6789"}],
    "office": "Durham",
    "projects": null
    },
    {"empNr": 468,
    "name": {"given": "Dave", "family": "Dunk"},
    "phone": [{"work": null}, {"home": null}],
    "office": "varies",
    "projects": null
    }
  ] }
```


16.4 Other Markup Languages

As the name suggests, markup languages, like XML, work by adding additional markings of some kind to the original information to indicate its function or purpose. Historically, the leading application for markup was in the publishing industry, where it was (and still is) used to show the treatment to be applied to the author's text to turn it into a properly formatted layout for printing. In a database context, we might consider markup as metadata.

A language close in spirit to JSON is *YAML*. This originally stood for "Yet Another Markup Language", but other interpretations of the name have emerged subsequently. YAML is a superset of JSON with some additional features, and utilities are available to convert back and forth between YAML and JSON. The main motivation for YAML was to provide a more "human-readable" format than JSON, especially avoiding extensive use of brackets (rounded, square, or curly). Nesting can be indicated by indenting items in a similar way to the Python language using spaces. This reduces the possibility of common keyboard errors.

Some brief features of YAML include the following:

- Items are nested by indenting child nodes below their parent.
- Key-value pairs are separated by a colon, but quotes are not required.
- New items in a list are indicated by a hyphen.
- Scalar values including strings, numbers, and Booleans can be used freely.
- Quotes are not needed for strings but can be used if needed (for example, if a number is required to be treated as a string).
- Comments can be indicated by a hash character (comments are stripped out when converted to JSON).
- Additional commands are provided to control the layout of text in output (e.g., suppressing newlines).

The following shows an abbreviated version of our movie data example expressed in YAML.


```
---
# This is a YAML document
movies:
  - movieNr: 1
    title: Cosmology
    released: '2006'
    director: Lee Lafferty
  - movieNr: 2
    title: Kung Fu Hustle
    released: '2004'
    director: Stephen Chow
    stars:
      - name: Stephen Chow
        gender: M
  - movieNr: 5
    title: The Da Vinci Code
    released: '2006'
    director: Ron Howard
    stars:
      - name: Tom Hanks
        gender: M
      - name: Ian McKellen
        gender: M
      - name: Audrey Tautou
        gender: F
...
```


YAML is a popular choice for relatively short documents where data structure needs to be explicitly represented. Two examples are configuration files containing application data and the interchange of structured messages. It is not widely used for large data collections. Its ease of use is partly due to automated decisions about the intention behind a data value, which can lead to unexpected consequences. For example, some parsers may interpret the unquoted value **No** as the Boolean value **false**, which may not have been the original intention.

The most familiar markup language is probably *HTML*, used to define the structure of web pages. Unlike XML, the tags in HTML are predefined and have specific meanings defined in the HTML standard. In the early days of web usage, HTML was also used as a rather weak method of identifying some aspects of data structure, but this was clearly unsatisfactory. Like XML, HTML can seem clumsy and awkward for authors, and several alternatives have been developed. Some examples include the following.

- The cutely named *Markdown* is a lightweight language that is used to add formatting indicators to textual documents. The marked document can be used in various applications to generate publishing formats such as PDF and convert to other formats such as a complete static website. However, it is obviously limited in its capability to express data structures.
- *AsciiDoc* is a plain text equivalent to the XML-based DocBook format used for structured documents. Documents can be created simply using a text editor and then rendered to a desired final format, such as PDF, HTML, and many others.
- *MediaWiki* is another lightweight language designed for ease of use. As the name suggests, it was developed for the creation of Wiki pages and is the basis for Wikipedia. As with other languages of this type, its documents can be re-processed into many other formats, including HTML.

The main purpose of most of these languages is to define the structure of documents, not the structure of data. Although there is a certain amount of overlap, there is no pressing need to use these languages for data description when better alternatives are available. We will therefore not explore them any further in this chapter.

16.5 XML, JSON, and ORM

A question worth considering is whether object-role modeling (ORM) can be used to design an XML schema or a JSON schema, in much the same way that it can be used to design a relational schema. The answer is certainly "Yes", but there are some additional considerations that are not relevant in the relational case.

As already mentioned, an ORM model can be represented in XML, and this indeed is the default file format used by the NORMA tool. The general structure can be outlined as follows:

```
<ORMModel>
   <Objects>
      <Object id = 123 ...> </ObjectType>
      <Object id = 456 ...> </ObjectType>
                 ...
      </Objects>
   <Facts>
     <Fact id = 321 ...>
        <FactRole id = 789 ...> </FactRole>
        <FactRole id = 987 ...> </FactRole>
   </Fact>
   ...
   </Facts>
  <Constraints> ... </Constraints>
  ...
  <ORMDiagram> ... </ORMDiagram>
</ORMModel>
```

Each item has a unique identifier (UUID), and these are used to relate items together. For example, the object playing a fact role can be referenced from the role entity using its UUID.

Although this structure captures all the ORM model details, it does not provide a good template for a data model. Some of the content, such as the diagram information, has no relevance in defining a data structure, and items in the *.orm file such as object types and fact types are represented in ways that are intended to facilitate modeling operations, not data transfer. Instead, what is required is a mapping from the ORM model to an XML schema that accommodates the relevant data.

Mapping from an ORM model to a relational schema was discussed in Chapter 11. There are only a few possible options in the Rmap procedure, such as how to treat subtypes. The procedure is largely automatic, and two different developers mapping the same ORM model will generally produce the same relational schema. However, the situation is different for XML.

The first issue is that XML obliges us to represent the data as a hierarchy, even if no natural hierarchy exists in the ORM model. This is a consequence of the nesting of elements within elements, which is intrinsic to XML. Consider the tiny ORM model shown in Fig. 16.5.

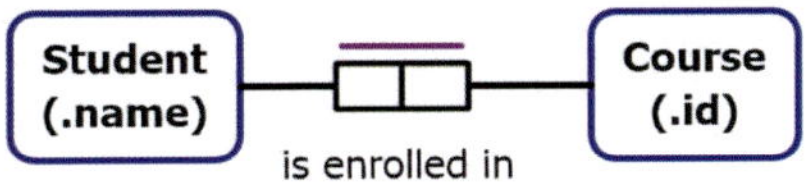

FIG. 16.5 Student-Course relationship.

One option would be to nest Courses under Students.

```
<Model>
    <Student name = “Alice” ...>
        <Course id = “CS101” ...> ...</Course>
        <Course id = “CS104” ...> ...</Course>
    </Student>
    <Student name = “Bob” ...>
        <Course id = “CS101” ...> ...</Course>
    </Student>
</Model>
```

But nesting of Students under Courses could be equally valid.

```
<Model>
    <Course id = “CS101" ...>
        <Student name = “Alice” ...> ... </Student>
        <Student name = “Bob” ...>... </Student>
    </Course>
    <Course id = “CS104” ...>
        <Student name = “Alice” ...> ... </Student>
    </Course>
</Model>
```

We could also consider something closer to a relational style, with a Student list, a Course list, and a list of linking Enrolment elements.

```
<Model>
      <Students>
            <Student name = "Alice" ...> ... </Student>
            <Student name = "Bob" ...>... </Student>
      </Students>
      <Courses>
            <Course id = "CS101" ...> ...</Course>
            <Course id = "CS104" ...> ...</Course>
      </Courses>
      <Enrolments>
            <Enrolment student = "Alice" course = "CS101"> ... </Enrolment>
            <Enrolment student = "Alice" course = "CS104"> ... </Enrolment>
            <Enrolment student = "Bob" course = "CS101"> ... </Enrolment>
      </Enrolments>
</Model>
```

Which of these would be preferable cannot be inferred from the ORM model because the choice depends on external factors such as the intended use of the data and patterns of access.

A second issue is that XML provides two options for storing data: as the value of an element attribute or as the content of an element. The examples above show student name as an attribute, but the following would be an equally valid alternative:

```
<Student>
  <Name>Alice</Name>
  ...
</Student>
```

There is no firm rule for deciding between these options, although various guidelines have been suggested, such as reserving attributes for metadata only. Attributes have some limitations that would make elements preferable in some cases, for instance, data that is multivalued or has some inner structure, but for straightforward scalar values either approach could be used.

Although we have been describing the situation with XML, similar issues arise with JSON, which also offers choices about how to represent the data. What all this means is that two different developers mapping the same ORM model may produce very different XML schemas (or JSON schemas).

To produce an XML rendering of a relational structure, we can follow a systematic procedure that is straightforward to automate.

Each relational table will map to an element type in XML, which can have a plural tag name (e.g., <Employees>). Each of these major element types will be nested under a suitably named root node. These major elements will contain a list of subelements of the relevant type.

Rows in a table will map to subelements with a singular tag name (e.g., <Employee>). nested under their parent element (e.g., <Employees>). Relational columns map to attributes of their related subelements. A UUID attribute can be added to each subelement for easy identification and cross-referencing.

The result is an XML file similar to the "relational" style in the examples above. Even if this is not the preferred format, it may still be a good first step. Once we have an XML file, it can be transformed into alternative XML structures using appropriate code, perhaps using the XML Stylesheet Language for Transforms (XSLT).

If we need to produce an XML file directly from an ORM schema, we can adapt the approach above by adding some extra steps. First, we follow the Rmap procedure to identify the relational structure corresponding to the ORM schema. Next, we apply the procedure above to this relational structure to define the XML schema elements. Finally, we need to add any required constraints from the ORM schema to the XML schema. This last step is not necessary when producing a data transfer file from a database because we can assume that the appropriate constraints will already have been applied to the data.

Mapping to a JSON schema can follow a similar procedure. Tables become JSON objects, rows become lists of nested objects within the parent types, and column values appear as key-value pairs within the nested object.

16.6 Summary

XML is a member of a family of markup languages based on the Standard Generalized Markup Language (SGML). Other languages based on SGML include HTML, though this has now diverged somewhat from its original basis. XML was designed to be easy to parse, initially in the context of web applications, but it is now well established as a general-purpose tool for describing data structure independent of any specific data container. It has become a built-in feature of many applications, though its presence may not be immediately obvious. As examples, the current formats of Microsoft Office files (such as *.docx, *.pptx, and *.xlsx) are based on XML, and the NORMA tool used to create and manage object relational models uses XML in its *.orm file format.

A key feature of XML is its extensibility. Unlike HTML, it does not assume predefined tag names, so a schema can be designed to meet any required scenario. If warranted, the schema can be documented in machine-readable form so that an XML document can be validated against the schema definition to ensure that it is consistent. The original schema definition method, inherited from SGML, used DTDs, which are expressed in their own language. Modern usage has moved to the expression of the schema in XML itself, using XML Schema Definition (XSD) or the simpler alternative Relax NG.

The XML definition is controlled by the World Wide Web Consortium (W3C). Several other W3C standards exist within the XML family. In addition to XML Schema, we have already discussed XPath and XQuery in some depth. These allow XML data to be queried in a way that corresponds roughly to the way that the **select** statement is used in querying SQL tables. Other W3C standards include the following:

XSLT (Extensible Stylesheet Language Transformations) is an XML-based language used to convert an XML document into a different XML schema or into a completely different format such as PDF.

XLink (XML Linking Language) provides methods for creating internal and external links within XML documents and associating metadata with the links.

XPointer is based on XPath and supports addressing into the internal structures of XML documents. This can be used to create something equivalent to a foreign key in a relational database.

XHTML is a somewhat stricter version of HTML with a syntax conforming to XML.

SVG (Scalable Vector Graphics) is a language for describing two-dimensional vector graphics that allows smooth scaling to different display resolutions.

There are many other XML-based standards produced by W3C and other bodies, perhaps running into hundreds of examples, as well as an uncountable number of private applications. This makes XML probably the most well-established method of defining structured data outside of a database system.

XML is a text format, and although it can be manipulated using text processing operations, it is more effective to use processing operations that are sensitive to the data structure being expressed. Database systems first started using XML as a text format, but recognition of its value led to its incorporation as a native data type in the SQL standard. Most database vendors have added features to store XML data, operate on XML data, and convert backward and forward between XML and relational table structures, sometimes including proprietary extensions that are not part of the official SQL standard.

JSON is a slightly younger language than XML. It was originally developed as a relatively lightweight communication language between web servers and browsers, using standard browser facilities instead of auxiliary plug-ins. This made JavaScript an obvious starting point for the definition of a markup language, and so JavaScript objects became the basis for the notation.

A major factor in the popularity of JSON is its simplicity and ease of processing compared to XML. It is a more compact format that reduces the bandwidth required for the transmission of data; this is an important factor given the reasons for its creation. It is also a better fit with web-related technology in general, relevant, for example, in application development. Against that, it is somewhat less expressive than XML. It has no support for namespaces, and it is unsuitable in applications where precise definition of data types is important. However, its adoption as the primary structuring principle in document databases and its incorporation into SQL as a native data type show that it is sufficiently robust to meet the demands of a wide range of real-world scenarios.

XML and JSON will continue to coexist for the foreseeable future because one is not a direct replacement for the other. It's tempting to caricature XML as heavyweight and JSON as lightweight solutions to data structuring outside of a data store, but this blurs some of the important features of each. It's certainly true that there is a large middle ground where either could be employed effectively, and any choice between them would be made on factors other than their data structuring capabilities.

Chapter Notes

The SQL standard has been adopted by the International Standards Organization (ISO) as ISO/IEC 9075. Copies can be ordered from ISO or via national standards organizations such as ANSI. Because of its size (several thousand pages in total), the standard is split into several parts. Part 2 (Foundations) and Part 14 (SQL/XML) are the most relevant to this chapter.

The relevant XML standards are being progressed through the W3C. Full details, including the current versions of the standards themselves, can be found at *www.w3c.org*. The standards documents are not particularly easy to read, but a large amount of more readable tutorial information is available. XQuery especially is still relatively unfamiliar to many database practitioners, but it seems likely to become progressively more important in utilizing heterogeneous data. Useful introductions can be found in Walmsley (2015) and Melton and Buxton (2006).

JSON arrived as a data format later than XML, but it is now established as an essential component in data management. A short introduction to JSON can be found in Basset (2015). Because of its origins in JavaScript, its standardization falls under ECMA, the same body responsible for the specification of the JavaScript language (now sometimes referred to as ECMAScript). A comprehensive description of JavaScript is given in Flanagan (2020). Copies of standards and further details can be found at *www.ecma-international.org*. JavaScript and JSON are covered in standards ECMA-262 and ECMA-404, respectively. The latter has also been adopted as an ISO standard (ISO/IEC 21778). JSON Schema is still a work in progress, but the latest status can be found at *www.json-schema.org*. For further details on JSONPath, see Gössner and Bormann (2021).

XML and JSON functionality is now included in major relational databases but because of variations between different implementations, the vendor's documentation is often the most useful resource. Most commercial systems provide online help that can be searched in various ways.

17

NoSQL and Other Nonrelational Databases

Chapter Outline

17.1 The Growth of the NoSQL Movement

Following the publication of E.F. Codd's ground-breaking relational paper in 1970 (Codd, 1970), it was not until the 1980s that commercial relational database systems became widely available. Since that time, they have become the dominant data storage model. Other data resources are used at smaller scales, such as spreadsheets, text files, and so on, but for large-scale data, relational systems have reigned supreme.

The NoSQL movement started to emerge in the 2010s in response to perceived shortcomings in the relational approach (Leavitt, 2010). It's worth noting straight away that "NoSQL" is not a particularly good choice of name. The crucial point is not really an issue with the structured query language (SQL). In fact, some NoSQL vendors try very hard to provide languages that are as similar as possible to SQL. Instead, NoSQL approaches rely on data models that are nonrelational to provide their distinctive features. There are some suggestions that "NoSQL" should be interpreted as "Not Only SQL", but this would include every kind of database, rendering it completely meaningless.

The driving forces behind the movement did not result from a failure of the relational model as such but from new challenges and opportunities exposed by the general evolution of technology. The following trends are especially relevant:

- The cost of data storage has fallen dramatically. In the 1980s, there was a preoccupation with minimizing storage requirements because of the expense of the hardware involved. Today, consumers are commonly offered services that include gigabytes of free data storage space, which would have been unthinkable in earlier times. The decreasing incentives to minimize storage have contributed to the continual increase in database sizes.

Information Modeling and Relational Databases. https://doi.org/10.1016/B978-0-443-23790-4.00005-0

- There is now a much greater understanding of the value of data. Even something as simple as a validated relationship between a person and their email address has a cash value. This has motivated organizations to collect as much data as possible, even if there appears to be no immediate use for the information.
- Relational databases evolved at a time when "client-server" was the dominant information system architecture. This naturally led to databases that served a limited internal pool of interests. Most organizations now recognize the advantages of being able to transfer data between disparate systems, both internally within the organization and externally with customers, suppliers, partners, and so on. This has been facilitated by interchange formats such as eXtensible Markup Language (XML) and JavaScript Object Notation (JSON), as discussed in Chapter 16.
- Since its public appearance in 1990, the Internet has grown by several orders of magnitude. Fast and reliable network connections are now widespread, and connectivity is taken for granted in most of the world. Many websites (and the databases that support them) have millions of users rather than the hundreds that would be considered a large user population in the 1980s.
- The day-to-day operations of most organizations are now reliant on some kind of information technology (IT) platform that facilitates data exchange within and without the organization. Growth in general data usage has been explosive. To a large extent, this has been fueled by today's extensive use of smartphones. In other words, most people carry a computer and communication device with them all the time and do not need to be tethered to a static business system. For example, the publication of street maps has virtually collapsed because most people will now get the relevant map (accessing a remote database) via their phone.

As these trends have evolved, they have exposed database systems to new challenges and opportunities. Rising to these has raised questions about the best way to organize data resources, perhaps by deviating from the relational model that has been the standard for so long.

Before diving into more detail about NoSQL, it will be useful to review some selected database characteristics to provide a context for discussion to follow. For this chapter, we'll assume that some important factors that might be relevant to choosing a database solution are equally balanced among all candidates. These include features such as security, ease of installation, documentation, vendor reputation, support, and so on. Our focus here will be on aspects that are important in discussing the distinction between relational and nonrelational databases.

The Relational Model

The obvious place to start is with the relational model. Many NoSQL proponents describe this as defining a particular type of database system, but that's not quite correct. What is being modeled is the structure of information, not the structure of a system. Relational databases evolved to meet the needs of relational structures, not the other way around. A relational model captures relationships within a collection of information in a natural and logical way, as explained in detail in earlier chapters of this book. The expression of a relational model is usually referred to as the relational *schema*.

One feature deserves special attention. A relational schema captures the intrinsic structure of the information but has nothing to say about how the information will be (or can be) used. The neutral standpoint guides attention to what the information *is*, rather than what it *does*. Depending on the point of view, this can be seen as either a strength or a weakness.

If we assume that the schema has been properly normalized, then we can be sure that unnecessary duplication has been avoided, as each piece of information occurs only once. This provides the maximum flexibility when we consider how the information might be used because it is not optimized in a direction that favors any specific type of usage. We might subsequently choose to depart from a pure normalized model to optimize performance in a certain direction. Sadly though, optimizing one factor in a given situation normally deoptimizes other factors, so having a neutral position is a useful starting point.

The most common reason for denormalizing a relational schema is to improve performance in a specified usage scenario. Values from different relations can be combined to give any desired result, but this involves join operations that are relatively expensive in computational terms. The joins can be avoided by combining the relations involved, in effect prejoining them rather than waiting for a query. However, although this reduces the cost of queries, it makes updates more expensive because of the need to manage duplicate information that the denormalization has introduced.

Determining the best policy requires knowledge of the anticipated data traffic. To take some examples, an online catalogue might have infrequent updates but large numbers of queries, whereas a logging application might have many updates but only rare queries. Concerns about performance figure largely in NoSQL databases, as we will see later.

Many NoSQL platforms announce the lack of a schema as a major benefit. Defining a database schema of any kind may indeed involve a considerable amount of work, such as identifying the data to be accommodated, understanding its relevance to the business, arbitrating between competing demands, and so on. The websites of NoSQL vendors often include examples that show how easy it can be to create arbitrary data structures without thinking about these kinds of issues.

However, the same could be true for relational databases. We could create a database, and then add as many tables as we like containing whatever columns we like without any reference to a schema. The standard approach (get the schema right before proceeding) reflects some hard-won experience in building real-world systems.

There are several reasons why defining a data schema might be important for a potential enterprise system. Defining what should or should not fall within the remit of the system may require accommodations between points of view within the enterprise. It is better to resolve these kinds of issues in advance before investing time and effort that may turn out to be wasted.

Applications using the database will always need to know how the information is structured. The idea of writing an application that can handle its required data in absolutely any structure or format is completely unrealistic. In the real world, application programs must be written to work with some defined structure.

A schema also has value as a convenient summary of the information deemed important by the enterprise. This makes it easier to form a shared culture among those responsible for the database and its applications.

Distributed Databases

Distributed systems play an essential role in modern computing. A distributed database system stores its data on more than just a single node, using multiple physical or virtual machines at the same time. The distributed machines have one shared state and operate concurrently.

Distributed systems offer three important features: scalability, availability, and fault tolerance. There is some overlap between these concerns, as we will see shortly, but scalability especially is one of the key motivators for the NoSQL movement.

The two main approaches to scalability can be summarized as *scale-up* or *scale-out*. The first of these uses progressively more powerful servers as demands increase. The second splits the database into different parts that can be mounted on an increasing number of servers as demands increase.

One way of scaling up is to migrate the database to a more powerful machine, but there are obvious limits to this. An alternative is to use a cluster of servers that shares a common data store. This obviously introduces additional complications and will sooner or later also come to a point where further extension is not feasible.

NoSQL databases focus more on the scale-out strategy, which can cope with databases far larger than would be possible by scaling-up. The key step is to identify a suitable method of partitioning the database. Partitioning could be done in two ways, vertically or horizontally. Consider the example shown in Fig. 17.1.

Employee:

empNr	*name*	*photo*
1	Alice	alicepic.jpg
2	Bob	bob2023.jpg
3	Carol	carol1.jpg
4	Dave	dave.jpg

FIG. 17.1 Simple employee data.

Vertical partitioning effectively assigns different table columns to different servers. This can provide some advantages. First, relatively slow-moving data can be separated from more dynamic data. Slow-moving data is a good candidate for an application to cache in a lower-performance store. Second, sensitive data can be stored in a separate partition with additional security controls. Finally, vertical partitioning can reduce the amount of concurrent access that's needed. A possible outcome is shown in Fig. 17.2.

EmpName:

empNr	*name*
1	Alice
2	Bob
3	Carol
4	Dave

EmpPic:

empNr	*photo*
1	alicepic.jpg
2	bob2023.jpg
3	carol1.jpg
4	dave.jpg

FIG. 17.2 Vertically partitioned employee data.

Vertical partitioning can be used within the same server. Horizontal partitioning (also known as *sharding*) is used to split data over several servers. This introduces an additional difficulty because applications now need to know which server contains the relevant partition. This can be done in several ways, for example:

- By range: Each server contains a known range of key values.
- By lookup: An auxiliary table contains a cross-reference between keys and servers.
- By hash: A hashing function is used on the key to identify the relevant server.

The distribution strategy may also consider other factors such as geography. If data traffic is primarily driven by localized concerns, it can be directed to servers that are physically close to where the information is created or consumed. Fig. 17.3 shows a simple horizontal partitioning, splitting the data over two servers.

Employee: (Server1)

empNr	*name*	*photo*
1	Alice	alicepic.jpg
2	Bob	bob2023.jpg

Employee: (Server2)

empNr	*name*	*photo*
3	Carol	carol1.jpg
4	Dave	dave.jpg

FIG. 17.3 Horizontally partitioned employee data.

The CAP Theorem

The CAP theorem was first introduced by Brewer in 2000 and is a widely used concept in distributed systems design (Gilbert and Lynch, 2002). At its core, the CAP theorem is about understanding the trade-offs that must be made when building distributed systems. Specifically, it states that it is impossible for a distributed system to simultaneously provide all three of the following guarantees:

- Consistency: Every node in a distributed system sees the same data at the same time, implying that every read operation must return the value of the most recent write of the data.
- Availability: Every request made to the system will receive a response in a reasonable time frame, regardless of whether a node in the system has failed or is experiencing issues.

- Partition tolerance: A distributed system should continue to function even when network partitions occur, due to nodes in the system becoming separated from each other due to network failures or other issues.

In other words, the CAP theorem proposes that it's only possible to have two of the three guarantees at any given time. This means that a highly available and partition-tolerant system must sacrifice consistency; a highly consistent and partition-tolerant system must sacrifice availability; and a system with high availability and consistency must sacrifice partition tolerance.

If the database is not partitioned, then it can be both consistent and available (sometimes known as a CA database). This is exactly the case in the traditional database with a large central server (i.e., a single-node system). If the database is partitioned, then it can prioritize consistency (CP) or prioritize availability (AP). Consider a distributed system, as shown in Fig. 17.4; just two nodes are sufficient to make the point.

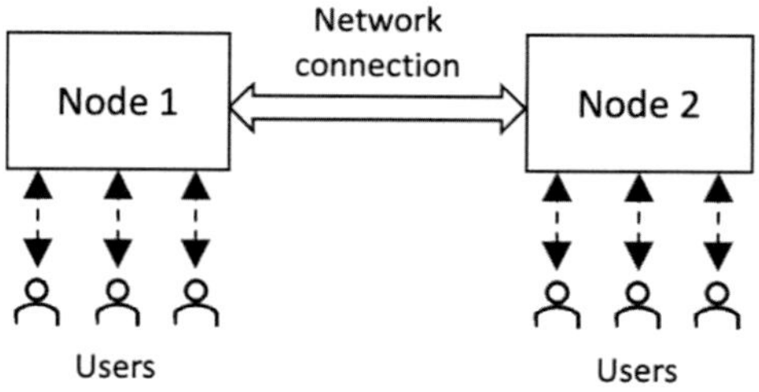

FIG. 17.4 A simple distributed system.

Updates made on one node can be transferred to the other node, with some latency, over the network connection. Both sets of users can eventually see consistent data. But if there is a failure, such as a loss of the network connection between Node 1 and Node 2, then there are two possibilities:

(a) Prevent access to one node. This assures consistency on the working node but removes availability for users on the other node (the CP option).
(b) Allow access to both nodes. The system remains available for all users, but updates on one node cannot reach the other, thus losing consistency (the AP option).

This example is obviously trivial, but it illustrates the balance that must be considered in a distributed system. There is no single right answer, as the choice depends on the purpose that the application is intended to serve. A financial system may prefer consistency, but a social media platform may prefer availability. Different distributed systems may explore different compromises in trading off some loss of consistency against some loss of availability.

ACID and BASE

Traditional databases usually conform to a set of principles known by the acronym ACID from the terms Atomic, Consistent, Isolated, and Durable (Gray and Reuter, 1993).

- *Atomic*: All operations in a transaction must either be completed successfully or not at all. If any operation fails during the execution of a transaction, the entire transaction is aborted and rolled back to its previous state.
- *Consistent*: The database must transition from one consistent state to another after each successful transaction. All data integrity constraints must be maintained before and after each transaction.
- *Isolated*: Concurrent transactions do not interfere with each other's execution. Each transaction should appear as if it is being executed independently without affecting other ongoing transactions.
- *Durable*: Once a transaction has completed successfully ("committed"), any changes made to the database should persist even in the event of a system failure.

The ACID viewpoint strongly values consistency, making it well-suited for traditional transactional database systems where reliable data is crucial. The rise of distributed systems has promoted interest in a different perspective, known by the acronym BASE, from the terms Basically Available, Soft state, and Eventually consistent (Pritchett, 2008).

- *Basically available* means that the system should always be available for requests, even if the system is experiencing some sort of failure or outage.
- *Soft state* refers to the fact that different copies of the data do not need to be in a completely consistent state at all times.
- *Eventually consistent* means that differences in data will eventually become reconciled, given enough time.

The BASE viewpoint strongly values availability over consistency by allowing temporary inconsistencies between replicas of the data. This is obviously a looser requirement than ACID, but it is more relevant for distributed systems.

Databases can ensure consistency by employing various techniques and mechanisms to synchronize data across multiple nodes in a distributed system. Some common methods used by CP databases include the following:

- Consensus protocols: CP databases often utilize consensus protocols, such as Paxos or Raft, to achieve agreement among nodes about the order and validity of data updates. These protocols ensure that all nodes in the system reach a consistent state by requiring a majority or unanimous agreement before committing changes.
- Two-phase commit: In a distributed transaction, a two-phase commit protocol can be used to coordinate the commit or rollback of changes across multiple nodes. It ensures that all participating nodes agree to commit the transaction or abort it, guaranteeing consistency across the system.
- Quorums and replicas: CP databases typically use replication to maintain multiple copies of data across different nodes. Quorums are used to determine the minimum number of nodes that need to agree on a read or write operation to ensure consistency. By requiring a certain number of replicas to participate in operations, CP databases can ensure that all replicas are consistent before returning a result.
- Synchronization and locking: CP databases may use synchronization mechanisms, such as distributed locks or optimistic concurrency control, to prevent concurrent

access and modifications to shared data. These techniques ensure that conflicting changes are serialized and applied in a consistent order.

- Data versioning and conflict resolution: CP databases often employ data versioning or timestamping to track changes and resolve conflicts when concurrent updates occur. By assigning unique versions or timestamps to data, conflicts can be detected and resolved based on predefined rules or policies.

Of course, achieving consistency can introduce latency and may reduce availability, especially during network partitions or failures. The specific mechanisms and trade-offs employed by a CP database depend on its design and the level of consistency required by the application.

Availability

Availability in distributed systems refers to the ability of the system to respond to client requests, even when some parts of the system have failed. This is important because one of the main reasons for using a distributed architecture is to increase availability and fault tolerance.

In a distributed database, several methods can be employed to provide high availability, ensuring that the system remains operational and accessible even in the presence of failures or network partitions. Here are some common techniques used to achieve high availability:

- Replication, which involves creating multiple copies of data across different nodes in the distributed database. By having replicas in different locations, the system can continue serving read and write requests even if some nodes fail. Replication also improves fault tolerance, as a copy of the data is available even if one or more nodes become unavailable.
- Data partitioning, or sharding, involves dividing the dataset into smaller subsets and distributing them across multiple nodes. Each node is responsible for a specific partition of the data. By distributing the data, the system can handle a larger volume of requests in parallel, improving availability and scalability.
- Load balancing techniques distribute incoming requests across multiple nodes in a distributed database, ensuring that the workload is evenly distributed. This helps prevent any single node from becoming overwhelmed with requests, improving availability by avoiding bottlenecks and optimizing resource utilization.
- Failure detection and recovery mechanisms are crucial for high availability. Nodes can periodically check the health of other nodes and detect failures. Upon failure detection, the system can take appropriate actions, such as promoting replicas to replace failed nodes or redistributing the data. Automatic failure detection and recovery minimize downtime and ensure continuous availability.
- Redundancy and Failover: Redundancy involves having multiple nodes or components that can serve the same function. In case of a failure, failover mechanisms can redirect

requests to healthy nodes or activate standby replicas. Redundancy and failover strategies provide fault tolerance and improve availability by minimizing the impact of failures on system operations.

By combining these techniques, distributed databases can achieve high availability by ensuring continuous access to data and uninterrupted operation, even in the face of failures or network disruptions. The specific methods used depend on the database system's design, requirements, and the desired level of availability.

NoSQL Benefits

When comparing relational database systems from various suppliers, it's clear that they have many commonalities. This is hardly surprising given that they are all based on the same fundamental logical concept. The same cannot be said for NoSQL databases. Each NoSQL database is designed to fill a specific niche, resulting in a wide range of variations among different systems. As we will see later, these can be classified into broad families, but they all share a common set of motivations for their existence. These fall into three main categories: coping with large amounts of data, improving performance, and enabling a more agile database development process.

As we've already noted, there is a consistent trend toward ever-bigger databases. For practical reasons, this means implementing a "scale-out" policy to support the database on multiple computers, raising all the distributed system issues discussed above.

Distribution is not so easy with a relational database, and many of the NoSQL approaches are aimed at providing ways of scaling using relatively cheap commodity hardware, usually by partitioning the database in some way. This also makes a cloud computing architecture a good fit for Big Data systems, and all cloud service providers offer managed NoSQL solutions.

NoSQL systems usually provide options for adjusting the trade-off boundary between consistency and availability, so that the system is as consistent as possible and as available as possible within the conditions of the intended type of application. A particular application area may offer special features that can be exploited to improve either consistency or availability within that given context.

Although the falling cost of hardware has made minimizing storage space less important, it can still become significant with very large volumes of data. Some NoSQL approaches provide ways of compressing data and storing sparse data in an efficient way, both to reduce storage space and to increase performance. However, this must be set against the increased storage required for the duplicate data that inevitably arises without normalization.

Dealing with large databases also raises issues of performance. This is always a concern with any kind of database, but an approach that might give an acceptable performance with a small database may be totally unacceptable in a larger system. The need to scale out in a distributed system also provides opportunities for improving performance. For example, the load on a server can be reduced by directing a proportion of the queries to other servers that hold copies of the same data.

In relational systems, it's quite common to denormalize the database to increase performance for specific types of queries, principally by avoiding the penalty of joins between tables. In most NoSQL systems, this is a standard feature, although it would be misleading to call the structure "denormalized" because it may never have been normalized in the first place.

Of course, knowledge of anticipated traffic is essential to maximizing performance, especially the relative frequency of reading from the database versus writing to the database. NoSQL systems are typically aimed at scenarios with a large proportion of reads but few writes. Database records may be entered with some form of bulk loading of relatively static data. This also eases the problems of consistency: if there are no writes there is no way for the database to become inconsistent!

Another efficiency bonus claimed for NoSQL databases is the simplification (and reduced cost) of system management. This can be due to built-in automation of various administrative tasks that traditionally required special expertise and complex manual procedures. Examples include automated sharding, data distribution and replication, easier deployment, and easier system configuration.

Agile development methodologies have gained popularity in all areas of system development. They promise rapid delivery of an initial working system, the ability to respond promptly to changes as they occur during development, and risk reduction through the early identification of problems.

Some NoSQL products define a database schema as a burdensome overhead and promote the lack of a schema as a major agility benefit. The usual example quoted is the ad hoc addition of new data elements to an existing database. However, this is not as distinctive as it may seem. It's also straightforward to extend a relational database using commands such as:

```
create table MyNewTable ..., and
alter table MyOldTable add myNewColumn ...
```

The situation is more difficult when we consider deleting or modifying an existing structure. The impact of this depends on whether the database is intended to support a single application (such as a mobile phone app) or multiple applications (sometimes called an integration database).

The latter case is more typical of business systems. The two styles are illustrated in Fig. 17.5.

In an application database, the developer of the application and the developer of the database are typically the same person. This makes it easy for changes in the application to be reflected in corresponding changes to the database structure and vice versa. However, things may not be so simple for an integration database because database changes to suit one application may require changes to other applications. Without some kind of schema, it becomes very difficult to manage database integrity. The lack of any authoritative

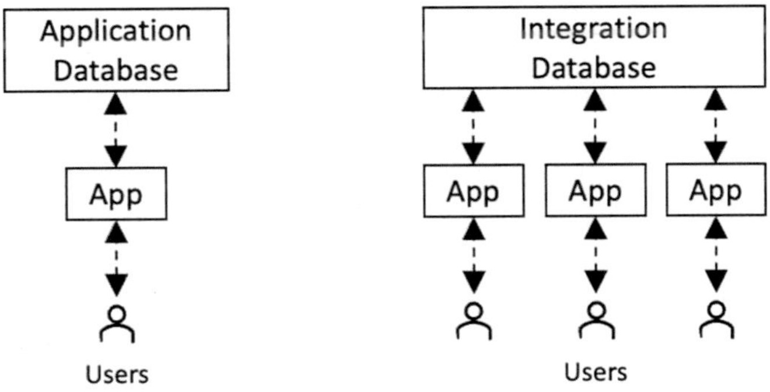

FIG. 17.5 Application database versus integration database.

reference to the data structure can also make general application development unnecessarily difficult and error prone.

NoSQL databases often store data in a form that directly matches its external usage, for example using the structure of a business document. This is typically encoded in the same way as the interchange formats described in Chapter 16, with JSON being a common choice. Interfacing between object-oriented application code and a relational database is usually accomplished through some kind of object-relational mapping layer, and it's possible to eliminate this so-called "impedance mismatch" in some NoSQL systems.

The definition of a relational table requires that a specific data type is allocated to each column. NoSQL databases are often considered in situations involving unstructured or ad hoc data or where data has high variability, and so dynamic typing may be preferable.

Potential Drawbacks of NoSQL Databases

NoSQL databases have gained popularity due to their flexibility and scalability for certain use cases. However, they also have some drawbacks that need to be considered. Here are some common drawbacks associated with NoSQL databases:

- Lack of a standardized query language: Unlike relational databases that use SQL as a standardized query language, each NoSQL database has its own unique query language or application programming interface (API). This can make it difficult to switch between different NoSQL databases or integrate them with existing systems. The lack of standardization leads to smaller user communities compared to SQL databases, which may make it more challenging to find support and resources.
- Data duplication and denormalization: NoSQL databases typically denormalize data to improve performance and enable horizontal scalability. This means that data may be duplicated across multiple records or collections, leading to increased storage requirements. More importantly, maintaining data consistency across denormalized structures can involve extra complexity.
- Lack of data integrity constraints: Unlike relational databases, which enforce data integrity through constraints such as primary keys, foreign keys, and unique

indexes, NoSQL databases often lack built-in mechanisms for ensuring data consistency and enforcing relationships among data entities. This makes developers responsible for handling data integrity at the application level.
- Lack of mature tooling and ecosystem: Although NoSQL databases have been around for some time, they generally have a less-mature tooling and ecosystem compared to relational databases. This can result in smaller community support, limited development frameworks, and fewer third-party integrations.
- Learning curve and developer expertise: Working with NoSQL databases may require developers to learn new concepts, query languages, and data modeling techniques. This can increase the learning curve and the need for specialized expertise, especially when migrating from a relational database environment.
- Limited transactional consistency guarantees: Some NoSQL databases sacrifice strong consistency guarantees in favor of high availability and partition tolerance, following the CAP theorem. This means that in certain failure scenarios, trade-offs are made that can lead to eventual consistency or data staleness. There is also an increased risk of inconsistent data due to the eventual consistency model used by some NoSQL databases.

It's important to note that the impact of these drawbacks can vary depending on the specific use case and requirements of the application. NoSQL databases excel in certain scenarios, such as handling large amounts of unstructured or semistructured data, real-time data processing, and highly scalable systems, but they may not be the optimal choice for all types of applications.

In conclusion, NoSQL databases offer a range of advantages that make them an attractive option for organizations dealing with large volumes of diverse data. Their flexibility, scalability, and high performance make them well-suited for big data applications, the Internet of Things (IoT), and real-time analytics. However, it's essential to carefully consider the specific requirements of the organization and the application area before choosing a NoSQL database.

17.2 Key-Value Stores

Key-value data stores are a type of NoSQL database that is becoming increasingly popular due to flexibility and scalability. Unlike traditional relational databases, key-value stores do not use a fixed schema for most of the data. Instead, data is stored as a collection of key-value pairs, where the key is used to uniquely identify the data and the value is the actual data. This type of data store is particularly useful for applications that require fast, efficient data access and retrieval.

There are many commercial key-value data stores available in the market today, in addition to others that are not promoted as key-value stores but offer similar capabilities. Redis is an in-memory data structure store that is used for database caching, message brokering, and real-time analytics. It is known for its speed and versatility and is often used as a distributed cache. RiakKV is a high-availability distributed key-value data store that is designed to be fault-tolerant and scalable. Apache HBase is an open-source, distributed key-value store built on top of the Hadoop Distributed File System (HDFS). It is designed for large-scale, random read/write access to Big Data and is modeled after Google's BigTable approach. Amazon DynamoDB is a fully managed key-value data store offered by Amazon Web Services (AWS). It offers features such as data replication, encryption at rest, and flexible indexing options.

The main feature of key-value stores is their conceptual simplicity. They build on the well-known technique of *hashing* to provide a consistent way of accessing data values that may not necessarily have a regular format. The values are held in a specialized structure, the internal details of which can be temporarily ignored by users. Values are associated with a unique key that provides a means of locating the values within the structure. A hash function is used that, given a key, provides access to the related value(s), either directly or indirectly using some form of pointer.

Such structures are supported in many programming languages. The terminology varies across different languages and such structures may be described as Hash Maps, Hash Tables, Dictionaries, Associative Arrays, and so on. The terminology is also not used consistently, so a precise definition is only relevant in the context of a specific language. For example, Python uses the term *dictionary* to refer to a structure containing key-value pairs. The items are ordered by key value in more recent versions of Python, but that's not necessarily true of equivalent structures in other languages. JavaScript can simply use JSON-style Objects for the same purpose. Despite differences in syntax, all these programmatic approaches support a broadly equivalent range of operations, such as:

- Insert item (key, value)
- Update value (key, new value)
- Delete item (key)
- Retrieve value (key)
- Test for existence (key)

Most languages allow the types of values that can be stored to be anything expressible in that language, ranging from simple types (number, string, etc.) to items that may have their own internal structure, such as objects, lists, and perhaps even further dictionaries. Although such structures can be used within a program, they can't be directly accessed from outside the program and must be exported in some form, such as JSON, if required externally.

Data within a program will obviously disappear when it terminates, and so key-value stores provide equivalent functionality in persistent storage. This also makes the data accessible from many different programs (assuming suitable language bindings).

We can approximate a key-value database using a relational database with a single table defined in a similar way to the following SQL fragment:

```
create table myKV (
  theKey varchar(20) primary key,
  theValue BLOB
)
```

Here, the key is assumed to be a string of up to 20 characters, but any SQL data type could be used. The value here is a BLOB datatype that is used to contain undifferentiated binary data that is not parsed by the database. BLOB may be replaced by other types such as varbinary( ) in some SQL dialects.

Key-value stores differ from relational databases by focusing solely on the efficient storage and retrieval of complete units of data. The internal structure of a data unit can be ignored by the data store so it can be treated as a single item. This removes the need to support complex queries, as the only essential operations needed are writing and reading individual data units. The data store is no longer responsible for understanding the internal values, data types, and structure of the information. That responsibility is now devolved to applications that use it. Any changes to the internal data structure, adding or removing values, have no impact on the data store, though they obviously will within any associated application.

In addition to making storage and retrieval more efficient, this flat structure also greatly simplifies the design and implementation of a data store because its schema can be trivially simple. However, it's important to remember that the need for a detailed schema has not gone away completely–it has simply been moved from the data store to the application.

Although key-value stores excel at rapid retrieval of data units, some also offer specific ways of operating on data values inside a unit, such as appending to a string, pushing an element to a list, and getting the member with the highest ranking in a sorted set. If the content of the data unit is encoded in some structured form such as JSON or XML, then facilities can be provided to query the content in the appropriate way.

Key-value stores are well suited to being distributed, and it's probably not a coincidence that their rise in popularity has occurred alongside the emergence of cloud computing as a viable platform. A data store can be scaled almost indefinitely by partitioning horizontally across many servers, each of which handles its own subset of the keys and values. Resilience and performance can be improved by replication so that multiple copies can be held on different servers and accessed through an appropriate load-balancing mechanism.

Distribution does not apply in all situations; some key-value stores are designed to be held in memory to provide the fastest possible speed of retrieval. Solid-state disks are about 100 times faster than older hard disks, but RAM is orders of magnitude faster still. Distribution is much more problematic for an in-memory store, but hybrid solutions can be found. For example, an in-memory store can be used to provide a lookup between an application-based key and a server-based key that can then be used to access a distributed key-value store.

In all cases, some thought needs to be given to the design of the key. Obviously, this needs to provide unique values and must be generated consistently throughout the life of the data store. It's a good idea to make the key as compact as possible, consistent with the need to fully identify the related data unit. Making the key somewhat human-readable can be useful for troubleshooting, providing it does not impact efficiency. In the case of distributed systems, the key must also indicate the partition to be accessed. This can be done by making the key more complex, in much the same way that a file can be uniquely identified on a computer through its file path. The key will almost certainly be indexed for rapid retrieval. In some situations, it may also be useful to provide a secondary index if values can be sought in different ways.

As already noted, there is a great deal of diversity among NoSQL databases, and this certainly applies to key-value stores. Individual systems have their own individual features, but there are some common characteristics associated with most popular key-value stores.

- Simplicity: Key-value stores typically offer a simple data model where data is stored as key-value pairs, making them easy to understand and use. They do not require a complex schema or complex table structures and can be easily integrated into existing applications.
- High performance: Key-value stores are designed for high-speed data access and low-latency operations. This is partly because they use simple data structures and avoid complex joins or indexing. They also tend to minimize disk I/O, using in-memory caching and optimizing data structures and algorithms for fast key-based lookups.
- Scalability: Key-value stores are easy to scale horizontally by distributing data across multiple nodes. They support sharding and partitioning techniques to handle large datasets and can grow gracefully as data volumes increase over time. This makes them a good choice for applications that need to handle a large volume of data and require high availability.
- Replication and fault tolerance: Key-value stores often provide replication mechanisms to ensure high availability and fault tolerance. They store redundant copies of data across multiple nodes or partitions and can provide automatic rebalancing of services in case of node failures.
- Flexible data models: Most key-value data stores can store a wide range of data types, including structured, semistructured, and unstructured data. This makes them useful for applications such as storing user profiles, sensor data, session data, and IoT applications that require a variety of types and formats.

Clearly, key-value stores offer some advantages that can be exploited in certain types of applications, but we must set these off against features that may make a key-value store less suitable in other circumstances.

- Ad hoc or complex queries are often not supported in the data store, such as searching for data with a specific feature or connections between data across different key-value pairs. Any such operations must be carried out in application code external to the data store.
- There are limited opportunities for operating on the data stored as the value of a key-value pair. Data retrieval is typically aimed at the efficient retrieval of the value as a unit. This also potentially makes modifications to some element within the data less efficient because the whole value must be handled as a complete data chunk.

- Unlike relational databases, there is no standard for the structure of key-value stores. There is little opportunity for the transfer of expertise from one key-value product to another because each uses its own proprietary functions and methodologies.
- Key-value stores are generally not well equipped to provide the kind of support for the ACID properties that are usually associated with relational systems. Transactions are generally limited to operations on a single data unit.
- There is limited scope for the validation of data or the application of constraints within the data store. Again, such operations must be carried out by the application code before the data is stored or after it is retrieved as a unit.

Identifying the right key-value store, or even deciding if a key-value store is appropriate, will depend heavily on the requirements of each individual case. Key-value stores are very well suited to applications that require rapid access to an individual unit of complex data from a very large pool of similar units. Some typical use cases include the following:

Caching. Data that is frequently accessed can be stored in a key-value data store for faster retrieval. This can help to improve application performance and reduce latency. For example, an e-commerce website may use a key-value data store to cache the product information for popular products so that the website can be served quickly without querying the traditional database every time.

Session management. A key-value data store can be useful in applications where a user's session needs to be maintained across multiple requests or across multiple servers. Values such as log-in information and shopping cart contents can be held for quick retrieval of the information when needed.

Storing user preferences. User preferences can be stored as key-value pairs, allowing for easy retrieval and updating of user preferences. This can be useful in applications where users need to be able to customize their experience or where user preferences need to be shared across multiple devices.

Content management. Digital content is now commonplace. This includes text, images, videos, documents, and other multimedia assets. An application may need to collect diverse kinds of media into a collection that can be stored and retrieved as a single unit, which is ideally suited to a key-value store.

17.3 Column-Oriented Databases

The main difference between column-oriented databases and relational databases is how data is organized and stored rather than the logical structure of the data. A relational database stores information in tables, organized into rows and columns. A table can be thought of as information about a particular type of entity, with each row representing an individual entity of that type and the columns containing values that are relevant to that entity type. A row is often described as a record, and the related column values as fields of that record. The main focus is on managing records, so it is legitimate to describe this type of database as row-oriented.

If we need to query for specific data, the database must scan through all the rows in the table to find the relevant information. This can be aided by indexes but for large databases with millions of records, it can be slow and inefficient.

In contrast, column-oriented databases store data in columns, with each column representing a single attribute or field in a table. This means that the values for a particular attribute type are stored together in one column, regardless of which records they belong to.

When the database is queried for specific data, the database only needs to scan the relevant columns, rather than every row in the table. This can be much faster and more efficient, especially when dealing with large datasets and operations such as aggregation.

Several column-oriented database systems are available, including the following: Amazon Redshift has excellent integration with other AWS products. It provides dynamic scaling and automates many system management activities. Several open-source systems are available. PostgreSQL is a well-known row-based database, but it also has a column store option that is useful when you want to use only one database for both column and row store. MariaDB is another popular open-source database that supports both row and column storage. Apache Cassandra is a column-oriented database designed to handle large amounts of data across many commodity servers.

Person:

id	*name*	*diet*	*weight*
1	Alice	glutenfree	68
2	Bob		92
3	Carol		
4	Dave	vegan	66
5	Edna		78

FIG. 17.6 Dietary data.

Fig. 17.6 shows a simple example. This does not represent any specific database product; it's intended only to illustrate the concept. Assume that we have information about individuals and their dietary preferences. A relational table might look something like the following (with empty cells corresponding to NULL in relational terminology).

If we imagine this serialized into physical storage, it might look something like the following:

1, Alice, glutenfree, 68, 2, Bob, NULL, 92, 3, Clare, NULL, NULL, 4, Dave, vegan, 66, 5, Edna, NULL, 78

Fig. 17.7 shows how the same information might look in a column-oriented database, assuming explicit identifiers.

Name:

1	Alice
2	Bob
3	Carol
4	Dave
5	Edna

Diet:

1	glutenfree
4	vegan

Weight:

1	68
2	92
4	66
5	78

FIG. 17.7 Column-oriented dietary data.

The example in Fig. 17.7 is laid out in separate tabular form for comparison with the relational version, but it is also possible to visualize the name, diet, and weight "tables" as rows in a column-oriented database, as shown in Fig. 17.8. From this perspective, it's possible to see that rows in this representation can contain a variable number of columns.

rowKey	values
name	(1, Alice), (2, Bob), (3, Clare), (4, Dave), (5, Edna)
diet	(1, glutenfree), (4, vegan)
weight	(1, 68), (2, 92), (4, 66), (5, 78)

FIG. 17.8 Dietary data expressed in row form.

In this case, serializing the data into physical storage might look something like the following:

```
1, Alice, 2, Bob, 3, Clare, 4, Dave, 5, Edna, 1, glutenfree, 4, vegan, 1, 68, 2, 92, 4, 66, 5, 78
```

At this simple level, we can't see the specific detail of the data layout in persistent storage. This depends very much on the specific database being considered. Even so, it's obvious that in one case, a retrieval will find colocated information for an entity and the other will find colocated information for an attribute. Another way of saying this might be that row-oriented retrieval gets values for the attributes of an entity, but column-oriented retrieval gets values for the entities of an attribute.

Consider a query to find the average person's weight from this information. The relational case would require the retrieval of each row and the `weight` column value within the row before calculating the average value. In the column-oriented case, we simply need to retrieve the `weight` row in the table shown in Fig. 17.8 to get all the necessary values for the calculation. Because data retrieval from persistent storage is the most time-consuming part of the operation, this shows that the column-oriented approach has the potential for much better performance on certain types of queries. Performance is probably the greatest single motivator for this type of database, especially as the database becomes larger. In a column-oriented database, increasing amounts of data result in rows becoming bigger, not a bigger number of rows, so these are sometimes referred to as "wide tables".

It's important to remember that the distinctive feature of a column-oriented database is the way that the information is stored physically. From a conceptual viewpoint, we could still think of the database in the traditional manner, but we would find some types of operation more efficient and others less efficient than in a standard normalized relational database. In some cases, column-oriented databases can be queried using standard SQL, making the "NoSQL" label a little bit pointless.

The advantages of column orientation are only valid if queries operate on one column or a small number of columns. Queries involving several columns get less benefit from the localization of information from a single column. If multiple columns are frequently used together, it's possible to add another level of organization by grouping columns into column families. For example, columns related to individual customer details (such as name, address, date of birth, etc.) could be grouped into a "Personal" column family. This can be used to ensure that the relevant columns are colocated in the database.

Column orientation provides additional opportunities for efficient storage of information. All the data from one row-oriented column (visualized as a row in a column-oriented database) is of the same type, opening possibilities for applying various types of compression to the values. An obvious instance of this is when the column contains many duplicate entries such as URLs or product names. In some cases, it may be possible to compress a column sufficiently to fit in memory for direct access, dramatically improving response time. Compression is useful because column-oriented databases are often used with very large datasets. There is also no need to allow for NULLs. An empty cell in a row-oriented database simply does not need to exist in its column-oriented equivalent.

The design of a column-oriented schema needs to consider not only the information to be stored but also how the database will be used and how it might evolve in the future. Adding a new column to the database is very easy–it can just be appended to the existing columns without impacting any existing queries. Other types of schema changes may be more problematic. Changes to query patterns can also reduce the effectiveness of the database if its design has been optimized for an anticipated scenario that no longer applies. It is also more difficult to design effective indexing schemes in the column-oriented case.

Column-oriented databases are very much skewed toward improving the efficiency of read operations (**select** in SQL). Operations such as **insert** (a new entity), **update** (an attribute of an existing entity), and **delete** (an existing entity) are generally slower because of the way the information is spread in physical storage. Also, read operations are improved only for certain types of queries. Consider the following two SQL queries:

```
select avg(weight) from Person
select * from Person
```

In a column-oriented database, the first of these will be faster but the second will be slower than the row-oriented equivalent (obviously, the differences will only be meaningful in large datasets). Generally, row-oriented operations or those involving many columns will be slower in a column-oriented database, making them unattractive for transactional applications.

Joins between tables in a column-oriented database are more complex than in the row-oriented case and are likely to have a lower performance. An alternative strategy is to consider the usage patterns of the database and define a schema that "prejoins" the relevant tables. This obviously requires additional controls to ensure the integrity of the duplicate data that will inevitably result, but the new table can then be treated in the same way as any other table with respect to a columnar arrangement. The risk here is that this optimization will make the database perform very badly if the usage patterns do not turn out to be as expected.

A major advantage of column-oriented databases is their ability to deal with very large datasets. Their structure lends itself very well to partitioning. An obvious first step is to partition the database vertically so that different columns (or perhaps column families) can be mounted on different servers. There is no need for each of these to be of the same type. Columns that are accessed less frequently can be mounted on less-powerful servers.

Within each vertical partition, there are options for further partitioning in the horizontal direction. A column can be split into shards, which, again, can be mounted on separate servers. Some additional effort is required to direct queries to the relevant shard, but this provides the potential for a huge expansion of capacity.

An additional measure to increase performance is replication: copies of the same data can be stored on multiple servers that can be accessed independently. This requires some kind of automated load-balancing mechanism to share queries evenly among the servers. A concern in replication is maintaining consistency between different copies of (what should be) the same data. However, as already noted, column-oriented databases are most applicable in situations with rare updates but frequent queries. Because updates are scarce in such cases, the consistency problem may be easily manageable in the right type of application.

The following characteristics summarize the features that are likely to be important in an application requiring a column-oriented database:

- Large amounts of data are involved, corresponding to millions of rows in more traditional row-oriented tables.
- Efficient use of storage is relevant, for example, in dealing with large amounts of sparse or repetitive data.
- Database operations are dominated by queries, with few (if any) changes to data once loaded.
- Queries typically use aggregation functions such as avg.(), count(), min(), max(), and sum().
- Each individual query involves only a small number of columns.

If we want to find a generic term to describe applications of column-oriented databases, the most appropriate is probably "analytic", often referred to by the acronym OLAP (online analytic processing). Column orientation is not appropriate for transactional systems, often referred to by the acronym OLTP (online transactional processing). Examples of typical applications include the following:

Data warehouses and data marts: These are commonly used to extract insights from existing data. The input is usually created (or updated) by extracting data from transactional systems and reorganizing it into a more appropriate format. Information is added not on a row-by-row basis, but by bulk loading, which can be quite efficient.

Scientific research: Scientific research applications can involve complex queries across large datasets, making column-oriented databases a natural fit. Researchers can use these databases to analyze data quickly and efficiently, allowing them to focus on their research rather than on managing data.

Financial trading: Financial trading applications often require real-time access to large volumes of data. Column-oriented databases are specifically designed to handle data of this type, allowing traders to access information quickly and accurately.

Analysis of historical data: Stored data can be modeled and analyzed using various statistical and logical methods and techniques to gain insights of various types, such as descriptive, exploratory, predictive, diagnostic, and prescriptive viewpoints.

17.4 Document Databases

In Chapter 3, we discussed how reports and other documents can be analyzed and expressed in terms of elementary fact types. Modeling these fact types in ORM, along with the relevant constraints, leads to a model that can form the basis for a database schema. Document databases take a different approach by storing documents directly using their original structure, so, as the name suggests, a database of this type consists of collections of documents. Because no preanalysis of documents is necessary, it seems possible to add any kind of document to any collection without any kind of preparation. In principle, this promises a great deal of flexibility, but in practice, some degree of control is required to have any hope of making the database manageable over its lifetime.

Several document database systems are available, including the following: MongoDB is one of the most popular document databases. It stores data in flexible, JSON-like documents and allows for dynamic schemas. Couchbase is a distributed NoSQL database that offers high performance and scalability. It's commonly used for applications requiring real-time data updates, caching, and mobile synchronization.

CouchDB is an open-source document database focused on ease of replication and conflict resolution. Document databases are available from cloud service providers such as Amazon DocumentDB and Google Firestore. Finally, MarkLogic is a NoSQL database that supports flexible schema design and stores data in XML or JSON documents. It's commonly used for applications that require complex data modeling and querying. The examples in this chapter are based on MongoDB, though other document database systems have broadly similar capabilities to the ones we will discuss.

Documents are typically expressed using the structural capabilities of JSON, previously described in Chapter 16. Internally, the information storage may adopt BSON, a binary form of JSON that incorporates additional metadata to make data scans faster and can also reduce storage requirements, but we will ignore this detail.

A document database is formed from *collections*. Each collection contains *documents* of a similar kind: for example, Products, Orders, Customers, and so on. Each document can contain several *fields*, but the data in a field need not be single-valued as it would be in a normalized relational database–a field can contain a structure that, in turn, can contain other structures and so on. There is a rough correspondence between these concepts and their relational equivalents, as shown in Table 17.1.

Table 17.1 Comparison of RelationalDB and Document DB concepts.

Relational DB	Document DB
Table	Collection
Row	Document
Column	Field

To illustrate some features of this type of structure, the following example shows two documents from a hypothetical Contracts collection:

```
Contracts: [
    {
        _id: ObjectId("6481d941f682789c3b8d71ef"),
        contractRef: 9876,
        activeDate: "Jan 1 2024",
        customerId: "DE876",
        staffAllocation: [ { person: "Alice123", hours: 15 },{ person: "Bob456", hours: 35 } ]
    },
    {
        _id: ObjectId("64820117f682789c3b8d71f0"),
        contractRef: 8765,
        startDate: "Jan 12 2024",
        customerId: 2024002,
        staffAllocation: {

        marketing: [ { person: "Carol999", hours: 5 } ],

         operations: [ { person: "Bob456", hours: 5 }, { person: "Dave666", hours: 30 } ]
        },
        contractType: "FP"
    }
]
```

Six points to be noted for this example may be summarized as follows:

1. The _id field has been automatically generated to provide a unique identifier for each document. For simplicity, we'll ignore this feature in subsequent examples.
2. The names of the same fields in different documents need not be consistent. The activeDate field in the first document and the startDate in the second document may refer to the same business concept, but they do not have the same field name.
3. The data types for the same field need not be the same in different documents. The customerId field is alphanumeric in the first document, but numeric in the second document.
4. Fields can be multivalued and can also contain other documents, as with staffAllocation in the two documents.
5. The internal structure of each document in a collection need not be the same. The staffAllocation field is a list in the first document, but an object (in the JSON sense) that contains multiple lists in the second document.
6. Different documents in the same collection need not contain the same fields. The contractType field appears in the second document but is absent in the first document.

This kind of flexibility is sometimes claimed as a great advantage of document databases over the relational alternative, where each row in a table (equivalent to each document in a collection) has a fixed structure with defined data types for each column (equivalent to each field in a document). While this flexibility makes it easy to add documents without worrying about any kind of schema, it makes other operations on such an uncontrolled database an almost impossible task to manage. Even something as simple as retrieving the value of a particular field could require an examination of the collection to determine all variants of the field name to write a sufficiently robust query. In the example above, listing all the dates when contracts began would need to query for both startDate and activeDate (and any other name variants that might have been used).

It is possible to add a *validator* to a collection to constrain the documents it contains to an acceptable structure. This can include checks for mandatory fields, data types, value ranges, and so on. In effect, a validator can provide something like a schema for an individual relational table.

Write operations for a document are atomic, as the document is either written in full or not at all. Writes involving multiple documents can be transactional, but this involves some additional effort. A session must be created for each update, and then the sessions must be combined so that all the updates are completed together or not at all. This is the same whether the collections are physically stored together on the same server or distributed over multiple servers. Such operations have a big impact on performance, and the general recommendation is to adjust the schema to avoid this situation if possible. For example, instead of a transactional update to collection A and to collection B, it may be possible to define a new AB collection so that the update can be handled by a single write operation.

Linking Data

Inserting, retrieving, and managing individual documents is relatively straightforward, but in any realistic database, there will almost certainly be a need to link different documents together. This can be seen as the equivalent of foreign key references in a relational database. There are two ways of achieving this in a document database. To provide some examples, assume that we have (highly simplified) documents relating to students and their grades in university courses.

The first approach to linking courses and students with customers is embedding. In general, if Document1 needs to be linked to Document2, this can be achieved by nesting one document within the other. In this example, we'll assume that the Course document contains the Student details.

```
{
  "Courses": [
      {"_id": "DB102",
      "courseName": "Databases",
      "creditValue": 15,
      "Students": [
            {"studentId": "12345", "name": "Abby Akbar", "grade": 3.4},
            {"studentId": "67890", "name": "Bill Briggs", "grade": 3.2}
            ]
      },
      {"_id": "SW101",
      "courseName": "Python",
      "creditValue": 10,
      "Students": [
            {"studentId": "12345", "name": "Abby Akbar", "grade": 3.8}
            ]
       }
  ]
}
```

It would also be possible to embed a list of Courses into a Student document, using the standard JSON structuring. Which of these would be preferable depends on the anticipated usage of the data. In this case, although we have two different kinds of information (Courses and Students), we end up with only one document collection: either Courses (as above) or Students, depending on how we decided to embed the documents.

It's clear that embedding the data in this way will result in redundancy, corresponding to denormalization in a relational database. Given the falling cost of hardware, the need for additional storage space might be acceptable. More serious is the problem of maintaining consistency. In the example above, if we realize that Abby Akhbar's name may have been misspelled, we must locate all the Course documents that may include her name to make the necessary changes. If this is not done rigorously, the result is an unreliable data store that is very difficult to manage. If, instead, we had embedded Courses into Students, an update to the Course details would entail accessing all Student documents.

The alternative approach to *embedding* is *referencing*. This concept is much closer to the way that foreign key references are used in a relational database. Using the same example as above, we could put Course documents and Student documents into different collections, as in the following example.

```
{
  "Courses": [
        {"_id": "DB102",
        "courseName": "Databases",
        "creditValue": 15,},
        {"_id": "SW101",
        "courseName": "Python",
        "creditValue": 10,}
  ]
}

{
  "Students": [
        {"_id": "12345",
        "name": "Abby Akbar",
        "courses": [ {"courseId": "DB102", "grade": 3.4},
             "courseId": "SW101", "grade": 3.8} ]
        },
        {"_id": "67890",
        "name": "Bill Briggs",
        "courses": [ {"courseId": "DB102", "grade": 3.2} ]
        }
  ]
}
```

Here, the two collections contain a list of Course documents and a list of Student documents, respectively. The relationships between students and courses are recorded by references from Student documents to the identifiers of Course documents. This avoids the data redundancy of the previous example, but introduces other issues. For instance, if we want to obtain the total value of credits gained by a student, we have to read additional Course documents to get the necessary information. This is roughly the equivalent of a join in a relational system.

The relationship between Courses and Students is many-to-many (so in the example above, a Student document needs to contain references to multiple Course documents). We could resolve this by inventing a new Course-Student document, equivalent to a linking table in a relational system. However, this could be rejected as an unnatural artifice. Thinking of a paper-based equivalent in the real world, it's quite plausible that there could be Course documents and Student documents, but a Course-Student document would be unlikely.

In designing a document database, the choice between embedding and referencing depends on finding the right balance between several competing factors.

First, we need to identify the significant document types. If we are starting from scratch, we can develop an object-role model (ORM) to clarify the information structure relevant to our concerns. From this, we can use relational mapping (as discussed in Chapter 11) to derive a set of candidate document types that could be represented in the appropriate JSON format. Alternatively, we may want to base our design on preexisting documents, respecifying these in JSON format if they have some other form (CSV, XML, ...).

The next set of decisions is highly dependent on the expected usage of the documents. Broadly speaking, we want a document to be as big as it needs to be but not bigger. This may indicate that some candidate documents should be split to improve performance (there may be a database-imposed size limit on individual documents, but this is usually large enough not to be a concern). In thinking about relationships between documents, we need to know the relative frequency of the usual Create, Read, Update, Delete (CRUD) operations on the data. It's quite common for Read operations to dominate, but that's not always the case.

If we consider two document types, A and B, that are related in some way, here are some of the kinds of scenarios that might be relevant.

- If A and B are often queried together, consider embedding A in B, or B in A (whichever gives the optimum performance).
- Long lists are inefficient, so if embedding B in A results in long lists of Bs, consider referencing instead.
- If B is rarely queried in isolation, consider embedding B in A.
- If B is embedded in A, but B is associated with many other document types, consider referencing instead.
- If B is embedded in A, and B is updated much more frequently than A, consider referencing instead.

Partitioning

Partitioning in a document database usually refers to horizontal partitioning. Something akin to vertical partitioning could be achieved by splitting a document type to make two new document types, each containing a different subset of the original document fields, but this is an intrinsic part of the original database design. Horizontal partitioning can be based on any part of a document, but the choice is determined by the way that the database is expected to be used. Each partition, or "shard", can hold a subset of the documents in a collection and is physically distinct from other shards. A crucial decision is the selection of the field or fields to be used as the basis for distribution. As with other aspects, this will depend on the expected pattern of usage of the database. For example, if queries often focus on data within a given year, it may be appropriate to assign a separate shard for each year and use a relevant date field as the basis for distribution.

Each document in a collection is assigned a *shard key* value based on the relevant field(s). The shard key should be chosen to distribute data evenly across the shards to prevent hotspots and ensure balanced performance. Each shard in the database maintains metadata that maps ranges of shard key values to specific shards. When an operation is performed, the database client includes the shard key value in the query or command sent to the cluster. This is handled by a specialized router that examines the shard key, determines which shards need to be involved in executing the operation, and then formulates a plan for executing the query across the relevant shards.

Assuming a query involving several shards, the router sends the parts of the query to the appropriate shards based on the shard key range for which the shard is responsible. Each shard locally executes its part of the query and returns partial results to the router containing the subset of the overall results specific to that particular shard. Once the router receives results from all shards, it combines them to form a complete result set. This may involve sorting, merging, and applying any additional query operations (e.g., filtering) to match the original query criteria. Finally, the router sends the merged result set back to the client application, providing a unified view of the query results as if it had been executed on a single nonsharded database.

The collection and merging of distributed results obviously impose a performance penalty, so the choice of shard key needs to be considered carefully. Ideally, the partitioning should be designed so that most queries can be addressed to a single shard within the cluster.

Indexing

As with Relational databases, document databases can use indexing to improve the performance of queries by creating data structures that allow for efficient data retrieval. As with other aspects of schema design, it's important to have a clear strategy for indexing. Although indexing can improve query performance, it will also have an impact on write performance and will increase storage requirements.

The application(s) using the database should be examined for the type of queries used and their relative frequency. Based on the number and type of operations, the appropriate index types can be identified. Several types are available, including:

- Single Field Index: The most basic type of index where queries are based on a single field.
- Compound Index: Created on multiple fields, combining them into a single index.
- Multikey Index: Separately indexes elements of an array.

There are also other types of indexes for more specialized applications, such as Text, Geospatial, and Hashed indexes.

Indexes are created using an operation to specify the collection and fields to be indexed, along with any additional options that might be required, such as creating the index in the background to mimimize disruption. For example, we could create a Single Field index on a courseName field in a Courses collection as follows:

```
db.Courses.createIndex( { courseName: 1 } )
```

Here, the digit 1 indicates that the index should be in ascending order; descending order would be indicated by a – 1 parameter.

It's also important to handle updates and deletes carefully in document databases. Unlike relational databases, document databases do not enforce referential integrity or cascading deletes. It's up to the developer to ensure that updates and deletes are performed correctly, without leaving orphaned data or breaking relationships.

Referential integrity is a concept commonly associated with relational databases, where relationships among tables are maintained and enforced using foreign keys. In a document database, which follows a NoSQL data model, referential integrity is not natively provided in the same way. However, referential integrity-like patterns and mechanisms can be employed to ensure data consistency and relationships among documents. Embedding and Referencing can provide some degree of integrity, as described above. Additional measures can include:

- Transactions, allowing multiple operations on different documents atomically to ensure data consistency across the affected documents. While transactions do not enforce referential integrity, they can help maintain data consistency among related documents.
- Triggers and Hooks, which allow custom logic to run before or after specific database operations, such as insert, update, or delete. These can be used to implement custom referential integrity checks and ensure that related documents remain consistent during data modification operations.
- Application code, external to the database, can be augmented with additional functionality to ensure that data modifications conform to the desired relationships and integrity constraints before persisting them to the database.

Database Queries

Because of their NoSQL status, document databases do not include the familiar **select** statement from SQL. However, they do provide equivalent facilities in their own query language. We will use MongoDB as an example (https://www.mongodb.com) and its MQL query language. The language allows similar functionality through functions that allow operations such as selection, projection, filtering, sorting and aggregation. Here, we will look at a few simple examples to give an impression of the scope available.

Selection is handled through a **find**( ) function, which contains the query criteria expressed as a JSON object. For example, we find relevant documents in a Courses collection as follows:

```
db.Courses.find( { creditValue: { $gt: 12 } } )
```

In this example, $gt is a query operator used to specify "greater than", so the function would return Courses with a credit value greater than 12. Other common query operators include **$eq, $ne, $gt, $lt, $in, $exists**, etc.

Projection allows fields to be included or excluded in the query results using projection. A parameter is used to specify the fields to be included (1) or excluded (0).

```
db.Courses.find( { }, { courseName: 1, _id: 0 } )
```

In this example, the names of courses would be returned, but the associated _id field would be suppressed.

Query results can be sorted based on one or more fields. The **sort( )** method specifies the fields to sort and the sort order, either ascending (1) or descending (−1).

```
db.Courses.find( ).sort( { creditValue: -1, courseName: 1 } )
```

Aggregation provides the ability to perform data transformations and calculations. The aggregation function can be applied in a simple way, as in the following.

```
db.Courses.aggregate( [ $avg { "$creditValue" } ] )
```

This provides the average creditValue for all Courses. However, the aggregate function is much more powerful than this because it can contain a "pipeline", a list of operations arranged so that the output of one operation becomes the input to the next operation. This allows complex operations combining grouping, summing, sorting, averaging, and so on. A typical sequence might be:

```
$project –> $match –> $group –> $sort
```

nested within the **aggregate**( ) function.

In general, we can see that the functionality available is similar to that provided by relational systems, though realized differently. The attractions of the document model include flexibility in the representation of data and the possibilities for partitioning. Against this must be set issues such as the need to maintain the consistency of replicated data and the complexity of transactions involving multiple documents.

Document databases are an obvious choice where the data to be stored can be visualized in terms of groups of values that share a common theme or purpose. This is certainly the case where the database is being used to store the electronic equivalent of traditional business documents such as statements, application forms, specifications, and so on. Other application areas where document databases are popular include:

Content management systems (CMS), used to store and manage various types of material such as articles, images, videos, and metadata used in support of websites and other applications.

Management of catalogs, also directories or listings such as business directories, real estate listings, and online marketplaces, which can benefit from the ability to store diverse data structures within a single database.
Storing user profiles and preferences to enable personalized experiences in applications such as social networks, online forums, and streaming platforms.
Managing healthcare information such as patient records, medical histories, test results, and treatment plans that form well-bounded units of information.
E-commerce platforms, used to manage product catalogs, user profiles, customer reviews, and order histories while providing flexibility to accommodate changing product attributes and relationships.

17.5 Graph Databases

Graph databases use a property graph model to represent and store data. The basic graph structure consists of nodes and relationships (edges). Each node represents an entity, object, or data point, and each relationship represents a connection or association between two nodes.

Nodes are the fundamental units of a property graph. Each node can have one or more properties that store data attributes or characteristics associated with the node. For example, in a social network graph, a node could represent a person, and node properties could include attributes related to that person.

Relationships define the connections or associations between nodes. They represent the links or interactions between entities. Relationships can also have properties that provide additional information about the connection. For instance, in a social network graph, relationships might represent "friendship" connections between people, and relationship properties could include the date the friendship was established.

Properties are key-value pairs associated with both nodes and relationships. They store specific attributes or data values related to the corresponding node or relationship. For example, a node representing a person might have properties such as "name", "age", and "email".

Graph databases offer several advantages over relational databases, especially when dealing with highly connected and complex data structures. In a relational database, it's possible to imagine a connection between a pair of tables by following the relationships between intermediate tables. However, if the two tables are widely separated (i.e., there are many intermediate tables), then the query becomes complex, and the cost of the multiple joins can be significant. Traversing relationships in a graph is a direct operation, which makes querying for highly connected data very fast. As the size of the graph grows, graph databases maintain consistent performance, making them scalable solutions for applications with expanding relationships.

Graph databases typically have a flexible schema, allowing nodes and relationships to have different properties without strict constraints. This adaptability can be beneficial in scenarios where the data model evolves over time, as changes can be made without altering the entire schema, which can be cumbersome in a relational database. Relationships between data points are stored directly, making the data model more intuitive and easier to work with when the relationships are the primary focus of the application. They also represent hierarchical data structures (which are common in some domains) in an efficient way.

The schema flexibility in graph databases makes them well suited for scenarios where data models are subject to change, relationships are dynamic, and applications require adaptability and agility in data representation. However, a more flexible schema implies fewer constraints in the database, and so structure and constraints may need to be defined through the application logic, rather than through a centralized schema definition enforced by the database itself.

In terms of database products, Neo4j is often considered the market leader, partly due to its early adoption and strong ecosystem. Amazon Neptune also has a notable presence due to its integration with other AWS services. Other graph databases such as ArangoDB, JanusGraph, and Stardog have their own niches and user bases, and some are able to integrate with other types of NoSQL systems such as document and column-oriented databases.

To provide some concrete examples of graph database features, we'll use Neo4j (www.neo4j.com). Its underlying graph processing engine is optimized for traversing and navigating graphs efficiently, making graph traversals and pattern-matching operations fast and scalable. It uses a language called Cipher to write graph queries and to carry out the usual CRUD types of operation. An Enterprise Edition provides advanced facilities for commercial operation, but we will focus on the facilities provided by the Community Edition.

Unlike other types of NoSQL databases, graph databases do not lend themselves to horizontal partitioning. The entire database resides on a single server, so as the database grows the server must be scaled vertically. Performance can be improved by providing a server cluster so that different read operations can be directed to different servers in the cluster to share the load, but updates are made through a master that then replicates changes to other servers in the cluster. Within a server, Neo4j supports ACID transactions to ensure data integrity and consistency.

Although there are some high-level similarities between relational and graph databases, the differences are considerable at the detail level. In the relational case, the most fundamental element is a table, whereas in the graph case, it is a node. However, these correspond to different kinds of concepts. A relational table relates to a particular *type* of object, but a graph node relates to an *instance* of a type. The notion of a node type certainly exists in graph databases; in Neo4j, it is referred to as a "label".

All nodes with the same label can be assumed to have similar properties. If a property value is unknown (or not relevant), a relational database will hold a NULL in the corresponding cell of the table. In the graph database equivalent, the property will simply not exist (i.e., NULLs are not stored).

There is an equivalent dichotomy when we consider the relationships between objects. In the relational database case, this is defined at the table level with foreign key references, perhaps to "join" or "link" tables for many-to-many relationships. This cements the relationship at the *type* level. In a graph database, relationships are defined between nodes at the *instance* level, which does not depend on the multiplicity of *types*, so relationships between instances are simply relationships.

This has implications when we consider the kinds of relationships that can be accommodated in a database. In the relational case, all instances (rows in a table) must comply with the same relationship structure, but in the graph case, different nodes with the same label could have different relationships. Comparing this back to a relational system, it's as though each row in a table could have its own individual foreign key references that differ from other rows in the table. The graph approach provides great flexibility to express complex relationships, but complicates the documentation needed to allow subsequent explorations of those relationships.

Another difference is the approach to data typing. SQL defines a range of data types, and all values in a table column are constrained to be of one specific type such as **int**, **char**, **float**, and so on (or perhaps null if a value is absent). In the community edition of Neo4j, the properties of nodes and relations are not preconstrained to a particular data type. Values have a type (string, integer, and so on) but properties do not, so, for example, a property may hold an integer value that later gets updated to a string value.

Graph Database Queries

Data querying in a graph database involves retrieving information from the graph by specifying patterns of nodes and relationships that match the desired data. Graph databases typically have their own query languages optimized for graph traversal and pattern matching, and Neo4j provides a language called Cypher.

Queries in a graph database often involve matching specific patterns of nodes and relationships. Cypher can be used to specify patterns, filter nodes based on properties, and include or exclude relationships based on certain criteria. Data can be retrieved using the **match** keyword, which is roughly the equivalent of **select** in SQL. The following query returns all the properties for a particular Customer node.

```
match ( c:Customer {customerID:"ABCDE"} ) return c
```

Indexing can be used to allow faster lookup of nodes and relationships based on specified properties, although this has the usual effect of slowing updates.

Cypher queries can include conditions based on node and relationship properties. This allows data to be filtered based on the values of attributes associated with nodes and relationships in a similar way to SQL, for example, by using a **where** clause. Results can also be organized in a similar way to SQL, for example, by using an **order by** clause. Aggregation functions are available to summarize data across nodes and relationships, again in a similar way to SQL using functions such as **count( )**, **sum( )** and **avg( )**.

A particular strength of graph databases is the ease of graph traversal, which involves navigating from one node to another through relationships to find data that matches certain criteria or patterns. Paths between nodes can be specified as a fixed length or a variable length with specified limits. A library of functions is provided to implement many graph algorithms, such as breadth-first and depth-first search, shortest path, page rank, and others. Queries of this kind are much harder to implement in a relational database.

Before data can be entered into a relational database, it must first be populated with tables, which, in turn, will have been created in accordance with some schema. A graph database does not require this kind of preparation, as nodes and edges can simply be added as needed. In a sense, we could say that the schema of a graph database emerges from the create, update, and delete operations carried out on it.

This should not be taken too far. A completely ad hoc approach can lead to a database that does not give reliable results and is impossible to maintain. One of the important benefits of a schema is that it defines a vocabulary for a given application area so that all participants can communicate in a consistent way. For example, we could have the following facts in a graph database:

```
Holmes – WATCHED -> Watson
Holmes – OBSERVED -> Watson
Holmes – NOTICED -> Watson
Holmes – RECOGNIZED -> Watson
```

Each of these could be a plausible record of some event. Listed together in this way, the problem is obvious, but in practice, the individual entries could be made by different people at different times who may not be aware of other data that has been entered. Can we query the database to confirm that Holmes SAW Watson? Not according to this data!

Populating a Graph Database

In SQL, objects are created by an **insert** statement that adds the relevant values to a table. Relationships are created in the same way by inserting the destination key in the source table, so no separate insert is required. In a graph database, nodes and relations are separate items and so are created separately.

The following example shows the general style of node creation:

```
merge ( :Customer { customerId: "ABCDE" } )
```

This creates a node with a label (i.e., node type) of Customer with a given property. The individual node can be identified by using the customerId property in any subsequent queries. The keyword **create** can be used in place of **merge**, the difference being that if a matching node already exists (in this case, a Customer node with the same customerId), **merge** will do nothing but **create** will introduce a duplicate node. A constraint can be added to enforce uniqueness of a property value. Adding the constraint automatically creates an index on the unique property.

```
create constraint cust_id_uniq for (c:Customer)
require c.customerID is unique
```

In a relational database, the absence of a value is indicated by a null in the relevant table cell. In a graph database, null has a slightly different interpretation. Properties do not have null as a value; null just indicates that the property doesn't exist. This means that nonexistence and null are equivalent for properties of nodes and relationships. A property can be constrained to be not null, which in Neo4j is equivalent to saying that it must exist.

```
create constraint cust_existence for (c:Customer)
require c.customerID is not null
```

The combination of uniqueness and existence constraints on a property provides something roughly equivalent to a primary key in a relational database.

The same **merge** or **create** keywords are used to insert relations. The Cypher notation allows the direction of the relation to be specified, as in the following example:

```
merge (:Customer {customerID:"ABCDE"}) -[:PURCHASED]-> (:Order {orderID:12345})
```

This will create the specified Customer and Order nodes together with a directed PURCHASED relationship between them. Adding a relationship between two *existing* nodes requires a slightly different strategy. We first find the relevant nodes by querying the database, and then use references to those nodes in creating the relationship, as in the following example.

```
match ( c:Customer { customerID:"ABCDE" } )
match ( o:Order { orderID:12345 } )
merge ( c ) - [ :PURCHASED ] -> ( o )
```

For large datasets, batch insertion methods can be used to insert multiple nodes and relationships in a single transaction. Utilities are also provided for the bulk import of data from external sources, such as CSV files or other databases.

Updating and Deleting

One of the most common update operations is modifying the properties of nodes and relationships. The relevant nodes or relationships can be found through a query, and the update applied to the results of the query. As with creation, batch update methods can be used to update multiple nodes and/or relationships in a single transaction. The following example shows a simple Cypher update data in a graph database:

```
match ( c:Customer { companyName: "Tamale Spice" } )
set c.country = "Mexico"
```

Here, the variable `c` is bound to a specific Customer, and its `country` property is then set to the specified value. If the property value already exists, it will be replaced by the new value.

If the property does not exist, it will be created. An existing property can be deleted completely by removing it. This is equivalent to setting the relevant table cell to null in a relational database.

```
match ( c:Customer { companyName: "Tamale Spice" } )
remove c.country
```

A complete node (or set of nodes) can be deleted along with all associated properties using appropriate filter conditions.

```
match ( c:Customer )
where c.country = "Mexico"
delete c
```

This will have the effect of deleting all data for Customers located in Mexico, and of course:

```
match ( x ) delete x
```

could delete all nodes in the database, so it's important to get the filtering conditions correct.

Comparison With Relational Databases

Table 17.2 lists some comparisons between the two database types (taking Neo4j as representative of graph databases).

Table 17.2 Comparison of relational and graph databases.

Relational database	Graph database
Tables	Labels
Rows	Nodes
Columns	Node or Relation properties
Primary keys	Uniqueness constraint on node properties
Foreign keys	Relations
"Join" or "Link" tables	Relations: columns become properties of the relation
Static data typing	Dynamic data typing

To illustrate the differences in structure between the two approaches, we can look at a typical customer-order-product scenario. This is similar to the well-known "Northwind" example database, but only a few properties are shown here to make the model more compact.

Fig. 17.9 shows a fragment of the table structure. The OrderDetails join table is used to resolve the many-to-many relationship that exists between Orders and Products.

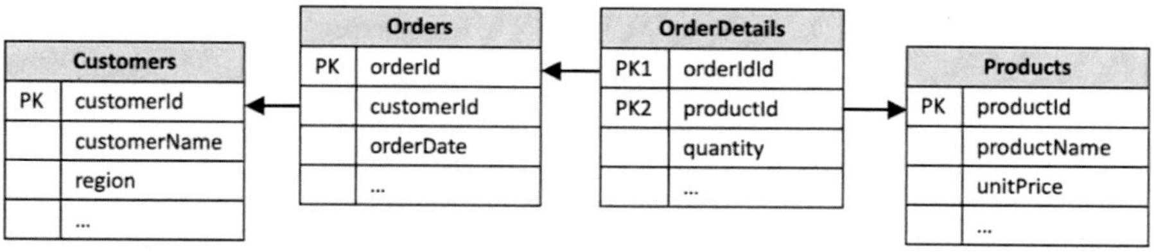

FIG. 17.9 Customer-order-product relational structure.

The graph database equivalent of this structure that contains the same information is shown in Fig. 17.10. The properties associated with each element are shown for clarity. Note that there is no need for an OrderDetails node type between Orders and Products. In the graph model, each Order-Product combination is a separate relation, and the associated quantity value is simply a property of that relation.

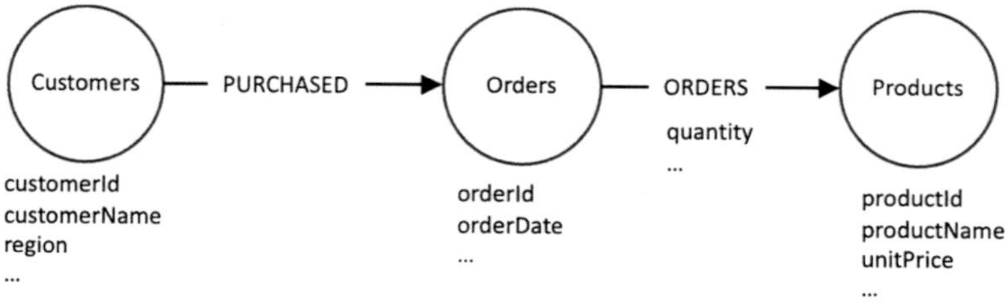

FIG. 17.10 Customer-order-product graph structure.

Assume that we need to know the names of all customers who have ordered the product "Tofu". In the relational case, we can express this in SQL using the following query:

```
select distinct customerName from Customers
join Orders on Customers.customerId = Orders.customerId
join OrderDetails on Orders.orderId = OrderDetails.orderId
join Products on OrderDetails.productId = Products.productId
where Product.productName = "Tofu"
```

Here, the **distinct** keyword is used to avoid duplicates if the same customer has ordered the product more than once. The equivalent query in Neo4j can be expressed in the Cypher language as follows:

```
match ( c:Customer ) - [ :PURCHASED ] -> ( :Order ) - [ :ORDERS ] -> ( p:Product )
where p.productName = "Tofu"
return distinct c.companyName
```

In the Cypher query, the variables c and p are bound to nodes with labels Customers and Products, respectively. No bindings are required for Orders nodes or the two relationships.

Mapping ORM to a Graph Database

Chapter 11 discussed the mapping of an ORM model to a relational structure and presented a systematic procedure (Rmap) to produce a redundancy-free relational design. We'll now consider the possibility of an equivalent approach for a graph database.

Fig. 17.11 shows a simple ORM model (constraints omitted, as discussed later) and a comparable graph model showing node labels. Leaving aside the obvious differences in graphical notation, it's clear that the two models are very similar in structure.

At this simple level, it appears that an ORM model could potentially be taken as a direct definition of a graph schema. Unfortunately, things become more complicated when we get into the detail. The first difficulty is that graph databases can only represent binary relationships.

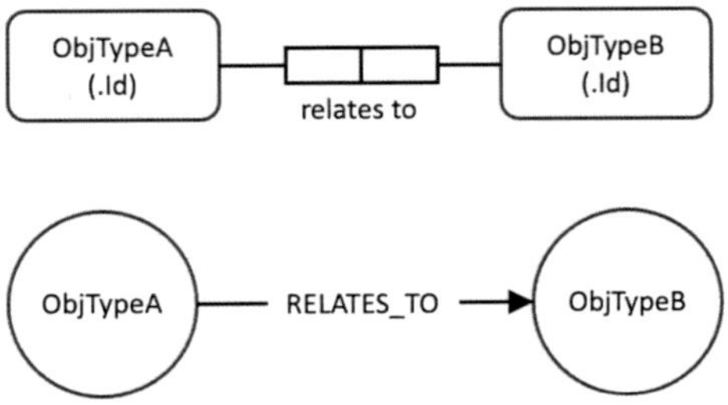

FIG. 17.11 Comparison of simple ORM and graph structures.

An ORM fact type of any other arity must be mapped to something structurally different in a graph model.

A unary fact type can be mapped to a graph property (normally a node property) or a binary relationship in a fairly straightforward way. Ternary or higher fact types must be remapped to binary relationships, perhaps involving multiple binary relationships for one ORM fact type. This may require additional constraints to preserve the intent of the original ORM model.

However, this raises the next difficulty, which is that graph databases generally provide very limited constraint options. In Neo4j, the only options are uniqueness and nonnull constraints (although a few more are available in the Enterprise Edition). It may be possible to work around these by taking care in the framing of create or update statements to preserve an appropriate structure, but for many ORM constraints, the only realistic solution is to implement the constraints in the application, not in the database.

A final difference in capability is the notion of subtypes. It is certainly possible to express subtyping notions in graph form, for example:

```
( :Species ) – [ :IS_MEMBER_OF ] -> ( :Genus ) – [ :IS_MEMBER_OF ] -> ( :Family ) ...
```

but the relationship is not distinguished in any way from other types of relationships. Similar considerations about absorption, separation, and partitioning arise when deciding if subtypes can be combined under one label or if different node labels are preferable. Any constraints involved in an ORM subtype relationship face the same issues as n-ary relationships and may need to be enforced in the related application code.

Despite these difficulties, the similarities between an ORM schema and a graph schema remain strong, and it's possible to outline a systematic procedure for ORM-to-graph mapping, which we could call Gmap. This could run on similar lines to the Rmap procedure, following a series of steps to coerce the result into the appropriate form.

This is best illustrated with a concrete example. Chapter 11 discussed the application of the Rmap procedure to several models, including the TV channel model reproduced here as Fig. 17.12. This can provide the basis for a useful comparison between the two approaches. This model shows a number of features that must be accommodated in any possible Gmap procedure.

Any one-to-one fact type can be resolved by combining the two object types so that one becomes a property of the other. Which object type should be considered the "parent" is usually obvious from the model, but in some cases an arbitrary choice must be made. In this model, we can make Frequency a property of Channel and ProgramTitle a property of Program. This is essentially the same strategy as in the Rmap procedure.

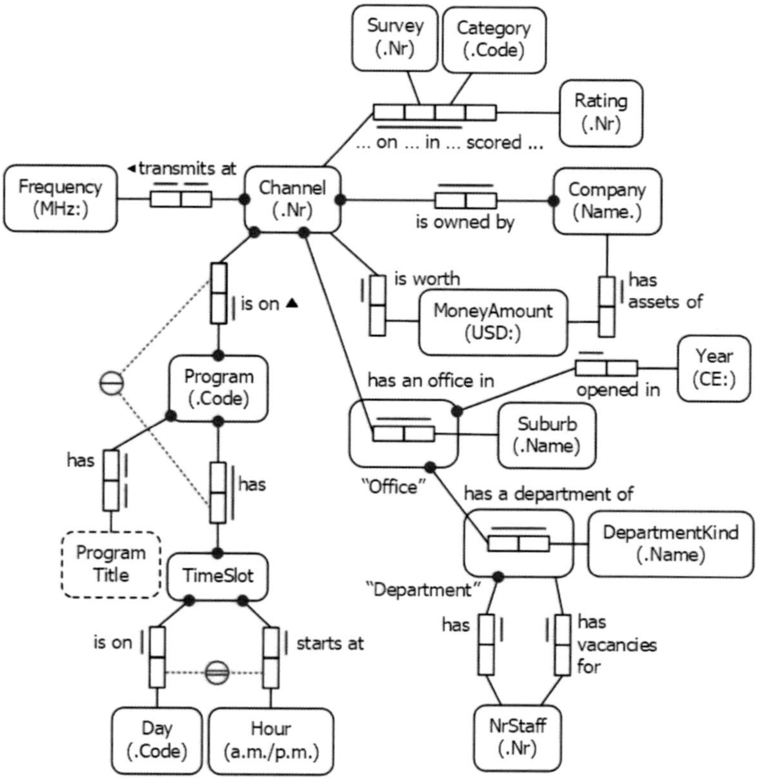

FIG. 17.12 An ORM schema for TV channels (repeated from Fig. 11.30).

The TimeSlot object type in the ORM model has a composite identifier: Day and Hour. Composite identifiers may be difficult to replicate in a graph database, so it might be necessary to introduce a surrogate identifier. In this case, we could form a timeSlotCode by concatenating Day and Hour, but it's not necessary because the TimeSlot is only relevant in the context of a particular Program.

Fact types with spanning uniqueness constraints will become relations in the graph model. For ternary or higher constraints, a choice must be made as to the best labels to use as the two ends of the relation because relations can only be binary in the graph model. The other roles will become properties of the relation. For the quaternary fact type in this model, we can set a relationship between Survey and Channel, with Category and Rating as properties.

Objectified fact types also map to relations, with any functionally dependent fact types indicating properties of the relation. This will apply to "Office" and "Department" in this case. A functional dependency between two objectified fact types introduces a further complication, to be discussed shortly.

Fact types with functional roles attached to the same object type can be grouped into the same node type (label), becoming properties of the "parent". In this model, the MoneyAmount object type can become a "worth" property of a Channel node and an "assets" property of a Company node.

The result so far is summarized in Fig. 17.13. In this figure, tentative node labels are shown in bold, and properties of nodes and relations are shown in italics to make the distinctions clearer.

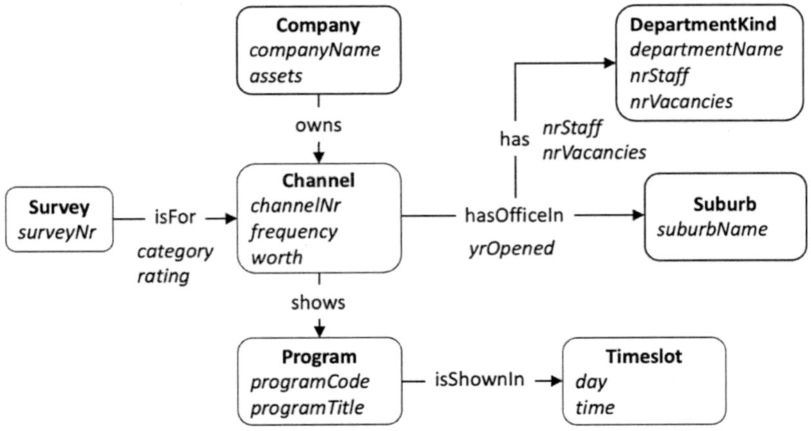

FIG. 17.13 Initial modification of the TV channel model.

We can immediately see a problem. A relation must be between two nodes and cannot be between a relation and a node. We can resolve the problem by moving the DepartmentKind relationship to Suburb. Recall that a graph schema is never created explicitly, as it is simply a high-level summary of labels for nodes and their relationships created in the database.

We can have multiple nodes with the same Suburb label but with different names. This allows a channel to be located in multiple suburbs, and multiple channels to be located in the same suburb. Similarly, a suburb can have multiple department kinds, and the same department kind can be in multiple suburbs. Fig. 17.14 shows a small example.

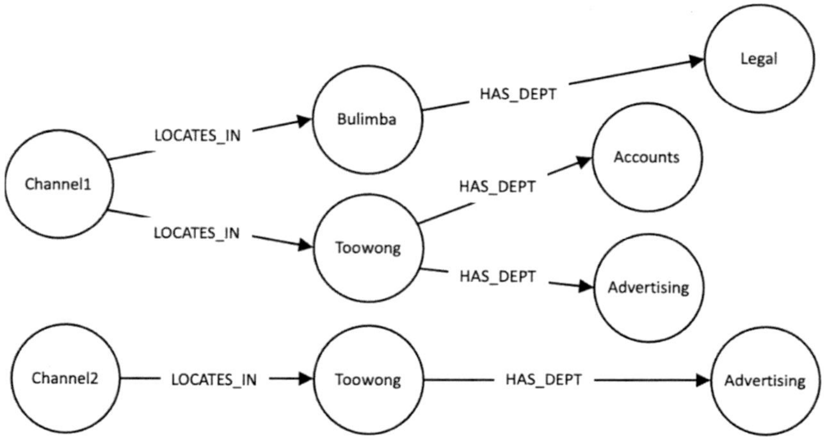

FIG. 17.14 Example of Channel, Suburb, and Department nodes.

The final graph schema is shown in Fig. 17.15. All the data items from the original ORM model are recorded as properties of nodes or relations and all fact types are covered explicitly as relations or implicitly by structure.

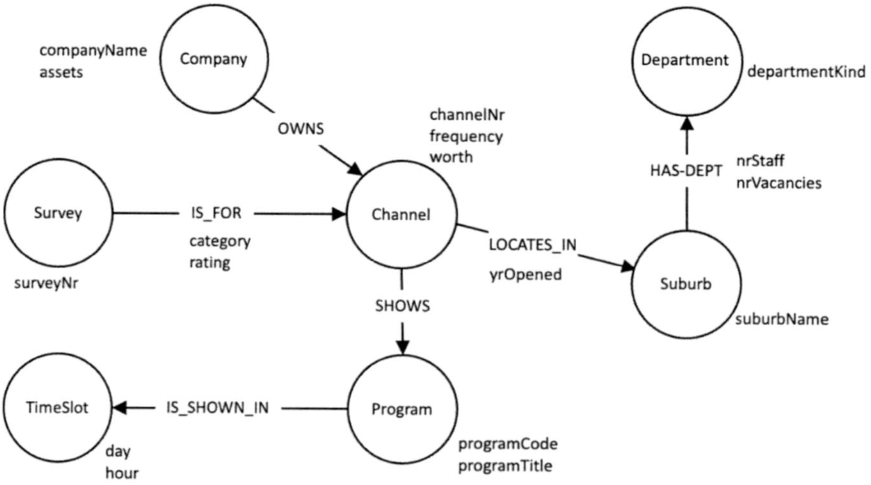

FIG. 17.15 Graph schema for TV channel model.

We also need to consider the constraints specified in the ORM model. This is potentially problematic because of the very limited constraint options available in the graph database equivalent. Some workaround possibilities exist through careful framing of statements that modify the structure, but this may not always be possible.

Mandatory constraints can be enforced to some extent by requiring that, for example, a property of a label is not null, but this can be a nuisance if the property value is not known when the relevant node is created.

Uniqueness constraints are available in Neo4j and can be used where appropriate. In the model above, the identifying properties for Company, Channel, Survey, and Program can be defined as unique. This means that, for example, there will only be one node with the label "Channel" and a ChannelNumber property equal to "1". However, this should not be done in the case of Suburb, Department, or TimeSlot because these labels can have multiple nodes with the same property, as can be seen in Fig. 17.14. These nodes can be distinguished from each other by having different instances of their relations with other nodes.

The external uniqueness constraint between the Channel role and the TimeSlot role is more difficult. We need to ensure that a TimeSlot can only relate to one Program for a given Channel, although the same TimeSlot can be used by other channels for their own programs. There is no constraint in Neo4j to express this, but we could specify the creation of the `IS_SHOWN_IN` relation between a specific Program and a specific TimeSlot using the following strategy:

1. Use a query to find the Channel that shows the specified Program.
2. Use a query to find any Program shown by that Channel in the specified TimeSlot.
3. If the result is null, create the `IS_SHOWN_IN` relation between Program and TimeSlot.

ORM has a very rich constraint language, and it may not be worthwhile to try to write complex Cipher code to capture the full intent of the ORM model. In such cases, it may be better to delegate the constraint to the application code instead of trying to enforce it in the database.

It's clear from this example that there is a close affinity between ORM and graph databases, which is perhaps not surprising because an ORM model is itself a graph. Graph databases offer a useful alternative to relational systems in applications that are focused more on relationships than data processing. They share many of the same issues on scalability as relational systems, but can provide high availability by replicating servers.

Translating an ORM model into a graph database can be done in a systematic way that preserves most of the intent of the ORM model, though some creativity may be required in dealing with constraints. For a specific graph database system, it would be feasible to automate a Gmap procedure to generate a schema design, possibly including some code (in the appropriate language) to deal with issues such as constraints.

Probably the most important feature of graph databases is their ability to follow links through a connected network of data, perhaps extending over long chains of connections.

Many common problems can be expressed in terms of searches through data space, and there are well-known algorithms to perform this in an efficient way. These capabilities make graph databases a popular choice in application areas such as the following:

- Modeling and querying social network data and other interpersonal relationships.
- Creating recommendations for users of a system by analyzing their preferences and activities.
- Identifying patterns of fraudulent behavior by analyzing connections between entities.
- Building knowledge graphs that represent semantic relationships between entities.
- Analyzing connections between individuals, organizations, and events in criminal investigations.

17.6 Other Nonrelational Databases

A conceptual schema may be mapped to various logical data models. The Rmap procedure discussed in Chapter 11 assumes that the application platform is a centralized, relational DBMS. It is a fairly simple task to specify other procedures for mapping to other types of DBMSs, as we've seen earlier in this chapter.

Although relational DBMSs have long been dominant in the commercial marketplace, many alternatives exist. These include older systems based on the hierarchic or network data model as well as newer alternatives such as those discussed previously. While purely relational DBMSs suit most business applications, they may not be the ideal match for some specialized applications such as computer-aided software engineering (CASE) tools, computer-aided design tools (e.g., VLSI design, mechanical engineering, architecture), document processing, spatial databases, expert systems, scientific databases (e.g., genetic engineering), and communications management. Note that most of these applications involve complex objects and probably account for at most 10% of database applications.

Many reasons are cited for dissatisfaction with *purely* relational DBMSs for complex applications. They may be too slow–they don't perform well with complex objects mainly because they require too many table joins. They often model objects in an unnatural way (e.g., information about a single object such as a person may be spread over several tables–this gets worse with complex objects). They are dependent on value-based identifiers. They don't facilitate reuse (e.g., they have no direct support for subtyping). They require access to a procedural language for difficult tasks (e.g., special rules and behavior), which leads to an "impedance mismatch" with the declarative, set-based, relational query language. They might not support binary large objects (BLOBs) such as images (icons, pictures, maps etc.), sound tracks, video, and so on.

Object Orientation

An *object database* (ODB), or *object-oriented database* (OODB), incorporates various object-oriented features. Historically, ODB research drew upon related developments in four areas: programming languages, semantic data models, logical data models, and artificial intelligence.

With programming languages, the need was seen for user-definable abstract data types and for persistent data. The object-oriented programming paradigm began with Simula in 1962 and Smalltalk in 1972. Nowadays, many object-oriented programming languages exist (e.g., Eiffel), and many traditional programming languages have been given object-oriented extensions (e.g., C++ and C#).

Some object-oriented features were taken from semantic data modeling, which as we know models reality in terms of objects and their relationships and includes notions such as subtyping (e.g., ORM, UML, and extended entity-relationship (ER) modeling). Various ideas were also borrowed from work on logical data models (network, hierarchic, relational, and especially the nested relational model). Finally, some concepts were adapted from artificial intelligence, where structures such as frames are used for knowledge representation.

What features must a DBMS have to count as "object-oriented?" There is no commonly agreed-upon answer. A classic paper ("The OODBMS Manifesto") on this issue was presented at the first international conference on object-oriented and deductive databases (Atkinson et al., 1990). To distinguish ODBMSs from OO programming languages, five essential features of a DBMS were identified, and then object-oriented features were added, of which eight were considered essential and five optional (see Table 17.3). Let's look briefly at the essential object-oriented features in this proposal.

Table 17.3 Feature list in the "OODBMS manifesto" (Atkinson et al., 1990).

DBMS features	*Essential OO features*	*Optional OO features*
Persistence	Complex objects	Multiple inheritance
Secondary storage	Object identity	Type checking and inferencing
Concurrency	Encapsulation	Distribution
Recovery	Types or classes	Design transactions
Ad hoc query facility	Inheritance	Versions
	Overriding and late binding	
	Computationally complete	
	Extensibility	

Complex objects are built from simpler ones by constructors. The manifesto proposed that these constructors should include at least set, list, and tuple, and be orthogonal (they can be applied in any order, recursively). For example, one object might be a list of sets of sets of tuples. Constructors are not orthogonal in the relational model (only sets of tuples of atomic values are allowed) or the nested relational models (e.g., the top-level construct must be a set). The notion of complex objects is considered important because it allows us to model a complex structure in a direct, natural way.

The basic idea behind *object identity* is that objects should be identified by system-generated object identifiers (OIDs) rather than by the values of their properties. This is in sharp contrast to the relational model, where for instance tuples are identified by the value of their primary key.

Among other things, OIDs can help keep track of objects whose external, value-based identification may change with time. This may occur because of simple renaming. For example, television channel 0 becomes channel 10, or a woman changes her family name upon marriage. More drastically, the reference scheme itself may change (e.g., a student identified by a student number becomes an employee identified by an employee number).

Object identifiers in the OO sense partly overcome this problem because they are rigid identifiers (i.e., they refer to the same object throughout time). They are system generated, nonreusable, immutable, and typically hidden. In ODB systems, they might be implemented as surrogates (logical identifiers or autoincremented counters, mapped by indexes to physical addresses), as typed surrogates (which makes migration between types awkward), or as structured addresses.

Encapsulation involves bundling the operations and data of an object together, with normal access to the object being through its operational interface, with implementation details hidden. For example, hiring, firing, and promoting employees are regarded as operations on Employee and are encapsulated with it. Implementations of operations are often called "methods".

Encapsulation includes the idea that, as in a conceptual schema, objects should be classified in terms of *types* or *classes*. Moreover, some form of *inheritance* mechanism should be provided (e.g., so that a subtype may inherit the data and operations of its supertype(s)). A subtype may have a specialized version of a function with the same name as one of its supertype functions. In this case, the specialized version will typically *override* the more general version when the operation on the subtype is invoked.

For example, the display procedure for a colored, equilateral triangle may differ from the general display procedure for a polygon. Various overriding options are possible.

The requirement for *computational completeness* means that any computable function can be expressed in the data manipulation language (if necessary, by calling programming languages). The *extensibility* requirement means that users may define their own types, and the system should support them just like its built-in types.

Partly in response to the OODB manifesto, a group of academic and industrial researchers proposed an alternative "third-generation DBMS manifesto" (Stonebraker et al., 1990). Here, they referred to hierarchic and network systems as first generation, and relational systems as second generation. Under this scheme, third-generation DBMSs were seen as the next generation.

The manifesto specified three basic tenets and 13 detailed propositions to be adhered to by the next-generation DBMSs (see Table 17.4). This proposal essentially argued for extending existing relational systems with object-oriented features and effectively laid the groundwork for today's object-relational database systems, which we'll discuss shortly.

Table 17.4 Next-generation DBMS features proposed by Stonebraker et al. (1990).

Basic tenets	*Detailed propositions*
1. Besides traditional data management services, next-generation DBMS will provide support for richer object structures and rules.	1.1 Next-generation DBMS must have a rich type system. 1.2 Inheritance is a good idea. 1.3 Functions, including database procedures, methods, and encapsulation, are a good idea. 1.4 Unique identifiers for records should be assigned by the DBMS only if a user-defined primary key is not available. 1.5 Rules (triggers, constraints) will become a major feature in future systems. They should not be associated with a specific function or collection.
2. Next-generation DBMS must subsume previous-generation DBMS.	2.1 Essentially all programmatic access to a database should be through a nonprocedural, high-level access language. 2.2 There should be at least two ways to specify collections, one using enumeration of members and one using the query language to specify membership. 2.3 Updatable views are essential. 2.4 Performance indicators have almost nothing to do with data models and must not appear in them.
3. Next-generation DBMS must be open to other subsystems.	3.1 Next-generation DBMS must be accessible from multiple high-level languages. 3.2 Persistent X for a variety of Xs is a good idea. They will all be supported on top of a single DBMS by compiler extensions and a (more or less) complex run-time system. 3.3 For better or worse, SQL is intergalactic dataspeak. 3.4 Queries and their results should be the lowest level of communication between a client and a server.

Although supporting several of the object-oriented features, the group argued against implementing them in such a way as to negate key advances made by the relational approach. For example, all facts in a relational system are stored in tables. In contrast, some facts in hierarchic, network, and object database systems may be specified as links between structures, requiring navigation paths to be specified when the data is accessed.

Hence, OO queries typically require specification of access paths. Although path expressions are often more compact than relational queries, their reliance on existing navigation links can make it difficult to perform ad hoc queries efficiently.

As an example, consider the UoD schematized in Fig. 17.16. Students either hold a degree or are currently seeking one (or both). The exclusion constraint forbids students from reenrolling in a degree that they already hold. Subjects are recorded only for current students, as shown by the subset constraint. Because this is not an equality constraint, students may enroll in a degree before choosing subjects.

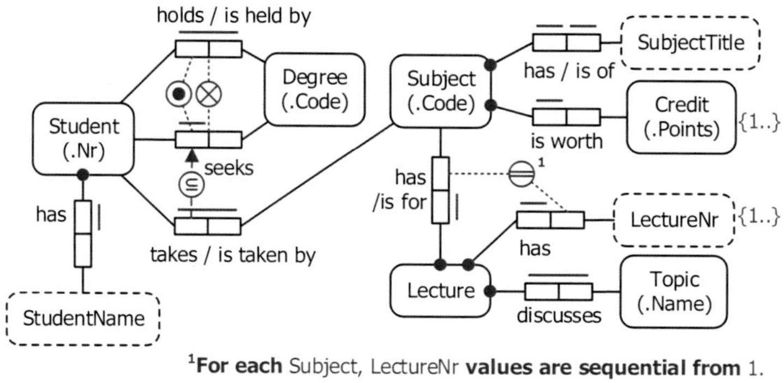

FIG. 17.16 An ORM schema about university records.

This aspect of the schema may alternatively be modeled by introducing the subtype CurrentStudent, to which the optional takes role is attached, using the definition **Each** CurrentStudent **is a** Student **who** seeks **some** Degree.

Each subject is identified by its subject code, but also has a unique title. For some subjects, a lecture plan may be available. This lists the topics discussed in the various lectures. For example, lecture 10 for CS114 might discuss relational projection, relational selection, and natural joins. The footnoted textual constraint qualifies (and implies) the external uniqueness constraint, requiring lectures to be numbered from 1 within each subject.

The conceptual schema of Fig. 17.16 (minus some constraints) might be specified in an ODB schema roughly as follows, using a syntax based on the Object Definition Language (ODL), as specified in Cattell and Barry (2000).

```
class Student {
    attribute unsigned long studentNr;
    attribute string studentName;
    attribute set<string> degreesHeld;
    attribute string currentDegree; };

class CurrentStudent extends Student {
    relationship set<Subject> takes
        inverse Subject::is_taken_by; };

class Subject {
    attribute string subjectCode;
    attribute string title;
    attribute unsigned short credit;
    relationship set<CurrentStudent> is_taken_by
        inverse CurrentStudent::takes;
    relationship list<Lecture> has
        inverse Lecture::is_for; };
```

```
class Lecture {
    attribute unsigned short lectureNr;
    relationship Subject is_for
        inverse Subject::has;
    attribute set<string> topics; };
```

Although readable, this OO schema is incomplete and needs additional code for the extra constraints in the conceptual schema. We could display this on a UML class diagram, with the extra constraints in notes (e.g., exclusive-or, pair exclusion, and subset constraints). In general, OO schemas are best developed as abstractions of ORM schemas.

Notice that the OO classes correspond to the major object types abstracted from the conceptual schema. Hence, they provide a useful basis for building screen forms for the application. Encapsulation involves adding generic operations as well as type-specific operations (e.g., a graduate operation might be added to Student).

One problem of the ODB approach is that it mixes too many levels together–an OO schema includes conceptual, logical, and physical elements. It also involves redundant specification of associations. For example, consider the fact type Student takes Subject. In the OO schema, this is specified twice: once on CurrentStudent and again on Subject.

In specifying the OO schema, inverses were used to declare bidirectional object links between CurrentStudent and Subject as well as between Subject and Lecture. This enables queries in either direction to be specified using path expressions. This versatility is lost if the link is made unidirectional. Notice that no such navigational problem can exist in the relational model, although of course many queries will require joins. While joins might slow things down, there is no restriction on their use because they do not require links to be set up beforehand. In contrast, the ODB approach obtains its efficiency by "hardwiring" in the links to be used in queries. This makes it difficult to optimize ad hoc queries.

The problem of relying on predeclared access paths in the schema itself to achieve efficiency is exacerbated when the schema evolves. We may then have to reset navigation pathways to optimize them for the new situation rather than simply relying on the system optimizer to do the job, as in current relational systems.

This is not to say that ODBs are a bad idea, or that complex objects should not be modeled as such at the implementation level. Many complex structures are awkward to model in relational terms. However, we should be able to specify such structures in a clean way, without resorting to low-level mechanisms.

Unlike ODBs, relational databases are based on a single data model formally based on predicate logic. Moreover, relational DBMSs are now dominant, and many are being extended to address several of the deficiencies mentioned earlier. Such extended relational database systems are usually called *object-relational database* (ORDB) systems. These are essentially relational DBMSs extended by adding support for many OO features, such as extra data types (spatial, image, video, text ...), constructors (arrays, sets, ...), and inheritance. Some of the major commercial systems support extended data types by allowing modules to be plugged in. Such modules may be developed by the vendor or a third party. These modules are variously called "relational extenders", "datablades", or "data cartridges".

The SQL standard has been progressively extended significantly to include several object-oriented features (e.g., user-defined types and functions, encapsulation, support for oids, subtyping, triggered actions, and computational completeness) as well as deductive features (e.g., recursive union).

Given the massive installed base of relational DBMSs, the ongoing extensions to these products, the cost of porting applications to a new data model, and the widespread adoption of the SQL standard, it may well be that the next generation of DBMSs will evolve out of current relational products. Just as Hinduism absorbed features from other religions that threatened its existence, the relational model can probably absorb the interesting features of the object-oriented faith without being replaced by it.

It's debatable whether all the transformations taking place in SQL-based DBMSs are actually desirable. While the use of extended data types (spatial, video, etc.) is clearly a step forward, the use of constructors (arrays, multisets, etc.) is questionable in many cases because they can make it harder to design efficient databases for ad hoc queries and updates, and they complicate the task of writing good optimizers.

For a detailed critique of the way in which many OO features have been grafted onto the relational model, see Date (2000) and Date and Darwen (1998). At any rate, relational and object-relational systems currently dominate commercially, and other contenders such as object database systems are struggling to gain any significant market share.

SQL includes arrays, row types, and multisets. One challenge then is to model such collections and provide mapping algorithms to implement them efficiently. Various constructors (e.g., for sets, bags, sequences, and schemas) have been added to some versions of ORM and ER. In addition, both ORM and UML allow some kinds of collections to be expressed as mapping annotations on the conceptual schema. Although the use of collection types in conceptual models can facilitate mapping to equivalent implementation structures, it is typically much harder to model complex objects directly at the conceptual level, and they are extremely awkward to validate by verbalization and population. Complex objects also tend to be overused, and simpler solutions are overlooked. For further discussion on collection types, see Chapter 10.

Nowadays, many applications are built using a third-generation language (3GL) such as C# or Java to specify an object model for the transient, in-memory storage, and using SQL to specify a relational or object-relational database for the persistent storage. As data must be moved between the object model and the relational model, considerable effort is spent on the *object-relational mapping*. Using a tool such as NORMA to generate both object and relational models from ORM models simplifies the task of creating an object-relational mapping process.

Deductive Database Systems

Logic programming languages such as Prolog are declarative languages based on first-order predicate logic. Facts and derivation rules (including recursive rules) may be declared concisely, with built-in backtracking support.

Datalog combines features from relational database systems and a safe subset of logic programming languages. Unlike Prolog, queries in Datalog are guaranteed to execute in a finite time. Datalog has been used for natural language processing and as a query language for deductive database systems.

Deductive databases offer elegant and powerful ways of managing complex data in a declarative way, especially for information that is derived by use of recursion. This gives them a strong rule enforcement mechanism with built-in backtracking and excellent support for recursive rules. For example, the isanAncestorOf relation can be derived from a base parent relation and the following two rules in Horn clause notation:

X isanAncestorOf Y :- X isaParentOf Y. (basis clause)
X isanAncestorOf Y :- X isaParentOf Z, Z isanAncestorOf Y. (recursive clause).

Here, the ":-" between the head and body of the rule means "**if**", and the comma "," in the body of the rule means "**and**". Individual variables in the head of the rule are assumed to be universally quantified, and variables introduced in the body are assumed to be existentially quantified, so the rules are equivalent to the following formulae in predicate logic:

∀x,y (x isanAncestorOf y if x isaParentOf y)
∀x,y [x isanAncestorOf y if ∃z(x isaParentOf z and z isanAncestor of y)]

In contrast, SQL-92 cannot express recursive queries at all. SQL:1999 introduced a recursive union operator, but its syntax is more complex and its execution does not enjoy built-in backtracking.

Various deductive database systems based on or influenced by Prolog or Datalog have been developed, such as LogiQL (Halpin and Rugaber, 2015) and Soufflé (https://github.com/souffle-lang). LogiQL was developed by LogicBlox, but is no longer available since the acquisition of LogicBlox by Predictix in December 2014. Soufflé is an open-source, parallel, logic programming language that includes both an interpreter and a compiler that targets parallel C++.

An extremely powerful programming language that supports deductive database features as well as knowledge graphs is under development by RelationalAI (https://relational.ai), with a likely release date around mid-2024. The system, called *RAI,* is cloud-based, and its language is called *Rel,* as its data structures are centered on relations (sets of ordered tuples).

The recursive rule for ancestorhood ∀x,y [x isanAncestorOf y if ∃z(x isaParentOf z and z isanAncestor of y)] may be expressed in Rel by means of the following definition: def isanAncestorOf (x, y): exists((z) | isaParentOf(x, z) and isanAncestorOf(z, y)).

As an indication of the power of Rel, a team of RelationalAI programmers has written a program in Rel that automatically transforms ORM models captured in the free, public-domain version of the NORMA tool into structures in Rel. This transformation is able to map extremely complex ORM models, and can capture essentially all alethic constraints in ORM. At the time of writing, deontic constraints are not supported.

The rest of this section provides a brief overview of the main logical features of the Rel language, and uses examples to illustrate how constraints may be formulated. Table 17.5 summarizes the basic *logical operators* between propositions supported in the Rel language. Multiple syntaxes are supported for each operator. Here, "*implies*" is used for the *material implication* operator of propositional logic, so "*p* implies *q*" is equivalent to "not *p* or *q*".

This differs significantly from the English meaning of "if *p* then *q*" because a false proposition materially implies any proposition whatsoever, and a true proposition is materially implied by any proposition. For example, the proposition "Earth is a star" materially implies the proposition "Earth has 10 moons", and the proposition "Mars is a star" materially implies the proposition "Mars is a planet".

Table 17.5 Summary of basic logical operators in the Rel language.

Symbol	*Meaning*
not	it is not the case that
and	and
or	inclusive or
implies	implies
xor	exclusive or
iff	if and only if

We'll use the ORM schema in Fig. 17.17 to illustrate how mandatory role and uniqueness constraints may be formulated in Rel. Constraint formulations start with "**ic**" (for "integrity constraint"), followed by a parenthesized list of the individual variables involved and then the detailed constraint expression in braces "{}".

Assuming the ORM fact type Country has CountryName has been declared in Rel using the binary predicate hasCountryName, the ORM *mandatory role* constraint **Each** Country has **some** CountryName may be expressed as follows:

```
ic (c) { Country(c) implies hasCountryName(c, _) }
```

The head variable c is universally quantified, and the underscore character "_" denotes the anonymous variable, so the constraint expression is equivalent to the following predicate logic formula:

$$\forall x\ [\text{Country}(x) \rightarrow \exists y(x\ \text{hasCountryName}\ y)].$$

The *internal uniqueness constraint* on Country's role in Country has CountryName may be declared by declaring the predicate to be *functional* as follows:

ic () requires function(hasCountryName)

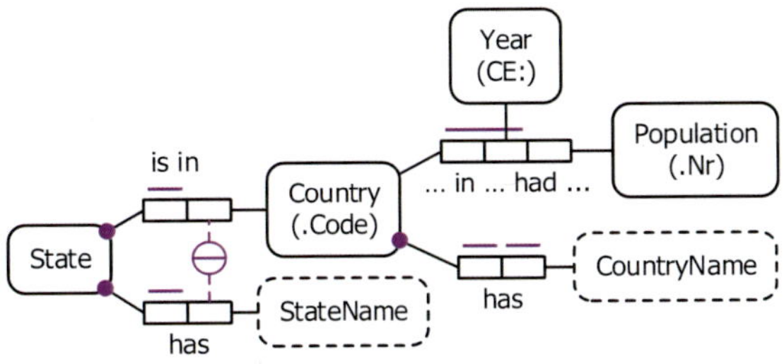

FIG. 17.17 Some uniqueness and mandatory role constraints.

The uniqueness constraint on CountryNames's role in Country has CountryName may be declared thus:

ic (c1, c2, n) requires (hasCountryName(c1,n) and hasCountryName(c2,n) implies c1 = c2

Assuming the ternary fact type has been declared, its *composite uniqueness constraint* may be declared as follows:

ic (c, y, p1, p2) requires
(inYearHadPopulation(c, y, p1) and inYearHadPopulation(c, y, p2)) implies p1 = p2

The *external uniqueness constraint* in the composite reference scheme for State may be declared thus:

ic (s1, s2, c, sc) requires
(isInCountry(s1,c) and hasStateName(s1,sc) and
isInCountry(s2,c) and hasStateName(s2,sc))
implies s1 = s2

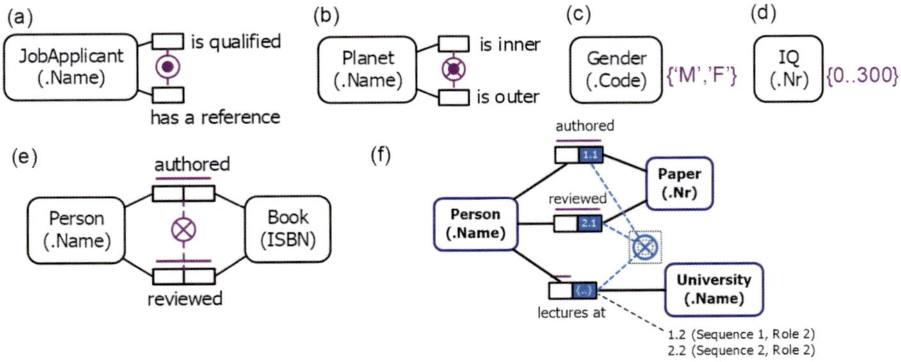

FIG. 17.18 Some more constraints.

The ORM schema in Fig. 17.18 includes several other kinds of constraints. The *inclusive-or constraint* in example (a) ensures that each job applicant is either qualified or has a reference (or both). This may be expressed in Rel thus:

```
ic (p) requires JobApplicant(p) implies (isQualified(p) or hasaReference(p))
```

The *exclusive-or constraint* in example (b) ensures that each planet in our solar system is an inner planet or an outer planet but not both. This may be expressed in Rel thus:

```
ic (p) requires Planet(p) implies (isAnInnerPlanet(p) xor isAnOuterPlanet(p))
```

The *value constraint* in example (c) restricts the value of GenderCode in the underlying reference type Gender has GenderCode. The *value constraint* in example (d) restricts IQ values to the range 0..300. These constraints may be formulated thus:

```
ic (g, gc) requires hasGenderCode(g,gc) implies (gc = "M" or gc = "F")
ic (x, y) requires hasIQNr(x,y) implies (x >= 0 and x <=300)
```

The *exclusion constraints* in examples (e) and (f) say that nobody can author and review the same book, and that nobody can review a paper authored by someone who lectures at the same university where he/she lectures. The diagram in (f) for the exclusion constraint over a join path is a screenshot from the NORMA tool. The constraints may be formulated thus:

```
ic (p, b) requires reviews(p, b) implies not authors(p, b)
ic (p1, p2, u, pr) requires
        (authors(p1, pr) and lecturesAt(p1, u) and reviews(p2, pr)) implies not lecturesAt(p2,u)
```

The ORM schema in Fig. 17.19 includes some more constraints. The *equality constraint* in example (a) ensures that a patient has a diastolic blood pressure reading if and only if the patient has a systolic blood pressure reading. This may be expressed in Rel using an anonymous variable as follows:

```
ic (p) requires hasDiastolicBP(p, _) iff hasSystolicBP(p,_)
```

The *internal frequency constraint* in example (b) ensures that each reviewer is assigned at most three papers to review. It may be formulated thus:

```
ic (r, n) requires isAssignedNrPapers(r, n) implies n <= 3
```

The *composite frequency constraint* in example (c) ensures that no student may enroll more than twice in the same course. This may be formulated as follows. Here, "def" prepends the definition or derivation rule for the inCourseHasNrEnrollments relation, which uses the built-in count function. The constraint makes use of the derived ternary relation.

```
def inCourseHasNrEnrollments(s, c, n):
n = count[(e): isanEnrollmentByStudent(s, e) and isanEnrollmentInCourse(e, c)]
ic (s, c, n) requires inCourseHasNrEnrollments(s, c, n) implies n <=2
```

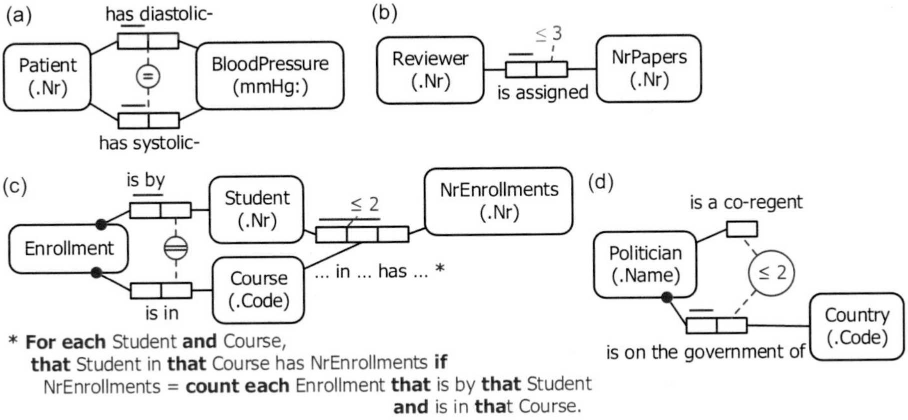

FIG. 17.19 Some equality and frequency constraints.

The *external frequency constraint* in example (d) ensures that for each country, at most two politicians are both in its government and are a coregent. This constraint may be formulated using the count function or instead simply thus:

```
ic (p1, p2, p3 c) requires
(isaCoregent(p1) and servesOnGovtOf(p1, c) and isaCoregent(p2)
and servesOnGovtOf(p2, c) and isaCoregent(p3) and servesOnGovtOf(p3, c))
implies p1 = p2 or p1 = p3 or p1 = p3
```

Ring constraints are easy to specify in Rel. The *reflexive* ring constraint that people know themselves may be expressed as: ic(p) requires Person(p) implies knows(p,p). The *irreflexive* ring constraint that nobody a parent of himself/herself may be expressed as: ic (p) requires not isaParentOf(p, p). The *symmetric* constraint on siblinghood may be specified thus: ic (p1,p2) requires isaSiblingOf(p1, p2) implies isaSiblingOf(p2, p1). The *asymmetric* ring constraint that nobody is a parent of himself/herself may be expressed thus: ic (p1,p2) requires isaParentOf(p1, p2) implies not isaParentOf(p2, p1).

The *transitive* nature of the isOlderThan relation may be formulated as follows: ic (p1, p2, p3) requires (isOlderThan(p1, p2) and isOlderThan(p2, p3)) implies isOlderThan(p1, p3). The *intransitive* constraint on the direct precedence relation may be formulated as follows: ic (e1,e2,e3) requires (directlyPrecedes(e1, e2) and directlyPrecedes(e2, e3)) implies not directlyPrecedes(e1, e3).

The ancestorhood relation may be defined as follows: def isanAncestorOf(p1, p2): isaParentOf(p1, p2) or isaParentOf(p1, p3) and isanAncestorOf(p3, p2). The *acyclic constraint* on ancestorhood may be specified thus: ic (p) requires not isanAncestorOf(p, p).

The pair-*subset constraint* in Figure 17.20(a) ensures that each student who passed a course also enrolled in that course. This may be expressed in Rel thus: ic (s, c) requires passedCourse(s, c) implies enrolledInCourse(s, c).

The *subset constraint over a join path* in Fig. 17.20 ensures that each advisor who serves in a country must speak at least one language that is used in that country. This constraint may be expressed as follows:

```
ic (a, c) requires servesInCountry(a, c) implies
(speaksLanguage(a, l) and isaLanguageUsedBy(l, c))
```

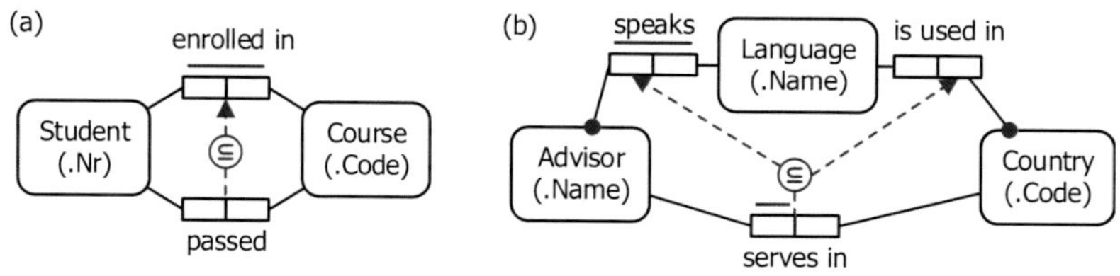

FIG. 17.20 Some subset constraints.

The simple examples give some idea of the expressive power of Rel for modeling ORM constraints. Large, highly complex ORM models can also be mapped to Rel structures.

Other Database Varieties

Specialized database varieties are sometimes invoked to deal with certain types of applications. *Spatial databases* require efficient management of spatial data, such as maps (roads, land, etc.), two-dimensional designs (circuits, town planning, etc.), and three-dimensional designs (visualization of medical operations, molecular structures, flight paths, etc.). They provide built-in support for spatial data types (points, lines, polygons, etc.), spatial operators (overlap, contains, intersect, etc.), and spatial indexes (R-trees, quad trees, etc.). This allows efficient formulation and processing of spatial queries, such as, How many houses are there within 5 km of the proposed shopping center? Which flights fly over the greater LA area? Which diagrams are similar to diagram 37?

Previously, special systems were used for spatial data while a relational DBMS was used for alphanumeric data. The current trend is to manage both standard and spatial data in the one system. A typical application of a geographic information system (GIS) might proceed as follows: standard data on traffic accidents is entered; the road maps are displayed to highlight the accident sites; and this is now used to determine regions where extra precautions need to be taken (e.g., radar traps).

Historically, GIS vendors adopted three main approaches: hybrid (georelational-objects in a spatial file store are given identifiers that can be referenced in relational tables); integrated (spatial and nonspatial data are stored in relational tables); and object-oriented (all data is stored in an ODB). More recently, many spatial database applications as well as other applications using nonstandard data have been implemented using object-relational technology.

Although time has only one dimension, unlike space's three dimensions, the efficient management of *temporal information* is no easy task. If historical rather than snapshot records need to be maintained about objects, time will feature largely in the modeling.

As discussed in Chapter 10, a variety of approaches may be adopted for modeling time. In some cases, we simply include object types such as Time and Period on the conceptual schema and map these like other object types. Distinctions may be needed between transaction time (when the system records a fact) and valid time (when the fact is true in the UoD being modeled). Often, we need to make use of temporal relations (such as before and after) and temporal operations (e.g., to compute an interval between two time points).

The SQL standard currently includes only basic support for temporal data, and sometimes an ordinary relational database does not allow temporal aspects to be implemented efficiently. For such applications, special DBMSs known as "temporal database systems" are sometimes used; these provide in-built support for automatic time stamping and the various temporal operators.

Most work on temporal databases focuses on maintaining relevant histories of application objects through time, with the assumption that the conceptual schema itself is fixed. Moving up one level, the problem becomes more complicated if we allow the conceptual schema itself to change with time. This is one aspect of the schema evolution problem.

Moving up another level, we might allow the conceptual metaschema itself to change with time (e.g., we might decide at a later stage to allow constructors for complex object types in our conceptual schema language). The management of such higher-order evolution has been addressed in research on evolving information systems. This topic provides one motivation for the discussion on metamodeling in Chapter 18.

Centralized and distributed databases were discussed earlier in this chapter. A further dimension is offered by *federated databases*. These deal with situations where there is a need for data sharing between several different database systems, possibly heterogeneous (e.g., some relational, some column-oriented, some key-value, and so on). In this framework, each individual system maintains its local autonomy and communicates with other systems on an as-needs basis. As the heterogeneity problem requires translation between different data models, the control of federated systems is nontrivial.

Interest in federated databases has been accelerated by the advent of cloud computing. Most organizations find that management of a collection of heterogeneous databases can be a headache, but this can be greatly eased by outsourcing functionality to a cloud service provider. Three levels of service are typically offered:

Infrastructure as a Service (IaaS) provides basic processing, storage, and networking functionality, based on dedicated or virtual hardware. This essentially replicates an in-house data center but within the cloud service. Users remain responsible for standard management functions.

Platform as a Service (PaaS) transfers responsibility for managing infrastructure resources to the cloud service provider. Users are freed from concerns such as operating system updates, hardware maintenance, and so on. Additional capacity can be obtained easily and incrementally from the service provider, allowing the user to focus on installing and maintaining applications.

Software as a Service (SaaS) includes all the previous features but also takes on responsibility for the software applications, including licensing, software updates, and so on. In this situation, the cloud service provider controls the application and users simply pay for access.

A cloud-based approach opens several other important possibilities. One is the idea of a *serverless database*. A database can be defined in the usual way, but the server hardware is absorbed into the broad cloud capability. This provides all the functionality of the traditional installation of a database installed on a server, plus some additional advantages. These include greater resilience, automatic scaling to meet demand, simpler resource management, and pay-as-you-go pricing.

Organizations commonly store large amounts of data in a data warehouse to make it available for further processing. In a cloud context, this idea can be extended into the concept of a *data lake*. The name is inspired by the analogy of a lake full of unprocessed water that can be drawn on and treated in various ways to make it suitable for consumption. A data lake is a repository for large amounts of data stored in its original "raw" form. This differs from a data warehouse in having no schema, as any necessary schema can be created on demand based on the requirement to be satisfied. Software tools can be used to develop the necessary functionality, such as creating a new SQL database from CSV files and spreadsheets in the data lake.

Various file formats have been created to cater for specific conditions and more will undoubtedly be defined in the future. Avro is an example of a row-oriented format that is useful for compressed data. Each record in the file contains binary data plus a JSON header that defines how the binary data can be parsed to extract individual fields. Parquet is an example of a column-oriented format that is useful for large datasets with nested values. Values from the same column are stored together and organized into row groups for efficient processing. Again, the file includes metadata to allow data to be located efficiently.

Creating a database in a cloud environment can be made easy through *low-code development*. Instead of writing code (perhaps in SQL) to define the structure, the creator can simply select from a range of options. The detailed operations to build the necessary structure can be automated and hidden from the user. Useful fields such as the user that created a record, the date it was last modified, and so on, can be automatically added to tables. An example of this approach is the Microsoft Dataverse, which is a cloud-based data platform. This allows users to create data-related applications without the need for technical knowledge using Microsoft's Power Platform tools.

As discussed in Chapter 16, XML and JSON are notations that support a systematic approach to describing structured data, and either could potentially provide the foundation for a database system. As we have already seen, JSON provides the basis of the document database approach. Some examples of *XML databases* have been developed, but these have not been widely adopted. The general usage of XML is in the form of individual

XML documents, particularly for the exchange of structured data between disparate systems. XML is now encompassed as a data type within SQL, and processing of XML data can be handled within a relational framework, reducing the appeal of a separate database for XML alone.

Although both XML and JSON are textual notations, the text is not well suited to human visualization of information in a business domain. In both cases, schemas are best developed by first modeling in a high-level language such as ORM and then mapping the model to the required lower-level notation. For a discussion of mapping ORM to XML schema, see Bird et al. (2000).

Artificial Intelligence

Since the phrase was first coined in the 1950s, artificial intelligence (AI) has been promoted as a breakthrough that could revolutionize technology. Early approaches to AI were dominated by the idea that knowledge could be expressed in terms of logical expressions that used data values through a systematic procedure. Although some successes were reported, crafting such systems proved to be difficult, and the resulting systems were generally not sufficiently robust to deal with real-world data.

In the 1990s, attention started to turn to "connectionist" approaches. Instead of a human-centric analysis of data to identify logical relationships, systems were trained on data from the application domain. This developed an internal model of patterns in the training data that could then be deployed against new data. One feature of this approach is that it may not be easy to explain how a system has arrived at a particular result. But this is no different to the way that humans operate. For example, most native speakers of a language produce results that can be judged as grammatically correct, even though the person may not be able to explain the grammatical "rules" that are being followed.

A major leap forward was created when these ideas were applied to very large datasets. This provided a solution to the robustness problem: how to deal with a wide range of situations. Instead of painstakingly trying to analyze each situation that can arise, a very large dataset can contain examples of all situations that can arise. Sophisticated learning algorithms can be applied to the data to build a capability sometimes referred to as "general AI". The large amount of computing required is supported by very large databases (including the kinds discussed earlier in this chapter) and special-purpose processors to carry out the many repetitive operations in an efficient way.

Although general AI holds tremendous promise, it has not yet become a regular feature of database creation and management. All the major vendors are investing heavily in its application, in addition to many start-up companies with a focus on specific features. Here, we'll limit ourselves to describing some of the areas where AI has obvious relevance, and in some cases is already being applied.

Schema design. Produce an initial schema design from a natural language description of a given domain. In the relational case, strict normalization could be enforced.

Query design. For a given database type and schema, write a query in the appropriate technical language to return results that have been expressed in natural language.

Application design. Produce a design based on a given database schema for an application described in natural language. This could include both desktop and mobile variants.

Database Management. Automation of routine operations that are normally carried out by a database administrator (DBA). These can include tasks such as index creation and updating, query tuning for evolving usage, performance assessment and optimization, and more.

Data cleansing. Historical data is often plagued with problems related to the original data capture that can obstruct subsequent analysis. Training on typical problems can be used to avoid or repair these issues.

Code creation. Technical code (in some given language) can be created from a description of the intent in nontechnical language. This is already operational as a "copilot" functionality in some development environments.

Schema migration. As the usage of a database evolves, it may become apparent that the schema originally envisioned is no longer appropriate. Migration to an alternative schema is a complex process that can be aided by automation.

Advanced analytics. The business-relevant information in the data can be processed to uncover patterns and trends that are not obvious through traditional queries.

The kinds of AI applications mentioned above often rely on a response to a prompt defining the necessary focus for the AI. Creating the appropriate prompt is at present a skill, but a skill that itself will surely become amenable to AI in time. The one area that cannot be determined by AI is what is, or is not, relevant to the user (often referred to as the UoD in an ORM context), and the user's view of how the objects in the UoD are related. In some sense, the UoD could be seen as the ultimate prompt that cannot be reduced to anything else and cannot be defined without user knowledge. Clearly, an AI agent must understand the UoD to be effective, but how can that information be supplied? An obvious solution is provided by an ORM schema. It is certainly possible to describe a UoD through an informal discourse, but such descriptions are plagued by well-known ambiguities. Creating an ORM schema can sometimes be a painful process for a business (internal politics, unclear direction, etc.), but the eventual prize is an unambiguous description of the business intent. Once this is adopted by the business, it can be used as the basis for many facets of subsequent AI processing.

17.7 Summary

Relational databases first appeared in the 1970s and became the dominant form of data management during the 1980s and 1990s. Apart from any intrinsic technological merits, their wide adoption was greatly aided by the similarities among the offerings of different vendors. This made investment in the technology a low-risk option, both in terms of deployment by companies and careers for individuals. Since their introduction, there have been many proposals for alternative approaches, but the technology and its associated language, SQL, have continued to evolve. The SQL standard has expanded to include features from many competing approaches to meet the perceived demand for alternatives.

However, there is one intrinsic feature of a relational database that cannot easily change, and that is its use of a client–server model with a single server providing resources to many clients. Increasing the size of the database requires increasing the size of the server, often at a disproportionally higher cost. Spreading a database over multiple servers is not practical unless the data can be partitioned into noninteracting subsets. When relational databases first appeared, this was not a serious problem because data volumes were generally low. In recent years, large datasets have become common, and this is one of the main driving factors behind the growth of the NoSQL movement.

A secondary factor has been the emergence of many niche applications where there is some advantage to be gained by a specialized architecture. In a sense, this is a reaction against the application-neutral approach of relational systems. Optimization of one factor is not always an advantage because it always involves deoptimization of other factors, and so seeking increased performance usually involves narrowing the applicability of the solution.

Today, many organizations use a mixture of database technologies, usually including one or more relational databases but supplemented by other approaches for selected application areas. This has been greatly facilitated by the growth of cloud-based facilities that reduce the barriers to deploying and managing a range of diverse data resources. We can expect this trend to continue for the foreseeable future.

18

Other Modeling Aspects and Trends

Chapter Outline

18.1 Introduction

Section 2.4 included an overview of the information systems life cycle. So far, we've focused on the stages that deal with the conceptual and logical design of the data. We've seen how to perform conceptual data modeling using object-role modeling (ORM), entity-relationship modeling (ER), or unified modeling language (UML), and applied the Rmap algorithm to map a conceptual schema to a relational database schema. We've discussed database manipulation using structured query language (SQL) and extensible markup language (XML), modeled business processes and states, and applied conceptual schema transformations and lower-level optimizations to improve the efficiency of relational designs.

This chapter discusses other aspects and recent trends related to modeling and querying information systems. We'll be covering a lot of ground, so the treatment of each topic is necessarily brief. You can use the references in the chapter notes to dive deeper into the areas examined.

Section 18.2 provides a brief introduction to data warehousing and online analytical processing. Section 18.3 discusses some very high-level languages for querying information systems. Section 18.4 outlines some ways to perform schema abstraction, enabling modelers to focus on various aspects by hiding other details.

Information Modeling and Relational Databases. https://doi.org/10.1016/B978-0-443-23790-4.00010-4

Section 18.5 discusses further design aspects, such as coordinating data and process models, and user interface design. Section 18.6 outlines some recent work on ontologies and the semantic web, especially the Web Ontology Language (OWL). Section 18.7 provides an introduction to metamodeling, in which schemas themselves are treated as instances of a higher-level metaschema. Section 18.8 furnishes a summary that completes this final chapter.

18.2 Data Warehousing and OLAP

Most commercial information systems are built to support heavy volumes of transactions by many users on a daily basis. Examples include banking, insurance, and order processing systems. These *Online Transaction Processing (OLTP)* systems typically require quick throughput for their largely predefined range of transactions, especially update transactions. To improve performance, historical data is often archived once it reaches a certain age, reducing the size of the datasets used for daily operations.

A single organization may have several OLTP systems (e.g., purchasing, sales, inventory, customer records), possibly implemented using different kinds of database management systems (DBMS) or other software applications, and the coupling between such systems may be weak or even nonexistent.

Over time, businesses became aware that the collective information contained in their various systems had great potential for analyzing market trends and improving their business processes. However, their OLTP systems were unsuitable for executives to perform this task, given the poor performance and complex interface for ad hoc analysis queries (e.g., aggregated multitable joins and nested correlated subqueries).

Moreover, insufficient integration or history made some queries simply impossible. Partly to address the problem of integrating data from prerelational systems for analysis purposes, IBM proposed the notion of an "information warehouse". Although performance problems delayed acceptance of this idea for some years, a later proposal for a "data warehouse" by Bill Inmon (1993) was enthusiastically embraced by the business community, and nowadays most large companies have a data warehouse.

A *data warehouse* is an enterprise-wide, integrated, historical database of information extracted from individual data sources for the purpose of supporting analysis of the business by management. During analysis, it is read-only. Because the patterns and trends sought from the analysis tend to evolve slowly, and some imprecision is acceptable, updates are performed in bulk according to an agreed schedule (e.g., weekly).

The construction of a data warehouse for a large enterprise can be a lengthy task. To exploit the benefits as soon as possible, the data warehouse is often built iteratively, one subject area at a time. As subject areas are added to the data warehouse, they may be used to load data marts (see Fig. 18.1).

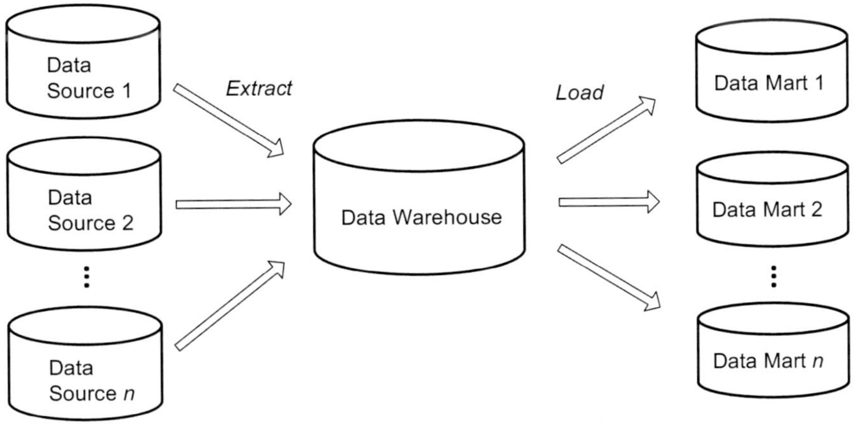

FIG. 18.1 Data is extracted into the data warehouse then loaded into data marts.

A *data mart* is a smaller "departmental warehouse" focused on one subject area, often containing more summarized and less detailed data than the data warehouse. For end users who perform their analysis within one subject area, a data mart provides a simpler model adapted to their needs, and the smaller data volume often leads to better performance.

Many different approaches to data warehousing exist. Sometimes, data marts are built first and used to incrementally construct the data warehouse. However, if an analytical query may span multiple subject areas, it is critical that an overall enterprise architecture be in place to make the appropriate connections.

It has been argued that the data warehouse should be implemented as a fully normalized relational model, based directly on the enterprise data model, with no summary data, postponing any denormalization and aggregation to the data marts loaded from the warehouse (e.g., Moody and Kortink, 2000). Current practice, however, usually does incorporate denormalized and summarized data in the data warehouse (Silverston, 2001a). A nice overview of alternative approaches is provided by Jukik (2006).

Before extracting data to the data warehouse, the business analysis needs should be determined, and the relevant data sources identified. The data sources may include operational databases as well as spreadsheets and legacy systems. Any details in the data sources irrelevant to the analysis needs should be removed. For example, the phone number and street address of a store are probably not of interest, but its city probably is. The remaining data now needs to be transformed to remove errors (applying integrity rules). For example, a genderCode field with many "F" and "M" entries might include a few instances of "'D", "G", "J", or "N". Assuming that these are typographical errors, can you guess the intended letter? (*Hint:* look at the keyboard.)

We must also ensure that all facts of the same type are represented in the same way. As a trivial example, a customer's birthdate might be recorded as a character string in a "DOB" field in one source, and elsewhere in a "birthdate" column based on a date data type. Once the data is "cleansed", it is transformed into a uniform representation in the data warehouse (typically a relational database).

To facilitate the analysis of historical trends, appropriate temporal data should be included at the desired granularity (e.g., daily, weekly, or monthly). For example, suppose an operational data source stores the following fact type: Employee manages Store. Over time, a store may have different managers. To retain history of these changes, when store management facts are loaded into the data warehouse, the load date may be inserted into the key to populate the historical fact type Employee on Date managed Store. Basic aspects of temporal modeling were discussed in Section 10.3.

To improve query performance, data marts (and usually the data warehouse) often contain derived data and denormalized structures. As a simple example of derived data, suppose an operational source stores the fact type Customer was born on Date. For demographical analysis, we may be interested in how product preferences are influenced by the age of customers. In this case, it may be more appropriate in a data mart to store the age, or even just the age group, of the customer rather than their birthdate, using a fact type such as Customer on Date belonged to AgeGroup. The snapshot dates are inserted when the fact type is updated incrementally.

A more typical example incorporating derivation and denormalization is a data mart for analyzing Sales trends. An ORM conceptual model of such a mart (simplified) is shown in Fig. 18.2. In this universe of discourse (UoD), the company makes sales from many stores located in various cities within the same country (e.g., the United States). A city is identified by combining its name and state (e.g., Portland, Oregon, differs from Portland, Maine).

States are grouped into regions (e.g., Oregon and Washington belong to the Northwest region). Items sold have a unique code and title, and belong to a category (e.g., Developer Tools). A line item is identified by its line number on a given invoice, and records the sale of an item as well as the quantity and unit price for that item. The line total for each line item is derived by multiplying the quantity by the unit price and is then stored.

To support sales analysis over time, the month number, quarter number, and year for each date are derived and stored. Although calendar years are used here, we could use fiscal years instead or as well. In Fig. 18.2, fact types that are derived and stored are marked "**", but for simplicity the derivation rules are omitted.

Using the default Rmap algorithm, this conceptual schema maps to the following six table schemes: *LineItem*(invoiceNr, lineNr, itemCode, quantity, unitPrice, lineTotal); *Invoice* (invoiceNr, saleDate, storeNr); *Item*(itemCode, itemTitle, category); *Store*(storeNr, stateCode, cityName); *StateLocation*(stateCode, region); *TimeDimension*(saleDate, saleYear, QuarterNr, MonthNr). However, to improve query performance it is decided to denormalize the relational schema to four table schemes, as shown in Fig. 18.3.

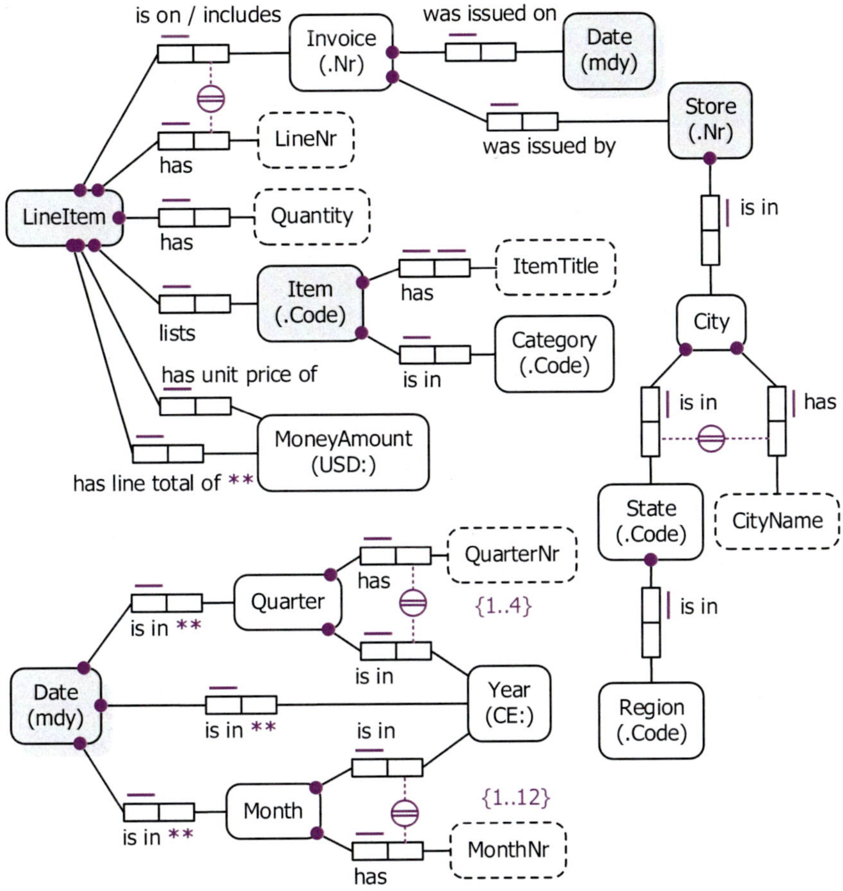

FIG. 18.2 A conceptual schema for the Sales data mart (derivation rules omitted).

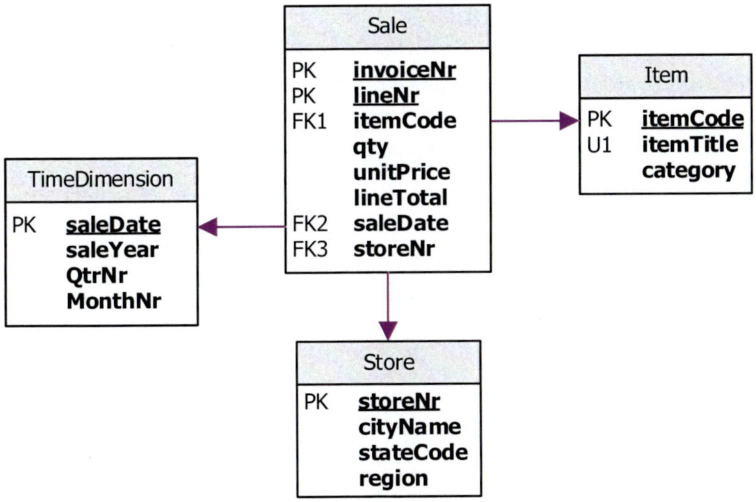

FIG. 18.3 A denormalized, relational star schema for the model in Fig. 18.2.

The Item and TimeDimension tables are normalized, but the Sale and Store tables are not. The Sale table is denormalized to 1NF because saleDate and storeNr are functionally dependent on invoiceNr, which is just part of the primary key. The Store table is denormalized to 2NF because region is functionally dependent on stateCode, a nonkey attribute.

The decision to denormalize in this way is indicated by annotating the conceptual schema, as shown in Fig. 18.2. Here, the *key object types* are *shaded*. An object type is a key object type if and only if its preferred identification scheme is used as the primary key of a table. Graphically, each key object type forms the root of a tree, where each node is an object type and each edge is a functional (*n*:1 or 1:1) predicate.

For example, from the Store object type, we run down functional chains to the leaf object types CityName and Region, gathering all the fact types on the way to group them into a single table based on the identifier for Store.

A functional chain stops if it runs into a key object type (or a leaf or a nonfunctional predicate). For example, starting at LineItem, we gather up all its functional fact types as well as those for Invoice, but we cannot proceed past Date, Store, or Item because these are key object types. This leads to the Sale table in Fig. 18.3.

The denormalized Sale and Store tables contain embedded functional dependencies (e.g., stateCode → region), but there is no need to enforce these because they have already been enforced in the operational tables from which the data mart is derived. Because the operational tables are used for base updates, and not the data mart, it is acceptable to denormalize the data mart in this way. Reducing the number of tables eliminates the need for many joins, leading to faster queries.

The schema in Fig. 18.3 is composed of a central table (Sale) linked by foreign key connections to outer tables (Item, Store, and TimeDimension). This is called a *star schema* because the central table may be viewed as the center of a star pattern, with its outer tables becoming "points of the star". In data warehousing terminology, the *central table* is called a *fact table* and the outer tables are *dimension tables*. Because all tables contain facts, the term "fact table" is rather inappropriate here. Moreover, ORM uses the term "fact table" to mean an atomic fact table. To avoid confusion, we'll use the more descriptive "central table" instead of the more popular "fact table" in this context.

Some approaches require the primary key of the central table to include all the keys of its dimension tables. This would require the (invoiceNr, lineNr) key of the Sale table in Fig. 18.3 to be expanded to the superkey (invoiceNr, lineNr, itemCode, saleDate, storeNr). From a purely logical standpoint, this is not required because joins can be made on nonkey attributes.

If some dimension tables are themselves used as central tables for other stars, the overall schema is called a "snowflake schema". A set of star schemas with shared dimension tables is sometimes called a "galaxy".

Data warehouses and data marts are used for *online analytical processing (OLAP)* and *data mining*. The term OLAP was introduced by E.F. Codd to describe interactive analysis of dimensioned and aggregated data for decision support (Codd et al., 1993). Data mining involves deeper analysis of the data, typically using sophisticated statistical techniques and complex algorithms for detecting patterns. Nowadays, many tools for OLAP and data mining are in use. Either topic deserves a book in itself. The remainder of this section provides a brief overview of OLAP.

There are three main approaches to OLAP. Each uses base data as well as aggregated data (e.g., sales figures might be summed and grouped at various levels). Multidimensional OLAP (*MOLAP*) stores both base and aggregated data in multidimensional structures rather than tables. Relational OLAP (*ROLAP*) stores both base and aggregated data in relational tables. Hybrid OLAP (*HOLAP*) stores base data in relational tables and aggregated data in multidimensional structures. Some DBMSs, such as Microsoft SQL Server, support all three kinds of OLAP.

The multidimensional structures used for OLAP are popularly known as *cubes*. In geometry, a cube is a three-dimensional box structure. In OLAP theory, a cube can have as many dimensions as you like. Cubes provide an intuitive way to visualize and browse data as well as fast access to aggregate data. Let's consider a simple example.

With reference to the star schema in Fig. 18.3, suppose we wanted to list the number of units sold in the years 2020 through 2022 for each geographic region and each item category. As an exercise, you might like to formulate the SQL query for this. You need to perform the natural join of the four tables; group by saleYear, region, and category; and compute sum(qty). An extract of a possible result is shown in Table 18.1, assuming only two categories (SW=Software, HW=Hardware) and four regions (N=North, S=South, E=East, W=West). The fact type underlying this table is the quaternary Year in Region had sales of items of Category in NrUnits. The full table display of the result would include 24 rows (eight for each year); only the first nine rows are shown here.

Table 18.1 Units sold.

Year	*Region*	*Category*	*NrUnits*
2020	N	SW	10000
2020	N	HW	500
2020	S	SW	12000
2020	S	HW	330
2020	E	SW	7500
2020	E	HW	440
2020	W	SW	12000
2020	W	HW	...350
2021	N	SW	14500
...	...	...	...

An alternative way of displaying the data using a cube is shown in Fig. 18.4. Here, the independent variables (year, region, and category) appear as the *dimensions* making up the edges of the cube, and the values for the dependent variable or *measure* (units sold) appear in the relevant cells. Only the software sales figures are shown here in the 12 cells making up the front half of the cube. You can imagine seeing the hardware sales figures by rotating the cube to see its other side.

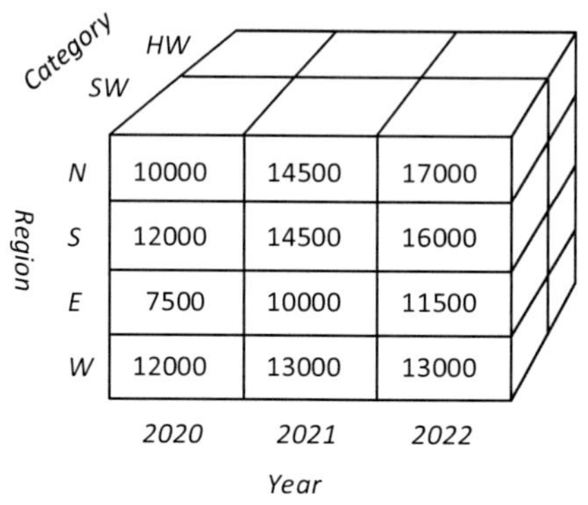

FIG. 18.4 Cube depiction of units sold (only the software figures are shown here).

OLAP cubes can be much more complex than this example. For any set of independent dimensions, there may be more than one measure (e.g., units sold and revenue). These different measures are often collectively referred to as the *Measures dimension*. Moreover, each independent dimension typically has a *hierarchy* of *levels*. For example, a Location dimension might have regions at its top level, composed of states at the second level, with cities at the third level, and stores at the fourth level. Similarly, a Time dimension may be decomposed into Years, then Quarters, then Months, and then Days. Finally, the Item dimension may be decomposed into categories and then items.

So, our data mart example can be used to construct a cube with three independent, hierarchical dimensions (Location, Time, and Item) and one dependent dimension for UnitsSold and Revenue measures. The neat thing about the cube structure is that it enables aggregate values for the measures to be efficiently stored and accessed for all levels of the hierarchies.

When analyzing a cube, you can choose to consolidate or *roll up* these aggregates (e.g., roll up sales figures for cities to regional sales figures). You can also *drill down* to a finer level of granularity. Moreover, you can *slice and dice* the cube whichever way you like by taking a subcube of it. For example, if you restrict the item category in Fig. 18.4 to software, you get the front slice of the cube.

Because of such advantages, star schemas in data marts are often used to create a variety of cubes for easy analysis. SQL Server provides wizards to simplify the task of cube creation and also supports a *Multidimensional Expression (MDX)* language for querying cubes.

In MDX, the **select** statement specifies a result set that contains a subset of multidimensional data returned from a cube. The select-clause specifies the query axes (0 to 128 axes are allowed), the from-clause determines which cube will be queried, and an optional where-clause determines which members or tuples to use on the slicer axis to restrict the data returned.

As a simple example, assume SalesCube is based on the hierarchical Location, Time, and Item dimensions discussed earlier. The following MDX query will result in a two-dimensional grid that lists the software units sold and revenue in Washington State in each of the first three quarters of 2020.

```
select { [Measures].[UnitsSold], [Measures].[Revenue] } on columns,
       { [Time].[2020].[Q1] : [Time].[2020].[Q3] } on rows
from SalesCube
where ( [Item].[SW], [Location].[N].[WA] )
```

In MDX, braces are used to delimit sequences, and square brackets may be used to delimit identifiers or values. The dot notation is used to move down a hierarchy. Notice that Time is drilled down to quarters and Location is drilled down to states. A colon ":" is used to indicate a range (like ".." in English). The result grid from the above query has two columns for units sold and revenue, and three rows for the first three quarters of 2020.

Different OLAP tools work differently. SQL:1999 onward includes support for cubes and roll up, so it is hoped that some OLAP features will become more standardized in the future. A full coverage of MDX would require another chapter or two, so it is not pursued further here. For an extensive treatment of MDX for SQL Server 2022, see MDX Language Reference (MDX) - SQL Server | Microsoft Learn.

MDX queries can be very complex, and Microsoft SQL Server provides a very high-level language called *English Query* that can be used by end users to generate MDX queries. Natural language-based queries are discussed in the next section.

18.3 Conceptual Query Languages

An information system may be modeled at any of four levels: conceptual, logical, physical, and external. In principle, it may also be *queried* at any of these levels. In practice, however, although conceptual modeling is widely used, it is comparatively rare for systems to be queried at the conceptual level. Instead, queries are typically formulated either at the external level using forms, at the logical level using a language such as SQL, or at the physical level using a programming language.

At the external level, Query By Form (QBF) enables users to enter queries directly on a screen form by entering appropriate values or conditions in the form fields. This form-based interface is well suited to simple queries where the scope of the query is visible on a single form and no complex operations are involved. However, this cannot be used to express complicated queries. Moreover, saved QBF queries may rapidly become obsolete as the external interface evolves. For such reasons, QBF is too restrictive for serious work.

For relational databases, SQL and Query By Example (QBE) are more expressive. However, complex queries and even queries that are easy to express in natural language (e.g., who does not speak more than one language?) can be difficult for nontechnical users to express in these languages. Moreover, an SQL or QBE query often needs to be changed if the relevant part of the conceptual schema or internal schema is changed, even if the meaning of the query is unaltered. Finally, relational query optimizers ignore many semantic optimization opportunities arising from knowledge of constraints.

Query languages for object-oriented and object-relational DBMSs suffer similar problems. Their additional structures (e.g., sets, arrays, bags, and lists) often lead to greater complexity in both user formulation and system optimization. For example, object query language (OQL) extends SQL with various functions for navigation as well as composing and flattening structures, thus forcing the user to deal directly with the way the information is stored internally (Cattell and Barry, 2000). At the physical level, programming languages may be used to access the internal structures directly (e.g., using pointers and records), but this very low-level approach to query formulation is totally unsuitable for end users.

Several commercial *SQL query builder tools* are now available for users to easily create SQL queries without any knowledge of SQL. They offer drag-and-drop or point-and-click interfaces that allow nontechnical users to build SQL queries quickly. Some also provide a charting environment for saving SQL query results as charts, reports, and dashboards. Details on several of these may be accessed at 9 Best SQL Query Builder & Generator Tools (2023) (guru99.com) and www.activequerybuilder.com/.

Some commercial tools allow users to enter queries directly in English, and then translate these into database queries in a language such as SQL or MDX. Early tools of this type often suffered from problems with ambiguity and expressibility as well as the correctness of the SQL code generated. Most of the natural language input tools we've looked at require you to spend considerable effort setting up a dictionary to relate conceptual terms to the underlying database structures, and most of them do an imperfect job at handling natural language input. This is not really surprising, given the inherent ambiguity and complexity of natural language.

One approach to address the ambiguity problem is taken by *English Query*, which was included many years ago in Microsoft SQL Server, but is no longer available. When you enter a question in English, it's rephrased to indicate the interpretation taken. Sometimes, more than one rephrasing is given, and you can pick the one that conveys your intended meaning.

Of course, there are limits on what kinds of questions can be understood, and you need to spend time setting up the connections between your semantic model and the database structures. For example, if you want to make a join between tables other than a foreign key to primary key reference, you need to explicitly add this beforehand. Also, the SQL that's generated might not be as efficient as you might construct yourself.

However, once you've done all this work, you can make it possible for nontechnical end users to query the database in plain English. Although common, predictable queries are best made available through prewritten procedures, English Query was especially useful for ad hoc queries. For example, with reference to the data mart schema in the previous section, and assuming adequate setting up, the question "Which region has the most stores?" might generate the following SQL code:

```
select top 1 with ties dbo.Store.region as "Region", count(*) as "count"
from dbo.Store
where dbo.Store.region is not null
group by dbo.Store.region
order by 2 desc
```

In addition to returning the region(s) with the most stores, this query returns the number of stores in those regions, which we might not want. Moreover, the **where** clause and the **order by** clause are both redundant. But these are small issues for the nontechnical user who is incapable of formulating the SQL query. For OLAP applications, English Query was also used to generate MDX queries. For example, referencing the SalesCube discussed in the previous section, the question "List the three items with the most unit sales" might result in the restatement "Show the three items with the highest total unit sales" and the following MDX query:

```
select { Measures.[UnitsSold] } on columns,
        topcount([Item].[ItemCode].members, 3, Measures.[UnitsSold] ) on rows
from [SalesCube]
```

As these examples illustrate, free input in natural language has a lot of potential for opening up databases to anybody who can write. As language recognition technology matures, it will also open up databases to anybody who can speak. However, we still have a long way to go before this becomes a totally reliable technology.

Several *conceptual query languages* have been proposed to allow users to formulate queries directly on the conceptual schema itself. Most of these language proposals are academic research projects, with at best prototype tool support, and are typically based on ER schemas (e.g., ERQL, Super, Hybris) or deductive models (e.g., CBQL). These are often challenging for naïve users, and their use of attributes exposes their queries to instability because attributes may evolve into entities or relationships as the application model evolves.

These problems of ambiguity and onerous dictionary preparation can be eliminated in one stroke by taking a different approach: *conceptual-schema-based queries*. Here, the user issues queries by selecting paths through the existing conceptual schema rather than using free text input, so no ambiguity can arise. As discussed later, if an ORM conceptual schema is used, there is no need to predeclare any join paths because ORM object types are the semantic domains on which all joins are based. Because ORM is attribute-free in its base conceptual model, it also avoids the instability problems of attribute-based queries, which require reformulation when attributes evolve into relationships (see later example). Moreover, ORM's sole data structure is the relationship type, providing a simple, sentence-based framework. For such reasons, we believe that ORM conceptual query technology has the greatest potential of any existing approach.

The first significant ORM-based query language was Reference and Idea Language (RIDL), a hybrid language with both declarative and procedural aspects (Meersman,

1982; Meersman et al., 1984). RIDL is a textual language for defining, populating, and querying models in NIAM, an early version of fact-oriented modeling. It was developed at the Control Data research laboratory at the University of Brussels, and was the first truly conceptual query language ever implemented. Let's look at some RIDL examples. Consider a military conceptual schema with the following fact types: Officer has Rank; Officer was born in Year. In RIDL, these fact types would normally be declared as Officer having Rank; Officer born-in Year. To find which sergeants were born before 1950, the following RIDL query may be used:

```
LIST Officer (having Rank 'sergeant' AND born-in Year < 1950)
```

In RIDL, reserved words are in uppercase. To illustrate the procedural side of RIDL, the following query lists pairs of officers of the same rank who were born before 1950:

```
FOR EACH o1, o2 IN Officer born-in Year < 1950 DO
    IF Rank OF o1 = Rank OF o2
        THEN LIST o1, o2
    END-IF
END-FOR
```

The original RIDL software mapped RIDL queries into internal statements that were directly executable on some early Control Data computers. In later years, the data definition part of RIDL was implemented graphically in the RIDL* tool, but the query component was not supported.

Another ORM query language is Language for Information Structure and Access Descriptions (LISA-D). Although very expressive, it is technically challenging for end users, and currently lacks tool support (ter Hofstede et al., 1996).

Like ORM, the Object-Oriented Systems Modeling (OSM) approach avoids the use of attributes as a base construct. An academic prototype was developed for the graphical query language OSM-QL based on this approach (Embley et al., 1996). For any given query, the user selects the relevant part of the conceptual schema, and annotates the resulting subschema diagram with restrictions to formulate the query. Negation is handled by adding a frequency constraint of "0", and disjunction is introduced by means of a subtype-union operator. Projection is accomplished by clicking on the relevant object nodes and then on a mark-for-output button.

Another ORM query language is *ConQuer* (the name derives from "CONceptual QUERy"). ConQuer is more expressive than OSM-QL, and easier for novice users. Its tool implementations transform conceptual queries into SQL queries for a variety of back-end DBMSs, and do not require the user to be familiar with the conceptual schema or the ORM diagram notation. The first version of ConQuer was released in InfoAssistant (Bloesch and Halpin, 1996). Feedback from this release led to the redesign of both the language and the user interface for greater expressibility and usability, resulting in a tool called ActiveQuery (Bloesch and Halpin, 1997). Typical queries can be constructed by just clicking on objects with the mouse and adding conditions. The ActiveQuery technology was later acquired by Microsoft.

A basic understanding of the ConQuer technology will help provide insights into a conceptual query environment. The ConQuer language and its tool support were designed for expressibility, clarity, simplicity, semantic stability, and semantic relevance. These design principles were explained in Section 3.1. A ConQuer query can be applied only to an ORM schema. Using a software tool, an ORM schema may be entered directly, or instead reverse engineered from an existing logical schema (e.g., a relational or object-relational schema). While reverse engineering is automatic, some refinement by a human improves the readability (e.g., the default names generated for predicates are not always as natural as a human can supply).

Although ConQuer queries are based on ORM, users don't need to be familiar with ORM or its notation. A ConQuer query is set out in textual (outline) form (basically as a tree of predicates connecting objects) with the underlying constraints hidden because they have no impact on the meaning of the query.

With ActiveQuery, a user can construct a query without prior knowledge of the schema. On opening a model, the user is presented with an object pick list. When an object type is dragged to the query pane, another pane displays the roles played by that object in the model. The user drags over any relationships of interest. Clicking an object type within one of these relationships displays its roles, and the user may drag over those of interest, and so on. In this way, users may quickly declare a query path through the information space without any prior knowledge of the underlying data structures. Users may also initially drag across several object types. The structure of the underlying model is then used to automatically infer a reasonable path through the information space, which is called a *point-to-point query.*

Items to be displayed are indicated with a check mark "✓". These check marks may be toggled on/off as desired. The query path may be restricted in various ways by the use of operators and conditions. As a simple example, suppose a company has branch offices in several cities around the world. Now, consider the query, List each employee who lives in the city that is the location of branch 52. This may be set out in ConQuer thus:

```
Q1   ✓Employee
           └─ lives in City
                    └─ is location of Branch 52
```

This implicit form of the query may be expanded to reveal the reference schemes (e.g., EmployeeNr, BranchNr), and an equals sign may be included before "52". Because any ConQuer query corresponds to a qualified path through an ORM schema, where all the object types and predicates are well defined, the meaning of the query is essentially transparent. ActiveQuery also generates an English verbalization of the query in case there is any doubt. This ensures *semantic clarity.*

Because ORM object types are semantic domains, they act as semantic "glue" to connect the schema. This facilitates not only strong typing but also query navigation through the information space, enabling joins to be visualized in the most natural way (recall Section 4.6). The underlying ORM schema fragment for query Q1 is shown in the shaded path within Fig. 18.5, along with the identification scheme for City.

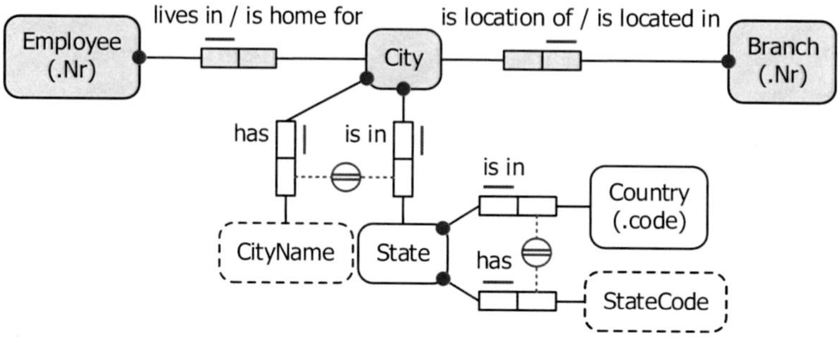

FIG. 18.5 Only the shaded path is relevant to query Q1.

Notice how City is used as a join object type for this query. If attributes are used instead, the query formulation becomes more cumbersome. If composite attributes are allowed, we might use **List** Employee.empNr **where** Employee.city = Branch.city **and** Branch.branchNr = 52. If not, we might resort to **List** Employee.empNr **where** Employee.cityName = Branch.cityName **and** Employee.stateCode = Branch.stateCode **and** Employee.country = Branch.country **and** Branch.branchNr = 52. Apart from awkwardness, both of these attribute-based approaches violate the principle of *semantic relevance.* Because the identification scheme of City is not relevant to the question, the user should not be forced to deal explicitly with it.

Even if we had a tool that allowed us to formulate queries directly in ER or OO models, and this tool displayed the attributes of the current object type for possible assimilation into the query (similar to how ActiveQuery displays the roles of the highlighted object type), this would not expose immediate connections in the way that ORM does. For example, inspecting Employee.city does not tell us that there is some connection to Branch.city.

At the relational level, we have two tables, *Employee*(empNr, countryCode, stateCode, cityName, ...) and *Branch*(branchNr, countryCode, stateCode, cityName, ...). Because there are no foreign key references to relate the city attributes in these tables, free text query tools rely on the join connection being explicitly entered into the dictionary beforehand, which is a risky assumption. The only way to automatically reveal all join possibilities is to use the domains themselves as a basis for connectedness, and this is one of the distinguishing features of ORM.

Because ConQuer queries are based on ORM, they continue to produce the desired result as long as their meaning endures. In other words, you never need to change a ConQuer query if the English meaning of the question still applies. In particular, ConQuer queries are not impacted by typical changes to an application, such as the addition of new fact types or changes to constraints or the relative importance of some feature. This ensures *semantic independence* (i.e., the conceptual queries are independent of changes to underlying structures when those changes have no effect on meaning). This results in greater *semantic stability* than attribute-based approaches.

As a simple example, suppose that in our model each employee is identified by an employee number and has exactly one gender and at most one title. In the real world, an employee may have many titles (e.g., "Dr.", "Prof.", "Sir"), but we decide to record at most one of these.

Now suppose we want to list the titled employees and their gender. An ORM schema and ConQuer query for this are shown in Fig. 18.6(a). There is an implicit **and** between the two branches stemming from Employee. The rest of the figure shows an equivalent UML class diagram, relational schema, and SQL query. Not only is the SQL code subconceptual (nulls are an implementation detail), but it is unstable because a simple change to a conceptual constraint on the title relationship requires the code to be changed as well.

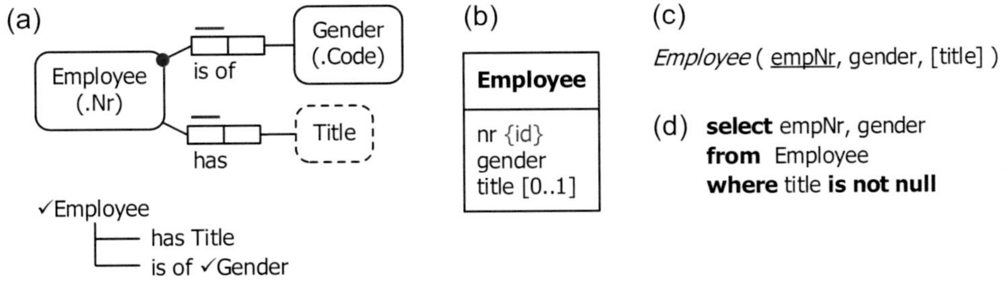

FIG. 18.6 Schemas and queries to list titled employees and their gender.

For example, suppose we change our business rules to allow the recording of many titles for the same employee, and we also decide to record any required orders for listing titles (e.g., the combination "Prof. Dr." should be listed that way, not "Dr. Prof."). From an ORM viewpoint, this new situation is depicted in Fig. 18.7(a). The uniqueness constraint on the title association is relaxed from *n*:1 to *m*:*n*, and a title precedence fact type is added. These schema changes have no impact on the ConQuer query, whose meaning relates just to the shaded fact types, not to their constraints and not to the additional fact type. Because the shaded path has unchanged semantics, the original ConQuer query means the same, so it may be reused.

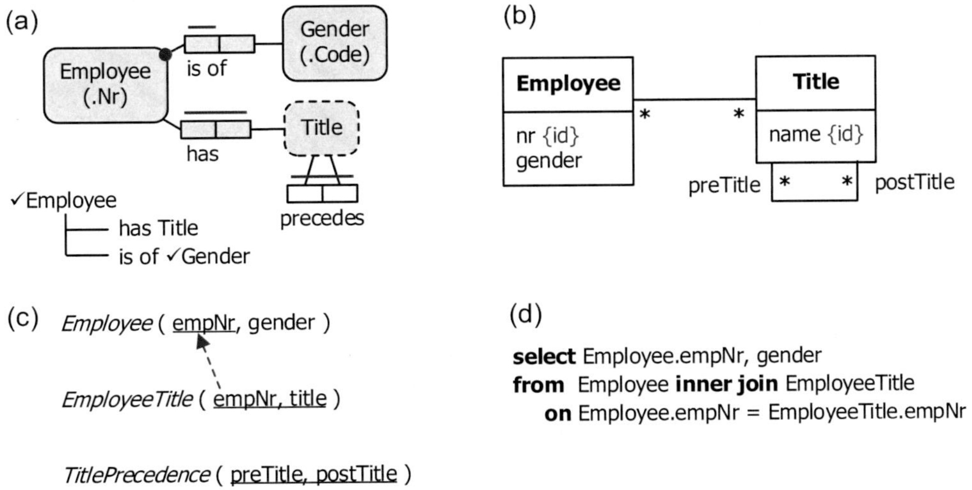

FIG. 18.7 ORM has greater semantic stability than attribute-based approaches.

For attribute-based approaches, the business change requires major changes to both the model and the query. The rest of Fig. 18.7 shows the new UML schema, the new relational schema, and the new SQL query. Substantial changes are needed because facts about titles are now stored in an association rather than an attribute. Another case forcing facts modeled with title and even gender attributes to be remodeled with associations was discussed in Section 10.1 to record restrictions of some titles (e.g., "Sir") to specific genders. In contrast, the ConQuer query stays valid; all that changes is the SQL code that gets automatically generated from the query.

An OO query approach is often more problematic than an ER or relational approach because of the many extra choices on how facts are grouped into structures, and the user is exposed to these structures in order to ask a question. Moreover, these structures may change drastically to maintain performance as the business domain evolves.

In the real world, UoD changes often occur, requiring changes to the database structures. Even more work is required to modify the code for stored queries. If we are working at the logical level, the maintenance effort can be very significant. We can minimize the impact of change to both models and queries by working in ORM at the conceptual level and letting a tool look after the lower-level transformations.

The ActiveQuery tool generates *semantically optimized* SQL queries wherever possible using knowledge of ORM constraints to produce more efficient code. For example, if the title role were actually mandatory, the SQL generated by the previous ConQuer query would, in both cases, simplify to **select** empNr, gender **from** Employee. The tool can also generate different code for different back ends to cater for their different support for the SQL standard, sometimes generating chains of queries over temporary tables to enable the query to run at all.

The simple examples just given illustrate how ConQuer achieves semantic clarity, relevance, and stability. Let's look briefly at its semantic strength or expressibility. The language supports the usual comparators (=, <, **in**, **like**, etc.), logical operators (**and**, **or**, **not**), and bag functions (**count**, **sum**, etc.) as well as a modal operator (**possibly**) for conceptual left outer joins. Subtype/supertype connections appear as "**is**" predicates.

For example, assuming that Manager is a subtype of Employee in the ORM model, the following query Q2 asks, List each manager and their cars (if any) where the managers do not work at branch 52.

```
Q2  ✓Manager
        └ is Employee
                ├ possibly drives ✓Car
                └ not works at Branch 52
```

The language is fully orthogonal and supports arbitrary correlation, using subscripted variables (e.g., $Person_1$, $Person_2$) when necessary to distinguish different instances of the same object type. As a simple correlation example, consider the query, "Who supervises an employee who lives in the same city as the supervisor but was born in a different country from the supervisor?" Assuming birth and supervision fact types in the ORM model, query

Q3 shows one way to express this in ConQuer.

```
Q3  ✓Employee1
          ├─ lives in City1
          ├─ was born in Country1
          └─ supervises Employee2
                        ├─ lives in City1
                        └─ was born in Country2 <> Country1
```

When an object type appears more than once, ActiveQuery automatically appends subscripts to distinguish the occurrences. You can equate instances by equating their subscripts (e.g., $City_1$). More generally, you can use comparators to compare instances (e.g., $Country_2$ <> $Country_1$). Try this in SQL. It's not that hard, but you have to admit it's easier in ConQuer!

As a final example, suppose the ORM model includes the *m:n* fact type Employee has Rating. Now, consider the query, "List each branch and those of its employees whose maximum individual rating exceeds the average of his/her branch" (remember that an employee may have many ratings). Query Q4 shows how to do this in ConQuer, illustrating the use of bag functions in for-clauses.

```
Q4  ✓Branch
          └─ employs ✓Employee
                        └─ achieves Rating
                                   └─ max(Rating) for Employee >
                                      avg(Rating) for Branch
```

At the relational level, two tables are needed: *Employee*(empNr, branchNr, ...); Achieves (empNr, rating). The SQL for this query is shown here. This is tricky, especially the final correlation between branch numbers, and even experienced SQL programmers might have difficulty with it. The equivalent ConQuer query would be easy for most end users who have even minimal experience with the language. This gives some indication of the potential of a language such as ConQuer for empowering end users.

```
select max(X1.branchNr), X2.empNr
from Employee as X1, Achieves as X2
where X1.empNr = X2.empNr
group by X2.empNr
having max(X2.rating) >
   (select avg(X4.rating)
    from Employee as X3, Achieves as X4
    where X3.empNr = X4.empNr
      and X3.branchNr = X1.branchNr)
```

ActiveQuery allows you to define derived predicates and store these definitions. These derived predicates (or "macros") can then be used just like base predicates in other queries. A subtype may be thought of as a derived object type, with its definition provided by a ConQuer query. For example, the population of the subtype MalePatient can be obtained by executing the following ConQuery query: **Find each** Patient **who** is of Gender 'M'. Given its generality, ConQuer could be adapted to provide a very high language for specifying derivation rules.

A constraint may be viewed as a check that a query searching for a violation of the constraint returns the null set. Hence, constraints may also be expressed in terms of queries. Various higher-level constructs can be provided in the language to make it more natural than the not-exists-check-query form provided in SQL. Although there is no room here to go into detail, it should be clear that this approach is quite powerful.

ORM's rich graphical constraint notation still needs to be supplemented by a textual language to provide complete coverage of the kinds of constraints and derivation rules found in business domains. While ORM's graphical constraints can be verbalized automatically, input textual languages such as ConQuer could be used to define other business rules mapped automatically to SQL. Although the ActiveQuery tool is currently no longer available, work on the NORMA tool is under way to extend the FORML textual language for ORM with the expressive power of ConQuer. This should result in a very higher-level textual language for both conceptual modeling (including fact types, constraints, and derivation rules) and conceptual queries.

18.4 Schema Abstraction Mechanisms

At times, we might feel almost swamped by the level of detail captured on an ORM schema diagram, or even on a UML, ER, or relational schema. Nevertheless, such detail is important when developing, transforming, or mapping schemas. The ORM notation has been crafted to facilitate these tasks. As we've seen, the diagrams may be verbalized naturally and populated with fact instances, their object-role-based notation allows many constraints to be expressed intuitively, and their object types reveal the semantic domains that glue the schema together. All this helps the modeler to get a complete and correct picture, and to transform the model in a rigorous way with formal control of information loss or gain.

However, once a schema has been developed, we may at times wish to hide some of the information to obtain a quick overview or to focus on some aspects. This is particularly the case if the schema is large and complex. Hence, there is a need for *abstraction mechanisms* by which unwanted details may be removed from immediate consideration. This section outlines a few of the more useful ways of doing this. The chapter notes provide references for further study in this regard.

One obvious abstraction strategy is *modularization*. Here, the complete schema is "divided up" into a number of conveniently sized modules or subschemas. One way to do this is to use different schemas for different modeling *perspectives*. For example, UML uses class and object diagrams for a data view, state charts and activity diagrams for behavioral views, and so on. In addition, you might use a purely conceptual schema for analysis and annotated or modified conceptual schemas for a design view.

Within a given modeling perspective, the global schema is still likely to be too large to conveniently inspect in one go, even though wall-sized schemas can be very useful at times for seeing far-reaching connections. One trivial way to divide a global schema diagram is to overlay a grid to partition the schema into separate cells.

However, it is typically much more useful to allow modules to overlap so that, for example, the same object type might appear in more than one module. This technique also allows greater flexibility in basing modules on semantic groupings. One useful technique is to divide the global schema into *reusable subschemas*, which can be imported into other schemas or projects at a later time. The most popular technique of all is to divide a given schema into *pages*. Each page should be given a meaningful name and include a reasonable number of tightly coupled modeling elements.

Various means may be used to specify connections between modules, such as hyperlinks, annotations, and border overlaps (cf. a directory of road maps). Electronic browsing opens up greater possibilities than hard-copy browsing (e.g., scrolling, zooming in and out, hypertext navigation), but printed documentation on standard pages needs to be catered for as well.

With large applications, the original schema might be developed by different teams as separate modules to be later integrated. Additional care is then required for the *schema integration* process. In particular, global consistency must be ensured, either by agreeing on a uniform treatment of terminology, constraints, and rules or by specifying appropriate translations. Ideally, a basic architectural framework will be agreed upon early in the development, and all the modelers will attach the same meaning to words by resolving any synonym and homonym problems.

For example, if the same object type is called "Subject" in one module and "Course" in another, then one of these will be chosen as the standard term and the other term replaced by it. If different object types or fact types in different modules are given the same name or reading, one of these must be changed. As long as they uniquely refer, alternative names and readings may be stored as aliases.

Global identification schemes and constraints should be agreed upon. Where necessary, contextual identification may be supported via subtyping, and textual constraints can be added to strengthen global constraints for a given context (e.g., restricted mandatory role and restricted uniqueness constraints). For some applications (e.g., federated databases), partial integration between relevant modules may be favored instead of global integration.

Another useful abstraction mechanism is provided by *display toggles,* or *layering.* For instance, the display of role names and selected classes of constraint or derivation rule may be toggled off when they are not of immediate interest to obtain a simpler picture. On an ORM schema, for example, we might toggle off value lists, ring constraints, frequency constraints, set-comparison constraints, textual constraints, and derivation rules to focus on uniqueness and mandatory role constraints. Each constraint class may be thought of as a constraint layer, with the modeler choosing which layers to view at any given time. An extreme option would be to toggle off all constraints.

Subtype display can be suppressed in many ways. We can hide the subtype defining rules or graphic constraints (e.g., exclusion and totality). We can hide just some of the subtypes, preferably indicating somehow that others exist (e.g., with an ellipsis or one of the UML annotations discussed in Chapter 9). We can also collapse subtypes into their top supertype. Depending on the model type, various other modeling elements can be hidden (e.g., reference schemes or attributes). When additional layers of detail are needed, their display can be toggled on again. Although this layering concept can be adapted to hard copy (e.g., by using a series of overlaid transparencies or progressively detailed printouts), it should ideally be exploited interactively with a computer-aided software engineering (CASE) tool.

Yet another abstraction mechanism is *object-type zoom*. Here, the modeler selects an object type of interest to have the display focused on that object type and its immediate *neighborhood*. The object type is displayed with all its fact types. By specifying a *logical radius* for the zoom (e.g., in predicate units), the neighborhood may be expanded to include fact types of the object types in the first-level zoom, and so on.

ActiveQuery uses this mechanism to progressively open up the universe to the user, displaying a neighborhood of one-predicate radius for the currently highlighted object type. The NORMA tool also provides a context window for this purpose, allowing expansion up to three predicate lengths.

A simple example will help illustrate some of the basic ways of abstracting. Fig. 18.8 depicts an ORM schema for a small application. Here, movies are identified by numbers, although the combination of their title and director is also unique. Some movies are based on another. For example, the western *The Magnificent Seven* was based on an early Japanese movie, *The Seven Samurai*.

When available, figures about the gross revenues of a movie are recorded, and the net profit too if this is known. The only people of interest in this application are movie directors and movie stars. We record their country of birth and, if known, their birthdate. The country in which a movie was made is noted as well as any export countries for it.

As a mental exercise, imagine toggling off the display of the constraint types on Fig. 18.8 one at a time (ring, then set-comparison, then mandatory role, then uniqueness) to visualize abstraction by peeling off constraint layers.

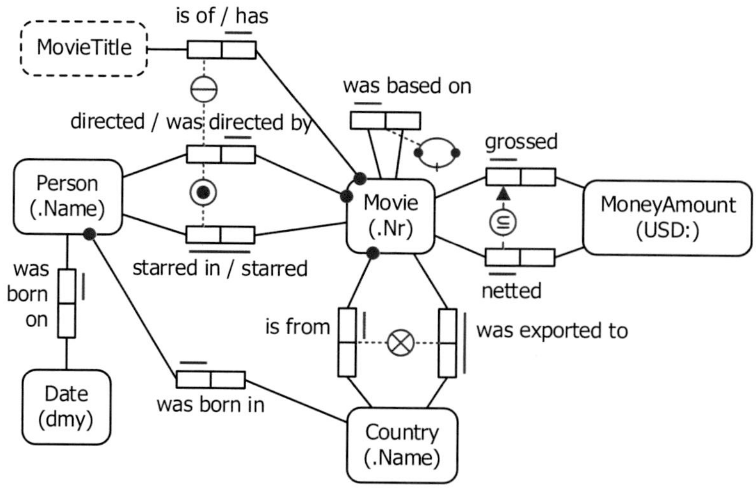

FIG. 18.8 A small but detailed ORM schema.

Fig. 18.9 depicts a zoom on the object type Person, with a logical radius of 1, so that only the fact types directly connected to Person are shown. To highlight the zoom object type, a thicker line is used for it. Visualize for yourself a zoom on Country; this would display three fact types, one of which is also contained in the Person zoom.

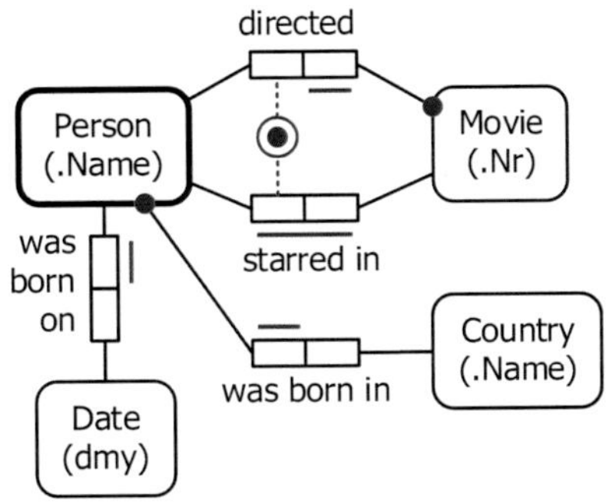

FIG. 18.9 Zooming in on the Person object type.

Another abstraction mechanism especially useful with large ORM schemas is that of *refinement levels* based on *major object types.* This can be used in conjunction with *attribute abstraction* to provide an ER schema or UML class diagram.

When developing a conceptual schema, it is useful to treat all object types equally because the relative importance of an object type is based on all the parts it plays in the global schema. Once the schema is complete, however, this information is available and can be used to determine which are the most important, or *major object types.* A procedure for deciding the major object types on ORM schemas was developed by Campbell and Halpin (1993) and was used as a basis for generating default screen forms for the external user interface (because application screens basically enable users to perform operations on major objects).

Once the major object types are decided, an ORM diagram can be lifted to a higher level of abstraction. For example, we may choose to display only the *major fact types* (i.e., those fact types in which at least two roles are played by a major object type). With large schemas, this diagram itself may be subjected to the same procedure, yielding a higher level abstraction, and so on, until the top level view is obtained. As long as the fully detailed schema is accessible, this bottom-up abstraction may be reversed, allowing the top-level view to be successively refined down to the detailed bottom level.

The detailed schema diagram in Fig. 18.8 has six object types: Movie, MovieTitle, Person, MoneyAmount, Country, and Date. Intuitively, these are not all equally important in this domain. Which are the major object types? Apart from using your intuition, you can use the constraint patterns to help decide. In particular, mandatory role and uniqueness constraints are relevant. If an object type has an *explicit* mandatory role, it is major. Usually, an object type playing a functional role is also major, but this is not always true (e.g.,

MovieTitle would still be a minor object type even if its role was functional; its role is still only implicitly mandatory).

There are some finer points to determining major object types, but this is enough to get the basic idea. Based on the previous reasoning, Fig. 18.8 has only two major object types: Movie and Person. Fig. 18.10 provides one overview of the application by displaying only the major fact types.

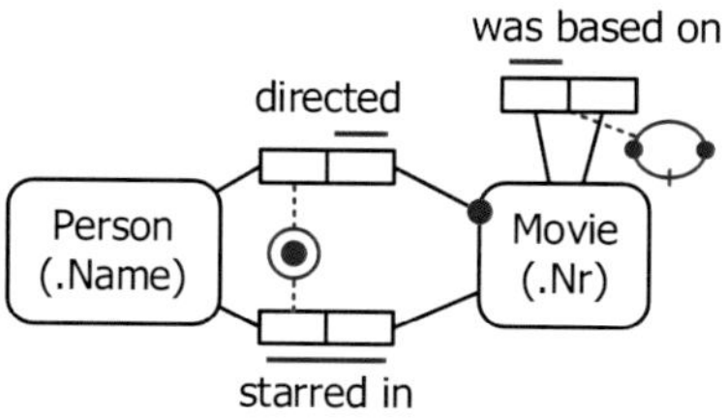

FIG. 18.10 Only the major fact types are shown.

Instead of being hidden, minor fact types may, if desired, be viewed in terms of *attributes* of the major object types. Although this attribute viewpoint hides structural information (e.g., domains and certain constraints), the ER or UML diagram obtained provides a compact picture that is useful for quick overviews. Fig. 18.11 shows a UML class diagram for the current example. Unlike most ER notations, UML allows the attribute domains to be specified if desired, and the missing constraints (here disjunctive mandatory, subset, exclusion, external uniqueness, and asymmetric constraints) could be documented in notes, although at the price of losing compactness.

Notice that country information is displayed here in terms of attributes (single and multivalued) rather than as associations with a Country class. This is because Country is a minor object type in the original ORM schema. If instead, Country played an explicit mandatory role or a functional role not already absorbed, it would become major, and we would display it as a class. This is usually the case (e.g., we normally store both the code and name of countries).

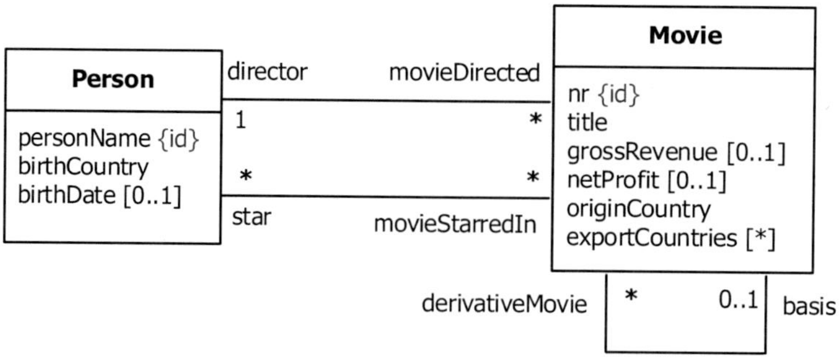

FIG. 18.11 A UML class diagram abstracted from Fig. 18.8.

Attribute-based diagrams can be made more compact by hiding all their attributes. For example, Fig. 18.12 shows just the major fact types in both UML and Barker ER notations. Again, some ORM constraints are missing (inclusive-or and asymmetric constraints). By using ORM to actually generate such diagrams, or detailed UML and ER diagrams, we can retain the benefits of ORM (e.g., for validation and completeness) and still have compact views when desired.

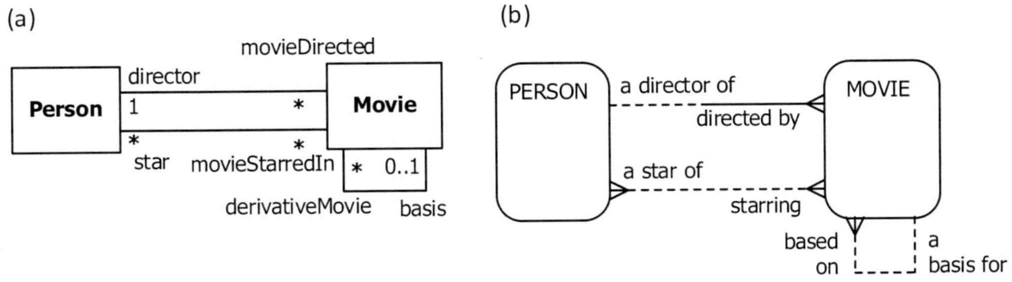

FIG. 18.12 Major fact types displayed in (a) UML and (b) Barker ER.

18.5 Further Design Aspects

We have seen how a data model may be specified at the conceptual level (e.g., with an ORM schema) and the logical level (e.g., with a relational schema). The overall information systems life cycle includes other tasks, as discussed in Chapter 2. Chapter 15 discussed how to specify business processes using UML and business process model and notation (BPMN). This section provides a brief overview of other design tasks such as relating processes to data, organizing large models, and specifying the user interface.

Suppose we need to design an information system for an automated teller machine (ATM) to be operated by a credit union. As part of the requirements analysis phase, we determine that four main functions of the system are to enable clients to make deposits, withdrawals, and transfers (e.g., from a savings account to a checking account) as well as obtain printouts of account balances. In UML, each of these functions may be treated as a use case and included in a use case diagram to provide an overview of the processes supported by the system. We might then flesh out these functions in a set of activity diagrams in UML or workflow diagrams in BPMN.

Suppose that when a client inserts a credit card into the ATM, the system checks whether the card is one of its own and, if so, accesses the client's identifier (client number) from the card. The same account could be used by many clients (e.g., a joint account for husband and wife), and each client has a distinct client number. In this way, the system can track which client actually makes any given transaction.

Suppose in diagramming the ValidateClient and GetBalance processes, we see that the combined process of validating a client and printing a requested balance involves two input data flows (clientNr, accountType) from the Client and one output data flow (balance) to the client. If we consider the credit union's account database system to be separate from the ATM, it may be thought of as another actor that actually provides the balance details in response to a request (e.g., an SQL query) sent from the ATM.

In such higher-level views, little or no detail is provided on the structure of the data store or on how the process manages to find the relevant data. Because such details are so far undefined, process diagrams often fail to provide an adequate basis for modeling the data perspective. Despite such informality, they can be of use in clarifying the UoD and for identifying functions to be included in the application's screen menus. Indeed, ordinary cartoons with intuitive icons can be quite useful for communication between modeler and domain expert in the early stages of requirements analysis.

In principle, a process model could be successively refined to elementary transactions that expose the underlying fact types. However, this approach is simply too time consuming for data intensive systems and is often unstable because business processes tend to change far more rapidly than the underlying data. To expedite the data modeling process, we should seed it with data use cases and apply a modeling procedure such as the ORM conceptual schema design procedure (CSDP). If examples already exist, use these. If not, sit down with the domain expert and generate them. These could be in the form of tables, forms, graphs, or whatever. By now, you should have the verbalization skills to do CSDP step 1 properly on any of these varieties, and because clients find it easier to work at the instance level, this is the safest way to go. Moreover, by doing this you've already started prototyping external screens and reports the clients want.

Process models should be formally connected to the data models. This is best done after the conceptual and logical data models are determined. Although skilled modelers can refine process models into data models, this approach tends to be error-prone. For example, suppose we bypass conceptual data modeling, instead refining the process models mentioned earlier for validating clients and getting balances into the following table scheme and SQL procedure. Here, "&1" and "&2" are input parameters or placeholders. When the procedure is executed, actual values are input to these parameters.

```
D1:  Account ( clientNr, accountType, balance )

P1:  select balance from Account
     where clientNr = &1
        and accountType = &2
```

Can you spot a potential problem with this design? It's easiest to see the problem if you do a conceptual schema for D1. This is depicted in Fig. 18.13. Each account has one or more users and has exactly one account type (e.g., L=loans, C=checking, S1=main savings, S2=other savings). That's fine. The problem lies in the identification scheme for Account. Although the combination of clientNr and accountType does identify an account in this domain, this should not be used for preferred reference because (clientNr, accountType) pairs do not relate to accounts in a 1:1 fashion.

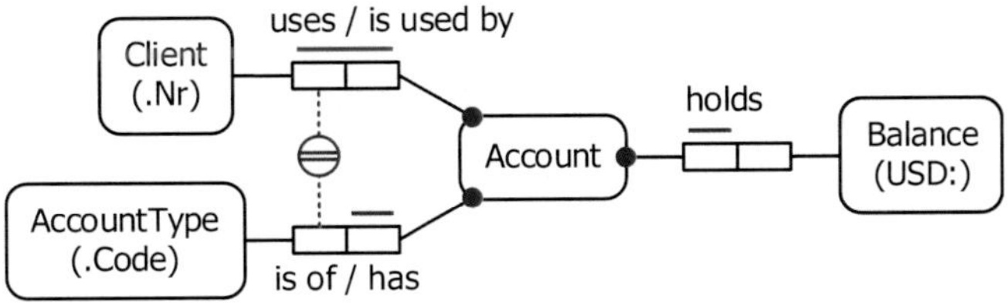

FIG. 18.13 The preferred reference scheme for accounts is unsafe.

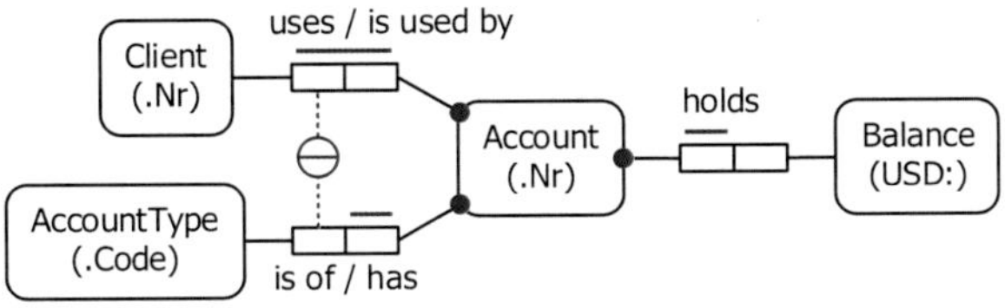

D1': *Account* (<u>accountNr</u>, accountType, balance)

AccountUse (<u>accountNr, clientNr</u>)

P1':
```
select balance
from   Account inner join AccountUse
  on   Account.accountNr = AccountUse.accountNr
       and clientNr = &1 and accountType = &2
```

FIG. 18.14 The preferred reference scheme for accounts is now safe.

For example, suppose your spouse's clientNr is 1001, yours is 1002, and you share a main savings account (type= "S1"). This joint account may be referenced by (1001, "S1") and by (1002, "S1"). This will not do for preferred reference. For instance, the Account table used for D1 will accept the population {(1001, "S1", 9000), (1002, "S1", 0)}. But this means that the same account has two different balances, which is nonsense. The obvious solution is to introduce another preferred identifier (e.g., accountNr), as shown in Fig. 18.14. This figure also includes the new relational schema (D1′) and the new procedure for computing the balance (P1′).

This example is simplified (e.g., Exercise 6.4 discussed how balances might be derived from account transactions) but it does illustrate how conceptual modeling facilitates a correct data design, and that a detailed process specification is of little use unless the data structure on which the process operates is defined accurately.

In practice, the use of accountNr for preferred reference should be obvious by inspecting other data examples associated with this use case or a related one for the accounting system (e.g., a balance statement). As discussed in Section 5.3, a reference scheme within a given context might not be suitable for the global application. Bear this in mind when discussing data use cases with domain experts.

Although a rigorous process model is best built on top of a data model, an overview of the processes can be useful as a precursor to data modeling, especially if the application is large or only vaguely understood. It is often helpful to get a clear picture of the functions of the application first.

Starting with the overall objectives, the main functions and processes of the application may be specified systematically using a variety of diagrams. This kind of analysis helps us divide the domain into coherent modules of manageable size and facilitates the task of specifying information examples where none exist. At this point, CSDP step 1 may be applied to each module.

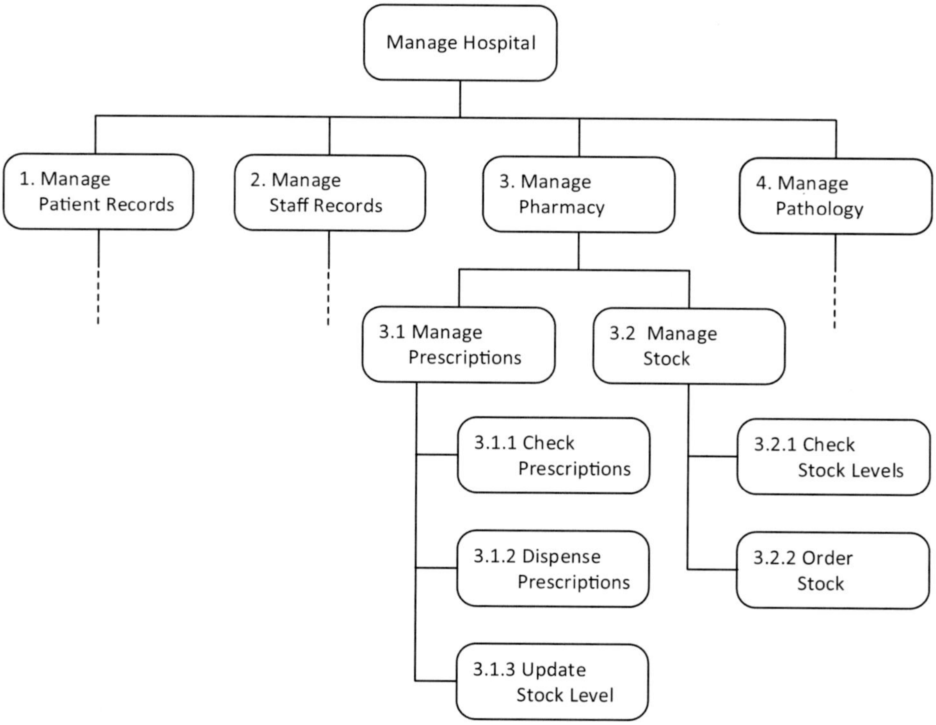

FIG. 18.15 A function tree for a hospital information system.

Functional requirements may be laid out in top-down form using a *function tree*. For example, Fig. 18.15 lists functions of a hospital information system. The four main functions are numbered 1..4. Each of these is refined or decomposed into a number of lower-level functions, numbered 1.1, and so on.

For example, function 3 is refined to functions 3.1 and 3.2. This may be refined further, for example, function 3.1 is refined to functions 3.1.1, 3.1.2, and 3.1.3. To help fit the tree on a single sheet, this third level has been laid out vertically rather than horizontally. Tools such as the Org-Chart solution in Microsoft Visio allow a very flexible layout of such diagrams. The other three main functions may be refined in a similar way. Other kinds of diagrams may be used to show dependencies between functions, which can help in deciding how to group functions into processes.

One common way of specifying information flows between processes is by means of *data flow diagrams*. Although invented long before UML use case diagrams, which have a somewhat similar purpose, data flow diagrams tend to be more useful in practice (Jeyaraj and Sauter, 2007) and are still popular.

By decomposing the processes, a data flow diagram at one level may be refined into several lower-level diagrams, and so on. As a precursor to this refinement, we might specify a very higher-level view in the form of a *context diagram*.

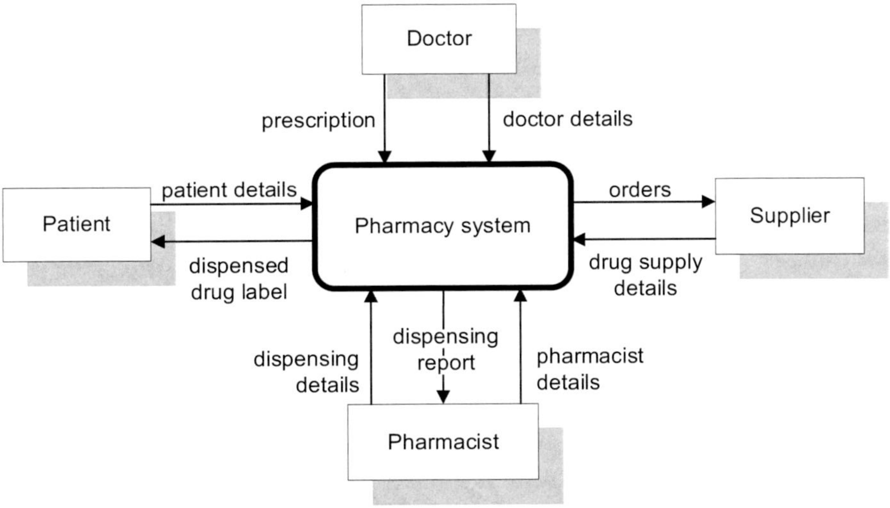

FIG. 18.16 A context diagram for the pharmacy subsystem.

A context diagram provides an overview of the interaction between the information system and the environment, treats the system as a single process, and ignores any data stores. A context diagram for the pharmacy module of our hospital application is shown in Fig. 18.16. External agents appear as shadowed boxes. Material flows are not shown, but some data flows may be associated with material flows (e.g., drug supply details).

At a lower level of detail, Fig. 18.17 shows a data flow diagram for a fragment of the pharmacy system. This could be further refined (e.g., decompose the order placement process into processes for computing totals, printing order forms, etc.).

However, because such refinements are time consuming, somewhat arbitrary, and lack the safeguards of population checks, there is usually little point in specifying low-level data flows. In general, then, once the requirements of the application are clear and modules of manageable size are identified, modeling efforts are best focused on obtaining a set of information examples and applying the CSDP. Once the conceptual schema is specified, the logical schema is obtained by applying a mapping procedure, such as Rmap.

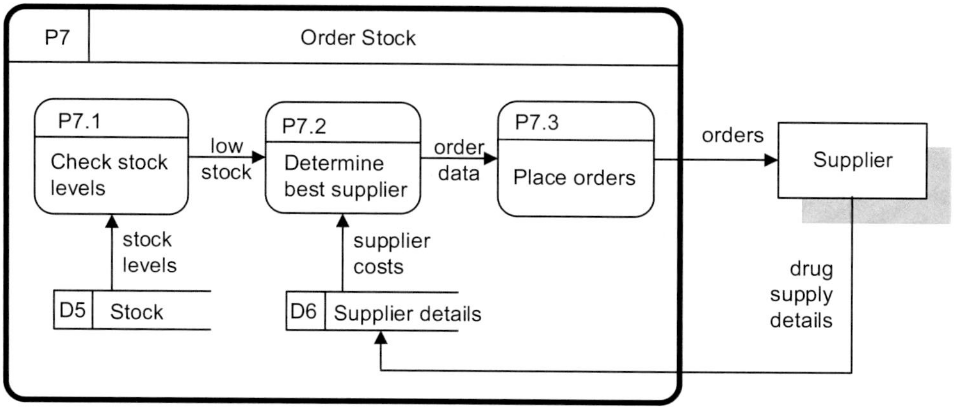

FIG. 18.17 A data flow diagram for part of the pharmacy subsystem.

The external schemas may now be developed for the various categories of user. It is through the interaction of users with the external interface that the original functions specified in the requirements analysis phase are ultimately executed. Hence, it is important to design the external interface in a way that naturally reflects these functions.

Fundamentally, all that happens when users interact with an information system is that elementary facts are added, deleted, retrieved, or derived. In principle, we could specify the process perspective in terms of these two update and two query operations. But this is too fine a level of granularity for humans to work with efficiently.

We need a way of grouping these elementary operations into larger units. It is often natural to think of operations being performed on significant *objects*, such as patients and stock items. The major object types abstracted from an ORM schema (see previous section) correspond closely to these object types, and hence provide a good basis for defining operations. The more important classes or entity types in UML class diagrams or ER diagrams provide another good choice for this.

A generic set of external update operations may now be generated with three varieties: add object, delete object, and modify object. Some generic query operations may be generated (e.g., list all facts about an object), but specific queries require an appropriate selection of the relevant fact types (stored or derived).

Users perform such operations through an *external interface*, typically *screen* forms or menus. For each screen, we need to consider *content* (which fact types and operations to include), *format* (how the screen is displayed), and *use* (how the operations are invoked by the user). As discussed, default decisions about screen content may be based on operations on the major object types. For example, if Patient is a major object type for the hospital system, a screen is required for it with at least the operations add Patient, modify Patient, delete Patient, show Patient.

Unlike conceptual and logical tables, forms may overlap in their content. This redundancy at the external level conveniently allows users to perform an operation whenever relevant, without having to switch screens. For example, consider the *m:n* fact type Doctor treated Patient. When working with a Doctor screen, we might query which patients were treated by that doctor. When at a Patient screen, we might query which doctors treated that patient. In the conceptual and logical schemas, facts of this type are stored only once. As long as the external screens have been properly related to the underlying data structures, this overlap is safe. For a discussion on the use of ORM schemas to generate screen forms and transactions, see Campbell and Halpin (1993).

Although there is only one (global) conceptual schema for an application, there may be many external schemas. Different user groups might have different *authorization* levels that can be specified in *access rights tables*. This may be done at the conceptual level (e.g., for each fact type, which groups have read, add, delete rights) or at the logical level (e.g., for each table or column, which groups have read, insert, delete, update rights).

Regarding access to options on screen forms, it is generally best to display to users only those options that they are authorized to select. As usual, access rights to screen options may be soft coded as data in tables rather than hard coded in procedures. This avoids having to recompile the forms when access rights are changed, and simplifies reuse of common code. As part of the overall security, passwords are required, and users are given the power to change their passwords.

Apart from access rights, users might also differ in their *ability* level. Designing application screens to optimize the way various users can perform their tasks is an important aspect of *human-computer interface (HCI) design*.

Many kinds of interfaces can be used. For example, dialog with the user may use natural language, short questions, command lines, menus, forms, tables, diagrams, images, or sound. Input devices include keyboard, mouse, joystick, pen, and microphone.

In general, the user interface should be *easy to learn* and *easy to use*. To achieve this, the interface should be consistent, simple, structured, efficient, and adapted to the user. Let's examine each of these in turn.

Consistency of interface implies that all the components use the same basic conventions for display format and operations. This promotes positive transfer (learning how to use one screen helps you with the others) and avoids negative transfer. Consistency may be achieved at different levels, such as within the one screen, within the application, among applications from the same company, and with external standards. For example, many Windows applications are being developed nowadays in conformity with Microsoft's user interface guidelines.

Simplicity of interface implies it is intuitive to use, with minimal effort for users. Each screen should provide the right amount of relevant information (e.g., limit the number of menu options seen at once). Don't clutter the screen. To allow many operations directly from the one screen, remove all but the main options from the permanent screen display, allowing the user to toggle extra options on as required. For example, use pull-down menus, pop-up windows (e.g., for look-up codes), and status-line help (and pop-up detailed help). To facilitate navigation within and between screens, consider the use of arrow keys for movement, and allow windows and menus to be cascaded or tiled.

Efficiency of interface entails that options may be selected quickly and intuitively. A *graphical user interface* (*GUI*) provides a more intuitive means of operation, especially where graphics or sophisticated layouts are involved; here, it is important that "what you see is what you get" (WYSIWYG). However, key-based shortcuts (e.g., hot keys) are sometimes faster and should be provided for the experienced user.

Menu design has a major impact on efficiency. Menus come in various shapes and sizes (full-screen, pull-down, pop-up, ring, button bars, etc.). They provide most, if not all, of the application's operations. Options on the same menu should be semantically related.

Navigation between menus or screens requires planning (state charts can help here), especially for web-based applications that use hyperlinks to navigate from one place to another. Some navigation patterns are shown in Fig. 18.18. Hyperlink access is typically cyclic.

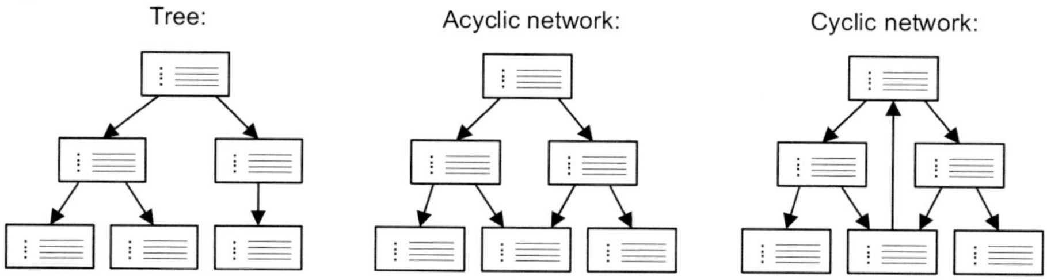

FIG. 18.18 Some ways of navigating between menus or screens.

In addition, the following principles should be observed; Order menu options in decreasing frequency of use. Vary color and fonts, but don't overdo it. For options, use either intuitive icons or terse, clear textual descriptions that are mostly in lowercase. Allow selection of options by mouse clicks or hot keys, and support undo operations.

Adapt the interface to the user as follows. Show only what is relevant to the user. Provide simple procedures for casual users and shortcuts for experienced users. Let users configure the environment to their own taste. Provide levels of context-sensitive help: short messages on status lines, detailed help in pop-up windows. Trap errors as soon as possible and give short and long error messages. Provide online tutorials, user manuals, technical support, and training courses.

If you use some software packages that you find particularly intuitive, have a close look at the interface design and see if you can adapt its good features to your own (as long as you don't violate any copyright in doing so!). Once you develop your prototype interface, test it on typical users and use their feedback to finesse it.

That's a quick run through several topics that could easily fill a book on their own. Indeed, whole books are devoted to interface design; many refinements exist for modeling processes, including some very technical approaches that attempt to formally integrate these models with data models. Some interesting work in this regard is referenced in the chapter notes. Various methods and tools are also in common use for associated tasks such as project management (e.g., Gantt charts and PERT charts), testing, performance measures, documentation, and versioning.

18.6 Ontologies and the Semantic Web

Etymologically, the term "ontology" means the study of existence. In philosophy, ontology is the branch of metaphysics concerned with the fundamental nature of being, addressing deep questions such as "Do nonphysical things exist?", "Does an object remain identical to itself when it changes?", and so on. The informatics community later adopted the term "ontology", typically using it to mean a conceptual model about some business domain where the model facilitates sharing information about that domain by conforming to some standard set of constructs.

With the rise of the Internet, millions of documents are now readily accessible over the World Wide Web. In 2001, to simplify sharing or combining information in different web documents, Sir Tim Berners-Lee, the founder of the Internet, proposed a new vision for a *semantic web* as an extension of the current web "in which information is given well-defined meaning, better enabling computers and people to work in cooperation" (Berners-Lee et al., 2001). The main idea is to add global identifiers and structure to the documents (e.g., using uniform resource identifiers and embedded tags) to reveal the underlying semantics of what is to be shared in a way that is accessible to automated agents.

In 1999, the World Wide Web Consortium (W3C) produced the first version of the *Resource Description Framework* (RDF) language as a standard on top of XML to capture metadata for web resources (e.g., who authored a web document). Other work on adding automatically accessible meaning to language via markup tags was conducted in the United States by the Defense Advanced Research Products Agency (DARPA), which produced the DARPA Agent Markup Language (DAML). With similar objectives, researchers in the European Union (EU) developed the Ontology Inference Layer (OIL). In 2001, joint work between the United States and the EU incorporated many concepts from OIL into DAML, resulting in the DAML+OIL language, which was submitted in late 2002 to the W3C.

In 2004, the W3C adopted *RDF Schema* (RDFS), which extended RDF to a simple ontology language, and also recommended the *Web Ontology Language* (OWL), which incorporates many aspects from DAML+OIL and RDFS, has a cleaner formal semantics based mainly on description logics, and is now the most popular language for developing ontologies on the semantic web. Fig. 18.19 summarizes the dependencies between layers, where upper layers depend on lower layers.

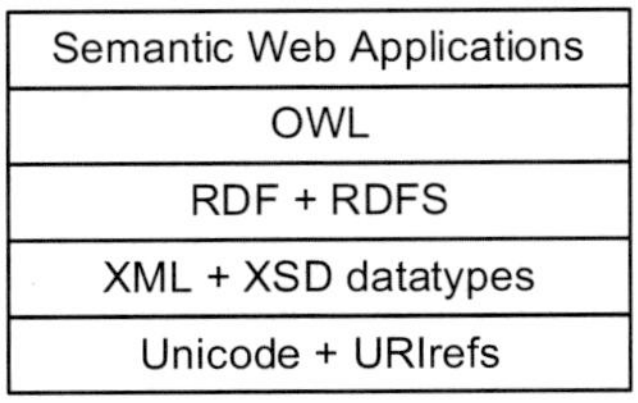

FIG. 18.19 Foundational layers for the semantic web.

RDF and RDFS

RDF models are basically directed graphs, where each node is either a resource or a literal, and each directed arc represents a binary relationship between the nodes. A resource is basically an entity instance or a logical predicate.

RDF 1.0 identifies resources by *Uniform Resource Identifier (URI)* references. A URI is a *Uniform Resource Locator* (URL) or a *Uniform Resource Name* (URN). A URL (e.g., http://www.orm.net) includes a network access type (e.g., http for hypertext transfer protocol) plus a network address to locate the resource. A URN (e.g., urn:isbn:1-55860-672-6) includes a name space (e.g., ISBN for International Standard Book Number) plus a name that identifies that resource within that name space. Unlike a URL, a URN does not provide a way to locate the resource on the web. Both URLs and URNs are global identifiers, so when used in different documents they refer to the same resource.

A URI reference (or URIref) is a URI, optionally followed by a fragment identifier, for example, the URI reference https://www.w3.org/TR/rdf-primer/#basicconcepts consists of the URI https://www.w3.org/TR/rdf-primer/, a "#" separator, and the fragment identifier basicconcepts. Character strings used for URIrefs are composed of Unicode characters.

Fig. 18.20 shows a simple RDF 1.0 model depicted as a graph. Here, we show resource nodes as named, soft rectangles; literals as named, hard rectangles; and relationships as named arrows from the subject node to the object node. The graph is intended to convey the semantics that there is a book identified by the ISBN 1-55860-672-6 that is authored by a person named "Terry Halpin".

RDF captures the information as simple *statements*, each represented as a *triple* (subject, predicate, object). The subject and predicate are resources, so are identified by a URIref. The object is either a resource or a literal. Literals may be untyped (as here) or typed (by pairing the value with a data type that itself is identified by a URIref to a standard XML Schema data type). Each predicate includes a predicate reading, for example, authored_by. Because the same predicate reading means different things in different contexts, a URIref is required to assign a unique meaning to a predicate.

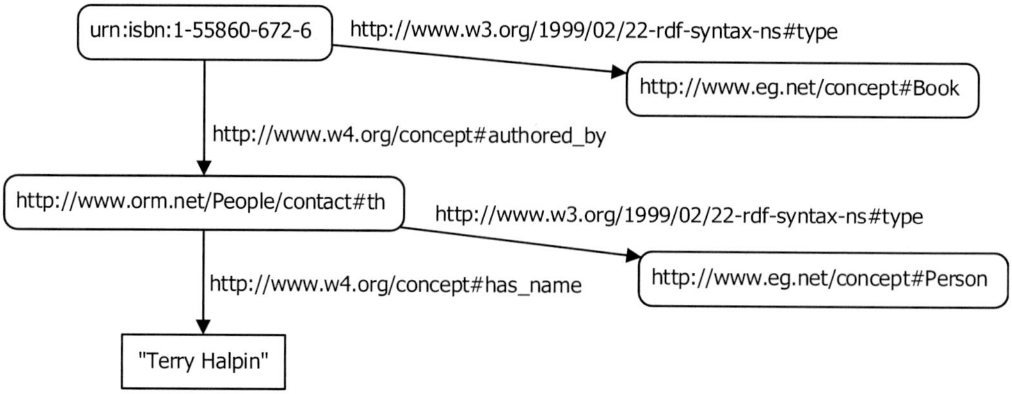

FIG. 18.20 A simple RDF 1.0 model.

The resource nodes on the left of Fig. 18.20 denote an individual book and an individual person. However, the resource nodes on the right denote the types Book and Person. The type predicate relates an instance to a type, and means "is an instance of". RDF also allows descriptions of resource collections (e.g., sets and lists), and RDF graphs can be serialized into XML format using RDF/XML.

As the example in Fig. 18.20 shows, RDF graphs allow nodes that are instances or types, and allow arcs that represent relationships between instances, or relationships between instances and types. RDF also allows relationships between types and types (e.g., Woman is_a Person). RDF even allows you to state that a type is an instance of itself (e.g., Class is_of_type Class), which can lead to formal problems such as Russell's paradox. Moreover, no special semantics is assigned to any predicate, so the is_an_instance_of relationship and the is_a_subtype_of relationships are treated the same as the has_name relationship (apart from different strings used for the URIrefs). So, there is no ability to perform inferences based on knowledge of special properties for certain predicates (e.g., transitivity of is_a_subtype_of).

RDFS builds on RDF by providing inbuilt support for classes and subclassing. For example, resources can be typed as classes (using rdfs:Class) and the predefined predicate rdfs:subClassOf is defined to be transitive. For a given predicate, the sets of instances playing the subject and object roles are called the *domain* and *range*, respectively. RDFS allows predicates to be typed by restricting their domain (using rdfs:domain) and range (using rdfs:range), respectively. Membership in a collection can also be specified using the rdfs:member property.

Despite these improvements, RDFS is still both too inexpressive (e.g., it cannot constrain associations to be other than *m:n*, and it cannot express complex properties by conjoining existing properties) and too expressive (e.g., like RDF, it allows you to state that Class is an instance of itself, and its underlying logic is undecidable).

In 2014, RDF was upgraded to RDF 1.1, and RDFS was upgraded to RDFS 1.1. Both these standards identify resources by *Internationalized Resource Identifiers (IRIs)*. While URIs use ASCII characters, IRIs use the Universal Coded Character Set (Unicode/ISO/IEC 10646) to cater for non-English characters.

For a detailed primer on RDF 1.1, see https://www.w3.org/TR/rdf11-primer/. The RDFS 1.1 standard is accessible at http://www.w3.org/TR/2014/REC-rdf-schema-20140225/. In 2023, the W3C published a working draft of RDFS 1.2 (see https://www.w3.org/TR/rdf-schema/.

OWL and Its Sublanguages

Like RDFS, OWL is restricted to binary fact types (no *n*-aries or unaries). OWL has a simpler formal semantics than RDFS, disallowing some freedoms in RDFS while at the same time adding the ability to capture further constraints. OWL is used to specify the ontology schema (called a terminological box or *Tbox*) for a business domain, and RDF/XML is used to mark-up conforming instance data (the fact population is called an assertional box or *Abox*).

Because information on the web is often incomplete, OWL adopts the *open world assumption* (the failure to find or infer some proposition does not imply that proposition is false). OWL was endorsed as a W3C recommendation in February 2004. In October 2009, a major upgrade to the language, known as *OWL 2*, was recommended by the W3C, and the previous version was renamed *OWL 1*. In December 2012, a second edition of OWL 2 was recommended, and is still the current version (see https://www.w3.org/TR/owl-overview/).

OWL 1 has three flavors: OWL Lite, OWL DL, and OWL Full. OWL Lite is a subset of OWL DL, which itself is a subset of OWL Full. *OWL Lite* is used for basic classification hierarchies where all the constraints are simple. It is based on the description logic *SHIF(D)*, has a low formal complexity, and is decidable. OWL Lite can express only simple constraints, and supports only simple inferences (e.g., those based on transitivity). *OWL DL* is much more expressive while still being decidable, and it is based on the description logic *SHOIN(D)*. OWL DL includes all the constructs of OWL Full but restricts their use. OWL DL is a fragment of first-order logic, so does not allow a class to be declared an instance of another class in the same model. *OWL Full* is more expressive, allowing all RDF expressions, including allowing a class to be an instance of itself. OWL Full is undecidable, going beyond even first-order logic. In spite of its undecidable nature, OWL Full still cannot express some important constraints that are expressible in ORM (e.g., acyclicity).

OWL 2 introduced three new sublanguages as subsets of OWL DL: OWL 2 EL, OWL 2 QL, and OWL 2 RL. *OWL 2 EL* is designed to provide good performance for basic reasoning tasks on large ontologies and is based on the *EL* family of description logics that provides existential quantification but not universal quantification. *OWL 2 QL* is designed to provide good performance with basic queries on large ontologies, where those queries can be implemented by rewriting them into queries in a standard relational query language such as SQL. *OWL 2 RL* is designed to provide good performance for managing ontologies with rule languages such as rule-extended database technologies operating directly on RDF triples.

Cardinality Restrictions in OWL 1

By default, all relationship types are optional and *m:n*. OWL Lite allows fact roles to be declared mandatory and fact types to specified as *n*:1, 1:*n*, or 1:1 using minCardinality and maxCardinality restrictions. Fig. 18.21 explains the meaning of these restrictions by relating them to an equivalent representation in ORM. The ORM fact type *A R B* is diagrammed as an RDF graph, where the subject and object nodes are assumed to be classes. The predicate *R* is directed from *A* to *B*.

OWL itself has no graphic notation, so the fact type and its restrictions are declared textually (see example later). Restricting *R* with a minCardinality of 1 means that each instance of *A* plays in that relationship type with at least one instance of *B*. In ORM terms, the first role of *R* is mandatory. In OWL DL and OWL Full, a minCardinality above 1 may be specified (in ORM this adds a frequency constraint to the mandatory constraint).

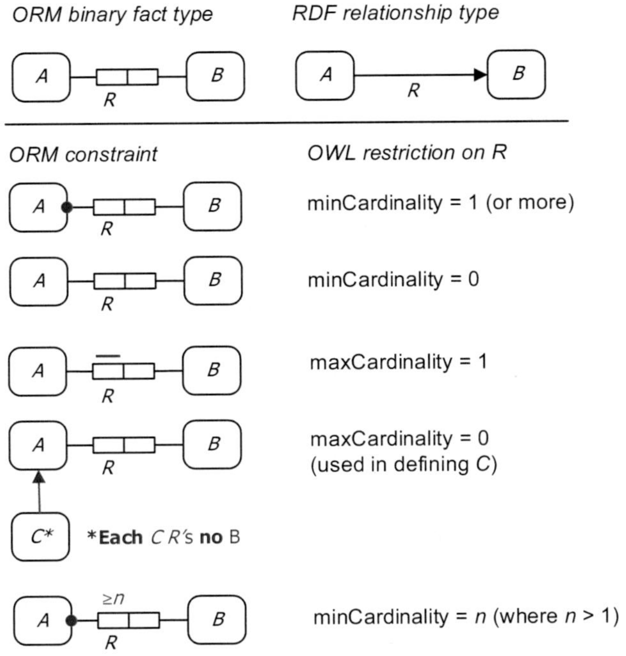

FIG. 18.21 Some correspondences between ORM and cardinality restrictions in OWL.

Assigning a minCardinality of 0 means that the role is optional. Restricting the maxCardinality to 1 means that each instance of *A* plays in that relationship type with at most one instance of *B*. In ORM terms, the first role of *R* has a simple uniqueness constraint. In OWL DL and OWL Full, a maxCardinalty above 1 may be specified (in ORM this replaces the uniqueness constraint by a frequency constraint).

Restricting the maxCardinality to 0 means that each instance of the class (let's call this class *C*) being discussed (perhaps in a query) plays in that relationship type with no instance of *B*. In ORM terms, this could be handled explicitly by a subtype definition as shown, or simply included as a condition in an ORM query. For example, we might define a teetotaller as a person who does not drink alcohol.

As an example of some cardinality declarations, the following OWL code declares that the hasFaxNr predicate is optional and *n*:1.

```
<owl:Restriction>
  <owl:onProperty rdf:resource="#hasFaxNr"/>
  <owl:minCardinality rdf:datatype="&xsd;nonNegativeInteger">0</owl:minCardinality>
</owl:Restriction>
<owl:Restriction>
  <owl:onProperty rdf:resource="#hasFaxNr"/>
  <owl:maxCardinality rdf:datatype="&xsd;nonNegativeInteger">1</owl:maxCardinality>
</owl:Restriction>
```

As a shortcut when the minimum and maximum cardinalities are the same, the value may be assigned to the cardinality property. So, setting owl:cardinality to 0 means forbidden, and setting it to 1 means exactly one.

OWL Syntaxes

From now on, our discussion of OWL relates to OWL 2, Table 18.2 lists the five concrete syntaxes proposed for expressing an OWL 2 ontology in textual form. RDF/XML is the only syntax that is required to be supported by OWL 2 conformant tools, but it is awkward to read and onerous to write. OWL/XML is also designed for processing by XML tools, so it too is painful to read and write. The Functional Syntax is designed to expose the formal structure of an OWL ontology, and is readable by technical users.

For general readability purposes, the *Manchester syntax* (see https://www.w3.org/TR/owl2-manchester-syntax) and the *Turtle* (Terse RDF Triple Language) syntax (see https://www.w3.org/TeamSubmission/turtle/) are far easier for non-technical people to use than the other three syntaxes. Though popular, Turtle is not a product of the OWL working group. The Manchester syntax is typically a little easier to read than Turtle syntax, and is structured in a way to facilitate specifying a large list of OWL statements in one go.

Table 18.2 Concrete syntaxes for OWL 2.

OWL Syntax	*Status*	*Purpose*
RDF/XML	Required	Interchange between tools
OWL/XML	Optional	Facilitate processing by XML tools
Manchester Syntax	Optional	Facilitate reading/writing OWL DL ontologies
Functional Syntax	Optional	Expose the formal structure of ontologies
Turtle	Optional, and unofficial	Facilitate reading/writing RDF triples

As a simple example of all five syntaxes, the statements in Table 18.3 can be used to assert that Biden is a politician (i.e., the individual thing referenced in the current document as Biden is an instance of the class referenced in the current document as Politician) and that each politician is a person (i.e., the Politician class is a subclass of the Person class).

Overall, we consider the Manchester Syntax to be the most convenient of the five syntaxes for communicating typical ontologies to nontechnical people. From now on, we'll typically phrase the OWL examples in terms of Manchester syntax. We now discuss some basic features of OWL 2 using Manchester syntax.

Namespace prefixes may be declared as abbreviations, using a prefix command of the form Prefix: *namespacePrefix*: *IRI*. The prefixes "rdf:", "rdfs:", "owl:", and "xsd:" are predefined. Each user-defined prefix used in an ontology document must have exactly one prefix declaration in the ontology document, such as Prefix: owl: Prefix: gup: https://www.galacticUni.edu/programs#. If no prefix is supplied, the current document is assumed to be the namespace.

Table 18.3 Example statements in OWL 2 syntaxes.

OWL Syntax	*"Biden is a politician"; "Each politician is a person"*
RDF/XML	<Politician rdf:about="Biden"/> <owl:Class rdf:about="Politician"> <rdfs:subClassOf rdf:resource="Person"/> </owl:Class>
OWL/XML	ClassAssertion> <Class IRI="Politician"/> <NamedIndividual IRI="Biden"/> </ClassAssertion> <SubClassOf> <Class IRI="Politician"/> <Class IRI="Person"/> </SubClassOf>
Manchester Syntax	Individual: Biden Types: Politician Class: Politician SubClassOf: Person
Functional Syntax	ClassAssertion(:Politician :Biden) SubClassOf(:Politician :Person)
Turtle	:Biden rdf:type :Politician . *aliter:* :Biden a :Politician . :Politician rdfs:subClassOf :Person .

Literals are delimited by double quotes (e.g., "64"), and may be typed (e.g., "64"^^xsd:integer). An individual may be declared by prepending "*Individual*:" to its name, optionally followed by a declaration of properties that apply to it. For example, the classes to which the individual belongs may be listed after "*Types*:". An individual may belong to many classes at once, such as Individual: Biden Types: Politician, MalePerson.

A class may be declared after "*Class*:", and declared as a subclass of some (typically other) class using "*SubClassOf*", such as Class: Politician SubClassOf: Person. In OWL, subclassing is reflexive (so each class is a subclass of itself) and transitive (so if *A* is a subclass of *B*, and *B* is a subclass of *C*, then *A* is a subclass of *C*). This differs from the subtyping relationship in ORM, which for efficiency reasons is used only for proper subtyping (and hence is irreflexive) that is direct (and hence intransitive).

A binary fact that relates individuals may be specified by declaring the subject as an individual and then listing the predicate and object of that statement after "*Facts*:". One or more facts may be listed using comma separators. For example, we could state that Terry is the husband of Norma and that Terry is the father of Selena as follows: Individual: Terry Facts: isHusbandOf Norma, isFatherOf Selena. Manchester syntax allows you to use "*some*" for the *existential quantifier*. So, we can state that Terry is father of somebody thus: Terry isFatherOf some Person.

A domain of a binary predicate is a class to which its subject must belong, and a range of a binary predicate is a class to which its object must belong. In Manchester syntax, a predicate may be declared as an *ObjectProperty*, and its domain(s) and ranges(s) specified after "*Domain*:" and "*Range*:" respectively. For example:

```
ObjectProperty: isFatherOf
    Domain: MalePerson
    Range: Person
```

A predicate may be assigned multiple domains and ranges, in which case the subject must belong to the intersection of the predicate's domains and the object must belong to the intersection of the predicate's ranges.

Some OWL 2 Taxonomy

Fig. 18.22 shows a UML class diagram for several concepts in OWL 2. This diagram is based on https://www.w3.org/TR/owl2-syntax/. What we called "resources" in our discussion of RDF are called "*entities*" in OWL. OWL entities may be named individuals, classes, properties (binary predicates), or data types. This differs from the use of the term "entity" in ORM and ER, where all entities are nonlexical individuals (not types or classes).

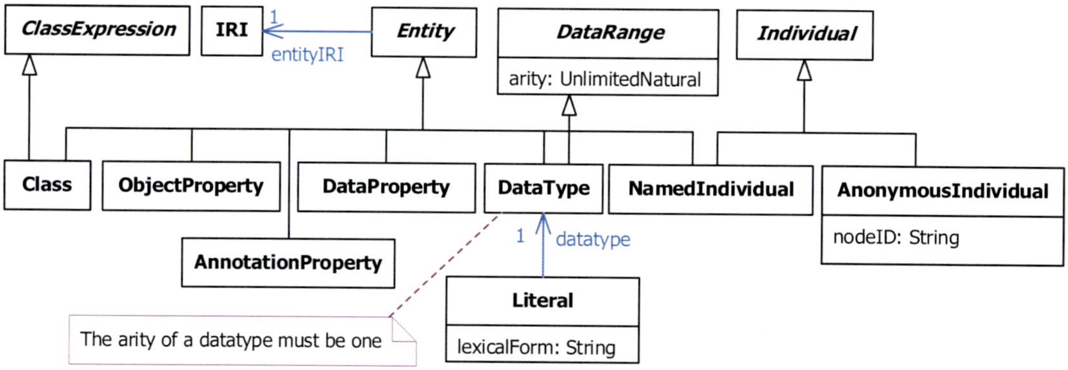

FIG. 18.22 Class diagram of basic OWL 2 concepts (based on the OWL 2 specification).

In OWL, entities are identified by IRIs, but unlike some approaches, OWL does *not* adopt the *Unique Name Assumption* (UNA). So, in OWL the same entity may be assigned different IRIs, even within the same document. Hence, the multiplicity constraint of 1 on entityIRI in Fig. 18.22 should instead be 1..* (1 or more). As discussed earlier, OWL includes the SameAs predicate to indicate when two different IRIs denote the same entity (e.g., DownUnder SameAs Australia).

Literals (data values such as names or numerals) are not treated as individuals. This practice differs from formal logic, where literals do count as individuals. Only OWL individuals (not literals) may be members of OWL classes.

OWL predicates are directed, binary relationship types that map subject individuals (instances) to object individuals. OWL includes a predefined, universal class called *owl:Thing* that includes all individuals. It also predefines the empty class *owl:Nothing*, which has no members. By default, all OWL predicates have an unrestricted domain and range (equal to owl:Thing).

Object properties are binary predicates that relate entities to entities. *Data properties* are binary predicates that relate entities to literals. For example:

```
ObjectProperty: wasBornIn
DataProperty: hasName
Individual: Einstein
   Facts: wasBornIn Germany, hasName "Albert Einstein"^^xsd:string
```

The following extreme OWL predicates ("properties") are predefined. The *owl:topObjectProperty* is the class of all object properties, and hence is a binary predicate relating all individuals in the model to all individuals in the model (its population is the Cartesian product owl:Thing × owl:Thing). Every object property is a subproperty of this property. The *owl:topDataProperty* is the class of all data properties, and hence is a binary predicate relating all individuals to all literals. Every data property is a subproperty of this property. The *owl:bottomObjectProperty* is the empty object property, and hence is a subproperty of every object property. The *owl:bottomDataProperty* is the empty data property, and hence is a subproperty of every data property.

Annotation properties are binary predicates that provide informal documentation annotations about ontologies, statements, or IRIs. For instance, the *rdfs:comment* annotation property is used to provide a comment, as in this example:

```
Class: Moon
Annotations: rdfs:comment "Natural satellite of a planet or other celestial body."
```

Other predefined annotation properties are shown below:

- The *rdfs:label* annotation property is used to provide a human-readable label.
- The *rdfs:seeAlso* annotation property is used to provide an IRI with another IRI such that the latter provides additional information about the former.
- The *rdfs:isDefinedBy* annotation property is used to provide an IRI with another IRI where the latter provides a definition of the former.
- An annotation with the *owl:deprecated* annotation property and the value equal to "true"^^*xsd:boolean* is used to specify that an IRI is deprecated.
- The *owl:versionInfo* annotation property is used to provide an IRI with a string that describes the IRI's version.
- The *owl:priorVersion* annotation property specifies the IRI of a prior version of the containing ontology.
- The *owl:backwardCompatibleWith* annotation property specifies the IRI of a prior version of the containing ontology that is compatible with the current version of the containing ontology.

- The *owl:incompatibleWith* annotation property specifies the IRI of a prior version of the containing ontology that is incompatible with the current version of the containing ontology.

OWL tools typically support *ontology parsing*—the process of converting an ontology document written in a particular syntax into an OWL ontology. Depending on the syntax used, the ontology parser may need to know which IRIs are used for which kinds of entity (e.g., is a predicate an object property, data property, or annotation property?). For example, is wasBornIn an object property or a data property? This typing information can be extracted from declarations—axioms that associate IRIs with entity types (e.g., ObjectProperty: wasBornIn).

Comparing Individuals, Classes, Datatypes, and Predicates

OWL predefines the *owl:sameAs* and *owl:differentFrom* predicates for "=" and "≠" respectively, when they apply between individuals (so their range and domain is owl:Thing). These predicates are used to declare whether two IRIs in the same or different documents refer to the same individual or to different individuals, for example,

```
Individual: Elvis SameAs: TheKing
    DifferentFrom: Donovan
```

OWL predefines the *owl:equivalentClass* predicate to equate two classes or class expressions. In Manchester syntax, this is called "*EquivalentTo*", for example,

```
Class: Human
Class: Person EquivalentTo: Human.
```

In Manchester syntax, EquivalentTo may also be used to define a *data range*. For example, the datatype RatingNr may be confined to {1..7}, thus Datatype: RatingNr EquivalentTo: integer[>=1, <=7].

In Manchester syntax, the DisjointWith predicate is used to declare that two classes (or class expressions) are mutually exclusive, for example,

```
Class: Cat DisjointWith Dog.
```

The predefined class *DisjointClasses* is a class whose members are pairwise exclusive (i.e., each pair of member classes is disjoint). In Manchester syntax, "*DisjointClasses*:" prepended before two or more class descriptions indicates they are mutually exclusive (so have no instances in common). This provides an efficient way to declare that many classes are pairwise disjoint.

In ORM, top-level entity types are assumed to be mutually exclusive. For example, in the ORM schema in Fig. 18.23(a), if Person, Car, and Dog are not subtypes, then they are assumed to be disjoint. In OWL, however, this disjointness needs to be explicitly stated, as shown in the first statement of Fig. 18.23(b).

(a)

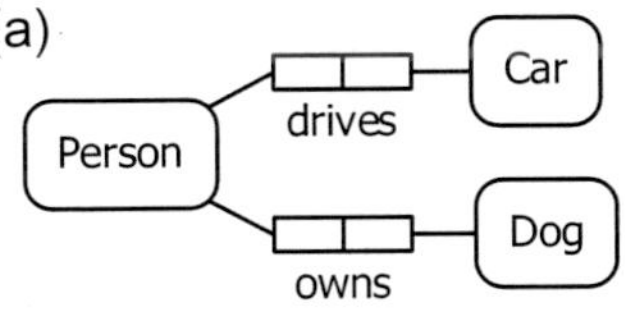

(b)

```
DisjointClasses: Person, Car, Dog
ObjectProperty: drives
  Domain: Person
  Range: Car
ObjectProperty: owns
  Domain: Person
  Range: Dog
```

FIG. 18.23 Declaring some fact types in (a) ORM and (b) OWL (Manchester syntax).

The single statement using DisjointClasses is equivalent to the lengthier approach of using DisjointWith to declare disjointness between each pair. The Domain and Range specifications restrict the predicates to the object types indicated.

Given the following statements:

```
Individual: Adam Types: Man
Individual: Eve Types: Woman
DisjointClasses: Man, Woman
```

we may infer:

```
Individual: Adam DifferentFrom: Eve
```

In Manchester syntax, the *SubPropertyOf* predicate may be used to declare that all (subject, object) instances of one predicate must be instances of another. Once a resource is declared, you can list all its properties in one go. For example, you could declare the fatherhood predicate along with its superproperty, domain, and range together, thus:

```
ObjectProperty: isFatherOf
    SubPropertyOf: isaParentOf
    Domain: MalePerson
    Range: Person
```

In Manchester syntax, the *DisjointWith* predicate may be used to indicate that the populations of binary predicates must be *mutually exclusive*. This is equivalent to a pair-exclusion constraint in ORM. For example, the exclusion constraint that no person authored and reviewed the same book may be specified as follows:

```
ObjectProperty: authored
  Domain: Person
  Range: Book
ObjectProperty: reviewed
  Domain: Person
  Range: Book
  DisjointWith: authored
```

If more than two predicates are mutually exclusive, it's more efficient to simply list them after the *DisjointProperties* header, for example,

ObjectProperty: gotFirstPlaceIn
ObjectProperty: gotSecondPlaceIn
ObjectProperty: gotThirdPlaceIn
DisjointProperties: gotFirstPlaceIn, gotSecondPlaceIn, gotThirdPlaceIn

If two predicates must always have the same instance populations, they are said to be (materially) equivalent. This is the same as declaring a subPropertyOf relationship in both directions. In Manchester syntax, equivalence between predicates is declared by listing the predicates after the *EquivalentProperties* header.

For example, an ORM pair-equality constraint to ensure that people drive a car if and only if they own the car (this constraint actually applies in some taxi companies) may be specified thus: EquivalentProperties: owns, drives.

Another use of equivalent properties is to support synonyms for predicate readings in the same document (or different documents), for example, EquivalentProperties: worksFor, isEmployedBy. A third use of equivalent properties is to specify derivation rules for derived predicates.

Functional, Inverse, Mandatory, and Key Relationships

By default, all OWL predicates are optional and many-to-many (*m:n*). Although the predefined predicates for equating and distinguishing individuals have unrestricted domains and ranges, user-defined predicates typically have at least one domain and at least one range explicitly declared. This is always the case for relationships declared in information modeling approaches such as ORM, ER, and UML.

Fig. 18.24(a) depicts two ORM fact types. The fact type Politician was born in Country is a many-to-one (*n*:1) association (each politician was born in at most one country, and many politicians may be born in the same country). Both roles are optional as our knowledge might be incomplete. The fact type Politician heads/is headed by Country is 1:1 (each politician heads at most one country, and each country is headed by at most one politician).

Fig. 18.24(b) depicts the ORM schema in OWL using Manchester syntax. OWL confusingly uses the term "property" to mean predicate. Hence, an *n*:1 predicate is called a *functional property*. A functional predicate is declared as an object property, and then its functional nature is declared by including *Functional* as one of its *characteristics*. Knowledge of functional properties enables OWL tools to make useful inferences. For example, given:

Individual: Terry
    Facts: wasBornIn TheLuckyCountry, wasBornIn Australia

we may deduce:

TheLuckyCountry SameAs Australia.

ORM treats an elementary fact type as a set of one or more typed predicates. In Fig. 18.24 (a), the head of government fact type has a forward predicate reading "heads" as well as an *inverse* predicate reading "is headed by" directed from Country to Politician (a slash "/" separates forward and inverse readings).

(a)

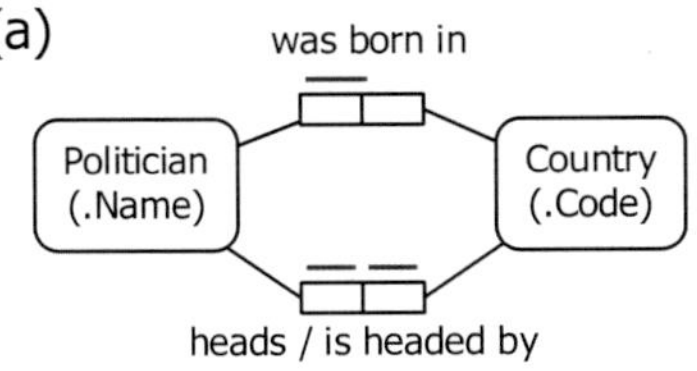

(b) ObjectProperty: wasBornIn
Domain: Politician
Range: Country
Characteristics: Functional
ObjectProperty: heads
Domain: Politician
Range: Country
Characteristics: Functional, InverseFunctional
InverseOf: isHeadedBy

FIG. 18.24 *n*:1 and 1:1 associations in (a) ORM and (b) OWL Manchester syntax.

OWL allows you to declare one predicate to be the inverse of another using the owl: inverseOf predicate. In Manchester syntax, you prepend "*InverseOf:*" to the inverse predicate, for example, ObjectProperty: isaChildOf InverseOf: isaParentOf.

Functional and inverse functional declarations apply *globally* to the predicate in all contexts. OWL cardinality constraints, however, apply only locally to the domain classes being introduced in a local context. This local nature of cardinality restrictions differs markedly from the usual approach in information modeling.

OWL does not allow an inverse to be declared for a data property, partly because it *never allows a literal to be the subject of a predicate.* This is unfortunate because we may wish to talk about literals.

Consider the data property Person misspelt Word. OWL does not allow us to declare its inverse Word wasMisspeltBy Person. Moreover, OWL does not let us assert the fact type Word is a synonym for Word. To work around such problems in OWL, you have to cheat, for example by treating Word as an entity type rather than as a value type (literal type).

Consider the fact types Person is in Country and City is in Country. In information modeling approaches such as ORM, ER, and UML, the two is in predicates would be treated as different predicates/relationships/associations. However, in OWL we may treat these instead as *different occurrences of the same predicate,* simply by assigning them the same IRI.

OWL does not allow a data property to be declared inverse functional. In practice, many data properties are either 1:1 or 1:*n*, thus requiring a uniqueness constraint on the role played by the literal. Although we can't declare uniqueness constraints on roles played by literals as inverse functional characteristics, we can specify such constraints in OWL 2 using *HasKey* declarations.

In Fig. 18.24(a), politicians are identified by their name and countries by their country codes. The ORM diagram depicts this compactly using parenthesized reference modes, but internally these are expanded to the injective (mandatory, 1:1) reference relationships Politician has PoliticianName and Country has CountryCode.

Unfortunately, the HasKey feature only partly addresses the need to specify identification schemes (simple or compound) for entity types because *HasKey captures only the inverse functional nature of the data property.*

The name choice "HasKey" for this feature is somewhat misleading, as it is weaker than the notion of "key" in relational databases. To fully capture the identification scheme, we need to explicitly declare the data property to be both functional and mandatory.

In Manchester syntax, the *mandatory role constraint* that each Politician has a name is expressed by using the *SubClassOf* property to declare that Politician is a subclass of those things that play the subject role in the hasPoliticianName predicate with a min*Cardinality* of 1 (or more), as follows:

```
Class: Politician
  SubClassOf: hasPoliticianName min 1.
```

A maxCardinality of 1 may be declared in a similar way using max 1, which is another way to declare that the predicate is functional. If minCardinality = maxCardinality, you can set both at once using exactCardinality. For example, the 1:1 reference scheme for Politician may be declared in OWL 2 as follows:

```
Class: Politician
   SubClassOf: hasPoliticianName exactly 1.
   HasKey: hasPoliticanName
 DataProperty: hasPoliticianName
   Domain: Politician
   Range: PoliticianName
```

As a composite identification scheme example, consider the ORM schema in Fig. 18.25(a). Here, each building is identified by its building number, and each room is identified by combining its local room number with its building.

(a)

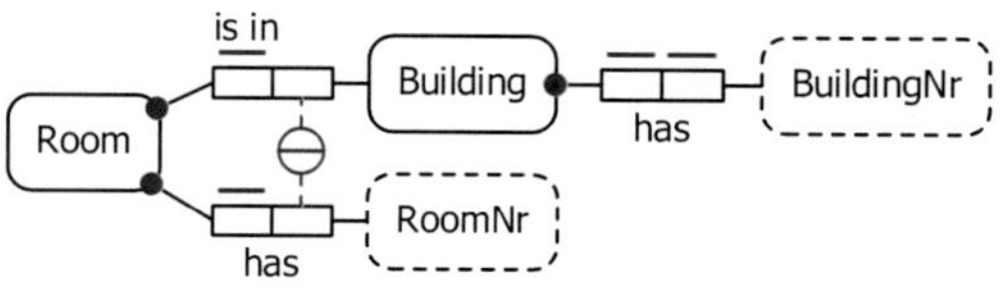

(b)

```
Class: Room
  SubClassOf: isInBuilding exactly 1.
  SubClassOf: hasRoomNr exactly 1.
  HasKey: isInBuilding, hasRoomNr
ObjectProperty: isInBuilding
  Domain: Room
  Range: Building
DataProperty: hasRoomNr
  Domain: Room
  Range: RoomNr
Class: Building
  SubClassOf: hasBuildingNr exactly 1
  HasKey: hasBuildingNr
DataProperty: hasBuildingNr
  Domain: Building
  Range: BuildingNr
```

FIG. 18.25 An ORM schema where Room has a composite reference scheme.

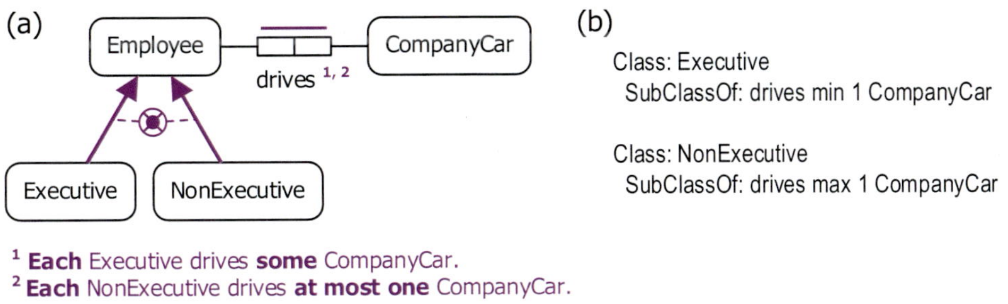

FIG. 18.26 Restricted mandatory role constraint and restricted uniqueness constraint.

Now, consider the ORM schema in Fig. 18.26(a). In the fact type Employee drives Car, the drives predicate is optional and *m:n*. The two textual constraints in FORML add a *restricted mandatory role constraint* (see constraint footnote 1) and a *restricted uniqueness constraint* (see constraint footnote 2). Here, the drives predicate is functional only with respect to NonExecutive, not Employee. We should *not* declare drives as a FunctionalProperty because that would then apply to all employees (and any other subjects of the drives predicate in an OWL version of the model). These restricted constraints may be declared in OWL using qualified cardinality restrictions, as shown in Fig. 18.26(a) using Manchester syntax.

In practice, it is very rare to encounter explicit declarations for minCardinality = 0 and minQualifiedCardinality = 0 because these apply by default, so they may be ignored. However, setting the maximum cardinality to zero does have some use. The setting *maxCardinality = 0* means that each instance of the subject class expression *A* is related via the predicate *R* to no object instance.

The setting *maxQualifiedCardinality = 0* means that each instance of the subject class expression *A* is related via *R* to no instance of the specified range *B*. These restrictions are used mainly in defining subclasses, and the qualified version is typically used in such cases. Fig. 18.27(a) and Fig. 18.27(b) summarize the corresponding ORM patterns. Fig. 18.27(c) provides a concrete example, and Fig. 18.27(d) shows the OWL code.

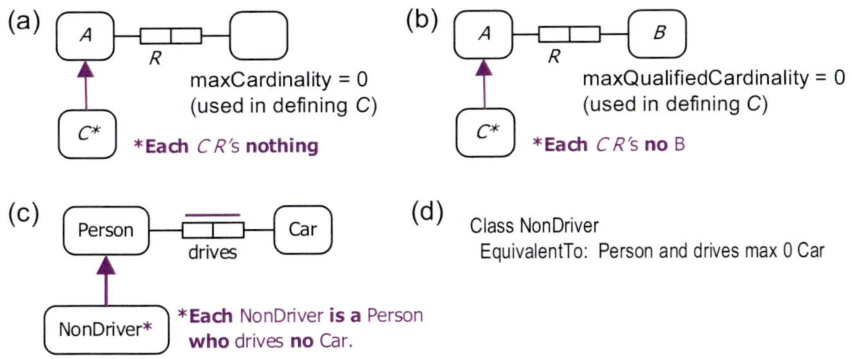

FIG. 18.27 Use of maxCardinality = 0 and maxQualifiedCardinality = 0 for subtyping.

ExactCardinality restrictions other than 1 are allowed. For example, the following OWL 2 code may be used to state that Asimov authored exactly 506 books.

```
ObjectProperty: authored
  Domain: Person
  Range: Book
Individual: Asimov
  Types: Person
       authored exactly 506 Book
```

Database systems typically identify countries by country codes or country names. Many country names contain embedded spaces (e.g., "Czech Republic") so can't be used directly as IRIs (though modified names using underscores or hyphens for spaces could be).

In practice, many OWL ontologies use meaningless, automatically generated identifiers as IRIs just as some relational database designers prefer artificial, surrogate keys for their table schemes. Human-readable labels can be associated with IRIs using rdfs: label annotation properties, and OWL software tools such as Protégé (http://protege.stanford.edu/) can be configured to display labels instead of IRIs to make the ontologies more readable.

Constraints specified in data models are *not* translated into actual *constraints* in OWL, but rather logical *propositions* that are understood in terms of open world semantics where some individuals may be unnamed. For example, the subclass restriction "hasPoliticianName min 1" for Politician simply says that each politician has a politician name—it doesn't require the system to know what that code is.

Union, Intersection, and Complement

Recall that the *union* of sets *A* and *B* is the set of all elements in *A* or *B* or both. The union operator is supported in all versions of OWL, except for OWL Lite, and may be applied to classes or data ranges (e.g., datatypes or other collections of literals). The complex class expression or data range expression resulting from a union may be used anywhere a simple class or data range may be used. The union of two or more classes contains all the individuals that occur in at least one of those classes. In Manchester syntax, the union operator appears as "or", placed in infix position. This is an *inclusive-or* operator. For example, we may define a promising student as a student who is either bright or industrious (or both) as follows:

```
Class: PromisingStudent EquivalentTo: BrightStudent or IndustriousStudent
```

This implies that BrightStudent and IndustriousStudent are subclasses of PromisingStudent. Inferences may be drawn about membership in the superclass. For example, given Individual: Hermione Types: BrightStudent, IndustriousStudent, we may infer Individual: Hermione Types: PromisingStudent.

A class *partition* may be declared in OWL by adding a *disjoint classes* declaration to the union operation. In Manchester syntax, to declare two or more classes that are disjoint (i.e., mutually exclusive), simply prepend "DisjointClasses:". Adding this to an inclusive-or operator results in an *exclusive-or* operator. For example, we may define a Party as either a person or an organization but not both, thus:

```
DisjointClasses: Person, PrivateCompany, PublicOrganization
Class: Party EquivalentTo: Person or Organization
```

A partition of more than two classes may be specified in a similar way, for example,

```
DisjointClasses: Person, PrivateCompany, Organization
Class: Party EquivalentTo: Person or PrivateCompany or PublicOrganization
```

As an alternative for declaring partitions, OWL supports the *disjoint union* operator. In Manchester syntax, this is rendered by using "DisjointUnionOf:". For instance, the partition just discussed may be declared more concisely thus:

```
Class: Party
  DisjointUnionOf: Person, PrivateCompany, PublicOrganization
```

Recall that the *intersection* of sets *A* and *B* is the set of all elements that occurs in both *A* and *B*. In OWL, the intersection operator may be applied to classes or data ranges, and the complex class expression or data range expression resulting from an intersection may be used anywhere a simple class or data range may be used. In Manchester syntax, the intersection operator appears as "and", placed in infix position. For example, the final line of the following OWL code defines the class StudentEmployee as the *intersection* of Student and Employee, which are themselves subclasses of Person.

```
Class: Person
Class: Student SubClassOf: Person
Class: Employee SubClassOf: Person
Class StudentEmployee EquivalentTo: Student and Employee
```

Contrast this example with the following declaration, which is based on an example from the official OWL 2 Primer. Here, the class Grandfather is simply asserted, not defined. Instead, it is simply constrained to be included in (rather than being identical to) the intersection of the Man and Parent classes. Not all men who are parents are grandparents.

```
Class: Grandfather SubClassOf: Man and Parent
```

Recall that the *complement* of set *A* is the set of all elements in the universal set but *not* in the set *A*. In OWL, the complement operator may be applied to classes or data ranges, and the complex class expression or data range expression resulting may be used anywhere a simple class or data range may be used.

The complement of a class contains each individual that is not a member of that class. In Manchester syntax, the complement operator appears as "**not**". For example, the class NonHuman may be defined as the complement of the class Human as follows:

```
Class: Human
Class: NonHuman EquivalentTo: not Human
```

The definition below of the class NonDriver is another example:

```
Class: Person
Class: Driver EquivalentTo: Person and drives min 1 Car
Class: NonDriver EquivalentTo: Person and not Driver
```

All the examples so far have used classes. In OWL 2, the union, intersection, and complement operations may also be applied to data ranges. For example:

```
Datatype: StringOrInteger EquivalentTo: xsd:string or xsd:integer
Datatype: PassGradeNr EquivalentTo: GradeNr and not FailGradeNr
```

Reflexive, Irreflexive, Symmetric, Asymmetric, and Transitive Predicates

In OWL, all properties are binary predicates. If the domain and range are the same or compatible, one may compare subject and object instances. In ORM, such predicates are called *ring predicates*, and constraints that apply specifically to them (or more generally, pairs of compatible roles) are called *ring constraints*.

ORM includes graphical notation for several kinds of ring constraints, some of which are simply defined (e.g., irreflexive, asymmetric) while others are recursively defined (e.g., acyclic, strongly intransitive). OWL 2 supports only the following five kinds of restrictions on ring predicates: reflexive, irreflexive, symmetric, asymmetric, and transitive. Of these, OWL 1 supported only symmetric and transitive. Unlike ORM, OWL allows restrictions on ring predicates to be applied only to object property expressions, not data property expressions.

A relation R is *reflexive* if and only if every individual thing in the domain of R bears the relation R to itself. If the domain is owl:Thing, the only predicate likely to be reflexive is the identity relation "=".

OWL allows object property expressions to be declared reflexive by characterizing them as *reflexive properties*. In Manchester syntax, we declare "Reflexive" as a characteristic. For example, if each person knows himself/herself, the "knows" predicate may be declared to be reflexive as follows:

```
ObjectProperty: knows
 Domain: Person
 Range: Person
 Characteristics: Reflexive
```

Given this, the following inferences may be made:

```
Individual: Einstein
  Types: Person"       →  Individual: Einstein Facts: knows Einstein.
Individual: Terry
  Facts: knows Tony"   →  Individual: Terry Types: Person
                          Individual: Tony Types: Person
```

As an example with a union domain, the following declaration of the isAsOldAs predicate asserts that each Person, each Cat, and each Dog is as old as itself.

```
ObjectProperty: isAsOldAs
  Domain: Person or Cat or Dog
  Characteristics: Reflexive
```

A ring predicate R is *irreflexive* if and only if no individual can bear the relation R to itself. For example, the parenthood relation is irreflexive because nothing can be a parent of itself. OWL object property expressions are declared irreflexive by characterizing them as *irreflexive properties*. To do this in Manchester syntax, we declare "Irreflexive" as a characteristic. For example, the isaParentOf predicate is declared irreflexive as follows. Any assertion that someone is his/her own parent would then render the OWL ontology inconsistent.

```
Class: Person
ObjectProperty: isaParentOf
  Domain: Person
  Range: Person
  Characteristics: Irreflexive
```

A ring predicate R is *symmetric* if and only if $\forall x \forall y(xRy \rightarrow yRx)$; that is, for each individual x and y (not necessarily distinct), if xRy then it is also the case that yRx. For example, the siblinghood relation is symmetric because if one person is a sibling (brother or sister) of another, then the second person is a sibling of the first.

OWL object property expressions are declared symmetric by characterizing them as *symmetric properties*. In Manchester syntax, we declare "Symmetric" as a characteristic, for example, isaSiblingOf may be declared to be symmetric in OWL as shown below. For completeness, the irreflexive nature of the predicate is also declared.

```
Class: Person
ObjectProperty: isaSiblingOf
  Domain: Person
  Range: Person
  Characteristics: Symmetric, Irreflexive
```

In OWL, if a predicate is declared to be symmetric, and we assert a fact instance of it, then its converse may be inferred. For example, if we now add the declaration.

Individual: Romulus Facts: isaSiblingOf Remus

we may infer

Individual: Remus Facts: isaSiblingOf Romulus.

In ORM, constraining a predicate to be symmetric means that the asserted fact population must be symmetric. For example, if we assert that Romulus is a sibling of Remus, then we must also assert that Remus is a sibling of Romulus (both facts are stored). In OWL, however, you can assert just one of these facts and have the other fact inferred. To do that in ORM, you would provide a derivation rule for inferring the converse rather than constraining the fact type to be symmetric.

For example, in ORM you could declare the fact type Person is a sibling of Person to be *semiderived*, meaning that some of its instances may be asserted and some may be derived, and use the following rule to derive the converse facts: Person_1 is a sibling of Person_2 **if** Person_2 is a sibling of Person_1. In this case, you would constrain the siblinghood fact type to be irreflexive, but not symmetric.

A ring predicate *R* is *asymmetric* if and only if, for each individual *x* and *y* (not necessarily distinct), if *xRy* then it is not the case that *yRx*. In other words, if the relationship applies, then its converse cannot apply. For example, the parenthood relation is asymmetric because if one person is a parent of another, then the second person cannot be a parent of the first.

OWL object property expressions are declared asymmetric by characterizing them as a*symmetric.* To do this in Manchester syntax, we declare "Asymmetric" as a characteristic. For example, isaParentOf predicate may be declared to be asymmetric in OWL, as shown below.

```
Class: Person
ObjectProperty: isaParentOf
  Domain: Person
  Range: Person
  Characteristics: Asymmetric
```

Note that if a predicate is asymmetric, it is automatically irreflexive as well (consider the definition of asymmetry for the case where $x = y$). Although asymmetry implies irreflexivity, the converse does not apply. For example, isaSiblingOf is irreflexive but not asymmetric.

If the above asymmetric constraint has been declared and we assert that Ann is a parent of Bill, then any attempt to add the fact that Bill is a parent of Ann would render the OWL ontology inconsistent.

A ring predicate *R* is *transitive* if and only if, for each individual *x*, *y*, and *z* (not necessarily distinct), if *xRy* and *yRz* then it is also the case that *xRz*. For example, the ancestorhood relation is transitive because if one person is an ancestor of another, and the second person is an ancestor of a third person, then the first person is an ancestor of the third person.

Ignoring reincarnation, the ancestorhood predicate is actually acyclic (an object can never cycle back to itself by applying one or more ancestorhood facts). Unlike ORM, OWL does not support acyclicity constraints. However, OWL does allow object property expressions to be declared transitive by characterizing them as transitive. To do this in Manchester syntax, we declare "Transitive" as a characteristic. For example, isanAncestorOf predicate may be declared to be transitive, as shown below. For completeness, the asymmetric nature of the predicate is also declared.

```
Class: Person
ObjectProperty: isanAncestorOf
  Domain: Person
  Range: Person
  Characteristics: Transitive, Asymmetric
```

In OWL, declaring the ancestorhood predicate to be transitive does not require that all its instances must be asserted. For example, if you declare the predicate to be transitive, and assert that Ann is an ancestor of Bill and that Bill is an ancestor of Chris, then the OWL engine can infer that Ann is an ancestor of Chris.

Enumerated Types in OWL 2

Recall that OWL allows you to define a new class or datatype by specifying its *intension* (i.e., meaning) in terms of operations on classes or datatypes that are already declared. A type definition that specifies the necessary and sufficient conditions for an item to be a member of the type is said to be an *intensional definition* of the type. For example, intensional definitions for StudentEmployee and PassGradeNr may be declared in Manchester syntax as shown below, assuming that the types referenced (Student, Employee, GradeNr, FailGradeNr) have been previously declared.

```
Class: StudentEmployee
  EquivalentTo: Student and Employee
Datatype: PassGradeNr
  EquivalentTo: GradeNr and not FailGradeNr
```

The set of all possible instances of a type is said to be the *extension* of the type. If a class or data range is composed of a known, finite set of items, it may be defined by *enumerating* (i.e., listing) those items. For example, while EvenDigit could be defined intensionally as $\{x \mid x$ is a digit that is divisible by 2$\}$, it could also be defined extensionally as a member of {0, 2, 4, 6, 8}. A type that is defined by enumerating its extension is said to be an *enumerated type*, and its definition is said to be an *extensional definition*. In Manchester syntax, the usual set notation is used to specify the extension, using braces as set delimiters and commas as element separators, such as.

```
Class: Weekday EquivalentTo: { Monday, Tuesday, Wednesday, Thursday, Friday }
Datatype: EvenDigit EquivalentTo: {0, 2, 4, 6, 8}
```

"Value" Restrictions on Predicates

OWL uses the term "*value*" loosely to mean either an individual entity or a literal. Given this understanding, a named class or (unnamed) class expression may be defined as the subject of a predicate whose range is restricted to include one of the following three options: (1) some specific value; (2) some value from a specified class or data range; (3) only values from a specified class or data range. We consider these cases in turn.

In Manchester syntax, the set of all elements that relates via a binary relation *R* to the value *v* is expressed simply as "*R* value *v*". As an example of option (1), if we treat Gender as an enumerated class of two individuals (male and female), the following OWL code defines MalePerson as the class of people who have male gender.

```
ObjectProperty: Person hasGender
  Domain: Person
  Range: Gender
Class: Gender
  EquivalentTo: { male, female }
Class: MalePerson
  EquivalentTo: Person and hasGender value male
```

In Manchester syntax, the set of all elements that relates via *R* to some value in a specified subrange *D* is expressed simply as "*R* some *D*". As an example of option (2), the following OWL code defines CarDriver as the class of people who drive a car.

```
ObjectProperty: drives
  Domain: Person
  Range: Vehicle
Class: Car
  SubclassOf: Vehicle
Class: Driver
  EquivalentTo: Person and drives some Car
```

In Manchester syntax, the set of all elements that relate via *R* to some value in a specified subrange *D only* (if at all) is expressed simply as "*R* only *D*". As an example of option (3), given the previous declarations for the drives predicate and Car subclass, the following OWL code defines CarOnlyDriver as the class of people who drive *only* vehicles that are cars. This subtype includes those people who drive no vehicles at all.

```
Class: CarOnlyDriver EquivalentTo: Person and drives only Car
```

Property Chains in OWL 2

OWL 2 allows you to derive a new object property fact by composing two or more object property facts together in a chain, where the entity in the object position of one fact (other than the last fact) is also the subject of the following fact. In relational database work, this corresponds to performing an equijoin between two elementary relations and then projecting on the first and last attribute in the chain. In OWL, the equijoin operation is known as a *composition* operation, and an expression that composes two or more predicates is called a *property chain*.

To illustrate this notion, consider a business domain in which each person is born in at most one city (e.g., some people might instead be born in a village), and each city is located in exactly one country. One way of specifying this schema in OWL using Manchester syntax is as follows:

```
ObjectProperty: wasBornInCity
  Domain: Person
  Range: City
  Characteristics: Functional
ObjectProperty: isLocatedInCountry
  Domain: City
  Range: Country
Class: Person SubClassOf: wasBornInCity max 1
Class: City SubClassOf: isLocatedInCountry exactly 1
```

Now, suppose we are also interested in knowing which countries people are born in. In ORM, we could add the semiderived fact type Person was born in Country, allowing some instances to be asserted and others derived using the derivation rule Person was born in Country **if** Person was born in **some** Country **that** is located in **that** Country.

In OWL 2, we can specify the derivation rule by declaring the wasBornInCountry predicate to be a subproperty of the predicate formed by projecting on the start and end of the property chain that composes the wasBornCountry and isLocatedInCity predicates. In Manchester syntax, the property chain is specified using the symbol "o" as the *composition operator* placed between adjacent predicates in the chain, and listed after the header "SubPropertyChain:". Our current example may be specified in Manchester syntax by adding the following OWL code:

```
ObjectProperty: wasBornInCountry
  Domain: Person
  Range: Country
   SubPropertyChain: wasBornInCity o isLocatedInCountry
```

The subproperty declaration is just a simple implication that applies to any satisfying models. OWL is not concerned with choices about how to actually implement implications. A property chain may compose multiple occurrences of the same predicate. For example, grandparenthood facts may be derived by composing two parenthood facts, as shown.

```
ObjectProperty: isaGrandparentOf
   SubPropertyChain: isaParentOf o isaParentOf
```

If knowledge of parenthood facts is incomplete, we may wish to allow some grandparenthood facts to be simply asserted by adding the following declarations.

```
ObjectProperty: isaGrandparentOf
   Domain: Person
   Range: Person.
```

Property chains composed from more than two predicate occurrences are also allowed. For example, the following declaration enables great-grandparenthood facts to be inferred from three relevant parenthood facts. For instance, given the facts that Edward VII is a parent of George V, George V is a parent of George VI, and George VI is a parent of Elizabeth II, we may derive the fact that Edward VII is a great-grandparent of Elizabeth II.

```
ObjectProperty: isaGreatGrandparentOf
   SubPropertyChain: isaParentOf o isaParentOf o isaParentOf
```

As an example of a property chain with three different predicates, the following declarations capture the derivation that a person drives a car made by a specific company if that person drives a car that is of a carmodel made by that company.

```
ObjectProperty: drives
   Domain: Person
   Range: Car
ObjectProperty: isOfCarModel
   Domain: Car
   Range: CarModel
ObjectProperty: isMadeBy
   Domain: Car
   Range: CarModel
ObjectProperty: drivesCarMadeBy
   SubPropertyChain: drives o isOfCarModel o isMadeBy
```

In dealing with property chain axioms, OWL includes formal restrictions to *avoid circular reasoning* that could lead to problems of decidability. We now consider an example. In general, an uncle is either a brother of a parent or the husband of an aunt. *Unclehood by birth* may be defined as follows: given persons *x* and *y*, *x* is a birth uncle of *y* if *x* is a brother of a parent of *y*. Now, suppose we also define brotherhood as follows: given persons *x* and *y*, *x* is a brother of *y* if *x* is a birth uncle of a child of *y*. These derivation rules may be specified in OWL as follows.

```
ObjectProperty: isaBirthUncleOf
      SubPropertyChain: isaBrotherOf o isaParentOf
ObjectProperty: isaBrotherOf
      SubPropertyChain: isaBirthUncleOf o isaChildOf
```

Note that the derivation rule for brotherhood calls the birth unclehood predicate, which itself depends on brotherhood. So, the derivation rules in combination involve a vicious circle. In OWL, this property chain combination is illegal.

Some Differences Between OWL Ontologies and Typical Databases

In OWL 2, a named individual is identified by an IRI. OWL 2 also allows unnamed or *anonymous individuals* (corresponding to blank nodes in RDF), which may be referenced within a given ontology by a local *nodeID* that starts with an underscore, for example, _:x or _:p1. The following example asserts the existence of a woman president of Taiwan, using the nodeID _:p1 to denote this person.

```
Individual: _:p1
  Types: Woman
  Facts: isPresidentOf Taiwan
```

In relational databases, individuals are normally identified by simple or compound names, so existential assertions like those above are typically not supported, although we can store specific facts that imply them, such as Tsai Ing-wen is a woman and president of Taiwan.

From a logical perspective, IRIs correspond to individual constants rather than names in the usual sense they have in relational databases. To illustrate how OWL's treatment of IRI-based identification schemes differs from the value-based identification schemes used in database systems, consider the following declarations:

```
DataProperty: hasCountryCode
  Domain: Country
  Range: xsd:string
Class: Country
  SubClassOf: hasCountryCode exactly 1
Individual: Australia
  Types: Country
  Facts: hasCountryCode "AU"
Individual: Germany
  Types: Country
```

Here, we asserted that Germany is a country without explicitly providing a country code for it. This would not be allowed in an information model or relational database because our declaration that each country has a country code is implemented as an integrity constraint, forbidding a country without a country code.

In contrast, OWL allows this because the declaration that each country has a country code is *interpreted simply as a proposition, not as a constraint.* This proposition may be formalized in logic as $\forall x$(Country $x \rightarrow \exists y$ x hasCountryCode y). Because we asserted that Germany is a country, the system infers that Germany has a country code, which is currently unknown. This is allowed in OWL, which adopts open world semantics and permits unnamed individuals.

Note also that if we add the data property hasCountryName, OWL has no way to choose one of the hasCountryCode or hasCountryName relationships for the preferred reference scheme, as any distinction between primary and alternate keys is lost.

InverseFunctional declarations apply to any kind of individual (named individuals, anonymous individuals, and individuals whose existence is implied by existential quantification). However, the uniqueness aspect of *HasKey declarations applies only to named individuals* (individuals that are explicitly named by an IRI). Suppose we declare the object property wasBornInCountry and the data property hasCountryCode in the usual way and then add the following declarations.

```
Class: Country
  SubClassOf: hasCountryCode exactly 1
  HasKey: hasCountryCode
```

We can now assert that Julia Gillard (a former prime minister of Australia) was born in some country that has the country code "GB" without knowing the name of that country, as follows:

```
Individual: JuliaGillard
  Facts: wasBornInCountry _:c1
Individual: _:c1
      Facts: hasCountryCode "GB"
```

Now, suppose we discover that the United Kingdom has the country code "GB" (chosen by ISO based on the long name "The United Kingdom of Great Britain and Northern Ireland"). Using TheUK as an intuitive IRI for the United Kingdom, we can now assert that the UK has the country code "GB" as follows:

```
Individual: TheUK
      Facts: hasCountryCode "GB"
```

One might expect that it is correct to infer the fact that Julia Gillard was born in the United Kingdom, which may be expressed thus:

```
Individual: JuliaGillard
  Facts: wasBornInCountry TheUK
```

However, this fact does not follow! This is because the HasKey declaration for Country applies only to named individuals (those with IRIs explicitly asserted) while allowing that Julia Gillard may be born in many unnamed countries.

Negated Facts and Unary Facts in OWL

Similar to the way OWL uses "not" as a complement operator for classes and data ranges, OWL enables you to explicitly assert the *negation* of a proposition by declaring a *NegativePropertyAssertion*. Manchester syntax uses the logical **not** operator before the relevant predicate object pair in the Facts slot for the individual. Consider the ORM fact type Scientist was born in Country. This is used purely to record which scientists are born in which country. Suppose we know that Newton was born in England, and that although we don't know where Einstein was born, we do know he wasn't born in England. OWL allows us to assert both the positive and negated facts as follows:

```
ObjectProperty; wasBornIn
  Domain: Scientist
  Range: Country
  Characteristics: Functional
Individual: Newton
  Facts: wasBornIn England
Individual: Einstein
  Facts: not wasBornIn England
```

Contrast this approach with the closed world approach of typical databases where any fact that is neither asserted nor derived is assumed to be false. The previous example negates an object property. Data properties may be negated in a similar way. For example, treating IQs as numbers, we may assert that Einstein's IQ is not equal to 100 as follows:

```
DataProperty: hasIQ
  Domain: Scientist
  Range: xsd:nonNegativeInteger
  Characteristics: Functional
Individual: Einstein
  Facts: not hasIQ 100
```

OWL adopts the open world assumption, so if a fact is neither asserted nor implied, it is unknown. Hercule Poirot, Captain Hastings, and Inspector Japp are characters from Agatha Christie novels. *Unary facts* (e.g., Poirot smokes) and their *negations* (e.g., Hastings does not smoke) have to be remodeled in OWL as binary relationships that relate the subject to an object that is either a Boolean value (true or false) or a status value with two possibilities (e.g., "Smoker", "NonSmoker").

For example, the following OWL declarations assert that Poirot smokes, Hastings doesn't smoke, and leave it *unknown* whether Japp smokes. In ORM, this example would be modeled using the unary fact type Person smokes under open world with negation semantics.

```
DataProperty: isaSmoker
  Domain: Person
  Range: xsd:boolean
```

```
  Characteristics: Functional
Individual: Poirot
  Facts: isaSmoker true
Individual: Hastings
  Facts: isaSmoker false
Individual: Japp
  Types: Person
```

Join Semantics of External Uniqueness Constraints

External uniqueness constraints with *inner join* semantics may be declared in OWL simply by using compound HasKey properties on the referenced class. Recall Fig. 18.25, where each room was identified by combining its local room number with its building, and this OWL code was used for the composite reference scheme.

```
Class: Room
  SubClassOf: isInBuilding exactly 1
  SubClassOf: hasRoomNr exactly 1
  HasKey: isInBuilding, hasRoomNr
```

Capturing *external uniqueness constraints with outer join semantics* in OWL requires more work. Typically, this will involve *introducing a subclass for the instances that don't play the relevant optional role and then declaring a HasKey property for that subclass.* To ensure that the HasKey declarations are affective, *IRIs must be included for the referenced instances.*

Fig. 18.28 reproduces an example from Section 10.1. The external uniqueness constraint has *outer join semantics,* with the added proviso that *nulls produced in the outer join are treated as actual values.* Hence, in addition to being identified by its course code, each course that is offered by a department may be referenced by the combination of its course title and its department, for example, (Mechanics, PY) and (Mechanics, MA). Each course offered by no department may be identified either by its code or its title.

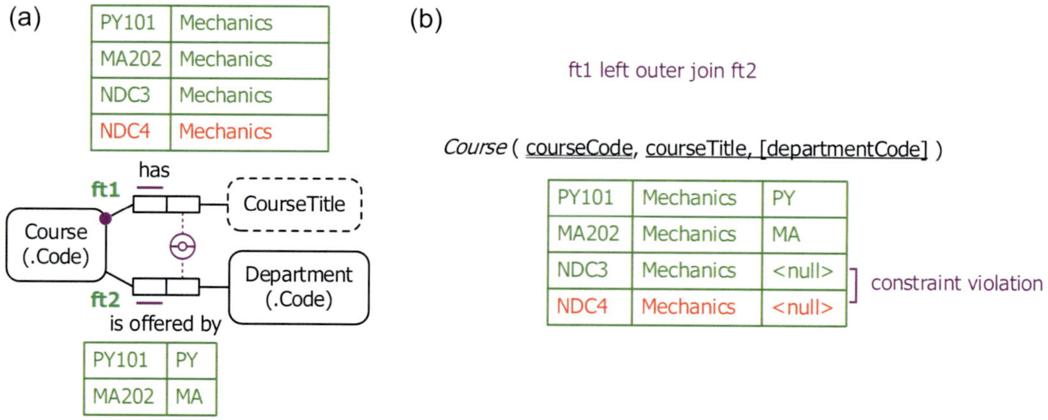

FIG. 18.28 (a) An external UC with outer join semantics; (b) the relational schema.

Adding course NDC4 with the same title "Mechanics" as course ND3 would violate the external uniqueness constraint. Assuming the object properties and data properties are declared in the usual way, the outer join semantics of the external uniqueness constraint may be captured in Manchester syntax as shown below. The subclass NonDepartmentalCourse is added for courses offered by no department, and hasCourseTitle is declared as a HasKey property for it.

```
Class: Course
  SubClassOf: hasCourseTitle exactly 1 xsd:string, hasCourseCode exactly 1 xsd:string
  HasKey: hasCourseCode

Class: Department
  SubClassOf: hasDepartmentCode exactly 1 xsd:string
  HasKey: hasDepartmentCode

Class: NonDepartmentalCourse
  EquivalentTo: Course and (isOfferedBy max 0 Department)
  HasKey: hasCourseTitle
```

If we add the test data shown in Fig. 18.28, this will violate the HasKey condition that nondepartmental courses can be identified by their titles because two nondepartmental courses, NDC3 and NDC4, would have the same title "Mechanics".

Mapping ORM to OWL

Recent research efforts have investigated mapping from conceptual models in ORM, UML, and ER to OWL as well as other description logic-based languages. The direct formal semantics of OWL 2 is based on the description logic SROIQ (D). Much of ORM can be mapped cleanly to these description logics, facilitating the use of efficient DL-based theorem provers for consistency checking of ORM schemas, and interoperability with other semantic web applications. Unlike OWL 2, ORM has a clean, first-order formalization, so problems about types being instances are avoided. However, some ORM structures and constraints cannot be captured in OWL. This has both good and bad consequences.

On the plus side, ORM is capable of capturing, in a natural fashion, several complex aspects of reality that commonly occur in business domains but cannot be captured in popular description logics. Hence, ORM enables higher-fidelity models of typical business domains.

On the negative side, ORM is so expressive that its underlying logic extends a little beyond decidable fragments of first-order logic. This makes it difficult to achieve efficient automated checking of aspects such as schema satisfiability (e.g., in very large schemas with millions of constraints, checking that none of the constraints contradict one another).

This is yet another example of the trade-off between expressibility and tractability. The more expressive the language, the harder it is to efficiently verify the logical consistency of models expressed in it. Because we ought to ensure that our models accurately reflect the business domains they are intended to model, it seems reasonable to capture the intended semantics in a language such as ORM, apply efficient automated checking where possible, and use heuristic techniques for the rest.

Examples of ORM features that can't be translated into SROIQ (D), and hence OWL DL, include uniqueness and frequency constraints spanning multiple roles (of the same or different relationships), set-comparison constraints involving multirole arguments, acyclic ring constraints, and deontic constraints. The $\mathcal{DLR}_{ifd}$ description logic adds support for identifier and functional dependency constraints, so it can capture internal uniqueness constraints spanning multiple roles as well as external uniqueness constraints.

Unlike ER and UML, ORM is attribute-free and hence is easier to relate to RDF, RDFS, OWL, and description logics, which also represent all facts in the form of relationships. Although RDF and OWL use binary relationships only, it is a trivial matter to transform unary and *n*-ary fact types into binary form, as discussed in earlier chapters. So, there is a fairly obvious mapping of facts and fact types in binary ORM to these semantic web languages.

The only aspects lost in the mapping are those ORM constraints and derivation rules that have no counterpart in the logic underlying the target language. These can be mapped over as unambiguous verbalizations in ancillary notes.

An atomic fact type in ORM offers an ideal semantic unit for sharing semantics between different models. Such a unit is constraint-free, allowing different models to apply different constraints without altering the semantics of the fact type. For example, the fact type Person drives Car might be optional and *m:n* in one model, but mandatory and *n*:1 in another model. Using a fact-based approach instead of an attribute-based approach facilitates this kind of exchange. This is one reason why the OMG's SBVR approach for exchanging semantics of business rules is also fact-oriented.

18.7 Metamodeling

Modeling involves making models of business domains. *Metamodeling* involves making models of models—this time the things being modeled are themselves models. Just as recursion is one of the most elegant and powerful concepts in logic, metamodeling is one of the most beautiful and powerful notions in conceptual modeling. This section uses a simple example to convey the basic idea.

A database holds fact instances from a business domain while its conceptual schema models the *structure* of the domain. Fig. 18.29 recalls our basic view of an information system where the database conforms to the rules laid down in the conceptual schema. Essentially, a DBMS is a system for managing various databases; for each database that models some UoD, it checks that each database state agrees with the structure specified in the conceptual schema for that UoD.

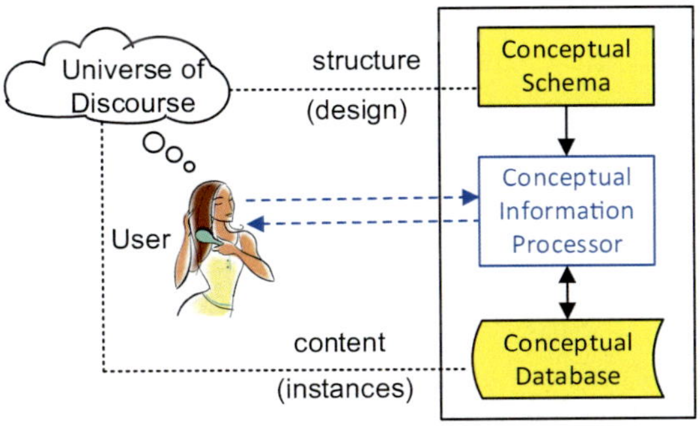

FIG. 18.29 The database must conform to the structure of the conceptual schema.

Among other things, a conceptual modeling tool is a system for managing conceptual schemas. Each valid schema diagram in this book may be thought of as an output report from this system. The trick then is to treat a schema as an instance of this higher-level system. As long as we can verbalize the diagrams into atomic facts, we can use the CSDP to develop a conceptual schema for such conceptual schemas. We would then have a conceptual *metaschema* (schema about schemas).

Suppose that Table 18.4 is part of an output report from a movie database. Other reports from the database provide further information (e.g., people's birth places).

Table 18.4 One report extracted from a movie database.

MovieNr	*Title*	*Director*	*Stars*
1	Wilderness	Tony O'Connor	
2	Sleepy in Seattle	Anne Withanee	Ima Dozer Anne Withanee
3	Wilderness	Anne Withanee	Paul Bunyip

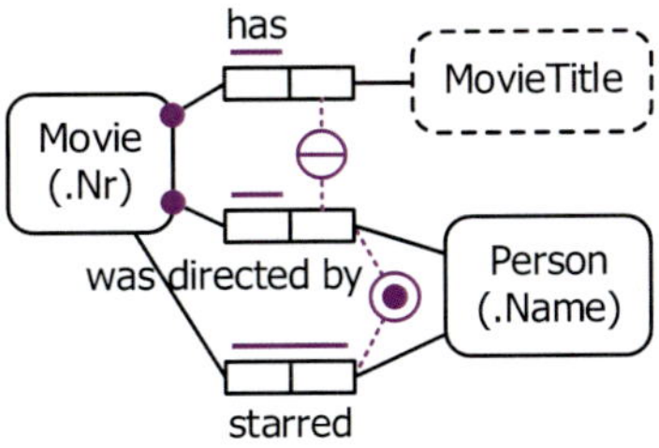

FIG. 18.30 A conceptual schema for Table 18.4.

The conceptual subschema for this report is shown in Fig. 18.30. The inclusive-or constraint on Person is shown explicitly because Person plays other roles in the global schema (e.g., as in Fig. 18.8 from an earlier section).

In this case, the information system architecture in Fig. 18.29 still applies, but the user is an information modeler, the database holds a conceptual schema, the UoD is about conceptual schemas, and the conceptual information processor ensures that only valid conceptual schemas are placed in the database by checking that they satisfy the structure specified in the metaconceptual schema.

Rather than developing a complete metaschema for ORM conceptual schemas, let's confine our discussion here to simple examples such as Fig. 18.30, ignoring nesting, subtyping, derivation, and all constraints other than uniqueness and mandatory role constraints. If you're new to metamodeling, it may seem a bit strange at first. As a challenge, perform CSDP step 1 using Fig. 18.30 as a sample report.

Metamodeling is like ordinary modeling, except the thing being modeled is itself a model. You might begin by describing Fig. 18.30 roughly, such as "It has two entity types (Movie and Person) and one value type (MovieTitle)". The first role of the "was directed by" predicate is mandatory and has a uniqueness constraint, and so on. This verbalization conveys the information, but we need to express it in terms of atomic facts. For example, using "ObjectKind" to mean "kind of object type", we might say: "The ObjectType named 'Movie' is of the ObjectKind named 'Entity'".

In previous chapters, we saw that the same UoD may be modeled in different ways. This applies here too because there are many different ways in which the information in Fig. 18.30 can be verbalized as atomic facts. For example, consider the information that both roles of the predicate called "starred" are spanned by the same uniqueness constraint. How would you say this in atomic facts?

With diagrammatic applications like this, you often find that you want to talk about an object (such as a constraint) but it *hasn't got a name* on the diagram. You would naturally identify it to somebody next to you by *pointing* to it, but this won't help you convey the information over the telephone to someone.

In such cases, it is often convenient to introduce an artificial name, or *surrogate*, to identify the object. This is done in Fig. 18.31, where each constraint is given an identifying constraint number. For convenience, we've also introduced role numbers and predicate numbers (although we could have identified roles by their positions in predicates, and predicates by their expanded fact type readings).

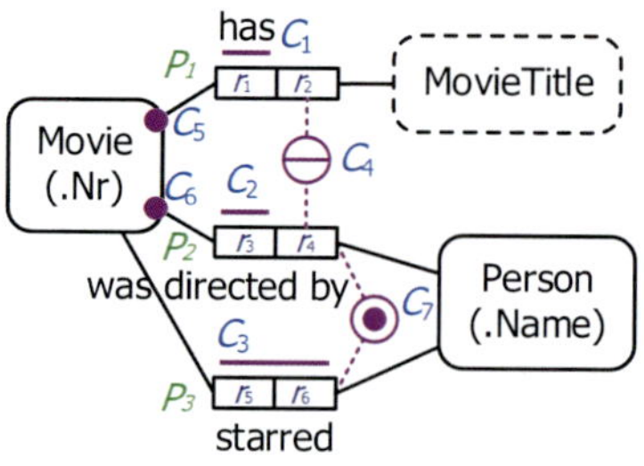

FIG. 18.31 Surrogates are added to identify constraints, roles, and predicates.

If readings are used to identify predicates, we need to expand them with the object type names (e.g., to distinguish the starred predicate in Movie starred Person from that in Play starred Person, or the runs predicate in Horse runs Race from that in Person runs Barbershop). In these examples, the two "starred" predicates have the same meaning, whereas the two "runs" predicates have different meanings. We could extend the metamodel to capture this distinction, but for simplicity we ignore this issue here.

The metaschema shown in Fig. 18.32 is only one of many possible solutions. As an aid to understanding, it is populated with the database, or set of facts, that corresponds to the conceptual schema shown in Fig. 18.31. Here, "UC" and "MR" abbreviate "uniqueness constraint" and "mandatory role", respectively.

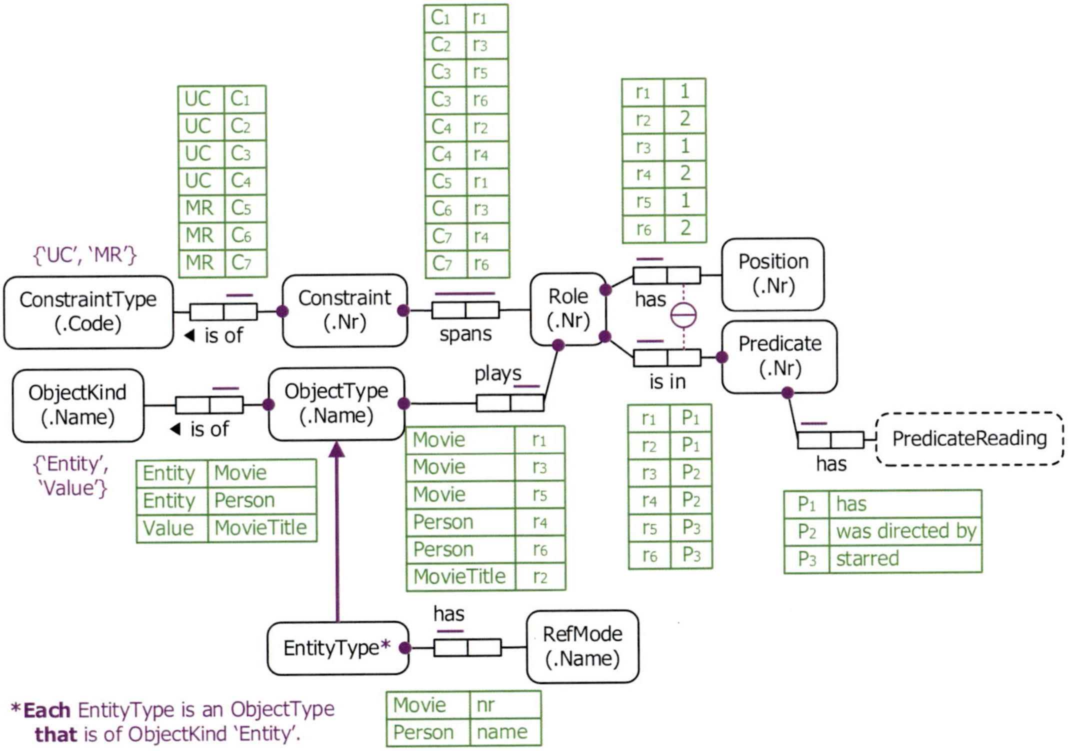

FIG. 18.32 A conceptual metaschema, populated with the schema of Fig. 18.31.

Whether a uniqueness constraint is internal or external is derivable by checking whether its roles come from the same predicate. Whether a mandatory constraint is simple or disjunctive (inclusive-or) is derivable by checking whether it spans multiple roles. For simplicity, the diagram omits these derivations.

Recall that other constraints are ignored in this discussion. For simplicity, only one reading is stored for each predicate, and nesting is ignored. This solution also ignores the implicit reference types implied by reference modes. If you developed an alternative metaschema, don't forget to do a population check.

The metaschema in Fig. 18.32 actually contains features not present in our original example (e.g., subtyping and other kinds of constraints). So it is not rich enough to capture itself. As a nontrivial exercise, you may wish to extend the metaschema until it can capture any ORM schema. For instance, you can capture subtype links by adding the fact type ObjectType is a subtype of ObjectType. To test a full ORM metaschema, you should be able to populate it with itself.

Metamodeling is not restricted to conceptual schemas. Any well-defined formalism can be metamodeled. Apart from being used to manage a given formalism, metamodels can also be developed to allow translation between different formalisms (e.g., ER, ORM, and UML). This is sometimes referred to as metametamodeling.

Because of ORM's greater expressive power, it is reasonably straightforward to capture data models in UML or ER in ORM. Although less convenient, it is possible to work in the other direction as well. To begin with, UML's graphic constraint notation can be supplemented by textual constraints in a language of choice (e.g., OCL). Moreover, the UML metamodel itself has built-in extensibility that allows many ORM-specific constraints to be captured within a UML-based repository.

For example, the ORM model in Fig. 18.33(a) contains four constraints, $C_1..C_4$. While the uniqueness constraints are easily expressed in UML as multiplicity constraints, the subset and exclusion constraints have no graphic counterpart in UML. The UML metamodel fragment shown in Fig. 18.33(b) extends the standard UML metamodel with constraintNr, constraintType, and elementNr attributes, and SetCompConstraint as a subtype along with the argLength attribute.

The full UML metamodel is vast, so we've included only the fragment relevant to the example. The attribute constraintType stores the type of constraint (subset, exclusion, mandatory, etc.). SetCompConstraint denotes set comparison constraint (subset, equality, or exclusion), and argLength is the argument length or number of roles (association ends) at each end of the constraint.

This metamodel fragment is probably easiest to understand in ORM.

Fig. 18.33(c) shows an ORM metamodel fragment with sample population based on Fig. 18.33(a). The four ORM constraints may now be stored as the object-relation shown in Table 18.5. The subset (SS) and exclusion (X) constraints have their argument length recorded. The actual arguments of these two constraints may now be derived by "dividing" the role lists by this number. Thus, the arguments of the subset constraint are the simple roles r_4 and r_2, whereas the arguments of the exclusion constraint are the role pairs (r_1, r_2) and (r_3, r_4). The constraint type may now be used to determine the appropriate semantics.

Although this simple example illustrates the basic idea, transforming the complete ORM metamodel into an extended version of UML is complex. For example, as the UML metamodel fragment in.

Fig. 18.33(b) indicates, UML associations must have at least two roles (association ends), so artificial constructs must be introduced to deal with unaries.

The following exercise includes several questions to hone your metamodeling skills.

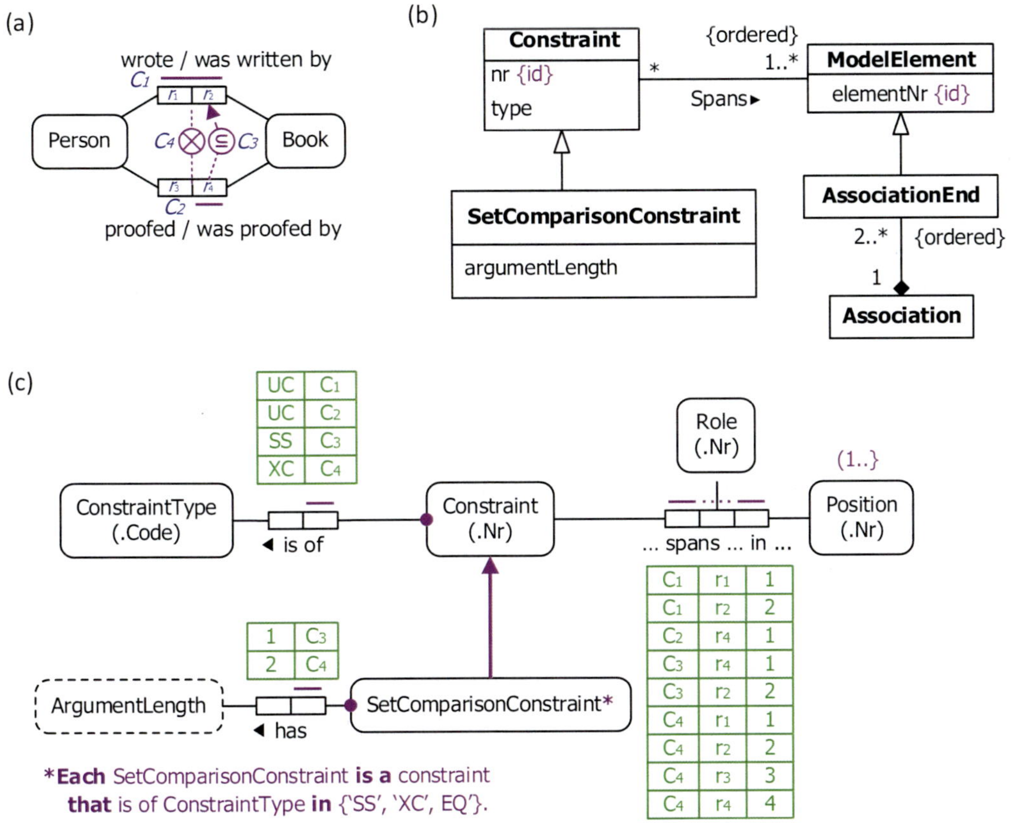

FIG. 18.33 Modeling ORM set-comparison constraints in ORM and extended UML.

Table 18.5 Metatable for storing ORM constraints.

Constraint:

constraintNr	*constraintType*	*roles*	*argLength*
C_1	UC	r_1, r_2	
C_2	UC	r_4	
C_3	SS	r_4, r_2	1
C_4	X	r_1, r_2, r_3, r_4	2

Exercise 18.7

1. Devise a conceptual metaschema (in ORM, UML, or ER) to store simple SQL schemas where each table has at most one primary key (possibly none). A primary key may be composite (multicolumn), but each foreign key references a simple (single column) primary key. Tables are identified by simple names. Columns are ordered by position, and may be optional or mandatory. Each column in a primary key is mandatory, but it is possible that all columns in a keyless table are optional. Ignore domains and all other constraints (e.g., uniqueness constraints on column sets

other than primary keys). For example, your metaschema should be able to store the following schema:

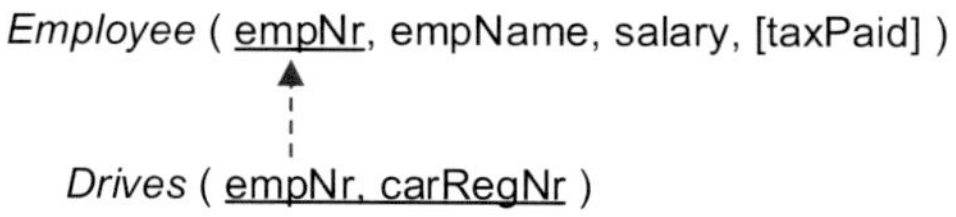

2. Map your answer to Question 1 to a relational metaschema.
3. A sample UML class diagram is shown.

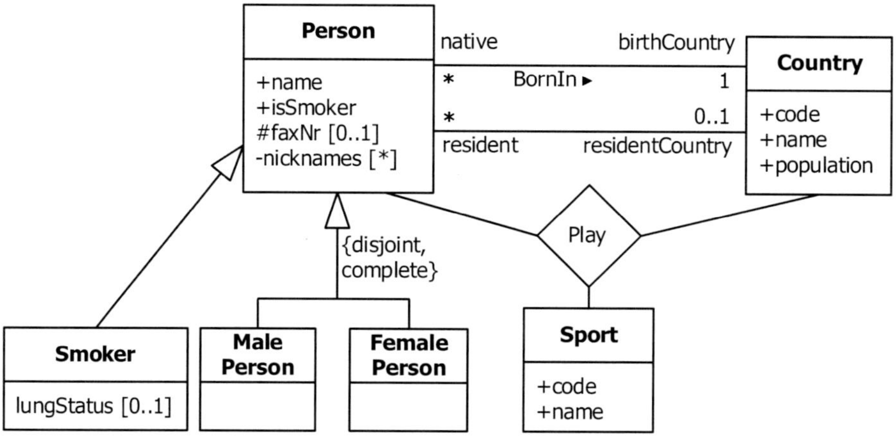

Specify a metaschema in ORM notation for simplified UML class diagrams that are restricted to the following features. Classes have attributes only (e.g., no operations). Attributes have visibility (+=public, #=protected, −=private, ~=package) and multiplicity (*=0 or more, 0.0.1=0 or 1, 1..*=1 or more, 1=exactly 1). Associations may be binary or *n*-ary. No association classes or qualified associations are allowed. Association ends have multiplicity (*=0 or more, 0..1=0 or 1, 1..*=1 or more, 1=exactly 1). Role names are mandatory (whether or not they are displayed). The only constraints that may be added to a subclassing scheme are disjoint and complete. No derived attributes or derived associations are allowed. No notes are allowed. No aggregation is allowed.

No other graphic constraints (e.g., {ordered}, {xor}, {subset}) or textual constraints are allowed. Ignore our extensions (e.g., {P}, {U1}). Ignore data types. Ignore instance populations and presentation aspects (e.g., layout, or whether the display of a feature is suppressed). Your metaschema should be able to store a single class diagram like the one shown.

4. Specify a metaschema in ORM notation for Barker ER (as discussed in Section 8.2). Your metaschema should be able to store a single Barker ER diagram like the one shown.

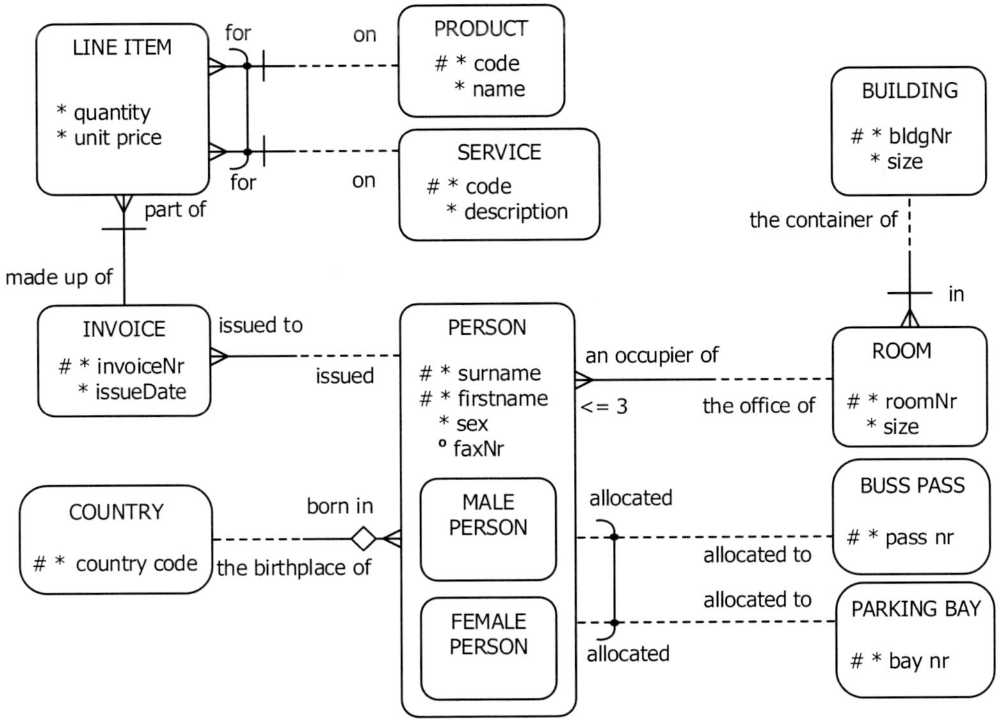

5. Specify a metaschema in ORM for *deterministic, finite-state automata*. Two examples of a deterministic automaton are depicted in the transition graphs shown. Each state appears as a named circle. The starting state has a no-input (unlabeled) arrow attached to it. Each automaton has one or more accepting states, depicted as named double-circles. Transitions between states are depicted as labeled arrows from a state to the next state, where the label names the action (or disjunction of actions) causing the state transition.

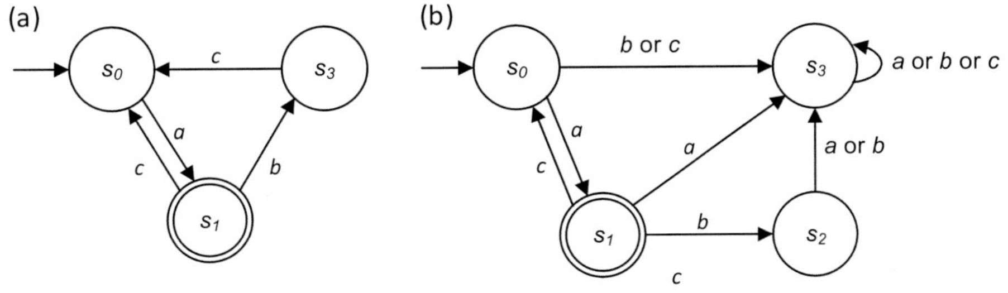

A sequence of transitions is acceptable (well formed) only if it ends in an accepting state. An accepting state cannot be the start state. The examples have only one accepting state, but your metaschema should allow for multiple accepting states. A finite-state automaton is deterministic if and only if for each action applied to a state, there is only one possible resulting state. Your metaschema should be able to store the information content of single transition graphs like the examples shown.

6. In a *nondeterministic* finite-state automaton, a given action performed on a given state may result in one of many different transitions (we cannot know in advance which transition option will occur for any specific occurrence of the action being applied to the state). For instance, in the following example the action *a* performed on state s_0 might cause a transition to state s_1 at one time and to state s'_1 at another time. Modify your answer to (a) to cater for nondeterministic automata.

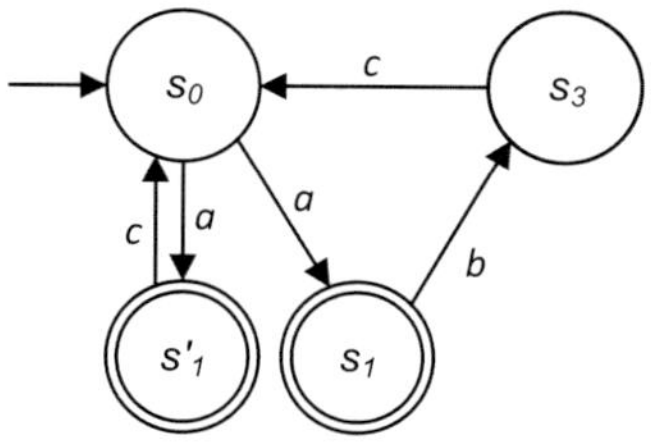

18.8 Summary

OLTP systems are used for typical database applications involving frequent updates and queries. A *data warehouse* is updated only on a scheduled basis from online sources and is used in read-only mode for enterprise-wide analysis of trends by management. *Data marts* are departmental subsets derived from the data warehouse but focused on a single subject area. Data marts are often denormalized for performance reasons and are often based on a "*star schema*" structure with one central "fact" table, referenced by dimension tables. Data warehouses and data marts are used for *OLAP* and data mining. Often a star schema is used to construct a multidimensional structure, or *cube*, containing base and aggregated measures that can be accessed across various dimensions, including time.

Just as information systems can be modeled at different levels, so can they be queried at different levels. Conceptual query languages allow queries to be formulated either in natural language (e.g., English Query) or by selecting and constraining paths on a conceptual schema (e.g., ConQuer). Conceptual query tools automatically transform these higher-level queries into queries in a lower-level language such as SQL. Conceptual queries based on an attribute-free model such as ORM are far more stable when the schema evolves.

With large, complex schemas, there is a need for *abstraction mechanisms* to hide details that are not of immediate relevance. Modularization divides the global schema into conveniently sized subschemas. Constraint layers and textual rules may be toggled off and on as desired. We may zoom in on, or out from, a selected object type, specifying how much of its neighborhood is to be displayed.

Major object types may be identified by their importance in the global schema (e.g., due to having an explicit mandatory role), and the display suppressed for "minor" fact types (those without two roles played by major object types). This view may be refined further by focusing on its major fact types, and so on. Such bottom-up abstraction may be reversed to give top-down refinement.

Minor fact types may also be displayed in terms of attributes of major object types. In this case, ER diagrams and UML class diagrams may be generated as abstractions of ORM diagrams. While such abstractions are good for compact overviews, detailed ORM diagrams are best for developing, transforming, and evolving conceptual schemas.

Although the data model is the foundation of any information systems application, there is also a need to specify *process models* to indicate the permissible *operations* on the data and to design the *external interface.* For such purposes, various diagrams and notations may be used to supplement data model diagrams.

In early stages, informal diagrams such as cartoons can help clarify the UoD, and *function trees* and UML *use case diagrams* may be used to provide overviews of the system requirements. *Data flow diagrams* may also be used to picture information flows among processes, agents, and information stores. However, at some stage a formal connection needs to be made between processes and data. For this purpose, *data use cases* may be used to determine the data model to which processes may be bound (e.g., by defining *operations on the major object types*). This can also be used to generate the content (operations) of default screen forms for the external interface. Some default decisions can be made about the format and use of these screens.

Design of the HCI typically requires careful planning by humans. Access rights for different user groups need to be controlled. Users may also differ in their ability level. To make the *user interface* (UI) easy to learn and use, the interface should be consistent, simple, structured, efficient, and adapted to the user. In most cases, a GUI is preferable. The *menu* design needs to provide efficient navigation, and various levels of online and offline help are required.

In information modelling, the term "*ontology*" is used for a conceptual model that is specifically designed for sharing with other models. The term "*semantic web*" refers to an enhanced web for the future where documents on the Internet are structured in such a way as to enable automated agents to extract their semantics. Semantic web applications are typically built on top of *OWL*, which itself is built on top of RDF and RDF Schema, using XML Schema for the tagged textual representation and data types.

RDF breaks down information into atomic units that can be represented as simple (subject, predicate, object) triples denoting a binary relationship, using *URIrefs* to provide global identifiers for the resources that instantiate the subject, predicates, and object nodes in a triple. An object node may also be instantiated by a literal value (either untyped or typed).

RDF and RDFS are weakly expressive (very few constraints can be captured) but also too relaxed in allowing problematic assertions (e.g., a class may be an instance of itself). Because information on the web is often incomplete, OWL adopts the *open world assumption*. OWL Lite and OWL DL are based on description logics that lie in a decidable fragment of first-order logic that is efficient to compute. OWL Full has the same looseness as RDFS.

In 2009, a major upgrade to the OWL language, known as *OWL 2*, was recommended by the W3C. In 2012, a second edition of OWL 2 was recommended, and is still the current version (see https://www.w3.org/TR/owl-overview/). OWL 2 can express most aspects of ORM models, but it is not rich enough to capture several ORM features, such as set-comparison constraints involving multirole arguments, acyclic ring constraints, and deontic constraints.

OWL models may be expressed textually in various syntaxes (e.g., RDF/XML, OWL/XML, Manchester syntax, Functional syntax, and Turtle) but lacks a graphic notation. To fully exploit OWL, it seems best to capture the ontology first in a conceptual language such as ORM and then map to OWL or a more appropriate description logic, noting those constraints that are lost in the mapping.

By treating conceptual schemas as sample database states, the CSDP may be used to develop a conceptual *metaschema*. This may be used by a CASE tool to ensure that only valid conceptual schemas are entered. The activity of *metamodeling*, or making models of models, also helps to clarify and translate between different formalisms.

Chapter Notes

The literature on data warehousing and OLAP is substantial. Date (2000, Chapter 21) includes a clear overview of decision support. Inmon (1993), Kimball (1996), and Silverston (2001a,b) provide practical advice and examples of data warehousing. Jukik (2006) provides a useful overview of alternative strategies for modeling data warehouses. The gendercode example of data cleansing was suggested by Scot Becker—thanks Scot! For an extensive treatment of MDX for SQL Server 2022, see MDX Language Reference (MDX) - SQL Server | Microsoft Learn. For detailed coverage of Microsoft's OLAP architecture, see Logical Architecture (Analysis Services - Multidimensional Data) | Microsoft Learn.

A brief review of conceptual query prototypes, such as ERQL, Super, Hybris, and CBQL, is given in Bloesch and Halpin (1997). Some other graphical database query languages are surveyed in Ozsoyoglu and Wang (1993). Details on several query builders for SQL may be accessed at 9 Best SQL Query Builder & Generator Tools (2023) (guru99.com) and www.activequerybuilder.com/.

The RIDL language was originally used in conjunction with Control Data's IAST tool for NIAM, which was the first true CASE tool for information modeling (Meersman and Van Assche, 1983). A full description of the RIDL language is given in Meersman et al. (1984), and further discussions of RIDL may be found in Meersman (1982) and Verheijen and van Bekkum (1982). For further discussion of ConQuer, see Bloesch and Halpin (1996, 1997) and Halpin (1998).

For more details about schema abstraction within ORM, see Campbell and Halpin (1994a), Campbell et al. (1996), and their various references. A procedure for generating default external forms from major object types is discussed in Campbell and Halpin (1993).

McLeod (2000) offers useful suggestions for extending use case and activity diagrams in UML. For practical advice on traditional ways to model the process aspects of information systems, see Barker and Longman (1992) and Barker (1990a,b). Some promising ways to integrate data and process are discussed by Shoval and Kabeli (2000), Halpin and Wagner (2003), Dietz and Halpin (2003), and Morgan (2007).

Extensive discussions of HCI design are provided by Shneiderman (1992) and Dix et al. (1993). Extensive documentation on Microsoft products and technologies is accessible at https://learn.microsoft.com/en-gb/docs/. Enter "user interface guidelines" in the search field for Microsoft's recommend UI guidelines.

Specifications for RDF, RDFS, and OWL may be found online at www.w3.org. For OWL 2, see https://www.w3.org/TR/owl-overview/. A standard reference on description logics is Baader et al. (2003). For discussion of mapping ORM to OWL and description logics, see Keet (2007). For an example of an ORM-based ontology tool, see Trog et al. (2006). Our discussion of ontological modeling and the semantic web is largely drawn from a 15-part series of articles by Halpin on ontological modeling (see www.orm.net/resources). This series includes many OWL examples in Turtle syntax as well Manchester syntax.

For a comparison of metamodels for ORM, Barker ER, and UML, see Halpin (2004b). A metamodel for UML 2 is contained in the online UML specifications accessible at https://www.omg.org/spec/UML. Click on the abstract syntax sections of the most recent version to see diagrams for various parts of the metamodel.

ORM Glossary

This glossary lists key terms and symbols used in Object-Role Modeling (ORM), and briefly explains their meaning. A concise explanation of other technical terms may be found in the chapter summaries. Further details on technical terms may be accessed by using the Index.

Alethic constraint: Constraint that holds necessarily for all states of the model.

Arity: Number of roles in a relationship (unary =1, binary = 2, ternary = 3, etc.).

Asserted fact: Fact that is simply asserted, rather than being derived from others; also called a primitive fact or base fact.

Asserted subtype: Subtype that is simply asserted (not defined by a subtype definition).

Assertable role: Role is a fact type that is either asserted or semiderived.

Association: Relationship type, usually involving at least two roles.

Atomic fact: Either an elementary fact or an existential fact.

Base fact: Fact that is primitive (not derived from others). Also called an asserted fact.

Compound fact type: Fact type that is equivalent to a conjunction of smaller fact types.

Conceptual schema: Conceptual view of the UoD structure. A design that specifies the object types, fact types, subtyping, constraints, derivation rules, and what states and transitions are possible.

Conceptual schema design procedure (CSDP):

0. Divide the UoD into manageable subsections
1. Transform familiar examples into elementary facts, and apply quality checks
2. Draw the fact types, and apply a population check
3. Check for entity types that should be combined, and note arithmetic derivations
4. Add uniqueness constraints, and check arity of fact types
5. Add mandatory role constraints, and check for logical derivations
6. Add value, set comparison (subset, equality, exclusion), and subtype constraints
7. Add other constraints and perform final checks
8. Integrate the subschemas into a global conceptual schema

Constraint: Restriction on possible or permissible states (static constraint) or transitions (dynamic constraint).

Compositely identified object type: Either a coreferenced object type or a nested object type.

Coreferenced object: Object that is identified by means of two or more reference types in combination; hence its identification scheme involves an external uniqueness constraint.

Database: Variable set of related fact instances.

Deontic rule: An obligation, that is, a rule that ought to be obeyed (but might be violated).

Derivation rule: Rule that declares how one fact type may be derived from others.

Derived fact: Fact that is derived from other fact types using a derivation rule.

Derived subtype: Subtype that is derived from other object types using a subtype definition.

Elementary fact: Assertion that an object has a property, or that one or more objects participate in a relationship, where the fact cannot be split into simpler facts with the same object types without information loss. Application of an atomic predicate to a sequence of objects.

Entity: Object that is referenced by relating it to other objects (e.g., the Country that has CountryCode 'AU'); not a value; typically, an entity may undergo changes over time; an entity is either atomic or nested (i.e., an objectified relationship); at the top level, entities are partitioned into primitive entity types, from which subtypes may be defined.

Existential fact: Assertion that an object exists (e.g., there is a Country that has CountryCode 'AU'); also called a reference.

Fact: Proposition that is taken to be true by the relevant business community, where the proposition is elementary or existential (rather than being a constraint or derivation rule).

Fact role: Role in an elementary fact type.

Fact type: Kind of fact, including object terms and either a predicate or existential quantifier.

Flatten: Restate without nesting.

Functional fact type: Fact type with a functional role.

Functional role: Role with a simple uniqueness constraint.

Generalization: Forming a more general case from one or more specific cases; the inverse of specialization.

Independent object: Object that may exist without participating in any elementary fact.

Independent object type: A primitive object type that may include instances that are simply asserted to exist. The disjunction of elementary fact roles played by an independent object type is optional.

Instance: An individual occurrence (one specific member of a type).

Mandatory role: Role that must be played by all instances in the population of the object type playing the role; also called a total role.

Modality: Mode in which a proposition is expressed. In ORM 2, modalities are either alethic (expressing necessities or possibilities) or deontic (expressing obligations or permissions).

Nested object: Relationship that plays some role (also called an objectified relationship).

Object: Thing of interest; an object may be an entity or a value.

Objectification: Treating a relationship as an object; also called nesting. Strictly speaking, objectification in ORM 2 distinguishes the object formed by the objectification from the original relationship, and hence involves situational rather than propositional nominalization.

Object-Role Modeling (ORM): Conceptual modeling method that pictures a business domain in terms of objects playing roles; it provides graphical and textual languages for verbalizing and querying information as well as various design and transformation procedures.

Population: Set of instances present in a particular state of the database.

Predicate: Proposition with object-holes in it, e.g., "… works for …".

Primitive object type: Object type that is not a subtype of another object type.

Reference: Relationship used as the preferred way to reference or identify an object (or to provide part of the identification).

Reference mode: Mode or manner in which a single value references an entity; used to abbreviate simple reference schemes, such as (.Code), (kg:).

Referential role: Role in a reference (existential fact). Also known as an existential role.

Relationship: Property or association involving one or more objects.

Rigid subtype: Subtype whose instances must remain in that type for their whole lifetime (e.g., Person).

Rmap: Relational mapping procedure.

Role: Part played by an object in a relationship (unary, binary, ternary, etc.).

Role subtype: Subtype whose instances may leave that type during their lifetime (e.g., Child).

Semiderived fact type: Fact type, some of whose instances may be derived from others while some other instances may be simply asserted.

Semiderived subtype: Subtype, some of whose instances may be derived using a derivation rule while some other instances may be simply asserted.

Subtype: Object type that is properly contained in another object type (e.g., Woman is a subtype of Person).

Type: Set of possible instances.

Uniqueness constraint (UC): Repetition is not allowed in the role or role sequence spanned by the constraint; a uniqueness constraint on a single predicate is an internal UC, and a uniqueness constraint over roles from different predicates is an external UC.

Universe of Discourse (UoD): Business domain (the aspects of the world that we want to talk about).

Value: Unchangeable object that is identified by a constant. In this book, a value is either a character string or a number; sometimes called a label.

The following symbol glossary covers the main graphical symbols in ORM 2. It is based on the standard ORM 2 symbol glossary copyrighted to Terry Halpin, and is used by permission.

Construct and Examples	*Description/Notes*
Entity type 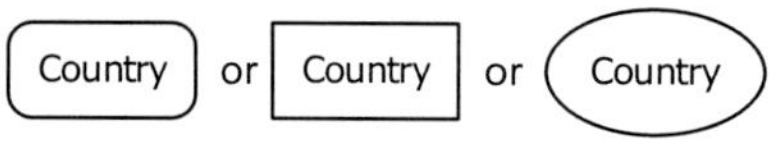	Named soft rectangle, named hard rectangle, or named ellipse. The soft rectangle shape is the default.
Value Type 	Named, dashed, soft rectangle (or hard rectangle or ellipse).
Entity type with popular reference mode 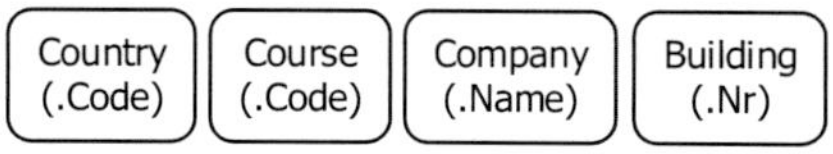	Abbreviation for injective reference to value type, such as 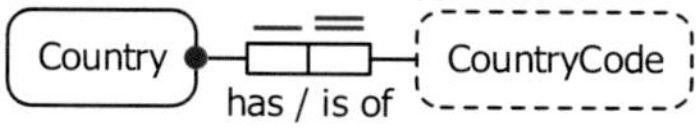
Entity type with unit-based reference mode 	Abbreviation for reference type, such as 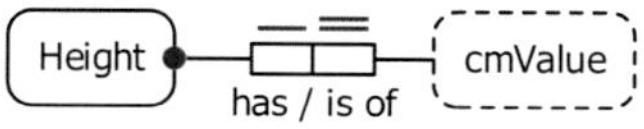 Optionally, the unit type may be displayed (as shown opposite).
Entity type with general reference mode 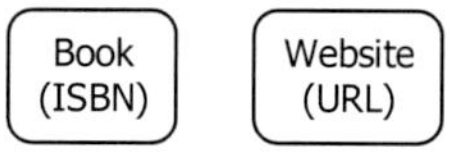	Abbreviation for reference type, such as 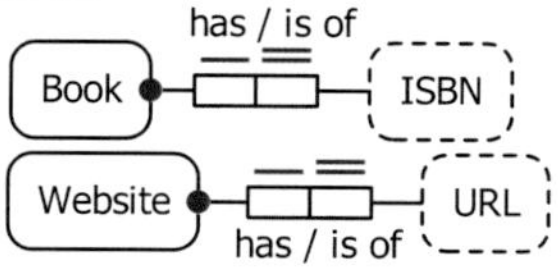
Independent Object Type 	Instances of the type may exist, without playing any elementary fact roles.
External Object Type 	This notation is tentative, and is not supported by NORMA.
Predicate (unary, binary, ternary, etc.) 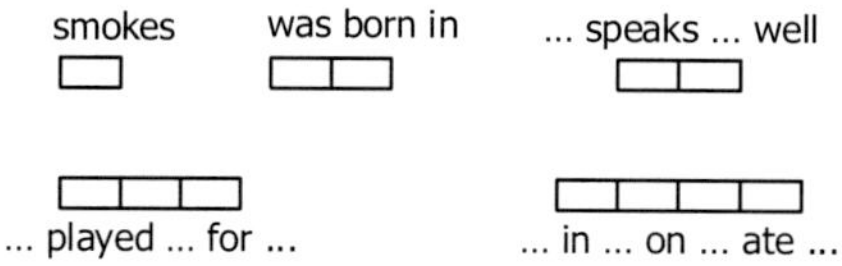	Ordered set of 1 or more role boxes with at least one predicate reading. If shown, object place-holders are denoted by "...". If placeholders are not shown, unaries use prefix notation. Binaries use infix notation. Ternaries, etc., use mixfix notation.
Duplicate type or predicate shape 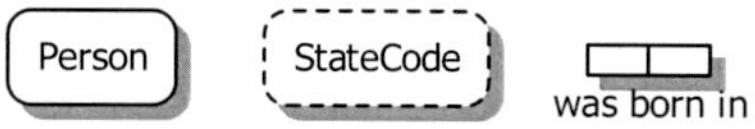	If an object type or predicate shape is displayed more than once (on the same page or different pages), it is shadowed.

Construct and Examples	*Description/Notes*
Unary fact type 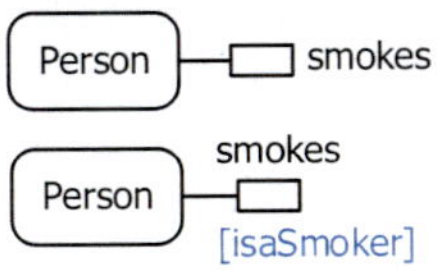	Attaching a role box to an object type shape means that only instances of that object type may play that role. A role name may be added in square brackets.
Binary Fact Type 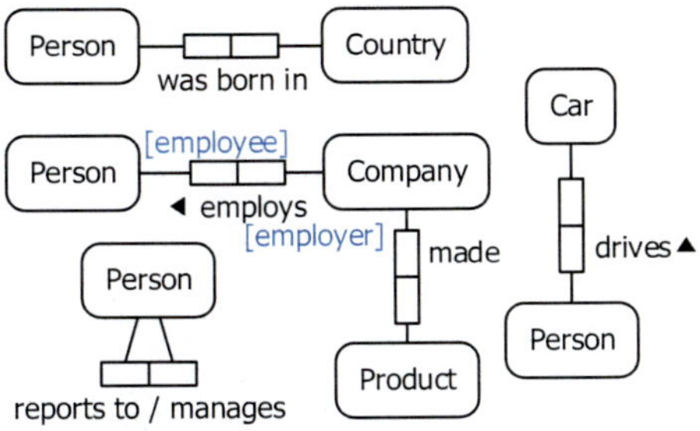	By default, predicate readings (binary or longer) are read left to right or top to bottom. An arrow tip is used to display a different reading direction. Role names may be displayed in square brackets beside their role. Forward and inverse readings for binaries may be shown together, separated by "/".
Ternary fact type 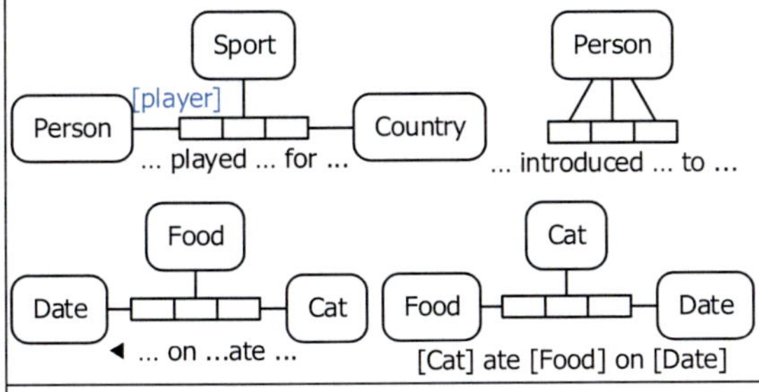	Role names may be added in square brackets. Arrow tips are used to reverse the default left-right or top-down reading order. Reading orders other than forward and reverse are shown using named placeholders.
Quaternary fact type 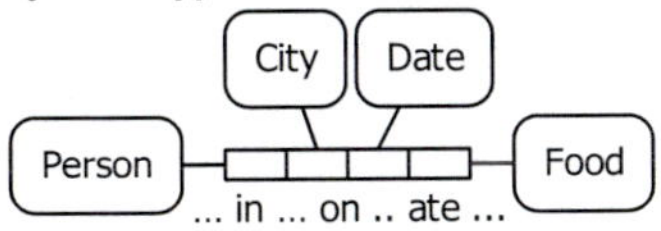	The above notes for the ternary case apply here also. Fact types of higher arity (number of roles) are also permitted.
Objectification (a.k.a. nesting) 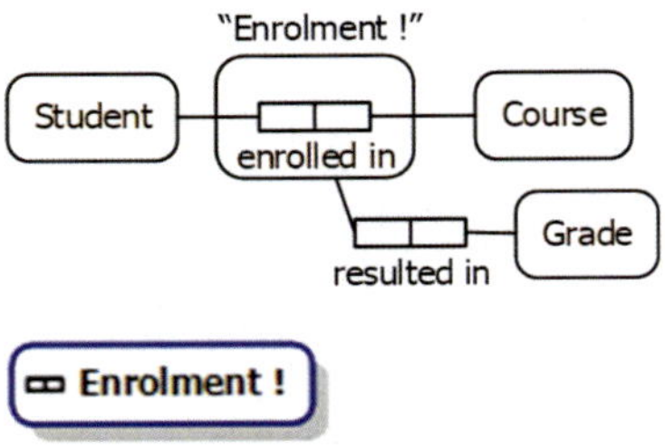	The enrolment* fact type is objectified as an entity type whose instances can play roles. In this example, the objectification type is independent, so we can know about an enrolment before the grade is obtained. Objectification object types may also be displayed without their defining components, using an object type shape containing a small predicate shape, as shown. *Australian spelling used here.

Construct and Examples	*Description/Notes*
Internal uniqueness constraint (UC) on a unary fact type 	These examples are equivalent. By default, fact types are assumed to be populated with sets of facts (not bags of facts), so no whole fact may be duplicated.
Internal UCs on a binary fact type 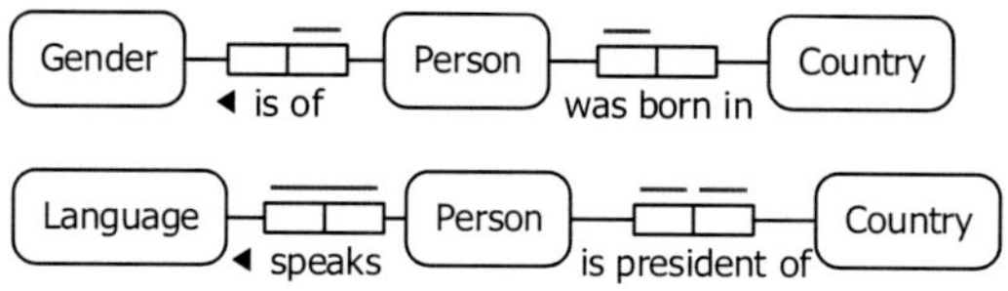	The examples show the four possible patterns: 1:*n* (one-to-many); *n*:1 (many-to-one); *m:n* (many-to-many); 1:1 (one-to-one).
Internal UCs on a ternary fact type 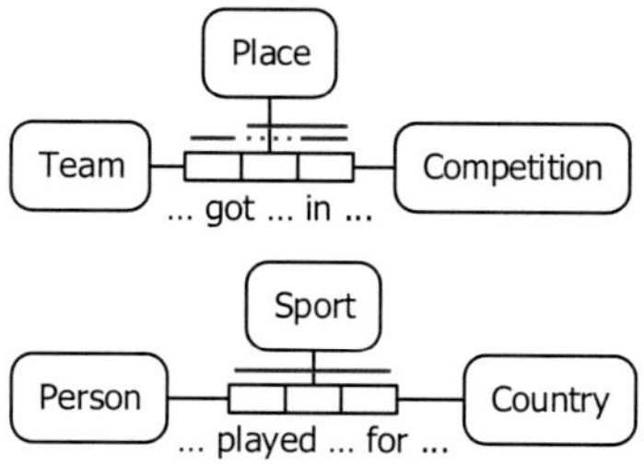	The first example has two two-role UCs: the top UC forbids ties. The other UC ensures that each team got at most one place per competition (a dotted line excludes its role from the UC). The second example has a spanning UC (many-to-many-to-many). For an *n*-ary ($n > 2$) fact type to be atomic, each UC on it must span at least n-1 roles.
Simple mandatory role constraint 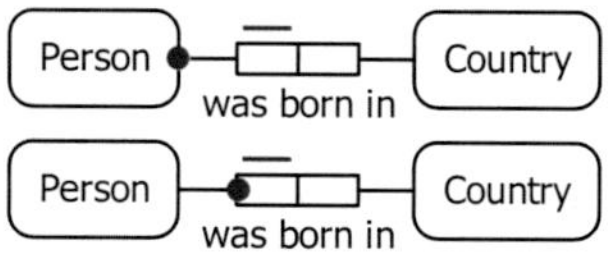	The example constraint means that each person was born in some country. The mandatory role dot may be placed at either end of the role connector.
Inclusive-or constraint 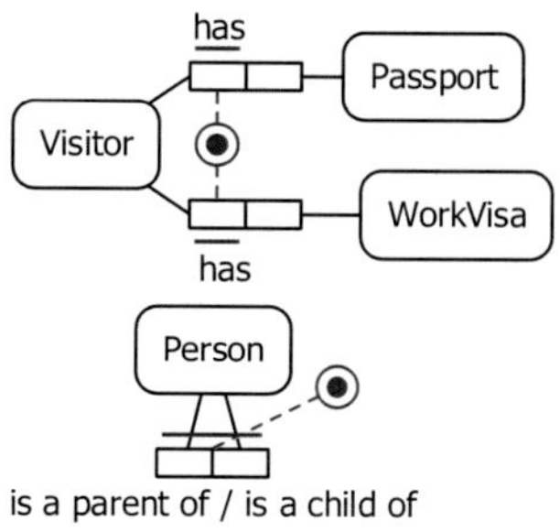	An inclusive-or constraint is also called a disjunctive mandatory role constraint. The constraint is displayed as a circled dot connected to the constrained roles (or the junction of adjacent roles hosted by the same object type). The first constraint means that each visitor referenced in the model must have a passport or work visa (or both). The second example means that each person is a parent of a person or a child of a person (or both).

<table>
<tr><th>Construct and Examples</th><th>Description/Notes</th></tr>
<tr><td>

Preferred internal UC

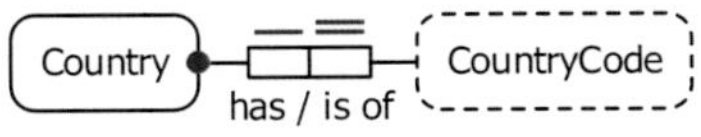

</td><td>A double bar on a UC indicates that it underlies the preferred reference scheme.</td></tr>
<tr><td>

External UC (inner-join semantics)

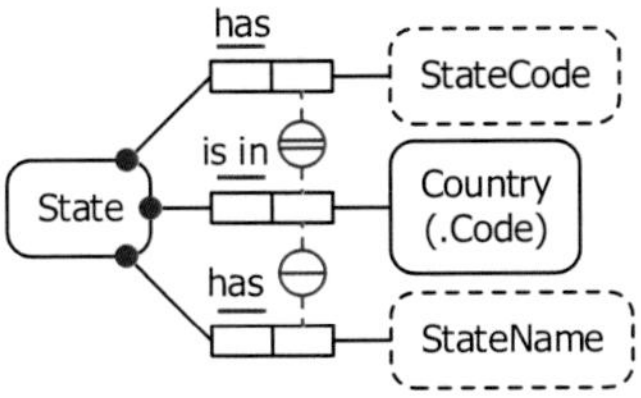

</td><td>A double-bar indicates that the constrained roles provide the preferred reference for the object type at the other end.
Here, each state is primarily identified by combining its country and state code. Each combination of country and state name also applies to only one state.</td></tr>
<tr><td>

External UC (outer-join semantics)

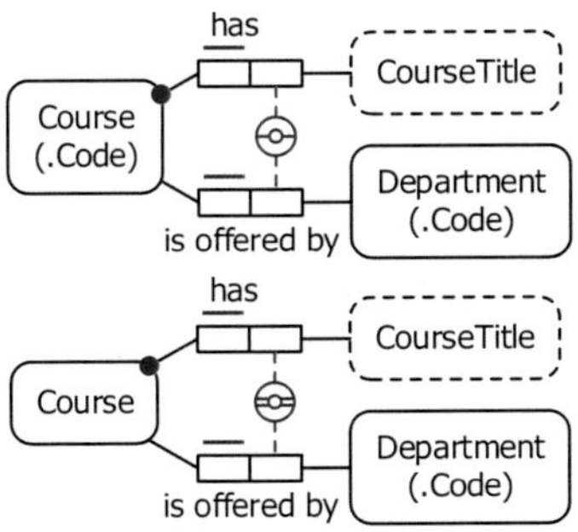

</td><td>An inner "o" through the uniqueness bars indicates that the external UC has outer join semantics, with the added proviso that nulls produced in the outer join are treated as actual values (see Section 10.1 for further discussion).
A double uniqueness bar indicates the UC is used in the preferred reference scheme.
The inner "o" notation is not yet supported by the NORMA tool.</td></tr>
<tr><td>

Object type value constraint

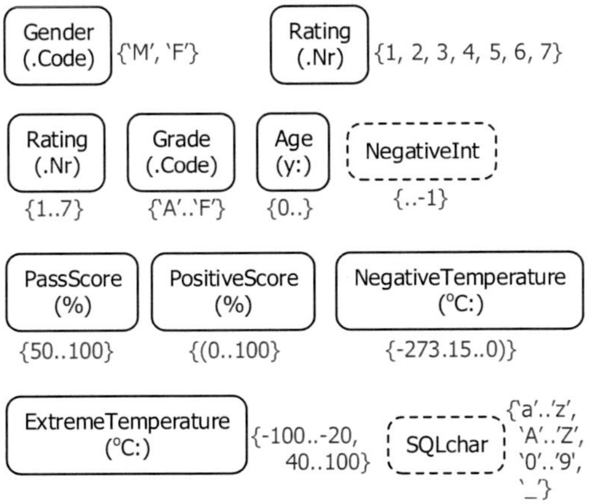

</td><td>The allowed values may be specified as a list of discrete values and/or value ranges. The first two examples specify an enumerated list of values.
Ranges are inclusive of end values by default.
Round brackets exclude an end value.
Square brackets explicitly declare inclusion, such as the constraint on PositiveScore may also be specified as {(0..100]}.
Multiple combinations of value constraints may also be specified.</td></tr>
</table>

Construct and Examples	*Description/Notes*
Role value constraint 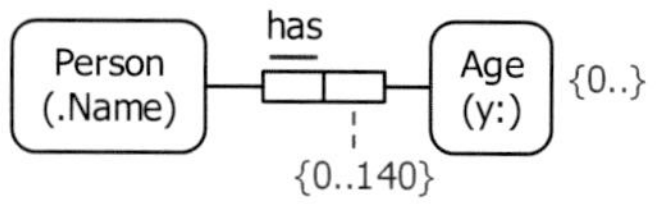	As for object type value constraints, but connected to the constrained role. Here, an age of a person must be at most 140 years.
Subset constraint 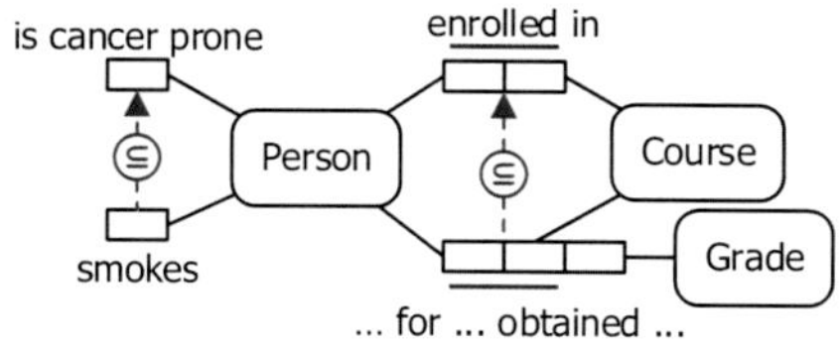	The arrow points from the subset end to the superset end (e.g., if a person smokes then that person is cancer prone). The role sequences at both ends must be compatible. A connection to the junction of two roles constrains that role pair.
Join subset constraint 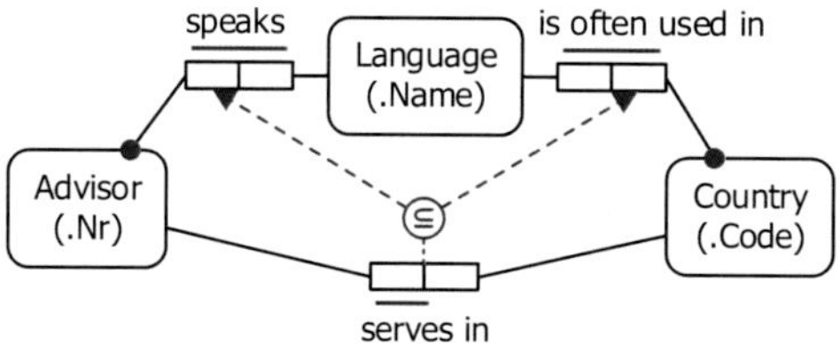	The constrained role pair at the superset end is projected from a role path that involves a conceptual join on Language. The constraint declares that if an advisor serves in a country, then that advisor must speak a language that is often used in that country.
Exclusion constraint 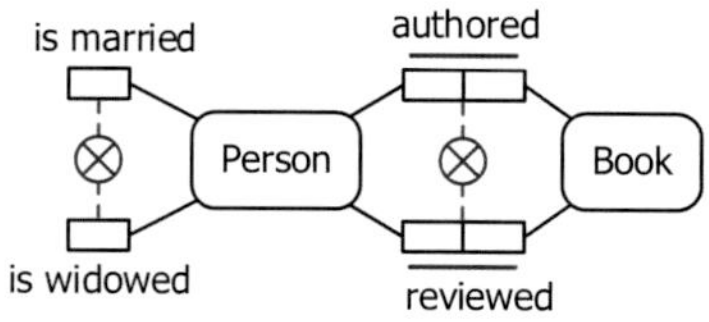	These exclusion constraints mean that no person is both married and widowed, and no person reviewed and authored the same book. Exclusion may apply between two or more compatible role sequences, possibly involving joins.
Exclusive-or constraint 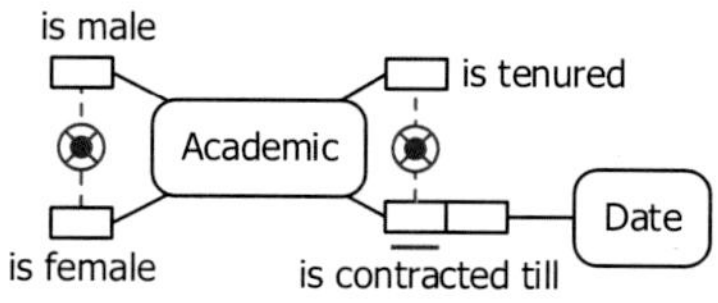 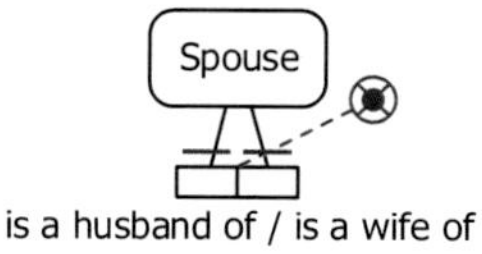	An exclusive-or constraint is simply the conjunction of an inclusive-or constraint and an exclusion constraint. The first two examples say that each academic is male or female but not both, and is tenured or contracted until some date but not both. The third example says each spouse is a husband or wife of a spouse but not both. The 1:1 nature of the fact type assumes monogamy.

Construct and Examples	*Description/Notes*
Equality constraint 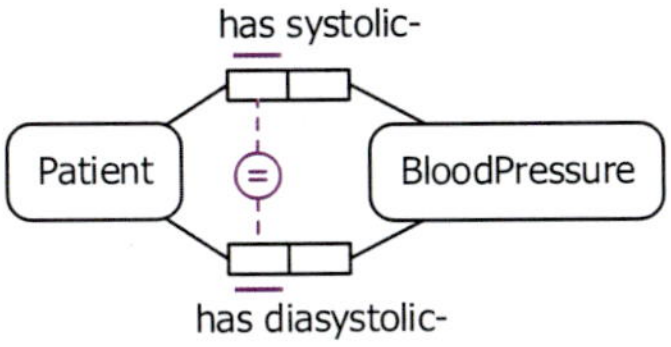	This constraint means that a patient's systolic BP is recorded if and only if his/her diastolic BP is recorded. An equality constraint may apply between two or more compatible role sequences, possibly involving joins.
Derived fact type, and derivation rule 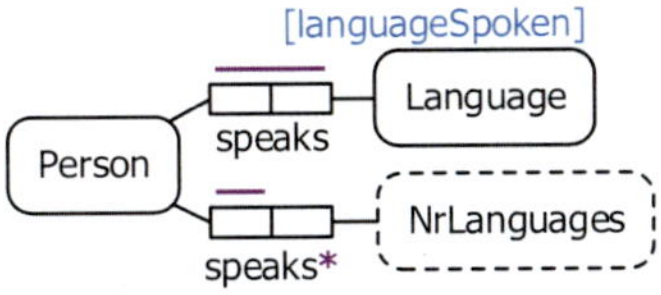 ***For each** Person, nrLanguages = **count**(languageSpoken).	A fact type is either asserted, derived, or semiderived. A derived fact type is marked with an asterisk "*". A derivation rule is supplied. A double asterisk "**" indicates derived and stored (eager evaluation).
Semiderived fact type, and derivation rule 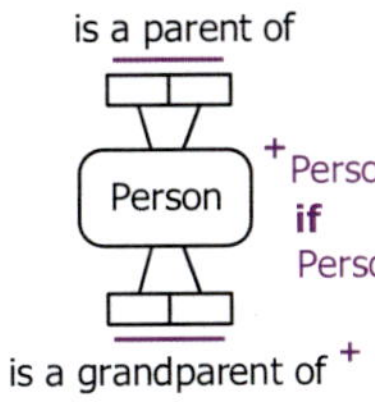 +Person$_1$ is a grandparent of Person$_2$ **if** Person$_1$ is a parent of **some** Person$_3$ **who** is a parent of Person$_2$.	A fact type is semiderived if some of its instances may be derived, and some of its instances may be simply asserted. It is marked by "+" (half an asterisk). "++" indicates semiderived and stored (eager evaluation for derived instances).
Subtyping 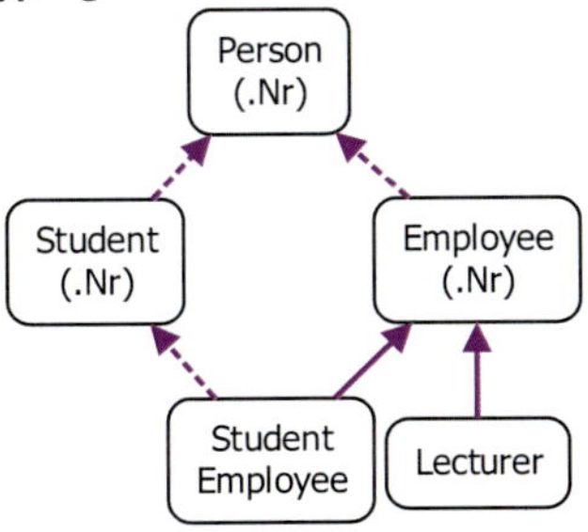	All subtypes are proper subtypes. An arrow runs from subtype to supertype. A solid arrow indicates a path to the subtype's preferred identifier (e.g., here, student employees are primarily identified by their employee number). A dashed arrow indicates the supertype has a different preferred identifier.
Subtyping constraints 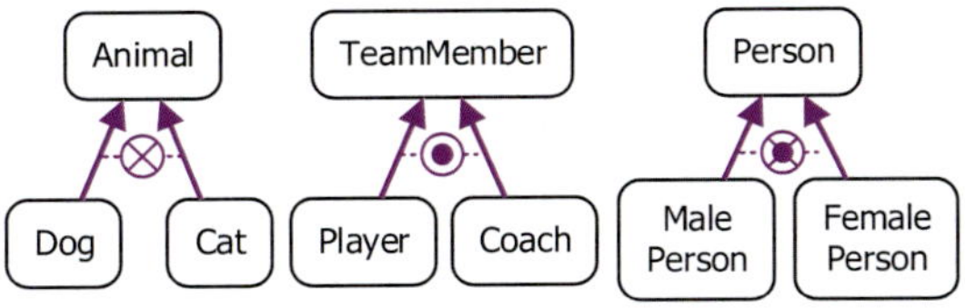	⊗ indicates that the subtypes are mutually exclusive. ◉ indicates that the supertype equals the union of the subtypes. The combination (xor constraint) indicates that the subtypes partition the supertype (exclusive and exhaustive).

Construct and Examples	*Description/Notes*
Subtype derivation status 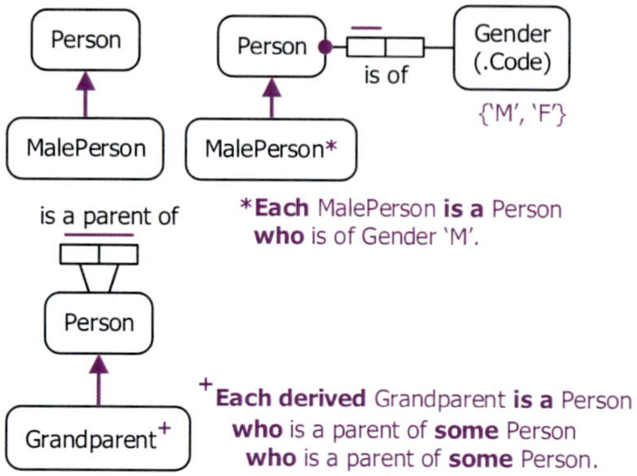	A subtype may be asserted, derived (denoted by "*"), or semiderived (denoted by "+"). If the subtype is asserted, it has no mark appended and has no derivation rule. If the subtype is derived or semiderived, a derivation rule is supplied.
Internal frequency constraint 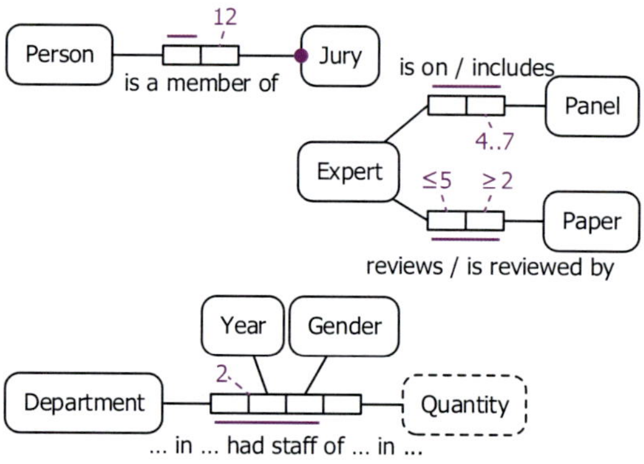	This constrains the number of times an occurring instance of a role or role sequence may appear in each population. Here, each jury has exactly 12 members; each panel that includes an expert includes at least four and at most seven experts; each expert reviews at most five papers; each paper that is reviewed is reviewed by at least two experts; and each department and year that has staff numbers recorded in the quaternary appears there twice (once for each gender).
External frequency constraint 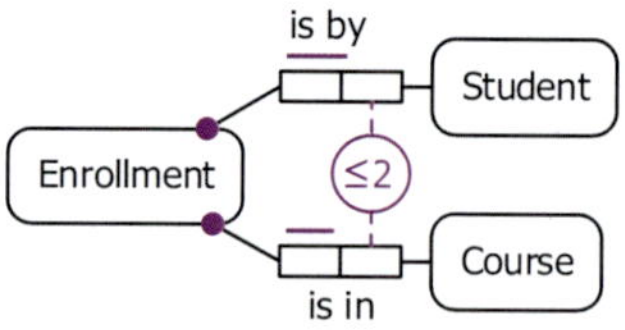	The example constraint has the following meaning. In this context, each combination of student and course relates to at most two enrolments (i.e., a student may enroll at most twice in the same course).
Value-comparison constraint 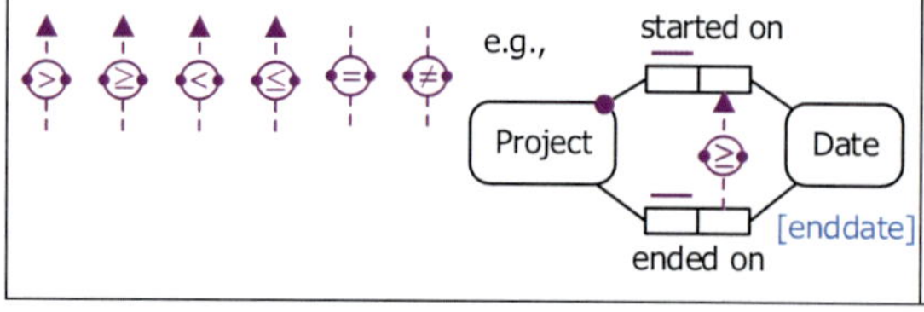	The example constraint verbalizes as: **For each** Project, **existing** enddate >= startdate.

Construct and Examples	*Description/Notes*
Object cardinality constraint 	The example constraints ensure that at any given time there is at most one president and either zero or at least five and at most 15 members of the UN Security Council.
Role cardinality constraint 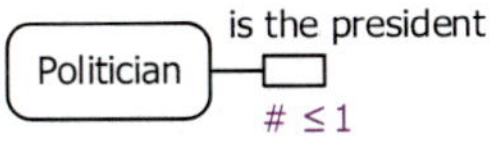	The example constraint ensures that at most one politician is the president (at any given time).
Ring constraints 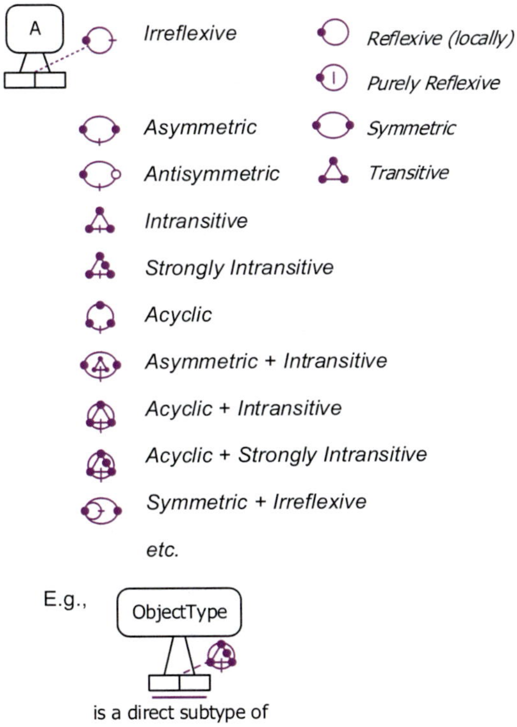	A ring predicate R is locally reflexive if and only if, for all x and y, xRy implies xRx. For example, "knows" is locally but not globally reflexive. Reflexive, symmetric, and transitive properties may also be enforced using semiderivation rather than by constraining asserted fact types. The example constrains the subtyping relationship in ORM to be acyclic (no cycles can be formed by a chain of subtyping connections) and strongly intransitive (no object type A can be both a direct subtype of another type B and an indirect subtype of B, where indirect subtyping means there is a chain of two or more subtyping relationships that lead from A to B). Ring constraints may be combined only if they are compatible, and one is not implied by the other. ORM tools ensure that only legal combinations are allowed.
Objectification display options 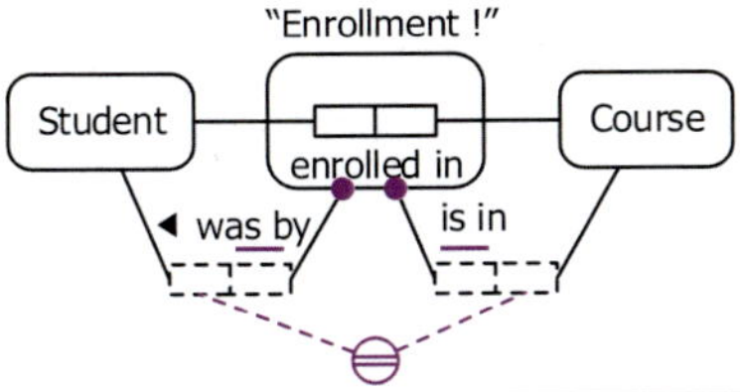	Internally, link fact types connect objectified associations to their component object types. By default, display of link fact types is suppressed. If displayed, dashed lines are used.

Construct and Examples	*Description/Notes*
Deontic constraints 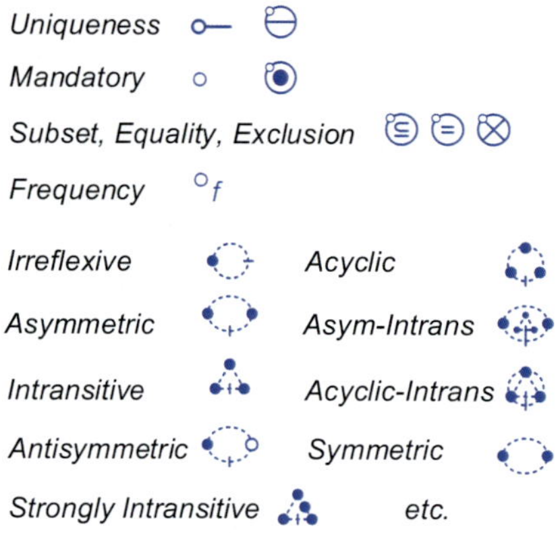 E.g., 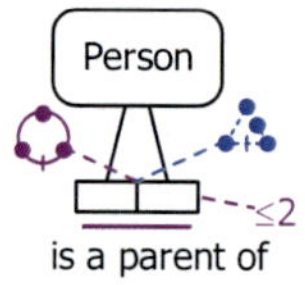 	Unlike alethic constraints, deontic constraint shapes are colored blue rather than violet. Most include "o" for "obligatory". Deontic ring constraints use dashed lines. In the parenthood example, the alethic frequency constraint ensures that each person has at most two parents, the alethic ring constraint ensures that parenthood is acyclic, and the deontic ring constraint makes it obligatory for parenthood to be strongly intransitive.
Textual constraints 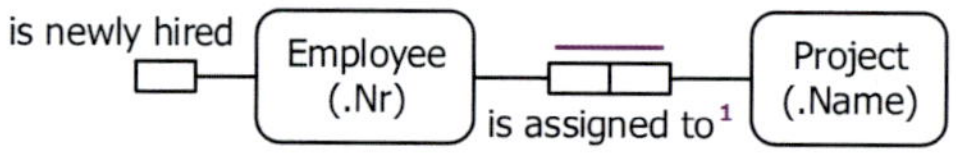 [1] **Each** Employee **who** is newly hired is assigned **at most one** Project.	First-order constraints with no graphic notation may be expressed textually in the FORML 2 language. This example uses footnoting to capture a restricted uniqueness constraint.
Derived constraints E.g. 	Derived constraints are colored green. This notation is supported in the Visio ORM stencil, but is not currently supported in NORMA.

UML Glossary

This glossary lists key terms and symbols used in UML data models (*class diagrams*), as discussed in Chapter 9.

Aggregate: Class that is the "whole" in a whole-part relationship.
Association: Relationship type, involving two or more association roles.
Association class: An association that is also a class; an objectified association in ORM.
Association end: Corresponds to a role in ORM.
Attribute: Property of an object; either single-valued or multivalued.
Changeability: How an attribute or role value may change: unrestricted, readOnly, addOnly, removeOnly.
Class: Type of object (cf. ORM entity type).
Collection property string: Annotation to an attribute or association role indicating the kind of collection to be used. For example, {ordered}, {nonunique}, and {ordered, nonunique} specify an ordered set, bag, and sequence, respectively.
Composition (*composite aggregation*): Each part belongs to at most one whole.
Data value: Scalar constant; data type instance; self-identifying (cf. ORM value).
Derived element: An attribute or association that is computed or inferred from others.
Generalization: Forming a more general class (a superclass) from one or more classes.
Link: Relationship instance—a fact in ORM.
Multiplicity: Number of instances to which the same object may relate in an attribute or association.
Object: A thing that may typically change its state; a class instance; identified by an oid (cf. ORM entity instance).
Oid: Object identifier; system generated constant that references an object.
Qualified association: Association where one class relates to another via a qualifier.
Shared aggregation: A part may belong to more than one whole.
Specialization: Forming one or more specific classes (subclasses) from a more general class.
State: The properties of an object at a point in time.
Subset constraint: Restricts the population of one association to be a subset of another.
Xor constraint: Exclusive-or constraint between roles (association ends).

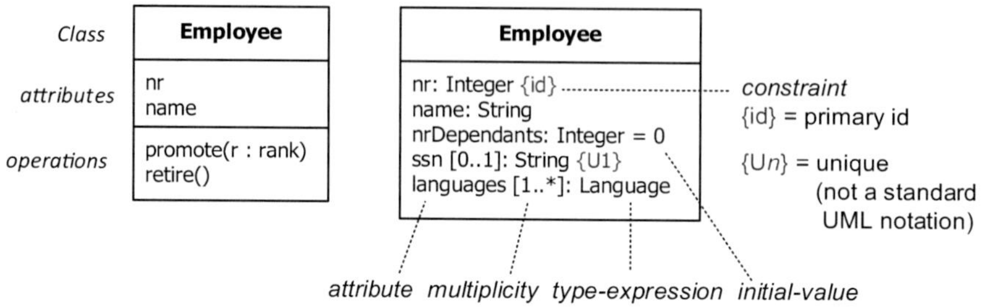

Binary association

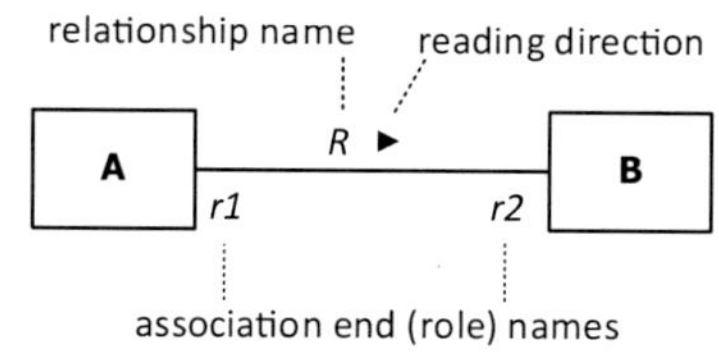

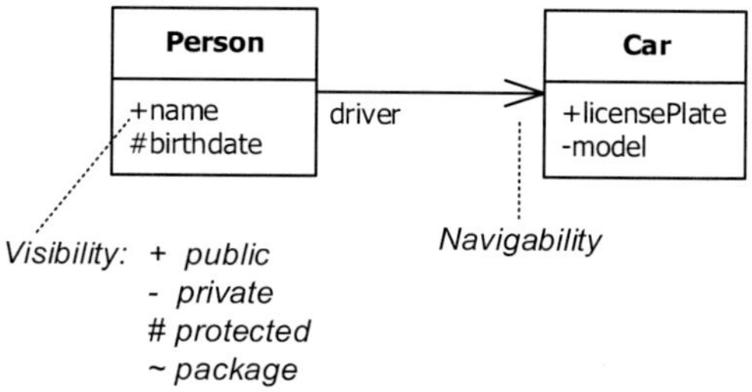

Multiplicities:

*	0 or more
0..1	0 or 1
1..*	1 or more
1	exactly 1
n..m	at least *n* and at most *m*

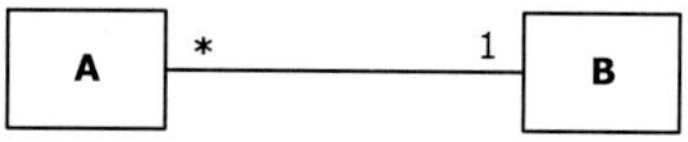

Each *A* plays with exactly one *B*.
Each *B* plays with 0 or more *A*'s

default attribute multiplicity = 1

default association role multiplicity = *

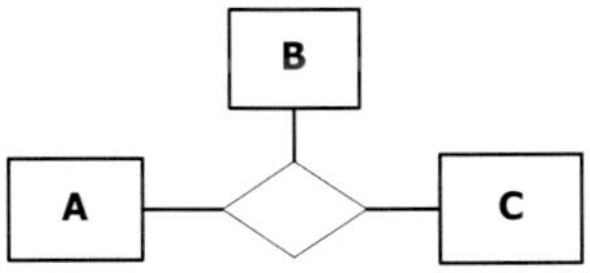

Ternary association

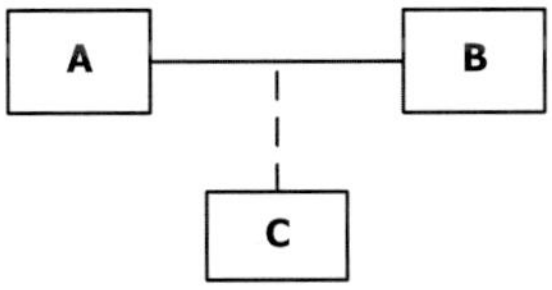

Association class

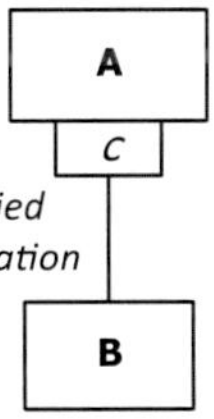

Qualified association

Exclusive-or constraint

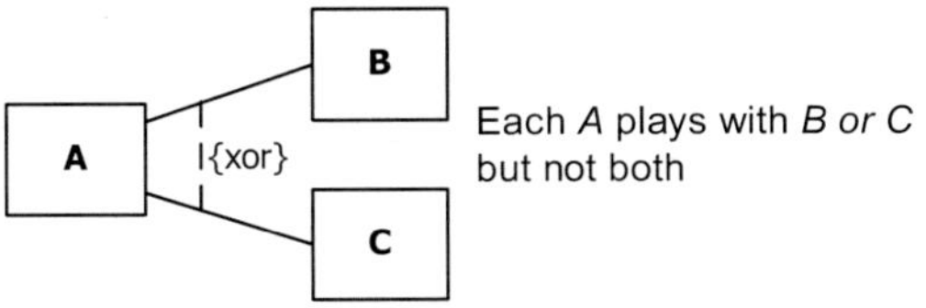

Subset constraint

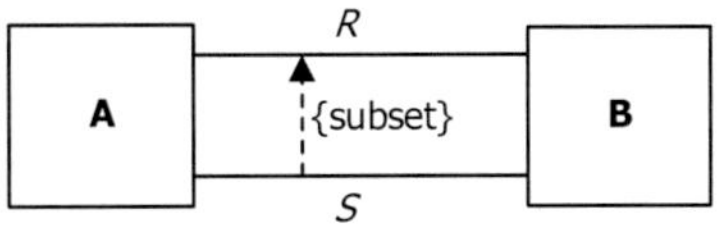

pop(S) is a subset of pop(R)

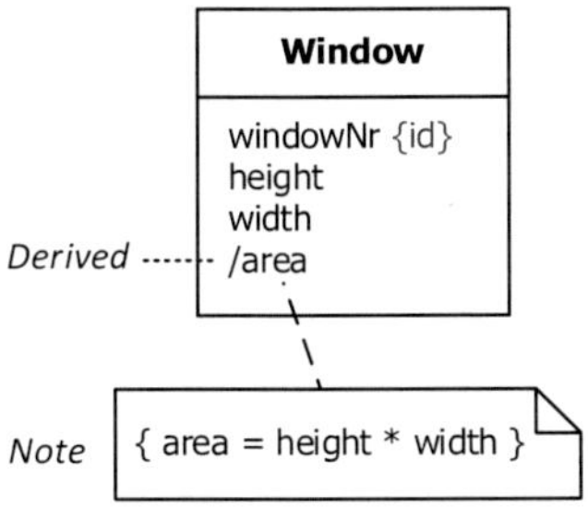

Subtyping

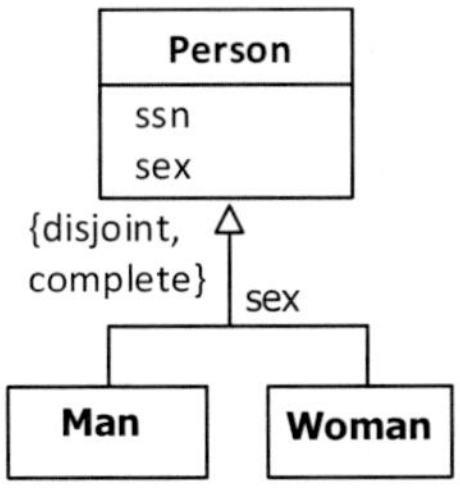

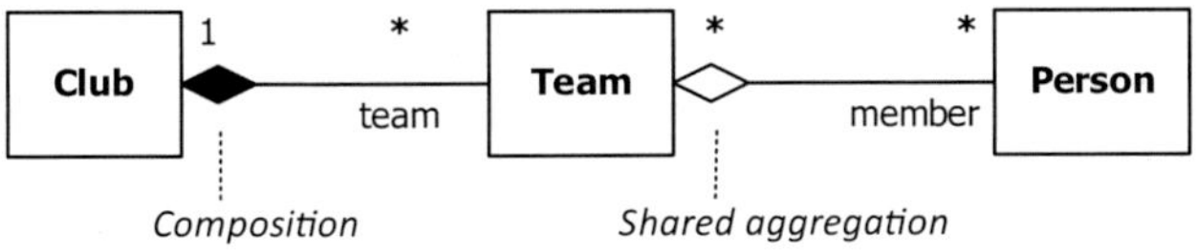

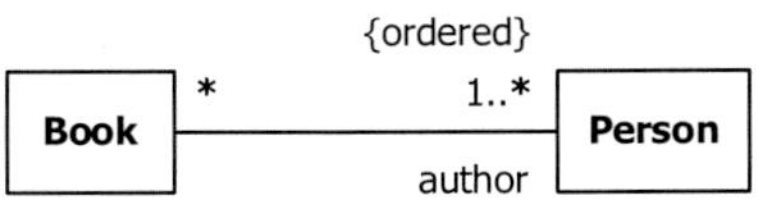

Authors are *ordered*

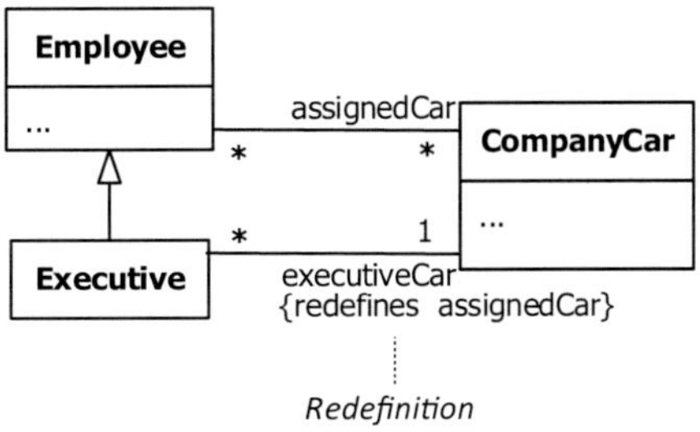

ER Glossary

This glossary lists the key symbols used in the Barker ER notation, the Information Engineering notation for ER, and the IDEF1X notation, as discussed in Chapter 8.

Barker ER notation

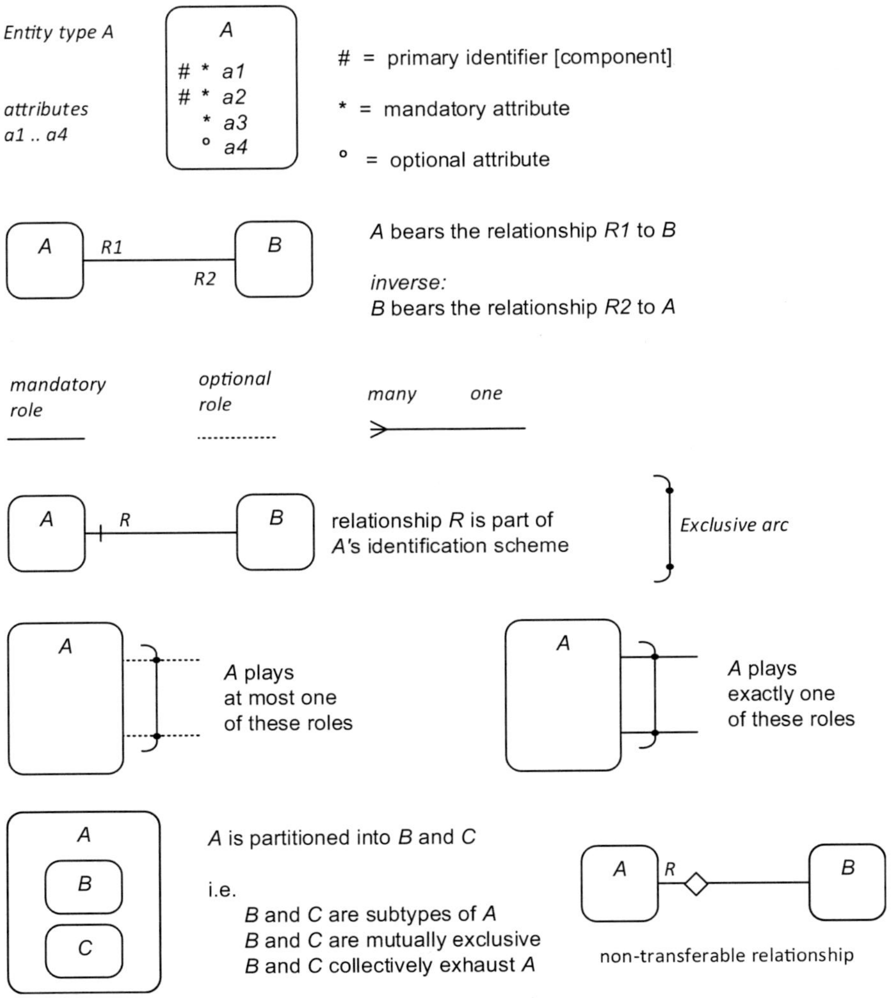

Information Engineering notation

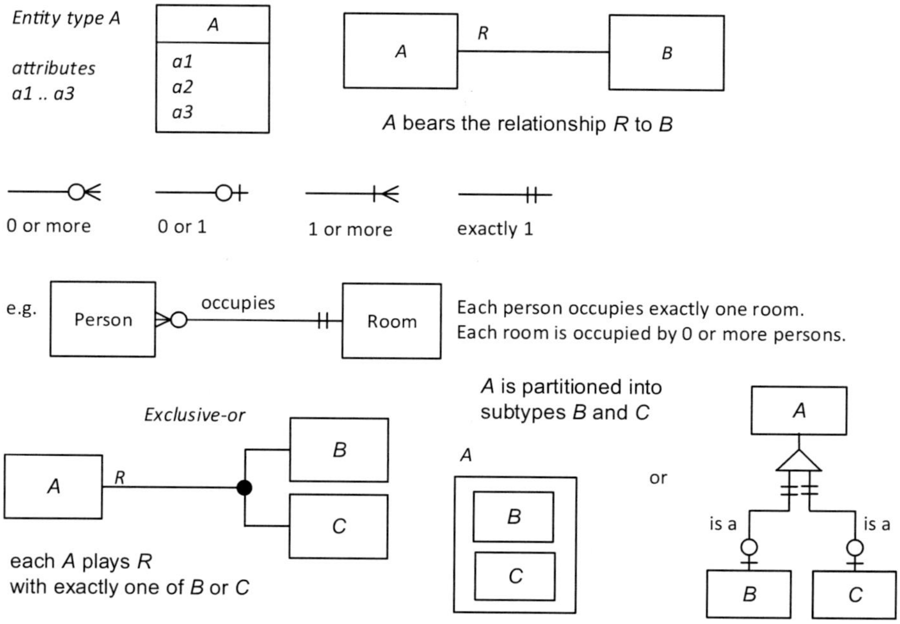

IDEF1X notation

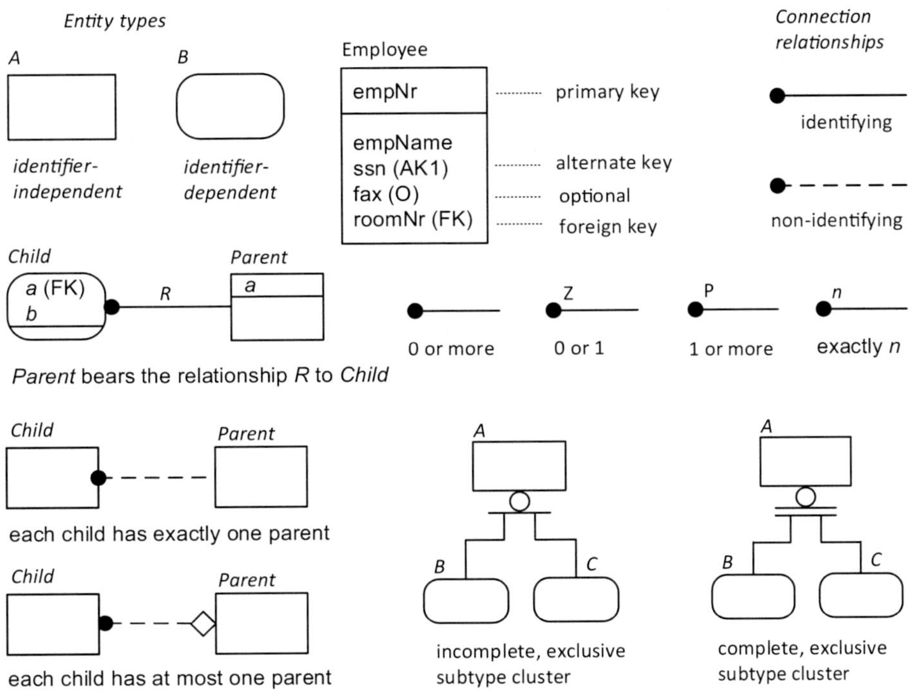

Useful websites

https://www.elsevier.com/books-and-journals/book-companion/9780443237904. Morgan Kaufmann website for this book; includes online appendices.

Fact-oriented modeling (general)

www.orm.net. Terry Halpin's ORM site.
www.ormfoundation.org. ORM Foundation site
https://geverest.umn.edu/. Gordon Everest's website. Includes links to ORM lectures.
https://everestdm.blogspot.com/. Gordon Everest's blog on data modeling.
www.pna-group.com. PNA Group: Dutch consultancy for fact-oriented modeling.

Fact-oriented modeling tools

NORMA (ORM2 tool). https://ormsolutions.com/ or https://sourceforge.net/projects/orm. For details on installing and using NORMA, see the Resources page at www.orm.net.
NORMA Pro has additional features such as multipage relational schemas and formal support for derivation rules. For details, contact Matt Curland at mcurland@live.com.
Most of the ORM schema diagrams in this book were drawn using the Visio stencil and template available at www.orm.net/ORM2_Draw2v25.zip. These files are free, but they require the standard or professional version of Microsoft Visio to run.
CaseTalk (FCO-IM tool). https://www.casetalk.com.
iKnow tools based on CogNIAM. www.pna-group.com.
ActiveFacts (ORM 2 tools). http://dataconstellation.com/ActiveFacts/.
Boston. https://www.factengine.ai/.

Business rules, data management, and data architecture frameworks

www.brcommunity.com/index.php. Business Rules Journal.
www.dama.org/. DAMA International.
https://en.wikipedia.org/wiki/Zachman_Framework. Zachman Framework.
https://zachman-feac.com. Zachman International and the FEAC Institute.
www.opengroup.org/togaf/. The Open Group Architecture Framework.
http://committee.iso.org/sites/jtc1sc7/home/projects/flagship-standards/isoiec-10746.html. The ISO RM-ODP viewpoints framework.

IDEF1X, SQL, XML, RDF, JSON, and OWL

www.iso.org/standard/60614.html. Latest version of IDEF1X.
www.wiscorp.com/SQLStandards.html. Free downloads of older SQL standards, etc.
http://sqlzoo.net. Basic tutorials on SQL, XML, etc.
www.w3.org/RDF/. W3C website on Resource Description Framework (RDF).
www.w3.org/2001/sw/wiki/RDFS. W3C website on RDF Schema.
www.w3.org/TR/owl2-overview. W3C website on OWL 2.
www.w3.org/XML/. XML and related standards.
www.iso.org/standard/76584.html. Latest version of the SQL standard (ISO/IEC 9075). Press the Preview button for informative details.
www.ecma-international.org/. JavaScript and JSON standards.

UML, OCL, MDA, and SBVR

www.uml.org. The OMG website on UML.
https://www.omg.org/spec/UML/2.5.1.The latest version of the UML specification.
https://www-st.inf.tu-dresden.de/ocl/. Portal on the Object Constraint Language.
https://www.omg.org/mda/. Model-Driven Architecture.
https://www.omg.org/spec/SBVR/1.5. SBVR specification on the OMG website.
https://en.wikipedia.org/wiki/Semantics_of_Business_Vocabulary_and_Rules. Wikipedia article on SBVR.

Workflow modeling

www.workflowpatterns.com. A compendium of resources on workflow patterns.
www.workflowpatterns.com/yawl/. Free YAWL tool and docs on workflow modeling.
www.oasis-open.org. The original home of the BPEL standard.
www.omg.org. The latest information on BPMN and related standards can be found here.

Bibliography

Entries marked with an asterisk "*" are accessible online at www.orm.net.

Abrial, J., 1974. Data semantics. In: Klimbie, J., Koffeman, K. (Eds.), Data Base Management. North-Holland, Amsterdam, pp. 1–60.

Aho, A., Beeri, C., Ullman, J., 1979. The theory of joins in relational databases. ACM Trans. Database Syst. 4 (3), 297–314.

Atkinson, M., Bancilhon, F., DeWitt, D., Dittrick, K., Maier, D., Zdonik, S., 1990. The object-oriented database system manifesto. In: Kim, W., Nicolas, J.-M., Nishio, S. (Eds.), Proc. First Int. Conf. on Deductive and Object-Oriented Databases (DOOD). Elsevier, Dec., 1989, pp. 40–57.

Baader, F., Calvanese, D., McGuinness, D., Nardi, D., Patel-Schneider, P. (Eds.), 2003. The Description Logic Handbook. Cambridge University Press.

Bakema, G., Zwart, J., van der Lek, H., 1994. Fully communication oriented NIAM. In: Nijssen, G.M., Sharp, J. (Eds.), NIAM-ISDM 1994 Conf. Working Papers, Albuquerque, NM, pp. L1–35.

Balsters, H., Halpin, T., 2007. Modeling data federations in ORM. In: Meersman, R., Tari, Z., Herrero, P., et al. (Eds.), On the Move to Meaningful Internet Systems 2007: OTM 2007 Workshops. Springer LNCS 4805, Vilamoura, pp. 657–666.

Balsters, H., Halpin, T., 2008. Formal semantics of dynamic rules in ORM. In: Meersman, R., Tari, Z., Herrero, P., et al. (Eds.), On the Move to Meaningful Internet Systems 2008: OTM 2008 Workshops. Springer LNCS 5333, Monterrey, Mexico, pp. 699–708.

Balsters, H., Halpin, T., 2016. Formal semantics of dynamic constraints and derivation rules in ORM. Int. J. Inform. Syst. Model. Des. 7 (2). IGI Global.

Balsters, H., Carver, A., Halpin, T., Morgan, T., 2006. Modeling dynamic rules in ORM. In: Meersman, R., Tari, Z., Herrero, P., et al. (Eds.), On the Move to Meaningful Internet Systems 2006: OTM 2006 Workshops. Springer LNCS 4278, Montpellier, pp. 1201–1210.

Barbier, F., Henderson-Sellers, B., Opdahl, A., Gogolla, M., 2000. The whole-part relationship in UML: a new approach. In: Siau, K., Halpin, T. (Eds.), Unified Modeling Language: Systems Analysis, Design and Development Issues. Idea Group, Hershey, PA.

Barker, R., 1990a. CASE*Method: Entity Relationship Modelling. Addison-Wesley, Wokingham.

Barker, R., 1990b. CASE*Method: Tasks and Deliverables. Addison-Wesley, Wokingham.

Barker, R., Longman, C., 1992. CASE*Method: Function and Process Modelling. Addison-Wesley, Wokingham.

Basset, L., 2015. Introduction to JavaScript Object Notation. O'Reilly Media.

Batini, C., Ceri, S., Navathe, S., 1992. Conceptual Database Design: An Entity-Relationship Approach. Benjamin Cummings, Redwood City, CA.

Bennett, S., McRobb, S., Farmer, R., 2006. Object-Oriented Systems Analysis and Design, third ed. McGraw Hill, Berkshire.

Bentley, J., 1998. Little languages. In: More Programming Pearls. Addison-Wesley, Reading, MA.

Berners-Lee, T., Hendler, J., Lassila, O., 2001. The semantic web. Scientific Am. 2001.

Bird, L., Goodchild, A., Halpin, T., 2000. Object-role modelling and XML schema. In: Conceptual Modeling—ER2000, Proc. 19th ER Conf. Springer LNCS 1920, pp. 309–322.

Blaha, M., Premerlani, W., 1998. Object-Oriented Modeling and Design for Database Applications. Prentice Hall, New Jersey.

Bloesch, A., Halpin, T., 1996. ConQuer: a conceptual query language. In: Proc. ER'96: 15th Int. Conf. on Conceptual Modeling. Springer LNCS 1157, pp. 121–133.*

Bloesch, A., Halpin, T., 1997. Conceptual queries using ConQuer-II. In: Proc. ER'97: 16th Int. Conf. on Conceptual Modeling. Springer LNCS 1331, pp. 113–126.*

Booch, G., Rumbaugh, J., Jacobson, I., 1999. The Unified Modeling Language User Guide. Addison-Wesley, Reading, MA.

Bruce, T., 1992. Designing Quality Databases: Practical Information Management and IDEF1X. Dorset House, New York.

Bubenko, J., 2007. From information algebra to enterprise modelling and ontologies—a historical perspective on modelling for information systems. In: Krogstie, J., Opdahl, A., Brinkkemper, S. (Eds.), Conceptual Modelling in Information Systems Engineering. Springer, Berlin, pp. 1–18.

Burlton, R., 2001. Business Process Management: Profiting From Process. Sams Publishing.

Campbell, L., Halpin, T., 1993. Automated support for conceptual to external mapping. In: Brinkkemper, S., Harmsen, F. (Eds.), Proc. 4th Workshop on Next Generation CASE Tools. Univ. Twente Memoranda Informatica 93-32, pp. 35–51.

Campbell, L., Halpin, T., 1994a. Abstraction techniques for conceptual schemas. In: Proc. 5th Australasian Database Conf., Christchurch (Jan.). World Scientific, Singapore.

Campbell, L., Halpin, T., 1994b. The reverse engineering of relational databases. In: Proc. 5th Workshop on Next Generation CASE Tools, Utrecht (June).

Campbell, L., Halpin, T., Proper, H., 1996. Conceptual schemas with abstractions: making flat conceptual schemas more comprehensible. Data Knowl. Eng. 20 (1), 39–85.*

Carver, A., Halpin, T., 2008. Atomicity and normalization. In: Halpin, T., Krogstie, J., Proper, H. (Eds.), Proc. EMMSAD2008. CEUR, Montpellier. June 2008.

Carver, A., Halpin, T., 2013. Reference scheme reduction on subtypes in ORM. In: Demey, Y.T., Panetto, H. (Eds.), On the Move to Meaningful Internet Systems 2013: OTM 2013 Workshops. Springer LNCS 8186, pp. 457–466.

Cattell, R., 1991. Object Data Management. Addison-Wesley, Reading, MA.

Cattell, R., Barry, D. (Eds.), 2000. The Object Data Standard: ODMG 3.0. Morgan Kaufmann, San Francisco.

Celko, J., 1999. Data and Databases: Concepts in Practice. Morgan Kaufmann, San Francisco.

Celko, J., 2000. Joe Celko's SQL for Smarties: Advanced SQL Programming, second ed. Morgan Kaufmann, San Francisco.

Celko, J., 2004. Joe Celko's Trees and Hierarchies in SQL for Smarties. Morgan Kaufmann, San Francisco.

Celko, J., 2005. Joe Celko's SQL Programming Style. Morgan Kaufmann, San Francisco.

Chamberlin, D., 1998. A Complete Guide to DB2 Universal Database. Morgan Kaufmann, San Francisco.

Chamberlin, D., Boyce, R., 1974. In: SEQUEL: a structured english query language. Proc. 1974 ACM SIGMOD Workshop on Data Description, Access and Control, Ann Arbor, MI.

Chawdhry, P., 1992. NIAMEX—a NIAM compiler and EXPRESS pre-processor. In: Proc. EXPRESS Users Group Conf., Oct., 1992.

Chen, P., 1976. The entity-relationship model—toward a unified view of data. ACM Trans. Database Syst. 1 (1), 9–36.

Choobineh, J., Mannino, M., Tseng, V., 1992. A form-based approach for database analysis and design. CACM 35 (2), 108–120.

Codd, E., 1969. Derivability, Redundancy, and Consistency of Relations Stored in Large Data Banks. IBM Research Report RJ599, August 19, 1969.

Codd, E., 1970. A relational model of data for large shared data banks. CACM 13 (6), 377–387.

Codd, E., 1990. The Relational Model for Database Management: Version 2. Addison-Wesley, Reading, MA.

Codd, E., Codd, S., Salley, C., 1993. Providing OLAP (On-line Analytical Processing) to User-Analysts: An IT Mandate. Arbor Software Corporation. Accessible online at www.arborsoft.com/essbase/wht_ppr/coddToc.html.

Connolly, T., Begg, C., Strachan, A., 1999. Database Systems, second ed. Addison-Wesley.

Control Data, 1982. IAST: Information Analysis Support Tools. Reference Manual, Control Data Publication no. 60484610.

Creasy, P., 1989. ENIAM—a more complete conceptual schema language. In: Proc. 15th VLDB Conf., Amsterdam.

Crockford, D., 2006. The Application/JSON Media Type for JavaScript Object Notation (JSON). Available at: https://www.ietf.org/rfc/rfc4627.txt.

Curland, M., Halpin, T., 2010. The NORMA software tool for ORM 2. In: Soffer, P., Proper, E. (Eds.), CAiSE Forum 2010, LNBIP 72. 2010. Springer-Verlag, Berlin Heidelberg, pp. 190–204.

Curland, M., Halpin, T., 2012. Enhanced verbalization of ORM models. In: Herrero, P., Panetto, H., Meersman, R., Dillon, T. (Eds.), On the Move to Meaningful Internet Systems 2012: OTM 2012 Workshops. Springer LNCS 7567, pp. 399–408.

Curland, M., Halpin, T., Stirewalt, K., 2009. A role calculus for ORM. In: Meersman, R., Herrero, P., Dillon, T. (Eds.), On the Move to Meaningful Internet Systems 2009: OTM 2009 Workshops. Springer LNCS 5872, pp. 692–703.

Cuyler, D., Halpin, T., 2003. Metamodels for object-role modeling. In: Proc. EMMSAD'03: 8th IFIP WG8.1 Int. Workshop on Evaluation of Modeling Methods in Systems Analysis and Design, Velden, Austria (June).

Czejdo, B., Elmasri, R., Rusinkiewicz, M., Embley, D., 1990. A graphical data manipulation language for an extended entity-relationship model. IEEE Comput. 1990, 26–37.

D'Atri, A., Sacca, D., 1984. Equivalence and mapping of database schemas. In: Proc. 10th Int. Conf. on Very Large Databases, VLDB, Singapore, pp. 187–195.

Date, C., 1998. The birth of the relational model: parts 1–3. In: Intelligent Enterprise. vol. 1. Miller Freeman, Inc., San Mateo, CA. Oct–Dec issues.

Date, C., 2000. An Introduction to Database Systems, seventh ed. vol. 1 Addison-Wesley, Reading MA.

Date, C., Darwen, H., 1997. A Guide to the SQL Standard, fourth ed. Addison-Wesley, Reading, MA.

Date, C., Darwen, H., 1998. Foundation for Object/Relational Databases: The Third Manifesto. Addison-Wesley, Reading, MA.

Date, C., Fagin, R., 1992. Simple conditions for guaranteeing higher normal forms in relational databases. ACM TODS 17 (3), 465–476.

Date, C., Darwen, H., Lorentzos, N., 2014. Time and Relational Theory—Temporal Databases in the Relational Model and SQL. Elsevier-Morgan Kaufmann, ISBN: 9780128006313.

De Troyer, O., 1991. The OO-binary relationship model: a truly object-oriented conceptual model. In: Advanced Information Systems Engineering: Proc. CAiSE-91. Springer LNCS 498, Trondheim.

De Troyer, O., 1993. On Data Schema Transformation. PhD Thesis, Katholieke Universiteit of Brabant, Tilburg.

De Troyer, O., Meersman, R., 1995. A logic framework for a semantics of object oriented data modeling. In: Proc. OOER'95: Object-Oriented and Entity-Relationship Modeling. Springer LNCS 1021, Gold Coast Australia, pp. 238–249.

De Troyer, O., Meersman, R., Verlinden, P., 1988. RIDL* on the CRIS case: a workbench for NIAM. In: Olle, T., Verrijn-Stuart, A., Bhabuta, L. (Eds.), Computerized Assistance during the Information Systems Life Cycle: Proc. CRIS-88. North-Holland, Amsterdam.

Delaney, K., 2000. Introducing indexed views. In: SQL Server Magazine, No. 15. Duke Communications, pp. 32–37. www.sqlmag.com.

Denning, P., Metcalfe, R. (Eds.), 1997. Beyond Calculation: The Next Fifty Years of Computing. Springer-Verlag, New York.

Dietz, J., Halpin, T., 2003. Combining DEMO and ORM—an investigation of mutual benefits. In: Proc. EMMSAD'03: 8th IFIP WG8.1 Int. Workshop on Evaluation of Modeling Methods in Systems Analysis and Design, Velden, Austria (June).

Dix, A., Finlay, J., Abowd, G., Beale, R., 1993. Human-Computer Interaction. Prentice Hall, New York.

Elmasri, R., Navathe, S., 1994. Fundamentals of Database Systems, second ed. Benjamin/Cummings, Redwood City, CA.

Embley, D., 1998. Object Database Management. Addison-Wesley, Reading, MA.

Embley, D., Kurtz, B., Woodfield, S., 1992. Object-Oriented Systems Analysis. Prentice Hall, Englewood Cliffs, NJ.

Embley, D., Wu, H., Pinkston, J., Czejdo, B., 1996. OSM-QL: A Calculus-based Graphical Query Language. Tech. Report, Dept. of Comp. Science, Brigham Young Univ., Utah.

Eriksson, H., Penker, M., 2000. Business Modeling with UML. John Wiley.

Everest, G., 1976. Basic data structure models explained with a common example. In: Proc. Fifth Texas Conference on Computing Systems, Austin TX, IEEE Computer Society Publications, Long Beach, CA, pp. 39–45.

Everest, G., 1986. Database Management: Objectives, System Functions, and Administration. McGraw-Hill.

Everest, G., 1994. Experiences teaching NIAM/OR modeling. In: Proc. 2nd NIAM-ISDM Conf., Albuquerque, NM.

Everest, G., 2020. Recognizing and Treating "Tableitis" (tay-bul-eye-tusd) A Widespread Condition Suffered by Professional Data Modelers, Commonly Called "TABLE THINK". TDAN. Accessible at: https://tdan.com/recognizing-and-treating-tableitis/26060.

Fagin, R., 1977. Multivalued dependencies and a new normal form for relational databases. ACM Trans. Database Syst. 2 (3).

Fagin, R., 1979. Normal forms and relational database operators. In: Proc. 1979 ACM SIGMOD int. Conf. on Management of Data, Boston.

Fagin, R., 1981. A normal form for relational databases that is based on domains and keys. ACM Trans. Database Syst. 6 (3), 387–415.

Falkenberg, E., 1976. Concepts for modelling information. In: Nijssen, G. (Ed.), Modelling in Database Management Systems. North-Holland Publishing, Amsterdam.

Falkenberg, E., 1993. DETERM: deterministic event-tuned entity-relationship modeling. In: Entity-Relationship Approach—ER'93. Springer LNCS 823, pp. 230–241.

Falkenberg, E., Oei, J., 1994. Meta-model hierarchies from an object-role modeling perspective. In: Halpin, T., Meersman, R. (Eds.), Proc. First Int. Conf. On Object-Role Modeling (ORM-1). Magnetic Island, Australia, pp. 218–227.

Falkenberg, E., Hesse, W., Lindgreen, P., Nilson, B., Oei, J., Rolland, C., Stamper, R., Van Assche, F., Verrijn-Stuart, A., Voss, K., 1998. A Framework of Information System Concepts: The FRISCO Report (Web Edition). IFIP. Available online by anonymous ftp://ftp.leidenuniv.nl/pub/rul/fri-full.zip.

Fillmore, C., 1968. The case for case. In: Bach, E., Harms, R. (Eds.), Universals in Linguistic Theory. Holt, Rinehart and Winston, New York, pp. 1–88.

Finkelstein, C., 1989. Introduction to Information Engineering. Addison-Wesley, Reading, MA.

Finkelstein, C., 1992. Information Engineering: Strategic Systems Development. Addison-Wesley, Reading MA.

Finkelstein, C., 1998. Information engineering methodology. In: Bernus, P., Mertins, K., Schmidt, G. (Eds.), Handbook on Architectures of Information Systems. Springer-Verlag, Berlin, Germany, pp. 405–427.

Finkelstein, C., Aiken, P., 2000. Building Corporate Portals with XML. McGraw-Hill, New York.

Fitting, M., 2000. Databases and higher types. In: Lloyd, J., et al. (Eds.), Computational Logic–CL2000. Springer-Verlag, pp. 41–52. LNAI 1861.

Flanagan, D., 2020. JavaScript—The Definitive Guide, seventh ed. O'Reilly Media.

Fowler, M., with Scott, K, 1997. UML Distilled. Addison-Wesley.

Gilbert, S., Lynch, N., 2002. Brewer's conjecture and the feasibility of consistent, available, partition-tolerant web services. ACM SIGACT News. Assoc. Comput. Mach. (ACM) 33 (2), 51–59.

Gorman, M., 1998. Great News, The Relational Data Model is Dead!. http://www.wiscorp.com/SQLStandards.html#sql1999readings.

Gössner, S., Bormann, C., 2021. JSONPath—XPath for JSON. Available at: www.ietf.org/archive/id/draft-goessner-dispatch-jsonpath-00.html.

Gray, J., Reuter, A., 1993. Distributed Transaction Processing: Concepts and Techniques. Morgan Kaufmann, ISBN: 1-55860-190-2.

Greenwald, R., Stackowiak, R., Stern, J., 2013. Oracle Essentials, fifth ed. O'Reilly Media, Inc.

Guarino, N., Welty, C., 2002. Evaluating ontological decisions with onto clean. CACM 45 (2), 61–65.

Guizzardi, G., 2005. Ontological Foundations for Structural Conceptual Models. CTIT PhD Thesis Series, No. 05-74, Enschede, The Netherlands.

Guizzardi, G., Wagner, G., Guarino, N., van Sinderen, N., 2004. An ontologically well-founded profile for UML conceptual models. In: Persson, A., Stirna, J. (Eds.), Proc. 16th Int. Conf. on Advanced Inf. Sys. Engineering, CAiSE2004. Springer LNCS 3084, pp. 112–126.

Haack, S., 1978. Philosophy of Logics. Cambridge University Press, London.

Habrias, H., 1993. Normalized object oriented method. In: Encyclopedia of Microcomputers. vol. 12. Marcel Dekker, New York, pp. 271–285.

Halpin, T., 1989a. Venn diagrams and SQL queries. Aust. Comput. J. 21 (1), 27–32.

Halpin, T., 1989b. A Logical Analysis of Information Systems: Static Aspects of the Data-Oriented Perspective. PhD Thesis, University of Queensland.

Halpin, T., 1989c. Contextual equivalence of conceptual schemas. In: Proc. Advanced Database Systems Symposium, Inform. Proc. Society of Japan, Kyoto, pp. 47–54.

Halpin, T., 1990a. Conceptual schemas and relational databases. In: Srinivasan, B., Zeleznikov, J. (Eds.), Databases in the 1990s: Proc. 1st Australian Database Conf. World Scientific, Singapore, pp. 45–56.

Halpin, T., 1990b. Conceptual schema optimization. In: Proc. 13th Australian Computer Science Conf. Monash University, Melbourne.

Halpin, T., 1991a. Optimizing global conceptual schemas. In: Srinivasan, B., Zeleznikov, J. (Eds.), Databases in the 1990s: 2-Proc. 2nd Australian Database Conf. World Scientific, Singapore.

Halpin, T., 1991b. A Fact-oriented approach to schema transformation. In: Proc. MFDBS-91 Int. Conf. on Mathematical Fundamentals of Database and Knowledge Base Systems. Spinger LNCS 495, Rostock.

Halpin, T., 1993. What is an elementary fact? In: Proc. 1st NIAM-ISDM Conf., Utrecht.*

Halpin, T., 1995. Subtyping: conceptual and logical issues. In: Ross, R. (Ed.), Database Newsletter. vol. 23(6). Database Research Group, Inc, pp. 3–9.*

Halpin, T., 1998. Conceptual queries. In: Ross, R. (Ed.), Database Newsletter. vol. 26(2). Database Research Group, Inc, Boston. (March/April 1998).*

Halpin, T., 1998–99. UML data models from an ORM perspective: parts 1–10. In: Journal of Conceptual Modeling. InConcept, Minneapolis, USA.*

Halpin, T., 1999. Data modeling in UML and ORM revisited. In: Proc. EMMSAD'99: 4th IFIP WG8.1 Int. Workshop on Evaluation of Modeling Methods in Systems Analysis and Design, Heidelberg, Germany (June).*

Halpin, T., 2000a. Integrating fact-oriented modeling with object-oriented modeling. In: Siau, K., Rossi, M. (Eds.), Information Modeling for the New Millennium. Idea Group, Hershey, PA.

Halpin, T., 2000b. Modeling collections in UML and ORM. In: Proc. EMMSAD'00: 5th Int. Workshop on Evaluation of Modeling Methods in Sys. Analysis and Design, Kista, Sweden.*

Halpin, T., 2000c. An ORM metamodel. In: Journal of Conceptual Modeling. vol. 16. InConcept, Minneapolis, MN.*

Halpin, T., 2000d. An ORM metamodel of Barker ER. In: Journal of Conceptual Modeling. vol. 17. InConcept, Minneapolis, MN.*

Halpin, T., 2001. Supplementing UML with concepts from ORM. In: Siau, K., Halpin, T. (Eds.), Unified Modeling Language: Systems Analysis, Design and Development Issues. Idea Group Publishing, Hershey, PA, pp. 168–185.

Halpin, T., 2002a. Join constraints. In: Halpin, T., Krogstie, J., Siau, K. (Eds.), Proc. Seventh CAiSE/IFIP-WG8.1 International Workshop on Evaluation of Modeling Methods in Systems Analysis and Design, Toronto, Canada, pp. 121–131.

Halpin, T., 2002b. Information analysis in UML and ORM: a comparison. In: Siau, K. (Ed.), Advanced Topics in Database Research. vol. 1. Idea Publishing Group, Hershey, PA, pp. 307–323.

Halpin, T., 2004a. Information modeling and higher-order types. In: Grundspenkis, J., Kirkova, M. (Eds.), Proc. CAiSE'04 Workshops. vol. 1. Riga Tech. University, pp. 233–248.

Halpin, T., 2004b. Comparing metamodels for ER, ORM and UML data models. In: Siau, K. (Ed.), Advanced Topics in Database Research. vol. 3. Idea Group, Hershey, PA, pp. 23–44.

Halpin, T., 2005a. Objectification. In: Castro, J., Teniente, E. (Eds.), Proc. CAiSE'05 Workshops. vol. 1. FEUP, pp. 519–532.

Halpin, T., 2005b. ORM 2. In: Meersman, R., Tari, Z., Herrero, P., et al. (Eds.), On the Move to Meaningful Internet Systems 2005: OTM 2005 Workshops. Springer LNCS 3762, Cyprus, pp. 676–687.

Halpin, T., 2005c. Constraints on conceptual join paths. In: Krogstie, J., Halpin, T., Siau, K. (Eds.), Information Modeling Methods and Methodologies. Idea Group, Hershey, PA, pp. 258–277.

Halpin, T., 2005d. Higher-order types and information modeling. In: Siau, K. (Ed.), Advanced Topics in Database Research. vol. 4. Idea Publishing Group, Hershey, PA, pp. 218–237.

Halpin, T., 2005e. Information modeling in UML and ORM: a comparison. In: Khosrow-Pour, M. (Ed.), Encyclopedia of Information Science and Technology. vol. 3. Idea Group, Hershey, PA, pp. 1471–1475.

Halpin, T., 2006a. Business rule modality. In: Latour, T., Petit, M. (Eds.), Proc. CAiSE'06 Workshops. Namur University Press, pp. 383–394.

Halpin, T., 2006b. Object-role modeling (ORM/NIAM). In: Handbook on Architectures of Information Systems, second ed. Springer, Heidelberg, pp. 81–103.

Halpin, T. 2007-2009, 'Temporal modeling: parts 1–7', Bus. Rules J., 8 and 9. Accessible at: www.BRCommunity.com and Modeling Issues page of www.orm.net.

Halpin, T., 2007a. Modality of business rules. In: Siau, K. (Ed.), Research Issues in Systems Analysis and Design, Databases and Software Development. IGI Global, Hershey, PA, pp. 206–226.

Halpin, T., 2007b. Subtyping revisited. In: Pernici, B., Gulla, J. (Eds.), Proc. CAiSE'07 Workshops. vol. 1. Tapir Academic Press, pp. 131–141.

Halpin, T., 2007c. Fact-oriented modeling: past, present and future. In: Krogstie, J., Opdahl, A., Brinkkemper, S. (Eds.), Conceptual Modelling in Information Systems Engineering. Springer, Berlin, pp. 19–38.

Halpin, T., 2008. Objectification of relationships. In: Siau, K. (Ed.), Advanced Topics in Database Research. IGI Global, Hershey, PA.

Halpin, T. 2009-2013, 'Ontological modeling: parts 1–15', Bus. Rules J., 10–14. Accessible at www.BRCommunity.com and Modeling Issues page of www.orm.net.

Halpin, T., 2009a. Enriched conceptualization of subtyping. In: Halpin, T., Krogstie, J., Proper, H. (Eds.), Innovations in Information Systems Modeling: Methods and Best Practices. IGI Global, Hershey, PA, pp. 1–16.

Halpin, T., 2009b. Object-role modeling. In: Liu, L., Ozsu, M.T. (Eds.), Encyclopedia of Database Systems. Springer-Verlag, Berlin.

Halpin, T., 2010. Object-role modeling: principles and benefits. In: International Journal of Information Systems Modeling and Design. vol. 1(1). IGI Global, pp. 32–54.

Halpin, T. 2014-2018, 'Logical data modeling: parts 1–14', Bus. Rules J., 15–19. Accessible at www.BRCommunity.com and Modeling Issues page of www.orm.net.

Halpin, T., 2015a. A comparative review of information modeling in UML and ORM. In: Khosrow-Pour, M. (Ed.), Encyclopedia of Information Science and Technology, third ed. Information Science Reference, Hershey, PA, USA, pp. 1622–1630.

Halpin, T., 2015b. Object-Role Modeling Fundamentals. Technics Publications, New Jersey.

Halpin, T., 2016. Object-Role Modeling Workbook. Technics Publications, New Jersey.

Halpin, T., 2019. Reference scheme modeling. In: IGI Global (Ed.), New Perspectives on Information Systems Modeling and Design. Hershey, pp. 227–254.

Halpin, T., 2020. Objectification and Atomicity. Accessible at Modeling Issues page of www.orm.net.

Halpin, T., Curland, M., 2006. Automated verbalization for ORM 2. In: Meersman, R., Tari, Z., Herrero, P., et al. (Eds.), On the Move to Meaningful Internet Systems 2006: OTM 2006 Workshops. Springer LNCS 4278, Montpellier, pp. 1181–1190.

Halpin, T., Curland, M., 2011. Enriched support for ring constraints. In: Meersman, R., Dillon, T., Herrero, P. (Eds.), On the Move to Meaningful Internet Systems 2011: OTM 2011 Workshops. Springer LNCS 7046, Hersonissou, pp. 309–318.

Halpin, T., Curland, M., 2013. Recent enhancements to ORM. In: Demey, Y.T., Panetto, H. (Eds.), On the Move to Meaningful Internet Systems 2013: OTM 2013 Workshops. Springer LNCS 8186, pp. 467–476.

Halpin, T., Girle, R., 1981. Deductive Logic, second ed. Logiqpress, Brisbane.

Halpin, T., Proper, H., 1995a. Database schema transformation and optimization. In: Papazoglou, M. (Ed.), Proc. OOER'95. Springer LNCS 1021, pp. 191–203.*

Halpin, T., Proper, H., 1995b. Subtyping and polymorphism in object-role modeling. Data Knowl. Eng. 15, 251–281.*

Halpin, T., Ritson, P., 1992. Fact-oriented modelling and null values. In: Srinivasan, B., Zeleznikov, J. (Eds.), Proc. 3rd Australian Database Conf. World Scientific, Singapore.

Halpin, T., Rugaber, S., 2015. LogiQL A Query Language for Smart Databases. CRC Press, Boca Raton, FL.

Halpin, T., Vermeir, D., 1997. Default reasoning in information systems. In: Database Applications Semantics. Chapman and Hall, London, pp. 423–441.

Halpin, T., Wagner, G., 2003. Modeling reactive behavior in ORM. In: Conceptual Modeling—ER2003, Proc. 22nd ER Conference, Chicago, October 2003. Springer LNCS.

Halpin, T., Weber, G., 2012. Form-and-fact based modeling. In: Bider, I., et al. (Eds.), BPMDS 2012 and EMMSAD 2012. Springer LNBIP 113, pp. 285–299.

Halpin, T., Wijbenga, J., 2010. FORML 2. In: Bider, I., et al. (Eds.), Enterprise, Business-Process and Information Systems Modeling. Springer-Verlag, Berlin Heidelberg, pp. 247–260. LNBIP 50.

Halpin, T., Evans, K., Hallock, P., MacLean, W., 2003. Database Modeling with Microsoft® Visio for Enterprise Architects. Morgan Kaufmann, San Francisco.

Halpin, T., Carver, A., Owen, K., 2007. Reduction transformations in ORM. In: Meersman, R., Tari, Z., Herrero, P., et al. (Eds.), On the Move to Meaningful Internet Systems 2007: OTM 2007 Workshops. Springer LNCS 4805, Vilamoura, pp. 699–708.

Hammer, M., 1997. Beyond Reengineering: How the Process-Centred Organization Is Changing Our Work and Our Lives. Collins.

Hammer, M., Champy, J., 1993. Reengineering the Corporation: A Manifesto for Business Revolution. Harper Business Books.

Hammer, M., McLeod, D., 1981. Database description with SDM: a semantic database model. ACM Trans. Database Syst. 6, 351–386.

Harel, D., Politi, M., 1998. Modeling Reactive Systems With Statecharts: The Statemate Approach. McGraw-Hill.

Harris, R., 1996. Information Graphics: A Comprehensive Illustrated Reference. Management Graphics, Atlanta, GA.

Havey, M., 2005. Essential Business Process Modeling. O'Reilly Media, Inc.

Hay, D., 1996. Data Model Patterns: Conventions of Thought. Dorset House, New York.

Hay, D., 1999a. There is no object-oriented analysis. In: DataToKnowledge Newsletter, vol. 27(1). Business Rule Solutions, Inc., Houston, TX.

Hay, D., 1999b. Object orientation and information engineering: UML. In: Reiner, R. (Ed.), The Data Administration Newsletter. No. 9 (June 1999), available online at www.tdan.com.

Hay, D., 2003. Requirements Analysis. Pearson Education, New Jersey.

Hay, D., 2006. Data Model Patterns: A Metadata Map. Morgan Kaufmann.

Heath, I., 1971. Unacceptable file operations in a relational database. In: Proc. 1971 ACM SIGFIDET Workshop on Data Description. Access and Control, San Diego, CA.

Hohenstein, U., Engels, G., 1991. Formal semantics of an entity-relationship-based query language. In: Kangassalo, H. (Ed.), Proc. 9th ER Conf. Elsevier Science, Amsterdam.

Inmon, W., 1993. Building the Data Warehouse. John Wiley and Sons, New York.

Jacobson, I., Booch, G., Rumbaugh, J., 1999. The Unified Software Development Process. Addison-Wesley, Reading, MA.

Jarke, M., Mylopoulos, J., Schmidt, J., Vassiliou, Y., 1992. DAIDA: An Environment for Evolving Information Systems. ACM Trans. Inform. Syst. 10 (1), 1–50.

Jeyaraj, A., Sauter, V., 2007. An empirical investigation into the effectiveness of systems modeling and verification tools. CACM 50 (6), 63–67.

Jukik, N., 2006. Modeling strategies and alternatives for data warehousing projects. CACM 49 (4), 83–88.

Keet, C., 2006. Part-whole relations in object-role models. In: Proc. OTM Workshops 2006. Springer LNCS 4278, pp. 1118–1127.

Keet, C., 2007. Mapping the Object-Role Modeling language ORM2 into Description Logic Language DLR_{ifd}. KRDB Research Centre TR KRDB07-2, Faculty of Computer Science, Free University of Bozen-Bolzano, Italy.

Kendall, K., Kendall, J., 1988. Systems Analysis and Design. Prentice Hall, Englewood Cliffs, NJ.

Kent, W., 1977. Entities and Relationships in Information. In: Nijssen, G. (Ed.), Proc. 1977 IFIP Working Conf. on Modelling in Data Base Management Systems. North-Holland Publishing, Nice, France, pp. 67–91.

Kent, W., 1983. A simple guide to five normal forms in relational database theory. CACM 26 (2), 120–125.

Kent, W., 1991. A rigorous model of object reference, identity, and existence. J. Obj.-Orient. Program. 1991, 28–36.

Kent, W., 2000. Data and Reality, second ed. The 1st Books Library.

Khoshaflan, S., Copeland, G., 1990. Object identity. In: Zdonik, S., Maier, D. (Eds.), Readings in Object-Oriented Database Systems. Morgan Kaufmann, San Mateo, CA, pp. 37–46.

Kimball, R., 1996. The Data Warehouse Toolkit. John Wiley and Sons, New York.

Kobayashi, I., 1990. Transformation and equivalence among predicate systems. In: Proc. French-Japanese Seminar on Deductive Databases and Artificial Intelligence. INRIN, Sophia-Antipolis.

Kulkarni, K., 2006. Overview of SQL:2003. www.wiscorp.com/SQL2003Features.pdf.

Kurzweil, R., 2005. The Singularity is Near: When Humans Transcend Biology. Penguin Group, New York.

Lassen, K., van der Aalst, W., 2006. WorkflowNet2BPEL4WS: a tool for translating unstructured workflow processes to readable BPEL. In: BETA Working Paper Series, WP 167. Eindhoven University of Technology, Eindhoven.

Leavitt, N., 2010. Will NoSQL databases live up to their promise? IEEE Comput. 43 (2), 12–14.

Levesque, H., 1984. A fundamental trade-off in knowledge representation and reasoning. In: Proc. CSCSI-84, London, Ontario, pp. 141–152.

Ling, T., Tompa, F., Kameda, T., 1981. An improved third normal form for relational databases. ACM Trans. Database Syst. 6 (2), 329–346.

Mark, L., 1988. The Binary Relationship Model. Tech. Report UMIACS-TR-88-67, University of Maryland.

Martin, J., 1993. Principles of Object Oriented Analysis and Design. Prentice Hall, Englewood Cliffs, NJ.

Martin, J., Odell, J., 1998. Object-Oriented Methods: A Foundation, UML Edition. Prentice Hall, Upper Saddle River, NJ.

McCreary, D., Kelly, A., 2014. Making Sense of NoSQL. Manning Publications.

McLeod, G., 2000. Beyond use cases. In: Proc. EMMSAD'00: 5th IFIP WG8.1 Int. Workshop on Evaluation of Modeling Methods in Systems Analysis and Design, Kista, Sweden.

Meersman, R., 1982. The RIDL Conceptual Language. Research Report, Int. Centre for Information Analysis Services, Control Data Belgium, Brussels.

Meersman, R., Van Assche, F., 1983. Modeling and manipulating production databases in terms of semantic nets. In: Proc. 8th IJCAI Conference. Morgan Kaufmann.

Meersman, R., De Troyer, O., Ponsaert, F., 1984. The RIDL User Guide. Control Data Corp, Brussels.

Mellor, S., Balcer, M., 2002. Executable UML: A Foundation for Model-Driven Architecture. Addison-Wesley.

Melton, J., Buxton, S., 2006. Querying XML: XQuery, XPath, and SQL/XML in Context. Morgan Kaufmann.

Melton, J., Simon, A., 2002. SQL: 1999 Understanding Relational Language Components. Morgan Kaufmann.

Milner, R., 1999. Communicating and Mobile Systems: The Pi Calculus. Cambridge University Press.

Mok, W., Embley, D., 1996. Transforming conceptual model to object-oriented database designs: practicalities, properties and peculiarities. In: Proc. ER'96: 15th Int. Conf. on Conceptual Modeling. Springer LNCS 1157, pp. 309–324.

Moody, D., Kortink, M., 2000. From enterprise models to dimensional models: a methodology for data warehouse and data mart design. In: Proc. DMDW'00: Int. Workshop on Design and Management of Data Warehouses, Report 25. Swiss Life IT RandD, Zurich.

Morgan, T., 2006. Some features of state machines in ORM. In: Meersman, R., Tari, Z., Herrero, P., et al. (Eds.), On the Move to Meaningful Internet Systems 2006: OTM 2006 Workshops. Springer LNCS 4278, Montpellier.

Morgan, T., 2007. Business process modeling and ORM. In: Meersman, R., Tari, Z., Herrero, P., et al. (Eds.), On the Move to Meaningful Internet Systems 2007: OTM 2007 Workshops. Springer LNCS 4805, Vilamoura.

Muller, R., 1999. Database Design for Smarties. Morgan Kaufmann, San Francisco.

Mylopoulos, J., 1998. Information modelling in the time of the revolution. Inform. Syst. 23 (3/4), 127–155 (Pergamon, Exeter).

Nijssen, G., 1976. A gross architecture for the next generation database management systems. In: Nijssen, G. (Ed.), Proc. 1976 IFIP Working Conf. on Modelling in Data Base Management Systems. North-Holland Publishing, Freudenstadt, Germany, pp. 1–24.

Nijssen, G., 1977. Current issues in conceptual schema concepts. In: Nijssen, G. (Ed.), Proc. 1977 IFIP Working Conf. on Modelling in Data Base Management Systems. North-Holland Publishing, Nice, France, pp. 31–66.

Nijssen, G., 1994. A general analysis procedure: recent advances in universal informatics. In: Proc. 2nd NIAM-ISDM Conf., Albuquerque, NM.

NIST, 1993. Integration Definition for Information Modeling (IDEF1X). FIPS Publication 184, National Institute of Standards and Technology. www.sdct.itl.nist.gov/~ftp/idef1x.trf.

Odell, J., 1998. Advanced Object-Oriented Analysis and Design using UML. Cambridge University Press, and SIGS Books, New York.

Olle, T., Hagelstein, J., Macdonald, I., Rolland, C., Sol, H., van Assche, F., Verrijn-Stuart, A., 1991. Information Systems Methodologies: A Framework for Understanding, second ed. Addison-Wesley, Wokingham, England.

OMG, 2019. Semantics of Business Vocabulary and Business rules (SBVR) Specification. Available online at: https://omg.org/spec/SBVR.

Orlowska, M., Zhang, Y., 1992. Understanding the fifth normal form (5NF). Austral. Comput. Sci. Comm. 14, 631–639.

Ouyang, C., van der Aalst, W., Dumas, M., ter Hofstede, A., 2006. From BPMN process models to BPEL web services. In: Pro 4th International Conference on Web Services (ICWS). IEEE Computer Society, Chicago.

Ozsoyoglu, G., Wang, H., 1993. Example-based graphical database query languages. Computer 26 (5), 25–38.

Petri, C.A., 1962. Kommunication mit Automaten. Ph.D. Thesis, Technischen Hoschule Darmstadt.

Petrounias, I., Loucopoulos, P., 1994. Time dimension in a fact-based model. In: Proc. ORM-1 Conference. Dept of Computer Science, University of Queensland.

Pohl, K., 1994. The Three Dimensions of Requirements Engineering: A Framework and its Applications. In: Information Systems. Pergamon, Exeter, pp. 243–258.

Pritchett, D., 2008. Base: an acid alternative. Queue 6 (3), 48–55.

Proper, H., 1994. A Theory for Conceptual Modelling of Evolving Application Domains. PhD Thesis, University of Nijmegen.

Redmond, E., Wilson, J.R., 2012. Seven Databases in Seven Weeks. Bookshelf, The Pragmatic.

Rissanen, J., 1977. Independent components of relations. ACM Trans. Database Syst. 2 (2), 317–325.

Ritson, P., Halpin, T., 1993a. Mapping one-to-one predicates to a relational schema. In: Orlowska, M., Papazoglou, M. (Eds.), Advances in Database Research: Proc. 4th Australian Database Conf. World Scientific, Singapore, pp. 68–84.

Ritson, P., Halpin, T., 1993b. Mapping integrity constraints to a relational schema. In: Proc. 4th ACIS, Brisbane (Sep.), pp. 381–400.

Rochfeld, A., Morejon, J., Negros, P., 1991. Inter-relationship links in E-R model. In: Kangassalo, H. (Ed.), Proc. 9th ER Conf. Elsevier Science, Amsterdam.

Ross, R., 1998. Business Rule Concepts. Business Rule Solutions, Inc, Houston, TX.

Rumbaugh, J., Jacobson, I., Booch, G., 1999. The Unified Modeling Language Reference Manual. Addison-Wesley, Reading, MA.

Rummler, G., Brache, A., 1995. Improving Performance: How to Manage the White Space in the Organization Chart, second ed. Jossey-Bass Publishing.

Russell, N., van der Aalst, W., ter Hofstede, A., 2016. Workflow Patterns: The Definitive Guide. MIT Press, Cambridge, MA.

Sadalage, P., Fowler, M., 2013. NoSQL Distilled. Addison-Wesley.

Seidman, C., 2000. Creating horizontally partitioned views. In: SQL Server Magazine, No. 14. Duke Communications, pp. 51–53.

Senko, M., 1975. Information systems: records, relations, sets, entities and things. Information Systems 1 (1), 3–13. Jan. 1995, Pergamon Press.

Shneiderman, B., 1992. Designing the User Interface, second ed. Addison-Wesley, Reading, MA.

Shoval, P., Kabeli, J., 2000. Functional and object-oriented analysis and design of information systems—an integrated methodology. In: Proc. EMMSAD'00: 5th IFIP WG8.1 Workshop on Evaluation of Modeling Methods in Systems Analysis and Design, Kista, Sweden.

Shoval, P., Shreiber, N., 1993. Database reverse engineering: from the relational to the binary relational model. Data Knowl. Eng. 10, 293–315.

Shoval, P., Gudes, E., Goldstein, M., 1988. GISD: A graphical interactive system for conceptual database design. Inform. Syst. 13 (1), 81–95.

Siau, K., Halpin, T. (Eds.), 2000. Unified Modeling Language: Systems Analysis, Design and Development Issues. Idea Group, Hershey, PA.

Silverston, L., 2001a. The Data Model Resource Book. vol. 1 John Wiley, New York.

Silverston, L., 2001b. The Data Model Resource Book. vol. 2 John Wiley, New York.

Simsion, G., Witt, G., 2005. Data Modeling Essentials, third ed. Morgan Kaufmann.

Snodgrass, R., 1990. Temporal databases: status and research directions. SIGMOD Rec. 19 (4), 83–89.

Snodgrass, R., 2000. Developing Time-Oriented Database Applications in SQL. Morgan Kaufmann, San Francisco.

Stonebraker, M., Rowe, L., Lindsay, B., Gray, J., Carey, M., Brodie, M., Bernstein, P., Beech, D., 1990. Third generation database system manifesto. ACM SIGMOD Rec. 19 (3).

Sundblad, S., Sundblad, P., 2002. Design Patterns for Scalable Microsoft.NET Applications. Sundblad and Sundblad ADB-Arkitektur AB, Uppsala, Sweden. Electronic version available online at: www.2xsundblad.com/.

Syverson, B., Murach, J., 2023. Murach's SQL server 2022 for developers. Mike Murach & Associates Inc, Fresno, CA.

Taylor, A., 2018. SQL for Dummies, nineth ed. For Dummies (A Wiley Brand), New York.

ter Hofstede, A., 1993. Information Modeling in Data Intensive Domains. PhD Thesis, University of Nijmegen.

ter Hofstede, A., van der Weide, T., 1994. Fact orientation in complex object role modeling techniques. In: Halpin, T., Meersman, R. (Eds.), Proc. 1st Int. Conf. on Object-Role Modeling (ORM-1), Magnetic Island, Australia, pp. 45–59.

ter Hofstede, A., Proper, H., van der Weide, T., 1992. Data modeling in complex application domains. In: Loucopoulos, P. (Ed.), Proc. CAiSE'92: Fourth Int. Conference on Advanced Information Systems Engineering. Springer LNCS 593, pp. 364–377.

ter Hofstede, A., Proper, H., van der Weide, T., 1993. Formal definition of a conceptual language for the description and manipulation of information models. Inform. Syst. 18 (7), 489–523.

ter Hofstede, A., Proper, H., van der Weide, T., 1996. Query formulation as an information retrieval problem. Comput. J. 39 (4), 255–274.

Thalheim, B., 1984. Deductive basis of relations. In: Proc. MFSSS-84. Springer LNCS 215, Berlin Heidelberg, pp. 226–230.

Thalheim, B., 1994. Fundamentals of Entity-Relationship Modelling. Springer, Heidelberg.

Thomsen, E., Spofford, G., Chase, D., 1999. Microsoft OLAP Solutions. John Wiley, New York.

Trog, D., Vereecken, J., De Leenheer, P., Meersman, R., 2006. T-Lex: a role-based ontology engineering tool. In: Meersman, R., Tari, Z., Herrero, P., et al. (Eds.), On the Move to Meaningful Internet Systems 2006: OTM 2006 Workshops. Springer LNCS 4278, Montpellier.

van Bommel, P., van der Weide, T., 1992. Reducing the search space for conceptual schema transformations. Data Knowl. Eng. 8, 269–292.

van der Aalst, W., van Hee, K., 2004. Workflow Management: Models, Methods, and Systems. MIT Press.

van Griethuysen, J. (Ed.), 1982. Concepts and Terminology for the Conceptual Schema and the Information Base. ISO TC97/SC5/WG3, Eindhoven.

Verheijen, G., van Bekkum, J., 1982. NIAM: an information analysis method. In: Information Systems Design Methodologies: a comparative review, Proc. IFIP WG8.1 Working Conf. North Holland Publishing, Noordwijkerhout, The Netherlands, pp. 537–590.

Vermeir, D., 1983. Semantic hierarchies and abstractions in conceptual schemata. Inform. Syst. 8 (2), 117–124.

Vincent, M., 1994. Semantic Justification of Normal Forms in Relational Database Design. PhD Thesis, Monash University.

Vincent, M., Srinivasan, B., 1993. Redundancy and the justification for fourth normal form in relational databases. Int. J. Found. Comput. Sci. 4 (4), 355–365.

Vincent, M., Srinivasan, B., 1994. A note on relation schemes which are in 3NF but not in BCNF. Inform. Process. Lett. 48, 281–283.

Walmsley, P., 2015. XQuery, second ed. O'Reilly Media, Inc.

Warmer, J., Kleppe, A., 2003. The Object Constraint Language, second ed. Addison-Wesley, Boston, MA.

Wintraecken, J., 1990. The NIAM Information Analysis Method: Theory and Practice. Kluwer, Deventer, The Netherlands.

Zachman, J., 1987. A framework for information systems architecture. IBM Syst. J. 26 (3), 276–292.

Zaniolo, C., 1982. A new normal form for the design of relational database schemas. ACM Trans. Database Syst. 7 (3), 489–499.

Zhao, A., 2021. SQL Pocket Guide, fourth ed. O'Reilly Media, Sebastopol, CA.

Index

Note: Page numbers followed by *f* indicate figures and *t* indicate tables.

D

F